Handbook of
Digital Human
Modeling

Research for Applied
Ergonomics and Human
Factors Engineering

Human Factors and Ergonomics

Series Editor

Gavriel Salvendy

HANDBOOK OF DIGITAL HUMAN MODELING

Research for Applied Ergonomics and Human Factors Engineering

EDITED BY

VINCENT G. DUFFY

CRC Press
Taylor & Francis Group
Boca Raton   London   New York

CRC Press is an imprint of the
Taylor & Francis Group, an **informa** business

CRC Press
Taylor & Francis Group
6000 Broken Sound Parkway NW, Suite 300
Boca Raton, FL 33487-2742

Library of Congress Cataloging-in-Publication Data

Handbook of digital human modeling : research for applied ergonomics and human factors
    engineering / edited by Vincent Duffy.
        p. cm. -- (Human factors and ergonomics)
    Includes bibliographical references and index.
    ISBN-13: 978-0-8058-5646-0 (hardcover : alk. paper)
    ISBN-10: 0-8058-5646-3 (hardcover : alk. paper)
    1. Human engineering--Handbooks, manuals, etc. 2. Digital computer simulation--Handbooks, manuals, etc. 3. Human-machine systems--Handbooks, manuals, etc. 4. Human mechanics--Computer simulation--Handbooks, manuals, etc. I. Duffy Vincent G. II. Title. III. Series.

TA166.H2735 2009
620.8'20113--dc22   2008014185

Visit the Taylor & Francis Web site at
http://www.taylorandfrancis.com

and the CRC Press Web site at
http://www.crcpress.com

Contents

Foreword

With the rapid introduction of highly sophisticated computers, (tele)communication, service, and manufacturing systems, a major shift has occurred in the way people use technology and work with it. The objective of this book series is to provide researchers and practitioners a platform where important issues related to these changes can be discussed, and methods and recommendations can be presented for ensuring that emerging technologies provide increased productivity, quality, satisfaction, safety, and health in the new workplace and information society.

With the rapidly expanding use of computer technology, computer-aided human digital modeling emerged as a promising methodology in the 1980s. The objective is to ensure that the anthropometric and physical characteristics of the human are considered in the design of products and services. By so doing, manufacturers ensure that the products they design and produce meet the requirements of the customer. The *Handbook of Digital Human Modeling* provides comprehensive coverage of the theory, tools, and methodologies to effectively achieve this objective.

The 56 chapters of this handbook are written by 113 contributing authorities from Canada, China, France, Germany, Hong Kong, the Netherlands, Poland, Spain, Sweden, Taiwan, the United Kingdom, and the United States, and they provide a wealth of international knowledge and guidelines, which is crucial as customers are more and more concerned about the usability and comfort of the products and services they use. The *Handbook of Digital Human Modeling* includes 435 figures and 54 tables, which provide easy to read and easy to comprehend materials for both researchers and practitioners. The 2,597 references cover the entire breadth and depth of the discipline and provide information for further in-depth study. The chapters provide an in-depth treatment of digital human modeling, and this handbook is an excellent source to consult before embarking on any human digital modeling research, project, or application.

Gavriel Salvendy
Purdue University and Tsinghua University, P.R. China
Series Editor

Preface

In my view, the growing body of literature makes it challenging for newcomers to identify the key elements of digital human modeling (DHM) quickly. One of the objectives of this book was to enable practicing engineers and researchers the opportunity to find the key elements in one updated resource that provides context on the independent works in relation to one another. As you read through these chapters, you will get a sense of the history and a compilation of the research and applications that contributed to the development of this field. The breadth of coverage across the *Handbook* will give some perspective to readers about current and potential future applications. The references within the chapters facilitate further study that can provide additional insight into each of the areas of coverage that contribute to the whole of the digital human modeling field.

The community of researchers and practitioners in this area recognizes the significance of using digital human modeling to consider digital representations of the human inserted into a simulation or virtual environment to facilitate predictions of safety and performance in design. Such models typically include a visualization with math or science in the background and are intended to reduce or eliminate the need for physical prototypes. This *Handbook* provides human factors specialists and ergonomists the knowledge and appropriate engineering tools that might suggest possible modifications to improve new designs.

This *Handbook* also enables engineers from many disciplines to quantify acceptability and risk in design in terms of the human factors and ergonomics aspects. The current gap between capability to correctly predict outcomes and set expectations for new and existing products and processes affects human-system performance, market acceptance, product safety, and satisfaction at work. Tools identified in this *Handbook of Digital Human Modeling* can help us evaluate product design and work design while reducing the need for physical prototyping. Up-to-date research and practice related to technology, computing, optimization, human factors, and ergonomics are included. The chapters focus in detail on such matters as advanced manufacturing, aerospace, automotive, data visualization and simulation, defense and military systems, design for impaired mobility, health care and medical applications, information systems, and product design, among others.

The Scientific Advisory Board is comprised of 13 members from industry and leading universities with expertise in modeling, product development, and human factors and ergonomics. The authors of these chapters are all leading authorities in their respective fields and represent a cross-disciplinary perspective. Each member of this distinguished group belongs to at least one of the following professional societies:

American Society for Engineering Education (ASEE)
American Society of Mechanical Engineers (ASME)
Association for Computing Machinery (ACM)
Human Factors and Ergonomics Society (HFES)
Institute of Electrical and Electronics Engineers (IEEE)
Institute of Industrial Engineers (IIE)
The Ergonomics Society
National Academy of Engineering (NAE)
Society of Automotive Engineers (SAE)

Additional demonstration materials are available through the CRC Web site exclusively for the *Handbook of Digital Human Modeling*. These include a never-before-released UGS-Jack help manual developed by H. Onan Demirel at Purdue University. This supplement is entitled, *User Manual and*

Examples: Tecnomatix Jack 5.0, and is available at http://www.crcpress.com/e_products/downloads/download.asp?cat_no=ER564X

I would like to thank our chief scientist, colleague, and friend, Gavriel Salvendy, for his guidance throughout the process. I would also like to thank the members of the digital human modeling community for their support for the project and contributions to this handbook. My work with the graduate students and the DHM community inspired a lot of the effort on the manuscript. The assistance of Amanda Cripps, Jason Lim, and Lindsay Squillace invaluable in finalizing the manuscript as it neared production. Anne Duffy at Lawrence Erlbaum Associates and Cindy Carelli, Cathy Giacari, and Jill Jurgensen at Taylor & Francis were instrumental in enabling the book to be brought to the readers. I would like to add a special thanks to my wife Colleen and children Gavin, Brendan, Ashlynn, Caitrin, and Arianna for their patience. My parents William and Arlene, and grandparents Arthur and Anne, provided all the right guidance in earlier endeavors and encouragement during this one. I hope you enjoy reading the book. Certainly, I enjoyed the opportunity to compile the great work of these authors within this volume.

Vincent G. Duffy
Purdue University
West Lafayette, Indiana

Scientific Advisory Board

About the Editor

Vincent G. Duffy is a faculty member in the College of Engineering at Purdue University. He has participated on national research committees including the Research Associateship Program of the National Academies in the United States, and international research collaborations such as the European Research Consortium on Informatics and Mathematics (ERCIM). His research supports the development of new science and applications in digital human modeling and simulation, human factors, safety engineering, and ergonomics. Elements of the work have been transferred to automotive design, manufacturing, health care, and other service engineering initiatives.

Contributors

Karim Abdel-Malek
Center for Computer Aided Design
University of Iowa
Iowa City, Iowa

Jan M. Allbeck
Computer and Information Science
University of Pennsylvania
Philadelphia, Pennsylvania

Vinnie Ahuja
Center for Ergonomics
University of Michigan
Ann Arbor, Michigan

Thomas Alexander
Research Institute for Communication,
Information Processing, and Ergonomics (FKIE)
Wachtberg, Germany

Dean H. Ambrose
3D Group
Pittsburgh, Pennsylvania

Susan Archer
MAAD Operations
Alion Science and Technology
Boulder, Colorado

Thomas J. Armstrong
Center for Ergonomics
University of Michigan
Ann Arbor, Michigan

Jasbir Arora
Center for Computer Aided Design
University of Iowa
Iowa City, Iowa

Sergio Ausejo
Center for Technical Study and Investigation
of Gipuzkoa (CEIT)
University of Navarra
San Sebastian, Spain

Richard W. Backs
Department of Psychology
Central Michigan University
Mount Pleasant, Michigan

Norman I. Badler
Computer and Information Science
University of Pennsylvania
Philadelphia, Pennsylvania

Timothy M. Bagnall
MAAD Operations
Alion Science and Technology
Boulder, Colorado

Steve Beck
Center for Computer Aided Design
University of Iowa
Iowa City, Iowa

Cecilia Berlin
Product and Production Development
Chalmers University of Technology
Gothenburg, Sweden

Wolfram Boucsein
Department of Physiological Psychology
University of Wuppertal
Wuppertal, Germany

Heiner Bubb
Institute of Ergonomics
Technical University of Munich
Munich, Germany

Brad Cain
RDDC Research Defense Agencies
Defense R&D Canada
Toronto, Ontario, Canada

Barrett S. Caldwell
School of Industrial Engineering
School of Aeronautics and Astronautics
Purdue University
West Lafayette, Indiana

Keith Case
Department of Mechanical and Manufacturing
Engineering
Loughborough University
Loughborough, United Kingdom

Don B. Chaffin
College of Engineering
University of Michigan
Ann Arbor, Michigan

Jim Chiang
Human Simulation Engineer
Work in Progress Ergonomics
Windsor, Ontario, Canada

Jaewon Choi
Center for Ergonomics
University of Michigan
Ann Arbor, Michigan

Robert R. Conatser, Jr.
Department of Biomedical Sciences
Ohio University
Athens, Ohio

Daniel DeLaurentis
School of Aeronautics and
Astronautics
Purdue University
West Lafayette, Indiana

Jack T. Dennerlein
Department of Environmental Health
Harvard School of Public Health
Boston, Massachusetts

Yingzi Du
Department of Electrical and Computer
Engineering
Indiana University–Purdue
University
Indianapolis, Indiana

Vincent G. Duffy
College of Engineering
Purdue University
West Lafayette, Indiana

Tania Dukic
Human Factors
Swedish National Road and Transport
Research Institute (VTI)
Gothenburg, Sweden

David S. Ebert
School of Electrical and Computer
Engineering
Purdue University
West Lafayette, Indiana

Stephen R. Ellis
NASA Ames Research Center
Moffett Field, California

Hongbing Fang
Department of Mechanical Engineering and
Engineering Science
University of North Carolina at Charlotte
Charlotte, North Carolina

Julian J. Faraway
Department of Mathematical Sciences
University of Bath
Bath, United Kingdom

Laura Frey Law
Carver College of Medicine
University of Iowa
Iowa City, Iowa

Florian Fritzsche
Institute of Ergonomics
Technical University of Munich
Munich, Germany

Afzal Godil
Information Technology Laboratory
National Institute of Standards and
Technology (NIST)
Gaithersburg, Maryland

Christina Godin
Research and Development
Human Simulation Products
Siemens PLM
Ann Arbor, Michigan

Ravindra S. Goonetilleke
Department of Industrial Engineering and
Logistics Management
Hong Kong University of Science and
Technology
Clear Water Bay, Hong Kong

Brian F. Gore
Human Systems Integration Division
SJSURF/NASA Ames Research Center
Moffett Field, California

Anand K. Gramopadhye
College of Engineering
Clemson University
Clemson, South Carolina

Jerzy Grobelny
Institute of Organization and
Management
Wroclaw University of Technology
Wroclaw, Poland

Diane E. Gyi
Department of Human Sciences
Loughborough University
Loughborough, United Kingdom

Lars Hanson
Department of Design Sciences
Lund University
Lund, Sweden

Riender Happee
Faculty of Mechanical, Maritime, and
Materials Engineering
Delft University of Technology
Delft, The Netherlands

Christoph M. Hoffmann
Computer Science Department
Purdue University
West Lafayette, Indiana

Dan Högberg
School of Technology and Society
University of Skövde
Skövde, Sweden

John N. Howell
Department of Biomedical Sciences
Ohio University
Athens, Ohio

Jeffrey A. Hudson
General Dynamics AIS
Wright-Patterson Air Force Base, Ohio

Peter W. Johnson
Department of Environmental and Occupational
Health Sciences
University of Washington
Seattle, Washington

Monica L. H. Jones
Assembly Ergonomics
Ford Motor Company
Dearborn, Michigan

Waldemar Karwowski
Department of Industrial Engineering and
Management Systems
University of Central Florida
Orlando, Florida

Hoo Sang Ko
College of Engineering
Purdue University
West Lafayette, Indiana

David A. Kobus
Pacific Science and Engineering
San Diego, California

Andrea Laake
Center for Computer Aided Design
University of Iowa
Iowa City, Iowa

Michael LaFiandra
Human Research and Engineering Directorate
U.S. Army Research Laboratory
Aberdeen Proving Ground, Maryland

Dan Lämkull
Manufacturing Engineering
Volvo Car Corporation
Gothenburg, Sweden

Mark R. Lehto
Department of Industrial Engineering
Purdue University
West Lafayette, Indiana

Zhizhong Li
Department of Industrial Engineering
Tsinghua University
Beijing, China

Yan Liu
Department of Biomedical, Industrial,
and Human Factors Engineering
Wright State University
Dayton, Ohio

John F. Lockett III
Human Research and Engineering
Directorate
U.S. Army Research Laboratory
Aberdeen Proving Ground, Maryland

Jia Lu
Center for Computer Aided Design
University of Iowa
Iowa City, Iowa

Jun-Ming Lu
Department of Industrial Engineering and
Engineering Management
National Tsing Hua University
Hsinchu City, Taiwan

Ameersing Luximon
Institute of Textiles and Clothing
Hong Kong Polytechnic University
Hung Hom, Hong Kong

Lennart Malmsköld
General Assembly
Saab Automobile AB
Trollhättan, Sweden

Tim Marler
Center for Computer Aided Design
University of Iowa
Iowa City, Iowa

Russ Marshall
Department of Design and Technology
Loughborough University
Loughborough, United Kingdom

Aninth Mathai
Center for Computer Aided Design
University of Iowa
Iowa City, Iowa

Sandra A. Metzler
SEA, Limited
Columbus, Ohio

Pierre Meunier
RDDC Research Defense Agencies
Defense R&D Canada
Toronto, Ontario, Canada

Rafal Michalski
Institute of Organization and Management
Wroclaw University of Technology
Wroclaw, Poland

Gilles Monnier
Altran AIT
Caluire, France

Douglas R. Morr
SEA, Limited
Columbus, Ohio

Mark Morrissey
Safework-Dassault Systems
Montréal, Quebec, Canada

Peter B. Muller
Potomac Training Corporation
Lansdowne, Virginia

Arne Nåbo
Saab Automobile
Trollhättan, Sweden

Jennifer McGovern Narkevicius
Jenius LLC
California, Maryland

Shimon Y. Nof
College of Engineering
Purdue University
West Lafayette, Indiana

Amirali Noorinaeini
Department of Industrial Engineering
Purdue University
West Lafayette, Indiana

Julie Cowan Novak
School of Nursing
Purdue University
West Lafayette, Indiana

Roland Örtengren
Product and Production Development
Chalmers University of Technology
Gothenburg, Sweden

Aernout Oudenhuijzen
Department of Human Performance
TNO Defence, Security, and Safety
Soesterberg, The Netherlands

John E. Owen
U.S. Naval Air Systems Command
Patuxent River, Maryland

Woojin Park
Mechanical, Industrial, and Nuclear
Engineering Department
University of Cincinnati
Cincinnati, Ohio

Matthew B. Parkinson
Engineering Design and Mechanical
Engineering
Pennsylvania State University
University Park, Pennsylvania

Amos Patrick
Center for Computer Aided Design
University of Iowa
Iowa City, Iowa

Voicu Popescu
Computer Science Department
Purdue University
West Lafayette, Indiana

J. Mark Porter
Department of Design and Technology
Loughborough University
Loughborough, United Kingdom

Robert W. Proctor
Department of Psychological Sciences
Purdue University
West Lafayette, Indiana

Kathryn Rapala
Indianapolis Patient Safety Coalition and
School of Nursing
Purdue University
West Lafayette, Indiana

Leah Reeves
Center for Neurotechnology Studies
Potomac Institute for Policy Studies
Arlington, Virginia

Sandy Ressler
Information Technology Laboratory
National Institute of Standards and Technology
(NIST)
Gaithersburg, Maryland

Kelly A. Rossi
Situation Awareness Inc.
Arlington, Virginia

Sajay Sadasivan
College of Engineering
Clemson University
Clemson, South Carolina

Robert A. Sargent
MAAD Operations
Alion Science and Technology
Boulder, Colorado

Dylan D. Schmorrow
Office of Naval Research
U.S. Navy
Arlington, Virginia

Thomas B. Sheridan
Departments of Mechanical Engineering and
Aeronautics-Astronautics
Massachusetts Institute of Technology
Cambridge, Massachusetts

Ruth E. Sims
Department of Design and Technology
Loughborough University
Loughborough, United Kingdom

Allison Stephens
Assembly Ergonomics
Ford Motor Company
Dearborn, Michigan

Don Stredney
Biomedical Sciences and Visualization
Ohio Supercomputer Center
Columbus, Ohio

Paris F. Stringfellow
College of Engineering
Clemson University
Clemson, South Carolina

Steve Summerskill
Department of Design and Technology
Loughborough University
Loughborough, United Kingdom

Colby Swan
Center for Computer Aided Design
University of Iowa
Iowa City, Iowa

Jules Trasbot
Renault SA
Guyancourt, France

Karl Van Orden
Naval Health Research Center
San Diego, California

Mao-Jiun Wang
Department of Industrial Engineering and
Engineering Management
National Tsing Hua University
Hsinchu City, Taiwan

Xuguang Wang
Biomechanics and Impact Mechanics
Laboratory (INRETS)
National Institute for Transport and
Safety Research
Bron, France

John F. Wiechel
SEA, Limited
Columbus, Ohio

Robert L. Williams II
Department of Mechanical Engineering
Ohio University
Athens, Ohio

Jac Wismans
TNO Automotive Safety Solutions
Delft, The Netherlands
and
Eindhoven University of Technology
Eindhoven, The Netherlands

Ting Xia
Center for Computer Aided Design
University of Iowa
Iowa City, Iowa

Motonori Yamaguchi
Department of Psychological Sciences
Purdue University
West Lafayette, Indiana

Jingzhou Yang
Department of Mechanical Engineering
Texas Tech University
Lubbock, Texas

Gregory F. Zehner
711th Human Performance Wing
U.S. Air Force Research Laboratory
Wright-Patterson Air Force Base, Ohio

Ming Zhou
Altair Engineering, Inc.
Irvine, California

Part 1

Foundations of Digital Human Modeling

1

Introduction

Vincent G. Duffy

Digital human modeling can be considered a digital representation of the human inserted into a simulation or virtual environment to facilitate prediction of safety and/or performance (Demirel & Duffy, 2007a). These include some visualization as well as the math or science in the background (Demirel & Duffy, 2007b). Digital human modeling (DHM) is intended to reduce or eliminate the need for physical prototypes in new product and process design, and enables engineers of various disciplines to incorporate ergonomics science and human factors engineering principles earlier in the design process. These methods can provide real cost savings as is evidenced in various chapters throughout this DHM handbook. The book is divided into five parts.

Part 1: Foundations of DHM
Part 2: Modeling Fundamentals
Part 3: Evaluation and Analysis
Part 4: Applications
Part 5: Current Implementation and the Future of DHM

The chapters within these sections are intended to outline various DHM issues with a common focus.

This field, and interest in it, is growing. It will be important to determine how to best incorporate this subject into university curriculums at both the undergraduate and graduate levels so that the needs of industry can be best met now and in the future. Product and process designers in various engineering domains have experienced some limitations in predicting usability, safety, satisfaction, and human performance without a more substantial contribution from the practitioners with background and experience with the sciences of human factors and ergonomics (Luximon et al., 2001; Duffy, 2003; Chan et al., 2004). Practitioners in human factors and ergonomics have sometimes been limited in their contribution to team-based design efforts when their different languages have not always mapped well (Duffy, 2005). The engineering disciplines typically work in teams and utilize computer-aided software tools. The engineering curriculums and the graduating engineers have typically not had a great deal of formal training in the human factors and ergonomics areas.

Commercially available tools for computer-aided design incorporate some of the analysis tools specifically designed to support the physical aspects of an ergonomic analysis (Du et al., 2005; Du & Duffy, 2007). The effective utilization of these tools is still primarily an engineering function and requires some depth of knowledge and perspective in human factors and ergonomics in order to know especially the capabilities, limitations, and best practices. The development of these tools has more than a 50-year history, and more recent DHM efforts developed out of the tradition of computer-aided ergonomics and safety from the 1980s. You will also see references within these chapters to some early digital human modeling efforts that precede the 1980s, which have evolved into digital human modeling, a growing community of practitioners and researchers committed to providing the math and science as well as the visualizations to provide additional insight about key elements of design for the decision makers in the business process (Duffy et al., 2004).

The Society of Automotive Engineers has held an annual conference in this area for the past 10 years. This conference has drawn participants from all over the world, and recent hosts are Germany, France, and Sweden. The Human Factors and Ergonomics Society (HFES) formed a Technical Group in Human Performance Modeling (HPM) with overlapping interests and dedicated sessions at the annual conferences. The Institute of Industrial Engineers (IIE) is beginning a DHM-related track in this area, and the Human Computer Interaction International (HCI International) community recently held a successful DHM conference, called the International Conference on Digital Human Modeling. The first was held in Beijing, China, in 2007.

At present, not all product design can be simulated, and some elements of product and process design are still difficult to represent in virtual environments (Duffy, 2005, 2007; Fuhua et al., 2002). It has become apparent that when a product or process requires more interactivity, there is still a greater need for physical prototyping. However, when a product or process requires less interactivity, or some activity of a repeated nature, there are more opportunities for a full simulation. Currently available tools will be outlined and described from both the theoretical and applications perspectives within this handbook. Some comparison and contrasting of previously available digital human models will give the practitioner additional insight into trade-offs that may be faced in design implementation. Key elements of common interest to these different professional societies, industry employers, and government agencies are represented in various chapters within this book, including modeling requirements and fundamentals, historical perspectives, and modern methods such as virtual interactive design, which are highlighted in the first section of the book on foundations of DHM.

Some elements of this handbook focus on the current tools; others focus on the development of the tools to give insight into the processes by which new analysis tools can be designed and derived (Duffy et al., 2004, 2007). Examples are also shown in some of the references of this chapter. Together these demonstrate how some of the models may be utilized, as well as limitations of current models that can be addressed through further research. There are some aspects of the digital human models, particularly the cognitive aspects, that are not currently well integrated into the computer-aided engineering tools available to most practicing engineers. Authors of related chapters who represent both researchers and practitioners provide insight into the vision of what digital human modeling has available and what it may be in the future.

A wide range of research and perspectives contributed to the development of this handbook. For example, human factors and ergonomics-related research demonstrated perception-based empirical model development in virtual and real environments that contributes to safety design (Duffy & Chan, 2002; Duffy et al., 2003). Fundamental contributions include the application of Stevens' Power Law for predicting hazard perception, a demonstration of the relationship between hazard perception in virtual and real, and theoretical concepts such as dual coding to explain the impact of virtual training accidents on improved safety decisions (Duffy, Ng, & Ramakrishnan, 2004). These led to the development of a testbed that had virtual reality integrated with live motion capture and computer-based ergonomics analysis capabilities (Li & Duffy, 2006) and enabled a real-time product design assessment for people with limited mobility. This virtual interactive product evaluation demonstrated a methodology suitable for both virtual and real industrial environments. The interested reader could review additional literature included in the reference section of this and many other chapters within this book.

Other research was intended to demonstrate how the cognitive and physical aspects of human factors and ergonomics can be considered in the same research testbed to enable more integrated predictions of performance and safety in product and process design (Carruth & Duffy, 2005; Kukula et al., 2007). This kind of testbed that includes both the cognitive and physical aspects can influence the development of new models for digital human modeling, and is particularly important as the nature of work is changing from more physical, to incorporating more of cognitive aspects. A series of experiments demonstrated how thermography can be used as a non-intrusive indicator of change in mental workload (Or & Duffy, 2007; Wang & Duffy, 2007).

Research participants included drivers in a driving simulator and an instrumented car. Results demonstrated differences in nose temperature and mental workload were illustrated by introducing a secondary task during simulated driving and through different driving modes. Measures of comfort were monitored through optical motion capture and analyzed in terms of joint angles. These results demonstrated the potential for using such physiological parameters in driving and provide a significant scientific contribution toward efforts to combine cognitive and physical modeling capabilities in the same virtual interactive design testbed. Potential applications of this research include real-time, unobtrusive, and automated mental workload assessment for human-system interaction during product development and interface design.

The lessons learned from these types of projects have provided the vision for future work and challenges to the digital human modeling community (Duffy, 2005; Duffy et al., 2005; Boone-Seals & Duffy, 2005; Tong et al., 2005; Wu et al., 2005a and b; Duffy, Or, & Lau, 2006; Duffy, Duffy, & Yen, 2007; Sutherland & Duffy, 2007). Additional opportunities for current and future digital human modeling contributions lie in computationally driven modeling, data-driven modeling, and efforts to validate that can help, in an iterative way, to inform both data-driven and computationally driven modeling efforts. These will be outlined in various chapters including chapters on data-driven modeling and simulation, as well as those on computational approaches contained in the second section of this handbook on modeling fundamentals. Engineers who utilize these models, as in any engineering model formulation, should be aware of the conditions in which the models are developed, the intended use, and limitations described throughout chapters in Part 3 of this handbook on evaluation and analysis.

The *Handbook of Digital Human Modeling* is intended, ultimately, to facilitate more reliable human-machine modeling—address the validation challenges of predictive digital human models and simulations now in order to minimize or eliminate the need for full-scale prototypes and reduce design costs in the future (Vice-Cappelli & Duffy, 2005). Engineers, ergonomists, and human factors specialists will see a broad spectrum of applications and can incorporate their previous knowledge to future human-system design efforts in the automotive and manufacturing industries, military, health care and aerospace. The authors' perspectives on current implementation and the future of digital human modeling are outlined in the final part of this handbook.

References

Boone-Seals, A., and Duffy, V.G. 2005. Toward a model for reducing medication administration errors, *Ergonomics*, 48 (9), 1151–1168.

Cappelli, T., and Duffy, V.G. 2007. Motion capture for job risk classifications incorporating dynamic aspects of work, *SAE 2006 Transactions Journal of Passenger Cars—Electronic and Electrical Systems*, 2006-01-2317, 1069–1072.

Carruth, D., and Duffy, V.G. 2005. Using the Virtual Build test bed in cognitive modeling for remote manipulation, *HCII 2005, 9th Conference on Human-Computer Interaction International*, July 22–27, 2005, Las Vegas, Nevada, CD-ROM.

Chan, A.H.S., Kwok, W.Y., and Duffy, V.G. 2004. Using AHP for determining priority in a safety management system, *Industrial Management and Data Systems*, 104 (5), 430–445.

Demirel, H.O., and Duffy, V.G. 2007a. Applications of human digital modeling in industry. In *Digital human modeling*, ed. V.G. Duffy, LCNS 4561, 824–832. Berlin: Springer-Verlag.

Demirel, H.O., and Duffy, V.G. 2007b. Digital human modeling for Product Lifecycle Management. In *Digital human modeling*, ed. V.G. Duffy, LCNS 4561, 372–381. Berlin: Springer-Verlag.

Du, C.J., Williams, S.N., Duffy, V.G., Yu, Q., McGinley, J., and Carruth, D. 2005. Using computerized ergonomic analysis tools for assembly task assessment, *HAAMAHA 2005 Conference Proceedings*, July 18–21, San Diego, CA, CD-ROM.

Du, J.C., and Duffy, V.G. 2007. Development of a methodology for assessing industrial workstations using optical motion capture integrated with digital human models, *Occupational Ergonomics*, 7 (1), 11–25.

Duffy, C.M., Duffy, V.G., and Yen, B.P.C. 2007. Toward a hybrid model for usability resource allocation in industrial software product development, *Human Factors and Ergonomics in Manufacturing*, 17 (3), 245–262.

Duffy, V.G. 2003. Effects of training and experience on perception of hazard and risk, *Ergonomics*, 46 (1), 114–125.

Duffy, V.G. 2005. Impact of force feedback on a Virtual Interactive Design assessment, *HAAMAHA 2005 Conference Proceedings*, July 18–21, San Diego, CA, CD-ROM (plenary presentation).

Duffy, V.G. 2005. Digital human modeling for applied ergonomics and human factors engineering, *Cyberg 2005*, Sept. 15–Oct. 15, 2005, CD-ROM. (on-line conference)

Duffy, V.G. 2007. Using the Virtual Build methodology for computer-aided ergonomics and safety, *Human Factors and Ergonomics in Manufacturing*, 17 (5), 413–422.

Duffy, V.G., and Chan, H.S. 2002. Effects of virtual lighting on visual performance and eye fatigue, *Human Factors and Ergonomics in Manufacturing*, 12 (2), 193–209.

Duffy, V.G., Jin, M., Eksioglu, B., Yu, Q., Kang, J., and Du, C.J. 2005. Virtual design optimization tool for improved interaction with hybrid automation, *HCII 2005, 9th Conference on Human-Computer Interaction International*, July 22–27, 2005, Las Vegas, Nevada, CD-ROM.

Duffy, V.G., Ng, P.P.W., and Ramakrishnan, A. 2004. Impact of a simulated accident in virtual training on decision making performance, *International Journal of Industrial Ergonomics*, 34 (4), 335–348.

Duffy, V.G., Or, C.K.L., and Lau, V.W.M. 2006. Perception of safe robot speed in virtual and real industrial environments, *Human Factors and Ergonomics in Manufacturing*, 16 (4), 369–383.

Duffy, V.G., Wu, F., and Ng, P.P.W. 2003. Development of an Internet virtual layout system for improving workplace safety, *Computers in Industry*, 50 (2), 207–230.

Duffy, V.G., Yen, B.P.C., and Cross, G.W. 2004. Internet marketing and product visualization (IMPV) system: development and evaluation in support of product data management, *International Journal of Computer-Integrated Manufacturing*, 17 (1), 1–15.

Fuhua, L., Duffy, V.G., and Su, C.J. 2002. Developing virtual environments for industrial training, *Information Sciences*, 140, 153–170.

Kukula, E., Elliot, S., and Duffy, V.G. 2007. The effects of human interaction on biometrics system performance. In *Digital human modeling*, ed. V.G. Duffy, LCNS 4561, 904–914. Berlin: Springer-Verlag.

Li, K., and Duffy, V.G. 2006. Universal accessibility assessments through Virtual Interactive Design, *International Journal of Human Factors Modelling and Simulation*, 1 (1), 52–68.

Luximon, A., Duffy, V.G., and Zhang, W. 2001. Performance, satisfaction and mental workload in voice assisted interface, *International Journal of Industrial Ergonomics*, 28, 133–142.

Or, C.K.L., and Duffy, V.G. 2007. Development of a facial skin temperature-based methodology for non-intrusive mental workload measurement, *Occupational Ergonomics*, 7 (2), 83–94.

Sutherland, J., and V.G. Duffy 2007. Validating optical motion capture assessments of the dynamic aspects of work. In *Digital human modeling*, ed. V.G. Duffy, LCNS 4561, 197–204. Berlin: Springer-Verlag.

Szczerba, J., Duffy, V.G., Geisler, S., Rowland, Z., and Kang, J. 2007. A study in driver performance: alternative human-vehicle interface for brake actuation, *Society of Automotive Engineers SAE 2006 Transactions—Journal of Engines*, 2006-01-1060, 605–610.

Tian, R., Duffy, V.G., and McGinley, J. 2007. Effecting validity of ergonomics analysis during virtual interactive design. In *Digital human modeling*, ed. V.G. Duffy, LCNS 4561, 988–997. Berlin: Springer-Verlag.

Tong, J., Duffy, V.G., Cross, G., Tsung, F., and Yen, B.P.C. 2005. Evaluating the industrial ergonomics of service quality for online recruitment websites, *International Journal of Industrial Ergonomics*, 35, 697–711.

Vice-Cappelli, T., and Duffy, V.G. 2005. Toward utilization of motion capture for dynamic ergonomic assessment, *HCII 2005, 9th Conference on Human-Computer Interaction International,* July 22–27, 2005, Las Vegas, Nevada, CD-ROM.

Wang, L., Duffy, V.G., and Du, Y. 2007. A composite measure for the evaluation of mental workload. In *Digital human modeling,* ed. V.G. Duffy, LCNS 4561, 460–466. Berlin: Springer-Verlag.

Wu, L.T., Duffy, V.G., Kang, J., Carruth, D., and Rowland, Z. 2005a. Validation of the Virtual Build methodology for automotive manufacturing, *HCII 2005, 9th Conference on Human-Computer Interaction International,* July 22–27, 2005, Las Vegas, Nevada, CD-ROM.

Wu, T., McGinley, J., Duffy, V.G., and Liu, L. 2005b. Application and validation of a mechanical motion capture-based industrial ergonomics assessment system, *Proceedings of the Society of Automotive Engineers, Conference on Digital Human Modeling for Design and Engineering,* SAE-DHM-2005-01-2733) June 14–16, 2005, Iowa City, Iowa.

2

Some Requirements and Fundamental Issues in Digital Human Modeling

Don B. Chaffin

2.1 Introduction

The use of digital human models to improve certain ergonomic attributes in a proposed design is not a new concept. Ryan and Springer at Boeing Aircraft in the late 1960s developed a digital human model that could be used to assess pilot reach requirements for people of varied anthropometry (Ryan & Springer, 1969). It had an exertion optimization method for predicting reach postures. About the same time, Chaffin, Kilpatrick, and Hancock (1970) described the development of a seated digital human model. It accepted data about the locations and orientations of objects that needed to be manipulated, and when combined with a list of tasks to be performed by a person, it produced an MTM prediction of performance times. It also provided a graphical illustration of a simple 3D avatar that contained an empirically driven optimization IK that displayed various predicted postures required to reach the objects and tools being used. The University of Michigan's 3D Static Strength Prediction Program was developed in the 1970s to run in batch mode on a mainframe computer. It was later released in 1984 to run on early PCs, and it included a 3D stick figure avatar that could be manipulated by the users to allow them to simulate people of varied anthropometry, when performing high exertion tasks. Evans and Chaffin (1985) described how workplace and task information could be integrated into the UM 3DSSPP along with a wire mesh 3D avatar to perform ergonomics assessments. In a similar development at the University of Pennsylvania, starting in the late 1970s, Jack was being created, and by the late 1980s it had an enfleshed 3D avatar and a sophisticated set of inverse kinematics routines to allow a user to easily manipulate simple movements and object-grasping tasks (Badler et al., 1993). Also, by the late 1980s the U.S. Air Force was using derivatives of the earlier Boeing model, entitled COMBIMAN and CREWCHIEF, to evaluate pilot and maintenance tasks. These programs included a simple 3D avatar (McDaniel, 1990), and provided access to a database of anthropometric and human strength attributes related to the tasks of interest. Similarly, in the United Kingdom during the 1980s, Case, Porter, and Bonney (1990) were developing a sophisticated 3D avatar and workplace/vehicle CAD system, referred to as SAMMIE, which allowed reach and sight line evaluations.

The point of this brief summary describing some of the early DHM models is to remind the reader that various types of digital human models have been around for over 35 years. Unfortunately, their acceptance and use have not been rapidly assimilated into organizations to improve the ergonomic design of most of the hardware and software systems used today. This is despite the fact that the benefits when using such a technology have been well acknowledged over the last decade in books by Badler et al. (1993), Peacock and Karwowski (1993), and Chaffin (2001).

The goal of this chapter is to review some of the fundamental issues that appear to be affecting the rather slow adoption of digital human modeling by various organizations. One proposed reason for this phenomenon is that the DHM technologies often may not be compatible with the needs of specific user groups. This proposition serves in this review as the basis for a second discussion about requirements for evaluating present DHM technologies, which in turn leads to speculation about future developments in the field.

2.2 Why Use DHM Technologies to Meet Ergonomics Goals?

To begin to answer this question, one should consider the three different approaches (numbered 1, 2, and 3 in fig. 2.1) that are typically used when attempting to include ergonomics in the design of a system. Though this diagram is a very simplified depiction of how organizations operate when designing new products for end users, it does depict the DHM approach (#3) as only one means of addressing complex ergonomics issues during (and hopefully early in) the design process.

The diagram in Figure 2.1 assumes that the managers, system designers, and/or marketing personnel are sophisticated enough to recognize that there is the need in any system to specify user interface goals for performance and safety when considering a new product or service. Traditional design goals might include statements, such as: all of our future customers should be able to get in and out of our new vehicles without adverse discomfort or risk of injury, or all of our workers should be able to operate our new materials handling system without risk of overexertion injury, while also providing a significant improvement in productivity. It also is assumed that the organization has someone on staff, or has access to a consultant who is a competent human factors/ergonomics specialist. This person often is required to review initial, proposed design requirements, and then establish a precise specification regarding the physical, perceptual, and cognitive tasks that would be required when operating or maintaining the proposed new system to meet the stated human performance and safety goals for the system. Once the human task requirements are agreed upon by all concerned, then the ergonomics expert has some choices to make regarding the process to determine if and how the intended user population will be accommodated by the proposed design. The traditional approach (labeled #1 in fig. 2.1) is to consult the literature, including ergonomics guidelines, human factors databases, ergonomics and human factors standards and codes of practice, and so on. Fortunately today there are a number of web-based organizations that can be consulted to assist in finding sources of information about various population performance

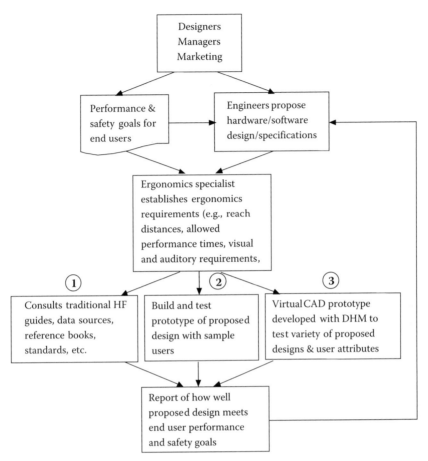

FIGURE 2.1 Three different approaches numbered 1, 2, and 3 are typically used to include ergonomics information when attempting to meet stated end-user human performance and safety goals. (Adapted from Gabriel, R. F., *What Engineers and Managers Need to Know about Human Factors*, SAE International, Warrendale, PA, 2003.)

capabilities and risk factors. When this approach is used, however, it becomes challenging to carefully interpret these empirical, discrete, and/or very general design recommendations, and to apply these correctly to a particular design under consideration. Most often this requires a very knowledgeable person or group of people who are true specialists in ergonomics and human factors. The fundamental problem in this traditional, literature-oriented approach was well stated by David Meister several years ago (Meister, 1993):

> Although there are many numbers in behavioral journals, these have not been incorporated in what I call a predictive database. We have descriptive databases … , but we lack a database to which the design engineer can relate, because behavioral databases are solely that, and are not linked to the physical aspects of design.

Commenting further on this issue as it relates to the design of aircraft, Smith and Hartzell (1993) stated:

> Unlike CFD (Computational Fluid Mechanics) and FEA (Finite Element Analysis), human factors engineering requires not one or two scientific disciplines, but over a dozen. Good human factors engineers must not only deal with the acoustic, life support, vision, anthropometry,

motor, and cognitive demands placed on the intended operators, but also must address input from avionics engineers related to the physical design, function and operation of the crew station.

Unfortunately, it appears we are graduating very few engineers (probably less than 10%) who even have a first course in human factors and ergonomics (Chaffin, 2005). This is despite surveys of practicing engineers and designers that indicate that they would like to know much more about ergonomics so they can include this information when proposing new designs (Broberg, 1997). This lack of understanding and sensitivity to potential human-hardware interface issues during the design of a new system may also mean that even the most fundamental questions about how operators and maintenance people will be affected by the design are not realized until the design either is finished completely, or worse yet, is being used by workers or consumers with complaints and/or injuries being reported.

The limitations noted above with the traditional literature-based approach have led many organizations over the past 60 years to require that several physical mockups of a proposed new design be built and tested by a sample of representative end users. This approach is designated #2 in Figure 2.1. Though such an empirical approach is conceptually good, it has some major limitations. First and foremost is the time and cost involved in designing and building testable prototypes. Given the need for organizations today to quickly develop and sell new and better products and services, if an organization is delayed for even a few months in reaching the marketplace compared to competitors, it may mean the failure of the product to sell as expected. Providing a new product or service that meets changing customer expectations, and being in the marketplace at the right time, are primary and daunting challenges for many organizations today.

Another limitation with using the prototype testing approach to ergonomics is that the sample of people chosen to test the prototype may not truly represent the intended end user group. Further, if the testing organization designated to conduct the prototype evaluation is not sufficiently trained in ergonomics and human factors, the types of statistical data that they report may not directly relate to the human performance and safety goals of interest to the organization. In reality, often physical prototypes are tested to meet engineering goals (e.g., can a new braking system on a proposed car stop the car in a given distance, or is the manufacturing equipment capable of running at a certain speed for a year without producing defects?). Comparable human performance goals would be: Can 99.9% of drivers actuate the brakes in a prescribed situation that assures the car will stop within a certain distance, or can a large, diverse population of trained workers set up, operate, and maintain the new machine in a way that it meets the quality and production standards expected by management? It is proposed in this chapter that it is these latter types of questions that must be answered early in the design process, and that reliance on DHMs (approach #3 in fig. 2.1) has a great deal of potential to supply such information to engineers and managers in a timely manner.

In general, what is being suggested in this handbook is that there needs to be a means for an engineer/designer to become immersed in probable human interface issues during the initial phases of a new design, not after much of the design has been frozen. In the past such an immersion in human behavior was very difficult to achieve without physical prototypes that engineers could try out, as well as observing others trying to perform various important tasks. But today there is the ability to depict both a proposed physical design and a fully enfleshed avatar in such a manner that the designer can simulate a large variety of human tasks. This technology that we refer to as digital human modeling, or simply DHM, has been enabled in the past 15 years by the advent of several supporting technologies, including: (1) greatly increased computing speeds that enable extremely complex and realistic 3D graphics to be shared among managers and engineers, (2) human-looking 3D avatars that represent diverse and statistically well defined anthropometric groups of people, and (3) greatly improved I/O structures within DHM software that allow designers/engineers to easily manipulate the actions of the avatar within common CAD environments.

2.3 What Conditions Today Are Influencing DHM Development?

From the preceding depiction of the design process used by many organizations, it should be apparent that there are several conditions that support more emphasis on DHM in the design of new systems. Some of these are:

- To address human interface issues that are being acknowledged by many marketing and safety groups as a primary reason for producing a new design. In this context, Gabriel (2003) discusses many of the benefits listed in Table 2.1 that result from the inclusion of human factors in new designs.
- To provide shorter design concept-to-market times that are often required to meet rapidly changing market conditions.
- To reduce prototype building and testing costs and time associated with customer-specified human performance and safety requirements.

Digital human modeling can help organizations meet these goals, provided the underlying human models are robust and accurate. The DHM approach to ergonomics problem solving also can provide the following additional benefits.

Because DHMs can be used as part of existing CAD and CAE systems, their interactive nature allows designers and engineers to explore what-if scenarios with immediate feedback about the potential ergonomics outcomes as part of their normal design and engineering activities. This is potentially one of the largest benefits to an engineer or designer for two reasons.

First, by linking ergonomic models of population capabilities (e.g., strengths, range-of-joint-motion, visual acuity, hearing acuity, postural discomfort levels, normative motion trajectories, response times, etc.) with the DHM avatar, the designer is provided a graphical means to make first-order approximations about important ergonomic outcomes related to specific design attributes of interest.

TABLE 2.1 Potential Benefits Often Stated from Adoption of Pro-Active Ergonomics Early in Designs

Reduced Frequency and Severity of Accidents
Improved Job Performance
Fewer Days Lost to Injuries
Improved User Acceptance
Improved Efficiency
Improved Effectiveness
Reduced Design and Manufacturing Costs
Increased Standardization
Shorter Training Time
Enhanced Marketing
Improved Public Relations
Better Readiness
Faster Turnaround Time
Protection from Litigation
Greater Job Satisfaction
Less Employee Turnover
Less User Fatigue

Source: Adapted from Gabriel, 2003.

Second, because a DHM often requires only a few seconds of CPU time, it becomes an interactive ergonomics teaching aid to improve an engineer/designer's repertoire of tools that can be used in future projects, thus addressing the needs expressed by many designers to learn more about ergonomics to improve their future designs (Broberg, 1997).

Digital human modeling during design also meets the higher virtuous desire to not harm our fellow human beings, but rather to improve human conditions for future generations. In other words, it allows designers to become more human-centered when creating a new product or service.

2.4 What Organizational and Technical Conditions May Be Inhibiting the Faster Adoption of DHM Technologies?

There are several different circumstances that appear to be inhibiting a faster rate of adoption of digital human modeling by various organizations.

Lack of trained DHM personnel. This is a large problem. As mentioned earlier in this chapter, if less than 10% of engineers and designers have even one course in human factors or ergonomics, organizations may not be able to find qualified ergonomics professionals, who also are capable of both designing hardware and software systems and correctly using a particular DHM tool. At present this author knows of only a few engineering or design programs that require graduates to be competent in the use of even one of the DHM software products that are now available.

The DHM tool that would best be able to help an engineer or designer to solve a particular human performance or safety problem does not interface well within the existing CAD tools used in an organization. This is a major issue for many organizations. It can mean that graphic and human simulation data files may not work in a seamless manner, requiring time-consuming manual entry of relevant data before performing even elementary human simulations.

The existing DHM tools are limited in their ability to simulate the infinite number of complex human motions and behaviors that may comprise a particular task of interest. Though human motion prediction modeling has been a major research topic within various academic circles for the past 10 years, as described in papers by Gleicher (1997), Faraway (1997), Allard et al. (1998), Zatsiorsky (1998), Nigg et al. (2000), Zhang and Chaffin (2000), Jagacinski and Flach (2003), and Park et al. (2006), having robust models imbedded in DHM applications that are capable of predicting many different and common types of motions for different groups of people is another matter (Chaffin, 2002).

To simulate many ergonomic conditions accurately requires a certain amount of specific human performance and safety requirement data to be provided as input to the DHM. An example could be: to perform a manual task successfully (e.g., actuate a parking brake), a 30N force is required in a certain direction. Unfortunately, human performance requirements may not be documented by engineers and designers in their initial designs unless a human factors specialist is involved (as is assumed in the earlier fig. 2.1). The lack of critical, human requirement information early in a design means that the user of a DHM application must guess at a requirement to simulate the capabilities of a group of people. If provided enough time, the designer/engineer may then run repeated simulations with varied estimates to test the sensitivity of the initial guess. This procedure not only takes additional simulation time, but it introduces variance and uncertainty in the results of the simulations. Such variance is used by some managers as a reason for not attempting to perform DHM simulations early in the design process, but rather delay until prototypes are constructed, and sample groups of people can be used to improve the human performance estimates. Stephens (2006) acknowledges this situation and proposes that it is often necessary to use the DHM along with physical mockups and a motion capture system to refine initial designs.

Though merging human performance and risk prediction models with DHM avatars has occurred in various ways over the years, it is very important that this coupling be done in a manner that is seamless to the user. At the present time, most DHM applications require that the human performance model of interest be called from an external file after the DHM avatar has been used to simulate potential postures or motions in a given CAD rendered environment. There are two problems with this process.

First, the motions and postures that were initially simulated may need to be altered to achieve the human performance goals being modeled. For instance, if the motions and postures being simulated involved the movement of a heavy object, but the hand force requirements were not included in the initial motion simulations, subsequent biomechanical analyses will be in error because of the well-documented and strong dependency between postures and strength capabilities (Chaffin & Erig, 1991).

The user may not fully realize that there exists a strong dependency between the motion or posture predictions in a simulation, and the human performance model outputs. This is true for perceptual models as well as biomechanical models. For instance, a simulated whole-body motion of a reaching activity may place the head in a certain location in space without considering the ability of a person to see an important visual target. Since such dependencies are common, many DHM simulations of complex tasks will be in error if these are not acknowledged and managed well by the user.

There is a lack of human performance and risk prediction models that are valid and accurate for a diverse set of task simulations and demographic groups. The reality is that predicting the capabilities of specific groups of people when performing a variety of physical, perceptual, and cognitive tasks is very difficult to achieve. Many excellent research groups have been performing studies to produce such models, as described in many of the chapters to follow. Considering the variety of tasks that people perform in their daily activities, and the diverse population groups that may be of interest, developing valid and robust models is challenging to say the least. This means that at best the present DHM models that access existing human performance and risk prediction models allow only first approximations of human behavior and capabilities, with error estimates that can be very large. This lack of precision in predicting a specific outcome, when acknowledged by DHM developers, can produce a lack of confidence and interest in adopting the general technology, even though some human activities may be well simulated.

2.5 What Are Some of the Technical Challenges to Improve Future DHM Simulations?

The preceding sections of this chapter have alluded to several important technical challenges to those developing and using DHM simulation methods. The following delineates these further.

Human Model Accuracy. The accuracy of DHM simulations when attempting to predict what people can perform safely must be much better studied and documented to allow users to understand the limitations of the underlying models. Today there is an effort to improve human avatar graphics and movement simulation, but this is not sufficient for future DHM applications. Human motion accuracy is a minimum expectation for a good DHM model these days, but this is only sufficient if one is interested in how well certain population groups "fit" into a specified environment. If the user wishes to know how much time might be required to perform a specific task, or whether the person will suffer a certain amount of discomfort for instance, then posture and motion prediction accuracy is only part of what is needed. Much better models are needed for predicting: performance time, discomfort, fatigue, mental workload, perceptual acuity, biomechanical tissue stresses, and other human performance attributes. Some of these human attribute models exist today and are described later in this book, and some are even directly accessible by existing DHM simulation methods, but they often have limited validity or apply to a small group of people or tasks, thus restricting their general use when designing a system to accommodate large numbers of diverse users.

Coupling of DHM Avatar with CAD. Coupling the DHM avatar with the physical environment is still very challenging. For instance, how do people grasp and move objects? Simulating the movement of a full coffee cup or a box that is full of breakable objects is quite different from moving a block of cement.

The differences begin with how the hand grasps an object. Once this is set, the normal motion trajectory for the hand and object must be predicted. Of course this may be different if the trajectory is impeded. Consideration also must be given as to how the object may be oriented when it arrives at its destination. What happens if there is limited clearance for the object, hand, forearm, arm, shoulder, torso, and so on, at some point in the motion? Does the person need to keep the object in sight throughout the motion, or only at the beginning or end of a motion? How does the motion change if the object is heavy? Some of these questions can be addressed by improved human motion prediction methods, but at times the user of the DHM must be provided a means to easily alter the tasks and environment to determine the affects on ergonomic indices of interest (Badler et al., 2005).

Integration of Avatar and Human Performance Models. Achieving a seamless integration of graphical information about a population (as represented by avatars within the CAD environment), with a large variety of valid and accurate human performance prediction models, is very important. The speed and memory capacity of present computers can allow many different computational models to run simultaneously. When a user of a DHM needs to know whether a certain size person can see an object, or push or pull on a lever, or decide among several alternative tasks, the DHM should be able to respond appropriately and quickly, without having to ask for a great deal of additional information. Such integration must include not only biomechanical and physiological models of population capabilities, but also perceptual and cognitive load prediction models. For instance, it may be enough to have the DHM simulate the motions of various drivers when reaching from the steering wheel to the radio, but only if the designer of a vehicle is interested in how many people can easily perform such a reach. On the other hand, what would be the case if the designer also needed to know how much cognitive load and possible distraction may occur when a simulated person is reaching to tune the radio or an iPod using different types of controls and displays? The integration of cognitive and physical models of people should provide such insights, but today such model integration is just starting to be discussed.

Task Planning. Providing high-level task descriptors as inputs to a DHM is very important in achieving wider acceptability and use of the technology among disparate user communities. Such high-level command inputs should allow users who are not human factors specialists to effectively simulate complex human-machine interactions, and have results that are consistent and accurate. Discussions of how some of the emerging task planning algorithms can work are presented elsewhere in this handbook. An example of a high-level task specification might be: "Have the avatar step over to shelf A, get part number 45, and attach it to subassembly 22 using the nut runner." As one can readily imagine, there are an infinite number of motion scenarios that might be simulated to accomplish this type of a task. The DHM must be capable of quickly simulating a number of feasible motion scenarios, and then select and display a few that are most likely, based on human motion data and models of normative behaviors of people doing similar tasks. In the process of making such selections, issues such as the following need to be considered: Are one or both hands needed to grasp and carry the part and position it on the subassembly? Does the worker need to see how the part must be oriented to attach it? How much torque needs to be provided to fasten the part? Where are the fasteners? If the DHM is to address these types of issues, it may well require a biomechanical analysis of the strength of the population in grasping, lifting, and carrying the part in different ways, an anthropometric model of different sized workers to determine reach and sight lines, a visual acuity model to assure that the lighting is adequate to see how the part fits onto the subassembly, and a cognitive model to predict the mental loads and response times associated with various methods of performing the task within the required standard time allotted to complete the job. The DHM also has to decide if additional information/data are needed and query the user for this, depending on the goals of the simulation. For instance, if the designer is concerned about limiting injurious stresses on the upper extremity or low back in the preceding example, the peak hand forces needed to grasp, lift, and carry the object may be needed, as well as the hand forces to operate the nut runner. A future DHM algorithm must be able to make initial estimates and then ask for more information if the design goals are not adequately met.

2.6 Summary

This chapter attempts to make the argument that there are three different but inter-related issues that have caused DHM technologies to be slowly adopted by various engineering and design organizations over the last 35 years. First is the existence of organizational structures that often do not recognize the need for human factors and ergonomics expertise early in the design process. If the need is recognized, and this is happening now more than in the past, then these organizations need to hire ergonomics and human factors specialists that have cross-training and experience in engineering and design. This leads to the second issue: Academic organizations are not providing many engineers and designers with even elementary human factors and ergonomics training. The third and perhaps the most relevant issue in the context of this text is the fact that DHM simulations are not very robust and valid in predicting, for many different common tasks, what populations of workers or consumers can effectively perform, and with a specified degree of safety.

This presentation lists a number of social and organizational conditions that would appear to support the rapid future expansion of DHM methods and models to meet ergonomic goals. Among these are factors such as the continuing shortening of "concept-to-market times," the realization that human-hardware interfaces are very important, the desire to reduce the very high cost of building and testing physical prototypes, and the need to provide a computational tool to assist engineers and designers in their natural desire to create designs that meet people's expectations and capabilities. It is argued that even if an organization has good access to a skilled human factors or ergonomics expert, this may not be sufficient to assure that the right type of human-hardware interface information is provided to engineers and designers in a timely manner. Further, it is proposed that the DHM approach to design is very powerful in that the designer or engineer can get immersed in various ergonomics issues, and by running what-if scenarios, actively learn how people are affected by proposed design specifications.

Unfortunately, though it would appear that many organizations are ready to use DHM in design, the models that exist today and are described in this handbook may not meet the expectations of an organization. Some of this relates to the inability of these DHM systems to accurately and quickly portray how a variety of people might perform a job or complex perceptual-motor task. Such lack of validity in the underlying models used in the DHM provides the basis for much of the skepticism often expressed by upper management. Compounding this problem is the fact that the human performance prediction models and the DHM avatar models may not be highly integrated, causing slow and laborious transfers of data, which can result in delayed reports and increased analysis errors.

The good news is that many different research groups are working to produce more accurate and robust human performance models. Also, the computational speed and memory available in today's computers enable a number of different types of human performance and DHM avatar models to run simultaneously, providing more seamless transfer of geometric, human, and task data, as well as sharing common operational instructions that control the models when running simulations. There also is great progress being shown in combining lower level control of the avatar and human performance models by high-level task planning algorithms, though much remains to be developed and tested in this effort.

As is described in many of the chapters to follow, though human performance prediction models and avatar control models are becoming very sophisticated, there is still much to be done. Not only must the human performance prediction models be improved and validated, but these must be integrated into commonly used CAD systems in a manner that will assure the use of the resulting DHM technology by many different designers and engineers, who for the most part control the environment in which we work and play.

References

Allard, P., A. Cappozzo, A. Lundberg, and C. L. Vaughan, eds. 1998. *Three-dimensional analysis of human locomotion*. West Sussex: J. Wiley and Sons Ltd.

Badler, N. I., J. Allbeck, S. J. Lee, R. J. Rabbitz, T. T. Broderick, and K. M. Mulkern. 2005. New behavioral paradigms for virtual human models. *SAE DHM Conference,* Iowa City.

Badler, N. I., C. B. Phillips, and B. L. Webber. 1993. *Simulating humans: Computer graphics animation and control.* New York: Oxford University Press.

Broberg, O. 1997. Integrating ergonomics into the product development process. *Int J Ind Ergon* 19: 317–27.

Case, K., J. M. Porter, and M. C. Bonney. 1990. SAMMIE: A man and workplace modeling system. In *Computer-aided ergonomics,* ed. W. Karwowski, A. M. Genaidy, and S. S. Asfour, 47, London: Taylor & Francis.

Chaffin, D. B., ed. 2001. *Digital human modeling for vehicle and workplace design.* Warrendale, PA: Society of Automotive Engineers.

Chaffin, D. B. 2002. On simulating human reach motions for ergonomics analyses. *HFEM,* 12(3): 1–13.

Chaffin, D. B. 2005. Engineers with HFE Education—Survey Results, *ETG Newsletter of the Human Factors and Ergonomics Society,* 3: 2–3.

Chaffin, D. B., and M. Erig. 1991. Three-dimensional biomechanical static strength prediction model sensitivity to postural and anthropometric inaccuracies. *IIE Trans* 23(3): 215–27.

Chaffin, D. B., K. E. Kilpatrick, and W. M. Hancock. 1970. A computer-assisted manual work-design model. *AIIE Transactions* 2(4): 347–54.

Evans, S. M., and D. B. Chaffin. 1985. A method for integrating ergonomic operator information within the manufacturing design process. *Proceedings of the IIE Conference* (December), Chicago, Illinois.

Faraway, J. J. 1997. Regression analysis for a functional response. *Technometrics* 39(3): 254–61.

Gabriel, Richard F. 2003. *What engineers and managers need to know about human factors.* Warrendale, PA: SAE International.

Gleicher, M. 1997. Motion editing with spacetime constraints. *Proceedings of the 1997 symposium on Interactive 3D graphics,* Providence, Rhode Island, 139ff.

Jagacinski, R. J., and J. M. Flach. 2003. *Control theory for humans: Quantitative approaches to modeling performance.* Mahwah, NJ: Lawrence Erlbaum.

McDaniel, J. W. 1990. Models for ergonomic analysis and design: COMBIMAN and CREWCHIEF. In *Computer-aided ergonomics,* ed. W. Karwowsi, A. M. Genaidy, and S. S. Asfour, 138–56. London: Taylor & Francis.

Meister, David. 1993. Non-technical influences on human factors. *CSERIAC Gateway* 4(1): 1–2.

Nigg, B. M., B. R. Macintosh, and J. Mester, eds. 2000. *Biomechanics and biology of movement.* Champaign, IL: Human Kinetics.

Park, W., D. B. Chaffin, B. J. Martin, and J. J. Faraway. 2006. A computer algorithm for representing spatial–temporal structure of human motion and a motion generalization method. *J Biomech* 38: 2321–29.

Peacock, B., and W. Karwowski. 1993. *Automotive ergonomics.* Washington, DC: Taylor & Francis.

Ryan, P. W., and W. E. Springer. 1969. Cockpit geometry evaluation final report. Vol. V, *JANAIR Report 69105,* Washington, DC: Office of Naval Research.

Smith, B. R., and E. J. Hartzell. 1993. A³I: Building the MIDAS touch for model-based crew station design. *Gateway* 4(3): 13–14.

Stephens, A. M. 2006. The truck that Jack built: digital human models and their role in the design of work cells and product design. *SAE International Conference, Digital Human Modeling for Design and Engineering,* July 2–6, Lyon, France.

Zatsiorsky, V. M. 1998. *Kinematics of human motion.* Champaign, IL: Human Kinetics.

Zhang, X., and D. Chaffin. 2000. A three-dimensional dynamic posture prediction model for simulating in-vehicle seated reaching movements: development and validation. *Ergonomics* 43(9): 1314–30.

3

A Scientific Perspective of Digital Human Models: Past, Present, and Future

Heiner Bubb and
Florian Fritzsche

3.1 Introduction

The oldest credible evidence about the existence of human beings is the discovery of prehistoric artifacts, resembling either tools or the artistic expression of human shapes. Therefore, many scientists think humans have an essential need to re-create our own image. In fact, we can find impressive examples of these kinds of activities in every culture. "The human being is the measure of all things" can be seen as a core principle of the Greek/European worldview. Since antiquity, especially in arts and sciences, we have tried to answer the question of ideal proportions concerning the human body. Well known is the "theory of proportions" presented by Vitruv (1st century AD), which was a basis of the anthropometric studies of Leonardo da Vinci (1485/90, see fig. 3.1). Today's art schools continue these deliberations and even modern computer dummies are influenced by them (e.g., RAMSIS; Geuss, 1995). Not only by way of the external shape, but in other properties and abilities as well, humans try to reproduce themselves. Well-known are the humanoid automatons in the 18th and 19th century (e.g., the mechanical flautist of Jacques de Vaucanson, 1737); and the human mind has been an object of reproduction attempts (i.e., the first automatic calculators of Schickard, 1623; Pascal, 1645; Leibniz, 1673; Babbage, 1822, 1833; and Zuse, 1937). The modern android robots combine a humanoid shape with the cognitive abilities of the

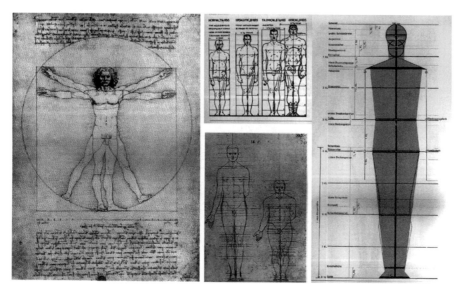

FIGURE 3.1 Vitruv's canon of human proportions after da Vinci, about 1490, and further development in the area of artists.

computer. For example, in 1973 a long-term research program began at Waseda University, focusing on the robot Wabot-1. Starting in 1986, Honda worked at the so-called E-Series, from which the Asimo was created. Recently, a common field of international research has been the development of android robots with the ability to walk (e.g., Johnnie of the Technical University of Munich; Lohmeier et al., 2002; see fig. 3.2).

All of these artistic, philosophical, and scientific efforts have one common goal: The human mind wants to gain knowledge about the core of its own essence. It uses different ways to approach this goal, but in the end, this ambition has no immediate purpose. The human mind simply wants to know.

The way we perceive our modern world now has changed tremendously. Since the Age of Enlightenment (which was mostly influenced by the discoveries of Galileo, the philosopher Descartes, and changes in society due to the beginning of the industrial revolution), everything had to be set in numbers. Therefore, it became increasingly important to measure everything about human life. It is well known that this new point of view is not only the basis for the emerging of an industrial culture but also for the development of a modern nation. Besides a lot of other new developments, most nations introduced compulsory military service, making it necessary to systematically measure the body height and weight of young men. Consequently, we have sound knowledge of average body heights, weights, and proportions today. The beginning of industrial mass production required detailed knowledge about the human body in measures and proportions. Therefore, the field of anthropometry began. It can be seen as a new branch of the original scientific discipline of anthropology, which studied the human being as a whole. Essentially independent from anthropometry, the so-called constitution research was developed especially in the psychology field. At first, based on rather vague visual estimations of body proportions, those were defined as pyknomorph/hyperplastic and endomorph/mesomorph. Not only were different nutritional behaviors connected with these body shape types (Huter, 1880, in Conrad, 1963) but also psychological behavior was determined. Kretschmer (1961, in Conrad, 1963) tried to predict the suitability of an individual for specific jobs based on these findings. Upon the foundation of the metric-index of Strömgren (1937), Conrad (1963) developed a two-dimensional schematic with the axis hypo-/hyperplastic and pykno-/leptomorph, in which a localization and categorization of an individual subject was possible by the use of anthropometric data.

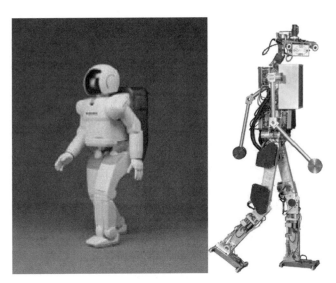

FIGURE 3.2   Examples for android robots. Left: Asimo of Honda. Right: Johnnie of Technical University of Munich.

A branch of precise and distinct, natural-science-oriented anthropometry, especially the measurement equipment of Martin (1914), must be mentioned at this point: It prefers the distance measurements between well-defined bone points. Measurements of human body proportions were carried out around the world on the basis of this or similar methods. The results were published in tables, which contained the corresponding measures classified according to the so-called percentile. As real-world applications of these percentile tables were not successful, templates were developed to represent the measures of these tables as adequately as possible. The SAE template is of extraordinary importance, especially in connection with automotive developments. In Germany the so-called Kiel mannequin was developed, which is available as a male and female template of various percentiles (see fig. 3.3).

A problem of "static" paper is its limitations in illustrating movement. Therefore, moving areas are defined. Of particular importance are the gripping spaces, which are published for different percentiles and depend on gender. For example, in SAE J 287, the gripping space for the driver of a car is defined (see fig. 3.4).

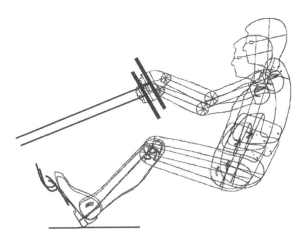

FIGURE 3.3   Kiel-mannequin after DIN 33 416.

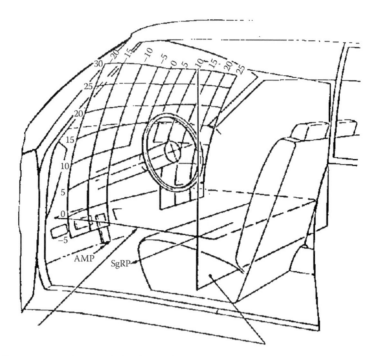

FIGURE 3.4   Gripping space for cars defined according to SAE J 287.

All these movement areas must be related to reference points such as the shoulder point, the hip point, and the so-called "Seat Reference Point" (SRP). A serious problem is how to find these reference points in reality. Particularly difficult scenarios arise in connection with upholstered seats. For this reason, a lot of measuring tools were developed just to find the SRP. Of great importance, especially in conjunction with the development of vehicles, is the SAE-SRP-machine. It was recently modernized and equipped with a representation of the human back. It defines the so-called H-point (hip point), which serves as the basis for the construction of the eye-ellipse as well as the above-mentioned gripping space. For further design processes, the SAE template is fitted with the H-point. Therefore, the position of the template in the surroundings is defined. The Kiel-mannequin yields more answers to more detailed and difficult questions. Being quite similar for tasks concerning larger body shapes, the Kiel-mannequin is used instead of the SAE template. Especially in Germany, it is still used in the area of car development, in addition to its use as computer-animated, so-called soft dummies.

As a part of anthropometry, the scientific measurement and description of body forces must also be mentioned. In respect to the wide distribution of physical abilities, it makes sense to create percentile specifications of maximum forces. Without going into detail about the problems with defining maximum forces (see Rohmert, 1993), two fundamentally different approaches may be mentioned here: On one side, the force specifications are given as body related. This means that the force is related to a reference point (e.g., shoulder or hip point, etc.); the result in this case can be presented (depending on the body posture) in the form of so-called isodyns (e.g., Rohmert, 1966). On the other side, there are externally applied forces during different kinds of activities (e.g., turning a wheel, locking a lever, pushing a pedal, and so on) to be measured (Rühmann & Schmidtke, 1991). Of course, not every conceivable application can be considered during such investigations. In order to compensate for this deficiency, Burandt (1978) developed a body-related prediction procedure, which was originally published as tables and calculation specifications. It allows the prediction of maximum forces to be applied permanently for a large area of applications depending on gender, age, and state of training respectively.

In the 1960s, new possibilities became available due to advancements in computer technology. Computer technology was applied to represent the complexity of the human appearance in a totally new manner. Also, the different kinds of interactions between humans and their surroundings could be simulated more realistically than ever before. However, the available data and the knowledge based on existing experiments were not sufficient to immediately satisfy all demands for human models. At first the existing paper versions of human models were simply transformed into computer programs. In this way, computerized variants of the SAE template, the Kiel-mannequin, and the like were created.

In the early 1960s, true human computer models started to be developed. The limitations of templates with regard to physiological movement simulation, anthropometric scalability, and the correct computation of anthropometric data correlations in the last few years have been more or less eliminated by modern computer technology. Since the 1980s, modern 3D CAD technologies have been developed and established—and this has led to a need for 3D aids to design a 3D human model. The term "digital human model" (DHM) was established for computer-based models of human appearance and behavior.

3.2 An Overview of Historic Developments

3.2.1 Developments in North America

The first real development of human computer models had its origin in the American air and space research field. As early as in 1960, a human model was developed by Simon and Garner in order to simulate the mass inertia within a zero-gravity environment. This first model was improved continuously and eventually led to Hanavan, which then was the starting point for further developments. In the following text, the very wide development in this area of modeling will only be attended to briefly. A very comprehensive development in this area is the CVS system (Crash Victim Simulator), which started in 1963 at the Cornell Aeronautic Laboratories (CAL) in the United States as a 2D system primarily for aviation applications. In 1966, the system was adapted for the automotive industry, initiated by the "Alliance of Automobile Manufacturers" (AAM) in the form of ROS. Between 1967 and 1974, the system was under further development by General Motors for the simulation of accidents. Simultaneously, the Calspan Corp., which emerged from CAL, developed a 2D approach to a real 3D-simulation system (the software system CALSPAN-3D-CVS), which allowed cinematic analysis under dynamic conditions. This system was applied to analyze safety systems (such as airbags, belts, and so forth) during collisions between cars, pedestrians, and motorcycles (Hickey et al., 1985).

Aside from this branch of dynamic modeling, simultaneously a branch of anthropometric models was established. In 1967, William A. Fetter, a Boeing employee, developed the anthropometrical computer model First Man that could substitute for templates (see fig. 3.5).

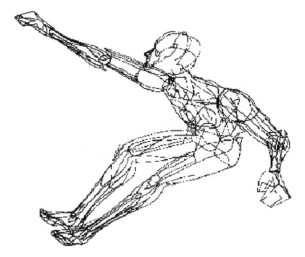

FIGURE 3.5 The man model First Man.

Based upon the anthropometric dimensions of a scalable 50th percentile man, it was used to check the reach and accessibility in aircraft cockpits. Between 1969 and 1977, based on this development, the Second Man, Third Man, and Fourth Man followed. As a parallel project, Boeing presented in 1970 the man model Boeman-I. The development had started in 1968 already and was aimed at a comprehensive simulation of the human body. This goal was supposed to have been reached within the CGE project in six steps within one year. However, after the third year, the project was stopped. The result was Boeman-II. After stopping the CGE project, the software was sold to the Aerospace Medical Research Laboratory where the improved Combiman was presented in 1972–1973. Base upon the results of the Boeman development, Boeing used this know-how to develop the CAR program (Crewstation Assessment of Reach) in 1976. Applied in this model was the first effort to compensate for the incorrect addition of percentile values, where software designs of human models were inaccurate until then. In the actual version of CAR-IV, the anthropometric data are generated by a Monte-Carlo simulation on the basis of 400 fictitious individuals. This model was repeatedly evaluated by scientific institutions and now it represents a valuable tool for the industry, especially for analysis.

In 1978 the Chrysler Corporation started developing the man model CYBERMAN (Cybernetic Man-Model). This model is a 3D wire model consisting of 3D splines and ellipses. The emphasis of its application lies in the design of dashboards, seats, and the geometry of seat belts. Although CYBERMAN was on the market for a certain time, it was not distributed widely (Dooley, 1982).

Also in 1978, N. I. Badler started the continuous development of man models. BUBBLEMAN was a simplified volume model consisting of balls, which was used to analyze moving processes. The knowledge gained from this system was transferred into the model TEMPUS in 1985, which was developed for NASA at the Johnson Space Center. The anthropometric modeling was based upon the program CAR, but the main focus in its development was the simulation of movements. It already allowed for extensive animation of complex movements in real time, like sitting on a chair, grabbing a flight stick, and such. TEMPUS is the predecessor for the modern, worldwide man model JACK (see fig. 3.6). JACK is not a genuine ergonomic tool. By opening interfaces, however, users have the opportunity to integrate their own methods and procedures and then visualize those with JACK.

The main field of application for the mannequin JACK is animation and visualization in vehicle design and architecture. The model consists of 39 body elements; visualization takes place using area segment imagery with textures. The anthropometric database is derived from the Human Solutions Library, NASA data, and from in-house studies, surveys, and tests. Posture and movement

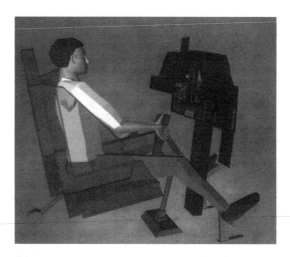

FIGURE 3.6 The man model JACK.

simulation enables interactive movement of the entire mannequin in real time. Animation is based on methods used in robotics but without the priority of smooth, realistic movements. A variety of modules are available in the form of analysis tools for factory planning and vehicle development. The system is available either as a stand-alone version or integrated in Unigraphics. JACK is used by John Deere, British Aerospace, Caterpillar, Volvo, and General Dynamics as well as various universities.

During the 1980s, the computer man model SAFEWORK (see fig. 3.7) was developed at the Ecole Polytechnique in Montreal. It covers three modules: anthropometry, movement, and analysis. Additionally it offers the possibility of displaying surroundings. SAFEWORK was conceived for workplace designs in factory planning as well as in product designing. The anthropometric data-base is based on U.S. Army data and encompasses an anthropometric body type generator that can create statistical test samples. The posture and movement simulation enables the movement of short chains in the model's arms, legs, and torso by means of inverse kinematics. Vision simulation, fixed accessibility areas, a joint-dependent comfort evaluation, maximum force calculation, and a center of gravity analysis are available as analysis tools. SAFEWORK is available solely in the form of an integration unit with CATIA by DassaultSystems. SAFEWORK is used by Chrysler, Boeing, and various universities and academies.

The newest development is the project Virtual Soldier, initiated and supported partly by the U.S. Army TACOM project (Digital Humans and Virtual Reality for Future Combat Systems (FCS)) and by the Caterpillar Inc. project Digital Human Modeling for Safety and Serviceability. It is called Santos, and, according to Abdel-Malek and colleagues (2006; see fig. 3.8), it promises to be the next generation of virtual humans. It is an anatomically correct human model with more than 100 degrees of free-dom. The ongoing project is to develop a system in which an avatar's motion is controlled by a variety of human performance measures. Those are incorporated in a unique, optimization-based approach to posture and motion prediction. Focus was set to an accurate musculoskeletal system, incorporated within the virtual human. The system will allow for real-time interaction and analysis of muscle forces within a package that is designed for widespread use. Also included in this package is a hand model with a 25 degree of freedom movement range.

FIGURE 3.7 The man model SAFEWORK.

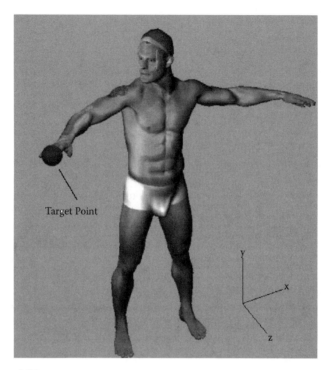

FIGURE 3.8   Men-model Santos.

3.2.2 European Developments

The European developments basically followed the same directions as their American colleagues. At the University of Nottingham, the development of the computer man model SAMMIE (System for Aiding Man-Machine Interaction Evaluation, see fig. 3.9) was started in 1967. This system consisted of 3D polygons. The 5th, 50th, and 95th percentile male model was available. Change of surface measures due

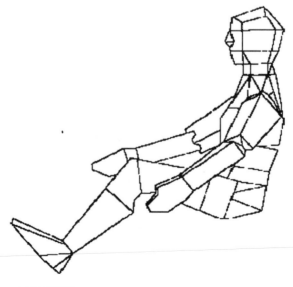

FIGURE 3.9   The man model SAMMIE.

to clothing could be considered. The body posture was generated by the definition of target points for hands and feet. A detailed simulation of the line-of-sight originating from three key points (left eye, right eye, and the so-called Cyclops point of view) was possible. Additionally, SAMMIE allowed for the evaluation of mirror sight with consideration of plane, concave, and convex mirror surfaces. SAMMIE was the first computer man model that was commercially available worldwide. It was used for the development of industrial working places, automobiles, and utility vehicles.

In the early 1980s, an extensive development of computer models started in many different institutions throughout continental Europe. In France, based upon the international databank "Ergodata," at the Laboratoire d'Anthropologie Appliquée et d'Ecole Humaine in Paris, the man model ERGO-MAN was created in 1984. It is a 3D wire-, plane- (three-angular), and volume-model, which allows for analysis of moving areas for the arms and legs. Additionally, detailed studies for fingers and the spinal column are possible. The software platform serves the CAD system EUCLID-IS. Besides the analysis of car interiors, further applications existed in the aerospace industry (HERMES) and the railway industry (Laboratoire d'Anthropologie Appliquée et d'Ecole Humaine, 1988; Hickey et al., 1985). Another development was done by INRETS, in cooperation with the French automotive industry, which led to the non-commercially available man model Man3D. It is especially used to calculate gripping areas in cars.

In Germany, in the mid-1980s, within the scope of a BMFT-program came the creation of computer model FRANKY. Based upon the data and the ideas that led to the Jenik-Bosch templates at the TH-Darmstadt, the system HEINER was developed. At the same time, development of system ANYBODY began by Lippmann. The last model to be named is OSCAR, a development of a Hungarian management-consulting firm. ANYBODY was a 3D static model (available as a man, woman, and child model) consisting of approximately 1,000 points, connected by splines. The 53 body elements could be animated by the input of joint angles. The model had a very flexible percentile-oriented databank. Sight simulation was possible as well. The system was totally integrated in the CAD software Cadkey from CADStar. The successor ANTHROPOS was presented in 1989 (see fig. 3.10). It was totally re-designed as a volume model with a lot of additional functions. Now aim-oriented movements were possible and a very detailed spinal column could be integrated into the animation.

FIGURE 3.10   The man model ANTHROPOS (1989).

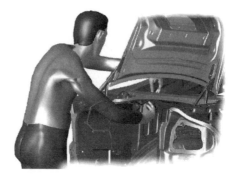

FIGURE 3.11 The man model ANYSIM. Left: in the original form (1993). Right: in the actual RAMSIS integrated form, now called eM-Human.

Whereas in the United States man modeling normally aims at a reach functionality, in Germany the development concentrated on functional specialities. Systems were developed, which answered special questions concerning high precision. An example is ERGOMAS, respectively ERGOMAN, which were especially created for the design of workplaces and production lines. A further example for this purpose is ANYSIM (see fig. 3.11), a development of iwb (Technical University of Munich), on which the system CHAMP (Siemens AG) is based; it is a powerful tool for work planning when integrated with MTM analysis. This model is under continuous development. Meanwhile, this system is offered commercially by Tecnomatix. The man models are substituted by RAMSIS (see below) and are now commercialized under the name eM-Human.

A development with a special rating is the RAMSIS mannequin (Realistic Anthropological Mathematical System for Interior Comfort Simulation, see fig. 3.12). It was developed to aid in the design of ergonomic interiors of vehicles between 1987 and 1994 under contract to, and in cooperation with, the entire German automotive industry, the company TECMATH (now Human Solution), and the Institute of Ergonomics at the Technical University of Munich. This is undoubtedly the most extensive industrial project ever done for the development of a mannequin. Apart from the CGE project, this was one of the most powerful industrial projects in regards to developing a computer man model. In order to eliminate the lack of transformability of literature data to man models, a closed model-adequate process sequence

FIGURE 3.12 The man model RAMSIS (1995).

from the data investigation all the way up to the simulation of body types and postures was developed and realized. Systems for contactless measurements of anthropometric data and posture measures, as well as comfort analysis, were developed. The program system RAMSIS offers the advantage to define a certain task only once; the evaluation with different body types works automatically, whereas the system provides the feature to automatically adjust areas and vary sizes. Based upon detailed investigations, the integrated comfort analysis allows us to compare different design variants and to optimize these in respect to comfort. After an extensive evaluation phase in 1995, the system was commercially offered worldwide.

The mannequin RAMSIS was developed for vehicle design. The anthropometric database encompasses a physical typology that takes body measurement correlation, desired percentile models, an international database with more than 10 global regions, and a secular growth model into account. Posture and movement simulation is carried out by a task-related animation, enabling a forecasting of the most probable posture. More than 80 functions for the analysis of vehicles and vehicle interiors are available (vision and mirror simulation, seat simulation, accessibility limits, comfort, belt analyses, etc.). The system is available as a stand-alone version for UNIX and Windows, as a CAD integration unit with CATIA, and as a programming library for independent applications. RAMSIS is used by more than 75% of all car manufacturers (Audi, BMW, Mercedes-Benz, General Motors, Ford, Porsche, Volkswagen, Seat, Honda, Mazda ...), by aircraft manufacturers (Airbus), and by manufacturers of heavy machinery (e.g., Bomag).

A very new development is ANYBODY, presented by Rasmussen et al. (2003). The aim of this model is not primarily on the external appearance, but on an accurate modeling of the muscles and their connection to the skeleton (see fig. 3.13). Also in Europe, dynamic models have been developed. In 1987, for the calculation system ADAM (a multi-body simulation system common in the automotive industry) a processor was developed in order to simulate the changed posture due to gravitational behavior of men in accelerated systems as well as to calculate the arising forces and moments in such. The model, consisting of 15 body elements, had the correct dimensions of a man, a woman, and a child. It used the data system of NASA (Waldhier, 1989). In 1988 the firm TNO in The Netherlands developed the dynamic crash analyzing system MADYMO-3D (Mathematical Dynamic Models; see fig. 3.14) on the basis of the 2D MADYMO model, which was very common in the European automotive industry. It does not directly simulate human properties but the physical properties of the real crash dummy HYBRID III. Apart from calculations about the accelerations and forces during a crash, the connection to finite-element systems (e.g., PISCES-3DELK) enables a simulation of a collision with finite element method (FEM) structures like, for instance, the collision with an airbag.

3.3 Scanner-Based Modeling

With the possibilities of modern computer technology, anthropometric measuring became automated as well. In this field we separate data-capturing methods based on optical procedures (silhouette procedure and laser scanning) and on measurements of atomic processes (magnetic resonance imaging). The optical scan methods became important. Using the so-called body scanning methods, only points of the subject's surface are obtained (see fig. 3.15). This is in opposition to the models described above. Therefore, the importance of these points in the first approach cannot be defined automatically. Usable anthropometric data can only be received when, by a second procedure, meaning is appointed to these measured coordinates. Two different procedures can be differentiated: the landmark method and the modeling method.

In the case of the *landmark method*, before measuring the body, points of interest on the subject's skin surface are marked by hand. With a second procedure, it must be ensured that these landmarks can be found in the data field of the scan as well. Then, in general, the same is to be done here (on the data level) as in the case of traditional anthropometric measurement: The distances between the interesting landmarks are calculated. The first anthropometric data gained during the CAESAR Project, by scanning

FIGURE 3.13 A model of the human body comprising more than 300 muscles developed in the AnyBody Body Modeling System.

approximately 7,000 U.S., 2,000 Dutch, and 1,000 Italian citizens (Robinette et al., 2002), is based on this procedure. Meanwhile, different procedures have been developed in order to find landmarks automatically (Kohno et al., 2005). Procedures have been developed by which landmarks on the head and at the feet are found automatically. However, the goal of these procedures is less to win anthropometric data than to obtain geometric data for certain applications (like the individual adaptation of helmets or the automatic production of custom-tailored shoes; Preatoni et al., 2005).

In connection with the modeling procedure, an existing DHM, or parts of it, is changed in such way that it corresponds to these data as well as possible. In the case of the silhouette procedure, only the external contour of the subject needs to correspond to the contour of the DHM. This method, for instance, is applied in connection with the PCMAN measuring tool (Seitz et al., 2000). When 3D data, obtained by body scanners, are processed by a computer-based optimizing procedure, the mean distance between the points that define the landmarks at the surface of the man model and the landmarks at measured scan coordinates are minimized. This adaptation of the model according to the scan data

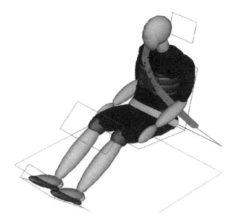

FIGURE 3.14 The dynamic man model MADYMO-3D.

cannot be successful with the first approach. Therefore, silhouettes, respectively scans of different body postures (e.g., upright posture and deflected extremities, defined according to ISO 20685), must be taken. Using an iterative process, in which alternating data are processed, the measured body dimensions and joint angles, which define the posture of the human "mold," are used to "rock" the man model into the given desired position until a break-off criterion is reached (i.e., 99% similarity). Using such an optimizing algorithm, a mathematically defined so-called snake can be adapted to the scan data to mark the spinal column (De Wilde et al., 2003). The parameters of this snake, for instance, are used to support and to objectify the medical diagnosis of spinal column damage in medical applications.

3.4 The Five Lines of the DHM Development

Depending on the application field, various man models are to be differentiated. As seen with describing the historic development, on one side the anthropometric man models are of special importance to allow representation of the variability of human sizes. On the other side, there are the biomechanical

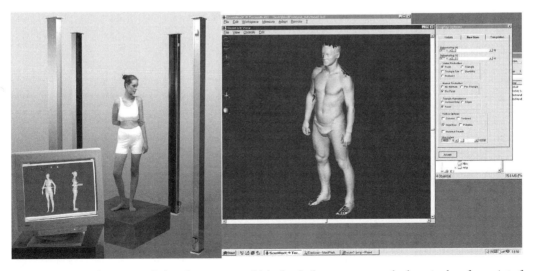

FIGURE 3.15 Left: an example for a laser scanner. Right: by the laser scanner method received surface point of the subject.

models that enable us to simulate physical dynamic behavior. The first types are mainly used to design space-critical working conditions, especially in vehicles, whereas the second types are mainly used to calculate the unintended dynamic behavior under crash conditions. In a third application field, models are used to simulate working processes (depending on the geometry of the working place) and to calculate the necessary working time (often under use of the MTM methods). These models are complemented by so-called voxel models, which represent position and form of internal organs. A completely different category of man models are the cognitive models that represent human behavior in decision-making and control situations.

Altogether, more than 150 man models exist worldwide that have been developed for workplace design, product design, safety evaluation, and documentation of planning results. All of these models, more or less, belong to one of the mentioned categories, although there might be crossovers in features from other categories. That means a classification can never be absolute. Nevertheless, the scientific problems in the different categories shall be described in the following.

3.5 Anthropometric Models

According to the example, the templates of the category *anthropometric models* try to represent the different sizes and shapes of the human body in order to design human-orientated products. The main application field of these models is the design of space-saving cabins or cubicles, in which a passenger or operator has to accomplish a certain task. Therefore, such models are mainly developed and applied by the aviation and automotive industry.

An anthropometric model is characterized by the exterior skin model that gives it a realistic appearance as well as by the interior skeleton model. The task of this interior model is to represent all posture and motion functions of the human body using as few joints as possible (see fig. 3.16). However, due to recent improvements in computing capabilities, this restriction disappears more and more, favoring a realistic representation of the functionality of the skeleton. Nowadays even the skeleton can be

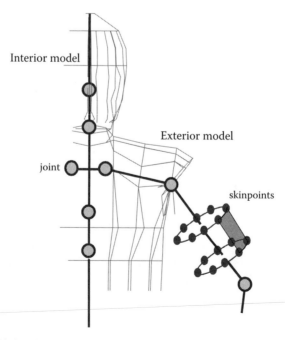

FIGURE 3.16   Interior (skeleton) and exterior (skin) model of a digital human model.

displayed in a realistic form. The interior model serves as a framework for the exterior features connected to it. Whether those are statically or elastically connected is determined by mathematical algorithms. The main task of these mathematical algorithms is to enable the model to show harmonic and realistic transitions and deformations of the skin during the animation. Various research projects focus on how to calculate the deformability of the skin and the subcutaneous muscle and fatty tissue. One way is by applying FEM in order to correctly simulate, for instance, the contact area to seats and similar difficult tasks (e.g., Schmale et al., 2002; Verver, 2004).

A special challenge for anthropometric human models is the correct representation of body parts, focusing on mutability and reach spheres. A simple transfer of values derived from anthropometric tables generated by conventional measuring methods is, in fact, not allowed. Nevertheless, most of the models developed in the past used anthropometric measures from tables which had to be adapted to the necessities of DHMs. In fact, conventional anthropometric data are derived from bone-to-bone measurements and therefore need to be modified for existing models. The main problem with using statistical analysis of the percentilization is that changing data about joints and the external surface results in a loss of the individual connections between the measures. To clarify, a 95-percentile man in body height may not automatically be a 95-percentile in leg length, sitting height, arm length, and so on. The first model considering this problem was *SAFEWORK*. In order to solve the problem, it uses correlation coefficients that were derived from the percentiles published in the table values. In its first approach, the human model RAMSIS used the real correlations between these values on the basis of a survey conducted in the former German Democratic Republic during the 1980s. Now, even for different ethnic populations, correct data are available. A correct consideration of the correlation between the anthropometric measures results in different phenotypes of the DHM; the principal procedure in how to achieve this is discussed in the following.

With respect to the anthropometry concerning body height as a key dimension (based on the results of the statistic analysis of Flügel et al., 1986), two auxiliary key dimensions were chosen to describe the morphology of a person in the RAMSIS method:

The waist circumference (corpulence) was chosen as a key dimension for horizontal measurements. The ratio of the sitting height and length of the legs was chosen as a key dimension for proportion.

By the dimensions body height, corpulence, and proportion, a 3D space can be defined. The dimension body height was divided into the following five groups: very small, small, medium, tall, and very tall. The mean values of body height in this group correspond to the 5th, 25th, 50th, 75th, and 95th percentile. Projecting one of the groups into a system of coordinates with the axes corpulence and proportion, every subject will be represented by a dot. The body measures of a normal-weighted/medium-legged person are calculated by taking the average values of body measurements of all persons within a given circle. The radius of this circle is dimensioned that way so that 60% of all persons are included. The other groups are developed by defining sectors within the remaining area of the coordinate system. According to this method, a 3D typology box is filled with this data. In total, 45 different body shape types per gender are created, which can be used for ergonomic layout purposes (see fig. 3.17). The center of interest in each box of the anthropometric cube of Figure 3.17 can now be connected to each other. Using this method allows the creation of every intermediate type between the original types.

Actually, much effort is done worldwide to simulate and calculate muscular strength. There are different methods in competition. One approach is to reproduce the anatomically correct properties of muscles und their tendons on the skeleton. The modeling of AnyBody was designed according to this idea. It is a software tool for modeling and analysis of the musculoskeletal system based on inverse dynamics. It presumes the musculoskeletal system to be a mechanism of rigid segments connected by joints and actuated by hill-type muscles. Using the correct simulation of the anatomic muscle constellation, optimal body postures for various tasks can be found (see fig. 3.18; Rasmussen et al., 2003).

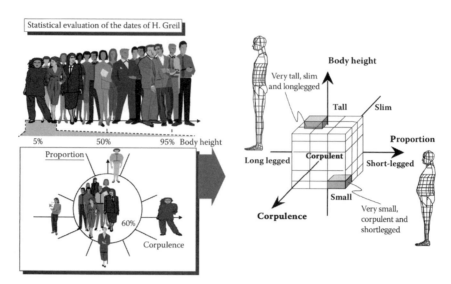

FIGURE 3.17 Anthropometric typology (example: RAMSIS).

Another approach in this area consists of the experimental research of maximum strength in the different body joints. The idea is to define a 3D volume for each of these local joint forces, which describe the actual maximum torque able to be applied in the current posture (Schwarz, 1997; Schaefer et al., 2000). If the posture changes, these visualized 3D volumes change their geometry

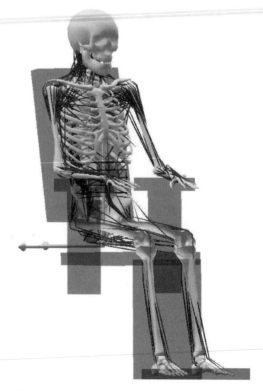

FIGURE 3.18 Modeling the sitting position by the muscle model of Rasmussen et al. (2003).

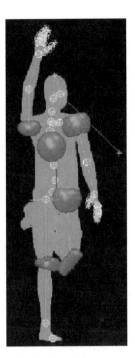

FIGURE 3.19　3D body representing the local moment in a certain body posture.

depending on the maximum amount of force possible (see fig. 3.19). Special measurement equipment must be developed and extensive experimental effort is necessary in order to define the shapes and the statistical distribution of these 3D volumes, which consequently depend on individual properties. The approach assumes that a human being wants to minimize the joint strain when taking a desired posture. This approach has been formulated within a *Mathematica* and RAMSIS simulation (FOCOPP = Force-Controlled Posture Prediction-Model; Seitz et al., 2005). An integral part of this approach is an accurate physical description of test subjects. The physical characterization includes anthropometry, masses of body parts, center of gravity, angle-dependent passive torques, and maximum torques (M_{max}) for each joint of the body. Equation 3.1 shows the core procedure of the FOCOPP approach where i is the index for joints, $M_{passive}$ is the retracting moment that pulls the joint in a neutral posture, and $M_{affective}$ is the actual applied moment.

$$\min \phi = \sum_{i=1}^{n} \left(\frac{M_{i,passive}}{M_{i,max}} \right)^2 + \left(\frac{M_{i.affective}}{M_{i,max}} \right)^2$$

(3.1)

Through experimental verification, this model is well confirmed. Furthermore, the individual load and discomfort based upon the verified correlation between load and discomfort assessment can be predicted (Zacher & Bubb, 2005; Schaefer & Zacher, 2006).

New application demands of DHMs also include the modeling of movements, which becomes increasingly important. Various approaches are known to solve the problem. The simplest approach is animation, the reconstruction of motion received from a computer-supported motion-tracking system. In contrast to animation, the simulation approach tries to find the correct motion (based upon a model) by using knowledge about the task, and the start and the target position. This can be realized by inverse kinematics using optimization criteria that include the application of multi-body systems or on the basis of modeling experimental observations. As Chaffin (2002) concludes, using regression methods promises good results. Faraway (1997) has developed a functional regression model for this purpose that uses the following form:

$$\Theta(t) = \beta_0(t) + C_x\beta_x(t) + C_y\beta_y(t) + C_z\beta_z(t) + C_xC_y\beta_{xy}(t) + C_yC_z\beta_{yz}(t)$$
$$+ C_zC_x\beta_{zx}(t) + C_x^2\beta_x^2(t) + C_y^2\beta_y^2(t) + C_z^2\beta_z^2(t) + D$$

where

$\Theta(t)$	=	the predicted joint angle over time
C_x, C_y and C_z	=	target coordinates
$\beta(t)$	=	parametric functions to be estimated
D	=	a demographic variable (e.g., age, stature, gender, etc.), which could modify the prediction

Especially utilizing the Memory-Based Motion Simulation (MBMS) algorithm (Park et al., 2002), good correlation between observed motion and simulated motion could be obtained.

A new approach is to define so-called leading body elements, as found by experiments of Arlt (1999). Cherednichenko et al. (2006) enhanced this model by investigations about "complex motions" in the context of entering a car. As already found by Arlt, Cherednichenko experimentally defined the leading body element by the fact that the motion of it is always exactly in one plane. It is remarkable that only one leading body element exists at a time. It appears to be a testimony to the restrictions of the active brain's capacity. However, in the case of a complex motion, several leading body elements are activated in a sequence according to a motion script. It was found that posture and motion of the remaining body elements can be calculated by the application of the already mentioned FOCOPP model, considering the condition of gravity and the obstacles restricting the motion. When the FOCOPP model is connected to a discomfort evaluation, even a virtual evaluation of motion discomfort seems to be possible in the future.

3.6 Models for Production Design

Anthropometric models are also often used to design workplaces. As in this case, the exact anthropometric modeling sometimes is of secondary interest; often more simplified models are used. However, in connection with *factory planning*, motion behavior is to be modeled also and especially the necessary time demand for it. Most of these models calculate the time demand after the MTM or work-factor methods. As an example, the 3D-Human Simulation of Tecnomatix Technology, should be mentioned (Geyer & Rösch, 2002), applying the MTM method. The concept of Manufacturing Process Management (MPM) realized by those simulations enables multiple users to collectively plan manufacturing processes and to broadcast this information to everyone within the organization. After the assembly processes including the required resources are defined, a 3D layout can be realized. The 3D model helps to increase the planning quality and to avoid planning errors. After a coarse plan of the layout structure and material flow, the details of the workplace design are planned in 3D. MPM as well will ensure the implementation of the work practices and workplaces. Starting from outlined 3D workplace blueprints to electronic work instructions, all necessary data can be generated electronically (see fig. 3.20).

However, Dukic et al. (2002, see also Laring et al., 2005) mentions that the computer mannequin is a capable tool in verifying ergonomics and it allows the detection of many problems prior to building pre-production samples. His results demonstrate that computer mannequin users and others involved in the virtual manufacturing process are very important for the quality of the final product in addition to the manufacturing equipment that is being used. Stephens and Godin (2006)

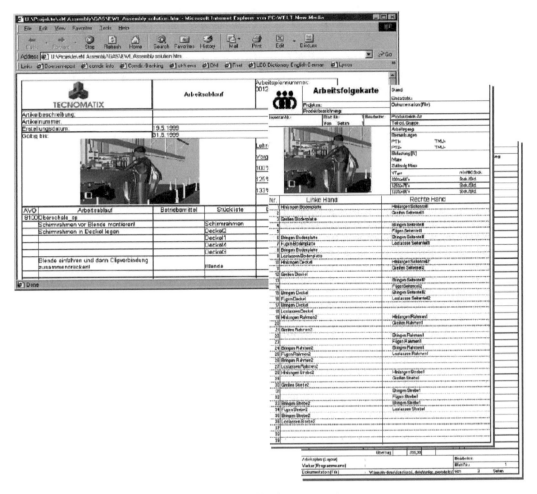

FIGURE 3.20   Electronic work instruction generated by the eHuman-System.

state that digital human models, used together with motion capture technology and biomechanical evaluations, are all critical in achieving an accurate ergonomic analysis. Figure 3.21 shows the difference between the manual posture prediction realized by a trained user and the real posture received by a motion capturing system. Once a human simulation is generated, the power of the digital human model is not just in the graphical representation. The ability to see a digital human interact with its proposed digital workstation is powerful and should not be underestimated. The significance of using a DHM lies in the amount of biomechanical information available about many joints in the body at any given moment within the task. Compared to those years of specific joint data collection on a single static posture, the biomechanical information available on the thousands of postures from an entire task simulation is more information than an ergonomist has ever had before. For instance, the spinal compression and shear forces occurring on the spine allow for predictions of back injury risks (Jäger et al., 2001).

The differences between manually determined postures and real postures, found in working situations, reveal the importance of an improvement of posture prediction systems. Rider et al. (2003) has shown how MBMS could be successfully applied to redesigning work stations.

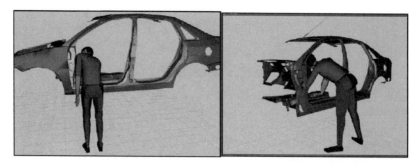

FIGURE 3.21 Left: posture prediction by a trained user. Right: same operation as left; however, posture received by a motion capturing system. (From Stephens, A., and C. Godin, *The Truck That Jack Built*, SAE International, Warrendale, PA, 2006. With permission.)

3.7 Biomechanical Models

Besides the geometric representation of human properties, a special domain of the computer applications used in ergonomics is the calculation of dynamic behavior. From the beginning, the application of computers in human modeling was geared to "real" *biomechanical models*. The models have been developed in respect to realistic masses, inertia, spring, and damping elements that represent the body parts connected to each other by joints. The anthropometrical properties (like segment length, masses, and inertia of a body part as well as the position of a joint and the center of gravity) are taken from empirical data (e.g., photogram-metrical methods, measurement of acceleration of individual segments after quick-release, radio medical methods), where a homogeneous density of the tissue is assumed. The mathematical models that describe these measurements in the form of regression calculation depend on a variety of basic anthropometrical data (Hatze, 1980; Saziorski et al., 1984; de Leva, 1996). The human body is treated as a mechanical system comprised of a multitude of rigid elements that are modeled within a multi-body system (MBS) software. Thus, the models are governed by Newton's laws of motion, which are automatically generated by the MBS software. This can be done by two different mechanical principles: the principle of d'Alambert (realizing "force equilibrium" between static and dynamic forces); and the principle of Hamilton (general integral principle of mechanics), which leads to the Euler-Lagrange differential equations, used by SIMPACK and alaska (in the form of generalized coordinates). Besides these MBS models there are also human models or submodels developed as finite element models (FEM).

There are two ways to apply these equations of motion: In the inverse dynamics, one can analyze forces and moments that act within the human model when the external forces and kinematic data are measured. The movement is usually analyzed by video so that the segment accelerations can be calculated to control the model. Problems can arise with measurement inaccuracies, such as when the markers used for the motion analysis move with the skin in dynamic situations. By smoothing the trajectories or using a closed-loop control to keep the model along the measured movement, these inaccuracies can be minimized. In direct dynamics, the rigid bodies of the model start at an initial condition and at each time step the trajectory is calculated depending on the activation or gross torques in the joints and contact forces implemented in the model. Especially in highly dynamic movements with ground contact, the rigidity of the models is an oversimplification and leads to unrealistic physiological forces and moments within the simulation. This is due to the fact that the model is abruptly accelerated during ground contact whereas the soft tissue of a human (muscles, fatty tissue, or organs) slows this movement in reality. Therefore Ruder and Gruber established wobbling masses. In order to achieve better results, the soft tissue masses were connected to the model's skeletal structure in a nonlinear visco-elastic condition (Gruber et al., 1997). Musculoskeletal models (as described in section 3.1) can also be considered a form of biomechanical models with more detail.

FIGURE 3.22 Biomechanical model MADYMO (left) and DYNAMICUS (right).

Biomechanical models are used in a variety of applications, like safety issues in traffic (e.g., crash or whiplash simulations; Golinski & Gentle, 2002). Another application is the study of impacts on passengers during a traffic accident (Keppler, 2003). In order to simulate crash tests, *MADYMO* was used as simulation software. Therefore, instead of a human body the crash dummy had to be modeled based upon the Hybrid III. The main results of the simulation are the injury criteria (e.g., head injury criteria: HIC). These results help evaluate and improve the crash performance during the design process. The transmission of vibrations to the human body (so-called whole body vibration: WBV) is also an important task to evaluate with the goal to improve the quality and comfort of car seats. Verver used the *MADYMO* human model and calculated the tension on the spine (Verver & Van Hoof, 2002). Pankoke calculated transfer functions with the FEM model *CASIMIR*, using anthropometrical data and postures of RAMSIS through an interface (Pankoke et al., 2002). Also the contact area between the human and the car seat is subject to simulation. Johnson Controls and the Institute of Mechatronics at the University of Chemnitz have developed the simulation tool COSYMAN in order to calculate the SRP point and pressure distribution (Schmale et al., 2002). For medical applications it is possible to examine problems of the kinematic behavior and strains within the knee joint (Lehner & Wallrapp, 1999). Within the field of ergonomics it is important to simulate strains in the spine and other joints during working tasks (Fritz, 2000). Also, ergonomic designs for ingress and egress (Rasmussen & Christensen, 2005) have advanced through the use of biomechanical models. In the field of sports science, human models are used to analyze the performance of top athletes and simulate new techniques or materials to improve their performance (King & Yeadon, 2006; IfM, 2006; Härtel et al., 2006). To cover these fields, countless individual models and problem-solvers have been developed in addition to the commercially available products. The reader may also refer to Chapters 5, 10, and 29, which discuss the Virtual Soldier Project. The simulation software ADAMS, developed for calculating mechanical problems, has a plug-in of *LifeMOD*, which is a human modeler. Other MBS solvers are Dymola/Modelica and DADS.

3.8 Anatomical Models

Rather late in the development of *anatomic models*, the so-called voxel models have been established in the area of medical information technology. The Voxel-Man project dates back to about 1985, when a research group at the University Medical Centre Hamburg–Eppendorf made first experiments for 3D visualization of tomographic volume data, with application in surgery planning and simulation. Voxel models are characterized by a specific arrangement of cubic volume elements, the so-called voxel (a portmanteau of the words "volumetric" and "pixel") that represent kind, position, and size of internal

FIGURE 3.23   Voxel model HUGO.

organs. After an originally very rough modeling using coarse volumes, today with increasing computer power, very precise representations of these organs are possible (see fig. 3.23). In the first approach voxel models are rigid (i.e., it is impossible to shift the position of internal organs and, correspondingly, to deform the skin depending on a change of body posture). In the area of medical technology, such voxel models play an increasingly important role for different purposes, from education to the representation of the patient's individual conditions in order to prepare and observe a surgery. An ergonomic application field of use for voxel models is the more realistic calculation of the impact of radiation. In this case, the goal is to consider the effects of shading radiation by any object or by the body itself. For instance, the GSF (located at Neuherberg, Germany) developed a whole family of voxel models from children to adults for this purpose (Petoussi-Henss et al., 2002). The automotive industry is also interested in these voxel models. In analyzing traumatic head injuries, caused by car crashes, fast response treatment is strongly related to the patient's recovery and long-term condition. For this reason, a drive recording system was proposed: capable of predicting the amount of damage to occupants and the time of transfer of the injured person to an appropriate hospital after a serious accident (Nishimoto et al., 2001). For this purpose, among other things, a detailed human-head model is important to determine the precise cause of damage for impact-injury analysis. Ejima et al. (2004) developed such a model on the basis of the voxel approach.

A totally new kind of application, using voxel models, was presented by Gopinath et al. (2005). As proposed, it uses multiple synchronized infrared video streams in order to obtain real-time tracking data of a subject and to use this data to drive an animated avatar. Basically, this approach is reconstructing a 3D object using the 2D silhouettes obtained from multiple camera feeds, applying the so-called voxel carving method. They adopted a straightforward approach that checks for each voxel (that is part of a volume of interest) if it is consistent with all the obtained 2D silhouettes. In this system, the person is known to remain inside a predetermined space of interest. With this procedure a method was developed, where, using the known volume and shape of a person, changes during the movement can be captured automatically; a method that could substitute the body scanning technologies and landmark-based motion tracking systems.

In future developments, a combination of the anthropometric and the voxel models could be realized. First steps in this direction are already being taken. For example, the functional joint to joint distances in many models can be visualized by bone pictures, and the muscle representation by the Anybody model already shows similarities to the very detailed voxel models. With further development, these models will be linked together with movable anthropometric models. This will enable the user to calculate the changed position of organs depending on an actual body posture. As the approach of Gopianth et al. shows, such a link could also be realized related to an individual subject. Such enlarged models could not only open new possibilities for medical diagnostics but also for the ergonomic applications in product design and layout of production assemblies.

3.9 Cognitive Models

Totally different from the models described above are *cognitive human models*. Cognitive models simulate the information flow inside the human operator between the input (represented by the sensory organs) and the output (given by the activation of muscles). Roughly, we distinguish between so-called cause-effect models and probability models. Independent from this differentiation, a common fundamental aspect in all approaches of cognitive modeling is the task analysis. This means, for every task the sequence and dependability between the single steps are to be described. Bayssie et al. (2003) presented an example of a mathematical, descriptive model: It simulates human activity using the case of flight manoeuvres with the goal to improve the safety of a given system when human operators are in scope.

Cause-effect models assume an absolutely clear relationship between information input and information output, normally defined with a mathematical formula. Very early approaches are control-oriented human models, developed originally without any computer support. The general idea behind it is that the operator is controlling a process (e.g., following the course of the road by a car or handling an aircraft) working like an automated control unit. If one assumes certain properties (like delayed reaction, excessive reaction due to sudden alterations, or smoothing movement delays caused by the inertia of the hand-arm system), the human reaction to deviations between the expected nominal value and the observed actual value can be calculated within certain limits. The most important model of this kind, which is still used, is the model of Tustin (1944), the so-called paper-pilot. This model is to be seen in connection with the research results of McRuer (e.g., Magdaleno & McRuer, 1971; McRuer & Krendel, 1974) in the 1960s. According to these results, the human operator recognizes the dynamic properties of the machine and adapts his own in such a way that the overall properties of the closed loop—consisting of man and machine—remain similar within certain boundaries. Today's computer programs are able to calculate these models and to simulate the behavior in real time. A huge diversity of applications and further developments of this fundamental approach are published in the meantime.

Probability models try to judge the probability of a successful action—the mathematical complement to the so-called human error—depending on external conditions. The probability tables of Swain and Guttmann (1983) are a famous example of this kind of model. As the deliberations of Rasmussen (1987) about the three levels of cognitive behavior (i.e., skilled-based, rule-based, and knowledge-based behavior) shows and the rough allocation of successful performance on these levels, the GEMS model (Generic Error Modeling System) of Reason (1990) provides the scientific background to this approach. The HCR approach (Human Cognitive Reliability model) of Hannaman and Spurgin (1984) considers only the available time in relation to the necessary time to calculate the error probability. In fact, time plays an important role for understanding human behavior. For instance, Gore and Milgram (2006) presented a time estimation model in order to predict human performance in complex environments.

In addition to these original approaches, various psychologically oriented decision models (the calculation of so-called internal models (chunks) and further representations of cognitive behavior) are an object of scientific research today. A model widely accepted was presented by Anderson (1983, 1996). Most models of human cognition provide a symbolic level of description. Pure sub-symbolic descriptions, like neural network architectures, are difficult to apply on more complex tasks and therefore are mostly used with modeling human perception. Nevertheless, Bouslimi et al. (2005) could show that such an approach gives a good prediction of driver actions on car control (steering wheel and pedal actions). Modeling human cognition on a symbolic level means to model knowledge and rules of cognition as a set of chunks that can be interpreted. The common architectures of cognition used today differ mainly on the symbolic level on the definition of chunks and on the framework on how these chunks interact in the process of cognition. Anyhow, if a cognitive user model and a model of the task environment are given, theoretically the probability of any possible evolution of the overall system could be calculated. Of course, for complex models only estimates can be calculated in reality. Winkelholz et al. (2003) developed a method to heuristically validate cognitive models with the help of variable length MARKOV chains. This method's focus is on the validation of action sequences rather than the validation of time consumption. This method was applied successfully in the investigation of visual attention in multiple target search tasks on computer-based radar displays. Similar to the McRuer approach in the framework of control theory (however, he assumed a continuous behavior), they assumed that the whole system, task environment and cognitive model, can be described as the transition between discrete states.

With the visual perception being the most important input channel within the human cognitive system, its modeling is subject to many scientific experiments. Winkelholz and Schlick (2006) presented a Bayesian approach to modeling visual-spatial cognition within a symbolic framework. The subject's task was to memorize spatially distributed objects on electronic information displays under time pressure. The developed model was validated through laboratory experiments with 30 subjects. The validation tests showed a good correlation among model prediction and empirical data.

Another approach was carried out by Schweigert (2003). He investigated the glance behavior during car driving. The result showed that scanning glances last on average 0.4 seconds. With scanning especially the edges of the road, other traffic participants and traffic signs are recognized. Processing means that special—so-called—areas of interest (AOs, e.g., the instruments, the mirror, and the display of the car but also objects in the environment) are fixated. The processing of glances needs, on average, twice the time of scanning glances. While scanning, internal models are stimulated, which are stored as a general concept in the form of a structural engram within the long-term memory. The modeling of this behavior follows the ideas of Anderson.

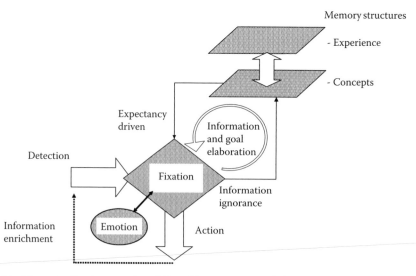

FIGURE 3.24 Elements of an integrated modeling of human cognitive behavior concerning reaction times, decision making, and human error.

A combination between the cause-effect models and probability models was presented by Sträter (2005). His model represents a system that basically can be programmed into a computer. The structure of this model is shown in Figure 3.24. Information coming from the outside world is redetected by the sensory system. In this perspective, the modeling of, for example, glance behavior or visual-spatial cognition is of interest in this connection. By a dynamic information and goal elaboration process this information is imprinted inside the memory structure, represented in the model as concepts and experience. Expectancy driven, a certain action is induced whereas emotions can also influence this process. Dörner (1999) proved that even emotions can be represented by a computer simulation. When a certain event changes the perception of the external world, it causes "information enrichment," by which (depending on the memory structure) changed or new actions may result. This, in relation on the outside world situation, can lead to wrong actions.

3.10 Future Development

The future in the development of digital human models will most likely end in the integration of the various, already existing models into one system adaptable to all required tasks. A first proposal in this direction was presented by Gore and Jaris (2005), introducing an MS-Windows-based version integrating the anthropometric character JACK with MIDAS (Man-machine Integration Design and Analysis System; Hart et al., 2001) validating perceptual and attention mechanisms. As shown in the chapter above, today various concepts already exist for such integrations. The combination of anthropometric and biomechanical approaches is known to be able to predict correct movements in certain application areas (e.g., Cherednichenko et al., 2006). Further into the future, the integration of anatomic human models in anthropometric models and additionally the connection with biomechanical models will allow predicting the shift of internal organs depending on body posture and especially during body movements. The connection of such models with cognitive models will then enable the user to rationalize work processes already within the design phase. It will be possible to determine (during the design phase) what the effects are on the reading frequency and on the probability of erroneous reading when positioning an instrument a certain way. Figure 3.25 illustrates the interaction of such a hybrid man model with a simulated task by the example of car driving.

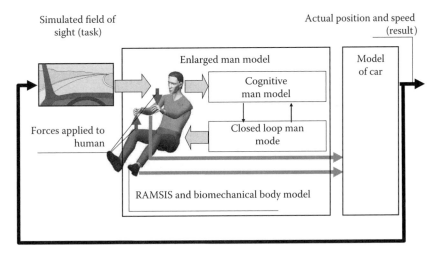

FIGURE 3.25   Total structure of a biomechanical-anthropometric-cognitive human model with the goal to adapt car properties to common human abilities.

3.11 Conclusion

The development of digital human models began in the 1960s. Although more than 150 ergonomic mannequins are known today (Hickey et al., 1985; Porter et al., 1993; Seidl, 1997) only a few systems have been able to establish themselves successfully on a global industrial scale in the last few years. Three ergonomic mannequins have become established in professional applications, controlling more than 95% of the market. The main field of application of the mannequin *JACK* (see fig. 3.6) is animation and visualization in vehicle design and architecture. Numerous modules are available in the form of analysis tools for factory planning and vehicle development. The system is available either as a stand-alone version or integrated in Unigraphics. *SAFEWORK* (see fig. 3.7) was conceived for workplace design in factory planning and product design. The anthropometric database is based on U.S. Army data and encompasses an anthropometric body-type generator that can create statistical test samples. The posture and movement simulation enables the movement of short chains in the model's arms, legs, and torso by means of inverse kinematics. Vision simulation, fixed accessibility areas, a joint-dependent comfort evaluation, maximum force calculation, and a center of gravity analysis are available as analysis tools. *SAFEWORK* is available solely in the form of an integration unit with CATIA by Dassault Systems. The mannequin RAMSIS (see fig. 3.12) was developed for vehicle design. The anthropometric database encompasses a physique typology that takes body measurement correlation, desired percentile models, an international database with more than 10 global regions and a secular growth model into account. Posture and movement simulation is carried out by task-related animation, enabling the forecasting of the most probable posture. More than 80 functions for the analysis of vehicles and vehicle interiors are available (vision and mirror simulation, seat simulation, accessibility limits, comfort, belt analyses, etc.). The system is available as a stand-alone version for UNIX and Windows, a CAD integration unit with CATIA, and a programming library for independent applications.

Presently, five main lines of the application can be observed: anthropometric models especially developed for product design, models for the production line design, biomechanical models that realize dynamic properties, anatomic models that represent the internal construction of the human body, and cognitive models. It can be expected that future developments will consist of an integration of these five development lines.

References

Abdel-Malek, K., Yang, J., Marler, T., Beck, S., Mathai, A., Zhou, X., Patrick, A., and Arora, J. (2006): Toward a new generation of virtual humans. *International Journal of Human Factors Modeling and Simulation*. Vol. 1, No. 1, 2–39.

Anderson, J. R. (1983): *The architecture of cognition*, Harvard University Press, Cambridge, Mass.

Anderson, J. R. (1996): *Kognitive psychologie*, Spektrum Akademischer Verlag, Heidelberg.

Arlt, F. (1999): *Untersuchung zielgerichteter Bewegungen zur Simulation mit einem CAD-Menschmodell*, Dissertation am Lehrstuhl für Ergonomie der Technischen Universität München.

Bayssie, L., Chaudron, L., Le Blaye, P., Maille, N., and Sadok, S. (2003): Human activity modeling for flight safety. Paper 2003-01-2211 in the Proceedings of the SAE-DHMC.

Bouslimi, W., Kassaagi, M., Lourdeaux, D., and Fuchs, P. (2005): Modeling driver behavior in a straight-line emergency situation. Paper 2005-01-2700 in the Proceedings of the SAE-DHMC.

Burandt, U. (1978): *Ergonomie für Design und Entwicklung*, Verlag, Dr. Otto Schmidt KG, Köln.

Chaffin, D. B. (2002): Simulation of human reach motion for ergonomic analyses. *SAE Digital Human Modeling Conference*, Munich. VDI-Berichte Nr. 1675, 9–23.

Cherednichenko, A., Assmann, E., and Bubb, H. (2006): Computational approach for entry simulation. Paper 2006-01-2358 in the Proceedings of the SAE-DHMC.

Conrad (1963): *Der Konstitutionstyp*, 2. Auflage, Springer Verlag, Berlin.

de Leva, P. (1996): Adjustments to Zatsiorsky-Seluyanov's segent inertia parameters. *Journal of Biomechanics*, Vol. 29, No. 9, 1223–1230.

De Wilde, T., Haex, B., Forausberger, C., Denis, K., Huysmans, T., and Vander Sloten, J. (2003): A contact-free optical measuring system for the acquisition of anatomical structures in 3D. *Proceedings of the 15th Triennial Congress of the International Ergonomics Association*, Seoul, Korea, August 24–29, 2003.

Dooley, M. (1982): Anthropometric modeling programs—A survey. *IEEE Computer Graphics and Applications*, 2:17–25.

Dörner, D. (1999): *Bauplan der Seele*, Rowohlt, Hamburg.

Dukic, T., Rönnäng, M., Ortengren, R., Christmansson, M., and Johansson Davidsson, A. (2002): Virtual evaluation of human factors for assembly line work: A case study in an automotive industry. *SAE Digital Human Modeling Conference,* Munich. VDI-Berichte Nr. 1675, 129–150.

Ejima, S., Nishimoto, T., Yuge, K., Tomonaga, K., Murakami, S., and Takao, H. (2004): Three-dimensional human-head model using VOXEL approach developed for head-injury analysis. *SAE Digital Human Modeling for Design and Engineering Symposium*, Rochester, MI. Paper 2004-01-2135.

Faraway, J. J. (1997): Regression analysis for functional response. *Technometrics*, 39, 3, 254–262.

Flügel, B., Greil, H., and Sommer, K. (1986): *Anthropologischer Atlas*. Edition Woetzel, Frankfurt.

Fritz, M. (2000): Simulating the response of a standing operator to vibration stress by means of a biomechanical model. *Journal of Biomechanics*, 33, 795–802.

Geuss, H. (1995) Entwicklung eines anthropometrischen Meßverfahrens für das CAD-Menschmodell RAMSIS, PhD thesis, TU München, 1995.

Geyer, M., and Rösch, B. (2002): 3D-human simulation and manufacturing process management. *SAE Digital Human Modeling Conference*, Munich. VDI-Berichte Nr. 1675, 121–128.

Golinski, W. Z., and Gentle, R. (2002): Whiplash injury assessment—A biomechanical FE model approach. *SAE Digital Human Modeling Conference*, Munich. VDI-Berichte Nr. 1675, 431–443.

Gopinath, A., Carroll, J. J., and MacIntosh, D. (2005): Human body modeling and tracking using 3D voxel data from infrared images. *SAE Digital Human Modeling Conference*, Iowa City, IA. Paper 2005-01-2725.

Gore, B. F., and Jarvis, P. A. (2005): New integrated modeling capabilities: MIDAS' recent behavioral enhancements. *SAE Digital Human Modeling Conference*, Iowa City, IA. Paper 2005-01-2701.

Gore, B. F. and Milgram, P. (2006): The conceptual development of a time estimation model to predict human performance in complex environments. *SAE Digital Human Modeling Conference*, Warrendale, PA. Paper 2006-01-2344.

Gruber, K., Ruder, H., Denoth, J., and Schneider, K. (1997): A comparative study of impact dynamics: Wobbling mass model versus rigid body models. *Journal of Biomechanics*, 31, 439–444.

Hannaman, G. W., and Spurgin, A. J. (1984): *Systematic human action reliability procedure (SHARP)*. EPRI-NP-3583, Electric Power Research Institute, Palo Alto, CA (USA).

Hart, S. G., Dahn, D., Atencio, A., and Dalal, K. M. (2001): Evaluation and application of MIDAS v2.0 in *Proceedings of the 2001 Aerospace Congress*, SAE Paper No. 2001-01-2648, Warrendale, PA.

Härtel, T., Hildebrand, F., and Knoll, K. (2006): Methods of simulation and manipulation for the evaluation of figure skating jumps, In E. F. Moritz and S. Haake (Eds.): *The Engineering of Sport 6, Volume 2, Developments for Disciplines*, Springer Science+Business Media, New York, 179–184.

Hatze, H. (1980): A mathematical model for the computational determination of parameter values of anthropometric segments. *Journal of Biomechanics*, 13, 833–843.

Hickey, D., Pierrynowski, M. R., and Rothwell, P. (1985): Man-modeling CAD-programs for workspace evaluations. Vortragsmanuskript, University of Toronto, Canada.

IfM (2006): Tätigkeitsbericht des Institutes für Mechatronik e.V. an der Technischen Universität Chemnitz.

Jäger, M., Luttmann, A., Göllner, R., and Laurig, W. (2001): The Dortmunder—Biomechanical model for quantification and assessment of the load on the lumbar spine. *SAE Digital Human Modeling Conference*, Arlington, VA. Paper 2001-01-2085.

Keppler, V. (2003): Biomechanische Modellbildung zur Simulation zweier Mensch-Maschinen-Schnittstellen. Dissertation an der Universität Tübingen.

King, M. A., and Yeadon, M. R. (2006): A comparison of activation timing profiles for single and double layout somersaults. *Journal of Biomechanics*, 39, 5445.

Kohno, Y., Yahara, H., Fukui, Y., Mochimaru, M., and Kouchi, M. (2005): Automatic landmarking in 3D human head scans. *SAE Digital Human Modeling Conference*, Iowa City, IA. Paper 2005-01-2731.

Krendel, E. S., and McRuer, D. (1959): The human operator as a servo-system element. *Journal of the Franklin Institute*, 267, 381–403.

Laboratoire d'Anthropologie Appliquee et d'Ecole Humaine Paris (1988): Produktinformation ERGO-MAN, Paris.

Laring, J., Christmansson, M., Dukic, T., Sundin, A., Sjöberg, H., Örtengren, R., Hanson, L., Lämkull, D., Davidsson, A., Falck, A-C., and Klingstam, P. (2005): Simulation for manufacturing engineering (ViPP). *SAE Digital Human Modeling Conference*, Iowa City, IA. Paper 2005-01-2696.

Lehner, S., and Wallrapp, O. (1999): 3D-simulation of the human knee-joint, in *Proceedings of the 10th Conference of European Society of Biomechanics* (G. V. d. Perre, ed.), Leuven.

Lohmeier, S., Loeffler, K., Gienger, M., Ulbrich, H., and Pfeiffer, F. (2002): Computer system and control of biped "Johnnie." *Proceedings of the 2004 IEEE International Conference on Robotics and Automation*, New Orleans, LA.

Magdaleno, R. E., and McRuer, D. T. (1971): Experimental validation and analytical elaboration for models of the pilot's neuromuscular subsystem in tracking tasks. NASA CR-1757.

Martin, R. (1914). *Lehrbuch der anthropologie*. Verlag Gustav Fischer, Jena.

McRuer, D. T., and Krendel, E. S. (1974): Mathematic models of human pilot behaviour. *AGARDograph* No. 188, Advisory Group for Aerospace Research and Development, Neuilly sur Seine, France.

Nishimoto, T., Arai, Y., Nishida, N., and Yoshimoto, K. (2001): Development of high performance drive recorders for measuring accidents and near misses in the real automobile world. *ISAE Review*, 22, 311–317.

Pankoke, S., Balzulat, J., Wölfel, H. P. (2002): Vibrational comfort with CASIMIR and RAMSIS using a finite-element model of the human body. *SAE Digital Human Modeling Conference*, Munich. VDI-Berichte Nr. 1675, 493–503.

Park, W., Chaffin, D. B., and Martin, B. J. (2002): Memory-based motion simulation. *SAE Digital Human Modeling Conference*, Munich. VDI-Berichte Nr. 1675, 255–270.

Petoussi-Henss, N., Zankl, M., Fill, U., and Regulla, D. (2002): A family of human voxel models for various applications. *SAE Digital Human Modeling Conference*, Munich. VDI-Berichte Nr. 1675, 91–103.

Porter, J. M., Case, K., Freer, M. T., and Bonney, M. C. (1993): Computer-aided ergonomics design of automobiles. In B. Peacock and W. Karwowski (Eds.): *Automotive ergonomics*, Taylor & Francis, 43–78.

Preatoni, E., Andreoni, G., Squadrone, R., and Forlani, C. (2005): Assessment of 3D surface anthropometry methods for mass customized shoes. *SAE Digital Human Modeling Conference*, Iowa City, IA. Paper 2005-01-2724.

Rasmussen, J. (1987): The definition of human error and a taxonomy for technical system design. In J. Rasmussen et al. (Eds.): *New technology and human error*, Wiley & Sons, New York.

Rasmussen, J., and Christensen, S. T. (2005): Musculoskeletal modeling of egress with the anybody modeling system. *SAE Digital Human Modeling Conference*, Iowa City, IA. Paper 2005-01-2721.

Rasmussen, J., Dahlquist, J., Damsgaard, M., de Zee, M., and Christensen, S. T. (2003): Musculoskeletal modeling as an ergonomic design method. *Proceedings of the 15th Triennial Congress of the International Ergonomics Association*, Seoul, Korea, August 24–29, 2003.

Reason, J. (1990): *Human error.* Cambridge University Press, Cambridge.

Rider, K. A., Park, W., Chaffin, D. B., and Reed, P. (2003): Redesigning workstations utilizing motion modification algorithm. *SAE Digital Human Modeling Conference*, Montreal. Paper 2003-01-2195.

Robinette, K., Blackwell, S., Daanen, H., Boehmer, M., Fleming, S., Brill, T., Hoeferlin, D., and Burnsides, D. (2002): *Civilian American and European surface anthropometry resource (CAE-SAR), final report, volume I: Summary.* AFRL-HE-WP-TR-2002-0169, Air Force Research Laboratory, Wright-Patterson AFB, OH, and Society of Automotive Engineers International, Warrendale, PA.

Rohmert (1966): Maximalkräfte von männern im *Bewegungsraum der Arme und Beine*, Forschungsbericht Nr. 1616 des Landes Nordrhein-Westfalen, Köln.

Rohmert (1993): Biomechanische grundlagen, in Schmidtke, *Ergonomie*, 3. Auflage, Carl Hanser Verlag München, Wien.

Rühmann, H., and Schmidtke, H. (1991): Körperkräfte des menschen. Dokumentation Arbeitswissenschaft. Band 31.

Saziorski, A., Arunin, A. S., and Selujanow, W. N. (1984): *Biomechanik des menschlichen bewegungsapparates.* Sportverlag, Berlin.

Schaefer, P., Rudolph, H., and Schwarz, W. (2000): Digital man models and physical strength—A new approach in strength simulation. *SAE Digital Human Modeling Conference*, Dearborn, MI. Paper 2000-01-2168.

Schaefer, P., and Zacher, I. (2006): On the way to autonomously moving manikins empowered by discomfort feelings, *16th World Congress on Ergonomics*, Maastricht.

Schmale, G., Stelzle, W., Kreienfeld, T., Wolf, C.-D., Härtel, T., and Jödicke, R. (2002): COSYMAN: A simulation tool for optimization of seating comfort in cars. *SAE Digital Human Modeling Conference*, Munich. VDI-Berichte Nr. 1675, 301–311.

Schwarz, W. (1997): 3D Video Belastungsanalyse—Ein Neuer Ansatz zur Kraft—und Haltungsanalyse, VDI Fortschrittberichte Reihe 17 Nr. 166.

Schweigert, M. (2003): *Fahrerblickverhalten und Nebenaufgaben.* Dissertation an der Technischen Universität München.

Seidl, A. (1997). Computer-menschmodelle in der ergonomie. In Schmidtke, H., *Handbuch der ergonomie* (Part A – 3.3.3). Bundesamt für Wehrtechnik und Beschaffung. Hanser Verlag.

Seitz, T., Balzulat, J., and Bubb, H. (2000): Anthropometry and measurement of posture and motion. *International Journal of Industrial Ergonomics*, 25, 447–453.

Seitz, T., Recluta, D., and Zimmermann, D. (2005): FOCOPP—An approach for a human posture prediction model using internal/external forces and discomfort. *SAE Digital Human Modeling Conference*, Iowa City, IA. Paper 2005-01-2694.

Stephens, A., and Godin, C. (2006): The truck that Jack built: Digital human models and their role in the design of work cells and product design. *SAE Digital Human Modeling Conference*, Lyon, France. Paper 2006-01-2314.

Sträter, O. (2005): *Cognition and safety—An integrated approach to system design and assessment.* Ashgate Aldershot, Burlington, Singapore, Sydney.

Strömgren (1937): Über anthropometrische Indizes zur Unterscheidung von Körperbautypen, Zeitschrift Neurologie, *Psychiatrie*, 159.

Swain, A. D., und Guttmann, H. E. (1983): Handbook of human reliability analysis with emphasis on nuclear power plant applications. NUREG/CR-1278, Sandia Laboratories, Albuquerque, NM 97185.

Tustin, A. (1944): An investigation on the operator's response in manual control of a power-driven gun. Metropolitan-Vickers Electrical Co. Ltd., C.S. Memorandum No. 169, Sheffield, England.

Verver, M. M. (2004): *Numerical tools for comfort analysis of automotive seating.* Technische Universiteit Eindhoven, Proefschrift, Eindhoven University Press.

Verver, M. M., and van Hoof, J. (2002): Vibration analysis with MADYMO human models. *SAE Digital Human Modeling Conference*, Munich. VDI-Berichte Nr. 1675, 447–455.

Verver, M. M., and van Hoof, J. (2003): Human modeling for automotive seating comfort. *IEA Proceedings*.

Waldhier, T. (1989): Menschmodellierende Verfahren zur rechnergestützten Arbeitsplatzgestaltung— Eine Übersicht. Marktanalyse am Lehrstuhl für Arbeitswissenschaft und Betriebsorganisation der TH Karlsruhe.

Winkelholz, C., Schlick, C., and Motz, F. (2003): Validating cognitive human models for multiple target search tasks with variable length markov chains. *SAE Digital Human Modeling Conference*, Montreal. Paper 2003-01-2219.

Winkelholz, C., and Schlick, C. (2006): A symbolic theory of visual-spatial cognition. In Proceedings of the IEA-Congress 2006 in Maastricht. Maastricht: Elsevier.

Zacher, I., and Bubb, H. (2005): Ansatz zur Entwicklung eines Diskomfortmodells für Bewegungen. In Proceedings of the Spring University, Balatonfüred, Hungary.

4

Historical Perspectives on Human Performance Modeling

Thomas B. Sheridan

4.1 Introduction

When did modeling human performance begin? To provide a specific answer would beg the question, What is meant by modeling? Modeling, in general, means to represent objects or events with words, symbols, images, song, dance, or other means of expression. From that perspective, people have been modeling human performance since the dawn of human intelligence. Almost all such representations that we humans make are qualitative, taking the form of conjectures, theories, descriptions, or stories. Mostly these are expressed in words—terms of law, politics, morality, religion, physics, art, or just ordinary idiom.

This chapter will narrow the scope to quantitative models and focus mostly on normative quantitative models of human behavior. Further, it will omit the large class of models of learning. The remaining terrain is still quite wide, so the chapter will discuss models that seem to have been both popular and widely applicable.

It is useful to make some generic distinctions. A *descriptive model* uses mathematics to characterize after-the-fact some set of data from experimental measurements in the form of a relation between independent and dependent variables. A *predictive model* seeks to predict the results of future measurements where the independent variables are similar but actual values are different from those used to develop the model. A *normative* model states what the dependent variables *should be,* given both the independent variables, and some (hopefully stated) assumptions about the mechanism of cause and effect. Critics argue that the normative model almost never fits the observed human behavioral events very well, and they are correct. One tries to improve the fit by adjusting the model parameters. When the fit is good, one can infer that the "structural mechanism" assumptions of the normative model are appropriate to characterize the cause-effect relation underlying human performance, and in this case use the normative model to predict human performance.

Quantitative descriptive models commonly use correlation and curve fitting and by extrapolation can become predictive models. There are elegant descriptive techniques such as factor analysis; the Brunswik Lens Model (Kirlik, 2005) embodies a structure of stages of correlation and is being actively pursued as a descriptive model of judgment. Because these approaches involve no cause and effect structure or mechanism they are not strictly normative models. Nor are taxonomies, such as classes of errors. Rasmussen's (1976) distinctions between skill, rule, knowledge, and level of automation do not qualify as quantitative normative models. Theories of naturalistic decision making (Zsambok & Klein, 1997) do not qualify as quantitative normative models. Nor do the techniques of statistical analysis such as null hypothesis tests or analysis of variance.

Normative models have existed in the physical sciences since before Newton. For human performance, it may be asserted that engineering development of physical weapons systems during World War II led to normative engineering theories that eventually, after the war, were adopted by psychologists and applied to model human behavior. Theories that underpinned the development of control systems for aircraft, ships, and guns, and that formed the basis for radar and sonar detection of enemy aircraft, were seen by psychologists to have analogies to human control and signal detection. Probability-cost decision theories that so-called *operations research* mathematicians used to develop military tactics were considered by psychologists to apply to individual human decision-making. An exception was information theory, which had its origins in Bell Laboratories well after the war. Recently, rule-based (both crisp and fuzzy) models evolved from the development of computers and software, and enabled "production rules" to be executed (Laird et al., 1987).

Discussed below are four classes of quantitative normative models, in more or less the order in which they were developed: (1) feedback control models, (2) expected value decision models, (3) information communication models, and (4) rule-based models. Some purely qualitative models that imply causative structure but are still begging for normative quantification are then mentioned.

It should be noted that the first three models predate digital computing, though of course all four are now implemented that way.

4.2 Feedback Control Models

4.2.1 Origins

Feedback control is an old idea. It may have been invented by mechanical designers of such devices as the flush toilet and the fly-ball governor for steam engines. We are all familiar with the flush toilet. In the fly-ball governor, as speed increases, spring-loaded balls are subjected to greater centrifugal force and move outward against springs, at the same time reducing the steam flow and eventually stabilizing the speed. MIT's Norbert Wiener, so-called father of Cybernetics (1964), is perhaps the best-known developer of control theory. The first documented set of lectures on automatic control engineering (as a subject in its own right) was given at Cambridge University (UK) by MacLellan in 1946–1947. An early American text was James, Nichols, and Phillips (1947). Figure 4.1 illustrates the basic idea of feedback control. Symbols G_c and G_p represent differential equations. A critical property of the closed loop is that if phase delay goes to near 180 degrees, the normally negative feedback becomes positive feedback, and the system becomes unstable. Barring this unhappy circumstance increasing gain of G_c causes output x to more and more closely track input r and bring error e to zero. This occurs when steering a car, for example.

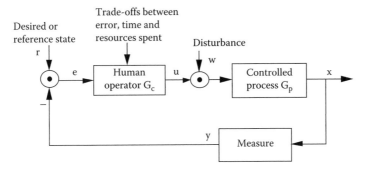

FIGURE 4.1 The feedback control paradigm.

4.2.2 Application of the Quasi-Linear Model to Human Performance

An early application of control theory to human tracking was by Tustin in the United Kingdom on "gun laying" in aircraft. Elkind (1956), then McRuer and Krendel (1957), and later McRuer and Jex (1967) were leaders in a multi-year Air Force project to develop the simplest (*quasi-linear*) differential equations to describe the human pilot. The objective was to engineer stable control loops for various pilot-aircraft combinations. The remarkable research finding by McRuer and his colleagues was that whatever the aircraft dynamics, within a surprisingly broad range, the pilot adapted dynamic equations to make the pilot-aircraft open loop closely resemble a simple integrator.

4.2.3 Nonlinear Control Applications

Other researchers soon took models beyond the quasi-linear assumption. Some observed a tendency of the human to be discontinuous in response, much as a "sampled-data" controller (Bekey, 1962). Other studies were mounted to explore how the equation parameters change with time in learning or fatigue (Sheridan, 1960) or how well the human could control a system where the control signal was a constant negative or a constant positive or zero, so-called *bang-bang-off control* (Pew, 1966). Young and Stark (1963) pioneered application of nonlinear control models to eye movements and the vestibular system.

Most of the studies assumed *compensatory tracking* (i.e., the human controller can only try to null an observed error). Modeling *pursuit tracking* (the human observes both error and reference input) was found to be more difficult, as the human element had two input signals rather than only error. There was also interest in *preview tracking* (the human can observe the desired path a certain distance ahead before actually getting there). Sheridan (1966) showed how a dynamic programming algorithm could define a normative model for the latter. The community of human factors professionals interested in applying control models met annually from roughly 1960 to 1985, different universities and government agencies hosting this "Annual Manual," which continues to this day in Europe.

4.3 Expected Value Decision Models

4.3.1 Origins

Typically the criterion (*objective function*) for making a decision is *expected value*—that is, take whatever decision will yield the greatest product of probability times benefit (in dollars, energy, time, safety, or a weighted combination of these). These ideas date to the *felicific calculus* of the 18th-century English utilitarian philosophers Jeremy Bentham and John Stuart Mill. (The visitor to the main entrance of London's University College will find Professor Bentham's embalmed and fully clothed corpse in a glass box, all in keeping with Bentham's will—allegedly ready to be carted to faculty meetings and be declared "present but not voting.")

A further normative criterion was provided by the 19th-century Italian economist Vilfredo Pareto, who articulated the social norm (*Pareto optimality*) that any policy change should never leave some people worse off, even though it will make things better for other people.

Reverend Thomas Bayes, another 18th-century Englishman, showed that if one knows the probability of some data D contingent upon a given cause C, one may work backwards to infer C from observed D, assuming knowledge of prior probabilities of C and D: $P(C|D) = P(D|C) P(C)/P(D)$. This simple expression, also known as Bayes' theorem, becomes particularly useful when one forms a ratio of two such expressions for two different candidate causes C_1 and C_2, in which case the common P(D) drops out. Then, given *likelihood ratio* $P(D|C_1)/P(D|C_2)$ and the *prior odds ratio* $P(C_1)/P(C_2)$ one can determine, after the occurrence of D, the *posterior odds ratio*, namely $P(C_1|D)/P(C_2|D)$. By repeating this calculation as new and different data become available (and presumably having as a prior for each D the correct P(D|C) based on experiments with C, the posterior odds ratio converges to the true value. Because of this convergence the initial prior odds ratio need only be a rough guess: One can even start with 50-50 provided there are enough new data.

It took many more years to provide a quantitative basis for relative worth or *utility*. Von Neumann and Morganstern (1944) showed how to elicit relative worth by means of a lottery judgment: an experimental subject specifies at what value C possessing of some objective variable (e.g., C in dollars) with 100% certainty is equivalent (i.e., the judge is indifferent) to having a 50-50 chance of X dollars or nothing. The underlying assumption of expected value means that utility $U(C) = 0.5 U(X)$ in this case. One can substitute C for the initial X and repeat the experiment, thus tracing out a utility function of that physical variable at $U(X) = 1$, then 0.5, then 0.25, and so on for many values of D.

An important decision model within the human factors community is the signal detection model. This model evolved during World War II from the need to discriminate signals from noise in radar (enemy aircraft detection) and sonar (enemy submarine detection). It draws on both Bayes' theorem and on subjective expected utility in its derivation and assumes overlapping probability densities of two types of events, classically, noise N and signal-plus-noise SN; fig. 4.2).

These probabilities are further assumed to be functions of some not-necessarily-observable *discrimination evidence variable* e (e.g., related to signal strength relative to noise) so that the likelihood ratio $P(e|SN)/P(e|N)$ increases monotonically with e. The important question is then: above what critical e^* to decide SN, and below that e^* to decide N? e^* sets probabilities of true positive TP or "hit," true negative TN, false positive or "false alarm" FA, and false negative or "miss" MS. These are represented by areas in Figure 4.2 and the coordinates in Figure 4.3. It can be seen that p(MS) is 1 − p(TP), and p(TN) = 1 − p(FA) so that there are only two independent variables.

Each of the curves in Figure 4.3 is known as a receiver operating characteristic (ROC), indicating the hit versus false alarm trade-off available to the decision maker, depending on what e^* criterion is selected (presumably on the basis of relative costs of miss and false alarm). The power of this model is

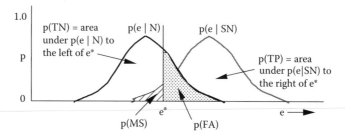

FIGURE 4.2 Overlapping probability densities assumed for two hypotheses in signal detection theory.

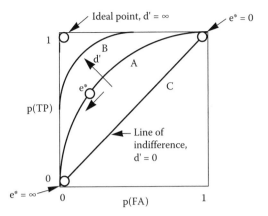

FIGURE 4.3 Curves of the relative operating characteristic (ROC).

that the discrimination parameter d′ (greater as the probability densities are more separated) is orthogonal to the criterion e*, also known as β.

4.3.2 Applications of Decision Models to Human Performance

The person best known for using Bayes' theorem as a normative model for human decision behavior was Edwards (1962). Edwards was famous for his "probabilistic information processor," wherein many hypothetical Cs can be considered by a computer, and the human judge need only put in the appropriate likelihood ratios. He made the controversial suggestion that battlefield decisions be determined this way. He and his students got permission to instrument a game at a Las Vegas casino to study Bayesian decision-making in real life.

There is much evidence that as humans gather evidence they are not able to converge on the underlying hypothesis nearly as fast as the ideal Bayesian process. Much of this work is attributed to Kahneman and Tversky (1982). Bayesian algorithms are now widely used as decision aids to reveal underlying causes and anticipate failures.

Green and Swets (1966) are best known for applying signal detection theory to problems of psychophysics. They demonstrated that so-called absolute thresholds are not fixed; the experimental subject "detects the signal" as a function of the relative payoffs for hit, miss, false alarm, and true negative. More recently Swets and others have applied this model to medical diagnosis, where the relative rewards and costs are critical for both patient and caregiver.

Economist Herbert Simon introduced the notion of *satisficing*, meaning that decision makers invest only limited effort in finding an optimum resolution, but at some level are satisfied that their decision is "good enough for now." Humans are limited in their ability to weigh in their minds between multiple "worth variables," but computer decision aids can help. For example, if many items were available that had different initial costs and maintenance costs (both dark and light circles in fig. 4.4) a computer can readily identify the items on the resulting Pareto frontier (dark circles only) and discover which alternatives maximize subjective utility (where "best" utility indifference curve contacts the Pareto frontier). This specifies the theoretical optimal decision.

Attention allocation has long been of interest to psychologists, but without much quantitative underpinning. A classic experiment by Senders (1964) and colleagues measured allocation of visual attention to multiple aircraft instruments each varying with a different bandwidth (highest frequency component). His normative basis was the Nyquist criterion, which states that if any signal can be sampled at twice its greatest bandwidth no information is lost. Senders also had subjects drive a car along the

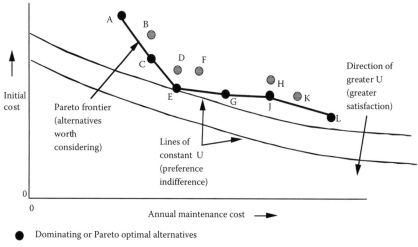

FIGURE 4.4 Intersection of Pareto optimal frontier (dark circles) with curves of constant utility (utility indifference curves).

streets of Cambridge, Massachusetts, with varying traffic density, wearing a helmet that allowed a quick glance only when he actuated it. In both cases he compared glance frequency to highway-traffic bandwidth and found a good correspondence.

4.4 Information Communication Models

4.4.1 Origins

Information theory is commonly attributed to Claude Shannon (1949), then at Bell Telephone Laboratories, later at MIT. The basic idea is that the information content in a message is its surprise value, the inverse of its probability, and has nothing to do with its meaning. This itself was a surprising (novel) idea. Shannon defined average information H(x) over a set of messages i is $\Sigma_{ij} P(x_i) \log_2[1/P(x_i)]$. He also showed that information transmitted H(x:y) from a set x_i to a set y_j is $\Sigma_{ij} P(x_i y_j) \log_2[P(x_i \mid y_j)/P(x_i)]$. McGill (1954) extended Shannon's ideas to multi-dimensions, and showed its close equivalence to analysis of variance.

4.4.2 Applications of Information Models to Human Performance

Psychologists quickly adapted Shannon's idea to modeling human performance, and started measuring human visual-motor capacity in various tasks such as reading, piano playing, and so on. In the United Kingdom, Hick (1952) showed how the time it takes to make a simple selection (e.g., which of N lights is lit) is proportional to $\log_2 N$. Fitts (1954) showed that the time it takes to make a precise movement (e.g., tap a pencil from A inches away to within a tolerance B) is proportional to $\log_2(A/B)$. Both experiments suggest a constant rate of information processing.

Miller (1956) showed that human memory capacity in making absolute judgments along a single stimulus continuum (e.g., how many different pitches can be identified correctly, on average, after brief training) is constant at about 7 (slightly more for vision and hearing variables, less for taste, touch, and smell). This capacity increases logarithmically with the number of stimulus dimensions (e.g., pitch and loudness). The logarithmic property of absolute judgments is not itself a surprise considering the Weber-Fechner law for differential thresholds and the logarithmic physical-to-subjective-magnitude scales that psychophysics already had established. Nowadays the Shannon information measure is not so much used by human factors professionals, which is curious given its earlier popularity.

4.4.3 Information Value

A final information topic to be mentioned is Howard's (1966) underappreciated notion of *information value*. This term refers to the expected payoff to act knowing the truth of any specific message from a set of messages, otherwise having to act knowing only their probabilities. (This comes closer to the *meaning* issue that Shannon avoided with his measure.) The idea is as follows: If message x_i is known exactly, then a rational decision maker takes action u_j (selects j) to maximize payoff V for each occurrence of x_i, in each instance yielding $\max_j[V(u_j \mid x_i)]$. Then average payoff over a set of x_i is $V_{avg} = \Sigma_i p(x_i)\{\max_j[V(u_j \mid x_i)]\}$. If x_i is known only by a probability density, $p(x_i)$, then the best a rational decision-maker can do is to adjust u_j once, to be the best in consideration of the whole density function $p(x_i)$. In this case the average reward over a set of x_i is $V'_{avg} = \max_j\{\Sigma_i p(x_i) V(u_j \mid x_i)\}$. *Information value*, then, is the difference V^*_{avg} between the gain in taking the best action given each specific xi as it occurs, and the gain in taking the best action in ignorance of each specific x_i, that is, knowing only $p(x_i)$: $V^*_{avg} = V_{avg} - V'_{avg}$.

4.5 Rule-Based Models

All models may be said to be rule-based. But the term rule-based usually refers to discrete explicit "if … is the case, then do… , else do…" rules.

4.5.1 Network Models

Network models include decision trees and Markov networks that specify conditions of transition (often with accompanying probability) to a next state given a present state. Computer flow-charts are network models where the boxes can represent any input-output transformation function couched in mathematics. But because many such structures are so arbitrary I do not see them as normative in the sense earlier defined.

4.5.2 Production Rule-Based Models

Production rules are typically "if… then… else …" statements represented in software. When put together with working memory about what has been learned so far, plus an *interpreter* that decides which rules to invoke on a next selection-execute cycle, the computer can solve problems, recognize patterns, and show "intelligence." It is the computer science behind "expert systems" and "intelligent agents." Production rules were the basis of the original Mycin drug diagnosis expert system. More recently Soar and related "cognitive architectures" (for problem solving) were begun by Alan Newell (1990). Soar remains an active project at several universities (Laird, 1987). A problem with production rule models is that they are not analytical in the sense of closed-form mathematics. One can run the programs with systematic parameter changes or in Monte Carlo fashion, but cannot easily optimize.

4.5.3 Fuzzy Rule-Based Models

Fuzzy set theory, invented by Lotfi Zadeh (1965), is particularly applicable to modeling human performance, because it is built on the idea of overlap of ideas, and particular overlap of words of natural language. Any given "state of the world," a set of values for multiple continuous physical variables characterizing some object or event, can be specified as *memberships* or *relevance weights* on each of a given set of discrete *fuzzy variables* as illustrated in Figure 4.5. Independently of the memberships of any particular object or event, fuzzy production rules can be stated in terms of these fuzzy variables (e.g., "If aircraft

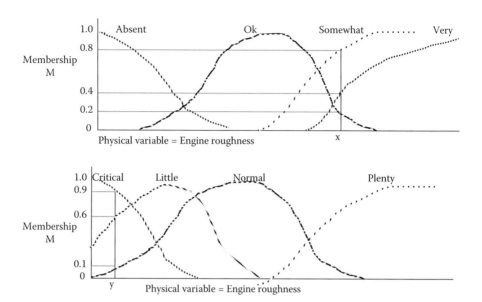

FIGURE 4.5 Translation of continuous state into memberships on fuzzy variables to determine relevance of rules and take action.

engine is SOMEWHAT rough or VERY rough, and there is LITTLE remaining fuel,—or if fuel is CRITICAL, then land the aircraft." This rule may be parsed (using the different brackets to convey how the rule applies to the object or event being considered:

If { [(somewhat rough) OR (very rough)] AND [(little fuel)] } OR {(fuel critical)} -->> Land,

then the relevance M of that rule for the state in Figure 4.5 is

$$max \{min [max (0.8), (0.4)], [(0.6)]\}, \{(0.9)\} = 0.9.$$

With more rules the same procedure is applied. The approach models decision and action according to discrete production rules for a continuous physical situation.

4.6 Some Popular Qualitative Models Begging for Normative Quantification

At the end of this very limited review of the history of models of human performance, particularly normative models, some caveats are offered concerning popular qualitative models deserving of effort to make them quantitative.

The problem of attention has particularly benefited from Wickens' (1980) resource theory, but *attention allocation* can also be viewed as a knowledge and decision problem. In that vein, and because of its criticality to driving, piloting, and operating other systems in time critical situations, it deserves further normative modeling beyond the Senders model based on just bandwidth (Sheridan, 2005).

Situation awareness (SA) is a very popular idea, and a procedure for interrupting a simulation and asking questions is a practical experimental measurement procedure (Endsley, 1995), but a normative model is lacking. There seems to be a close tie between SA and attention.

Mental workload garnered great attention in the late 1970s after the airline pilots union contested the airlines over 2- versus 3-person flight crews, and subsequently mental workload rating scales were developed by NASA (Hart, 1984) and the Air Force. But no normative model yet exists.

There has been discussion of *supervisory control* since 1960, and again there are taxonomies and "frameworks" (Sheridan, 1992) but yet nothing exists that could be called a normative model.

Finally there is that most elusive notion of *mental model* (a hypothetical model in the head that comprehends how some variables in the world interact and can effect real-time simulations). This may be no different from "thinking" and problem solving, which the Soar and AI community is trying to cope with.

References

Bekey, G.A. (1962). The human operator as a sampled-data system. *IEEE Trans. Human Factors in Electronics*, HFE3, 43–51.

Edwards, W. (1962). Dynamic information theory and probabilistic information processing. *Human Factors*, 4, 59–73.

Elkind, J.I. (1956). *Characteristics of Simple Manual Control Systems*. Cambridge, MA: MIT Lincoln Laboratory Tech. Rep. 111.

Endsley, M.R. (1995). Measurement of situation awareness in dynamic systems. *Human Factors* 37 (1), 65–84.

Fitts, P.M. (1954). The information capacity of the human motor system in controlling the amplitude of movement. *J. Exper. Psychol.*, 47, 381–391.

Green, D.M. and Swets, J.A. (1966). *Signal Detection Theory and Psychophysics*. New York: Wiley.

Hart, S.G. and Sheridan, T.B. (1984). Pilot workload, performance and aircraft control automation. In *Proceedings of NATO/AGARD Conference 371, Human Factors Considerations in High Performance Aircraft*, Nuilly sur Seine, France.

Hick, W.E. (1952). On the rate of gain of information. *Quart. J. Experimental Psychology*, 4, 11–26.

Howard, R.A. (1966). Information value theory. *IEEE Trans. on Systems Science and Cybernetics*, SSC-2, 22–26.

James, H.M., Nichols, N.B. and Philips, R.S. (1947). *Theory of Servomechanisms*. New York: McGraw Hill. (See Chapter 15, *Modeling Response Time and Accuracy for Digital Humans*.)

Kahneman, D. and Tversky, A. (1982). The psychology of preferences. *Scientific American*, 246 (1), 160–173.

Kirlik, A., Ed. (2005). *Adaptive Perspectives on Human-Technology Interaction*. New York: Oxford.

Laird, J., Newell, A. and Rosenbloom, P. (1987). SOAR: An Architecture for General Intelligence. *Artificial Intelligence*, 33.

March, J.G. and Simon, H.A. (1958). *Organizations*. New York: Wiley.

McGill, W. (1954). Multi-variate information transmission. *Psychometrika*, 19, 97–116.

McRuer, D.T. and Jex, H.R. (1967). A review of quasi-linear pilot models. *IEEE Trans. Human Factors in Electronics*, HFE-4, 3, 231–249.

McRuer, D.T. and Krendel, E.S. (1957). Dynamic response of human operators. WADC-TR 56-524, US Air Force.

Miller, G.A. (1956). The magical number 7, plus or minus 2: some limits on our capacity for processing information. *Psych. Review*, 63, 81–97.

Newell, A. (1990). *Unified Theories of Cognition*. Cambridge, MA: Harvard University Press.

Pew, R.W. (1966). A model of human operators in a 3-state with velocity augmented displays. *IEEE Trans. Human Factors in Electronics*, HFE-7, 2, 77–83.

Rasmussen, J. (1976). Outlines of a hybrid model of the process plant operator. In T. Sheridan and G. Johannsen (Eds.), *Monitoring Behavior and Supervisory Control* (pp. 371–384). New York: Plenum.

Senders, J.W., Elkind, J.I, Grignetti, M.C. and Smallwood, R.P. (1964). An investigation of the visual sampling behavior of human observers. NASA–CR–434. Cambridge, MA: Bolt, Beranek and Newman.

Shannon, C.E. (1949). Communication in the presence of noise. *Proceedings of the IRE, 37*, 10–22.

Sheridan, T.B. (1960). Experimental analysis of time variation of the the human operator's transfer function. *Proc. 1st Intl. Federation of Automatic Control, Moscow* (pp. 1681–1686). London: Butterworths.

Sheridan, T.B. (1966, June). Three models of preview control. *IEEE Trans. Human Factors in Electronics,* HFE–6.

Sheridan, T.B. (1992). *Telerobotics, Automation and Human Supervisory Control.* Cambridge, MA: MIT Press.

Sheridan, T.B. (2005, June). Attention and its allocation. *Proc. Applied Attention: From Theory to Practice* (festschrift for C. Wickens). Champaign-Urbana, IL: University of Illinois.

Von Neumann, J. and Morganstern, O. (1944). *Theory of Games and Economic Behavior.* Princeton, NJ: Princeton University Press.

Wickens, C.D. (1980). The structure of attentional resources. In R. Nickerson (Ed.), *Attention and Performance* (VIII, pp. 239–257). Hillsdale, NJ: Erlbaum.

Wiener, N. (1964). *God and Golem, Incorporated.* Cambridge, MA: MIT Press.

Young, L.R. and Stark, L. (1963). Variable feedback experiments testing a sampled data model for eye tracking movements. *IEEE Trans. Human Factors in Electronics*, HFE-4, 1, 38–51.

Zadeh, L.A. (1965). Fuzzy sets. *Information and Control*, 8, 338–353.

Zsambok, C.E. and Klein, G.A. (1997). *Naturalistic Decision-Making.* Mahwah, NJ: Erlbaum.

5

Physics-Based Digital Human Modeling: Predictive Dynamics

Karim Abdel-Malek
and Jasbir Arora

5.1 Introduction

Digital humans are characters that live inside the virtual world. They can be as strong as we program them to be; they can run for miles without fatiguing; and they may collide with objects without feeling pain and causing injury. Physics-based modeling and simulation not only provide for the science behind making the simulation as realistic as possible but also provide a method to predict human motion. This is a significant departure from current methods that require user manipulation of the digital mannequin.

Typical methods for evaluating safety and ergonomics using digital humans have been based on simple cartoonish avatars that must be manipulated in the digital environment. An expert ergonomist, for example, would have to manipulate the avatar's shoulder and elbow joints to simulate a reach, and the user would then evaluate the resulting posture and deem it good or no good.

Computer games have rapidly advanced in the last decade to a high level of sophistication. Realism is paramount. While gaming developers have created simplified physics-based mathematical models to characterize motion, collision, trajectories of projectiles, and to some extent human gait, these models remain simplistic in nature due to the computational constraints mandated by gaming computers.

The need for a uniform, systematic, and computationally capable strategy for a physics-based approach to digital human modeling is significant. Because digital human models have been developed for the purpose of static manipulation of the mannequin, the level of sophistication has been relatively low. As digital human model (DHM) environments have become part of the design cycle in a manufacturing environment—used for safety analysis, integrated into product life cycle management solutions, and employed to simulate soldiers—it has become abundantly clear that static analysis alone is not enough. The DHM is no longer a static mannequin, but rather a character on the move that has force actions and reactions with inertia and resulting dynamics.

In this chapter, we present a comprehensive methodology for creating digital humans that can predict their motion while subject to the laws of physics. This method has been developed over the past 8 years, drawing upon expertise in robotics, kinematics, multi-body dynamics, optimization, human performance, and ergonomics (Abdel-Malek et al., 2001, 2004a, 2004b, 2004c, 2006a, 2006b; Farrell, 2005; Horn, 2005; Yang et al., 2004, 2006a, 2006b, 2007; Kim et al., 2004, 2005; Xiang et al., 2007; Chung et al., 2007). It is based on PhD dissertations (Yang, 2003; Yu, 2001; Marler, 2004; Mi, 2004; Kim, 2006; Wang 2006), many MS theses, and the work of 35 researchers at the University of Iowa led by the authors.

5.2 Predicting Human Motion

Predicting how humans move is no easy task. In this chapter, we introduce a new theory in the prediction of human motion called *predictive dynamics*. Its inventors will tell you that it is not new because it is based on Newton's laws of motion published in his *Philosophiae Naturalis Principia Mathematica* in 1687. However, it is new in the sense that it is the only method known to date that predicts realistic human motion while conforming to the laws of physics.

5.2.1 What Is Predictive Dynamics?

Predictive dynamics is a term coined to characterize a new methodology for predicting human motion while considering the dynamics of the human and the environment. Whereas kinematics is the study of motion (position, velocity, and acceleration) without forces and torques, dynamics is the study of motion with all external and internal forces taken into consideration.

For every motion affected by physics, there are laws that govern that motion. These laws have undergone the test of time, have sent people to the moon, and have been implemented into every computer that governs motion. Equations that represent motion are called the equations of motion (EoM). For large redundant systems such as the human body, these equations become very sophisticated nonlinear differential equations subject to algebraic constraints, hence the term often used to represent these equations is *differential algebraic equations* (DAEs).

For a sophisticated system of segments such as the human being, which is made up of joints and rigid links (in this case we assume rigidity of our bones), the formulation for multi-body dynamics becomes large and complex. Solving the consequent system of equations is almost impossible. Indeed, for structural systems with limited number of degrees of freedom (DOF), integrators have been developed to solve the problem iteratively. For high degrees of freedom, however, integrators come to a halt.

Recent results have demonstrated significant promise for resolving the problem of predicting human motion while taking into consideration external forces, obstacles, physiology, and most importantly the equations of motion. This method, which we call predictive dynamics, provides a way to address the issue of predicting human motion in a general manner. Our recent results have shown that this method is applicable to gait prediction, lifting movements, pushing and pulling movements, climbing, and many others.

5.2.2 How Does Predictive Dynamics Work?

Predictive dynamics works based on a special optimization formulation. Consider a general optimization problem, for which there are three main ingredients: (1) a set of design variables, which in our case are the joint profiles (i.e., joint angles as a function of time) and the torque profiles at each joint; (2) multiple cost functions to be optimized, which are human performance measures that represent functions that are important to accomplishing the motion (e.g., energy, speed, joint torque); and (3) constraints on the motion (e.g., collision avoidance, joint ranges of motion). This general optimization problem is readily solved using existing optimization code. The field of optimization is mature, and many such codes exist, have been verified and validated, and have been tested with many different complex problems.

Solving the above problem predicts human motion. It has been shown that for static postures (i.e., predicting final human postures to reach an object), this method is very successful. The Virtual Soldier Research Program has put forth a great deal of effort over the past few years to make sure that this method is indeed valid, produces humanlike results, and is easily implemented. We are happy to report that the method passed the tests with flying colors.

Now we add the issue of dynamics. We are interested in seeing how human motion is predicted for scenarios that involve dynamic influences including but not limited to external loads, obstacles, and running. In general, we consider any case where a human segment is undergoing motion that warrants the consideration of masses and moments of inertia. Predictive dynamics can incorporate such general cases.

5.2.3 Why Data-Driven Human Motion Prediction Does Not Work

We firmly believe that the data-driven approach to human motion prediction is the wrong approach. Thousands of experiments are typically done to capture a few motions. These motions are then compiled into large tables with many parameters. The data are then analyzed and modeled as a nonlinear or functional regression model that should, in principle, predict motion. There are many obvious problems with this method:

- Difficulties in collecting the data for varying anthropometries. This includes the changes in masses, moments of inertia, muscle performance, and many other parameters for each person.
- Difficulties in managing a large number of parameters in a functional or nonlinear regression algorithm. A large number of parameters means a sophisticated and less accurate model.
- Difficulties in predicting postures and motions for reaches that have obstacles. For each obstacle, the experiments must be repeated.
- Difficulties in predicting motions where dynamics (external forces and loads) play an important role.

After the apple fell from the tree onto Newton's head, he proceeded to measure a few more, came up with the general theory, and finally devised a rigorous mathematical formulation for all falling objects and, furthermore, for all objects in motion. He did not measure every apple on every tree to come up with a theory that works and is the fundamental basis for all motion in our universe.

Recording every motion for thousands of people and for thousands of different scenarios does provide a good way to study motion and to validate motion predicted with various methods. However, it has no value for the prediction of motions beyond static postures.

5.3 Kinematics

The objective of this section is to establish a systematic method for representing human anatomy and to develop mathematical methods for kinematic analysis as the human body undergoes motion. Kinematic analysis in this context means the study of motion characterized by the position, velocity, and acceleration of human segmental links as opposed to dynamics, which is the study of forces and torques affecting or resulting from the motion.

Throughout this chapter, we shall consider the various segments of the body as individual rigid bodies that are connected via joints. Human modeling techniques have rapidly evolved in recent years, driven by the need for safety, security, and better ergonomics, as well as the need for avatars to perform tasks that could not be performed in the real world. Perhaps the most influential force behind this fast pace is the gaming industry, where avatars are extensively used to interact and respond in real time. Similarly, in the movie industry, digital characters are used to replace actors, and it is difficult in some cases to differentiate the real from the virtual. In general, human models have been represented as stick figures, skeletons, mesh surfaces, profiles, and mannequins.

Consider the motion of a person's arm from one position to another, where only the elbow joint is changed. As a result of this simple motion, the hand is also moved in space to a final configuration defined by a position and an orientation (also called a "pose"). In order to characterize the motion of these segmental links and their associated joints, it is necessary to establish a systematic approach for specifying coordinate systems defined by **xyz** on each link, and to establish a method for relating any two such coordinate systems.

Let us assume for a moment that we are able to specify values of the joint variables for the shoulder complex, the elbow, and the wrist, and we want to know the final position of the hand (fig. 5.1). In this case, we specify a vector **x** that describes the position of the hand with respect to another coordinate system (e.g., the foot). This chapter addresses exactly this issue. It will provide a rigorous method for formulating a set of equations that have the joint variables as their parameters. If the final hand position is required, variations in the joint variables are substituted into the equation, and the final position is readily obtained.

We shall also introduce a method for modeling human joints, as simple or as complex as necessary, that represents the resulting interaction between any two segmental links. The simplest form of these joints is the rotational joint (e.g., the elbow joint). The combination of a number of simple joints can become complex in nature but still be represented using this straightforward approach.

The Denavit–Hartenberg (DH) method was created in the 1950s to systematically represent the relationship between two coordinate systems but was not extensively used until the early 1980s with the introduction of computational methods and hardware that enabled the necessary calculations. The method is currently used to a great extent in the analysis and control of robotic manipulators. This method has also been successful in addressing human motion, in particular toward a better understanding of the mechanics of human motion.

It is important to distinguish the difference between a rigid body and a flexible body. A rigid body is one that cannot deform (we typically consider bone as non-deforming, at least for the moment). However,

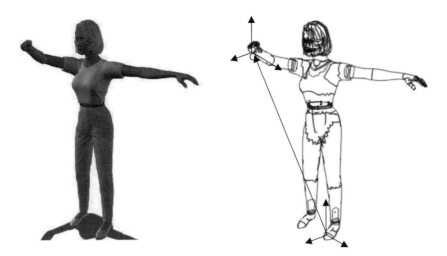

FIGURE 5.1   A vector **x** representing the position of the hand with respect to the foot.

a flexible body (or deformable object) is one that undergoes relatively large strains when subjected to a load (e.g., soft tissue). For the approach presented in this section, only rigid body motion is assumed at all times. Indeed, for ergonomic design considerations, rigid body motion is adequate to address most problems. Muscle interaction and deformation will be addressed in later chapters.

The DH modeling method is suitable for addressing the motion of kinematic structures that are arranged in series. The DH method will be used to perform analysis on the human body in this chapter and to predict postures and perform ergonomic analysis in later chapters. A posture is defined in this text as the configuration of a series of segmental links in the human body.

The human body is indeed arranged in series, where each independent anatomical structure is connected to another via a joint. Consider, for example, that there exists a main coordinate system located at the waist. From that coordinate system, one may be able to draw a branch by identifying a rigid link, connected through a joint to another rigid link, connected to another link, until you reach the hand. Each finger also comprises a number of segmental links connected via joints. Similarly, starting from the waist, one may follow the connection to reach the head, the other hand, the left foot, and the right foot. We shall refer to one such chain as a branch. For example, Figure 5.2 depicts the modeling of a human into a number of kinematic branches. A hand can be represented by five branches, one for each finger.

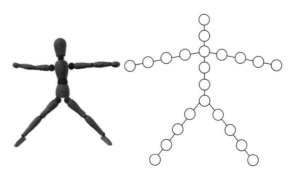

FIGURE 5.2   Modeling of a human using a series of rigid links connected by joints.

Because we consider gross human motion, detailed modeling of the joint connection is less important at this stage. Nevertheless, because in many cases accurate biomechanical modeling of a joint is needed, in this chapter we will present a more elaborate method for representing the kinematic interactivity within the joint while taking into consideration muscle action, ligaments, and other effects that are not considered in the DH representation method.

The general translational and rotational motion of an object will first be presented, followed by a standardization of a method for embedding coordinate systems (also called *triads*) in each segmental link. We will then develop a formulation that relates any two segmental links in this chain.

5.3.1 Basic Transformation Matrices

The concept of a basic transformation matrix is fundamental to the study of human motion. This $^A\mathbf{T}_B$ matrix is called the *transformation matrix* and is read as the transformation from B to A. It can be seen as acting on a rigid body causing a transformation (i.e., changing its configuration). On the other hand, and this is the concept that will be used throughout this text, it can be seen as an operator acting on a vector $^B\mathbf{x}$, which is resolved in the B-coordinate system. This is similar in action to the rotation matrix but includes the translation as well.

The transformation matrix is indeed a partitioned matrix where the upper left corner submatrix is the rotation matrix and the upper right vector is the position vector.

$$^A\mathbf{T}_B = \begin{bmatrix} \text{rotation} & \text{translation} \\ \text{matrix} & \text{vector} \\ \hline \mathbf{0} & 1 \end{bmatrix} \tag{5.1}$$

$$^A\mathbf{T}_B = \begin{bmatrix} ^A\mathbf{R}_B & ^A\mathbf{p} \\ \hline \mathbf{0} & 1 \end{bmatrix}. \tag{5.2}$$

Basic homogeneous transformation matrices for rotation about the x and y axes, respectively, are given by

$$\mathbf{T}_{x,\alpha} = \begin{bmatrix} 1 & 0 & 0 & 0 \\ 0 & \cos\alpha & -\sin\alpha & 0 \\ 0 & \sin\alpha & \cos\alpha & 0 \\ 0 & 0 & 0 & 1 \end{bmatrix} \tag{5.3}$$

and

$$\mathbf{T}_{y,\phi} = \begin{bmatrix} \cos\phi & 0 & \sin\phi & 0 \\ 0 & 1 & 0 & 0 \\ -\sin\phi & 0 & \cos\phi & 0 \\ 0 & 0 & 0 & 1 \end{bmatrix}. \tag{5.4}$$

where the rotation matrices are those of rotation about x and y, respectively.

A pure translation matrix, called a *basic transformation matrix* for translation, can be written by specifying no rotations (i.e., an identity matrix for the rotation matrix) but by specifying the coordinates

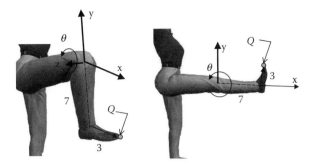

FIGURE 5.3   Rotation of the lower limb around one axis.

of the three elements in the last column. A transformation matrix for translation along the x axis by a units can be written as

$$\mathbf{T}_{x,a} = \begin{bmatrix} 1 & 0 & 0 & a \\ 0 & 1 & 0 & 0 \\ 0 & 0 & 1 & 0 \\ 0 & 0 & 0 & 1 \end{bmatrix}. \tag{5.5}$$

5.3.1.1 Example

In this example, we seek to represent the position vector of a point specified on the foot as the lower limb undergoes a rotational motion at the knee. A point Q on the foot is shown in Figure 5.3 and is represented by the vector $\mathbf{x}_Q = [3\ -7\ 0\ 1]^T$ with respect to the coordinate system located at the knee. The knee is constrained to allow motion only about the axis $\mathbf{z}$ by an angle θ. It is necessary to calculate the final position of the foot point Q as the joint rotates (i.e., as a function of θ).

The lower limb rotates about the axis $\mathbf{z}$; thus the pure rotational homogeneous matrix is

$$\mathbf{T}_{z,\theta} = \begin{bmatrix} \cos\theta & -\sin\theta & 0 & 0 \\ \sin\theta & \cos\theta & 0 & 0 \\ 0 & 0 & 1 & 0 \\ 0 & 0 & 0 & 1 \end{bmatrix}. \tag{5.6}$$

Note the translation vector is zero. To determine the location of Q at any joint displacement θ, we multiply $\mathbf{x}_Q$ by $\mathbf{T}_{z,\theta}$ to calculate the rotated vector $\mathbf{x}'_Q$:

$\mathbf{x}'_Q = \mathbf{T}_{z,\theta}\mathbf{x}_Q = [3\cos\theta + 7\sin\theta\ \ 3\sin\theta - 7\cos\theta\ \ 0\ \ 1]^T$. With this expression, it is possible to calculate the position of the lower limb at any specified value of θ. For example, let $\theta = 90°$: then the rotated limb's new position is $\mathbf{x}'_Q(90°) = [7\ \ \ 3\ \ \ 0\ \ \ 1]^T$, which is shown in Figure 5.3 as the lower limb is extended.

5.3.2 Directed Transformation Graphs

Consider the coordinate frames depicted in Figure 5.4. Each graph from one frame to another represents a transformation matrix and is denoted by the **T**-matrix. The direction of the arrow indicates subscript and superscript, respectively, of the **T**-matrix (i.e., the transformation from frame 0 to frame 1 is denoted by $^0\mathbf{T}_1$). A graph from frame 1 to frame 2 is represented by $^1\mathbf{T}_2$.

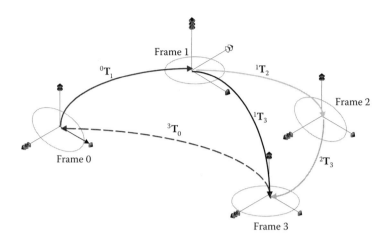

FIGURE 5.4   Directed transformation graphs.

Applying a sequence of transformations such as $^0\mathbf{T}_1$ followed by $^1\mathbf{T}_2$ yields a graph from frame 0 directly to frame 2 represented by the transformation matrix

$$^0\mathbf{T}_2 = {}^0\mathbf{T}_1\,{}^1\mathbf{T}_2. \tag{5.7}$$

Similarly, applying another transformation from frame 2 to frame 3 can be represented by a directed graph from frame 0 to frame 3 and be characterized by

$$^0\mathbf{T}_3 = {}^0\mathbf{T}_1\,{}^1\mathbf{T}_2\,{}^2\mathbf{T}_3. \tag{5.8}$$

A graph from frame 3 to frame 0 is represented by $^3\mathbf{T}_0$; therefore, the multiplication of a transformation matrix by its inverse yields the identity matrix

$$^0\mathbf{T}_3\,{}^3\mathbf{T}_0 = \mathbf{I}. \tag{5.9}$$

It is also noted here that a transformation between any two frames can be obtained independent of the route followed; that is, the transformation from frame 0 to frame 2 can be obtained by

$$^0\mathbf{T}_2 = {}^0\mathbf{T}_1\,{}^1\mathbf{T}_2 \tag{5.10}$$

or by

$$^0\mathbf{T}_2 = {}^0\mathbf{T}_3\,{}^3\mathbf{T}_2 \tag{5.11}$$

where the transformation $^3\mathbf{T}_2 = (^2\mathbf{T}_3)^{-1}$.

The order in which matrices are multiplied is important since matrix multiplication is not commutative; in general, $^A\mathbf{T}_B\,{}^B\mathbf{T}_C \neq {}^B\mathbf{T}_C\,{}^A\mathbf{T}_B$. In fact, great care must be given to the order of multiplication. Consider two coordinate systems, $X_1Y_1Z_1$ being the world coordinate system, and $X_2Y_2Z_2$ being the body reference frame. Two rules must be followed in applying the order of multiplication of transformation matrices. These rules are given without proof.

- A transformation taking place with respect to the world reference frame $(X_1Y_1Z_1)$ necessitates the *pre-multiplication* of the previous transformation matrix by an appropriate basic homogeneous transformation matrix.
- A transformation taking place with respect to the body's own reference frame $(X_2Y_2Z_2)$ necessitates the *post-multiplication* of the previous transformation matrix by an appropriate basic transformation matrix.

5.3.2.1 Example

A digital human lives in a computer-aided engineering environment. This digital human will be asked to grasp and move some objects. To enable the digital human to identify objects in the workspace, a virtual camera is embedded in its head and will function as eyes. This camera will determine the position and orientation of an object in space and will return a homogeneous transformation matrix. The virtual camera senses the position of a joystick **J** shown in Figure 5.5. The virtual camera's coordinate system is represented by $\mathbf{c}_1$, $\mathbf{c}_2$, and $\mathbf{c}_3$, as shown in Figure 5.5.

The camera identifies the position and configuration of the shoulder and the joystick. The homogeneous transformation matrix $^C\mathbf{T}_J$ of the joystick J as seen by the camera C is

$$^C\mathbf{T}_J = \begin{bmatrix} 0 & 1 & 0 & 5 \\ 0 & 0 & 1 & 24 \\ 1 & 0 & 0 & -15 \\ 0 & 0 & 0 & 1 \end{bmatrix}.$$

The transformation matrix of the coordinate system embedded in the shoulder with respect to the camera is given as

$$^C\mathbf{T}_S = \begin{bmatrix} 0 & 0 & 1 & 10 \\ 1 & 0 & 0 & 0 \\ 0 & 1 & 0 & 15 \\ 0 & 0 & 0 & 1 \end{bmatrix}.$$

It is necessary to

1. Sketch the unit basis vectors $\mathbf{j}_1$, $\mathbf{j}_2$, and $\mathbf{j}_3$ in the correct orientation at the origin O_j of the joystick J
2. Sketch the unit basis vectors ($\mathbf{s}_1$, $\mathbf{s}_2$, $\mathbf{s}_3$) in the correct orientation at the origin O_S of the shoulder S
3. Calculate the coordinates of the joystick origin O_j relative to the shoulder S
4. Determine the transformation matrix $^S\mathbf{T}_H$ for the hand H as seen by the shoulder, when the hand is positioned to grasp the cube, as shown in Figure 5.6

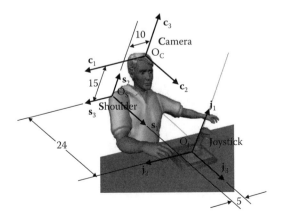

FIGURE 5.5 A digital human whose eyes are replaced by a camera.

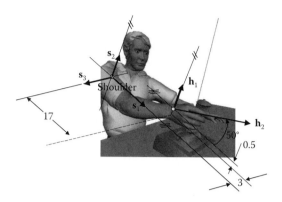

FIGURE 5.6   Orienting the hand to grasp an object.

Since the rotation matrix $^C\mathbf{R}_J$ extracted from $^C\mathbf{T}_J$ is given by $^C\mathbf{R}_J = \begin{bmatrix} 0 & 1 & 0 \\ 0 & 0 & 1 \\ 1 & 0 & 0 \end{bmatrix}$, the vector basis at O_C is given by $\begin{bmatrix} \mathbf{j}_1 & | & \mathbf{j}_2 & | & \mathbf{j}_3 \end{bmatrix} = \begin{bmatrix} 0 & | & 1 & | & 0 \\ 0 & | & 0 & | & 1 \\ 1 & | & 0 & | & 0 \end{bmatrix}$, which can readily be drawn at O_J with respect to the video camera coordinate system (see fig. 5.7). Similarly, the unit basis $\begin{bmatrix} \mathbf{s}_1 & | & \mathbf{s}_2 & | & \mathbf{s}_3 \end{bmatrix}$ is extracted from $^C\mathbf{T}_S$ and plotted as shown in Figure 5.7.

The coordinates of the joystick origin relative to the shoulder coordinate frame S (i.e., seeking the vector $^S\mathbf{p}_{SJ}$) can be either read from Figure 5.7 as $^S\mathbf{p}_{SJ} = [24\ 0\ -5]^T$ or calculated numerically from $^S\mathbf{T}_J$, which can be written as

$$^S\mathbf{T}_J = {}^S\mathbf{T}_C\,{}^C\mathbf{T}_J.$$

Since only $^S\mathbf{T}_C$ is given, its inverse is computed as

$$^S\mathbf{T}_C = \begin{bmatrix} ^C\mathbf{T}_S \end{bmatrix}^{-1} = \begin{bmatrix} 0 & 1 & 0 & 0 \\ 0 & 0 & 1 & 15 \\ 1 & 0 & 0 & -10 \\ 0 & 0 & 0 & 1 \end{bmatrix}.$$

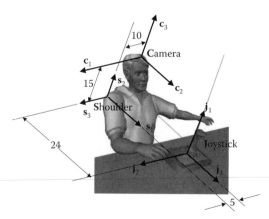

FIGURE 5.7   Solution to points 1 and 2.

Calculating $^S\mathbf{T}_J$ yields

$$
^S\mathbf{T}_J = \begin{bmatrix} 0 & 1 & 0 & 0 \\ 0 & 0 & 1 & 15 \\ 1 & 0 & 0 & -10 \\ 0 & 0 & 0 & 1 \end{bmatrix} \begin{bmatrix} 0 & 1 & 0 & 5 \\ 0 & 0 & 1 & 24 \\ 1 & 0 & 0 & -15 \\ 0 & 0 & 0 & 1 \end{bmatrix} = \begin{bmatrix} 0 & 0 & 1 & 24 \\ 1 & 0 & 0 & 0 \\ 0 & 1 & 0 & -5 \\ 0 & 0 & 0 & 1 \end{bmatrix},
$$

and extracting $^S\mathbf{p}_{SJ} = [24\ 0\ -5]^T$ yields identical results.

The transformation matrix $^B\mathbf{T}_F$ can be first obtained by identifying the unit basis vectors with respect to the shoulder coordinate frame as

$$
^S\mathbf{R}_H = \begin{bmatrix} \mathbf{h}_1 & \vdots & \mathbf{h}_2 & \vdots & \mathbf{h}_3 \end{bmatrix} = \begin{bmatrix} 0 & 0.64 & -0.76 \\ 1 & -0.76 & 0 \\ 0 & 0 & -0.64 \end{bmatrix} \text{ and } ^S\mathbf{p}_{SH} = [17\ 0.5\ -3]^T.
$$

Because the method for determining the configuration of a rigid body with respect to a second coordinate frame is now well established through this systematic approach, it is only natural to extend this method for use in human modeling, for the purpose of simulating human motion to perform tasks in the virtual world. However, logistics regarding the embedding of coordinate frames in each rigid body (a link) are to be developed, particularly when many segmental links and joints are needed. Similarly, a systematic representation of a homogeneous transformation matrix between two consecutive links should also be developed because coordinate frames are typically arbitrarily oriented.

5.3.3 Determining the Position of a Multi-Segmental Link: Forward Kinematics

The forward kinematics problem is characterized by determining the final position and orientation of a link (e.g., anatomical landmark on the hand) with knowledge of the joint variables. One can think of the forward kinematics (sometimes called direct kinematics) as a black box that contains the necessary calculations for accepting joint coordinates as input and producing position and orientation parameters as output for any of the segmental links in the chain. This black box approach is depicted in Figure 5.8.

Consider the arm of a person constrained to move on the surface of a table. Assume that this arm is represented by only two joints, characterized by two variables, q_1 and q_2. The question that will be addressed throughout this chapter is as follows:

Given a displacement of $q_1 = 15°$ and $q_2 = 30°$, what is the final position of the hand?

In order to answer this question, it is necessary to formulate an equation that contains the two independent variables q_1 and q_2 as its parameters and that evaluates to a position. Because two parameters are involved, we would also need two independent coordinate systems, which we shall denote by frame 0 and frame 1, shown in Figure 5.9.

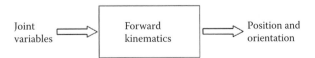

FIGURE 5.8 Understanding forward kinematics.

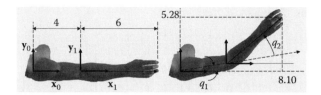

FIGURE 5.9   The upper limb with two embedded coordinate systems.

In this simple example, it can be seen from the geometry that the x- and y-values of the hand with respect to the first coordinate system x_0 and y_0 are

$$x = 4\cos q_1 + 6\cos(q_1 + q_2)$$

$$y = 4\sin q_1 + 6\sin(q_1 + q_2)$$

Therefore, to calculate the final position of the hand, we substitute for the values of q_1 and q_2, which yields the final position of the hand as $\mathbf{x} = [8.10 \quad 5.28]^T$.

From this simple example, it is evident that if the number of DOF becomes large, and the orientation of each joint with respect to another is spatial rather than planar, the formulation of the x, y, and z equations becomes complicated. If the orientation of the hand is required, further complexity is introduced. Therefore, we need to develop a systematic methodology for

- Locating coordinate systems on each segmental link in a consistent manner
- Calculating the relation between any two segmental links
- Characterizing the position and orientation of a distal link on the kinematic chain with respect to another link on the same or a different chain

5.3.4 The Denavit–Hartenberg Representation Method

In order to obtain a systematic method for describing the configuration (position and orientation) of each pair of consecutive segmental links, a method was proposed by Denavit and Hartenberg (1955). We shall utilize the DH method to address human kinematics.

The DH method is based upon characterizing the configuration of link i with respect to link i–1 by a (4×4) homogeneous transformation matrix representing each link's coordinate system. If each pair of consecutive links represented by their associated coordinate system (fig. 5.10) is related via a matrix, then using matrix chain-rule multiplication, it is possible to relate any of the segmental links (e.g., the hand) with respect to any other segmental link (e.g., the shoulder).

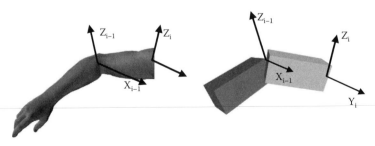

FIGURE 5.10   Joint coordinate systems between two segmental links.

We shall always refer to **T** as having the following vectors:

$$\mathbf{T} = \begin{bmatrix} n_x & s_x & a_x & p_x \\ n_y & s_y & a_y & p_y \\ n_z & s_z & a_z & p_z \\ 0 & 0 & 0 & 1 \end{bmatrix} = \left[\begin{array}{c|c|c|c} \mathbf{n} & \mathbf{s} & \mathbf{a} & \mathbf{p} \\ \hline 0 & 0 & 0 & 1 \end{array} \right]. \tag{5.12}$$

A single general transformation matrix between any two coordinate systems is needed. A general motion (translation and rotation) between any two configurations can be identified by four consecutive transformations, taking the rigid body from one configuration to its final destination. These transformations are illustrated in Figure 5.11 and are listed as follows:

- A rotation of α angle about the OX axis
- A translation of a units along the OX axis
- A translation of d units along the OZ axis
- A rotation of θ angle about the OZ axis

Since all transformations are about the world coordinate system, the transformation matrices are pre-multiplied. Remember that a basic homogeneous transformation matrix characterizes each transformation. The final transformation can be written as

$$\mathbf{T}(d, \theta, \alpha, a) = \mathbf{T}_{z,\theta} \mathbf{T}_{z,d} \mathbf{T}_{x,a} \mathbf{T}_{x,\alpha}$$

$$= \begin{bmatrix} \cos\theta & -\sin\theta & 0 & 0 \\ \sin\theta & \cos\theta & 0 & 0 \\ 0 & 0 & 1 & 0 \\ 0 & 0 & 0 & 1 \end{bmatrix} \begin{bmatrix} 1 & 0 & 0 & a \\ 0 & 1 & 0 & 0 \\ 0 & 0 & 1 & 0 \\ 0 & 0 & 0 & 1 \end{bmatrix} \begin{bmatrix} 1 & 0 & 0 & 0 \\ 0 & 1 & 0 & 0 \\ 0 & 0 & 1 & d \\ 0 & 0 & 0 & 1 \end{bmatrix} \begin{bmatrix} 1 & 0 & 0 & 0 \\ 0 & \cos\alpha & -\sin\alpha & 0 \\ 0 & \sin\alpha & \cos\alpha & 0 \\ 0 & 0 & 0 & 1 \end{bmatrix}$$

$$\mathbf{T}(d, \theta, \alpha, a) = \begin{bmatrix} \cos\theta & -\cos\alpha\sin\theta & \sin\alpha\sin\theta & a\cos\theta \\ \sin\theta & \cos\alpha\cos\theta & -\sin\alpha\cos\theta & a\sin\theta \\ 0 & \sin\alpha & \cos\alpha & d \\ 0 & 0 & 0 & 1 \end{bmatrix} \tag{5.13}$$

The resulting homogeneous transformation matrix of Equation 5.13 is an important element in developing the DH representation. For any two rigid bodies, this transformation matrix characterizes the configuration (position and orientation) of one with respect to the other in terms of the four important

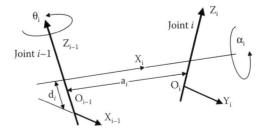

FIGURE 5.11   The relation between two coordinate systems with four parameters.

parameters (*d*, *θ*, *α*, *a*). Therefore, if it is possible to expand this result to the representation of rigid bodies connected in a serial chain, their multiplication will yield a transformation matrix relating any two links in the chain. We will expand on this issue in the following discussion on how to establish a systematic method for embedding coordinate systems.

5.3.5 The Kinematic Skeleton

In order to establish a systematic method for biomechanically modeling human anatomy, it is necessary to establish a convention for representing segmental links and joints. We can represent human anatomy as a sequence of rigid bodies (links) connected by joints. Of course, this serial linkage could be an arm, a leg, a finger, a wrist, or any other functional mechanism. Joints in the human body vary in shape and form. The complexity offered by each joint must also be modeled, to the extent possible, to enable a correct simulation of the motion. The degree by which a model replicates the actual physical model is called the level of *fidelity*.

Perhaps the most important element of a joint is its function, which may vary according to the joint's location and physiology. The physiology becomes important when we discuss the loading conditions of a joint. In kinematics, we shall address the function in terms of the number of DOF associated with its overall movement. Muscle action and ligament and tendon attachments at a joint are also important and contribute to the function.

For example, consider the elbow joint, which is considered a hinge or one DOF rotational joint (e.g., the hinge of a door) because it allows for flexibility and extension in the sagittal plane (fig. 5.12) as the radius and ulna rotate about the humerus. We shall represent this joint by a cylinder that rotates about one axis and has no other motions (i.e., one DOF). Therefore, we can now say that the elbow is characterized by one DOF and is represented throughout the book as a cylindrical rotational joint, also shown in Figure 5.12.

On the other hand, consider the shoulder complex (fig. 5.13). The glenohumeral joint (shoulder joint) is a multiaxial (ball and socket) synovial joint between the head of the humerus (5) and the glenoid cavity (6). There is a 4:1 incongruency between the large round head of the humerus and the shallow glenoid cavity. A ring of fibrocartilage attaches to the margin of the glenoid cavity, forming the glenoid labrum. This serves to form a slightly deeper glenoid fossa for articulation with the head of the humerus.

There are a number of methods that can be used to model this complex joint (fig. 5.14). One such method (Maurel et al., 1996) is to consider the shoulder girdle (considering bones in pairs) as four joints that can be distinguished as (1) the sterno-clavicular joint, which articulates the clavicle by its proximal end onto the sternum, (2) the acromio-clavicular joint, which articulates the scapula by its acromion onto the distal end of the clavicle, (3) the scapulo-thoracic joint, which allows the scapula to glide on the

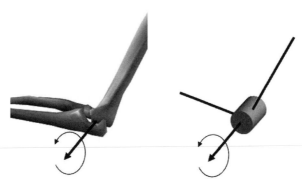

FIGURE 5.12 A one-DOF elbow.

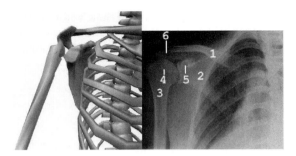

FIGURE 5.13 The shoulder joint: (1) clavicle, (2) body of scapula, (3) surgical neck of humerus, (4) anatomical neck of humerus, (5) coracoid process, and (6) acromion.

thorax, and (4) the gleno-humeral joint, which allows the humeral head to rotate in the glenoid fossa of the scapula.

Another method takes into consideration the final gross movement of the joint (Abdel-Malek et al., 2001), as abduction/adduction (about the anteroposterior axis of the shoulder joint), flexion/extension, and transverse flexion/extension (about the mediolateral axis of the shoulder joint). Note that these motions provide for three rotational DOF with axes that intersect at one point. This gives rise to the effect of a spherical joint typically associated with the shoulder joint.

5.3.6 Establishing Coordinate Systems

In order to obtain a systematic method for generating the (4×4) homogeneous transformation matrix between any two links, it is necessary to follow a convention in establishing coordinate systems on each link. This can be accomplished by implementing the following rules. It should be emphasized that a suitable *home configuration* must first be established before these rules are applied. A home configuration denotes the start configuration of the serial chain (segmental links). It is customary to start from a well-known position that the user identifies as the home configuration.

The rules for establishing coordinate frames at each link are as follows:

- Name each joint, starting with 1, 2, ... up to n DOF.
- Embed the $\mathbf{z}_{i-1}$ axis along the axis of motion of the ith joint.
- Embed the $\mathbf{x}_i$ axis normal to the $\mathbf{z}_{i-1}$ (and of course normal to the $\mathbf{z}_i$ axis).

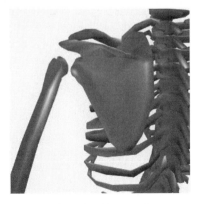

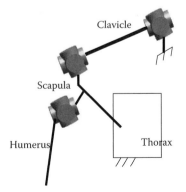

FIGURE 5.14 A model of the shoulder complex.

- Embed the y_i axis such that it is perpendicular to the x_i and z_i subject to the right-hand rule. However, on the kinematic skeleton, it is customary to not show the y_i axis to avoid cluttering the drawing because it is not needed for determining the DH parameters.

The location of the origin of the first coordinate frame (frame 0) can be chosen anywhere along the z_0 axis. In addition, for the nth coordinate system, it can be embedded anywhere in the nth link, subject to the four rules outlined above. In order to generate the matrix relating any two links, four parameters are needed. The four parameters are as follows:

- θ_i is the joint angle, measured from the x_{i-1} to the x_i axis about the z_{i-1} (the right-hand rule applies). For a prismatic joint, θ_i is a constant; it is basically the angle rotation of one link with respect to another about the z_{i-1}.
- d_i is the distance from the origin of the $(i-1)$th coordinate frame to the intersection of the z_{i-1} axis with the x_i axis along the z_{i-1} axis. For a revolute joint, d_i is a constant. It is basically the distance translated by one link with respect to another along the z_{i-1} axis.
- a_i is the offset distance from the intersection of the z_{i-1} axis with the x_i axis to the origin of the ith frame along the x_i axis (the shortest distance between the z_{i-1} and z_i axes).
- α_i is the offset angle from the x_{i-1} axis to the z_i axis about the x_i axis (right-hand rule).

Careful attention must be given when the following cases occur:

- When two consecutive axes are parallel, the common normal between them is not uniquely defined (i.e., the direction of x_i must be perpendicular to both axes); however, the position of x_i is arbitrary.
- When two consecutive axes intersect, the direction of x_i is arbitrary.

The four values for the DH parameters θ_i, d_i, a_i, α_i are typically entered into a table known as the DH. A 10-DOF model will have a table with 10 rows. Each row is used to generate the homogeneous transformation matrix.

5.3.6.1 Example

Consider the lower limb shown in Figure 5.15. For the purpose of this example, a total of four DOFs are used to model the limb.

FIGURE 5.15 A simple model of the lower limb.

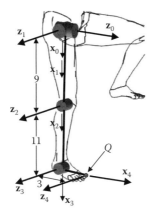

FIGURE 5.16 Establishing coordinate systems on a 4-DOF model of the lower limb.

- Determine the coordinate systems and sketch the figure.
- Determine the DH table.
- Write down the matrices relating any two consecutive coordinate systems.
- Write down the matrix relating the first link to the coordinate system embedded in the foot.

If the lower limb has moved to $q_1 = 0$, $q_2 = 90°$, $q_3 = -90°$, and $q_4 = 0$, calculate the final orientation of the foot and the final coordinates of point Q located at the tip of the foot.

- Coordinate systems are established as shown in Figure 5.16.
- The DH parameters are shown in the DH table.
- Substituting each row of the DH table into Equation 5.13 yields the four transformation matrices as follows:

$$
{}^0\mathbf{T}_1 = \begin{bmatrix} \cos q_1 & 0 & \sin q_1 & 0 \\ \sin q_1 & 0 & -\cos q_1 & 0 \\ 0 & 1 & 0 & 0 \\ 0 & 0 & 0 & 1 \end{bmatrix}
$$

$$
{}^1\mathbf{T}_2 = \begin{bmatrix} \cos q_2 & -\sin q_2 & 0 & 9\cos q_2 \\ \sin q_s & \cos q_2 & 0 & 9\sin q_2 \\ 0 & 0 & 1 & 0 \\ 0 & 0 & 0 & 1 \end{bmatrix}
$$

$$
{}^2\mathbf{T}_3 = \begin{bmatrix} \cos q_3 & -\sin q_3 & 0 & 11\cos q_3 \\ \sin q_3 & \cos q_3 & 0 & 11\sin q_3 \\ 0 & 0 & 1 & 0 \\ 0 & 0 & 0 & 1 \end{bmatrix} \qquad (5.14)
$$

$$
{}^3\mathbf{T}_4 = \begin{bmatrix} -\sin q_4 & -\cos q_4 & 0 & -3\sin q_4 \\ \cos q_4 & -\sin q_4 & 0 & 3\cos q_4 \\ 0 & 0 & 1 & 0 \\ 0 & 0 & 0 & 1 \end{bmatrix}
$$

- To relate the first frame to the fourth frame, the following multiplication of homogeneous transformation matrices must be carried out:

$$^0\mathbf{T}_1\,^1\mathbf{T}_2\,^2\mathbf{T}_3\,^3\mathbf{T}_4 = {^0\mathbf{T}_4}, \quad (1)$$

where the resulting (4×4) matrix is

$$^0\mathbf{T}_4 = \begin{bmatrix} n_x & o_x & a_x & p_x \\ n_y & o_y & a_y & p_y \\ n_z & o_z & a_z & p_z \\ 0 & 0 & 0 & 1 \end{bmatrix}, \tag{5.15}$$

where

$n_x = -\cos q_1 \sin(q_2 + q_3 + q_4)$
$n_y = -\sin q_1 \sin(q_2 + q_3 + q_4)$
$n_z = \cos(q_2 + q_3 + q_4)$
$S_x = -\cos q_1 \cos(q_2 + q_3 + q_4)$
$S_y = -\sin q_1 \cos(q_2 + q_3 + q_4)$
$S_z = -\sin(q_2 + q_3 + q_4)$
$a_x = \sin q_1$
$a_y = -\cos q_1$
$a_z = 0$
$p_x = \cos q_1\left(-3\sin(q_2 + q_3 + q_4) + 9\cos q_2 + 11\cos(q_2 + q_3)\right)$
$p_y = \sin q_1\left(-3\sin(q_2 + q_3 + q_4) + 9\cos q_2 + 11\cos(q_2 + q_3)\right)$
$p_z = 9\sin q_2 + 11\sin(q_2 + q_3) + 3\cos(q_2 + q_3 + q_4)$

- To determine the orientation of the fourth coordinate system with respect to the hip, we substitute $q_1 = 0$, $q_2 = 0$, $q_3 = 0$, and $q_4 = 0$ into $^0\mathbf{T}_4$, and the orientation is identified as $\mathbf{n} = [0 \quad 0 \quad 1]^T$, $\mathbf{s} = [-1 \quad 0 \quad 0]^T$, and $\mathbf{a} = [0 \quad -1 \quad 0]^T$.

In order to determine the position of point Q, defined with respect to the fourth coordinate system, we shall use the extended vector equation as

$$\begin{bmatrix} ^0\mathbf{v} \\ 1 \end{bmatrix} = {^0\mathbf{T}_4(\mathbf{q})} \begin{bmatrix} ^4\mathbf{v} \\ 1 \end{bmatrix}.$$

The position of a point on the foot is given by $\mathbf{v}_Q = [0 \quad 0 \quad 0]^T$. For the initial posture of the lower limb, the joints are $q_1 = 0$, $q_2 = 0$, $q_3 = 0$, and $q_4 = 0$.

In order to calculate the new posture given the change in joint variables, we substitute into $^0\mathbf{T}_4$

$$^0\mathbf{T}_4(0,\ 90°,\ -90°,\ 0) = \begin{bmatrix} 0 & -1 & 0 & 11 \\ 0 & 0 & -1 & 0 \\ 1 & 0 & 0 & 12 \\ 0 & 0 & 0 & 1 \end{bmatrix}$$

The position is calculated as

$$\mathbf{v}_Q\,(0,\ 90°,\ -90°,\ 0) = [12 \quad 0 \quad -11],$$

and the orientation is calculated as $\mathbf{n} = [1 \quad 0 \quad 0]^T$, $\mathbf{s} = [0 \quad 0 \quad 1]^T$ and $\mathbf{a} = [0 \quad -1 \quad 0]^T$, as shown in Figure 5.17.

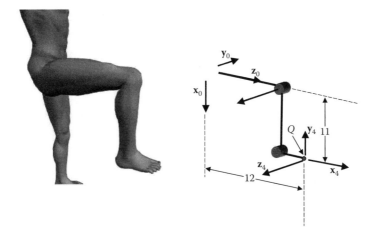

FIGURE 5.17   A 4-DOF model of the lower limb.

5.3.6.2 Example of a 15-DOF Model of the Torso-Spine-Shoulder-Arm

Consider a model of the human body comprising the torso, spine, shoulder, arm, and wrist with a total of 15 DOFs as shown in Figure 5.18. A point Q is embedded in the hand.

Sketch the coordinate systems using the DH representation method.

Enter the DH parameters into a table.

Determine the position and orientation of the 15th coordinate system (the hand) with respect to the waist (0th coordinate system) in the home (starting) configuration and determine the position of point Q.

For the following set of joint coordinates, calculate the final position of point Q

$$\mathbf{q} = [-5 \quad -4 \quad -3 \quad -3 \quad -4 \quad -4 \quad 0.3 \quad 0.2 \quad 30 \quad 20 \quad -30 \quad 0 \quad 10 \quad 20 \quad -10]^T.$$

The coordinate systems are located along each joint. The DH table is shown in Table 5.2.

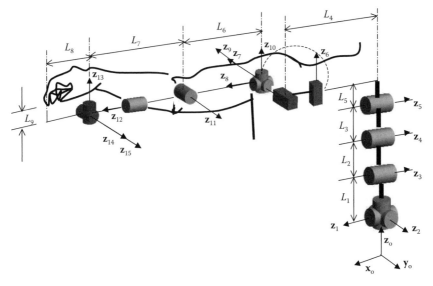

FIGURE 5.18   A 15-DOF model of the torso, spine, shoulder, arm, and wrist.

TABLE 5.1 DH Parameters for the Lower Limb

Joint	θ_i	α_i	a_i	d_i
1	q_1	0	0	$\pi/2$
2	q_2	0	9	0
3	q_3	0	11	0
4	$q_4 + \pi/2$	0	3	0

Each row of the DH table is entered into the 4×4 homogeneous transformation matrix of Equation 5.13. The 15 matrices are multiplied, and the resulting $^0\mathbf{T}_{15}$ matrix is

$$^0\mathbf{T}_{15} = \begin{bmatrix} 1 & 0 & 0 & 14.5 \\ 0 & 0 & 1 & 0 \\ 0 & -1 & 0 & 9 \\ 0 & 0 & 0 & 1 \end{bmatrix}.$$

The position of point Q is located at $^{15}\mathbf{v}_Q = [2.5 \quad -0.5 \quad 0]^T$ with respect to the 15th coordinate system. This same point resolved in the 0th coordinate system is $^0\mathbf{v}_Q = [17 \quad 0 \quad 9.5]^T$.

For the set of joint variables,

$$\mathbf{q} = [-5 \quad -4 \quad -3 \quad -3 \quad -4 \quad -4 \quad 0.3 \quad 0.2 \quad 30 \quad 20 \quad -30 \quad 0 \quad 10 \quad 20 \quad -10]^T.$$

The $^0\mathbf{T}_{15}$ matrix is

$$^0\mathbf{T}_{15} = \begin{bmatrix} 0.349 & 0.889 & -0.296 & 10.914 \\ -0.304 & 0.406 & 0.861 & -4.992 \\ 0.886 & -0.211 & 0.412 & 14.901 \\ 0 & 0 & 0 & 1 \end{bmatrix}.$$

The new position for point Q is $^0\mathbf{v}_Q = [11.342 \quad -5.957 \quad 17.222]^T$.

TABLE 5.2 The DH Table for the 15-DOF Human Model

	θ_i	d_i	α_i	a_i
1	$\pi/2 + q_1$	0	$\pi/2$	0
2	$\pi/2 + q_1$	0	$\pi/2$	0
3	q_3	0	$\pi/2$	$L_1 = 3$
4	q_4	0	0	$L_2 = 2$
5	q_5	0	0	$L_3 = 23$
6	$\pi/2 + q_6$	$-L_4 = -3.5$	$\pi/2$	0
7	$-\pi/2$	$L_5 + q_7 (L_5 = 1)$	$\pi/2$	0
8	$\pi/2$	q_8	$\pi/2$	0
9	q_9	0	$-\pi/2$	0
10	$-\pi/2 + q_{10}$	0	$-\pi/2$	0
11	q_{11}	0	$-\pi/2$	$L_6 = 5$
12	$-\pi/2 + q_{12}$	0	$-\pi/2$	0
13	$-\pi/2 + q_{13}$	$L_7 = 6$	$-\pi/2$	0
14	$-\pi/2 + q_{14}$	0	$-\pi/2$	0
15	q_{15}	0	0	0

5.3.6.3 Variations in Anthropometry

Because the DH method is dependent upon a well-defined set of parameters as entered into the DH table, it is indeed straightforward to vary these parameters as the anthropometric model is changed. Therefore, variations in anthropometry are readily performed in a computer simulation by changing the values of the DH parameters.

5.4 Laws of Physics—Equations of Motion

Equations of motion are mathematical formulas that describe the behavior of a system (e.g., the motion of a rigid body under the influence of a given force) as a function of time. While there are many methods to obtain the equations of motion, they are typically characterized by a system of differential equations that govern the motion of the system. The two most used methods for creating the equations of motion are either according to Newton's second law or according to the Euler-Lagrange equations. The solution to the equations of motion is the predicted (calculated) motion.

As an introduction to the equations of motion, consider the rotating leg modeled as one DOF and shown in Figure 5.19.

The equation characterizing this motion is given by:

$$\tau = I\ddot{q} + c\dot{q} + mg\cos(q),$$

where τ is the torque required to accomplish the motion, I is the moment of inertia of the leg (combined inertia for all elements of the model in this example), and $\ddot{q}$, $\dot{q}$, q represents the angular acceleration, angular velocity, and joint angle of the leg rotating about its degree of freedom.

Solutions to this equation are readily available, whether in closed form because of the simplicity of the mathematics or through numerical integration.

The problem quickly becomes more difficult if more than one DOF is used, and becomes substantially more difficult (almost impossible) if constraints on the motion are enforced.

The field of multi-body dynamics has evolved in recent years; several well-established methodologies have been developed and demonstrated for the creation of the equations of motion. Multi-body dynamics is concerned with finding rigorous and systematic methods for modeling systems that undergo motion, as well as finding their solutions. A review of this field is provided by Shabana (2001).

For a human model with open-loop chains, we continue using the DH representation method to develop the equations of motion. The mathematics used to arrive at the equations of motion is presented in a comprehensive manner by Fu et al. (1987). It is based on the Lagrange-Euler formulation for calculating the potential and kinetic energy.

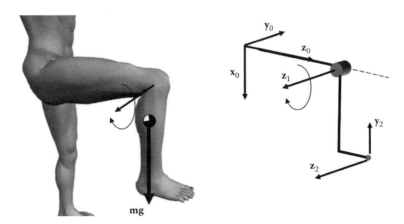

FIGURE 5.19 Lower-extremity model.

The typical form of the equations of motion (Kim et al., 2006) is given by

$$\boldsymbol{\tau} = \mathbf{M}(\mathbf{q})\ddot{\mathbf{q}} + \mathbf{V}(\mathbf{q},\dot{\mathbf{q}}) + \sum \mathbf{J}_i^{\mathrm{T}} \, m_i \mathbf{g} + \sum \mathbf{J}_k^{\mathrm{T}} \, \mathbf{F}_k + \mathbf{K}\big(\mathbf{q} - \mathbf{q}^N\big), \tag{5.16}$$

where τ_i is the generalized torque of joint-I, $\mathbf{J}_i^T$ is the transpose of the Jacobian matrix, and $\mathbf{v}_i(\mathbf{q}, \dot{\mathbf{q}})$ is the Coriolis and centrifugal vector:

$$\mathbf{V}_i(\mathbf{q},\dot{\mathbf{q}}) = \sum_{k=1}^{n}\sum_{l=1}^{n}\sum_{j=\max(i,k,l)}^{n} Tr\left(\frac{\partial^0 \mathbf{T}_j(\mathbf{q})}{\partial q_k}\mathbf{I}_j\left[\frac{\partial^0 \mathbf{T}_j(\mathbf{q})}{\partial q_i}\right]^T\right)\dot{q}_k\dot{q}_l \,. \tag{5.17}$$

$i, k, l = 1, 2, \ldots, n$, M_{ik} is the (i, k) element of the mass-inertia matrix $\mathbf{M}(\mathbf{q})$ such that

$$M_{ik}(\mathbf{q}) = \sum_{j=\max(i,k)}^{n} Tr\left(\frac{\partial^0 \mathbf{T}_j(\mathbf{q})}{\partial q_k}\mathbf{I}_j\left[\frac{\partial^0 \mathbf{T}_j(\mathbf{q})}{\partial q_i}\right]^T\right) \tag{5.18}$$

$\mathbf{I}_j$ is the inertia matrix and is written as follows:

$$\mathbf{I}_i = \begin{bmatrix} \dfrac{-I_{xx}+I_{yy}+I_{zz}}{2} & -I_{xy} & -I_{xz} & m_i\overline{x}_i \\[2mm] -I_{xy} & \dfrac{I_{xx}-I_{yy}+I_{zz}}{2} & -I_{yz} & m_i\overline{y}_i \\[2mm] -I_{xz} & -I_{yz} & \dfrac{I_{xx}+I_{yy}-I_{zz}}{2} & m_i\overline{z}_i \\[2mm] m_i\overline{x}_i & m_i\overline{y}_i & m_i\overline{z}_i & m_i \end{bmatrix}. \tag{5.19}$$

m_i is the mass of link-i; $(\overline{x}_i, \overline{y}_i, \overline{z}_i)$ is the location of the center of gravity of link-i, expressed in terms of ith coordinate frame (local frame); and $I_{xx}, \ldots, I_{xy}, \ldots$ are the moments/products-of-inertia of link-i with respect to the ith coordinate system. $\mathbf{K}$ is a diagonal matrix of elastic constants given as follows:

$$\mathbf{K} = \begin{bmatrix} k_1 & & \mathbf{0} \\ & \ddots & \\ \mathbf{0} & & k_n \end{bmatrix}_{n\times n} \tag{5.20}$$

Note that all of the variables are functions of time, that is, $\boldsymbol{\tau} = \boldsymbol{\tau}(t)$, $\mathbf{F}_k = \mathbf{F}_k(t)$, and $\mathbf{q} = \mathbf{q}(t)$, with $k = 1, 2, \ldots, n$.

5.4.1 Example

Consider the example of the leg shown in Figure 5.19 with the following characteristics:
Mass = 0.5 kg, Length = 0.4 m.

The inertia about A is calculated as $I_A = \dfrac{ml^2}{3} = 0.0267(kg \cdot m^2)$.

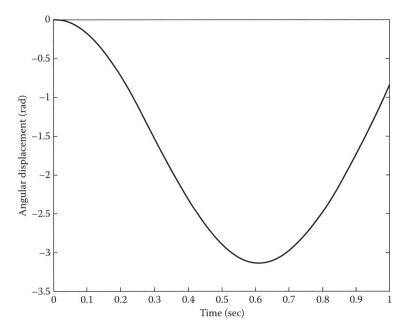

FIGURE 5.20 Joint angle from direct integration.

The equation of motion for this one-DOF system is:

$$I_A \ddot{q} + mg\frac{l}{2}\cos q = \tau.$$

The solution to this equation using traditional integration methods is shown in Figure 5.20.

The objective of this section is simply to introduce the field of multi-body dynamics to the reader and to reiterate that these equations are only part of representing human motion. Solving them alone does not yield a predicted human motion but one that is only subject to the laws of physics. Other influences to be considered are the task to be executed, the physical and world constraints, and other physiological and psychological factors. Indeed, EoM alone are not enough to predict human behavior. Solutions to these equations are repeatable; there is no human aspect to solving them. Driving the motion for humans requires methodology that is able to predict behavior in such a way that the same person may accomplish the same task in many different ways.

This gives rise to the so-called human performance measures, which are used not only to evaluate the state of the human model at any instant in time, but are also used in an optimization formula to drive the motion.

5.5 Cost Functions

In this section, we address the development of simple human performance measures that enable the mathematical evaluation of a cost function. These cost functions must be considered drivers of a system toward a realistic human motion.

The basic plot relies on obtaining a real number that evaluates the task, where each task comprises several cost functions. Each cost function must evaluate to a number and must be mathematically defined. Once this is achieved, it is then possible to formulate an optimization algorithm that iteratively evaluates the task. One must think of a cost function as the driving force behind the task—that is, why a task is to be accomplished.

Cost functions will then be used in the optimization formulation to arrive at a solution that optimizes the human performance measure characterized by the cost function.

Various human performance measures provide the objective functions for the optimization formulation. Details concerning these performance measures are provided by Yang et al. (2004). In this section, we present an overview. The most fundamental of these functions is *joint displacement*, which is given as follows:

$$f_{\text{JointDisp}}(\mathbf{q}) = \sum_{i=1}^{n} w_i \left(q_i - q_i^N \right)^2 . \tag{5.21}$$

q_i^N is the *neutral position* of a joint, and $\mathbf{q}^N$ represents the overall neutral posture. The neutral posture is selected as a relatively comfortable posture, typically a standing position with arms at one's sides. A weight w_i is introduced to stress the importance of a particular joint. Currently, these weights are determined by trial and error. Generally, with joint displacement, the avatar gravitates toward the neutral position.

Energy is perhaps the most important cost function that drives human motion (Kim, 2006). Humans tend to perform tasks while minimizing their energy. Energy is related to the torque exerted at every joint and is therefore proportional to muscle energy expended to execute a task.

Effort is similar to joint displacement, but $q_i^{initial}$ replaces q_i^N and represents the avatar's initial position or starting position. Thus, when effort is used as the objective function, the avatar gravitates to its starting configuration, no matter what that configuration may be. This performance measure is most significant when a series of target points are selected, with the posture changing from point to point.

Using *delta-potential-energy* as an objective function provides an alternative approach for determining the weights in Equation 5.21. Various segments of the avatar's body are treated as lumped masses. The total change in the potential energy for the masses is minimized. When determining the change in potential energy, the initial configuration is always set as the neutral position described above. In this way, the masses of the different sections of the body essentially provide inherent weights for the motion of the different segments.

Finally, *discomfort* is modeled as another variation on joint displacement. Again, the avatar moves toward the neutral position. However, this function incorporates three facets of comfort: (1) the tendency to move toward a generally comfortable position, (2) the tendency to avoid postures with which joint angles are pushed to their limits, and (3) the idea that people strive to reach or contact a point using one set of body parts at a time. Typically, in terms of upper-body motion, one first tries to reach a point using one's arm. If that is unsuccessful, only then does one bend the torso. Finally, if necessary, the clavicle is extended. Note that the intent in developing this performance measure is not necessarily to quantify discomfort. Rather, the function is designed to be proportional to discomfort. Only its relative values (from one posture to another), not its absolute values, are significant.

5.6 Constraints

Constraints are used to account for the world's physical limitations as well as our own. We shall examine a number of constraints that are to be used in the formulation for predicting human motion. A large number of constraints exist and are typically developed for a given task. Constraints for a simple posture prediction, for example, would include joint limits only, while constraints for walking would include joint limits, collision with the ground, reaction force constraints, and foot slippage.

We will focus on the most important constraints and enumerate others.

5.6.1 Joint Angle Limits

Each joint in the human model has limits for movement. These limits are given as:

$$q_i^L \leq q_i \leq q_i^U$$
$$(1 \leq i \leq n_{dof})$$

(5.22)

where q_i is the ith joint angle, q_i^U, q_i^L represents the upper and lower limits of the ith joint angle, and n_{dof} is the number of DOFs.

This bounded condition for joint limits yields two inequality HPMs, called joint angle limits (JAL), for each joint i:

$$g_i^{JAL,1} = q_i^L - q_i$$
$$g_i^{JAL,2} = q_i - q_i^U$$
$$(1 \leq i \leq n_{dof})$$

(5.23)

5.6.2 Foot Slippage

The friction force between the foot and the ground is large enough that the foot never slips. Under this no-slippage condition, the in-plane components of velocity and acceleration of the foot point in contact with the ground are zero.

5.6.3 Ground Penetration

During walking the foot points that are in contact with ground should stay on the ground and the other foot points should be above the ground. This condition gives one equality and one inequality constraint.

5.6.4 Dynamic Equilibrium

Zero movement point (ZMP) is a concept (turned into a constraint) related to dynamics and control of legged locomotion. It specifies the point at which dynamic reaction force at contact of the foot with the ground does not produce any movement—the point where total inertia force equals 0 (zero).

5.6.5 Initial and Final Position Constraints

Theoretically, any point on the body can be constrained to a position at the initial and final time frames.

5.7 Optimization: An Efficient Computational Framework

The field of optimization has matured to a level at which it exists in every walk of life, from the stock market to our digital cameras. Various numerical methods and approaches have been developed over the years, and most exist in commercial programs available on the market. Once an optimization problem is formulated, it can be addressed using an appropriate solver. We will assume that the reader has a good understanding of this field. A well-established reference was written by the second author (Arora, 2004).

5.7.1 Predictive Dynamics for a Simple Example

Consider again the example of the lower limb presented in the section "Laws of Physics—Equations of Motion." We will now address the same problem using predictive dynamics.

The mass moment of inertia about point A is

$$I_A = \frac{ml^2}{3} = 0.0267(kg \cdot m^2),$$

where m is mass, and l is the dimension of the lower leg. The EoM is

$$I_A \ddot{q} + mg\frac{l}{2}\cos q = \tau .$$

5.7.2 Optimization Formulation

We formulate the predictive dynamics problem for this one-DOF problem of the lower limb as follows:

Find: joint profiles

$$\mathbf{x} = \{q_{t=0},\, q_{t=\Delta t},\, \dots,\, q_{t=n\Delta t}\}$$

Minimize: cost function

$$\int_0^\tau \tau^2 dt \approx \sum_i \tau_i^2 \Delta t$$

$$\tau = I_A \ddot{q} + mg\frac{l}{2}\cos q$$

Subject to: constraints

$$q_{t=i\Delta t} = \tilde{q}_{t=i\Delta t} \qquad 0 \le i \le n$$

Where $\tilde{q}_{t=i\Delta t}$ is tje specified joint angle at $t = i\Delta t$

Providing this problem to a numerical optimization solver yields the results shown in Figure 5.21.

Compared with the results of the example in the section "Laws of Physics—Equations of Motion" and solved using traditional methods, it is clear that predictive dynamics offers a viable alternative to the traditional step-integration method for dynamics. This is simply another approach to solving traditional multi-body dynamics problems while considering cost functions and using optimization.

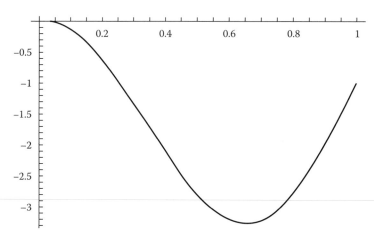

FIGURE 5.21   Joint angle from optimization.

5.8 Applications of Predictive Dynamics

Almost every task can be modeled with predictive dynamics. Setting up the cost function, EoM, and constraints has become much simpler as we better understand human motion and as we continue to experiment with modeling various tasks.

5.8.1 Example of an Upper-Body Task

In this section, we summarize a complete numerical example of a human torso-arm model that is assigned the task of pulling a load with a rope using a one-arm and torso motion. The model is shown in Figure 5.22.

The optimization formulation for this problem is outlined as follows:

Find: joint profiles

$$\mathbf{x} = \{q_{t=0}, q_{t=\Delta t}, ..., q_{t=n\Delta t}\}$$

Minimize: cost function

$$f_{torque} = w_1 |\tau_1| + w_2 |\tau_2| + ... + w_n |\tau_n|$$

$$\boldsymbol{\tau} = \mathbf{M}(\mathbf{q})\ddot{\mathbf{q}} + \mathbf{V}(\mathbf{q}, \dot{\mathbf{q}}) + \sum_i \mathbf{J}_t^T m_t \mathbf{g} + \sum_k \mathbf{J}_k^T \begin{bmatrix} -\mathbf{F}_k \\ -\mathbf{M}_k \end{bmatrix} + \mathbf{K}(\mathbf{q} - \mathbf{q}^N)$$

Subject to: constraints:

—Physical constraints

$$q_i^L \le q_i \le q_i^U$$
$$\tau_i^L \le \tau_i \le \tau_i^U \quad (i = 1, ..., n)$$

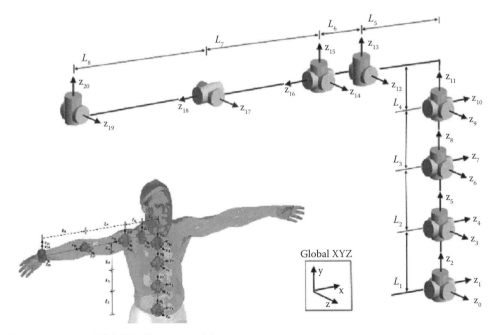

FIGURE 5.22 A 21-DOF digital human model.

—Path constraints

$$\| \mathbf{x}(\mathbf{q}(t)) - \mathbf{p}(t) \| \leq \varepsilon$$

$0 \leq t \leq t_f$ ε Is a small number

—Minimum jerk constraints

$$C = \frac{1}{2} \int_0^{t_f} \left((\frac{d^3 x}{dt^3})^2 + (\frac{d^3 y}{dt^3})^2 + (\frac{d^3 z}{dt^3})^2 \right) dt$$

where the constraints are as follows:

- Physical constraints limiting rotational angles and torques at each joint
- Path constraints that require the end effector to follow a specified path $\mathbf{p}(t)$
- Minimum jerk constraint to enforce the end effector to minimize jerk denoted by C (i.e., the third derivative of acceleration)

The motion of pulling a rope is predicted using the right arm with light load (0.01 N) and heavy (270 N) constant tension from a given initial position (–75, 40, 0) (cm) to a given final position (–32, 40, 0) (cm) of the right hand in the global Cartesian frame. The rope is horizontal on the right side of the digital human. Imagine a digital human pulling a rope where a counterweight is hanging at the other end of the rope. Figure 5.23 shows the resulting predicted motions of one-arm rope-pulling with 0.01 N and 270 N forces. The numbers of iterations were 3051 and 3485, respectively.

The optimum joint kinematic profiles (joint variables, joint velocities, and joint accelerations) are obtained every 0.04 sec in both cases. The predicted profiles for the 21 joints are shown in Figure 5.24. Each joint moves smoothly, but the overall curve shapes are quite different for the different magnitudes of pulling forces.

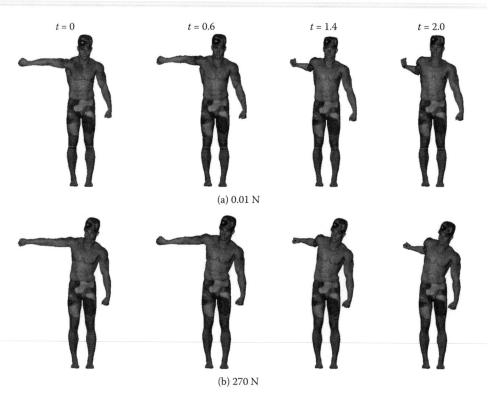

$t = 0$ $t = 0.6$ $t = 1.4$ $t = 2.0$

(a) 0.01 N

(b) 270 N

FIGURE 5.23   Predicted motion of one-arm rope-pulling.

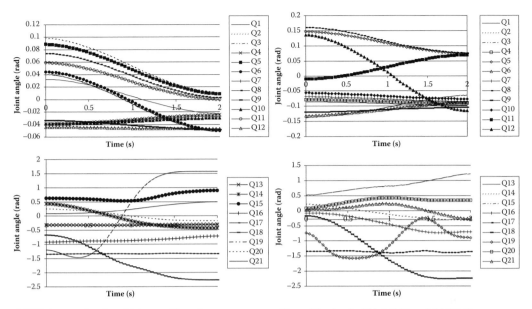

FIGURE 5.24 Predicted joint profiles of one-arm rope-pulling with a load of (a) 0.01 N and (b) 270 N.

5.8.2 Validation

It is imperative that modeling and simulation results be validated through experimental testing. While one cannot validate every aspect of the predicted motion, there are many well-established methods for designing experiments that produce results against which models can be validated. Typical methods for validation include initial and final joint angle comparative studies, joint profile comparative studies, as well as video-based studies that compare the task execution for realistic behavior.

5.8.3 Other Predictive Dynamics Examples

We enumerate a number of tasks that have been met successfully with modeling and simulation.

5.8.3.1 Simple Whole-Body Motion

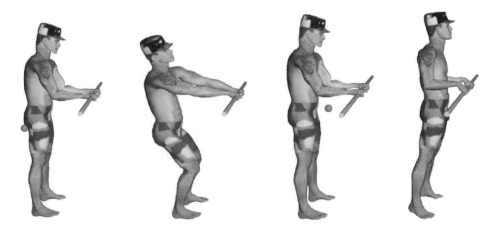

FIGURE 5.25 Simulation result of simple whole-body motion.

FIGURE 5.26   Simulation result of climbing.

5.8.3.2 Climbing

This task is inherently complex because of the large number of DOF and the requirement to have the world coordinate system embedded in the hip as a moving target (requiring additional DOF).

5.8.3.3 Walking and Running

Several papers have been presented for the prediction of a gait cycle.

FIGURE 5.27   Snapshot of Santos running with arm motion.

5.8.3.4 Lifting

FIGURE 5.28 Simulation of lifting.

5.8.3.5 Complete Tasks

Tasks such as diving, shouldering a weapon, aiming, and throwing a grenade have been successfully modeled and predicted using predictive dynamics.

FIGURE 5.29 Simulation of tasks using predictive dynamics.

5.8.3.6 Other Tasks

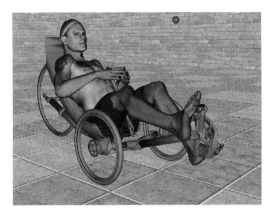

FIGURE 5.30 Simulation of Santos in other tasks.

5.9 Conclusions

This chapter offers a comprehensive methodology and associated mathematics for predicting human motion. The approach is task based; every task is modeled using a consistent and systematic method for kinematic modeling while obeying the laws of physics and taking into consideration a driver function (cost).

Benefits of Predictive Dynamics:

- Presents a rigorous and systematic method for predicting human motion
- Takes into consideration human performance measures characterized by cost functions that drive the motion
- Accounts for the laws of physics such that objects have mass and inertia
- Uses the mature field of optimization to numerically solve for a motion
- Accounts for the world's physical limitations, represented by constraints

References

Abdel-Malek, K., Yu, W., Tanbour, E., and Jaber, M. (2001) "Posture prediction versus inverse kinematics," *Proceedings of the ASME Design Automation Conference*, Pittsburgh, PA.

Abdel-Malek, K., Mi, Z., Yang, J., and Nebel, K. (2004a) "Optimization-based layout design," *4th International Symposium on Robotics and Automation ISRA*, Queretaro, Mexico, August 25–27, 2004.

Abdel-Malek, K., Yang, J., Mi, Z., Patel, V., and Nebel, K. (2004b) "Human upper body motion prediction," *Conference on Applied Simulation and Modeling (ASM)*, Rhodes, Greece, June 28–30, 2004.

Abdel-Malek, K., Yu, W., Yang, J., and Nebel, K. (2004c) "A mathematical method for ergonomic-based design: placement," *International Journal of Industrial Ergonomics*, 34(5), 375–394.

Abdel-Malek, K., Mi, Z., Yang, J., and Nebel, K. (2006a) "Optimization-based trajectory planning of human upper body," *Robotica*, 24(6), 683–696.

Abdel-Malek, K., Yang, J., Marler, T., Beck, S., Mathai, A., Patrick, A., and Arora, J. (2006b) "Towards a new generation of virtual humans," *International Journal of Human Factors Modelling and Simulation*, 1(1) 2–39.

Arora, J. (2004) *Introduction to Optimum Design*, New York: Elsevier.

Chung, H.J., Xiang, Y., Mathai, A., Rahmatalla, S., Kim, J., Marler, T., Beck, S., Yang, J., Arora, J., Abdel-Malek, K., and Obusek, J. (2007) "A robust numerical formulation for the prediction of human running," *Proceedings of SAE Digital Human Modeling for Design and Engineering*, Seattle, WA.

Denavit, J. and Hartenberg, R.S. (1955) "A kinematic notation for lower-pair mechanisms based on matrices," *ASME Journal of Applied Mechanics*, 22, 215–221.

Farrell, K. (2005) "Kinematic human modeling and simulation using optimization-based posture prediction," MS Thesis, University of Iowa.

Fu, K.S., Gonzalez, R.C., and Lee, C.S.G. (1987) *Robotics: Control, Sensing, Vision, and Intelligence*. New York: McGraw-Hill.

Horn, E. (2005) "Optimization-based dynamic human motion prediction," MS Thesis, University of Iowa.

Kim, J., Abdel-Malek, K., Mi, Z., and Nebel, K. (2004) "Layout design using an optimization-based human energy consumption formulation," *SAE Digital Human Modeling for Design and Engineering Symposium*, Rochester, MI, June 2004.

Kim, J., Abdel-Malek, K., Yang, J., and Nebel, K. (2005) "Optimization-based dynamic motion simulation and energy consumption prediction for a digital human," *Journal of Passenger Car-Electronic and Electronical Systems*, 114.

Kim, J., Abdel-Malek, K., Yang, J., and Marler, T. (2006) "Prediction and analysis of human motion dynamics performing various tasks," *International Journal of Human Factors Modelling and Simulation*, 1(1), 69–94.

Kim, J.H. (2006) "Dynamics and motion planning of redundant manipulators using optimization, with applications to human motion," PhD dissertation, University of Iowa.

Marler, T. (2004) "A study of multi-objective optimization methods for engineering applications," PhD dissertation, University of Iowa.

Maurel, W. (1996) "3D biomechanical modeling of the human upper limb joints, muscles, and soft tissues," PhD dissertation, EPFL, Switzerland.

Mi, Z. (2004) "Task-based motion prediction," PhD dissertation, University of Iowa.

Piegl, L. and Tiller, W. (1995) *The NURBS Book*, Berlin: Springer-Verlag.

Shabana, A.A. (2001) *Computational Dynamics*, 2nd ed., New York: John Wiley & Sons.

Wang, Q. (2006) "A study of alternative formulations for optimization of structural and mechanical systems subjected to static and dynamic loads," PhD dissertation, University of Iowa.

Xiang, Y., Chung, H.J., Mathai, A., Rahmatalla, S., Kim, J., Marler, T., Beck, S., Yang, J., Arora, J.S., and Abdel-Malek, K. (2007) "Optimization-based dynamic human walking prediction," *Proceedings of the Digital Human Modeling Conference*.

Yang, J. (2003) "Swept volumes: theory and implementation," PhD dissertation, University of Iowa.

Yang, J., Marler, R.T., Kim, H., Arora, J.S., and Abdel-Malek, K. (2004) "Multi-objective optimization for upper body posture prediction," *10th AIAA/ISSMO Multidisciplinary Analysis and Optimization Conference*, Albany, NY, August 30, 2004.

Yang, J., Abdel-Malek, K, Marler, R.T., and Kim, J. (2006a) "Real-time optimal reach posture prediction in a new interactive virtual environment," *Journal of Computer Science and Technology*, 21(2), 189–198.

Yang, J., Marler, T., Beck, S., Kim, J., Wang, Q., Zhou, X., Pena Pitarch, E., Farrell, K., Patrick, A., Sinokrot, T., Potratz, J., Abdel-Malek, K., Arora, J., and Nebel, K. (2006b) "New capabilities for the virtual-human Santos," *SAE 2006 World Congress*, April 3–6, 2006, Cobo Center, Detroit, Michigan.

Yang, J., Kim, J.H., Abdel-Malek, K., Marler, T., Beck, S., and Kopp, G.R. (2007) "A new digital human environment and assessment of vehicle interior design," *Computer-Aided Design*, 39, 548–558.

Yu, W. (2001) "Placement of manipulators for maximum functionality," PhD dissertation, University of Iowa.

6

Workplace Methods and Use of Digital Human Models

Allison Stephens and
Monica L. H. Jones

6.1 Introduction

Digital human models (DHMs) are used to predict postures and ergonomic stresses associated with automotive assembly tasks. The engineering community uses DHMs to evaluate assembly feasibility and identify product and process compatibility issues. More specifically, DHMs have increased the ability to determine risk and acceptability of designs very early in the product development cycle. They also afford the unique opportunity of completing ergonomic assessments while interacting with 3D computer-aided designs (CAD). Ultimately, DHMs reduce the number of assumptions typically associated with traditional ergonomic assessment tools. The use of DHMs has coincided with a significant reduction in engineering design time and time to market. Even still, the full potential of digital human modeling has yet been realized.

The primary advantage of using DHMs for ergonomic and safety risk assessment is the infinite biomechanical, physiological, and anthropometric data available at every instance of simulated operation. DHMs have increased degrees of freedom and capacity in comparison to traditional quasi-static biomechanical models. DHMs also enable direct interaction between the mannequin and 3D CAD product data that results in improved "assembly representative" posturing.

The challenge lies in developing efficient applications of a DHM. The motivation or goal of completing virtual assessments, visualization versus feasibility assessments, is critical to determining how

DHM technology can best be applied. The following is a narrative example of how DHMs are integrated into virtual manufacturing processes at Ford Motor Company. In this context, the primary function of both DHMs and virtual manufacturing is to facilitate assembly and ergonomic feasibility evaluations.

6.2 A Short History of Applied Ergonomics within Automotive Manufacturing

The integration of ergonomics as a core element of manufacturing engineering has been a slow process. Historically, in the absence of standards and specifications for human performance capability, assembly feasibility issues were identified by manufacturing engineers who assessed risk based on experience.

The practice of ergonomics has traditionally involved the systematic evaluation of current physical production parts. More specifically, measurements of forces, joint angles, and time studies are completed from observing an operator on the assembly plant floor. Thus, product design and workstation modifications involved a costly, reactive, and highly iterative process.

Today, ergonomic, product, and manufacturing communities of practice are integrated. It is important to note that assembly ergonomics requirements are starting to be included as part of the engineering manufacturing requirements. Ergonomic specialists provide in-depth ergonomic assessments and specifications, and offer support to drive change into the product development processes and justify product design and facility changes.

The introduction of biomechanical static strength prediction models has enabled the ergonomic assessment of assembly postures and installation efforts. One of the more prominent software tools for performing strength evaluations is the University of Michigan's 3D Static Strength Prediction Program (3DSSPP). Prior to DHMs, such ergonomic assessment tools were stand-alone software packages, each with its own set of constraints and assumptions. They often involved a time-consuming, iterative process of progressively entering higher hand loads or different postures to determine an appropriate engineering criterion in the design for manufacturing process.

The advancements in virtual technology and DHMs have enabled a more data-driven, multi-faceted, proactive approach to ergonomics. The major benefit of DHM software is that ergonomic assessments are based on a DHM mannequin that interacts with its virtual environment. In the manufacturing phase of the product life cycle, DHMs facilitate the virtual evaluation of workstation layout, workflow simulation, assembly accessibility, human reach, clearance and strength capability studies, and safety analysis.

6.3 Workflow Methods

Decisions regarding the use of DHMs and virtual simulation technologies are dependent upon the specific needs of an engineering community. At Ford Motor Company, the approach has been to integrate assembly ergonomics within the manufacturing engineering objectives. DHMs have therefore become a tool for manufacturing engineering to assess both assembly and ergonomic feasibility within an assembly workstation.

Current workflow methods involve assembly and ergonomic feasibility studies performed throughout a hierarchical process. These methods are graduated from: (1) initial high-level feasibility evaluations performed during virtual build event; (2) to preliminary assembly and ergonomic checkpoints completed by the manufacturing engineers during the digital preassembly; and (3) lastly, a more in-depth ergonomic assessment completed by ergonomists.

6.3.1 Virtual Manufacturing Process

The virtual manufacturing (VM) process is the simultaneous compatibility of product and process. Dynamic simulations are performed to verify product design and manufacturing engineering processes (ergonomics, assembly, facility, and tooling). The objectives of virtual manufacturing include: early

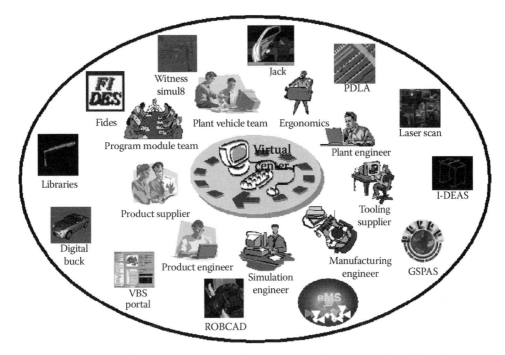

FIGURE 6.1 Virtual manufacturing process—seamless data between integrated and multidisciplinary teams.

identification of ergonomic and safety risk hazards, work element allocation, cost reduction, improved quality, and decreased time to deliver new product to the customer. To this end, virtual manufacturing is a strong enabler of engineering cultural change. It has facilitated process manufacturing involvement and input further upstream in the product development cycle. Thereby, the fundamental objective of virtual manufacturing is to drive a strategy of process-driven product development. The virtual technology is also leveraged to generate alternatives for both product and process in the virtual environment and to drive decisions with data.

Seamless data flow and integrated, multidisciplinary teams are critical to the success of VM processes. Cross-functional teams include representation from: plant engineering, program management, manufacturing engineering, product and tooling suppliers, product design engineering, ergonomics, simulation engineer, and assembly line operators.

Dynamic virtual simulations integrate product (part), manufacturing (assembly operation) process, and facility 3D CAD data resources. Product (parts) data are the pieces that make up the complete vehicle. The product (parts) tree lists all of the parts that make up the final product as a hierarchical model that depicts how the parts relate to one another in the completed vehicle assembly. Manufacturing (assembly operation) process trees contain the hierarchy of the completed manufactured product. The top level of this hierarchy represents the finished product and each individual process contains the actions included within a given workstation in order to manufacture the product. Resource trees are the factory facilities that perform the operations on the parts. These include assembly lines, stations, zones, work-cells, tools, fixtures, and operators. Each work station will have a list of operations assigned to it. The resource tree lists the stations, processes, and corresponding resources (tools, fixtures, and operators) used during the manufacturing process.

6.3.2 Digital Pre-Assembly Process (Product and Process Validation)

Digital pre-assembly (DPA) is a set of CAD-based checks that evaluate geometric compatibility for adjacent systems, subsystems, or components in vehicle position. The goal of the DPA process is to achieve total vehicle geometric capability prior to physical prototype builds. There are different types of DPA checks, including function, craftsmanship, service, ergonomic/assembly feasibility and systems package, and human factors. The manufacturing and ergonomic DPA checks are performed by manufacturing engineers and ergonomists. These checks consist of feasibility, ergonomics, assembly process, and process control. The manufacturing DPA is the evaluation of the product from a manufacturing point of view and evaluation of the manufacturing process and facilities—that is, that the product can be produced (at the desired quality level) in the intended manufacturing facility within the allocated cost and at the required line speed. Among the attributes evaluated are the cycle line layout (manufacturing and facility layout), assembly sequence, dynamic clash (assembly load path), and ergonomics (posture, hand clearance, and strength).

6.3.3 Manufacturing Feasibility Assessments

Contributing factors to assembly and ergonomic feasibility include manufacturing assumptions, dynamic clash (assembly path interference), assembly sequence, ergonomics, facility, and tooling assumptions. Each simulation is reviewed and assessed using basic engineering tools as required. Manufacturing engineers are able to rotate, pan, and zoom to obtain the best view of the simulation. Users are able to take point-to-point measurements and cut sections through the components of a study. Collision detection tools can be used to easily identify collisions between parts when trying to evaluate part locations, paths, robot/fixture locations while playing simulation or in static conditions. Obstacle avoidance can be configured to take a static, dynamic, or conservative approach to evaluating part and human clearance conditions.

6.3.4 Preliminary Ergonomic Assessments

The goal of DHMs in dynamic simulations is to evaluate all simulations for reach/posture and hand clearance. As a preliminary filter, statically postured DHM models are included with dynamic assembly simulations. These static posture models are used by manufacturing engineering to discern the ergonomic DPA checks.

FIGURE 6.2 Human collision detection—manufacturing feasibility simulation.

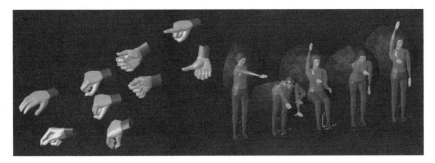

FIGURE 6.3   Preliminary ergonomic filters.

6.3.5 REACH Assessments

To address the reach requirement the simulation must quantify if the installation of the part falls within the ergonomic reach requirements. Installation includes routing, stowing, loading, or any other time the operator must touch/handle or use a tool to secure the component. Static postures have been created to simplify postures that may be adopted during the manufacturing process. Each posture has a reach envelope attached. This envelope defines a preliminary filter to assess the reach capability of the 5th percentile female.

Based upon the manufacturing assumptions, simulation engineers are required to select the appropriate posture and position within the dynamic simulation. The decision process to determine the appropriate static reach posture is defined by a set of predetermined postures, which are classified based upon assembly conditions.

Manufacturing engineers assess the simulations to determine whether or not the environment matches manufacturing assumptions. Considerations include: the workstation height, environment, all the required parts are properly sequenced and positioned, tools or assistive devices required for assembly, and the static posture (operator) positioned properly relative to the vehicle.

The reach posture filter decision criterion is as follows. Pass: If the center of the part and/or any tool required to secure the component fall within the reach envelope. Fail: If the center of the part does not fall within the reach envelope. If the assessment fails the ergonomic reach filter, a more in-depth ergonomic evaluation is required.

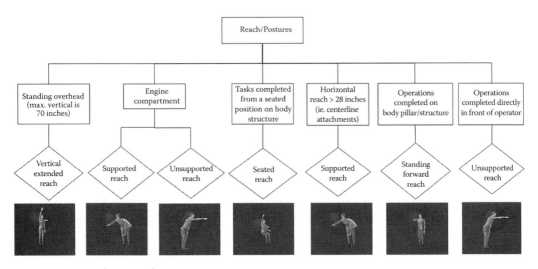

FIGURE 6.4   Reach posture decision tree.

6.3.6 HAND Clearance Assessments

The DPA check for hand clearance evaluates if the 95th percentile male has the necessary hand/finger clearance to perform the task. Tasks include, but are not limited to, hand starts, connections, part load, routing, stowing, and any other time the operator must touch/handle the part.

Based upon the manufacturing assumptions simulation engineers are required to select the appropriate hand posture and position within the dynamic simulation. The decision process to determine the appropriate static hand posture is defined by a set of predetermined postures, which are classified based upon specific part commodities.

To assess the hand clearance, the manufacturing engineers must first assess the clearance of the hand posture relative to the surrounding product data. Therefore a full 360-degree rotation about the part must be reviewed to identify any collisions or interactions with product data. Engineers must also be aware that in addition to the clearance for the hand, there must be sufficient package space for the upper extremity to reach the hand posture.

The hand clearance filter decision criterion is based upon the following. Pass: Line to line there are no collisions or intersection points between the hand and the product geometry. Fail: If there is collision between the hand posture and product data.

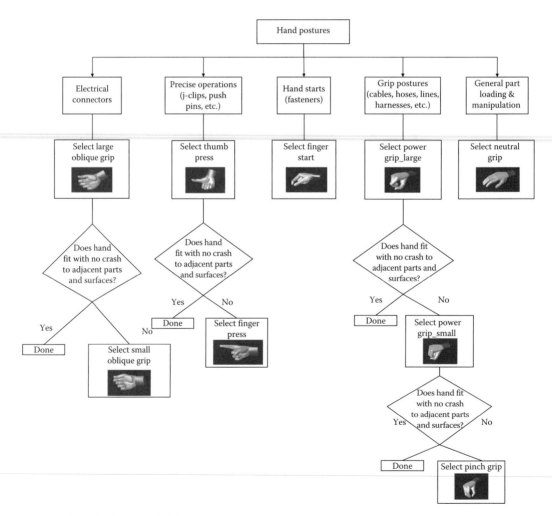

FIGURE 6.5 Hand posture decision tree.

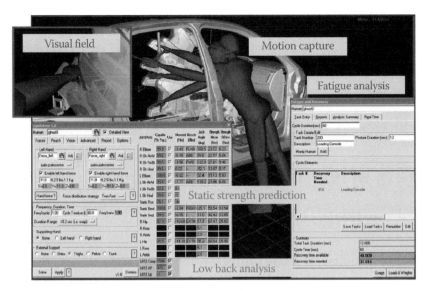

FIGURE 6.6 Integrated ergonomic analysis within a stand-alone DHM solution.

6.3.7 In-Depth Ergonomic Analysis

In-depth ergonomic analysis is required for all assessments that fail the preliminary static human posture filters. A more detailed ergonomic analysis considers complex posturing of the mannequin, interaction of the DHM mannequin with the CAD geometry, part weight, whole-body biomechanical strength capability, grip strength and posture, and energy expenditure and fatigue analyses, which will influence the acceptability of a product design or manufacturing process.

Often the DHM integrated within the PLM (Product Lifecycle Management) solution lacks the sophistication and enhancements required to complete a comprehensive ergonomic evaluation. To this end, there is commercially available DHM software that integrates ergonomic analysis capabilities. The appropriate DHM solution depends upon the motivation for the analysis. If it is in-depth ergonomic evaluations, the biomechanical foundation that the DHM is built upon is imperative. It is recommended that one evaluate and validate the biomechanical model and assumptions that constrain a given DHM solution prior to implementing.

6.4 Workflow Methods for Stand-Alone DHM

The use of a stand-alone DHM involves unique work methods to ensure that cycle line layout and manufacturing assumptions are maintained. First, users are required to transfer product geometry data from the PLM data management software to the stand-alone DHM software. It is critical that the virtual environment be accurately recreated and represented in the stand-alone software. This can be a cumbersome process due to file format compatibility issues.

The time required to create DHM simulations within a stand-alone system varies dependent upon the complexity of the posture prediction. Subsequently, the number of joints or degrees of freedom of the DHM that require manipulation to obtain an "assembly representative" posture is directly correlated with the amount of time to create a simulation. For an ergonomic evaluation, the end posture is a vital output of the simulation. Minor difference in degrees of flexion can dramatically affect the injury risk evaluation. Posturing the DHM mannequin is especially critical for those tasks that have failed the preliminary ergonomic filter.

There are three methods to posture a DHM for the purpose of an in-depth ergonomic analysis: (a) manual manipulation of mannequin, (b) posture-prediction algorithms, and (c) motion capture.

6.4.1 Manual Manipulation of Mannequin

In an effort to derive an "assembly representative" end posture for the purpose of ergonomic evaluation, most DHMs enable the user to manipulate the individual body segments connected by joints. As the user moves an individual body segment on the mannequin, the DHM uses real-time inverse kinematics to determine the position of linked segments and joints. Typically, a DHM can be manipulated by manually posturing its head, eyes, shoulders, torso, center of mass, pelvis, arms, feet, or its entire body. The human body is capable of moving in an almost infinite number of ways, which is the challenge for computer-aided ergonomics.

Depending on the user's experience in posture prediction and ergonomics, manual manipulation can be an iterative, laborious, and inconsistent process. Even when given the same goal, users may realize very different movement strategies based on experience, knowledge of human biomechanics, and the assembly process. This is illustrated in the two different postures evaluated for the 50th percentile operator performing an automotive assembly task (see fig. 6.7).

6.4.2 Posture-Prediction Algorithms

Posture prediction capability, algorithms that enable the estimation of human joint angles for a given posture, offer a time-saving alternative to manual posturing. There are two main approaches currently utilized for motion prediction: empirical statistical modeling and inverse kinematics or biomechanics. The first approach uses anthropometrical data and motion patterns collected in the laboratory that are statistically analyzed to form a predictive regression model of posture with rule-based adjustments to accommodate the infinite motions possible. The second approach uses common inverse kinematics characterization to represent mathematically feasible postures. Inverse kinematics and optimization are used to assess objective functions, such as joint limitations, physiology or cost functions, and are evaluated prior to selecting a final posture for ergonomic analysis.

Validation of all approaches to posture-prediction algorithms is the subject of great debate in the academic arena. Independent of posture-prediction algorithms, it is the final static posture that is critical for the assessment of assembly ergonomics. Therefore, work methods detail that whenever possible digital photos or videos of comparative assembly operations should be referenced upon predicting extreme or awkward postures to ensure accurate representation of assembly relevant postures. Likewise, postures should be predictive for both the 5th and 95th percentile subjects to account for anthropometric differences.

FIGURE 6.7   Two different posture strategies result from two different simulation engineers manually manipulating the mannequin to simulate the same automotive assembly task.

6.4.3 Motion Capture

Motion Capture (MOCAP) technology provides an alternative option for generating dynamic human motion within a virtual environment. Specifically, this technology allows an operator to become immersed within a virtual environment. Body markers are used to track and transpose the human subject onto the DHM, ultimately driving the mannequin. This approach can improve the simulation time, provide more accurate posture predictions, and capture the variability in movement represented on the assembly plant floor. Subsequently, MOCAP work methods detail that test subjects of different anthropometry, 5th and 95th percentile workers, be captured to identify a variety of postural strategies adopted to complete an assembly task. Physical props are also used to enable a subject to interact more realistically within a given virtual environment. If a simulated posture involves bracing or positioning relative to geometric CAD data, it is reasonable to assume that physical barriers or hard-points provide necessary feedback to the subject.

Therefore, it is generally accepted that all postures obtained from motion capture technology are realistic and accurate.

Upon obtaining an accurate and "assembly representative" posture, the DHM provides all the inputs necessary for a comprehensive ergonomic assessment. One of the major considerations in assembly ergonomics is determining the acceptable hand load(s) for a given task, with the task being defined by workstation and product packaging constraints. Determining the acceptable hand forces will often involve performing a strength evaluation. These evaluations are based on the force and joint moment requirements for the given task and an available strength database for comparison. The Static Strength Prediction tool in Classic Jack (Siemens, UGS, Germany) is the University of Michigan's 3D Static Strength Prediction Program (3DSSPP). This is one of the more prominent software tools to evaluate

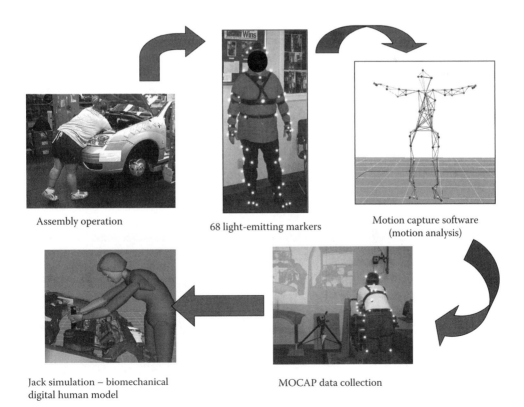

Assembly operation

68 light-emitting markers

Motion capture software
(motion analysis)

Jack simulation – biomechanical
digital human model

MOCAP data collection

FIGURE 6.8 Motion capture data collection process.

the percentage of a worker population that has the strength to perform a task based on posture, exertion requirements, and anthropometry.

Most commercially available DHM software integrates available and documented ergonomic assessment tools. Classic Jack's Task Analysis Toolkit provides tools for: low back spinal force analysis, NIOSH (National Institute for Occupational Safety and Health) lifting analysis, predetermined time analysis, rapid upper limb assessment, metabolic energy expenditure, manual handling limit, fatigue/recovery time analysis, and working posture analysis.

6.5 Ergonomic Specifications and Requirements

DHMs also provide a logical solution to perform many evaluations of strength capability to develop commodity-specific specifications. Posturing of a DHM mannequin via manual manipulation, posture-prediction algorithms or MOCAP within a virtual environment enables the derivation of commodity-specific strength capability assessments. Each of the three methods of posturing the DHM mannequin offers a level of precision and flexibility. Thus, the development of specifications is based on comprehensive evaluations of the 5th to the 95th percentile anthropometric distribution and subsequent postural strategies adopted upon interaction with 3D CAD vehicle data. Surrogate video and plant data of assembly operators installing these components on current model vehicles are also to ensure that the mannequin is manipulated to an "assembly representative" posture prior to determining a corresponding strength evaluation. This process of simulating postures using the DHM can be reiterated for all of the feasible, 'worst-case,' and potential postures adopted to install component parts.

Commodity-specific specifications are subsequently derived based upon the following criteria. (1) All human-part interactions must be within an acceptable operator reach zone. This is based on the maximum reach of a 5th percentile female and is dependent on posture. (2) Sufficient clearance must be provided around all manually installed commodities for the 95th percentile male hand to access and install the commodity. (3) The force exerted by an operator during the installation/manipulation of all commodities must be deemed acceptable to 75% of the female population (which includes 99% of the male population).

The verification of posture/reach and hand clearance assessments has been outlined above. However, there are many challenges associated with virtually assessing insertion efforts and manipulation of flexible parts. As stated, DHMs are primarily used to posture the human and determine an associated acceptable strength capability requirement. The remainder of strength assessments and installation requirements are verified using physical parts.

6.6 Future Direction

Current DHMs are being used to assist ergonomists and engineers when making educated decisions on product geometry for future vehicle models. However, there are some limitations associated

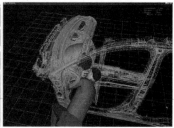

FIGURE 6.9   Evaluate and assess all human variability prior to developing a commodity-specific specification.

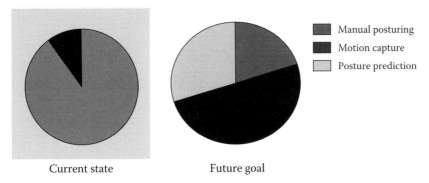

FIGURE 6.10 Future ergonomic and virtual manufacturing work methods.

with the functionality and the integration of comprehensive DHMs within commercially available PLM solutions. Manual manipulation remains the most common method for posturing a DHM mannequin into an "assembly representative" posture. Thus, the workflow methods associated with DHMs continue to be cumbersome and dependent upon the expertise of the user. This being said, Ford Motor Company has realized some advantages to using Motion Capture as an alternative posturing tool or work method for assembly and ergonomic feasibility evaluations and specification development.

The latest trend in academic and technical research is focused on motion prediction. The hope is that the user will indicate the start and end points of a movement and the DHM will adopt a posture that is based upon validated, data-driven, posture-prediction models. Such models offer time savings in simulation setup, consistency of movement patterns between users, and accuracy of virtual ergonomic and assembly feasibility assessments.

New research is also focusing on interpretation of movement. Recent efforts are committed to understanding the limits and constraints associated with cumulative fatigue and dynamic movement. Currently dynamic simulations or key-frame animations are built from manual posturing key frames and interpolation. To date, most ergonomic assessment tools are also constrained to static, terminal end postures. Yet during the execution of automotive assembly operations, humans rely on momentum and dynamics to perform tasks. The goal of calculating and predicting injury risk in a dynamic environment is both challenging and exciting!

References

Chaffin, D.B., Anderseson, G.B.J., and Martin, B.J. (2006). *Occupational Biomechanics.* Wiley, New York, Toronto.

Classic Jack—Commercially Available Digital Human Modeling Software from UGS (Siemens, UGS, Germany).

Godin, C.A., Chiang, J., Stephens, A., and Potvin, J. (2006, July). Assessing the Accuracy of Ergonomic Analyses When Human Anthropometry Is Scaled in a Virtual Environment, Society of Automotive Engineering (SAE at Digital Human Modeling Conference, Lyon, France).

Regents of the University of Michigan (2000). University of Michigan 3DSSPP 4.21.

7

Virtual Environments and Digital Human Models

Thomas Alexander
and Stephen R. Ellis

7.1 Virtual Environments

Virtual reality (VR) and virtual environments (VE) subsume methods and technologies for a natural and potentially intuitive human-computer interaction (Burdea & Coiffet, 1994; Kalawsky, 1993; Ellis, 1995; Brooks, 1999). They have become valuable tools for a broad spectrum of applications in different domains. The spectrum ranges from visualization for product development and scientific data exploration to simulation-based education and training (Bullinger et al., 1997; Seidel & Chatelier, 1997; Alexander et al., 2004; Stanney, 2002).

There is a close link between virtual environments and digital human modeling: A digital human model (DHM) simulates the body shapes, postures, reach ranges, and motions of a user population, and VE technology can be applied to visualize the results in a natural way (Bullinger et al., 1997). In fact, one of the first applications of 3D computer graphics was the visualization of human postures and reach envelopes (Fetter, 1961, 1964). For a comprehensive understanding of further connections between both topics, it is necessary to specify and define key terms related to VE.

7.2 Background and Basic Definitions

The term *virtual reality* (VR), originally introduced as a marketing buzzword by Jaron Lanier (2006) in the late 1980s, comprises new methods, procedures, and technologies for multi-modal human-computer interaction with 3D data representing inhabitable environments. In these systems, users' movements and actions are processed to activate body-referenced or body-worn displays to enable perception and experience of the environments as if they were real (Ellis, 1994; Brooks, 1999). Consequently, the experience of these VEs is not limited to technology but also has psychological, social, and philosophical dimensions (Reeves & Nass, 1996; Slater et al., 1999).

The basic concept of VE derives from a continuous technological development during the past decades. Pioneers like Ivan Sutherland (1965), Myron Krueger (1977), who developed the first prototypes, and later Michael McGreevy and his colleagues (Fisher et al., 1986) who elaborated the concept, suggested a wide variety of possible applications. These demonstrations were helpful to present the concept, but neither the original nor most of the 1980s' developments had sufficient computer power or simulation fidelity for a practical application. Probably the earliest high-fidelity implementations of the general concept of personal simulators were head-worn flight simulators implemented in the mid-1980s (Furness, 1986; CAE, 1991). With growing graphic rendering power of commercial personal computers, the technology has now become available to a broader audience. Today, academic research, system developers, and industrial customers can adapt it for practical applications (Carr & England, 1995; Stanney, 2002; Alexander & Ellis, 2005).

A strict interpretation of the term *virtual reality* implies generating and presenting a comprehensive reality with all its facets. This is conceptually and technically impossible. Therefore, generating and presenting VR concentrated on the special aspects of reality that were relevant for the particular settings. For this reason, the term *virtual environments* (VE) was adopted in the scientific community (Ellis, 1991, 1993). The term is intended to cover the intersection of technological capabilities, characteristics of the user, and the specific applications, and to relate the technology to its origins in robotics, simulation science, stereoscopic visual displays, and interactive computer graphics.

Many different approaches for defining VE exist. Their common elements include computer-generated, synthetic environments, associated displays and interaction technology, a natural, intuitive perception, and the support of active exploration. Ellis (1995) defined VE as a synthetic, interactive,

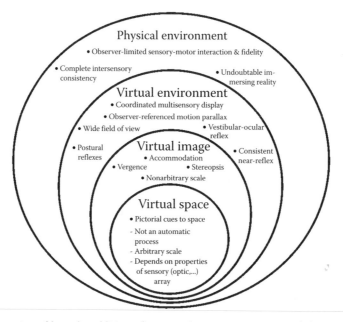

FIGURE 7.1 Illustration of how the addition of more and more sensory cues and dimensions of interaction transform the displays like 2D pictures, which depict virtual spaces through the principles of perspective to fully immersing environments that are approximations of true physical environments (Ellis, 1995).

illusory environment, perceived when a user wears or inhabits an appropriate apparatus, providing a coordinated presentation of sensory information that can convince the users that they are located in a place other than their current physical location. An alternative definition characterizes VE as the experience of being in a synthetic environment and the perceiving and interacting through sensors and effectors, actively and passively, with it and the objects in it, as if they were real. Virtual environment technology allows the user to perceive and experience sensory contact and interact dynamically with such contact in any or all modalities. (Werkhoven et al., 2001)

This definition circumscribes the technical, objective aspects of VE as multimodality properties of hard- and software components, as well as user-related, subjective aspects as perception, interaction, and experience.

Thus, the impression of a VE is the outcome of a system of two components: technology and the user. Consequently, the characteristics of both components have to be considered to optimize system performance and usability.

7.3 VE Technologies

VE serves as a medium for a physiologically close and intensive human-computer interaction. This requires special hardware and software that produce an artificial environment allowing interaction in terms of human perceptual, cognitive, and motor skills. The main output modalities of a VE system are visual, acoustic, and haptic. Further modalities (e.g., kinesthetic or olfactory) are rarely included.

Because of the dominance of the visual modality, the visual display of a VE system is crucial (Heineken, 2002). An optimal visual representation has to match the visual and color resolution of the eye, facilitate a 360° field-of-regard, and provide rendering update without noticeable flickering and transport delay effects (Carr & England, 1995). Depth cues like perspective, occlusion, and motion parallax facilitate depth perception as in reality. Most of the systems include stereoscopic presentation capabilities, which enhance visual depth perception significantly, especially for short distances (IJsselsteijn et al., 2001; Alexander et al., 2003). Inaccuracies of visual displays (e.g., a limited resolution) or inconsistencies of the presented stimuli (e.g., a delay between visual and acoustic feedback) often lead to degraded depth perception and introduce additional visual stress (Roscoe, 1991; Mon-Williams & Wann, 1998; McCandless et al., 2000; Hofmann et al., 2001). There are many different VE-displays that facilitate a 3D stereoscopic presentation. An extensive summary is available online (www.stereo3d.com). These displays can be categorized into three groups.

A simple, though practicable alternative is to use conventional *cathode ray tube (CRT) monitors*, or, more recently, flat panel display technologies (e.g., plasma, LCD, TFT). Images for the left and the right eye are calculated and rendered separately. The separation of the two stereoimages may be handled by shutter glasses, which darken alternately and synchronously with the rendering to present left and right eye stereoimages to the users' eyes (Hodges & McAllister, 1993; Lipton, 1993). Autostereoscopic or lenticular displays are an alternative to shutter glasses (Delaney, 2005). They consist of a pixel-based display with a special overlaid mask or lenticular optical element that separates the left and right eye images so that no special glasses are required to perceive a stereoscopic view. When the physical size of these monitors is limited, such displays are referred to as desktop VE. The latest model of displays of this sort has been introduced for the HDTV format and computer monitor displays with similar aspect ratios by Phillips (2004).

Projection technologies can be used to present wall-sized stereoscopic displays. A rear projection system, for instance, allows close-up interaction with virtual stereoscopic objects because many of the occlusive interference effects of the user's shadow can be avoided. Such systems, however, cannot present virtual objects between the user's hand and eye. Common projection-based devices are projection tables or projection walls, sometimes with multiple projectors (Krueger, 1983; Krueger & Froehlich, 1994; Bolas et al., 2004). Projection rooms are more space-intensive solutions with a large field-of-view and fields-of-regard. They consist of multiple projection walls, a projection floor, and projection ceiling.

Projection rooms facilitate a comprehensive presentation of a computer-generated scenario (Crux-Neira et al., 1993).

In comparison to projection systems, *individual systems* require smaller spaces. The most common individual displays are head-mounted displays (HMD), which are worn like glasses or, more precisely, helmets (Sutherland, 1965; Furness, 1986; Fisher et al., 1986; Bungert, 2006). HMDs consist of miniaturized displays and optics, presenting an individual image to the user. With a typical, closed HMD, users feel completely present in the VE and are not able to perceive real, visual stimuli in their physical environment.

The decision about which display to use depends on the type of application, task requirements, and budget. Conventional monitors are cost efficient, but offer the smallest field of view. Projection rooms and HMDs have capabilities to present a wide field-of-regard, but they are technologically more challenging and may require larger investments of money and space. But only the latter two, fully immersing options provide a strong sense of presence in an alternative environment.

There are different techniques for auditory presentation. One straightforward technique is the use of monaural sound without the simulation of spatial location. Most current VE systems feature stereo or pseudo-spatial sound. The acoustic models supporting them have some capabilities for directional sound. But these only support a rough spatial localization of sound sources. For more precise auditory localization, more sophisticated models utilizing filtering with head-related transfer functions must be used (Wightman & Kistler, 1989). Such virtual auditory displays include various spatial parameters, such as reflection, damping, and room resonances (Begault, 1994; Vince, 1995; Shilling & Shinn-Cunningham, 2002). Acoustic presentation can be implemented by means of headphones, speakers, or arrays of speakers for spatial sound.

Haptic feedback is essential for most manipulative tasks. Several different approaches for providing haptic feedback in VE exist (Adelstein et al., 1996; Burdea, 1996). A simple, though effective solution is the use of a simplified real-world model as a prop, with objects or parts of the environment modeled by wooden mock-ups (Schiefele, 2000). This enables the user to perceive haptic feedback as soon as a virtual object is hit. Other approaches use mechanical linkages with electromagnetic, electrohydraulic, or pneumatic actuators. The user holds a grip or pen connected to the system and perceives inertial forces (Adelstein & Rosen, 1992; Massie & Salisbury, 1994; Burdea, 1996; Biggs & Srinivasan, 2002). An alternative to this approach are gloves that are connected to a wire system. In this case actuators shorten or lengthen the wires depending on the touch force (Sturman & Zeltzer, 1994). As interaction is an essential part of a VE, means for user input have to be provided as a basis for system control, function allocation, and determining the appropriate system response.

VE systems adopt various input devices (Bowman et al., 2005). Some installations include conventional input devices like mice or keyboards. For a more natural interaction, however, there are special input devices with six degrees of freedom available. Several of them are based on an extended control stick concept: The grip of the device can be shifted like a control stick and rotated around the three spatial axes, allowing the manipulation of position and orientation of viewpoint or objects. These devices can also be augmented with haptic actuators, making them active, haptic displays as well (Burdea, 1996).

A common method for detecting and measuring user actions is motion tracking. Motion tracking may be used to capture biomechanical data or to control digital human models interactively (Landauer et al., 1997). Motion tracking systems are based on inertial measures, time-delay measures of ultrasonic or electromagnetic signals, or optical image processing (Welch & Foxlin, 2002). They allow measurement of position and orientation by the use of fiduciary or body-mounted markers, or interaction devices, or the user's limbs (Foxlin, 2002; Bowman et al., 2005).

Devices often mentioned in connection with VE are specially equipped gloves, so defined as data gloves (Sturman & Zeltzer, 1994; Bowman et al., 2005). These devices are used for manipulations with several degrees of freedom, for pointing, and for gesture input. Gloves incorporate sensors for measuring hand position and bending of each finger (Zimmermann et al., 1987; Kramer & Leifer, 1994). They can control hand and finger actions of digital human models within a VE or for subsequent gesture recognition for function allocation.

Speech recognition uses another modality than the previously described manipulative input or gestures. It is often applied to control global functions. Natural speech or command speech allows a reliable allocation of special functions without interfering with gestures or manipulation actions. However, conceptual and technical issues have to be considered when using speech recognition, ranging from position of the microphone to additional cognitive workload with natural speech input especially associated with inadvertent command execution. Speech input can be combined with other modalities, such as gestures, to provide better spatial information (Oviatt et al., 1997).

7.4 Related Human Factors Issues

The inclusive human-computer interaction within a VE, the extent of computer-generated stimuli, and the broad application area require the consideration of multiple relevant human factors issues at the conceptual, functional, and technical level (Brooks, 1988; Ellis et al., 1991; Carr & England, 1995; Bullinger et al., 1997; Alexander et al., 2007).

The conceptual base of a VE system is the interaction metaphor. This and related metaphors on lower levels can be used to compare a new system with previously established systems (Tourangeau & Sternberg, 1981; Laurel, 1991, 1993). By defining an appropriate metaphor the interaction with the VE relates to real manual procedures during task processing.

A common metaphor is virtual fly- or walk-through. The user controls a virtual camera in the computer-generated environment by modifying its position and orientation. The metaphor is frequently used for group presentations with one user controlling the camera and guiding the group. A modification of this metaphor is frequently used for showcases or workbenches of virtual designs and projects, which can be examined as if they were real. In this case, the exploration range is limited to a smaller physical surrounding. Consequently, only procedures for exploration of nearby space are integrated. By adding functionality for manipulation these metaphors can also be applied for rapid engineering and rapid prototyping (Bullinger et al., 1997).

These baseline metaphors are supported by special interaction methods. They refer to functionality known from conventional desktop computer systems (e.g., cut and paste, drag and drop) and combine them with the new interaction technologies (Kalawsky, 1993; Turk, 2002; Bowman et al., 2005).

In addition to the conceptual design of a VE system, there are functional and technological requirements. Display and system characteristics have to take into account the perceptual capabilities of the human user. For a VE this relates especially to the resolution, field-of-view, frame-rate, delay times, and system latency (Ellis et al., 2002). Each of these factors affects performance and might induce negative aftereffects (Frank et al., 1983; Kennedy et al., 1989; Pausch et al., 1992; Kolasinski, 1995; Mania et al., 2004). The same applies for cue conflicts between modalities (Potel, 1998; Kennedy et al., 1988). Additionally, characteristics of the task have to be considered because a mismatch between a user's expectation and a system's response leads to degraded performance.

A severe potential, negative effect of VE is cybersickness, which is related to simulator sickness. It results from inconsistencies between these different technological, task-specific, and user-dependent characteristics. The various effects of simulator sickness can be clustered into nausea, oculomotoric, and disorientation effects with a range from hardly measurable effects to massive health problems preventing the application of systems (Frank et al., 1983; Kennedy et al., 1988; Pausch et al., 1992; Kolasinski, 1995; Potel, 1998; Stanney & Hash, 1998; Young et al., 2007).

Further relevant multifactor phenomena in connection with VE are immersion and presence. Immersion generally refers to the technological basis for experiencing presence (Slater & Usoh, 1994; Slater et al., 1996; Slater, 1999). It defines the technological requirements for addressing multiple sensory modalities; presenting stimuli in a wide area; extinguishing real stimuli; presenting manifold; high-resolution, and high-quality information, including the body of the user; and presenting matching cues and stimuli. An alternative definition of immersion is given by Witmer and Singer (1998) and Stanney et al. (2003). The authors define immersion as the psychological state of the user being included in an environment, and perceiving and experiencing it interactively.

This understanding leads to the concept of presence. Virtual presence describes the feeling of being included in the virtual scene and experiencing it as if it were real (Sheridan, 1992; Ellis, 1996; Biocca, 1997). It is the synthesis of all external and internal stimuli, and thus the outcome of a wide-reaching integrative process. Heeter (1992) differentiates among personal presence (i.e., the individual feeling of being present in a VE), social presence (i.e., the feeling that other humans are present in the VE), and environmental presence (i.e., the interactivity of the computer-generated scene itself). In addition to its technical basis, presence is strongly affected by the task and user characteristics (Welch et al., 1996; Schumie et al., 2001; Regenbrecht & Schubert, 2002).

7.5 Present Applications of VEs

The potential of VE systems to present realistic, interactive environments made them promising tools for various applications. Since a list of every application would be too lengthy, just several broad application areas are included in the following discussion (cf. Kalawsky, 1993; Pimentel & Teixeira, 1993; Burdea & Coiffet, 1994; Ellis et al., 1995; Bullinger et al., 1997; Stanney, 2002; Bowman et al., 2005; Alexander et al., 2004).

In the broader sense VE serves as a communication medium (Ellis, 1991). They are capable of visualizing complex, often abstract data in an intuitively understandable way. This makes VE a valuable tool for scientific analyses. In general, a high-end computing system performs analyses and provides the data for the VE system. The VE system visualizes these data streams and facilitates an interactive exploration of the data. By allocating available functions, the displayed information can be modified or analyses can be redone with modifications.

An important similar application area is design. VEs make future product designs applicable within the early phases of the development process, supporting decisions about further design steps and product optimization. This application is subsumed by the term virtual mock-up. Like conventional mock-ups, virtual mock-ups can also be used to assess empirical data. By integrating simulator models of the future products the VE systems offer functionality for operational evaluation and testing. This way VE strikes a balance between design, mock-ups, and prototypes. Though they are technologically not capable of completely replacing real mock-ups, they can help to reduce the total number of mock-ups needed for a design. Since only a small fraction of all the physical components of a product may be required and only electronic CAD-data of the future product may be available, virtual mock-ups are easier to modify than traditional (wooden or clay) ones. Even extensive design changes can be realized in a short amount of time. This efficiency supports rapid engineering and prototyping, and it reduces time and cost. At present several automotive companies (e.g., GM, Ford, Fiat, Chrysler-Benz) have established VE centers and use them for the design of their future products.

After final product prototypes have been defined, the computer-based models used to develop them in the VE can be applied as training tools. This leads back to the basis of VE in simulation and training. In contrast to traditional vehicle simulators, most of which are restricted to single vehicle types only, VEs are variable to simulate different vehicle types. This development is possible because of the large range of computer-generated stimuli. Yet, technological shortcomings, especially in haptic feedback, have to be considered for practical use and VE-based product testing. Therefore, today's VE is not the general solution and must be utilized in conjunction with conventional training methods. New training concepts using a mix of constructive (compare applications in design), virtual, and live training seems to be most promising.

A final application area is teleoperation or telepresence. In this application the user gets the feeling of being physically present at a different physical environment. The environment might be scaled (e.g., laparoscopic surgery) or distant (e.g., dangerous or cost-effective locations). By applying VE technology, the distant remote system is controlled in a natural way. This simplifies navigation and orientation significantly. Cyberspace is another telepresence application referring to a virtual communication platform for social activities (Gibson, 1984). Participants meet each other online for virtual exchange of ideas or collaborative work. In this case, digital representations of the users typically are available to facilitate a more natural communication.

7.6 Digital Human Models within Virtual Environments

Digital human models (DHM) and virtual environments (VE) have been developed and used mostly independently. DHMs primarily originated from work in anthropometry and biomechanics. Their primary application fields were workplace and product design. VE derived from computer graphics and initially provided visualization and training tools. But with recent development digital human representations are now included in VE. Moreover, VE has become a promising medium for workplace and product design, and both DHM and VE activity has begun to overlap.

7.6.1 Tools for Product and Vehicle Design

One of the main application areas of DHMs and VEs is product and vehicle design review. But DHMs and VEs have had different development paths. DHMs have evolved from physical drawing templates used in design work into sophisticated digital tools for computer-aided design or CAD. Thus, they have been integrated into modern design processes. Although still a separate research topic, VE has become a powerful tool for visualization in product design and engineering. Consequently, many companies have built up VE centers and use them for the presentation and design of future products.

7.6.2 Presentation

In general, the inclusion of DHMs into VE enhances realism for the observer. Designs and virtual products simply look more realistic when humans are involved (Plantec, 2004). However, the degree of modeling and accuracy of these realistic-looking virtual humans varies from simple, 2D representations to 3D agents with intelligent-appearing behaviors. A simple approach is to include photographs and photographically based textures into the virtual CAD design. By building a database with different postures from different viewing angles, it is possible to achieve higher face-validity. However, this approach is not easily extensible and requires a large photograph database. But it becomes more adaptable if connected to more complex human shape models. In this case pictorial information is used as texture of the corresponding geometric properties of the human shape model. This approach is similar to that of projects in 3D surface anthropometry, which also combine pictorial and 3D information (Robinette et al., 2002; Robinette & Daanen, 2003). There are several approaches in computer graphics for pasting pictorial information (e.g., photos of real persons) to 3D human shape models (Foley et al., 1995) and for generating realistic-looking virtual humans (Magnenat-Thalmann & Thalmann, 2000; Prendinger & Ishizuka, 2004; Magnenat-Thalmann & Thalmann, 2004). The results often look realistic, though they are still limited to single human templates. Improved realism can come from increasing the number of human models representing individuals and giving them interactive behavioral models (e.g., to create the appearance of realistic crowds) (O'Sullivan et al., 2002; Ulicny & Thalmann, 2002).

The main application purpose of these virtual human models is product presentation and marketing. Potential products can be presented as if they were already in use. This presentation is considered to have impact on basic management decisions at early conceptual phases. However, despite the photorealistic appearance, the validity of body dimensions and posture is often not guaranteed. Instead, more sophisticated models with valid anthropometric and biomechanical results are used for more advanced design processes.

7.6.3 Exocentric and Egocentric Analysis Tools in Design

Applications for ergonomic workplace and product design are the key applications of DHMs (SAE DHMC, 1998ff; Landau, 1999; Chaffin, 2001). DHMs serve as tools for analyzing sight, reach, posture, comfort, fatigue, and so on, for a user population, an application that was pioneered by William Fetter at Boeing Aircraft in 1960 (Fetter, 1961, 1964). In contrast to the first DHMs, which were simplified digitized versions of anthropometric percentile drawing templates, today's commercial anthropometric models are far more sophisticated (cf. Delmia Safeworks, 2006; HumanSolutions Ramsis, 2006; UGS Jack, 2006; Virtual Soldier Research, 2006). They apply complex algorithms for describing and modeling human body shape, anthropometric dimensions, and their variability. Modeling behavior and its variability is still a research topic, often relying on biokinematic or biomechanical models, which facilitate the simulation of simple goal-directed movements. More comprehensive human models, which also consider higher levels of human behavior such as tactical and strategic thinking, are still research topics and not totally integrated into design processes (Alexander & Ellis, 2005).

By applying VE technology, some of the deficits in behavior modeling can be finessed. Instead of modeling higher behavioral levels, the intelligence of the designers may be utilized in that they may be enabled to control the DHM interactively from an exocentric or an egocentric perspective.

In case of an *exocentric* perspective, the VE can serve as an extension of conventional CAD. First, the designer defines the anthropometric dimensions based on the target user population and generates the appropriate DHM. Afterwards, the initial posture of the DHM and procedural tasks (e.g., reaching) may be defined. Finally, the DHM is positioned in the design and the analysis is carried out. During the analysis in the VE, the designer is able to change the viewpoint of the virtual camera interactively by natural body and/or head movements. This natural exploration supports the identification of deficits and deeper analysis. By utilizing large-scale projection displays, this kind of analysis can also involve collaborative design reviews by groups (Bullinger et al., 1997; Abdel-Malek, 2006).

The *egocentric* analysis facilitates initial operational analyses. After defining the anthropometric dimensions of the DHM, the designer is able to map and link the design eyepoint of the DHM with the real viewpoint (Badler et al., 1993; Slater & Usoh, 1994; Delaney, 1998; Emering et al., 1998; Kalra et al., 1998; Sturman, 1998; Abdel-Malek, 2006). The designer then can experience the surrounding as if it were real. By mapping actual body posture to DHM posture, it is also possible to control the DHM interactively: the model acts as a virtual puppet on a string. In this way, first operational analyses are feasible with the designer performing the task. Subsequently, operational deficits can be identified.

A third approach is the combination of both exocentric and egocentric analysis. In this case the egocentric analysis is followed by another exocentric evaluation. In contrast to the purely exocentric analysis, captured postures and movements of an egocentric analysis may be recorded and replayed. The designer is able to analyze the recordings interactively by changing the viewpoint of the virtual camera. In this way designs can be evaluated under operational conditions from different viewpoints. This approach enhances the overall quality of the design process and reduces the risk of corrections at later design phases.

Each of these approaches can be integrated into the design process considering the actual conceptual and technological capabilities of the VE system. A continuous adjustment is needed as VE is still a topic of ongoing development.

7.6.4 Telepresence and Teleoperation

Combinations of DHMs and VE also may advance applications in telepresence and teleoperation. Meetings, for example, may be conducted in a virtual space in which participants are required to be present as so-called avatars for more nature communication, including extralinguistic cues. A simple icon (a "you are here" symbol) and verbal communication are often not sufficient because real conversation requires many nonverbal cues (Argyle, 1974; Knapp & Hall, 2002). By including appropriately animated virtual avatars, the communication between users is enhanced (Cassel et al., 2000; Gerhard et al., 2002; Schroeder, 2002; Rist, 2003; Pfleger & Alexandersson, 2004).

This amount of detail might not be required for a simple communication, but it is promising for applications in distributed education and training. In such systems trainees and instructors would not have to be physically present at the same location but could conduct courses remotely. DHM with VE systems would accordingly enhance training efficiency in situations in which the trainee is remote but needing specialized, interactive training and help (e.g., maintenance training, operational training, or military tactical training).

Other applications of DHM are oriented toward using video streaming with limited bandwidth (Roehl et al., 1997; Capin et al., 1999; Alexander & Gärtner, 2000; Thalmann & Vexo, 2003; H-Anim, 2006). In case of broadband communication, uncompressed video streaming is sufficient to transmit communication data for teleconferencing. With more restricted bandwidths channels compression algorithms have to be applied and video streams need to be preprocessed to minimize transferred data. For such situations human posture information in a DHM can be used as a preprocessing technique requiring changes of posture to be transmitted. By applying DHM and VE technology, it is possible to decompress the information again and to create the illusion of physical presence within an education environment.

7.6.5 Intelligent Agents and Avatars

Another general application of DHMs in VEs is for realistically populating virtual worlds created on the Internet (e.g., Hilton et al., 1999; Shao & Terzopoulos, 2006; Pelechano & Badler, 2006). Actors in such worlds can include autonomous intelligent agents as well as user-controlled avatars (Allbeck & Badler, 2002; Plantec, 2004; Delleman, 2006). In an ideal setting, a user would not be able to tell the difference between human controlled and autonomously acting DHMs (Delleman, 2006).

7.6.6 Populating Virtual Worlds

A typical characteristic of the real world is that most areas are populated. Scenarios totally lacking any human beings seem to be unreal. This feature disturbs the user's sense of immersion and presence. Therefore, inclusion of human representations acting like real humans is desired for computer-generated scenarios in VE. But simple static geometric representations are insufficient, because the avatars appear more like storefront mannequins than real humans. They would be more appropriately represented as active DHMs (Allbeck & Badler, 2004; Prendinger & Ishizuka, 2004; Magnenat-Thalmann & Thalmann, 2004).

Thus, in addition to the user's appearance, the behavior of the virtual humans has to appear naturally (Ellis et al., 2004). Moreover, visual appearance of the DHM and its motion has to be consistent. Realistic motion of virtual humans within the VE is especially critical because human users are very sensitive to inaccuracies. From lifelong experience we have gained detailed knowledge about gestures, facial expressions, and so on, so that small inconsistencies are instantly noticeable. Furthermore, the human users often process motion to infer emotional states, intentions, and goals of acting entities. Accordingly, slight inaccuracies in motion modeling might easily lead to incorrect inferences about future actions and goals.

Inferring and understanding the semantic content of a situation in a VE involves hierarchical classifications of virtual humans, their motions, and different behaviors. In addition to model appearance and low-level movements, it is also important to build up a consistent, integrated story. This so-called virtual storytelling is essential for the reasonable communication of information within the DHM (Aylett & Cavazza, 2001; Darcy et al., 2003).

As stated before in *Related Human Factors Issues*, virtual humans affect presence, especially social presence. Slater (2002) observed natural reactions in VE users while delivering a public speech in front of a virtual auditorium or facing other social situations. Other authors observed the same and concluded that social presence has an impact on overall presence as well (e.g., Zhao, 2003).

7.6.7 Virtual Tutors in Education

Because one of the major applications of VE is education and training, a further step would be to include DHMs in these training scenarios. As described in the previous section, this includes populating the training scenarios with virtual humans, but also integrating virtual tutors into the training session (Loftin et al., 2004; Gratch & Marsella, 2005; Macedonia & Rosenbloom, 2001).

Training is based on special educational and pedagogical objectives achieved by educational means. Consequently, the computer-generated scenario has to be considered a part of the overall pedagogical concept so that training objectives are met. Trainees may need to experience a high degree of presence so that they behave as if they were in the physical world. There are approaches integrating DHMs for practicing and training communication and social skills. In one case, a checkpoint control was simulated, featuring virtual soldiers and passersby (Loftin et al., 2004). Trainees needed to learn how to check ID or secure their teammates. The passerby and teammates were simulated and controlled by artificial intelligence. They reacted to the trainee's actions instantly. A further recent installation simulates a highly emotional situation during a military peace-keeping mission (Gratch & Marsella, 2005). In this case the behavior of a crowd of people was simulated. Further installations work on using VEs and DHMs in order to train medical surgery personnel (Lok, 2006; dstl, 2006). By integrating virtual tutors as a primary point-of-contact, educational scenarios may be personalized. Such virtual tutor systems can be used for visualizing and training procedural tasks, such as in maintenance or manual operation of machines (Johnson, 1997; Rickel & Johnson, 2000). Installations of virtual tutor systems included sensorimotor systems to monitor communication and react to environmental stimuli. If the trainee addressed the agent, it reacted and gave recommendations like a real tutor.

7.7 Short Outlook: Anthropomorphic Interaction Agents

Adding advanced autonomous behavior to DHMs would enhance the field of applications of DHMs in VEs significantly. As described earlier, the designer has to define special posture and task interactively. In most extant systems the DHM acts as a puppet with no further intelligence. Initial approaches exist to provide artificial intelligence to the model so that it is able to identify objects and infer goals. This reduces the required input to a task (e.g., "sit") and an object (e.g., "driver's seat"). Moreover, adverbs can be added to specify the action (e.g., "fast"). One example for an implementation of this procedure is parameterized action representation (PAR) (Badler et al., 1993).

Another potential application of DHMs in VE is the integration of an intelligent conversational agent (Cassell et al., 2000; Aylett & Cavazza, 2001). VEs have previously been characterized as highly interactive, enabling a natural human-computer interface. Because of the reference and close link to reality, there is always the problem of including natural system responses. The worry would be that artificial system responses with an integrated standard graphical user interface might hamper immersion and presence. An alternative approach is the inclusion of anthropomorphic conversational agents for system interaction. In this case the user would be able to communicate with the system as with a human being (Plantec, 2004). Although this application may appear to be coming directly from a science fiction novel, there are several installations in the area of electronic commerce that already use anthropomorphic interfaces to help distribute their products via the Internet (VirtualHumans, 2006).

Of course, the inclusion of intelligent conversational agents is a large topic with many relevant aspects. They involve behavior modeling on various levels, including appearance, anthropometry, and biomechanics, but also incorporate emotional, psychological, and sociological factors. The resulting anthropomorphic intelligent agents would then become a "real" virtual human for various applications.

References

Abdel-Malek, K., Yang, J., Marler, R.T., Beck, S., Mathai, A., Zhou, X., Patrick, A., Arora, J. (2006): Towards a New Generation of Virtual Humans: Santos. *International Journal of Human Factors Modeling and Simulation,* 1(1), pp. 2–39.

Adelstein, B.D., Ho, P., Kazerooni, H. (1996): Kinematic Design of a Three Degree of Freedom Parallel Hand Controller Mechanism. *Fifth Annual Symposium on Haptic Interfaces for Virtual Environment and Teleoperator Systems. Proceedings, Dynamic Systems and Control,* DSC-Vol. 58, American Society of Mechanical Engineers, New York, pp. 539–546.

Adelstein, B.D., Rosen, M.J. (1992): Design and Implementation of a Force Reflecting Manipulandum for Manual Control Research. In: Kazerooni (ed.): *Advances in Robotics.* American Society of Mechanical Engineers, New York, 1992, pp. 1–12.

Alexander, T., Ellis, S.R. (2005): Linguistics in Motion: Structuring Human Motion Behavior. *Proceedings of the SAE Digital Human Modeling Conference in Iowa City.* Society of Automotive Engineers (SAE), Warrendale, MI.

Alexander, T., Gärtner, K.-P. (2000): Design of a Human Model for Internet Telecooperation. In: Landau (ed.): *Ergonomic Software Tools in Product and Workplace Design.* IfAO Institut für Arbeitsorganisation, Dortmund, Germany.

Alexander, T., Goldberg, S.R., Magee, L., Rasmussen, L., Borgval, J., Lif, P., Delleman, N., Smith, E., Cohn, J. (2004): *Virtual Environments for Intuitive Human-System Interaction. National Research Activities in Augmented, Mixed and Virtual Environments.* RTO-TR-HFM-121-Part-I. NATO Research & Technology Organization, Neuilly-sur-Seine, France.

Alexander, T., Goldberg, S.R., Magee, L., Rasmussen, L., Borgval, J., Lif, P., Delleman, N., Smith, E., Cohn, J. (2007): *Virtual Environments for Intuitive Human-System Interaction. Human Factors Considerations in the Design, Use, and Evaluation of Augmented and Virtual Environments.* RTO-TR-HFM-121-Part-II. NATO Research & Technology Organization, Neuilly-sur-Seine, France.

Alexander, T., Winkelholz, C., Conradi, J. (2003): Depth Perception and Visual After-effects at Stereoscopic Workbench Displays. *Proceedings of the IEEE Virtual Reality 2003 Conference,* Los Angeles. IEEE Press, Los Alamos, CA.

Allbeck, J., Badler, N.I. (2002): Embodied Autonomous Agents. In: Stanney (ed.): *Handbook of Virtual Environments.* Erlbaum, Mahwah, NJ.

Allbeck, J., Badler, N.I. (2004): Creating Embodied Agents with Cultural Context. In: R. Trappl and S. Payr (eds.): *Agent Culture: Designing Virtual Characters for a Multi-Cultural World.* Erlbaum, New York.

Argyle, M. (1974): *Bodily Communication.* Taylor & Francis, London.

Aylett, R., Cavazza, M. (2001): Intelligent Virtual Environments—a State-of-the-Art Report. *Proceedings of Eurographics 2001* in Manchester, UK. ACM.

Badler, N.I., Hollick, M., Granieri, J. (1993): Real-time Control of a Virtual Human Using Minimal Sensors. *Presence* 2(1), pp. 82–86.

Badler, N.I., Phillips, C., Webber, B. (1993): *Simulating Humans: Computer Graphics Animation and Control.* Oxford University Press, New York.

Begault, D. (1994): *3D Sound for Virtual Reality and Multimedia.* Academic Press, Boston.

Biggs, S.J., Srinivasan, M.A. (2002): Haptic Interfaces. In: Stanney (ed.): *Handbook of Virtual Environments.* Erlbaum, Mahwah, NJ.

Biocca, F. (1997): The Cyborg's Dilemma: Progressive Embodiment in Virtual Environments. *Journal of Computer-Mediated Communication,* 3(2).

Bolas, M., McDowall, I., Corr, D. (2004): New Research and Explorations into Multiuser Immersive Display Systems, *IEEE Computer Graphics and Applications,* 24(1), pp. 18–21.

Bowman, D.A., Kruijff, E., LaViola, J.J., Poupyrev, I. (2005): *3D User Interfaces.* Addison-Wesley, Boston.

Brooks Jr., F.P. (1999): What's Real About Virtual Reality?, *IEEE Computer Graphics and Applications,* 19(6), pp. 16–27.

Brooks, F.P. (1988): Grasping reality through illusion—interactive graphics serving science. *Proceedings of the SIGCHI conference on human factors in computing systems.* ACM, New York, 1–12.

Bullinger, H.-J., Brauer, W., Braun, M. (1997): Virtual Environments. In: Salvendy, G. (ed.): *Handbook of Human Factors and Ergonomics,* John Wiley & Sons, New York, pp. 1725–1759.

Bungert, C. (2006): *HMD/VR Helmet Comparison Chart.* Internet document retrieved from http://www .stereo3d.com/hmd.htm#chart

Burdea, G. (1996): *Force and Touch Feedback for Virtual Reality.* John Wiley & Sons, New York.

Burdea, G., Coiffet, P. (1994): *Virtual Reality Technology.* John Wiley & Sons, New York.

CAE Electronics (1991): Product Literature. CAE Electronics, Montréal, Canada CAE Electronics, Montreal, Canada.

Capin, T.K., Pandzic, I.S., Magnenat-Thalmann, N., Thalmann, D. (1999): *Avatars in Networked Virtual Environments.* John Wiley & Sons, New York.

Carr, K., England, R. (eds.) (1995): *Simulated and Virtual Realities.* Taylor & Francis, London.

Cassell, J., Sullivan, J., Prevost S. (eds.) (2000): *Embodied Conversational Agents.* MIT Press, Boston.

Chaffin, D.B. (ed.) (2001): *Digital Human Modeling for Vehicle and Workplace Design.* Society of Automotive Engineers (SAE), Warrendale, PA.

Crux-Neira, C., Sandin, D., DeFanti, T. (1993): *Surround-Screen Projection-Based Virtual Reality: The Design and Implementation of the CAVE,* Computer Graphics (SIGGRAPH '93 Proceedings), pp. 135–142.

Darcy, S., Dudgedale, J., El Jed, M., Pallamin, N., Pavard, B. (2003): *Virtual StoryTelling: A Methodology for Developing Believable Communication Skills in Virtual Actors.* Springer Lecture Notes in Computer Science, Berlin.

Delaney, B. (1998, September/October): *The Mystery of Motion Capture.* IEEE Computer Graphics and Applications, pp. 14–19.

Delaney, B. (2005, May/June): *Forget the Funny Glasses.* IEEE Computer Graphics and Applications, pp. 14–19.

Delleman, N. (2006): Agents and Avatars. *Proceedings of the 2006 Digital Human Modeling Conference for Design and Engineering* in Lyon, France. Warrendale: SAE.

Delmia (2006): *Delmia Human.* Internet document retrieved from http://www.delmia.com/gallery/pdf /DELMIA_V5Human.pdf in 12/2006.

dstl (2006): *Datasheet Interactive Trauma Trainer.* Internet document retrieved from http://www.hfidtc. com/PDF/ITT.pdf in 12/2006.

Ellis, S.R. (1991): Nature and Origin of Virtual Environments: A Bibliographical Essay. *Computer Systems in Engineering,* 2, 4, pp. 321–347. Reprinted in *Readings in Human Computer Interaction, 2nd ed.,* Toward the Year 2000, Baeker, R.M., Grudin, J., Buxton, W.A.S. and Greenberg, S. (eds.) Morgan-Kaufman, San Francisco, 1995.

Ellis, S.R. (1993): Review of Virtual Reality by Howard Rheingold. *Ergonomics,* 36, 6, pp. 743–746.

Ellis, S.R. (1994): What are Virtual Environments? *IEEE Computer Graphics and Applications,* 14, 1, pp. 17–22.

Ellis, S.R. (1995): Virtual Environments and Environmental Instruments. In: K. Carr and R. England (eds.): *Simulated and Virtual Realities.* Taylor & Francis, London, pp. 11–51.

Ellis, S. R. (1996): Presence of Mind: A Reaction to Sheridan's Musings on Telepresence. *Presence* 5, 2, pp. 247–259.

Ellis, S.R., Kaiser, M.R, Grunwald, A.J. (eds.) (1991): *Pictorial Communication in Virtual and Real Environments.* Taylor & Francis, London.

Ellis, S.R., Slater, M., Alexander, T. (eds.) (2004): *Conference Proceedings: Intelligent Motion and Interaction within Virtual Elements.* University College, London, United Kingdom, September 2003. NASA Conference Proceedings 213468, NASA Ames Research Center, Moffett Field, CA.

Ellis, S.R., Wolfram, A., Adelstein, B.D. (2002): Large Amplitude Three-Dimensional Tracking in Augmented Environments: A Human Performance Trade-Off between System Latency and Update Rate. *Proceedings of HFES*, pp. 2149–2154.

Emering, L., Boulic, R., Thalmann, D. (1998, January/February): *Interacting with Virtual Humans through Body Actions*. IEEE Computer Graphics and Applications, pp. 8–11.

Fetter, W.A. (1961): *Computer Graphics, Aircraft Applications*: Document No. D3-424-I, Boeing Airplane Company, Wichita Division, 1961.

Fetter, W.A. (1964): *Computer Graphics in Communication*, McGraw-Hill, New York, p. 52.

Fisher, S.S., McGreevy, M., Humphries, J., Robinett, W. (1986): Virtual Environment Display System. ACM 1986 Workshop on 3D Interactive Graphics. Chapel Hill, NC. ACM.

Foley, J.D., van Dam, A., Feiner, S.K. (1995): *Computer Graphics*. Addison-Wesley, Amsterdam.

Foxlin, E. (2002): Motion Tracking Requirements and Technologies. In: Stanney (ed.): *Handbook of Virtual Environments*. Erlbaum Associates Publishers, London, pp. 163–210.

Frank, L.H., Kennedy, R.S., Kellog, R.S., McCauley (1983): Simulator Sickness: A Reaction to a Transformed Perceptual World 1. Scope of the Problem. *Proceedings of the Second Symposium of Aviation Psychology*, Ohio State University, Columbus, OH.

Furness, T.A. (1986): The Supercockpit and Its Human Factors Challenges. *Proceedings of the 30th Annual Meeting of the Human Factors Society*. Dayton, OH, pp. 48–52.

Gerhard, M., Moore, D.J., Hobbs, D.J. (2002): An Experimental Study of the Effect of Presence in Collaborative Virtual Environments. In: Earnshaw, R., Vince, J. (eds.): *Intelligent Agents for Mobile and Virtual Media*. Springer, London.

Gibson, W. (1984): *Neuromancer*. Ace Books, New York.

Gratch, J., Marsella, S. (2005): Some Lessons for Emotion Psychology for the Design of Educational Agents, *Journal of Applied Artificial Intelligence* (special issue on Educational Agents—Beyond Virtual Tutors), 19, pp. 215–233.

H-Anim (2006): *ISO/IEC FCD 19774:200x Information Technology—Computer Graphics and Image Processing*—Humanoid animation (H-Anim).

Heeter, C. (1992): Being There: The Subjective Experience of Presence. *Presence: Teleoperators and Virtual Environments,* 1(2).

Heineken, E. (2002): Wahrnehmungspsychologische Grundlagen. In: Schmidtke (ed.): *Handbuch der Ergonomie*. Bundesamt für Wehrtechnik und Beschaffung, Koblenz, Germany.

Hilton, A., Beresford, D., Gentils, T., Smith, T., Sun, W. (1999): Virtual People: Capturing Human Models to Populate Virtual Worlds. *Proceedings of the Computer Animation*, Washington, DC. IEEE Computer Society, Los Alamos, CA.

Hodges, L.F., McAllister, D.F. (1993): Computing Stereoscopic Views. In McAllister (ed.): *Stereo Computer Graphics and Other True 3D Technologies*, University Press, Princeton, Princeton, NJ, pp. 71–89.

Hofmann, J., Jäger, T.J., Deffke, T., Bubb, H. (2001): Measuring an Illusion: The Influence of System Performance on Size Perception in Virtual Environments. In: Fröhlich, Deisinger, and Bullinger (eds.): *Proceedings of the Workshop Immersive Projection Technology and* Virtual Environments. Wien, Springer Computer Science.

Human Solutions (2006): *Fahrzeugindustrie – Human Solutions*. Internet document retrieved from http://www.human-solutions.com/automotive_industry/ ergonomic_simulation_de.php in 12/2006.

IJsselsteijn, W., De Ridder, H., Freeman, J., Avons, S.E., Bouwhuis, D. (2001): Effects of Stereoscopic Presentation, Image Motion, and Screen Size on Subjective and Objective Corroborative Measures of Presence. *Presence*, 10, pp. 298–311.

Johnson, R. (1997): STEVE, An Animated Pedagogical Agent for Procedural Training in Virtual Environments. *ACM SIGART Bulletin*, 8 (1), pp. 16–21.

Kalawsky, R.S. (1993): *The Science of Virtual Reality and Virtual Environments*, Addison-Wesley Publishers, New York.

Kalra, P., Magnenat-Thalmann, N., Moccozet, L., Sannier, G., Aubel, A., Thalmann, D. (1998): Real-Time Animation of Realistic Virtual Humans. *IEEE Computer Graphics and Applications*, pp. 42–56.

Kennedy, R.S., Hettinger, L.J., Lilienthal, M.G. (1988): Simulator Sickness. In: Crampton (ed.): *Motion and Space Sickness* (Ch. 15). CRC Press, Boca Raton, FL.

Kennedy, R.S., Lane, N.E., Berbaum, K.S., Lilienthal, M.G. (1989): Simulator sickness in US Navy Flight Simulators. *Aviation, Space and Environmental Medicine*, 60, pp. 10–16.

Knapp, M.L., Hall, J.A. (2002): *Nonverbal Communication in Human Interaction*. Wadsworth Publishing.

Kolasinski, E.M. (1995): *Simulator sickness in virtual environments*. Technical Report 1027, U.S. Army Research Institute for the Behavioural and Social Sciences, Alexandria, VA.

Kramer, J., Leifer, L.J. (1994, January): A Survey of Glove-Based Input. *IEEE Computer Graphics and Applications*, 14 (1), pp. 30–39.

Krueger, M. (1983): *Artificial Reality*. Addison-Wesley, Reading, MA.

Krüger, W., Fröhlich, B. (1994): The Responsive Workbench. *IEEE Computer Graphics and Applications*, pp. 1215.

Landau, K. (1999): Ergonomic Software Tools in Product and Workplace Design. IFAO, Darmstadt, Germany.

Landauer, J., Blach, R., Bues, M., Rosch, A., Simon, A. (1997): Toward Next Generation Virtual Reality Systems. *Proceedings of the International Conference on Multimedia Computing and Systems* (ICMCS'97), p. 581ff.

Lanier, J. (1984): *Biography of Jaron Lanier*. Internet document retrieved 2006 from http://www.jaronlanier.com/

Laurel, B. (1991): *The Art of Human-Computer Interface Design*. Addison-Wesley, Reading, MA.

Laurel, B. (1993): *Computers as Theatre*. Addison-Wesley, Boston.

Lipton, L. (1993): Composition for Electrostereoscopic Displays. In: McAllister (ed.): *Stereo Computer Graphics and Other True 3D Technologies*. University Press, Princeton, NY, pp. 11–25.

Loftin, R.B., Scerbo, M.W., McKenzie, F.D., Catanzaro, J.M. (2004): Training in Peacekeeping Operations Using Virtual Environments. *IEEE Computer Graphics and Applications*, 24 (4), pp. 18–21.

Lok, B. (2006): Teaching Communication Skills with Virtual Humans. *IEEE Computer Graphics and Applications*.

Macedonia, M., Rosenbloom, P.S. (2001): Entertainment Technology and Virtual Environments for Training and Education. In: M. Devlin, R. Larson, and J. Meyerson (eds.), *The Internet and the University: 2000 Forum*. Educause, Boulder, CO, pp. 79–95.

Magnenat-Thalmann, N., Thalmann, D. (eds.) (2000): Deformable Avatars. *Proceedings of Avatars 2000 Workshop* in Lausanne. Kluwer Academic, Boston.

Magnenat-Thalmann, N., Thalmann, D. (eds.) (2004): *Handbook of Virtual Humans*. John Wiley & Sons, London.

Mania, K., Adelstein, B., Ellis, S.R., Hill, M. (2004): Perceptual Sensitivity to Head Tracking Latency in Virtual Environments with Varying Degrees of Scene Complexity. *Proceedings of ACM Siggraph Symposium on Applied Perception in Graphics and Visualization*, ACM Press, pp. 39–47.

Massie, T.H., Salisbury, J.K. (1994): The PHANTOM Haptic Interface: A Device for Probing Virtual Objects. *Proceedings of the ASME Winter Annual Meeting, Symposium on Haptic Interfaces for Virtual Environment and Teleoperator Systems*, Chicago, IL.

Mon-Williams, M., Wann, J.P. (1998): Binocular Virtual Reality Displays: When Problems Do and Don't Occur. *Human Factors*, 40, pp. 42–49.

O'Sullivan, C., Cassell, J., Vilhjálmsson, H., Dingliana, J., Dobbyn, S., McNamee, B., Peters, C., and Giang, T. (2002): Levels of Detail for Crowds and Groups. *Computer Graphics Forum*, 21(4), pp. 733–742.

Oviatt, S.L., DeAngeli, A., Kuhn, K. (1997): Integration and Synchronization of Input Modes during Multimodal HCI. *Proceedings of Conference on Human Factors in Computing Systems CHI 1997*. ACM Press, New York.

Pausch, R., Crea, T., Conway, M. (1992): A Literature Survey for Virtual Environments: Military Flight Simulator Visual Systems and Simulator Sickness. *Presence*, 1, pp. 344–363.

Pelechano, N., Badler, N.I. (2006): Modeling Crowd and Trained Leader Behavior during Building Evacuation. *IEEE Computer Graphics and Applications*, 26 (6).

Pfleger, N., Alexandersson, J. (2004): Modeling Non-Verbal Behavior in Multimodal Conversational Systems. In: Wahlster (ed.): Special Journal Issue "Conversational User Interfaces", IT—Information Technology, 46 (6). Oldenbourg Wissenschaftsverlag, München, Germany.

Phillips (2004): *Phillips Showcases 3D Display Technology at SID 2004* http://www.research.philips.com /newscenter/archive/2004/3d-display-sid.html

Pimentel, K., Teixeira, K. (1993): *Virtual Reality: Through the New Looking Glass.* Intel/Windfrest/ McGraw-Hill, New York.

Plantec, P. (2004): *Virtual Humans.* AMACOM, New York.

Potel, M. (1998): Motion Sick in Cyberspace. *IEEE Computer Graphics and Applications*, 18 (1), pp. 16–21.

Prendinger, H., Ishizuka, M. (eds.) (2004): *Life-like Characters: Tools, Affective Functions and Application.* Springer, Berlin.

Reeves, B., Nass, C. (1996): *The Media Equation. How People Treat Computers, Televison, and New Media Like Real People and Places.* CSLI Publications, Stanfort, CA.

Regenbrecht, H., Schubert, T. (2002): Real and Illusory Interactions Enhance Presence in Virtual Environments. *Presence: Teleoperators and Virtual Environments*, 11 (4).

Rickel, J., Johnson, W.L. (2000): Task-Oriented Collaboration with Embodied Agents in Virtual Worlds. In: Cassell, Sullivan, and Prevost (eds.), *Embodied Conversational Agents*, MIT Press, Boston.

Rist, T. (2003): From Virtual Presenters to Interactive Role Plays with Multiple Conversational Characters. *Künstliche Intelligenz*, 17 (4), pp. 18–23.

Robinette, K., Blackwell, S., Daanen, H., Boehmer, Fleming, S., Brill, T., Hoeferlin, D., Burnsides, D. (2002): *Civilian American and European Anthropometry Resource (CAESAR), Final Volume I: Summary,* AFRL-HE-WP-TR-2002-0169, Force Research Laboratory, Human Effectiveness Directorate, Crew System Interface Division, Street, Wright-Patterson AFB OH 45433-7022 Society of Automotive Engineers International, Commonwealth Drive, Warrendale PA, 15096.

Robinette, K.M., Daanen, H. (2003): Lessons Learned from CAESAR: A 3-D Anthropometric Survey. In: Pikaar, R.N., Koningsveld, E.A.P., and Settels, P.J.M. (eds.): *Meeting Diversity in Ergonomics. Proceedings of the 16th World Congress on Ergonomics.* Elsevier.

Roehl, B., Couch, J., Reed-Ballreich, C., Rohaly, T., Brown, G. (1997): *Late Night VRML 2.0.* Ziff-Davis Press, New York.

Roscoe, S.N. (1991): The Eyes Prefer Real Images. In: Ellis, S.R., Kayser, M.K., Grunwald, A.C. (eds.) (1991): *Pictorial Communication in Virtual and Real Environments.* Taylor & Francis, London.

SAE (1999ff): *Proceedings of the Digital Human Modeling Design and Application Conference 1999–2006.* Society of Automotive Engineers (SAE), Warrendale, MI.

SAE DHMC (1998ff): Proceedings of the SAE-Digital Human Modeling Conference Series. Society of Automotive Engineers (SAE), Warrendale, PA.

Schiefele, J. (2000): *Realization and Evaluation of Virtual Cockpit Simulation and Virtual Flight Simulation.* Shaker-Verlag, Aachen, Germany.

Schroeder, R. (ed.) (2002): *The Social Life of Avatars: Presence and Interaction in Shared Virtual Environments.* Springer, London.

Schumie, M., van der Straaten, P., Krijn, M., van der Mast, C. (2001): Research on Presence in VR: A Survey. *Journal of Cyberpsychology and Behavior*, 4 (2).

Seidel, R., Chatelier, P. (eds.) (1997): *Virtual Reality, Trainings's Future?* Plenum Press, New York.

Shao, W., Terzopoulos, D. (2006): Environmental Modeling for Autonomous Virtual Pedestrians. *SAE Transactions Journal of Passenger Cars: Electronic and Electrical Systems*, 114 (7), pp. 735–742.

Sheridan, T. (1992): Musings on Telepresence and Virtual Presence. *Presence*, 1 (1).

Shilling, R.D., Shinn-Cunningham, B. (2002): Virtual Auditory Displays. In: Stanney (2002): *Handbook of Virtual Environments*. Erlbaum, Mahwah, NJ.

Slater, M. (1999): Measuring Presence: A Response to the Witmer and Singer Presence Questionnaire. *Presence*, 8 (5).

Slater, M. (2002): Do Avatars Dream of Digital Sheep? Virtual People and the Sense of Presence. *Proceedings of the IEEE VR 2002 Conference* in Orlando, FL. IEEE Computer Society, Los Alamos.

Slater, M., Linakis, M., Usoh, M., Kooper, R. (1996): Immersion, Presence, and Performance in Virtual Environments: An Experiment with Tri-Dimensional Chess. *Proceedings of the ACM Symposium on Virtual Reality Software and Technology* (VRST 96), pp. 163–172.

Slater, M., Pertaub, D., Steed, A. (1999): Public Speaking in Virtual Reality: Facing and Audience of Avatars. *IEEE Computer Graphics and Applications*, 19 (2), pp. 6–9.

Slater, M., Usoh, M. (1994): Body Centred Interaction in Immersive Virtual Environments. In: Thalmann and Thalmann (eds.): *Artificial Life and Virtual Reality*. John Wiley and Sons, New York, pp. 125–148.

Stanney, K.M. (ed.) (2002): *Handbook of Virtual Environments*. Erlbaum, Mahwah, NJ.

Stanney, K.M., Hash, P. (1998): Locus of User-Initiated Control in Virtual Environments: Influences on Cybersickness. *Presence* 7 (5), pp. 447–459.

Stanney, K., Mollaghasemi, M., Reeves, L., Breaux, R., Graeber, D. (2003): Usability Engineering of Virtual Environments (VE): Identifying Multiple Criteria that Drive Effective VE System Design. *International Journal of Human-Computer Studies*, 53.

Stanney, K., Mourant, R., Kennedy, R. (1998): Human Factors Issues in Virtual Environments: A Review on Literature. *Presence*, 7 (4).

Sturman, D.J. (1998, January/February): Computer Puppetry. *IEEE Computer Graphics and Applications*, pp. 38–45.

Sturman, D.J., Zeltzer, D. (1994): A Survey of Glove-Based Input. *IEEE Computer Graphics and Applications,* 14 (1), pp. 30–39.

Sutherland, I. (1965): *The Ultimate Display*. International Federation of Information Processing, 2, 506ff.

Thalmann, D., Vexo, F. (2003): *MPEG-4 Character Animation*, KI - Künstliche Intelligenz Special Issue Embodied Conversational Agents, 17 (4).

Tourangeau, T., Sternberg, R.J. (1981): Adaptness in Metaphor. *Cognitive Psychology*, 19, pp. 27–55.

Turk (2002): Gesture Recognition. In: Stanney (ed.): *Handbook of Virtual Environments*. Erlbaum, Mahwah, NJ, pp. 223–237.

UGS (2006): *Human Performance: Jack*. Internet document retrieved from http://www.ugs.com/products/tecnomatix/human_performance/jack/ in 12/2006.

Ulicny, B., Thalmann, D. (2002): Towards Interactive Real-Time Crowd Behavior Simulation. *Computer Graphics Forum*, 21 (4), pp. 767–775.

Vince, J. (1995): *Virtual Reality Systems*. Addison-Wesley, Reading, MA.

Virtual Soldier Research (2006): The Virtual Soldier Research Program. Internet document retrieved from http://www.digital-humans.org/ in 12/2006.

VirtualHumans (2006): *Virtual Humans Project*. Internet document retrieved from http://www.virtual-humans.de/ in 12/2006.

Welch, G., Foxlin, E. (2002): Motion Tracking: No Silver Bullet, but a Respectable Arsenal. *IEEE Computer Graphics and Applications* (Special Issue on Tracking) 22 (6), pp. 24–38.

Welch, R., Blachmon, T., Liu, A., Mellers, B., Stark, L. (1996): The Effects of Pictorial Realism, Delay of Visual Feedback, and Observer Intractivity on the Subjective Sense of Presence. *Presence: Teleoperators and Virtual Environments*, 5 (3).

Werkhoven, P. (ed.): *What Is Essential for VR Systems to Meet Military Human Performance Goals?* HFM-MP-058. NATO Research & Technology Organization, Neuilly-sur-Seine, France.

Wightman, F.L., Kistler, D.J. (1989): Headphone Simulation of Free-Field Listening I: Stimulus Synthesis. *Journal of the Acoustical Society of America*, 85, pp. 858–867.

Witmer, B.G., Singer, M.J. (1998): Measuring Presence in Virtual Environments: A Presence Questionnaire. *Presence*, 7 (3), pp. 225–240.

Young, S.D., Adelstein, B.D., Ellis, S.R. (2007): Demand Characteristics in Assessing Motion Sickness in a Virtual Environment: or Does Taking a Motion Sickness Questionnaire Make You Sick? *IEEE Transaction in Visualization and Computer Graphics*, 13 (3), pp. 422–428.

Zhao, S. (2003): Toward a Taxonomy of Copresence. *Presence*, 12 (5), pp. 445–455.

Zimmermann, T., Lanier, J., Blanchard, C., Bryson, S., Harvill, Y. (1987): A Hand Gesture Interface Device. *Proceedings of CHI 1987, Human Factors in Computing Systems and Graphics Interface*, ACM Press, pp. 189–192.

Part 2

Modeling Fundamentals

8
Methods, Models, and Technology for Lifting Biomechanics

Michael LaFiandra

8.1 Executive Summary

Many soldier tasks require lifting heavy objects while in unusual postures. The percentage of soldiers who can complete specific tasks and the potential for injury to these soldiers resulting from completing these tasks is unknown. There is no easy way to directly measure the internal and external forces generated by a human subject while completing lifting tasks in unusual postures; however, these forces are likely associated with the potential for injury. Nevertheless, biomechanical models of lifting can be used to design the task in a way that is more suited to the soldier, can estimate the percentage of soldiers who can complete a specific task, and can be used to estimate the internal and external forces exerted on the soldier while completing a task. Redesigning the task to be more suited to the soldier can reduce the potential for injury, make the soldier more efficient, and increase the percentage of soldiers that can safely complete the task.

Depending on the data available for input and the information required as output, forces can be estimated with differing levels of fidelity from various types of models. The purpose of this chapter is to review the current state of biomechanical models with special attention given to lifting models of the lower back. Because some of the models and techniques used to model the lower back are generalizable, they may be highly useful in estimating forces in a variety of lifting tasks such as reloading ammunition in an M1A1 main battle tank, moving or lifting sandbags, or lifting a litter into an evacuation vehicle. Additionally, these models can be applied to other tasks, including walking, running, riding in a vehicle, and so on. Therefore, knowledge of these models and their appropriate application can be extremely valuable.

The first part of this chapter will serve as a primer on the types of lifting models. The three basic categories of lifting/lower back models are: (1) guideline models, which are typically used to better suit the task to the lifter; (2) models that are useful in estimating the external forces applied to the body and joint reaction forces and torques; and (3) models that estimate the contribution of muscles to the internal forces during a task. This will provide the background needed to further understand the advantages and disadvantages of each of the software packages designed to assist with calculating the potential for low-back pain that are described in the second part of the chapter.

Each category of models has advantages and disadvantages; however, there are some underlying limitations common to the three categories of lifting models. Lacking is an investigation into how forces are distributed between two or more people lifting a single object. Also absent is an understanding of the effects of physical fatigue on biomechanics, more specifically the effects of repeated lifts on the kinematics and kinetics of lifting. Similarly, some of the guideline models account for energy expenditure; however, none of them explains the effects of fatigue on the metabolic cost of repeated lifts.

8.2 Introduction

Many soldier tasks require lifting heavy objects while in unusual postures. The percentage of soldiers who can complete a specific task and the potential for injury to these soldiers are unknown; both are likely affected by the amount of force required to lift the object. Directly measuring the bone-on-bone forces or the forces generated by individual muscles is very difficult and highly invasive. However, ergonomic models of lifting can be used to design the task in a way that is better suited to the soldier, and biomechanical models can be used to estimate the internal and external forces exerted on the soldier while performing a task. Knowledge of internal and external forces is valuable in estimating the percentage of soldiers who are capable of completing a task and evaluating the potential for injury resulting from performing a task. Redesigning the task to be more suited to the soldier can reduce the potential for injury, make the soldier more efficient, and increase the percentage of soldiers who can safely complete the task.

Many biomechanical lifting models can be generalized to estimate forces for any lifting task, such as reloading ammunition in an M1A1 main battle tank, moving or lifting sandbags, and so on. Because biomechanical models and techniques are generalizable, these models may be highly useful in estimating forces in a variety of tasks, including walking, running, riding in a vehicle, lifting a litter into an evacuation vehicle, and so on, and are therefore extremely valuable. There are three basic categories of lower back lifting models: (1) ergonomic models; (2) models that estimate external forces, also called link segment models; and (3) models that account for muscle force, involving the use of electromyography (EMG).

The purpose of this chapter is to review the current state of ergonomic and biomechanical models with special attention given to lifting models of the lower back. Lifting models of the lower back are emphasized because repetitive lifting is associated with low back discomfort and other musculoskeletal disorders. The Bureau of Labor Statistics reports that nearly 34% of all nonfatal occupational injuries, 65% of back pain, and 71% of back injuries in 2001 were related to musculoskeletal disorders (Bureau of Labor Statistics, 2003).

Ergonomic models, such as the National Institute of Occupational Safety and Health (NIOSH) equation (Waters, Putz-Anderson, & Garg, 1994), the design criteria outlined by Mital, Nicholson, and

Ayoub (1997) or the Snook tables (Snook & Ciriello, 1991) are the easiest to use and most general. The Snook tables are included in the *Manual Material Handling Guidelines* put forth by Liberty Mutual Insurance (Liberty Mutual, 2004). The high level of generalizability of ergonomic models is due in part to the ease with which data for the model's inputs can be obtained. The required input to these models includes parameters such as the weight of the object to be lifted, the height the object needs to be lifted, body weight and sex of the subject lifting the object, and number of times the object needs to be lifted in a given period. These models do not provide highly specific information (such as the force exerted on a specific joint) but are useful in designing a workspace or task. A defining difference between ergonomic models and the other two categories is that ergonomic models typically include factors other than biomechanical factors, including environmental stressors and physiological factors.

Link segment models (LSMs), such as presented in Kingma, de Looze, Toussaint, Klijnsma, and Bruijnen (1996), Chaffin, Andersson and Martin (1999), or Nordin and Frankel (1980) are typically used to estimate external forces applied to the body and resulting joint reaction force and torque. These models usually require precise measurements of segment position (i.e., upper and lower leg, upper and lower arm, trunk, etc.), and require more precise information on subject anthropometrics than the ergonomic models require. The more precise measurements require greater control over the data collection than is required when collecting information for the ergonomic models. However, LSMs typically produce information of a higher fidelity than guideline models. For instance, LSMs can produce information on the external forces applied to the body and on specific joints during a task. Information from LSMs can be used to calculate the amount of work performed by the subject, and when oxygen consumption data are collected in concert with segment position data, efficiency can be determined. LSM can be either static or dynamic. Static LSMs are used to estimate forces for a single instant or snapshot in time and subsequently assume the effects of velocity and acceleration are negligible. In contrast, dynamic LSMs are used to estimate forces throughout a movement, account for effects of velocity and acceleration, and estimate mechanical energy, work, power, and momentum. LSMs are typically less generalizable than guideline models.

Finally, the forces generated by individual muscles and their contribution to the net force can be estimated using biomechanical models that include a measure of EMG, such as presented in Marras and Sommerich (1991), Chung, Song, Hong, and Choi (1999), and Dolan et al. (2001). Including a measure of muscle force to estimate the internal forces increases the fidelity of the information obtained. Typically, these models also need a measure of the external forces applied to the body, so LSM may be used. Measuring EMG allows the researcher to take into account the amount of electrical activity in muscles, which can be used to estimate the force generated by individual muscles. Moment arms for each muscle are estimated by magnetic resonance imaging (MRI) or anthropometric tables, allowing the torque to be calculated from the force estimates. Because intra-abdominal pressure (IAP) may contribute to the internal forces during a lift, models that estimate internal muscle forces may also include a measurement of IAP. Because of the amount of instrumentation required to collect EMG data (or EMG + IAP data), biomechanical analysis using these models is typically restricted to data collected in highly controlled environments. Results of models that employ EMG (or EMG + IAP) are typically more specific to the individuals the data were measured on and not necessarily as generalizable to novel tasks or different subjects as results from the guideline models.

The first part of this chapter serves as a primer on the types of lifting models. The three basic type or categories of lifting/lower back models are: (1) guideline models, which are typically used to better suit the task to the lifter; (2) models that are useful in estimating the external forces applied to the body and joint reaction force and torque; and (3) models that estimate the internal forces during a task. This section provides the background needed to further understand the advantages and disadvantages of each of the software packages designed to assist with calculating the potential for low-back pain that is described in the second part of the chapter.

Following the primer on the types of lifting models and their advantages or disadvantages is a review of the commercial software that is available to perform these analyses. Additionally, some types of analysis, such as EMG, may need custom software to be written. Software packages discussed include NexGen's Ergomaster and Mannequin, University of Michigan's 3-Dimensional Static Strength Prediction Program (3DSSPP), the Air Force's Articulated Total Body Model, The Boeing Human Model System, Safework, Jack, Digital Biomechanics, and Santos.

8.3 Types of Models and Review of Current Research

Lifting models can be divided into three categories. These are ergonomic, link segment, and EMG models (Table 8.1). This section describes each category and gives examples of representative models.

8.3.1 Ergonomic (Guideline) Models

Guideline models evaluate tasks based on specific criteria. Additionally, the purpose of some guideline models is to design tasks that allow workers to complete tasks safely. One advantage of guideline models is that they typically include factors other than solely biomechanics, such as environmental stressors. Some of these models are generalizable to many different tasks. Also, the data for the inputs required for the model are relatively easy to collect, do not require a large amount of instrumentation, and usually can be collected at a job or test site. Additionally, the output is easily interpreted. However, because these models are highly generalizable, they do not typically provide highly specific information. For instance, the NIOSH equation provides a recommended weight limit (RWL) and the design criteria output from the model of Mital, Nicholson, and Ayoub (1997) is used as input for designing a workstation. The ergonomic/guideline models do not provide estimates of the forces exerted on the body during a task.

The three guideline models that will be presented in this review (Table 8.2) are the NIOSH equation (Waters, Putz-Anderson, & Garg, 1994), the design criteria outlined by Mital, Nicholson, and Ayoub (1997), and the updated Snook tables (Snook & Ciriello, 1991). The NIOSH equation sets the RWL as a function of specific task characteristics, such as the height a box needs to be lifted, the number of times the box needs to be lifted in a specific time period, and the types of handles on the box being lifted. A complete list of the factors and how they are calculated is included in the following sections. This model is ideally suited for evaluating a workstation that is already in use.

In contrast to the NIOSH equation, the design criteria outlined by Mital et al. (1997) was focused more on establishing standards that should be considered for improving manual material handling tasks or for designing new manual material handling tasks. The criteria include the subject's anthropometry, age, gender, training, the environment, the task duration, any protective equipment, and so on, and include factors that can help account for unusual postures, such as lying, kneeling, and sitting. In contrast to the NIOSH equation, which was developed specifically for lifting while standing, the guide contains a lookup table for each factor. Essentially, each factor is measured, the correct lookup table is referenced, and safe limits for that specific subject and activity are determined.

Both the NIOSH equation and the design criteria outlined by Mital et al. (1997) differ from the Snook tables in that the Snook tables are designed specifically to provide guidance on pushing and pulling. Additionally, the Snook tables provide a quantitative value indicating the percent of the population that can complete a task.

8.3.2 NIOSH Equation

The advantages of the NIOSH lifting equation are that it is easily employed, accounts for factors such as the quality of the handholds and frequency of lifting, and is not highly dependent on data that are difficult to collect. While the mechanics are taken into account, joint reaction force, joint torque, required

TABLE 8.1 Comparison of Types of Models

Model	Factors	Advantages	Limitations
Ergonomic	Biomechanics Psychophysical Physiological	Established Guidelines. Include factors other than solely biomechanics. Input data are easily collected. Output can be compared to specific criteria.	Assumes everyone completes the motion similarly No specific information (no joint forces) Not always able to be tailored to novel tasks. Some of these types of models do not give quantitative values.
Link Segment	Biomechanics—External Forces Dynamics 22–60% greater than static models.	Task-specific results. Quantitative data provided. Higher resolution information determined (i.e., Joint Reaction Force and Torque). Lifting kinematics has been shown to be an important indicator of work-related low back disorders. LSMs account for this. Accounts for differences between people. If data are collected for a prolonged period, these models can account for effects of fatigue. If a dynamic model, accounts for changes in force throughout a motion. When combined with metabolic cost data, efficiency can be determined and tasks can be optimized.	Pure biomechanics does not account for other factors such as metabolic cost, psychophysical factors, and environment. Instrumentation needed. Results not always easily interpreted (there is no set criterion like those seen in ergonomic models). More expertise needed to collect the data. Typically data need to be collected in a lab. No internal (muscle) forces. Additional measure of internal force (obtained through EMG models) needed to calculate net force. Cannot estimate force distribution over different structures in spine.
EMG	Biomechanics—Internal Forces	Can estimate force distribution over different structures. Provide information on neural activity and underlying mechanisms of overuse injuries. Higher resolution information than ergonomic models (such as force exerted by individual muscles). Quantitative data provided. Account for differences in anthropometrics. Typically used in conjunction with a measure of external force to obtain net forces.	Pure biomechanics does not account for other factors such as metabolic cost, psychophysical factors, and environment. More instrumentation than LSM needed. Assume force can be estimated based on muscle activity. Results difficult to interpret. Limited accuracy due to inherent variability of EMG signal. More complex than LSM. Estimate muscle locations and distances from axis via MRI, requiring even greater instrumentation and special equipment. Data collected through fine wires or surface electrodes. Additional measure of external force (obtained through LSM) needed to calculate net force.

TABLE 8.2   Comparison of Ergonomic Models

Specific Model	Advantages	Disadvantages
NIOSH Lifting Equation	Easily employed. Not dependent on data that are difficult to collect. Account for factors such as quality of handholds, etc. Provide Lifting Index (LI), a quantitative value that can be compared to LI's for other tasks to determine which task is least difficult.	Results not scaled to anthropometrics or sex. Based on Standard Lifting task. No accounting for walking or taking steps (assumes feet are stationary). Not a biomechanics model, therefore no forces calculated.
Mital	Very comprehensive, accounts for many factors other than solely biomechanics (includes psychophysical and physiological measures). Guidelines included for unusual postures, varying environment (temperature), and protective equipment.	No quantitative value produced, difficult to compare tasks. No accounting for walking or carrying. Some of the factors (such as handling technique) are difficult to quantify. No specific information provided—only guidelines. Not a biomechanics model, therefore no forces calculated.
Snook	Psychophysical approach—account for more than solely biomechanics. Comprehensive tables on lifting provided, including effect of sex. Quantitative value provided. Easy to use.	Not a biomechanics model, therefore no forces calculated. No unusual postures. No varying environment. No effect of protective clothing or equipment.

muscle force, and internal and external work are not explicitly calculated. The disadvantages are that the estimates are not scaled for anthropometrics, and therefore according to this model, the 5th percentile female and 95th percentile male have the same recommended weight limit for lifting. Also, the model was developed based on a standard lifting task (such as lifting a box off the floor and placing it on a table); it is unclear how this model would be applied to nonstandard lifting tasks (such as reloading ammunition in an M1A1 tank). Also, the model does not include a factor to account for walking while carrying an object, and assumes that if there is axial rotation there is no pivoting around the foot or stepping. Additionally, while the model provides information on the potential for low back pain, information on other body parts is not provided.

The remaining part of this section summarizes the applications manual for the revised NIOSH lifting equation written by Waters et al. (1994). In 1981, NIOSH developed analytic procedures and a lifting equation for calculating a RWL for two-handed, symmetrical lifting and an approach for controlling the hazards of low back injury from manual lifting. In 1991, the NIOSH equation was adapted to include updated information on the physiological, biomechanical, psychophysical, and epidemiological aspects of manual lifting. The revised NIOSH lifting equation reflects methods for evaluating asymmetrical lifting tasks and lifts of objects with less than optimal couplings between the object and the worker's hands (Waters et al., 1994).

The revised NIOSH lifting equation (Equation 8.1) is a multifactorial equation, incorporating aspects of lifting that include the position of the load's center of mass, the horizontal and vertical distance the load needs to be moved, the number of times the load needs to be lifted, and the type of handles or way of lifting the object. All the information necessary to employ the revised NIOSH lifting equation is provided. For examples of the revised NIOSH lifting equations, refer to Waters et al. (1994).

$$RWL = LC * HM * VM * DM * AM * FM * CM \qquad (8.1)$$

where:

Term	Description	Accounts for:
RWL	Recommended Weight Limit	
LC	Load Constant	51 lbs. for U.S. Customary or 23 kg for metric
HM	Horizontal Multiplier	Horizontal distance between the lifter and the object
VM	Vertical Multiplier	Distance between knuckle height* and the handles
DM	Distance Multiplier	Vertical distance the object needs to be lifted
AM	Asymmetric Multiplier	Amount of axial twisting required to complete the task
FM	Frequency Multiplier	Number of times the lift needs to be performed
CM	Coupling Multiplier	Type of handles the object has

* Knuckle height is assumed to be 30 inches for a worker of average height

The RWL is the weight, under a specific set of task conditions, that nearly all healthy workers could lift over a substantial period of time without the increased risk of developing low back pain. For more information on the terms of the equations and an explanation of how to determine each, consult Waters et al. (1994).

The lifting index (Equation 8.2) provides a relative estimate of the physical stress associated with a manual lifting job.

$$LI = \frac{Load\ Weight}{RWL}$$

(8.2)

The greater the lifting index, the smaller the fraction of workers capable of safely sustaining the level of activity (fig. 8.1). Comparisons between two jobs or alternate methods of the same lifting task can be

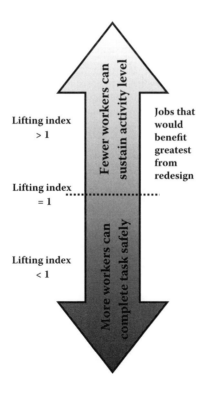

FIGURE 8.1

made using the lifting index. An LI near 1 indicates the load weight is nearly equal to the RWL. If $LI \approx 1$, the individual parameters that comprise the RWL can be used to identify specific job-related problems. Jobs with $LI > 1$ would benefit the most from redesign.

8.3.3 *Guide to Manual Materials Handling*

Mital, Nicholson, and Ayoub (1997) developed a *Guide to Manual Materials Handling*. In contrast to the NIOSH equation, which was designed to evaluate and rate manual material handling tasks that were already employed, the work of Mital et al. (1997) is focused on establishing criteria that should be considered for improving manual material handling tasks or for designing new MMH tasks. Also, while the NIOSH equation produces a quantitative value based on measurable quantities that could be used to compare different lifting tasks, the guidelines outlined by Mital et al. (1997) are more general and do not provide a single number that can be interpreted as a measure of goodness of the lifting task being investigated.

In addition to providing a list of factors that should be considered in designing an MMH task, Mital et al. (1997) also provide information to assist in designing specific tasks. For example, given the box size, frequency of lift, height of lift, type of lift (one or two handed), and target population percentile that may have to complete the task, a recommended weight that can be lifted over an 8-hour period can be estimated. This weight can then be adjusted for specific heat stresses, hand/handle couplings, load placements, and postures. In contrast to the NIOSH equation, the *Guide to Manual Materials Handling* is applicable to more tasks than solely lifting, including tasks, such as pushing and pulling, carrying and moving, while in unusual postures.

The MMH task design criteria are listed below (Table 8.3).

The guide contains a lookup table for each factor. Essentially, each factor is measured or decided on, the correct lookup table is referenced, and safe limits for that specific subject and activity are determined. Three of the most important parts of this process are developing correct and accurate lookup tables, correctly measuring each factor, and referencing the correct table. The most difficult of these parts is developing correct and accurate lookup tables.

According to Mital et al. (1997), there are three approaches to understanding MMH abilities: biomechanical, physiological, and psychophysical.

The biomechanical approach focuses on the mechanical stresses on the lower back and the rest of the body, relying on the compression and shear forces generated at the L5/S1 disk interface and pressure generated in the spinal column. The ultimate compressive strength of the spinal column can be estimated based on gender, age, lumbar level, cross section, and structure (Equation 8.3) (Jager, Luttmann, & Lauring, 1991).

$$CS(kN) = (7.65 + 1.18G) - (0.502 + 3.82G)A + (0.035 + 0.127G)C - 0.167L - 0.89S \tag{8.3}$$

TABLE 8.3 Factors to Consider When Designing an MMH Task

Physique/Anthropometry/Strength	Physical Fitness/Spinal Mobility
Age/Gender	Psychophysical Factors/Motivation
Training and Selection	Effects of Static Work
Posture/Handling Techniques	Loading Characteristics
Handles/Coupling	Repetitive Handling
Asymmetrical Lifting/Load Asymmetry	Confined Environments/Spatial Constraints
Safety Aspects	Protective Equipment
Handling in a Hot Environment	Task Duration
Work Organization	

where:

Term	Description	Example
CS	Compressive strength	
G	Gender	0 = female, 1 = male
A	Age in decades	30 years = 3
L	Lumbar level	0 = L5/S1
C	Cross-sectional area	
S	Structure	0 = disk, 1 = vertebra

This is only an estimated value. True measurement of compression and shear forces on the spine during different activities has been the focus of substantial research, requiring more sophisticated biomechanical models, such as LSM or EMG models discussed later in this chapter.

According to Mital et al. (1997), the physiological approach focuses more on the effects of repetitive lifting and the ability of the muscles to contract under varying loads. Not being able to complete the task due to physiological constraints is more related to lack of muscle strength than to a structural failure (as is the case in the biomechanical approach). Not only does the contracting muscle need to be strong enough, metabolites need to be delivered and waste needs to be removed. The inability to deliver oxygen or remove carbon dioxide would prevent the handler from continuing the task. The physiological approach focuses on metabolic energy expenditure. Minimizing energy expenditure is typically considered a favorable design criterion.

The psychophysical approach refers to the perceived physical strain. It is based on the notion that tasks that are perceived to be more difficult will be more difficult. The final workload is a function of the amount of work a worker can sustain without undue strain or discomfort. Mital et al. (1997) provide details on how to quantify the effects of psychophysical strain on the worker, and how to design tasks to minimize these effects.

8.3.4 Snook Tables

Snook and Ciriello (1991) determined the maximum acceptable weights and forces that workers should be exposed to during manual materials handling (MMH). Liberty Mutual Insurance has incorporated these data into their *Manual Material Handling Guidelines* (available at http://libertymmhtables .libertymutual.com/CM_LMTablesWeb/pdf/LibertyMutualTables.pdf). Essentially, Snook and Ciriello (1991) conducted four studies on the ability of workers to complete lifting, lowering, carrying, pushing, and pulling tasks, using the psychophysical approach. Their work is mentioned here because many of the ergonomic software packages incorporate the Snook tables to estimate either the maximum loads that can be lifted, lowered, carried, pushed, or pulled, or the percent of the population that can complete certain tasks.

In their series of experiments, Snook and Ciriello asked subjects to complete realistic lifting, lowering, and carrying tasks. Independent variables were task frequency (ranging from 12 per minute to 1 per 8 hours), box width (distance between the hands), hand distance (distance from front of body to hands), presence of handles, box weight, and starting and ending height. Subjects were blind to box weight. Pushing and pulling tasks were also evaluated using a special treadmill that was powered by the subject; subjects were blind to the amount of force they were exerting. The subject on the treadmill would push or pull against a stationary bar that had a load cell and measured the force exerted.

Because this research involved the psychophysical approach, the subjects would start the task with either a very light or very heavy box (or a pulling/pushing task that required either a very large or very small amount of force). Subjects were allowed to adjust the weight/force until they arrived at a workload they could sustain for a long period of time (the period of time required by the task). The final weight is set as the criterion for that specific frequency and lifting task. The result of Snook and Ciriello's (1991) work is a list of tables that can be referenced when developing a manual materials handling task.

8.3.5 Link Segment Models

Link segment models (LSMs) are used to calculate joint reaction forces and torques due to external forces applied to the body. External forces result from contact between the body and either the ground or an object. All LSMs are based on several assumptions: (1) each segment has a fixed mass; (2) the location of each segment's COM remains constant relative to the endpoints of the segment during the motion; (3) the joints are considered hinged, or ball and socket; (4) the moment of inertia of the segment (around the segment's COM) remains constant during the motion; and (5) the length of each segment remains constant during the movement (Winter, 1990).

Forces and torques are estimated based on the mass properties of each link and the acceleration the link is experiencing. The mass properties of each link (mass, center of mass location, moment of inertia, and so on) of each link are estimated based on the subject's body mass, segment lengths, and anthropometric tables. LSMs can be either static (based on an instant in time or snapshot of a motion) or dynamic (analyze the entire motion, including velocity and acceleration).

For static LSMs force is estimated based solely on the acceleration due to gravity, the mass of each of the segments, and the weight of the object being lifted. Many commercially available software packages for biomechanics analysis estimate joint forces and torques based on static analysis. The advantage of static analysis is that it allows for a fast estimate of the forces at a specific point in time with a minimal amount of instrumentation. For instance, NexGen's Ergomaster Lift Analyzer has a 2D biomechanical prediction tool that allows a user to load a digital picture of a subject performing the task. The user then clicks on the subject's joints in the picture, and enters the weight of the object being lifted and the subject's body mass. The software returns the joint reaction forces and torques of the ankle, knee, hip, L5/S1, and so on.

However easy to use, static models are limited in the aspect that they only analyze an instant in time during an activity and do not include forces due to the motion of the subject and/or the object. Consequently, factors such as linear or angular velocity and acceleration are ignored. These factors are important for calculating the dynamic forces exerted on the body during a movement. Dynamic force calculation is important because during movement, when the body is required to accelerate or decelerate an object that is being lifted, the forces experienced by the body are much greater and subsequently the potential for injury is higher. For instance, Marras and Sommerich (1991a) indicate the forces determined with dynamic models are 22% to 60% greater than those determined with static models. Also, without taking into account the motion of the subject and object, other factors such as momentum, mechanical energy, work, power, rotational inertia, and so on are ignored.

Dynamic LSMs on the other hand are ideal for measuring changes in force throughout a motion (such as lifting an object). The calculation of forces using a dynamic LSM is based on the notion that the observed motion of the link is caused by the forces exerted on that link, known as inverse dynamics. Consequently by measuring the motion of the link, the forces and torques that resulted in that motion can be calculated.

To solve an inverse dynamics problem, the motion of each of the link segments needs to be known for every instant in time. In live data collection, this can be accomplished with the use of a motion capture system. Also, many simulation products provide this information. Acceleration is the second derivative (with respect to time) of change in position. Force is then calculated by multiplying acceleration by the mass of each segment. The appropriate gravity vector is accounted for. Also, the motion capture system will provide information necessary to calculate changes in angular position (and angular acceleration) of each segment during the motion, information which then can be used (when multiplied by moment of inertia of the segment) to obtain joint torque. Knowledge of position, velocity, acceleration, mass, and moment of inertia provides the information necessary to estimate momentum, angular momentum, mechanical energy, work, and power, which are factors that cannot be calculated by using a static LSM, but are important in quantifying human movement and the potential for injury.

One disadvantage of LSMs compared to guideline models is the greater amount of instrumentation and expertise required to use them. This is especially true for dynamic link segment models. Additionally, while there are some motion capture systems that can be used outdoors, the standard for motion capture data collection is in a laboratory environment. This may limit the applicability of these models. Also, while the guideline models are designed to be applied to a specific population, link segment models (in particular, those involving live subject data collection), require the recruitment of subjects that are representative of specific populations. This may be advantageous in that specific training can be administered (which the guideline models do not fully account for), or disadvantageous in that subjects representing a specific population may be difficult to recruit.

However, a clear advantage is that the data obtained from LSMs are more precise than the data obtained from guideline models. For instance, while guideline models rely on everyone performing a motion similarly, LSMs are based on motion capture data. Information on within-subject differences in the way a motion is completed could be useful in determining the repeatability of an action or how the action changes as a function of time, fatigue, and so on.

Additionally, the forces and torques determined from the LSMs can be input into injury prediction models or, when available, compared against known thresholds for injury. Also, knowledge of the internal and external work performed by a subject (which is easily calculated with a dynamic LSM) can be used with metabolic cost data to give an estimate of mechanical efficiency (Winter, 1990). LSMs are lacking relative to EMG and EMG/IAP models in that LSMs do not account for internal forces, such as muscle force. LSMs account for external force and net joint torque, but EMG is needed to estimate the amount of force each muscle is contributing to the net joint torque.

8.3.6 Example Link Segment Model

Kingma et al. (1996) presented a full-body 3D dynamic linked segment model. The objective of their study was to validate a general-purpose 3D dynamic linked segment model, incorporating all body segments. The unique component of their model was the use of full 3D data for all the body segments. Many 3D link segment models only use two markers on each segment, allowing only planar angles to be calculated. In contrast, Kingma et al. used rigid braces attached to the segments of interest. The advantage of using rigid braces is that each brace had several markers attached to it, thereby allowing for a calculation of flexion/extension, abduction/adduction, and internal/external rotation of the segment. Other advantages were that the use of the braces increases the number of possible applications of the model and results in a reduction of systematic as well as random errors as compared to the use of skin markers.

Seven subjects were asked to perform a lifting task. Subjects started the lift in an asymmetrical squatted position holding a small 5-kg dumbbell in their right hand in front of their left knee. Subjects were asked to lift the dumbbell; the end position was with the right shoulder at 90 degrees of abduction and straight legs. The Kingma model contained 14 segments: two feet, two lower legs, two upper legs, two upper arms, two forearms (including hands), a pelvis, a trunk (including neck), a head, and a dumbbell. The braces that had the markers affixed to them were attached to each segment. Motion capture data were collected at 60 Hz via a Vicon motion capture system. The task took place on a force plate. Ground reaction force was low-pass filtered with an analog filter at a cutoff frequency of 30 Hz and was sampled at 60 Hz.

The joint center position, center of mass position, and inertia of each body segment were defined with respect to anatomical axis systems (McConville, Churchill, Kaleps, Clauser, & Cuzzi, 1980). This process required the measurement of 75 anthropometric quantities to be plugged into regression equations defined in McConville et al. (1980). Joint center positions were defined for the ankle, knee, hip, lumbro-sacral joint, neck, shoulder, and elbow. The three principal moments of inertia were derived from regression equations using anthropometric measures and a rotation matrix provided by McConville et al. (1980) and were used to transform the segment inertia in a way that describes the inertia of the body segment with respect to the anatomical axis.

During a segment calibration, the position of markers on the braces as well as markers on anatomical landmarks were recorded. This allowed for an estimation of the location of the anatomical landmarks based on the location of the braces. After the segment calibration, the anatomical axis system could be reconstructed, allowing for the orientation of the three anatomical axes of each body segment as well as the position of their common origin to be calculated. Then, for each instant in time during the movement, the segment center of mass, joint centers, and inertia tensor for each segment were calculated. Finally, the LSM used an inverse dynamics model to calculate the reactive forces and torques at all intersegmental joint centers.

An interesting aspect of the Kingma et al. study is that it included a validation of the LSM model for estimating forces. Reactive forces exerted on the L5/S1 joint were estimated in two ways: (1) the bottom-up approach started the inverse dynamics analysis at the feet; and (2) the top-down approach started the inverse dynamics analysis at the hands. In the bottom-up analysis, the force exerted on the feet is known (from the force plate); the remaining joint reaction forces can be calculated joint by joint, accounting for the joint reaction force of the next most distal joint. For the top-down approach, the load applied to the hand by the load is estimated based on the acceleration of the load and the load's mass (Force = Mass * Acceleration). Joint reaction forces for the next most proximal joint can be calculated in the same manner as for the bottom-up approach. If the model is valid, using either approach should yield the same joint reaction force for the same joint. For example, regardless of whether or not the top-down or bottom-up approach is used, the joint reaction force at the right knee should be the same.

Part of their validation included a comparison of the ground reaction force calculated by the top-down approach to the ground reaction force measured from the force plates. Another part was a comparison of the top-down to bottom-up estimates of force exerted on L5/S1. The estimated ground reaction forces were 11% to 50% greater than the measured ground reaction forces. Kingma et al. attributed this to the fact that the kinematic data used to estimate the ground reaction force were noisier than the data recorded directly from the force plates. The flexion/extension torques measured by top-down and by bottom-up were well correlated ($r = 0.96$ or higher) for each trial. Twisting torques showed a slightly lower correlation (median r value = 0.933) and lateral flexion/extension showed the lowest correlation of the three movements ($r = 0.706$). While these correlation coefficients may be acceptable, the model suffers from the same limitation as other estimators of low back forces, namely the fact that there is no direct measure of the force exerted on the lower back. Consequently, validity studies lack a gold standard to which results can be compared. However, the model presented by Kingma et al. (1996) is an important example of the application of a LSM to the study of low back movement, partly because it demonstrates the utility of using LSMs for studying unique movements.

8.3.7 Electromyography (EMG) Models

EMG models include a measure of the electrical activity of a muscle and estimate of the force generated by that muscle. The net force exerted during an activity is the sum of the internal (muscle) and external forces during the activity. Therefore, models that account for internal force offer the advantage of higher fidelity of information. In fact, LSMs that only calculate external force have been shown to consistently predict lower peak moment values than models that include EMG (difference of ~21%), possibly due to different anthropometric assumptions, differing degrees of smoothing and filtering of raw data, and differences in the model's ability to assess the affects of antagonistic muscle activity (Dolan et al., 2001). The observation that antagonistic muscle activity may contribute to the differences observed between external force models and muscle force models is important. LSMs alone do not have a way of determining the amount of antagonistic muscle activity, while EMG models do.

According to Kingma et al. (1998), a fundamental trade-off between LSM and EMG models is that LSMs cannot estimate the distribution of force over different structures in the spine, while EMG models can make this estimate but are limited in accuracy due to inherent variability of EMG signals. In addition, EMG models are typically more complex than LSMs, requiring greater expertise and more instrumentation. Because EMG models aim to account for internal force in order to calculate net force, additional instrumentation is required to gather external force data. This is usually accomplished via a motion capture system and an LSM; however, other techniques have been used (Marras & Sommerich, 1991a). While EMG data are more difficult to collect, the information EMG provides on muscle activity may offer insight into the underlying mechanisms of overuse injuries.

The application of the electrodes and the techniques used to collect and analyze the EMG data are more complex than the techniques used to collect and analyze the motion data needed for the LSMs. For instance, EMG is collected through the use of either fine wire electrodes that are inserted into the muscle or through surface electrodes that are placed on the skin. EMG models are based on the assumption that a given amplitude of electrical activity in the muscle EMG consistently corresponds to the magnitude of force being generated by the muscle, a notion that was supported by Lippold's (1952) finding that during isometric or constant velocity contractions, the integrated EMG signal is related to muscle force. EMG readings are highly sensitive to electrode placement; therefore, moving the electrode off the center of the muscle belly will result in a vastly different EMG amplitude reading than if the electrode is on the muscle belly. Because of this sensitivity, every time EMG electrodes are placed on the body, a calibration reading, called maximum voluntary contraction (MVC), must be taken. This can be difficult when investigating 10 muscles, as done by Marras and Sommerich (1991b).

Because the distance of the line of action of each muscle from the axis of rotation will affect the moment or torque that muscle can generate, information on the distance between the muscle and the axis of rotation is needed. Sometimes this information is estimated from published tables, while at other times highly detailed models of anthropometry are used. For instance, magnetic resonance imagery (MRI) has been used to help estimate the distance between the muscle and the axis. While this may increase the accuracy of the model, using MRIs also increases the amount of instrumentation required in order to collect all the data.

Another factor that affects the internal forces exerted on the spine during lifting is intra-abdominal pressure (IAP) (Daggfeldt & Thorstensson, 2003). Daggfeldt and Thorstensson (2003) developed a model and tested subjects at various trunk flexion angles ranging from −45° to 20°. Their model demonstrated the fraction of the total torque that can be attributed to IAP during flexion and extension to be about 9% when trunk flexion was 45° (the most flexed position tested) and about 13% when trunk extension was 20° (the most extended tested). While introducing IAP as a factor for estimating the internal forces may be beneficial, the model presented by Daggfeldt and Thorstensson (2003) was not evaluated using dynamic motion and required a large amount of instrumentation. Also, because data for the model were collected with subjects lying on their sides on a spin table that could provide a measure of the external force generated by the trunk, the applicability of this to real-life tasks may be limited. This suggests that the advantages of being able to calculate the 9% to 13% of the trunk torque that could be attributable to IAP may be outweighed by the instrumentation required to do so. In contrast, Marras and Sommerich (1991a) state that IAP has a minimal effect on trunk torque, and do not include IAP in their models.

8.3.8 Example EMG Model

Marras and Sommerich (1991a & b) introduced an EMG model of the loads on the lumbar spine. Their model is intended to be a research tool to investigate the effects of constant velocity trunk motion and trunk asymmetric position on the loading and torque production capabilities of the trunk. Additionally, the model is a tool that allows someone to evaluate the collective influence of the trunk musculature on spine loading as the trunk moves under constant laboratory conditions. The input, output, constraints on data collection, assumptions, and limitations of this model are similar to many EMG models; subsequently, the Marras and Sommerich model is presented in this chapter as a representative EMG model.

As with other EMG models, the Marras and Sommerich model assumes EMG amplitude is directly related to muscle force (during constant velocity trunk motion) and consequently requires isokinetic conditions. Three isokinetic conditions were tested (10, 20, and 30 degrees per second), and velocity was controlled by the use of a KIN/COM isokinetic dynamometer. Subjects were placed in the dynamometer, which is designed to constrain movement velocity. Controlling movement may affect the way the subject completes the task and the resulting forces and torques generated by the subject. Consequently, isokinetic conditions and the use of a KIN/COM is a severe limitation compared to the LSM or ergonomic models. EMG was sampled during the movement from the *latissumus dorsi, erector spinae, internal and external oblique* and *rectus abdominus*. EMG data were high-pass filtered at 80 Hz, and low-pass filtered at 1000 Hz, and digitized with an AD converter. The model can be expressed by Equation 8.4.

$$Force = gain * \frac{EMG}{EMG_{max}} * Vratio * L - Sfactor * area$$

(8.4)

where *gain* represents the muscle force per unit area and was determined separately for each subject. *EMG* represents the EMG reading at any particular point during the movement. EMG_{max} represents the EMG reading during a maximal voluntary contraction. *Vratio* represents the velocity modulation factor, which is a ratio the numerator of which is the average normalized EMG response for a particular muscle with respect to trunk angle and external torque production, at an angular trunk extension velocity of zero. This value is in a database at Ohio State University. In the denominator of the ratio is the average EMG response (from the same database), for the same angle and torque conditions but performed at the speed at which the trial of interest was conducted. The *L – Sfactor* represents the length-strength modulation factor, determined by equations derived from Brobeck (1973) and Chaffin and Andersson (1984), as well as pilot data collected at University of Ohio. The *area* term is the muscle cross-sectional area, calculated from the subject's torso depth and breadth and definitions from Shultz et al. (1982). The model output is shear and compressive forces exerted on the spine, including the contribution due to muscle force.

One limitation of this model is the high level of difficulty an average researcher would have in calculating the inputs. For instance, *gain* is subject-specific and was determined by trial and error, and the *Vratio* requires information that is in a database at the University of Ohio. As with the LSM, the validation of this model is limited by the lack of a gold standard with which to compare the force values calculated from the model. Additionally, the requirement of isokinetic movement and the consequent requirement for the use of a KIN/COM isokinetic dynamometer severely limit the applicability of this model to tests that would take place outside the laboratory, and thereby the generalizability of the model across subjects and tasks. However, it should be noted that the model was not designed to be used outside the laboratory; it was designed to be a research tool.

8.4 Team Lifting

Given the awkward and heavy lifting conditions that may be encountered by soldiers, of particular interest to the Army may be the notion of team lifting. The NIOSH guidelines (Waters et al., 1994), Snook tables (Snook & Ciriello, 1991), and the *Guide to Manual Material Handling* (Mital et al., 1997) focus on individual people lifting, and do not discuss team lifting in detail. Team lifting offers a unique problem in that the distribution of the load and the lifting capabilities may differ between the team members.

There is evidence that supports the notion that the lifting ability of the team is less than the sum of the lifting abilities of the individuals on the team. Karwowski and Mital (1986) asked teams of two or three males to perform two isokinetic lifting tasks evaluating dynamic back extension and dynamic lift strength. In both dynamic lifting conditions, the teams of two or three males could only lift an average maximum of 68% and 58% (respectively) of the sum of the individual maximums of the team members. A similar trend was observed in isometric conditions, but to a lesser extent. The teams of two and three males were asked to perform four standard isometric strength measures: arm, stooped back, leg, and composite. On average, isometric strength was 94% (team of two) and 90% (team of three) of the sum of the maximums of the individuals. However, this trend of team lifts equaling less than the sum of individual lifts was not observed in isometric leg strength for the two-person team or isometric arm strength for the three-person team.

Similar trends but to a greater extent have been observed in females (Karwowski & Pongpatana-suegsa, 1988) performing the same isometric and isokinetic tasks. Aside from isometric arm strength, the maximum lifting strength of teams of two or three females were 83.3% and 83.9% (respectively) less than the sum of the individual team members' strength. In addition, the isokinetic strengths for teams of two or three females account for about 68% and 68.4% of the sum of the individual team members' strengths. Furthermore, Sharp, Rice, Nindl, and Williamson (1997) showed that mixed-gender teams lifted 80% of the sum of the individual team member's ability, while same-gender teams could lift as much as 90%.

The cause for the reduction in strength during team lifting is unclear. Marras et al. (1999) propose a biomechanical model of two-person lifting in an attempt to explain the differences observed between the sum of the individual team member's maximum lifting capability and the team's maximum lifting ability. In contrast to previous studies that focused on determining the maximum amount a team can lift, Marras et al. (1999) investigated the effect of lift symmetry (symmetric vs. asymmetric lifts) and of number of team members (one or two people lifting) on spine compression, and anterior/posterior shear force and lateral force about L5/S1.

Marras et al. (1999) asked teams to lift a load (45.4 kg) that was twice the mass of the load to be lifted by the individuals (22.7 kg). Subject trunk motions were recorded with a device that measures 3D torso motion called a lumbar motion monitor. EMG data were collected bilaterally from the *erector spinae, latissmus dorsi, internal abdominal oblique, external abdominal oblique,* and *rectus abdominus.* The EMG data were normalized with respect to the EMG level during an MVC for that specific muscle. During the team lifts, data were collected simultaneously for both individuals.

Spine compression was significantly lower for two-person lifts under sagittally symmetric conditions than for individual lifts. Marras et al. (1999) attribute this to the observed decrease in sagittal plane movements, increase in lateral plane movements, and differences in the kinematics observed during the lifts. For instance, a decrease in hip flexion and hip flexion acceleration during two-person lifts was noted. Additionally, two-person lifts resulted in significantly lower sagittal plane moments and greater 3D trunk velocities and accelerations than one-person lifts. In general, when two-person lifting occurred, the subjects would bend less from their hips and more from the back in the sagittal plane.

Dennis and Barrett (2002) conducted a similar experiment as Marras et al. However, Dennis and Barrett asked subjects to lift a box instrumented with force transducers in the handles. Individuals lifted boxes 15, 20, and 25 kg in mass; two-person teams lifted boxes 30, 40, and 50 kg in mass. Data from the force transducers and kinematic data were collected and used to estimate L4/L5 torque and joint reaction force in the sagittal plane. Compression and shear forces at L4/L5 were calculated using a LSM.

Similar to Marras et al. (1999), Dennis and Barrett (2002) showed average torque and compression force at L4/L5 were 18% to 20% less during the team lifting than during the individual lifting. The horizontal and vertical hand forces were analyzed to provide insight into the observed differences in torque and compression force. Based on these analyses, Dennis and Barrett (2002) concluded the reduction in torque about L4/L5 was due to an *increase* in the horizontal hand force (which acted to reduce the L4/L5 extensor torque by 21 Nm) and a *decrease* in the moment arm of the vertical torque (which acted to reduce the L4/L5 extensor torque by approximately 22 Nm).

8.5 Review of Current Software

This second part of the chapter provides information on specific software packages available for ergo-nomic assessment or estimating forces exerted on the low back. Because many laboratories write custom software for specific needs, this list is not exhaustive, but provides an overview on some of the more well-known software packages. After the information for each software package is a web address where much of the information was obtained; additional information can be found at the website.

8.5.1 Pocket Ergo

Pocket Ergo is a Pocket PC-based software program that contains office and industrial ergonomic assess-ment tools and guidelines. It is designed for quick and easy ergonomic analyses to improve ergonomics, quality, and productivity of a workplace by using standard and accepted tools for assessing ergonomic risk. Pocket Ergo includes guideline models such as the revised NIOSH lifting equation, a push/pull force calculator, a carry force calculator, and the rapid upper limb assessment tool (RULA). RULA is an upper limb ergonomics evaluation tool that is in the form of a worksheet. Assessment results are savable, and can be exported into Microsoft Word. More information concerning Pocket Ergo can be obtained at http://www.thehumansolution.com/pocketergo.html

8.5.2 NexGen's ErgoMaster

NexGen ErgoMaster (Version 3.05.0003) is a suite of ergonomics analysis tools. The suite includes a discomfort survey, lift analyst, biomechanics analyst, task analyst, posture analyst, and workstation analyst. The software is easily navigable between the different analyses. Generated reports are import-able into Microsoft Word and include the analysis results, pictures (of the subject completing the task), subject information, and a glossary. Multiple activities can be recorded for each subject.

The discomfort survey asks the subject to rate the level of discomfort for each body part on a scale from 1 to 10 during a specific activity.

The lift analyst includes two tools, a manual material handling assessment tool, and a 2D biomechan-ical prediction tool. In the manual material handling tool, a user enters information about the task to be completed, and the software provides information on how to complete the task most safely. The second tool is a 2D biomechanical prediction. In this tool, a user loads a digital picture of the subject perform-ing the task. Based on this picture, a 2D static evaluation of the forces exerted on the body is conducted. The limitation here is that the analysis is static; therefore forces due to the motion of the object or the subject are not accounted for. Also, the analysis is limited to the sagittal plane, and therefore does not take into account any motion in the transverse or coronal planes, such as twisting of the torso.

The biomechanics analyst is a shortened version of the University of Michigan Static Strength Pre-diction Program. The version included with NexGen's ErgoMaster (Version 3.05.0003) is a 2D ver-sion of the software described below (University of Michigan's 3D Static Strength Prediction Program (3DSSPP)).

The task analyst is comprised of a task assessment tool, a RULA tool, and a tool/product assessment tool. The task assessment and tool/product assessment tools are similar to the materials handling assess-ment in that a user is asked a series of questions, and the software provides information on how to com-plete the task more safely. The RULA tool asks the same questions as the RULA worksheet. Based on the answers to the questions, the RULA returns a relative risk of upper limb injury.

The posture analyst has two tools: the posture assessment tool and a dimensional assessment tool. Both tools are very similar. Essentially, a user loads a picture of a subject completing a task. The user clicks on the joints; and for the posture assessment tool, the software returns the joint angles. For the dimensional assessment tool, the software returns the distance between two clicks.

The last module is the workstation assessment and video display assessment. These are very similar to the material handling assessment, in that the user answers questions about the subject and the task. Based on the answers to those questions, the software provides information on how to set up the workstation more safely.

Information on ErgoMaster is available at NexGen's website: http://www.nexgenergo.com/ergonomics/ergomast.html

8.5.3 NexGen's MannequinPro

In addition to ErgoMaster, NexGen has a line of software called Mannequin, the most advanced of which is MannequinPro. This software is different from ErgoMaster in that it lets you create accurate, 3D digital humans (anthropometry based on 1988 Anthropometric Survey (ANSUR) database and NASA-STD3000) and measure the impact of ergonomics on products, equipment, and facilities. Computer-aided design (CAD) files of products can be imported into the software. Joint ranges of motion of the human models are similar to human joint ranges of motion. The software includes 3D predictions of joint forces and torques and the revised NIOSH equation. The mannequin can move through a space, lifting and carrying objects. The software allows for frame-by-frame animation of specific activities, such as walking, and allows for simulating lifting, pushing, and pulling by adding forces and torque in any direction on any body part.

On the surface, the ability for animation appears to be beneficial; however, there does not seem to be any embedded control of the digital human; the digital human's motion seems to be programmed by the user. Further, there is no information as to whether or not the trajectories of the digital human's body segments accurately represent the way a live human would move under the same conditions. Additionally, it is unclear if the model used to calculate forces is a static or dynamic link segment model.

More information on MannequinPro is available at NexGen's website: http://www.humancad.com/products/mqpro.html

8.5.4 University of Michigan's 3D Static Strength Prediction Program (3DSSPP)

3DSSPP software predicts static strength requirements for tasks such as lifts, presses, pushes, and, pulls. The program provides an approximate job simulation that includes posture data, force parameters, and male/female anthropometry. Output includes the percentage of men and women who have the strength to perform the described job, spinal compression forces, and data comparisons to NIOSH guidelines. Because the analysis is 3D, the user can analyze torso twists and bends (an advantage over NexGen's ErgoMaster) and make complex hand force entries. Analysis is aided by an automatic posture generation feature and 3D human graphic illustrations.

3DSSPP can be used as an aid in the evaluation of the physical demands of a prescribed job or in evaluating proposed workplace designs and redesigns prior to the actual construction or reconstruction of a workplace or task. One limitation of 3DSSPP is this use of static LSMs to calculate forces (as opposed to dynamic models). Consequently, the software assumes the effects of acceleration and momentum are negligible. The analysis of static postures allows for information on body segment angles, hand locations, and hand force magnitude and direction, L5/S1 disc compression, percentage of the population capable of completing the task, center of gravity location, and body segment weights. Additionally, parameters such as the resultant moments produced by the load and the body weight about the main reference axes at each joint, resultant moments and forces at the spinal segments, and resultant moment produced by the load and body weight for each joint articulation are calculated. Specific information is provided on L5/S1 compression and shear forces and estimated ligament strain and an in-depth lower back analysis that considers the effect of additional muscle action in the torso.

Essentially, the software appears to use a static LSM to calculate external forces. Additionally, the model appears to include estimates of muscle locations (and therefore estimates of distances between the muscle force and the axes of rotation), and the relative contribution of each muscle to a given posture. This information is used to estimate the muscle force required to maintain a specific posture.

In an effort to address the limitations of predicting forces based on a static LSM, the software allows the user to repeatedly conduct static evaluations of a subject at various postures throughout the motion. An easy way to do this is built into the software. However, this methodology only addresses part of the issue of using a static model (only analyzing a snapshot of the motion), not addressing the fact that static models assume forces and moments due to motion are negligible.

More information can be obtained at http://www.engin.umich.edu/dept/ioe/3DSSPP/

8.5.5 Safework Pro and Delmia Human

Safework Pro and Delmia Human are two software suites that allow users to load a virtual human into a 3D space and perform ergonomic evaluations. They are grouped in this chapter because there is a link between them that allows a user to create mannequins with realistic anthropometrics and joint centers of rotation in Safework, and use the direct and inverse kinematics and other aspects of Delmia to make the simulated human move through the environment.

Available analysis ranges from static posture analysis to analysis of dynamic movement, and interaction with the environment. Ergonomic evaluations include analysis based on the NIOSH equation, and allows for a determination of the effects of lifting, lowering, pushing, pulling, and carrying. Input variables to the model are anthropometrics, and initial and final posture; output variables include RWL (from the NIOSH equation), and maximum lifting/lowering weight. Tools are included that allow for balance calculation, the RULA, the envelope the upper limb can reach, and an estimation of metabolic cost (based on the GARG equation; Garg, Chaffin, and Herrin, 1978). There is an animation tool that assists the user in visualizing the motion of the virtual human and the interaction with the environment. Additionally, there is a vision analysis feature, postural analysis, comfort angle analysis, collision detection, a virtual reality feature, and a clothing module. Control of the motion of the human figure can be accomplished through either inverse kinematics or direct kinematics. When using inverse kinematics, the desired position of an end effector is established and the joint angles required to reach that position are determined. When using direct kinematics the position and orientation of the end effector is a function of joint variables.

The virtual human's anthropometrics are based on 103 variables from the 1988 ANSUR database, but individual parameters of the virtual human's anthropometrics are adjustable. Additionally, the software allows for the use of boundary mannequins to represent specific sub-populations. The virtual human has 100 independent links and 148 degrees of freedom, including a fully articulated hand, spine, shoulder, and hip model. Also included are limits of joint mobility and coupled range of motion between joints.

Delmia Human Task Simulation allows users to create, validate, and simulate activities for simulated humans. For instance, simulated humans may walk to a specific location in the environment, move from one target posture to another, pick up an object, and walk to another location. The simulated human would have anthropometric characteristics entered by the user, who would also determine the path and trajectories.

Safework Pro and Delmia Human are limited by the absence of a biomechanics toolbox or module. It does include an inverse kinematics tool; however, there is no way of calculating forces and torques in the software (i.e., no inverse dynamics). The features of Safework allow for interaction between the virtual human and the environment to be studied; however, there is no way of calculating forces internal and external to the virtual human. This is a severe limitation in Safework's applicability to investigating the biomechanics of lifting. For more information visit the Safework website (http://www.safework.com) or the Delmia Human website (http://www.delmia.com/gallery/pdf/DELMIA_V5Human.pdf)

8.5.6 Jack

Jack is a visual simulation tool that enables users to create virtual environments by modeling them directly or importing CAD data, inserting human figures, assigning tasks to the virtual humans, and obtaining information about the interaction between the digital human and the environment. Jack provides a high-fidelity human model with accurate joint limits, a fully defined spine and flexible anthropometric scaling. Additionally, the digital human can walk, balance, reach, grasp, bend, and lift. Jack allows for the evaluation of strength requirements, visibility, multi-person activity, reach, grasp, and manipulation of tools or objects, foot pedal operation, and injury risk assessment.

In the Jack environment, the simulated human interacts with objects that can be imported from various CAD software (supported formats include JT, IGES, VRML, STL, DXF, etc.). Jack is a real-time visual simulation system based on inverse kinematics. The digital humans created by the software are based on an anthropometry database derived from the ANSUR 88, National Health and Nutrition Examination Survey (NHANES) III, and others. Jack digital people have 69 segments, including a realistic 17-segment spine and 16-segment hands. Joint and strength limits derived from NASA studies are imposed on the digital humans.

To use Jack, a user would import the CAD models and build the virtual environment, insert the simulated human, assign or animate a task, and utilize human performance models to analyze the ergonomics of the task sequence. Software output includes still images and animations, low back spinal force analysis (however, the methodology for calculating forces is not defined), output from the NIOSH equation, metabolic energy expenditure estimates (methodology not defined), static strength predictions, reach envelopes, safety zones, seating accommodation reports, and visibility information.

Additionally, there is the Jack Motion Capture Toolkit, which is a set of tools that allows a user to assign movement to the simulated human via motion capture. This is different from the inverse kinematics methodology employed by Safeworks. With this functionality, markers are placed on landmarks of the person (such as the elbow, knees, shoulders) performing the action and tracked by the motion capture hardware. The Motion Capture Toolkit allows data on the position of the live human in the virtual environment to be collected and presented to the user and the live human (via virtual reality goggles) in real time. The advantage of using motion capture data to move the human through the virtual environment is that it allows for an understanding of true human movement (in contrast to movement programmed by the user) and the complex postures or movement nuances associated with completing a novel task. Additionally, movements can be recorded and used for later analysis (including calculation of external forces) or training. With Jack and Motion Capture, design teams can place real subjects into the virtual environment to see how people will perform assigned tasks or interact with product designs. For example, by tracking a real person doing the task, a design team can simulate and analyze the motions of the human driving a truck or servicing a machine. The real-time human performance analysis tools available in Jack can be run simultaneously to rapidly evaluate if the observed motions will put a user at elevated risk of injury.

The views from Jack's eyes can be output to a head mounted display so that the subject sees what Jack sees, and can investigate the virtual environment. Reaching for objects, looking into bins, even picking up virtual tools and working with them can be done. All Jack functionality including collision detection and human performance models are available during an immersive session, so sophisticated analyses can be performed very rapidly, including clearance investigations of a person with tools, reachability, visibility, and injury risk assessment.

For more information on Jack and the Motion Capture Toolbox, refer to the UGS website: http://www.ugs.com/products/tecnomatix/human_performance/jack/

8.5.7 Digital Biomechanics

Digital Biomechanics is a dynamic simulation tool that incorporates active control of the simulated human, built-in anthropometry, and realistic (CAD models) equipment models. The purpose of Digital

Biomechanics is to model the effect of equipment on a digital human engaged in actual tasks, including walking, running, crawling, and a virtual obstacle course. The use is to analyze the effects of prototype designs on human performance.

The process for using Digital Biomechanics is to (1) select a human model, (2) load equipment and attach it to the human model, (3) select soldier tasks to create virtual obstacle course, (4) simulate and record human and equipment performance, (5) analyze soldier and equipment performance, and (6) vary the equipment and repeat the task.

Digital Biomechanics is physics based; therefore, the simulated models obey the same physical laws as live subjects would. Anthropometry for the simulated human can be extracted from the 1988 ANSUR database. Equipment models can be constructed from CAD data. Equipment can be either statically or dynamically attached to the simulated human. With static attachments no forces are exchanged between the equipment and the simulated human, the equipment acts solely to affect the mass properties, volume, etc. of the human. In a dynamic attachment, forces are exchanged between the equipment and the simulated human. The dynamic simulation and control allows the simulated human to perform Soldier tasks like combat rolls and rush maneuvers. Equations of motion calculate the response of the digital human to external forces due to the acceleration due to gravity and due to collisions between segments or objects. The output of the model is a comprehensive biomechanical dataset that allows the user to evaluate the equipment design. Data can be viewed in Digital Biomechanics or exported to MATLAB®. The limitations of Digital Biomechanics include the lack of a capability that allows the user to easily use motion capture data as input to the model. While validation of Digital Biomechanics is ongoing, evidence has not been provided demonstrating the motion and forces generated by the simulated human accurately represents the motion and forces generated by a live human, especially in novel conditions. However, this is a limitation of any physics-based simulation system that does not utilize motion capture data of live humans performing actions in the novel conditions. http://www.bostondynamics.com

8.5.8 AnyBody Technology

AnyBody Technology software is capable of modeling on multiple scales, from the subsets of the musculoskeletal system to the entire body, and can compute forces and mechanical energy and estimate metabolic cost, and allows for dynamic or static models. This software differs from similar software packages in that individual muscles may be modeled using AnyBody Technology, and therefore information can be obtained on the effects of each muscle on a given movement. Muscle recruitment is estimated by an inverse dynamics optimization technique. A limitation of the system is that the motion needs to be prescribed; however, the company does indicate a motion capture interface is in development. As with any model, validation is highly important. AnyBody Technology is a spin off of a program at Aalborg University in Denmark, and has a substantial reference list of conference proceedings and peer-reviewed articles validating certain aspects of their software. While this is more validation than many other software packages offer, it is important to realize that validation is important for each movement simulated, especially when simulating novel tasks. That is, while the simulated muscles move the simulated limbs, information is lacking on how accurately the resulting motion captures the motion of a real human under the same conditions. Additionally, the methods used to estimate metabolic cost based on human movement are not clear. Another limitation is the lack of a library of digital humans (i.e., 1988 ANSUR database) to select simulated subjects from. http://www.anybodytech.com

8.5.9 Santos

Santos is a product of the University of Iowa Virtual Soldier Research Program. Santos differs from other biomechanics simulation tools in that it is based on Newtonian physics *and* incorporates an optimization-based approach to motion prediction. There are other biomechanics simulation packages that are physics based (i.e., Digital Biomechanics from Boston Dynamics); however, Santos is the only package that incorporates an optimization-based approach to motion prediction. Santos lists optimizing factors such as joint displacement, effort, minimizing discomfort, change in potential energy, visual acuity, and visual displacement. Evaluation of these parameters occurs on a frame-by-frame basis.

A significant component of Santos is the ability to predict posture in real-time. In order to accomplish this, the Virtual Soldier Research Program at the University of Iowa developed the direct human optimization posture prediction (D-HOPP) tool. The problem of posture prediction is constrained primarily by requiring a specified end-effector (i.e. a fingertip) to contact a specified point, line, or plane. The end-effector positions in Cartesian space are determined from the joint angles by using the Denavit–Hartenberg (DH) method, a kinematics technique stemming from the field of robotics. Joint limits are imposed as constraints and are based on anthropometric data (anthropometric database was not specified). Human performance measures that represent physically significant quantities, such as energy, discomfort, etc., provide the objective functions.

Motion prediction is also based on D-HOPP, using the characteristics of the joint angle curves (control points of B-Spline representing joint angle curves) as the design variables. Constraints on objective functions for motion are similar to those for posture prediction (such as joint displacement, effort, discomfort, change in potential energy, visual acuity, and visual displacement), although they are evaluated at each time step. This unique approach allows for functionality to be added by including different constraints. For instance, to calculate torques at joints, equations of motion and torque limits are used as additional constraints. In contrast to other forms of simulation (forward dynamics), the approach used for Santos does not require numerical integration (which is often computationally expensive). Additionally, the use of D-HOPP negates the need for pre-recorded data (or virtual reality data as used with Jack) to drive motion of the digital human.

The novel approach to predicting human motion used for Santos addresses some of the problems associated with many of the current simulation tools, such as the need for realistic human data to drive the position of the digital human. However, other limitations may be relevant. For instance, correctly choosing the optimization criteria poses a significant problem. Optimality criteria affecting one motion may differ from the optimality criteria affecting other motions. There is no discussion as to how the optimality criteria for Santos were determined. In order for the approach to be validated, a strong understanding of all the factors that affect human motion is needed. Additionally, unclear is the weighting of the optimality criteria, and how those weights were determined. Also missing is a discussion of how new objects are added to the environment and how Santos interacts with equipment. http://www.digital-humans.org/santos/

8.6 Discussion

The correct model and software to use for a biomechanical analysis of lifting depend on the data available for input, the ability to collect those data, and how rigorous the user wants the output to be (Appendix 1: Comparison of commercially available ergonomics software packages that can be used for lifting analysis). Many of the software packages commercially available, such as Pocket Ergo, NexGen's ErgoMaster, and University of Michigan's 3DSSPP, incorporate Ergonomic/Guideline models and have the advantage of requiring input data that are relatively easy to collect. Additionally, these software packages are somewhat easy to use, are highly generalizable, and are appropriate for assisting in designing or evaluating tasks that require lifting. Because many of these software packages use either the NIOSH lifting equation or static LSMs, the effects of velocity and acceleration of the load being lifted are ignored during analysis, and should be minimized in the task that is being designed or evaluated. Consequently, the software is limited in application to tasks that can be decomposed into a series of representative static postures. An additional limitation is the analysis is typically not very specific to an individual person.

While these limitations are important to consider, the value of software packages for ergonomic assessment should not be underestimated. Software such as Pocket Ergo, NexGen's ErgoMaster, and University of Michigan's 3DSSPP are highly valuable in providing methodology for quickly assessing a task and determining what percentage of a population can complete the task. One aspect of these software packages that make them valuable is the fact that they are based on accepted models (i.e., the NIOSH lifting equation, the Snook tables, etc.) and provide output that is comparable across different tasks or situations. For instance, Pocket Ergo can be used to determine the RWL and LI of a given lifting task. If the RWL is too low or the LI is too high, the task could be re-designed (for instance, possibly changing the coupling between the hands and the object or reducing the height the object needs to be lifted), reevaluated using Pocket Ergo again, and the new RWL and LI can be compared to the RWL and LI from the original task. This process can be repeated until acceptable values of the RWL and LI can be determined. When used in a broader context, such as evaluating many different lifting tasks a Soldier may encounter while servicing a vehicle, the most strenuous task can be determined, and if appropriate, redesigned.

Additionally, NexGen's ErgoMaster can be used to estimate joint forces and includes discomfort surveys and recommendations for improving tasks that are already designed. The idea is to use the output from ErgoMaster to redesign tasks if needed. Further, NextGen's Mannequin Pro includes human figure models and can import data from CAD files. The imported data are used to develop an environment that the human figure can interact with. This tool is useful for evaluating lifting tasks before actual prototypes are constructed. For instance, consider a new vehicle that may be being designed. The vehicle requires a Soldier to lie on his back underneath it in a specific manner while accessing a part of the vehicle that is being repaired or maintained, such as the vehicle's alternator or fuel pump. CAD drawings of the vehicle's undercarriage and alternator or fuel pump can be loaded into NexGen's Mannequin Pro. A digital human can be imported into the environment and configured in a posture that allows him access to the part of the vehicle that needs repair. Based on the weight of the part that needs repair, the posture needed to access the part that needs repair, and the anthropometry of the Soldier, NexGen's Mannequin Pro can calculate the forces needed to complete the task. These forces can then be used to determine what percentage of the population can complete the task. It should be noted that the documentation for NexGen's Mannequin Pro does not indicate whether the model used for calculating forces is static or dynamic.

If the task is dynamic and the effects of velocity and acceleration cannot be assumed to be negligible, software such as Jack, Digital Biomechanics, or Santos could be used. Jack and Digital Biomechanics are similar to NextGen's Mannequin Pro in that they allow for CAD data to be used to establish an environment for the simulated human figure to interact with. Jack has the capability of using motion capture data of a live subject performing the task to control the motion of the simulated human. However, Jack does not include a calculation of the forces required to complete a task. Once a task is constructed in Jack, representative postures (body joint angles) can be determined and used with the University of Michigan's 3DSSPP to determine the force required to maintain each posture. Additionally, there is a Task Analysis Toolkit that can be used with Jack that integrates ten ergonomics analysis tools including the University of Michigan's Strength Prediction Program, the NIOSH Lifting Equation, and a Low Back Spinal Force Estimator into Jack. The limitation of this using Jack in concert with these other models is that the other models are static, so all the limitations of static models would apply.

The ability to use motion capture data to control the virtual human in Jack allows a live subject performing the task to get feedback on task performance and to gather data needed for the calculation of dynamic forces. For example, a CAD model of a vehicle can be imported and a simulated human model of the correct anthropometry can be constructed in Jack. A live human can then perform tasks in a motion capture laboratory; the motion capture data can be used to control the motion of the simulated human in Jack. Used in concert with virtual reality goggles, the live human can see the simulated environment from the simulated human's perspective. This technology allows the live human to move in the real environment, but to have those movements copied by the simulated human in the Jack environment. The live human can obtain the needed postures for performing specific tasks on the vehicle in the simulated environment. Once specific postures are obtained, static LSMs (such as 3DSSPP or the Task Analysis Toolkit) can be used to determine the forces and torques required to maintain the posture and the percent of the population capable of completing the task. Additionally, the motion capture data can be used with a dynamic LSM to calculate the dynamic forces used to obtain the posture. The advantage of this approach is that the motion data used to calculate the dynamic forces are actual motion capture data from the live human.

Digital Biomechanics and Santos allow for analysis of dynamic forces without the need for live humans to perform the task. Both Digital Biomechanics and Santos use control algorithms to control the simulated human. These models may be appropriate to use if suitable live subjects are not available. However, both Digital Biomechanics and Santos are limited in that while they account for the motion of the body segments during the task, information is lacking as to whether the motion observed in the simulation accurately portrays the actual motion of a human subject performing the task. For instance, a live human lifting a 10-pound box may lift it differently than a live human lifting a 50-pound box. Information is lacking as to whether or not the simulations can accurately capture the differences in *movement* between lifting a 10-pound box and lifting a 50-pound box. It is important to capture the differences in movement between these tasks because the movement will affect the forces required to complete the task.

Motion capture of a live subject and the use of control algorithms to control a simulated human represent two ways of gathering kinematic data on a subject (live or simulated) completing a task. A third method, known as the Human Motion Simulator (HUMOSIM), is being developed at the University of Michigan. HUMOSIM is a database of motion capture data that can be used with programs such as Jack to simulate human motion. The advantage of HUMOSIM is that the motions are based on live humans performing specific tasks. HUMOSIM is still in development and offers a promising alternative to solely using control algorithms to control an avatar.

8.7 Concluding Remarks

The level of specificity of output data needed will dictate what model to use in which situations. There is a trade-off between model specificity and generalizability of the output. The ergonomic/guideline models are the least specific and offer the most generalizable output. On the other end of the scale are the EMG and EMG + IAP models that are highly specific but not generalizable.

If the user's goal is to design a lifting task Soldiers can complete, one of the Ergonomic/Guideline models will provide the information needed. The NIOSH lifting equation and Snook tables are highly accepted and in many cases will suffice. The advantage of the Mital guideline is the inclusion of non-standard postures and psychophysical factors. However, ergonomic/guideline models do not give information on the forces exerted by the subject while completing the task.

If the user's goal is to estimate the external forces exerted by the subject to complete the task, a LSM model is needed. Before a human subject or Soldier is asked to complete a task, one of the ergonomic/guideline models should be used to estimate and minimize the risk of injury to the subject. Once this is done, to get a more precise estimate of the forces and torques required to complete the task, a dynamic LSM based on motion capture data of the subject performing the task should be used. If the task is based on prototype equipment that does not exist, a mockup of the equipment may suffice. Additionally, a visual simulation (Jack) of the equipment may provide the information necessary to collect the appropriate data. More sophisticated simulations, such as Digital Biomechanics, and Santos, do not need human subjects and show great potential. Digital human models can be programmed to complete specific tasks, and the forces and torques required to complete those tasks can be estimated without actually collecting data on human subjects.

Finally, if the user's goal is to estimate the internal and external forces exerted by the subject to complete the task, an LSM model and an EMG model are needed. These models relate highly precise measurements of muscle activity and segment motion to the forces and torques required to complete a task. The limitation of applying these models is the amount of instrumentation, expertise, and data analysis required. Also, data for many of these models need to be collected in a laboratory.

Appendices

APPENDIX 1   Comparison of Commercially Available Ergonomics Software Packages That Can Be Used for Lifting Analysis

Software Package	Tools/Model Summary	Input/Output	Limitations
Pocket Ergo	Ergonomics software designed to run on pocket PC for field evaluations.	*Output* NIOSH lifting equation Push/Pull and Carry Force Calculator (Snook tables)	See limitations associated with NIOSH equation
NexGen's ErgoMaster	Discomfort Survey Lift Analyst Biomechanics Analyst Task Analyst Posture Analyst Workstation Analyst	*Input* Depends on tool used, typically questions or worksheet to be filled out by user Anthropometry estimated by body weight and segment lengths from digital photographs. *Output* Guidelines based on answers to questions	Restricted to 2D analysis Forces during lifting calculated in sagittal plane only, static link segment model
NexGen's ManneQuin Pro	NIOSH Equation Human Figure modeling Environment based on CAD data	Anthropometry is from 1988 ANSUR database and/or NASA STD3000 *Output* 3D joint force and torque	Frame-by-frame animation, however no information on how the Digital Human Model moves No indication as to whether the model is static or dynamic

APPENDIX 2   Commercially Available Biomechanics Modeling Software That Can Be Used for Lifting Analysis

Software Package	Tools/Model Summary	Input/Output	Limitations
3DSSPP	3D Biomechanics tool Predicts static strength requirements for tasks	*Input* Anthropometry, Position of joints and hands, Direction of forces applied to hands, Magnitude of forces, Allows use of anthropometry representing Male/Female; 5th, 50th, and 95th percentile, plus data entry within these ranges. *Output* NIOSH equation, Percentage of population capable of completing a task, Body segment angles, hand locations, force magnitude and direction, L5/S1disk compression	Static Model
Safework Pro and Delmia Human	Load Digital Human into 3D environment and perform ergonomic evaluation Human Figure Model with 100 independent links and 148 degrees of freedom.	*Input* Anthropometrics, initial and final postures, Anthropometry is from 1988 ANSURR database, and allows the use of boundary mannequins.	No Biomechanics Model, therefore no force calculation

(continued)

	Controlled through Inverse Kinematics or Direct Kinematics. The environment can be based on CAD data.	*Output*: NIOSH equation, Recommended Weight Limit, envelope of upper limb reach, estimate of metabolic cost, determination of effects of lifting, lowering, pushing, pulling, and carrying (Snook tables), postural analysis, comfort angle analysis, collision detection, virtual reality feature	
Jack	Visual simulation tool that allows motion capture data to drive the motion of the digital human in real time. Motion capture data can also be used to calculate forces (dynamic). Human Figure Model controlled through inverse kinematics or motion capture. Model has 69 segments, 17 spine segments, 16 hand segments, and joint strength limits derived from NASA studies. The environment can be based on CAD data.	*Input* anthropometry, Anthropometry from 1988 ANSURR database, NHANES III, CPSC Children. *Output* Evaluation of fit, visibility, multi-person activity, reach, grasp, and manipulation of tools or objects NIOSH equation, metabolic energy estimate, static strength predictions, reach envelopes, seating accommodation	The use of motion capture data and virtual reality is good, but having a subject tethered to a source for the virtual reality may inhibit natural motion. However, this appears to allow for very realistic motion of the digital human.
Digital Biomechanics	Physics-Based Dynamic simulation tool designed to measure the effects of equipment (backpacks, helmets) on human performance Human Figure Model, the control was not discussed. Joint Limits (range of motion) from peer-reviewed literature. The environment can be based on CAD data.	*Input* Anthropometry, task type, equipment model/type of attachment. Anthropometrics from 1988 ANSURR database. *Output* Comprehensive biomechanical dataset including joint positions and accelerations, forces, torques, etc.	Does not allow for easy importing of motion capture data in real-time to allow the Digital Human to react to the environment.
AnyBody Technology	Physics-based simulation tool Allows modeling to the level of individual muscles Estimates joint forces, mechanical energy, and metabolic cost.	*Input* Link segments, COM locations, etc. prescribed motion, applied external loads *Output* Biomechanical dataset including joint positions and accelerations, forces, torques, estimates of mechanical work, metabolic cost and efficiency.	Does not have a library of digital humans such as the 1988 ANSUR database
Santos	Product of the University of Iowa, Virtual Soldier Research Program. Physics-based simulation tool, where motion is based on optimality criteria Human Figure Model controlled through optimality criteria Only one model is constructed, no varying anthropometrics.	Input Optimality criteria and constraints on the motion, such as joint displacement, effort, discomfort, change in potential energy *Output* Objective functions (a measure of how good the optimization is) joint displacement, effort, discomfort, change in potential energy	Still in development, many tasks still being researched.

References

Brobeck, J. R. (Ed) (1973) *Best and Taylor's Physiological Basis of Medical Practice* (9th Ed.). Baltimore: Williams & Wilkins.

Bureau of Labor Statistics (2003) *Bureau of Labor Statistics.* Retrieved November 2006 from http://www .bls.gov/iif/oshwc/osh/case/ostb1154.pdf

Chaffin D. B., and Andersson, G. B. J. (1984) *Occupational Biomechanics.* New York: John Wiley and Sons.

Chaffin, D. B., Andersson, G. B., and Martin, B. J. (1999) *Occupational Biomechanics* (3rd Ed.). New York: John Wiley and Sons.

Chung, M. K., Song Y. W., Hong, Y., and Choi, K. I. (1999) A novel optimization model for predicting trunk muscle forces during asymmetric lifting tasks. *International Journal of Industrial Ergonomics 23,* 41–50.

Daggfeldt, K., and Thorstensson, A. (2003) The mechanics of back extensor torque production about the lumbar spine. *Journal of Biomechanics 36,* 815–825.

Dennis, G. J., and Barrett, R. S. (2002) Spinal loads during individual and team lifting. *Ergonomics 45* (10), 671–681.

Dolan, P., Kingma, I., DeLooze, M. O., van Dieen, J. H., Toussaint, H. M., Baten, C. T. M., and Adams, M. A. (2001) An EMG technique for measuring spinal loading during asymmetric lifting. *Clinical Biomechanics 16 S1 S17–S24.*

Garg, Arun, Chaffin, Don B., and Herrin, Gary D. (1978) Prediction of metabolic rates for manual materials handling jobs. *American Industrial Hygiene Association Journal,* 39(8), 661–674.

Gordon, C. C., Churchill, T., Clauser, C. E., Bradtmiller, B., McConville, J. T., Tebbetts, I., and Walker, R. A. (1989) *1988 Anthropometric Survey of U.S. Army Personnel: Methods and Summary Statistics* (Natick Technical Report, TR-89/044).

Jager, M., Luttmann, A., and Lauring, W. (1991) Lumbar load during one handed bricklaying. *International Journal of Industrial Ergonomics 8*(3), 261–277.

Karwowski, W., and Mital, A. (1986) Isometric and isokinetic testing of lifting strength of males in teamwork. *Ergonomics 29*(7), 869–878.

Karwowski, W., and Pongpatanasuegsa, N. (1988) Testing of isometric and isokinetic lifting strengths of untrained females in teamwork. *Ergonomics 31*(3), 291–301.

Kingma, D., Toussaint, K., and Bruijnen (1996) Validation of a full 3D dynamic linked segment model. *Human Movement Science (15),* 833–860.

Kingma, I., van Dieen, J. H., de Looze, M., Toussaint, H. M., Dolan, P., and Baten, C. T. M. (1998) Asymmetric low back loading in asymmetric lifting movement is not prevented by pelvic twist. *Journal of Biomechanics 31,* 527–534.

Liberty Mutual Insurance (2004) *Liberty Mutual Manual Materials Handling Guidelines.* Retrieved November 2006 from http://libertymmhtables.libertymutual.com/CM_LMTablesWeb/pdf/ LibertyMutualTables.pdf

Lippold, O. C. J. (1952) The relationship between integrated action potentials in human muscle and its isometric tension. *Journal of Physiology 117,* 492–499.

Marras, W. S., Davis, K. G., Kirking, B. C., and Granta, K. P. (1999) Spine loading and trunk kinematics during team lifting. *Ergonomics 42* (10), 1258–1273.

Marras, W. S., Granta, K. P., and Davis, K. G. (1999) Variability in spine loading model performance. *Clinical Biomechanics 14,* 505–514.

Marras, W. S., and Sommerich, C. M. (1991a) Three-dimensional motion model of loads on the lumbar spine: I. Model structure. *Human Factors 33* (2), 123–137.

Marras, W. S., and Sommerich, C. M. (1991b) Three-dimensional motion model of loads on the lumbar spine: II. Model validation. *Human Factors 33* (2), 139–149.

McConville, J. T., Churchill, T. D., Kaleps, I., Clauser, C. E., and Cuzzi, J. (1980) Anthropometric relationships of body and body segments moments of inertia. Air Force Aerospace Medical Research Laboratory, Wright Patterson Air Force Base, Ohio AFAMRL-TR-80–119.

Mital, A., Nicholson, A. S., and Ayoub, M. (1997) *A Guide to Manual Materials Handling* (2nd Ed.). New York: Taylor & Francis.

Nordin, M., and Frankel, V. (1980) *Basic Biomechanics of the Musculoskelatal System* (2nd Ed.). Baltimore: Williams & Wilkins.

Sharp, M. A., Rice, V. J., Nindl, B. C., and Williamson, T. L. (1997) Effects of team size on the maximum weight bar lifting strength of military personnel. *Human Factors 39* (3), 481–488.

Shultz, A., Andersson, G. B., Haderspeck, K., Ortengren, R., Nordin, M., and Bjork, R. (1982) Analysis and measurement of lumbar trunk loads in tasks involving bends and twists. *Journal of Biomechanics 15,* 669–675.

Snook, S. H., and Ciriello, V. M. (1991) The design of manual handling tasks—revised tables of maximum acceptable weights and forces. *Ergonomics 34* (9), 1197–1213.

Waters, T. R., Putz-Anderson, V., and Garg, A. (1994) *Applications Manual for the Revised NIOSH Lifting Equation.* US Department of Health and Human Services DHHS (NIOSH) Publication No. 94–110.

Winter (1990) *Biomechanics and Motor Control of Human Movement* (2nd Ed.). New York: John Wiley and Sons.

9

Data-Based Human Motion Simulation Methods

Woojin Park

9.1 Abstract

Human motion simulation has been researched as a basic function of computer-aided ergonomic design software programs as it is an ideal means for providing motion and posture data necessary for virtual ergonomic analyses. Various motion simulation methods have been developed during the last few decades; yet, at this time, the human motion simulation technology, as a whole, does not seem to have reached the level of sophistication necessary to support its wide use in design practice. An understanding of the existing simulation methods may guide us in developing new methods that better facilitate computer-aided ergonomic design. The objective of this chapter is to provide a review of existing human motion simulation methods. In particular, this chapter focuses on the data-based human motion simulation approach, which has been explored by recent research studies. The data-based approach utilizes pre-existing motion data to predict novel motions. In this approach, a large human motion dataset, obtained from motion capture experiments, is statistically summarized as prediction equations or is analyzed to identify the rules that the human motor system employs to resolve the kinematic redundancy during motion planning. Also, some recent data-based simulation methods predict human motions by retrieving relevant sample motions from a motion database and systematically modifying them. It has been shown that such a retrieve-and-modify approach has the potential to achieve obstruction avoidance in a robust and rapid manner. Five different data-based human motion simulation methods are reviewed in this chapter. Future research directions are suggested.

9.2 Introduction

Human motion and posture simulation (hereinafter simply referred to as "human motion simulation" unless distinction is necessary) aims to accurately predict natural human motion trajectories or postures for given input simulation scenarios (Chaffin, 2005; Hsiang & Ayoub, 1994; Jung et al., 1995). An input simulation scenario is typically specified as a brief description of the task and performer. The output is a whole-body motion-time trajectory (typically, a set of joint angle-time trajectories) or a posture (a set of joint angles) that corresponds to the input scenario. The output motion trajectory is computed by a mathematical model, algorithm, or empirically developed equations. It is then visualized via anthropometrically correct, articulated digital human figures within a computer-aided design (CAD) environment.

In ergonomics and computer-aided design, human motion simulation has been researched as a basic function of ergonomic human CAD software programs because it is an ideal means for providing motion and posture data necessary for computer-aided ergonomic design. Computer-aided ergonomic design involves conducting various virtual ergonomic analyses, including reach, visibility, postural discomfort, biomechanical muscular strengths, and low back stress analyses. These analyses require supply of accurate human motion-time trajectories or postures as the input data. Conventional means for providing motion and posture data, such as manual posturing or motion capture, are unsuitable for addressing this need. Manual posturing based on manual entry of joint angles or direct manipulation of human figures is time-consuming and requires a high level of skill from the users; yet, it does not guarantee accurate estimations of natural human postures. Conducting motion capture experiments is not only time-consuming but it also requires participation of real human subjects and construction of experimental setups in the real world, and thus is not suitable to be used during computer-aided design activities. Human motion simulation, if done accurately and rapidly, can address the need for posture and motion data during virtual ergonomic analyses in a time- and cost-effective manner (Chaffin, 2005).

Human motion simulation is generally considered a difficult problem to solve. The human body is a kinematically redundant linkage system that consists of many joint degrees-of-freedom (DoFs). Due to the kinematic redundancy, the human body can perform a manual task based on an infinite number of possible motion trajectories (Flash, 1990; Kawato, 1996). A healthy person seems able to rapidly resolve the kinematic redundancy and make a choice out of the infinite number of possible motion trajectories. The way such choice is made does not appear completely random. When asked to perform identical manual tasks repeatedly, an individual generally exhibits consistency in the adopted motion patterns across different motion trials.

Despite this consistency, however, some intra- and inter-individual variability also seems to exist. When performing identical manual tasks, different individuals, or even a single individual across multiple motion trials, may adopt substantially different motion patterns, which are known as alternative movement styles or techniques (Burgess-Limerick & Abernethy, 1997a and b; Zhang et al., 2000; Park et al., 2005a). Despite advances in human motion research, how the human motor system resolves the kinematic redundancy and what causes the variability in the movement technique are not well understood. This represents a major difficulty in human motion simulation. Developing a motion simulation method is essentially an attempt to develop a model of the human motor system. It involves developing a proposition as to how the human motor system resolves the inherent kinematic redundancy, and then implementing it as a computer program.

For the last few decades, various human motion simulation methods have been developed. They vary substantially in terms of the underlying propositions and computational techniques employed. The existing human motion simulation methods, nonetheless, may be classified into two broad categories: the data-based and non-data-based approaches. The data-based approach utilizes pre-existing motion data to predict novel motions; in this approach, typically, a large human motion dataset, obtained from motion capture experiments, is statistically summarized as prediction equations (Faraway, 1997) or is analyzed to identify the rules employed by the human motor system to resolve the kinematic redundancy during motion planning (Zhang et al., 1998; Zhang & Chaffin, 2000). Also, some recent data-based simulation methods predict novel human motions by retrieving relevant motion samples from a motion database and systematically modifying them according to new simulation scenarios (Park et al., 2004; Monnier et al., 2006; Wang et al., 2006; Chateauroux et al., 2007). This retrieve-and-modify approach could be utilized to accomplish rapid and robust obstruction avoidance during posture prediction (Park et al., 2006). The non-data-based methods, on the other hand, do not utilize pre-existing motion data; instead, they employ some theoretically plausible criteria or principles to resolve the kinematic redundancy during motion and posture planning. The non-data-based methods typically solve constrained optimization or constraint satisfaction problems to simulate motions or postures (Nubar & Contini, 1961; Hogan, 1984; Uno et al., 1989; Hsiang & Ayoub, 1994; Chang et al., 2001).

The existing simulation methods have achieved various successes and represents significant progress in the human motion simulation research. However, at this point in time, the human motion simulation technology, as a whole, does not seem to have reached the level of sophistication necessary to support its wide use in design practice. An important step toward enhancing the state of the art is to gain a full understanding of the existing simulation methods. Such understanding may guide us in developing new methods that better facilitate computer-aided ergonomic design. Therefore, the objective of this chapter is to provide a review of existing human motion simulation methods. In particular, this chapter focuses on the data-based human motion simulation approach, which has been explored by recent research studies. Five data-based motion simulation methods are reviewed in this chapter. Future research directions are suggested.

9.3 Data-Based Motion Simulation Methods

Five data-based human motion simulation methods are described in what follows. They are: the functional regression model for reach prediction, the data-driven differential inverse kinematics model for reach prediction, the memory-based motion simulation model, the data-based motion simulation method for ergonomic vehicle design, and the memory-based posture planning model.

9.4 Functional Regression Model for Reach Motion Prediction

Faraway (1997) developed a statistical functional regression model for predicting human reach motions. The model aimed to rapidly and accurately predict multi-segment human reach motion trajectories based on a set of input covariates, such as the performer attributes (age, gender, stature, etc.) and the target location. Functional regression is a statistical modeling technique that relates smooth functional response, $y(t)$, to some known covariates, x, by a linear combination of parameter functions, $\beta(t)$. $\beta(t)$ are estimated to fit the response with minimum error.

The functional regression model for reach motion prediction was developed based on a large set of real human reach motion data. The reach motion data were part of the motion database created by the University of Michigan Human Motion Simulation (HUMOSIM) Laboratory. In one motion capture study conducted by the HUMOSIM researchers, 38 subjects of varying age and anthropometry performed reach motions while seated in simulated vehicle interiors and industrial workplaces. A total of over 7,000 reach motions were recorded using a hybrid motion capture system that utilizes both the optical and the electromagnetic motion capture technology. Each recorded motion was represented as a set of multiple joint angle-time trajectories. The functional regression model was fitted to the joint angle-time trajectory data. Given input data, such as the performer's stature, age, and gender, and the reach target location, the regression model predicts the average joint angle-time trajectories, and also the corresponding angle-time confidence envelopes. The model takes the following form:

$$\theta(t) = \beta_0(t) + C_x \beta_x(t) + C_y \beta_y(t) + C_z \beta_z(t) + C_x C_y \beta_{xy}(t) + C_y C_z \beta_{yz}(t)$$
$$+ C_z C_x \beta_{zx}(t) + C_x^2 \beta_{x^2}(t) + C_y^2 \beta_{y^2}(t) + C_z^2 \beta_{z^2}(t) + D$$

One desirable aspect of the functional regression modeling technique is that it allows evaluating how well the model fits the empirical human motion data. The above quadratic model was shown to account for nearly 80% of the variation in one set of reach motion data (Faraway, 1997).

One implicit assumption of the Faraway functional regression model is that joint angle-time trajectories can be planned directly from a task description (i.e., the target location) without prior planning of other kinematic variables (e.g., the hand position trajectory). Without prior planning of hand motion trajectory, the model did not guarantee the end-effector's arrival at the intended target location at the final posture. This is due to the statistical errors inherent in regression-based predictions and the variability in individuals' body segment link lengths. To resolve this problem, a final posture rectification method based on inverse kinematics was developed (Faraway et al., 1999). The final posture rectification method was shown to improve the accuracy of joint angle predictions at the final posture.

9.5 Data-Driven Differential Inverse Kinematics Model for Reach Prediction

Zhang et al. (1998) and Zhang and Chaffin (2000) developed an optimization-based differential inverse kinematics scheme for modeling and predicting 3D seated reach motions. The method aimed to identify the hand-to-joint mapping strategy in the velocity domain. A model-guided data-fitting process was developed.

In this approach, a seated human worker performing reach motions was modeled as a four-link open kinematic chain (torso, clavicle, upper arm, and lower arm). The linkage model had a total of seven DoF $\theta = [\theta_1 \cdots \theta_7]^T$. The hand position of the linkage model $\mathbf{P} = [x \quad y \quad z]^T$ is determined as a function of the joint DoF:

$$\mathbf{P} = [f_1(\theta_1, \cdots, \theta_7) \quad f_2(\theta_1, \cdots, \theta_7) \quad f_3(\theta_1, \cdots, \theta_7)]^T = \mathbf{f}(\theta)$$

Differentiating the above forward kinematics function with respect to time, the following relationship between the hand linear velocity $\dot{\mathbf{P}}$ and the joint angular motion rates $\dot{\theta}$ is derived:

$$\dot{\mathbf{P}} = \dot{\mathbf{f}}(\theta) = \frac{\partial \mathbf{f}}{\partial \theta} \dot{\theta} = \mathbf{J}(\dot{\theta})\dot{\theta}$$

In the above equation, $\mathbf{J}(\theta)$ is the 3×7 manipulator Jacobian matrix. It governs the mapping from the joint angular motion rates $\dot{\theta}$ to the hand linear velocity $\dot{\mathbf{P}}$. The inverse of the above equation, which maps the hand velocity $\dot{\mathbf{P}}$ to the joint angular motion rates $\dot{\theta}$, can be derived using a weighted pseudo-inverse of the Jacobian matrix $\mathbf{J}$:

$$\dot{\theta} = \mathbf{W}^{-1}\left[\mathbf{J}\mathbf{W}^{-1}\right]^{\#} \dot{\mathbf{P}}$$

In the above equation, # represents the matrix pseudo-inverse. $\mathbf{W}$ is a 7×7 matrix that contains weighting factors: $\mathbf{W} = diag\,(w_1 \cdots w_7)$. The choice of the weighting factor values $w_1 \cdots w_7$ determines the way a given hand velocity vector $\dot{\mathbf{p}}$ is translated into the joint angular motion rates $\dot{\theta}$. The weighting factor values, $w_1 \cdots w_7$, were selected through an optimization-based data-fitting process; they were determined such that the resulting joint angular motion trajectories have the minimum deviations from the joint angle trajectories of real human reach motions. A global optimization technique, simulated annealing, was utilized to determine the optimal set of weighting factor values. The weighting factor values determined by the data-fitting process are considered interpretable; they are viewed as representing an inter-joint motion apportionment strategy.

The differential inverse kinematics method by Zhang et al. (1998) and Zhang and Chaffin (2000) assumes that hand motion trajectory is planned prior to the planning of joint angular motion trajectories. Due to this hierarchy of motion planning, the differential inverse kinematics method seems to facilitate simulating certain types of motions (e.g., tracing predetermined paths in the task space using an end-effector), which may be difficult to simulate using other simulation methods.

9.6 Memory-Based Motion Simulation Model

The memory-based motion simulation (MBMS) model was developed by the researchers at the University of Michigan Human Motion Simulation (HUMOSIM) Laboratory (Park et al., 2004, 2005a and b). This model is based on the idea that human motion samples can be systematically modified to generate motions for new scenarios.

The MBMS model is comprised of four basic components: the motion database, the root motion finder, the motion variability analyzer, and the motion modification algorithm (fig. 9.1). The motion database stores samples of real human motions, which serve as templates for simulating novel motions. These motion samples are obtained from laboratory motion capture experiments, in which a group of subjects performs a set of well-defined, discrete, goal-directed motions specific to a particular application domain. For each sample motion recorded, the kinematic motion data are stored in the database, with the motion scenario associated with the motion. A motion scenario consists of motion task descriptors (motion type/category, initial conditions, hand load weight, motion task goals, etc.) and motion performer characteristics (gender, height, weight, etc.).

Given an input simulation scenario, the root motion finder searches the motion database to retrieve the motion samples that closely match the input scenario. The motions retrieved by the root motion finder are called "root motions" and serve as templates for simulating motions for the given input simulation scenario. The root motions retrieved for an input simulation scenario may significantly differ from one another in the overall movement patterns and may represent different movement techniques. It is important that a motion simulation method identifies and predicts the range of possible alternative motion behaviors as opposed to predicting a single representative motion because different movement techniques could result in different biomechanical, physiological, and psychophysical consequences. Given a set of root motions retrieved for an input simulation scenario, the motion variability analyzer analyzes and graphically summarizes the variability in the movement technique within the root motion set.

The motion variability analyzer first quantitatively represents each root motion's underlying movement technique using the joint contribution vector (JCV) index (Park et al., 2005a). A JCV computed for a root motion is a vector whose elements represent the contributions of individual body joint angle rotations to the task goal achievement (generally, hand target acquisition). Once JCVs are computed for all the root motions, then the motion variability analyzer graphically summarizes the dissimilarities between the root motions in movement technique as a map on a low-dimensional space using the multidimensional scaling (MDS) method (Johnson & Wichern, 1998). Based on the graphical summary of dissimilarity relationships between the root motions, a user can rapidly identify alternative movement techniques and further reduce the initially retrieved root motion set to a more manageable, smaller subset that contains all the alternative movement techniques without repetitions of similar root motions. An example of the motion variability analysis is provided in Figure 9.2.

The input simulation scenario for this simulation example was as follows: the performer is 175 cm tall; from an initial standing position, the digital human figure was asked to transfer a box from one shelf to another using both hands. The initial and the final positions of the right hand center position are shown in Figure 9.2a. Given the input simulation scenario, the root motion finder searched the HUMOSIM motion database and retrieved five root motions for the scenario. The motion variability analyzer then computed JCVs for the five root motions and depicted their proximity relationships on a 2D space using the MDS method (fig. 9.2b). The proximity relationships between the five root motions

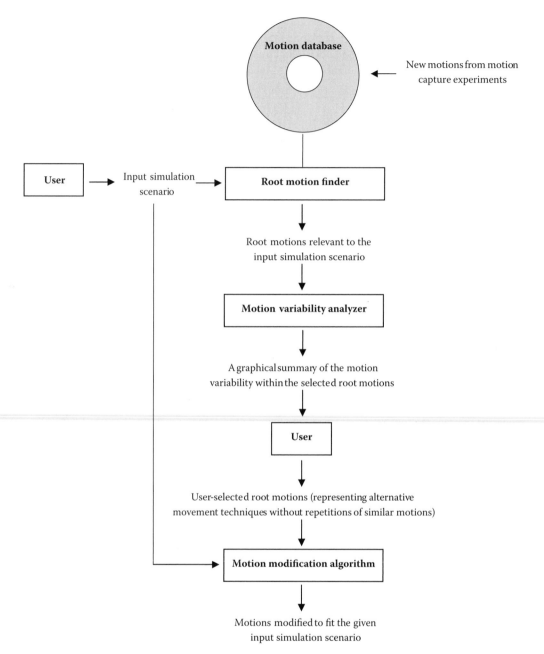

FIGURE 9.1 Memory-based motion simulation process. (From Park, W., et al., "Memory-Based Human Motion Simulation for Computer-Aided Ergonomic Design," *IEEE Transactions on Systems, Man, and Cybernetics, Part A: Systems and Humans*, accepted 2007. With permission.)

suggested that the five root motions may represent two different movement techniques: one represented by root motions 2–5 and the other by root motion 1. To verify this classification, the five root motions were visually inspected. Stick figure animations of the five root motions are provided in Figure 9.2c. The visual inspection of the root motions was in agreement with the classification based on the MDS plot.

Once a subset of retrieved root motions is selected by a user, the selected root motions are modified by the motion modification (MoM) algorithm to exactly fit the input simulation scenario. For a given root motion, the MoM algorithm first analyzes the motion using the symbolic motion structure

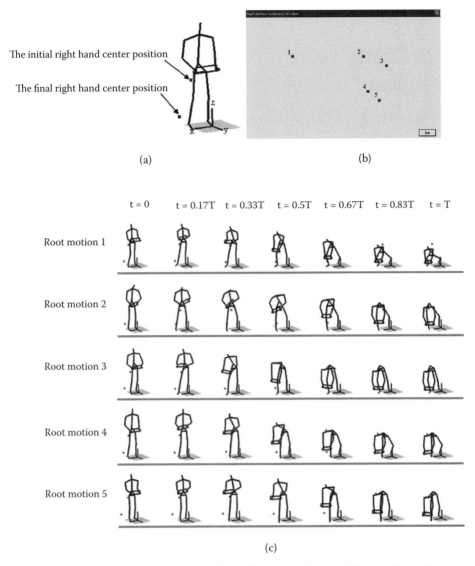

FIGURE 9.2 An example of motion variability analysis. (From Park, W., et al., "Memory-Based Human Motion Simulation for Computer-Aided Ergonomic Design," *IEEE Transactions on Systems, Man, and Cybernetics, Part A: Systems and Humans*, accepted 2007. With permission.)

representation (SMSR) algorithm (Park et al., 2005b). The SMSR algorithm identifies a motion's fundamental spatial-temporal structure in the joint angle-time domain. The SMSR algorithm analyzes each joint angle-time trajectory of a root motion into a sequence of geometric motion primitive segments. Three types of motion primitives are utilized: a monotonically increasing (symbol U), a monotonically decreasing (symbol D), or a stationary segment (symbol S). By concatenating symbols according to their order in time, the spatial-temporal *structure* of a joint angle-time trajectory is represented as a symbolic string (fig. 9.3). The structure of a multi-joint motion is then represented as a set of symbolic strings.

A root motion, whose structure is identified by the SMSR algorithm, can be parametrically generalized to produce an infinite number of realistic motion variants (Park et al., 2005b). To generate a variant of a root motion, segment boundary points of the root motion identified by the SMSR algorithm are first relocated to new locations in the angle-time space, and then individual motion segments of the original joint angle trajectories are shifted and proportionally rescaled to fit the new segment boundary points

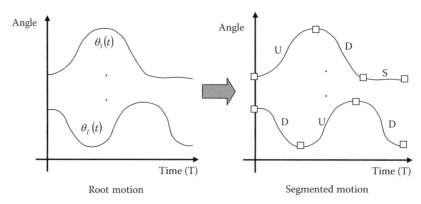

FIGURE 9.3 Motion segmentation by the symbolic motion structure representation (SMSR) algorithm. The hollow squares represent the segment boundary points. The spatial-temporal structures of the two joint angle-time trajectories are represented symbolically as UDS and DUD. (From Park, W., et al., "Memory-Based Human Motion Simulation for Computer-Aided Ergonomic Design," *IEEE Transactions on Systems, Man, and Cybernetics, Part A: Systems and Humans*, accepted 2007. With permission.)

(fig. 9.4). Thus, segment boundary point locations in the joint angle-time space are the parameters for generalizing a root motion. The new segment boundary point locations are subjected to a motion structure maintenance constraint: the new segment boundary point locations must not breach the spatial-temporal structure of the original root motion.

Once a root motion's structure is revealed by the SMSR algorithm, the motion simulation problem becomes a problem of relocating the segment boundary points in the angle-time space such that the resulting motion variant (a modified motion) exactly satisfies the input simulation scenario. In general, for a given input simulation scenario, there exists an infinite number of motion variants that satisfy the scenario. A minimum dissimilarity principle was proposed as a criterion for selecting the most desirable variant: Among the possible variants, the motion that is minimally dissimilar with the root motion is selected. A two-step optimization procedure was developed based on this idea (Park et al., 2004): First, the terminal segment boundary points at the beginning and end of the root motion (which represent the initial and final postures), which are relocated in the angle-time space. In other words, the initial and final postures are modified.

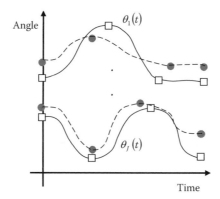

FIGURE 9.4 Deriving a motion variant from a root motion. The hollow squares represent the segment boundary point locations of a root motion. The filled-in circles represent new segment boundary point locations. The solid and the dashed curves represent the joint angle trajectories of the original root motion and those of a newly derived motion variant. (From Park, W., et al., "Memory-Based Human Motion Simulation for Computer-Aided Ergonomic Design," *IEEE Transactions on Systems, Man, and Cybernetics, Part A: Systems and Humans*, accepted 2007. With permission.)

A numerical optimization scheme was developed, which finds the new initial and final postures that satisfy the input simulation scenario with the minimum angle changes from the initial and final postures of the root motion. Then, the motion trajectory between the initial and final postures is modified to fit the newly determined initial and final postures. This is done by relocating the nonterminal segment boundary points. An optimal solution was analytically derived that minimizes the deviation of the new joint angle trajectory from the root joint angle trajectory in the domain of the first time-derivative of the joint angle displacement. An example of motion modification is presented in Figure 9.5. The root motion is shown in Figure 9.5a. It was a forward, downward reach performed by a 191-cm-tall male. The root motion was modified for six new scenarios: six new target locations, which were 30 cm away from the original in different directions (upward, downward, forward, backward, and sideways). Motion modification results are presented in Figure 9.5b to g.

The prediction accuracy of motion simulations based on the MoM algorithm was empirically evaluated. The MoM algorithm was found to be able to accurately predict both seated target reach and one-handed whole-body load-transfer motions with errors comparable to the inherent variability in repeated human motions (Park et al., 2004).

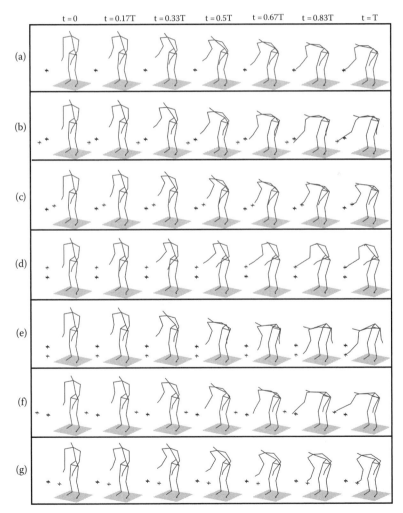

FIGURE 9.5 A forward, downward reach (a) and its variants generated for new target locations (b)–(g). The thick crosshairs represent the original target location. The thin crosshairs represent the new target locations. (From Park, W., et al., *Journal of Biomechanics*, 38, 2321–29, 2005. With permission.)

9.7 Data-Based Motion Simulation Method for Ergonomic Vehicle Design

Researchers at the French National Institute for Transport and Safety Research (INRETS) developed a data-based motion simulation method for simulating various human motions during man-vehicle interactions (Monnier et al., 2006; Wang et al., 2006; Chateauroux et al., 2007). This method intends to simulate not only simple, discrete motions (for example, a hand target reach inside a vehicle) but also complex motions, such as object manipulations and vehicle ingress/egress motions.

The motion simulation process consists of three steps: the construction of a structured motion database, the extraction of a referential motion from the database, and the adaptation of a referential motion to a given input simulation scenario (Monnier et al., 2006). Thus, the method is based on the retrieve-and-modify concept, similar to the memory-based motion simulation model.

The motion database stores typical human motions during man-vehicle interactions; a range of different vehicle configurations is considered during such motion data collection. For example, the motion database constructed by Monnier et al. (2006) stored ingress/egress motions of 23 healthy subjects for four different vehicle configurations. For each motion, the following data were stored in the motion database:

Position- and orientation-time trajectories of body parts
Joint angle-time trajectories
Environment parameters (vehicle geometry configuration)
Motion performer characteristics (gender, stature, age, etc.)
Motion characteristics (motion styles and key frames)

When an input simulation scenario is given, one or multiple referential motions that closely match the input simulation scenario are selected and modified to fit the input simulation scenario. An input simulation scenario is specified as a set of end-effector position and orientation constraints defined as different time points (referred to as key frames). Multiple end-effectors (e.g., hands, feet, pelvis, etc.) can be considered in defining constraints. Such end-effectors involved in setting up constraints at key frames are referred to as control points. The use of multiple control points and multiple key frames in defining constraints enables expressing complex man-vehicle interactions as input simulation scenarios.

The position and orientation time-trajectories of the end-effectors are modified to meet given constraints while preserving the shapes of the original motion trajectories. A motion modification method similar to the methods proposed by Park et al. (2000) and Zhang (2002) is employed to do so. Then, joint angle-time trajectories are modified to fit the new position and orientation time trajectories of the end-effectors based on motion adaptation techniques (Choi & Ko, 2000; Klein & Huang, 1983). The modified motion is used to predict the discomfort experienced by the motion performer during the motion.

The motion database can be easily expanded without altering the structure of the entire motion simulation system. Chateauroux et al. (2007) expanded the database developed by Monnier et al. (2006) by adding motion samples performed by elderly and disabled subjects.

9.8 Memory-Based Posture Planning Model for Obstruction Avoidance

The existing human motion simulation models generally lack the capability of simulating human obstruction avoidance during goal-directed manual tasks. This compromises the utility of digital human models for ergonomics, as many design problems involve interactions between humans and obstructions. To address this problem, a novel memory-based posture planning (MBPP) model was developed that plans reach postures that avoid obstructions (Park et al., 2006).

The human motor system is able to rapidly plan and execute target reaches even when obstructions exist in the task space. In addition to this, the human ability of obstruction avoidance is also characterized by its robustness; that is, in performing target reaches, the human motor system seems able to avoid obstructions of various configurations in a robust manner. The MBPP model attempts to achieve both the robustness and the rapidity of human obstruction avoidance by utilizing multiple posture memories called cells. In this new memory structure, the task space is partitioned into many small square-like (or cube-like if the task space is 3D) regions called cells. For a human figure of a particular size and a particular linkage structure, each cell is linked to a memory specific to the human figure, which stores various alternative feasible postures of the human figure that locate the hand within the cell. Postures stored in a cell are assumed to respect physiological and physical constraints, such as the joint range of motion and the static body balance constraint, and are assumed to approximate the entire range of possible postures for reaching the cell. An example of partitioning the 2D task space into cells is provided in Figure 9.6. Four postures stored in a cell (shaded) are depicted; only four are presented for the clarity of presentation.

The MBPP model also utilized the sensor approach by Badler et al. (1994) to detect a collision between a human figure and an obstruction configuration. In this approach, an obstruction configuration is represented as a combination of multiple geometric primitive objects (e.g., boxes and spheres). A finite number of sensors (point objects) are attached to a human figure; and detection of collision between a human figure and an obstruction is accomplished by testing whether or not collision occurs between the sensors and the obstruction. An example of a sensor attachment scheme adopted for a sagittal-plane human model is shown in Figure 9.7.

When a reach posture planning problem is given in terms of a target position and an obstruction configuration for a human figure, the MBPP model first identifies the cell that contains the target and retrieves all the postures in it. Among the postures, only collision-free ones are selected as candidate

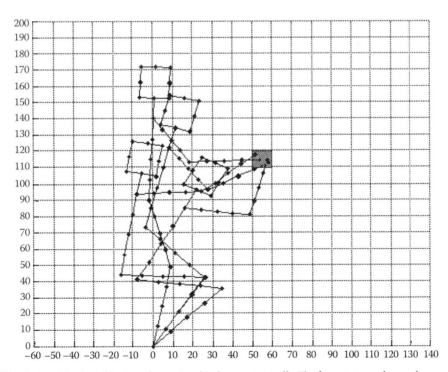

FIGURE 9.6 Partitioning of the two-dimensional task space into cells. The four postures shown above are stored in the memory associated with the shaded cell. Only four postures are shown for clarity of illustration. (From Park, W., D. Singh, and B. J. Martin, *Ergonomics*, 49, 1565–80, 2006. With permission.)

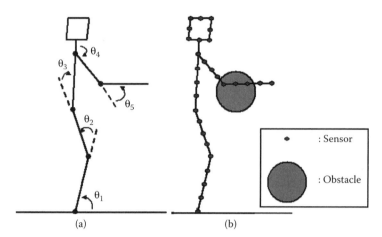

FIGURE 9.7 A simple kinematic linkage model representing the human body in the sagittal plane (a) and a sensor placement scheme (b). (From Park, W., D. Singh, and B. J. Martin, *Ergonomics*, 49, 1565–80, 2006. With permission.)

solutions. Since the postures stored in the cell approximate the entire range of possible postures for reaching or locating the hand near the given target position, this mapping from a target to a cell's member postures enables a rapid global posture search. Thus, collision-free candidate postures are likely to be found if the target position is actually reachable for the human figure in the presence of the obstruction configuration (if no collision-free posture is found, then it is concluded that the target position is unreachable for the human figure in the presence of the obstruction configuration). Each collision-free candidate posture is locally adjusted to exactly meet the target acquisition constraint; only slight adjustments would be required since each candidate posture's hand position is already near the target location. The local posture adjustment is based on a simple gradient-based, iterative posture update scheme.

The success of the MBPP model as a robust posture planner depends on whether each cell contains many alternative postures that approximate the entire range of possible postures for reaching the cell, and thus enables a rapid global posture search when solving a reach posture planning problem. For a human figure, such cells are pre-computed by an iterative random posture generation and registration process. The random posture generation and registration process prevents repetitions of similar postures using a criterion on the minimum spacing between postures in the joint space, and thus maintains the number of postures stored in a cell at the necessary minimum. To provide each cell with enough postures approximating the entire range of reach postures, the random posture generation and registration process needs to be repeated until cells are saturated with postures (in other words, until additional posture registrations rarely occur).

Some examples of posture planning based on the MBPP model are presented in Figure 9.8. A 10 cm × 10 cm square-shaped target, whose center is located 55 cm and 95 cm away from the ankle joint in the horizontal and vertical directions, respectively, was considered. A total of six obstruction configurations, including no obstruction condition, were considered. For each obstruction configuration, postures were planned for the 95th percentile U.S. male figure representation; the human figure was a 2D, sagittal-plane linkage model. The number of feasible reach postures found and the time required to plan and display the postures are shown for each example. For all six examples, posture planning and visualization was near real time. The maximum time required was 0.66 seconds.

9.9 Summary and Future Directions

Five different data-based motion simulation methods have been reviewed in this chapter. The commonality between these methods is that they utilize pre-existing human motion (or posture) data in order to

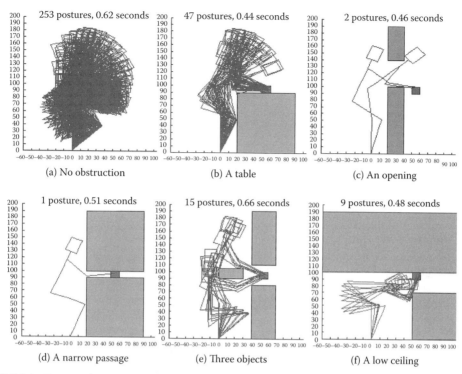

FIGURE 9.8 Posture planning examples for six obstruction configurations. (From Park, W., D. Singh, and B. J. Martin, *Ergonomics*, 49, 1565–80, 2006. With permission.)

predict motions for newly given input simulation scenarios. The pre-existing motion data are typically motion capture recordings of real human performances, but also can be artificially generated as shown in the memory-based posture planning model.

The functional regression model (Faraway, 1997) statistically summarizes a large set of real human reach motion data in the form of simple prediction equations. For a given input simulation scenario, the prediction equations compute the average joint angle-time trajectories and also determine percentile confidence envelopes of joint angle-time trajectories. The functional regression modeling approach allows evaluating how well the regression equations fit the sample data. The data-driven differential inverse kinematics method (Zhang et al., 1998; Zhang et al., 2000) identifies the velocity domain hand-to-joint mapping strategies of human reach motions, based on a model-guided data-fitting process. This method assumes that hand motion trajectory is planned prior to the planning of joint angle-time trajectories. The functional regression model and the data-driven differential inverse kinematics method are capable of rapidly predicting motions, as motion prediction using them does not require a complex optimization or iterative search process. Also, they both reduce a large motion dataset into a smaller, manageable set of parameters or parameter functions, which is desirable in terms of efficient utilization of computer memory and data portability.

The memory-based motion simulation model (Park et al., 2004; Park et al., 2005a and b) and the data-based motion simulation method for ergonomic vehicle design (Monnier et al., 2006; Wang et al., 2006; Chateauroux et al., 2007) retrieve and modify real human motion samples to predict motions for new simulation scenarios. The motion modifications are conducted such that newly derived motions meet the constraints specified in the input simulation scenario while preserving the spatial-temporal structures of the original motions. These methods utilize a large-scale motion database, which could be viewed as a model of the human memory containing motor programs (Schmidt & Lee, 1999). They also have the following desirable characteristics: First, they allow simulating categorically different motions (e.g., reach, lifting/lowering, pushing/pulling, walking, carrying, vehicle ingress/egress, etc.) based on a single model. Second, they can continually learn new motion behaviors by simply registering new motion samples to the motion database. Third, they are able to predict a range of possible alternative motion behaviors for a given input simulation scenario. The memory-based motion simulation model provided a quantitative method for exploring alternative movement techniques for motion tasks. The data-based motion simulation method for vehicle design provided a scheme for simulating complex motion patterns, which involve multiple end-effectors interacting with the environment at various points in time.

The memory-based posture planning model (Park et al., 2006) plans collision-free reach postures for a human figure in the presence of obstructions, based on a unique posture memory structure. This model does not rely on motion-capture recordings of real human performances but utilizes a memory of computer-generated human postures. The use of pre-computed postures enables human-like rapidity and robustness in planning collision-free reach motions in the presence of obstructions. Although the model is currently limited to posture prediction, it may serve as a starting point for developing dynamic motion simulation methods that are capable of obstruction avoidance.

When compared with non-data-based motion simulation methods (Nubar & Contini, 1961; Hogan, 1984; Uno et al., 1989; Hsiang & Ayoub, 1994; Chang et al., 2001), which do not rely on real human motion data but utilize theoretically plausible criteria, strategies, or principles to predict human motions, data-based methods seem to have some inherent advantages: First, data-based simulation methods utilizing real human motion data are likely to predict natural human motions more accurately than non-data-based methods. This is because they have access to more information on natural human motion patterns than non-data-based methods; data-based methods aim to closely mimic natural human motion patterns contained in pre-existing motion data, and therefore, do not require a wild guess regarding the criterion, strategy, or principle underlying the human motion planning. Second, data-based methods utilizing real human motion data would better facilitate understanding and simulating the intra- and inter-individual motion variability than non-data-based methods. This is because the use of large motion datasets provides an opportunity to observe and analyze such variability. Without observing or analyzing empirical human motion data, it would be difficult to artificially generate the inter- and intra-individual motion variability. Third, pre-existing motion and posture datasets could be utilized so as to facilitate achieving computational efficiency and robustness in solving complex simulation problems, as shown in the memory-based posture planning model.

With the advantages above, data-based motion simulation methods have a great potential as practical motion simulation tools for ergonomic design applications. However, they also have relative weaknesses compared with non-data-based methods: First, these methods (except for the memory-based posture planning model) require construction of a large human motion database through motion capture experiments, which are generally time-consuming and costly. Second, the motion simulation methods relying on real human motion data do not seem to facilitate understanding the principles underlying human motion planning; to gain insights into the principles underlying the formation of human motion patterns, a hypothesis-driven, non-data-based motion simulation models approach might be more adequate. This is because, as mentioned earlier, data-based simulation methods aim to mimic real human motion behaviors focusing on "what do human motions look like" but are not designed to shed light on the question "why do humans move in certain ways?" One exception to this may be the data-driven differential inverse kinematics methods (Zhang et al., 1998; Zhang & Chaffin, 2000); although this method is based on data-fitting, the parameter values determined by the data-fitting process are considered "interpretable" due to the unique structure of the underlying hand-to-joint mapping model. Such interpretability of empirically determined parameters may assist forming and testing hypotheses on the kinematic motion strategy of human reach motions.

Based on the current review of data-based motion simulation methods, future research topics were identified. They are:

Optimal motion/posture sampling scheme: Motion capture experiments are generally costly and time-consuming. Thus, when constructing a motion database or dataset for data-based human motion simulation, it is desirable to reduce the number of sample motions as much as possible. However, at the same time, a motion dataset must be large enough to sufficiently capture the human motion variability. The following questions need to be addressed: How can a motion capture experiment be designed optimally such that recordings of similar motion patterns, and thus, redundancy in the dataset can be minimized, and thus, the total number of motion samples recorded can be minimized? How can it be determined whether or not a motion dataset includes most of the possible alternative motion behaviors for a given task?

Integration of multiple motion databases and datasets: Since motion capture experiments are time-consuming and costly, it would be difficult for one research group to construct a large, comprehensive motion database. Nonetheless, a large-scale motion database could be obtained easily by combining multiple motion databases or datasets constructed independently by different research groups. Such database integration will greatly enhance the utility of the data-based human motion simulation approach. Independently developed motion databases, however, may significantly differ from one another in the motion data format and the human linkage model. Thus, a method for handling such differences would be needed.

Handling of various constraints during motion simulation: The current data-based motion simulation methods do not seem able to handle constraints describing individuals' physical capabilities (e.g., compromised muscular strengths, reduced joint ranges of motion, obesity, etc.) or task and environmental conditions (e.g., heavy hand load weights, repetitive and prolonged tasks, obstructions in the task space, low feet-floor friction, sloped floor, etc.). Data-based motion simulation satisfying such constraints may be accomplished by collecting real human motion data from the experimental conditions corresponding to the constraints; however, such approach will require a prohibitively large number of motion capture recordings and therefore would be highly costly and time-consuming. Thus, developing methods for accurately adapting sample motions according to various types of constraints will enhance the utility of the data-based human motion simulation approach.

Simulation of long sequences of human motion behaviors: The data-based motion simulation methods currently do not support combining one-time, discrete movements to represent long motion sequences. The lack of a tool for composing long motion sequences limits the utility of human motion simulation. To develop a motion sequence composer, the following questions must be addressed: What is the minimum set of discrete motion categories for representing typical human activities in workplace or product use? How should a motion database be structured in order to support simulating sequences of human motion behaviors? How can two discrete motions be joined or blended seamlessly?

References

Badler, N., Bindiganavale, R., Granieri, J., Wei, S., Zhao, X., 1994. Posture interpolation with collision avoidance. In *Computer Animation' 94*, 25–28 May 1994, Geneva, Switzerland (Los Alamos, CA: IEEE Computer Society Press), 13–20.

Burgess-Limerick, R., Abernethy, B., 1997a. Qualitatively different modes of lifting. *International Journal of Industrial Ergonomics* 19, 413–417.

Burgess-Limerick, R., Abernethy, B., 1997b. Toward a quantitative definition of manual lifting postures. *Human Factors* 39, 141–148.

Chaffin, D. B., 2005. Improving digital human modelling for proactive ergonomics in design. *Ergonomics* 48, 478–491.

Chang, C., Brown, D. R., Bloswick, D. S., Hsiang, S. M., 2001. Biomechanical simulation of manual lifting using spacetime optimization. *Journal of Biomechanics* 34, 527–532.

Chateauroux, E., Wang, X., Trasbot, J., 2007. A database of ingress/egress motions of elderly people. SAE Technical Paper 2007-01-2493. Society of Automotive Engineers, Warrendale, PA.

Choi, K. J., Ko, H. S., 2000. Online motion retargeting. *Journal of Visualization and Computer Animation* 11, 223–235.

Faraway, J. J., 1997. Regression analysis for functional response. *Technometrics* 3, 254–261.

Faraway, J. J., Zhang, X., Chaffin, D. B., 1999. Rectifying postures reconstructed from joint angles to meet constraints. *Journal of Biomechanics* 32, 733–736.

Flash, T., 1990. The organization of human arm trajectory control. In *Multiple Muscle Systems: Biomechanics and Movement Organization*, J. Winters and S. Woo, Eds. (New York: Springer-Verlag).

Hogan, N., 1984. An organizing principle for a class of voluntary movements. *Journal of Neuroscience* 4, 2745–2754.

Hsiang, S. M., Ayoub, M. M., 1994. Development of methodology in biomechanical simulation of manual lifting. *International Journal of Industrial Ergonomics* 19, 59–74.

Johnson, R. A., Wichern, D. W., 1998. *Applied Multivariate Statistical Analysis* (Upper Saddle River, NJ: Prentice-Hall).

Jung, E. S., Kee, D., Chung, M. K., 1995. Upperbody reach posture prediction for ergonomics evaluation models. *International Journal of Industrial Ergonomics* 16, 95–107.

Kawato, M., 1996. Trajectory formation in arm movements: minimization principles and procedures. In *Advances in Motor Learning and Control*, H. N. Zelaznik, Ed. (Champaign, IL: Human Kinetics).

Klein, C. A., Huang, C. H., 1983. Review of pseudo inverse control for use with kinematically redundant manipulators. *IEEE Transactions on Systems, Man, and Cybernetics* 13, 245–250.

Monnier, G., Renard, F., Chameroy, A., Wang, X., Trasbot, J., 2006. A motion simulation approach integrated into a design engineering process. SAE Technical Paper 2006-01-2359. Society of Automotive Engineers, Warrendale, PA.

Nubar, Y., Contini, R., 1961. A minimal principle in biomechanics. *Bulletin of Mathematical Biophysics* 23, 377–391.

Park, W., Chaffin, D. B., Martin, B. J., 2000. Development of an angle-time-based dynamic motion modification method. SAE Technical Paper 2000-01-2156. Society of Automotive Engineers, Warrendale, PA.

Park, W., Chaffin, D. B., Martin, B. J., 2004. Toward memory-based human motion simulation: development and validation of a motion modification algorithm. *IEEE Transactions on Systems, Man, and Cybernetics, Part A: Systems and Humans* 34, 376–386.

Park, W., Martin, B. J., Choe, S., Reed, M. P., Chaffin, D. B., 2005a. Representing and identifying alternative movement techniques for goal-directed manual tasks. *Journal of Biomechanics* 38, 519–527.

Park, W., Chaffin, D. B., Martin, B. J., Faraway, J. J., 2005b. A computer algorithm for representing spatial-temporal structure of human motion and a motion generalization method. *Journal of Biomechanics* 38, 2321–2329.

Park, W., Singh, D., Martin, B. J., 2006. A memory-based model for planning target reach postures in the presence of obstructions. *Ergonomics* 49, 1565–1580.

Schmidt, R. A., Lee, T. D., 1999. *Motor Control and Learning: A Behavioral Emphasis* (Champaign, IL: Human Kinetics).

Uno, Y., Kawato, M., Suzuki, R., 1989. Formation and control of optimal trajectory in human multijoint arm movement—minimum torque-change model. *Biological Cybernetics* 61, 89–101.

Wang, X., Chevalot, N., Monnier, G., Trasbot, J., 2006. From motion capture to motion simulation: an in-vehicle reach motion database for car design. SAE Technical Paper 2006-01-2362. Society of Automotive Engineers, Warrendale, PA.

Zhang, X., 2002. Deformation of angle profiles in forward kinematics for nullifying end-point offset while preserving movement properties. *Journal of Biomechanical Engineering* 124, 490–495.

Zhang, X., Chaffin, D. B., 2000. A three-dimensional dynamic posture prediction model for in-vehicle seated reaching movements: development and validation. *Ergonomics* 43, 1314–1330.

Zhang, X., Kuo, A. D., Chaffin, D. B., 1998. Optimization-based differential kinematic modeling exhibits a velocity-control strategy for dynamic posture determination in seated reaching movements. *Journal of Biomechanics* 31, 1035–1042.

Zhang, X., Nussbaum, M. A., Chaffin, D. B., 2000. Back lift versus leg lift: an index and visualization of dynamic lifting strategies. *Journal of Biomechanics* 33, 777–782.

10

Computational Approaches in Digital Human Modeling

Tim Marler, Jasbir
Arora, Steve Beck, Jia Lu,
Aninth Mathai, Amos
Patrick, and Colby Swan

10.1 Introduction

The human body is a multiscale machine composed of subsystems that range from DNA to relatively large joints like the hip. Studying and understanding this kind of machine requires a variety of disciplines, and for this reason, the field of digital human modeling (DHM) has the potential to draw on an enormous breadth of computational expertise and foster collaboration. However, DHM has not attracted the wide variety of expertise that it could, possibly because many researchers may not realize that their work is applicable to DHM. Consequently, the purpose of this chapter is to demonstrate the expanse of computational fields that can be leveraged when modeling humans and to highlight key themes that should be considered as the DHM field grows. This is not a survey, and not every possible approach is discussed. Rather, a sample of various approaches is presented in the context of ongoing work, primarily at the University of Iowa's Virtual Soldier Research (VSR) lab.

10.1.1 Key Concepts for Computational DHM

In general, the idea of developing computational approaches in DHM entails developing models of the human or specific aspects of the human, and the purpose of such models is simply to replace the artifact or process being modeled. Using a model is cheaper, faster, and more insightful. Models of any sort should be versatile enough such that they can be modified and re-run to present different cases and scenarios. Of course, it can be helpful to analyze a single occurrence with a single set of parameters. However, the more versatile a model is, and the more scenarios and conditions a single model can represent, the better. Given this preferred versatility, which is not necessarily available with most current DHM tools, computational models of a human should be able to predict an outcome based on a set of input parameters that may change. In this way, models can be used to formulate and test hypotheses; they can be used to answer questions. Based on the fundamental deductive scientific method, a hypothesis is formed in order to answer questions; a model is developed based on the hypothesis and associated assumptions; and the model is tested and refined. At the end of this process, a question embodied in the hypothesis has been answered. These steps should always be considered when developing a computational approach for DHM. Although a human model can provide a tool for evaluating products, using a model that cannot be varied in order to test hypotheses concerning the human itself defeats a substantial purpose of actually developing models in the first place.

When creating a computational model, the question arises as to what exactly is being modeled, and this question depends primarily on the hypothesis being tested. Essentially, any model, short of representing the complete universe, has boundary conditions or constraints, and such constraints must be defined clearly, either explicitly or implicitly. Humans constantly interact with and depend on the environment around them, and one must determine *a priori* to what extent such interaction will be considered. In addition, rarely does anyone model every aspect of the human, so one must also consider which components of the human should be modeled in order to answer the pertinent questions or conduct the relevant studies. In short, any approach to human modeling must allow for variable boundary conditions and constraints.

After the questions or problems have been identified, and after the appropriate limits of the model have been determined, one must address the issue of appropriate model fidelity or complexity. Fidelity is closely tied to the hypothesis and the boundary conditions. With regards to computational models, there is typically a trade-off between fidelity and computational costs. An absolutely complete and high-fidelity human model is not yet feasible, so fidelity and accuracy generally cater to the application. Traditionally, most digital humans have been used to study gross motion and to model coarse interaction with various products (automobiles, furniture, etc.). In fact, the discussion in this chapter assumes this is the predominant application, although broader potential applications are recognized. Given this application, fidelity has been measured primarily with respect to the number of degrees of freedom (DOF) used to represent the skeleton.

In general, the pursuit of a human model that involves every muscle, every bone, and every organ, all coordinated seamlessly and operating in real time with no computational restrictions, has resulted in multiscale human models (systems of systems). Although the ideal human model is not yet feasible, results from a large-scale model can provide input to more refined models representing specific joints, skin, internal organs, clothing, external items in the environment with which a human must interact, and so on. Each individual model has varying degrees of fidelity and computational speed.

As with any set of multiscale models, connectivity is critical. As separate human-modeling efforts mature, there must always be the potential for subsystems, models, and/or tools to link with and communicate with each other. If the field of DHM is to grow as efficiently as possible, with the ultimate goal of modeling the complete human, then there must be connectivity between components. Of course, one can argue that boundary conditions may be applied in such a way that the whole body is not needed to answer every question, and certainly limited models have served and will continue to serve with significance. However, the actual human body is not composed of independent systems, so to impose such independence is to introduce inaccuracy.

Closely related to the idea of fidelity or model complexity is realism. In the context of this chapter, *realism* refers to the appearance of the human model and to the visual results of a simulation. Two key points should be considered with respect to realism. First, the overall appearance of any human model is critical, not just as a matter of esthetics but as a matter of interactivity. Certainly, visual appearance is important for demonstrations and basic simulations, but it can also play a more substantial role. A significant application for whole-body human models is training simulation, where a real human ultimately interacts with a virtual human. Consequently, the virtual human must look as real as possible. Otherwise, the interactivity is incomplete. Secondly, simulated results must look realistic with regard to posture and motion. Models must be validated and must have a certain level of quantifiable accuracy. In addition, one must be able to see cause and effect with variations simulations. When parameters in a human model or in a modeling scenario are changed, one should be able to see realistic changes in simulated motion or posture. Obtaining accurate numerical results is not enough.

The most significant capabilities for whole-body human models involve predicting posture or motion for multiple joints concurrently. Most other human-modeling capabilities either stem from or feed into these capabilities. Ideally, simulating posture and motion should operate in real time, given the reasons for developing a model and the presumed desire to see cause and effect. Consequently, all other capabilities should eventually function in real time as well.

10.1.2 Overview of the Chapter

The remainder of this chapter presents a series of examples of computational approaches in DHM. Methods involving the following topics are discussed: model visualization, posture prediction, dynamic motion prediction, clothing, physiology, and muscles. Because predicting posture and motion are especially important, more time is spent with these topics. With each topic, various fields or areas of expertise that are leveraged are summarized, thus demonstrating how different fields can be applied to DHM. Then, a brief explanation of the associated computational model and/or method is provided. The intent is simply to give an overview of the different approaches, while technical details are provided in many of the other chapters. Finally, the concepts discussed above are highlighted in the context of the specific topics.

10.2 Modeling the Human Form

Before developing methods for simulating posture, motion, or other aspects of human operations, it is necessary to develop a method for visualizing the human and the output from human simulations. This visual aspect of human modeling draws heavily on the field of computer graphics, which in turn includes computer science, solid modeling, and advances in the movie and 3D gaming industries. Here, we briefly present some of the aspects involved in visualizing a human model, based on the work of Yang et al. (2005) and Abdel-Malek et al. (2006). Then, we present an example of a whole-body skeletal model, which provides the basis for further discussions of computational methods.

10.2.1 Visual Model

An avatar's movement is computed for joints in real time and is, therefore, not a result of traditional three-dimensional (3D) animation techniques like key-framing, inverse kinematics (IK), or the use of

FIGURE 10.1 Wire frame view of a virtual human. (From Abdel-Malek, K., et al., *International Journal of Human Factors Modelling and Simulation*, 1, 2–39, 2006. With permission.)

IK constraints. However, traditional 3D modeling, texturing, and animation rigging techniques are used to prepare the avatar for deployment. One can think of the actual 3D model as the "skin" of the avatar. This skin is comparable to an infinitely thin but hollow shell that defines the avatar's shape (fig. 10.1).

Once the shape of the avatar is created, shaders (a compilation of effects that dictate how a 3D surface responds to light) and textures (2D images that are projected onto or wrapped around 3D surfaces) are used to provide the avatar's shape with the visual queues necessary to create the illusion of human skin (fig. 10.2).

By themselves, 3D models, shaders, and textures can provide convincing renderings that suggest human form, but this form cannot move until it is bound to a hierarchical joint structure. This hierarchical joint structure, referred to as an IK skeleton, is a series of interdependent local coordinate systems strategically positioned at locations within the 3D model to suggest shoulders, elbows, knees, and so on. The skeleton is shown in the context of the complete model, as in Figure 10.2. The avatar's skin is bound to the IK skeleton, allowing the avatar to move in a human way.

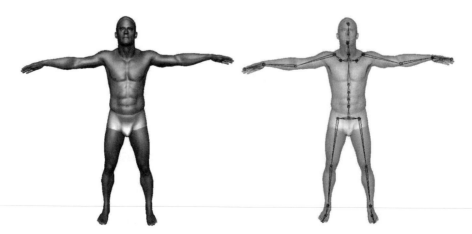

FIGURE 10.2 Textured view of a virtual human with underlying skeletal model. (From Abdel-Malek, K., et al., *International Journal of Human Factors Modelling and Simulation*, 1, 2–39, 2006. With permission.)

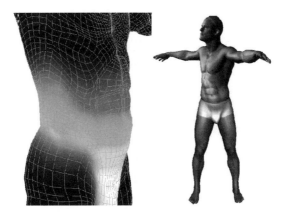

FIGURE 10.3　Skin with adjusted weight. (From Abdel-Malek, K., et al., *International Journal of Human Factors Modelling and Simulation*, 1, 2–39, 2006. With permission.)

Computational methods are used to determine how exactly the joints move. To visually simulate the elasticity of human skin as the joints are exercised, the amount of movement in the area of skin around a particular point must be defined when that point in the skin moves during joint rotation. This is done with a traditional animation technique called *skin weighting*, which addresses the aesthetic issue that would otherwise cause a 3D model to tear or break at the joints when rotated. Typically, this technique is accomplished subjectively through interactive tools—much like using a can of spray paint—that allow 8-bit gray-level values to represent how specific regions of the skin are anchored to specific joints. The higher the gray-level value, the greater the effect a given anchor has on that region. Figure 10.3 shows the skin weight associated with an anchor below the first spine joint. Note the high gray-level values around the groin area, which cause the geometry in that skin region to be completely immoveable. Alternatively, the decreasing gray-level values above the groin area allow the skin increasing elasticity the further it is from the groin region.

Successful weighting of the skin over the IK skeleton using this technique provides the illusion that the 3D mesh that defines the avatar's shape has skin-like properties when the joints are exercised.

10.2.2　Computational Model

Given the visual model discussed above, a computational model is used when simulating posture or motion. Figure 10.4 illustrates one such skeletal model, where each cylinder represents a DOF. Computational models can then be used to calculate movement in the DOFs. Note that the two components of the human model, the skin and the skeleton, can be altered. In fact, all of the currently available human modeling tools have slightly different skeletal structures with different external appearances.

10.2.3　Key Concepts

When creating the outer skin for a digital human, the primary concern is realism. In fact, modeling skin that moves in a realistic fashion can be just as important as skeletal motion, depending on the application. In addition to realism, the idea of fidelity also pertains to the skin. The skinning process discussed above concerns visualization only, but skin can constitute a computational model that calculates deformation resulting from applied loads. In fact, skin deformation is a component of the broader topic of human surface deformation that occurs when one sits down for instance. Including such high fidelity requires not only deformation in the skin but also in muscle and fat tissue, possibly even in internal organs.

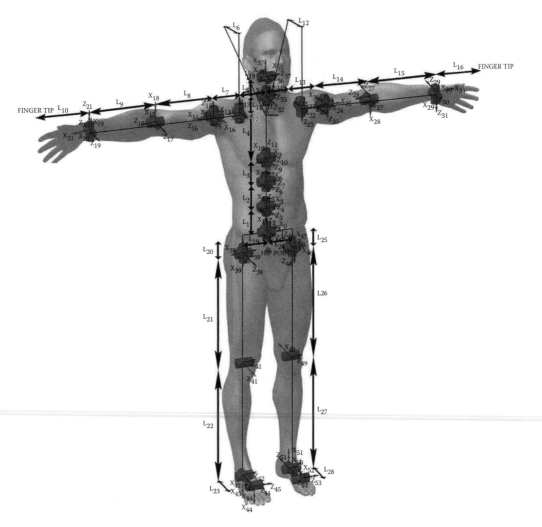

FIGURE 10.4 Computational skeletal model.

Determining an appropriate level of fidelity is also critical when building a human skeleton. Not every DOF for the human body is considered, especially with respect to the spine and neck. However, the primary applications for current digital human models tend to involve gross whole-body motion. For such purposes, a complete spine (24 vertebrae with 72 DOF) is not necessary and would increase computational demands. It is a matter of whether one is interested in modeling the spine itself or just modeling how the spine effects the overall motion of the body. In addition to the spine, the shoulder is another area with varying degrees of fidelity. Technically, a human shoulder model can have as many as 14 DOFs (Lu et al., 2005), but this level of fidelity is not necessary for realistically simulating overall human posture or motion.

10.3 Predicting Posture

Given a visual and skeletal structure, this section summarizes a variety of computational approaches for predicting human posture and focuses on a relatively recent optimization-based approach from

VSR. These approaches can involve the fields of motion-capture, inverse kinematics, robotics, and optimization.

10.3.1 Methods

A common approach to posture prediction actually draws from reaching-motion prediction. It involves using prerecorded motion (from motion capture systems), anthropometric data, and functional regression models in terms of joint angles (Faraway, 1997; Zhang & Chaffin, 1997; Chaffin et al., 2000; Chaffin, 2002). This provides analytical expressions for joint-angles as a function of time, from which individual postures are extracted. This approach is not always accurate with respect to constraints in Cartesian space (Chaffin, 2002). Faraway et al. (1999) provide a modification whereby a posture is selected that is as close as possible to the initial data-based prediction, while satisfying restrictions in Cartesian space modeled with traditional inverse kinematics. The translation between joint space and Cartesian space is completed using traditional inverse kinematics (using a pseudo-inverse of the Jacobian matrix).

The pseudo-inverse approach is one of the most common methods to inverse kinematics (Liegeois, 1977; Klein & Huang, 1983; Jung et al., 1995). Essentially, a set of joint angles representing a posture is determined iteratively using the pseudo-inverse of the Jacobian matrix, which represents the derivatives of the *end-effector* position with respect to the joint angles. An end-effector is any point of interest on a human or robot but traditionally represents the end or tip of a kinematic chain. During each iteration, an optimization algorithm is run to minimize the deviation of the resolved posture from a predetermined reference posture. Eventually, the algorithm converges on a final posture (set of joint angles). Zhang et al. (1998) incorporate optimization in a weighted pseudo-inverse approach whereby the weights determined such that the predicted motion approximates empirical data (prerecorded motion). Reed et al. (2000) also combine the use of optimization and experimental data by using an optimization prediction model with three DOFs to find a posture that approximates the data most accurately.

Zhao and Badler (1994) provide one of the earliest approaches that directly incorporate optimization for posture prediction. A gradient-based unconstrained optimization routine is used to minimize an objective function formed by the weighted sum of components that model various factors, such as the position of the end-effector (a specified point, line, or plane) or the orientation of the hands. Limits on the joint angles are incorporated as constraints. The work is demonstrated using a 22-DOF full-body virtual human. Riffard and Chedmail (1996) use a similar approach and determine the optimum placement of the torso and the optimum posture of a seven-DOF arm, using simulated annealing, which is a global optimization method. Equations for target contact, collision avoidance, vision, body orientation, and torque are combined in a weighted sum to form the objective function. Aside from limits on the design variables, these approaches do not involve constraints in the optimization formulation. What constitute constraints conceptually are included in the objective function and minimized along with other criteria.

Researchers at VSR have developed an optimization-based approach to posture prediction using constrained optimization with the model shown in Figure 10.7 (Mi et al., 2002; Farrell et al., 2005; Yang et al., 2006). The design variables for this problem are joint angles, measured in units of radians. q_i is a joint angle and represents the rotation of a single revolute joint with respect to a local coordinate system. There is one joint angle for each DOF. $\mathbf{q} = [q_1, \ldots, q_n]^T \in R^n$ is the vector of joint angles in an n-DOF model and represents a specific posture. $\mathbf{x(q)} \in R^3$ is the position vector in Cartesian space that describes the location of the end-effector as a function of the joint angles, with respect to the global coordinate system. For a given set of joint angles q, $\mathbf{x(q)}$ is determined using the Denavit–Hartenberg (DH) method (Denavit & Hartenberg, 1955). This method essentially allows one to work with either joint space or Cartesian space.

With this approach, the constraints are modeled independently. The first fundamental constraint, called the *distance* constraint, requires the end-effector to contact a target point $\mathbf{x}^{target\ point}$. In addition, each joint angle is constrained to lie within predetermined limits. q_i^U represents the upper limit for q_i, and q_i^L represents the lower limit. These limits are derived from anthropometric data. Additional constraints can be used depending on the task being modeled.

The basic benchmark performance measure represents joint displacement (Jung et al., 1994; Mi et al., 2002). This performance measure is proportional to the deviation from the *neutral position*, which is selected as a relatively comfortable posture, typically a standing position with arms at one's sides. q_i^N is the neutral position of a joint. Because some joints articulate more readily than others, a weight w_i is introduced to stress the relative stiffness of a joint. The final joint displacement is given as follows:

$$f_{JointDisplacement}(\mathbf{q}) = \sum_{i=1}^{n} w_i \left(q_i - q_i^N\right)^2 \tag{10.1}$$

Additional performance measures can also be used, such as discomfort (Marler et al., 2005a), visual displacement (Marler et al., 2006), and potential energy (Marler, 2005). In addition, multiple performance measures can be combined, representing the idea that more than one physically significant quantity may drive human posture (Yang et al., 2004; Marler et al., 2005b).

The optimum posture for the system shown in Figure 10.2 is then determined by solving the following problem using any number of algorithms for constrained optimization:

Find: $\mathbf{q} \in R^{DOF}$

to minimize: Performance Measure($\mathbf{q}$)

subject to: $distance = \left\| \mathbf{x}(\mathbf{q})^{end\text{-}effector} - \mathbf{x}^{target\,point} \right\| \leq \varepsilon$ (10.2)

$q_i^L \leq q_i \leq q_i^U; \; i = 1, 2, K, DOF$

where ε is a small positive number that approximates zero. Some results from this approach are shown in Figure 10.5 with a 35-DOF upper-body model. Legs are not considered computationally, although they are shown visually.

The formulation in Equation 10.2 is based on kinematics only. However, Horn (2005) extends this approach to include static equilibrium and to include global DOFs (position and orientation of the hip point). The consequent conceptual formulation is given as follows:

FIGURE 10.5 Kinematic posture prediction. (From Farrell, K., R. T. Marler, and K. Abdel-Malek, *SAE 2005 Transactions Journal of Passenger Cars—Mechanical Systems*, 114(6), 2891, 2005. With permission.)

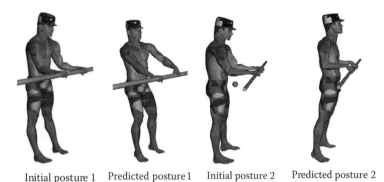

Initial posture 1   Predicted posture 1   Initial posture 2   Predicted posture 2

FIGURE 10.6   Posture prediction with static equilibrium.

Find: joint angles, global DOFs, and reaction forces at eth hands and feet
to minimize: sum of torque-squared
subject to:

      (a) joint-angle limits
      (b) reaction-force limits
      (c) hand and foot end-effector locations   (10.3)
      (d) hand and foot orientations
      (e) hip location
      (f) static equilibrium for global DOFs at the hip point

Results of this work are shown in Figure 10.6, with a 55-DOF full-body model (including six global DOFs for the hip point), where the sphere indicates a target point for the hip. With Posture 1, the hip of the avatar begins at the target point with an initial posture specified by the user.

10.3.2 Key Concepts

All of the approaches to posture prediction have their pros and cons. One convenient advantage to the approach formulated in Equation 10.2 is the ability to test hypotheses easily, and as we will see, this advantage carries over to motion prediction with a similar approach. Different types of postures associated with different tasks are driven by different performance measures, and by testing different performance measures, one is able to study what governs human posture. In addition, with the optimization formulation, it is relatively easy to represent boundary conditions for the model by including various types of constraints that are necessarily satisfied. This approach to posture prediction is computationally fast, and operates in real time. Finally, it is possible to alter model parameters (constraint limits, performance measures, constraint characteristics representing different physical entities, etc.) and see the consequences in the simulation (the predicted posture).

10.4 Predicting Dynamic Motion

The approaches discussed for posture prediction can be extended to incorporate changes in time and thus dynamic quantities. In this section, computational methods for predicting human motion while considering applied and reaction loads are summarized. The development of methods for human motion prediction leverages advances in the fields of robotics, controls, and of course dynamics.

10.4.1 Methods

Much of the initial work with motion prediction (both kinematic and dynamic) is based on robotic controls with simple three-DOF robots (Kahn & Roth, 1971; Shin & Mckay, 1986). These methods determine what torques need to be applied to joints in order to yield a specified motion (i.e., have the end-effector follow a specified parameterized path) while minimizing operation time and/or energy loss. Typically, some type of *path* descriptor is provided for the end-effector, where a path refers to Cartesian position-history for a single point. Then, the first stage of the problem involves *trajectory planning*, where the term trajectory refers to a time history, and this stage entails determining the time histories for the joint angles and angular velocities of each robotic joint. Then, *trajectory tracking* involves determining actuation that results in the motion determined in the trajectory planning stage.

One of the earliest works with motion prediction, outside the field of classical controls, is provided by Lin et al. (1983). With this approach, first, a series of positions and corresponding orientations are specified for an end-effector. Each set of specifications is transformed to a set of joint angles, and this yields a series of points in joint-angle space that represents a discretized path. Time histories for the joint angles in a seven-DOF robot are represented with splined cubic polynomials, and each polynomial is made to fit the series of joint-angle values determined above. Thus, the joint angles are essentially constrained by the above-mentioned specified points. The polynomials are written in terms of angular acceleration and in terms of time intervals. The angular acceleration of a joint is determined by solving a system of linear equations based on stipulations for continuity in the time histories of the joint angular velocity and acceleration. Note that this process of solving a system of equations can be cumbersome for larger models with a relatively high number of DOFs. The time intervals provide the design variables for an optimization problem. The sum of the time intervals for each spline segment is minimized, thus maximizing the speed of operation for the robot. Limits (constraints) are placed on the angular velocity, acceleration, and jerk for each joint. Saramago and Steffen (1998) use the same approach and apply it to a three-DOF and a six-DOF robotic arm. However, in this case mechanical and potential energy are determined, and generalized forces are determined based on Lagrangian mechanics. In addition, mechanical energy is also incorporated in the objective function.

Drawing on the work of Gilbert and Johnson (1985), Chen (1991) outlines the first use of B-splines for general robotic motion prediction and includes obstacle avoidance capabilities. Compared to using splined cubic polynomial, one of the advantages of using B-splines is inherent continuity. A general objective function is presented in terms of time, joint angles, angular velocity, and a general vector of control parameters, such as applied joint torques. The joint angles are represented by B-splines, and the B-spline coefficients provide the design variables for an optimization problem. The approach is tested on basic two- and three-DOF robots. Wang et al. (1999) use a similar approach with a six-DOF robot and incorporate a recursive formulation for inverse dynamics (including derivatives). In both cases, equations of motion are modeled as optimization constraints.

Saramago and Steffen (2000) expand on this approach. They also use the DH method for kinematic analysis. Only the initial and final values for the joint angles are provided as input; no intermediate points are stipulated. Thus, no path is specified for the end-effector. The objective function is a function of the total traveling time for the robot and of generalized forces at the joints. The generalized forces are functions of angular velocity and angular acceleration, which can be written as functions of the B-spline coefficients. These coefficients provide design variables for the optimization problem along with the total traveling time for the robot. The joint angles are determined from the B-spline coefficients using a system of linear equations that incorporates user-specified initial and final angular velocities. The coefficients are then used to determine angular velocities and accelerations, which are used in the objective function and in the constraints. Saramago and Ceccarelli (2002, 2004) also use B-splines in this way, and consider externally applied loads. Again, the method is tested on a six-DOF robot.

Alternatively, Zhang et al. (1998) essentially solve a system identification problem to determine the time histories for the joint angles in a seven-DOF robot. Parameters in an approximating function for a joint angle are determined by minimizing the error between the approximating function and experimental data.

Much optimization-based research has been conducted in context of modeling biped robotic walking, which requires one to pay special attention to modeling stability (Chevallereau & Aousin, 2001; Saidouni & Bessonnet, 2003). Mu and Wu (2003), however, develop an approach that does not involve optimization. A system of equations is developed for a biped robot, based on constraints that characterize gait in the sagittal plane. This system is solved explicitly for foot and hip trajectories with a three-DOF robot and does not consider dynamics.

Lo et al. (2002) present one of the first methods for applying optimization to dynamic human motion prediction. With this work, the authors also provide one of the first examples of dynamic human motion prediction. Where dynamics had long been considered with robotics, it had not yet been included with motion prediction in the context of DHM. Joint angles, angular velocity, and angular acceleration are written in terms of B-splines, and the B-spline coefficients provide the design variables. The objective function is the integral of the Euclidean norm of joint torque over time. Torque is determined based on equality constraints that model the equations of motion. The equations of motion are determined using recursive dynamics based on a Newton-Euler formulation, which is written in terms of Cartesian space. As with previous work, the initial and final values for the joint angles (initial and final postures) are provided as constraints. In addition, joint angles and joint torques are subject to limits. This approach is applied to a human model where the number of articulated DOFs is adjusted depending on the task being completed, with a maximum of 12 DOFs. The authors consider the following basic tasks: arm lifting, leg lifting, whole-body weight lifting, chin-ups, dip-downs, and pushups. Note that although Lo et al. (2002) describe a formulation for a standard constrained optimization problem, they present their work as an optimal control technique. The fields of controls and optimization are interrelated. Technically, an optimal control problem is a problem in which the optimization design variables are continuous (i.e., a curve over time), although they may be discretized for computational implementation.

Mi (2004) and Abdel-Malek et al. (2004) focus on the kinematics of human motion with a 15-DOF upper-body human model. They use B-splines and the DH method to relate Cartesian space to joint-angle space. The objective function and the constraints are functions of joint angles, which are directly related to the B-spline coefficients, which in turn provide the optimization design variables. The objective function is similar conceptually to Equation 10.1. The end-effector is constrained to remain on a parametric path that is determined by minimizing the jerk of the end-effector during the motion (Flash & Hogan, 1985). Note that although using minimum jerk to determine the path for a human end-effector is common, Alexander (1997) uses metabolic energy for path (trajectory) planning and suggests that most currently available models for human paths, including the minimum jerk model, have significant deficiencies. Because the work of Mi (2004) and Abdel-Malek et al. (2004) concerns kinematic human motion, additional mathematical terms are included in the objective function to refine the motion, making it more realistic. These terms are necessary to ensure consistency in the joint angle profiles, by reducing the amount of reversal in the sign of the angular velocity. In addition, they enforce smoothness by reducing the magnitude of the angular acceleration. Finally, they reduce the magnitudes of the initial and final angular velocities.

Kim et al. (2006) extend the above-mentioned kinematic model and include equations of motion. Whereas Lo et al. (2002) incorporate the equations of motion as explicit constraints, Kim et al. (2006) enforce them implicitly to calculate torque as part of the objective function, which is an extensive model of metabolic-energy usage. The authors also extend the equations of motion to include joint stiffness (as torsional springs). Initial and final postures are not modeled as constraints but are predicted implicitly as part of the B-spline joint-angle profiles. The approach is used with a 21-DOF upper-body human model.

The work of Kim et al. (2006), Xiang et al. (2007), and Chung et al. (2007) provide the most recent computational approach to human motion prediction. The design variables are again control points for B-splines. The objective function is the sum of the torque-squared, similar to that which is used by Lo et al. (2002). The constraints, however, are modified. It is found that the path constraints discussed above can be overly restrictive, and ideally an elegant motion-prediction method should provide such a path implicitly. Consequently, these constraints are omitted. The equations of motion are determined using a recursive

Lagrangian formulation, which is written in terms of joint space (not Cartesian space). This formulation is especially convenient when using the DH notation. As with Kim et al. (2006), although the equations of motion are not formulated as explicit constraints in the optimization formulation, they are satisfied, and torque is calculated accordingly. This approach is applied to a 55-DOF whole-body human model, and the final conceptual formulation for modeling walking is given as follows (Xiang et al., 2007):

Find: B-spline control points
to minimize: sum of torque-squared
subject to:

      a) joint-angle limits
      b) approximate torque limits
      c) feet cannot penetrate the ground
      d) soft impact (zero velocity for feet at impact)   (10.4)
      e) stability: zero moment point (ZMP) should remain within the foot
         supporting region (FSR)
      f) specified step length
      g) constant forward pelvic velocity
      h) knee-flexion angle at mid-swing
      i) symmetry between legs
      j) coupling between leg and arm motion

Note that the constraint regarding the knee is one of six commonly accepted determinants used to quantify gait (Ayyappa, 1997), and the specific range of values for the knee angle is obtained from the literature. Results from Equation 10.4 are shown in Figure 10.7, with output for the ground reaction force (GRF) and for knee torque when different loads are applied.

10.4.2 Key Concepts

With respect to the variety of fields that can contribute to human modeling, the area of predictive dynamic motion is an excellent example of cross-pollination. There is a long history of work with modeling robotic motion, initially with no consideration of DHM. However, clearly, progress made with robotics has been applicable to human models.

The computational methods discussed in this section for predicting motion are able to represent cause-and-effect visually, not just numerically. For instance, the simulated gait in Figure 10.10 changes if the weight of the backpack is altered.

The various methods for predicting human motion also provide excellent examples of the significance of model constraints. With earlier approaches, it was necessary to constrain end-effectors to a specified path, which was predicted separate from the overall motion of the body. Later approaches were able to incorporate this behavior (of the end-effector) implicitly. The question of whether or not to constrain the path of the end-effector sheds light on the idea of *forced constraints*. Forced constraints are constraints that are imposed to enforce preconceived results rather than to represent physically significant quantities or boundary conditions that then yield a realistic simulation. *Physical constraints* are used to represent the latter. Producing realistic motion is relatively easy if artificial constraints are overused. However, the intent is not just to reproduce realistic motion, but to model the human and what drives realistic motion, and there is a subtle difference between these two goals. When conducting research to develop human models, artificial constraints are often necessary, but ideally they are eventually removed. As an example, in the constraint that stipulates that feet con not penetrate the ground represents a boundary condition. Given the purpose of the human model, it is not necessary to model the interaction between a foot and the ground surface, so this constraint is imposed. The stability constraint, however, represents a preconceived idea of appropriate walking motion.

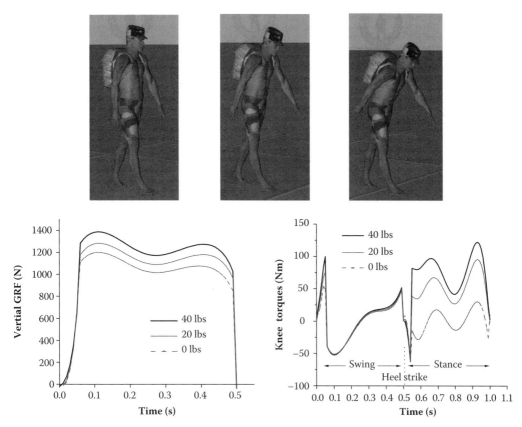

FIGURE 10.7 Dynamic motion prediction for walking. (From Xiang, Y., et al., "Optimization-Based Dynamic Human Walking Prediction," *SAE Human Modeling for Design and Engineering Conference*, SAE International, Warrendale, PA, 2007. With permission.)

In general, two types of physical constraints must be used. *General constraints*, such as the joint-angle limits, are used to model the human, and they apply to any task. *Task constraints*, such as foot-ground penetration and soft impact, are used to model a specific task. Task constraints change depending on what type of task is being modeled or studied, and every task typically entails different challenges. For instance, modeling running presents the difficulty of modeling the impact of the foot hitting the ground, which is not a significant issue when modeling walking. Modeling throwing presents substantial redundancy; there are many significantly different ways to throw an object.

While the constraints allow one to model various scenarios and various aspects of the human, the performance measures (objective functions in the optimization problem) allow one to study what drives or governs human motion or behavior. To date, most performance measures have involved some form of energy or joint torque. However, many different measures or combinations of measures can be considered, as the area of motion prediction grows.

10.5 Modeling Clothing

Clothing can substantially affect predicted posture or motion, especially when heavy suits are worn such as space suits or protective garb. Rahmatalla et al. (2006) show experimentally that the stability of human motion can be significantly affected by clothing restrictions, which eventually causes wearers to change their strategies for accomplishing physical tasks. This effect can be modeled on two levels. Either one can try to model the effects of the clothing by incorporating modified joint limits in the simulation, or one can model clothing directly. The difficulty with the former approach is determining reasonable values for such limits, especially when considering flexible materials. The latter approach, of course, represents a higher fidelity model and is discussed in this section. Modeling clothes directly involves the fields of computational mechanics, finite element analysis, and materials.

Most frameworks for modeling clothing require three interacting components: (1) fabric mechanics/dynamics modeling, (2) collision detection and contact computation, and (3) digital human modeling. The first component essentially breaks down into two alternative frameworks: (1) particle-based methods in which the fabric is directly discretized into a system of springs and masses, and (2) surface-based methods in which the fabric is treated as an elastic continuum. The second component constitutes a large area with a variety of potential methods and is beyond the scope of this chapter. Finally, the third component is provided by predicted motion discussed above. In this section, computational approaches for modeling the first component (the mechanics/dynamics of cloth) are reviewed.

10.5.1 Particle-Based Methods

Particle-based methods treat fabrics as a discrete dynamic system composed of mass points or particles that interact through a system of interconnected membrane and bending springs. The coupled equations of motion for all of the "particles" are integrated in time using explicit and implicit algorithms. Simple as they are, such particle methods can generate visually realistic clothing animations and thus have been widely applied in computer graphics and movie-making.

A representative work in particle methods is the mass-spring cloth model proposed by Provot (1995) wherein the fabric is modeled as an array of mass particles interconnected by linear springs of three different types: structural, shear, and flexion (fig. 10.8), which characterize the in-plane stretching, in-plane shear, and out-of-plane bending behaviors, respectively. Structural springs connect a particle with its direct neighbors along two perpendicular axes, which are usually aligned with warp and weft yarn fabric directions, while shear springs connect a particle with its neighbors in the diagonal directions. Flexion springs also act along the two perpendicular yarn axes but each connects every other particle. Cloth drape is solved by an explicit time integration of the system. Since the step size of the integrator is inversely related to the spring stiffnesses, compliant springs were used, which resulted in some unrealistic overstretching of the fabric model. To address this issue, Provot (1995) proposed a heuristic method to adjust particle positions to account for overstretched springs. An extension of the mass-spring model was proposed by Choi and Ko (2002), who considered fabric buckling and included it in the formulation of the bending springs. In their model, a bending spring is treated as a buckling column with both ends pinned, and a nonlinear force-compression relation was derived.

Breen et al. (1994a and b) incorporated experimental fabric mechanics results into their particle modeling framework that accounts for four different types of mechanical interactions between particles: (1) repulsion, (2) stretching, (3) bending, and (4) in-plane shear. Each type of particle interaction is governed using an independent nonlinear energy function. Draping configurations of fabrics were computed by minimizing the total potential energy of the whole system. The specific energy functions

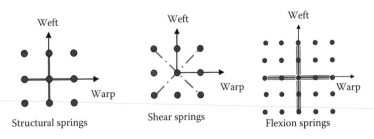

FIGURE 10.8   Mass-spring cloth model.

of bending and in-plane shear were based on experimental data obtained from Kawabata Evaluation System for Fabrics (KES-F) (Kawabata, 1980). As the internal forces between particles are computed from the spatial derivatives of the energy functions, the model can be reformulated as a generalized mass-spring model with nonlinear internal forces. The particle-based models of Eberhardt et al. (1996) were similar to those of Breen et al. (1994a, 1994b) with the exception that the internal energy functions were differentiated symbolically, yielding Lagrangian equations of motion for each particle. The resulting differential equations were solved by a Runge-Kutta method with adaptive step-size control. The strain energy functions for bending and in-plane shear were based on experimental data, and fabric hysteresis was included by constructing piecewise linear approximation to experiment curves.

The particle-based modeling works cited above and numerous other similar works not referenced here are notable for being able to create visually realistic and wrinkled draping configurations of clothing and fabrics. Relatively simple as they are, and yet able to capture the complicated draping configurations of fabrics, particle methods yield helpful insights on issues of fabric modeling. Nevertheless, particle-based modeling of fabrics does have a major shortcoming, and this is that the fabric is directly discretized into a system of lumped masses and springs. In essence, the fabric is treated as a fishnet where the strings and their spacings do not correspond to the yarns that comprise the fabric. As the spatial fishnet discretization of a garment is refined, the masses and spring stiffnesses must be adjusted accordingly, and this is not necessarily a trivial matter when dealing with fabric patches of irregular shape and size. The vague physical meaning of spring stiffnesses in particle methods also makes it difficult to translate between fabric spring forces in such models and the actual stress level in the fabric being modeled. Ad-hoc assumptions can and have been made to answer these types of questions, but particle methods lack the rigor of continuum-based methods wherein the relation between forces in the fabric model and fabric stresses and strains is handled both rigorously and straightforwardly.

10.5.2 Continuum Surface-Based Methods

Unlike particle-based methods, surface-based methods take the local equilibrium of a continuum as the point of departure. Models are derived using well-established computational techniques, such as approximation of spatial derivatives using finite difference methods, or approximation of the solution space using linearly independent nodal basis functions as in the finite element method. Continuum surface-based methods are generally more rigorous than particle-based methods, both mathematically and mechanically, since the relation between the continuum properties and forces/displacements in the discretized model is based on exact spatial integration. While continuum surface-based methods have a more rigorous foundation than particle-based methods, the implementation of continuum surface-based methods is more involved. Furthermore, since continuum surface-based methods are continuously performing spatial integrations of stresses and strains, such methods are more computationally intensive than particle-based methods, which do not require any spatial integration at all.

While particle-based methods have been more prevalent in mathematical clothing modeling over the past two decades, a number of works have utilized continuum surface methods. To create animations of deformable bodies in computer graphics, Terzopoulos et al. (1987) started with the local form of the Lagrangian equation of motion for a hyperelastic medium, which was then approximated over a regular mesh using finite difference operators. A set of second-order ordinary differential equations was obtained and solved by implicit time integration. This framework was extended by Terzopoulos and Fleischer (1988) to allow for inelastic material behaviors. This framework was later extended by Carignan et al. (1992) for cloth and garment simulation on virtual humans.

Collier et al. (1991) showed that fabric drape can be predicted using nonlinear shell finite element models. The draping of a circular piece of cotton plain-weave fabric was modeled with four-node bilinear plate-shell elements and the results were compared with experimental drape results (Chu et al., 1950; Cusick, 1968). The fabric was modeled first with isotropic and then orthotropic linear elasticity by Collier et al. (1991), and it was discovered that orthotropic elasticity with very high shear compliance is more appropriate for plain-weave fabrics. Three input parameters were needed for the orthotropic model, the tensile moduli in the two yarn family directions, which were measured using KES-F system, and the Poisson's ratio, for which literature values were used. An interesting effect reported was that the computed shape of the draped fabric was sensitive to the Poisson's ratio. A number of similar studies were reported in the mid-1990s by Gan et al. (1995) and by Chen and Govindaraj (1995), who modeled the fabric with continuum-degenerated shell elements. The constitutive relationship used to represent the fabric was orthotropic linear elasticity in which the Young's moduli and the shear moduli were obtained by KES-F and the Poisson's ratio was determined from tensile tests. Chen and Govindaraj (1995) found that the elastic fabric moduli obtained in the low strain range of Kawabata tests gave realistic drape configurations.

Deviating from continuum-degenerated shell elements, Eischen et al. (1996) modeled fabric draping with the geometrically exact resultant shell theory described by Simo and Fox (1989) and Simo et al. (1989, 1990). An isotropic elastic material model with a nonlinear moment/curvature relationship derived from the KES-F system was used. Quasi-static simulations of fabric drape and handling were performed, and an arc-length controlled solution technique was utilized to treat instabilities due to fabric buckling. The contact between fabrics and rigid surfaces was considered, and the nonpenetration constraint was enforced by a penalty method.

Researchers at VSR have developed a clothing-modeling approach involving nonlinear shell theory based on continuum mechanics and dynamics. As noted previously, such a framework is both mathematically and mechanically rigorous and provides a clear relationship between fabric stress-strain relations and the forces/displacements in the discretized model. If the intent were simply to generate somewhat realistic videos of clothing draped on animated virtual humans, then a particle-based fabric modeling approach would be more appropriate in favor of its relative simplicity and computational efficiency. The collision and contact module enforces the impenetrability constraint between the clothing system and digital human models, and computes the mutual contact forces between the two. Putting the entire framework together, the clothing is first draped onto the human model's form. Then, as the human form goes through prescribed motions, the contact forces between the draped clothing and the model are computed. The contact forces between the wearer and the clothing are then integrated over space and time to quantify the resistance that the clothing exerts on the wearer during the specific activity considered.

Figure 10.9 shows the results of a simulation that predicts how pants deform as one steps over an obstacle, and Figure 10.10 indicates the consequent additional torque necessary to bend the knee with two different material thicknesses. Further details on this type of computation can be found in Man and Swan (2007).

10.5.3 Key Concepts

Computational models for clothing highlight a variety of issues that concern DHM in general. Compared to the skeletal model, the scale of the clothing model is much smaller. The behavior of clothing

FIGURE 10.9 Simulation of pants interacting with lower body striding and then stepping over an obstacle.

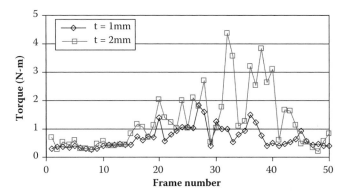

FIGURE 10.10 Computed resistance torques exerted by two pairs of cotton pants of different fabric thicknesses about the right knee.

fabrics on the human scale is derived from interactions of yarns and fibers on much smaller scales, which can be studied with actual fabric weaves and finite elements, as suggested by Figure 10.11 (Man, 2006). This demonstrates the necessity for viewing a human model as a multiscale model.

The idea of selecting an appropriate level of fidelity surfaces in the choice between continuum models, which provide a relatively high level of fidelity, and particle-based models, which sacrifice fidelity for computational efficiency. However, in either case, clothing models tend not to function in real time. Collision detection in particular can be computationally demanding.

Modeling clothing separately from the human body requires that the two models be integrated or linked. This connectivity is achieved by passing energy, joint torques, and joint-angle limits from the clothing model to the motion-prediction mode. The clothing model can provide the energy necessary to move clothing a specified distance, which can be added to an energy performance measure, such as the one developed by Kim et al. (2006). The joint torque necessary to deform the cloth, as shown in Figure 10.9, can be added to the equations of motion in a dynamic motion model. Finally, if collision detection methods are used not just to calculate interference between the cloth and the body, but also between various sections of cloth, then new joint-angle limits can be determined and used in posture or motion prediction.

10.6 Modeling Physiology

Whereas the clothing model provides input to the motion-prediction mode, physiological processes can be simulated using output from motion prediction. The energy performance measure used to predict motion can be used to calculate various physiological indices. This section provides an overview of these indices and how they relate to energy consumption. Many of the relationships used here are available

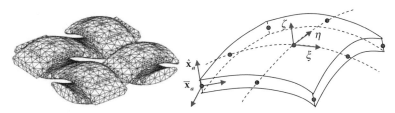

FIGURE 10.11 Finite element unit cell of plain-weave fabric on the millimeter scale and shell finite elements used to model clothing on the human scale.

in the literature, and this section demonstrates how various pre-existing mathematical models can be coupled and applied to DHM.

This section focuses on models for oxygen uptake and heart rate. However, one can also determine body temperature from energy consumption (Givoni & Goldman, 1972) and determine physiological strain index from heart rate (Moran et al., 1998). Throughout the description of the model, necessary assumptions are highlighted, given the potential complexity of physiological models.

10.6.1 Methods

A process has been developed at VSR whereby existing models for various physiological quantities have been linked and incorporated in a digital human (Mathai, 2005; Yang et al., 2005). The overall process is summarized as follows. As work is done, blood circulation increases to deliver more oxygen to the muscles, which in turn increases the ventilation rate to supply oxygen to the blood. However, the details of this process can be prohibitively complex when it comes to practical computational models, so two primary assumptions have been made. First, it is assumed that the energy consumed by the human is known, determined from motion prediction as described above. Second, physiological fatigue or exhaustion is not considered.

First, we discuss a method for calculating oxygen uptake. The net effect of work to the human body is the contraction of muscles, assuming of course that the energy expended by the body while doing work is available to us from any one of the sources mentioned. In order for the muscles to contract, they require energy. The energy is provided by the mitochondria of the cells in the form of adenosine tri-phosphate (ATP) molecules, and to produce the ATP, the cells require oxygen. Thus, the energy required to contract the muscles can be translated into a physiologically significant quantity based on the consumption of oxygen. This leads to the definition of total body oxygen uptake, or ventilatory oxygen uptake, which is defined as the amount of oxygen utilized from the inspired air to perform work (Froelicher & Myers, 2006). Energy can be converted into oxygen uptake (VO_2) equivalents. It has been found from experimental studies that 1 liter of oxygen corresponds to about 5 Kcal of energy liberation (Åstrand & Rodahl, 1970). It is assumed that this oxygen uptake provides the energy for the entire body, not just the energy required to perform a specific activity. The oxygen uptake includes not just the energy required by muscles, but also by all other physiological processes.

Oxygen uptake bears a linear relationship with the work rate, with moderate activity (Barstow & Mole, 1991), where moderate activity is activity below the lactic acidosis threshold (LAT), which is approximately 60% of the maximum heart rate. The relationship between oxygen uptake and the work rate becomes nonlinear above the LAT, and the models presented in this section are valid only below the LAT. Oxygen uptake is related to work rate as follows:

$$Vo_2^{req} = Vo_{2rest} + Vo_{2unloaded} + \alpha \cdot WR \qquad (10.5)$$

where Vo_2^{req} is the steady-state value of the oxygen uptake in L/min, Vo_{2rest} is the baseline value of oxygen uptake in L/min, $Vo_{2unloaded}$ is the value of oxygen uptake at the start of an activity in L/min, α is the oxygen cost of work in L min^{-1} W^{-1}, and WR is the work rate of the current activity in W. The Vo_{2rest} represents all the oxygen required by the body to function at rest; it represents the energy consumed by the body under normal circumstances (resting conditions). The $Vo_{2unloaded}$ is the oxygen uptake value of the human body while pedaling on a bicycle ergometer with no load, and this value is almost equal to the Vo_{2rest}. α is the slope of the regression line relationship between the oxygen uptake and the work rate.

When the body moves from rest to an increased work level, the muscles need more oxygen to maintain the new work level, and the body's VO_2 level increases. This new VO_2 is the Vo_2^{req} derived from Equation 10.5. However, this is the VO_2 at the steady state, and in reality the VO_2 does not jump to the steady state but follows an exponential path (Özyener et al., 2001; Barstow & Mole, 1991). The same is true when the body moves from a state of activity to a state of rest; the VO_2 drops toward its level at rest with logarithmic decay.

In order to model the exponential rise and fall of the VO_2 response to activity, the energy consumed every second has to be sampled. A time interval of 1 second is appropriate, because it is significantly shorter than the duration of a single breath and much shorter than the time constants involved in modeling VO_2 (Lamarra et al., 1989). The energy consumed per second is also called the work rate. The change in VO_2 per unit time is given by Equation 10.6 for rising physical activity and Equation 10.7 for drops in physical activity (Özyener et al., 2001).

$$\Delta VO_2 = A_1 \left(1 - e^{(-t-TD)/\tau_1}\right) \tag{10.6}$$

$$\Delta VO_2 = A_1 \left(e^{(-t-TD)/\tau_1}\right) \tag{10.7}$$

TD is the time delay term in seconds, τ_1 is the time constant in seconds, and t is the elapsed time in seconds. A_1 is the amplitude and is related to the work rate as follows (Barstow & Mole, 1991):

$$A_1 = -75 + 11.5 \times WR \tag{10.8}$$

WR is the work rate of the current activity in W. The values of the constants used in Equation 10.6 and Equation 10.7 are summarized in the literature (Barstow & Mole, 1991; Engelen et al., 1996; Bearden & Moffatt, 2001; Özyener et al., 2001).

The oxygen uptake does not increase infinitely, and the maximum value to which it can increase is called the maximal oxygen uptake (VO_{2max}). VO_{2max} is part of the description of the physical capabilities of a digital human, and its value has to be set. VO_{2max} is related to the physical fitness, age, mass, and other physical descriptors of a human. The simulated VO_2 can not increase to levels higher than the VO_{2max} and can never fall to values below the VO_{2rest} (baseline VO_2 value).

Given the oxygen uptake, it is possible to model the heart rate. The relationship between heart rate (HR) and VO_2 is linear (Wilmore & Haskell, 1971). As activity levels increase, the heart rate increases as well. The increase in heart rate is in response to the increased need for oxygen to generate ATP by the muscles involved in activities. By beating faster, the heart pumps more blood to the muscles. Although there are additional functions of increased blood flow, they are relatively minor compared to the needs of the muscles and thus may be neglected. There are three mechanisms by which the body supplies increased oxygen demands to the working muscles: arterio-venous oxygen difference; increased stroke volume; and increased heart rate (Åstrand & Rodahl, 1970).

The arterio-venous oxygen difference is the difference between the oxygen content of the blood flowing through the arteries and the blood flowing through the veins. During an increased work rate, there is a hemoconcentration of blood. This translates into an increase in the oxygen-carrying capacity of the blood. The blood becomes more viscous and increases the transportation capacity not only for oxygen but also for carbon dioxide. The relationship between the arterio-venous oxygen difference and the oxygen uptake is linear and can be correlated to the percentage of maximum oxygen uptake as follows (Stringer et al., 1997):

$$C(a-vD_{O_2}) = 5.72 + 0.105 \times \%Vo_{2max} \tag{10.9}$$

$C(a-vD_{O_2})$ is the arterio-venous oxygen difference in mL/100mL, and $\%Vo_{2max}$ is the percent of maximum oxygen uptake in L/min.

The stroke volume is the volume of blood ejected into the main artery by each ventricular beat of the heart. As the energy requirement in the muscles increases, the stroke volume increases. In order to relate stroke volume and oxygen uptake, a regression equation is derived from data in the literature and is given as follows (Mathai, 2005):

$$y = (32.553 + 14.291 Ln(x)) \times {}^{q_{max}}\!/\!100 \tag{10.10}$$

y is the stroke volume in mL, x is the percentage of maximum oxygen uptake, and q_{max} is the maximum stroke volume in mL.

With a final form of oxygen uptake, Fick's mass balance equation is used to relate the cardiac output to the oxygen uptake, as follows (Margaria, 1976):

$$Q = \frac{Vo_2 \times 100}{C(a - vD_{O_2})} = f \times q \tag{10.11}$$

where Q is the cardiac output in L/min, $C(a - vD_{O_2})$ is the arteriovenous oxygen difference in mL/100mL, Vo_2 is the oxygen uptake in L/min, f is the stroke volume in L, and q is the heart rate in beats/minute. Equation 10.11 can be solved for heart rate as follows:

$$q = \frac{Vo_2 \times 1000}{\left(f \times C(a - vD_{O_2}) \Big/ 100 \right)} \tag{10.12}$$

10.6.2 Key Concepts

A primary consideration in modeling any physiological component is scope. What exactly will be modeled and what assumptions will be made (what constraints will be placed on the model)? The human body can be viewed as a complex machine that is constantly working. Thousands of chemical reactions and physical processes occur constantly, and almost all these processes are linked and form control loops. Extensive research has been done on these individual processes and their effects on the body. Thus, in order to model physiology in a digital human, one must define the scope of the model. While it is not totally impossible to model all the reactions occurring in the body and their interaction with each other, such a model would be extremely complex and computationally intensive. It is possible to delve beyond the key physiological indices and consider the performance of individual internal organs, but considering the primary interest in gross whole-body movement and performance, this level of fidelity is not necessary.

The various models described in this section can certainly operate independently, simply providing feedback concerning the digital human. However, there is the potential for relating such models to the potential for injury or to the tendency for exhaustion.

10.7 Modeling Muscles

More so than many other aspects of digital humans, the area of muscle-modeling is extremely broad, with a variety of ongoing research efforts investigating different aspects on different levels. When modeling muscles, one has many options with respect to desired fidelity, varying from modeling the different components of a single muscle, to simply modeling the resultant force from a set of muscles. In this section, we sample a few efforts surrounding muscle models, in order to demonstrate the different directions this area can take. Course models for muscle force are reviewed, and then a relatively new, more detailed approach for determining muscle stress and deformation is summarized. Ultimately, muscle-modeling can involve the fields of biochemistry, optimization, controls, finite-element analysis, differential equations, and others.

10.7.1 Muscle Force

When simply modeling the forces exerted by various muscles, there are two components to consider: (1) the line-of-action for the force, and (2) the magnitude of the force (muscle activation). First, we discuss some methods for addressing the first component.

Seireg and Avikar (1973) present an early *straight-line approach*, whereby straight lines represent the lines-of-action for the muscular tensile forces. This model is adequate for some muscles in certain orientations, but it is not robust enough for a real-time model that allows the user to orient the limb interactively to any position within the prescribed joint limits. For example, considering the biceps, a straight line running from the insertion to the origin may be acceptable for orientations when the elbow is bent. However, if the forearm is fully extended, the line-of-action then passes through the distal end of the humerus. Many muscles, such as the deltoid, can never be modeled as a straight line, regardless of the orientation. Because of the curved nature of the deltoid muscle, representing the line-of-action as one or more straight lines is insufficient.

Jensen and Davy (1975) present an alternative that is referred to here as the *centroid approach*. It does not assume that the force runs in a straight line between known attachment points. Instead, it assumes that the lines-of-action are along the centroid of the transverse cross section of the muscle. The unit vector of the action line predicted by this model is significantly different from the line-of-action predicted by the straight-line approach (An et al., 1981; Mikosz et al., 1988). While this method creates a more accurate line-of-action, it only represents muscles in a fixed orientation.

A third approach, called the *via point approach*, provides a compromise between the simplicity of the straight-line approach and the accuracy of the centroid approach. This method (Charlton & Johnson, 2001) uses a line-of-action that has one end attached to the origin point and the other attached to the insertion point. However, instead of a simple straight line, the line is allowed to wrap around a number of obstacles through the use of *via points*, which represent underlying muscles and bones. Typically, these via points are fixed to a bone during movement. While this is appropriate for muscles that only cross one joint with one degree of rotational freedom, fixed via points become a problem when more complex wrapping occurs. For example, consider the force line shown in Figure 10.12, which represents the lower line-of-action for the trapezius. This line originates in the thoracic region of the spine, and wraps around the thorax and the posterior surface of the scapula. The movement of the scapula and shoulder affects the shape of this muscle. If one were to use fixed via points, the line-of-action would not slide over the medial edge of the scapula during movement of the shoulder and would result in an inaccurate approximation of the muscles' centroid line.

Patrick and Abdel-Malek (2007) present a *floating via point approach* for modeling and visualizing in real time muscles that wrap and slide around bones, for the whole body, as shown in Figure 10.13. The bones are approximated as spheres and cylinders, and floating via points indicate contact points for the lines of action. The difference between floating via points and fixed via points is that the floating points are not fixed to a bone but are allowed to slide across the surface of the obstacle. This approach makes two assumptions about the action lines. The first is that the action line or lines of a muscle can be modeled as a frictionless elastic string wrapping around prescribed obstacles. The second is that spheres and cylinders can represent the underlying anatomical structures that the muscle must wrap around. The lines being wrapped will then be used to approximate the force lines generated by the muscles.

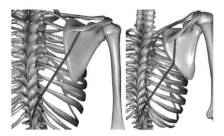

FIGURE 10.12   An action line of the trapezius with a fixed via point on medial edge of the scapula. (From Patrick, A., and K. Abdel-Malek, "A Musculoskeletal Model of the Upper Limb for Real Time Interaction," *SAE Human Modeling for Design and Engineering Conference*, SAE International, Warrendale, PA, 2007. With permission.)

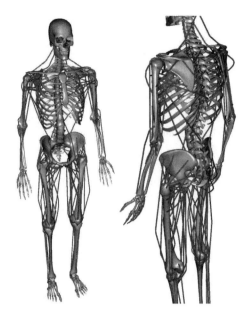

FIGURE 10.13   Whole-body muscle model.

10.7.2 Muscle Activation

Given the line-of-action for a muscle, one must then determine the magnitude of the muscle force. Bones act as the levers, and the joints between them serve as fulcrums. Motion is created by the muscles pulling on the bones to create torque about a joint. Since muscles can only act in tension, each joint requires one set of muscles to create a positive torque and a separate set to create negative torque. Assuming joint torque is constrained or known (possibly determined from dynamic motion prediction), optimization can be used to calculate the actual muscle force (muscle activation).

There are a variety of subtly different optimization-based approaches to determining muscle activation (Pedotti et al., 1978; An et al., 1981, 1984; Happee, 1994). In general, most of the recent approaches involve the following conceptual formulation:

> Find: activation level or force for each muscle
> to minimize: a function of activation level or forces that depend on activation
> subject to: (10.13)
> > (a) equations of static equilibrium
> > (b) activation limits or force limits

Raikova (1992, 1996) solve this problem analytically for an elbow model using necessary optimality conditions. However, this approach becomes intractable with more extensive models. Glitsch and Baumann (1997) solve this problem computationally for lower extremities, and experiment with different objective functions based on muscle force and muscle stress. Chadwick and Nicol (2000) use a two-stage optimization process for the elbow and wrist, where Equation 10.13 is solved in the first stage, and in the second stage joint and ligament forces are minimized. Considering just the elbow, Li et al. (2006) also consider both the muscle forces and joint forces, but they include the location of the rotational center for the joints as design variables.

A common basic approach is outlined as follows. Figure 10.14 presents an example that represents the medial head of the bicep and is used to demonstrate the muscle-modeling approach. The bicep originates on the scapula and inserts below the elbow on the radius. When the bicep muscle is activated,

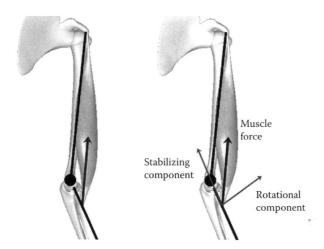

FIGURE 10.14 Force from bicep. (From Abdel-Malek, K., et al., *International Journal of Human Factors Modelling and Simulation*, 1, 2–39, 2006. With permission.)

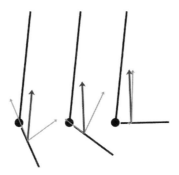

FIGURE 10.15 Effect of joint angle. (From Abdel-Malek, K., et al., *International Journal of Human Factors Modelling and Simulation*, 1, 2–39, 2006. With permission.)

it creates a force, and the vector representing this force is assumed to intersect the two endpoints of the muscle. This force creates a torque about the elbow by acting on a moment arm. As shown in Figure 10.14, the force can be resolved into two components: (1) the rotational component, which is normal to the moment arm, and (2) the stabilizing component, which is parallel to the moment arm. The rotational component is the portion of the force that actually contributes to joint torque. Note that as the joint angle changes, so does the rotational component (fig. 10.15). Thus, the resulting torque is not only a function of the muscle force, but also a function of the current joint angle.

For this one-muscle example, the muscle force needed to achieve a desired torque at a specific joint angle is found with $\mathbf{T} = \mathbf{r} \times \mathbf{F}$, where $\mathbf{T}$ is torque, $\mathbf{r}$ is a vector pointing from the joint to the point where the muscle acts on the bone, and $\mathbf{F}$ is the force. However, when more muscles are considered, the solution is nontrivial, as shown in Figure 10.16.

In this case torque is found with $\mathbf{T} = \mathbf{r}_b \times \mathbf{F}_b + \mathbf{r}_{bc} \times \mathbf{F}_{bc} + \mathbf{r}_{br} \times \mathbf{F}_{br}$, where b denotes bicep, bc denotes the brachialis, and br denotes the brachioradialis. Finding the muscle forces to achieve a desired torque at a specified angle becomes an indeterminate problem, because there is one equation with three unknowns ($\mathbf{F}_b$, $\mathbf{F}_{bc}$, and $\mathbf{F}_{br}$). To solve this problem, optimization must be used.

The design variables for this problem are muscle activation, where the activation value for each muscle ranges between 0 and 1, where 0 indicates no muscle activation, and 1 indicates a fully activated muscle. Muscle force can then be defined as $F = a * F_{max}$, where a is the activation and F_{max} is the maximum

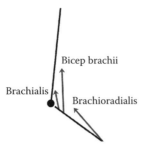

FIGURE 10.16   Three muscles acting on the lower arm. (From Abdel-Malek, K., et al., *International Journal of Human Factors Modelling and Simulation*, 1, 2–39, 2006. With permission.)

theoretical force the muscle can generate. F_{max} is proportional to the muscle's physical cross-sectional area. Torque is represented as $\mathbf{T} = \mathbf{r} \times (a\mathbf{F}_{max})$. Only the torque being generated about the rotational axis of the elbow is considered. Then, $T = raF_{max}$, where r is a constant and F_{max} is the rotational component of the vector F_{max}. This problem can now be solved with the following optimization formulation, where n is the number of muscles:

$$\text{Find: } a_1, a_2, \text{K}, a_n$$

$$\text{to minimize: } J = \sum_{i=1}^{n} a_i^2$$

$$\text{subject to: } T = \sum_{i=1}^{n} r_i(a_i F_{i\,max})$$ (10.14)

$$0 \le a_i \le 1$$

Note that by simply adding more constraints, this formulation can be extended to additional joints.

The problems in Equations 10.13 and 10.14 are *static optimization* problems, in that they are solved for a single set of variables. However, muscles can also be modeled using *dynamic optimization*, where the design variables are all continuous functions of one or more other parameters (Davy & Audu, 1987; Anderson & Pandy, 1999, 2001). With this approach, the design variables are functions of time. These models output a pattern of muscle excitation and associated motion that optimizes the specified performance measure. Equations of motion are solved using numerical integration and are implemented as constraints in the dynamic optimization problem. Consequently, this approach couples the problem of modeling muscles and motion. Unfortunately, dynamic optimization and numerical integration can be computationally expensive. This approach is applied to a 23-DOF lower-body model for walking and for vertical jumping.

In order to couple muscle models with motion simulation without overbearing computational demands, Thelen et al. (2003) present a method of combining static optimization with feedforward and feedback controls, and they apply it to a three-DOF lower-body model for pedaling. The equations of motion are integrated numerically and are solved using output from the static optimization problem, which is similar to Equation 10.13. The process is essentially nested in a control loop that requires initial kinematic data as input.

10.7.3 Muscle Stress and Displacement

Once the forces exerted by (and thus applied to) the muscles are known, it is possible to force data input to higher fidelity muscle-models. The same could be done with more detailed joint models. The isolated muscle models can be for fatigue analysis, or for stress and displacement analysis, which is discussed in this section.

Several major methods have been proposed for physically based deformable modeling: mass-spring models, finite element methods, finite volume models (Teran et al., 2003), and other low-degree approximated continuum models. There are essentially two issues in deformable body modeling: how to parameterize a body and how to bring in physical behavior to the system.

Biomechanical analysis of muscles is an important task in the development of a digital human system. The stress level within the muscle can provide information concerning the propensity for injury. However, stress analysis with traditional finite element methods is usually computationally intensive with a dense mesh resulting from the irregular shapes of muscles. Considering the variety of muscle shapes and functional differences, simulations with oversimplified or idealized muscle geometry and biomechanics can be unsatisfactory, especially when trying to examine a group of cooperating muscles in action.

Zhou and Lu (2005) developed an approach that overcomes the difficulties of computational intensity of the standard finite element method in an interactive virtual environment. The method is based on the combination of non-uniform rational B-spline (NURBS) geometric representation with the Galerkin finite element methods. NURBS geometry is used to represent the complex geometry of the muscles. The actual initial muscle shape can be extracted from medical image data such as the Visible Human Data Set (U.S. National Library of Medicine). The NURBS surface is then extended to a NURBS solid, which keeps the geometric information of a surface while extending to the inner volume. The NURBS solid representation can be written as follows:

$$\mathbf{P}(u,v,w) = \sum_{i=0}^{n}\sum_{j=0}^{m}\sum_{k=0}^{l}\mathbf{P}_{i,j,k}R_{i,j,k}(u,v,w) \tag{10.15}$$

Using the isoparametric mapping technique, the NURBS geometric shape functions are used as the interpolation functions of finite elements. Therefore, the displacement mapping can be written as follows:

$$\mathbf{d}(u,v,w) = \sum_{i=0}^{n}\sum_{j=0}^{m}\sum_{k=0}^{l}\mathbf{d}_{i,j,k}R_{i,j,k}(u,v,w) \tag{10.16}$$

where $\mathbf{d}_{i,j,k}$ is the displacement of the control point $\mathbf{P}_{i,j,k}$.

Based on this mapping, the deformation gradient and other needed quantities for the finite element computation are calculated. Because muscles are generally subjected to large deformations during human motion, a fully nonlinear formulation for the analysis of large-strain motion is used.

The internal structure of the skeletal muscle is comprised of soft tissues with various material properties. More important, skeletal muscles are active, and the properties of the active tissue are altered upon activation. With this work, the muscle model is described by a hyperelastic, anisotropic constitutive equation that includes an active component.

The hyperelastic strain energy function is assumed to be the sum of two parts: energy function of the passive ground substance and active fibrous structure. The passive ground substance consists of connective tissue, water, and so on, and is modeled as isotropic neo-Hookean material. The muscle fiber strain-energy is assumed to be a function of the muscle fiber stretch and the muscle activation level. The muscle fiber directions are generally distributed with different patterns among different muscles. In this work, isocurves of the NURBS solid are used to model the muscle fibers.

Motion simulation provides at discrete points of time the joint positions and rotations as well as the activation level in individual muscles. The displacement boundary-conditions for the muscles can be

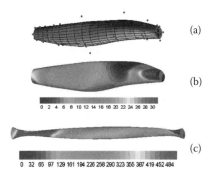

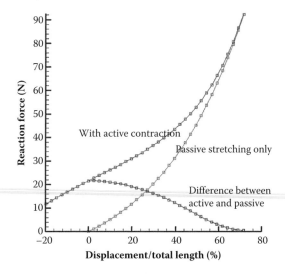

FIGURE 10.17   Stress analysis of muscle: (a) initial rest geometry, (b) active contraction, (c) large passive defor-mation. (From Zhou, X., and J. Lu, *SAE 2005 Transactions Journal of Passenger Cars—Mechanical Systems*, 114(6), 2921–29, 2005. With permission.)

FIGURE 10.18   Force-extension curve. (From Zhou, X., and J. Lu, *SAE 2005 Transactions Journal of Passenger Cars—Mechanical Systems*, 114(6), 2921–29, 2005. With permission.)

determined from the position of the bones. Stress analysis is then conducted using the NURBS-FEM. Both passive and active motions of muscles are simulated.

Using the pseudo-static simulation of muscle deformation at discrete points of time, the muscle force-extension is shown in Figure 10.17 with different length changes. The curves in Figure 10.18 show the muscle behavior, which qualitatively matches what is reported in the literature (Zajac, 1989).

10.7.4 Key Concepts

The computational approaches with muscles demonstrate how multiscale models can be coordinated. Dynamic motion simulation feeds joint torques to a model that determines lines-of-action for forces, which are then used to determine muscle activation and force. The muscle forces can then be used to determine stress and deformation. Such forces could also be used in complex muscle fatigue models. Fatigue is discussed in a separate chapter.

In general, the more refined or higher fidelity a model becomes, the more computationally demanding it becomes. Although the NURBS-FEM approach is an improvement over traditional finite element anal-ysis, it cannot yet function in real time, even if applied to only one or two muscles. In addition, modeling muscle fatigue for all of the muscles in the body simultaneously in real time is far from feasible today.

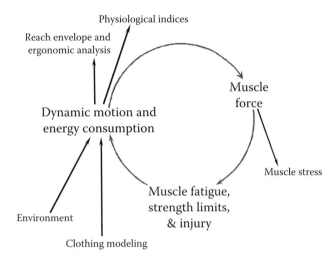

FIGURE 10.19   A system of integrated systems.

10.8 Conclusions

This chapter has provided a sample of the variety of computational efforts that are involved in creating a digital human model. Despite the exciting momentum that currently exists in the field of DHM, there is much opportunity for collaboration and cross-pollination in order to develop a realistic, high-fidelity, complete digital human model that operates in real time. The intermediate steps in obtaining that goal clearly entail progress with a system of systems that must be integrated, as shown in Figure 10.19. Most of the components in Figure 10.19 have been discussed in this chapter, but there are many others that can be included and that are discussed throughout the remainder of the text.

The most challenging aspect in Figure 10.19 is making the connection back to motion prediction, from fatigue or strength models. Ideally, a digital human model will not simply report that a task is too difficult. Rather, it will predict how the motion changes to accommodate difficulties. The work with dynamic optimization (Davy & Audu, 1987; Anderson & Pandy, 1999, 2001) provides one approach to making this connection. In addition, it is possible to provide strength limits, in the form of joint torque limits, as constraints in Equation 10.4. Such limits could even change over time, thus approximating muscle fatigue.

One of the primary considerations that is not addressed in this chapter is cognitive modeling. Given the current state of the art for DHM, it is the most significant boundary condition for human models. It can potentially affect all of the components of Figure 10.16. However, in general, the field of cognitive modeling is not at the point where a cognitive model can dictate how one will move or fatigue, without substantial amounts of input, in the form of both data and preconceived rules. Nonetheless, substantial progress is being made in this direction and is discussed in other chapters.

In summary, there are many fields that can contribute to DHM, in an effort to produce a more accurate and autonomous digital human. All models are data driven to some extent, but robust and general models that can respond to different scenarios with minimal input and maximum autonomy provide the most helpful tools in evaluating humans and evaluating items that humans interact with.

References

Abdel-Malek, K., Yang, J., Marler, T., Beck, S., Mathai, A., Zhou, X., Patrick, A., and Arora, J. (2006), "Towards a New Generation of Virtual Humans," *International Journal of Human Factors Modelling and Simulation*, 1 (1), 2–39.

Abdel-Malek, K., Yang, J., Mi, Z., Patel, V. C., and Nebel, K. (2004), "Human Upper Body Motion Prediction," *Conference on Applied Simulation and Modeling (ASM)*, June, Rhodes, Greece.

Alexander, R. McN. (1997), "A Minimum Energy Cost Hypothesis for Human Arm Trajectories," *Biological Cybernetics*, 76, 97–105.

An, K. N., Hui, F. C., Morrey, B. F., Linscheid, R. L., and Chao, E. Y. (1981), "Muscles across the Elbow Joint: A Biomechanical Analysis," *Journal of Biomechanics*, 14 (10), 659–669.

An, K. N., Kwak, B. M., Chao, E. Y., and Morrey, B. F. (1984), "Determination of Muscle and Joint Forces: A New Technique to Solve the Indeterminant Problem," *Journal of Biomechanical Engineering*, 106 (4), 364–367.

Anderson, F. C., and Pandy, M. G. (1999), "A Dynamic Optimization Solution for Vertical Jumping in Three Dimensions," *Computer Methods in Biomechanics and Biomedical Engineering*, 2, 201–231.

Anderson, F. C., and Pandy, M. G. (2001), "Dynamic Optimization of Human Walking," *Journal of Biomechanical Engineering*, 123, 381–390.

Åstrand, P., and Rodahl, K. (1970), *Textbook of Work Physiology*, McGraw Hill, New York.

Ayyappa, E. (1997), "Normal Human Locomotion, Part 1: Basic Concepts and Terminology," *Journal of Prosthetics and Orthotics*, 9 (1), 10–17.

Barstow, T. J., and Mole, P. A. (1991), "Linear and Nonlinear Characteristics of Oxygen Uptake Kinetics during Heavy Exercise," *Journal of Applied Physiology,* 71 (6), 2099–2106.

Baruch, G., and Goldman, R. F. (1972), "Predicting Rectal Temperature Response to Work, Environment and Clothing," *Journal of Applied Physiology,* 32 (6), 812–822.

Bearden, S. E., and Moffatt, R. J. (2001), "V_{O2} and Heart Rate Kinetics in Cycling: Transitions from an Elevated Baseline," *Journal of Applied Physiology*, 90, 2081–2087.

Breen, D. E., House, D. H., and Wozny, M. J. (1994a), "A Particle-based Model for Simulating the Draping Behavior of Woven Cloth," *Textile Research Journal*, 64 (11), 663–685.

Breen, D. E., House, D. H., and Wozny, M. J. (1994b), "Predicting the Drape of Woven Cloth Using Interacting Particles," *Computer Graphics (Proc. SIGGRAPH)*, 365–372.

Carignan, M., Yang, Y., Thalmann, N. M., and Thalman, D. (1992), "Dressing Animated Synthetic Actors with Complex Deformable Clothes," *Computer Graphics*, 26 (2), 99–104.

Chadwick, E. K. J., and Nicol, A. C. (2000), "Elbow and Wrist Joint Contact Forces during Occupational Pick and Place Activities," *Journal of Biomechanics*, 33, 591–600.

Chaffin, D. B. (2002), "On Simulating Human Reach Motions for Ergonomic Analysis," *Human Factors and Ergonomics in Manufacturing*, 12 (3), 235–247.

Chaffin, D. B., Faraway, J. J., Zhang, X., and Woolley, C. (2000), "Stature, Age, and Gender Effects on Reach Motion Postures," *Human Factors*, 42 (3), 408–420.

Charlton, I. W., and Johnson, G. R. (2001), "Application of Spherical and Cylindrical Wrapping Algorithms in the Musculoskeletal Model of the Upper Limb," *Journal of Biomechanics*, 34, 1209–1216.

Chen, B., and Govindaraj, M. (1995), "A Physically Based Model of Fabric Drape Using Flexible Shell Theory," *Textile Research Journal*, 65 (6), 324–330.

Chen, B., and Govindaraj, M. (1996), "A Parametric Study of Fabric Drape," *Textile Research Journal*, 66 (1), 17–24.

Chen, Y.-C. (1991), "Solving Robot Trajectory Planning Problems with Uniform Cubic B-Splines," *Optimal Control Applications and Methods*, 12, 247–262.

Chevallereau, C., and Aousin, Y. (2001), "Optimal Reference Trajectories for Walking and Running of a Biped Robot," *Robotica*, 19, 557–569.

Choi, K., and Ko, H. (2002), "Stable but Responsive Cloth," *Computer Graphics (Proc. SIGGRAPH)*, 604–611.

Chu, G. C., Cummings, C. L., and Teixeira, N. A. (1950), "Mechanics of Elastic Performance of Textile Materials, Part V: A Study of the Factors Affecting the Drape of Fabrics—The Development of a Drape Meter," *Textile Research Journal*, 20, 539–548.

Chung, H. J., Xiang, Y., Mathai, A., Rahmatalla, S., Kim, J., Marler, T., Beck, S., Yang, J., Arora, J., and Abdel-Malek, K. (2007), "A Robust Formulation for Prediction of Human Running," *SAE Human Modeling for Design and Engineering Conference*, June, Seattle, WA, Society of Automotive Engineers, Warrendale, PA, SAE paper number 2007-01-2490.

Collier, J. R., Collier, B. I., O'Toole, G., and Sargand, S. M. (1991), "Drape Prediction by Means of Finite-Element Analysis," *Journal of the Textile Institute*, 82 (1), 96–107.

Cusick, G. E. (1968), "The Measurement of Fabric Drape," *Journal of the Textile Institute*, 59 (6), 253–260.

Davy, D. T., and Audu, M. L. (1987), "A Dynamic Optimization Technique for Predicting Muscle Forces in the Swing Phase of Gait," *Journal of Biomechanics*, 20 (2), 187–201.

Delp, S. L., Loan, J. P. (1995), "A Graphics-Based Software to Develop and Analyze Models of Musculo-skeletal Structures," *Computers in Biology and Medicine*, 25 (1), 21–34.

DeLucam, C. J., and Forrest, W. J. (1973), "Force Analysis of Individual Muscles Acting Simultaneously on the Shoulder during Isometric Abduction," *Journal of Biomechanics*, 6, 385–393.

Denavit, J., and Hartenberg, R. S. (1955), "A Kinematic Notation for Lower-Pair Mechanisms Based on Matrices," *Journal of Applied Mechanics*, 22, 215–221.

Dvir, Z., and Berme, N. (1978), "The Shoulder Complex in Elevation of the Arm: A Mechanism Approach," *Journal of Biomechanics*, 11, 219–255.

Eberhardt, B., Weber, A., and Strasser, W. (1996), "A Fast, Flexible, Particle-System Model for Cloth Draping," *IEEE Computer Graphics and Applications*, 16 (5), 52–59.

Eischen, J. W., Deng, S., and Clapp, T. G. (1996), "Finite-Element Modeling and Control of Flexible Fabric Parts," *IEEE Computer Graphics and Applications*, 16 (5), 71–80.

Engelen, M., Porszasz, J., Marshall, R., Wasserman, K., Maehara, K., and Barstow, T. J. (1996), "Effects of Hypoxic Hypoxia on Oxygen Uptake and Heart Rate Kinetics during Heavy Exercise," *Journal of Applied Physiology*, 81 (6), 2500–2508.

Faraway, J. J. (1997), "Regression Analysis for a Functional Response," *Technometrics*, 39 (3), 254–262.

Faraway, J. J., Zhang, X. D., and Chaffin, D. B. (1999), "Rectifying Postures Reconstructed from Joint Angles to Meet Constraints," *Journal of Biomechanics*, 32, 733–736.

Farrell, K., Marler, R. T., and Abdel-Malek, K. (2005), "Modeling Dual-Arm Coordination for Posture: An Optimization-Based Approach," *SAE 2005 Transactions Journal of Passenger Cars—Mechanical Systems*, 114-6, 2891, SAE paper number 2005-01-2686.

Flash, T., and Hogan, N. (1985), "The Coordination of Arm Movements: An Experimental Confirmed Mathematical Model," *The Journal of Neuroscience*, 5 (7), 1688–1703.

Froelicher, V. F., and Myers, J. (2006), *Exercise and the Heart*, 5th ed., Saunders Elsevier, Philadelphia.

Gan, L., Ly, N. G., and Steven, G. P. (1995), "A Study of Fabric Deformation Using Nonlinear Finite Elements," *Textile Research Journal*, 65 (11), 660–668.

Gilbert, E. M., and Johnson, D. W. (1985), "Distance Functions and Their Application to Robot Path Planning in the Presence of Obstacles," *IEEE Journal of Robotics and Automation*, 1, 21–30.

Givoni, B., and Goldman, R. F. (1972), "Predicting Rectal Temperature Response to Work, Environment and Clothing," *Journal of Applied Physiology*, 32 (6), 812–822.

Glitsch, U., and Baumann, W. (1997), "The Three-Dimensional Determination of Internal Loads in the Lower Extremity," *Journal of Biomechanics*, 30 (11/12), 1123–1131.

Happee, R. (1994), "Inverse Dynamic Optimization Including Muscular Dynamics, a New Simulation Method Applied to Goal Directed Movements," *Journal of Biomechanics*, 27 (7), 953–960.

Horn, E. N. (2005), "Optimization-Based Dynamic Human Motion Prediction," M.S. Dissertation, University of Iowa, Iowa City, IA.

Jackson, K. M., Joseph, J., and Wyard, S. J. (1977), "Sequential Muscular Contraction," *Journal of Biomechanics*, 10, 97–100.

Jenesn, R. H., and Davy, D. T. (1975), "An Investigation of Muscle Lines of Action about the Hip: A Centroid Line Approach VS the Straight Line Approach," *Journal of Biomechanics*, 8, 103–110.

Jung, E. S., Choe, J., and Kim, S. H. (1994), "Psychophysical Cost Function of Joint Movement for Arm Reach Posture Prediction," *Proceedings of the Human Factors and Ergonomics Society 38th Annual Meeting*, October, Nashville, TN, Human Factors and Ergonomics Society, Santa Monica, CA, 636–640.

Jung, E. S., Kee, D., and Chung, M. K. (1995), "Upper Body Reach Posture Prediction for Ergonomic Evaluation Models," *International Journal of Industrial Ergonomics*, 16, 95–107.

Kahn, M. E., and Roth, B. (1971), "The Near-Minimum-Time Control of Open-Loop Articulated Kinematic Chains," *Journal of Dynamic System, Measurement, and Control*, 93 Ser G (3), 164–172.

Kawabata, S. (1980), *The Standardization and Analysis of Hand Evaluation*, The Textile Machinery Society of Japan, Osaka.

Kim, J. H., Abdel-Malek, K., Yang, J., and Marler, R. T. (2006), "Prediction and Analysis of Human Motion Dynamics Performing Various Tasks," *International Journal of Human Factors Modelling and Simulation*, 1 (1), 69–94.

Klein, C. A., and Huang, C.-H. (1983), "Review of Pseudoinverse Control for Use with Kinematically Redundant Manipulators," *IEEE Transactions on Systems, Man, and Cybernetics*, SMC-13 (3), 245–250.

Lamarra, N., Ward, S. A., and Whipp, B. J. (1989), "Model Implications of Gas Exchange Dynamics on Blood Gases in Incremental Exercise," *Journal of Applied Physiology*, 66 (4), 1539–1546.

Li, G., Pierce, J. E., and Herndon, J. H. (2006), "A Global Optimization Method for Prediction of Muscle Forces of Human Musculoskeletal System," *Journal of Biomechanics*, 39, 522–529.

Liegeois, A. (1977), "Automatic Supervisory Control of the Configuration and Behavior of Multibody Mechanism," *IEEE Transactions on Systems, Man and Cybernetics*, SMC-17 (2), 868–871.

Lin, C.-S., Chang, P.-R., and Luh, J. Y. S. (1983), "Formulation and Optimization of Cubic Polynomial Joint Trajectories for Industrial Robots," *IEEE Transactions on Automatic Control*, AC-28 (12), 1066–1074.

Lo, J., Huang, G., and Metaxas, D. (2002), "Human Motion Planning Based on Recursive Dynamics and Optimal Control Techniques," *Multibody System Dynamics*, 8, 433–458.

Lu, T.-W., Lin, Y. -S., Kuo, M. -Y., Hsu, H.-C., and Chen, H.-L. (2005), "A Kinematic Model of the Upper Extremity with Globally Minimized Skin Movement Artifacts," *The XXth Congress of the International Society of Biomechanics and 29th Annual Meeting of the American Society of Biomechanics*, Cleveland, OH.

Man, X. (2006), "A Mathematical and Computational Multiscale Clothing Modeling Framework," Ph.D. Dissertation, University of Iowa, Iowa City, IA.

Man, X., and Swan, C. C. (2007), "A Mathematical Modeling Framework for Analysis of Functional Clothing," *Journal of Engineered Fibers and Fabrics* (in press).

Margaria, R. (1976), *Biomechanics and Energetics of Muscular Exercise*, Clarendon Press, Oxford.

Marler, R. T. (2005), "A Study of Multi-objective Optimization Methods for Engineering Applications," Ph.D. Dissertation, University of Iowa, Iowa City, IA.

Marler, R. T., Farell, K., Kim, J., Rahmatalla, S., and Abdel-Malek, K. (2006), "Vision Performance Measures for Optimization-Based Posture Prediction," *SAE Human Modeling for Design and Engineering Conference*, July, Lyon, France, Society of Automotive Engineers, Warrendale, PA.

Marler, R. T., Rahmatalla, S., Shanahan, M., and Abdel-Malek, K. (2005a), "A New Discomfort Function for Optimization-Based Posture Prediction," *SAE Human Modeling for Design and Engineering Conference*, June, Iowa City, IA, Society of Automotive Engineers, Warrendale, PA.

Marler, R. T., Yang, J., Arora, J. S., and Abdel-Malek, K. (2005b), "Study of Bi-Criterion Upper Body Posture Prediction Using Pareto Optimal Sets," *IASTED International Conference on Modeling, Simulation, and Optimization*, August, Oranjestad, Aruba, International Association of Science and Technology for Development, Canada.

Mathai, A. J. (2005), "Towards Physiological Modeling in Virtual Humans," M.S. Thesis, University of Iowa, Iowa City, IA.

Maurel, W., and Thalmann, D. (1999), "A Case Study on Human Upper Limb Modeling for Dynamic Simulations," *Computer Methods in Biomechanics and Biomedical Engineering*, 2 (1), 65–82.

Maurel, W., Thalmann, D., Hoffmeyer, P., Beylot, P., Gingins, P., Karla, P., and Thalmann, N. (1996), "A Biomechanical Musculoskeletal Model of Human Upper Limb for Dynamic Simulation," *EGCAS '96, 7th Eurographics Workshop on Computer Animation and Simulation '96* (Springer Poitiers 1996), 121–136.

Mi, Z. (2004), *Task-Based Prediction of Upper Body Motion*, Ph.D. Dissertation, University of Iowa, Iowa City, IA.

Mi, Z., Yang, J., Abdel-Malek, K., Mun, J. H., and Nebel, K. (2002), "Real-Time Inverse Kinematics for Humans," *Proceedings of the 2002 ASME Design Engineering Technical Conferences and Computer and Information in Engineering Conference*, 5A, September, Montreal, Canada, American Society of Mechanical Engineers, New York, 349–359.

Mikosz, R. P., McKersie, R., Berg, R., Snitovsky, P., Lobick, J., Logue, S., Andersson, G. B. J., and Andriacchi, T. P (1988), "Muscle Lines of Action in the Lower Extremity: A Centroid Line Approach vs. the Straight Line Approach," *Proceedings of the 12th American Society of Biomechanics*, Urbana-Champaign, IL, 30–31.

Moran, D. S., Avaraham, S., and Kent, B. P. (1998), "A Physiological Strain Index to Evaluate Heat Stress," *American Journal of Physiology*, 44, R129–R134.

Mu, X., and Wu, Q. (2003), "Synthesis of a Complete Sagittal Gait Cycle for a Five-Link Biped Robot," *Robotica*, 21, 581–587.

Özyener, F., Rossiter, H. B., Ward, S. A., and Whipp, B. J. (2001), "Influence of Exercise on the On- and Off-Transient Kinetics of Pulmonary Oxygen Uptake in Humans," *Journal of Physiology*, 533 (3), 891–902.

Patrick, A., and Abdel-Malek, K. (2007), "A Musculoskeletal Model of the Upper Limb for Real Time Interaction," *SAE Human Modeling for Design and Engineering Conference*, June, Seattle, WA, Society of Automotive Engineers, Warrendale, PA, SAE paper number 2007-01-2488.

Pedotti, A., Krishnan, V. V., and Stark, L. (1978), "Optimization of Muscle-Force Sequencing in Human Locomotion," *Mathematical Biosciences*, 38 (1–2), 57–76.

Provot, X. (1995), "Deformation Constraints in a Mass-Spring Model to Describe Rigid Cloth Behavior," *Proc. Graphics Interface*, 147–154.

Rahmatalla, S., Kim, H., Shanahan, M., and Swan, C. C. (2006), "Effect of Restrictive Clothing on Balance and Gait Using Motion Capture and Dynamic Analysis," *SAE 2005 Transactions Journal of Passenger Cars-Electronic and Electrical Systems*, SAE paper number 2005-01-2688.

Raikova, R. (1992), "A General Approach for Modeling and Mathematical Investigation of the Human Upper Limb," *Journal of Biomechanics*, 25 (8), 857–867.

Raikova, R. (1996), "A Model of the Flexion-Extension Motion in the Elbow Joint—Some Problems Concerning Muscle Forces Modelling and Computation," *Journal of Biomechanics*, 29 (6), 763–772.

Reed, M. P., Manary, M. A., Flannagan, C. A. C., and Schneider, L. W. (2000), "Comparison of Methods for Predicting Automobile Driver Posture," *SAE 2000 Digital Human Modeling for Design and Engineering Conference and Exposition*, June, Detroit, MI, SAE International, Warrendale, PA, SAE paper number 2000-01-2180.

Riffard, V., and Chedmail, P. (1996), "Optimal Posture of a Human Operator and CAD in Robotics," *Proceedings of the 1996 IEEE International Conference on Robotics and Automation*, April, Minneapolis, MN, Institute of Electrical and Electronics Engineers, New York, 1199–1204.

Saidouni, T., and Bessonnet, G. (2003), "Generating Globally Optimized Sagittal Gait Cycles of a Biped Robot," *Robotica*, 21 (2), 199–210.

Saramago, S. F. P., and Steffen, V. S. (1998), "Optimization of the Trajectory Planning of Robot Manipulators Taking into Account the Dynamics of the System," *Mechanism and Machine Theory*, 33 (7), 883–894.

Saramago, S. F. P., and Steffen, V. S. (2000), "Optimal Trajectory Planning of Robot Manipulators in the Presence of Moving Obstacles," *Mechanism and Machine Theory*, 35, 1079–1094.

Saramago, S. F. P., and Ceccarelli, M. (2002), "An Optimum Robot Path Planning with Payload Constraints," *Robotica*, 20, 395–404.

Saramago, S. F. P., and Ceccarelli, M. (2004), "Effect of Basic Numerical Parameters on a Path Planning of Robots Taking into Account Actuating Energy," *Mechanism and Machine Theory*, 39, 247–260.

Seireg, A., and Avikar, R. J. (1973), "A Mathematical Model for the Evaluation of Forces in the Lower Extremities of the Musculoskeletal System," *Journal of Biomechanics*, 6, 313–326.

Simo, J. C., and Fox, D. D. (1989), "On a Stress Resultant Geometrically Exact Shell Model. Part I: Formulation and Optimal Parameterization," *Computer Methods in Applied Mechanics and Engineering*, 72, 267–304.

Simo, J. C., Fox, D. D., and Rifai, M. S. (1989), "On a Stress Resultant Geometrically Exact Shell Model. Part II: The Linear Theory; Computational Aspects," *Computer Methods in Applied Mechanics and Engineering*, 73, 53–92.

Simo, J. C., Fox, D. D., and Rifai, M. S. (1990), "On a Stress Resultant Geometrically Exact Shell Model. Part III: Computational Aspects of the Nonlinear Theory," *Computer Methods in Applied Mechanics and Engineering*, 79, 21–70.

Stringer, W. W., Hansen, J. E., and Wasserman, K. (1997), "Cardiac Output Estimated Non Invasively from Surgery," *Journal of Applied Physiology*, 82 (3), 908–912.

Teran, J., Blemker, S., Ng-Thow-Hign, V., and Fedkiw, R. (2003), "Finite Volume Method for the Simulation of Skeletal Muscle," *Proceedings of the 2003 SIGGRAPH/Eurographics Symposium on Computer Animation*, 68–74.

Terzopoulos, D., Platt, J., Barr, A., and Fleischer, K. (1987), "Elastically Deformable Models," *Computer Graphics (Proc. SIGGRAPH)*, 21 (4), 205–214.

Terzopoulos, D., and Fleischer, K. (1988), "Modeling Inelastic Deformation: Viscoelasticity, Plasticity, Fracture," *Computer Graphics*, 22 (4), 269–278.

Thelen, D. G., Anderson, F. C., and Delp, S. L. (2003), "Generating Dynamic Simulations of Movement Using Computed Muscle Control," *Journal of Biomechanics*, 36, 321–328.

Vignes, R. (2004), "Modeling Muscle Fatigue in Digital Humans," M.S. Thesis, University of Iowa, Iowa City, IA.

Wang, C.-Y. E., Timoszyk, W. K., and Bobrow, J. E. (1999), "Weightlifting Motion Planning for A Puma 762 Robot," *Proceedings of the 1999 IEEE International Conference on Robotics and Automation*, May, Detroit, MI, Institute of Electrical and Electronics Engineers, New York, 480–485.

Wilmore, J. H., and Haskell, W. L. (1971), "Use of the Heart Rate-Energy Expenditure Relationship in the Individualized Prescription of Exercise," *American Journal of Clinical Nutrition*, 24, 1186–1192.

Xiang, Y., Chung, H. J., Mathai, A., Rahmatalla, S., Kim, J., Marler, T., Beck, S., Yang, J., Arora, J. S., and Abdel-Malek, K. (2007), "Optimization-Based Dynamic Human Walking Prediction," *SAE Human Modeling for Design and Engineering Conference*, June, Seattle, WA, Society of Automotive Engineers, Warrendale, PA, SAE paper number 2007-01-2489.

Yang, J., Marler, T., Beck, S., Abdel-Malek, K., and Kim, H.-J. (2006), "Real-Time Optimal-Reach Posture Prediction in a New Interactive Virtual Environment," *Journal of Computer Science and Technology*, 21 (2), 189–198.

Yang, J., Marler, R. T., Kim, H., Arora, J. S., and Abdel-Malek, K. (2004), "Multi-objective Optimization for Upper Body Posture Prediction," *10th AIAA/ISSMO Multidisciplinary Analysis and Optimization Conference*, August, Albany, NY, American Institute of Aeronautics and Astronautics, Washington, DC.

Yang, J., Marler, R. T., Kim, H. J., Farrell, K., Mathai, A., Beck, S., Abdel-Malek, K., Arora, J., and Nebel, K. (2005), "Santos: A New Generation of Virtual Humans," *SAE 2005 World Congress*, April, Detroit, MI, Society of Automotive Engineers, Warrendale, PA.

Zajac, F. E. (1989), "Muscle and Tendon: Properties, Models, Scaling, and Application to Biomechanics and Motor Control," *Critical Reviews in Biomedical Engineering*, 17, 359–411.

Zhang, X., and Chaffin, D. B. (1997), "Task Effects on Three-Dimensional Dynamic Postures during Seated Reaching Movements: An Investigative Scheme and Illustration," *Human Factors*, 39 (4), 659–671.

Zhang, X., Kuo, A. D., and Chaffin, D. B. (1998), "Optimization-Based Differential Kinematic Modeling Exhibits A Velocity-Control Strategy for Dynamic Posture Determination in Seated Reaching Movements," *Journal of Biomechanics*, 31, 1035–1042.

Zhao, J., and Badler, N. I. (1994), "Inverse Kinematics Positioning Using Nonlinear Programming for Highly Articulated Figures," *ACM Transactions on Graphics*, 13 (4), 313–336.

Zhou, X., and Lu, J. (2005), "Biomechanical Analysis of Skeletal Muscle in an Interactive Digital Human System," *SAE 2005 Transactions: Journal of Passenger Cars: Mechanical Systems*, 114-6, Society of Automotive Engineers, Warrendale, PA, SAE paper number 2005-01-2709.

11

Impact Simulation and Biomechanical Human Body Models

Riender Happee
and Jac Wismans

11.1 Abstract

Computational human body models are widely used for automotive crash-safety research and design and as such have significantly contributed to a reduction of traffic injuries and fatalities. Models based on crash dummies are effectively used in the vehicle development process, but crash dummies differ significantly from the real human body and, moreover, crash dummies are only available for a limited set of body sizes. Models of the real human body offer in this aspect some promising advantages including the prediction of injury mechanisms and injury criteria. Topics presented in this chapter include multibody versus FE modeling, human body geometry, human body material modeling, and applications. An outlook on developments in this field concludes the chapter.

11.2 Introduction

Biomechanical human body models are widely used to simulate human motion and to relate motion to body internal loads and neuromuscular control strategies. In automotive impact, three main types of biomechanical human body models are used:

- Crash dummies
- Computational models of crash dummies
- Computational models of the real human body

This chapter addresses computational models of the real human body for impact and briefly discusses models of crash dummies. For an introduction into crash dummies, the reader is referred to AGARD (1996).

Biomechanical models have been developed using multibody and finite element techniques, using forward or inverse dynamics and defining one, two, or three dimensions (see Tables 11.1 and 11.2). Multibody is used primarily to analyze motion and forces while finite elements also enable analysis of local material deformations. Inverse dynamic simulation has been used extensively to analyze recorded human motion and has contributed to a better understanding of human movement control (e.g., Van Ingen Schenau et al., 1992; Koopman et al., 1995; Van der Helm et al., 2002; Damsgaard et al., 2006). Both 2D and 3D multi-camera motion data are routinely applied to analyze voluntary motion. However, the need for human kinematic input data is a serious limitation in product design, where virtual prototyping is increasingly successful and physical prototypes are built only in the latest design stages. For impact obviously volunteer kinematic data cannot be obtained in potentially injuriously loading levels. This makes forward simulation the leading solution in impact research and development. However, in forward simulation of human motion, muscular activation has to be defined as input signal (see Table 11.1, bottom row). Horst et al. (1997, 2001) investigated effects of muscular activity by exploring assumptions regarding muscle pretension and reflex-induced activation. Only a few models aim to predict activation using neuromuscular feedback models (see Cappon et al., 2007), and most human models for impact simulation simply include some additional joint resistance to account for muscular activity.

TABLE 11.1   Common Representation of the Human Body in Multibody and Finite Element Models

Tissue/Property	Multibody Representation	Finite Element Representation
Inertial properties of all tissues	Lumped into rigid bodies (or occasionally flexible bodies)	Deformable FE Bones of minor importance are often switched to rigid
Stiffness of joint structures such as ligaments, capsules etc.	Spring-damper elements Kinematic joints constraining motions regarded as minor or irrelevant	Spring-damper elements Deformable FE
Contact surfaces	Ellipsoids, cylinders, planes 3D surfaces (membrane elements rigidly supported on bodies)	Surface, volume, and line elements
Contact surface compliance resulting from deformation of underlying tissues (fat, muscle, organs) and materials such as seat foams	Non-linear contact stiffness and damping	Described by underlying deformable FE
Active muscle forces	Passive joint stiffness and damping models "tuned" to match volunteer data; this is the status of most dummies and human models for impact Active elements applying moments at joint level Muscles as line elements, applying tensile forces to bodies	Generally ignored Sometimes modeled using multibody options
Muscle activation levels	Estimated using inverse analysis Optimization of muscular activation sequences such that some mathematical performance criterion is optimized Arbitrary assumptions regarding muscle pretension and reflex-induced activation Predictive modeling of motor feedback	

TABLE 11.2 Forward and Inverse Dynamic Simulation

	Multibody	Finite Element
Forward dynamic simulation	Body kinematics and loads are calculated as a result of applied external loads (and sometimes muscle forces).	Nodal kinematics, stresses, and strains are calculated as a result of applied external loads (and sometimes muscle forces).
Inverse dynamic simulation	Joint moments and muscle forces are calculated using body kinematics and external loads.	

11.2.1 Multibody Models

Bodies in a multibody formulation can be connected by various joint types, which constrain the number of degrees of freedom between the bodies. External forces generated by so-called force-interaction models cause the motion of the joint-connected bodies in a multibody model. Examples of force-interaction models in a multibody model for crash analyses are the model to account for an acceleration field, spring-damper elements, restraint system models, and contact models. Multibody models in which the complete human body is simulated for the purpose of crash analyses are often referred to as crash victim simulation (CVS) models, human body gross-motion simulators, or whole body response models. One of the early human CVS models was published by McHenry (1963). The model that represents the human body together with restraint system and vehicle is 2D and has 7 degrees of freedom. McHenry compared model calculations with experimental data and was able to show quite good agreement for quantities like hip displacements, chest acceleration, and belt loads. The results of this model were so encouraging that since then several, more sophisticated, models have been developed. The most well known are the 2D 8-body MVMA-2D (Robbins et al., 1974), the 3D CAL3D allowing up to 20 elements (Fleck et al., 1974), and MADYMO 2D/3D allowing an arbitrary number of bodies, both rigid and flexible ones and a range of joint types (Koppens et al., 1993; MADYMO, 2006). MADYMO, which was developed and is supported by TNO Automotive in the Netherlands, has gone through an extensive further development and validation program and includes a finite element part for crash analyses. The multibody module of MADYMO calculates the contribution of the inertia of bodies to the equations of motion; the other modules calculate the contribution of specific force elements such as springs, dampers, muscles, interior contacts, and restraint systems. Flexible bodies can be specified to simulate 3D deformations with a few degrees of freedom (Koppens et al., 1993). Surfaces that are 3D can be supported on rigid and flexible bodies, and contact algorithms provide options to simulate nonlinear surface compliance in a realistic manner (Verhoeve et al., 2001). Special models are available for vehicle dynamic applications including tire models and a control module that offers the possibility to apply loads to bodies based on information extracted from sensors.

Apart from MADYMO, the above models are hardly used anymore. One exception is a special version of the CAL3D program called the ATB (Articulated Total Body) program developed by the Air Force Aerospace Medical Research Laboratory in Dayton, Ohio, for aircraft safety applications (Ma et al., 1995).

11.2.2 Finite Element Models

In a finite element model, the system to be modeled is divided in a number of finite volumes, surfaces, or lines representing an assembly of finite elements. These elements are interconnected at a discrete number of points: the nodes. In the displacement-based finite element formulation, which is applied in practically all major finite element software packages, the motion of the points within each finite element is defined as a function of the motion of the nodes. The state of stress follows from the deformations and the constitutive properties of the material modeled. Three of the currently most often used software packages for crash simulation using the finite element method are LSDYNA 3D, RADIOSS, and PAMCRASH. These packages are based on the public domain version of the DYNA explicit finite element program that was developed at the Lawrence Livermore Laboratories during the 1970s (Hallquist, 1976). This is the reason why the theoretical concepts of these crash codes do not differ very much from each other. For spatial discretization, the available elements include shell elements, solid elements, beam elements, and membrane elements. A large number of material models are available, such as those describing elastic material behavior, elastic-plastic material behavior with isotropic hardening and failure, crushable foam with failure, orthotropic material behavior, and strain-rate-dependent material behavior.

11.2.3 Hybrid Modeling

The above crash codes allow the inclusion of rigid bodies and are able to simulate some of the specific features of multibody crash simulation programs. In the MADYMO crash simulation program that originally was developed as a multibody code, most of the capabilities of the finite-element-based crash codes are provided in the integrated FE module. Moreover, external interfaces between MADYMO and the FE-based crash codes are available allowing integrated multibody finite element simulations, further referred to as hybrid simulations. Both the multibody method and the finite element method offer their specific advantages and disadvantages in case of crash analyses. The multibody approach is particularly attractive due to its capability of simulating, in a very efficient way, complex kinematical connections present in the human body and in parts of the vehicle structure, such as the steering assembly and the vehicle suspension system. The finite element method offers the capability of describing (local) structural deformations and stress distributions, from which it becomes possible to study, for instance, injury mechanisms in the human body parts. Usually much larger computer times are required to perform a finite element crash simulation than a multibody crash simulation, making the finite element method less attractive for optimization studies involving many design parameters. Extreme material deformations, as occur in impact conditions, challenge the robustness of finite element solutions.

11.3 Available Human Models

Human body models for crash analyses can be subdivided into models of crash dummies and models of the real human body. For a long time the focus has been on crash dummy modeling based on the need, in particular, from the design departments in the automotive industry, for well-validated design tools that can reduce the number of prototype tests with crash dummies in order to shorten and optimize the development process of a new car model and its safety features. For most of the current crash dummies today, databases (often well-validated) for the various crash codes are available and continuous activities take place in various organizations and user groups to further improve such databases, as well as to develop databases for new crash dummies. Detailed FE dummy models are available with 100,000 to 200,000 or more elements (e.g., Schuster et al., 2004). Multibody dummy models using flexible bodies and facet surfaces provide an efficient and predictive alternative (Verhoeve et al., 2001). Experimental databases with thousands of components and assembly tests and hundreds of full dummy tests are being used in the modeling and validation process, and standards are being defined to objectively "rate" the quality of dummy models by systematic comparison of model responses to experimental data (Hovenga et al., 2005). Stochastic dummy models are being introduced to cover the inherent variability of dummy hardware (Dalenoort et al., 2005; Rutjes et al., 2007).

This chapter will further focus on models of the real human body. A model of the real human body is much more difficult to develop than a model of a physical crash dummy. This type of model offers improved biofidelity (human-likeness) compared to crash dummy models and allows the study of aspects like body size, body posture, muscular activity, and post-fracture response. Furthermore they potentially allow analysis of injury mechanisms on a material level. In the next sections three important aspects of human body models will be discussed—namely, dealing with human body geometry (anthropometry), modeling human body tissues, and validating human models for impact loading.

11.3.1 Anthropometry

One of the challenges in modeling the real human body compared to crash dummy modeling is how to deal with the large variations in human body sizes. In case of a multibody approximation, several methods exist, allowing the generation of an arbitrarily sized human body model. Two of the most widely used methods will be mentioned here. The first one was developed in the early 1980s and is available through the software package GEBOD (Baughman, 1983). This software generates geometric and inertia properties for a 15-segment ATB or MADYMO multibody model. Computations for the geometrical parameters and mass distribution are based on a set of 32 body measurements to be specified by the user or generated by GEBOD using regression equations on the basis of body height and/or weight for both adult males and females. Also for children, regression equations are available. A major limitation of GEBOD is the approximation of body segments by simple geometrical volumes.

The second method is based on the use of software packages for ergonomic analyses like the RAMSIS software (Geuß, 1994; Seidl, 1997). RAMSIS allows the generation of human models with a wide range of anthropometry parameters. The RAMSIS model describes the human body as a set of rigid bodies connected by kinematic joints, and the skin is described as a triangulated surface. RAMSIS provides a detailed geometric description of the body segments based on extensive anthropometric measurements on various civilian populations including automotive seated postures. Segment mass and centers of gravity are derived in RAMSIS using this realistic geometric description. RAMSIS provides a mathematical prediction for the increase of the average body height of the entire population during a given time period (secular growth). Anthropometric studies have shown that the body dimensions of each individual can be classified according to three dominant and independent features: body height, the amount of body fat, and body proportion (i.e., the ratio of the length of the limbs to the length of the trunk). Using this classification scheme RAMSIS describes the entire population in a realistic way. This method takes into account the correlations between body dimensions that are disregarded in GEBOD. For instance, tall persons typically have long legs combined with a comparably short trunk. A translator has been developed to convert RAMSIS models into MADYMO models (Happee et al., 1998b). The resulting database contains joint locations, joint ranges of motion, segment masses and centers of gravity, and a triangulated skin connected to various body segments. Inertia properties are derived by integration over segment volume assuming a homogeneous density. This conversion into MADYMO can be performed for any anthropometry specified in RAMSIS. The methodology can also be applied to scale crash dummies. Happee et al. (1998a) created 30 different models by scaling male and female Hybrid III dummy models toward various RAMSIS anthropometries for the purpose of evaluating restraint systems in a frontal collision. The models accounted for human variance with respect to length, corpulence, and the proportion of seated height to standing height.

The CAESAR project (Robinette et al., 2002) provides a next generation of anthropometric data. Data were gathered in North America, the Netherlands, and Italy, and two different 3D scanning technologies were used. More than 13,000 3D scans were provided and 4,431 subjects were measured. CAESAR provides 3D surface (skin) geometries as well as traditional measures for standing and seated subjects.

Finite element full body human models have been developed with 50,000 to 80,000 elements (Robin, 2001; Maeno & Hasegawa, 2001; Iwamota et al., 2002; Yang et al., 2003; Murakami et al., 2004; Haug et al., 2004). For such models information is needed on the structures within the human body (fig. 11.2). Public data from the VISIBLE HUMAN PROJECT (1994) has been used for several models. One of the efforts to achieve consistent full body detailed information from one specific seated subject is the work done in the European HUMOS-1 project (Robin, 2001). For this purpose the method of physical slicing of a human cadaver in driving position was chosen. The slices were photographed and afterwards digitized. The cadaver approximated a 50th percentile human male. Alternative methods to achieve the detailed human body anthropometry are based on MRI techniques. A limitation of these models, although they are very detailed, is that they represent one unique human body size—namely, the one cadaver body that was actually measured. In the European project (HUMOS-2), data were acquired on 64 volunteers and on 24 PMHS (postmortem human subjects) using low-dose stereo x-rays on volunteers, and were used to develop a generic scaling method for FE models (Serre et al., 2006a).

11.3.2 Tissue Characterization

Simulations with human body finite element models require constitutive descriptions for the various materials that constitute the human body (such as bone, cartilage, ligament and tendon, muscular tissue, and various organs such as brain, heart, lung, etc.). Accurate representations of the mechanical behavior of the various components are essential for reliable predictions of injury in impact situations (see Haug et al., 2004, for an overview of tissue properties, and Wismans et al., 2005, for discussion).

In general, properties of biological tissues are viscoelastic (i.e., their response is rate dependent and shows stress relaxation at constant strain level), nonlinear, and anisotropic due to the specific microstructure (for example, consisting of an arrangement of collagen fibers). A full characterization of the constitutive behavior considers the behavior in various deformation modes (shear, uniaxial tension, compression, biaxial deformation, etc.) and complex loading paths (e.g., reverse loading). Furthermore, the use of constitutive models for biological materials in impact biomechanics simulations requires a characterization of these materials at high strain rates. The viscoelastic characteristics are typically determined in (small strain) oscillatory experiments, stress relaxation experiments, and constant strain rate tests at varying strain rates. Prior to the characterization experiments, specimens are often preconditioned (e.g., Meyers et al., 1998; Funk et al., 2000). The frequency range that can be addressed in oscillatory experiments is often limited by the capabilities of the experimental setup. Characterization experiments are often conducted *in vitro*, on small specimens, either in compression/extension (Miller, 2001) or, more commonly, in shear (e.g., Brands et al., 2000). *In vivo* experiments are sometimes carried out by indentation of organs (e.g., Gefen & Margulies, 2004).

The mechanical properties of living tissues may depend on age and gender. Furthermore, properties vary largely between different subjects. Therefore, mechanical experiments on biological materials show a large scatter in obtained results. Additional scatter in data is due to varying test conditions (both physical and mechanical), handling and treatment, and postmortem time (since most tests are conducted *in vitro*). Furthermore, regional differences within an organ or within the body may exist. Tests are conducted either on human cadaver material or on tissues from animal donors. The advantage of the latter is that material can often be tested at shorter postmortem times or even *in vivo*. Consequently, the properties of biological materials as reported in the literature vary widely. For example, the mechanical properties reported for brain tissue vary by orders of magnitude (e.g., Nicolle et al., 2004).

The properties of skeletal muscles can be separated in an active and a passive component. The active response may become important in low-speed accidents such as for example rear-end collisions producing whiplash disorders. One-dimensional phenomenological models are often used to describe the response of skeletal muscles. The Hill muscle model (Hill, 1970) is frequently used for bar elements that simulate the active and passive response of skeletal muscles.

Injury will develop if the mechanical response (e.g., strains, stresses, etc.) of the biological material attains a level at which either the structural integrity of the materials is affected or functionality is reduced. The latter may be the result of physiological processes that occur after the impact, at time scales that are much larger than the time scale of the loading conditions.

11.3.3 Validation

Validation is the process of assessing the reliability of a simulation model in comparison to one or more reference tests with human subjects. Very important in this process is that the reference tests, often also referred to in literature as biofidelity tests, are not the same tests as used for determination of model input data. If results from different tests are available, usually so-called biofidelity corridors are defined by way of envelopes of resulting time histories.

Human models have been validated for frontal, lateral, and rear impact (e.g., Happee et al., 1998a, 2000; Lange et al., 2005) as well as pedestrian loading (e.g., Hoof et al., 2003) using volunteer tests for low-severity loading and PMHS tests for higher severity loading. Accident reconstructions provided additional support for the validity of human models (e.g., Lange et al., 2006). Recently, the validity range of the MADYMO multibody human model has been extended to vertical vibration transmission (Verver et al., 2003).

FE human models have also been validated using bone segment testing (Haug et al., 2004), and some progress validating soft tissue responses has been made using marker and ultrasound techniques.

While human models have been validated extensively for kinematics, accelerations, and compliance (force deflection), the next step is to demonstrate injury prediction capabilities of human body models, which is an area still in its infancy. Some promising results in this respect have been achieved, among others, for long bones where initiation of bone failure was predicted based on yield stress and plastic strain (e.g., Rooij et al., 2003, 2004).

11.4 Applications

Human models are used in the vehicle design process, in accident reconstruction, and in biomechanics research.

11.4.1 Design (CAD) of the Crash Response of Vehicles and Safety Devices

Current regulations and consumer tests prescribe hardware testing with crash dummies representing car occupants in frontal, lateral, and rearward impact. Simulation of these tests is instrumental in vehicle development and subsequently hardware prototyping is postponed to the later design cycles. Full vehicle structural deformation with one or more dummies is effectively simulated using full FE

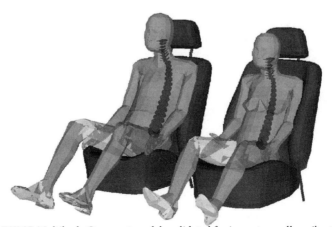

FIGURE 11.1 MADYMO Multibody Occupant models validated for impact as well as vibration analysis.

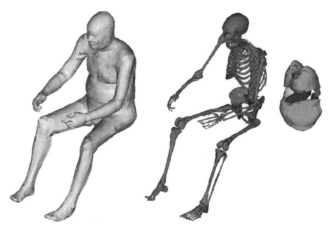

FIGURE 11.2 MADYMO Finite Element Occupant model in automotive seating position based on the HUMOS-1 project (left: soft tissues; middle: skeleton; right: organs).

structural models. However, such full FE vehicle simulations easily require 8 hours to calculate even if multiple CPUs are used for parallel computation. For efficient optimization, submodels of the occupant compartment with dummy, airbag, belt, and the most important interior parts are effectively used (fig. 11.3). Such occupant safety simulations typically require less than one hour to calculate and thereby allow efficient optimization and verification of design robustness through stochastic analysis (Rutjes et al., 2007). Dummy model simulations are sometimes complemented by human model simulation for load cases where suitable dummies do not exist. These include evaluation of risks of airbag-induced injuries (fig. 11.4).

Current pedestrian safety evaluation procedures prescribe loading of the vehicle front with subsystems representing the head and the leg segments (fig. 11.5). Full body human models in various sizes are used to verify whole body kinematics as well as the effects of combined loading from different body parts (Hoof et al., 2003; Yazuki, 2005).

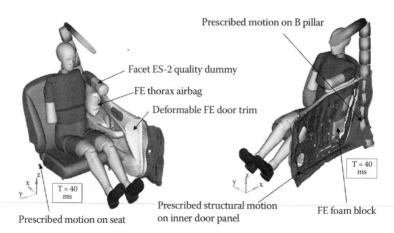

Prescribed motion on B pillar

Facet ES-2 quality dummy

FE thorax airbag

Deformable FE door trim

T = 40 ms

T = 40 ms

Prescribed motion on seat

Prescribed structural motion on inner door panel

FE foam block

FIGURE 11.3 Side impact occupant safety analysis with multibody ES2 crash dummy model. Vehicle structural deformation is prescribed based on a separate FE analysis (Rutjes et al., 2007).

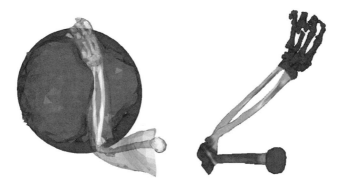

FIGURE 11.4 Airbag-arm impact simulated with FE arm model (Rooij et al., 2003).

11.4.2 Accident Reconstruction and Biomechanics Research

An increased usage of computer models can be observed in the area of accident reconstruction. Three-dimensional dynamic reconstruction enables evaluation of the physical response of vehicles and humans in complex scenarios. Reconstructions are being performed by industry aiming to enhance safety by learning from real accidents, from parties involved in legal cases, and from research institutes. Challenges remain in dealing with the usually large number of unknown accident parameters and the lack of experimental data available for validation for the case under consideration. Development of a code of practice with guidelines for usage of models in accident reconstruction is highly recommended.

Reconstructions generally focus on derivation of vehicle and occupant kinematics matching observed tire traces, vehicle damage, and injury locations. A next step will be to match injury levels with simulation results. A range of European and U.S. research projects now gather real accident data to reconstruct with human models. These include frontal, side, rear, and rollover cases, as well as pedestrian, cyclist, and motorcycle accidents. Early studies already show promising results with current human models (Liu & Yang, 2002; Lange et al., 2006; Serre et al., 2006b; Yao et al., 2006). In the longer term, when larger sets of well-documented cases become available, these may be used to systematically validate human models with real-world accident data, and to develop human-model-based injury criteria and thresholds.

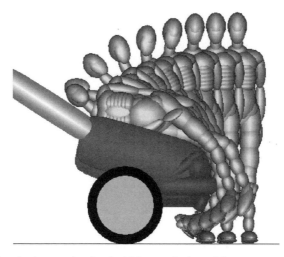

FIGURE 11.5 Car-pedestrian impact simulated with human body model.

Human models are also used in the development of new test procedures and crash dummies. Full body human simulations have been used to predict body kinematics in car to pedestrian impact and have provided guidelines for simplified test procedures with headform devices. Human models are used to define optimal test conditions for experiments with volunteers and cadavers, where a careful pre-study allows achievement of desired loading levels while remaining below injury levels in volunteers, and covers a desired range of loading levels in cadavers. Simulation has also been used to derive body internal deformations and loads from biomechanical data, to investigate injury mechanisms, to investigate potential effects of muscle tone, to scale data from subjects of different body size, and to transfer animal data and cadaver data into data representative of living humans.

While current crash dummies are still designed to meet biofidelity requirements directly based on volunteer and cadaver tests, it is expected that human models will be increasingly used as a reference in dummy development and validation.

11.5 Outlook

Models of the human body can be subdivided into models of crash dummies and models of the real human body. Many models of crash dummies have been developed, and extensive series of validation studies have been conducted with rather impressive results. Also in the field of real human body models promising results have been achieved. An important advantage of real human body models is that they allow the study of the effect of body size, posture influence, as well as muscular activity. Furthermore, human models can benefit with limited delay from new scientific knowledge on injury mechanisms and injury criteria obtained through biomechanical research. In case of a crash test dummy–based design strategy, usually a long period elapses before new findings actually can be implemented in crash dummy hardware. Road maps issued by IRCOBI (2006) and APSN (2006) foresee that virtual testing with human models will be included into regulated test procedures when the benefits of human models are substantiated and standardized models become available.

Current dummies and human models are developed mainly to evaluate severe injuries in high-speed impact. However, the very large cost to society of low-severity injuries warrants further research into the applicable injury mechanisms, injury criteria, and injury tolerances. Also, effects of aging on biomechanical response and injury tolerance require further study. Detailed human models may predict injury at tissue level, but much more knowledge of tissue responses is needed, and robust formulations are needed to achieve numerical stability under severe tissue compression. Validation of human models is still focusing on global measures like contact force, and displacement and accelerations measured with sensors strapped to the body. Detailed measurement techniques are needed to match the level of detail of current human models. In addition, reconstruction of real-world accidents can be of value to prove and enhance the injury predictive value of human models.

Another challenge lies in the development of innovative safety systems—for instance, adapting the operation of airbags and belts toward occupant size and position and/or activating these systems prior to the actual crash. While current restraints act in an open loop fashion, a further injury reduction potential has been shown for real-time controlled restraint systems adapting continuously to occupant position, restraint loads, and vehicle acceleration. Innovative design methods and tools are needed to develop such complex systems and to prove reliability in real life with real people. Herein it will be essential to further develop models predicting the effects of muscle activity prior to and during impact (Cappon et al., 2007).

References

AGARD (1996). Advisory Report 330 Anthropomorphic Dummies for Crash and Escape system testing, AGARD/AMP/WG21, 1996. ISBN 92-836-1039-3. NATO UNCLASSIFIED.

APSN (2006). Secondary Safety Research Action Plan issued by the Advanced Passive Safety Network (APSN). www.passivesafety.com.

Baughman, L.D. (1983). Development of an Interactive Computer Program to Produce Body Description Data. University of Dayton Research Institute, Ohio, USA, Report No. AFAMRL-TR-83-058, NTIS doc. no. AD-A 133 720.

Brands, D.W.A., Bovendeerd, P.H.M., Peters, G.W.M., and Wismans, J.S.H.M. (2000). The Large Shear Strain Dynamic Behaviour of In-Vitro Porcine Brain Tissue and a Silicone Gel Model Material. Proceedings of the 44th Stapp Car Crash Conference.

Cappon, H., Mordaka, J., Van Rooij, L., Adamec, J., Praxl, N., and Muggenthaler, H. (2007). A Computational Human Model with Stabilizing Spine: A Step Towards Active Safety. SAE World Conference, April 2007, Paper SAE-2007-01-117.

Dalenoort, A., Griotto, G., Mooi, H., Baldauf, H., and Weissenbach, G. (2005). A Stochastic Virtual Testing Approach in Vehicle Passive Safety Design: Effect of Scatter on Injury Response, SAE 2005 World Congress and Exhibition, Paper No. 2005-01-1763, April 2005.

Damsgaard M., Rasmussen, J., Christensen, S., Surma, E., and de Zee, M. (2006). Analysis of Musculoskeletal Systems in the AnyBody Modeling System. *Simulation Modelling Practice and Theory* 14 (8), 1100–1111.

Fleck, J.T., Butler, F.E., and Vogel, S.L. (1974). An Improved Three-Dimensional Computer Simulation of Motor Vehicle Crash Victims. Final Technical Report No. ZQ-5180-L-1, Calspan, (4 Vols.).

Funk, J., Hall, G., Crandall, J.R., and Pilkey, W. (2000). Linear and Quasi-linear Viscoelastic Characterization of Ankle Ligaments. *Journal of Biomechanical Engineering* 122, 15–22.

Gefen, A., and Margulies, S.S. (2004). Are In Vivo and In Situ Brain Tissues Mechanically Similar? *Journal of Biomechanics* 37, 1339–1352.

Geuß H. (1994). Entwicklung eines anthropometrischen Mebsystems für das CAD-Menschmodel Ramsis. Ph.D. thesis, München University.

Hallquist, J.O. (1976). A Procedure for the Solution of Finite-Deformation Contact-Impact Problems by the Finite Element Method. Report UCRL 52066, Lawrence Livermore Laboratory, Livermore, California.

Happee, R., Haaster, R., Michaelsen, L., and Hoffmann, R. (1998a). Optimisation of Vehicle Passive Safety for Occupants with Varying Anthropometry, ESV Conference 1998, Paper 98-S9-O-03.

Happee, R., Hoofman, M., van den Kroonenberg, A.J., Morsink, P., and Wismans, J. (1998b). A Mathematical Human Body Model for Frontal and Rearward Seated Automotive Impact Loading. Proceedings of the 42nd Stapp Car Crash Conference, Tempe, AZ.

Happee, R., Ridella, S., Nayef, A., Morsink, P., de Lange, R., Bours, R., and van Hoof, J. (2000). Mathematical Human Body Models Representing a Midsize Male and a Small Female for Frontal, Lateral and Rearward Impact Loading, IRCOBI Conference.

Haug, E., Choi, H.-Y., Robin, S., and Beaugonin, M. (2004). Human Models for Crash and Impact Simulation. *Handbook of Numerical Analysis*, Vol. XII, Editor P.G. Ciarlet, Elsevier.

Hill, A.V. (1970). *First and Last Experiments in Muscle Mechanics*. Cambridge.

Hoof, J., Lange, R., and Wismans, J.S.H.M. (2003). Improving Pedestrian Safety Using Numerical Human Models. *Stapp Car Crash Journal* 47.

Horst, M.J., Bovendeerd, P.H.M., Happee, R., Wismans, J.S.H.M., and Kingma, H. (2001). Simulation of Rear End Impact with a Full Body Human Model with a Detailed Neck: Role of Passive Muscle Properties and Initial Seating Posture. International Technical Conference on the Enhanced Safety of Vehicles (ESV), Amsterdam, the Netherlands.

Horst, M.J., Thunnissen, J.G.M., Happee, R., and Haaster, R.M.H.P. (1997). The Influence of Muscle Activity on Head-Neck Response during Impact. Stapp Conference, SAE 973346.

Hovenga, P.E., Spit, H.H., Uijldert, M., and Dalenoort A.M. (2005). Improved Prediction of Hybrid-III Injury Values Using Advanced Multibody Techniques and Objective Rating. SAE World Conference, SAE Paper 2005-01-1307.

IRCOBI (2006). Future Research Directions in Injury Biomechanics and Passive Safety Research. Issued by the International Research Council on the Biomechanics of Impact (IRCOBI).

Iwamoto, M., Kisanuki, Y., Watanabe, I., Furusu, K., Miki, K., and Hasegawa, J. (2002). Development of a Finite Element Model of the Total Human Body for Safety (THUMS) and Application to Injury Reconstruction. IRCOBI Conference, Munich, Germany, 31–42.

Koopman, B., Grootenboer, H.J., de Jongh, H.J. (1995). An Inverse Dynamics Model for the Analysis, Reconstruction and Prediction of Bipedal Walking. *Journal of Biomechanics* 28 (11), 1369–1376.

Koppens, W.P., Lupker, H.A., and Rademaker, C.W. (1993). Comparison of Modeling Techniques for Flexible Dummy Parts. Stapp Conference 1993, SAE933116.

Lange, R., Happee, R., Rooij, L., and Liu, X.J. (2006). Validation of Human Pedestrian Models Using Laboratory Data as Well as Accident Reconstruction. Expert Symposium on Accident Research (ESAR), Hannover.

Lange, R., Rooij, L., Mooi, H., and Wismans, J. (2005). Objective Biofidelity Rating of a Numerical Human Occupant Model in Frontal to Lateral Impact. *Stapp Car Crash Journal* 49, 457–479.

Liu, X.J., and Yang, J.K. (2002). Development of Child Pedestrian Mathematical Models and Evaluation with Accident Reconstruction. *Traffic Injury Prevention* 3 (4), 321–329.

Ma, D., Louise, A., Obergefell, L.A., and Rizer, A.L. (1995). Development of Human Articulating Joint Model Parameters for Crash Dynamics Simulations. SAE World Conference, Document Number: 952726.

MADYMO Theory Manual, Version 6.3.2 (2006). TNO, the Netherlands.

Maeno, T., and Hasegawa, J. (2001). Development of a Finite Element Model of the Total Human Model for Safety (THUMS) and Application to Car-Pedestrian Impacts. ESV Conference, Paper No. 494.

McHenry, R.R. (1963). Analysis of the Dynamics of Automobile Passenger Restraint Systems. Proceedings of the 7th Stapp Car Crash Conference, 207–249.

Meyers, B.S., Woolley, C.T., Slotter, T.L., Garrett, W.E., and Best, T.M. (1998). The Influence of Strain Rate on the Passive and Stimulated Engineering Stress-Large Strain Behavior of the Rabbit Tibialis Anterior Muscle. *Journal of Biomechanical Engineering* 120, 126–132.

Miller, K. (2001). How to Test Very Soft Biological Tissues in Extension? *Journal of Biomechanics* 34, 651–657.

Murakami, D., Kitagawa, Y., Kobayashi, S., Kent, R., and Crandall, J. (2004). Development and Validation of a Finite Element Model of a Vehicle Occupant. SAE World Conference, SAE 2004-01-0325.

Nicolle, S., Lounis, M., and Willinger, R. (2004). Shear Properties of Brain Tissue over a Frequency Range Relevant for Automotive Impact Situations: New Experimental Results. *Stapp Car Crash Journal* 48.

Robbins, D.H., Bowman, B.M., and Bennett, R.O. (1974). The MVMA Two-Dimensional Crash Victim Simulation. Proceedings of the 18th Stapp Car Crash Conference, 657–678.

Robin, S. (2001). HUMOS: Human Model for Safety: A Joint Effort Towards the Development of Refined Human-like Car Occupant Models. 17th International Technical Conference on the Enhanced Safety of Vehicles, Paper No. 297.

Robinette, K.M., Blackwell, S., Daanen, H.A.M., Fleming, S., Boehmer, M., Brill, T., Hoeferlin, D., and Burnsides, D. (2002). Civilian American and European Surface Anthropometry Resource (CAE-SAR), Final Report, Volume I: Summary, AFRL-HE-WP-TR-2002-0169, United States Air Force Research Laboratory, Human Effectiveness Directorate, Crew System Interface Division, 2255 H Street, Wright-Patterson AFB OH 45433-7022 and SAE International, 400 Commonwealth Dr., Warrendale, PA 15096.

Rooij, L., Bours, R., Hoof, J., Mihm, J.J., Ridella, S.A., Bass, C.R., and Crandall, J.R. (2003). The Development, Validation and Application of a Finite Element Upper Extremity Model Subjected to Air Bag Loading. *Stapp Car Crash Journal* 47.

Rooij, L., Hoof, J., McCann, M.J., Ridella, S.A., Rupp, J.D., Barbir, A., Made, R., and Slaats, P. (2004). A Finite Element Lower Extremity and Pelvis Model for Predicting Bone Injuries due to Knee Bolster Loading. SAE Digital Human Modeling Conference, Paper No. 2004-01-2130.

Rutjes, N., Hassel, E., Happee, R., and Cappon H. (2007). Evaluation and Improvement of Side Impact Occupant Safety Using Optimization and Stochastic Analysis. SAE World Conference. Paper SAE-2007-01-0365.

Schuster, P., Franz, U., Stahlschmidt, S., Pleschberger, M., and Eichberger, A. (2004). Comparison of ES-2re with ES-2 and USSID Dummy, Considerations for ES-2re Model in FMVSS Tests. LS DYNA User Conference, Bamberg Germany.

Seidl, A. (1997). RAMSIS—A New CAD Tool for Ergonomic Analysis of Vehicles Developed for the German Automotive Industry. SAE International Congress, Detroit, February 1997 Paper No. 970088.

Serre, T., Brunet, C., Bruyere, K., Verriest, J.P., Mitton, D., Bertrand, S., Skalli, W., Bekkour, T., and Kayvantash, K. (2006a). HUMOS (Human Model for Safety) Geometry: From One Specimen to the 5th and 95th Percentile. SAE Digital Human Modeling Conference, SAE-DHM 2006-01-2324.

Serre, T., Masson, C., Perrin, C., Chalandon, S., Llari, M., Cavallero, C., Py, M., and Cesari, D. (2006b). Pedestrian and Cyclist Accidents: A Comparative Study Using In-Depth Investigation, Multibody Simulation and Experimental Test. IRCOBI Conference, Madrid.

Van der Helm, F.C.T., Schouten, A.C., de Vlugt, E., and Brouwn, G.G. (2002). Identification of Intrinsic and Reflexive Components of Human Arm Dynamics during Postural Control. *Journal of Neuroscience Methods* 119 (1), 1–14.

Van Ingen Schenau, G.J., Boots, P.J.M., De Groot, G., Snackers, R.J., and Van Woensel, W.W.L.M. (1992). The Constrained Control of Force and Position in Multijoint Movements. *Neuroscience* 46, 197–207.

Verhoeve, R., Kant, R., and Margerie, L. (2001). Advances in Numerical Modeling of Crash Dummies. ESV Conference.

Verver, M.M., Hoof, J., Oomens, C.W.J., Wouw, N., and Wismans, J.S.H.M. (2003). Estimation of Spinal Loading in Vertical Vibrations by Numerical Simulation. *Clinical Biomechanics* 18 (9), 800–811.

Visible Human Project (1994). National Library of Medicine. Visible Human Database. 8600 Rockville Pike, Bethesda, MD 20894 (http://www.nlm.nih.gov/research/visible).

Wismans, J., Happee, R., and Dommelen, J.A.W. (2005). Computational Human Body Models. Iutam Conference.

Yang, K.H., Beillas, P.D., Zhang, L., Lee, J.B., Shah, C.S., Hardy, W.N., Demetropoulos, C.K., and King, A.I. (2003). Advanced Human Modelling for Injury Biomechanics Research. Document Number: SAE Digital Human Modelling Conference, SAE-DHM-2003-01-2223.

Yao, J., Yang, J., and Otte, D. (2006). Investigation of Brain Injuries by Reconstructions of Real World Adult Pedestrian Accidents. IRCOBI Conference, Madrid.

Yazuki, T. (2005). Using THUMS (Total Human Model for Safety) for Pedestrian Safety Simulation. Aachener Kolloquium Fahrzeug- und Motorentechnik.

12

Development of Hand Models for Ergonomic Applications

Thomas J. Armstrong,
Jaewon Choi, and
Vinnie Ahuja

12.1 Introduction

Development of digital human models of the hand has much to offer those concerned with design of equipment that is to be produced, used, or serviced by people. First it is necessary to understand to what end, where, and how people interact with equipment. Ergonomics provides a foundation for understanding how people interact with equipment in a given environment. These interactions can be used to determine how models can be crafted to solve specific problems.

12.2 Ergonomics

The Human Factors and Ergonomics Society provides a working definition for ergonomics:

> The Society furthers serious consideration of knowledge about the assignment of appropriate functions for humans and machines, whether people serve as operators, maintainers, or users in the

system. And, it advocates systematic use of such knowledge to achieve compatibility in the design of interactive systems of people, machines, and environments to ensure their effectiveness, safety, and ease of performance. (HFES 2007)

A key concept is "the system." Figure 12.1 shows a simple human-machine system that contains one operator and one machine. The system has specific objectives, such as entering data into a computer or joining two parts with threaded fasteners using a hand tool. System objectives also include quality and safety concerns, such as acceptable error rates or risk of musculoskeletal disorders.

To achieve the system objectives, it is necessary for the operator to gather information, process information, and operate controls. Information may come from visual, auditory, or tactile pathways. Tactile feedback requires coupling between the hand and the grip object. Many of the actions require coupling between the hand and controls, tools, or materials. The performance of the system will be affected by how well the hand and the grip object fit and by the force required to manipulate the grip object.

The human-machine system is an important tool for identifying how the hand is used to achieve system objectives. Interactions between the human and machine can then be described as a list. Each step can then be inspected to identify sources of required information and which involve use of the hands. Similarly, each step can be inspected to see how the hands are used to grasp, hold, or manipulate one or more machine components. For example, the objective of a job may be to attach an electrical connector to the motherboard of a computer. Step 1: The worker must first visually locate the loose end of the connector and the correct location on the motherboard. Step 2: The worker must reach for the connector. Step 3: The worker must grasp the connector. Step 4: The worker must move the free end of the connector to the motherboard. Step 5: The worker must reposition the hand on the connector so to have sufficient clearance for the hand. Step 6: The worker must position the free end of the connector on the motherboard. Step 7: The worker must exert sufficient force to join the connectors. The position of the hand is important because it affects the worker's strength and dexterity. It also affects the clearance between the hand and obstructions in the workspace. Tactile feedback also affects how the worker grips and manipulates the grip object.

Biomechanical models provide a basis for understanding and predicting many aspects of human-machine performance. Specific examples include: gathering information, reaching for grip objects, and manipulating grip objects. Gathering information may require operators to position their heads where they can see or hear or to position their hands where they can feel. Tactual stimulation is a biomechanical process that entails deformation of tissue when it comes into contact with another surface. Reaching requires positioning the body or body part near a grip object where it can be touched or grasped. It also may involve moving around or through obstructions in the work environment. Grasping objects involves applying force with the hand so that it can be supported against another part of the hand or another object in the work environment. Manipulation involves moving an object from one location or orientation to another. In the course of moving the object it may be necessary to reposition the hand to avoid contact with another object or overcome excessive load forces on the hand. Exertions of the body

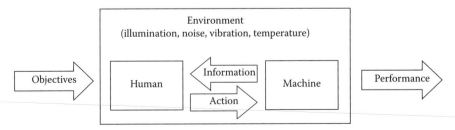

FIGURE 12.1 The human-machine system is an important tool for determining the reach and strength required for performance to achieve the system objectives.

produced internal forces on joints, tendons, and nerves. Repeated exertions or very high forces may cause chronic or acute injury.

Whole body models based on kinematics and biomechanics are used to evaluate a worker's ability to see and reach objects in the work environment (Chaffin et al., 2006). Whole body models also are used to assess a worker's ability to lift or push. Hand models have not yet reached the same level of functionality as that of whole body models. Designers still rely heavily on population studies of hand size and strength. Kinematic and hand biomechanical models show great promise for the future.

12.3 The Hand

12.3.1 Anthropometry

There have been a number of anthropometric studies of the hand. These include static dimensions such as hand length, breadth, joint thickness, and joint breadth (Garrett, 1971; Roebuck, 1995). Some studies have examined functional anthropometric dimensions such as inside grip diameter, distance between the center of grip or center of pinch and the wrist, and the space required for the hand to hold and use some tools (Churchill et al., 1957; Clauser et al., 1988). Although these data provide important design benchmarks, they lack the flexibility and robustness needed to generalize them from one design problem to another.

12.4 Hand Posture

12.4.1 Hand Strength

Hand strength is affected profoundly by the position of the hand. Power grip is the position of maximum strength (Landsmeer, 1962; Napier, 1962; Long et al., 1970). In power grip, the fingers and thumb are wrapped around the handle in opposite directions and the palmar surface of the hands and fingers are in continuous contact with the grip object or another part of the hand. Studies have shown that maximum strength is produced for handles 1.25 to 1.50 inches in diameter. Hammers, axes, tennis rackets, and baseball bats are grasped using power grip postures.

Pinch grip is a position of maximum control (Landsmeer, 1962; Napier, 1962; Long et al., 1970). The grip object is held between the tips of one or more fingers and the tip or side of the thumb or against another grip object. Pinch grip strength is only 15 to 20 percent that of power grip. Pinch grip is typically used for manipulation of surgical or dental tools, to assemble small parts, to write, to use a mouse or a keyboard. In some cases workers may shift between pinch and power grip depending on the force requirements of a task. For example, a pinch grip may be used to start a screw but a power grip may be used to tighten it. These categories of hand posture provide important insights into how people can be expected to perform a given task using a given grip object.

There is a growing body of literature about kinematic and biomechanical models. These models were created to develop a basic understanding of how the hand works, to understand the cause of various pathologies, and to design medical interventions. With the availability of powerful desktop computers, development of digital hand models for evaluation and design of equipment that is held or manipulated with the hand is now feasible.

12.5 Kinematic Hand Models

All hand models utilize a kinematic chain to represent the digits of the hand. Each phalange and metacarpal is represented as a separate link. A system of links described by Buchholz and Armstrong (1992) is shown in Figure 12.2. The length of the links and the range of joint motions correspond to those of the hand. Buchholz and Armstrong (1992) proposed a model that can be used to predict the length of each

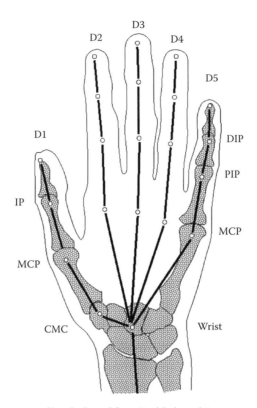

FIGURE 12.2   Link representation of hand adapted from Buchholz and Armstrong (1992).

link as a function of hand length. The outer surface of the hand can be represented as a geometric object, such as an ellipsoid, or as an array of points or nodes (Buchholz & Armstrong, 1992; Hui & Wong, 2002; Choi & Armstrong, 2006). Geometric shapes can be described mathematically, which facilitates manipulations of the hand; however, the resulting hand does not look lifelike. Also, grip objects may have irregular surfaces that are difficult to describe mathematically. Surface representation using points makes it possible to capture irregularities that make the computer image look like a hand.

Posture can be predicted using a kinematic chain based solely on fit, based on biomechanics, based on behavior, or based on some combination of all three. Fitting the hand to the grip object entails first placing the grip object near the hand and then closing the fist until each segment of the hand is in contact with the object. The resulting posture will be affected by the size, shape, orientation, and position of the grip object. It also will be affected by how the hand is closed and by mechanical constraints on the object. Figure 12.3a shows a simple 2D rendering of a forearm, palm, and finger through the third digit that will be used to illustrate some of the issues associated with posture prediction using kinematic models. Each segment is depicted with an arc around each end of the phalange and tangent lines representing the dorsal and the palmar surfaces. The surface is depicted using 200 equally spaced points. For the purposes of the following illustrations, a 50th percentile male hand length of 196.3 mm was selected based on Garrett (1971). Link lengths for the first, second, third, and fourth segment (palm and first, second, and third phalange) were calculated as 88, 52, 33, and 21 mm based on Buchholz and Armstrong (1992) and the 50th percentile hand length. A grip object with a 38-mm-diameter circular cross-section is depicted using 100 equally spaced points about the circumference of the cylinder. The following examples illustrate the effects of the algorithm used to close the fist, the position of the grip object, and the physical constraints on the grip object.

In Figure 12.3a, the grip object is positioned adjacent to the first segment of the hand (the palm) 20 percent of the hand length from the wrist joint. The fingers are flexed one at a time starting with the

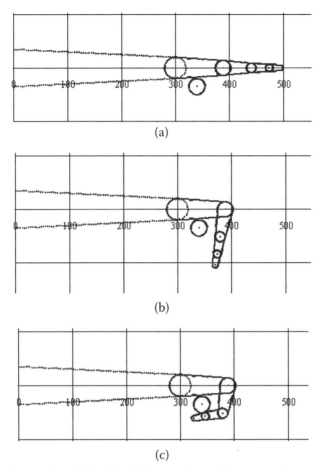

FIGURE 12.3 Fiftieth percentile male hand length (196.3 mm), HL (Garrett, 1971), average finger length proportions (Buchholz & Armstrong, 1992), 38-mm-diameter object, diameter located 20% HL from wrist. Fingers are rotated one at a time until contact is made. Final hand posture: MCP = –100°, PIP = –70°, DIP = 0°.

MCP joint, followed by the PIP joint, and finally the DIP joint until they make contact with the grip object. This algorithm is summarized as follows:

1. Identify joint i distal to grip object
2. Increment joint θ_i by $\Delta\theta$
3. If no contact and $\theta_i \leq \theta_{imax}$ then go to step 2
4. Increment i
5. If i $\leq$ number of joints, i_{max}, then go to step 1
6. Stop

The joints are numbered 1 to 4, which corresponds to the MCP, PIP, and DIP joints. There are a number of algorithms for detecting contact (Lin & Gottschalk, 1998; Bourg, 2002), but for the purposes of this example, contact was determined by visual inspection of the graphical output of the simulation.

The first joint (MCP) rotates to 100°—the limits of its range-of-motion—without making contact (fig. 12.3b). The second joint (PIP) rotates to 70° when contact occurs for the fourth segment (fig. 12.3c). The third joint (DIP) does not rotate because contact occurs distal to the third joint. The grip object appears to be supported between the palmar surfaces of the first and fourth segments of the hand.

12.6 Posture Prediction: Multiple Joints Together

In the preceding examples, the fingers were flexed one joint at a time. People don't move this way. They move their joints together at the same time. We now consider a second example in which the fingers move together at the same rate until contact is achieved between an external object and a distal segment of the hand. Figure 12.4 shows the same hand size and proportions and the same grip object size and location as shown in Figure 12.3. In this case the joint angles are incremented using the following algorithm.

1. Identify joint i distal to grip object
2. Increment joints i to i_{max} by $\Delta\theta$
3. If no contact and $\theta_i \le \theta_{imax}$ then go to step 2
4. Stop

   In this case, contact occurs when all of the joints have reached 78° and the tip of the finger touches the grip object. The posture shown in Figure 12.4 is quite different from the previous result shown in Figure 12.3 even though the hand and grip object are the same size and the grip object is in the same location. The only difference is the procedure for closing the fist. In reality the fingers tend to move together, although the relative rates may vary depending on the size, shape, and location of the grip object and the purpose of the grasp. There also may be inter-subject variations when everything else is the same.

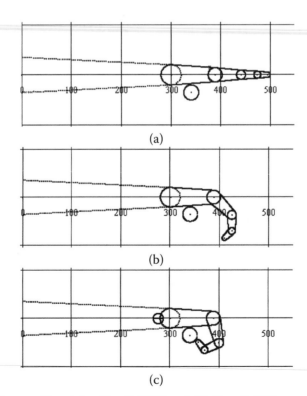

(a)

(b)

(c)

FIGURE 12.4 Fiftieth percentile male hand length (196.3 mm), HL (Garrett, 1971), average finger length proportions (Buchholz & Armstrong, 1992), 38-mm-diameter object, diameter located 20% HL from wrist. Fingers are rotated simultaneously at the same rate until contact is made. Final hand posture: MCP = PIP = DIP = 78°.

12.7 Posture and Object Placement

The predicted posture of the hand also may vary according to where the object is positioned. Figure 12.5 shows a 50th percentile hand length with a 3.8-cm-diameter object placed at 20, 30, and 60 percent of the hand length. The fingers are all rotated at the same rate until contact occurs. In Figure 12.5a, the first contact occurs when the end of the finger contacts the object. In Figure 12.5b, the palmar surfaces of the proximal and distal phalange both contact the object at the same time. Even though the same joint angles are obtained, the contact between the grip object and the hand is quite different. In Figure 12.5c, there is contact between the object and the proximal phalange before any movement occurs. All of the rotation in this case is at the PIP and DIP joints, and rotation stops when the distal phalange contacts the grip object. The posture shown in Figure 12.5c is quite different from that shown in Figures 12.5a and 12.5b.

12.8 Object Constraints

The postures obtained in Figures 12.3 to 12.5 all assume that the object is fixed so that finger rotation stops when contact occurs between the grip object and the hand. Such a scenario might correspond to someone grasping a railing or the rung of a ladder. In many cases the object is not fixed in space. This scenario might involve someone picking up a drinking glass, a bar of soap, a part, or a tool. In this case, initial contact between the grip object and the hand causes the object to move until the grip object contacts a second fixed object or another part of the hand. Figure 12.6 shows a 50th percentile hand length hand grasping a 3.8-cm-diameter object. The object is initially located at 60 percent of hand

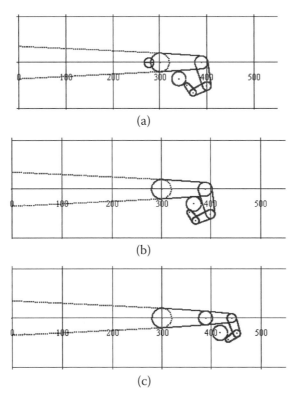

(a)

(b)

(c)

FIGURE 12.5 Fiftieth percentile male hand length (196.3 mm), HL (Garrett, 1971), average finger length proportions (Buchholz & Armstrong, 1992), 38-mm-diameter object, placed at 20%, 30%, and 60% HL. Joints rotate at same rate together until contacting object. (Final hand postures a: $\Theta_1 = \Theta_2 = \Theta_3 = -78°$; b: $\Theta_1 = \Theta_2 = \Theta_3 = -78°$; and c: $\Theta_1 = 0°$, $\Theta_2 = \Theta_3 = -72°$.)

length. All of the joint angles are assumed to rotate at the same rate. The object is in contact with the proximal phalange at the beginning of the grasp, so it begins to move as the hand closes. The PIP and DIP joints stop rotating when the distal phalange makes contact with the grip object. Finally the MCP joint stops rotating when the object comes into contact with the palm. The resulting posture shown in Figure 12.5c is quite different from that shown in Figure 12.6e.

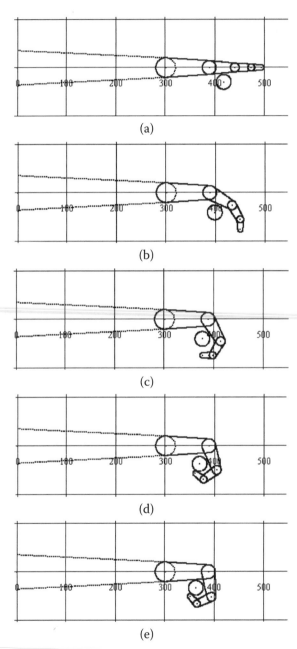

FIGURE 12.6 Fiftieth percentile male hand length (196.3 mm), HL (Garrett, 1971), average finger length proportions (Buchholz & Armstrong, 1992), 38-mm-diameter object placed at 60% HL. Joints rotate at same rate together until contacting object. Object free to rotate with fingers until it contacts proximal segment of the hand. (Final hand posture $\Theta_1 = -83°$, $\Theta_2 = -72°$ and $\Theta_3 = -72°$.)

12.9 Finger Rotation and Hand Movement

It was previously shown that predicted hand postures are affected by the relative rates in interdigit joint rotations. In the second algorithm, it was assumed that the joints all rotate at the same rate. Opening and closing the fist is a complex process involving biomechanical, neural, and cognitive processes (Jeannerod & Prablanc, 1983; Jeannerod, 1986; Cole & Abbs, 1986; Walsh, 1994; Kamper et al., 2002; Kamper et al., 2003). Figure 12.7 shows hand aperture and hand distance as functions of time during a 300-mm reach for a 60-mm-diameter cylindrical handle starting from a relaxed position. Hand aperture is the distance between the tip of the index finger and the thumb. The aperture increases to 150 mm before decreasing to 50 mm to complete the grasp. Figure 12.8 shows a plot of the interdigit joint angles during this reach. The relaxed starting position of the MCP, PIP, and DIP joints are 37°, 28°, and 9°, respectively. It can be seen that the MCP, PIP, and DIP joint extend −9°, −2°, and −3°, respectively, and then close 14°, 37°, and 27°, respectively, to complete the grasp. Movements about the MCP joint have a much greater effect on aperture than those about the DIP joint because of the distance between the tip of the finger and the joint. This is particularly true for pinch type exertions.

Models have been proposed for describing finger motions using principal component analysis, hyperbolic tangent functions, and spline functions (Santello et al., 2002; Braido & Zhang, 2004; Li & Tang,

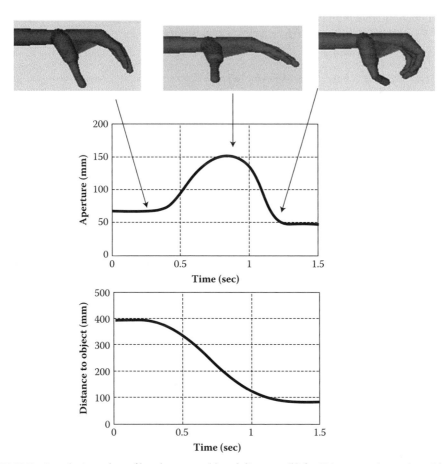

FIGURE 12.7 Sample time-plots of hand aperture (a) and distances (b) for 300 mm reach starting with relaxed hand to grip of 60-mm-diameter cylindrical handle. (Aperture is defined as the distance between the tip of the thumb and the tip of the index finger.)

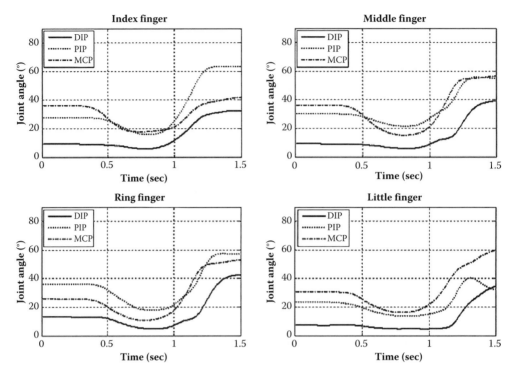

FIGURE 12.8 MCP, PIP, and DIP joint angles for index, middle, ring, and little finger versus time for 300 mm reach to grasp a 60 mm handle.

2007). These models are empirical and are not easily adjusted for differences in object size and shape, object location, or task variables. Models are needed that integrate control and learning with kinematics and biomechanics. Movements vary within and among people so model predictions should be expressed probabilistically—not deterministically.

12.10 Optimizing Object-Hand Fit

An alternative to closing the fist is calculation of joint angles based on assumed contact between the hand and grip object. Lee and Zhang (2005) proposed the use of a procedure that minimized the sum of distances from the finger joints to the object surface. This approach eliminates the need to consider relative finger rotation. It also eliminates the need for repeated use of a contact algorithm each time one or more joints are incremented; however, the biological or behavioral rationale for minimizing the sum of the distances is not clear. Also, it is not clear how it would work for non-cylindrical handles for pinch grip or other nonpower grip postures.

12.11 Three-Dimensional Models

The preceding examples were based on 2D simulations of one finger for simplicity. Many real-world problems require 3D models with multiple fingers. The 3D simulations are much more computationally intensive than 2D problems—especially as routines are added to detect contact and to refine surfaces. Still such computations and simulations are within the capability of contemporary desktop computers. Also, there are a number of tricks that can be employed to find approximate distances between the

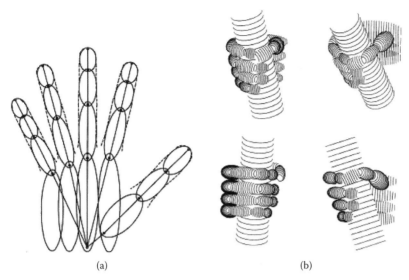

(a)   (b)

FIGURE 12.9 Three-dimensional kinematic hand model. (From Buchholz, B. PhD thesis. Ann Arbor, University of Michigan, 1989.)

hand and the grip objects (Lin & Gottschalk, 1998; Bourg, 2002). It is possible to start out with a small number of points and add points as the hand gets closer to the grip object. $\Delta\theta$ can be adjusted according to the distance between the hand and the grip object. It is also possible to employ search strategies that reduce the number of required iterations. Yet another strategy employs the use of geometric surfaces to describe approximate shape of hand and grip object parts. Distances and contact between geometric shapes can be computed faster than for all of the surface points.

Buchholz (1989) and Buchholz and Armstrong (1991 and 1992) described a 3D multifinger model. They utilized a stick figure surrounded with ellipsoids to represent the fingers and palm as shown in Figure 12.9a. This implementation was well suited for computers of the mid-1980s because contact between two surfaces is not computationally intensive. The ellipsoid model provided good predictions of hand posture using the first algorithm described above for closing the fist (see fig. 12.3) for power grip postures (see fig. 12.9b; Buchholz, 1989; Buchholz & Armstrong, 1992). Limitations of this implementation included the unrealistic look of the hand, difficulty depicting irregular-shaped grip objects using geometric objects, and the time required to display graphical images.

Choi and Armstrong (2006) utilized Buchholz's link lengths to implement a kinematic model of the hand. They used an array of points to depict the surface of the hand (see fig. 12.10a). The locations of the surface points were calculated using truncated cones and data from Buchholz and Armstrong (1991). OpenGL graphic functions were used to fill in the surface to enhance its lifelike look (see fig. 12.10b). The fingers were all rotated together at the constant rate to close the fist as described in the second algorithm above. A contact algorithm was implemented in which the distance between all points was computed. Contact occurred when the distances decreased to a predetermined distance. Model predictions were compared with observed postures for six subjects ranging from small females to large males gripping 26-, 60-, and 114-mm-diameter cylinders. The overall correlation, r^2, between predicted and observed interdigit joint angles was 0.72 (see fig. 12.10c), but varied from 0.47 for the DIP joint of the little finger to 0.93 for the MCP joint of the middle finger. Joint angles were found to be very sensitive to cylinder diameters (see fig. 12.11)—angles increased with decreased cylinder diameter—which is consistent with expectations and previous investigators (Lee & Rim, 1990). As shown above for the 2D model, joint angles were sensitive to position of the grip object in the hand for the 3D model. It was also shown that the angle of the object in the hand affects the angle of the interdigit joints. The effect of skin deformation was investigated

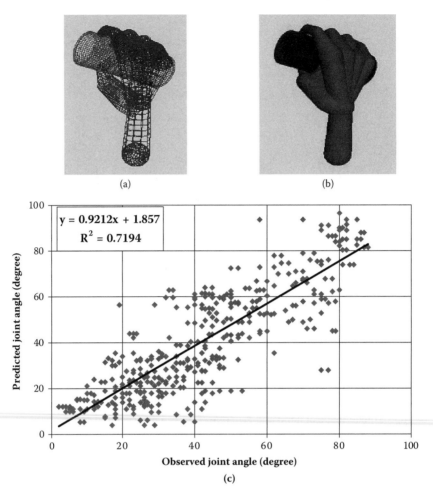

FIGURE 12.10   (a) The hand is represented as a series of links surrounded by an array of points. (b) The surface is filled in to make the hand look more lifelike using OpenGL graphic functions. (c) A comparison of observed versus predicted joint angles for gripping cylindrical handles (Choi & Armstrong, 2006).

by modifying the thickness of the finger segment—reducing the thickness 1 mm corresponds to 1 mm of skin deformation. Deformations as much as 20 percent resulted in only a few degrees difference compared to no deformation. The truncated cones model looks better than the ellipsoid model, but can be improved using scanned images of real hands. This implementation of the model is suitable for contemporary desktop or laptop computers, but efficient contact algorithm will make it run faster.

12.12 Space Requirements

Performance of the human-machine system can be adversely affected if there is not sufficient space for the hand to grasp, manipulate, and use grip objects, such as controls, tools, and parts. Choi et al. (2007) examined the use of a kinematic model to predict the space required to install flexible hose onto a flange. Figure 12.12 compares two views of a real hand grasping a rubber hose with two views based on a simulation using a kinematic model. A rectangle is constructed around the hand to determine the vertical and horizontal space required for the hand gripping the hose at fixed intervals along the length of the hose. Experiments were conducted to compare the space requirements based

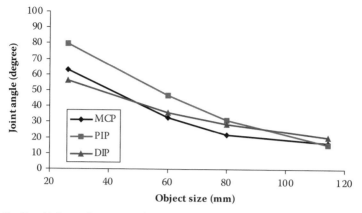

FIGURE 12.11 Predicted joint angles versus cylinder diameter (Choi & Armstrong, 2006).

on observations with those based on simulation. Light emitting diodes were placed on the dorsal side of each joint. These positions were then tracked as subjects installed the hose onto a flange. The data were transformed to a coordinate system that corresponded to the hose. The points on the hand were stratified into groups that corresponded to their position along the hose (see fig. 12.12a). The data in each stratum were then searched to find the minimum and maximum horizontal and vertical values. In the simulation, a cylindrical object of the same diameter and length of the hose was

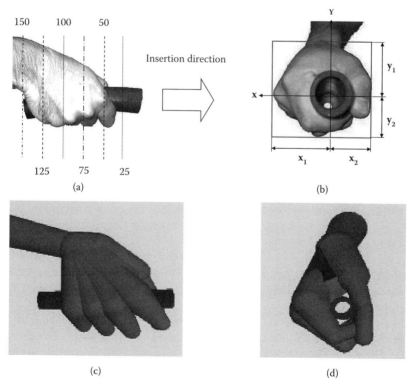

FIGURE 12.12 Space is required for the hand to attach a flexible hose to a flange, a and b. The required space can be predicted using a kinematic model of the hand, c and d (Choi et al., 2007).

TABLE 12.1   Comparison of Simulations and Observations

Distance from the End	Study	Horizontal Axis		Vertical Axis		Area (mm²)
		X_1 (mm)	X_2 (mm)	Y_1 (mm)	Y_2 (mm)	
25	*	34.2	27.1	12.5	37.4	2216
	**	36.4	35.3	12.5	35.3	2946
50	*	34.2	50.9	49.6	37.6	7415
	**	40.5	38.8	43.7	38.9	6549
75	*	38.1	50.9	49.6	44.5	8375
	**	41.7	54.6	53.7	49.1	9887
100	*	38.1	86.0	86.7	48.6	16798
	**	48.8	64.2	62.5	60.1	13846
125	*	36.9	89.3	90.2	49.8	17673
	**	50.1	67.0	72.9	65.9	16248
150	*	31.8	89.3	90.2	49.8	16950
	**	48.0	85.6	87.8	67.5	20748

*: Simulation, **: Grieshaber et al. (2006).

placed at the same location that the subjects were observed to hold the hose and the fist was closed until contact occurred between the fingers, thumb, and hose. The surface points were then stratified in the same ranges as the observed points. The points were then searched for maximum vertical and horizontal values. One of the complications of these experiments was the tendency of subjects to twist or wiggle the hoses as they are installed—which increases the space required for the hand. The observed movements of the hose were measured so that the simulated hand could be moved the same way. Table 12.1 compares the observed and simulated results. The observed area for the first strata was 25 percent greater than that observed; from their error decreased to 9 percent for the strata at the wrist. Although the relative difference is high for the first strata, the absolute difference is smallest for this location.

With refinements, kinematic models will be powerful design tools for predicting grip postures for given tasks and grip objects. Models are needed that capture the effects of tasks and grip object variables as well as subject behaviors.

12.13 Biomechanical Models

The models discussed thus far are based solely on kinematic or the spatial relationships between parts of the hand and grip objects. Kinematic models are important; first, they can be used to examine the fit between the hand and objects in the work environment. Second, kinematic models can be used to predict possible postures that can be used to hold a given grip object. Posture has a profound impact on hand strength (Blackwell et al., 1999; Yan & Downing, 2001). Third, kinematic models provide a framework that is necessary for biomechanical models that describe the relationship between dynamic and static forces acting on the hand. Inverse biomechanical models are used to predict muscle forces based on observed movements (Buchner et al., 1988; Sancho-Bru et al., 2001). Forward biomechanical models are used to predict hand movements and external forces based on muscle forces and movements (Chao et al., 1976; Armstrong, 1982; Lee & Rim, 1990; Esteki & Mansour, 1997; Valero-Cuevas et al., 1998; Li, Zatsiorsky, & Latash, 2000; Sancho-Bru et al., 2003). Exertions of the hand involve multiple muscles that produce moments about the same joint. As a result, computation of muscle forces requires simplifying assumptions or optimization algorithms. At present, these models provide a basic understanding of how the hand works, which provides insights for the design and use of grip objects (Sancho-Bru et al., 2003). Skin deformation and friction are important biomechanical considerations of grip (Johannson & Westling, 1984; Buchholz et al., 1988; Frederick & Armstrong, 1995; Seo & Armstrong, 2006). Current biomechanical models have not yet reached a level of functionality where they are suitable for use in a desktop environment by designers not familiar with the underlying biomechanics. Future studies using inverse biomechanics to describe and model behavior patterns are needed for development of forward biomechanical models that can be used for design of grip objects and tasks.

12.14 Biomechanical Stresses

Work-related musculoskeletal disorders (WMSDs) of the hand-wrist and forearm are a leading cause of worker disability and compensation (NRC, 1999; NRC & IOM, 2001; BLS, 2006). Biomechanical models provide a basis for understanding the mechanisms of WMSDs and for the evaluation and design of jobs. Exertions and movements of the hand involve the transmission of force from muscles in the forearm through the finger flexor and extensor tendons. The flexor digitorum profundus (FDP) and flexor digitorum superficialis (FDS) muscles are the major force producing for grip and pinch; although the extensor and intrinsic muscles also may be subjected to repetitive loads to stabilize the hand. Figure 12.13 shows how forces produced by the FDP are transmitted through the carpal tunnel to the fingers via the flexor tendons. In addition to tensile stresses, compressive stresses are produced between tendons and adjacent anatomical structures. In addition, shear stresses are produced on the surface of tendons and tendon sheaths as the tendons move back and forth through the carpal tunnel. Studies have shown that repetitive exertions, high forces, and extreme postures can be associated with WMSDs such as hand and wrist tendonitis and carpal tunnel syndrome. These studies include biomechanical analyses that show

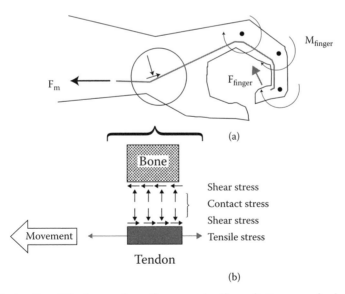

FIGURE 12.13 Contraction of the forearm finger flexor muscles (flexor digitorum profundus and superficialis) produces tendon tension, which produces moments about the interphalangeal joints, which close the fist about external objects or another part of the hand (a). Tension on the finger flexor tendons causes them to rub against adjacent anatomical surfaces (b).

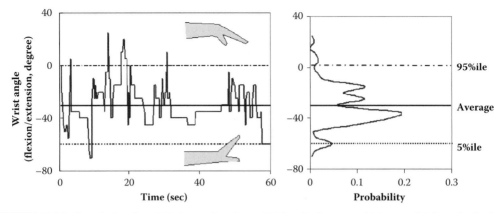

FIGURE 12.14 Sample time-based (4 observations/second) (a) and a frequency histogram (b) for wrist flexion/extension in a medium repetition job.

correspondence between the location of stress concentrations and injuries of tendons and nerves (Armstrong & Chaffin, 1979; Moore et al., 1991; Moore & Garg, 1994; Moore, 2002). These also include *in vivo* studies of tendon loads (Brain et al., 1947; Smith et al., 1977; Goldstein et al., 1987; Kursa et al., 2005). Finally, the relationship between the risk factors (repetitive exertions, high forces, and extreme postures) and WMSDs has been shown epidemiologically (Silverstein et al., 1987; Marras & Schonmarklin, 1993; Moore & Garg, 1994; Latko et al., 1999; Leclerc, 2001; Gell et al., 2005). These studies support the use of computer models to evaluate and control the risk of WMSDs. Choi and Armstrong (2007) re-examined jobs studied by Latko et al. (1999) to investigate the relationship between tendon excursions at the wrist and risk of WMSDs. A time-based analysis of wrist posture (see fig. 12.14) was performed as described by Armstrong et al. (2003). Average angular velocities and accelerations were calculated using the time plots of wrist posture. (Tendon excursions, $\Delta x = r \, \Delta\theta$ where r is a constant and $\Delta\theta$ is the angle of joint rotation.) Armstrong and Chaffin (1978) developed a model for the constants used to predict FDP and FDS tendon excursions based on joint angles and hand size. Choi and Armstrong (2007) computed and integrated the absolute value of tendon excursions to be expressed in meters excursion per hour so that they could be compared among jobs with different WMSD risk levels (fig. 12.15). It can be seen that increasing angular velocity, acceleration, and tendon excursions all correspond with increasing risk of WMSDs.

Choi's study used time-based observations of job videos to determine wrist posture. Additionally it only considered postures of the wrist. Future models will make it possible to predict finger motions as a function of object and task attributes. These same models will make it possible to predict tendon excursions

TABLE 12.2   Mean and Standard Deviation of Velocity, Acceleration, and Tendon Excursions

WMSD Risk Group	Flexion/Extension		Radial/Ulnar Deviation		Tendon Excursion	
	Velocity (°/s)	Acceleration (°/s²)	Velocity (°/s)	Acceleration (°/s²)	FDP (m/hr)	FDS (m/hr)
High	51.4 ± 20.0	338.0 ± 157.8	19.1 ± 7.9	118.3 ± 49.6	42.9 ± 19.3	49.7 ± 21.9
Medium	25.5 ± 9.3	155.5 ± 44.8	12.7 ± 3.3	87.0 ± 19.7	18.9 ± 8.9	22.3 ± 10.1
Low	7.9 ± 3.5	60.4 ± 29.1	2.46 ± 1.4	18.7 ± 11.5	5.7 ± 2.7	6.7 ± 3.2

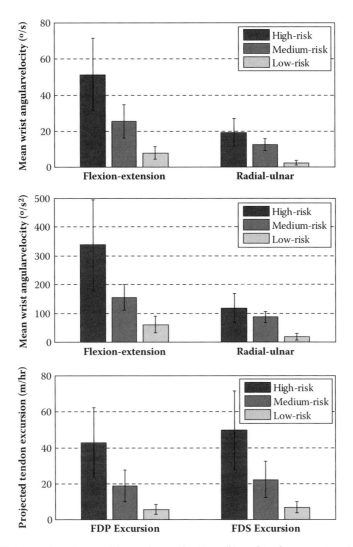

FIGURE 12.15 Wrist angular velocities (a), angular accelerations (b), and tendon excursions (c) are all related to risk of hand-wrist work-musculoskeletal disorders (Choi & Armstrong, 2007).

based on object and task attributes. In addition to eliminating the time required for time-based observations, it also will make it possible to compare alternative job designs proactively.

12.15 Summary

The ability to grasp and manipulate grip objects and the risk of musculoskeletal disorders are necessary considerations of equipment design. Population studies of hand size, strength, behavior, and injury patterns are valuable design resources. However, it may not be possible to generalize data collected for one population and one set of conditions to others. Kinematic and biomechanical hand models show great potential for overcoming this limitation, but have not yet reached a level of functionality where they can be used by equipment designers in a desktop environment. Studies of hand use and behavior will lead to more powerful and robust models and use of models for design purposes.

References

Armstrong, T. (1982). Development of a biomechanical hand model for study of manual work. *Anthropometry and biomechanics: Theory and application*. R. Easterby, K. H. E. Kroemer and D. B. Chaffin. New York, Plenum Press: 183–92.

Armstrong, T. J., D. B. Chaffin (1978). An investigation of the relationship between displacements of the finger and wrist joints and the extrinsic finger flexor tendons. *J Biomech* 11(3): 119–28.

Armstrong, T. J., D. B. Chaffin (1979). Some biomechanical aspects of the carpal tunnel. *J Biomech* 12(7): 567–70.

Armstrong, T. J., W. M. Keyserling, D. C. Grieshaber, et al. (2003). Time based job analysis for control of work related musculoskeletal disorders. 15th triennial Congress of the International Ergonomics Association, Seoul, Korea.

Blackwell, J. R., K. W. Kornatz, E. M. Heath (1999). Effect of grip span on maximal grip force and fatigue of flexor digitorum superficialis. *Appl Ergon* 30(5): 401–5.

BLS (2006). *Workplace injuries and illnesses, 2005*. Washington, DC, United States Department of Labor, Bureau of Labor Statistics. http://stats.bls.gov/news.release/pdf/osh.pdf.

Bourg, D. M. (2002). *Physics for game developers*. Sebastopol, CA: O'Reilly and Associates, Inc.

Braido, P., X. Zhang (2004). Quantitative analysis of finger motion coordination in hand manipulative and gestic acts. *Hum Mov Sci* 22(6): 661–78.

Brain, W. R., A. D. Wright, M. Wilkinson (1947). Spontaneous compression of both median nerves in the carpal tunnel. *The Lancet* 1(March): 277–82.

Buchholz, B. (1989). A kinematic model of the human hand to predict its prehensile capabilities. PhD thesis, Biomedical Engineering, Ann Arbor, University of Michigan.

Buchholz, B., L. J. Frederick, T. J. Armstrong (1988). An investigation of human palmar skin friction and the effects of materials, pinch force and moisture. *Ergonomics* 31(3): 317–25.

Buchholz, B., T. J. Armstrong (1991). An ellipsoidal representation of human hand anthropometry. *Hum Factors* 33(4): 429–41.

Buchholz, B., T. J. Armstrong (1992). A kinematic model of the human hand to evaluate its prehensile capabilities. *J Biomech* 25(2): 149–62.

Buchholz, B., T. J. Armstrong, S. A. Goldstein (1992). Anthropometric data for describing the kinematics of the human hand. *Ergonomics* 35(3): 261–73.

Buchner, H. J., M. J. Hines, H. Hemami (1988). A dynamic model for finger interphalangeal coordination. *J Biomech* 21(6): 459–68.

Chaffin, D. B., G. Andersson, B. J. Martin (2006). *Occupational biomechanics*. Hoboken, NJ: Wiley-Interscience.

Chao, E. Y., J. D. Opgrande, F. E. Axmear (1976). Three-dimensional force analysis of finger joints in selected isometric hand functions. *J Biomech* 9(6): 387–96.

Choi, J., T. Armstrong (2006). Sensitivity study of hand posture using a 3-dimensional kinematic hand model. 16th congress of the International Ergonomics Association, Maastricht, the Netherlands, Elsevier.

Choi, J., T. J. Armstrong (2006). Examination of a collision detection algorithm for predicting grip posture of small to large cylindrical handles. 2006 Digital Human Modeling for Design and Engineering Conference and Exhibition, Lyon, France, SAE.

Choi, J., T. J. Armstrong (2007). Assessment of the risk of MSDs using time-based video analysis. Sixth International Scientific Conference on Prevention of Work-Related Musculoskeletal Disorders, Boston, MA.

Choi, J., D. C. Grieshaber, T. J. Armstrong (2007). Estimation of grasp envelope using a 3-dimensional kinematic model of the hand. Human Factors and Ergonomics Society 51st Annual Meeting, Baltimore, MD.

Churchill, E., A. Kuby, et al. (1957). Nomograph of the hand and its related dimensions. Wright Patterson Air Force Base, Aero Medical Laboratory: 49.

Clauser, C. E., I. O. Tebbetts, et al. (1988). Measurer's handbook: U.S. Army anthropometric survey, 1987–1988. *Technical Report*, NATICK, TR-88/043, AD A202 721.

Cole, K. J., J. H. Abbs (1986). Coordination of three-joint digit movements for rapid finger-thumb grasp. *J Neurophysiol* 55(6): 1407–23.

Esteki, A., J. M. Mansour (1997). A dynamic model of the hand with application in functional neuromuscular stimulation. *Ann Biomed Eng* 25(3): 440–51.

Frederick, L. J., T. J. Armstrong (1995). Effect of friction and load on pinch force in a hand transfer task. *Ergonomics* 38(12): 2447–54.

Garrett, J. W. (1971). The adult human hand: some anthropometric and biomechanical considerations. *Hum Factors* 13(2): 117–31.

Gell, N., R. A. Werner, A. Franzblau, S. S. Ulin, T. J. Armstrong (2005). A longitudinal study of industrial and clerical workers: incidence of carpal tunnel syndrome and assessment of risk factors. *J Occup Rehabil* 15(1): 47–55.

Goldstein, S. A., T. J. Armstrong, D. B. Chaffin, et al. (1987). Analysis of cumulative strain in tendons and tendon sheaths. *J Biomech* 20(1): 1–6.

Grieshaber, D., T. Armstrong, M. Lau, J. Choi (2006). The effect of insertion method and required force on the grasp envelope during rubber hose insertion tasks. 16th congress of the International Ergonomics Association, Maastricht, the Netherlands, Elsevier.

HFES. (2007). About the Human Factors and Ergonomics Society. From http://www.hfes.org/web/AboutHFES/about.html.

Hui, K. C., N. N. Wong (2002). Hands on a virtually elastic object. *The Visual Computer* 18: 150–63.

Jeannerod, M. (1986). Mechanisms of visuomotor coordination: a study in normal and brain-damaged subjects. *Neuropsychologia* 24(1): 41–78.

Jeannerod, M., C. Prablanc (1983). Visual control of reaching movements in man. *Adv Neurol* 39: 13–29.

Johansson, R. S., G. Westling (1984). Roles of glabrous skin receptors and sensorimotor memory in automatic control of precision grip when lifting rougher or more slippery objects. *Exp Brain Res* 56(3): 550–64.

Kamper, D. G., E. G. Cruz, M. P. Siegel (2003). Stereotypical fingertip trajectories during grasp. *J Neurophysiol* 90(6): 3702–10.

Kamper, D. G., T. G. Hornby, W. Z. Rymer (2002). Extrinsic flexor muscles generate concurrent flexion of all three finger joints. *J Biomech* 35(12): 1581–89.

Kursa, K., E. Diao, L. Lattanza, D. Rempel (2005). In vivo forces generated by finger flexor muscles do not depend on the rate of fingertip loading during an isometric task. *J Biomech* 38(11): 2288–93.

Landsmeer, J. M. (1962). Power grip and precision handling. *Ann Rheum Dis* 21: 164–70.

Latko, W. A., T. J. Armstrong, A. Franzblau, S. S. Ulin, R. A. Werner, J. W. Albers (1999). Cross-sectional study of the relationship between repetitive work and the prevalence of upper limb musculoskeletal disorders. *Am J Ind Med* 36(2): 248–59.

Leclerc, A., M. F. Landre, J. F. Chastang, I. Niedhammer, Y. Roquelaure (2001). Upper-limb disorders in repetitive work. *Scand J Work Environ Health* 27(4): 268–78.

Lee, J. W., K. Rim (1990). Maximum finger force prediction using a planar simulation of the middle finger. *Proc Inst Mech Eng* [H] 204(3): 169–78.

Lee, S. W., X. Zhang (2005). Development and evaluation of an optimization-based model for power-grip posture prediction. *J Biomech* 38(8): 1591–97.

Li, Z. M., J. Tang (2007). Coordination of thumb joints during opposition. *J Biomech* 40(3): 502–10.

Li, Z. M., V. M. Zatsiorsky, M. L. Latash (2000). Contribution of the extrinsic and intrinsic hand muscles to the moments in finger joints. *Clin Biomech* (Bristol, Avon) 15(3): 203–11.

Lin, M., S. Gottschalk (1998). Collision detection between geometric models: A survey. Proc. of IMA Conference on Mathematics of Surfaces.

Long, C., 2nd, P. W. Conrad, E. A. Hall, S. L. Furler (1970). Intrinsic-extrinsic muscle control of the hand in power grip and precision handling: An electromyographic study. *J Bone Joint Surg Am* 52(5): 853–67.

Marras, W. S., R. W. Schoenmarklin (1993). Wrist motions in industry. *Ergonomics* 36(4): 341–51.

Moore, A., R. Wells, D. Ranney (1991). Quantifying exposure in occupational manual tasks with cumulative trauma disorder potential. *Ergonomics* 34(12): 1433–53.

Moore, J. S. (2002). Biomechanical models for the pathogenesis of specific distal upper extremity disorders. *Am J Ind Med* 41(5): 353–69.

Moore, J. S., A. Garg (1994). Upper extremity disorders in a pork processing plant: relationships between job risk factors and morbidity. *Am Ind Hyg Assoc J* 55(8): 703–15.

Napier, J. (1962). The evolution of the hand. *Sci Am* 207: 56–62.

NRC (1999). *Work-related musculoskeletal disorders: a review of the evidence.* Washington, DC: National Academy Press.

NRC and IOM (2001). *Musculoskeletal disorders and the workplace: Low back and upper extremities.* Washington, DC: National Academy Press.

Roebuck, J. A. (1995). *Anthropometric methods: Designing to fit the human body.* Santa Monica, CA: Human Factors and Ergonomics Society.

Sancho-Bru, J. L., A. Perez-Gonzalez, M. Vergara-Monedero, D. Giurintano (2001). A 3-D dynamic model of human finger for studying free movements. *J Biomech* 34(11): 1491–500.

Sancho-Bru, J. L., A. Perez-Gonzalez, M. Vergara, D. J. Giurintano (2003). A 3D biomechanical model of the hand for power grip. *J Biomech Eng* 125(1): 78–83.

Sancho-Bru, J. L., D. J. Giurintano, A. Perez-Gonzalez, M. Vergara (2003). Optimum tool handle diameter for a cylinder grip. *J Hand Ther* 16(4): 337–42.

Santello, M., M. Flanders, J. F. Soechting (2002). Patterns of hand motion during grasping and the influence of sensory guidance. *J Neurosci* 22(4): 1426–35.

Seo, N., T. J. Armstrong (2006). The effect of torque direction on hand-object coupling. The American Society of Biomechanics 30th meeting, Blacksburg, VA.

Silverstein, B. A., L. J. Fine, T. J. Armstrong (1987). Occupational factors and carpal tunnel syndrome. *Am J Ind Med* 11(3): 343–58.

Smith, E. M., D. A. Sonstegard, W. H. Anderson, Jr. (1977). Carpal tunnel syndrome: contribution of flexor tendons. *Arch Phys Med Rehabil* 58(9): 379–85.

Valero-Cuevas, F. J., F. E. Zajac, C. G. Burgar (1998). Large index-fingertip forces are produced by subject-independent patterns of muscle excitation. *J Biomech* 31(8): 693–703.

Walsh, B. S. (1994). Movement before contact. *The grasping hand.* C. L. Mackenzie and T. Iberall. New York, North-Holland. 104: 109–201.

Yan, J. H., J. H. Downing (2001). Effects of aging, grip span, and grip style on hand strength. *Research Quarterly for Exercise and Sport*, 71–77.

13

Foot Modeling and Footwear Development

Ameersing Luximon and
Ravindra S. Goonetilleke

13.1 Introduction

In the early years, foot coverings made of animal skin were used to protect human feet against harsh weather and rough terrain. As time went by, foot coverings became more elaborate, complex, and specialized. The design and construction of footwear, which were customized to each person, were indicative of the wearer's wealth and status. After the industrial revolution, footwear production was mechanized, thereby becoming an affordable commodity. With the emergence of mass-produced footwear, the focus shifted to lowering costs, and, as a result, a lot of footwear did not meet the basic criteria of fit and comfort. Recent technological innovations in digital modeling and the attraction of mass customization are enabling companies to focus on comfort and fit, while maintaining near mass-production efficiency. But, both the human foot and footwear design must be clearly understood to improve fit and comfort. This chapter discusses the human foot, traditional footwear production, mass customization of footwear, and recent advances in foot modeling to enable improvements in footwear design and development.

13.2 The Human Foot

The human foot (Morton, 1935) consists of 26 bones: 14 phalanges, five metatarsals, three cuneiforms (medial, lateral, and intermediate), the navicular, cuboid, calcaneus, and talus (fig. 13.1). A common division of the foot is the hindfoot (cuneiforms, navicular, cuboid, calcaneus, and talus), the midfoot (metatarsals), and the forefoot (phalanges). The inner side (or big toe side) of the foot is called the medial side, while the outer side (or little toe side) is called the lateral side of the foot.

Cartilage at the joints provides smooth surfaces for the movement of bones, while ligaments, connecting one bone to another, provide stable joints (Abboud, 2002). For example, the calcaneofibular ligament at the ankle joint is attached to the calcaneus and the fibula. At the ankle joint, the talus joins the foot to the tibia and fibula on the leg. The medial end of the tibia forms the medial malleolus, while the lateral side of the fibula forms the lateral malleolus. The alignment of the tibia (torsion) influences the position of the medial malleolus and in turn affects both posture and gait, while the fibula influences foot orientation and allows complex foot movements.

The calcaneus, also known as the heel bone, is the primary weight-bearing bone that is subjected to heavy impact during activities such as walking and running. The Achilles tendon, which attaches the gastrocnemius (calf) and soleus muscles to the rear (posterior) surface of the calcaneus bone, is the thickest and strongest tendon in the body.

The navicular bone lies between the talus and the three cuneiforms, while the plantar calcaneonavicular ligament (also called spring ligament) joins the calcaneus and the navicular and is an important ligament as it supports the arch of the foot. The tuberosity (small outward growth) on the medial side of the navicular can be detected by palpation; its height is the navicular height, which is highly correlated with arch height. The three cuneiforms make the transverse arch of the foot. The cuboid provides facets for the bases of the fourth and fifth metatarsal bones. The metatarsals are the five long bones of the foot, the first three metatarsals are joined to the cuneiform bones (lateral, medial, and intermediate), while the last two metatarsals are joined to the cuboid. The forefoot has 14 phalanges (phalangeal bones) that form the toes of the foot. The plantar (bottom of the foot) fascia consists of a network of ligaments that extend from the calcaneus to the metatarsals and the phalangeal bones. The plantar fascia provides the shape and structure for the whole foot and also influences the shape and height of the medial arch.

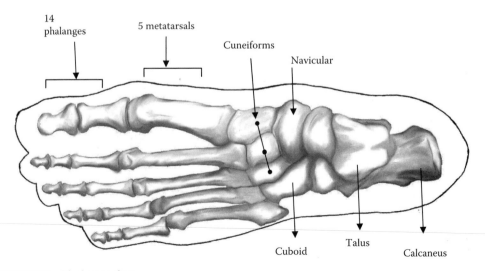

FIGURE 13.1 The human foot.

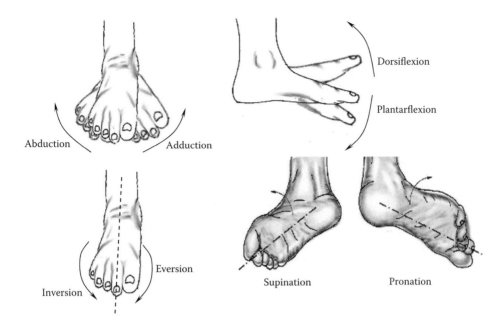

FIGURE 13.2 The foot movements.

The ankle joint, subtalar joint, midtarsal joint, metatarsal-phalangeal joints, and phalangeal-phalangeal joints are the main joints in the foot. The ankle joint is between the tibia, fibula, and talus bones and actually provides movement in only the vertical plane (hinge type of joint) and enables dorsiflexion and plantar flexion (fig. 13.2). Injuries to the ankle joint are very common among athletes (Wexler, 1998). The subtalar joint (talocalcaneal joint) is between the talus and calcaneus, and it allows inversion and eversion of the foot. Inversion is the movement of the sole of the foot away from the median plane (center of foot), while eversion is the movement of the sole toward the median plane (fig. 13.2). The metatarsal-phalangeal and the phalangeal-phalangeal joints act mainly as hinge joints and enable toe grasping. The adductor hallucis and the abductor hallucis muscles enable adduction and abduction of the foot. Abduction is the motion of the foot away from the midline of the body, while adduction is the motion toward the midline of the body. There tends to be pronation and supination of the foot during walking. Pronation is a combination of the three motions of eversion, abduction, and dorsiflexion. On the other hand, supination is a combination of inversion, adduction, and plantar flexion. Supination is a rolling motion to the outside edge or lateral side of the foot, while pronation is the rolling motion to the medial side of the foot. The bones, muscles, ligaments, and other tissues of the foot are enclosed in a protective layer, the skin. This complex structure allows the foot to perform its functions of locomotion and support. Due to the many structures in the foot, any changes to them while in static and dynamic situations result in differences in the internal and external shape of the foot.

13.3 Foot Variations

Foot size and shape changes depend on the load the foot bears, and these changes are primarily related to the static or dynamic state of a person. The static state includes standing on one foot or both feet. Of particular interest is standing on one foot (full weight-bearing), standing on both feet (semi-weight-bearing), sitting while the foot is resting (partial weight-bearing), and no weight bearing where the foot is not touching the ground. In 80% of any population, the changes in the two feet differ with weight bearing (Rossi, 1983). For example, Rossi (1983) reported that, in 85% of women, the length of one foot increased by 4 mm (1/2 a shoe size in the U.S. sizing system), while the length of the other increased

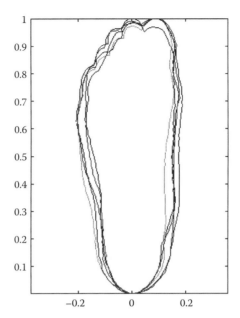

FIGURE 13.3   Static foot outlines (normalized by foot length).

by more than 8.7 mm with increases in load bearing. Differences in shape exist among persons as well. Figure 13.3 shows the differences in the foot outlines of five subjects when in a semi-weight-bearing position. Due to the complexities of the foot structure, there has been no clear model proposed to predict foot shape changes with increased weight bearing until recently (Xiong & Goonetilleke, 2007).

Most footwear research has focused on the pressure distributions under the foot (Luximon & Zhang, 2006). Equipment such as the Fscan mat (Tekscan Inc., Boston, MA, USA) and the Pedar system (NOVEL, http://www.novel.de) have been widely used to measure static as well as dynamic foot pressures, in spite of research that shows the inaccuracies of such systems (Lee et al., 2001). Other methods to measure pressure include the Harris mat, which is a foot impression system using ink that is employed largely in clinical settings. Plantar pressure distributions have been used to classify the arch type with the extremes being labeled as flat foot (*pes planus*) and high arched foot (*pes cavus*). In addition, comparisons of the pressure patterns of full and semi-weight-bearing conditions are indicative of the flexibility within the foot. Flexible feet make foot impressions that are similar to flat-foot impressions during full weight-bearing conditions while normal feet make foot impressions that are similar to the semi-weight-bearing patterns. Mildly flat feet tend to be "corrected" using foot orthotics, while those with serious cases must undergo surgery to restore the arch. Most children below the age of 3 years tend to have flat feet but they usually grow out of this condition with age. Whether distributing pressures increases the level of comfort is yet unknown as the effects of localizing forces and distributing forces seem to depend on the thresholds of discomfort and pain in the various locations of the foot (Goonetilleke, 1998). Apart from pressure distribution patterns, parameters such as the arch index, the arch angle, the navicular drop, the rear-foot angle, the valgus index, and the like, have been used to quantify the level of flat-footedness (Razeghi & Batt, 2002). The pressure distributions tend to differ with the amount of load bearing and by the height the heel is raised, commonly referred to as the heel height of a shoe. Figure 13.4 shows examples of the foot pressure distributions in full and half weight-bearing positions with no heel height and with a four-inch heel height. Generally, increases in heel height cause a shift in the pressure from the heel to the forefoot resulting in an increase in forefoot pressure.

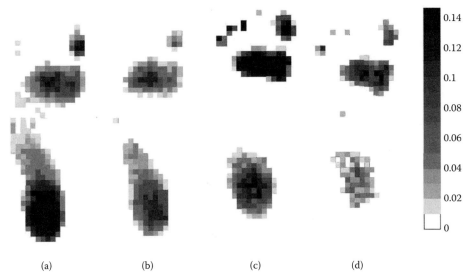

(a) (b) (c) (d)

FIGURE 13.4 Foot pressure distribution for differing heel height and differing weight bearing conditions: (a) full-weight, 0″ heel height; (b) half-weight, 0″ heel height; (c) full-weight, 4″ heel height; (d) half-weight, 4″ heel height.

When a person walks, runs, or jumps, the forces acting on the foot can be quite different and depend on the activity. The linear and angular acceleration, velocity, and the corresponding displacements impose different forces on different parts of the foot. It is not an easy task to quantify all of these forces for different activities and different individuals. However, numerous researchers have focused their attention on the walking cycle known as the human gait. The gait cycle begins when one foot touches the ground at the heel and ends when the same foot touches the ground again (fig. 13.5). The human gait has a stance phase and a swing phase. The stance phase involves 60% of the cycle and is the time during which the foot is in contact with the ground, while the swing phase is when the foot is off the ground. If one foot is in the swing phase, the other foot is in the stance phase and helps support the whole body. There are times when both feet are in the stance phase and are touching the ground. Generally, the load acting on the two feet is usually not the same for a number of different reasons. As a result, the foot shape and plantar pressure distributions between feet tend to be different at any given moment.

The body motion during gait and other activities can be evaluated using two-dimensional (2D) or three-dimensional (3D) motion capture systems, while foot pressure distributions and forces can be measured using pressure transducers and force platforms, respectively. Figure 13.5 shows an example of the forces acting on a foot when walking. The force on each foot can be as much as 1.2 times the body weight during walking and tends to be much higher during running or jumping. Hence, it is no surprise to have footwear-related injuries and illnesses when some types of footwear give a false sense of security to the wearer (Goonetilleke, 1992). With advances in technology, footwear manufacturers are now in a position to account for these large forces and torques through improved designs.

13.4 Shoe Manufacture

From a bag-like wrapping made of animal skin, footwear has now advanced to a level to include computer chips, as in the smart shoe released by Adidas in 2005. Footwear improvements have always lagged behind other technological advancements. The first intelligent load-bearing system was developed in 1992 (Gross et al., 1992) but its equivalent implementation in footwear is just starting, primarily because footwear fit and comfort have for the most part lacked in-depth study. Manufacturers have circumvented the issue by using gross shape approximations in the development of shoe lasts used in mass production

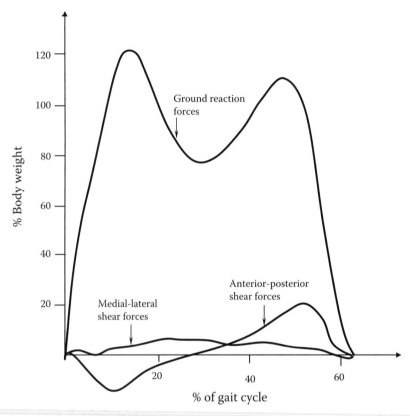

FIGURE 13.5   Forces on the foot when walking.

(fig. 13.6). The recent shift toward mass customization is providing a window of opportunity to improve subjective perceptions of footwear comfort and fit through improvements in our understanding of the foot-shoe interaction (Au & Goonetilleke, 2007) and the associated manufacturing processes.

13.5 History of Shoe Making

Footwear production has undergone many cycles. Prior to the industrial revolution, most footwear was custom-made using a wooden shoe last (fig. 13.7), which itself is a 3D mold. In 1807, William Young of Philadelphia was granted a shoe last patent and he is considered to be the pioneer last maker in the United States. After Thomas Blanchard's invention in 1819 for turning irregular forms, machines were used to produce lasts (Quimby, 1944). In 1820, the first last-making plant was established. The invention of the sewing machine by Lyman Reed Blake in 1858 enabled the soles of shoes to be attached to the uppers by machine. In 1867, Christian Dancel, who had migrated to New York City, had just invented a machine for sewing shoes. Charles Goodyear Jr. bought the rights to this machine and employed Dancel in his factory. Together, in 1874, they invented the Goodyear Welt machine for sewing boots and shoes, which is in use even today with some minor improvements.

By 1909, Boston became the center of shoe making and at that time a pair of shoes cost around $4 to $6. In 1914, Salvatore Ferragamo, a leading Italian footwear designer, moved to Boston where one of his brothers worked in a cowboy boot factory (Ricci, 1992). While working in the factory, he was appalled at the poor quality and aesthetically unpleasant shoes made there and convinced his brothers to move with him to California where he started making made-to-measure shoes for celebrities. With his fashion design background, Ferragamo attempted to find out why his shoes "pleased the eye yet hurt the foot."

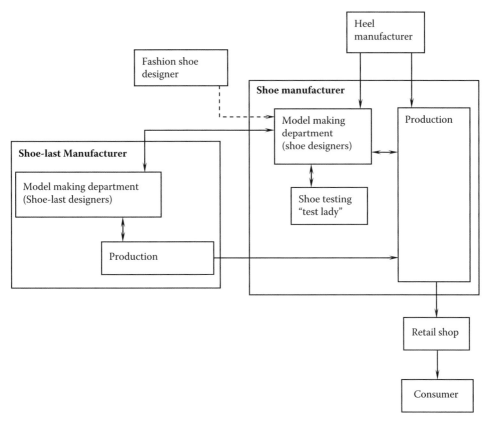

FIGURE 13.6　Simplified diagram to illustrate footwear manufacturing.

It was at that time that Ferragamo proceeded to study anatomy at the University of Southern California with the hope of improving the comfort of shoes. After 13 years in the United States, Ferragamo moved back to Italy and expanded his operations in the early 1950s.

Today, the high-density, high-intensity footwear-manufacturing region of the world is Hong Kong and South China, where the bulk of footwear is manufactured for the whole world. Hong Kong's footwear industry earned HK$46.8 billion in 2007 (Trade Development Council, 2007).

As mass-produced footwear became ubiquitous, there were others who were focused on measuring feet. In 1926, Charles F. Brannock, the son of a shoe industry entrepreneur, improved on the wooden RITZ stick and patented the Brannock device to measure feet. It is in use today in most retail stores around the world.

The first breakthrough in last making occurred in 1961 when the first plastic last (fig. 13.8b) was developed by the Sterling Last Company. Unlike a wooden last that changes shape with changes in temperature and humidity, plastic lasts retain their shape and are not as sensitive to environmental changes. Today, almost all production lasts are made of plastic or aluminum with the aluminum last being used for injection-molded shoes and boots. Wooden lasts, which can be expensive depending on the wood used, are still in use for making the master or designer's last.

13.6　Shoe Lasts

A shoe gets its shape from the shoe last, and so it is an important component that gives the shoe its comfort and fitting qualities. The last comes from the Anglo-Saxon word *laest*, which means a footprint

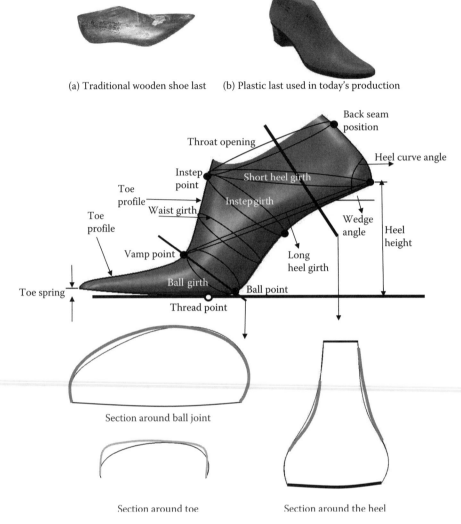

(a) Traditional wooden shoe last   (b) Plastic last used in today's production

FIGURE 13.7   Shoe last.

or a foot track. The shoe last has a relatively complex shape without any straight lines. Even though the American Footwear Manufacturers Association (AFMA) defines as many as 61 terms related to last dimensions, there is no direct mapping from the foot to the last. Some of the last dimensions are shown in Figure 13.7. When making a new shoe last, the last maker rarely starts from scratch. Instead, the last maker uses an existing last as a starting point and modifies it by cutting and replacing the toe shape (fig. 13.9), or adjusting the back part of the last to fit the heel (fig. 13.10). In order to change the wedge angle (to fit different heel designs with the same heel height) the last maker cuts the back part of an existing shoe last and adjusts the shape to get at the desired wedge angle. Then, new material is added to the last, sanded, and filed until the last looks right. Last makers believe that the back part of the shoe last is for comfort and the toe part is for design. They are primarily guided by their experience and skill. Typical last factories have tens of thousands of shoe lasts, most of which differ in just the toe shape as dictated by fashion. The other components of a shoe are the upper, outsole, insole, midsole, and the heel. Sports shoes have additional components in the outsole and midsole for extra cushioning.

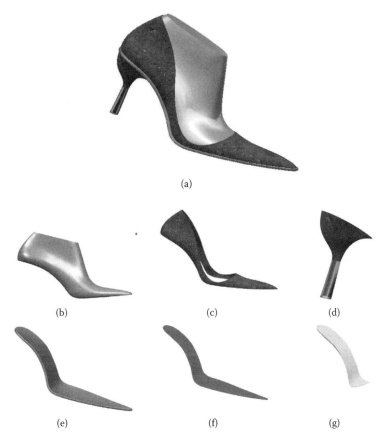

FIGURE 13.8 Components of a shoe: (a) complete shoe with shoe-last; (b) shoe-last; (c) shoe upper; (d) heel; (e) outsole; (f) insole; and (g) midsole.

13.7 Systems of Last Design

Last makers adopt differing guidelines depending on where they design and manufacture the last. An American last maker is guided by the six dimensions of ball girth, waist girth, instep girth, long heel girth, short heel girth, and the stick length. The stick length is influenced by the heel height, toe spring, and toe style. The degree of toe spring in a last depends on the heel height, shoe style, upper material,

FIGURE 13.9 Variations in toe shapes.

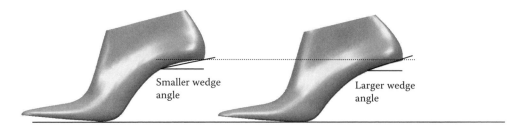

Smaller wedge
angle

Larger wedge
angle

FIGURE 13.10   Adjustment for heel designs.

and flexibility of the shoes. With a higher heel height, the toe spring tends to be lower while stiff upper material requires more toe spring. The amount of toe spring depends on the type of footwear as well. Walking shoes would require more toe spring than dress shoes, while ballet shoes do not have any toe spring. The back part of a last is somewhat a standard for any given last factory and is based on the heel height, shoe type, and shoe construction.

Most shoe factories use the AKA64/WMS system as the basis for design and modification (fig. 13.11). In 1964, shoe manufacturers in Germany initiated an extensive foot measurement study to improve shoe fit for children, which resulted in the development of the AKA64 last. This last is available in three widths and hence the term WMS: W (Weit = wide), M (Mittel = medium), S (Schmal = narrow). The AKA64 system is quite detailed and provides all necessary dimensions. For example, a ladies size 6B last would have the following dimensions (fig. 13.11a): $AI = 246$ mm, $HI = 15$ mm, $IO_1 = 37\%AI$, $AO_1 = 63\%AI$, Angle $E_1O_1I = 74°$, $E_1O_1 = 15\%$ Ball girth, $O_1D_1 = 23\%$ Ball girth, Angle $O_1E_1I_1 = 96°$, Angle $O_1D_1I_2 = 71°$, $D_1I = 20\%AH$, $AB = 1/6$ AH, $B_1B = BB_2 = 1/3$ $E_1D_1 + 1$, Angle $O_1BO = 6°$, Angle $OBB_1 =$ Angle $OBB_2 =$ Angle $O_1HH_2=O_1II_2 = 90°$ (Adrian, 1991). The angle O_1BO, also known as the flare angle, is a useful measure to account for foot shape. The design system, however, does not give any indication on how to make a curved last bottom, which determines the surface on which the foot is supported. As a result, differing shoe last manufacturers adopt their own systems and have developed their own templates.

Comparatively, the Chinese system (Xing et al., 2000) is more detailed and gives templates for heel curves and instep curves for differing heel heights and differing shoe constructions. For example, the Chinese shoe size 23 of width grade 1 having a heel height of 20 (fig. 13.11b) would have the dimensions: $AI = 242$ mm, $GI = 16.5$ mm, $AH = 225.5$ mm, $AG = 202.5$ mm, $AF = 174.9$ mm, $AE = 162.3$ mm, $AD = 141.5$ mm, $AC = 89.8$ mm, $AB = 36.9$ mm. The length AH represents the maximum foot length reduced by a rear foot allowance of 4.5 mm. The various points represent corresponding foot anatomical locations: G_1 represents the position of the big toe, while F_1 represents the position of the little toe. E_1 represents the position of the medial metatarsal joint, while D_1 represents the position of the lateral metatarsal joint. Position C_1 represents the position of the cuboid. Other dimensions include the width $G_1G = 30.6$ mm, width $FF_1 = 44.9$ mm, width $EE_1 = 33.3$ mm, width $DD_1 = 46.9$ mm, width $CC_1 = 35.4$ mm, width $B_1B = BB_2 = 26.7$ mm. Angle $ABB_2 =$ angle $ABB_1 = 90$. $OD1 = EE_1$. The ball girth is 216.5 mm. The instep girth is equal to 1.0185 times ball girth, the toe spring = 16, heel height = 20 mm, toe height = 16.5 mm. The Chinese system does not provide any information on how to design the last bottom. In spite of being relatively detailed, the Chinese system is not widely used due to its complexity and the lack of software to implement the last system. Coincidentally, the Chinese last template is very similar to that of the AKA64/WMS template.

Figure 13.12 shows the U.K., U.S., and Chinese measurement systems for a shoe last. The U.S. measurement system (Adrian, 1991) includes toe spring, heel height, wedge angle, height of the back of the heel, toe thickness, stick length, ball girth, waist girth, instep girth, width of ball, width of heel, short heel girth, long heel girth, and throat opening as shown in Figure 13.12a. The U.K. system based on SATRA (1982) has measures for heel pitch, toe spring, ball girth, big toe depth, little toe depth, little toe width, 70° width, instep girth, and effective last length. The Chinese measurement system includes toe

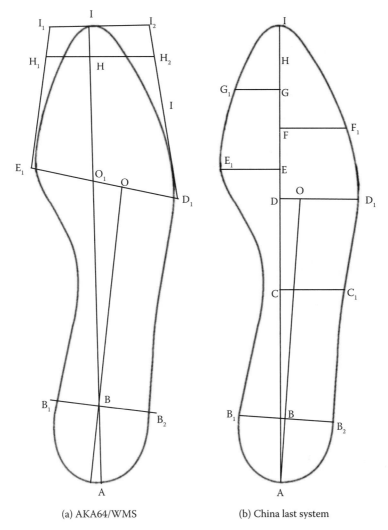

(a) AKA64/WMS   (b) China last system

FIGURE 13.11 AKA64/WMS and China last systems.

spring, total toe spring (when the ball point and heel point are touching the ground), heel height, heel top length, heel top height, last bottom length, last length horizontal, max last length, toe height, heel to toe length, last width, ball girth, instep girth, and short heel girth. All of these measurement systems are governed by dimensions rather than geometric shapes, resulting in large variations in the overall shape of a shoe last even with the same dimensions.

13.8 Shoe Sizing

A shoe sizing system is different from a measurement system. Sizing is basically creating a set of shoes of different sizes. In order to make the sizing simplistic, a combination of length and ball girth is used. A sizing system provides a guideline to create different sizes; however, the size marking is rarely exact (SATRA, 1993). Foot measuring devices (fig. 13.13), such as the Brannock, RITZ stick, and Scholl devices, have concentrated on measuring the foot length and the maximum width at around the MPJ joint area. The most widely used size systems are the U.K., U.S., Continental, Chinese, and Mondopoint systems.

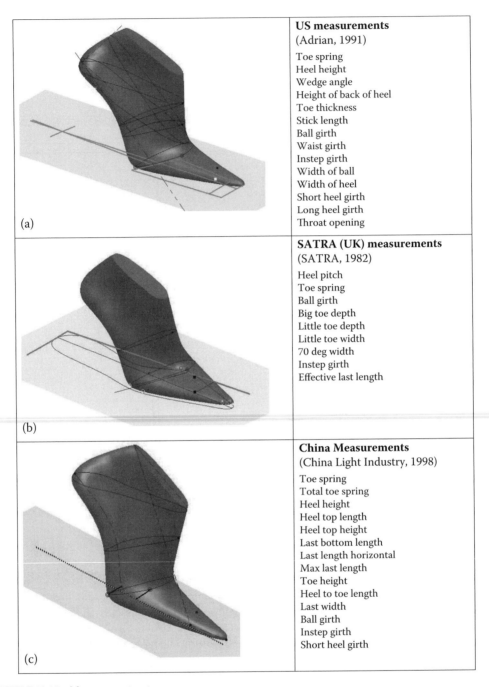

(a)	**US measurements** (Adrian, 1991) Toe spring Heel height Wedge angle Height of back of heel Toe thickness Stick length Ball girth Waist girth Instep girth Width of ball Width of heel Short heel girth Long heel girth Throat opening
(b)	**SATRA (UK) measurements** (SATRA, 1982) Heel pitch Toe spring Ball girth Big toe depth Little toe depth Little toe width 70 deg width Instep girth Effective last length
(c)	**China Measurements** (China Light Industry, 1998) Toe spring Total toe spring Heel height Heel top length Heel top height Last bottom length Last length horizontal Max last length Toe height Heel to toe length Last width Ball girth Instep girth Short heel girth

FIGURE 13.12   Measurement systems.

13.8.1 The English Size System

The U.K. footwear sizing system was the first foot sizing system. It is based on the English measurement units of feet and inches. One foot is equal to 12 in or 304.8 mm, while the inch is equal to 25.4 mm. In the English system, one size is ⅓ in (8.46 mm). After the year 1880, half sizes were introduced (½ sizes = 4.23 mm) to improve fit and, in the 20th century, quarter sizes were introduced but not adopted due to

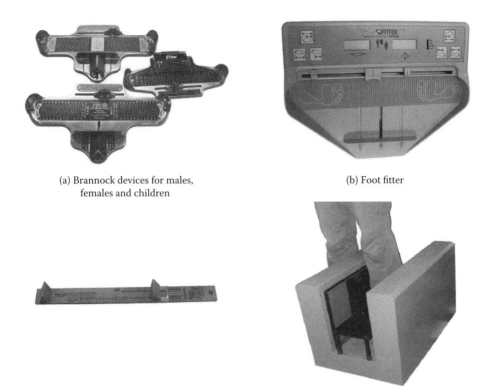

(a) Brannock devices for males, females and children

(b) Foot fitter

(c) Ritz Stick

(d) Foot scanner

FIGURE 13.13   Foot measurement devices.

the added production and inventory costs. The English size 0 is equivalent to a length of 4 in. Widths are expressed as A, B, C, D, E, F, G, H, I, J, and K, with A being the narrowest and K being the widest. The girth increase between any two neighboring designations is ¼ in.

13.8.2 The American Size System

The United States adopted the English sizing system, but instead of starting at 4 in, it starts at 3 and 11/12 in, hence the size designations of any given length are different between the U.S. and U.K. systems. A woman's shoe in the U.S. system would be 1½ sizes larger than the U.K. size while a men's shoe would differ by one full size. For example, a U.K. size 7 women's shoe will be equivalent to a U.S. size 8½. Similarly a U.K. men's size 8 shoe will be equivalent to a U.S. size 9. The American system has widths designated as AAAA, AAA, AA, A, B, C, D, E, EE, and EEE, where AAAA is the narrowest and EEE is the widest. The width difference between two neighboring designations is ¼ in.

13.8.3 The Continental Size System

Most European countries use the French system, also known as the Continental system or the Paris points. The ratio of length increase to girth increase is the same as in the U.S. and U.K. systems. However, the metric increment is not the same. In the French system, the length grade is ⅔ cm (6.66 mm) and is called the stitch. The stitch tape starts at 15 stitches and ends at 50 (10 cm–33.3 cm). The girth grade is ½ cm between sizes.

13.8.4 The Chinese Size System

In the Chinese system, the length interval is 5 mm and the width interval is 3.5 mm. The width grade ranges are denoted as 1, 1.5, 2, 2.5, 3, 3.5, and 4. The Chinese system is not as common as the others.

13.8.5 The Mondopoint or Metric System

The Mondopoint system, proposed by the International Standards Organization (ISO 9407, 1991), is based on foot length and foot width. In contrast to the other systems, the length and width of the foot that will fit the shoe is used. The shoe size is based on the foot length when wearing socks and measured in centimeters. The scale, however, is not continuous (i.e., it does not fit exact foot measurements), and each size increase is 10 cm with half-size increase at 5 cm. The shoe length is equivalent to the foot length plus 1 cm. The shoe size given as 24/95 represents a foot that fits the shoe having a length of 24 cm and a width equivalent to 95% of 24 cm (22.8 cm). Sometimes the shoe size is given as 240/95, representing a foot length of 240 mm and a width of 95 mm. The width grade is then 5 mm. The Mondopoint shoe sizing system is mainly used for sizing military boots.

13.9 Grading Systems

A grading system allows the generation of a complete size range from one master last. The three types of grading systems are the arithmetic grade, the geometric grade, and the proportional grade. In the geometric grade, the increment per size and/or width of any dimension is a specified percentage of the dimension. Proportional grading is a system in which the increments of all dimensions per size within a size run are a constant. In an arithmetic grade, the increment of a dimension per increase in foot size is a constant (Clarks, 1989). The arithmetic grade is currently used for sizing shoes. In the U.S. grading system, when the length increases by one size (⅓ in), the ball girth, waist girth, and instep girth increases by ¼ in (6.35 mm). The ball, waist, and instep girths reduce by ¼ in from a width of B to A and from A to AA. From a width size AA to AAA and from AAA to AAAA, the ball waist and instep girth reduces by 3/16 in. Similarly, from a width size B to C and from C to D, the ball waist and instep girth increases by ¼ in. From a width size D to E, from E to EE, and from EE to EEE, the ball waist and instep girth increase by 3/16 in. Grading tables are available to scale the heel, toe spring, ball to heel, shank, and other measures.

13.10 Process of Shoe Making

The components of a ladies' shoe are shown in Figure 13.8. First, the shoe model-making department finds the material and heel to design one pair of Continental size 36 (Adrian, 1991) or U.S. size 7 shoes. The last maker is either provided with a sample shoe and a sample heel or is told to use a similar last and match with a given heel. If a sample shoe is given, the model maker would copy the toe shape of the given shoe using thermoplastic material. The back part of the shoe last is then adjusted to match the design of the heel. The shoe last is thereafter sent to the manufacturer for making a sample shoe and then for testing. Fit testing is performed by a "test lady," who wears a size 36 shoe. The test lady will wear the shoe and convey the fit perceptions to the prototype maker who will then relay the information to the shoe-last designer. The shoe-last maker would then modify the shoe last and the cycle is repeated a few times before the last can be finalized. This process is rather inefficient and, with today's technology, researchers are now perfecting matching algorithms to generate "fit scores" based on a 3D scanned foot without involving a test lady.

After the sample shoe is made, the shoe factory may order different sizes of the shoe last using a specification sheet. The last specification sheet includes all the information on the shoe-last base model, such as the model number, model size, style of last, model number, shoe construction, heel grade, toe spring,

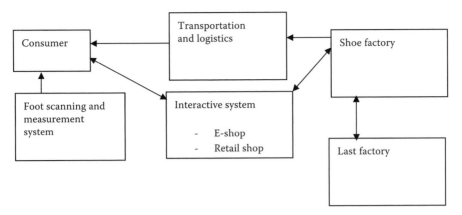

FIGURE 13.14 Simplified diagram to illustrate mass customization of footwear.

heel height, wedge angle, height of back, shank template (template for the arch curve), shank number, heel curve template, counter frame, toe thickness, ball girth, waist girth, instep girth, short heel girth, long heel girth, heel seat template, stick length, and length of bottom. The last specification sheet would also include information on sizing and grading and the last construction information related to the toe plates, shank plug, hinge, and turning models. The turning model specification enables the shoe manufacturer to use a few heels for different shoe sizes. For example, U.S. sizes in the range 3 to 4½ (3, 3½, 4, 4½) use the same back part to fit a model 4 heel, and U.S. sizes in the range 5 to 6½ (5, 5½, 6, 6½) all fit a model 6 heel. The last specification also includes grading information. The production lasts are made and used in the manufacture of the shoes.

13.11 Mass-Customized Shoes

Mass customization, a recent concept, is an attempt to make customized products or near customized products at near mass-production costs. In order to compete with Asian footwear manufacturers, the European Union funded a €16.7 million project to develop equipment, processes, and management tools from 2001 to 2003 in order to design and develop mass-customized footwear in Europe (Piller, 2002). A survey conducted on 420 European consumers reported that 63.1% of females and 61.7% of males encountered difficulties in finding appropriate shoes due to "design" and 58.8% of females and 51.5% of males have issues related to "fit" (Piller, 2002). The study also found that a majority of European consumers are interested in customized shoes (65% women, 55% men) and aesthetic customization was rated as less important than comfort and fit. Today, consumers are ready to pay 20% to 40% more for customized shoes compared to the equivalent mass-produced footwear. The process of mass customization is illustrated in Figure 13.14.

In spite of a clear indication that design is not the major interest to consumers, companies like Nike (www.nikeid.com) and Adidas (www.miadidas.com) are offering aesthetic customization while the shoes are manufactured on generic lasts. Comfort and fit are still neglected by footwear companies as they lack computer models to select or create custom lasts.

13.12 Foot Modeling

Foot modeling can be separated into 1D, 2D, 3D, and 4D modeling. Anthropometry, which involves measurement using calipers, tapes, rulers, and other types of measuring devices for linear and angular measurements, basically provides 1D data. Foot length, foot width, and foot height are insufficient to create 3D shapes. However, an anthropometric study together with statistical modeling can be used to provide some insights on 2D, 3D, and 4D foot shape information. Alternatively, equipment and techniques such as scanners, magnetic resonance imagery, ultrasound, and 3D motion analysis are all very useful in determining the 3D shapes of feet.

13.12.1 Anthropometry

Anthropometric studies of feet include Freedman et al. (1946), Rossi (1983), Goonetilleke et al. (1997), Luximon (2001), and Kouchi and Mochimaru (2002), to name a few. Due to space limitations, only the Freedman et al. (1946) study is discussed to demonstrate the usefulness and limitations of anthropometric modeling.

The Freedman et al. (1946) study included 27 dimensions and 7,559 U.S. male Army participants. The 27 dimensions included six height measures (height of big toe tip, toe height, ball height, outside ball height, plantar arch height, and dorsal arch height); four girth measures (ball girth, instep girth, diagonal ankle girth, and lower leg girth); seven length measures (foot length, ball length, fifth toe length, outside ball length, outside ball length diagonal, toe length, and ankle length); five width measures (breadth of three forward toes, foot breadth diagonal, foot breadth horizontal, breadth of instep, and heel breadth); foot flare (representing a normal abduction or adduction of the foot measured as a proportion or angle); and four shape contours (anterior curvature and orientation of toes, angular orientation of metatarsal heads, lateral foot contour, and posterior heel contour) derived from a cardboard template. Results indicated that the error between repeated measures can be large in some cases. For example, the replicated measures had differences of more than 6 mm in toe length. The statistical analysis of that study included mean, standard deviation, and correlation analysis, which are not insufficient for footwear development. Other analysis tools include regression analysis, multiple regression analysis, and multivariate techniques such as factor analysis (Jeffrey & Thurstone, 1955; Goonetilleke et al., 1997) and principal component analysis (Goonetilleke et al., 1997) to interpret anthropometric data.

Randall et al. (1951) performed multivariate analysis of the Freedman et al. data and concluded that foot length and foot width might not be the best measures for sizing feet and suggested that ball length and ball girth be used instead. In addition, they pointed out that arch height, foot flare, and the angular orientation of the metatarsals were critical for fit. In 1955, Jeffrey and Thurstone did factor analysis on the same Freedman et al. data and obtained 10 factors. The first factor was loaded on the length measures and was called bone length, and the second factor was loaded on foot flare. This shows the relative importance of flare compared to other measures. Similarly, Goonetilleke et al. (1997) indicated that the correlation ($r^2 = 0.43$) between foot length and width, and therefore sizing based on only foot length and foot width, might not be appropriate. It is also not surprising to have sizing based on length alone, as the width can be predicted using length measures. Goonetilleke and Luximon (1999) recognized the importance of flare and proposed a robust and accurate foot flare measurement that can match the shoe last. When looking at the shoe last design method, the foot flare measure influences the O_1BO angle in the AKA64/WMS system and the DAO angle in the Chinese last system (fig. 13.11). Currently, the angles are fixed at 6° and 5° in the AKA64/WMS and Chinese design systems, respectively, while the average measured foot flare angle is around 3.2° (Luximon & Goonetilleke, 1999). Holscher and Hu (1976) have alluded to the detrimental effects of inflared shoes that give rise to greater foot adduction. Anthropometric measures together with statistical analysis are useful for determining the measures and the mapping to use. Luximon and Goonetilleke (2003) proposed critical measures for footwear fitting including length and flare sizing. Dimensions alone are not indicative of the complex shape of the foot and therefore other types of evaluations have been conducted by many.

13.12.2 Foot Shape

The shape of the foot is critical in making footwear. Foot shape can be obtained in numerous ways. With the availability of digitization technologies such as laser scanners, foot shape can be obtained very quickly (figs. 13.15a and b). However, these scanners remain expensive, and simpler methodologies such as modeling the shape from a few dimensions and profiles have been proposed in the literature (figs. 13.15c and d). Foot shape modeling includes the generation and prediction of a footprint, foot outline (plantar view), and 3D foot shape. Foot outline and footprint can be obtained by tracing the foot outline with a pen or from 2D pictures of the foot plantar surface. Traditional methods for custom footwear make use of the footprint and foot outline together with foot girth measures. Static and dynamic footprints can be obtained with a Harris mat. Sensor-based computerized pressure measurement systems, such as the Fscan and RS scan systems, can also be used to measure foot plantar pressure in static as well as dynamic situations.

The Chinese last design system provides information on making custom shoe lasts using the foot outline and the footprint (Xing et al., 2000). The shoe-last bottom is designed to be larger than the footprint, while slightly smaller than the foot outline (Witana et al., 2004). Furthermore, footprint and footprint pressures have been widely used in clinical settings to evaluate foot health.

Luximon et al. (2003b) modeled the 2D foot outline using 18 landmarks to an accuracy of less than 1 mm. Then they obtained a set of 12 points such that the average absolute positive and negative absolute

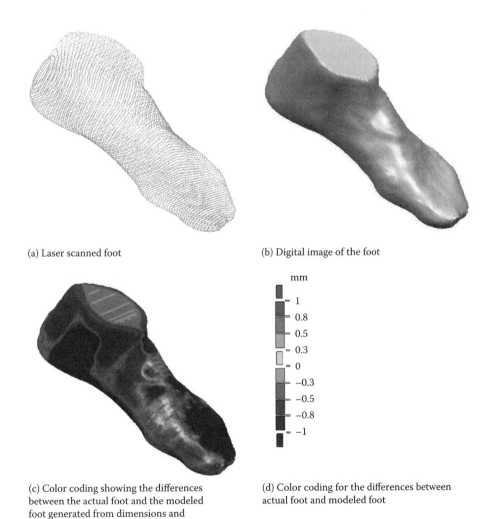

(a) Laser scanned foot

(b) Digital image of the foot

mm

1
0.8
0.5
0.3
0
−0.3
−0.5
−0.8
−1

(c) Color coding showing the differences between the actual foot and the modeled foot generated from dimensions and profiles

(d) Color coding for the differences between actual foot and modeled foot

FIGURE 13.15 The stages of digitized feet.

errors were on average 1.69 and 0.93 mm, respectively. This shows that relatively few points are sufficient to accurately determine the foot outline and having one-to-one mapping between feet and lasts will allow the generation of the outline of a shoe last as well. Studies such as Goonetilleke et al. (2000) and Witana et al. (2004) have proposed a method for quantifying footwear-fit based on dimensional differences between the foot and shoe last outline.

Regarding the 3D foot shape, Luximon and Goonetilleke (2004) used a parameterized approach to generate the 3D shape from a "standard" foot and the four parameters of foot length, foot width, foot height, and foot curvature to a mean accuracy of 2.1 mm for the left foot and 2.4 mm for the right foot. Using different prediction models and parameters, Luximon et al. (2005) developed a model to generate individual foot shapes to a mean absolute error of 1.36 mm for the left foot and 1.37 mm for the right foot using foot outline and foot height. When using foot outline and the foot profile, the mean absolute error was 1.02 mm for the left foot and 1.02 mm for the right foot.

Mochimaru et al. (2000) developed a technique using free-form deformation to transform foot shape data to the shoe last.

13.12.3 Finite Element Models

Finite element models have the ability to generate 3D foot shape changes over time to create a 4D model of the foot. Biomechanical models range from simple line diagrams to complex 3D dynamic models. Earlier foot models were 2D and included stress/strain analyses (Lemmon et al., 1997; Gefen, 2002) and in some cases a simplified 3D partial foot skeleton with connected bony structures (Chu et al., 1995). Simple biomechanical models were used to calculate the forces acting on the calcaneus and the MPJ joint when a force, F (weight of the body), is applied. The models were used to predict the stability of the medial arch.

Computational models, based on finite element (FE) methods, are increasingly popular these days due to increased computing power and the cost of experimental setup. In recent years, researchers have developed accurate 3D FE foot models (Chen et al., 1988; Gefen, 2002; Cheung & Zhang, 2005). Unlike previous models, new models capture the complete anatomical geometry of the bones, ligaments, and soft tissue. The biological structures, such as bones, ligaments, and tendons, are modeled using mechanical quantities, such as mass, density, stiffness, reaction forces, and so on. Using finite element simulation techniques, internal stresses are computed and predictions about foot function can be made. FE researchers believe that FE modeling, if properly conducted, could potentially make significant contributions to the understanding of the human foot and enable better footwear design.

Development of a FE model constitutes obtaining the 3D foot shape information including the bones. Magnetic resonance imaging (MRI) and computerized tomographic (CT) 2D images are acquired. Daniel et al. (2002) performed a CT scan at an interval of 1 mm on a fully loaded cadaver foot. By computing the 2D outlines of each CT image (286 images) and combining them, each bone can be generated in respect to the whole 3D foot shape. Similar techniques have been used by using MRI images at 2 mm interval scans (Cheung & Zhang, 2005). The 3D model of bones and 3D foot shape are converted into small elements, usually 3D tetrahedral elements. Using biomechanical properties, such as Young's modulus and Poisson's ratio, of each tissue and bone, the foot model can be simulated. For example, Cheung and Zhang (2005) have modeled 28 foot bones, 72 ligaments, the plantar fascia, and the cartilaginous structures between the phalanges using idealized homogeneous, isotropic, and linearly elastic properties while using hyperelastic properties of the encapsulated soft tissue. Once the foot model has been developed, it can be used to evaluate foot pressure on different surfaces, the effects of changes in the properties of the biomechanical properties, and dynamic as well as static predictions of foot pressures. Figure 13.16 shows an FE model with a plantar pressure distribution when standing.

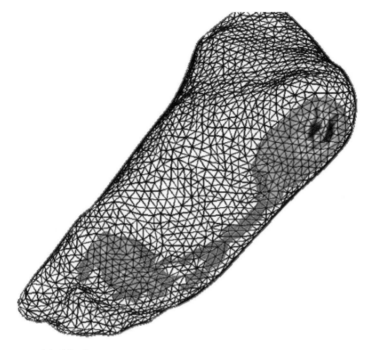

FIGURE 13.16 FE model of the foot.

FE models have disadvantages. FE models assume that the material properties are linear and boundary conditions do not have friction and slip. FE models are like a black box, and the output is purely dependent on the input data. FE models are usually validated using pressure measurements, and it is possible to generate similar pressure patterns by small adjustments in the biomechanical values. In addition, most FE models are built using one participant's foot. The parameters of the foot are usually from the existing literature and they may not be similar to the foot being investigated. FE models lack external validity. In spite of existing problems with FE models of the foot, FE modeling seems promising and more research is being conducted to improve its accuracy.

13.13 Conclusions

Foot modeling is useful in footwear design, since the human foot is a complex structure that is constantly changing due to foot dynamics and foot pressure. With the need to customize footwear, more research has been focused on capturing 3D foot shapes and on the development of foot shape models in order to generate customized shoe lasts. Customization requires understanding of the foot and its function, last development, and material characteristics. Custom-made shoes are not easy to develop, but the rapid advancement of technology is moving us closer toward the development of comfortable footwear.

Acknowledgments

This work was made possible due to numerous grants (HKUST6162/02E, 613205, 613406) from the Research Grants Council of Hong Kong.

References

Abboud, R.J., 2002. Relevant foot biomechanics. *Current Orthopaedics*, 16, 165–179.

Au, E.Y.L., Goonetilleke, R.S., 2007. A qualitative study on the comfort and fit of ladies' dress shoes. *Applied Ergonomics*, 38(6), 687–696.

Chen, J., Siegler, S., Schneck, C.D., 1988. The three-dimensional kinematics and flexibility characteristics of the human ankle and subtalar joint—Part II: Flexibility characteristics. *Journal of Biomechanical Engineering*, 110(4), 374–85.

Cheung, J.T., Zhang M., 2005. A 3D finite element model of the human foot and ankle for insole design. *Archives of Physical Medicine and Rehabilitation*, 86, 353–358.

Chu, T.M., Reddy, N.P., Padovan, J., 1995. Three-dimensional finite element stress analysis of the polypropylene, ankle-foot orthosis: static analysis. *Med Eng Phys*. 17, 372–379.

Clarks, Ltd. Training Dept., 1989. Manual of shoe making, Training Department Clarks.

Daniel, L.A.C., William, R.L., Eric, S.R., Bruce, J.S., Randal, P.C., 2002. A three-dimensional, anatomically detailed foot model: A foundation for a finite element simulation and means of quantifying foot-bone position. *Journal of Rehabilitation Research and Development*, 39(2), 401–410.

Freedman, A., Huntington, E.C., Davis, G.C., Magee, R.B., Milstead, V.M., and Kirkpatrick, C.M., 1946. Foot dimensions of soldiers (Third Partial Report Project No. T-13). Armored Medical Research Laboratory, Fort Knox, Kentucky.

Gefen, A., 2002. Stress analysis of the standing foot following surgical plantar fascia release. *Journal of Biomechanics*, 35, 629–637.

Goonetilleke, R.S., 1992. Aerobics Dance Exercise: A Survey of Physical Problems. World Research Forum, IDEA International Convention, July 19, 1992.

Goonetilleke, R.S., 1998. Designing to minimize discomfort. *Ergonomics in Design*, 6(3), 12–19.

Goonetilleke, R.S., Ho, C.-F., So, R.H.Y., 1997. Foot sizing beyond the 2-D Brannock method. *Annual Journal of IIE* (HK). (December 1997), 28–31.

Goonetilleke, R.S., Luximon, A., 1999. Foot flare and foot axis. *Human Factors*, 41, 596–607.

Goonetilleke, R.S., Luximon, A., Tsui, K.L., 2000. The quality of footwear fit: What we know, don't know and should know, Proceedings of the IEA 2000/HFES 2000 Congress (pp. 515–518), San Diego: CA.

Gross, C., Goonetilleke, R.S., Banaag, J., Nair, C., 1992. Feedback system for load bearing surfaces. U.S. Patent Number 5, 170, 364; December 8, 1992.

Holscher, E.C., Hu, K.K., 1976. Detrimental results with the common inflared shoe. *Orthopedic Clinics of North America*, 7, 1011–1018.

Hunt, K.D., 1993. The evolution of human bipedality: ecology and functional morphology. *Journal of Human Evolution*, 26, 183.

ISO 9407, 1991. Shoe sizes—Mondopoint system of sizing and marking, International Organization for Standardization, Geneva, Switzerland,

Jeffrey, T.E., Thurstone, L.L., 1955. A factorial analysis of foot measurements. Natick, MA: US Army.

Kouchi, M., Mochimaru, M., 2002. Japanese body dimensions data 1997–1998. Tokyo, Japan, Digital Human Research Center, National Institute of Advanced Industrial Science and Technology.

Lee, N.K.S., Goonetilleke, R.S., Cheung, Y.S., So, G.M.Y., 2001. MEMS Pressure Sensor Packaging Technology for Biomechanical Applications. *Microsystem Technologies*, 7(2), 55–62.

Lemmon, D., Shiang, T.Y., Hashmi, A., Ulbrecht, J.S., Cavanagh, P.R., 1997. The effect of insoles in therapeutic footwear: a finite-element approach. *J Biomech*. 30, 615–620.

Luximon, A., 2001. Foot shape evaluation for footwear fitting. PhD unpublished thesis (Hong Kong: Hong Kong University of Science and Technology).

Luximon, A., Goonetilleke, R.S., 2003. Critical dimensions for footwear fitting, IEA2003 conference, Seoul, Korea (CD-ROM).

Luximon, A., Goonetilleke, R.S., 2004. Foot shape modelling. *Human Factors*, 46(2), 304–315.

Luximon, A., Goonetilleke, R.S., Tsui, K. L., 2003a. A 3-D methodology to quantify footwear fit. In M.M. Tseng and F. Piller (eds.), *The customer centric enterprise—Advances in customization and personalization*. New York/Berlin: Springer 491–499.

Luximon, A., Goonetilleke, R.S., Tsui, K.L., 2003b. Foot landmarking for footwear customization. *Ergonomics* 46(4), 364–383.

Luximon, A., Goonetilleke, R.S., Zhang, M., 2005. 3D foot shape generation from 2D information. *Ergonomics*, 48(6), 625–641.

Luximon, A., Zhang, M., 2006. Foot biomechanics, *International encyclopedia of ergonomics and human factors*, 2nd ed., by W. Karwowski (ed.), Boca Raton, FL: CRC Press, Chapter 75, 333–337.

Mochimaru, M., Kouchi, M., Dohi, M., 2000. Analysis of 3-D human foot forms using the free form deformation method and its application in grading shoe lasts. *Ergonomics*, 43(9):1301–1313.

Morton, D.J. 1935. *The human foot*. New York: Columbia University Press.

Piller, F.T. 2002. The market for customized footwear in Europe—Market demand and consumer preferences, Euroshoe Project report (www.euroshoe.net).

Pivecka, J., Laure, S., 1995. The shoe last: practical handbook for shoe designers, Pivecka Jan Foundation, Slavicin, Czech Republic.

Quimby, H.R. 1944. *The story of lasts*. New York: National Shoe Manufacturers Association.

Randall, F.E., Munro, E.H., White, R.M., 1951. Anthropometry of the foot (Report No. 172). Environmental Protection Section, Quartermaster Climatic Research Laboratory, Lawrence, Massachusetts.

Razeghi, M., Batt, M.E., 2002. Foot type classification: a critical review of current methods. *Gait and Posture*, 15, 282–291.

Ricci, S., 1992. *Salvatore Ferragamo: The art of shoe 1898–1960*. New York: Rizzoli.

Rossi, W.A., 1983. The high incidence of mismated feet in the population. *Foot and Ankle*, 4(2), 105–112.

SATRA, 1982. Foot, Last and Shoe Measurement. SATRA Test Method FLM 1: Shoe and Allied Trades Research Association (SATRA) Footwear Technology Centre, UK.

SATRA, 1993. How shoes are made. Shoe and Allied Trades Research Association (SATRA) Footwear Technology Centre, UK.

Trade Development Council, 2007. Profiles of Hong Kong major manufacturing industries (http://www.tdctrade.com/main/industries/t2_2_13.htm).

Wexler, R.K., 1998. The injured ankle. *American Academy of Family Physicians*, 57(3), 474–490.

Witana, C.P., Goonetilleke, R.S., Feng, J., 2004. Dimensional differences for evaluating the quality of footwear fit. *Ergonomics*, 47(12), 1301–1317.

Xiong, S., and Goonetilleke, R.S., 2007. Foot shape changes upon weight-bearing. Unpublished report. Dept. of Industrial Engineering and Logistics Management. HKUST, Hong Kong.

Xing, D.H., Deng, Q.M., Ling, S.L., Chen, W.L., Shen, D.L., 2000. *Handbook of Chinese shoe making: Design, techniques and equipment*. Beijing: Press of Chemical Industry (in Chinese).

14

Shape and Size Analysis and Standards

Afzal Godil and
Sandy Ressler

14.1 Introduction

The field of anthropometry is the science of measurement of the human body from which comparisons and characterizations of the size and shape of the body in different postures can take place. The size and shape of human bodies are important in many applications, such as clothing design, machine design, transportation, in the medical/healthcare field, aircraft cockpit design, space suit design for astronauts, safety, biometrics, criminology, interface design for household/industrial products, and so on. The anthropometric data are some of the basic tools used for analysis and design requirements by human factors, ergonomics professionals, architects, interior designers, and industrial engineers. Anthropometry has its roots in physical anthropology, a discipline that focuses on how human body size and shape has varied with ethnic origins, gender, and climatic and altitude variations. Most of the anthropometric measurement methods were first used by physical anthropologists. This study required the development of two sets of tools: (1) measurement techniques to obtain data from humans; and (2) statistical techniques to transform the measurement data into summary data.

Anthropometric data have been collected as scalar values (one-dimensional data) with a measuring tape, anthropometers, calipers, and so on. There are measurement errors because of error in location of landmarks, variation in posture, and instrument errors because of orientation and position location. There are also variations in measurements from one measurer to the other. The other problem with traditional measurements is that they lack shape and spatial relationship information. The work of Robinette (1992) was first to characterize the human body shape by contours. With the advancement in 3D scanning technology, it became possible to scan the entire human body in a reasonable amount of time. The first large-scale use of the 3D scanning anthropometric survey was the CAESAR project (Robinette, 2000). These sets of 3D anthropometric data are very effective to create highly accurate 3D human body surface models.

14.2 Anthropometric Surveys

The history of anthropometry dates back to the Renaissance. Dürer's (Dürer, 1528) four books of human proportions illustrate the diversity of humans through different drawings. Most of the anthropometric measurements techniques, which are still used today, were developed by physical anthropologists. They figured out the different anatomical postures that would allow for repeatable measurements and locations of landmark points for measuring distances. They also developed different ways to summarize group data; the main concept was that of percentile, the number that is greater or equal to a given percentage of the population. The concept of percentile is first-order approximation. This methodology is followed in books by Tilley (2002) and Panero (1986) and is also used in many of the human modeling software. But no person is extreme and they differ in ways as shown by McConville and Robinette (1981). The other way is to use the boundary models, which represent actual people who are extreme in some respect, as opposed to the percentile person. This model involves the use of multivariate data analysis. One of the reasons given for the widespread use of percentile methodology is that they are easily listed in a tabular form.

14.3 Military Personnel Surveys

The study of man-machine relationships and interactions became important with the introduction of new types of war machines, like tanks, planes, submarines, and so on. Since then the U.S. Department of Defense has been interested in the field of anthropometry. The U.S. Defense Department has conducted over 40 anthropometric surveys of military personnel between 1945 and 1988, which also includes the ANSUR (1988) survey of men and women, with over 240 measurements. The number of individuals involved on these surveys is over 75,000 military personnel.

14.4 Civilian Surveys

There are a few civilian anthropometric surveys. The most complete survey of the U.S. population was the National Health Survey (1965) conducted by the Department of Health, Education and Welfare (HEW). This study involved over 7,500 civilians between 18 to 79 years of age.

14.5 CAESAR Database

The CAESAR (2006) (Civilian American and European Surface Anthropometry Resource) project has collected 3D scans, 73 anthropometry landmarks, demographics data, and traditional measurements data for each of 5,000 subjects. The objective of this study was to represent, in three-dimensions, the anthropometric variability of the civilian populations of Europe and North America and it was the first

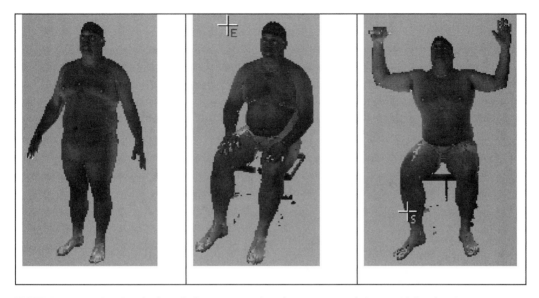

FIGURE 14.1 A CAESAR body with three postures (standing, sitting, and sitting with hands up).

successful anthropometric survey to use 3D scanning technology. The CAESAR project employs both 3D scanning and traditional tools for body measurements for people ages 18 to 65. A typical CAESAR body is shown in Figure 14.1.

The 73 anthropometric landmark points were extracted from the scans as shown in Figure 14.2. These landmark points are pre-marked by pasting small stickers on the body and are automatically extracted using landmark software. There are around 250,000 points in each surface grid on a body, and points are distributed uniformly.

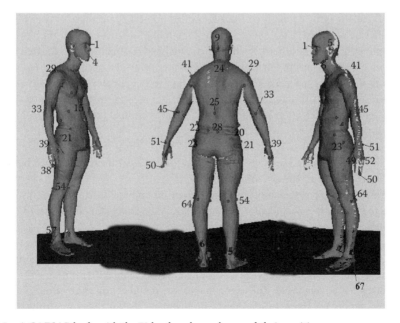

FIGURE 14.2 A CAESAR body with the 73 landmark numbers and their positions.

14.6 SizeUK and Other

SizeUK (2006) is the U.K. National Sizing Survey, where they have 3D scanned 11,000 subjects and have extracted 130 body measurements from each subject using two postures: standing and seated. The survey took place between July 2001 and February 2002.

Similarly, SizeUSA (2006) is the U.S. National Sizing Survey, where they have 3D scanned 10,000 people and have extracted 130 measurements from each subject using two postures: standing and seated. The survey was completed in September 2003.

14.7 Tools for Size/Shape Analysis

14.7.1 Tools for Virtual Models of Humans

There are a number of tools that use information from anthropometry measurements to create virtual models of humans also called digital mannequins. These models, which can be added to any virtual environment or workspace, can be assigned tasks to analyze reach, comfort, and performance. They are also widely used for ergonomics studies, design, and analysis of automobile interior spaces, aircraft cockpits, and for workspace studies. Some of this information can also help them design safer products and achieve cost savings by having fewer redesigns. Some of the most widely used tools are RAMSIS (2006), SAFEWORK (2006), Jack (2006), and ManneQuinPro (2006). All of these tools are based on anthropometric data based on a percentile person. SAFEWORK, however, departs from the percentile person by using the concept of boundary models, which are actual people who are extreme in some respect.

14.7.2 Visualization for Shape and Size Analysis

Visualization of the combined anthropometric data and the 3D scan data is also a powerful tool for data analysis. Figures 14.3, 14.4, and 14.5 show examples of the Visual Atlas developed at NIST by Ressler and Wang (2001, 2003). AnthroGloss is a visual 3D anthropometric landmark glossary usable over the web. Implemented using VRML, the virtual reality modeling language, users may easily locate and determine the names of these landmarks, which are visualized as small spheres located over the body. The goal is to create a 3D anthropometric glossary.

Two versions of the system have been implemented. Figure 14.3 illustrates a closeup of the head, and Figure 14.4 illustrates the entire standing male. Landmark names are displayed simply by moving the cursor over the spheres; no selection is needed. The second version is functionally identical and illustrates the landmarks for a male in a wheelchair.

Figure 14.5 shows a CAESAR viewer implemented in VRML. These interactive bodies are the equivalent of AnthroGloss; however, they are now generated automatically for each CAESAR body rather than manually constructed for a particular synthetic body.

Our current version includes the ability to toggle on or off body textures, landmarks, and contours. It also provides the ability to select a color for the entire body. Labels for the control slider change as appropriate to match the particular functionality selected.

The controls currently available to the user allow for the display of multiple (up to 10) bodies. The control panel operates on the "current" body indicated by a box surrounding the body. The current body is selected by simply clicking the body. Contour lines associated with sagittal, coronal, and transverse cutting planes can be displayed. The user can measure distances on the contour display by selecting start and end points on the contour lines. The distances are in the same units as the original CAESAR data.

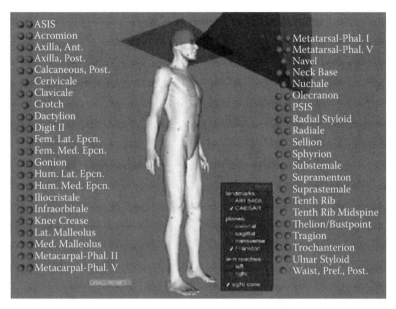

FIGURE 14.3 AnthroGloss screen layout with the Frankfort plane and sight cone indicators turned on.

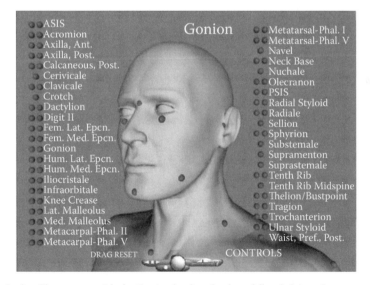

FIGURE 14.4 AnthroGloss screen with the Gonion landmark selected (head close up).

14.8 Compact Shape Descriptors of the 3D Human Body

The 3D scans of human bodies in the CAESAR human database contain over 250,000 grid points. To be used effectively for indexing, searching, clustering, and retrieval, this human body data require a compact representation. Pioneering work in content-based retrieval was done by Paquet and Rioux (2002, 2004). We at NIST also have developed shape descriptors for a human body and head. The next subsection presents more details.

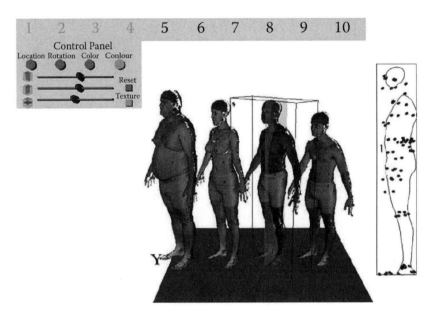

FIGURE 14.5 WEB3D version of AnthroGloss with multiple bodies.

14.9 Shape Descriptors for 3D Human Body

14.9.1 Shape Descriptors Based on CORD

The earliest work on creating a compact shape descriptor for human body representation for content-based retrieval was performed by Paquet and Rioux (2002, 2003). They performed content-based anthropometric data mining of 3D scans of humans by representing them with compact support feature vectors based on the concept of cords. A cord was defined as a vector from the center of the body to the center of a triangle grid on the surface. The distribution is then represented as a histogram. They also developed a virtual environment to perform visual data mining on the clusters and to characterize the population by defining archetypes. Paquet (2004) also introduced cluster analysis as a method to explore 3D body scans together with the relational anthropometric data as contained in the CAESAR anthropometric database.

14.9.2 PCA-Based Volumetric Shape Descriptor

Ben Azouz et al. (2002, 2004) analyzed human shape variability using a volumetric representation of 3D human bodies and applied a principal components analysis (PCA) to the volumetric data to extract dominant components of shape variability for a target population. Through visualization, they also showed the main modes of human shape variation.

14.9.3 Distance-Based Shape Descriptor

The shape descriptor described by Godil et al. (2003, 2005) uses vector d based on lengths mostly between single large bones. For descriptor vector purposes, we require lengths only between landmark points where their separation distance is somewhat pose independent. The reason it is not completely pose invariant is that distance is between landmark points which are on the surface body compared to

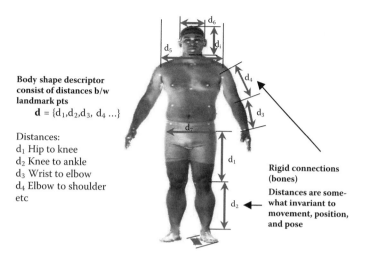

FIGURE 14.6 A distance-based body shape descriptor.

the distance between the center of the joint axis. This applies to points connected by a single large bone as shown in Figure 14.6. Thus, we form a descriptor vector of 15 distances, d, with d_1 wrist to elbow, d_2 elbow to shoulder, d_3 hip to knee, and so on. More details and shortcomings about this descriptor were described in the paper by Godil et al. (2003, 2005).

To test how well the distance-based descriptor performs, we studied the identification rate of a subset of 200 subjects of the CAESAR database where the gallery set contains the standing and the probe set contains the sitting pose of each subject. In this discussion, the gallery is the group of enrolled descriptor vectors and the probe set refers to the group of unknown test descriptor vectors.

The measure of identification performance is the rank order statistic called the cumulative match characteristic (CMC). The rank order statistics indicates the probability that the gallery subject will be among the top r matches to a probe image of the same subject. This probability depends upon both gallery size and rank. The CMC at rank 1 for the study is 40%.

The partial results from body-shape-based similarity retrieval for subject number 16270 are shown in Figure 14.7.

14.9.4 Fourier-Based Descriptor

The Fourier-based body shape descriptor, developed by Godil et al. (2003, 2005) is based on rendering the human body from the front, side, and top directions and creating three silhouettes of the human body as shown in Figure 14.8. The theory is that 3D models are similar if they also look similar from different viewing angles. The silhouette is then represented as R (radius) of the outer contour from the center of origin of the area of the silhouettes. These three contours are then encoded as Fourier descriptors, which are used later as features for similarity-based retrieval. The number of Fourier modes used to describe each silhouette is 16; hence each human body is described by a vector of length 48. This method is pose dependent, so only bodies of the same pose can be compared. The Fourier-based descriptor is then used with the L1 and L2 norm to create a similarity matrix.

14.10 Shape Descriptors Head Shape

We now describe three methods for creating descriptors based on the shape of the human head.

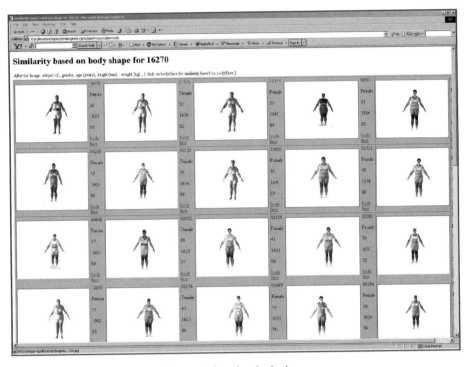

FIGURE 14.7   Similarity-based retrieval for 16270 based on body shape.

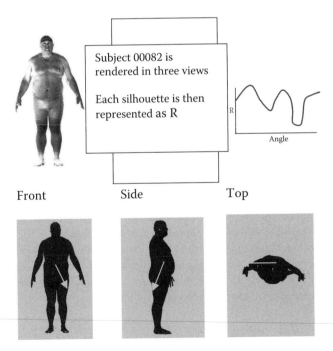

FIGURE 14.8   Subject 00082 is rendered in three silhouette views.

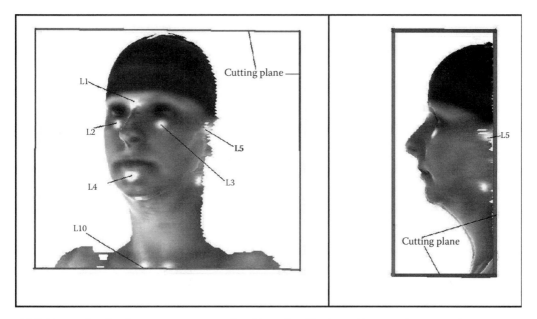

FIGURE 14.9 Landmark points 1, 2, 3, 4, 5, and 10. Vertical and horizontal lines are the cutting plane.

14.10.1 PCA-Based Facial Shape

Godil et al. (2006, 2004) applied principal component analysis (PCA) on the 3D facial surface and created PCA-based facial descriptors by cutting part of the facial grid from the whole CAESAR body grid using the landmark points 5 and 10 as shown in Figure 14.9.

Then we interpolate the facial surface information and color map on a regular rectangular grid whose size is proportional to the distance between the landmark points L2 and L3 (d = | L3 − L2 |) and whose grid size is 128 in both directions. For some of the subjects there are large voids in the facial surface grids. Figure 14.10 shows the facial surface and the new rectangular grid.

We properly positioned and aligned the facial surface and then interpolated the surface information on a regular rectangular grid. Next we perform principal component analysis (PCA) on the 3D surface

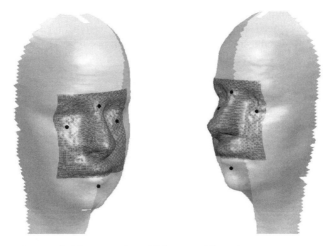

FIGURE 14.10 Shows the new facial rectangular grid for two subjects.

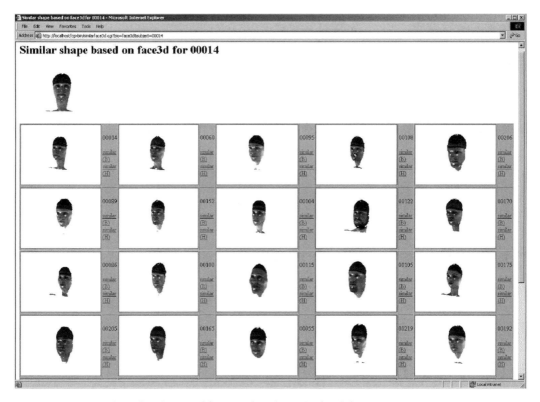

FIGURE 14.11   Similarity based retrieval for 00014 based on PCA facial shape.

and similarity-based descriptors are created. In this method the head descriptor is only based on the facial region. The PCA recognition method is a nearest neighbor classifier operating in the PCA subspace. The similarity measure in our study is based on L1 distance.

To test how well the PCA-based descriptor performs, we studied the identification between 200 standing and sitting subjects. The CMC at rank 1 for the study is 85%. More details about this descriptor are described in the paper by Godil et al. (2006, 2004).

The partial results from a head shape PCA-based similarity retrieval for subject number 00068 are shown in Figure 14.11 and for subject number 00014 are shown in Figure 14.12.

14.10.2 Spherical Harmonics-Based Descriptor

Godil and Ressler (2006) developed a spherical harmonic-based descriptor to represent a human head. The 3D triangular grid of the head is transformed to a spherical coordinate system by a least square approach and expanded in a spherical harmonic basis as shown in Figure 14.12. Since the CAESAR head grid has large voids in the top and also because of cutting the grid at the neck there is a circular hole. Since these holes are not filled properly, we have a convergence problem with 10% of the head grids. The main advantage of the spherical harmonics-based head descriptor is that it is orientation and position independent. In the near future we plan to fix this problem using a method that fills voids. The spherical harmonics-based descriptor is then used with the L1 and L2 norm to create similarity measure.

To test how well the spherical harmonics-based head descriptor performs, we studied the identification of the human head between 220 standing and sitting subjects. The CMC at rank 1 for the study is 94%.

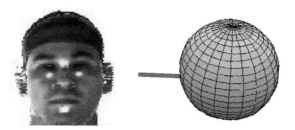

FIGURE 14.12   3D head grid is mapped into a sphere.

14.11 Extended Gaussian Images Descriptor

Retrieval based on head shape was performed by Ip and Wong (2002). Their similarity measure was based on extended Gaussian images of the polygon normal. They also compared it to an Eigenhead approach.

14.11.1 Clustering

Godil et al. (2006) have used the compact body and head descriptors for clustering. Clustering is the process of organizing a set of bodies/heads into groups in such a way that the bodies/heads within the group are more similar to each other than they are to other bodies belonging to different clusters. Many methods for clustering are found in various communities; we have tried a hierarchical clustering method. We then use dendrogram, which is a visual representation of hierarchical data, to show the clusters. The dendrogram tree starts at the root, which is at the top for a vertical tree (the nodes represent clusters) and Figure 14.13 shows the same with number of clusters at 30.

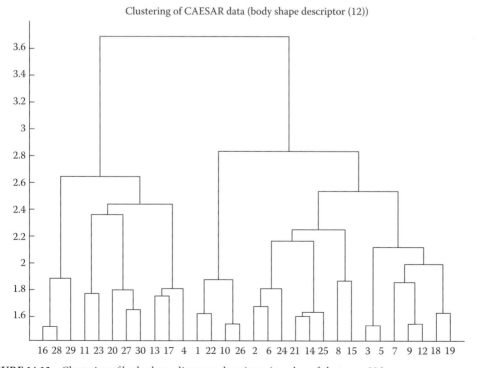

FIGURE 14.13   Clustering of body shape distances descriptor (number of clusters = 30).

14.12 Standards

In anthropometry, standard sets of measurements are needed to compare studies. However, anthropometric measurements differ from one application to another, depending on their intended use. Also, detailed standards are needed to define landmarks and provide traditional measurement naming in a standard way. There are, however, a few standards bodies that have tried to seek consensus on different naming conventions for landmark names and standard anthropometric database creation such as ISO, SAE 13, and WEAR.

Most countries have their own standards for ergonomics and garments/clothing and use different names and ways for classifying body shape, age, and size. Most of these standards are created by their own standards organization, such as ANSI (American National Standard Institute); BSI (British Standard Institute); DIN (Deutscher Normeausschuss); JIS (Japanese Industrial Standard); AFNOR (Association French Norme); and others. But because of globalization there is an imperative to create international standards. ISO (International Organization for Standardization), a worldwide federation of standard organization, produces industrial and commercial standards, the so-called ISO standards. International standards are created by a technical committee. The anthropometry related standards are prepared by the technical committee ISO/TC 19, Ergonomics, Subcommittee SC 3, Anthropometry and Biomechanics. One such standard is ISO 7250:1996, "Basic human body measurements for technological design"—the standard provides descriptions of different anthropometric measurements and different guidelines. Some of the standards deal with definitions of measurements and landmarks, while some deal with ergonomics, safety, and mannequins. There are also standards for 3D scanning methodologies for anthropometry and a number of standards related to garment and clothing designs. Because of space limitation we are not going to discuss any of the standards in detail, but the standards related to anthropometry, garments, ergonomics, and safety are listed in Table 14.1.

14.13 Conclusions and Future Trends

We believe that 3D human shape and size analysis will play a very import role in the field of digital human modeling, and in a number of applications such as clothing design, machine design, transportation, in the medical/healthcare field, interface design for household/industrial products, and so on. But there are still a number of important issues that need to be addressed:

The creation of a large publicly available Global Anthropometry database based on both 3D scanning and traditional tools for body measurements of people from around the world.

Developing shape descriptors that can efficiently represent the human body shape and size at different levels of detail and independent of the posture.

Developing techniques to locate different joint locations more precisely in the human body, such as in the hand, shoulder, neck, hip, and so on.

Developing techniques to locate the position of the different landmark locations on the body from the 3D scan of a person without pasting stickers on the body.

Developing better models to predict the range of motion for the different joints in the human body, which will have applications in reach analysis, comfort analysis, and so on.

Many of these topics described are receiving a lot of interest from researchers worldwide. We are optimistic that, because of this, there will be better digital human modeling tools that will help with the design, analysis, and ergonomics studies of different products. Some of this will also make these products more comfortable, safe, and cost efficient.

Finally, in spite of all the anthropometric (shape and size) tools, it is always critical to test the design or product with actual people. Human modeling software can be used for testing and design, but the results needs to be verified with actual user testing.

TABLE 14.1   Lists of ISO, ASTM Standards Related to Anthropometry, Garments, Ergonomics, Safety, etc.

	Organization and Standard Number	Title of Standards
1	AIRSTANDARD 61/83. Air Standardization Coordinating Committee. 1991	A Basis For Common Practices and Goals in the Conduct of Anthropometric Surveys
2	ASTM, Committee D-13.55	Standard Terminology Relating to Body Dimensions for Apparel Sizing. D 5219-97,West Conshohocken, PA: American Society for Testing and Materials (ASTM), Committee D-13.55 on Body Measurement for Apparel Sizing, October, 1997
3	ISO 7250:1996	Basic human body measurements for technological design
4	ISO/DIS 7250-1	Basic human body measurements for technological design—Part 1: Body measurement definitions and landmarks
5	ISO/NP 7250-2	Basic human body measurements for technological design—Part 2: Statistical summaries of body measurements from individual ISO populations
6	ISO 8559	Garments Construction and Anthropometric Surveys—Body Dimensions
7	ISO 3635	Size Designation of Clothes. Definition and Body Measurement Procedures
8	ISO 3636	Size Designation of Clothes-Men's and Boys' Outwear Garments
9	ISO 3637	Size Designation of Clothes-Women's and Girls' Outwear Garments
10	ASTM D 5586	Standard Tables of Body Measurements for Women
11	ASTM D 5219	Terminology Reverting to a Body Dimensions for Apparel Sizing
12	ASTM D4910	Standard Tables of Body Measurements for Infants
14	ISO 11226:2000	Ergonomics—Evaluation of static working postures
15	ISO 11226:2000/Cor 1:2006	
16	ISO 11228-1:2003	Ergonomics—Manual handling—Part 1: Lifting and carrying
17	ISO/FDIS 11228-2	Ergonomics—Manual handling—Part 2: Pushing and pulling
18	ISO/FDIS 11228-3	Ergonomics—Manual handling—Part 3: Handling of low loads at high frequency
19	ISO 14738:2002	Safety of machinery—Anthropometric requirements for the design of workstations at machinery
20	ISO 14738:2002/Cor 1:2003	
21	ISO 14738:2002/Cor 2:2005	
22	ISO 15534-1:2000	Ergonomic design for the safety of machinery—Part 1: Principles for determining the dimensions required for openings for whole-body access into machinery
23	ISO 15534-2:2000	Ergonomic design for the safety of machinery—Part 2: Principles for determining the dimensions required for access openings
24	ISO 15534-3:2000	Ergonomic design for the safety of machinery—Part 3: Anthropometric data
25	ISO 15535:2006	General requirements for establishing anthropometric databases
26	ISO 15536-1:2005	Ergonomics—Computer mannequins and body templates—Part 1: General requirements
27	ISO/FDIS 15536-2	Ergonomics—Computer mannequins and body templates—Part 2: Verification of functions and validation of dimensions for computer mannequin systems
28	ISO 15537:2004	Principles for selecting and using test persons for testing anthropometric aspects of industrial products and designs
29	ISO/TS 20646-1:2004	Ergonomic procedures for the improvement of local muscular workloads—Part 1: Guidelines for reducing local muscular workloads
30	ISO 20685:2005	3D scanning methodologies for internationally compatible anthropometric databases
31	ISO/IEC 19774:2006	Humanoid Animation (H-Anim)

References

Allen, B., Curless, B., Popovic, Z. Exploring the space of human body shapes: data-driven synthesis under anthropometric control. Proceedings of the Digital Human Modeling for Design and Engineering Conference, Rochester, MI. SAE International, 2004.

Allen, B., Curless, B., Popović, Z. The space of human body shapes: reconstruction and parameterization from range scans. SIGGRAPH 2003, San Diego, CA, July 2003.

Anthropometric Standardization Reference Manual. Lohman, T. G., Roche, A. F., and Martorell, R., Champaign, IL, Human Kinetics Books, 1988.

Anthropometric Survey of US Army Personnel: Summary Statistics Technical Report Natick/TR-89/044. Gordon, C. C., Churchill, T., Clauser, C. E., Bradtmiller, B., McConville, J. T., Tebbetts, I., and Walker, R. A. Natick, MA 01760-5000: United States Army Natick Research, Development and Engineering Center, (unclassified), 1989b.

ARD 50080 &AS5540 Anthropometric Dimensions for Creating Human Analogues, Yellow Springs, OH, Anthrotech, September 1999.

Ben Azouz, Z., Rioux, M., Lepage, R. 3D Description of the Human Body Shape: Application of Karhunen-Loève Expansion to the CAESAR Database. Proceedings of the 16th International Congress Exhibition of Computer Assisted Radiology Surgery, Paris, France, June 26–29, 2002.

Ben Azouz, Z., Rioux, M., Shu, C., Lepage, R. Analysis of Human Shape Variation using Volumetric Techniques, The 17th Annual Conference on Computer Animation and Social Agents (CASA2004), Geneva, Switzerland, July 7–9, 2004.

CAESAR: Civilian American and European Surface Anthropometry Resource web site: http://www.hec.afrl.af.mil/cardlab/CAESAR/index.html, Also at http://store.sae.org/caesar/, 2006.

Dürer, A. *Four Books on Human Proportions*, 1528.

Godil, A., Grother, P., Ressler, S. Human Identification from Body Shape. Proceedings of 4th IEEE International Conference on 3D Digital Imaging and Modeling, Banff, Canada, October 6-10, 2003.

Godil, A., Ressler, S. Retrieval and Clustering from a 3D Human Database Based on Body and Head Shape. SAE Digital Human Modeling Conference, Lyon, France, 2006.

Godil, A., Ressler, S., Grother, P. Face Recognition Using 3D Surface and Color Map Information: Comparison and Combination, SPIE's Symposium on Biometrics Technology for Human Identification, Orlando, FL, 2004.

Godil, A., Ressler, S. Similarity based Retrieval from a 3D Human Database. Poster Paper, SIGGRAPH 2005, Los Angeles, CA, 2005.

Ip, H.H.S., Wong, W. 3D Head Model Retrieval Based on Hierarchical Facial Region Similarity, Proceedings of 15th International Conference on Visual Interface (VI2002), Canada, 2002.

Jack, http://www.ugs.com/products/tecnomatix/docs/fs_tecnomatix_jack.pdf, 2006

ManneQuinPro, http://www.nexgenergo.com/ergonomics/mqpro.html, 2006

McConville, J.T., Robinette, K.M., White, R.M. An Investigation of Integrated Sizing for U. S. Army Men and Women (Natick/TR-81/033), U.S. Army Natick Research and Development Command, Natick, MA, 1981.

NASA Anthropology Research Project Staff, Anthropometric Source Book, Volume I. Anthropometry for Designers. NASA Reference Publication 1024, Houston, TX, NASA, 1978.

NASA Anthropology Research Project Staff, Anthropometric Source Book, Volume II. A Handbook of Anthropometric Data. NASA Reference Publication 1024, Houston, TX, NASA, 1978.

National Health Survey, by National Center for Health Statistics, Weight, Height, and Selected Body Dimensions of Adults, United States 1960-1962, PHS pub. no. 1000, series 11, no 8, June 1965.

Panero, J., Zelnik, M. *Human Dimension and Interior Space: A Source Book of Design Reference Standards*. London: Architectural Press, 1979.

Paquet, E. Exploring Anthropometric Data through Cluster Analysis. Digital Human Modeling for Design and Engineering (DHM). Oakland University, Rochester, MI. NRC 46564, June 2004.

Paquet, E., Rioux, M. Anthropometric Visual Data Mining: A Content-Based Approach, Submitted to IEA 2003—International Ergonomics Association XVth Triennial Congress. Seoul, Korea. NRC 44977, 2003.

Pheasant, S. *Bodyspace: Anthropometry, Ergonomics and the Design of Work*, 2nd ed. London: Taylor & Francis, 1996.

RAMSIS, http://www.human-solutions.com/automotive_industry/ramsis_en.php, 2006.

Ressler S. A Web-based 3D Glossary for Anthropometric Landmarks. Proceedings of HCI International 2001, New Orleans, LA, August 2001.

Ressler, S., Wang, Q. Using Web3D Technologies to Visualize and Analyze Caesar Data. Proceedings of XVth Triennial Congress International Ergonomics Association (IEA 2003), Seoul, Korea, August 2003.

Robinette, K.M. (1992). Anthropometry for HMD Design. Proceedings of the SPIE, Aerospace Sensing International Symposium and Exhibition.

Roebuck, J.A. Anthropometric Methods: Designing to Fit the Human Body, Human Factors and Ergonomics Society, 1992.

Roebuck, J.A., Kroemer, K.H.E., Thomson, W.G. *Engineering Anthropometry Methods*. New York: John Wiley & Sons, 1975.

SAFEWORK, http://www.safework.com, 2006.

SizeUK, UK National Sizing Survey (SizeUK), www.sizeuk.org, 2006.

SizeUSA, US National Sizing Survey (SizeUSA), http://www.tc2.com/what/sizeusa/index.html, 2006.

Tilley, A.R., Henry Dreyfuss Associates. *The Measure of Man and Woman: Human Factors in Design*. New York: Wiley, 2002.

15

Modeling Response Time and Accuracy for Digital Humans

Motonori Yamaguchi
and Robert W. Proctor

15.1 Introduction

Use of digital human models in computer-aided engineering has increased rapidly over the past 10 years (Chaffin, 2001; Sundin & Örtengren, 2006). Such models typically take into account anthropometric and biomechanical characteristics of humans, allowing a designer to develop a visual, physical model of the human, called a mannequin. The digital human can be placed in computer-generated environments of various types and programmed to carry out specific tasks that require interacting dynamically with products, systems, or workspaces. The designer receives direct visual feedback and summary data concerning whether the design is adequate to accommodate human users of various sizes, and can modify design properties until an acceptable degree of accommodation is attained. For example, Karim Abdel-Malek, director of the Virtual Soldier Research program at the University of Iowa, recently said of the digital human Santos, "We can ask Santos to change an oil filter on a dump truck or some similar task. As he goes about doing the job, we can query any part of his body functions, such as heart rate, temperature, muscle load, and others. At the same time, we can watch him work onscreen and observe any problems he might encounter" (quoted in Hudson, 2006).

For the most part, with Santos being an exception (Virtual Soldier Research, 2004), digital human modeling has focused primarily on the physical attributes of people (e.g., Sundin & Örtengren, 2006). Although there are many technical issues regarding the techniques used to generate human motion to carry out various tasks, the models are relatively successful at depicting the physical interactions of humans and products/systems. However, several authors have noted a need to model human cognitive characteristics as well (e.g., Badler, 2000). As an example, although Sundin and Örtengren (2006) focused their review of digital human models on simulation of physical attributes, they initially defined digital human modeling as involving cognitive functioning as well: "In the attempt to represent the complex human being digitally, functions being modeled include both physical and cognitive performance human aspects" (p. 1054). Bubb (2002) earlier made the point, "Modeling cognitive behavior will become more and more important as the technical design of machines demands a lot of human information processing abilities" (p. 252). Thus, there is a need to combine human cognitive models with anthropometric digital human models to simulate the full range of human performance.

Sundin and Örtengren's (2006) stated reason for focusing exclusively on physical simulation was simply, "Compared to the area of physical modeling, the area of cognitive and performance modeling is not as well known or developed" (p. 1055). However, although cognitive and performance modeling has not been integrated into physical digital human modeling and may not be well known to digital human modelers, a wealth of literature on cognitive and performance modeling exists outside of the domain of digital humans (Fisher, et al., 2006; Laughery, Lebiere, et al., 2006). This literature includes complete models of cognition and performance developed within unified cognitive architectures (e.g., Byrne, 2008), as well as more detailed models built to explain specific aspects of cognition (e.g., attention; Logan, 2004). Thus, cognitive and performance modeling is in fact relatively well developed.

An essential aspect of cognitive and performance modeling for digital humans in many situations is to model the speed and accuracy of their choice behavior. For example, Bubb (2002) stated, "Especially the opportunity to precalculate … the reaction behavior, for example, at what time what buttons are operated, can serve as forcing function for the appropriate movement of head, upper body, and extremities" (p. 263). Consequently, the present chapter focuses on ways to model response time (RT) and accuracy for digital humans in situations where rapid choices among alternative actions must be made. We review work on predictive models of RT and accuracy, with an emphasis on those models that distinguish information accumulation from thresholds for responding based on that information.

15.2 Stage Analysis of the Cognitive System

Since the 1950s, the human information-processing approach has been dominant in cognitive psychology (Proctor & Vu, 2006a). According to this approach, human cognition can be viewed as the product of processes that operate on information provided by the sensory systems. Modeling techniques developed in the context of the information-processing approach include Soar (Newell, 1990), ACT-R (Anderson & Lebiere, 1998), and EPIC (Meyer & Kieras, 1997). These techniques analyze the cognitive system into functionally distinct modules, such as the sensory, motor, and central processing systems, and simulate human performance through the dynamics of a general cognitive architecture. Thus, with the information-processing approach, human cognition is viewed as a dynamic system that consists of several subsystems, each of which has a specific function to interact with the environment.

The root of the techniques for analyzing the cognitive system into distinct subsystems can be traced back to the studies of Donders (1968/1969). Donders was concerned with measuring durations of mental processes. He reasoned that different tasks require distinct mental processes, and that RT should reflect the durations of the processes involved in those tasks. Suppose, for example, one performs two different tasks, Task 1 and Task 2. Task 1 is known to involve mental processes A, B, and C, whereas Task 2 involves mental processes A and B. Then, the duration of C can be obtained simply by subtracting the RT for Task 2 from that for Task 1. Thus, this technique is called the subtractive method (Sternberg, 1998).

The subtractive method has been applied not only to RT in behavioral studies but also to measures of brain activity in neural imaging studies (Newman, et al., 2001). With techniques such as functional magnetic resonance imaging (fMRI), brain activity occurring while people perform tasks can be monitored. It is by using the subtractive method that experimenters identify brain regions that are more active during one task than others and link the regions to cognitive processes uniquely involved in performing the task.

With the subtractive method, however, cognitive processes and functions of brain regions can only be inferred based on a separate analysis of the tasks that are compared. Disagreement about the component processes of tasks can lead to questions about the validity of conclusions drawn using the subtractive method. For instance, Donders' (1968/1969) original study provoked a criticism from Wilhelm Wundt, the founder of experimental psychology, on Donders' speculation about the mental processes involved in one of three tasks, called a c-reaction (or go/no-go task; Johnson & Proctor, 2004). For this task, an observer is to respond to one stimulus but withhold responding to another. Donders compared RT for c-reactions to that for a-reactions (or simple reaction tasks), in which an observer is always to make a response to a stimulus presentation, and considered the RT difference to reflect a stimulus-identification process. Wundt criticized Donders' assumption, and hence his interpretation, arguing that c-reactions also involve choice between responding or not responding (i.e., a response-selection process). This example illustrates the importance of task analysis for applying the subtractive method.

Sternberg (1969, 1998) elaborated on the subtractive method to provide a means for determining whether two variables affect task processing at a single stage or two distinct stages. Such variables include, for example, figure-ground contrast, number of alternative choices, stimulus duration, and stimulus-response compatibility. With Sternberg's method of additive factors analysis, additive effects of two variables indicate that the variables influence different stages in the processing sequence, whereas interactive effects suggest that the variables may influence the same stage.

Suppose the experimenter orthogonally manipulates two factors involved in a task. If the sum of the increase of RT for increases in difficulty of each variable alone is equivalent to the increase of RT for simultaneous increases in difficulty of the two variables, the effects are additive and indicate that the two factors may prolong separate processing stages. In contrast, if the sum of the increases of RT for increases in difficulty of each variable alone is less than the increase of RT for simultaneous increases in difficulty of the two variables, their effects are interactive and indicate that the factors may affect the same processing stage.

In contrast to Donders' subtractive method, the additive factors analysis involves manipulations of levels of multiple variables involved in a task. It presupposes that cognitive processes for a task are not qualitatively altered by manipulating levels of variables but that there are only increases or decreases of processing durations. On the other hand, additive factors analysis does not identify functions of processing stages directly but only indicates whether factors have a stage in common or not. Hence, the function of a processing stage has to be inferred from the variables that affect that stage. Nevertheless, application of additive factor analysis allows stages to be discovered that have not been recognized previously. By manipulating multiple factors, one can find factors that are additive to each other, and the number of those factors is equivalent to the minimum number of processing stages involved in the task. Then, one can assign descriptive labels to those processing stages based on the nature of the variables that interact.

Sanders (1998), a proponent of additive factors analysis, has used the method extensively and proposed six robust stages: preprocessing, feature extraction, identification, response selection, motor programming, and motor adjustment. According to his analysis, several factors modulate those processing stages. For instance, figure-ground contrast predominantly prolongs the preprocessing stage, whereas signal quality has a major influence on the feature-extraction stage. Moreover, stimulus-response compatibility seems to affect the latency and accuracy of the response-selection stage (see, e.g., Proctor & Vu, 2006b). In contrast, other factors, such as signal frequency and the number of alternative choices, tend to interact with several environmental factors, suggesting that they affect multiple stages of information processing. Thus, additive factors analysis is useful in modeling RT in complex situations because, by analyzing environmental factors involved in tasks, one can consider how manipulations of those factors should be treated in models of task performance.

Both the subtractive and additive factors methods assume that the cognitive processes involved in performing a task are arranged in a sequence of discrete stages. Thus, if the times for each individual processing stage can be estimated, total response time can be predicted simply by summing the times for the component processes. This is how predictions are derived using the model human processor, a simplified information-processing framework developed by Card, et al. (1983) of which most human factors and human-computer interaction professionals are aware: Time to perform basic tasks can be predicted by adding the times for component perceptual, cognitive, and motor operators determined from a task analysis.

However, the physical substrate of the cognitive system is the nervous system, which consists of a massive network of interconnected neurons where activities involve enormous parallelism. This structure suggests a need to consider models in which two or more cognitive processes proceed concurrently. Schweickert's (1978, 1984) latent network theory extends the additive factor analysis to apply to those situations. In particular, Schweickert demonstrated that when a network of cognitive processes allows some processes to be concurrent, the difference between RT for prolonging two processes and the summed RT for prolonging each process individually indicates whether the processes are performed concurrently or serially in the network. For instance, subadditive interaction of two variables, a smaller RT for the condition in which two variables are manipulated simultaneously than the sum of RTs for the conditions in which they are manipulated separately, indicates that these factors are processed concurrently.

Similarly, Dzhafarov and Schweickert (1995) developed a more general theory of testing RT components. As in Sternberg's additive factor analysis and Schweickert's latent network theory, RT is seen as reflecting properties of cognitive processes that are selectively influenced by specific environmental factors, such that, for example, additivity of two factors can be expressed as $RT(\alpha, \beta) \overset{d}{=} A(\alpha) + B(\beta)$, in which A and B are component times dependent on levels of factors α and β, respectively, and $\overset{d}{=}$ indicates the equivalence of the distributions of the left and right sides of the equation. In other words, RT can be decomposed into two component times whose relation is expressed by an algebraic operation, such as addition (+), minimum, or maximum. These operations are called decomposition rules. By applying Dzhafarov and Schweickert's decomposition tests (under certain assumptions), a unique decomposition rule can be recovered from the observed RT distribution. Once a decomposition rule can be identified, the architecture underlying the RT distribution can be interpreted.

Details of the latent network theory and the decomposition theory are beyond the scope of the present chapter. Interested readers can consult Schweickert (1978) and Townsend and Ashby (1983) for the latent network theory, and Dzhafarov and Schweickert (1995) and Van Zandt (2002) for RT decomposition theory. Also, applications of the decomposition theory and its statistical test are discussed by Cortese and Dzhafarov (1996) and Dzhafarov and Cortese (1996).

For modeling purposes, it is relatively easy to manipulate and add parameters arbitrarily to fit a model to observed RT patterns. However, a good fit of model outputs to experimental data obtained from human performance does not necessarily guarantee the validity of the model. Thus, it is important to construct RT models based on known facts about the behavior of the cognitive system (Proctor, 1986). Stage analysis offers a way to experimentally investigate what types of processing stages are involved in the task, how environmental factors affect such stages, and what structure they should be considered in the models. Consequently, stage analysis provides a fruitful basis for quantitative formulations and modeling of the cognitive system.

15.3 Accumulation Models of Response Times and Accuracy

As will be introduced in Chapter 32 of this volume, ACT-R (Anderson & Lebiere, 1998) is one of the cognitive modeling environments that has been applied to a wide variety of human decision tasks. ACT-R models are based on a general cognitive architecture, of which a central component is spreading activation throughout the memory network (Anderson, 1976, 1995). The speed and probability of retrieving a particular piece of knowledge are determined by the activation of a node in the network that represents that knowledge. The assumption is that a memory or chunk is retrieved only if its activation exceeds threshold. This mechanism of selection postulated in ACT-R resembles a more general family of mathematical models called sequential sampling models, which includes random walk models (Laming, 1968; Link & Heath, 1975; Stone, 1960), diffusion process models (Ratcliff, 1978; Ratcliff & Smith, 2004), counter models (LaBerge, 1962; Pike, 1973; Townsend & Ashby, 1983; Van Zandt, et al., 2000), and accumulator models (Usher & McClelland, 2001; Usher, et al., 2002; Vickers, 1970; Vickers, et al., 1971).

Whereas activation in ACT-R is concerned primarily with interconnectivity in the network space (the number of nodes connected to the target node and their associative strengths), sequential sampling models employ activation of responses that is dependent on a temporal function of the evidence accumulation process. In sequential sampling models, a decision is made when sufficient evidence in favor of one of the alternative choices is collected. The collection of evidence is represented as a temporal process of sampling from the available source of information (say, a stimulus representation) S that is associated with one of the alternative choices R. The decision process is terminated and response is made when the accumulating evidence U exceeds a pre-specified response criterion, or threshold, θ.

Sequential sampling models assume that the stimulus representations from which each sample is drawn (or communication paths through which the stimulus information is conveyed) are noisy, an assumption that has been adopted widely since development of information theory (Attneave, 1959; Shannon & Weaver, 1949) and signal detection theory (Swets, et al., 1964). This assumption of noisy representation provides a critical basis for the stochastic nature of the models. In classic signal detection theory, for instance, a stimulus representation is often assumed to have a normal distribution with mean μ that represents the true value associated with the stimulus and variance σ that is attributed to the background noise. A sample is represented as some psychological value X, which can arise from noise alone or the combined effect of the stimulus representation and the background noise. Thus, whether or not the sample comes from the stimulus representation S is probabilistic. Consequently, the confidence level of the decision is closely related to the likelihood ratio $p(x)/q(x)$, where $p(x)$ is the conditional probability of obtaining x from the combined effect of the stimulus representation and noise (correct sampling), and $q(x)$ is the conditional probability of obtaining x from noise alone (incorrect sampling).

Let us suppose a simple case of decision making, two-choice reaction tasks in which the stimulus S_A or S_B is presented and the associated choice R_A or R_B is selected. In a type of sequential sampling model, called the random walk model (Laming, 1968; Stone, 1960), the selection process is terminated and R_A is selected when $U_n > \theta$ or R_B is selected when $U_n < -\theta$. In other words, the random walk process is represented as the amount of accumulating evidence plotted against the time axis (see fig. 15.1). Though the model can be adapted to situations for which more than two choices are available by representing the accumulation process as a point moving in a multidimensional space (Laming, 1968), here we will focus on the cases of two-choice alternatives for simplicity.

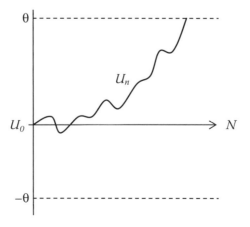

FIGURE 15.1 Illustration of a typical random walk process.

According to the random walk model developed by Laming (1968), RT corresponds to the time to attain either of the response criteria values, nh, where n is the number of samples and h is the time interval between two samplings. Generally, the random walk process is expressed as

$$U_n = U_0 + \sum_{i=1}^{n} v_i,$$

where U_0 is the initial state of the decision process and v_i is the increment of evidence at the ith sampling. The increment v is a specific value of a random variable V, corresponding to the confidence level and defined as the log-likelihood ratio, $v_i = \ln\{p(x_i \mid S_A)/p(x_i \mid S_B)\}$ (Laming, 1968; Stone, 1960), such that x is a specific value of a random variable X. Thus, x_i is the ith value of sampling. The conditional probability $p(x \mid S_A)$ is a probability of obtaining x given that the stimulus S_A is presented, and $p(x \mid S_B)$ is a probability of obtaining x given that the stimulus S_B is presented. The distribution of X is assumed to be symmetric with mean μ_A (μ_B) and variance σ_A^2 (σ_B^2), representing the stimulus S_A (S_B) (see fig. 15.2). The increment V is then given by

$$V = \ln \frac{\left(2\pi\sigma_A^2\right)^{-\frac{1}{2}} \exp\left\{-\left(X-\mu_A\right)^2 / 2\sigma_A^2\right\}}{\left(2\pi\sigma_B^2\right)^{-\frac{1}{2}} \exp\left\{-\left(X-\mu_B\right)^2 / 2\sigma_B^2\right\}}$$

Note that the variances σ_A^2 and σ_B^2 represent the background noise confounded with the actual stimulus representations μ_A and μ_B. Therefore, they are assumed to be equal, $\sigma_A = \sigma_B = \sigma$, and the expression can be rewritten as

$$V = \frac{\mu_A - \mu_B}{\sigma^2}\left(X - \frac{\mu_A + \mu_B}{2}\right).$$

It is important to note that V is expressed in terms of two concepts that are central to signal detection theory (Swets et al., 1964); the discriminability of two signals, $d' = (\mu_A - \mu_B)/\sigma$, and the optimal criterion for the detection process, $\beta = (\mu_A + \mu_B)/2$ (note that here we use the detection criterion β in the detection process that is different from the response criterion θ in the accumulation process). Thus, the above equation can be rewritten as

$$V = d'\frac{X - \beta}{\sigma}.$$

In effect, the increment of evidence V can be interpreted as the standardized distance of the observation X from the optimal criterion β weighted by the discriminability d' between the two stimuli. Thus, the increment will be greater when the observation is closer to either extreme of the partially overlapped distributions representing the stimuli. Hence, it is clear that in the random walk model, the decision process consists of a number of discrete detection processes over time.

It often appears that human decision making is biased to favor one choice over the other. A bias may reflect the situations, for example, in which one choice occurs more frequently than the other, more successes to accomplish one's goal have been attained with one choice than the other, or more costs are associated with one choice than the other. There are two ways to incorporate a decision bias into a random walk model. The first is a bias that is represented by the initial state of the decision process. More specifically, $U_0 > 0$ is a bias favoring choice R_A and $U_0 < 0$ choice R_B. Similarly, while adopting the initial state at the origin, response criteria for the two choices can be varied, such that $\theta_A = \theta + \alpha$ and $\theta_B = -\theta + \alpha$, to produce equivalent processes.

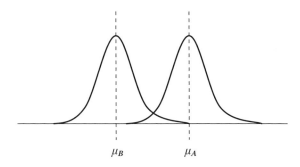

FIGURE 15.2 Hypothetical densities representing S_A (right) and S_B (left).

The second way to incorporate response biases is to allow the detection criterion β to shift. As in the above equation, placing the criterion for the detection process at the midpoint between the means of the two distributions, μ_A and μ_B, is optimal when all factors other than the means are identical for S_A and S_B. However, the optimal criterion can be shifted closer to either extreme as the probabilities and costs associated with S_A and S_B differ (Swets et al., 1964). Thus, if β shifts toward μ_B, the choice R_A is more likely chosen throughout the trials, whereas the choice R_B is more likely if β shifts toward μ_A. The general tendency for evidence to accumulate for one choice over the other is called the drift rate and reflected in the mean of increments within a trial.

In random walk models, the accuracy of a response on a particular trial is dependent primarily on the probabilities of correct and incorrect sampling p and q in the detection process. That is, if one sample is needed for response selection, the probabilities of correct response P and incorrect response Q are equal to p and q, respectively. On the other hand, when multiple samples are required to make a response, the response accuracy improves such that $P > q$ and $Q < q$. In other words, as the value of the response criterion θ increases and more samples are required, the latency and accuracy of responding also increase (Townsend & Ashby, 1983). In this way, the model is able to account for speed-accuracy trade-off (e.g., Vickers, 1970).

However, this property also yields a problem for random walk models. Namely, the models predict perfect accuracy if a sufficient time is allowed to select a response, which is not consistent with empirical observations (Ratcliff & Smith, 2004). To eliminate this problem, Ratcliff (1978) introduced an assumption that, depending on the relatedness of the alternative choices, the drift rate varies from trial to trial due to input perturbations. The between-trial variability is expressed as the standard deviation of the drift rate and termed the diffusion coefficient. Ratcliff's model is a version of diffusion process model, which is in turn a random walk model modified as a continuous time function t. Ratcliff's diffusion process model is considered one of the most powerful models for choice behaviors (Logan, 2004).

Another class of sequential sampling models is called the counter model (LaBerge, 1962; Pike, 1973; Townsend & Ashby, 1983; Van Zandt et al., 2000) and the accumulator model (Usher & McClelland, 2001; Usher et al., 2002; Vickers, 1970; Vickers et al., 1971). The major difference of counter models from random walk models is that, whereas random walk models assume a single accumulation process for multiple alternative responses, in which evidence for one response is taken as negative evidence against others, counter models assume independent and parallel accumulation processes, each of which corresponds to one of the response alternatives. According to a counter model developed by Pike (1973), the evidence accumulation process is represented as counting discrete signals. The counters increment 1 for the arrival of a signal, and response is made when the count for either choice accrues to a pre-specified response criterion. It is generally assumed that the arrival time of a signal is exponentially distributed, and thus the decision process can be viewed as a race between independent and parallel Poisson processes (Pike, 1973; Townsend & Ashby, 1983).

Similarly, the accumulator model assumes separate accumulators associated with alternative choices. On the other hand, whereas the counter model employs integer-valued increments of counts with exponentially distributed time intervals, increments of evidence in the accumulator model are a real-valued random variable. In a general version of the accumulator model (Vickers, 1970; Vickers et al., 1971), evidence is accumulated at equally spaced time steps and the amount of increment in an accumulator varies at each step. A more recent version of the accumulator model exploits a continuous temporal function of accumulation process (Usher & McClelland, 2001).

It is normally observed that increasing the number of alternative choices tends to slow the overall latency of responding. In such a condition, RT is modeled as a logarithmic function of the number of response alternatives, called Hick's law (Hick, 1952). The accumulator model can account for the observation with increase of response criterion θ, which reflects one's decision confidence (Vickers & Lee, 2000). In addition to this approach, Usher and McClelland's (2001) connectionist-based model devised lateral inhibition, the mechanism that an excitation of an accumulator produces inhibitory inputs to other accumulators. That is, the more response alternatives that are involved in the task, the more inhibitory inputs an accumulator receives. These models exemplify the flexibility and generality of the accumulator model in applying to various task settings.

Finally, Dzhafarov (1993) provided an argument on sequential sampling models that is potentially important in applying the models in complex task settings. According to his formal analysis, any stochastic accumulation model can be translated into a model that employs a strictly increasing, deterministic accumulation process and variable response thresholds, a type of model represented by Grice's (1968) variable criterion theory. In such models, the accumulation process is conceptualized as deterministic and the response criteria are represented as random variables. Brown and Heathcote (2005) recently demonstrated that Usher and McClelland's (2001) accumulator model can be modified as a similar deterministic model, which involved variable initial states as well as variable criterion values, and argued that the deterministic process model maintains performance at the same level as the original model and other stochastic accumulation models such as Ratcliff's diffusion process model. They agreed with Dzhafarov's conclusion that by using deterministic processes, sequential sampling models can be made simpler and more analytically tractable.

Whereas the modeling schema discussed in this section has been developed to account for simple choice behaviors, it should be emphasized that the schema has great flexibility to encompass a broad range of human cognitive tasks. For instance, Vickers and Lee (1998, 2000) constructed a neural network that takes an accumulator model as its basic unit. Similarly, Busemeyer and his colleagues (Busemeyer, 1985; Busemeyer & Goldstein, 1992; Busemeyer & Townsend, 1992, 1993) have developed the decision field theory of deliberative behavior in which a serial sampling process is embedded. These models provide examples of sequential sampling models to capture human performance in a more complex environment.

15.4 Conclusion

RT and accuracy of responses have been the two measurements of human cognitive performance that are most widely used in experimental psychology. Simulating these measurements is to reverse the procedures of theory construction. Thus, it is important in simulating RT and accuracy of human performance to be acquainted with assumptions involved in the procedures used to analyze experimental studies of human cognition. To this end, we reviewed modeling frameworks that have been studied and used extensively in psychological research and are of interest in simulating digital humans. It is not possible within a limited space to discuss more detailed mathematical properties associated with the modeling techniques. We thus recommend that interested readers consult respective references for further examination of the techniques.

As emphasized in this chapter, sequential sampling models have sufficient flexibility and can be useful components as the basis of more sophisticated cognitive models in digital human modeling.

Similarly, processing stage models offer a theoretical basis for cognitive processes and structures, based on which modelers can decide which parameters should be manipulated in given task settings. It was also emphasized that cognitive processes and functions in such models have to be specified by analyzing the task of interest. There are modeling tools (e.g., MicroSaint and GOMS) that adopt approaches to human performance based on task analysis (e.g., John, 2003; Laughery & Scott-Nash, 2000). These tools may be useful for the purpose of model specification.

Finally, though the decision process that sequential sampling models depict is a central aspect of all cognitive behaviors, digital human modeling needs to incorporate detailed perceptual and motor processes to simulate more complete human performance. Fortunately, theories and models of these processes are being continually developed in domains that are very active in psychological research. What is needed, then, is to choose, and perhaps combine, appropriate models that are suited to the relevant aspects of the tasks and environment in which human performance of digital humans is modeled. As in MIDAS and IMPRINT (e.g., Mitchell, 2003; Tyler, Neukom, et al., 1998), combined use of those different approaches may be valuable in the development of modeling environments for digital humans.

Acknowledgments

Preparation of the chapter was supported in part by Army Research Office Grant W9112NF-05-1-0153. We would like to thank Trisha Van Zandt for her valuable comments on an earlier draft of the chapter.

References

Anderson, J. R. (1976). *Language, memory, and thought.* Hillsdale, NJ: Lawrence Erlbaum.

Anderson, J. R. (1995). A simple theory of complex cognition. *American Psychologist, 51,* 355–365.

Anderson, J. R., and Lebiere, C. (1998). *The atomic components of thought.* Mahwah, NJ: Lawrence Erlbaum.

Attneave, F. (1959). *Applications of information theory to psychology.* New York: Holt, Rinehart, and Winston.

Badler, N. (2000). What's next for digital human models? Keynote, in *Proceedings of the SAE International Digital Human Modeling for Design and Engineering 2000 International Conference and Exposition,* June 6–8, Dearborn, MI.

Brown, S., and Heathcote, A. (2005). A ballistic model of choice response time. *Psychological Review, 112,* 117–128.

Bubb, H. (2002). Computer aided tools of ergonomics and system design. *Human Factors and Ergonomics in Manufacturing, 12,* 249–265.

Busemeyer, J. R. (1985). Decision making under uncertainty: A comparison of simple scalability, fixed-sample, and sequential-sampling models. *Journal of Experimental Psychology: Learning, Memory, and Cognition, 11,* 538–564.

Busemeyer, J. R., and Goldstein, W. M. (1992). Linking together different measures of preference: A dynamic model of matching derived from decision field theory. *Organizational Behavior and Human Decision Processes, 52,* 370–396.

Busemeyer, J. R., and Townsend, J. T. (1992). Fundamental derivations for decision field theory. *Mathematical Social Sciences, 23,* 61–104.

Busemeyer, J. R., and Townsend, J. T. (1993). Decision field theory: A dynamic-cognitive approach to decision making in an uncertain environment. *Psychological Review, 100,* 432–459.

Byrne, M. D. (2008). Cognitive architecture. In A. Sears and J. A. Jacko (Eds.), *The human-computer interaction handbook: Fundamentals, evolving technologies, and emerging applications* (2nd ed.; pp. 93–113). Boca Raton, FL: CRC Press.

Card, S. K., Moran, T. P., and Newell, A. (1983). *The psychology of human-computer interaction*. Hillsdale, NJ: Lawrence Erlbaum.

Chaffin, D. B. (2001). *Digital human modeling for vehicle and workplace design*. Warrendale, PA: Society of Automotive Engineers.

Cortese, J. M., and Dzhafarov, E. N. (1996). Empirical recovery of response time decomposition rules: II. Discriminability of serial and parallel architectures. *Journal of Mathematical Psychology*, 40, 203–218.

Donders, F. C. (1968/1969). On the speed of mental processes. In W. G. Koster (Ed.), *Acta psychologica, 30, attention and performance II* (pp. 412–431). Amsterdam: North-Holland.

Dzhafarov, E. N. (1993). Grice-representability of response time distribution families. *Psychometrika*, 58, 281–314.

Dzhafarov, E. N., and Cortese, J. M. (1996). Empirical recovery of response time decomposition rules: I. Sample-level decomposition tests. *Journal of Mathematical Psychology*, 40, 185–202.

Dzhafarov, E. N., and Schweickert, R. (1995). Decompositions of response times: An almost general theory. *Journal of Mathematical Psychology*, 39, 285–314.

Fisher, D. L., Schweickert, R., and Drury, C. G. (2006). Mathematical models for engineering psychology: Optimizing performance. In G. Salvendy (Ed.), *Handbook of human factors and ergonomics* (3rd ed.; pp. 997–1024). Hoboken, NJ: John Wiley.

Grice, G. R. (1968). Stimulus intensity and response evocation. *Psychological Review*, 75, 359–373.

Hick, W. E. (1952). On the rate of gain of information. *Quarterly Journal of Experimental Psychology*, 4, 11–26.

Hudson, J. (2006). Cashing in on virtual humans. *Wired News*, Feb. 22, 2006. Available at http://www.wired.com/news/technology/0,70253–0.html?tw=wn_technology_4. Downloaded Oct. 10, 2006.

John, B. E. (2003). Information processing and skilled behavior. In J. M. Carroll (Ed.), *HCI models, theories, and frameworks: Toward a multidisciplinary science* (pp. 55–101). San Francisco, CA: Morgan Kaufmann.

Johnson, A., and Proctor, R. W. (2004). *Attention: Theory and practice*. Thousand Oaks, CA: Sage.

LaBerge, D. (1962). A recruitment theory of simple behavior. *Psychometrika*, 27, 375–396.

Laming, D. R. J. (1968). *Information theory of choice-reaction times*. London: Academic Press.

Laughery, K. R., Jr., Lebiere, C., and Archer, S. (2006). Modeling human performance in complex systems. In G. Salvendy (Ed.), *Handbook of human factors and ergonomics* (3rd ed.; pp. 967–996). Hoboken, NJ: John Wiley.

Laughery, R., and Scott-Nash, S. (2000). *Progress in linking human performance models and human motion simulation models*. SAE Technical Paper Series 2000–01–2185. Warrendale, PA: SAE International.

Link, S. W., and Heath, R. A. (1975). A sequential theory of psychological discrimination. *Psychometrika*, 104, 175–182.

Logan, G. D. (2004). Cumulative progress in formal theories of attention. *Annual Review of Psychology*, 55, 207–234.

Meyer, D. E., and Kieras, D. E. (1997). A computational theory of executive control processes and human multiple-task performance: Part 1. Basic mechanisms. *Psychological Review*, 104, 3–65.

Mitchell, D. K. (2003). *Advanced improved performance research integration tool (IMPRINT) vetronics technology test bed model development*. Army Research Laboratory Technical Report ARL-TN-0208, September, 2003. Aberdeen Proving Ground, MD: U.S. Army Research Laboratory.

Newell, A. (1990). *Unified theories of cognition*. Cambridge, MA: Harvard University Press.

Newman, S. D., Twieg, D. B., and Carpenter, P. A. (2001). Baseline conditions and subtractive logic in neuroimaging. *Human Brain Mapping*, 14, 228–235.

Pike, R. (1973). Response latency models for signal detection. *Psychological Review*, 80, 53–68.

Proctor, R. W. (1986). Response bias, criteria settings, and the fast-same phenomenon: A reply to Ratcliff. *Psychological Review*, 88, 291–326.

Proctor, R. W., and Vu, K.-P. L. (2006a). The cognitive revolution at age 50: Has the promise of the human information-processing approach been fulfilled? *International Journal of Human-Computer Interaction*, *21*, 253–284.

Proctor, R. W., and Vu, K.-P. L. (2006b). *Stimulus-response compatibility principles: Data, theory, and application*. Boca Raton, FL: CRC Press.

Ratcliff, R. (1978). A theory of memory retrieval. *Psychological Review*, *85*, 59–108.

Ratcliff, R., and Smith, P. L. (2004). A comparison of sequential sampling models for two-choice reaction time. *Psychological Review*, *111*, 333–367.

Sanders, A. F. (1998). *Elements of human performance*. Mahwah, NJ: Lawrence Erlbaum.

Schweickert, R. J. (1978). A critical path generalization of the additive factor method: Analysis of a Stroop task. *Journal of Mathematical Psychology*, *18*, 105–139.

Schweickert, R. J. (1984). The representation of mental activities in critical path networks. *Annals of the New York Academy of Sciences*, *423*, 82–95.

Shannon, C. E., and Weaver, W. (1949). *The mathematical theory of communication*. Urbana, IL: University of Illinois Press.

Sternberg, S. (1969). The discovery of processing stages: Extensions of Donders' method. In W. G. Koster (Ed.), *Acta psychologica, 30, attention and performance II* (pp. 276–315). Amsterdam: North-Holland.

Sternberg, S. (1998). Discovering mental processing stages: The method of additive factors. In D. Scarborough and S. Sternberg (Eds.), *An invitation to cognitive science (Vol. 4: Methods, models, and conceptual issues*, pp. 703–863). Cambridge, MA: MIT Press.

Stone, M. (1960). Models for choice-reaction time. *Psychometrika*, *25*, 251–260.

Sundin, A., and Örtengren, R. (2006). Digital human modeling for CAE applications. In G. Salvendy (Ed.), *Handbook of human factors and ergonomics* (3rd ed., pp. 1053–1078). Hoboken, NJ: John Wiley.

Swets, J. A., Tanner, W. P., and Birdsall, T. G. (1964). Decision processes in perception. In J. A. Swets (Ed.), *Signal detection and recognition by human observers* (pp. 3–57). New York: John Wiley.

Townsend, J. T., and Ashby, F. G. (1983). *The stochastic modeling of elementary psychological processes*. London: Cambridge University Press.

Tyler, S. W., Neukom, C., Logan, M., and Shively, J. (1998). *The MIDAS human performance model*. In *Proceedings of the Human Factors and Ergonomics Society 42nd Annual Meeting* (Vol. 1, pp. 320–324). Santa Monica, CA: Human Factors and Ergonomics Society.

Usher, M., and McClelland, J. L. (2001). The time course of perceptual choice: The leaky, competing accumulator model. *Psychological Review*, *108*, 550–592.

Usher, M., Olami, Z., and McClelland, J. L. (2002). Hick's law in a stochastic race model with speed-accuracy tradeoff. *Journal of Mathematical Psychology*, *46*, 704–715.

Van Zandt, T. (2002). Analysis of response time distributions. In J. Wixted and H. Pashler (Eds.), *Steven's handbook of experimental psychology* (3rd ed.) *Vol 4: Methodology in experimental psychology* (pp. 461–516). New York, NY: John Wiley.

Van Zandt, T., Colonius, H., and Proctor, R. W. (2000). A comparison of two response time models applied to perceptual matching. *Psychological Bulletin and Review*, *7*, 208–256.

Vickers, D. (1970). Evidence for an accumulator model of psychophysical discrimination. *Ergonomics*, *13*, 34–58.

Vickers, D., Caudrey, D., and Willson, R. (1971). Discriminating between the frequency of occurrence of two alternative events. *Acta Psychologica*, *35*, 151–172.

Vickers, D., and Lee, M. D. (1998). Dynamic models of simple judgments: I. Properties of a self-regulating accumulator module. *Non-Linear Dynamics, Psychology and Life Sciences*, *2*, 169–194.

Vickers, D., and Lee, M. D. (2000). Dynamic models of simple judgments: II. Properties of self-organizing PAGAN (Parallel, Adaptive, Generalized Accumulator Network) model for multi-choice tasks. *Non-Linear Dynamics, Psychology and Life Sciences*, *4*, 1–31.

Virtual Soldier Research (2004). Santos: A 5th generation human model. Retrieved October 10, 2006, from http://www.digital-humans.org/santos/.

16

Psychophysiology in Digital Human Modeling

Richard W. Backs and
Wolfram Boucsein

16.1 The Case for Psychophysiology in Interactions with Virtual Agents

There is ongoing development in the modeling of interactions between virtual agents and their human counterparts to increase the verisimilitude of the interaction (André, Klesen, Gebhard, Allen, & Rist, 2000). In addition to making the appearance of virtual agents more humanlike (e.g., their face and body shape, skin texture, hairstyle, motion, and clothing), creating believable relationships between agents and humans constitutes a major challenge for the future (Magnenat-Thalmann & Thalmann, 2005). Various attempts have been made to introduce realistic high-level behavior in digital human models (DHM). Agents are equipped with features to appear intelligent, to express different emotions, and even generate personality (Egges, Kshirsagar, & Magnenat-Thalmann, 2004; Gratch et al., 2002).

Because almost all aspects of human interaction are modulated by emotion (Picard, Vyzas, & Healey, 2001), the expression of emotions is a key component needed for agents to appear intelligent or as having personality. Much current effort is being devoted to modeling an agent's appearance so that humans can reliably recognize the emotion that the agent expresses. So far, the standard for modeling emotional behavior is using facial expressions and gestures together with expressive speech (e.g., Kalra, Garchery, & Kshirsagar, 2004; Pelachaud & Bilvi, 2003; see also the volume by Gratch et al., 2006).

However, virtual agents as they appear today are still far from approaching the complexity of human emotional behavior, let alone human personality. One reason might be that an important feature of emotional behavior is missing in these agents—its psychophysiological component. From a psychophysiologist's point of view, emotions are expressed on three levels: subjective experience;

overt behavior; and physiological responses (Hugdahl, 1981). So far, modeling of emotional behavior in DHM has been dominated by the pure cognitive model provided by Ortony, Clore, and Collins (1988), which has not had much impact in psychology, and in which both the neurophysiological basis and psychophysiological concomitants of emotions are disregarded. Given this background, the physiological aspect that is central for emotion and which also includes behavioral arousal has not had much impact in DHM. Furthermore, the Ortony et al. model assigns emotions to distinct categories, which is also the case in the Ekman and Friesen (1975, 1976) facial expression model that is frequently used in DHM. As a consequence, the intensity aspect is currently missing in virtual agents' emotional behavior, which makes their emotional expression still rather stiff. Because the psychophysiological approach favors modeling of emotions as dimensional (see chapter 35 by Boucsein & Backs in this volume) instead of categorical, DHM might benefit from considering psychophysiological models for arousal and emotion.

Further, the veridical *expression* of emotion by a virtual agent is only half of the problem. Human interaction, whether between two people or within a group, is an inherently closed-loop process where the behavior of one person affects the behavior of others on multiple levels (e.g., cognitive, behavioral, emotional), only some of which are available to conscious awareness. Hatfield, Cacioppo, and Rapson (1994) refer to the process whereby two people synchronize their emotional responses as *emotion contagion*, where the emotional state of one person spreads to the other who begins to behave (through facial expression and gesture) and vocalize as though he or she is experiencing the same emotional state. This behavioral synchrony goes beyond these overt expressions of emotion to include synchrony between psychophysiological responses such as respiration and heart rate as well. This type of feedback loop between humans has also been called *physiological compliance* and can be understood in control theory terms similar to those used to model human-machine interactions (Smith & Smith, 1987). Physiological compliance has also been found when humans function as a team to control a virtual system, where team members who exhibit greater physiological compliance have superior team performance (Henning, Boucsein, & Gil, 2001; Henning & Korbelak, 2005). The relatively recent finding of mirror neurons that respond to the behavior and emotion (at least for negative emotions) of others is providing the basis for how these feedback systems function neurophysiologically (e.g., Frith & Frith, 1999; Gallese, Keysers, & Rizzolatti, 2004).

Although virtual agents do not yet have a physiology (Magnenat-Thalmann & Thalmann, 2005), we believe that they will need to in the future. That is, virtual agents will need to have the properties of expressing emotions via physiological responses and acquiring information about the human's physiology. Some work is already in progress with regard to facial expressions and vocalization (e.g., Deng, Bailenson, Lewis, & Neumann, 2006). However, the DHM community should also know about the possibilities of continuously recording the human's psychophysiological responses (and modeling the virtual agent's) during emotional (or affective) states because that will be of great use for establishing a more realistic conversational dialogue in virtual environments.

So far, the use of physiology in DHM has been restricted to models for medical training, such as blood pressure changes during surgery or the disturbance of breathing by an artificial pneumothorax (De Carlo et al., 1995). The following sections will review the current and future use of psychophysiological methods and concepts in this field.

16.2 DHM and Physiology in Medicine and Related Fields

Physiology can be modeled at many scales: from intracellular molecular processes that take microseconds, to cellular membrane processes that take milliseconds, to organ functions that can take from milliseconds to minutes, to developmental changes that can take decades. Numerous models at each level have been proposed over the years, and it is beyond the scope of the present chapter to review them. Instead, we will focus on one emerging approach that we believe will have the greatest potential application to DHM, the Physiome Project (www.physiome.org). The Physiome Project is unique in that:

> The Physiome Project will provide a framework for modelling the human body, using computational methods that incorporate biochemical, biophysical and anatomical information on cells, tissues and organs. The main project goals are to use computational modelling to analyse integrative biological function and to provide a system for hypothesis testing (Hunter & Borg, 2003, p. 237).

The Physiome Project is attempting to take empirically validated models of physiological processes that span many spatial and temporal scales and integrate them into a hierarchy of models (Hunter & Borg, 2003; Hunter, Robbins, & Noble, 2002). Models at each level in the hierarchy will use a common markup language and be linked with databases that provide parameters for the models. The primary purpose of the project is to aid integrative physiological research and clinical medical practice.

One example of how the models of the Physiome Project can impact clinical medicine can be seen in the Virtual Center for Renal Support (VCRS; Prado, Roa, Renia-Tosina, Palma, & Milan, 2002). The VCRS is an example of telemedicine where patients in remote locations are monitored physiologically and their data are transmitted over a network to the VCRS where a patient physiological image is maintained. This image can be tracked for changes that occur over time as disease progresses or after a change in medication. The patient physiological image can also be compared against a model of renal function from the Physiome Project where the effect of proposed therapies can be simulated using the patient physiological image.

Many areas of clinical practice from medicine (e.g., Srinivasan, Hwang, West, & Yellowless, 2006) to dentistry (e.g., Kleinert et al., 2007) to psychology (e.g., Kiss et al., 2004) are exploring the potential of virtual patients. Applications for virtual patients during clinical training include, but are not limited to, skill acquisition, skill maintenance, and skill assessment using a standardized patient to present cases that are objective and structured. Other potential uses of virtual patients include the modeling of patients that have special needs (e.g., autism, Down syndrome, etc.) and other low-incidence conditions that clinicians may not see during their training or may need to review when the need arises in their practice. These virtual special cases will also need the appropriate psychophysiology (cf. Edgar, Keller, Heller, & Miller, 2007), which may very well be different from the psychophysiology of "normal" virtual agents.

Unfortunately, there is currently no role for psychology in the Physiome model, but that does not mean that it could not be added as a higher level in the current hierarchy. What is needed is a mapping between psychological processes and physiological responses. The psychophysiological model presented in Boucsein and Backs (chapter 35, this volume) may be a first step toward an interface with the virtual agent's model (chapter 35, this volume).

16.3 DHM and Psychophysiology in Improving the Agent's Appearance

As mentioned, there is much current work underway to make the appearance of a virtual agent more lifelike. This section will present a few suggestions on how a psychophysiology for an agent may improve its appearance, especially with regard to the expression of emotion. We discuss in the Boucsein and Backs chapter elsewhere in this volume how psychophysiological measures can provide information about the intensity of emotional experience that may prove beneficial to virtual agents. In this section we limit our suggestions to a small subset of psychophysiological responses that could modulate the face of a virtual agent. Other work on the use of speech (e.g., Deng et al., 2006) and gesture (e.g., Smid, Zoric, & Pandzic, 2006) to convey emotion in virtual agents is beyond the scope of this chapter.

Ekman and Friesen (1975, 1976; see also Ekman & Rosenberg, 1997) developed the Facial Action Coding System (FACS) as a method of quantifying the emotional state of a person based upon the status of "action units" that correspond to whether a facial muscle (or small group of closely related muscles) is contracted or relaxed. The FACS method is used to identify which discrete emotion, or combination of emotions, a person is experiencing from the pattern of action units. For example, a smile that reflects happiness (also called a Duchenne smile) involves the contraction of both the zygomaticus major (action unit 12) and the orbicularis oculi, pars orbitalis (action unit 6), whereas a polite or insincere smile that may mask another emotion would involve only the zygomaticus major (Ekman & Friesen, 1982). FACS can also be used in a converse fashion as a guide for how to arrange the facial muscles to give the appearance of a felt emotion (e.g., Ekman, Levenson, & Friesen, 1983). DHM of virtual agents has already begun to take advantage of this phenomenon, which can even be extended to convey complex blends of emotions (Buisine et al., 2006). Therefore, we will not elaborate on this method further except to stress that the virtual agent needs a veridical representation of the underlying structure of the facial musculature for it to be maximally effective.

Ocular parameters appear to be unrecognized in DHM as a means to convey emotion, interest, and workload on the part of the virtual agent. Pupil dilation has long been known to increase with the interest value (Hess & Polt, 1960) and with the mental workload (Beatty, 1982; Hess & Polt, 1964) of a situation. Eye movements and blinking are also not random and can convey information about natural breaks in cognitive operations, affect, and fatigue (Fogarty & Stern, 1989; Stern, 1992; Stern et al., 1994). To our knowledge, none of these parameters has yet been used by the DHM community, and we believe that they deserve close inspection as fairly easy to implement ways to improve the virtual agent's appearance.

Other potentially useful psychophysiological responses may be skin temperature and respiration. Embarrassment and anger will both likely elicit changes in skin coloration in the face. Blushing with embarrassment (Edelmann, 1987; Keltner & Anderson, 2000) and flushing with increased skin temperature during anger (Ekman et al., 1983) may also be applied to virtual agents. Finally, respiration rate typically increases when heart rate increases and a subtle rising and lowering of the chest may be used to indicate faster respiration, with more pronounced chest movements indicating deeper respiration.

However, emotion expression by the virtual agent must not only appear natural, it must also occur in the correct context and at the correct time; therefore, a model for emotion in virtual agents that takes into account the reciprocal nature of agent-human interaction is critical (Gratch et al., 2006). Probably the most complete emotion model that meets these criteria and has the potential to integrate a psychophysiological component is the appraisal model of Gratch and Marsella (2004, 2005). Initially, the outcome of the appraisal process may guide how the virtual agent should reflect psychophysiological responses. However, to the extent that appraisal is a cognitive process it will not be able to capture psychophysiological changes elicited by emotion contagion or psychophysiological compliance that occur at a subconscious level as a part of social interactions. These processes still need more basic research in psychophysiology to inform DHM.

16.4 The Role of Emotions and Personality in Virtual Humans

In the process of making agents more humanlike, generating emotions and even creating a "personality" for agents has become an important issue in DHM. The ability to express emotions is regarded as a key feature for establishing believable agents (Bates, 1994). In contrast to the field of psychology, where a separation of cognitive and emotional phenomena prevails, emotional behavior is considered an important aspect of humanoid intelligence in the literature on DHM (e.g., Seif El-Nasr, Yen, & Ioerger, 2000). Contrarily, emotion is only a small (but growing) aspect of intelligence as discussed in the realm of psychology (cf. Goleman, 1995; Mayer, Salovey, & Caruso, 2000). Therefore, the present authors prefer discussing emotional behavior of virtual agents with respect to their believability, which is increased if their emotional behavior comes close to that of humans and can respond to the user's emotions. In the context of constructing more believable virtual agents, the term *personality* is frequently used in connection with, but not unequivocally separated from, emotion (e.g., André et al., 2000). In psychological terms, personality is what makes each individual unique, different from other human beings. Personality characteristics are considered relative stable traits, whereas emotions are states that change with situations.

It is beyond the scope of the present chapter to provide a complete review of works that include emotion in DHM. Instead, the state of the art will be demonstrated in a few examples. Recently, there have been several attempts to include emotions in the simulation process for creating virtual agents. For example, Seif El-Nasr et al. (2000) proposed an adaptive computational model of emotions to be incorporated into conversational agents. External events were evaluated by an emotional component, making use of expectations and event-goal associations taken from a learning component, to produce fuzzy-logic rules-driven emotional behavior. The model was successfully used to simulate "emotional" responses in a virtual pet. Gratch and Marsella (2001) combined the Ortony et al. (1988) approach of cognitive-appraisal-related emotions with three interactive behavior modes (withdrawal, partly engaged, and fully engaged in conversation) for the use in a virtual training environment. Marsella and Gratch (2002) applied their approach to a peacekeeping mission rehearsal exercise, extending their model to emotion-focused coping with distress. Egges et al. (2004) suggested a combination of Ortony et al.'s appraisal model with the five-factor personality model proposed by McCrae and John (1992) to determine the response of a virtual agent. The resulting emotion-modulated verbal response was combined with facial action patterns. Zhao, Kang, and Wright (2006) used the fear level of an agent in conjunction with a motion database to determine the threshold for a fire escape response. An approach that partly resembles what psychophysiologists would do has been used by Takács and Kiss (2003), who stated that virtual humans not only need to express fine-tuned emotions but also must be able to read the emotional responses of their human counterparts. Their goals were to make the users believe that the virtual character is paying personalized attention to him or her during their communication, getting access not only to declarative but also to procedural parts of the users' memory. They developed a controller for the virtual human's interface that uses the same internal representations as for the emotional facial expressions recorded from the interacting human.

In addition to facial expressions, psychophysiologists use a variety of measures to determine a current emotional state. Based on the psychological three-level model mentioned in the introduction, human emotions can be expressed verbally, through mimic and gesture, and physiologically. Establishing physiological concomitants of emotion has been one of the central topics in psychophysiology. Simple models of emotional behavior are mostly confined to at least two dimensions: valence and arousal (e.g., Larsen & Diener, 1992). However, there is ample evidence from neuroscience that arousal is not a one-dimensional phenomenon. Therefore, after an introduction to psychophysiological measures and before going into the psychophysiology of emotions and personality, we will provide a more refined arousal concept.

Autonomic nervous system (ANS) physiological phenomena are an essential part of human emotional states. Therefore, physiological responses will necessarily become a part of virtual agents as they become more humanlike. Psychophysiological responses could be implemented in DHM to indicate the internal emotional and arousal state of a virtual agent, very similar to biofeedback. For example, moving bars can be used to indicate changes in heart rate and electrodermal activity.

The ability to recognize and cope with human emotion will become essential for a well-functioning interaction between agent and the interacting human. Agents need to learn about the association between an event and emotions to develop rules for simulating the process of generating adequate emotions. Such a learning process can be performed by verbal rewards and punishments by the human who interacts with the agent (Seif El-Nasr et al., 1999). However, this is a very coarse method given the complexity of human emotions. Therefore, recording and evaluating psychophysiological concomitants of the participating human's emotional behavior is a much more sensitive method that will be used in the future.

16.5 Application of Psychophysiology to DHM

Psychophysiological methods have not as yet been applied in modeling arousal, emotion, and personality characteristics of virtual agents. Therefore, this final section will provide a framework for the future application of psychophysiology in DHM. Table 35.2 in the Boucsein and Backs chapter elsewhere in this volume provides an overview of psychophysiological measures in the order of their usability in the more general field of human-machine interactions, together with measures of their reliability (if available), their specificity for the different arousal systems (cf. fig. 35.1 and Table 35.1 in the Boucsein and Backs chapter) and examples for their applications in the field.

The most commonly used measure in psychophysiology is HR. Heart rate is greater for fear and anger than for happiness and disgust (Cacioppo, Berntson, Larsen, Poehlmann, & Ito, 2000). It accelerates with all kinds of physical activity, for example, in industrial workers (Vitalis et al., 1994), truck drivers (Vivoli, Bergomi, Rovesti, Carrozzi, & Vezzosi, 1993), or flight phases with increased handling activity (Roscoe, 1993; Hankins & Wilson, 1998). Heart rate variability decreases with mental workload in complex task environments such as simulated flight (Veltman & Gaillard, 1998) or driving through sharp road bends (Backs, Lenneman, Wetzel, & Green, 2003; Richter, Wagner, Heger, & Weise, 1998). Time pressure during human-computer interaction (HCI) can also enhance HR (Boucsein, 2000). Decreases of HRV can also appear as a consequence of frequently interspersed breaks during complex computer work (Boucsein & Thum, 1997) and as a consequence of night shift work (Boucsein & Ottmann, 1996). Short-term increases in HR that indicate a DR are seen in highly aversive situations (Klorman, Weissberg, & Wiesenfeld, 1977). Stimuli of moderate intensity induce an OR-typical initial deceleration, which is most pronounced in negatively toned stimulus material (Simons, Detenber, Roedema, & Reiss, 1999), whereas pleasant stimuli induce the greatest mid-interval acceleration following an initial deceleration (Lang, Greenwald, Bradley, & Hamm, 1993).

Increases in systolic BP appear as long-term consequences of highly demanding mental tasks such as air traffic controlling (Rose & Fogg, 1993), but also in data entry work if time pressure is a factor (Boucsein, 2000). The diastolic BP is higher when viewing angry compared to happy, sad, fearful, or relaxed faces (Schwartz, Weinberger & Singer, 1981). Pulse volume amplitude is a measure of OR if used as a phasic measure. Its modulation calculated as mean squares of successive differences indicates persisting interest in an interaction with a new product (Boucsein et al., 2002). More on the specificity of cardiovascular measures for engineering psychophysiology is given in Table 1.3 of Boucsein and Backs (2000).

Besides indicating the strength of muscular work (Vitalis et al., 1994), EMG is frequently used as an indicator of stress-related tension in specific muscles. For example, the trapezius EMG activity is increased during computer system breakdowns (Boucsein & Thum, 1997) and during increased mental workload through ambient noise (Hanson et al., 1993). Eyelid closure frequency is increased during complicated maneuvers in car driving (Richter et al., 1998) and piloting (Hankins & Wilson, 1998). A decrease of eyeblink rate while viewing a surgery film compared to neutral and threatening films (Palomba et al., 2000) can be interpreted as an increase in information uptake. Eye movements (in particular saccades) and eye fixations are frequently measured to determine the frequency, duration, and sequential analysis of looking at visual displays (Senders, 1983). Saccadic duration increases during the time spent on monitoring tasks (McGregor & Stern, 1996). The probability for fixating emotional stimuli is higher compared to neutral stimuli, indicating selective attention (Nummenmaa, Hyönä, & Calvo, 2006). A summary of the specificity of somatomotor measures for engineering psychophysiology is given in Table 1.5 of Boucsein and Backs (2000).

Affective facial expressions recorded by EMG reflect pleasant (zygomatic major) and unpleasant (corrugator supercilii) stimulus properties (Simons et al., 1999; Larsen, Norris, & Cacioppo, 2003), which was recently confirmed for emotional facial expressions displayed by an avatar (Weyers, Mühlberger, Hefele, & Pauli, 2006). Disgust is indicated by the levator labii muscle (Vrana, 1993; Schienle, Stark, & Vaitl, 2001). Corresponding facial EMG patterns are obtained in human-product interactions (Boucsein et al., 2002).

The frequency of non-specific electrodermal responses increases with an increase of driving demands on rural roads (Richter et al., 1998) and with the duration of computer system response times in the presence of time pressure (Boucsein, 2000). It decreases during involuntary breaks in HCI (Kuhmann, Boucsein, Schaefer, & Alexander, 1987) if time pressure is not a factor and after night work (Boucsein & Ottmann, 1996). Fast presented news on a pocket PC elicit more non-specific EDA than slowly presented ones (Kallinen & Ravaja, 2005). Slater et al. (2006) observed a characteristic non-specific EDA pattern as a response to whiteouts in an immersive virtual environment scenario. The amplitudes of spontaneous changes in EDA increase with the probability of an aversive event (Backs & Grings, 1985) and in user-hostile HCI (Muter, Furedy, Vincent, & Pelcowitz, 1993). They indicate the intensity dimension of emotion expressing faces (Johnsen, Thayer, & Hugdahl, 1995). Amplitudes in spontaneous EDA are higher in obstructive compared to conducive events in playing computer games (van Reekum et al., 2004). Skin conductance level increases with strain such as induced by prolonged waiting periods in HCI (Boucsein, 2000), but also with increased challenge during a computer game (Mandryk, Inkpen, & Calvert, 2006). A summary of the specificity of electrodermal measures for engineering psychophysiology is given in Table 1.4 of Boucsein and Backs (2000).

Body temperature increases during physical and environmental strain (Romet & Frim, 1987) and is an indicator of re-entrainment after crossing time zones (Gander, Myhre, Graeber, Andersen, & Lauber, 1989). More on the specificity of body temperature measures for engineering psychophysiology is given in Table 1.6 of Boucsein and Backs (2000). Skin temperature slopes behave differently in conducive and obstructive events during a computer game (van Reekum et al., 2004) and also change in a specific pattern during arousing, negative-emotion-inducing music compared to calm, positive-emotion-inducing music (McFarland, 1985). Furthermore, skin temperature decreases during emotional strain in computer work (Ohsuga, Shimono, & Genno, 2001) and during threatening personal questions, but increases during viewing happy film clips (Rimm-Kaufman & Kagan, 1996). Tension during car driving is also accompanied by a decrease in skin temperature (Min et al., 2002).

Respiration rate increases with increasing demands during simulated air traffic control (Brookings, Wilson, & Swain, 1996) and with increasing time pressure during computer work (Boucsein, 2000). Respiration becomes unstable as a consequence of boredom in computer work (Ohsuga et al., 2001).

Alpha activity in the EEG decreases with increased workload during normal flight performance (Sterman & Mann, 1995) but increases as a sign of hypovigilance during extended night flights (Gundel, Drescher, Maaß, Samel, & Vejvoda, 1995). Theta activity, on the other hand, increases during flight segments with high cognitive demands (Hankins & Wilson, 1998). Theta activity is also increased while listening to pleasant (as opposed to unpleasant) music (Sammler et al., 2007) and in positive emotional meditative states (Aftanas & Golocheikine, 2001). Combined indices derived from EEG bands such as beta/(alpha + theta) can be used for detecting hypovigilant states in biocybernetic systems to be used for adaptive automation (Scerbo, Freeman, & Mikulka, 2000). Relating both alpha and beta power to the sum of alpha and beta can be used to quantify the pleasantness/unpleasantness continuum during emotion-relevant imagination (Min, Chung, & Min, 2005).

The most popular ERP component is the P3 or P300. Its amplitude that follows irrelevant acoustic stimuli during a simulated flight decreases if the mental load in the flight task is increased, thus forming an indirect measure of effort demanded by the primary task (Sirevaag et al., 1993). The P300 increases if attention of radar operators is captured (Kramer, Trejo, & Humphrey, 1995). Specific P300 patterns can be found for emotional pictures (Diedrich, Naumann, Maier, & Becker, 1997). More information on the specificity of EEG- and EEG-derived measures for engineering psychophysiology is given in Table 1.2 of Boucsein and Backs (2000).

The contents of Table 35.2 in this volume will become applicable to DHM as it may appear in the future. Together with Figure 35.1 and Table 35.1 from Boucsein and Backs (this volume), it may serve as comprehensive framework for a model-based future use of psychophysiological methods in DHM. As a first step, physiological responses of the human interacting with an agent will be used to determine state of arousal and emotion, which will be fed into the system for the agent's information. In a later step, when the agent itself may have obtained the capacity to exert psychophysiological responses, those might be used to inform the interacting human about the agent's internal state.

Given the complexity of human arousal and emotions, a multivariate approach will be the most adequate one for determining the user's or the agent's current internal state. Multivariate psychophysiological

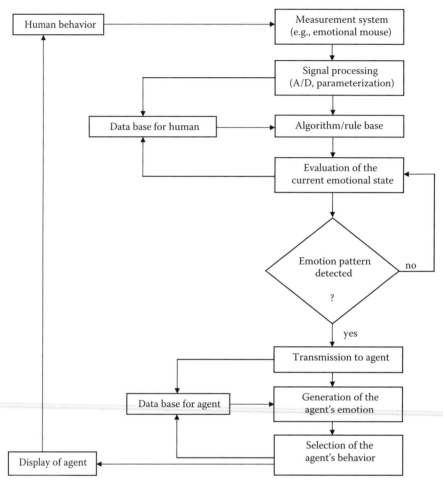

FIGURE 16.1   Flow chart for a system evaluating human emotion for appropriate emotional responses of a virtual agent (modified after Whang et al., 2003, Fig. 3). (Modified from Whang, M. C., J. S. Lim, and W. Boucsein, *Human Factors*, 45, 623–34, 2003.)

recording and evaluation has already been performed for the assessment of work phases with increased stress in a virtual building maintenance task (Ohsuga et al., 2001). One way to measure arousal and/or emotional behavior of a user online by means of psychophysiology is the so-called emotional mouse (Whang, 2007).

Computer systems that are capable of continuously monitoring psychophysiological data will require additional hard- and software to measure and classify human arousal and/or emotions. Figure 16.1 shows the flowchart of a system being able to perform psychophysiological recording of human emotions during the interaction with a virtual agent. An emotional mouse or other recording devices such as affective wearables (Picard & Healey, 1997) are continuously obtaining several key psychophysiological measures such as HR, EDA, peripheral blood flow, and skin temperature (Whang, 2007). A signal processing system is used for online parameterization. The signals are amplified, filtered, and digitized by a data acquisition system designed as small and light as possible, to be embedded in the computer main board or in a mobile computer system such as a personal digital assistant or cell phone (Whang, Lim, & Boucsein, 2003). For an undisturbed interaction between human and agent, wireless data communication and automatic noise detection/removal are recommended. The signals are pretreated in the recording device and further processed for extracting physiological parameters that are appropriate for emotion detection. The normalized data are referred to a special algorithm or rule base for evaluating the current emotional state of the human. Both measured and processed data are stored in a database, which is continuously updated and refined, including upgrading the algorithm or rule base. The adaptive nature of the emotion evaluation allows for a detection of emotional patterns in the human interacting with the agent.

If a certain pattern of emotion is detected, it will be transmitted to the virtual agent, which will then use its own database of emotional expressions (mimic, gestures, voice timbre, and later even its own psychophysiological pattern) to generate a believable emotional behavior. This will be displayed and viewed by the interacting human. Because of huge differences in the emotional behavior of different humans, the system needs a period of learning the psychophysiological patterns of each system user. This is performed by a subjective confirmation procedure, using an underlying dimensional emotion model (Whang, 2007).

References

Aftanas, L. I., and Golocheikine, S. A. (2001). Human anterior and frontal midline theta and lower alpha reflect emotionally positive state and internalized attention: High-resolution EEG investigation of meditation. *Neuroscience Letters, 310,* 57–60.

André, E., Klesen, M., Gebhard, P., Allen, S., and Rist, T. (2000). Integrating models of personality and emotions into lifelike characters. In A. M. Paiva (Ed.), *Affective interactions* (pp. 150–165). Berlin: Springer.

Backs, R. W., and Grings, W. W. (1985). Effects of UCS probability on the contingent negative variation and electrodermal response during long ISI conditioning. *Psychophysiology, 22,* 268–275.

Backs, R. W., Lenneman, J. K., Wetzel, J. M., and Green, P. (2003). Cardiac measures of driver workload during simulated driving with and without visual occlusion. *Human Factors, 45,* 525–538.

Bates, J. (1994). The role of emotion in believable agents. *Communications of the ACM, 37,* 122–125.

Beatty, J. (1982). Task-evoked pupillary responses, processing load, and the structure of processing resources. *Psychological Bulletin, 91,* 276–292.

Boucsein, W. (2000). The use of psychophysiology for evaluating stress-strain processes in human-computer interaction. In R. W. Backs, and W. Boucsein (Eds.), *Engineering psychophysiology. Issues and applications* (pp. 289–309). Mahwah, N.J.: Lawrence Erlbaum.

Boucsein, W., and Backs, R. W. (2000). Engineering psychophysiology as a discipline: Historical and theoretical aspects. In R. W. Backs, and W. Boucsein (Eds.), *Engineering psychophysiology. Issues and applications* (pp. 3–30). Mahwah, N.J.: Lawrence Erlbaum.

Boucsein, W., and Ottmann, W. (1996). Psychophysiological stress effects from the combination of night-shift work and noise. *Biological Psychology, 42,* 301–322.

Boucsein, W., Schaefer, F., Kefel, M., Busch, P., and Eisfeld, W. (2002). Objective emotional assessment of tactile hair properties and their modulation by different product worlds. *International Journal of Cosmetic Science, 24,* 135–150.

Boucsein, W., and Thum, M. (1997). Design of work/rest schedules for computer work based on psychophysiological recovery measures. *International Journal of Industrial Ergonomics, 20,* 51–57.

Brookings, J. B., Wilson, G. F., and Swain, C. R. (1996). Psychophysiological responses to changes in workload during simulated air traffic control. *Biological Psychology, 42,* 361–377.

Buisine, S., Abrilian, S., Niewiadomski, R., Martin, J. C., Deviliers, L., and Pelachaud, C. (2006). Perception of blended emotions: Corpus to expressive agent. In J. Gratch, M. Young, R. Aylett, D. Ballin, and P. Oliver (Eds.), *Intelligent virtual agents: 6th International Conference, IVA 2006 Proceedings* (pp. 14–27). Berlin: Springer-Verlag.

Cacioppo, J. T., Berntson, G. G., Larsen, J. T., Poehlmann, K. M., and Ito, T. A. (2000). The psychophysiology of emotion. In R. Lewis and J. M. Haviland-Jones (Eds.), *The handbook of emotion* (2nd ed., pp. 173–191). New York: Guilford Press.

DeCarlo, D., Kaye, J., Metaxas, D., Clarke, J. R., Webber, B., and Badler, N. (1995). Integrating anatomy and physiology for behavior modeling. In K. Morgan, R. M. Satava, H. B. Sieburg, R. Mattheus, and J. P. Christensen (Eds.), *Interactive technology and the new paradigm for healthcare* (pp. 81–87). Amsterdam: IOS Press.

Deng, Z., Bailenson, J., Lewis, J. P., and Neumann, U. (2006). Perceiving visual emotions with speech. In J. Gratch, M. Young, R. Aylett, D. Ballin, and P. Oliver (Eds.), *Intelligent Virtual Agents: 6th International Conference, IVA 2006 Proceedings* (pp. 107–120). Berlin: Springer-Verlag.

Diedrich, O., Naumann, E., Maier, S., and Becker, G. (1997). A frontal positive slow wave in the ERP associated with emotional slides. *Journal of Psychophysiology, 11*, 71–84.

Edelmann, R. J. (1987). *The psychology of embarrassment.* Chichester, UK: Wiley.

Edgar, J. C., Keller, J., Heller, W., and Miller, G. A. (2007). Psychophysiology in research on psychopathology. In J. T. Cacioppo, L. G. Tassinary, and G. Berntson (Eds.), *Handbook of psychophysiology* 3rd ed. (pp. 665–687). Cambridge, UK: Cambridge University Press.

Egges, A., Kshirsagar, S., and Magnenat-Thalmann, N. (2004). Generic personality and emotion simulation for conversational agents. *Computer Animation and Virtual Worlds, 15*, 1–13.

Ekman, P., and Friesen, W. V. (1975). *Unmasking the face: A guide to recognizing emotions from facial cues.* Oxford, UK: Prentice-Hall.

Ekman, P., and Friesen, W. V. (1976). *Pictures of facial affect.* Palo Alto, CA: Consulting Psychologists Press.

Ekman, P., and Friesen, W. V. (1982). Felt, false, and miserable smiles. *Journal of Nonverbal Behavior, 6*, 238–258.

Ekman, P., Levenson, R. W., and Friesen, W. V. (1983). Autonomic nervous system activity distinguishes among emotions. *Science, 221*, 1208–1210.

Ekman, P., and Rosenberg, E. L. (1997). *What the face reveals: Basic and applied studies of spontaneous expression using the Facial Action Coding System (FACS).* New York: Oxford University Press.

Fogarty, C., and Stern, J. A. (1989). Eye movements and blinks: Their relationship to higher cognitive processes. *International Journal of Psychophysiology, 8*, 35–42.

Frith, C. D., and Frith, U. (1999). Interacting minds: A biological basis. *Science, 286,* 1692–1695.

Gallese, V., Keysers, C., and Rizzolatti, G. (2004). A unifying view of the basis of social cognition. *Trends in Cognitive Sciences, 8,* 396–403.

Gander, P. H., Myhre, G., Graeber, C. R., Andersen, H. T., and Lauber, J. K. (1989). Adjustment of sleep and the circadian temperature rhythm after flights across nine time zones. *Aviation, Space, and Environmental Medicine, 60,* 733–743.

Goleman, D. (1995). *Emotional Intelligence.* New York: Bantam Books.

Gratch, J., and Marsella, S. (2001). Tears and fears: Modeling emotions and emotional behaviors in synthetic agents. In J. P. Müller (Ed.), *Proceedings of the 5th International Conference on Autonomous Agents, Montréal* (pp. 278–283). New York: ACM Press.

Gratch, J., and Marsella, S. (2004). A domain-independent framework for modelling emotion. *Cognitive Systems Research, 5,* 269–306.

Gratch, J., and Marsella, S. (2005). Lessons from emotion psychology for the design of life-like characters. *Applied Artificial Intelligence, 19,* 215–233.

Gratch, J., Okhmatovskaia, A., Lamothe, F., Marsella, S., Morales, M., van der Wolf, R. J., and Morency, L.-P. (2006). Virtual rapport. In J. Gratch, M. Young, R. Aylett, D. Ballin, and P. Oliver (Eds.), *Intelligent virtual agents: 6th International Conference, IVA 2006 Proceedings* (pp. 14–27). Berlin: Springer-Verlag.

Gratch, J., Rickel, F., André, E., Cassell, J., Petajan, E., and Badler, N. (2002). Creating interactive virtual humans: Some assembly required. *IEEE Intelligent Systems, 17,* 55–63.

Gundel, A., Drescher, J., Maaß, H., Samel, A., and Vejvoda, M. (1995). Sleepiness of civil airline pilots during two consecutive night flights of extended duration. *Biological Psychology, 40,* 131–141.

Hankins, T. C., and Wilson, G. F. (1998). A comparison of heart rate, eye activity, EEG and subjective measures of pilot mental workload during flight. *Aviation, Space, and Environmental Medicine, 69*, 360–367.

Hanson, E. K. S., Schellekens, J. M. H., Veldman, J. B. P., and Mulder, L. J. M. (1993). Psychomotor and cardiovascular consequences of mental effort and noise. *Human Movement Science, 12*, 607–626.

Hatfield, E., Cacioppo, J. T., and Rapson, R. L. (1994). *Emotion contagion.* New York: Cambridge University Press.

Henning, R. A., Boucsein, W., and Gil, M. C. (2001). Social-physiological compliance as a determinant of team performance. *International Journal of Psychophysiology, 40*, 221–232.

Henning, R. A., and Korbelak, K. T. (2005). Social-psychophysiological compliance as a predictor of future team performance. *Psychologia, 48*, 84–92.

Hess, E. H., and Polt, J. M. (1960). Pupil size as related to interest value of visual stimuli. *Science, 132*, 349–350.

Hess, E. H., and Polt, J. M. (1964). Pupil size in relation to mental activity during simple problem solving. *Science, 142*, 1190–1192.

Hugdahl, K. (1981). The three-systems model of fear and emotion: A critical examination. *Behaviour Research and Therapy, 19*, 75–85.

Hunter, P. J., and Borg, T. K. (2003). Integration from proteins to organs: The Physiome Project. *Nature Reviews: Molecular Cell Biology, 4*, 237–243.

Hunter, P., Robbins, P., and Noble, D. (2002). The IUPS human physiome project. *Pflügers Archiv: European Journal of Physiology, 445*, 1–9.

Johnsen, B. H., Thayer, J. F., and Hugdahl, K. (1995). Affective judgment of the Ekman faces: A dimensional approach. *Journal of Psychophysiology, 9*, 193–202.

Kallinen, K., and Ravaja, N. (2005). Effects of the rate of computer-mediated speech on emotion-related subjective and physiological responses. *Behaviour and Information Technology, 24*, 365–373.

Kalra, P., Garchery, S., and Kshirsagar, S. (2004). Facial deformation models. In N. Magnenat-Thalmann, and D. Thalmann (Eds.), *Handbook of virtual humans* (pp. 119–139). Chichester: John Wiley & Sons.

Keltner, D., and Anderson, C. (2000). Saving face for Darwin: The functions and uses of embarrassment. *Current Directions in Psychological Science, 9*, 187–192.

Kiss, B., Benedek, B., Szijarto, G., Csukly, G., Simon, L., and Takács, B. (2004). Virtual patient: A photo-real virtual human for VR-based therapy. *Studies in Health Technology and Informatics, 98*, 154–156.

Kleinert, H. L., Sanders, C., Mink, J., Nash, D., Johnson, J., Boyd, S., and Challman, S. (2007). Improving student dentist competencies and perception of difficulty in delivering care to children with developmental disabilities using a virtual patient module. *Journal of Dental Education, 71*, 279–286.

Klorman, R., Weissberg, R. P., and Wiesenfeld, A. R. (1977). Individual differences in fear and autonomic reactions to affective stimulation. *Psychophysiology, 14*, 45–51.

Kramer, A. F., Trejo, L. J., and Humphrey, D. (1995). Assessment of mental workload with task-irrelevant auditory probes. *Biological Psychology, 40*, 83–100.

Kuhmann, W., Boucsein, W., Schaefer, F., and Alexander, J. (1987). Experimental investigation of psychophysiological stress reactions induced by different system response times in human-computer interaction. *Ergonomics, 30*, 933–943.

Lang, P. J., Greenwald, M. K., Bradley, M. M., and Hamm, A. O. (1993). Looking at pictures: Affective, facial, visceral, and behavioral reactions. *Psychophysiology, 30*, 261–273.

Larsen, J. T., Norris, C. J., and Cacioppo, J. T. (2003). Effects of positive and negative affect on electromyographic activity over zygomaticus major and corrugator supercilii. *Psychophysiology, 40*, 776–785.

Larsen, R. J., and Diener, E. (1992). Promises and problems with the circumplex model of emotion. In M. S. Clark, (Ed.), *Review of personality and social psychology annual, Vol. 13: Emotion* (pp. 25–59). Thousand Oaks, CA: Sage.

Magnenat-Thalmann, N., and Thalmann, D. (2005). Virtual humans: Thirty years of research, what next? *The Visual Computer, 21,* 997–1015.

Mandryk, R. L., Inkpen, K. M., and Calvert, T. W. (2006). Using psychophysiological techniques to measure user experience with entertainment technologies. *Behaviour and Information Technology, 25,* 141–158.

Marsella, S., and Gratch, J. (2002). A step toward irrationality: Using emotion to change belief. *Proceedings of the 1st International Joint Conference on Autonomous Agents and Multiagent Systems, July 15-19, 2002* (pp. 334–341). Bologna, Italy.

Mayer, J. D., Salovey, P., and Caruso, D. (2000). Models of emotional intelligence. In R. Sternberg (Ed.), *Handbook of intelligence.* Cambridge, UK: Cambridge University Press.

McCrae, R. R., and John, O. P. (1992). An introduction to the five-factor model and its applications. *Journal of Personality, 60,* 175–215.

McFarland, R. A. (1985). Relationship of skin temperature changes to the emotions accompanying music. *Biofeedback and Self-Regulation, 10,* 255–267.

McGregor, D. K., and Stern, J. A. (1996). Time on task and blink effects on saccade duration. *Ergonomics, 39,* 649–660.

Min, B. C., Chung, S. C., Park, S. J., Kim, C. J., Sim, M. K., and Sakamoto, K. (2002). Autonomic responses of young passengers contingent to the speed and driving mode of a vehicle. *International Journal of Industrial Ergonomics, 29,* 187–198.

Min, Y. K., Chung, S. C., and Min, B. C. (2005). Physiological evaluation on emotional change induced by imagination. *Applied Psychophysiology and Biofeedback, 30,* 137–150.

Muter, P., Furedy, J. J., Vincent, A., and Pelcowitz, T. (1993). User-hostile systems and patterns of psychophysiological activity. *Computers in Human Behavior, 9,* 105–111.

Nummenmaa, L., Hyönä, J., and Calvo, M. G. (2006). Eye movement assessment of selective attentional capture by emotional pictures. *Emotion, 6,* 257–268.

Ohsuga, M., Shimomo, F., and Genno, H. (2001). Assessment of phasic work stress using autonomic indices. *International Journal of Psychophysiology, 40,* 211–220.

Ortony, A., Clore, G. L., and Collins, A. (1988). *The cognitive structure of emotions.* New York: Cambridge University Press.

Palomba, D., Sarlo, M., Angrilli, A., Mini, A., and Stegagno, L. (2000). Cardiac responses associated with affective processing of unpleasant film stimuli. *International Journal of Psychophysiology, 36,* 45–57.

Pelachaud, C., and Bilvi, M. (2003). Computational model of believable conversational agents. In M.-P. Huget (Ed.), *Communications in multiagent systems* (pp. 300–317). Berlin: Springer.

Picard, R. W., and Healey, J. (1997). Affective wearables. *Personal Technologies 1,* 231–240.

Picard, R. W., Vyzas, E., and Healey, J. (2001). Toward machine emotional intelligence: Analysis of affective physiological state. *IEEE Transactions on Pattern and Machine Intelligence, 23,* 1175–1191.

Prado, M., Roa, L., Reina-Tosina, J., Palma, A., and Milan, A. (2002). Virtual center for renal support: Technological approach to patient physiological image. *IEEE Transactions on Biomedical Engineering, 49,* 1420–1430.

Richter, P., Wagner, T., Heger, R., and Weise, G. (1998). Psychophysiological analysis of mental load during driving on rural roads: a quasi-experimental field study. *Ergonomics, 41,* 593–609.

Rimm-Kaufman, S. E., and Kagan, J. (1996). The psychological significance of changes in skin temperature. *Motivation and Emotion, 20,* 63–78.

Romet, T. T., and Frim, J. (1987). Physiological responses to fire fighting activities. *European Journal of Applied Physiology, 56,* 633–638.

Roscoe, A. (1993). Heart rate as a psychophysiological measure for in-flight workload assessment. *Ergonomics, 36,* 1055–1062.

Rose, R. M., and Fogg, L. F. (1993). Definition of a responder: Analysis of behavioral, cardiovascular, and endocrine responses to varied workload in air traffic controllers. *Psychosomatic Medicine, 55,* 325–338.

Sammler, D., Grigutsch, M., Fritz, T., and Koelsch, S. (2007). Music and emotion: Electrophysiological correlates of the processing of pleasant and unpleasant music. *Psychophysiology, 44,* 293–304.

Scerbo, M. W., Freeman, F. G., and Mikulka, P. J. (2000). A biocybernetic system for adaptive automation. In R. W. Backs, and W. Boucsein (Eds.), *Engineering psychophysiology: Issues and applications* (pp. 241–253). Mahwah, N.J.: Lawrence Erlbaum.

Schienle, A., Stark, R., and Vaitl, D. (2001). Evaluative conditioning: A possible explanation for the acquisition of disgust responses. *Learning and Motivation, 32,* 65–83.

Schwartz, G. E., Weinberger, D. A., and Singer, J. A. (1981). Cardiovascular differentiation of happiness, sadness, anger, and fear following imagery and exercise. *Psychosomatic Medicine, 43,* 343–364.

Seif El-Nasr, M., Ioerger, T. R., and Yen, J. (1999). PETEEI: A PET with evolving emotional intelligence. In E. Oren (Ed.), *Proceedings of the 3rd International Conference on Autonomous Agents* (pp. 9–15). New York: ACM Press.

Seif El Nasr, M., Yen, J., and Ioerger, T. R. (2000). FLAME—fuzzy logic adaptive model of emotions. *Autonomous Agents and Multi-Agent Systems, 3,* 219–257.

Senders, J. W. (1983). *Visual scanning processes.* Tilburg, The Netherlands: University of Tilburg.

Simons, R. F., Detenber, B. H., Roedema, T. M., and Reiss, J. E. (1999). Emotion processing in three systems: The medium and the message. *Psychophysiology, 36,* 619–627.

Sirevaag, E. J., Kramer, A. F., Wickens, C. D., Reisweber, M., Strayer, D. L., and Grenell, J. F. (1993). Assessment of pilot performance and mental workload in rotary wing aircraft. *Ergonomics, 36,* 1121–1140.

Slater, M., Guger, C., Edlinger, G., Leeb, R., Pfurtscheller, G., Antley, A. Garan, M., Brogui, A., and Friedman, D. (2006). Analysis of physiological responses to a social situation in an immersive virtual environment. *Presence, 15,* 553–569.

Smid, K., Zoric, G., and Pandzic, I. S. (2006). [HUGE]: Universal architecture for statistically based HUman GEsturing. In J. Gratch, M. Young, R. Aylett, D. Ballin, and P. Oliver (Eds.), *Intelligent virtual agents: 6th International Conference, IVA 2006 Proceedings* (pp. 107–120). Berlin: Springer-Verlag.

Smith, T. J., and Smith, K. U. (1987). Feedback control mechanisms of human behavior. In G. Salvendy (Ed.), *Handbook of human factors.* New York: John Wiley & Sons.

Srinivasan, M., Hwang, J. C., West, D., and Yellowless, P. M. (2006). Assessment of clinical skills using simulator technologies. *Academic Psychiatry, 30,* 505–515.

Sterman, M. B., and Mann, C. A. (1995). Concepts and applications of EEG analysis in aviation performance evaluation. *Biological Psychology, 40,* 115–130.

Stern, J. A. (1992). The eye blink: Affective and cognitive influences. In D. G. Forgays, T. Sosnowski, and K. Wrzesniewski (Eds.), *Recent developments in cognitive, psychophysiological, and health research.* Washington, DC: Hemisphere Publishing Corp.

Stern, J. A., Boyer, D., and Schroeder, D. (1994). Blink rate: A possible measure of fatigue. *Human Factors, 36,* 285–297.

Takács, B., and Kiss, B. (2003). The virtual human interface: A photorealistic digital human. *IEEE Computer Graphics and Applications, 23,* 38–45.

Van Reekum, C. M., Johnstone, T., Banse, R., Etter, A., Wehrle, T., and Scherer, K. R. (2004). Psychophysiological responses to appraisal dimensions in a computer game. *Cognition and Emotion, 18,* 663–688.

Veltman, J. A., and Gaillard, A. W. K. (1998). Physiological workload reactions to increasing levels of task difficulty. *Ergonomics, 41,* 656–669.

Vitalis, A., Pournaras, N. D., Jeffrey, G. B., Tsagarakis, G., Monastiriotis, G., and Kavvadias, S. (1994). Heart rate strain indices in Greek steelworkers. *Ergonomics, 37,* 845–850.

Vivoli, G., Bergomi, M., Rovesti, S., Carrozzi, G., and Vezzosi, A. (1993). Biochemical and haemodynamic indicators of stress in truck drivers. *Ergonomics, 36,* 1089–1097.

Vrana, S. R. (1993). The psychophysiology of disgust: Differentiating negative emotion contexts with facial EMG. *Psychophysiology, 30,* 279–286.

Weyers, P., Mühlberger, A., Hefele, C., and Pauli, P. (2006). Electromyographic responses to static and dynamic avatar emotional facial expressions. *Psychophysiology, 43,* 450–453.

Whang, M. C., Lim, J. S., and Boucsein, W. (2003). Preparing computers for affective communications: A psychophysiological concept and preliminary results. *Human Factors, 45,* 623–634.

Whang, M. C. (2007). The emotional computer adaptive to human emotion. In J. Westerink, M. Ouwerkerk, T. Overbeek, F. Pasveer, and B. de Ruyter (Eds.), *Probing experience: From academic research to commercial propositions.* Dordrecht, The Netherlands: Springer.

Zhao, Y., Kang, J., and Wright, D. K. (2006). Emotion-affected decision making in human simulation. *Biomedical Sciences Instrumentation, 42,* 482–487.

17

Mathematical Models of Human Text Classification

Amirali Noorinaeini
and Mark R. Lehto

17.1 Abstract

The human body of knowledge has been exponentially increasing since humanity first set foot on earth. Along with the growth in knowledge comes the difficulty to master it all, yet this obstacle has not caused any major problems for pushing the boundaries of knowledge even further. This is due to the growth of methods that help individuals learn only what is most relevant to their immediate interests. Text classification (TC) and information retrieval (IR) methods are of great importance for the information retrieval tasks as they help us search the complete body of knowledge available in the way we think will suit us most. In this chapter we briefly mention what was done in the past to help people find this material, introduce some heuristics that people use on a daily basis for such purposes, and finally discuss the mathematical models that are used to help with TC and IR.

Three specific mathematical methods are discussed in more detail after the common steps in preparing the knowledge base for the search purposes are introduced. These methods include Bayesian classifiers, singular value decomposition, and regression. These methods can be applied to both TC and IR with small changes. Research examples for each of these methods are also referred to in each section to help understand the implementation of such methods.

17.2 Introduction

In recent years it has become more feasible to extract useful information from the large and growing repository of text-based data most organizations possess, using automated or semi-automated data (or text) mining tools. As a result the fields of data mining (DM) and knowledge discovery in databases (KDD) have grown in leaps and bounds and many researchers have concluded that they have great future potential (Han & Kamber, 2000). A reasonable question that naturally arises when discussing these approaches is: "What did humans do before technology advances in this area became so dominant?" There were many ways that people attacked this problem, from asking the more knowledgeable, to developing abstracting reference books, to training individuals (librarians) to help with the problem. All of these methods used in the past provided a way to help find relevant material in a timely manner. So we are not facing a completely new problem, only the dimensions of the problem have grown, as have the methods to address them.

The need for knowledge classification is still great and despite all the technological advances in this area, the logic behind the methods of classification has not changed much. Currently most of the automated tools for classification and information retrieval use the same logic as a librarian would have used long ago, but with a new set of limitations; an automated tool at the time being is much faster, more precise and has more memory than any human who has ever lived, but it lacks the ability of processing and understanding the natural language the way people do. We will now point to the background on some models of human reasoning and the heuristics people use for categorization tasks. Interested readers can find more about these in the referenced literature.

From the very beginning of the 18th century until the early 19th century as the theory of probability—an extension of the logical calculus—was formed by scholars such as Thomas Bayes, Jakob and Nikolas Bernoulli, and Laplace, many researchers strongly believed in probability as the calculus of reason and rational choice. As probability theory improved and the extensive calculations needed for it became more apparent, such a view was abandoned and probability slowly became another mathematical tool of the natural sciences. In the mid-20th century the idea of a rational man's mental model as a probability calculator again resurfaced (de Finetti, 1937; Savage, 1954), this time in a milder manner. Since then many researchers have developed models that use probabilities (e.g., Bayesian) and statistical methods as a way to describe human decision-making processes. The most serious issue faced by this approach is that the number of calculations required becomes cumbersome even for computers (Martignon & Laskey, 1999). That's why many researchers have focused on the heuristics people use to perform different tasks, including categorization and classification tasks.

Many different heuristics have been identified that are similar to an automated text mining tool. In the area of estimation, for example, researchers have come to believe that people keep track of event frequencies in the past to help them with situations in the future (Hasher & Zacks, 1984). Others have concluded that people infer statistical qualities by using proximal cues that are correlated to them (Brunswik, 1955). Both of these methods (frequency count and estimation through correlation) are building blocks of almost all of the text mining tools. In the categorization area researchers have developed several models of the human categorization process that are actually simple statistical models. These models include Nosofsky's (1986) generalized context model (GCM), and Bayesian models (Blum & Langley, 1997; Domingos & Pazzani, 1996) among some more statistically complex models. Even very simple heuristics such as categorization by elimination (Berretty, et al., 1999) are applications of the classical two category classification problem.

The above arguments show that although people are often able to achieve near-perfect accuracy in categorization tasks (Ashby & Maddox, 1992), a text mining tool should not model the human performance, but rather use the methods which humans use simplified versions of. Also note that the text mining tools are still much weaker than people in taking advantage of the natural language properties in this task. Having noted the links between heuristics and text mining we have to briefly mention the differences between data mining and text mining as well.

Text mining is different from data mining in general, but the two areas have enough in common to share a lot of methods. In general it is text mining that uses the methods of data mining for either of the two main goals in text mining: classification and retrieval. The reason behind this is that data miners develop methods to do either of these two tasks on data other than text, whereas the text miners develop methods to transform text into acceptable input data for the methods developed by data miners.

The above argument shows that the main difference between data mining and text mining is the material they work on: Data mining deals with data in the form of numbers, while text mining deals with data in the form of text. The numerical data can be structured much easier than textual data. Also when working on texts, we are actually working on textual data coded into human language. Add the concepts of grammar and meaning to this and it should be obvious that text mining is a special case of data mining, but the two tasks have very different data to work on. From now on we will shift our focus to the text mining tools.

17.3 Text Mining Goals

Traditionally text mining has been used to help humans by reducing the level of expertise required, and the amount of time spent. We will now describe each of the two tasks that text mining tools may be designed and used for, and for each task we mention how they help humans.

A major task for text mining tools is the problem of text classification (TC). A well-known example of TC is spam filtering. These days it is almost impossible to find an email client (web based or machine installed) that does not have a spam detection feature. Since the amount of irrelevant emails sent to people is large and increasingly so, service providers need fast and reliable spam filters to help them reduce cost and increase customer satisfaction and reliability of service. The task at hand is to design an automated system that will distinguish desired emails from non-desired ones. From the tool designer's point of view the problem is to classify emails as spam or not. This is the simplest classification problem with only two categories, but yet the job is very hard as the variety of unwanted emails and their volume increases all the time.

This task can not be done manually due to privacy issues, let alone the impracticality of it due to the huge volume of emails. Using the text mining tools here is desirable as they reduce the required time greatly. On the other hand since the recognition of spam emails does not require high levels of expertise, the accuracy of the automated systems could theoretically be very high. There are other classification tasks that would help reduce the number of experts required in a field. For example a system that automatically files online insurance claims under different categories for further use by the insurance company, eliminates the need for human experts. The accuracy of such systems is still very high as well (Wellman, et al., 2007).

Another task in text mining is information retrieval (IR). The text mining tool in this case is responsible for finding the information within a corpus that is relevant to a user's specific query. Although this task may seem very different from TC, the two are very similar in nature. What the tool has to do in this case is classify the given query, and retrieve all the documents that fall in the same class as the query. Note that the task here is more complex because the query may not be as easily classifiable as the documents already in the corpus. Also most of the text miners now have features that rank the retrieved documents according to their relevance to the query, which makes the task somewhat more challenging.

The extent to which the information retrieval tools help with saving time and required expertise is evident for anyone who has used a web-based search engine. Imagine how much time and expertise one would need to prepare a list of all data-mining techniques if this task was to be done using all the data available in a small college library. With the use of Internet search engines this task can be done in a very short time using a much larger volume of information (all the resources available online).

Up to now we have used data mining and text mining almost interchangeably, while in reality text mining is partly under the domain of data mining and partly a different science. We will discuss this in the next section. Readers who are more interested in data mining and its application in DHM are referred to Chapter 37.

17.4 Text Mining Methods

There are two major approaches in text mining; one approach leans toward using the data coded in the text with respect to the language, grammar, and meanings as much as possible. This approach is commonly referred to as the linguistic approach. The linguists themselves differ with respect to how much of the mathematical methods they employ in their research. The other approach is concerned with using mathematical methods to extract as much information out of the text as possible. The researchers taking this approach may again use different levels of linguistic methods to refine or filter the text before applying the mathematical models.

The linguistic approach may seem the more reasonable path to take in the beginning because by ignoring much of the meaning, grammar, and structure of a language, not much remains to extract information from and use in the text-mining method. Although this argument is correct, it is extremely difficult to develop automated tools that understand text the way humans do, and therefore the linguistic methods are still far from perfect. On the other hand, several mathematical methods have been developed that perform very well, including Bayesian, singular value decomposition, and regression-based models. Application of these models requires several general steps to transform textual data into numerical form as expanded upon before.

17.5 Text Transformation

In this section we assume that a corpus has been chosen for text mining. Here we define *corpus* as a group of textual documents with at least one similar characteristic (belonging to the same area of research, having been published in the same newspaper, or simply having been chosen for testing the text-mining tool). Before the corpus can be read into the mathematical data-mining model, some steps should be taken. These steps are briefly discussed in this section.

17.5.1 Standardization

This step is necessary for all text-mining methods. In this step all the documents are transformed into the same format (e.g., XML). Other actions may also be done during standardization. For example a researcher may decide that the document titles are no different from the text, and therefore make all the text in one document a long string of text, while another may want to treat titles differently from the rest of the text and therefore skip this step in standardization.

17.5.2 Tokenization

During this necessary step, the stream of characters in each document is broken down into words or as some researchers call it, *tokens* (Weiss, et al., 2005). Although this seems like a simple task, in practice many problems may rise. The simplest problem is the existence of so many different delimiters. It is also common that some words don't have a delimiter between them. Another problem is in dealing with some characters that are sometimes delimiters and sometimes not. For example "_" between two words is a delimiter, but between two digits is usually not a delimiter. If the researchers are concerned with these issues, they should customize the tokenization process to best suit their corpus.

17.5.3 Lemmatization

Another step, which is not necessary, but is usually helpful, is to standardize the form of the tokens. This step is usually referred to as *lemmatization* or *stemming*. There are two major stemming steps: One is to unify all the different forms of a word, and the other is to transform all the words into their roots. An example of the first case is to transform all the plural nouns into their single form. The second type would try to replace all words by their common root. For example the noun "beginner" and the verb "begin" would both be transformed into the form "begin."

Lemmatization has two purposes: The first is to reduce the number of words that the model is dealing with, and the second is to increase the number of instances of each word. In theory, these two changes will help the mathematical models deal with a corpus that is more conforming to the assumptions of statistics tools (sparse datasets are not reliable for distributional statistics).

17.5.4 Dictionary Reduction

Several other steps are often taken to reduce the number of words used in the calculations. The reduction of the number of words or the reduction of the dictionary size not only helps with faster analysis of the data, but may also improve the results.

One of the most common steps in reducing the size of the dictionary is stopword removal. Stopwords are the words that very rarely have any predictive capability. Many articles and pronouns fall into this category and can be safely removed from the dictionary. Also removal of the infrequent words is usually helpful as they are in many cases typos or do not have enough predictive value due to their low frequency.

If the task at hand is to distinguish one class of documents from all the rest (e.g., filter spam emails out), creating a local dictionary that only contains the words in that class rather than a dictionary with all the words in all of the documents is a useful method in reducing the dictionary size. This problem is usually referred to as the binary prediction problem. Note that in order to distinguish 100 different categories or classes of documents, one can reduce it to solving 99 of such binary prediction problems, each of which needs a local dictionary.

17.5.5 Vector Generation

In order to describe the documents in a corpus without any deep linguistic analysis, researchers use the vector representation of documents. In this representation each of the word roots that are present in the corpus after the above steps have been done will be represented as a row. Other rows may also be added to the data. These rows could be the existence or absence of a feature or even the type of a feature. For example a row could indicate the category of the documents, while another could represent the presence or absence of a title for each document. On the other hand each document is represented as a column. This matrix is often referred to as the vector space model of the corpus.

Such a matrix can in theory represent the entire corpus, after development. Several different methods are commonly used for creating and filling in the matrix. The simplest method is to identify presence or absence of a word in a document. In this model if any of the forms of a previously identified root is present in a document at least once, the element in the matrix at the intersection of the relevant column and row will contain a value that is decided to show such an instance. Usually a "1" represents the presence of the word in the document and a "0" represents its absence. This representation is known as the binary vector space. The best feature of this representation is its simplicity, but because the frequency of the words in a document is ignored, it is very rarely used in practice.

Researchers use different methods to improve the amount of data captured in the vector space model, so later the data-mining models can use this richer matrix to do better classification or retrieval. One common method is the use of multiword features. This method is based on the fact that many times the co-occurrence of two words gives a better clue about the document than each of the words alone. For example the phrase "sports car" is very different from the use of these two words in a sentence such as "He parked his car at the sports center's parking lot." Having a multiword feature in the dictionary will help capture this difference later when applying data-mining techniques, although it could increase the dictionary size to a great extent.

A very common change from the binary vector space is to represent the frequency of a word happening in the document rather than checking for its presence or absence. A document that uses the word *fire* 10 times is much more likely to be related to a fire incident than the one using this word only once. Clearly the binary vector space lacks the required information to differentiate these two documents. Some variants of the frequency representation model have been used over time. One is to set thresholds for frequencies so as to decrease the complexity of the model. For example a "0" would represent any frequency below a predetermined value, say 4, a "1" would represent frequencies between 4 to 10, and a "2" would represent any frequency above 10.

One last improvement that will be introduced here is the normalization of the frequency scoring. Since the absolute frequency of a word in a document is not as important as its relative frequency in the entire corpus, most researchers normalize the frequency scores before using them. One common normalization is to divide the word frequency or as we'll call it from now on *term frequency* by some measure of its frequency within the entire corpus. The choices for the frequency of a word in the entire corpus are different, but all of them attempt to show the informative level of the term through checking its uniqueness within the corpus: "The less frequent a term is within the corpus, the more important it becomes in predicting the nature of a document that contains it."

Up to now we have discussed the preparation steps for text to be input into data-mining models. In the next sections we will introduce three different models, all of which are used in both IR and TC tasks with little change.

17.6 Mathematical Text-Mining Models

In the following section three main mathematical models have been chosen to be described in more detail: Bayesian methods, singular value decomposition methods, and regression methods. In choosing these methods we had two goals in mind: First, describe the main mathematical models that are closer to the human classifiers' methods than most others. Second, introduce the methods that can be easily incorporated into many text-mining algorithms as the main algorithm or a refining procedure within the algorithm. Other mathematical methods such as neural networks or clustering are in many cases as strong as the ones introduced in this section and may even sometimes perform better than these models, but their mathematical and statistical approaches are much farther from the human raters' methods than the ones introduced in this section.

17.7 Bayesian Classifiers

17.7.1 Naive Bayes Classifier

There are two main branches in Bayesian methods used in TC and IR: naive Bayes classification (NBC) and non-NBC. The main difference is that one approach makes an assumption on a conditional independency, which will be mentioned later, and the other approach tries to avoid this assumption. The concept of naive Bayes classifier will be discussed here, during which the difference between the two methods will be addressed. The focus will be on NBC as it is more often used.

The Bayesian classification approaches are mostly constructed on the concept of NBC. This is a statistical method based on the Bayes theorem. The most appealing characteristic of this approach is its simplicity. In order to understand this concept, a brief explanation is useful.

Let $Y = \{0, 1, ..., |Y|\}$ be the set of possible labels or classes for documents, and $W = \{w_1, w_2, ..., W_{|W|}\}$ be a dictionary of words. A document of N words is represented by the vector $X = (x_1, x_2, ..., x_n)$ of length N. The ith word in the document is $X_i \in W$. Note that N can vary for different documents, but for now let's assume this length is the same for all documents. Also note that in order to make the length N equal for all the documents, we just have to choose a dictionary as discussed in the earlier sections. The NBC model assumes that the class label Y is chosen from some prior distribution $p(Y = y)$, and each word X_i is drawn independently from some distribution $p(W = x_i | Y)$ over the dictionary. This is the conditional independence assumption, without which the following equation could not be derived:

$$p(X = x, Y = y) = p(Y = y) \prod_{i=1}^{n} p(W = x_i | Y = y) \tag{17.1}$$

NBC is a vector space model in which the similarity measure is defined as the posterior probability of a narrative belonging to a particular class given the words in the narrative as evidence. The conditional independence assumption is the main difference between NBC and non-NBC approaches. Here we are accepting that the set of words in a document are conditionally independent of one another, the condition being the specific category. Although it is obvious that this assumption is almost always wrong, much of the TC research based on Bayes theorem makes this assumption (Kim, et al., 2003; Lewis, 1998). There are two reasons behind this choice. One is that the alternative approaches are usually very complex and are computationally expensive, while usually making many other assumptions, which in most cases contribute to inaccuracy of the results as much as the conditional independence assumption. The other is the observed fact that the final results of NBC models are usually much better than the expectations, most times outperforming the non-NBC approaches (Lewis, 1998).

Now let there be a training set $S = \{x^{(i)}, y^{(i)}\}_{i=1}^{m}$ of m examples. The NBC uses S to calculate estimates $\hat{P}(X|Y)$ and $\hat{P}(Y)$ for the probabilities $p(X|Y)$ and $p(Y)$. The estimated probabilities are typically calculated in the usual vector space representation methods discussed earlier. To classify a new document, it is easy to see using Bayes rule that the estimated class posterior probabilities are

$$\hat{P}(Y = y_0 | X = x) = \frac{\hat{P}(X = x | Y = y_0)\hat{p}(Y = y_0)}{\sum_{y=1}^{|y|} \hat{p}(X = x | Y = y)p(Y = y)} \tag{17.2}$$

where

$$\hat{p}(X = x | Y = y) = \prod_{i=1}^{n} \hat{p}(W = x_i | Y = y) \tag{17.3}$$

It should be noted that Equation 17.3 could only be written if we made the conditional independence assumption. Equation 17.2 is the general form of the NBC, in which one can replace the estimates; the predicted class for the document is then just:

$$\text{arg. max}_{y \in Y} \hat{p}(Y = y | X = x) \tag{17.4}$$

In addition to the conditional independence assumption NBC has some other problems such as the maximum likelihood estimator problem with sparse data (Church & Hanks, 1990; Church et al., 1991). Most of these problems have been solved in practical applications, and are therefore not discussed here. Some of the models based on the NBC are the binary independent model (Lewis, 1998), integer-value approaches (Mosteller & Wallace, 1984; Margulis, 1993; Robertson & Walker, 1994) and multinomial models (Guthrie, et al., 1994; Chakrabarti et al., 1997) of which the last model has only been applied to TC because of a binding assumption that has not been solved for the IR purposes yet.

The naive Bayes model currently is rarely used alone to help with either TC or IR as it usually does not yield good results on corpuses that are not very subject specific. One example of the use of this method is Lehto's 1994 work. In this work Lehto used the naive Bayes and the fuzzy Bayes models to assign keywords to the texts and create automatic hyperlinks. The goal of this work was to compare the results with manual hyperlink creation. The models implemented in this version of the work were primary implementations of the naive Bayes discussed above, and the fuzzy Bayes, which will be discussed shortly. These models lacked some of the standardization steps and they only looked at partial evidence to establish similarity between different texts. For example the fuzzy Bayes model used only evidence from one word at a time.

Zhu and Lehto (1999) improved the models and performed experiments on the same dataset. The results were more satisfactory, showing comparable performance between the manual and automated hyperlink creation. The verification method for this claim consisted of two parts. The first was to compare the index terms suggested by the automated system against those suggested by the human raters and measure the hit rate (percentage of similar index terms) and the false alarm rates (percentage of index terms suggested by the automated tool, but not by human raters). The second experiment was to ask human subjects to use the index terms generated by either the human raters or the automated tool to perform some information retrieval tasks. Task performance was measured by task completion time and task accuracy measures.

17.7.2 Bayesian Networks

The Bayesian network is a newer technology for probabilistic representation and inference. What makes the Bayesian networks desirable for TC and IR is mostly its intuitive representation of uncertain relations and its efficient inference algorithms.

In the Bayesian network graph, each node represents a random variable and each arc which leads into the node represents a probabilistic dependence between the node and the node from which the arc is initiated. This leads to a representation of conditional independence relations among the variables in the network. Now suppose a user specifies one or more topics of interest and identifies some document features as evidence for the topics of interest. The Bayesian network retrieval task could be simplified into the following three steps:

Step 1. Build the network representing the query
Step 2. Score each document
    2.1 Extract the features from the document
    2.2 Instantiate the features in the retrieval network
    2.3 Calculate the posterior probability of the relevance
Step 3. Rank the document according to the posteriors

Most of the work done in this area is related to modeling of the query or step 1. After this step is done, all the other steps are very similar to what we have discussed up to now. In developing models for step 1, there are arguments about independence and conditional independence assumptions that need to be addressed. Deciding on these questions greatly affects the representation of the network constructed. For example, one problem could arise from the features of the query that are independent of each other, which will result in constructing nonrelated networks for each of the features in the query. Therefore, if a user runs a query including some synonyms or antonyms, the results of the query could be very poor.

Considering the many different approaches of modeling the network for a query and their irrelevance to understanding the overall method, we will not discuss them any further.

The results obtained from this method depend on the assumptions and the network constructed. One very important factor is the number of layers in the network. In the IR problems specifically, if all possible interdependencies are considered, the size of the network gets extremely large as the number of arcs grows exponentially. The reason is that between each two layers we would need to consider any possible combinations of the lower level to construct the higher level, and from each node we would need to draw arcs to virtually every other node in its higher and lower levels. This explosion of the network shows the huge number of computations needed to find the desired probabilities for a search. This task was almost impossible until 15 years ago because of the high cost of memory needed and the slow computation. Nowadays it's becoming more and more popular because of improvements in computer technologies. If a network with strong mathematical background is constructed, where the speed and cost constraints are met, the results from the Bayesian network are usually better than simple Bayes algorithms (Fung & Del Favero, 1995).

17.7.3 Fuzzy Bayes Models

Another approach to automated coding of narratives is using a fuzzy Bayes model (Lehto & Sorock, 1996; Wellman, et al., 2004). In this method the conditional probability of a category given the set of n words or word combinations is calculated using the following formula:

$$P(A_i \mid n) = MAX_j \frac{P(n_j \mid A_i)P(A_i)}{P(n_j)} \qquad (17.5)$$

where $P(A_i \mid n)$ is the probability of category A_i, given the set of n word combinations in the narrative. $P(n_j \mid A_i)$ is the probability of word n_j, given category A_i. $P(A_i)$ is the probability of category A_i, and $P(n_j)$ is the probability of word n_j in the entire keyword list. $P(A_i \mid n)$, $P(A_i)$, and $P(n_j)$ are all estimated using the relative word frequencies from the database.

Using Equation 17.5, the model selects the category that has the maximum probability for the set of keywords found in the narrative. To do so, the model will assign a probability to each category, using each of the words in that narrative. The model can also assign probabilities of categories to each narrative using combinations of any number of words in that narrative. Notice that this process will take a lot of computational effort. This is because the probability of any combination of words should be estimated using the frequency of that combination happening in the entire database. This model has proved to get very good results in practice. Zhu and Lehto (1999) showed that the fuzzy Bayes approach, unlike classic Bayes, does not assume conditional independence between words. Negative evidence (i.e., words that indicate an index term is not relevant) is considered indirectly. That is, each word or word combination increases the probability of alternative categories, rather than reducing the probability of a particular category. Wellman et al. (2004) included multiple-word combinations as a means of aggregating evidence in the fuzzy Bayes model. They used the fuzzy Bayes improved model that used pairs of words, triples, and even four-word combinations in calculating the best fitting classes for the accident narratives. Their results outperformed those of Chatterjee (1998), which will be discussed later.

Based on the findings of the above experiments, the fuzzy Bayes model was tested for use in the industry and accepted. This model is currently being used and constantly improved by Liberty Mutual for indexing their claims narratives (Corns, et al., 2007; Wellman et al., 2007). The model indexes the claims automatically and flags some that are not rated with high certainty. The flagged claims will later be indexed manually. The results show great benefits in using the model with respect to time, accuracy, and expense for Liberty Mutual.

The fuzzy Bayes model up to now seems to perform very well in both TC and IR tasks, especially for short narratives. The one drawback of this model is that its computational complexity grows with the length of the text, and therefore is not a good candidate for indexing long strings of text, such as articles or electronic books.

17.8 Singular Value Decomposition

The singular value decomposition (SVD) by itself is a mathematical concept. This concept is the building block of most of the latent semantic indexing (LSI) methods as a powerful technique for principal component analysis (PCA) for non-square matrices. Therefore, the discussion here will start with an introduction to SVD.

17.8.1 Mathematical Concepts of SVD

The SVD concept arises when the matrix of eigenvectors P of a given square matrix A is not a square matrix, implying that P cannot have a matrix inverse, and hence that A does not have eigen decomposition. This is not desirable for many applications in engineering, because being able to write the eigen decomposition of a matrix means that the simplest representation of the characteristics of that matrix is reached. The SVD concept is even more important as it can provide eigenvectors for a non-square matrix, A (m × n). Linear feature extraction using principal component analysis by SVD is very popular for TC (Husbands, et al., 2000; Kontostathis & Pottenger, 2002) as it avoids many of the drawbacks of other feature extraction techniques such as Fisher's linear discriminant analysis and Samon's projection (Chatterjee, 1998). The singular value decomposition of matrix A is in the form:

$$A = U\Sigma V^{T} \tag{17.6}$$

where U, Σ, and V are, respectively, m × m, m × n, and n × n. Matrices U and V are both orthogonal decompositions, and, therefore, Σ is a diagonal matrix. These three matrices are then used to make the best rank r approximation of A, which will help in reducing the computational efforts involving matrix A.

Integer number r is at most equal to the minimum of m and n. This is because as Σ is diagonal, for m > n the n + 1 to m rows of Σ will be zero. If n > m, the m + 1 to n columns of Σ will be zero. This means that they play no role in the computations and that they can be omitted. In practice, r is usually much smaller than the minimum of m and n.

This feature extraction technique is commonly used in engineering. When one is interested in the results of an equation for practical use, many times there is a willingness to sacrifice a lot of precision for computational effort and, as a result, computation time. The SVD can easily give a very good approximation of a matrix in a specified lower degree such as k, just by doing the following steps:

Omit the columns k + 1 to m from the matrix U
Omit both columns k + 1 to n and rows k + 1 to m from matrix Σ
Omit rows k + 1 to n from V^{T}
Multiply the new matrices in the same order as before

These four easy steps provide the best rank k approximation of A. The new matrix is called the truncated SVD matrix. The result of the multiplication is still an m × n matrix, but due to the omission of some rows and columns, it is different from A to some extent. Matrix U is m × k, matrix Σ is k × k, and matrix V is n × k. It should be noted that, in practice, step 4 rarely takes place, as the three truncated SVD matrices hold all of the information of the original matrix. One or a combination of two of these matrices will usually be input for computations involving the original matrix, depending on what features of matrix A are of interest.

17.8.2 Relation between SVD and IR

Not surprisingly, there is a way to represent a corpus based on the dictionary of words (the vector space model representation). Let's consider a matrix A with each column representing a different document and each row representing a term available in the whole set of documents. Now the element a_{ij} of the matrix is nothing but the number of times word i is found in document j. The only problem with this representation is that the resulting matrix A is going to become extremely big as the number of documents, and consequently words, increases. As noted earlier there are some preparation steps that can be taken to reduce the dictionary size and therefore the sparseness of the vector space representation. Almost always the resulting representation is still a sparse matrix. This is where the concept of SVD fits in. By applying the SVD technique to matrix A, we can then find the rank k approximation of this matrix. Since A is a sparse matrix, its rank k approximation is very close to the original in the sense of the information embedded in it, while it is less sparse.

As SVD extracts the features of the data at hand, it helps us in another way as well. It is known that the natural language text consists of a lot of noise, which implies that not every word or word combination is equally informative. Dimension reduction can help in eliminating the differences in word usage and eliminate some of the problems with synonyms (these steps can otherwise only be done using natural language processing, which is extremely difficult to automate).

17.8.3 Latent Semantic Structure of Data and the Use of SVD

When a dataset is categorized by human subjects, the semantic similarities between narratives play a very important role. This is a very important reason for human subjects' superiority to automated systems in categorization. It is also true that the more a model is capable of accounting for the semantic structure of a dataset, the better it performs. This is where SVD can help.

If the set of terms is arranged in the rows and the set of documents in the columns, it can be shown that there is a strong relation between matrix U of SVD and term-term relationships of a document set. Also there is a strong relationship between matrix V and document-document relationships (Berry, et al., 1995; Deerwester et al., 1990). The reason for this strong relationship lies in the construction of the SVD. Matrix U is actually the matrix of the eigenvectors of the square symmetric matrix $Y = XX^T$. Matrix V is, on the other hand, the matrix of eigenvectors of $Z = X^TX$. For the sake of simplicity, we will call matrix U and V the term vectors and document vectors, respectively. Figure 17.1 graphically shows the above discussions.

Now it is easy to note that matrix Y is nothing but the matrix of the cosine similarity measures, between all terms. Matrix Z, on the other hand, is the matrix of cosine similarity measures between all documents.

The matrix Σ is known as the matrix of singular values. It is a diagonal matrix, and its diagonal elements are monotonically decreasing in value. The interpretation of the meaning of matrix Σ is the

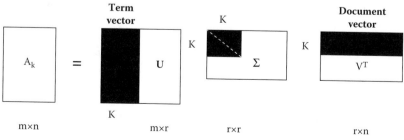

FIGURE 17.1 Graphical representation of SVD vectors (Chatterjee, 1998).

reason behind rank k approximations of matrix A. As matrix U and Σ are multiplied, one can see that each column of U is multiplied by its corresponding element in Σ. This means that the more to the left a column is in matrix U, the more weight is given to it through greater multipliers in Σ. It is based on this observation that one can say columns of U are arranged in order of the information they carry about the original matrix A. The same argument can be made for matrix V. Knowing this, one can conclude that omitting the right columns of U and V to get a lower rank approximation of A will result in the best approximation to the original matrix.

The explanation behind this fact is that rank k approximation of matrix A has omitted the noise from the original data. This is favorable in the area of IR as this lower dimension approximation of A contains higher quality information than the original vector space representation. This is primarily because two related narratives may be represented by completely different vectors in the original high-dimensional space. This may occur, for example, if the words used in each of the two narratives co-occurred frequently in other narratives in the corpus. Therefore, two narratives that are related semantically will be placed close to each other in the reduced space, while in the original space they will be placed further apart. Intuitively, since the number of dimensions, k, is much smaller than the number of unique terms, m, minor differences in terminology will be ignored by the representation in the reduced space. This is shown to be true in actual examples by Berry and Fierro (1996), Deerwester et al. (1990), and Berry et al. (1995).

Chatterjee (1998) used a hybrid SVD-neural networks model and tested its performance on a corpus of accident narratives. The results were very satisfactory and showed great promise for use of automated tools in indexing accident narratives. Earlier it was mentioned that Wellman et al.'s (2004) four-word string combination fuzzy Bayes model performed better than this model.

17.9 Regression Models

There have been two major regression approaches toward the text-mining tasks in general, and the TC specifically. These are known as logistic regression and linear least squares fit. The first approach is in its infancy in the area of IR because it has been highly computationally expensive, especially for the purpose of TC. Detailed discussions of logistic regression can be found in Hosmer and Lemeshow (2000), Minka (2001), and Komarek and Moore (2003a and b).

The linear least squares fit (LLSF) has been used for the purpose of text classification more often. Yang and Chute (1992, 1994) developed and used this model for the first time for text categorization and retrieval. This model has had a very high performance in both areas as reported by Yang and Liu (1999), but it has a disadvantage, which is its high computational cost. In order to use this method in practice, a weighted vocabulary set should be developed for the dataset this method will be used for. Then a subset of the extracted vocabulary will be used as a lower dimension representation of the document set. This preprocessing can be automated, using some statistical criteria such as word count, to decide which of the terms in the document set would be used. Until now, all of the models constructed using this approach have picked a subset of the available terms in the training set as the regression parameters. It would be very interesting to use some features of the document set as the parameters of the regression model as well, although no direct work in this area has been reported. Although not much research has been performed in this area other than the works of Yiming Yang, it is expected that this method will be used more often now that the computational power needed is at hand (Noorinaeini & Lehto, 2006).

Mixed models of regression and SVD have been used in text mining as well as other areas. Ghosh (2002) has used a two-step principal components analysis (PCA), one of which is the SVD technique, to find the most important gene expression profiles, which are then used in classifying tumors, using a regression model. Noorinaeini and Lehto (2006) developed three other hybrid SVD models encouraged by widespread use of the SVD models in text classification literature. The three models were SVD hybrids of Bayes and regression models, all of which used SVD as a preprocessing step for the other model. Although the results were satisfactory and the SVD-regression model performed even better than the simple fuzzy Bayes model, still the multi-word fuzzy Bayes outperformed all of these models.

17.10 Summary

In this chapter we discussed the increasing human need for text classification and information retrieval. We then briefly mentioned what was done to assist in the need for knowledge classification and retrieval prior to the emergence of automated methods. Some of the heuristics that people use, which have the same concept as classification and retrieval, were also noted.

In the later part of the chapter, we laid out the steps that are usually taken to code the textual data into numeric form, so it can be input into data-mining models for either classification or information retrieval. This was followed by the introduction of three mathematical inference models: the Bayesian, SVD, and regression models. All three models were shown to be effective, each having different strengths and weaknesses. We presented how these models can be used either alone or in combination through interrelated examples of previous research. The results showed that no particular model can be always superior, but rather the key to success is a single or combinatory model that best fits the task at hand.

References

Ashby, F.G., and Maddox, W.T. (1992). Complex decision rules in categorization: Contrasting novice and experienced performance. *Journal of Experimental Psychology: Human Perception and Performance*, 18, 50–71.

Berretty. P.M., Todd, P.M., and Martignon, L. (1999). Categorization by elimination: Using few cues to choose. In G. Gigerenzer, P.M. Todd, and the ABC Research Group (Eds.). *Simple Heuristics That Make Us Smart* (pp. 235–256). New York: Oxford University Press.

Berry, M.W., Dumais, S.T., and O'Brien, G.W. (1995). Using linear algebra for intelligent information retrieval. *SIAM Review*, 37(4), 573–595.

Berry, M.W., and Fierro, R.D. (1996). Low rank orthogonal decomposition for information retrieval applications. *Numerical Linear Algebra with Applications*, 3(4), 301–327.

Blum, A.L., and Langley, P. (1997). Selection of relevant features and examples in machine learning. *Artificial Intelligence*, 97(1–2), 245–271.

Brunswik, E. (1955). Representative design and probabilistic theory in a functional psychology. *Psychological Review*, 62, 193–217. [As referenced by Hertwig, Hoffrage, and Martignon, 1999].

Chakrabarti, S., Dom, B., Agrawal, R., and Raghavan, P. (1997). Using taxonomy discriminants and signatures for navigation in text databases. In M. Jarke, M. Carey, K.R. Dittrich, F. Lochovsky, P. Loucopoulos, and M.A. Jeusfeld (Eds.). *Proceedings of the 23rd VLDB Conference*, 446–455.

Chatterjee, S. (1998). *A Connectionist Approach for Classifying Accident Narratives*. Unpublished doctoral dissertation, Purdue University, West Lafayette, IN.

Church, K.W., Gale, W., Hanks, P., and Hindle, D. (1991). Using statistics in lexical analysis. In U. Zernik (Ed.), *Lexical Acquisition: Exploiting On-line Resources to Build a Lexicon* (pp. 115–164). Hillsdale, NJ: Erlbaum.

Church, K.W., and Hanks, P. (1990). Word association norms, mutual information and lexicography. *Computational Linguistics*, 16, 22–29.

Corns, H., Wellman, H.M., and Lehto, M.R. (2007). Development of an approach for optimizing the accuracy of classifying large numbers of claims narratives using a machine learning tool (TEXT-MINER). In *Proceedings of the 12th International Conference on Human-Computer Interaction*, 411–416.

De Finetti, B. (1937). La prevision: Ses lois logiques, ses sources subjectives. *Annales de l'Institut Henri Poincare,* 7, 1–68. (Reprinted in 1964 in English translation as Foresight: Its logical laws, its subjective sources. In H.E. Kyburg, Jr. & H.E. Smokler (Eds.), *Studies in Subjective Probability.* New York: Wiley.). [As referenced by Martignon and Laskey, 1999]

Deerwester, S., Dumais, S.T., Furnas, G.W., Landauer, T.K., & Harshman, R. (1990). Indexing by latent semantic analysis. *Journal of the American Society for Information Science,* 41(6), 391–407.

Domingos, P., and Pazzani, M. (1996). Beyond independence: Conditions for the optimality of the simple Bayesian classifier. In *Proceedings of the 13th International Conference on Machine Learning* (pp. 105–212). San Mateo, CA: Morgan Kaufmann.

Fung, R., and Del Favero, B. (1995, March). Applying Bayesian networks to information retrieval. *Communications of the ACM,* 38(3), 42–48, 57.

Ghosh, D. (2002). Singular value decomposition regression models for classification of tumors from microarray experiments. *Proceedings of Pacific Symposium on Bio Computing* (pp. 18–29). Lihue, HI.

Guthrie, L., Walker, E., and Guthrie, J. (1994). Document classification by machine: Theory and practice. *Proceedings of COLING 94: The 15th International Conference on Computational Linguistics,* 2, 1059–1063.

Han, J., and Kamber, M., (2000) *Data Mining: Concepts and Techniques.* San Francisco, CA: Morgan Kaufmann.

Hasher, L., and Zacks, R.T. (1984). Automatic processing of fundamental information: The case of frequency of occurrence. *American Psychologist,* 39, 1372–1388.

Hertwig, R., Hoffrage, U., and Martignon, L. (1999). Quick estimation: Letting the environment do the work. In G. Gigerenzer, P.M. Todd, and The ABC Research Group (Eds.). *Simple Heuristics That Make Us Smart* (pp. 209–234). New York: Oxford University Press.

Hosmer, D. W., and Lemeshow, S. (2000). *Applied Logistic Regression* (2nd ed.). New York: Wiley-Interscience.

Husbands, P., Simon, H., and Ding, C. (2000, October). On the use of singular value decomposition for text retrieval. *First SIAM Computational Information Retrieval Workshop.*

Kim, S., Seo H., and Rim, H. (2003, July). Poisson naive Bayes for text classification with feature weighting. *Proceedings of the 6th International Workshop on Asian Language* (pp. 33–40), Sapporo, Japan, ACL.

Komarek, P., and Moore, A. (2003a). Fast robust logistic regression for large sparse datasets with binary outputs. In C.M. Bishop and B.J. Frey (Eds.), *Proceedings of the Ninth International Workshop on Artificial Intelligence and Statistics,* Key West, FL.

Komarek, P., and Moore, A. (2003b, July). Fast logistic regression for data mining, text classification and link detection. Paper submitted to Neural Information Processing Systems Conference, La Jolla, CA.

Kontostathis, A., and Pottenger, W.M. (2002, December). *Detecting Patterns in the LSI Term-Term Matrix.* Paper presented at the workshop on the foundations of data mining and discovery, IEEE International Conference on Data Mining (ICDM'02).

Lehto, M.R. (1994). *Warnings & Safety Instructions: Electronic Hypertext Version 2.0.* Ann Arbor, MI: Fuller Technical Publications.

Lehto, M.R., and Sorock, G.S. (1996). Machine learning of motor vehicle accident categories from accident narratives. *Methods of Information in Medicine,* 35, 309–316.

Lewis, D.D. (1998). Naive (Bayes) at forty: The independence assumption in information retrieval. In N. Claire & C. Dellec (Eds.), *ECML'98: Tenth European Conference on Machine Learning* (pp. 4–15). Heidelberg, Germany: Springer Verlag.

Margulis, E.L. (1993). Modeling documents with multiple Poisson distributions. *Information Processing and Management,* 29, 215–227.

Martignon, L., and Laskey, K.B. (1999). Bayesian benchmarks for fast and frugal heuristics. In G. Gigerenzer, P.M. Todd, and The ABC Research Group (Eds.). *Simple Heuristics That Make Us Smart* (pp. 169–188). New York: Oxford University Press.

Minka, T. P. (2001). *Algorithms for maximum-likelihood logistic regression.* (Tech. Rep.Stat. No.758). Pittsburgh: Carnegie Mellon University.

Mosteller, F., and Wallace, D.L. (1984). *Applied Bayesian and Classical Inference* (2nd ed.). New York: Springer-Verlag.

Noorinaeini, A., and Lehto, M.R. (2006). Hybrid singular value decomposition: A model of text classification. *International Journal of Human Factors Modeling and Simulation,* 1(1), 95–118.

Nosofsky, R.M. (1986). Attention, similarity, and the identification-categorization relationship. *Journal of Experimental Psychology: General,* 115(1), 39–57.

Robertson, S.E., and Walker, S. (1994). Some simple effective approximations to the 2-Poisson model for probabilistic weighted retrieval. *Proceedings of the Seventeenth Annual International ACM-SIGIR Conference on Research and Development in Information Retrieval* (pp. 232–241). London: Springer-Verlag.

Savage, L.J. (1954). *The Foundations of Statistics.* Dover: New York. [As referenced by: Martignon and Laskey, 1999].

Weiss, S.M., Indurkhya, N., Zhang, T., and Damerau, F.J. (2005). *Text Mining: Predictive Methods for Analyzing Unstructured Information.* New York: Springer.

Wellman, H.M., Lehto, M.R., and Corns, H. (2007). Computer classification of injury narratives using a fuzzy Bayes approach: Improving the model.

Wellman, H.M., Lehto, M.R., Sorock, G.S., and Smith, G.S. (2004). Computerized coding of injury narrative data from the National Health Interview Survey. *Accident Analysis and Prevention,* 36, 165–171.

Yang, Y., and Chute, C.G. (1992, August). A linear least squares fit mapping method for information retrieval from natural language texts. *Proceedings of 14th International Conference on Computational Linguistics* (COLING-92) (pp. 447–453). Silicon Valley, CA: Morgan Kaufmann.

Yang, Y., and Chute, C.G. (1994). An example-based mapping method for text classification and retrieval. *ACM Transactions on Information Systems* (TOIS), 12, (3), 252–277.

Yang, Y., and Liu, X. (1999). A re-examination of text categorization methods. *Proceedings of 22nd Annual International SIGIR Conference* (pp. 42-49). London: Springer-Verlag.

Zhu, W., and Lehto, M.R. (1999). Decision support for indexing and retrieval of information in hypertext systems. *International Journal of Human-Computer Interaction,* 11(4), 349–371.

18

Modeling Task Administration Protocols for Human and Robot E-Workers

Hoo Sang Ko and
Shimon Y. Nof

18.1 Abstract

This chapter presents the design of task administration protocols (TAPs) for effective task allocation in distributed systems composed of human and robot e-Workers. Digital human models in such modeled environments must be able to operate under this type of protocol. Two e-Work logic models and tools are employed to model the activities and interactions of human and robot e-Workers: multi-agent systems and protocols. Protocols are designed to represent the rules of interactions for effective task allocation among agents. Coordination protocols have been developed for task allocation by managing some dependencies between tasks. As a result, it generates a dynamic alliance of agents, such as a task dependence network composed of task agents and resource agents, to complete the given tasks. In dynamic, distributed task environments, it is not feasible to devise a universal coordination protocol that performs effectively under all circumstances. Coordination protocols, however, address usually limited types of dependence. Furthermore, there are many complex situations requiring decisions beyond just coordination in real task environments. As a control mechanism, TAPs are designed to handle dependencies including coordination solutions. The suitable TAPs also manages variable dependencies.

This chapter explains the definition of TAPs and their objectives. As TAPs manage dependencies, types of dependencies between tasks are analyzed first. The appropriate protocol is determined based on the type of current dependence, which is dynamically decided by task priority evaluation function. In order to improve the performance under TAPs, decision parameters in TAPs are identified, for instance, space-related and time-related parameters. Improvement is envisioned by employing case-based reasoning with the concepts of (1) agent affinity and (2) time-out. Finally, a case study of TAPs for emergency task allocation is illustrated to show the rationale and logic of TAPs.

18.2 Background

This chapter presents the design of task administration protocols (TAPs) for effective task allocation in a network of autonomous entities where groups of humans and robots collaborate with each other to perform simple or complex tasks. Digital human models in such modeled environments must be able to also operate under this type of protocol. In the context of distributed systems composed of multiple humans and robots, a task can be seen as an input going into an entity in a network of distributed entities, then processed or served by the resources in the entity that has the skill and capability to perform it. Tasks requested from or required by the system can be performed through a process of distributing tasks to a cluster of nodes in the network of distributed entities and synthesizing the results to achieve the system's goals. For example, production in a plant involves many activities and collaboration of people, robots, machines, and equipment in the plant, each having its own skill and capability to perform subtasks in order to achieve the overall tasks (i.e., producing products).

There have been a number of research studies on the control of multiple collaborative robots. Architectures such as ALLIANCE (Parker, 1998), BLE (Werger & Mataric, 2000), and MURDOCH (Gerkey & Mataric, 2002) have shown the capability of autonomous control of multiple heterogeneous robot teams in order to achieve common goals with proper task allocation and role assignment. They are achieved through inter-robot communication and interaction, while the role of humans in such architectures is limited only to intervention, assistance, or initial planning. On the other hand, some research involved human-related issues in multiple robot systems and focused on human-machine interaction and interface design for the control of multiple robots (Jones & Snyder, 2001; Monferrer & Bonyuef, 2002). These researchers try to enhance operability of multiple collaborative robots through better information management and interface design. A common assumption of such research is that the performance of human operators usually deteriorates as the number of robots and the complexity of tasks increase. A way to solve the problem is to augment human operators' decision making by providing them with intelligent tools following certain rules of interactions.

18.3 Significance

e-Work is defined as any collaborative, computer-supported, and communication-enabled productive activities in highly distributed organizations of humans and/or robots or autonomous systems (Nof,

2003). According to the definition, activities of digital humans and robots in virtual environments can be seen as e-Workers' activities in e-Work environments. Two of the main e-Work models and tools are agents and protocols. By using these tools, virtual environments can be modeled as multi-agent systems, in which each agent represents a digital human or a robot thanks to agents' characteristics and capabilities (Wooldridge & Jennings, 1995). Besides, they will follow certain protocols, which represent the rules of interactions among agents and environments. In particular, in collaborative e-Work environments, digital human model simulations must follow collaborative control to yield valid results (Nof, 2007). The protocols function as a control mechanism that allocates tasks to distributed participants who must achieve their common goals by interacting and sharing information with each other in distributed systems. They greatly affect the performance of a group of humans and robots in distributed systems, since they decide how each individual entity coordinates tasks with other entities. In many cases of distributed, dynamic, decentralized environments, each entity has only local or limited information and its own goals, which may or may not conflict with other entities' goals. In order to accomplish overall goals with such limited information and their own goals, it is inevitable that an effective control mechanism must be provided to coordinate tasks by exchanging information among participants.

Many coordination protocols have been designed to achieve effective coordination by managing a given type of dependence (see section "Design of Task Administration Protocols") between tasks (Smith, 1980; Ferber, 1999; Fatima & Wooldridge, 2001). In real task environments, however, there are some situations that cannot be handled just by coordination protocols: (1) the type of dependence may change from time to time; and (2) there are many complex situations requiring decisions beyond just coordination, for example, handling an emergent problem that can cause a significant damage to the system. Thus, the protocol must be capable of identifying the current type of dependence between tasks and dealing with the current situation. Such a protocol, which assumes the responsibility of making decisions actively and triggering timely actions so that the overall performance can be improved, was originally defined by Huang and Nof (2000, 2002), and Anussornnitisarn et al. (2002) as a TAPs.

The next section introduces work related to coordination theory and coordination protocols. The rest of this chapter focuses on how to design TAPs for effective task allocation and for its digital human modeling applications by dealing with the following problems:

What are the definition and the objective of TAPs?
How can a task be defined and prioritized?
How can dependencies between tasks be classified for task administration?
Does a system under control of adaptive TAPs perform better than under general, non-TAPs coordination protocols?

Several examples illustrate how TAPs are vital to e-Work and e-Service environments, and to their digital human modeling implementations.

18.4 Coordination Theory and Coordination Protocols

18.4.1 Agent and Multi-Agent Systems (MAS)

An agent possesses several characteristics such as autonomy, interaction, reactivity, proactivity, reasoning/learning/adaptation, and cooperation/coordination (Wooldridge & Jennings, 1995). Due to such characteristics, multi-agent systems (MAS), a subfield of distributed artificial intelligence (DAI), are widely used in many domains, including multi-robot systems (Balch & Parker, 2002; Iocchi et al., 2003), supply chain management (Fung & Chen, 2005; Hu et al., 2001; Khoo & Yin, 2003; Ulieru & Cobzaru, 2005), agent-based manufacturing systems (ABMS; Huang & Nof, 2000; Shen & Norrie, 2001;

Frayret et al., 2004), to name a few. MAS can be applied to domains comprised of different organizations or participants with different, possibly conflicting goals (Stone & Veloso, 2000). It aims to provide foundations to construct dynamic, complex systems composed of multiple agents and mechanisms to coordinate the behaviors of independent agents. Every agent having proprietary information pursues its own objectives by communicating with other agents to obtain necessary information and makes a contribution to the coordinated solution of the overall problem. MAS, therefore, are appropriate to represent many application domains in heterogeneous, autonomous, distributed environments. For example, a network of exploratory robots may be composed of several robot agents and a human agent. In such a network, a human agent places an order (a task) to robot agents to explore a certain area, and selects a robot agent (bid selection) that is close to the area and has less work to do, in other words, proposes the best offer (bidding).

The distribution of tasks, or task allocation, is one of the major problems of MAS. It should decide who must do what, with what resources, based on the goals, the skills of the agents, and the constraints in the context (Ferber, 1999). Due to the decentralization and the dynamic nature inherent in the distributed systems, however, achieving effective task coordination among agents is a challenge. In the next section, some basic modes of task allocation will be reviewed.

18.4.2 Classification of Task Allocation Modes

MAS are used to solve complex problems or tasks by decomposing the tasks into small subtasks, each of which is handled by single agents having different capabilities and then combining those individual accomplishments. The main problem that needs to be addressed in MAS is task allocation, or the distribution of tasks among agents in such a way that the number of tasks completed successfully by their deadline is maximized while satisfying their desired level of quality.

Task allocation can be classified into four modes (Ferber, 1999; Lee & Lee, 2003), and the strengths and weaknesses of each task allocation mode are summarized in Table 18.1. Recently, decentralized modes are more preferred to centralized modes since they possess the following abilities:

Suitability for the nature of many systems: tasks are allocated in distributed, dynamic, decentralized environments (e.g., enterprise network, Internet, etc.)
Scalability: the number of participants may vary
Flexibility: network configuration can be changed
Fault-tolerance: robust to failure of participants

18.4.2.1 Contract Net Protocol (CNP)

CNP is a marketplace approach in which contracts are drawn up and allocated to other agents. This is a very popular and widely used method introduced by R. G. Smith (1980), and many coordination

TABLE 18.1   Comparison of Task Allocation Modes

	Centralized		Decentralized	
	Hierarchical	Heterarchical	Acq. Network	CNP
Advantage	Simple and fast	Coherence Flexibility	No bottleneck Fault-tolerance	Flexibility Simple and proven
Disadvantage	Inflexibility Fault-intolerance	Bottleneck Fault-intolerance	Coherence Scalability Latency	Communication overload Temporal ignorance Spatial ignorance

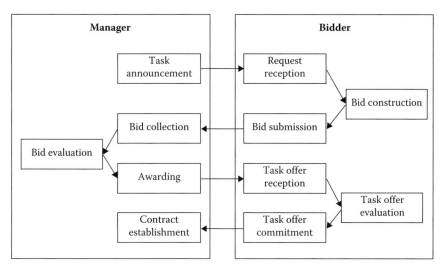

FIGURE 18.1 Stages of contract net protocol.

protocols in various domains have been created based on the framework of CNP. An agent that requires a service from other agents becomes a contractor or a manager, and the other agents capable of providing the required service become bidders in the market. The CNP takes four stages to allocate a task: task announcement (request for bids); bidding proposal; bid evaluation and awarding; and task commitment, as shown in Figure 18.1. CNP is very simple to use and easy to understand, but its performance can be seriously degraded by several factors in the environment such as the number of agents, communication overload, waiting time for bid collection, and so on.

Coordination protocols and TAPs can be under the contract net category. CNP, however, only provides a general framework for coordination protocols. Since TAPs is designed to handle situations requiring more than coordination, TAPs must be considered as a more expanded concept than the CNP framework. In the following sections, related research on coordination protocols based on CNP will be reviewed, and their limitations will be described.

18.4.3 Coordination Theory and Dependence Analysis

In coordination theory, *coordination* is defined as managing dependencies between activities (Malone & Crowston, 1994). The dependencies were classified into four types in their research:

Shared resource: multiple activities share some limited resource
Producer/consumer relationships: one activity produces something that can be used by another activity
Simultaneity constraint: some activities must (or not) occur at the same time
Task/subtask: an activity can be hierarchically decomposed into several subactivities

Given a few assumptions, all dependencies can be expressed as producer/consumer relationships (Gouaich, 2004). In other research, the types of dependencies were further classified into six types (Khanna et al., 1998):

- Flow dependence (data): a processor cannot execute if it does not receive material or information from its predecessor.
- Input dependence: a processor has to wait for a resource until another processor releases the resource.

- Output dependence: a processor cannot receive a task until the next processor is finished.
- Co-dependence: a task is separated and processed in a different processor simultaneously. The task cannot be sent to another processor until all the subtasks are completed.
- Flow dependence (control): a control signal allows other processors to execute their tasks.
- Anti-dependence: a change in a task affects other tasks in other processors.

The types of dependencies between tasks in coordination situations have been identified in the previous research. In real task environments, however, there are many complex situations requiring decisions beyond just coordination, such as handling urgent tasks that have higher priority first.

18.4.4 Selection of Coordination Mechanism

In the perspective of MAS, coordination is a fundamental capability of agents to make decisions on their own actions in the context of interactions with other agents in an environment (Durfee, 2001; Ferber, 1999). Durfee identified three properties that must be considered by a coordination strategy: agent population; task-environment properties; and solution properties. For each property, he defined three important dimensions that influence the applicability of specific coordination mechanisms, including heterogeneity, dynamics, distributivity, quality, etc. It is not possible to devise a coordination protocol that works under all circumstances as Durfee stated. To solve this problem, there has been some research on the selection of coordination mechanisms. A MAS in which agents can reason about the coordination process and then choose appropriate coordination mechanisms in the current situation was developed by Excelente-Toledo and Jennings (2004). In the MAS, the selection of coordination mechanism can be made at run-time during the coordination process. The researchers defined a generic template for a coordination mechanism composed of coordination technique components and meta-data attributes. In order to reason about choosing a particular mechanism, coordination mechanisms should be quantified with the two meta-data attributes. The quantification of coordination mechanisms, however, is not clear and is difficult to normalize. Generalized partial global planning (GPGP), a coordination framework in which local optimization solutions of agents are combined into a global solution, was developed by Decker and Lesser (1995). GPGP is composed of a set of modular coordination mechanisms and one of them is activated in a predefined situation during execution. Since only a fixed coordination mechanism can be activated in the given situation, the flexibility of mechanisms selection is limited.

An appropriate coordination protocol can be selected according to the type of dependence in the given environment. Four basic types of TAPs were developed to manage certain types of task dependencies by Anussornnitisarn (2003), as shown in Table 18.2. In the research, however, dynamic selection of an appropriate protocol, which is suitable for the current situation, was not addressed. It also failed to consider other types of dependencies more than coordination that must be dealt with in task administration.

TABLE 18.2 Task Classifications and Corresponding TAPs

Coordination Problem	Task Dependence	Constraints	TAPs
Managing shared resources	Input dependence, Output dependence	Resource, time	Resource allocation
Managing producer/consumer relationships	Flow dependence, Input dependence, Output dependence	Optimization, task, time, resource	Negotiation (bidding)
Managing simultaneity constraints	Flow dependence, Output dependence, Control flow dependence	Task, time, resource	Task synchronization
Managing task/subtask dependencies	Co-dependence, Anti-dependence	Conflict, task, time, resource	Conflict resolution and error recovery

Source: Anussornnitisarn, 2003.

18.4.5 Performance Enhancements of Coordination Protocols

An agent suffers from the following ignorance during the coordination process (Parunak, 1987):

Temporal ignorance: an agent cannot see the task announcements and the bids that will arrive later.

Spatial ignorance: an agent cannot see what other agents are doing.

Performance anomalies from such ignorance are caused by the fact that task allocation only depends on local knowledge of a manager (Walsh & Nahavandi, 2004). To overcome this ignorance, coordination protocols must be able to decide proper parameters for the following:

How far does a manager access bidders?

How long does a manager wait for bidding?

18.4.5.1 Case-Based Reasoning

If a manager tries to visit all the potential bidders, it will be time-consuming and too many messages will be created. Most coordination protocols based on CNP get into the trouble of message congestion due to broadcasting of tasks when the system is large. To overcome this inefficiency, some studies suggest reducing the scope of bidders and communication load on task coordination by some methods, such as instance-based learning (IBL; Deshpande et al., 2004) and case-based reasoning (CBR; Garland & Alterman, 2004; Wan et al., 2005; Lou et al., 2004; Weng & Zhu, 2005). In CBR, each agent stores the past successes in the individual case base. When a new problem emerges, the previous knowledge in the case base is reviewed and applied, so that it can save the overhead due to the task negotiation by coordination protocols. It is difficult, however, to determine how to define a case including time, how to assess the similarity between the cases in the case base and a new case, and how to maintain the case base when the size of the case base is increasing. How to apply CBR effectively will be discussed later in this chapter.

18.4.5.2 Time-Out Protocol

Time-out protocol has proved to be effective in several collaborative operations (Esfarjani & Nof, 1998; Williams et al., 2002; Anussornnitisarn et al., 2002; Liu & Nof, 2004). It has been developed to prevent a single task from occupying a shared resource excessively, or a single agent from delaying the manager by getting her wait too long for response. In order for the time-out protocol to be effective, the most suitable time-out threshold (t_o) must be determined by statistical analysis. A similar concept of time-out can be applied to decide time-related parameters of TAPs. A manager announces a task, and waits for bidding until every bidder sends a bid. A bidder, however, may not be interested in the task, so it may send a no interest message, or it may not respond to the task announcement. Besides, some bidders may not function due to certain reasons, such as failures or a broken communication channel. In these cases, the manager would waste excessive time waiting for the bids that are useless or not receivable. Therefore, an effective method to determine the appropriate cut-out timing or time-out threshold for a request session should be developed for time efficiency.

18.4.5.3 Protocol Adaptability

Many coordination protocols have been developed, most of which lack in protocol adaptability and only work well in a given, assumed task environment. In a study of coordination protocols, five types of coordination protocols were developed based on CNP (Reaidy et al., 2006). The protocols could be applied for different conditions, but their logics were pre-defined and, dynamic selection and adaptation of protocols were not considered. Limited work has addressed the dynamic selection of coordination mechanisms. The selection or adaptation of logics and parameters in the coordination protocol was considered only to a limited extent. For example, in bid calculation and selection during the coordination process, adaptation of parameters was partially addressed by calculating viability measures (Anussornnitisarn, 2003).

The impact of protocol adaptability in a testing local area network (TestLAN) was analyzed by Williams et al. (2003). In TestLAN, adaptation logic checks the changes of job parameters by four protocol adaptation methods: static system, event-based adaptability, event-batch triggered adaptability, and time-interval triggered adaptability. When the logic perceives the changes of job parameters over a certain level of thresholds, it automatically updates selected parameters of the protocol. This analysis showed the usefulness of parameter adaptation under varying operational conditions.

18.4.6 Priority-Based Task Allocation

Task priority assignment has been widely studied for task scheduling in the area of real-time systems. Preemptive-resume methods use priority of tasks for task scheduling; when a task with higher priority arrives, tasks with lower priorities are suspended and resumed when the higher priority task is completed. Algorithms such as rate monotonic algorithm have been developed for task priority assignment in real-time systems (Lehoczky et al., 1989; Karl et al., 1993). In this area, the priority of tasks is usually determined by time or frequency, and it is assigned in a centralized manner.

In the business context, priority of tasks in task allocation can be determined by managerial decision, by prior assignment based on task type, or by pricing (Malone & Crowston, 1994; Walsh & Wellman, 1998). In most coordination protocols, priority of tasks is determined by pricing. A coordination protocol that allocates tasks considering their priority has been developed by Fatima and Wooldridge (2001). In their work, a task has a priority element that represents how critical the task is. Tasks with higher priority get more funds, so the priority is reflected in funding. The limitation of such coordination protocols is that the priority of a task is predetermined and fixed by its price and deadline in the task description at the moment the task arrives. Thus, dynamic change of priority along time cannot be considered.

18.4.7 Limitations in Design of Coordination Protocols

Based on observations of the previous research, the limitations of coordination protocol are summarized as follows:

- Protocols for task allocation focus only on coordination, for instance, managing the four basic types of dependencies. In the real world, there are other types of dependencies beyond coordination; furthermore, there are dynamic changes in dependencies.
- Priority of a task is simply predefined, fixed, or even ignored. Priority should be re-evaluated dynamically whenever a new task arrives, since it determines the current type of dependence and the corresponding protocol.
- Consideration of logic and parameter adaptation in protocols has been limited, despite the evident value of such adaptation.

The above issues need to be addressed in design of TAPs, which will be explained in the next section.

18.5 Design of Task Administration Protocols

Considering the level of task administration, an agent basically has a three-layered structure as shown in Figure 18.2. The role of each component is described as follows:

Data processing component: This component is in charge of actual data processing. Physical data are exchanged through communication protocols.

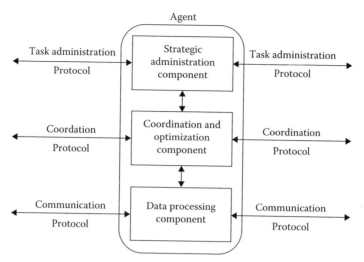

FIGURE 18.2   Basic structure of an agent.

Coordination and optimization component: This component handles the coordination process with other agents, for instance, for bid pricing, bid selection, awarding, rejection, and so on.

Strategic administration component: This component makes the strategic level decisions, for instance, it receives task information (T_i), decides the type of dependence (δ), sets the priority of tasks (PR_i), and then determines the appropriate protocol (P_j) and the value of its parameters ($PA(P_j)$).

18.5.1 Task Administration Protocols (TAPs) in a Coordination Network

A concept of a coordination network (Anussornnitisarn, 2003) is introduced to define the model of TAPs. A coordination network (Co-net) is a network of autonomous entities that enables collaboration among the entities. Each entity exchanges information and execution to achieve its objectives while it tries to maximize the system goals. The Co-net is defined as:

$$\text{Co-net} = \langle \pi, \tau, \alpha, \sigma \rangle \tag{18.1}$$

where π: a set of agents in Co-net;
τ: a set of tasks that can be processed in Co-net;
α: a set of activities performed by an agent to fulfill T_i;
σ: a set of control mechanisms used by each agent in π.

Malone and Crowston defined the coordination process as the process of managing dependencies between activities (1994). In the Co-net, dependence (δ) can be defined as:

$$T_i \times \pi(T_i) \times \alpha(T_i) \mid T_i \in \tau \rightarrow \delta \tag{18.2}$$

where $\pi(T_i)$: a set of agents who can process T_i;
$\alpha(T_i)$: a set of activities required to process T_i.

In the Co-net, a coordination protocol (CP) is designed to manage a particular dependence (δ) of any given task (T_i). A CP is defined as:

$$P_j = \langle \delta, I, R, PA, S \rangle \tag{18.3}$$

where P_j: a particular coordination protocol;
$\quad\quad\quad$ δ: a set of dependencies;
$\quad\quad\quad$ I: a set of initiators $(I \subset \pi)$;
$\quad\quad\quad$ R: a set of responders $(R \subset \pi$ and $I \cap R = \{\})$;
$\quad\quad\quad$ PA: a set of parameters of the coordination protocol;
$\quad\quad\quad$ S: a set of transition stages.

TAPs are defined as a set of protocols that assume the responsibility of making decisions actively and triggering timely actions so that it can improve the coordinated performance (Anussornnitisarn et al., 2002; Huang & Nof, 2000, 2002). According to the type of task dependence in a given situation, TAPs should select the appropriate protocol (P_j) and set proper parameter values $(PA(P_j))$ for the protocol. In the Co-net, therefore, TAPs is defined as:

$$\text{TAPs: } \sigma(T_i, \delta \mid \pi, \tau', \alpha) \rightarrow P_j = <\delta, I, R, PA, S> \tag{18.4}$$

where τ': a set of tasks that currently remain with an agent.
The effectiveness of protocols can be evaluated as follows:

$$f_{im}(\tau, \pi, P_j); 0 \le f_{im}(\tau, \pi, P_j) \le 1 \tag{18.5}$$

where $f_{im}(\tau, \pi, P_j)$: a measure of performance in the Co-net under a given coordination protocol or TAPs;
$\quad\quad\quad$ im: an index of measured factors.

The objective of TAPs is to select the best protocol (P_j) and its parameters $(PA(P_j))$ that maximize performance when a new task T_i arrives:

$$P_j = \arg\max_{P_j \in \sigma}(f_{im}) \tag{18.6}$$

18.5.2 Classification of Task Dependencies

In coordination theory, *coordination* is defined as managing dependencies between tasks, and four types of dependencies have been defined. In order to define the types of dependencies between tasks in task administration, including but not limited to coordination, Allen's interval temporal logic (Allen & Ferguson, 1994) is introduced. Allen originally defined 13 relations between two intervals of events. In this chapter, seven basic types of dependencies are defined to represent the dependencies between two tasks as shown in Table 18.3. In particular, interrupt dependence $(T_i \text{ i } T_j)$ is defined in the classification. $(T_i \text{ i } T_j)$ can be further classified into three subcategories: $(T_i \text{ ic } T_j)$, $(T_i \text{ ir } T_j)$, and $(T_i \text{ ik } T_j)$. Each dependence can be mapped to its corresponding dependence as defined by Khanna et al. (1998). Any dependence between tasks can be represented by one of basic dependencies or by their combinations.

Whenever a new task T_i arrives to an agent, the agent calculates the priority of T_i and previous tasks at time t by dynamic task priority evaluation function $(pf(T_i, t) = PR_i(t))$. If the priority of T_i is less than the priority of previous tasks, the previous dependence (δ) and the protocol (P_j) are maintained. Otherwise, the dependence is changed and another protocol (P_j') is activated to manage the new dependence (δ') as shown below:

$$\exists T_j \in \tau \, pf(T_i, t) < pf(T_j, t) \rightarrow \text{TAPs}(T_i, \delta \mid \pi, \tau', \alpha) \rightarrow P_j(\delta)$$
$$\exists T_j \in \tau \, pf(T_i, t) \ge pf(T_j, t) \rightarrow \text{TAPs}(T_i, \delta' \mid \pi, \tau', \alpha) \rightarrow P_j' = (\delta') \tag{18.7}$$

where $pf(T_i, t)$: task priority evaluation function that returns $PR_i(t)$.

TABLE 18.3 Classification of Task Dependencies

Dependence	Relation	Meaning	Dependence Analysis (Khanna et al., 1998)	Illustration
$T_i \leq T_j$	Before	T_i ends before T_j	Flow (data) Input/output	
$T_i \, o \, T_j$	Overlaps	T_j starts before T_i ends	Flow (control)	
$T_i \, d \, T_j$	During	T_i starts after T_j starts and ends before T_j ends	Flow (control)	
$T_i \, s \, T_j$	Starts	T_i and T_j start simultaneously	Co-dependence	
$T_i \, f \, T_j$	Finishes	T_i and T_j end simultaneously	Co-dependence	
$T_i = T_j$	Equals	T_i and T_j start and end simultaneously	Co-dependence	
$T_i \, i \, T_j = \{(T_i \, ic \, T_j), (T_i \, ir \, T_j), (T_i \, ik \, T_j)\}$	Interrupts (i∧ continues) (i∧ restarts) (i∧ killed)	T_j is stopped by T_i T_j continues when T_i ends T_j restarts when T_i ends T_j is removed by T_i	Anti-dependence	

For example, let's assume a robot is exploring a remote area. A task T_j, searching object 1, is being performed and a new task T_i arrives. In normal situations, for example in a case where T_i is to search object 2, the dependence will not change. In a special situation, such as in a case where T_i is to adjust the robot's camera settings, $PR_i(t)$ should be higher than $PR_j(t)$. In this case, even though T_j was being performed, T_i should interrupt T_j, which corresponds to the ($T_i \, i \, T_j$) dependence, thus a protocol that interrupts T_j and performs T_i should be activated. The priority of tasks at time t ($PR_i(t)$) will be evaluated by priority evaluation function ($pf(T_i, t)$).

18.5.3 Dynamic Task Priority Assignment

In general, a task can be defined as follows:

$$T_i = <TYPE_i, QTY_i, DL_i, Std\_cost_i, PR_i(t)> \tag{18.8}$$

where $TYPE_i$ is the type of T_i that requires a certain skill of a server;
   QTY_i is the amount involved in T_i that occupies the capacity of a server;
   DL_i is the latest time by which T_i must be performed by a server;
   Std_cost_i is the estimated cost of T_i;
   $PR_i(t)$ is the priority of T_i at time t.

$PR_i(t)$ represents the importance of T_i and the task that has higher priority should be performed first. Whenever a new T_i arrives to an agent, the agent calculates the priority of T_i ($PR_i(t)$) and previous tasks ($PR_j(t)$) assigned to itself, and then the tasks are sorted by priority. If there is a change in the ordered list of tasks ($PR_i(t) > PR_j(t)$), the type of dependence is changed and another protocol (P_j') is activated to manage the changed dependence (δ').

PR$_i$(t) is dynamically evaluated by task priority evaluation function (pf) at time t, which is defined by five factors:

$$pf(T_i,t)=w_1 s_i + w_2 \frac{QTY_i \cdot (Std\_cost_i - Std\_price_i)}{\sum_j QTY_j \cdot (Std\_cost_j - Std\_price_j)}$$

$$+ w_3\left(1-\frac{DL_i-t}{\sum_j DL_i-t}\right) + w_4\left(1-\frac{QTY_o^t}{QTY_o}\right) + w_5 \sum_k (1-p_k)$$

where w_n: weight of each factor; $0 \le w_n \le 1$; $\sum w_n = 1$; $n = 1, 2, ..., 5$;
        j: index of a task previously assigned to the agent;
        o: index of the current task;
        $s_i = 1$ if $TYPE_i = TYPE_o$; 0 otherwise;
        Std_price_i: estimated task price to perform T_i by a server;
        QTY_o^t: remaining quantity of T_o at t;
        k: index of tasks that will be decommitted due to T_i ($PR_i(t) > PR_k(t)$);
        p_k: penalty of decommitment of T_k; $0 \le p_k \le 1$.

The implication of the five factors is as follows:

If T_i is the same type of task as the current task, its priority is high since setup is minimized or not needed. In some case, however, it must be the opposite.
The more T_i's quantity and profit, the higher its priority.
A task close to its deadline has higher priority.
If the current task is not finished yet but almost done, it has very high priority.
If T_i will cause many tasks to be decommitted after assignment, its priority must be low.

18.5.4 Agent Model with TAPs

Collaboration between agents in the MAS is achieved by interactions through an appropriate protocol in the DCN. Figure 18.3 shows the generic structure of an agent. An agent is composed of the following components:

Communication protocol: handling physical data processing and exchange.
Coordination protocol: providing rules and procedures of information exchange for task coordination among agents.
Task administration protocol: choosing and applying the appropriate protocol and its parameters for incoming tasks and the types of dependencies in order to optimize the coordination performance.
Inference engine: deciding the priority of tasks and other performance-related parameters based on the knowledge in the case base and the current status of the agent.
Case base: storing useful knowledge to decide the best parameters for coordination protocol and to expedite the coordination process.
Coordination engine: making optimal decisions (e.g., bid pricing and bid selection) based on the current status of the agent and incoming tasks.
Task scheduler: scheduling tasks based on the decision made by the coordination engine.
Task/resource database: storing descriptions and status of tasks/resources.

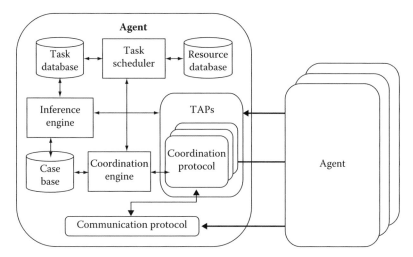

FIGURE 18.3 A generic structure of an agent with TAPs (Ko & Nof, 2007).

The general workflow under TAPs between a task agent (TA) and resource agents (RA) are depicted in Figure 18.4. Those activities between a TA and RAs are administered by TAPs. When a task T_i arrives at a TA, its inference engine decides the candidate list (CL_i) among RAs by using case-based reasoning, which will be described later. Then the TA announces T_i to the RAs in the candidate list. The inference engines in the RAs calculate the priority of T_i ($PR_i(t)$) and decide the type of dependence (δ). The information is sent to the TA. After getting the information, the TA selects the RAs that assign high $PR_i(t)$ and select an appropriate protocol (P_j) with proper parameters ($PA(P_j)$). The TA notifies the selected RAs of the selected protocol (P_j). Finally, the selected protocol (P_j) is activated and the coordination process begins. At the end of the coordination process, each TA and RA updates its status and information, which are to be stored in its own case base and task/resource database.

18.5.5 Performance-Related Decision Parameters of TAPs

When a TAPs is designed to coordinate tasks among a number of agents in MAS, there are four issues that affect the coordination performance significantly. The four issues should be reflected as the decision parameters of the TAPs model:

1. To whom does a task agent announce the task? (CL_i)

Most coordination protocols including CNP suffer from the performance degradation caused by the exponentially increasing number of messages when broadcasting task announcements. This problem could happen due to spatial ignorance; if the task agent knows about the status of other agents or it can decide possible candidates to perform the task. The search space and the number of messages incurred will be drastically reduced.

2. How long does a task agent wait for bidding from resource agents? (t_o)

A task agent (a manager) must open a request session for a certain duration of time to receive enough bids, but it cannot wait too long. The timing of closing a request session is difficult to decide due to temporal ignorance. Thus, an effective method to set the time-out value should be developed.

3. How does a resource agent evaluate the requests from task agents and select a task from a list of requests? (pricing, level of prudence)

A resource agent can decide how much it will bid for the task by a proper bid pricing algorithm while considering its schedule and capability. Viability can be used as one factor to adjust price according

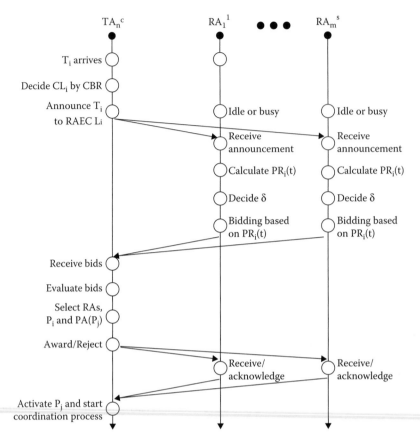

FIGURE 18.4 Workflow under TAPs between a task agent and resource agents (modified from Anussornnitisarn et al., 2005).

to its status. Another important factor that should be considered in such a bid pricing algorithm is the level of prudence, that is, the proportion of the previous contracts that will be reflected in the bid calculation. For example, if the level of prudence is zero, then the resource agent will calculate a bid without reserving any resource for the previous requests that are pending. In this case, the agent may be awarded the task, but it may suffer from too many contracts beyond its capability, which will decrease its credibility.

4. How is the best resource agent selected by a task agent to carry out the task? (weight of decision criteria)

In other words, how does a task agent evaluate bids proposed by resource agents? This decision can be made by an appropriate bid selection algorithm, in which the weight should be carefully selected according to the objective of the task agent (e.g., the cheapest price, the fastest delivery, etc.).

The four issues of TAPs, related parameters ($PA(P_j)$), solutions, and corresponding components in the agent structure are summarized in Table 18.4. This chapter only focuses on the first two issues, which will be described in the next section.

18.5.5.1 Case-Based Reasoning with Agent Affinity

As the number of agents and the number of tasks increase, the efficiency of coordination protocols is drastically degraded due to exponentially increasing messages. If the agent can find out who the possible

TABLE 18.4 Performance-Related Decision Parameters of TAPs and Solutions

Issue	Parameter	Solution	Component
Spatial ignorance	Candidate list (CL_i)	Case-based reasoning (CBR) with agent affinity ($F_{n,m}^i$)	Inference engine
Temporal ignorance	Time-out threshold (t_o)	Time-out protocol	Inference engine, task scheduler
Bidding	Viability, level of prudence	Bid pricing algorithm with viability (VRA_m^s)	Coordination engine
Evaluation	Weight of decision criteria	Bid selection algorithm with viability (VTA_n^c)	Coordination engine

candidates are for the given task from previous experience, the agent can limit the task announcement to only those candidates. This reduced set for search can be fed into the TAPs as a parameter in order to increase the performance of coordination and to achieve scalability by reducing communication overload. Case-based reasoning (CBR) is an efficient method to get such a reduced set.

CBR is a method for problem solving by looking for previous cases similar to the current problem. A case c_k can be defined as follows:

$$c_k = <T_i, RA_m^s, Bid_i^m, t_r, credibility_m, r> \qquad (18.10)$$

where   k: index of case; $1 \leq k \leq K$; K: the size of case base;
t_r: response time for the bidding after T_i announced;
$credibility_m$: the rate of successful commitment of RA_m^s; $0 \leq credibility_m \leq 1$;
r: the match rate of c_k used for inference.

It is difficult to extract knowledge from a case base by assessing the similarity between the previous cases and a new case. In order to make a candidate set of resource agents effectively, the concept of agent affinity is introduced. In physical or chemical science, affinity is defined as the force attracting objects or atoms to each other and binding them together in a system or a molecule. Affinity between two objects is proportional to the multiplication of mass of the two and inversely proportional to the square of the distance between the two. From the analogy, agent affinity can be defined as a degree of binding between a task agent and a resource agent and formularized as follows:

$$F_{n,m}^i = \frac{M_n M_m}{d^2}; d \propto \frac{1}{r} \cdot \frac{1}{credibility_m} \qquad (18.11)$$

where $F_{n,m}^i$: agent affinity for T_i between a TA_n^c and an RA_m^s;
$M_n = QTY_i$;
$M_m = Qqty_i^m$;
d: distance metric.

The formula shows that a larger quantity of T_i announced by the task agent, a larger quoted quantity for T_i proposed by the resource agent, and a shorter distance (a larger match rate) imply strong affinity between the two agents. The concept of agent affinity reflects preferences among agents; low affinity means the two agents have less mutual interests or hostility with each other. Each task agent keeps and maintains the affinity information obtained from its case base.

The inference engine selects the most similar cases by using agent affinity ($F_{n,m}^i$), makes the best candidate list for T_i (CL_i) of the resource agents found in the similar cases, and sets the list as a parameter of TAPs so that the TAPs can announce T_i only to the resource agents in the candidate list (CL_i). If the

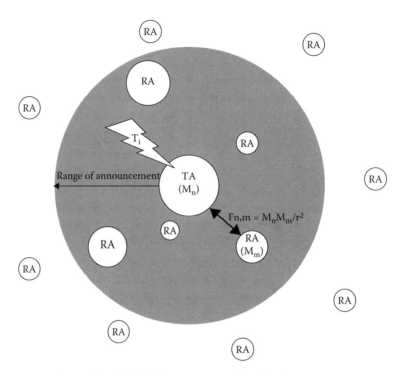

FIGURE 18.5 Setting the candidate list (CL_i) by using agent affinity ($F_{n,m}{}^i$).

length of the list is less than a certain threshold and thus not enough to get a meaningful result, the TAPs just broadcasts the task announcement.

The size of a case base (CB) is very small at the beginning. As task coordination proceeds, the CB expands fast, which may cause the time for searching CB to be very long. Besides, some cases in the CB may be changed or obsolete after coordination. Therefore, the CB should be maintained every time the coordination process is finished so that the size of CB should be limited and the CB can be adapted for the future sessions. The maintenance steps are as shown in Table 18.5.

Using CBR with agent affinity can be illustrated with the case study described later in this chapter. When an emergency situation (T_i) occurs at a location, the worker in charge of response to the emergency can request resource agents ($RA_m{}^s$) to send necessary resources. The worker can retrieve previous cases from CBR to find the $RA_m{}^s$ with high affinity ($F_{n,m}^i$), which means that they are close to the emergency location (d is small) or they have enough to provide many units of resources ($Qqty_i{}^m$) to the emergency location. Then the user can decide the candidate list (CL_i) among a number of resource agents and contact only the storages in the candidate list.

18.5.5.2 Time-Out Approach for TAPs

In a task queue of a TA, the task in the front may occupy too much time waiting for bids. The task may significantly degrade the protocol performance (e.g., execution time) as well as the system performance

TABLE 18.5 Case Base Maintenance Steps

1. Update r of all cases;
2. For all k, if $c_k(RA_m{}^s) == RA_m{}^s$, update $credibility_m$;
3. If length(CB) > K, sort CB by ($r * (1 - credibility_m)$);
4. Return the list of the first K cases.

(e.g., number of successful tasks) under control of the protocol by blocking other tasks waiting in the queue. When the waiting time is beyond a certain threshold (t_o), the task must start immediately or be preempted by the next task. Time-out threshold (t_o) for a request session should be determined for time efficiency so that a task agent can receive enough bids and it should not waste too much time waiting for the bids that may be useless or even not receivable.

Table 18.6 shows the time-out logic that decides the time-out threshold (t_o) by time-out conditioning in a request session of a task agent.

If t_s has not reached t_o, the protocol checks if it received enough bids to finish the session. Even though it has not received sufficient bids, it will terminate the session if t_w is too long and the number of bids is beyond a lower bound. When t_s becomes more than t_o, the protocol checks if the number of bids is greater than the lower bound. If it is not, then the protocol checks the task queue, and restarts the session with the same task if nothing is in the queue. If there are other tasks in the queue and the protocol has received at least one bid, it terminates the session. Otherwise, the task is sent back to the last slot in the queue if the DL_i of the task is still remaining.

The default value for t_o is application-specific; for example, if the application is a supply network, the default value of time-out may be hours to days; if the application is surveillance, it may be seconds to minutes. It depends on the time unit used in the application.

Time-out approach can be illustrated with the case study described later. When a user requests special equipment (T_i) to storages in other areas ($RA_m{}^s$), a certain storage may not reply quickly, or it cannot reply because it is damaged by another emergency situation. If the user waited enough ($t > t_o$) and got enough responses, the user must decide the best providers at the moment. If not and there are other tasks the user has to do, the user must stop waiting and start the next task.

18.5.6 Overview of TAPs Design

This section presented the definition of TAPs and their objective. TAPs are designed to manage dependencies (δ) including but beyond coordination. The dependencies (δ) between tasks were classified into several types. Dynamic task priority evaluation function ($pf(T_i, t)$) decides the type of dependence by

TABLE 18.6 Time-Out Protocol Logic

If (l(CB) too small), set t_o to a default value;

Otherwise, calculate μ and S and set $t_o = \mu + 2\sqrt{S}$;

Start a request session with the first T_i in Q;

If (($t_s \le t_o$) $\wedge$ ((n_b/n_r) $\ge U_t$), goto step 10;

If (($t_s \le t_o$) $\wedge$ ((n_b/n_r) $\ge L_t$) $\wedge$ ($t_w > 2\sqrt{S}$)), goto step 10;

If (($t_s > t_o$) $\wedge$ ((n_b/n_r) $\ge L_t$)), goto step 10;

If (($t_s > t_o$) $\wedge$ ((n_b/n_r) $\le L_t$) $\wedge$ (null? Q)), goto step 3;

If (($t_s > t_o$) $\wedge$ (n_b) $\wedge$ (not null? Q)), goto step 10;

Else move T_i to the end of Q and terminate;

Delete T_i from Q and terminate a request session.

where l(CB): length of CB;	t_o: time-out threshold;
μ: mean of t_r in similar cases;	S: variance of t_r in similar cases;
t_s: session time;	n_b: number of received bids;

n_r: number of resource agents who received the task announcement;

L_t: lower bound of termination threshold; $0 \le L_t \le 1$; (n_b/n_r) should be at least equal to L_t for termination;

U_t: upper bound of termination threshold; $U_t \approx 1$; a request session can be terminated when (n_b/n_r) reach U_t;

t_w: waiting time after a final bid; Q: task queue.

Source: Revised from Williams et al., 2002.

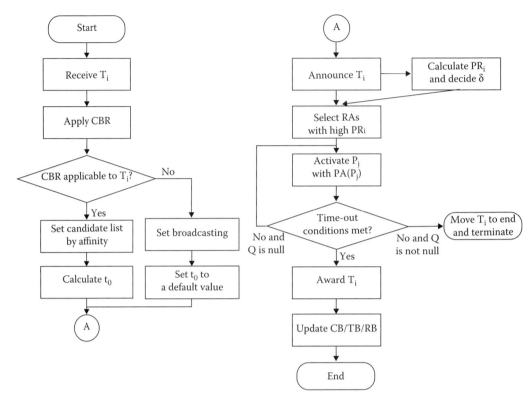

FIGURE 18.6 Overall task allocation process by TAPs with CBR and time-out (Ko & Nof, 2008).
Source: Modified from Williams, N. P., Y. Liu, and S. Y. Nof, *International Journal of Production Research*, 40, 4505–22, 2002.

task priority ($PR_i(t)$). In order to improve the performance under TAPs, the concepts of CBR with agent affinity ($F_{n,m}^i$) and time-out (t_o) are introduced. The overall process of task allocation by TAPs with CBR and time-out is depicted in Figure 18.6.

18.6 Case Study: Emergency Assignment by Human-Robot Team

A case study is introduced in this section in order to illustrate the feasibility of TAPs. For a simple illustrative purpose, there are only two operators on the field, at different locations. They need to operate mobile robots to perform certain tasks. There are two locations where the mobile robots stand by. Therefore, there are two TAs (workers) and two RAs (stand-by locations) in this example. Besides, there are two types of tasks: emergency and normal. The difference is that emergency tasks have more quantities and shorter deadlines than normal tasks. If an emergency task arrives, it may stop the previous task and be performed first. A TAPs should be designed and activated to decide effectively how many robots will be supplied from which location to which worker on a certain day based on the deadline and number of robots necessary for the operations. To sum up, the system can be specified as follows:

δ: ($T_i \leq T_j$) or (T_i ic T_j). For normal/routine operations, the type of dependence (δ) is usually $T_i \leq T_j$. For emergency operations, δ can be changed to T_i ic T_j according to the value of $PR_i(t)$. In that case, the current task T_j is interrupted by the new task T_i and continues its remaining task after T_i is completed when $PR_i(t) > PR_j(t)$.

$T_i = <TYPE_i, QTY_i, DL_i, Std\_cost_i, PR_i(t)>$
$TYPE_i$ = normal or emergency
QTY_i = Normal(m_1, s_1) (low) or Normal (m_2, s_2) (high)
DL_i = long or short
$Std\_cost_i \propto$ distance between TA_n and RA_m

A coordination and interruption-continuation protocol (CICP) was developed as a TAPs and the performance under CICP will be compared to the performance under a non-TAPs coordination protocol (CP). CP simply works in a first-come first-served (FCFS) manner, and it allocates the time in proportion to QTY_i. CICP also works based on FCFS, but it evaluates $PR_i(t)$ whenever a new T_i arrives at an RA. Thus, $PR_i(t)$ under CICP is variable while $PR_i(t)$ under CP is constant. If $PR_i(t)$ is higher than the current task, it can stop the current task and start the T_i with higher $PR_i(t)$. This information is sent back to TA, and then the TA selects the RA that suggests the best offer (i.e., the shortest QLT_i). Detailed specification of CICP is described in Figure 18.7. The simulation was developed with C++. Table 18.7 shows the parameters used for simulations.

For each of normal tasks and emergency tasks, the following hypothesis must be tested.

$$H_0: \mu_1 - \mu_2 = 0 \ \ vs. \ H_1: \mu_1 - \mu_2 > 0 \tag{18.12}$$

The results are shown in Figure 18.8, Figure 18.9, and Table 18.8. For normal tasks, it is obvious that H_0 is not rejected, which means the difference between TAR of CICP and TAR of CP is negligible. This is

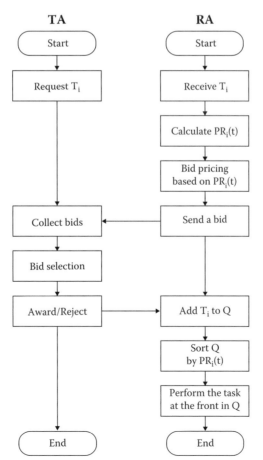

FIGURE 18.7 Coordination and interruption-continuation protocol (CICP) (Ko & Nof, 2008).

TABLE 18.7   Design of Experiments: Simulation Parameters

Parameter	Value
Number of RA	2
Number of TA	2
Quantity	Normal: $QTY_i = Normal(5, 1)$ per operation; Emergency: $QTY_i = Normal(20, 5)$ per operation
Interarrival time	Normal: Exponential(2) days; Emergency: Exponential(10) days
Deadline	Normal: $DL_i = 15$ days; Emergency: $DL_i = 5$ days
Resource capacity	5 robots/day
Weight of $pf(T_i, t)$	$w_2 = 0.4; w_3 = 0.6; w_1 = w_4 = w_5 = 0$
Simulation length	$T_E = 365$ days
Warm-up time	$T_0 = 30$ days
Replication	$r = 10$
Treatment	CP vs. CICP
Performance measure (f_{im})	Task allocation ratio: $TAR = 1 - \dfrac{\overline{T}}{|\tau|}$ where $|\tau|$: total number of tasks $\overline{T}$: number of un-allocated tasks

because there are enough resources for normal tasks. On the other hand, Table 18.8 shows H_0 is rejected for emergency tasks. This means that there is a significant difference for emergency tasks (10.91%), when resources become tight. As a result, both CP and CICP successfully perform all the normal tasks, but CICP performs better than CP when handling emergency tasks.

CICP was developed as a TAPs that can handle ($T_i \leq T_j$) and (T_i ic T_j) dependence. By using dynamic priority evaluation, CICP could decide the order of tasks according to their priority. As a result, the systems under the control of CICP performed better than the same systems under the control CP for emergency tasks which have higher $PR_i(t)$, while the performance difference of normal tasks was insignificant.

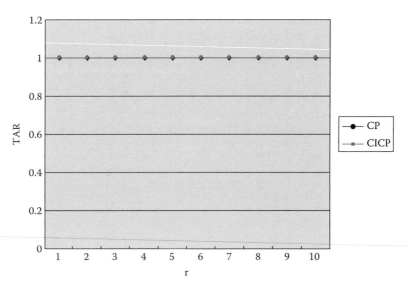

FIGURE 18.8   Task Allocation Ratio of normal tasks (Ko & Nof, 2008).

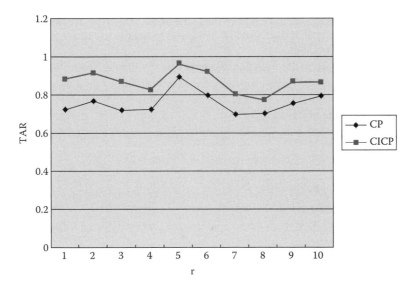

FIGURE 18.9 Task allocation ratio of emergency tasks (Ko & Nof, 2008).

18.7 Emerging Trends and Challenges

We have presented the design of TAPs and illustrated the case of TAPs design specifically for task allocation in distributed systems composed of human and robot e-Workers. As indicated earlier, such TAPs
are also needed to enable digital human models to operate effectively. Two of e-Work logic models
and tools are employed to model the activities and interactions of human and robot e-Workers: (1)
multi-agent systems are used to represent virtual environments in which each agent represents a digital
human or a robot; and (2) protocols are designed to represent the rules of interactions for effective task
allocation among agents. In this context, TAPs are designed as a control mechanism that can manage
dependencies between tasks including coordination solutions in the task environment. TAPs are composed of several types of basic protocols and it selects an appropriate protocol and parameters based on
the type of dependence. Seven types of basic dependencies were defined to represent the dependencies
between tasks. The type of dependence can vary and be dynamically determined by task priority, which
is evaluated by a priority evaluation function. By dynamic task priority assignment, TAPs can determine the order of tasks. It can also change the type of dependence based on their priority, even though
the priority may not be given. CICP was developed as a kind of TAPs to handle $(T_i \leq T_j)$ and $(T_i \text{ ic } T_j)$
dependence. The performance under TAPs and non-TAPs was compared in a case study and it showed
the feasibility of TAPs. The study is ongoing to address the following challenges:

Improving the performance of TAPs: Not only the system performance under TAPs, but also the
performance of TAPs itself must be considered. Adaptability of protocol parameters should be
developed to improve the performance of TAPs.

TABLE 18.8 t-Test for Task Allocation Ratio of Emergency Tasks

Treatment	Mean (μ)	Standard Deviation (σ)	Degree of Freedom (ν)	t	$t_{1-\alpha,\nu}$ ($\alpha = 0.05$)
CP	0.7580	0.05662	18	4.413	1.734

<u>Design of TAPs for managing other types of dependencies</u>: CICP only manages ($T_i \leq T_j$) and (T_i ic T_j). Other kinds of TAPs should be developed to handle situations occurring in other types of task dependencies.

<u>Adaptation of priority evaluation function</u>: According to the type of application, the terms and weights in the priority evaluation function should be customized since different applications have different criteria to evaluate task priority. Besides, it should be further modified to reflect other types of task dependencies.

References

Allen, J. F., and Ferguson, G. 1994. Actions and events in interval temporal logic. *Journal of Logic and Computation* 4 (5): 531–579.

Anussornnitisarn, P. 2003. Design of active middleware protocols for coordination of distributed resources. PhD dissertation, Purdue University, West Lafayette, IN.

Anussornnitisarn, P., Nof, S. Y., and Etzion, O. 2005. Decentralized control of cooperative and autonomous agents for solving the distributed resource allocation problem. *International Journal of Production Economics* 98 (2): 114–128.

Anussornnitisarn, P., Peralta, J., and Nof, S. Y. 2002. Time-out protocol for task allocation in multi-agent systems. *Journal of Intelligent Manufacturing* 13 (6): 511–522.

Balch, T., and Parker, L. E. 2002. *Robot Teams*. A K Peters.

Decker, K. S., and Lesser, V. R. 1995. Designing a family of coordination algorithms. *Proceedings of the First International Conference on Multi-Agent Systems*. San Francisco, CA.

Deshpande, U., Gupta, A., and Basu, A. 2004. Performance enhancement of a contract net protocol based system through instance-based learning. *IEEE Transactions on Systems, Man, and Cybernetics* 35 (2): 345–358.

Durfee, E. H. 2001. Scaling up agent coordination strategies. *IEEE Computer* 34 (7): 39–46.

Esfarjani, K., and Nof, S. Y. 1998. Client-server model of integrated production facilities. *International Journal of Production Research* 36 (12): 3295–3321.

Excelente-Toledo, C. B., and Jennings, N. R. 2004. The dynamic selection of coordination mechanisms. *Autonomous Agents and Multi-Agent Systems* 9 (1–2): 55–85.

Fatima, S. S., and Wooldridge, M. 2001. Adaptive task and resource allocation in multi-agent systems. *Proceedings of the Fifth International Conference on Autonomous Agents*. Montreal, Canada.

Ferber, J. 1999. *Multi-Agent Systems: An Introduction to Distributed Artificial Intelligence*, Addison Wesley Longman.

Frayret, J., D'amours, S., and Montreuil, B. 2004. Coordination and control in distributed and agent-based manufacturing systems. *Production Planning and Control* 15 (1): 42–54.

Fung, R. Y. K., and Chen, T. 2005. A multiagent supply chain planning and coordination architecture. *International Journal of Advanced Manufacturing Technology* 25 (7–8): 811–819.

Garland, A., and Alterman, R. 2004. Autonomous agents that learn to better coordinate. *Autonomous Agents and Multi-Agent Systems* 8 (3): 267–301.

Gerkey, B. P., and Mataric, M. J. 2002. Sold!: Auction methods for multirobot coordination. *IEEE Transactions on Robotics and Automation* 18 (5): 758–768.

Gouaich, A. 2004. Coordination and conversation protocols in open multi-agent systems. *Lecture Notes in Artificial Intelligence* 3071: 182–199.

Hu, Q., Kumar, A., and Zhang, S. 2001. A bidding decision model in multiagent supply chain planning. *International Journal of Production Research* 39 (15): 3291–3301.

Huang, C. Y., and Nof, S. Y. 2000. Autonomy and viability—measures for agent-based manufacturing systems. *International Journal of Production Research* 38 (17): 4129–4148.

Huang, C. Y., and Nof, S. Y. 2002. Evaluation of agent-based manufacturing systems based on a parallel simulator. *Computers and Industrial Engineering* 43 (3): 529–552.

Iocchi, L., Nardi, D., Piaggio, M., and Sgorbissa, A. 2003. Distributed coordination in heterogeneous multi-robot systems. *Autonomous Robots* 15 (2): 155–168.

Jones, H., and Snyder, M. 2001. Supervisory control of multiple robots based on a realtime strategy game interaction paradigm. *Proceedings of the IEEE International Conference on Systems, Man, and Cybernetics*. Tucson, AZ.

Karl, R. G., Lo, T. L., and St. Clair, D. C. 1993. Effects of nonsymmetric release times on rate monotonic scheduling. *Proceedings of the 1993 ACM Conference on Computer Science*. Indianapolis, IN.

Khanna, N., Fortes, J. A. B., and Nof, S. Y. 1998. A formalism to structure and parallelize the integration of cooperative engineering design tasks. *IIE Transactions* 30 (1): 1–15.

Khoo, L. P., and Yin, X. F. 2003. An extended graph-based virtual clustering-enhanced approach to supply chain optimization. *International Journal of Advanced Manufacturing Technology* 22 (11–12): 836–847.

Ko, H. S., and Nof, S. Y. 2008. Design of task administration protocols for priority-based task allocation. PRISM Research Memorandum, Purdue University, West Lafayette, IN.

Lee, K. M., and Lee, J-. H. 2003. Coordinated collaboration of multiagent systems based on genetic algorithms. *Lecture Notes in Artificial Intelligence: Intelligent Agents and Multigent Systems* 2891: 144–157.

Lehoczky, J., Sha, L., and Ding, Y. 1989. The rate monotonic scheduling algorithm: exact characterization and average case behavior. *Proceedings of the IEEE Real-Time Systems Symposium*. Santa Monica, CA.

Liu, Y., and Nof, S. Y. 2004. Distributed microflow sensor arrays and networks: Design of architectures and communication protocols. *International Journal of Production Research* 42 (15): 3101–3115.

Lou, P., Zhou, Z., Chen, Y. -P., and Ai, W. 2004. Study on multi-agent-based agile supply chain management. *International Journal of Advanced Manufacturing Technology* 23 (3–4): 197–203.

Malone, T. W., and Crowston, K. 1994. The interdisciplinary study of coordination. *ACM Computing Surveys* 26 (1): 87–119.

Monferrer, A., and Bonyuef, D. 2002. Cooperative robot teleoperation through virtual reality interfaces. *Proceedings of IEEE Sixth International Conference on Information Visualization*. London, UK.

Nof, S. Y. 2003. Design of effective e-Work: Review of models, tools, and emerging challenges. *Production Planning and Control* 14 (8): 681–703.

Nof, S. Y. 2007. Collaborative control theory for e-Work, e-Production, and e-Service. *Annual Reviews in Control* 31: 281–292.

Parker, L. 1998. ALLIANCE: An architecture for fault tolerant multi-robot cooperation. *IEEE Transactions on Robotics and Automaton* 14 (2): 220–240.

Parunak, H. V. D. 1987. Manufacturing experience with the contract net. In *Distributed Artificial Intelligence*, ed. M. N. Huhns, 285–310. Pitman.

Reaidy, J., Massotte, P., and Diep, D. 2006. Comparison of negotiation protocols in dynamic agent-based manufacturing systems. *International Journal of Production Economics* 99 (2006): 117–130.

Shen, W., and Norrie, D. H. 2001. Dynamic manufacturing scheduling using both functional and resource related agents. *Integrated Computer-Aided Engineering* 8 (1): 17–30.

Smith, R. G. 1980. The contract net protocol: high-level communication and control in a distributed problem solver. *IEEE Transactions on Computers* 29 (12): 1104–1113.

Stone, P., and Veloso, M. 2000. Multiagent systems: a survey from a machine learning perspective. *Autonomous Robots* 8 (3): 345–383.

Ulieru, M., and Cobzaru, M. 2005. Building holonic supply chain management systems: an e-logistics application for the telephone manufacturing industry. *IEEE Transactions on Industrial Informatics* 1 (1): 18–30.

Walsh, S. P., and Nahavandi, S. 2004. Applying a reconfigurable multi-agent scheduler to product distribution. *Proceedings of the 2nd International IEEE Conference on Intelligent Systems*. Varna, Bulgaria.

Walsh, W. E., and Wellman, M. P. 1998. A market protocol for decentralized task allocation. *Proceedings of the Third International Conference on Multi Agent Systems.* Paris, France.

Wan, W., Wang, X., and Liu, Y. 2005. Contract net protocol using fuzzy case based reasoning. *Lecture Notes in Artificial Intelligence* 3614: 941–944.

Weng, X. X., and Zhu, X. F. 2005. A distributed problem solving with contract net and case-based reasoning. *Proceedings of the International Conference on Control and Automation*, Budapest, Hungary.

Werger, B. B., and Mataric, M. J. 2000. Broadcast of local eligibility for multi-target observation. In *Distributed Autonomous Robotic Systems*, ed. Lynne E. Parker, George Bekey, and Jacob Barhen, 347–356. Springer-Verlag.

Williams, N. P., Liu, Y., and Nof, S. Y. 2002. TestLAN approach and protocols for the integration of distributed assembly and test networks. *International Journal of Production Research* 40 (17): 4505–4522.

Williams, N. P., Liu, Y., and Nof, S. Y. 2003. Analysis of workflow protocol adaptability in TestLAN production systems. *IIE Transactions* 35 (10): 965–972.

Wooldridge, M., and Jennings, N. R. 1995. Intelligent agents: Theory and practice. *The Knowledge Engineering Review* 10 (2): 115–152.

19

Visualization, Perceptualization, and Data Rendering

Don Stredney and
David S. Ebert

19.1 Abstract

Through our collaborations, we have developed systems that allow for the emphasis/de-emphasis of visual features to facilitate visual communication through interactive volume rendering (Svakhine, 2005). These renderings involve the 3D reconstruction of typical medical datasets acquired through discrete spatial sampling including, but not exclusive to, computed tomography (CT), magnetic resonance (MR), and serial microscopy. Through various representational techniques used to emphasize and de-emphasize features, a single data acquisition can be employed not only to represent both a myriad of contexts (e.g., anatomical atlas, surgical simulation), but also can be titrated to provide a representation commensurate with the proficiency of the user (e.g., schematic to complex) (Stredney, 2005). We present our parallel efforts that include creating illustrative renderings of volume reconstructions as well as interactive sessions for procedural surgical training. Although our focus here is to utilize these volume renderings to teach regional anatomy and procedural techniques such as surgery, these developments are extensible to the tasks of pre-operative assessment and pre-surgical and treatment planning.

19.2 Introduction

There are two purposes for representation that should be considered. The first is the drive to represent the data in the context of how it will be viewed. For instance, to maximize transfer from the simulation to real-world practice, it is useful to try, to the extent technically possible in real time, to represent the visual look faithfully as to what will be seen in the surgery—colors, contrast, lighting, etc., in a word, realism. On the other hand, the computer medium itself allows for unprecedented opportunities. Certainly, when using a particular medium, one wishes to exploit the unique capabilities it provides. Thus the ability to present data in ways that facilitate conception, rather than perception, is important. Motifs, such as color overlays, tinting, cross sections, and so on, can easily be introduced to facilitate comprehension. With a diverse toolkit allowing for a wide variance of representation, the issue then becomes when and how to best use such contrivances.

Most importantly, if the content is clearly understood, it is more likely to be extensively investigated by the user, thus promoting self-driven exploration and more comprehensive learning. We provide no aesthetic guidance, and certainly do not imply that a "best" solution exists for presenting visual information. Instead, we contend that useful solutions can be determined based on the context, including the task and user proficiency. Thus emerging systems will be able to more broadly engage and facilitate the user's understanding and automatically modify how the data are represented to optimize comprehension.

19.3 Methods

Our efforts have been focused on interactive surgical training, specifically in areas that involve the exacting complex anatomy of the temporal bone (Bryan, 2001; Stredney, 2003, 2005; Wiet, 2002, 2004). Subsequently, we have focused on two specific contexts that are required to provide useful simulations for training. The first includes an anatomical tutor scenario, where the user can localize pre-segmented structures, areas, and sub-areas requisite for comprehension of the regional anatomy associated with surgical intervention. A consummate internal representation of the regional anatomy is required for surgical training, especially in the recognition of anatomical and pathological variance. The second involves representing the data in the context of the surgery. This context includes the integration of aural, haptic, and visual stimuli to present a comprehensive emulation of the surgical procedure and its concomitant results. By creating a simulation environment that emulates surgery, our goal is to optimize performance through deliberate practice.

The anatomical tutor exploits direct volume rendering and per pixel shaders to create a lucid, comprehensible representation with unambiguous lighting cues to locate essential structures and regions. This provides a direct and definite representation. Through intuitive interactive movements, arbitrary orientation can be achieved as if turning an object in one's hand and the structural complexities and their relationship are easily perceived through structure from motion. Through the use of color, the user's attention is drawn to specific structures and regions to learn their location, boundaries, shape variance, and relationship with surrounding structures (see fig. 19.1).

In the surgical representation, the data are rendered to represent the structural anatomy as actually seen in the surgical context. This includes bright lighting from around the viewpoint, to emulate illumination from the microscope. This type of lighting diminishes the form shading, making the surfaces

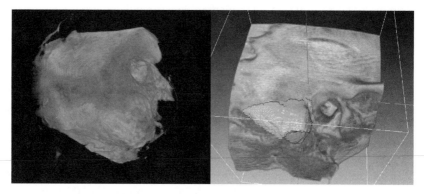

FIGURE 19.1 Left: CT data reconstructed. Right: Image from tutor showing segmented areas (i.e., mastoid process (green), mastoid tip (blue) and surrounding regional anatomy).

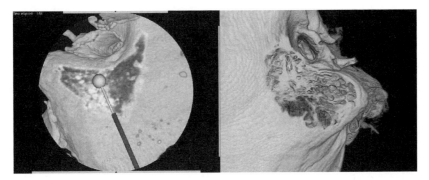

FIGURE 19.2   Left: Simulated surgical view with features de-emphasized to emulate surgical appearance. Right: Data, from a local trial with residents/medical students, with improved shading depicts how segmented data are used to color the data. The bluish area is the underlying venous structure, the sigmoid sinus.

appear slightly flat and insipid. In addition, the form is obscured with areas that represent surfaces obscured by resultant bleeding. These effects are employed to mirror the appearance of the surgical view, and especially to provide visual effects that have a tendency to mask and muddle the comprehension of the form.

Visual artifacts can plague real-time volume rendering exploiting the GPU. It is important to try and mitigate their effects, as they can distract the attention of the user. In Figure 19.3 we demonstrate the application of stochastic jittering to reduce visual artifacts caused by low slice frequency when employing direct volume rendering (Engel et al., 2006). These aliasing bands can be extremely distracting when the user is interactively exploring the form through arbitrary orientation.

Our volume illustration system is a combination of existing texture-based volume visualization algorithms, our illustrative volume visualization extensions, a high-level, more natural user interface, and a domain-specific illustration specification framework, as shown in Figure 19.4. More details on the implementation of this system can be found in Svakhine (2005).

One of the main issues in the design of such a system is providing high-level control of the many visualization parameters. As volume visualization systems incorporate more enhancement, rendering, and shading techniques, the number of user-adjustable parameters tends to increase dramatically and often limits their usefulness in medical domains and other domains. Therefore, we have created a new, more natural specification interface with intuitive and understandable parameters to enable users with various backgrounds to quickly generate informative visualizations.

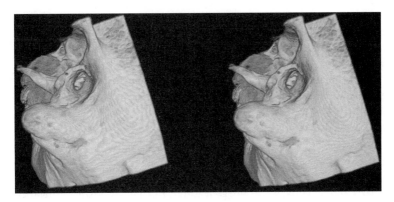

FIGURE 19.3   Left: Aliasing is seen in parallel bands. Right: Using stochastic jittering to reduce the artifacts caused by low slice frequency.

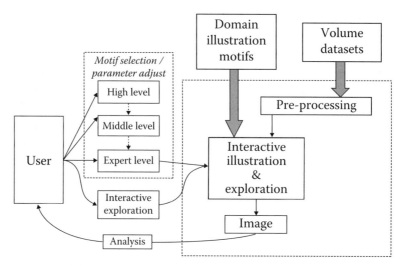

FIGURE 19.4 Design of the system architecture to allow enhanced, user- and task-adapted illustrative visualization.

Our interface incorporates three levels of interaction: an expert level, a mid level, and a high level. The expert user level is for software developers, experienced illustrators, and system builders. The mid-level interface is for illustrators and experienced end users who want to make adjustments once they understand the controls. The high-level interface is designed for the end user, and in our current application focus, it would be for medical students, surgeons, and surgical residents.

As mentioned previously, we have designed our illustration system on schema found in perception and subsequently exploited through illustrative techniques. One of the primary perceptual principles used by illustrators is visual focus: The most important information in the illustration will be rendered to facilitate orientation and to draw and hold the viewer's attention. The surrounding details are de-emphasized (subjugated) to provide context, while not overloading the viewer's attention. Therefore, a simple focal volumetric region is utilized to determine the spatial distribution of illustrative techniques to be applied, allowing different illustrative enhancements to be selectively applied through our high-level interface.

19.4 Results

Several example results from the system have been shown previously and illustrate our approach as well as the capabilities of illustrative visualization for medical training. Figure 19.5 shows the range of styles and informative representations that are achievable at interactive rates using the latest PC graphics cards for the Visible Woman foot CT dataset. Figure 19.6 shows further examples of applying these techniques to temporal bone datasets for education in radiology and surgical training, created by Lakshmi Ganapathy using our IVIS system.

19.5 Discussion

The adept emphasis and de-emphasis of image features, in context, provide critical clarity and focus for communicating information visually. Through skillful manipulation of features, interactive volume

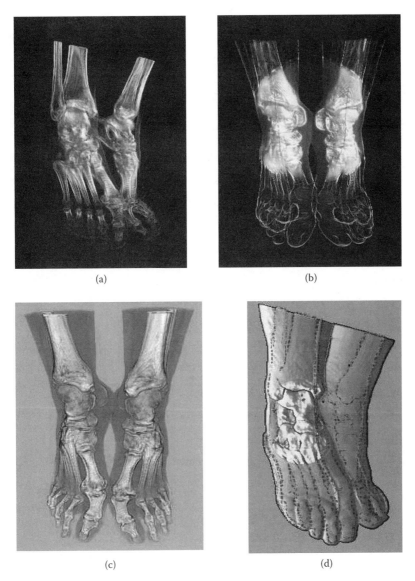

FIGURE 19.5 Visible Woman feet dataset illustrations, highlighting tarsal (ankle) joints. Three different styles are shown, with (b) utilizing attentive focus, (c) showing the effectiveness of depth highlighting using silhouette sketch styles, and (d) showing varying sketch styles to show hidden bone structures for surgery planning.

representations are emerging that engage the user and facilitate understanding of exacting structural form, intricate relationships, and complex interaction during procedural surgical techniques.

Acknowledgments

This research was funded in part from a grant from the National Institute on Deafness and Other Communication Disorders, of the National Institutes of Health, 1 R01 DC06458-01A1, by National Science Foundation grants 0121288 and 0328984, and from Adobe Systems.

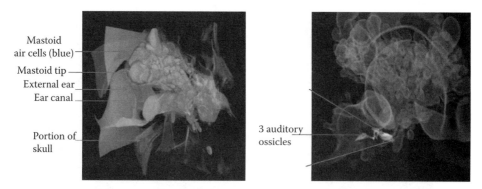

FIGURE 19.6   Two illustrations of the anatomy of the inner ear created using IVIS from patient temporal bone CT scans.

References

Bryan, J., D. Stredney, D. Sessanna, and G. J. Wiet. 2001. Virtual temporal bone dissection: A case study. *Proc. 12th IEEE visualization* 497–500.

Engel, K., M. Hadwiger, J. M. Kniss, C. Resk-Salama, and D. Weiskopf. 2006. *Real-time volume graphics.* Wellesley, MA: A. K. Peters.

Stredney, D., J. Bryan, D. Sessanna, and T. Kerwin. 2003. Facilitating real-time volume interaction. *Proc. medicine meets virtual reality 13*. Amsterdam: IOS Press, 329–35.

Stredney, D., D. S. Ebert, N. Svakhine, J. Bryan, D. Sessanna, and G. J. Wiet. 2005. Emphatic, interactive volume rendering to support variance in user experience. *Proc. medicine meets virtual reality 13*. Amsterdam: IOS Press, 526–31.

Svakhine, N., D. S. Ebert, and D. Stredney. 2005. Illustration motifs for effective medical volume illustration. *IEEE Computer Graphics and Applications* 25(3): 31–39.

Wiet, G. J., D. Stredney, D. Sessanna, J. Bryan, D. B. Welling, and P. Schmalbrock. 2002. Virtual temporal bone dissection: An interactive surgical simulator. *Otolaryngol Head Neck Surg* 127(1): 79–83.

Wiet, G. J., and D. Stredney. 2002. Update on surgical simulation: The Ohio State University experience. *Otolaryngol Clin North Am* 35(6): 1283–88.

Wiet, G. J., P. Schmalbrock, K. Powell, J. Bryan, D. Sessanna, and D. Stredney. 2004. Use of ultra-high resolution temporal bone data sets for surgical simulation. *Otolaryngol Head Neck Surg* 131(2): 218–19.

20

Computing Infrastructure and Methods for Visualizing Large-Scale Dynamic Simulations

Voicu Popescu and
Christoph M. Hoffman

20.1 Introduction

Finite element analysis (FEA) is a well-established tool for complex physical simulations in engineering and science. FEA has become so pervasive that an interest in simulation results reaches well beyond the laboratory and the engineering departments of industry. So common are simulations that they are often used in news programs to elucidate complex events and relationships. So valuable are simulations that they figure in forensic analysis, and so accurate are they that they can become the basis for legislation and statutes. This widespread interest in FEA necessitates an effective set of tools to communicate the results of the analysis.

Graphics and visualization are well-established tools for presenting sequences of events, whether real or imagined. By now a staple in the movie industry and at the core of computer games, visual presentation engages the premier channel of human cognition—sight. Humans can apprehend complex interrelationships when presented with sophisticated visuals. Thus, visualization is a uniquely powerful tool for disseminating the results of FEA.

The traditional applications in the movie and the gaming industries have resulted in sophisticated tools for creating visual presentations. However, visualization systems and FEA systems have evolved separately, and there are no simple combinations of the two that integrate the unique strengths of their individual capabilities. For instance, the two kinds of systems use vastly different representation of the geometry, making it difficult to integrate the results of one into the other system.

There are several tools for visualizing FEA results, including VTK (Schroeder, Martin, & Lorensen, 2003) and LS-Post (Livermore Software Corp., 2006). However, none of those tools offers an easy way to integrate environmental settings that contextualize the simulation results. Indeed, when setting up an FEA analysis, an early effort of the engineer or scientist is devoted to simplifying the set of phenomena and entities to be analyzed and simulated. Thus, environmental settings, in which these entities exist and in which the simulated phenomena take place, are eliminated up-front. As a consequence, the analysis problem is abstracted to a degree that robs it of the visual richness that could have significantly enlarged the set of people who understand and appreciate the results.

We describe here an approach to combining FEA systems and visualization systems to obtain the best of both worlds. Our approach *federates* the two types of systems, modifying neither, thus preserving the integration as each of the systems evolves and matures further. In this way, we do not obstruct the ongoing advances, either in the domain of FEA systems or in the domain of visualization and animation systems.

Our work has numerous applications. By restoring the visual richness of the simulation and its setting, decision making can be enhanced, virtual environments for training can be more effective, and policy makers and the public have a better understanding on which to base their views. Interdisciplinary teams, moreover, have a better language for understanding each other's work.

20.2 Strategy and Realism

The entities in a simulation should be recognized easily and associated effortlessly with their counterparts in the real world. Since the simplifications and abstractions employed by the FEA are often required for efficiency reasons, it must be possible later to re-introduce the real-world complexities of the settings, with little effort. Therefore, we consider the three numbered tasks shown in Figure 20.1.

Animation software has become so powerful that it can be difficult to discern whether a particular image or animation is a photograph/video or is a computationally constructed artifact (see fig. 20.2). Animation systems target moviegoers or gamers. They offer visual support for showing water, fire, smoke, and so on, even though these phenomena are not necessarily synthesized following the laws of physics.

Likewise, FEA software has become very powerful and accurate. Simulations involving millions of degrees of freedom are no longer uncommon. Applications range from structural engineering to fluid dynamics and biological systems. Complex systems, such as airplanes and jet engines, are built based on FEA. Anatomical deficiencies, such as arterial obstructions and heart valve performances, are analyzed using FEA. Disasters, such as explosions and crashes, are analyzed using FEA.

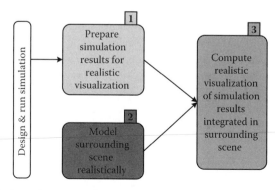

FIGURE 20.1 Main tasks in producing visualizations of simulation results.

FIGURE 20.2   Photorealistic rendering computed with state-of-the-art animation system. (© Proper Graphics, www.propergraphics.com. With permission.)

So, we seek to construct a link between FEA and visualization systems that is general, scalable, and can be reused. We avoid re-implementing any of the capabilities in either system, and construct the link so that it persists across software updates. In this way, we stay on the evolution curve of both systems and take advantage of the work of both the FEA development and the visualization development communities. The main challenge accomplishing this agenda is to bridge the two distinctly different ways of representing and conceptualizing geometric objects. Moreover, since visualization systems have particular ways of allocating the resources, we have to account for the fact that selecting from different ways to represent moving objects may have different impacts on the scalability of the translation from FEA system to visualization system.

Based on this approach, the three tasks identified in Figure 20.1 are carried out. Most of the work is performed automatically by an import module that creates an animation software 3D scene equivalent to the simulation results. The appearance of the simulation entities is refined using library and custom materials to finally produce a realistic and high-fidelity visualization by integration into the surrounding scene. The actual rendering is performed by state-of-the-art rendering algorithms implemented by the animation system.

20.3 Example

Our work in visualizing the 9/11 Pentagon attack provides an example of the approach we advocate (Popescu et al., 2003; Hoffmann et al., 2004; Popescu & Hoffmann, 2005). The work began with a simulation of the impact of the aircraft on the columns of the Pentagon ground floor. The simulation abstracted away the upper floors and focused on the role of the kinetic energy transferred in the impact from the liquid fuel mass to the building structure in the impact region. Only a small part of the Pentagon's structure was considered in the simulation, and similarly the aircraft structure was simplified to account primarily for the fuel in the tanks. Note that the representation of the plane for animation purposes is wholly unsuitable for FEA (see fig. 20.3).

The reintroduction of the deleted environmental structures and surrounding environs was part of the preparation for post-processing the simulation results. A relatively simple graphical model of the Pentagon was constructed and then embellished with textures derived from aerial photographs. The resulting 3D model and its surroundings are shown in Figure 20.4. The plane and its shadow have been added using native capabilities of the visualization software.

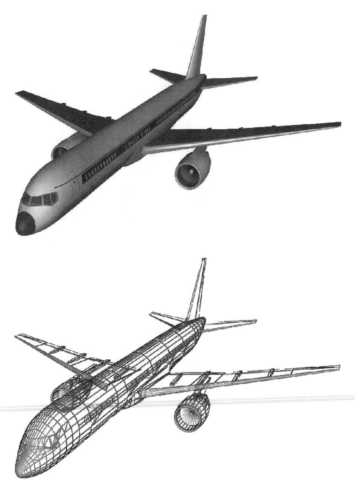

FIGURE 20.3   Aircraft model used for animation.

The difference between the customary visualization of FEA results, using software dedicated to this task, and using state-of-the-art animation software for the same purpose, with the help of our federation software, can be appreciated by the two images shown in Figure 20.5.

In particular, liquids are much better rendered in professional visualization systems in use for games and special effects, as seen in the side-by-side comparison given in Figure 20.6. So, once we have established a link that lets us import simulation results into animation software, then we can add the rich settings and render with great faithfulness the results of the FEA. This leads to compelling communication of the FEA.

20.4 Visualizing LS-DYNA Output in 3DS MAX

This section describes how to create an animation system 3D scene from a simulation results database. Even though the discussion is based on 3DS MAX, a state-of-the-art animation software system (Autodesk, 2006a), and LS-DYNA (Livermore Software Corp., 2006), the same approach can be applied to other pairs of animation and simulation systems.

Given the LS-DYNA output database, one could take several paths for creating a 3DS MAX scene. One option is to use the scripting language of 3DS MAX. In addition to the overhead of having to learn

FIGURE 20.4 Snapshots from the approach visualization.

the scripting language, this option comes with the severe disadvantage of poor scalability of the scripting language. It has been our experience that loading the Pentagon simulation via a script takes days. A second option is to translate the LS-DYNA database to a format that 3DS MAX can import, which has the disadvantages of introducing a labor-intensive intermediate step.

The third, and best, option is to import the simulation results directly into 3DS MAX with a custom plug-in. A plug-in is a software module that extends the functionality of a software system that uses an open architecture, like 3DS MAX. Once an LS-DYNA 3DS MAX plug-in is implemented, 3DS MAX naturally loads simulation results creating a 3D scene that can then be saved in 3DS MAX's native format.

Implementing a LS-DYNA 3DS MAX plug-in is a standard programming task that can be performed by a programmer with little or no expertise in simulation or visualization. The plug-in must load the simulation database, streamline the simulation data, and instantiate the 3DS MAX scene according to the simplified data.

The loading module of the plug-in asks the user to specify import parameters (e.g., materials, time steps, simplification parameters), parses the simulation database, and instantiates custom intermediate data structures that store the simulation results. Using custom intermediate data structures as opposed

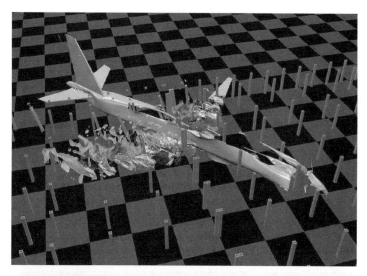

FIGURE 20.5 Comparison between visualizations produced with a state-of-the-art animation system (*top*) and with an FEA postprocessor (*bottom*).

to 3DS MAX data structures has the following important advantages. First, custom data structures will have only the required functionality, which brings efficiency. Second, the custom data structures will ensure that the front end of the plug-in is reusable for plug-ins that import LS-DYNA into other animation systems, for example into MAYA.

The simplification module discards all data that has little or no visualization relevance. This includes removing hidden faces of opaque geometry, removing position markers for nodes that move on a straight line, and removing material properties that do not affect rendering. If these simplification steps do not reduce the simulation scene to a manageable size, the scene can be further reduced by simplification (e.g., remove position markers for nodes that move approximately on a straight line).

The simplified custom intermediate data structures are translated into the actual 3DS MAX scene by instantiation. This is the step when, for example, beam and shell elements become 3DS MAX splines and editable meshes, and when element node positions are encoded as 3DS MAX vertex position controllers.

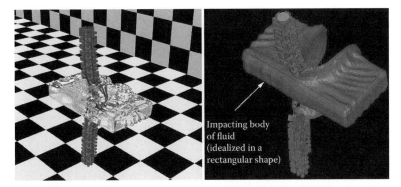

FIGURE 20.6 Simulation of impact between block of liquid and structural column visualized using animation system (*left*) and FEA postprocessor (*right*).

The position controllers are a scarce resource; encoding every vertex at every time step using position controllers would not give good performance for large simulations. So it is important that node trajectories be simplified as aggressively as possible.

The plug-in effectively implements a bridge between the worlds of simulation and animation. Once the simulation results are imported into 3DS MAX, one can use state-of-the-art material editing, lighting, camera animation, rendering, and scene management to produce high-quality visualizations that surpass what can be achieved in conventional simulation postprocessors.

20.5 Modeling the Surrounding Scene

Good simulation resource management requires that only the entities with direct relevance to the phenomena studied be included in the simulation. In the case of the Pentagon simulation, for example, the building was abstracted to only the first-floor concrete columns in the region of impact. In order to increase the eloquence of the visualization for the non-expert user, details that have no bearing on the simulation should be reinserted into the scene.

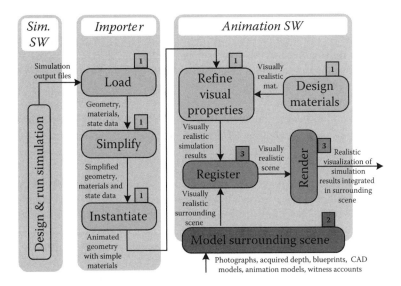

FIGURE 20.7 Visualization of simulation results by import into animation system.

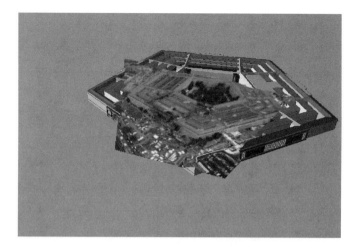

FIGURE 20.8 Photograph before (*top*) and after (*bottom*) matching to the geometric model of the Pentagon building.

The animation system is an ideal platform for modeling the scene surrounding the simulation and for combining it with the imported simulation results. Nonetheless, photorealistic 3D modeling of real-world scenes is a challenging task. A photorealistic model is constructed from two main ingredients: colorimetric data and geometric data.

High-resolution digital photography has become ubiquitous, and sensors mounted on satellites or aircraft can easily survey large scenes in detail. The Pentagon scene shown in Figure 20.4 was decorated with satellite imagery by texture-mapping (American Society of Civil Engineers, 2003). Texture mapping is the process of pasting an image onto geometry to provide high-resolution detail without increasing the geometric complexity.

Conventional texture mapping requires that each texture pixel (*texel*) samples the same scene area. This restriction is lifted in projective texture mapping, which allows using any photograph of a scene to provide photorealistic color. The camera that acquired the photograph is transformed into a projector that sprays the color samples and deposits them onto the faces of the geometric model.

Before a reference photograph can be used, one needs to estimate the position and orientation of the camera with respect to the scene when the photograph was taken. The six degrees of freedom of the camera can be computed *a posteriori* given the camera's focal length through a nonlinear optimization process that matches the photograph to the geometric model of the scene it sampled (see fig. 20.8). Projective texture mapping is a powerful means for adding realism to complex 3D scenes (see fig. 20.9).

FIGURE 20.9 Visualization of the scene after impact rendered using projective texture mapping.

While color data can be gathered efficiently from photographs, producing a geometric model of a complex 3D scene is far more challenging. The conventional approach of software-assisted, manual modeling produces good results but requires a substantial time investment and considerable artistic talent. Another, better approach is to build the geometric model from depth measurements acquired with a specialized device such as a time-of-flight laser range finder. This approach captures real-world scenes with great fidelity but it incurs large time, equipment, and operator expertise costs.

In the case of the Pentagon simulation, the aircraft model used for the approach scene (fig. 20.3) was obtained from a 3D gaming company, Amazing 3D Graphics, Inc. The Pentagon building was modeled in AUTOCAD using building blueprints (Autodesk, 2006b). The aircraft and building models were then imported into 3DS MAX where they were combined with the LS-DYNA simulation results. The results have been published by Popescu and Hoffmann (2005).

20.6 Summary and Conclusions

Finite element analysis is indisputably a core simulation technology that is relevant to science, technology, training, policy formulation, and so on. Because of its historical use, FEA has evolved into a sophisticated tool for simulating physical events based on first principles, but it has not paid equal attention to visualization of results which, in turn, is a core communications technology. The strong development of sophisticated computer tools in graphics and entertainment provides a rich environment for complementing the technical accomplishments of FEA with sophisticated, context-aware animations and visual presentations of the results. As we have illustrated, such animations provide eloquent presentations of technical insights.

In combining the strengths of FEA and animation systems, we have argued that a federation approach is best, because it is economical and allows future evolution of the separate systems without breaking the federating link between the systems. Based on the characteristics of the animation system, moreover, some consideration is required to achieve scalability, as FEA systems routinely deal with very large scenarios. In the example of the Pentagon simulation, these considerations included simplifying the trajectories of nodes, so as to reduce the number of controllers necessary for the animation, and eliminating geometric structures in the interior of opaque materials.

We conclude that our approach of combining sophisticated, state-of-the-art systems developed by separate communities is very useful to disseminate highly sophisticated technical insights to a general audience. Moreover, because our approach contextualizes the analysis, it confers a high degree of eloquence to this dissemination.

References

American Society of Civil Engineers. 2003. *The Pentagon building performance report*. Reston, VA: ASCE.

Autodesk, Inc. 2006a. 3D Studio Max (http://www.autodesk.com/3dsmax).

Autodesk, Inc. 2006b. AUTOCAD (http://www.autodesk.com/autocad).

Hoffmann, C., V. Popescu, S. Kilic, and M. Sozen. 2004. Modeling, simulation, and visualization: The Pentagon on September 11th. *IEEE Computing in Science and Engineering* 6(1): 52–60.

Livermore Software Corporation. 2006. LS-Post, LS-DYNA (www.lstc.com).

Popescu, V. 2003. High-fidelity visualization of large-scale simulations: Images and animations (http://www.cs.purdue.edu/cgvlab/projects/pentagon.htm).

Popescu, V., and C. Hoffmann. 2005. Fidelity in large-scale simulations. *Computer-Aided Design* 37: 99–107.

Popescu, V., C. Hoffmann, S. Kilic, M. Sozen, and S. Meador. 2003. Producing high-quality visualizations of large-scale simulations. *Proc. IEEE visualization*, 575–580.

Schroeder, W., K. Martin, and W. Lorensen. 2003. *The visualization toolkit: An object-oriented approach to 3d graphics*. Clifton Park, NY: Kitware.

21

Multi-Objective Optimization for Short Duration Dynamic Events

Hongbing Fang
and Ming Zhou

21.1 Introduction

With the ever-increasing complexities in engineering applications, it is almost impossible to obtain the exact analytical solutions. As a consequence, numerical methods have been widely employed taking advantage of modern supercomputers and parallel computing. Such a computational or simulation tool functions as a black box that generates a set of output (or responses) for a given set of input (or design variables) without knowing the explicit input-output relationships. By combining a simulation tool with some mathematical procedure, optimization can be carried out to improve the system design. This simulation-based design optimization has been seen in a wide range of engineering applications. However, for many large-scale simulations such as dynamic contact problems (e.g., crash), the computational cost is still high even with parallel computing and supercomputers. This makes it very difficult to perform simulation-based design optimization, because an optimization procedure typically requires hundreds or even millions of simulation runs. It is highly desirable to have explicit functions that are both inexpensive and accurate for the input-output relationships to replace the computational tool. This leads to the wide adoption of the metamodeling approach.

In the metamodeling approach, an unknown response function is approximated with a predefined function whose coefficients are to be determined using function values (obtained through numerical simulations) for some given input of design variables. Such an approximate function is called a response surface or metamodel. Since it has been used mostly for response approximations, the polynomial regression (PR) method is often called the response surface methodology (RSM) [1]. A PR model typically uses first- or second-order polynomials; therefore, it is not capable of representing a higher-order response over the entire (or global) design space. To overcome this deficiency of the global RSM, various local or successive methods have been developed [2–5]. When applied within the framework of an optimization process, local approximation enhancement is crucial as the final solution would be rendered meaningless if the functions involved are inaccurate. However, such local approach does have limitations in that (1) improvement of the local accuracy of the metamodel often comes at the cost of reduced overall fidelity, and (2) adding additional sampling points is associated with increased computational cost. Therefore, local adaptive PR methods may be impractical for computationally expensive problems such as vehicle crash simulations [6]. Moreover, local approaches are inappropriate for multi-objective optimization problems because the response regions of interest would never be reduced to a small neighborhood that was good for all the objectives typically conflicting with each other [7]. Despite its simplicity and limitation, PR methods including local approaches have been successfully used in many engineering applications.

Optimization is an inverse engineering process to find the set of input parameters (or design variables) that will generate the optimum (minimum or maximum) value of a given objective function. For most engineering applications, additional requirements often exist and are formulated as constraint functions that an acceptable design must satisfy. In addition, design variables typically are bounded and the values of these variables must be chosen within the design ranges. For such single-objective optimization problems, there exist a large set of well-established methods [8, 9] with a short list given as follows:

Gradient-based methods
Newton's method
Conjugate gradient method
Method of feasible direction
Reduced gradient method
Gradient projection method
Quasi-Newton method
Quadratic programming
Sequential quadratic programming
Non-gradient-based (direct search) methods
Simplex search method
Hooke-Jeeves pattern search method
Powell's conjugate direction method
Complex method
Random search method

For complex engineering systems, especially those composed of multidisciplinary subsystems, there are typically multiple objectives, and solutions to such problems are difficult to obtain [10]. Due to the conflicts among different objectives, a multi-objective optimization problem will have a set of trade-off solutions instead of a single optimum solution as in a single-objective problem. Although a multi-objective optimization problem can be converted into a single-objective problem (see the section "Solution Techniques" for details) and solve it using the aforementioned methods, only a single solution can be obtained and all other trade-off options are lost. To this end, it is preferable to obtain a set of trade-off solutions from which the final design can be selected by the designer.

Evolutionary algorithms [11–13] are particularly suitable for solving multi-objective optimization problems by simultaneously obtaining the trade-off solution set. These algorithms are typically expensive, and developing efficient algorithms for evolutionary computing is still an ongoing research area. An alternative approach is to assign a weight coefficient to each objective function and combine all of the objectives into one. Single-objective methods can then be used to obtain one solution, and multiple solutions can be obtained by varying the combinations of the weight coefficients. This weighted sum approach is efficient but becomes ineffective if the trade-off solutions do not form a convex surface [14]. Other approaches exist such as the analytic target cascading method [15–19] in which the objectives are prioritized into several hierarchical sub-problems. The highest-level sub-problem is first solved, and a target is set for the objective of the immediate lower-level sub-problem, which then tries to minimize the difference between the objective and the target. Feedback may also be provided from lower to upper level, in which case the upper-level sub-problem will be resolved. The process continues until a trade-off solution is obtained. This approach has been successfully used in large system problems [20–22]; however, it is similar to the single-objective approach in that only a single trade-off solution is obtained.

Short duration dynamic events such as crash and blast create great challenges to metamodeling and optimization of dynamic responses. First, a dynamic response is a function of time rather than a single value, along with the fact that multiple dynamic response histories may need to be considered simultaneously. Second, it is not obvious how to determine the design variables that have strong effects on the time-histories of multiple dynamic responses. Third, these dynamic responses could be highly nonlinear due to the abrupt change of structural status. These high nonlinearities make it difficult to solve the optimization problems where these responses are used as objective and/or constraint functions. Finally, the dynamic responses are not explicitly available and the high nonlinearities also impose difficulty to create metamodels with sufficient accuracy.

Recent studies [23–26] showed that alternative metamodeling methodologies such as Kriging [27] and radial basis functions (RBFs) [28] are better than the PR method for highly nonlinear responses. The Kriging method is also a regression method whose model accuracy can be estimated on design points. But unlike the PR method, they are suitable for highly nonlinear responses. The Kriging method, however, requires using an optimization process to determine its coefficients for each point on which the response is to be calculated. Therefore, the Kriging method is generally less efficient and its model accuracy depends on how well the coefficients are calculated by the optimization process. RBF models using newly developed basis functions [29] have been shown to give superior performance over traditional basis functions [30]. However, one should note that an RBF model is much more complicated than a PR model (see the section "Radial Basis Function (RBF)" for details). The number of basis functions in an RBF model is equal to the number of design points, with each basis function a complicated one due to the use of the Euclidian norm. This will inevitably increase the computational cost when RBF models instead of PR models are used in a multi-objective optimization problem.

Another alternative metamodel of global characteristics is the moving least square regression (MLSR) method [31]. In this approach a dynamic component is introduced to the PR model that uses a weighting factor as a decaying function of distance to a sampling point. As a result the response surface is capable of tracking local features of a highly nonlinear function. The MLSR method is particularly suitable for functions whose sampling points contain noise. However, unlike the static PR model, an MLSR model is an implicit function for which every function evaluation requires the solution of a least square problem involving a subset of the sampling points. Therefore, the PR method is still preferred wherever it is appropriate.

In the remaining portion of this chapter, metamodeling methodologies using the PR, MLSR, and RBF are first introduced. Some concepts of multi-objective optimization and solution techniques are then presented. Finally, two numerical examples are used to demonstrate the use of metamodeling in solving multi-objective optimization problems of short duration dynamic events. This is followed by a summary of the entire chapter.

21.2 Metamodeling Methodologies

Consider an input-output model given in the format

$$z = R(\boldsymbol{x}) \tag{21.1}$$

where $x = [x_1, x_2 \ldots x_m]$ is a vector of m input variables or design variables, and z is the output for a given input vector x from true response function $R(x)$. For most engineering applications, $R(x)$ is unavailable in explicit format but can be computed numerically for an input vector x. Such numerical computations are typically very expensive for complex engineering applications. Therefore, it is often desirable to construct an approximate function of $R(x)$ that is explicit, computationally less expensive, and accurate enough for the entire design space of the input variable x.

Before an approximate function can be constructed, the values of $R(x)$ need to be obtained at some sampling (or design) points, which are often determined by design of experiments (DOE) methods so that the design space can be well represented. Commonly used DOE methods include factorial design, central composite design, Latin hypercube sampling [1], and Taguchi orthogonal arrays [32]. With the known values of $R(x)$ corresponding to a set of design variables, the approximate function can be constructed using the PR, MLSR, or RBF methods.

21.3 Polynomial Regression (PR)

In this section, we present the PR method for constructing metamodels, parameters for assessing the accuracy of a PR model, and several examples of using the PR method to create metamodels.

21.3.1 Formulation

The PR method typically uses first-order (linear) or second-order (quadratic) polynomials to construct metamodels for the responses of interest. Let $f(x)$ be the approximation of the true response function $R(x)$; it can be written in the general form as

$$f(\boldsymbol{x}) = \beta_0 + \sum_{i=1}^{m} \beta_i x_i + \sum_{i=1}^{m} \beta_{ii} x_i^2 + \sum_{i=1}^{m-1} \sum_{j=i+1}^{m} \beta_{ij} x_i x_j \tag{21.2}$$

where the β's are unknown coefficients to be determined using the method of least squares. The four terms in Equation 21.2 represent a constant value, linear response, quadratic response, and paired interactions between any two variables, respectively. A quadratic model in its complete form consists of all of the terms in Equation 21.2, while an incomplete quadratic model does not have the fourth term. A linear model typically consists of the first and second terms.

For n sets of design variables x_k and the corresponding function values f_k ($k = 1, 2 \ldots n$), Equation 21.2 leads to n linear equations expressed as

$$f_1 = \beta_0 + \sum_{i=1}^{m} \beta_i x_{1i} + \sum_{i=1}^{m} \beta_{ii} x_{1i}^2 + \sum_{i=1}^{m-1} \sum_{j=i+1}^{m} \beta_{ij} x_{1i} x_{1j}$$

$$f_2 = \beta_0 + \sum_{i=1}^{m} \beta_i x_{2i} + \sum_{i=1}^{m} \beta_{ii} x_{2i}^2 + \sum_{i=1}^{m-1} \sum_{j=i+1}^{m} \beta_{ij} x_{2i} x_{2j} \tag{21.3}$$

$$\cdots$$

$$f_n = \beta_0 + \sum_{i=1}^{m} \beta_i x_{ni} + \sum_{i=1}^{m} \beta_{ii} x_{ni}^2 + \sum_{i=1}^{m-1} \sum_{j=i+1}^{m} \beta_{ij} x_{ni} x_{nj}$$

Equation 21.3 can be written in the matrix format as

$$\{f\}=[X]\{\beta\}$$ (21.4)

where $\{\beta\}$ is determined by the least square method as

$$\{\beta\}=\left([X]^T[X]\right)^{-1}\left([X]^T\{f\}\right)$$ (21.5)

21.3.2 Model Accuracy Assessment

Statistical analysis techniques can be used to evaluate the accuracy of a PR model. Major statistical parameters for determining the model accuracy are the F-statistic, coefficient of multiple determination (R^2), adjusted R^2 (R^2_{adj}), and root mean square error ($RMSE$). These parameters are not totally independent of each other; they are calculated as

$$F=\frac{(SST-SSE)/p}{SSE/(n-p-1)}$$ (21.6)

$$R^2=1-SSE/SST$$ (21.7)

$$R^2_{adj}=1-(1-R^2)\frac{n-1}{n-p-1}$$ (21.8)

$$RMSE=\sqrt{\frac{SSE}{n-p-1}}$$ (21.9)

where p is the number of non-constant terms in a PR model, SSE is the sum of square errors, and SST is the total sum of squares. The values of R^2 and R^2_{adj} range from 0 to 1.

In general, a model with good accuracy is indicated by large values of F-statistic, R^2 and R^2_{adj} and by small values of $RMSE$. When the number of design variables is large, it is more appropriate to look at R^2_{adj} instead of R^2, because R^2 always increases as the number of terms in the model is increased, while R^2_{adj} actually decreases if unnecessary terms are added to the model [1]. SSE and SST are calculated by

$$SSE=\sum_{i=1}^{n}\left(f_i-\hat{f}_i\right)^2$$ (21.10)

$$SST=\sum_{i=1}^{n}\left(f_i-\bar{f}\right)^2$$ (21.11)

where f_i is the true function value at the ith design point, $\hat{f}_i$ is the value calculated from the PR model at the ith design point, and $\bar{f}$ is the mean value of f_i ($i=1, 2 \dots n$).

In addition to these statistics, the accuracy of a PR model can also be measured by the prediction error sum of squares ($PRESS$) and R^2 for prediction ($R^2_{prediction}$) [1]. The $PRESS$ statistic and $R^2_{prediction}$ are calculated by

$$PRESS=\sum_{i=1}^{n}\left[f_i-\hat{f}_{(i)}\right]^2$$ (21.12)

$$R^2_{prediction}=1-PRESS/SST$$ (21.13)

where $\hat{f}_{(i)}$ is the predicted value at the ith design point using a PR model created by $(n-1)$ design points after excluding the ith point from the sample. Similar to R^2 and R_{adj}^2, $R_{prediction}^2$ ranges from 0 to 1 and a large value indicates good accuracy of the PR model.

21.3.3 Discussions on the PR Method

As an example of the PR method, the linear and quadratic PR models are created for the following response function

$$R(x) = \sin x + 0.25x^4 \tag{21.14}$$

Using five equally spaced sampling points on a range of $[0, 3]$, the two models are created and shown in Figure 21.1a with their functions given as follows.

$$f(x)_{Linear} = -0.695132 + 2.216115x$$

$$f(x)_{Quadratic} = 0.1172842 + 0.049672x + 0.7221477x^2 \tag{21.15}$$

The statistics parameters are calculated for both models and shown in Table 21.1. It can be seen that the quadratic model is a good approximation of $R(x)$ and is more accurate than the linear one.

The PR method has the advantage of being simple and efficient. In addition, the accuracy of a PR model can be assessed by just using the function values at the sampling points. However, the PR method also has limitations.

First, PR models typically do not pass through the design points (see fig. 21.1a), because they try to minimize the errors at all points. This means that the values given by the PR model on the sampling

TABLE 21.1 Values of Statistics Parameters for the PR Models

Model	F	Prob $> F$	R^2	R_{Adj}^2	RMSE	PRESS	$R_{Prediction}^2$
Linear	33.71	0.01	0.9183	0.8910	0.9053	4.956	0.8352
Quadratic	201.55	0.005	0.9951	0.9901	0.2725	0.016	0.9995

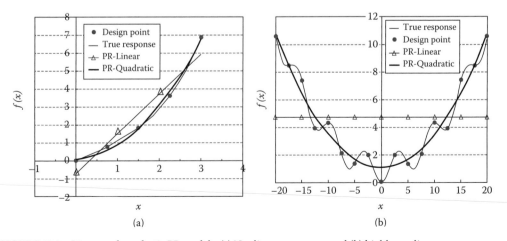

FIGURE 21.1 Linear and quadratic PR models. (a) Nonlinear responses; and (b) highly nonlinear response.

points are not accurate, even though the true values are already known from simulation. This feature could be appropriate for data analysis of physical experiments that involve random and bias errors. However, this feature is a drawback for simulation-based design, because all of the response values obtained from numerical simulation at any design points are considered true values.

Another limitation of the PR method is that PR models in the form of Equation 21.2 are not appropriate for highly nonlinear responses, as illustrated in Figure 21.1b. Although higher-order polynomials may be used for highly nonlinear responses, they do not work for some functions (e.g., the function in fig. 21.1b) and, in case they do, the order of the best polynomial has to be determined by trial and error. In addition, when the number of design variables becomes large, the terms in a high-order polynomial will quickly increase and the number of required sampling points becomes prohibitively large. For these reasons, high-order polynomials are seldom used and other methods suitable for highly nonlinear responses are typically sought (e.g., RBF, Kriging, and MLSR). A detailed discussion on the performance of high-order polynomial regression can be found in Reference 30.

21.4 Radial Basis Function (RBF)

The RBF has been traditionally used in applications with large sizes of data points. In recent years, RBF models have been used in applications with limited numbers of sampling points and found to have good accuracy.

21.4.1 Formulation of Non-Augmented RBF Models

An RBF model has the general form of

$$f(x) = \sum_{i=1}^{n} \lambda_i \phi(\|x - x_i\|) \qquad (21.16)$$

where n is the number of sampling points, x is a vector of design variables, x_i is the ith sampling point, $\|x - x_i\|$ is the Euclidean norm, ϕ is a basis function, and λ_i is the coefficient for the ith basis function. It can be seen from Equation 21.16 that the approximate function $f(x)$ is actually a linear combination of some basis functions with weight coefficients. Traditionally used basis functions are linear, cubic, thin-plate spline, Gaussian, multiquadric, and inverse-multiquadric. Newly developed basis functions include compactly supported functions [29, 33] that have a nice feature of being positive definite. These basis functions are summarized in Table 21.2.

Using function values at the n sampling points in Equation 21.16 leads to the following n equations

$$\begin{Bmatrix} f_1 \\ f_2 \\ \cdots \\ f_n \end{Bmatrix} = \begin{bmatrix} \phi(\|x_1 - x_1\|) & \phi(\|x_1 - x_2\|) & \cdots & \phi(\|x_1 - x_n\|) \\ \phi(\|x_2 - x_1\|) & \phi(\|x_2 - x_2\|) & \cdots & \phi(\|x_2 - x_n\|) \\ \cdots & \cdots & & \cdots \\ \phi(\|x_n - x_1\|) & \phi(\|x_n - x_2\|) & \cdots & \phi(\|x_n - x_n\|) \end{bmatrix} \begin{Bmatrix} \lambda_1 \\ \lambda_2 \\ \cdots \\ \lambda_n \end{Bmatrix} \qquad (21.17)$$

where coefficients $\{\lambda\}$ are obtained by solving Equation 21.17. Figure 21.2 shows the RBF models created for the two response functions in Figure 21.1 using the same sampling points. It can be seen that all of the RBF models are more accurate than the PR models.

TABLE 21.2   Commonly Used Radial Basis Functions

Name	Basis Function
Linear	$\phi(r) = r$
Cubic	$\phi(r) = r^3$
Thin-plate spline	$\phi(r) = r^2 \ln(cr), 0 < c \leq 1$
Gaussian	$\phi(r) = e^{-cr^2}, 0 < c \leq 1$
Multiquadric	$\phi(r) = \sqrt{r^2 + c^2}, \ 0 < c \leq 1$
Inverse multiquadric	$\phi(r) = 1/\sqrt{r^2 + c^2}, \ 0 < c \leq 1$
Wu's compactly supported (2,0)	$\phi_{2,0}(t) = (1 - t)^5 (1 + 5t + 9t^2 + 5t^3 + t^4), t = r/r_0$
Wu's compactly supported (2,1)	$\phi_{2,1}(t) = (1 - t)^4 (4 + 16t + 12t^2 + 3t^3)$
Wu's compactly supported (3,0)	$\phi_{3,0}(t) = (1 - t)^7 (5 + 35t + 101t^2 + 147t^3 + 101t^4 + 35t^5 + 5t^6)$
Wu's compactly supported (3,1)	$\phi_{3,1}(t) = (1 - t)^6 (6 + 36t + 82t^2 + 72t^3 + 30t^4 + 5t^5)$

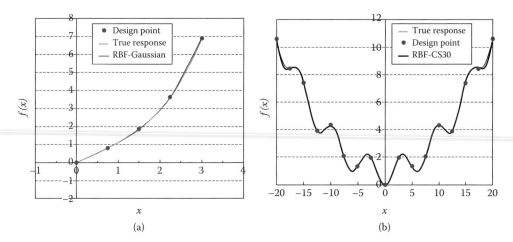

FIGURE 21.2   RBF metamodels of two nonlinear response functions. (a) RBF model by Gaussian function; and (b) RBF model by Wu's compactly supported function $\phi_{3,0}$.

21.4.2 Formulation of Augmented RBF Models

Although the RBF is good for highly nonlinear responses, it has been shown to have decreased accuracy for linear responses [26]. This is due to the use of highly nonlinear basis functions and the way the coefficients of an RBF model are obtained. To correct this deficiency, an RBF model can be augmented with a polynomial function given as

$$f(x) = \sum_{i=1}^{n} \lambda_i \phi \left(\|x - x_i\| \right) + \sum_{j=1}^{n} c_j p_j(x)$$

(21.18)

where $p(x)$ is a low-order (linear or quadratic) polynomial function, k is the total number of terms in the polynomial, and cj ($j = 1, 2 \dots k$) are the unknown coefficients. Equation 21.18 is underdetermined, because the number of parameters to be solved is $(n + k)$, which is more than the number of equations

(n) created with available data points. An orthogonality condition is thus imposed on coefficients $\{\lambda\}$ so that all of the unknown coefficients can be solved. The orthogonality condition is expressed as

$$\sum_{i=1}^{n} \lambda_i p_j(x_i) = 0, \quad for \; j = 1, 2 \ldots k \tag{21.19}$$

Combining Equations 21.18 and 21.19, we obtain ($n + k$) equations and the matrix form is given as

$$\begin{Bmatrix} f_1 \\ f_2 \\ \ldots \\ f_n \\ 0 \\ 0 \\ \ldots \\ 0 \end{Bmatrix} = \begin{bmatrix} \phi_{1,1} & \phi_{1,2} & \ldots & \phi_{1,n} & p_{1,1} & p_{2,1} & \ldots & p_{k,1} \\ \phi_{2,1} & \phi_{2,2} & \ldots & \phi_{2,n} & p_{1,2} & p_{2,2} & \ldots & p_{k,2} \\ \ldots & \ldots & \ldots & \ldots & \ldots & \ldots & \ldots & \ldots \\ \phi_{n,1} & \phi_{n,2} & \ldots & \phi_{n,n} & p_{1,n} & p_{2,n} & \ldots & p_{k,n} \\ p_{1,1} & p_{1,2} & \ldots & p_{1,n} & 0 & 0 & \ldots & 0 \\ p_{2,1} & p_{2,2} & \ldots & p_{2,n} & 0 & 0 & \ldots & 0 \\ \ldots & \ldots & \ldots & \ldots & \ldots & \ldots & \ldots & \ldots \\ p_{k,1} & p_{k,2} & \ldots & p_{k,n} & 0 & 0 & \ldots & 0 \end{bmatrix} \begin{Bmatrix} \lambda_1 \\ \lambda_2 \\ \ldots \\ \lambda_n \\ c_1 \\ c_2 \\ \ldots \\ c_k \end{Bmatrix} \tag{21.20}$$

where $\phi_{i,j} = \phi(\|x_i - x_j\|)$ and $p_{i,j} = p_i(x_j)$. Solving Equation 21.20 gives coefficients $\{\lambda\}$ and $\{c\}$ for the RBF function in Equation 21.18.

21.4.3 Discussions on RBF Models

An RBF model passes through sampling points exactly; in other words, an RBF model has no error on all of the sampling points. This feature makes RBF models more desirable than PR models for simulation-based designs. Furthermore, RBF models typically have good accuracy on off-design points as well. Figure 21.3 illustrates the RBF models for two highly nonlinear responses. Both RBF models are created with 121 (11×11) design points. The PR models created with quadratic polynomials using the same design points are also given.

Despite the abovementioned advantages, RBF models, particularly those using highly nonlinear basis functions, are more expensive to evaluate function values than PR models. In addition, the number of terms in an RBF model increases as the number of design points increases (see Equations 21.16 and 21.18). Furthermore, since the value of SSE is zero at sampling points, the accuracy of an RBF model cannot be assessed using statistics parameters given in Equations 21.6 to 21.9. Although the values of $PRESS$ and $R^2_{prediction}$ can be calculated at sampling points, they have been shown to be unable to give correct indication of model accuracy [34]. To have a good assessment of the accuracy of an RBF model, true function values at additional points (off-design points) are needed. The $RMSE$ values at off-design points are calculated by

$$RMSE = \sqrt{\frac{\sum_{i=1}^{n}(f_i - \hat{f}_i)^2}{n}} \tag{21.21}$$

where n is the number of off-design points, f_i is the true function value at the ith off-design point, and $\hat{f}_i$ is the function value calculated from the RBF model at the ith off-design point.

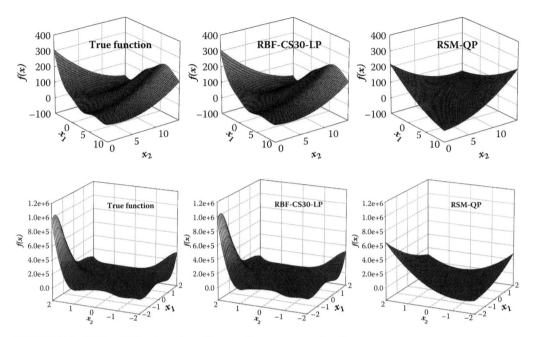

FIGURE 21.3   RBF and PR metamodels of two nonlinear response functions.

Note RBF-CS30-LP is an RBF model by Wu's compactly supported function $\phi_{3,0}$ and augmented by a linear polynomial. RSM-QP is the PR model created by a quadratic polynomial.

21.5 Moving Least Square Regression (MLSR)

The MLSR method can be regarded as an extension of the polynomial regression discussed in the section "Polynomial Regression (PR)." We give a brief introduction to the MLSR method in this section, because this method has also been widely used and it is particularly useful in noise filtering for dynamic responses. Two techniques are essential to this approach. First, weighing coefficients are introduced to the least square fitting scheme of the PR. Denoting [W] as a diagonal matrix of weighting factors for n sampling pointing, the least square method in Equation 21.5 becomes

$$\{\beta\}=\left([X]^{T}[W][X]\right)^{-1}\left([X]^{T}[W]\{f\}\right)$$

(21.22)

The weighting coefficient w_i indicates the relative importance of the information at the ith sampling point. The second element of MLSR is that the weighting coefficient w_i is formulated as a function that decays as a point x moves away from x_i. The term *moving* least square regression essentially means that the least square fitting moves dynamically with emphasis of information of sampling points close to the evaluation point x. As a consequence, the MLSR method can follow local hills and valleys of a highly nonlinear function. However, because the weights w_i are functions of x, the coefficients β are no longer constants. In other words, an MLSR model is an implicit function that is computationally more expensive than a traditional (static) PR model that has an explicit function expression. Typical weight decaying function includes Gaussian function and polynomial functions. The normalized distance from the ith sampling point to the evaluation point is calculated by

$$d_{i}=\frac{r_{i}}{R_{max}}$$

(21.23)

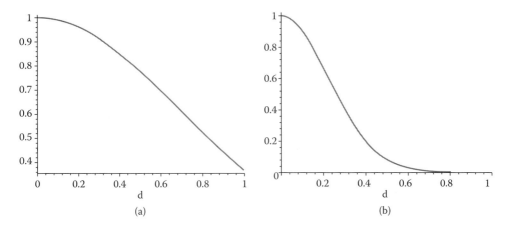

FIGURE 21.4 Gaussian weight functions. (a) $\theta = 1$; and (b) $\theta = 10$.

where r_i is the distance from the evaluation point to ith sampling point, R_{max} denotes the radius of the sphere of influence. The Gaussian weight decaying function is given as

$$w_i = exp\left(-\theta d_i^2\right) \tag{21.24}$$

where θ defines the closeness of fit. When θ is equal to zero, Equation 21.24 reduces to the traditional least square regression given by Equation 21.5. As parameter θ increases, the bias toward sampling points increases and hence the MLSR surface gets closer to sampling points. Figure 21.4 illustrates the effect of θ on the decaying rate of the weight function.

As an alternative, polynomial functions of various orders can be used as weight functions. Figure 21.5 shows two examples of polynomial weight functions.

21.5.1 Discussions on the MLSR Method

We now assess the capability of the MLSR method on approximating nonlinear responses. As aforementioned, the MLSR method uses weight decaying function to track local features; therefore, they are expected to generate better approximations than the static PR method. Figure 21.6 shows the MLSR

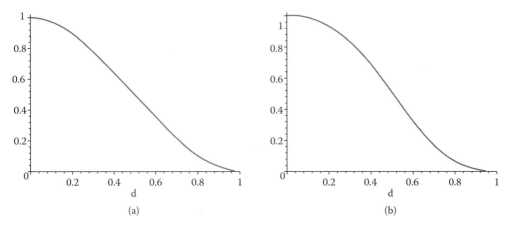

FIGURE 21.5 Examples of polynomial weight functions. (a) $w_i = 1 - 3d_i^2 + 2d_i^3$; and (b) $w_i = 1 - 10d_i^3 + 15d_i^4 - 6d_i^5$.

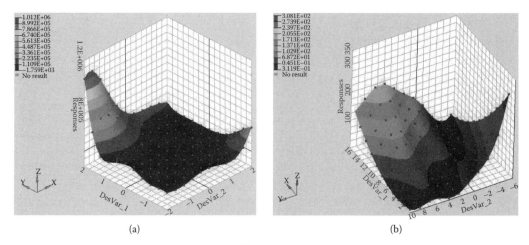

(a)   (b)

FIGURE 21.6 MLSR approximations of two highly nonlinear functions. (a) MLSR model of Goldstein-Price function; and (b) MLSR model of Branin-Rcos function.

models of two highly nonlinear responses in Figure 21.3. Both MLSR models were generated with the sample number of sampling points as those for the RBF models in Figure 21.3 (121 points). It can be seen that the two MLSR models are also good approximations of the two nonlinear responses.

The MLSR method was also used to generate the nonlinear response in Figure 21.2b but failed to produce a good approximation using the same set of sampling points. The reason is that MLSR is a regression-based method, and that it needs more samples to represent the small, detailed features of this highly nonlinear function. The MLSR method cannot "create" local features like RBF.

Besides its application in function approximation, the MLSR method has gained popularity in geometric modeling [35, 36]. A major strength of the MLSR method lies in its intrinsic capability of handling noisy inputs. The parameter "influencing radius R_{max}" provides a direct control on the size of features that are desired to be smoothed out in a surface reconstruction. The noise filtering capability is also a powerful feature for nonlinear dynamic events, since data obtained through either physical experiments or some numerical simulations often contain noises.

21.6 Responses of Short Duration Dynamic Events

Dynamic events that happened within short durations could have a large effect on humans and the surrounding structures. Such events include but are not limited to crashes, blasts, and earthquakes. Short duration dynamic events typically have severe adverse effects due to the abrupt changes of states on humans and structures. For example, in vehicle crashes, a human may be subject to a deceleration of more than 100 G within 40 ms after the initiation of the impact. The human and structural responses during a dynamic event are typically time dependent; therefore, a time history of response values exists for each of the responses of interest. Figure 21.7 shows a vehicle after an offset-frontal impact test and the corresponding time history of acceleration measured on the dummy chest in the impact direction [37].

Other responses such as the contact force on the dummy can also be measured in the crash test shown in Figure 21.7. Some responses, such as the vehicle's energy absorption history, may be of interest but cannot be measured from a test or a real event; such a history can be obtained from numerical simulation (see examples in the section "Numerical Examples").

Short duration dynamic events present challenges to the optimization of dynamic responses. First, these dynamic responses are typically highly nonlinear and will incur computational challenges to solve the optimization problems in which these responses are used as objective or constraint functions. Second, a dynamic response is a function of time and its values at several time instances (e.g., pick values)

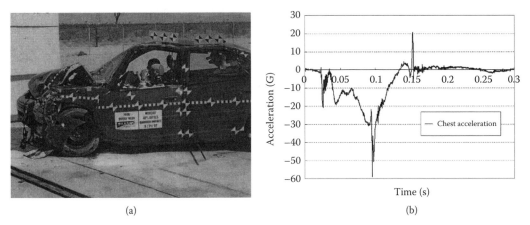

FIGURE 21.7 An offset-frontal impact test of a 1996 Dodge Neon [34]. (a) Test vehicle after impact; (b) acceleration history at the dummy's chest.

may be of interest, along with the fact that multiple dynamic responses may need to be considered. Third, it is difficult to select design variables that have strong effects on the time-histories of multiple dynamic responses. Finally, the dynamic responses are typically obtained through numerical simulations due to the complexity of the problem; there is typically no explicit function between a dynamic response and selected design variables. Therefore, challenges exist in both formulating the optimization problem for short duration dynamic events and in solving this type of multi-objective problems efficiently. In the next section, a brief summary is given on the methods on solving multi-objective optimization problems. Combining the metamodeling techniques presented in the section "Metamodeling Methodologies" and multi-objective optimization methodology, two numerical examples are given in the section "Numerical Examples" to demonstrate how this approach can be used to efficiently solve problems of short duration dynamic events.

21.7 Multi-Objective Optimization

21.7.1 Definitions and Formulations

An optimization problem is an inverse engineering process that finds the optimum set(s) of design variables that will minimize (or maximize) the objective functions. In an optimization problem with a single objective function, there exists a single set of design variables that is the optimum solution of the problem. Such a set of design variables is called the optimum solution. For an optimization problem with more than one objective function, there is typically no single optimum solution that will yield the minimum (or maximum) values of all of the objectives simultaneously. This is because there exist conflicts among the objective functions, and the optimum solution for one objective may not be the optimum solution for other objectives. An optimum solution in the context of multi-objective optimization, therefore, represents a trade-off among all of the objectives. A set of solutions representing the different levels of trade-offs among objective functions exists and such a set is formally called Pareto optimum set or Pareto frontier [38]. Figure 21.8 illustrates possible Pareto frontiers for a minimization problem of two objective functions.

All of the solutions in the Pareto optimum set (or Pareto frontier) are called nondominated solutions, because they are the trade-offs among objective functions and no solution is strictly better than any other solution in the Pareto set for all of the objectives. Solutions in the feasible domain but not in the Pareto set are called dominated solutions, because for each of such solutions, there is at least one solution in the Pareto set that is better on all of the objectives.

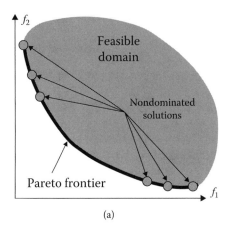
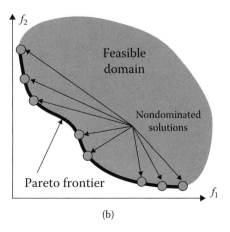

(a)   (b)

FIGURE 21.8   Pareto frontiers of a two-objective minimization problem. (a) Convex solution space; (b) solution space with concave regions.

A multi-objective minimization problem can be formulated as

$$\text{Min.} \quad F(x) = \left[f_1(x), f_2(x) \ldots f_M(x) \right]$$

$$\text{S.t.} \quad h_i(x) = 0, \ i = 1, 2 \ldots N_{EC}$$

$$g_j(x) \le 0, \ j = 1, 2 \ldots N_{IC} \tag{21.25}$$

$$X_k^L \le x_k \le X_k^U, \ k = 1, 2 \ldots N_{DV}$$

where x is a vector of N_{DV} design variables, $F(x)$ is a vector of M objective functions, $h_i(x)$ are the equality constraint functions, $g_j(x)$ are the inequality constraint functions, and X_k^L and X_k^U are the lower and upper bounds of the kth design variable x_k, respectively. If any of the objective functions, for example, $f_2(x)$, is to be maximized, Equation 21.25 can still be used by replacing the objective function with its negative form, for example, $-f_2(x)$ in this case.

21.7.2 Solution Techniques

A multi-objective optimization problem can be converted into a single-objective optimization problem by converting all but one objective into constraint functions. This is, however, undesirable and impossible in many situations, because it requires the knowledge of limits for those constraints. Furthermore, only a single solution can be obtained by this approach. It is desirable to obtain a set of solutions on the Pareto frontier and then select the final design(s) from it. To this end, obtaining a set of solutions on the Pareto frontier is the goal of solving a multi-objective optimization problem.

One popular approach is to use evolutionary algorithms (EAs), which is a family of algorithms including three major categories: evolutionary strategy; genetic algorithm; and genetic programming [11, 12]. An evolutionary algorithm starts with a randomly generated set of solutions (called *population*) that may include both feasible and infeasible solutions. The population is iteratively improved so that it will eventually become or be close to the Pareto frontier. One advantage of the EAs is that they do not require gradients for both the objective and constraint functions, because the gradients may be difficult to obtain for many numerical simulations. Another advantage of EAs is that they are typically not trapped in local optima for highly nonlinear objective functions. The drawback of the EAs is that it is computationally expensive due to the large number of function evaluations. To this end, it is still not feasible to directly combine an EA method with a numerical simulation procedure for complex engineering applications, even though the gradient information is not required from the simulation. In addition, when the solution set is close to the Pareto frontier, the genetic operations used in EAs do not allow for monotonic convergence and a forced stopping criterion has to be adopted after a given number of iteration.

Another approach that is commonly used is to assign each objective a weight coefficient and then combine all of the weighted objectives into a single objective function. A single solution can then be obtained using a single-objective optimization method (e.g., the sequential quadratic programming method). By varying these weight coefficients, different solutions can be obtained. By performing a Pareto optimality check on the set of these solutions, dominated solutions will be removed and the remaining will form a non-dominated solution set. The advantage of this approach is that the efficiency of gradient algorithms can be utilized. The disadvantage is that gradient algorithms require gradient values for the objective and constraint functions. Additionally, gradient algorithms are easily trapped into local optima for highly nonlinear objectives; therefore, a number of random starting points are needed for the optimization of a single solution. As a consequence, the computational costs are increased.

For both approaches, the metamodeling technique can be utilized to reduce the cost of evaluating objective and constraint functions. Improving the efficiency of multi-objective optimization algorithms is still an ongoing research topic.

21.8 Numerical Examples

In this section, we use two numerical examples to illustrate how to optimize the responses of short duration dynamic events. The first example is a 1996 Dodge Neon to be optimized for simultaneous weight reduction and improvement of safety performance considering two crash scenarios, offset-frontal, and side impacts. The second example is a minivan to be optimized for side impact and roof crashes.

21.8.1 Crashworthiness Optimization of a 1996 Dodge Neon

The objective of the tasks in this example is to optimize the design of a 1996 Dodge Neon to reduce its weight while improving its safety performance in offset-frontal and side impacts. Both types of impacts are standard vehicle crash testing specified by the U.S. Federal Motor Vehicle Safety Standards and Regulations [39]. Figure 21.9 shows the finite element models used in the simulations of both impacts. The finite element model of this vehicle was originally developed at the National Crash Analysis Center [40, 41]. The standard setup for the two tests is schematically illustrated in Figure 21.10.

In the offset-frontal impact (OFI), the vehicle impact onto a deformable barrier (made of aluminum foam) attached to a (rigid) concrete wall. During impact, the instrumental panel (and the attached driving wheel) intrudes toward the driver and may cause serious injury if the intrusion distance is large. In side impact (SI), the vehicle is hit by a moving deformable barrier from the driver side. Since the space between the drive and the side door is very small, the door's intrusion distance becomes even more

(a) (b)

FIGURE 21.9 Full-scale finite element models for impact simulations. (a) Offset-frontal impact; (b) side impact.

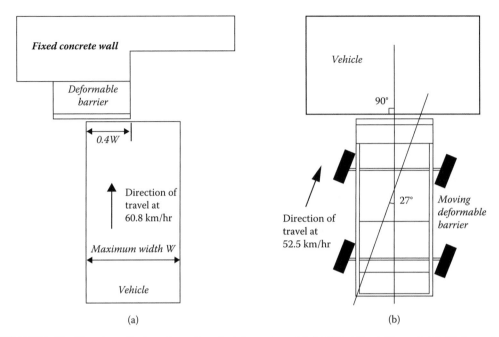

FIGURE 21.10   Plan view of test configurations (not drawn to scale). (a) Offset-frontal impact; (b) side impact.

critical than in OFI. In this example, the intrusion distances of the front panel in OFI and the driver-side door in SI, respectively, are selected as safety parameters to be minimized.

Most of the vehicle's components are made of sheet metals, and the thickness of these components can be used as design variables. Due to the large amount of components in the vehicle, only the most influential components should be selected. Unfortunately, there is no direct linkage between the two safety parameters (intrusion distances) and the components' thickness. Since the safety performance in such dynamic events is largely related to the energy absorbed by the structure, an analysis of the energy absorption history was performed based on crash simulations using the original design. Figure 21.11 illustrates the energy absorption histories for both OFI and SI.

It can be seen that a large amount of energy is absorbed within 60 ms for OFI and within 40 ms for SI. Components with the largest energy absorption at 60 ms and 40 ms for OFI and SI, respectively, were then

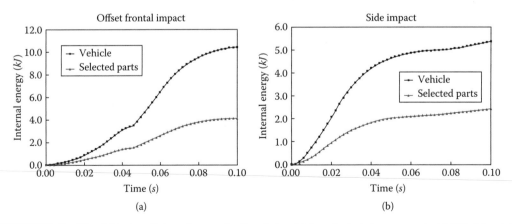

FIGURE 21.11   Time histories of internal energy for the whole vehicle and selected components. (a) Offset-frontal impact; (b) side impact.

FIGURE 21.12 Selected components for optimization.

identified. Additionally, components with large mass and small amounts of energy absorptions were also identified, because there is a third objective (weight reduction) to be minimized besides the two safety parameters. The rationale was that the reduction on weight of these components would not adversely affect the energy absorption of the vehicle. The final selected components are shown in Figure 21.12; the thickness of these components is represented by 13 design variables (due to structural symmetry). The total energy absorptions of the selected components in OFI and SI are also shown in Figure 21.11.

With the selected design objectives and components, this multi-objective problem is formulated as follows:

$$\text{Min. } F(x) = \left[f_1(x), f_2(x), f_3(x) \right]$$
$$\text{S.t. } g_1^0 - g_1(x) \le 0$$
$$g_2^0 - g_2(x) \le 0 \tag{21.26}$$
$$X_k^L \le x_k \le X_k^U, \ k = 1, 2 \dots 13$$

where $F(x)$ is a vector of three objective functions, the mass of selected components $f_1(x)$, the intrusion distance in OFI $f_2(x)$, and the intrusion distance in SI $f_3(x)$. The two constraint functions require that the energy absorptions at 40 ms for any new design ($g_1(x)$ for OFI and $g_2(x)$ for SI) should not be less than those of the original design (g_1^0 for OFI and g_2^0 for SI). The use of the two constraints is based on the consideration that a reduction on energy absorption at an earlier stage of an impact (e.g., 40 ms) may result in an increase on the peak acceleration on the occupant, which adversely affects the safety performance.

The mass of selected components can be directly represented using design variables with known initial mass. But functions $f_2(x)$, $f_3(x)$, $g_1(x)$, and $g_2(x)$ are not known and need to be created using the metamodeling methodology. The Taguchi orthogonal array L_{27} [32] was used to obtain the responses for the four functions at 27 design points. The PR models were then created with quadratic polynomials; they were shown to have good accuracies from the statistics parameters in Table 21.3.

With all objective and constraint functions now available, Equation 21.26 is formulated using the weight sum method as follows

$$\text{Min. } F(x) = W_1 f_1(x) + W_2 f_2(x) + W_3 f_3(x)$$
$$\text{S.t. } g_1^0 - g_1(x) \le 0$$
$$g_2^0 - g_2(x) \le 0 \tag{21.27}$$
$$W_i > 0, \ \sum W_i = 1, \ i = 1,2,3$$
$$X_k^L \le x_k \le X_k^U, \ k = 1, 2 \dots 13$$

TABLE 21.3   Statistical Parameters of Quadratic Polynomial Regression Models

Response Function	*F*-statistic	R^2	R^2_{adj}	*RMSE*
Intrusion distance for OFI, $f_2(x)$	13,616	0.99	0.99	0.41
Intrusion distance for SI, $f_3(x)$	10.9	0.99	0.90	0.03
Internal energy for OFI, $g_1(x)$	29.8	0.99	0.97	0.03
Internal energy for SI, $g_2(x)$	44.9	0.99	0.98	1.88

TABLE 21.4   Two Sample Optimum Solutions Compared with Original Design

Function	Response	Original Design	Optimum Design			
			Solution 1	Change	Solution 2	Change
	Mass (kg)	96.8	82.8	−14.5%	90.8	−6.2%
Objective	Intrusion distance in OFI (mm)	226	187	−17.3%	159	−29.6%
	Intrusion distance in SI (mm)	370	351	−5.1%	335	−9.5%
Constraint	Internal energy at 40 ms in OFI (kJ)	3.08	3.16	+2.6%	3.15	+2.3%
	Internal energy at 40 ms in OFI (kJ)	4.18	4.18	0%	4.26	+1.9%

where Wi are weight coefficients representing the relative importance of the objectives. The feasible sequential quadratic programming (FSQP) method [42], which was implemented into an integrated optimization framework, HiPPO [43], was used to solve the problem given in Equation 21.27. By varying the values of weight coefficients using a minimum value of 0.01 and an increment of 0.005, a total of 19,110 solutions were obtained and 1,795 solutions were found non-dominated after performing the Pareto optimality check. Table 21.4 gives two sample solutions from the non-dominated solution set along with the original values.

The results in Table 21.4 clearly show the conflicts between $f_1(x)$ and $f_2(x)$ and between $f_1(x)$ and $f_3(x)$. For example, if weight reduction changes from 14.5% (Solution 1) to 6.2% (Solution 2), the reduction on intrusion distances can be improved from 17.3% to 29.6% for OFI and from 5.1% to 9.5%. While the final decision is based on the designer's preference, these non-dominated solutions provide designers with useful information on the trade-offs of all of the objectives. FE simulations were performed using the optimum designs in Table 21.4, and the results show that the errors of the predicted optimum values using metamodels are all less than 3%.

21.8.2 Crashworthiness Optimization of a Minivan

This example presents the optimization of a minivan to improve its safety performance in a side impact (SI) and roof impact (RI) with simultaneous weight reduction. The intrusion distances in both impacts were used as the safety parameters (two objectives). Figure 21.13 illustrates the FE models of the minivan for both impacts along with the selected components whose thickness is the design variables in the optimization problem.

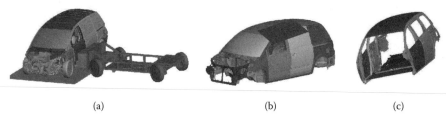

(a)   (b)   (c)

FIGURE 21.13   Finite element models of a minivan in side and roof impacts. (a) Side impact; (b) roof impact; (c) selected components for optimization.

TABLE 21.5   Comparison of Accuracy for PR and RBF Models

	Errors of PR Model			Errors of RBF Model		
Response	Max.	Min.	*RMSE*	Max.	Min.	*RMSE*
Intrusion distance for OFI	1.4%	−2.5%	5.72	0.9%	−1.5%	3.76
Intrusion distance for SI	0.0%	−6.5%	4.30	2.1%	−2.5%	1.83

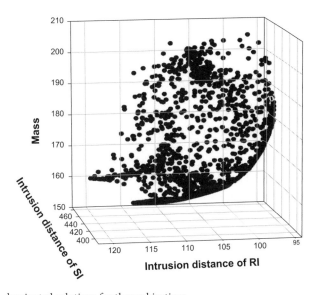

FIGURE 21.14   Non-dominated solutions for three objectives.

This example is similar to the previous one except that the response functions for the two intrusion distances use RBF models in this case, because PR models were found to be less accurate. The RBF models for the two responses were created using 51 design points determined by the Latin hypercube sampling method [1] for 20 design variables (representing the thickness of selected components). Model accuracy was evaluated at 20 validation points that were different from those design points, and the results are shown in Table 21.5.

The results in Table 21.5 show that the RBF models are more accurate than the polynomial ones. For example, the *RMSE* values indicate a 42% reduction of error for the first response and a 62% reduction for the second response. The optimization formulation of this example is similar to Equation 21.28 and a total of 2,901 non-dominated solutions were obtained. Figure 21.14 illustrates the trade-off surfaces formed by these solutions. As in the case of the previous example, strong conflicts exist between the objectives of mass reduction and the two intrusion distances.

21.9 Summary

Short duration dynamic events present several challenges to the optimization of such problems: a time history of responses instead of a single response function; implicit relationships of response functions and design variables; highly nonlinear responses; and high computational costs of numerical simulations. These challenges make it difficult to formulate and solve a multi-objective optimization problem of such an event.

In this chapter, we presented three metamodeling methodologies and discussed their advantages and disadvantages. We demonstrated how to combine metamodeling methodologies with multi-objective optimization techniques to obtain solutions to such optimization problems with affordable computational costs. The polynomial regression (PR) method, which has been widely used in many engineering applications, is simple and efficient but inappropriate for highly nonlinear responses. The radial basis functions (RBFs), which are traditionally used on applications with large sample sizes, have recently been shown to work well for highly nonlinear responses at small sample sizes. Nevertheless, the PR method is preferred over the RBF wherever the former is also appropriate, because an RBF model is more complicated and much more expensive to evaluate than a PR model. The solution techniques were shown to be effective through the two numerical examples of vehicle crashes.

References

1. Montgomery, D. C. (2001). *Design and Analysis of Experiments*, 5th Ed., John Wiley & Sons, Inc., New York.
2. Rodriguez, J., Renaud, J. E., and Watson, L. T. (1997). Trust region augmented Lagrangian methods for sequential response surface approximation and optimization. DETC'97 ASME Design Engineering Technical Conference, Sacramento, CA, DETC97/DAC3773.
3. Alexandrov, N. M., Dennis, J. E., Lewis, R. M., and Torczon, V. (1998). A trust region framework for managing the use of approximation models in optimization. *Structural Optimization*, 15, 16–23.
4. Wang, G., Dong, Z., and Aitchison, P. (2001). Adaptive response surface method: a global optimization scheme for approximation-based design problems. *Engineering Optimization*, 33, 707–734.
5. Rais-Rohani, M., and Singh, M. N. (2004). Comparison of global and local response surface techniques in reliability-based optimization of composite structures. *Structural and Multidisciplinary Optimization*, 26(5), 333–345.
6. Fang, H., Solanki, K., and Horstemeyer, M. F. (2005). Numerical simulations of multiple vehicle crashes and multidisciplinary crashworthiness optimization. *International Journal of Crashworthiness*, 10(2), 161–171.
7. Yang, B. S., Yeun, Y. S., and Ruy, W. S. (2002). Managing approximation models in multi-objective optimization. *Structural and Multidisciplinary Optimization*, 24(2), 141–156.
8. Arora, J. S. (2004). *Introduction to Optimum Design*, Elsevier Academic Press, San Diego, CA.
9. Reklaitis, G. V., Ravindran, A., and Ragsdell, K. M. (1983). *Engineering Optimization: Methods and Applications*, John Wiley & Sons, Inc., New York.
10. Papalambros, P. Y. (2002). The optimization paradigm in engineering design: promises and challenges. *Computer-Aided Design*, 34, 939–951.
11. Coello Coello, C. A. (1999). A comprehensive survey of evolutionary-based multi-objective optimization techniques. *Knowledge and Information Systems*, 1, 269–308.
12. Coello Coello, C. A., Van Veldhuizen, D. A., and Lamont, G. B. (2002). *Evolutionary Algorithms for Solving Multi-Objective Problems*, 1st Ed., Kluwer Academic Publishers, New York.
13. Zitzler, E., and Thiele, L. (1999). Multi-objective evolutionary algorithms: a comparative case study and the strength Pareto approach. *IEEE Transactions on Evolutionary Computation*, 3, 254–271.
14. Das, I., and Dennis, J. (1997). A closer look at drawbacks of minimizing weighted sums of objectives for Pareto set generation in multicriteria optimization problems. *Structural Optimization*, 14, 63–69.
15. Choudhary, R., Malkawib, A., and Papalambros, P. Y. (2005). Analytic target cascading in simulation-based building design. *Automation in Construction*, 14, 551–568.
16. Kim, H. M., Michelena, N. F., and Papalambros, P. Y. (2003). Target cascading in optimal system design. *Journal of Mechanical Design*, 125(3), 474–480.
17. Lassiter, J., Wiecek, M., and Andrighetti, K. (2005). Lagrangian coordination and analytical target cascading: solving ATC-decomposed problems with Lagrangian duality. *Optimization and Engineering*, 6(3), 361–381.

18. Michalek, J. J., and Papalambros, P. Y. (2005). An efficient weighting update method to achieve acceptable consistency deviation in analytical target cascading. *Journal of Mechanical Design*, 127(2), 206–214.

19. Michelena, N., Park, H., and Papalambros, P. Y. (2003). Convergence properties of analytical target cascading. *AIAA Journal*, 41(5), 897–905.

20. Kim, H. M., Rideout, D. G., Papalambros, P. Y., and Stein, J. L. (2003). Analytical target cascading in automotive vehicle design. *Journal of Mechanical Design*, 125(3), 481–489.

21. Kokkolaras, M., Fellini, R., Kim, H. M., Michelena, N. F., and Papalambros, P. Y. (2002). Extension of the target cascading formulation to the design of product families. *Structural and Multidisciplinary Optimization*, 24(4), 293–301.

22. Allison, J. T., Kokkolaras, M., and Papalambros, P. Y. (2005). On the use of analytical target cascading and collaborative optimization for complex system design. In: The 6th World Conference on Structural and Multidisciplinary Optimization, Rio de Janeiro, Brazil.

23. Fang, H., Rais-Rohani, M., Liu, Z., and Horstemeyer, M. F. (2005), A comparative study of metamodeling methods for multi-objective crashworthiness optimization. *Computers & Structures*, 83 (25–26), 2121–2136.

24. Jin, R., Chen, W., and Simpson, T. W. (2001). Comparative studies of metamodelling techniques under multiple modeling criteria. *Structural and Multidisciplinary Optimization*, 23(1), 1–13.

25. Hussain, M. F., Barton, R. R., and Joshi, S. B. (2002). Metamodeling: radial basis functions versus polynomials. *European Journal of Operational Research*, 138(1), 142–154.

26. Krishnamurthy, T. (2003). Response surface approximation with augmented and compactly supported radial basis functions. In: The 44th AIAA/ASME/ASCE/AHS/ASC Structures, Structural Dynamics, and Materials Conference, Norfolk, VA, AIAA-2003-1748.

27. Krige, D. G. (1951). A statistical approach to some basic mine valuation problems on the Witwatersrand. *Journal of the Chemical, Metallurgical and Mining Society*, 52(6), 119–139.

28. Hardy, R. L. (1971). Multiquadratic equations of topography and other irregular surfaces. *Journal of Geophysics*, 76, 1905–1915.

29. Wu, Z. (1995). Compactly supported positive definite radial function. *Advances in Computational Mathematics*, 4, 283–292.

30. Fang, H., and Horstemeyer, M. F. (2006). Global response approximation with radial basis functions. *Engineering Optimization*, 38(4), 407–424.

31. Lancaster, P. and K. Salkauskas (1981). Surfaces generated by moving least-squares methods. *Mathematics of Computation*, 37(155), 141–158.

32. Taguchi, G. (1993). *Taguchi Method—Design of Experiments.* Japanese Standards Association, ASI Press, Tokyo, Japan.

33. Wendland, H. (1999). On the smoothness of positive definite and radial functions. *Journal of Computational and Applied Mathematics*, 101, 177–188.

34. Fang, H., and Wang, Q. (2008). On the effectiveness of assessing model accuracy at design points for radial basis functions. *Communications in Numerical Methods in Engineering*, 24(3), 219–235.

35. Alexa, M., Behr, J., Cohen-Or, D., Fleishman, S., Levin, D., and Silva, C. T. (2001). Point set surfaces. *IEEE Visualization* 2001, 21–28.

36. Levin, D. (2003). Mesh-independent surface interpoloation. *Geometric Modeling for Scientific Visualization.* Springer-Verlag, 37–49.

37. KARCO Engineering (1997). Frontal barrier forty percent offset impact test. *Final Report for the National Highway Traffic Safety Administration*, No. KAR-97-13, Adelanto, CA.

38. Collette, Y., and Siarry, P. (2004). *Multiobjective Optimization: Principles and Case Studies*, Springer-Verlag, Berlin.

39. NHTSA. (1998). Federal Motor Vehicle Safety Standards and Regulations. U.S. Department of Transportation, National Highway Traffic Safety Administration, Washington, D.C.

40. Zaouk, A. K., Marzougui, D., and Bedewi, N. E. (2000). Development of a detailed vehicle finite element model, Part I: methodology. *International Journal of Crashworthiness*, 5(1), 25–35.

41. Zaouk, A. K., Marzougui, D., and Kan, C. D. (2000). Development of a detailed vehicle finite element model, Part II: material characterization and component testing. *International Journal of Crashworthiness*, 5(1), 37–50.

42. Lawrence, C. T., Zhou, J. L., and Tits, A. L. (1997). User's guide for CFSQP, Ver 2.5. Electrical Engineering Department and Institute for Systems Research, University of Maryland, College Park, MD.

43. Fang, H., and Horstemeyer, M. F. (2005). HiPPO: an object-oriented framework for general-purpose design optimization. *Journal of Aerospace Computing, Information, and Communication*, 2(12), 490–506.

22

Modeling the Role of Human Behaviors in a System-of-Systems

Daniel DeLaurentis

22.1 Abstract

This chapter explores the nature of optimization in a system-of-systems context in which human behavior plays a central role. An introduction to the emerging concept of system-of-systems is presented first to establish a foundation. Building from this foundation, the sources of complexity related to human participation in a system-of-systems are described and categorized, and the unique facets of optimization in this setting are identified. A methodology for modeling, analyzing, and optimizing is then outlined, with special attention placed upon how complexity might be better managed through structured accounting of hierarchy and scope dimensions. The examination of future air transportation architectures is introduced as a system-of-systems application that exposes many of the concepts described, especially the variety of roles played by human decision-making. In particular, making context and participant perspectives overtly visible to system users is emphasized—an important precondition for effective human system integration in a system-of-systems context. The ultimate objective of this research thrust, still far from achieved, is to better engineer robust capabilities in the various system-of-systems that exist in society.

22.2 Introduction

22.2.1 Characterizing a System-of-Systems

A system-of-systems (SoS) consists of multiple, heterogeneous, distributed systems that can and do operate independently but also assemble in networks to achieve a unique function. These networks occur within a hierarchy of levels and they evolve over time. While the terminology may be recent, the notion of a system-of-systems is not new. Examples abound in which services developed and delivered rely upon the interaction of otherwise independently operating systems that assemble to achieve a capability not otherwise reachable by the individual systems in isolation. Transportation, energy, health care, and defense are high-profile, large-scale domains in which such dynamics exist. (Challenges within these domains existed before the phrase system-of-systems entered common use and have been studied extensively through segregated fields of inquiry, but have rarely been examined holistically as a distinct problem class.)

In recent years, however, the formal study of SoS as a problem class has increased, driven largely by the defense and aerospace arenas and due to a radical shift in the government's procurement approach. Where government customers once issued detailed *requirements* for a specific platform system (e.g., a tank with certain speed, firepower, and armor thickness), they now ask instead for a broad set of *capabilities* that are needed over a significant time span (e.g., direct-fire capability on all terrain battlefields). As a result, the system developers in industry must determine the appropriate mix of systems and related interconnections to provide these capabilities. In particular, these industry stakeholders are being asked to provide capabilities that must integrate existing, new, and future systems that interface well with the individual end users and satisfy a global need. The U.S. Army's Objective Force, constituted by its Future Combat System and operating under a Network Centric Warfare paradigm, is the most widely known example.

Motivated by these defense system challenges, the most active group exploring methodologies for SoS is the systems engineering community. And there is a growing recognition in this community that systems engineering processes are not complete for the new challenges posed by the development of SoS [1]. Therefore, researchers are looking for new, more appropriate system characterizations and design and development frameworks. Rouse, for example, describes categories of implications of complexity on systems engineering approaches [2]. Sage and Cuppan present a working collection of traits for SoS that points to the possibility of a "new federalism"—an assertion of systems with flexible strength of relations—as a construct for dealing with the variety of levels of cohesion in SoS organization [3]. We describe later in this chapter an additional view with the objective of structuring complexity in the organization of an SoS.

While defense-related applications have driven the recent emphasis on system-of-systems, many of society's needs across diverse domains are met through such a construct. Unfortunately, since few were designed with an SoS mentality, many current SoS exhibit an inflexibility in response to both disruption (artificial or natural) and increases in service demand. Supply chains acquire excessive inventory and capacity if manufacturers and distributors ignore opportunities to collaborate on demand forecasts. Health care networks experience breakdowns of information flows and lose continuity of care as patients migrate among autonomous hospitals, outpatient clinics, nursing homes, and so on. Electric power blackouts illustrate consequences of emergent behavior in that modest breakdowns in component systems of energy grids can cascade quickly into major service disruptions. Well-publicized transportation delays from overcapacity operations at major airports prevent passengers and cargo from reaching their destinations in a timely, productive manner. Large urban areas seek sustainable development in the face of interactions between many of these service infrastructures—health, power, and transportation. We hypothesize that a common thread through all these challenges is the complexity that arises from their nature as an SoS. In the face of this complexity, the SoS tends to evolve in place through the un- or ill-coordinated development of constituent systems fielded at different time scales.

These civil/commercial SoS applications also highlight, to a larger degree, the importance of interactions between independent individuals and organizations. Since much of the present research toward SoS methods stems from the engineered systems community, SoS solutions are still conceived primarily as the appropriately organized mix of artificial autonomous systems. Less attention is being paid to the role of human behavior in these applications (and to adjacent research areas such as the socio-technical systems approach [4], which addresses the complexities and uncertainties that result when human and technical systems interact). (Combining this traditional approach to SoS solutions with human-technical system interactions leads to a more complete characterization of an SoS. This more complex depiction has striking similarities to the crafting of formulations for sustainable information societies, for example, as a mix of human and engineered systems that exhibit emergent behavior [5]).

Since most systems-of-systems have evolved into their present state with little strategic intent, and since methods and models have yet to reach a comprehensive capability, there is a clear need for tools. Customers for these tools are decision makers who need to understand the implications of various alternatives for SoS challenges. Thus, in the hope of better performance at the SoS level, there is a desire to take a more active role, a design role, which naturally leads to the need for optimization. However, optimization in an SoS context has several unique and important traits that require an approach that seeks to explore and influence as opposed to pin-point and select. Therefore, the goals of this chapter are to describe the emerging system-of-systems problem class, identify primary sources of complexity within this class (highlighting in particular the role of human systems in this context), and present a methodology that leads to relevant SoS optimization problem formulations.

In order to accomplish this, the perspectives from the human systems modeling community should be integrated with those of the complex systems engineering community. An example application in air transportation provides a venue for exploring these ideas.

22.3 SoS Traits and Sources of Complexity

22.3.1 Maier's Criteria for an SoS

A set of distinguishing traits for SoS problems has been proposed by Maier [6]. Maier's criteria, listed and described in Table 22.1, are: operational independence; managerial independence; geographic distribution; evolutionary behavior; and emergent behavior. The first three primarily describe the problem boundaries and mechanics of the interacting elements while the latter two describe overall behavior. Maier contends that if a majority of these criteria are met, one should treat the problem as an SoS. The one-sentence characterization for an SoS offered at the beginning of this chapter's introduction attempts to encapsulate these traits as well as additional aspects such as the heterogeneity of component systems and the multilevel structure of their networks. In particular, emergent behavior—the manifestation of behavior that develops out of complex interactions of component systems that are not present for the systems in isolation—presents a particular challenge. Emergent behavior is unpredictable. It is often nonintuitive, and can be manifested in a positive manner (e.g., a new capability arises) or negative manner (e.g., a new failure is created). The well-known phrase *unintended consequences* well encapsulates for many the emergence concept. A primary challenge in the study of SoS with greater effectiveness is to understand the mechanism of emergent behavior, develop cues to detect it, and create a methodology for managing it intelligently.

22.4 Sources of Complexity and Human Behaviors

There are implications for modeling clearly apparent from the unique SoS traits. In particular, the heterogeneity of participating systems, connectivity in networks, and emergence all introduce sources of *complexity*. Over the past two decades, numerous views of complexity have been adopted, including

TABLE 22.1   Maier's Criteria for a System-of-Systems and Related Observations

Trait	Synopsis	Observation
Operational & Managerial Independence	Constituent systems are useful in their own right and generally operate independent of other systems (i.e., with unique intent provided by the owner/operator).	- Intent for interoperability and global status of system coupling variables are not easily discerned. - At the SoS level, *no one is completely in charge.*
Geographic Distribution	Constituent systems are not physically co-located; but, they may communicate.	- Spatial dynamics are important for SoS behavior. - Communication rules/capabilities are critical.
Evolutionary Behavior	The SoS is never completely, finally formed, it constantly changes, and has a "porous" problem boundary; i.e., is a living system.	- Behavior and constitution is time varying. - Presence of multiple timescales, possibly spanning a generation in some applications.
Emergent Behavior	Properties appear in the SoS that arise due to system interactions but are not apparent (or predicted) from study of isolated systems.	- Drives non-monotonic character of evolution. - Not readily addressed in current design methods. - Unpredictable, but manageable?

internal system complexity, external complexity, computational (algorithmic) complexity, and so on. While complexity can be, and is, viewed in different manners, its most basic utility is as a comparative measure. A given system, observed at a particular scale, can be more or less complex than another system examined at the same scale. One measure for this complexity is the amount of information necessary to describe a system [7, 8]. Systems of greater complexity require more information to define them than systems of lesser complexity.

In the SoS context, the heterogeneity of constituent systems, the dynamic and uncertain connectivity that arises with operation at multiple levels, and the role of humans all increase the amount of information needed for comprehensive characterization. Indeed, the integration activities of humans and their differing perspectives on situations often bring about self-organization (an emergent property). A design methodology including optimization for SoS may be most challenged by complexity sources arising from these human behaviors. This view supports the findings in the recently released study conducted by the U.S. Air Force Scientific Advisory Board entitled "System-of-Systems Engineering for Air Force Capability Development" [9]. The board identified the role of the human as critical for the successful implementation of SoS in the field. Quoting from their summary:

> Whenever the Air Force generates a system-of-systems, interaction among the systems often includes human-to-human interactions. If the machine-to-machine aspect of SoS is weak, then it falls upon the humans to achieve the interaction. This can, and often does, create a very challenging environment for the human; sometimes leading to missed opportunities or serious mistakes. The lack of sound Human System Interface designs can exacerbate this. Coordinated situation awareness is difficult to manage if the individual systems miss or convey confusing or conflicting information to their operators.
>
> Assuming there are sound human systems interfaces for individual systems, the Air Force can greatly reduce the burden on the human-to-human coordination if effective intersystem interoperation is established. It is the objective of our study to suggest ways for improving intersystem interoperation at the hardware and software level while keeping in mind that sometimes the human-to-human interaction is the appropriate interface. An effective system-of-systems will promote collaborative decision making and shared situation awareness among the human operators. This occurs by addressing the need for common, consistent human-system interfaces.

Clearly evident within this quote is the heart of the Network Centric Warfare paradigm and the more generic SoS problem class—interoperability. As the advisory board highlights, achieving capabilities through system interoperability requires more than just developing integrated networks between all the engineered systems. Operational evidence is beginning to show that much information being relayed between all the human and robotic systems can "bog down" the overall performance and reduce the efficiency [10]. What is the optimal trade-off for cohesion between the human and engineered systems and the type of information layer that will satisfy the needs? In this vein, the sources of complexity in an SoS must be placed in context using a taxonomy to be presented in the section "Methodology Outline." Further, in the face of these sources, representing the structure of organization is key to managing complexity.

22.5 Optimization

The central motivation for the SoS concept was to bring together systems to achieve some desired capability not possible by the systems operating in isolation. We are interested in an SoS because we perceive that "the whole is greater than the sum of the parts"; proper connection and collaboration among systems can produce something new and useful. Thus, design (synthesis of feasible and viable alternatives) and optimization (selecting the best of all possible SoS implementations) is immediately relevant. However, optimization in an SoS context is especially difficult. There is a wider array of complications with which one must contend, such as shifting requirements, operating uncertainties, new systems emerging, old systems failing, participant perspectives, and so on. Additionally, conflicts may exist between optimization at an individual system's level and optimization at the SoS level; optimal operating conditions at one level may conflict with those at another level. Since the SoS likely hosts diverse participating stakeholders (at several levels of hierarchy), there will correspondingly be a multi-objective problem, and thus non-dominated solutions in which each participant might have to sacrifice some individual gain for the purposes of SoS gain. These Pareto optima represent goals for those seeking to establish the characteristics of a successful SoS. Such a formulation might also be cast via the model developed by Mistree [11], among others.

In the face of these complications, the goal of optimization in an SoS methodology must emphasize mapping versus pointing and decision support versus decision synthesis. We have often used the analogy of a minefield, in which knowing the regions one should *not* traverse is more important than finding a particular trajectory that seems optimal [12]. Particular expressions of optimization that will be explored later in this chapter emphasize the formulation of objective functions that reflect the SoS problem nature, such as:

Reachability—to what extent can new capabilities be achieved by the SoS under various levels of constraint relaxation?

Scalability—do such capabilities persist, or scale, as demands increase?

Robustness—over an ensemble of plausible scenarios [13], which types of operations, network connectivity, etc., in an SoS lead to preferred results?

Versatility—are unexpected performance gains possible via introduction of a new technology, policy, procedure, organizational design, and/or information flow?

Taken together, these objectives point to *sustainability*, and they correspond quite closely to the main features described by those seeking to develop sustainable ecosystems, both in the traditional natural/physical setting [5] and in the context of information [14].

Understanding complexity in and optimization of an SoS is an endeavor in its infancy. Researchers must develop a robust intellectual construct of SoS, its relation to adjacent concepts and methodologies, and the generic relevancy of progress in various application-specific efforts. While early work may be broad in its scope, it may serve to accelerate numerous more detailed tracks that will lead to tangible theories and methods that can be proven useful in application. In the limited space of this chapter, we hope to provide a glimpse of one methodology and its application for analysis of human-derived complexity in an SoS problem in one application area.

22.6 Methodology Outline

22.6.1 Framework

A framework defines and organizes the universal entities involved in a problem as well as the analyses that comprise the methodology. We have proposed and experimented with a three-phase framework shown in Figure 22.1. Each of the three phases is summarized in the following paragraphs.

The purpose of the *Definition Phase* of the framework is to characterize the customer's needs and the constitution of the SoS (systems, context, current barriers). Further, the *Abstraction Phase* maps the key entities, their organization in networks, and their evolution. Finally, the purpose of the *Implementation Phase* is to create a computational model for analysis of alternatives. In this latter phase, the execution of optimization as described in the section "Optimization" is pursued. The overarching objective of this framework is then to enable decision makers to discern, in a timely and cost-effective manner, whether various related configurations of operations, economic practices, infrastructure, policy, and technology (i.e., an SoS) together are effective over time.

22.6.2 Definition Phase: Scope and Structure

Of particular importance is the provision of structure with which the numerous and significant sources of complexity can be understood. For this purpose, a lexicon has been formed [12]. By definition, SoS problems have interconnectivity, often set in a hierarchical manner. The lexicon enumerates these interacting levels of various components with Greek letters. Starting from α, the levels proceed upward, leading to γ, δ, or higher depending on the complexity of the problem and the depth of analysis desired. Further, a β-level constituent represents a network of α-level entities; a γ-level constituent represents a network of β-level entities, and so on.

In addition to these hierarchical labels representing levels of complexity, a set of scope dimensions are defined. These dimensions highlight the trans-domain aspects of SoS, since not all entities within the levels are similar in their basic character. Thus, to properly distinguish constituent systems, they are categorized primarily into Resources, Operations, Policy, and Economics (ROPE). Each of these dimensions independently comprises the previously described levels, thereby completing the SoS lexicon. The categorization of the levels lends clarity in dealing with the different facets of the problem while maintaining the lucidity provided by the levels. The relationship between the categories and the levels can be

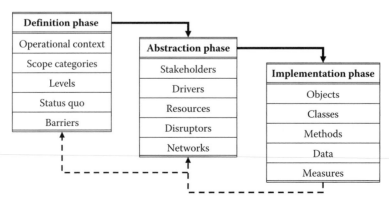

FIGURE 22.1   The three phases of the system-of-systems engineering framework.

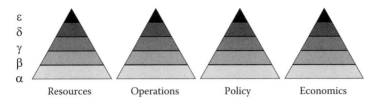

FIGURE 22.2 Unifying ROPE lexicon: scope dimensions and level hierarchy in an unfolded pyramid.

conceived as a pyramid, indicating that the number of systems decreases at higher levels. This pyramid representation is shown in Figure 22.2.

22.6.3 Abstraction Phase: Models

While a systematic representation of scope and structure is crucial, the ability to characterize the generic organization and workings of an SoS problem that is amenable to analysis is the next required step. This is embodied in the process of abstraction. Further, the abstraction of the SoS of interest must be crafted in a way that exposes appropriate level(s) and categorical scope dimensions. A taxonomy has been proposed to guide abstraction [15] consisting of a 3D space characterized by system type, autonomy, and connectivity, as illustrated in Figure 22.3.

Analysis methods must be appropriate for the *system types* (S-axis) that constitute the SoS. Some SoS consist predominantly of technological systems—independently operable mechanical (hardware) or computational (software) artifacts. Technological systems have no purposeful intent; these resources must be operated by, programmed by, or activated by a human. Other SoS consist predominantly of

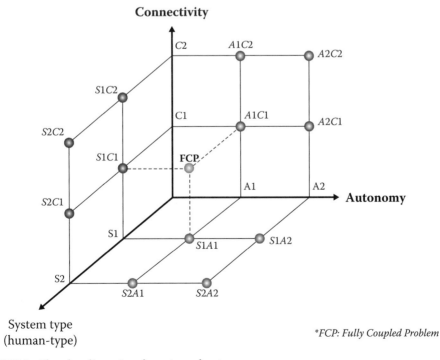

FIGURE 22.3 Three key dimensions for system-of-systems.

humans and human enterprise systems—a person or a collection of people with a definitive set of values. The second SoS dimension is the *degree of autonomy* (A-axis) over the entities by an authority. This relates to Maier's discussion of operational independence and managerial independence of systems within an SoS. (When arrayed by the degree of control/autonomy, a collection of systems with operational, but limited managerial, independence is sometimes referred to as a system-of-systems, while a collection of systems with little central authority is a federation of systems [3].) Finally, systems involved in an SoS are *interrelated and connected* (C-axis) with other (but likely not all) systems in the SoS. These interrelationships and communication links form a network. A key focus for design methods research in an SoS context lies in analysis and exploitation of interdependencies (i.e., network topology) in addition to the attributes of systems in isolation.

These dimensions serve as a taxonomy to guide the formulation and analysis of the SoS design problem. A particular SoS can therefore be considered as a "point" or "region" in the 3D space formed by the aforementioned dimensions as axes. Based on its location in this space and other indicators of particular problem structure, the approach and methodology necessary for analysis and design can be more intelligently selected. For instance, an SoS composed mostly of complex engineered systems with high autonomy (near point A2C2 in fig. 22.3), for example, a collection of unmanned aircraft, satellites, and unattended ground sensors linked by communication, would require modeling appropriate for autonomous engineered systems. Alternately, an SoS problem which lies at the S1 level or beyond demands modeling that treats directly the influence of human behaviors. These socio-technical systems require modeling approaches such as game theory and agent-based modeling that capture human decision-making. Further, in such cases when the taxonomy indicates that a particular SoS has a heterogeneous system type and thus a significant role of human behavior, complexity sources peculiar to these human behaviors must be addressed.

The following notional example, depicted in Figure 22.4 and using the previously introduced lexicon, illustrates the manifestation of several of these sources of complexity in an SoS. The α-level comprises the most basic components of the SoS. The result of the interactions at an α-level are felt at the corresponding β-level. Hence, emergence is detected at the β-level and higher. In addition, there is evolution at play in any given β-level during any interval of time. As shown in Figure 22.4, the different α-level components are nodes in a β-level network and the connectivity may shift over time (t_1, t_2, and t_3). There is an undirected relation between 1 and 2 at time t_1, while there is only a one-way relation between them at time t_2. Also, new entities appear and existing entities depart over time. Due to these changes among different α-level constituents, the performance of the β-level entities is altered by both evolution and emergence. New configurations lead to altered interactions between the α-level constituents (emergence) and this emergence in turn affects the course of the future makeup of the β-level (evolution). Thus, evolution and emergence are mutually influential and are manifested at each level.

Relative complexity between two SoS may differ for each level. One SoS may have simplistic organizational structure (lower complexity) at its β-level but its α-level systems are more complex than those of another SoS. This multilevel aspect is especially important from a computational complexity perspective [7]. Integration of high-fidelity analysis models across multiple layers of abstraction is impractical, and a more refined tack is required that is selective in which information is appropriate. This is well explained by Nobel Laureate Herbert Simon, who proposed pseudo-decomposable systems as a means to structure and manage complexity. Simon asserts that "resemblance in behavior of systems without identity of the inner systems is particularly feasible if the aspects in which we are interested arise out of the *organization* of the parts, independently of all but a few properties of the individual components" [8].

22.6.4 Models, Theories, and Tools

22.6.4.1 Modeling Connectivity: Network Topology Analysis and Network Science

The aspect of connectivity, as we have seen in the lexicon and taxonomy, is central to the traits of an SoS. Connectivity underpins the goal of interoperability in that it determines the interaction of two

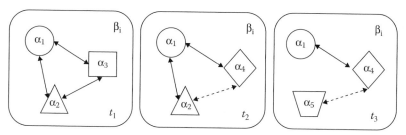

FIGURE 22.4 Notional example of a system-of-systems.

or more systems in a unique way. The recent developments in modern network science shed light on capturing the structure of connectivity and thereby help us understand the evolutionary mechanism and emergent behavior. The examination of network topologies and the dynamic processes taking place on networks enables the prediction of performance of the SoS as it evolves. Studies in vastly different domains—biology, computer science, sociology, etc.—have already produced tremendous insight in recent years [16, 17, 18].

Network science examines connectivity (links) between the entities (nodes) in a topology through various statistical properties, some of which are described here. For instance, the *degree* of a node is the sum of all links associated with a node, and the *degree distribution* represents the topology of a network. Further, *average shortest path* reflects the efficiency of propagating information across the network, whereas *average clustering coefficient* measures a network's cohesiveness or cliquishness. These properties (and others) are highly relevant to understanding the robustness, vulnerability, and overall efficiency of the network of systems [17]. Further, a study of the properties of the networks can help determine preferred patterns of the connectivity in an SoS, a topic we return to later in this chapter. We also highlight aspects of the air transportation study. The long-term intent is to build upon this knowledge base to create new, integrated network topological measures and apply them as design objectives to manage and improve the performance of the SoS.

22.6.4.2 Modeling Behavior via Agent-Based Simulation

While network modeling is useful in representing the connectivity at the different levels in an SoS, the heterogeneity of various systems involved—especially the individuals and organizations that have a stake in the SoS exhibited in the operations, economic, and policy dimensions—must also be addressed. To achieve this, it is necessary to employ modeling methods that reflect the competition and cooperation driving the stakeholder behavior and determine their actions to manipulate the resources within the SoS. This represents a departure point in the area of design theory, where the emphasis often lies on simply understanding the influence of human preference on a design. For SoS, we must incorporate human preference and behavior patterns explicitly inside the problem boundary along with the engineered systems that must be designed.

Agent-based modeling (ABM) has emerged as an approach of choice in this setting. ABM employs a collection of autonomous decision-making entities called *agents* imbued with simple rules of behavior that direct their interaction with each other and their environment. Agent functionality is quite flexible, as indicated by the eight different classifications shown in Figure 22.5 (generated based on discussion in [19]). In a complex ABM environment, a single agent can actually encompass more than one of these classifications. For example, a human modeled as a complex system can actually exhibit all eight characteristics in the course of daily routines. Typically, as models grow in complexity so will the number of characteristics that the constituent agents will exhibit and the types of interactions that they will have with other agents and their environment.

While the mathematical representation of individual agent rules is often quite simple, the resultant system-wide behavior is often more complicated and surprising, and thus instructive [20]. In ABM, the

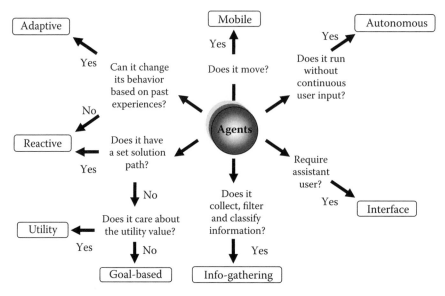

FIGURE 22.5 Classification of agent model behaviors.

ultimate goal is not to definitively prove an outcome, but to understand the processes and patterns that may appear in a modeled set of systems [21]. The mixture of human and technological components of an SoS exhibit complications that cannot be defined by physical laws. Thus, the use of ABM targets disparate interactions, which is essential in uncovering the emergent behavior at each level. Incorporating these results into the connectivity analysis would provide a holistic understanding of the connectivity amongst the heterogeneous systems involved in the SoS.

22.6.5 Integration and Implementation

22.6.5.1 Optimization Problem Formulation

We can, upon reexamining the simple example of Figure 22.4, establish a basic and generic formulation of optimization, which can subsequently be used to expose the several sub-classes of problems introduced in the section "Optimization." The components of this formulation reflect the lexicon put forth in the section "Methodology Outline" and are as follows:

Objective function: maximize capability achieved of the SoS
$\text{Max } J_{SoS} = F(J_{ij})_{i,j=\alpha}^n$ over x and subject to $\psi_{SoS} = (\psi_{ij})_{i,j=\alpha}^n$
Subject to constraints (x_{co}):   side constraints on x, and set of external performance constraints
Boundary conditions: initial conditions, x_0; desired final conditions, x_f
$\Psi = (x_0, x_f, x_{co})$
Design variables: $x = (x_r, x_h, x_p)$
x_r = engineered system (resource) capabilities
x_h = human/organizational system behaviors
x_p = policies that govern interoperability and restrict connectivity

where the subscripts i and j indicate the workings within a particular level ($i = j$) and the interplay between the levels i and j ($i \neq j$).

The availability of analytical relations to determine function evaluations for objectives and constraints is highly unlikely for most SoS problems. Further, a large number of degrees of freedom exist within a real-world SoS operation. Thus, implementation of modeling and simulation tools and techniques, especially those tailored for the analysis of connectivity and the diverse motivations for participating or influencing the SoS, must be integrated in a manner that allows multiscale analysis.

22.7 Exploration-Oriented Solution Approach

An approach for integrating the modeling approaches for SoS studies into an optimization setting is provided in Figure 22.6. The evolution of the network topology representing an SoS's connectivity is determined by the actions of stakeholders using ABM. Rule sets that represent possible sets of behaviors of stakeholders in the system can now be added to the arsenal of possible design or constraint parameters. This rule set (x_h) along with policies that shape them (x_p) can then be combined with resource capabilities establishing the design problem to be analyzed. Such problems must be evaluated over multiple scenarios that are of interest in a particular application domain. Optimization in this manner exemplifies that the focus in a SoS problem is not on point predictions as the ultimate measure of success, but instead on patterns of good solutions across an ensemble of possible scenarios.

In this approach, a network evaluator is employed as a quantitative tool to explore the connectivity of the constituents of the SoS and, in turn, evaluate the topology developed using ABM. The growth of the network topology may lead to well-known configurations, such as the exponential random network or scale-free network models [17], that directly enable characterization of some performance metrics on the networks. However, topologies belonging solely to either model might not be the best candidates for every problem. By incorporating knowledge derived known from the topology type (e.g., the random network is resistant to targeted disruption while the scale-free network is tolerant of random disruption [22]), the network evaluator assesses the overall performance of the networked systems, granting the designer a capability of shaping desirable network topologies. Tailored versions of objective functions and constraints can be developed and employed to assess whether or not a particular topology is competent to meet the design requirements. Also, in the presence of a variety of players (stakeholders) with

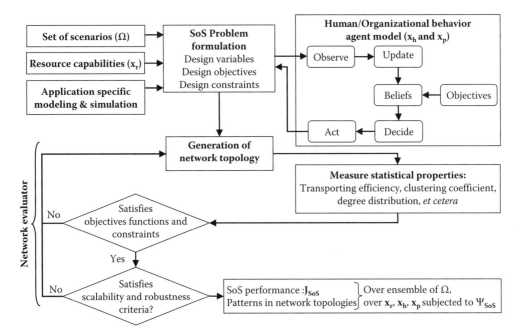

FIGURE 22.6 Exploration-oriented SoS optimization approach.

their own objectives, a multi-objective optimization naturally arises to generate topologies that concurrently optimize the individual systems and the operation network formed by them.

Finally, the framework must support important design-space exploration tasks to help impose checks and balances on qualified topologies to finally approve (or reject) them as the most suited solutions for an SoS problem. In particular, the aforementioned optimization problems of reachability, scalability, robustness, and versatility are emphasized in the approach:

Reachability—which behaviors, operating on which network topologies, allow achievable SoS capabilities within a range of possible engineered system capabilities?

Robustness—which SoS capabilities are vulnerable in the face of particular classes of network disruption, e.g., random and targeted?

Scalability—will a significant change in demand on the SoS produce a dramatic change in performance? Which interactions in lower level systems underlie this phenomenon?

Versatility—what "surprise synergies" arise when legacy systems are combined in new ways, or with new policies and procedures?

22.8 Insights from Air Transportation Example

National and global air transportations systems are undergoing profound transformation as the combined pressures of rising operational costs, increased demand, and growing delays stress the systems. In this brief section, through the air transportation example, we highlight aspects that explicitly tie together the concepts of human behavior and sources of complexity in an SoS. A more detailed report on the application of the framework and methodology described in Section 22.6 to air transportation is summarized in [23] and [24]. The overarching objective in this work is to identify robust architectures that could transform air transportation from its current strained state to one that better scales with shifts in demand and better recovers from disturbances. The information presented in Table 22.2 describes air transportation as an SoS using the lexicon introduced earlier in this chapter. The order numbers at each level indicate the relative number of entities that exist at that level.

Individuals, and the organizations in which they participate, influence the SoS through application of self-interests and perspectives (via operational and managerial independence). Dynamics from this influence take place at multiple levels and under multiple time scales. For example, humans can operate and implement policies for α-level systems. In air transportation, these roles include pilots, maintenance

TABLE 22.2 SoS Specification for Air Transportation

	Level	← System of Systems Dimensions →			
		Resources	**Operations**	**Economics**	**Policy**
Base Level	α ($\vartheta\ 10^6$)	Aircraft, Tower	Pilot/Crew Deployment, Maintenance Schedules	Fuel Price, Investments	Type Certification, Flight Procedures
↑ Network of Networks ↓	β ($\vartheta\ 10^4$)	Airport	Airline	Fuel Market, Labor/Union Costs	Airport Traffic Mgmt, Noise Policies
	γ ($\vartheta\ 10^2$)	Air Transportation System	Commercial Air Operations	Airline Industry	Air/Ground Safety, Accessibility
	δ ($\vartheta\ 10^1$)	National Transportation System	Operators of Total National Transportation System	Overall Transportation Forecasts/Market	National Transportation Policies
	ε ($\vartheta\ 10^0$)	Global Transportation System	Global Operators in the World Transportation System	WTO, Global Marketplace	Global Transportation System Policies

crew, inspectors, and so on. Further, actions at these lower levels of hierarchy tend to evolve on a short time scale; typically, they are near-term cognitive actions of a practical variety in response to the situation at hand and are operational in nature. Complexity at this level may arise here from ambiguous instructions, differing skill levels in distributed teams [25], goals and strategies, and detection [26] and diagnosis of system failures [27]. Humans can also participate as managers of α-levels systems, or parts of organizations at even higher levels. In air transportation, these roles include airlines, air traffic regulatory bodies (e.g., Federal Aviation Administration, FAA), labor unions, and so on. Actions of these organizations tend toward the longer-term influence based upon considered assessment of organizational goals; these decisions may establish well-intentioned policies that unintentionally restrict the existence of promising solutions.

The specification in Table 22.2 spans the hierarchy of systems across the dimensions of the air transportation enterprise. In fact, within it are numerous SoSs, for which particular boundaries can be drawn. The interactions of humans and human enterprises at various levels of an SoS exhibit the same kind of self-organizing behaviors described in the previous section on complexity. Interactions of this kind introduce a multiplicity of perspectives that arise from "institutional frames of reference and associated operational context scope that system users and system designers implicitly assume in their behavior and their designs" [28]. These perspectives form a backdrop for the operational assumptions adopted for the given system and enable (or prohibit) its integration with other systems in the system-of-systems. When systems come together (to exchange mass, energy, or information), the perspectives under which they were instantiated may conflict, requiring human intervention for resolution. These so-called trigger events force a reorganization of convergence protocols, assumptions under which the systems operate, the context in which they interact, or some combination of these. This reorganization is nothing more than a manifestation of the self-organization of a complex system. Humans effect this restructuring of the SoS to resolve this interoperability challenge, but require additional information (about unstated assumptions, perspective, and so forth) in order to respond effectively. As a result, this increase in information also increases the complexity of the overall SoS.

In an attempt to capture some of these interactions involving human participants at various levels of organization in the SoS, we have developed a National Transportation System Simulation (NTSS) to examine how transportation stakeholders (modeled as agents, such as service providers, infrastructure providers, regulators) act to evolve a network under various scenarios. Its basic architecture is displayed in Figure 22.7, and its genesis from the generic solution approach model shown in Figure 22.6 is readily apparent. Each agent goal is currently very simplistic, but additional beliefs and objectives can be implemented as needed for particular studies. The interactions between the various agents gradually build a new transportation network topology within the virtual world. Thus, new topologies evolve for the future; however, do they exhibit good network performance both in terms of capacity and robustness? Here the network analysis model plays a central role; it accepts a measure of the demand from a demand generator and each agent in the simulation acts on its self-interest driven by the predetermined goal. A network evaluator is employed to compare the simulated network to topologies that do exhibit preferred behaviors. For example, capacity and robustness can be encapsulated in the previously introduced concept of *scalability*—the characteristic that, as operations increase, network performance measures do not degrade. Results of comparative topology studies will be presented to understand which topologies (under which types of agent behavior) lead to scalability. Such findings can be used to guide technology and infrastructure investments for future air transportation systems. More specifically, the network evaluator will employ analysis of actual air transportation data and simulation of network growth using network theory.

Other researchers in the air transportation domain have pursued improved understanding of human behaviors and decision making within portions of the larger SoS dynamics. For example, traveler preference models (an α-level behavior) have been developed by Lewe et al. [29] and Xu et al. [30]. Further, models of β-level entities have been developed, such as MITRE's JetWise model of airlines [31]. Some of these studies have used agent-based modeling (ABM) to mimic the human factors and influences on the system-of-systems. ABM is from the same lineage as that of Complex Adaptive Systems [32], which

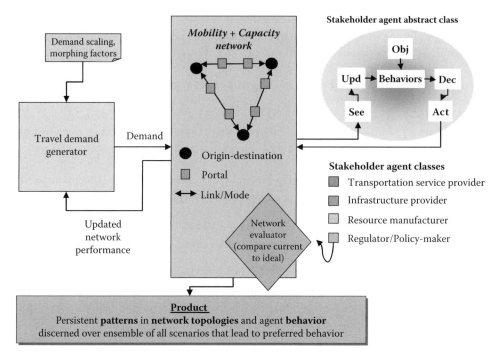

FIGURE 22.7   National Transportation System Simulation (NTSS).

represents a problem class with many of the same dynamic behaviors of SoS problems (e.g., emergence). For the particular case of air transportation, Donohue has pursued methodologies for a CAS-inspired approach [33].

22.9 Summary

A brief introduction to system-of-systems problems was presented, especially crafted as an entrée into dialogue with researchers in human system modeling. The sources of complexity related to human participation in an SoS were documented, couched in a lexicon and taxonomy for this problem class, and include heterogeneous behaviors, dynamic connectivity, and the unique perspectives (and thus interpretations) each human participant brings to an SoS. We observe that these sources reside at multiple levels and across different scope dimensions. And these peculiarities give rise to the need for unique formulations for optimization, including objectives such as reachability, scalability, and versatility. A challenge for effective design in a system-of-systems context is to make these perspectives transparent during interactions and to characterize performance on dynamic networks. In particular, a better understanding of the mixing between short- and long-term time scales for human influence on an SoS must be obtained.

Progress along these areas in the domain of air transportation research was described to lend more clarity to the general constructs via example. Clearly, however, a need remains for an increased ability to understand and represent the complex human behaviors involved in a human-technical SoS. We recommend collaborative explorations between engineered system architects and designers with human system modeling experts in order to increase the effectiveness of system-of-systems that may involve significant human presence in their constitution as described via the proposed taxonomy. Perhaps part of this exploration could even reexamine the rich heritage of this kind of cross-fertilization (e.g., cybernetics, general systems theory, and the like).

References

1. Keating, C., et al, "System of Systems Engineering," *Engineering Management Journal*, vol. 15, no. 3, 2003, pp. 36–45.
2. Rouse, W.B., "A Theory of Enterprise Transformation," *Systems Engineering*, vol. 8, no. 4, 2005.
3. Sage, A.P., and Cuppan, C.D., "On the Systems Engineering and Management of Systems of Systems and Federation of Systems," *Information, Knowledge, Systems Management*, vol. 2, 2001, pp. 325–345.
4. Ottens, M.M., Franssen, M. P. M., Kroes, P. A., and Poel, I. v. d., "Systems Engineering of Socio-technical Systems," Proceedings of INCOSE International Symposium, Rochester, NY, 2005.
5. Gustavsson, R., and Fredriksson, M., "Humans and Complex Systems: Sustainable Information Societies," *System Approaches and Their Application: Examples from Sweden*, edited by M.-o. Olsson & G. Sjostedt, Kluwer Academic Publishers, The Netherlands, 2004, pp. 269–289.
6. Maier, M.W., "Architecting Principles for System-of-Systems," *Systems Engineering*, vol. 1, no. 4, 1998, pp. 267–284.
7. Casti, J., *Connectivity, Complexity, and Catastrophe in Large-Scale Systems*, John Wiley & Sons, New York, NY, 1979, pp. 105–106.
8. Simon, H., *Science of the Artificial*, 3rd Ed., Massachusetts Institute of Technology Press, Cambridge, MA, 1996.
9. USAF Scientific Advisory Board, "System-of-Systems Engineering for Air Force Capability Development," Quick Look Study, October 23, 2004.
10. Talbot, D., "We Got Nothing Until They Slammed into Us," *MIT Technology Review*, November, 2004, pp. 36–45.
11. Mistree, F., et al, "Learning How To Design: A Minds-On, Hands On, Decision-Based Approach," 1995, URL: http://cohomology.princeton.edu/books/Computers/Learning_How_ to_Design.pdf
12. DeLaurentis, D.A., and Callaway, R.K., "A System-of-Systems Perspective for Future Public Policy," *Review of Policy Research*, vol. 21, no. 6, Nov. 2004.
13. Lempert, R., Popper, S.W., and Bankes, S.C., "Shaping the Next One Hundred Years: New Methods for Quantitative, Long-Term Policy Analysis," The RAND Pardee Center, Report MR-1626-CR, Santa Monica, CA, 2003.
14. Polzer, H., "Creating an Evolutionary Creating an Evolutionary Environment to Foster Capability-Environment to Foster Capability-Oriented Interoperability," presentation to the NDIA, [online database] URL: www.dtic.mil/ndia/2002interop/polzer.pdf [cited 15 April 2007].
15. DeLaurentis, D.A., and Crossley, W.A., "A Taxonomy-Based Perspective for System-of-Systems Design Methods," *2005 IEEE International Conference on Systems, Man and Cybernetics*, vol. 1, October 10-12, 2005, pp. 86–91.
16. Newman, M.E.J., "The Structure and Function of Complex Networks," *SIAM Review*, vol. 1, March 25, 2003, URL: http://arxiv.org/PS_cache/cond-mat/pdf/0303/0303516.pdf
17. Barabasi, A.L., and Albert, R., "Emergence of Scaling in Random Networks," *Science*, vol. 286, no. 5439, Oct. 1999, pp. 509–512.
18. Dorogovstev, S. N., Goltsev, A. V., and Mendes, J. F. F. "Ising Model on Networks with an Arbitrary Distribution of Connections." *Physical Review E*, 66:016104{1{016104{5, July 8 2002.
19. Coppin, B., *Artificial Intelligence Illuminated* (Chapter 19: Intelligent Agents), Jones and Bartlett Publishers, Sudbury, MA, 2004.
20. Bonabeau, E., "Agent-based Modeling: Methods and Techniques for Simulating Human Systems," *Proceeding of the National Academy of Sciences of the Unites States of America*, vol. 99, suppl. 2, May 14, 2002, pp. 7280–7287, URL: www.pnas.org/cgi/doi/10.1073/pnas.082080899
21. Bonabeau, E., "Agent-based Modeling: Methods and Techniques for Simulating Human Systems," *Proceeding of the National Academy of Sciences of the United States of America*, vol. 99, suppl. 2, May 14, 2002, pp. 7280–7287, URL: www.pnas.org/cgi/doi/10.1073/pnas.082080899

22. Albert, R., and Barabási, A.-L., "Statistical Mechanics of Complex Networks," *Reviews of Modern Physics*, 74, 2002, pp. 47–97.

23. DeLaurentis, D., and Fry, D.N., "Understanding the Implications for Airports of Distributed Air Transportation Using a Systems-of-Systems Approach," accepted in *Transportation Planning and Technology*, publication forthcoming in Summer 2007.

24. DeLaurentis, D., Han, E.-P., and Kotegawa, T., Establishment of a Network-based Simulation of Future Air Transportation Concepts. 6th AIAA Aviation Technology, Integration and Operations Conference (ATIO), Wichita, KS, 25–27 September 2006, AIAA 2006–7719

25. Caldwell, B., "Analysis and Modeling of Information Flow and Distributed Expertise in Space-Related Operations," *Acta Astronautica*, vol. 56, 2005, pp. 996–1004.

26. Patrick, J., James, N., and Ahmed, A., "Human Processes of Control: Tracing the Goals and Strategies of Control Room Teams," *Ergonomics*, vol. 49, 2006, pp. 12–13, pp. 1395–1414.

27. Rasmussen, J., and Rouse, W.B. (Eds.), *Human Detection and Diagnosis of Systems Failures*, New York: Plenum Press, 1981.

28. Polzer, H., DeLaurentis, D.A., and Fry, D.N., "Multiplicity of Perspectives, Context Scope, and Context Shifting Events," *International Conference on System of Systems Engineering*, San Antonio, Texas, April 2007, 16–18.

29. Lewe, J-H., "An Integrated Decision-Making Framework for Transportation Architectures: Application to Aviation System Design," Ph.D. Dissertation, Georgia Institute of Technology, Atlanta, GA, May 2005.

30. Xu, Y., Trani, A.A., and Baik, H., "Preliminary Assessment of Lower Landing Minima Capabilities in the Small Aircraft Transportation System Program," *Transportation Research Record: Journal of the Transportation Research Board*, no. 1915, 2005, pp. 1–11.

31. Liguori, P.A., and Niedringhaus, W.P., "The JetWise Model of National Airspace System Evolution: Model Description and Validations Results," *Transactions on Simulation*, May 1, 2003.

32. Waldrop, M. M., *Complexity: The Emerging Science at the Edge of Order and Chaos.* Simon & Schuster, 1992.

33. Donohue, G., "Air Transportation Is a Complex Adaptive System: Not an Aircraft Design," *AIAA International Air and Space Symposium and Exposition: The Next 100 Years* Dayton, OH, July 14–17, 2003, AIAA-2003-2668.

23

Verification and Validation of Human Modeling Systems

Aernout Oudenhuijzen,
Gregory F. Zehner, and
Jeffrey A. Hudson

23.1 Introduction

Today, human modeling systems (HMSs) are considered a basic element in crew station design. Consequently, HMSs not only need to be realistic, but their results must be reproducible and verifiable in order to be useful tools for design and evaluation purposes. Also, any study performed with digital human models must demonstrate its cost-effectiveness. However, the verification and validation process for the use of HMSs as tools for cockpit ergonomic assessment has not been thoroughly completed.

The aim of this report is to answer the following question: Are HMSs usable for accommodation studies on crew stations? The answer to this question was obtained through the present verification and validation study conducted jointly by the U.S. Air Force and TNO Human Factors. Oudenhuijzen and Hudson (2000) defined procedures for verification and validation of HMSs in the front crew station of the F-16D. The core methodology of the verification and validation effort was a comparison of results from a live subject field study to those obtained via HMSs in the digital realm. The verification and validation procedure consisted of two separate phases:

In the verification phase, HMS anthropometry was compared to the corresponding body measurements on eight subjects.

In the validation phase, the overall results of the HMS functions were compared to corresponding results of the field test with human subjects. Crew accommodation issues, such as fit, reach, and vision, were the primary focus of the validation effort.

Subsequent studies may focus on other aspects such as joint mobility, motion paths, etc.

The following HMSs were examined in this verification and validation project:

RAMSIS (version 3.5)
Jack (version 2.1)
Safework (PTC/Safework option)
BHMS (Boeing Human Modeling System) (version 3.5.2)
COMBIMAN (version 11, AutoCAD r13)

All five HMSs were tested following the defined procedure. The following sections report the data analysis, results, and conclusions.

23.2 Methods

This section will elaborate on the methods used for verification and validation. The field test data used in the analysis were obtained using a predetermined method (Zehner, & Hudson, 2002). These field test data will be compared to the corresponding data obtained digitally. However, a direct comparison would be unfair. Instead, the field test data include standard error margins and serve as a baseline for comparison to the digital study results.

23.2.1 Verification

The verification phase consisted of several steps. The first step was to generate the mannequins using the HMSs. Therefore, the input variables had to be selected from the 40 CAESAR variables (Robinette, 1997). Not all HMSs required the same input variables, however. For instance, some of the HMSs did not make use of the arm length measurements provided, while others needed input such as gender, demographic profile, age, weight, or somatotype.

The second step was to measure the anthropometric variables on the mannequins generated in the HMSs. This was done using the CAD tools provided with the HMSs and resulted in an extensive list of digitally obtained anthropometric data. Due to some of the HMS mannequin shapes it was not always possible to obtain certain anthropometric measurements. Next, the error margins, mentioned above, were determined. Two error margins were used for the verification (see fig. 23.1):

Error margin 1 (+/− S): consisted of the standard error (SE) of the mean of the measured subject anthropometric data

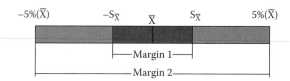

FIGURE 23.1   The two error margins used for the verification.

Error margin 2 (+/− 5%): was the 10% range of deviation around input variable (plus or minus 5% of live measured mean).

The last step was to determine mannequin deviation from the live subject measures using the two error margins.

23.2.2 Validation

The mannequins created were positioned in a CAD model of the forward F16D crew station (the same crew station was used in the field tests).

Following the procedure specified by Oudenhuijzen and Hudson (2000), the mannequins were positioned in the ACES II ejection seat model. Two sets of data were obtained using each of the seated HMS mannequins:

The successes or misses (including miss distances) for reaches to 14 controls;
F16 accommodation limits using the mannequins as "rulers" or "measuring devices."

Both sets of data were compared to the field test results.

23.3 Results

23.3.1 Verification

Table 23.1 shows the total percentage of all variables examined (see Table 23.2), in each HMS, which fell within the two error margins.

(If a variable fell within Margin 1, by definition, it also fell within Margin 2.)

Safework is the best performing HMS for both error margins, followed by RAMSIS and COMBI-MAN. Jack and BHMS have the worst (similar) values for both error margins.

Table 23.2 shows the absolute mean error percentage for all five HMSs for a set of measurements that are often used for anthropometric accommodation studies of crew stations. The gray cells represent variable measures required for mannequin generation, while the other cells represent values generated by the HMS during mannequin generation. Some cells are empty due to the mannequin's shape or postural restrictions (COMBIMAN could not assume a standing posture, therefore, standing measurements could not be obtained).

It is clear in Table 23.2 that percentage error is greater in measures that were not part of the original input variable list (unshaded cells). For example, all Safework errors are close to 0% except for the generated measurement of elbow height with an error of 2.1%. Another fact is that some HMSs actually *generate* error from the original measurement value that was given as part of the mannequin input list. This means that during the process of mannequin generation the original anthropometric values were somehow changed (i.e., the input does not equal the output). RAMSIS was the only HMS to include

TABLE 23.1 The Total Percentage of Examined Variables Falling in the Two Error Margins

	Margin 1	Margin 2
BHMS	20%	65%
COMBIMAN	24%	79%
Jack	17%	64%
RAMSIS	60%	80%
Safework	74%	94%

TABLE 23.2   The Absolute Mean Error Percentage for All 5 HMSs

		Safework	COMBIMAN	Jack	BHMS	RAMSIS
1	Acromial Height, Sitting	0.0%	2.7%	0.9%	4.3%	2.3%
10	Buttock-Knee Length	0.0%	3.4%	8.4%	0.8%	3.2%
12	Crotch Height	0.0%		8.4%	0.9%	
13	Elbow Height, Sitting (Rt.)	2.1%		7.5%	7.7%	6.7%
14	Eye Height, Sitting	0.5%	1.0%	1.2%	3.3%	3.6%
16	Foot Length	0.0%	1.0%	0.8%	0.4%	0.1%
18	Hand Length	0.0%	0.6%	4.0%	6.0%	5.6%
22	Hip Breadth, Sitting		9.0%	1.6%	7.9%	
25	Knee Height, Sitting	0.0%	5.8%	3.8%	0.8%	0.0%
27	Shoulder (Bideltoid) Breadth	0.4%	14.6%		5.7%	0.1%
28	Sitting Height	0.3%	2.2%	1.0%	2.7%	0.0%
29	Stature	0.0%		0.5%	0.4%	0.0%
33	Thumb Tip Reach	0.3%		4.7%	2.0%	
37	Waist Circ., Preferred					0.0%

waist circumference values in its listed output. The difference between the modeled waist circumference and that of the live subject is not known for the other HMSs.

23.4 Discussion

A large range in anthropometric fidelity was observed during mannequin generation in the five HMSs. For the anthropometric dimensions we examined, Safework was the best performing HMS, closely followed by RAMSIS. The difference between RAMSIS and Safework is attributed to differences in the measurement-input lists for constructing the digital models. Specifically, RAMSIS required measurement values that were not taken on the subjects. BHMS and Jack produced the largest discrepancies between subject measure input and mannequin measurements. COMBIMAN differences were slightly smaller than BHMS and Jack. However, COMBIMAN was the only HMS not able to provide data on standing subjects.

Some of the HMSs can produce output lists of the anthropometric measurements of their generated mannequins. There were instances where these values from the listings differed from the data actually measured on the mannequins using CAD tools. To make proper anthropometric verification comparisons in the future, it is advised to standardize on a set of critical anthropometric measurements.

Most of the anthropometric errors found in the digital human mannequins are the result of faulty anthropometric engines, which include the modification or use of original input measurement values, as well as in the regression equations to produce the non-inputted measures. The act of measuring anthropometric dimensions on mannequins in CAD packages could also add error. It is fairly easy for a trained person to palpate the acromion on a live subject, in order to obtain acromial height. However, this is not possible on a digital mannequin. This difficulty is compounded when, as in some cases, there is no HMS offering landmark locations. It would be very useful for a standard and predetermined set of landmarks to be available on all HMSs. Postural differences (e.g., in pelvis tilt) between the generated mannequin and the original subject during measurement could also result in discrepancies. Therefore, posture also needs to be standardized for verification purposes.

The factors mentioned thus far (i.e., input variables, anthropometric engines, standardization of landmarks, and posture) are all much more easily addressed than the issue of tissue compression. Tissue compression, or redistribution of subject soft tissue when moving between postures or in contact with other surfaces, is not taken into account by any of the models and has probably contributed error in this study.

When the intent of human modeling is to estimate and assess fit, vision, and reach in a workstation, as in cockpit accommodation, then this is a very important factor and its proper modeling must be assured.

Finally some HMSs mannequins were wearing clothing, while the subjects were not. The resulting additions to certain digital measurements, as a result of added clothing, had to be subtracted when measured in the digital realm (for example the boots in COMBIMAN and BHMS). However, in some cases the effect of the added clothing for some HMSs was not documented and therefore is unknown. It is possible that too much or too little has been subtracted from the digitally derived data.

23.4.1 Validation

The validation phase was comprised of two tasks; they were:

Determine reach capability for each mannequin or subject
Determine the overall physical accommodation of forward crew station in the F-16D related to both large and small pilot issues

For both of these tasks the results obtained with traditional methods were compared to those obtained in the digital realm with the HMSs.

23.5 Reach Task: Matched Reach Successes and Misses, a Qualitative Comparison

Table 23.3 reports the number of baseline reach successes for the field test subjects, combined, as well as the comparable successes (reaches to the same 14 controls) for their digital counterparts (combined by HMS). All differences in tallies, between the HMSs and baseline, under both Zone 1 and Zone 2 reach conditions, represent misses being incorrectly reported for baseline successes. There were some instances, however, where the opposite was reported (mannequin successes were assigned to baseline misses); these are marked with parentheses. Zone 1 errors are probably the result of poor arm length modeling or incorrect initial mannequin positioning.

All Zone 3 reaches to the 14 controls were successes for the live subjects, and were likewise modeled by all HMSs except for BHMS and COMBIMAN. They were not able to reach controls placed directly to the side and aft. However, the single reach algorithm in BHMS is not intended to reach controls placed aft of the frontal plane of the subject. This restriction resulted in high miss distances to three controls located aft of the shoulder plane under Zones 1, 2, and 3 reach conditions. The BHMS mannequin miss distances in question are constant within each subject, suggesting joint lockup or restricted motion regardless of increasing freedom (Zone 1 to Zone 3 reach conditions). COMBIMAN also had difficulty with these reaches: (1) The mannequin model for Art, the largest subject, missed Zones 2 and 3 to Canopy Jettison, and (2) mannequins for the four large males all missed their Zone 2 reaches to 100% oxygen.

TABLE 23.3 Correctly Matched Reach Successes

	Subject	Zone 1 Successes		Zone 2 Successes		Zone 3 Successes	
		Baseline	HMS	Baseline	HMS	Baseline	HMS
1	BHMS	86	51	102	73	112	88
2	Safework	86	83(+2)*	102	100(+4)*	112	112
3	RAMSIS TNO	86	77(+1)*	102	91	112	112
4	Jack	86	82(+1)*	102	100(+1)*	112	112
5	COMBIMAN	86		102	95	112	111

*A mannequin success was incorrectly assigned to a baseline miss.

TABLE 23.4   Correctly Matched Reach Misses

		Zone 1 Misses		Zone 2 Misses		Zone 3 Misses	
	Subject	Baseline	HMS	Baseline	HMS	Baseline	HMS
1	BHMS	26	26(+35)*	10	10(+29)*	0	(+24)*
2	Safework	26	24(+3)*	10	6(+2)*	0	0
3	RAMSIS TNO	26	25(+9)*	10	10(+11)*	0	0
4	Jack	26	25(+4)*	10		0	0
5	COMBIMAN	26		10	10(+7)*	0	(+1)*

*Mannequin misses incorrectly assigned to baseline successes.

It is important to note that, in this study, the amount of potential overreach, or excess reach past a control, was not addressed. For example, under Zone 2 reach conditions, a test subject may have been able to reach 10 cm beyond a control while the digital mannequin just missed it. If the mannequin's small miss is compared to the test subject's "success," the actual 10 cm difference in reach magnitude is ignored. However, Table 23.4 only tallies correctly matched reach misses (i.e., both test subject and mannequin had a miss to the same control), which better illustrates the portion of the data used in our quantitative comparison. For quantitative purposes, it is appropriate only to compare mannequin misses to test subject misses. The tallies enclosed in parentheses represent the number of the combined mannequin misses of the HMSs incorrectly assigned to test subject successes.

All HMSs are comparable in the number of correctly matched reach misses; however, they differ widely with respect to incorrectly matching mannequin misses to baseline successes (tallies in parentheses). BHMS had a great number of these errors, which occur when mannequins do not reach as far as their live subject counterparts, and are most likely the result of placing the shoulders too far aft in the cockpit.

23.6 HMSs Reach Misses versus Baseline Misses, a Quantitative Comparison

Table 23.5 reports mean absolute deviations (Mannequin – Baseline miss) for the HMSs by the 14 controls. The list of controls corresponds to the movement from the left side of the cockpit to the right. The mean deviations are calculated using absolute values for miss distance differences from each subject by control and reach zone. The means of these means, by HMS, are listed at the bottom. Note that a superscript 3 marks instances where there was only one value offered and it was not possible to calculate a mean.

In estimating the Zone 1 miss distances recorded in the field study, HMS accuracy is influenced mostly by a mannequin's (1) initial position in the seat, (2) reaching posture, and (3) anthropometric fidelity. Because live subjects would actually produce a distribution of miss distances, if asked repeatedly to reach toward the same control, it is appropriate to allow answers offered by the HMSs to fall within a reasonable range of the field test miss distances and still consider them "matches." Thus, deviations in Table 23.5 that are set in boldface type fell within 1 SE of repeatability test results, which were conducted under controlled lab conditions. Except for COMBIMAN, estimates for the Zone 2 miss distances were found by subtracting constants (calculated in the same repeatability experiment) from the HMS Zone 1 miss estimates. Hence, any deviation in a HMS Zone 2 estimate was also observed in its Zone 1 estimate and compounded by any discrepancy between the constant and the field test miss distance. Although COMBIMAN had a malfunctioning Zone 1 algorithm and could not be evaluated, it was one of two HMSs (BHMS is the other) that offered internal algorithms for conducting Zone 2 reach simulations. These COMBIMAN results are shown below.

TABLE 23.5   Mean Absolute Miss Distance Deviations (values in cm)

	1 BHMS		2 Safework		3 RAMSIS		4 Jack		5 COMBIMAN
Control & Direction	Zone 1	Zone 2	Zone 1	Zone 2	Zone 1	Zone 2	Zone 1	Zone 2	Zone 2
Left Down									
1 Canopy Jettison									
Left Oblique Down									
2 Throttle A–B	6.52*		0.70*						
3 Man. Pitch Override	4.57		1.11*		11.27		0.50		
4 Downlock Release	8.37		3.42		12.30		1.30		
Left Fwd Ahead									
5 Threat Warn Knob	3.30	6.05	1.97	2.98	4.35	2.84	4.84	8.34	1.68
6 Left MFD Cont.	2.82	3.06	1.76	1.78³	8.90	6.31	2.30	3.22	1.83
7 Master Caution	1.72		2.07		1.26		3.75		
Right Fwd Ahead									
8 Engine Fire	3.58*		1.68*		7.27		2.45		
9 Right MFD	2.17	3.83	3.25	3.27*	5.28	3.27	2.04	4.77	1.83
Right Oblique Down									
10 Liquid Oxygen	2.92	2.20	0.70	0.44*	12.34	15.73*	1.47	0.39*	2.57*
11 Stick									
12 Master Zeroize									
Right Down									
13 Air Source Knob									
Right Aft Down									
14 100% Oxygen									
Mean	**4.00**	**3.79**	**1.86**	**2.12**	**7.87**	**7.04**	**2.33**	**4.18**	**1.98**

Note: Values set in boldface type fall within the SE for repeatability test results.
*Single value presented since calculation of a mean was not possible.

For Zone 1 estimates, the direction of the reach to controls seemed to be related to accuracy. For BHMS and RAMSIS the best estimates were associated with controls placed directly forward, while Jack and Safework had their best miss estimates toward controls located to the oblique. Although not apparent in Table 23.3, there are trends linking mannequin size and accuracy. Jack appears to obtain more accurate estimates with the smaller mannequins, while both BHMS and RAMSIS are more accurate with the larger mannequins. The Safework results show no apparent pattern linking mannequin size and accuracy. As stated above, means of Zone 2 estimates are simply adjusted Zone 1 estimates. The repeatability experiment from which we calculated the adjustment constants also provided an interesting side note regarding the use of live subjects. Of the three small subjects that provided Zone 2 data, one of them reached incredibly better in the lab experiment than in the actual cockpit during the field study. Currently, we assume this is related to consistent coaching during the repeatability experiment. As a result, her adjustment constants actually added error to the Zone 2 estimates. However, in contrast, the smallest subject generated very good adjustment constants that paralleled her performance in the field study.

23.7 Accommodation Task

Using the digital mannequins in the HMSs, accommodation limits were estimated for the forward crew station of the F-16D. In Table 23.6, these values are compared to the USAF accommodation limits

TABLE 23.6   F-16D Accommodation Limits—Survey versus the HMSs (values in cm)

Related Anthropometry	Traditional Survey	BHMS	Safework	RAMSIS	Jack	COMBIMAN
Minimums						
Sitting Eye Height	76.7	71.6	72	72	76	76.9
a: Buttock-Knee Length	52.1[†]	59.1	62.3	62.3	62.8	53.3
b: Knee Height, Sitting	45.7[†]	51.3	54	54	53	47.5[††]
COMBOLEG (a + b)	97.8	110.4	116.3	116.3	115.8	100.8
Maximums						
Sitting Height*	99.4[†]	95.5	98	98	97	96.7
a: Buttock-Knee Length	68.8[†]	68.2	68.2	68.2	68	68.6
b: Knee Height, Sitting	62	63.2	63.2	63.2	63	61.5[††]

Notes: Values in gray cells fall within the SE of the mean for the repeatability field test. Calculation of these values includes 3.8 cm free space for canopy deformation during bird strike and a 7.4 cm for the thickness of a helmet. Italicized values were not reported through a COMBIMAN algorithm, but rather represent estimates obtained through visual inspection.

*This value is based on the absolute difference between the traditional and the digitally obtained accommodation limits.

[†] Estimate regressed from other leg dimensions.

[††] Estimate regressed from buttock-knee length.

obtained using traditional live subject methods during the USAF Cockpit Accommodation Survey. Estimates of the large subject maximum buttock-knee length limits are quite comparable for instrument panel clearance issues. The sitting height maximums, used to ensure head to canopy clearance, have a slightly larger difference. The small subject minimums have an even greater disparity in the minimum COMBOLEG value (about 14 cm difference between the HMSs and the field test), which is used to indicate adequate leg reach for rudder and brake authority. The minimum sitting eye height estimate, required to achieve the design eye angle, is, on average, 3.5 cm shorter than the traditional survey result. COMBIMANs results are very close to the field test results. This is not very remarkable since COMBIMAN offers algorithms based on field test results with subjects seated in an ejection seat. The other HMSs do not offer this function. This function offers an advantage for this study; however, this advantage may become a disadvantage, in bigger errors, when focusing at other crew systems with different seats and restraint systems.

23.8 Reverse Analysis for Small Mannequins

With respect to the results in the accommodation task, the mannequin's seated posture may be the best explanation for the disparity between most of the HMSs accommodation estimates and the traditional survey results. Relative to the live subject survey results, most of the human model estimates offer both a smaller minimum sitting eye height and a greater minimum COMBOLEG length. As an exercise, the pelvis of a small mannequin was rotated to better simulate a better live subject posture (figs. 23.2 and 23.3). This rotation moved the hip joints forward and the sacroiliac joint downward toward the seat pan. The resulting accommodation estimates, offered by this mannequin with modified initial posture, were affected. Specifically, there was a simultaneous reduction of minimum COMBOLEG required to reach rudder and brake and an increase in the minimum sitting eye height required. (As the head was pulled down with the pelvis rotation the eye height had to increase to maintain minimum vision.) An overly erect torso posture is also the most likely explanation for the larger mannequins and their overhead

FIGURE 23.2 Rudder miss with original mannequin placement.

FIGURE 23.3 Rudder miss (=0) with tilted pelvis and tissue deformation.

clearance estimates. A more relaxed posture would increase the mannequin estimates. This illustrates the importance of initial mannequin posture in generating realistic accommodation estimates in cockpit applications. Some discrepancy, however, can also be attributed to live subject tissue deformation that takes place when the posterior aspect of the thigh is pressed into the seat during rudder and brake actuation, which is not modeled in the HMSs.

23.8.1 Discussion: Validation Phase

Different errors were found in the examined HMSs. All of the models either (1) were not validated, (2) had previous validation studies that did not apply to crew station applications, or (3) had previous validation studies that could not be obtained due to confidentiality. There are a number of factors that contributed to the errors found. Anthropometric accuracy is one of the most important factors. However, it is still possible for an HMS to generate errors when starting with a mannequin that is anthropometrically perfect (i.e., Safework). Other factors contributing to error in the validation phase include: (1) joint location in the HMSs, (2) initial positioning and posturing of the mannequins, and (3) harness influence and subject flexibility and motivation. The influence of each of these separate aspects is unknown and should be investigated in successive studies.

While the anthropometry of an HMS may be correct, the proper initial positioning of the model into the ACES II seat, and the subsequent posturing during reaches against locked or unlocked reels, may not be fully correct. During the validation study the users had to visually guess at the correct initial position. For some models this included estimating spinal posture and the amount of lumbar kyphosis present in order to position the mannequin. The deformation of the buttocks and upper leg against the seat was also estimated. None of the models provided any aid in estimating initial mannequin positioning. For some of the HMSs the same lack of aid or guidance was present in selecting mannequin posture when reaching to controls. However, some of them do have a posture module, which helps the user to obtain a natural mannequin posture. The validity of the human posturing engine remains questionable, however. The difficulty of positioning and posturing the mannequin is compounded when the user has to manually model the influence of the harness restraint system. All of the guesses made by the users in the process of estimating initial mannequin posture and position certainly contribute to the errors reported for the validation phase.

In the reach validation phase the Zone 1 results had less error than Zone 2. With respect to the accommodation task, the mean deviation for maximum accommodation limits for tall subjects was 1.5 cm for sitting height, while the maximum leg length deviation was 0.8 cm (buttock-knee length). The mean deviations for minimum accommodation limits (related to small subjects) were much larger at 4.7 cm for sitting height and 9.5 cm for COMBOLEG.

For the best HMSs, the accuracy for maximum accommodation limits is within the standard error for the repeatability tests. However, the accuracy for the minimum limits is far outside the standard error of the repeatability tests for all HMSs. If the user will continue to be responsible for initial mannequin positioning and posturing, error can be reduced with the quantification of live subject tissue deformation and spine and pelvis orientation data. The observed HMS errors do not currently affect the selection of pilots for the Royal Netherlands Air Force. Arm length, which is directly associated with reach, is not used as a selection criterion. However, the HMS inaccuracy in modeling reaches may be too high for use in the U.S. Air Force accommodation assessment. Pilot selection in the U.S. Air Force uses span (the maximum distance from between the fingertips with the arms placed horizontally) as a selection criterion. It is recommended to the Royal Netherlands Air Force to follow the U.S. Air Force policy and to use span as a minimum pilot selection criterion in order to ensure the selected pilots reach to controls.

23.9 Conclusions and Recommendations

We conclude that in their present form, HMSs may yield inaccurate results when attempting to determine accommodation limits for an aircraft cockpit. However, HMS results can be drastically improved if platform-specific field studies are conducted to define initial body position and posture for the digital mannequins. Live subject data can also serve as a guideline for defining mannequin posture during reach, as well as for quantifying the effects of any restraint system or protective equipment, in the particular application environment. In addition, there is a glaring absence of the modeling of human tissue and seat cushion deformation in all HMSs. When this is properly modeled and included in the HMSs, the fidelity and the results will be greatly improved. Without these improvements, current HMSs can give misleading results and should be used with caution when determining cockpit accommodation limits.

There are currently three options for cockpit accommodation assessment: (1) a live subject study in a mockup or cockpit (traditional method), (2) using HMS mannequins in a CAD model, and (3) using the HMS approach guided by data from a live subject study. The cost of traditional assessment methods is quite high when the proper number of subjects is used, especially when combined with the cost of an accurate mockup for new aircraft applications. However, the distribution of data resulting from a live subject study includes overall variation due to uncontrolled factors. This is quite useful in that it suggests what will be observed in the final flying population. The accommodation results from an analysis using a few mannequins do not offer this, nor do a series of Monte Carlo style mannequin runs. The cost of using an HMS approach is much lower but, given the lessons learned in this study, the results may be inaccurate, which may incur expensive fixes. Option 3, a combination of modeling and live subject studies, may be the best alternative. In this option some of the cost advantages of using an HMS are realized while final accuracy is increased via improved position and posture data from live subject studies.

The accommodation test and the funds available will ultimately determine which option is best. Some tests do not require high accuracy, while others do. Sometimes it is not possible to use traditional live subjects if a mockup is not available or is too costly to construct.

References

Oudenhuijzen, A.J.K., & Hudson, J.A. (2000). Digital human modelling systems: Anthropometric verification and validation. Report 1: Test methods and instructions (Report TM-00-A007). Soesterberg, The Netherlands: TNO Human Factors.

Robinette, K.M. (1997). Minutes of First CAESAR Partner Meeting, September 16–17. Wright Patterson Air Force Base, OH.

Zehner, G.F., & Hudson, J.A. (2002). Body size accommodation in USAF aircraft (AFRL-HE-TR-2002-0118). Wright-Patterson Air Force Base, OH: Human Engineering Division, Human Effectiveness Directorate, Air Force Research Laboratory.

Part 3

Evaluation and Analysis

Digital Human Modeling: Evaluation Tools

Dan Lämkull,
Cecilia Berlin, and
Roland Örtengren

24.1 Introduction

Digital human modeling software (DHMS) enables early assessment of important design parameters of future products and production systems. Typically, humans affect the system performance, and in order to achieve the expected system efficiency, ergonomics needs to be considered in the design process in addition to the more technical or logistical matters. Hence, there is a call for ergonomics to be a natural part of the product and production system development process, also at virtual stages. Commercial DHMS support the consideration of ergonomics at virtual stages of the design process. Simulation is becoming an increasingly important approach in the design of products and production systems. Simulation and predictive models enable evaluation of the consequences of design choices before financial resources are committed to constructing either prototypes or actual systems; the rapid adoption of these tools pose an opportunity to integrate consideration of ergonomics into early design stages (Mathiassen et al., 2002).

DHMS are used primarily for industrial purposes before any physical objects are built (Sundin & Örtengren, 2006). The major reason for this is the expense involved in building mockups and prototypes for evaluation of space requirements, reach functionality, and ergonomics (Landau, 2000). Another reason is to facilitate a faster and more cost-efficient design process (Chaffin, 2001).

The use of evaluation tools is one step in a chronological work flow process referred to as the evaluation method. A generic digital human modeling work flow process has been described by Green (2000), who includes the use of evaluation tools in an action called *performing the analysis*. However, in most of the analysis human factors judgment comes into play and the assumptions and limitations of the modeling system being used must be taken into account. This action is, according to Green, analyzing and applying judgment to the results.

The functionalities of DHMS are broad and cover features like collision detection, clearance and reach verifications, posture and motion predictions, vision, and comfort. For special purposes there are manikins that represent inner organs (Petoussi-Henss et al., 2002), the structure of muscles and other tissues (Rasmussen & Toerholm Christensen, 2005), and effects of traumatic impacts (Nishimoto & Murumaki, 1998), while others focus on representing performance, human behaviors, and mental workloads (Ntuen, 1999; Zhang, 2003). Another area for manikins is evaluation of thermal comfort where a physiological finite element model represents the human body (McGuffin et al., 2002; Farrington et al., 2004) or seating comfort (Mergl et al., 2004; Verver, 2004; Siefert et al., 2006). Given all these features and application areas, a chapter describing all evaluation tools would be a vast creation. Thus, in this chapter we will only briefly describe the most commonly used evaluation tools for *physical work loads*, *body postures*, *reach abilities,* and *fields of vision* integrated in different DHMS. We try to do so in a way that is neither too specific to any particular DHMS nor so generic that all evaluation tools, more or less integrated in DHMS, are covered. The continuation of this chapter is structured according to the different tool sets that are available in the most commercial DHMS. We have chosen to divide the tool sets into three groups:

- Quantitative evaluation tools
- Semi-quantitative tools
- Tools for discomfort, anthropometry, human performance, and cognition

The main focus of this chapter is addressed to group one (quantitative evaluation tools), whereas groups two and three are less extensive.

24.1.1 Quantitative Evaluation Tools

These tools are used to evaluate working postures and physical workloads. All of these tools require as input a postured manikin (digital human model) representing an operator performing the task being analyzed. Typically, they give quantitative values as answers. Often, the tools are based on a multi-zone system (e.g., red, yellow, and green) to provide an easy indication of working conditions that are clearly hazardous, might be hazardous, or entail negligible risks. Very often, the connections between work and the risk of musculoskeletal disorders are complicated. The tools have been simplified, so as to improve the chances of their actually being put to practical use (Arbetarskyddsstyrelsen, 1998; Laring, 2004). They take in just a few aspects of one type of load at a time, and for this reason they cannot be used to give exact load limit values. But they furnish sufficient guidance to serve as the basis of a practical process of change, both at existing workstations and in the planning of new workstations and jobs. Due to their simplified nature, if the models are employed uncritically, this can result in both overestimates and underestimates of the actual risks. To arrive at a comprehensive assessment, more factors need to be taken into account and more accurate models used, which in turn calls for a thorough knowledge of ergonomics and of correct work technique. Other tools in this group give the percentage of men and women who have the strength to perform an evaluated job task according to, for instance, an allowable weight-lifting standard.

24.1.2 Semi-Quantitative Tools

These tools are used to visualize or analyze the manikin's interaction with its environment. The environment is made up of components (3D-CAD models) describing products and process equipment. Examples of such tools are field of vision, reach envelopes, and accessibility and clearance analysis. Most of the tools are available in the CAD software itself. We have chosen to call these tools semi-quantitative. Although some quantitative values are given by these tools, in most of the analyses human factors judgment comes into play and the assumptions and limitations of the modeling system being used must be taken into account, all in congruency with Green (2000).

24.1.3 Tools for Discomfort, Anthropometry, Human Performance, and Cognition

Finite element tools are already efficiently used to improve crash safety and these tools are now also being used for comfort evaluation. There are several studies published regarding the simulation of seat comfort (Verver & van Hoof, 2003; Marx et al., 2005; Pankoke & Siefert, 2007). Digital human models (DHM) are increasingly used in numerical simulations; with their quality improvement, simulations are becoming more realistic and precise studies can be performed, focusing on the whole body as well as specific body parts (Bidal et al., 2006). They can be used in many fields of application, such as comfort/discomfort evaluation.

When using DHMS to analyze vehicle interior design or workstation design, usually the smallest manikin (female: 5th percentile (stature)) and the largest manikin (male: 95th percentile (stature)) of the designated group are used. One aspect of the difficulty of using percentile values is that they are not additive, except the 50th percentile values (Robinette & McConville, 1981; Annis & McConville, 1990) (fig. 24.1). Since many databases of anthropometric values present data for male and female as 5th percentile, 50th percentile, and 95th percentile values, it is reasonable for a *non-specialist* to assume that such a "constant percentile human" exists, and that by designing from the 5th percentile female up to the 95th percentile male, the product or workstation design would accommodate 90% of the designated group. This is, however, *not* true for multivariate problems, such as vehicle interior design or workstation design (Roebuck et al., 1975; Porter et al.,

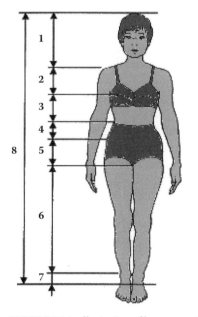

Variable	Measurement	5th Percentile	95th Percentile
1	Shoulder to vertex	270.5	327.9
2	Bust to shoulder	107.9	177.7
3	Waist to bust	134.2	217.8
4	Buttock to waist	137.8	216.7
5	Crotch to buttock	47.8	104.2
6	Ankle to crotch	578.4	710.9
7	Ankle height	92.3	132.9
	TOTAL (mm)	1368.9	1888.1
8	Stature	1525.0	1730.5

FIGURE 24.1 llustration of how percentile values are *not* additive (Robinette & McConville, 1981).

2004). Furthermore, Annis and McConville (1990) highlight the fact that gender, ethnicity, age, and occupation are sources of anthropometric variation, which can have significant effects on anthropometric data.

Human performance and cognitive models include the modeling and simulation of the performance and cognitive aspects of the human being (e.g. modeling of human computer interaction, graphic user interface, man-machine interaction, behavioral realism, and artificial intelligence). Early works in the area of human factors and engineering psychology were performed by Fitts and Jones (1947) and Mackworth (1948). Underlying reasons for these research contributions were several mysterious accidents and mistakes during World War II due to poor design. Systems with well-trained operators simply were not working—for instance, airplanes crashed into the ground with no apparent mechanical failures. Since the 1970s, human performance modeling has been a valuable tool in the human factors engineering toolkit. An extensive and recent overview of the history of these techniques has been given by Laughery et al. (2006). Also worth reading is Wickens (1992).

24.2 Evaluation Tools

24.2.1 Quantitative Evaluation Tools

24.2.1.1 NIOSH 81/91

The NIOSH (National Institute of Occupational Safety and Health) Lifting Equation is a tool used to identify, evaluate, or classify risks associated with lifting tasks. The original reference for this tool is the *Work Practices Guide to Manual Lifting* (NIOSH WPG, 1981). This version defined a compressive force of about 3400 N[1] on the spine segments L5–S1 as the criteria for establishing an action limit (AL) and the maximum permissible limit (MPL) for a task (Chaffin & Andersson, 1984). The calculated AL is the weight that can safely be lifted by 75% of the female and 99% of the male population. If the load exceeds the AL, administrative or engineering-controlled action is required in order to reduce the risk of injury. The MPL is the equivalent weight of a compressive force of 3 × AL on the spine, according to Chaffin and Andersson (1984).

NIOSH updated this lifting guide by issuing the revised 1991 NIOSH lifting equation. The new equation increased the types of task variables the tool could assess, such as asymmetrical lifting tasks and lifts of objects with sub-optimal hand-object couplings, and also provided guidelines for a larger variety of work durations and lifting frequencies than the 1981 equation (Waters et al., 1993).

The revised NIOSH lifting equation requires (as input) the weight of the object being lifted, the horizontal and vertical hand locations at key points in the lifting task, the frequency rate of the lift, the duration of the lift, the type of handhold on the object being lifted, and any angle of twisting. The NIOSH lifting equation calculates the recommended weight limit (RWL) and the lifting index (LI). The RWL is the recommended weight of the load that nearly all healthy workers could lift over a period of time (up to 8 hours) without an increased risk of developing lifting-related low back pain or injury. The LI is the ratio between the weight of the lifted object and RWL, and acts as a relative estimate of the physical stress associated with a manual lifting job. As the magnitude of the LI increases, the level of the risk for a given worker increases, and a greater percentage of the workforce is likely to be at risk for developing low back pain. From the NIOSH perspective, lifting tasks with a LI > 1.0 pose an increased risk for lifting-related low back pain and injury in the workforce. NIOSH considers that the goal should be to design all lifting jobs to achieve a LI of 1.0 or less.

---

[1] Specified in the original equation as a compressive force of 770 lbs. Here, translated into SI units.

Successful application of the NIOSH lifting equations assumes the following:

(From OSHA, 2006)

(1) Lifting task is two-handed, smooth, in front of the body, hands are at the same height or level, moderate-width loads (i.e., they do not substantially exceed the body width of the lifter), and the load is evenly distributed between both hands.
(2) Manual handling activities other than lifting are minimal and do not require significant energy expenditure, especially when repetitive lifting tasks are performed (i.e., holding, pushing, pulling, carrying, walking, or climbing).
(3) Temperatures (19–26° C) or humidity (35–50%) outside of the ranges may increase the risk of injury.
(4) One-handed lifts, lifting while seated or kneeling, lifting in a constrained or restricted workspace, lifting unstable loads, wheelbarrows, and shovels are not tasks designed to be covered by the lifting equation.
(5) The shoe sole to floor surface coupling should provide for firm footing.
(6) Lifting and lowering assumes the same level of risk for low back injuries.
(7) Using the Guidelines in situations that do not conform to these ideal assumptions will typically underestimate the hazard of the lifting task under investigation.

24.2.1.2 Garg Energy Model

The Garg energy model (Garg et al., 1978) was developed to calculate metabolic energy requirements for a job using a worker's physical characteristics and a description of the job as input data. The main aim expressed by the author was to provide a tool for planning manual work in such a way that workers can perform without excessive strain or fatigue. The model is based on the assumption that a job can be divided into tasks or activity elements, for which the individual energy expenditure requirements can be added together to determine the energy expenditure of the entire job. The average metabolic energy expenditure rate (expressed in kcal/minute) for the job is then predicted as the average (over time) of the sum of the energy requirements of the individual tasks, plus the energy required to maintain various body postures.

The energy expenditures of the tasks are calculated using prediction equations derived from empirical data. The information for each task needed to compute these energy requirements include: force exerted, distance moved, frequency, task posture, lifting technique for lifting tasks, and the time needed to perform the tasks. Gender and body weight, two worker factors, are also needed.

24.2.1.3 OWAS

OWAS (Karhu et al., 1977), short for Ovako Working Posture Analysing System, was developed as an observation tool at the Finnish steel manufacturing company Ovako. Its purpose was to identify and evaluate potentially harmful working postures observed among workers. The tool summarizes work postures in the form of a four-digit code, where each digit represents an assessment score for the observed postures of the back, arms, legs, and work load (force), respectively. Karhu et al. also employed a fifth-digit position to code which subtask that each posture occurred in sequentially. The method was later extended by adding an additional digit position, signifying the head posture, as in Kant et al. (1990). The higher the score, the greater the potential risk for harm (fig. 24.2).

As a rule, OWAS is carried out at regular time intervals during observation of a work task resulting in a sequence of OWAS assessment codes. The system thereby reveals the frequency and relative proportion of time spent in specific postures. The postures signified by the OWAS codes are then related to a four-level scale of harmfulness, where each level (a.k.a. "action category") indicates the urgency to correct harmful postures. These four categories are:

1. Normal postures—where no remedial action is needed
2. Postures that will need remedial action in the near future

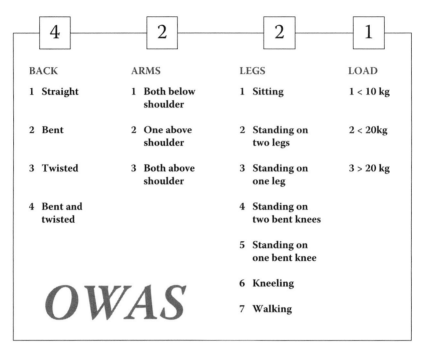

FIGURE 24.2   An example showing the different possible scores for each digit in the OWAS code. The code 4221 describes a posture where the observed person is working standing on two legs with a bent and twisted back, one arm above shoulder height, and handling a load less than 10 kg. The fifth-digit position, signifying the head, is not included in the example.

3. Postures that will need remedial action as soon as possible
4. Postures that immediately require remedial action

24.2.1.4 RULA

RULA (McAtamney & Corlett, 1993), short for Rapid Upper Limb Assessment, is a survey method developed for investigations of workplaces where work-related upper limb disorders are reported. RULA is a screening tool that assesses biomechanical and postural loading on the upper body, with particular attention to the neck, trunk, and upper limbs.

The procedure for RULA is to observe working postures and score them for each body segment, respectively. The scores are given based on the observed joint angles for each body segment. As in OWAS, higher scores imply greater risks of harmful postures. For specified extreme positions or conditions that increase the physical impact of a posture (such as bending or twisting), RULA can add "penalty" points to a posture score (fig. 24.3). Using three conversion tables developed by McAtmney and Corlett, RULA practitioners translate the scores into a final "Grand Score," which ranks the observed working posture on a four-level action list similar to that of OWAS. The Grand Score translates to a corresponding recommendation regarding how urgently the posture requires remedial action.

ACTION LEVELS, RULA:		
Level	**Score**	**Assessment**
Level 1:	1–2	Acceptable if not maintained or repeated for long periods
Level 2:	3–4	Further investigation needed, changes may be required
Level 3:	5–6	Investigation and changes required soon
Level 4:	7+	Investigation and changes required immediately

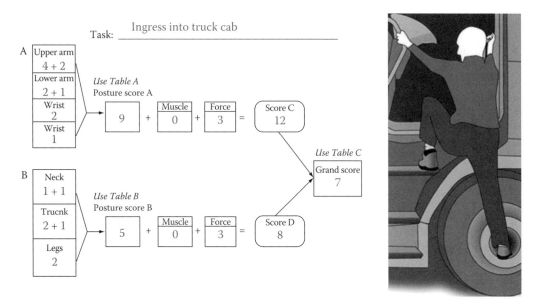

FIGURE 24.3 An example of what RULA scoring can look like. The scores are based on observed joint angles, and "penalty" scores are shown as additions.

As stated by McAtamney and Corlett, RULA is meant to be used as a preliminary screening tool or as part of a wider ergonomics assessment.

24.2.1.5 REBA

REBA (Hignett & McAtamney, 2000), Rapid Entire Body Assessment, is a survey/observation method based on the aforementioned RULA, where consideration has been taken of the body as a whole and focus has been shifted from upper limb evaluation to more unpredictable, transient postures. The assessment is carried out using a scoring procedure similar to that of RULA, with additional scoring to take consideration of coupling (meaning the physical interface between the human and a load force, for example, grip) and activity (in the sense that the observed work posture can be characterized as static or dynamic). REBA also introduces the concept of gravity-assisted upper limb position. As in RULA, a final score is generated that translates to an action level recommendation, in this case with five levels. These are slightly different from RULA, as follows:

ACTION LEVELS, REBA:

Level	Score	Risk Level	Action (including further assessment)
Level 0:	1	Negligible	None necessary
Level 1:	2–3	Low	May be necessary
Level 2:	4–7	Medium	Investigation and changes required soon
Level 3:	8–10	High	Necessary soon
Level 4:	11–15	Very high	Necessary now

24.2.1.6 Snook & Ciriello

The research on manual handling tasks carried out by Snook and Ciriello (1991), often referred to as Snook tables (and also released as Liberty Mutual Force Tables; see Liberty Mutual, 2004), is the aggregated result of several experiments carried out to determine the maximum acceptable weight of lift (or alternatively, the maximum frequency of lift) for different types of work tasks (lifting, lowering, pushing, pulling, and carrying). The tables from 2004 are based on experimental data from 119 industrial workers. The main aim of the tables is to "assist industry in the evaluation and design of manual handling tasks, and thereby contribute to the reduction of disability from low back pain" (Snook & Ciriello, 1991, p. 1212). This is done by establishing a maximum acceptable loading level for as large a population percentile as possible. The tables are sorted by gender, and data is given based on the type of task, the spatial requirements of the movement, and the frequency at which the task is carried out.

24.2.1.7 Burandt & Schultetus

There are several ways described in the literature to establish limits for human force exertions in material handling activities. Many of these are quite similar and are based on risk evaluations for work-related injuries in different companies. Tools for determination of maximum force limits have been developed for use in Volvo (both Volvo Group and Volvo Car Corp) (Munck-Ulfsfält et al., 1999, 2003; Volvo Corporate Standard, 2002) and Saab Automobile (Svensson & Sandström, 1995). These tools are also very similar to international standards and ordnances issued by standardization bodies and national boards of occupational safety and health (e.g., Arbetarskyddsstyrelsen, 1998).

One line of methods goes back to the principles developed by Burandt and Schultetus for use in Siemens. This tool was originally published as an internal Siemens document (Burandt & Schultetus, 1978), but also independently by Burandt (1978) and later by Schultetus et al. (1987).

The tool has been adopted in slightly modified form by several German organizations, for example, REFA and Mercedes-Benz and also the Bosch company who developed its own, somewhat simplified, version. The tool is described in a publication on workplace design and work organization by the Association of German Engineers (VDI, 1980). The principles behind the tool (based on Schultetus et al., 1987) are also outlined in BIA-Report 5/2004 (2004). This report is in German and can be accessed on the Internet. A description in English can be found in Schaub et al. (1997).

The tool yields a result "from practical experience acceptable" limit value for force moment or load respectively. The limit values are determined by the following influencing factors:

- Personal factors (gender, age, training state)
- Type of force exertion (static, dynamic)
- Force-exerting body part (finger-hand, arm-shoulder, leg) and type of exertion (force, moment, weight)
- Frequency and, depending on the way of working, also the duration of the force exertion
- Point of attack of the force (far, medium, or close; forward, diagonal, or to the side; height—head, shoulder, waist, or pelvis)
- Direction of force
- Position of the hand

The force and moment limits are calculated using the factorial expression:

$$F_{lim} = T \cdot P_1 \cdot P_2 \cdot P_3 \cdot F_{max}$$

where F_{max} is the maximum force that can be exerted under ideal conditions. T, P_1, P_2, P_3 are correction factors, the values of which depend on the working conditions. F_{lim} is the resulting limit value for the work situation under study. The same factorial equation is valid for both dynamic and static forces and moments. However, the equation for static forces in the hand–arm system is a little different since also the weight of the arm is included in the calculation.

The maximum force or moment is obtained from different tables depending on which part of the body is executing the force, either

- Hand–arm system
- Hand–finger system
- Legs (Burandt only, no data for moments)

Actual data on maximum force and moments can be taken, for example, from the method description or from an appropriate standard, such as the national German standard DIN 33411-5 (1999).

The procedure to follow when determining the limit values involves several steps and requires input from tables and diagrams. Therefore, the tool should rather be used in software form. Fortunately, software versions of the tool have been developed and are available as a module, such as in eM-Ramsis.

Regarding validity, it must be remembered that it is not scientifically validated. However, the tool has been tested in a large number of situations and been found to yield useful data by many independent users (BIA-Report 5/2004, 2004). Therefore, the tool has been adopted as a quasi-standard for industrial use in Germany. But, there are several questions that need to be answered before the tool can be accepted for general use. The origin and compilation of some of the data are not known, and it is not clear for which percentage of the working population the safe limit values are valid. The judgments of body constitution and training state, which influence the limit values in the range from minus 20% to plus 40%, create uncertainty as the judgments cannot be done without a solid experience of job design or occupational medicine.

24.2.1.8 3DSSPP

The 3DSSPP software, short for 3D Static Strength Prediction Program, predicts static strength requirements for tasks such as lifts, presses, pushes, and pulls, using a computation algorithm based on a 12-link biomechanical model. The model is primarily focused on the prevention of low back injuries (Chaffin, 1997). 3DSSPP allows input of body link angles, hand forces (magnitude and direction), and anthropometrical data, and generates a predicted percent of the population (with differentiation between men and women) who have sufficient strength to perform the job, as well as spinal compression forces and data comparisons to NIOSH guidelines. Analysis is aided by an automatic posture generation feature and three-dimensional (3D) human graphic representations. The developers have acknowledged some limitations; for example, the program specifically targets analysis of "very slow or static exertions" (Chaffin, 1997, p. 305) rather than dynamic movements, for which it tends to overestimate the population's capabilities. Moreover, the author's experience is that 3DSSPP tends to overestimate the minimum strength percentages for older populations, due to the anthropometric database it bases the calculations on.

Table 24.1 gives a summary of quantitative characteristics of the eight quantitative evaluation tools mentioned in this chapter.

24.2.2 Semi-Quantitative Tools

24.2.2.1 Field of Vision

According to Cushman et al. (1983), one important principle that should be utilized in workplace design is a clear line of sight for the operators. In a manufacturing context it is important that the assembly workers have a clear view of what they are doing, or the quality of the work will decrease (Rönnäng et al., 2004). Several authors have stated that visual requirements influence working postures (Corlett, 1981; Delleman, 1992; Haslegrave et al., 1995; Delleman & Dul, 2002). Dukic (2003) states that solving a visibility problem can have a considerable effect on the overall work situation. The possibility to accurately and by easy actions direct the field of vision area is of major importance for assessing workplace design (Hoekstra, 1993). DHMS provide the opportunity to see through the

TABLE 24.1   Summary of Quantitative Characteristics

Tool	Focus	Input	Output	Comment
OWAS	Body postures (observation)	Assessment score for torso, arms, legs, and weight load, summarized in a 4-digit OWAS code	Indication of how urgently the posture needs correction on a 4-level scale	Available and widely used in a stand-alone software called WinOWAS.
NIOSH Lifting Equation	Lifting tasks (equation)	NIOSH 81: horizontal and vertical load locations, vertical travel distance to destination, frequency rate of the lift, and maximum sustainable frequency of lift (from a table). NIOSH 91: Mass of the object being lifted, horizontal and vertical hand locations, frequency rate of the lift, duration, type of handhold, and asymmetry	NIOSH 81: Action limit (AL) in kg or lbs and maximum permissible limit (MPL) in kg or lbs. NIOSH 91: Recommended weight limit (RWL) in kg or lbs and lifting index (LI), which is the ratio between the RWL and the weight of the lifted object	Released in two versions, first as NIOSH 81; the one currently used is known as NIOSH 91. The output data differ between the two versions.
RULA	Body postures with specific attention to upper limbs (observation)	Scoring system for 7 body segments (based on joint angle ranges and twisting/bending), characteristics of muscle use and force involved in the posture.	Individual assessment for body segments as well as an indication of how urgently the posture needs correction on a 5-level scale.	Intended originally as an observation/screening tool specifically for work places reporting upper limb injuries.
REBA	Body postures (observation)	Scoring system for 6 body segments (based on joint angle ranges and twisting/bending), characteristics of load/force and coupling,* and level of activity when the posture occurs.	Individual assessment for body segments as well as an indication of how urgently the posture needs correction on a 5-level scale.	Very similar to RULA, takes more consideration of unpredictable body postures, coupling to loads and static/dynamic movements.
3DSSPP	Strength prediction (simulation)	Posture data, hand force parameters, male/female anthropometrical data	Percentage of males and females in the specified population who will be able to perform the task. Also calculates spinal compression forces and comparisons with NIOSH guidelines.	This tool is a stand-alone software that provides an approximate job simulation. Includes 3D human graphic illustrations to aid analysis.
Snook & Ciriello	Manual handling tasks—lifting, lowering, pushing, pulling, carrying (tables)	Task frequency, distance, height, duration, object size, nature of grip/handles, extended horizontal reach, combination tasks, gender of worker	Maximum acceptable weight of lift or maximum frequency of lift applicable to a population. Tables are sorted by gender and nature of the task.	
GARG energy model	Energy expenditure—physical strain or fatigue (equation)	For each sub-task: force exerted, distance moved, frequency, task posture, lifting technique, time needed. For the worker: gender and body weight.	Total energy expenditure of the job expressed as a metabolic rate (in Kcal/minute), as calculated from the individual tasks.	The aim of the model is to link energy requirements specifically to worker characteristics and a description of the job.
Burandt & Schultetus	Limit values for load or force moments (equation and tables)	Personal factors, type of force (static/dynamic), body part, point of attack of force and duration of force exertion, frequency and duration of force exertion, point of attack of force (distance, height), direction of force, and position of hand. Also, depending on the body part used, data on maximum force taken from tables.	A "from practical experience acceptable" limit for force moment or load.	This tool is based on "experience-based" data from tables, and is incorporated into German standards

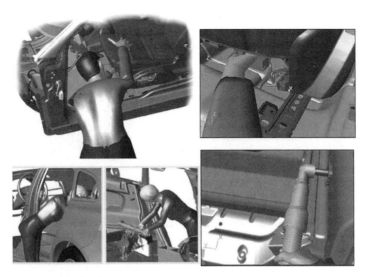

FIGURE 24.4 Examples of field of vision evaluations. DHMS provide the opportunity to see through the manikin's eyes, and a view obtained via this function is a very natural way for visual inspection of human workspace design (image courtesy of Volvo Car Corporation).

manikin's eyes, and a view obtained via this function is a very natural way for visual inspection of human workspace design (fig. 24.4). An extensive and recent overview of visual demand in manual task performance has been given by Dukic (2006).

In most DHMS, several types of vision simulation are provided. The most common vision simulations are binocular, ambinocular, monocular left, and monocular right. Visual characteristics are displayed as peripheral cones, central cones, blind spot cones, and central spot cones that permit the user to gain an insight into the operator's view. Visual fields are differently represented by various DHMS (Table 24.2).

From the idea to simply visualize what the operator will be able to see at a workplace, the vision module has become an expert's module requiring accurate knowledge concerning human vision. Another vision aspect would be that of realism in colors and contrasts. Although certain simulation software can produce photorealistic images including glare, shadows, reflections, and so on, it is not possible to render images of such a high quality in any DHMS. Furthermore, it is very often the case that the manikin user uses high-contrast colors and intense lighting to make it easier to detect and indicate collisions and measure clearance, although this generates an unrealistic visualization.

24.2.2.2 Reach Envelopes

"Let the smallest woman reach. Let the largest man fit" is an ergonomic rule of thumb used in workplace design (Konz, 1983). Maximum reach changes with individual parameters such as gender, age, physical training state, and obesity. External parameters such as wearables and safety equipment (e.g., helmets, life jackets, and gas and protective masks) also affect reach. Reach envelopes are either presented as statistical or functional data (Roe, 1993). Statistical data depend on segment length and joint mobility and do not consider external parameters. Applied concepts are, for instance, maximum reach (full arm length and full back flexion), convenient reach (full arm length), working area (lower arm length), and comfortable reach (application of comfort joint angles) (Hanson, 2004). Functional data use similar concepts as statistical data but in contrast consider external parameters.

Driver reach simulations were among the earliest uses of DHM in the design of vehicle interiors. The analysis is typically concerned with whether controls or other targets can be reached by a sufficient percentage of drivers (fig. 24.5). For industrial task analysis, tasks are evaluated by determining whether workers will be able to reach required hand locations. Seated reach capability, particularly in vehicles,

TABLE 24.2   Vision Module Characteristics of Three DHMS

Computer manikin	Vision module characteristics	Examples from software
Ramsis, version 3.7	Three areas of vision and acuity are defined: • Sharp sight area: opening angle +/−5° • Optimum sight area: opening angle +/−15° • Maximum sight area: opening angle +/−50°	
Jack, version 4.1	It is possible to define view cones, one for each eye. View cone's length (default value 200 cm). View cone's angle (default 40°). Angles and length have no limitations. Visual fields defined by: • Eye point: both, right or left • Type: achromatic, green, blue, yellow, red and blind spot. Achromatic field represents the peripheral vision.	
Safework, version 5.11	Four types of vision are defined: binocular, ambinocular (total field of vision seen simultaneously by both eyes), monocular left, monocular right and stereo. For each type of vision one can define the field of view characteristics (horizontal monocular, horizontal ambinocular, vertical top, and vertical bottom angles) Visual characteristics are displayed as peripheral cones, central cones, blind spot cones and central spot cones.	

Source: Dukic, 2006.

has been studied more than standing reach capability, probably because seated postures are generally more constrained (Parkinson & Reed, 2004).

Before the widespread use of DHM in CAD, reach analyses were primarily performed using the reach surfaces in the SAE J287 Standard (SAE, 2007). This standard provides surfaces within which 95 percent of a particular driver population are expected to be able to reach with different levels of torso restraint. Note that this is not the same as the average surface within which a person who is 95th-percentile by stature (or some other dimension) can reach, but rather reflects the combined effects of body size, posture, and other factors that affect reach capability. This approach provides the vehicle designer with important information for locating controls, but is not useful for simulating reaches with individual figure models (J287 does not indicate how or how far a person described by a particular set of body dimensions would reach).

24.2.2.3 Accessibility and Clearance Analysis

The size of hand clearance affects several parameters of performance. Over 40 years ago, Kama (1963, 1965) showed that the time taken to remove and replace a component decreased sharply as the aperture available for hand clearance increased. It has also been shown that the torque capability of the hand is

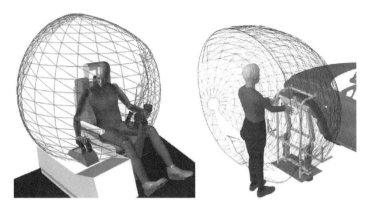

FIGURE 24.5 Examples of visualizations of reach envelopes. The left image shows the use of Boeing Human Modeling System's reach envelope to define pilot reach requirements (image courtesy of the Boeing company). The right image shows the maximum reach envelope of the manikin eM-Ramsis. The maximum reach envelope has been cut so it only displays the maximum reach volume in front of the standing manikin (image courtesy of Volvo Car Corporation).

affected by the clearance between the hand and any physical obstruction (Adams & Peterson, 1988). Small hand clearance is also a problem when working with the hands in spaces where sharp edges are present, such as car bodies made of joined metal sheets. Besides accessibility for hands and arms, manual assembly tasks also require accessibility for hand tools of different kinds (e.g., screw drivers, wrenches, lifting devices). In most DHMS several types of accessibility and clearance analysis tools are provided. The DHMS users establish specific rules that define what parts in the product design should be checked against one another as well as any spatial requirements that should be considered. The most common way is to create two collision lists that will check the distance between the parts (3D models) included in the two different lists. In the first list, for instance, the human body and any used hand tool could be included, and in the second list the surrounding parts could (preferably) be included. The frequency of the distance check (checks per second) is adjustable—the higher the frequency, the higher the workload for the computer becomes. Thus, it is important to use a frequency that is not too high (causing the computer to slow down) or too low (might miss a collision, where a collision occurs between two, or more, checks). The system can be set in different modes:

Alarm if parts in the different lists are violating a predefined specific clearance distance, either via a sound signal or by coloring the parts in a chosen color (often reddish)
Stop all motions if any part is violating a predefined specific clearance distance
Just record all violations in a stored and readable list

Figure 24.6 shows how a manikin can be used to assess clearance for the hands.

24.2.3 Comfort/Discomfort Tools

An extensive amount of work has been conducted involving discomfort that originates from factors outside the human body or unrelated to motion (Toftum et al., 2000; Hoppe, 2001; Fehren et al., 2003; Kaynakli & Kilic, 2005). Da Silva (2002) provides a thorough review of environmental factors (thermal conditions, air quality, noise, etc.) that affect comfort. However, this section focuses on discomfort issues related to the musculoskeletal system.

Comfort is a qualitative factor considered in the area of ergonomics. Several definitions of comfort and discomfort exist. Hertzberg (1972) states that people are only aware of discomfort, therefore he plainly defines comfort as absence of discomfort. Hedberg (1987) defines comfort as no pain and discomfort as pain. Both describe comfort as being related to physical ergonomics, albeit being aware

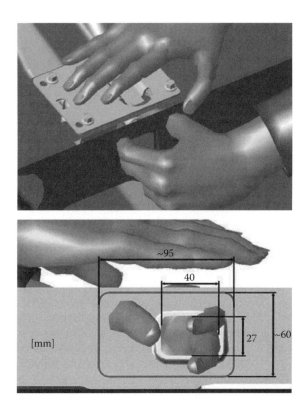

FIGURE 24.6 An example of a clearance/access study. To be able to carry out the connection task with the left hand, the hole must be enlarged from 40 × 27 mm to 95 × 60 mm. However, this might have an effect on the strength (image courtesy of Volvo Car Corporation).

of the fact that comfort, as well as discomfort, are aspects of more complex concepts. Helander and Zhang (1997) found that discomfort is related to physical characteristics of the environment. This is congruent with Allread et al. (1998) who have performed experiments to evaluate discomfort in manufacturing environments and found that overall total-body discomfort depends on external loads and on the nature of tasks that are being completed. Studies by Zhang et al. (1996) show that comfort and discomfort are not arranged on one single axis of a coordinate system. Comfort is mainly an aspect of pleasure related to design and subjective feelings. Discomfort is more related to biomechanical factors and physical fatigue, for instance, suffering, and can be assigned to objective criteria.

In several DHMS, comfort and discomfort tools are available. Some of the tools use joint angles (joint displacements) as variables (Hanson, 2004). These comfort/discomfort angle intervals are defined by work of several researchers (Rebiffe, 1969; Grandjean, 1980; Tilley & Dreyfuss, 1993; Krist, 1994; Porter & Gyi, 1998). Other tools use finite element models to predict static and dynamic behaviors of a human on a car seat, for instance, pressure distribution, H-Point position, and behavior under vertical vibrations (Marx et al., 2005; Pankoke & Siefert, 2007). Marler et al. (2005) state that discomfort is not the same as simple joint displacement, despite the semantic discrepancy in the literature. Instead they use a novel optimization approach to develop a new human performance measure for direct posture prediction.

24.3 Anthropometry

Anthropometry is the study of human body measurements (height, weight, waist circumference, proportions, etc.) and its biomechanical characteristics. Anthropometric data refer to a collection of physical dimensions of a human body. The aim of anthropometry is therefore to characterize the human body by a set of measurements (Paquet & Rioux, 2003). Most anthropometric variables conform quite closely to the normal (Gaussean) distribution (at least within reasonably homogeneous populations) (Pheasant, 1991). Hence, regular parametric statistics apply in most cases. Most available anthropometric information relies on data taken on military populations (e.g., NASA, 1978). However, civilian-based surveys do exist (e.g., Lewin, 1969; Abraham et al., 1979). Later surveys have benefited from recent body scanning technologies, enabling 3D descriptions of the body surface, e.g., CAESAR (Civilian American and European Surface Anthropometry Resource) and HQL's (Research Institute of Human Engineering for Quality Life) Japanese body size data. The CAESAR database contains anthropometric data (e.g., weight, height), demographic data (e.g., income, car brand), and 3D full-body scans concerning thousands of individuals in the United States, Canada, Italy, and the Netherlands (Robinette, 1998). HQL's database contains anthropometric data, 178 measurements of 19,000 males and 15,000 females (Japanese aged from 7 to 90 years) (HQL, 1997). Besides these examples of national and multinational anthropometric databases, an international collaborative effort to create a worldwide resource of anthropometric data for a wide variety of engineering application is in progress. The project is called WEAR (World Engineering Anthropometry Resource). In addition, WEAR is the primary project of a formal CODATA[2] [1]task group titled Anthropometric Data and Engineering. The objective of this task group is to identify and develop data models, software tools, theoretical constructs, and principles that support the development of an online worldwide information system for utilizing the latest anthropometry databases in concerned engineering environments.

When one or two body dimensions affect the design, it is relatively easy for the designer to assess and achieve the expected accommodation level of the product. However, when the product should fit a population of users and several body dimensions affect the design, the design task becomes very complicated due to human anthropometric diversity (Högberg, 2005). Consequently, approaches have been developed that permit the power of computers to be applied in a way that more realistically represents the wide diversity that occurs in a population (Roebuck, 1995). When using anthropometric data, it is important to make sure that the data are valid for the design issue at hand, and to know for example from what population the sample was drawn, how large the sample was, and how old the study is.

Several commercial DHMS contain advanced anthropometry tools. With these tools the user can create manikins with specific anthropometric measurements for quite a number of (more than 20) different body dimensions. The user can create individuals for accommodation studies in which a single dimension is changed at a time. Furthermore, today efficient body scanner systems for serial measurements are available. The systems enable the measurement of large groups of people and a database containing anthropometric data of hundreds of individuals can be created at a reasonable price and time.

24.4 Tools for Human Performance and Cognition

Examples of software dealing with human cognitive performance are: Fatigue Avoidance Scheduling Tool (FAST) (Hursh et al., 2004), Crew Station Design Tool (CSDT) (Walters et al., 2005), and Graph-Based Interface Language (GRBIL) tool (Archer et al., 2007). Scripting interface is a recent and major step toward constrained language-based commands for the digital human models. Examples of work related to scripting interface are Digital Human Model Testbed (Badler et al., 2005) and Task Simulation Builder (Raschke et al., 2005). Compared to the area of physical modeling, the area of cognitive and performance modeling is not as well known or developed, mainly due to its complexity and abstract nature (Sundin & Örtengren, 2006). Small user communities or the specificity and difficulty in developing

---

[2] CODATA, the Committee on Data for Science and Technology, is an interdisciplinary Scientific Committee of the International Council for Science (ICSU). CODATA is a resource that provides scientists and engineers with access to international data activities for increased awareness, direct cooperation, and new knowledge.

some of these tools are also mentioned as obstacles to human performance modeling (Badler et al., 2005). In this section, we do not describe the area of cognitive and performance digital human modeling any further. The main reason for this, in contrast to the tools used to evaluate working postures and physical workloads, is that these tools rarely require a manikin as input. Chapters in this handbook related to tools for human performance and cognition are Chapters 15 and 37.

24.5 Discussion

Most current commercial DHMS focus on the analysis of static postures (Locket et al., 2005; Raschke et al., 2005; Lämkull, 2006; Wegner et al., 2007). Tools available for static analysis include, for instance, balance maintenance capability, calculation of joint moments, joint-specific strength, low back compression, and shear force estimates. For many tasks, the inertial loads due to acceleration of body segments and/or external objects may contribute considerably to internal body forces and tissue stresses (Wagner et al., 2007). Due to the complexity of incorporating the dynamics of motion into analysis, most DHMS do not have the capability to include dynamic effects. The obstacles to implementing dynamic analysis tools in DHMS are vast. At present, it is not possible to simulate task motions in DHMS with sufficiently accurate velocities and accelerations. Thus, a dynamic analysis still requires motion data captured from a human operator performing the analyzed work task or motion. The costs, complexity, and time-consuming activities of acquiring motion capture data restrict the simulation activities of dynamic analysis to relatively few tasks.

DHMS users of a decade ago used manikins with approximately 50 degrees of freedom in contrast to some of the manikins of today with about three times the number of degrees of freedom. With the increased level of detail in human models comes added complexity in manually programming postures and motions. In response to user requirements, DHMS developers must provide more sophisticated manikins. In order to be feasible in today's product and manufacturing engineering processes, they must also provide automated tools for simulating human motions. Because of these developments, a dynamic set of assessments will eventually become possible within DHMS. In light of this, it is interesting to consider the many recent advancements in the movie animation and gaming industries. The quantity and quality of computer-generated movies and computer games have increased greatly over the past 10 years. The realism of humans represented in these models and their associated motions are attracting customers and driving new business opportunities. DHMS providers should investigate the technology used in these industries for application in commercial DHMS. With an improved DHM and an open architecture that will support the inclusion of a company's own internal standard time data, there would be increased opportunity for users to validate productivity improvements, perform ergonomic assessments, and utilize DHM for virtual training. Implementation of intelligent manikins that self-posture and move realistically would enable application of DHMS to manual manufacturing evaluations.

Industry has benefited from the application of DHMS for static posture assessments. So far, limited success has resulted from attempts to apply the technology to assess dynamic motion. However, biomechanical inverse-kinematics software developers have begun to take consideration of dynamic moment variations in joints, and the lack of dynamic assessment tools has been identified as a potential area for development of new applications of DHMS.

One example is AnyBody Technology (of Aalborg University in Denmark). In their initial work, it is pointed out that the inertia of moving body segments affects the joint moments, indeed, and in such a way that static analysis consistently tends to underestimate peak values of analyzed tasks (Wagner et al., 2007). However, they point out that they have not demonstrated that ergonomists are currently failing to diagnose dangerous jobs by not including dynamic effects. It is at this stage difficult for a task analyst or ergonomist to determine when dynamic analysis becomes necessary, as opposed to when static analysis will suffice, as there are no established dynamic criteria that help guide the choice between static or dynamic analysis. AnyBody Technology concludes that including dynamic applications in ergonomic task analysis will only be meaningful if such criteria are further developed and disseminated. Another example is HUMOSIM (Human Motion Simulation Laboratory at the University of Michigan). HUMOSIM develops models to predict and evaluate realistic human movements. These models can be used by commercially available DHMS to enable ergonomic analysis of products and workplaces. Yet another example of a software developer is the Virtual Soldier Research (VSR) program housed in the University of Iowa. VSR's objective is to create humanlike life in a virtual world—virtual humans that can walk, behave, talk, and answer questions. VSR's method capitalizes on a novel optimization-based approach to motion prediction (Abdel-Malek et al., 2006). Using this new approach, motion is governed by human performance measures, such as speed and energy, which act as objective functions in an optimization formulation. In addition, constraints on joint torques and angles are imposed quite easily. Predicting motion in this way allows one to use manikins to study how and why humans move the way they do, given a specific scenario. It also enables manikins to react to an infinite number of scenarios with substantial autonomy.

Software representations of humans are increasingly the major tool used to perform ergonomic analyses of products and workplaces during the design phase. Compared with other human factors tools that are more abstract, DHM provide a high level of apparent validity and have proven to be very valuable for communicating physical design issues, such as reach and clearance. However, the Complete Digital Human—a digital replica of a real human who can think, learn, teach, smell, hear, taste, see, move, become worried, happy, angry, tired, wounded, and so forth—is far from a reality. The Complete Digital Human would be much too ambitious for any individual or group to undertake at present. It must be built by a global community of developers who are able, and willing, to share and review each other's work. The challenges lie in the vast amount of highly complex information that already exists about the human body and mind, the rate at which new information is being developed, and the fact that the information is currently stored in diverse forms in different research databases. However, at the rate that the capabilities of computers, developers, and DHMS are increasing alongside each other, this research area is expected to flourish at an impressive rate. It may be that the Complete Digital Human is not so much an ambitious vision, but rather an attainable goal in the not-so-distant future.

Acknowledgments

Financial support for writing this chapter comes from the Swedish Agency for Innovation System (VIN-NOVA) and Volvo Car Corporation. This support is gratefully acknowledged.

References

Abdel-Malek, K., Yang, J., Marler, T., Beck, S., Mathai, A., Zhou, X., Patrick, A., and Arora, J. (2006). Towards a new generation of virtual humans. *International Journal of Human Factors Modeling and Simulation*, Vol. 1, No. 1, pp. 2–39.

Abraham, S., Johnson, C.I., and Najjar, M.F. (1979). Weight and height of adults 18–74 years of age, United States, 1971–1974. US Department of Health, Education and Welfare, Washington, DC. Series 11, no. 211.

Adams, S.K. and Peterson, P.J. (1988). Maximum voluntary handgrip torque for circular electrical connectors. *Human Factors*, Vol. 30, pp. 733–745.

Allread, W.G., Marras, W.S., and Fathallah, F.A. (1998). The relationship between occupational musculoskeletal discomfort and workplace, personal, and trunk kinematic factors. Proceedings of the Human Factors and Ergonomics Society 42nd Annual Meeting, Chicago, IL. Human Factors and Ergonomics Society, Santa Monica, CA, pp. 896–900.

Annis, J.F. and McConville, J.T. (1990). Applications of anthropometric data in sizing and design. *Industrial Ergonomics and Safety II, International Industrial Ergonomics and Safety Conference*, Montreal, Taylor & Francis, pp. 309–314.

Arbetarskyddsstyrelsen (Swedish National Board of Occupational Safety and Health). (1998). AFS 1998:1. *Arbetarskyddsstyrelsens föreskrifter om belastningsergonomi* (*Ergonomics for the Prevention of Musculoskeletal Disorders*).

Archer, S., Archer, R., and Matessa, M. (2007). Our GRBIL has a split personality. *Proceedings of the 2007 SAE Digital Human Modeling for Design and Engineering Conference*, Seattle, WA. SAE 2007-01-2505.

Badler, N., Allbeck, J., Lee, S-J., Rabbitz, R.J., Broderick, T.T., and Mulkern, K.M. (2005). New behavioral paradigms for virtual human models. *Proceedings of the 2005 SAE Digital Human Modeling for Design and Engineering Symposium*, Iowa City, IA. SAE 2005-01-2689.

BIA-Report 5/2004. (2004). *Untersuchung der Belastung von Flugbegleiterinnen und Flugbegleitern beim Schieben und Ziehen von Trolleys in Flugzeugen. Hauptverband der gewerblichen Berufsagenossenschaften*, Berufsagenossenschaftliches Institut für Arbeitsschutz, Sankt Augustin.

Bidal, S., Bekkour, T., and Kayvantash, K. (2006). M-COMFORT: Scaling, positioning and improving a human model. Proceedings of the 2006 Digital Human Modeling for Design and Engineering Conference and Exhibition, Lyon, France. SAE 2006-01-2337.

Burandt, U. (1978). Ermitteln zulässiger Grenzwerte für Kräfte und Drehmomente. Firmeninterne Schulungsunterlagen zur Arbeitsgestaltung. Siemens.

Burandt, U. and Schultetus, W. (1978). Ermitteln zulässiger Grenzwerte für Kräfte und Dremomente. Firmeninterne Schulungsunterlagen zur Arbeitsgestaltung. Siemens.

Chaffin, D.B. (1997). Development of computerized human static strength simulation model for job design. *Human Factors and Ergonomics in Manufacturing*, Vol. 7, No. 4, pp. 305–322.

Chaffin, D.B. (2001). *Digital Human Modeling for Vehicle and Workplace Design*. Warrendale, PA, Society of Automotive Engineers.

Chaffin, D.B. and Andersson, G. (1984). *Occupational Biomechanics*. New York, John Wiley & Sons.

Corlett, E.N. (1981). Pain, posture and performance. *Stress, Work Design and Productivity*. Corlett, E.N. and Richardsson, J. (Eds.). London, UK, Wiley, pp. 27–42.

Cushman, W.H., Nielsen, W.J., and Pugsley, R.E. (1983). Workplace design. *Ergonomic Design for People at Work 1*. Eastman Kodak (Ed.). Belmont, CA. Lifetime Learning Publications.

Da Silva, M.C.G. (2002). Measurements of comfort in vehicles. *Measurement Science and Technology*, Vol. 13, No. 6, pp. R41–R60.

Delleman, N.J. (1992). Visual determinants of working posture. *Computer Application in Ergonomics, Occupational Safety and Health*. Mattila, M. and Karwowski, W. (Eds.). Elsevier Science Publishers.

Delleman, N.J. and Dul, J. (2002). Sewing machine operation: workstation adjustment, working posture, and worker's perceptions. *International Journal of Industrial Ergonomics*, Vol. 30, pp. 341–353.

DIN 33411-5. (1999). Körperkräfte des Menschen, Teil 5: Maximale statische Aktionskräfte—Werte (11/1999). Bleuth, Berlin.

Dukic, T. (2003). Applying computer mannequins to evaluate the visual demands in manual tasks. Licentiate thesis. Chalmers University of Technology, Gothenburg, Sweden.

Dukic, T. (2006). Visual demand in manual task performance: Towards a virtual evaluation. Doctoral thesis. Chalmers University of Technology, Gothenburg, Sweden.

Farrington, R.B., Rugh, J.P., Bharathan, D., and Burke, R. (2004). Use of a thermal mannequin to evaluate human thermoregulatory responses in transient, non-uniform, thermal environments. *Proceedings of the 34th SAE International Conference on Environmental Systems* (ICES), July 19–22, Colorado Springs, CO. SAE 2004-01-2345.

Fehren, M., Thunnissen, J., and Nicol, K. (2003). Influence of package factors on driver's posture and conclusions for posture assessment and modeling. *Proceedings of the 2003 SAE Digital Human Modeling for Design and Engineering Conference and Exposition*, Montreal, Canada. SAE 2003-01-2224.

Fitts, P.M. and Jones, R.E. (1947). Analysis of factors contributing to 460 "pilot errors" experiences in operating aircraft controls (Memorandum Report TSEA 4-694-12, Aero Medical Laboratory). Wright Patterson AFB, OH, Harry Armstrong Aerospace Medical Research Laboratory. Reprinted in: *Selected Papers on Human Factors in the Design and Use of Control Systems.* Sinaiko, H.W. (Ed.), New York, Dover, 1961.

Garg, A., Chaffin, D.B., and Herrin, G.D. (1978). Prediction of metabolic rates for manual materials handling jobs. *American Industrial Hygiene Association Journal,* Vol. 39, No. 8, 0002-8894.

Grandjean, E. (1980). Sitting posture of car drivers from the point of view of ergonomics. *Human Factors in Transport Research, part 1.* Grandjean, E. (Ed.). London, UK, Taylor & Francis.

Green, R. (2000). A generic process for human model analysis. *Proceedings of the Digital Human Modeling Conference,* Munich, Germany. SAE 2000-01-2167.

Hanson, L. (2004). Human vehicle interaction: Drivers' body and visual behaviour and tools and process for analysis. Doctoral thesis. Lund University, Sweden.

Haslegrave, C.M., Li, G., and Corlett, E.N. (1995). Factors affecting posture for machine sewing tasks. *Applied Ergonomics,* Vol. 26, pp. 35–46.

Hedberg, G. (1987). Epidemiological and ergonomical studies of professional drivers. *Arbete och Hälsa 9.*

Helander, M.G. and Zhang, L.J. (1997). Field studies of comfort and discomfort in sitting. *Ergonomics,* Vol. 40, pp. 895–915.

Hertzberg, H.T.E. (1972). The human buttock in sitting: Pressures, patterns, and palliatives. *American Automobile Transactions,* Vol. 72, pp. 39–47.

Hignett, S. and McAtamney, L. (2000). Rapid entire body assessment (REBA). *Applied Ergonomics,* Vol. 31, No. 2, pp. 201–205.

Hoekstra, P.N. (1993). Computer aided anthropometric assessment: "Seeing what you are doing." *Proceedings of International Conference on Engineering Design* (ICED'93).

Högberg, D. (2005). Ergonomics integration and user diversity in product design. Doctoral thesis. University of Skövde, Sweden and Loughborough University, UK.

Hoppe, P. (2001). Different aspects of assessing indoor and outdoor thermal comfort. *Energy and Buildings,* Vol. 34, No. 6, pp. 661–665.

HQL (1997). Research Institute of Human Engineering for Quality Life. Japanese Body Size Data 1992–1994, Osaka (in Japanese).

Hursh, S.R., Balkin, T.J., Miller, J.C., and Eddy, D.R. (2004). The fatigue avoidance scheduling tool: Modeling to minimize the effects of fatigue on cognitive performance. *Proceedings of the 2004 SAE Digital Human Modeling for Design and Engineering Symposium,* Rochester, MI. SAE 2004-01-2151.

Kama, W.N. (1963). Volumetric workspace study: I. Optimum workspace configuration for using various screwdrivers. Report No. AMRL-TDR-63-68 (I). Aerospace Medical Research Laboratories, WPAFB, OH.

Kama, W.N. (1965). Volumetric workspace study: II. Optimum workspace configuration for use of wrenches. Report No. AMRL-TDR-63-68 (II). Aerospace Medical Research Laboratories, WPAFB, OH.

Kant, I., Notermans, J., and Borm, P. (1990). Observations of working postures in garages using the Ovako Working Posture Analysing System (OWAS) and consequent Workload Reduction Recommendations. *Ergonomics,* Vol. 33, pp. 209–220.

Karhu, O., Kansi, P., and Kuorinka, I. (1977). Correcting working postures in industry: A practical method for analysis. *Applied Ergonomics,* Vol. 8, No. 4, pp. 199–201.

Kaynakli, O. and Kilic, M. (2005). Investigation of indoor thermal comfort under transient conditions. *Building and Environment,* Vol. 40, No. 2, pp. 165–174.

Konz, S. (1983). *Work Design: Industrial Ergonomics.* Columbus, OH, Grid Publishing.

Krist, R. (1994). Modellierung des Sitzkomforts: Eine experimentelle Studie (Modeling of sit comfort: An experimental study). Doctoral thesis. Katolischen Universität Eichstätt, Germany.

Lämkull, D. (2006). *Computer Mannequins in Evaluation of Manual Assembly Tasks*. Licentiate thesis. Chalmers University of Technology, Gothenburg, Sweden.

Landau, K. (2000). *Ergonomics Software Tools in Product and Workplace Design: A Review of Recent Developments in Human Modeling and Other Design Aids*. Landau, K. (Ed.). Stuttgart, Germany, ERGON GmbH.

Laring, J. (2004). Ergonomic workplace design analysis: Development of a practitioner's tool for enhanced productivity. Doctoral thesis. Chalmers University of Technology, Gothenburg, Sweden.

Laughery, K.R., Lebiere, C., and Archer, S. (2006). Modeling human performance in complex systems. *Handbook of Human Factors and Ergonomics*. Salvendy, G. (Ed.). Hoboken, NJ, John Wiley & Sons, pp. 967–996.

Lewin, T. (1969). Anthropometric studies on Swedish industrial workers when standing and sitting. *Ergonomics*, Vol. 12, pp. 883–902.

Liberty Mutual (2004). Liberty Mutual Manual Materials Handling Tables. Retrieved September 21, 2007, from: http://libertymmhtables.libertymutual.com /CM_LM TablesWeb.

Lockett, J.F., Assmann, E., Green, R., Reed, M.P., Rascke, R., and Verriest, J-P. (2005). Digital human modeling research and development user needs panel. *Proceedings of the 2005 SAE Digital Human Modeling for Design and Engineering Symposium*, Iowa City, IA. SAE 2005-01-2745.

Mackworth, N.H. (1948). The breakdown of vigilance during prolonged visual search. *Quarterly Journal of Experimental Psychology*, Vol. 1, pp. 5–61.

Marler, T., Rahmatalla, S., Shanahan, M., and Abdel-Malek, K. (2005). A new discomfort function for optimization-based posture prediction. *Proceedings of the 2005 SAE Digital Human Modeling for Design and Engineering Symposium*, Iowa City, IA. SAE 2005-01-2680.

Marx, B., Amann, C., and Verver, M. (2005). Virtual assessment of seating comfort with human models. *Proceedings of the 2005 SAE Digital Human Modeling for Design and Engineering Symposium*, Iowa City, IA. SAE 2005-01-2678.

Mathiassen, S.E., Wells, R.P., Winkel, J., Forsman, M., and Medbo, L. (2002). Tools for integrating engineering and ergonomic assessment of time aspects in industrial production. *Proceedings of the 34th Annual Congress of the Nordic Ergonomics Society*, Kolmården, Sweden, pp. 579–584 (Vol. 2).

McAtamney, L. and Corlett, E. N. (1993). RULA: A survey method for the investigation of work-related upper limb disorders. *Applied Ergonomics*, Vol. 24, No. 2, pp. 91–99.

McGuffin, R., Burke, R., Huizenga, C., Hui, Z., Vlahinos, A., and Fu, G. (2002). Human thermal comfort model and mannequin. *Proceedings of the 2002 SAE Future Car Congress*, Arlington, VA. SAE 2002-01-1955.

Mergl, C., Anton, T., Madrik-Dusik, R., Hartung, J., Librandi, A., and Bubb, H. (2004). Development of a 3D finite element model of thigh and pelvis. *Proceedings of the 2004 SAE Digital Human Modeling for Design and Engineering Symposium*, Rochester, MI. SAE 2004-01-2132.

Munck-Ulfsfält, U., Falck, A., and Forsberg, A. (1999) Requirement specification for load ergonomics. Internal document at Volvo Car Corporation, Göteborg, Sweden.

Munck-Ulfsfält, U., Falck, A., Forsberg, A., Dahlin, C., and Eriksson, A. (2003). Corporate ergonomics programme at Volvo Car Corporation. *Applied Ergonomics*, Vol. 34, pp. 17–22.

NASA (1978). *Anthropometric Source Book*. National Aeronautics and Space Administration, Scientific and Technical Information Office, NASA Technical Report. TL 787 U5494 no. 1024, Webb Associates.

NIOSH WPG (1981). Work Practices Guide for Manual Lifting. NIOSH Technical Report No. 81-122. U.S. Department of Health and Human Services. National Institute for Occupational Safety and Health, Cincinnati, OH 45226.

Nishimoto, T. and Murumaki, S. (1998). Prediction on traumatic head injury by computer biomechanics. *Journal of the Society of Automotive Engineers of Japan*, Vol. 52. Tokyo, Japan.

Ntuen, C.A. (1999). The application of fuzzy set theory to cognitive workload evaluation of electronic circuit board inspectors. *Human Factors and Ergonomics in Manufacturing*, Vol. 9, pp. 291–310.

OSHA (2006). U.S. Department of Labor. Occupational Safety & Health Administration. OSHA Technical Manual (OTM), Directive number: TED 1-0.15A. Section VII: Ergonomics, Chapter 1. Back disorders and injuries.

Pankoke, S. and Siefert, A. (2007). Virtual simulation of static and dynamic seating comfort in the development process of automobiles and automotive seats: Application of finite-element-occupant-model CASIMIR. *Proceedings of the 2007 SAE Digital Human Modeling for Design and Engineering Conference*, Seattle, WA. SAE 2007-01-2459.

Paquet, E. and Rioux, M. (2003). Anthropometric visual data mining: A content-based approach. *Proceedings of the XVth Triennial Congress of IEA 2003*. International Ergonomics Association, Seoul, South Korea.

Parkinson, M.B. and Reed, P. (2004). Standing reach envelopes incorporating anthropometric variance and postural cost. *Proceedings of the 2007 SAE Digital Human Modeling for Design and Engineering Conference*, Seattle, WA. SAE 2007-01-2482.

Petoussi-Henss, N., Zankl, M., Fill, U., Heide, B., and Regulla, D. (2002). A family of human voxel models for various applications. *Proceedings of the Digital Human Modeling Conference*, Munich, Germany. SAE 2002-07-0030.

Pheasant, S. (1991). Anthropometry and the design of work places. *Evaluation of Human Work: A Practical Ergonomics Methodology*. Wilson, J. and Corlett, N. (Eds.). London, Taylor & Francis, pp. 455–471.

Porter, J.M., Case, K., Marshall, R., Gyi, D.E., and Sims, R.E. (2004). Beyond Jack and Jill: designing for individuals using HADRIAN. *International Journal of Industrial Ergonomics*, Vol. 33, pp. 249–264.

Porter, J.M. and Gyi, D.E. (1998). Exploring the optimum posture for driver comfort. *International Journal of Vehicle Design*, Vol. 19, pp. 255–266.

Raschke, U., Kuhlmann, H., and Hollick, M. (2005). On the design of a task based human simulation system. *Proceedings of the 2005 SAE Digital Human Modeling for Design and Engineering Symposium*, Iowa City, IA. SAE 2005-01-2702.

Rasmussen, J. and Toerholm Christensen, S. (2005). Musculoskeletal modeling of egress with the Any-Body Modeling System. *Proceedings of the 2005 SAE Digital Human Modeling for Design and Engineering Symposium*, Iowa City, IA. SAE 2005-01-2721.

Rebiffe, R. (1969). An ergonomic study of arrangement of the driving positions in motorcars. Symposium, London, England.

Robinette, K.M. (1998). The Civilian American and European Surface Anthropometry Resource (CAESAR) project. *Proceedings of Workshop on 3D Anthropometry and Industrial Products Design*, Paris, France, pp. 18.1–18.3.

Robinette, K.M. and McConville, J.T. (1981). An alternative to percentile models. Warrendale, Society of Automobile Engineers, SAE, Technical Paper 810217.

Roe, R.W. (1993). Occupant packaging. *Automotive Ergonomics*. Peacock, B. and Karwowski, W. (Eds). London, UK, Taylor & Francis, pp. 11–42.

Roebuck, J.A. (1995). *Anthropometric Methods: Designing to Fit the Human Body*. Santa Monica, CA, Human Factors and Ergonomics Society.

Roebuck, J.A., Kroemer, K.H.E., and Thomson, W.G. (1975). *Engineering Anthropometry Methods*. New York, John Wiley & Sons.

Rönnäng, M., Lämkull, D., Dukic, T., and Örtengren, R. (2004). Task-related field of view parameters. *SAE 2004 Transactions Journal of Aerospace and Proceedings of the 2004 SAE Digital Human Modeling for Design and Engineering Symposium*, Rochester, MI, SAE 2004-01-2195.

SAE (2007). SAE J287 (R) Driver Hand Control Reach. SAE Handbook (2007). SAE International, Warrendale, PA.

Schaub, K., Landau, K., Menges, R., and Grossman, K. (1997). A computer-aided tool for ergonomic workplace design and preventive health care. *Human Factors and Ergonomics in Manufacturing*, Vol. 7, pp. 269–304.

Schultetus, W., Lange, W., and Dorken, W. (Eds). (1987). *Praxis der ergonomie—Montagegestaltung*. TÜV Rheinland, Köln.

Siefert, A., Delavoye, C., and Cakmak, M. (2006). CASIMIR: Human finite-element-model for static and dynamic assessment of seating comfort. *Proceedings of the 16th Triennial Congress of the International Ergonomics Association*. Maastricht, the Netherlands.

Snook, S. H. and Ciriello, V. M. (1991). The design of manual handling tasks: Revised tables of maximum acceptable weights and forces. *Ergonomics*, Vol. 34, No. 9, pp. 1197–1213.

Sundin, A. and Örtengren, R. (2006). Digital human modeling for CAE applications. *Handbook of Human Factors and Ergonomics*. Salvendy, G. (Ed.). Hoboken, NJ, John Wiley & Sons, pp. 1053–1078.

Svensson, I. and Sandström, R. (1995). BUMS- Belastningsergonomisk Utvärderings Mall Saab, Produktion. Saab Automobile AB, Trollhättan, Sweden.

Tilley, A. and Dreyfuss, H. (1993). *The Measure of Man and Woman: Human Factors Design*. New York, Whitney Library of Design.

Toftum, J., Rasmussen, L.W., Mackeprang, J., and Fanger, P.O. (2000). Discomfort due to skin humidity with different fabric textures and materials. *American Society of Heating, Refrigerating, and Air-Conditioning Engineers' Transactions*, Vol. 106, pp. 521–529.

VDI. (1980). *Handbuch der Arbeitsgestaltung und Arbeitsorganisation*. VDI Verlag, Düsseldorf.

Verver, M. (2004). *Numerical Tools for Comfort Analyses of Automotive Seating*. Doctoral thesis. Technische Universiteit Eindhoven, The Netherlands.

Verver, M.M. and van Hoof, J. (2003). Development of a FEM pelvis model for analysis of pressure distribution. *Proceedings of the 2003 Digital Human Modeling Conference*, Munich, Germany. SAE 2003-01-2214.

Volvo Corporate Standard. (2002). Ergonomic guidelines with respect to strain. STD 8003, Vol. 2, No. 5, 2002–04.

Wagner, D., Rasmussen, J., and Reed, M. (2007). Assessing the importance of motion dynamics for ergonomic analysis of manual materials handling tasks using the AnyBody Modeling System. *Proceedings of the 2007 SAE Digital Human Modeling for Design and Engineering Conference*, Seattle, WA. SAE 2007-01-2504.

Walters, B., Bzostek, J., and Li, J. (2005). Integrating human performance and anthropometric modeling in the Crew Station Design Tool. *Proceedings of the 2005 SAE Digital Human Modeling for Design and Engineering Symposium*, Iowa City, IA. SAE 2005-01-2698.

Waters, T. R., Putz-Anderson, V., Garg, A., and Fine, L. J. (1993). Revised NIOSH equation for the design and evaluation of manual lifting tasks. *Ergonomics*, Vol. 36, No. 7, pp. 749–776.

Wegner, D., Chiang, J., Kemmer, B., Lämkull, D., and Roll, R. (2007). Digital human modeling requirements and standardization. *Proceedings of the 2007 SAE Digital Human Modeling for Design and Engineering Conference*, Seattle, WA. SAE 2007-01-2498.

Wickens, C.D. (1992). *Engineering Psychology and Human Performance*. New York, HarperCollins. ISBN 0-673-46161-0.

Zhang, L. (2003). Cognitive task network modeling and human task simulation visualization in 3D environment (-preliminary feasibility study). *Proceedings of the 2003 SAE Digital Human Modeling for Design and Engineering Conference and Exposition*, Montreal, Canada, SAE 2003-01-2202.

Zhang L., Helander, M.G., and Drury, C.G. (1996). Identifying factors of comfort and discomfort in sitting. *Human Factors*, Vol. 38, No. 3, pp. 377–389.

25

Discomfort Evaluation and Motion Measurement

Xuguang Wang

25.1 Introduction

Ergonomic evaluation of a product often requires building up a physical mockup or a prototype, having a group of experts or a representative sample of users to test it and to give their discomfort feeling. This is an expensive and time-consuming process, especially when the product design has to be modified and revalidated. Thanks to recent progress on motion capture, motion analysis, and motion modeling, simulating industrial application oriented complex motions is becoming possible (see, for instance, Monnier et al., 2003, 2006; Park et al., 2004; also Ausejo & Wang, 2009, and Monnier et al., 2009 in this handbook). However, the evaluation of the discomfort associated with such complex motions is another challenge for digital human modeling researchers. Two fundamental questions are posed:

How to measure the discomfort perceived by a subject, knowing that it is subjective and that there is no other measurement instrument than the subject himself?

How to define discomfort criteria based on biomechanical parameters, such as joint angles, joint forces, work, energy, muscle efforts, …

Much work has been published on the methods for postural classification and evaluation of postural stressfulness (Karhu et al., 1977; Keyserling, 1990; Genaidy et al., 1994; McAtamney & Corlett, 1993; Hignett & McAtamney, 2000; Kee & Karwowski, 2001). Some of them are implemented in digital human software, such as RULA (Rapid Upper Limb Assessment; McAtamney & Corlett, 1993) and OWAS (Ovako Working posture Analysis System; Karhu et al., 1977). However, most of these methods were initially developed for observation of working postures in industry. Only a very rough estimation of posture is usually required either from direct visual estimation or from recorded video. In addition, the postural evaluation criteria were decided by a group of ergonomics experts. Only a few recently proposed postural ranking systems are based on criteria other than experts' opinion. For instance, the LUBA proposed by Kee and Karwowski (2001) was based on subjects' perceived discomfort and maximum holding times of the static postures studied by Miedema et al. (1997).

Clearly, there is a need to develop motion-oriented discomfort predictive models intended for implementation in a digital human model. However, modeling the discomfort feeling of a task-oriented movement is not simple. First, the task involves, in most cases, the motion of more than one joint. Second, it is a dynamic process, and the associated discomfort depends not only on the final posture but also on the whole motion. Third, discomfort criteria should be generic enough so that they can be used to evaluate a large number of activities. Finally, they should also be quantitative and easily identified through experiments.

The purpose of this chapter is to overview currently available discomfort models for assessing motion and to outline some future research directions.

25.2 Comfort/Discomfort and Rating Methods

Comfort and discomfort were assumed by many researchers and practitioners as two opposites on a rating scale, like the general comfort rating scale (GCR) used by Shackel et al. (1969) or the overall seven-point discomfort rating scale by Corlett and Bishop (1976). Although this was questioned by a few investigators (e.g., Branton, 1969), it was not until a very recent date that Helander and his colleagues (Zhang et al., 1996; Helander & Zhang, 1997; Helander, 2003) clearly argued that discomfort and comfort are two different and complementary entities. Discomfort is primarily associated with physiological and biomechanical factors, whereas comfort is primarily associated with aesthetics. A practical consequence of this conceptual distinction is that comfort and discomfort need to be treated as different and complementary entities in ergonomic evaluation (De Looze et al., 2003). Another consequence is that they need to be evaluated separately. A unipolar is therefore preferred to evaluate discomfort. Moreover, it is believed that a biomechanical approach is appropriate only for investigating the discomfort associated with a task-oriented movement.

A subjective rating scale is commonly used for comfort/discomfort evaluation. Shen and Parsons (1997) tested and compared six commonly used rating scales for seated pressure discomfort. The category partitioning scale CP-50 was found to be the most reliable and valid for rating pressure intensity and perceived discomfort. Chevalot and Wang (2004) preferred a modified CP-50 scale (fig. 25.1) and found it more easily understandable by subjects. Instead of asking subjects to directly state a number from 0 to 50 (or higher) as used by Shen and Parsons, subjects were asked to first select the category that fits their feeling, then to refine their judgment by choosing a number from 1 to 10 within the selected category. The real scale from 0 to 50 (original CP-50) was hidden from the subject in order to give the priority to the category choice rather than a direct numerical rating.

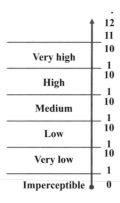

FIGURE 25.1 Slightly modified category partition CP-50 scale. (From Chevalot, N., and X. Wang, "Experimental Investigation of the Discomfort of Arm Reaching Movements in a Seated Position," SAE International Conference and Exposition of Digital Human Modeling for Design and Engineering, SAE International, Warrendale, PA, 2004. With permission.)

Discomfort as a response to be measured is clearly multidimensional. It may not have quantitative structure as questioned by Annett (2002). The rating scales of direct estimation may not possess interval or ratio properties. Stevens's free modulus estimation method offers an alternative to direct estimation. Ratio judgments methods are designed, by construction, to follow a ratio scale. Kee and his colleagues (Kee & Karwowski, 2001a and b, 2003; Chung et al., 2005) used free modulus estimation method to estimate perceived discomfort for evaluation of postural load. However, it should be noted that this method requires complex experimental procedure and careful data processing such as subject selection, training, data normalization and response anchoring, and so on (Han et al., 1999). Recently, conscious of possible non-numerical structure of the constructs like discomfort, Maurin (2003) developed an original method to indirectly estimate numerical responses from an ordered category scale. The method is based on the Thurstonian randomness of subjects' responses under the general frame of measurement theory. The basic idea behind this method is to consider the rating as a random variable which may obey a probability law. The proposed method was applied to the study of the comfort of pedal operation in relation with seat height, pedal inclination, pedal travel, and pedal resistance (Wang & Maurin, 2003) in case of mono-variables. The method was also extended to bivariate situations for studying reach discomfort (Maurin, 2006).

25.3 Methods for Modeling Discomfort Associated with Motion

It is well accepted that a body discomfort model should take into account the following three factors, as formulated in the conceptual cube model by Laring et al. (2002) for ergonomic workload analysis of complex tasks:

Postural strain (POSTURE)
External force exertion in pushing/pulling or in manual material handling (FORCE)
Static or repetitious loads (TIME)

Although most body discomfort assessment systems such as OWAS (Karhu et al., 1997), RULA (McAtamney & Corlett, 1993), and REBA (Hignett & McAtamney, 2000) are more oriented toward postural analysis, level of muscle forces and load type are explicitly or implicitly taken into account in recommendations. Ideally, a discomfort predictive model should include these three factors.

Up to now, most discomfort models are specific and their range of application is limited to experimental conditions. Two types of specific models can be distinguished, one based on design parameters and the other on biomechanical parameters. For design parameter-based models, investigators try to relate the perceived discomfort with relevant design parameters of a product. For instance, pressure distribution was found to be the most associated with the sitting (dis)comfort ratings from a review of 21 studies (De Looze et al., 2003). Giacomin and Quattrocolo (1997) attempted to correlate comfort feeling of car ingress/egress motion with several car configuration variables. The main advantage is that design parameters discomfort models can be directly used by designers to optimize the product design. But these models are completely independent of a human model and disconnected with associated motions. In addition, it is impossible to investigate the effects of a design parameter that is not included in the experimental design.

Biomechanical discomfort models are believed to be more appropriate for explaining perceived discomfort as they are based on body-related parameters, such as joint angle, torque, or even muscle forces. Discomfort is a subjective feeling about the body's interaction with the external world (e.g., extra-body excitation, tool use, working posture, etc.). Relevant biomechanical parameters can be a kind of translation by the body to this interaction. Jung and Choe (1996) developed a discomfort model for arm reaching posture based on joint angles. Wang et al. (2004) searched for a possible relationship between (dis)comfort rating and biomechanical parameters when depressing a clutch pedal. More recently, Dickerson et al. (2006) investigated the relationship between shoulder torque and perception of muscular effort in loaded reaches. Compared to specific discomfort models in terms of design parameters, biomechanical discomfort models help us understand possible sources of discomfort. But these models are still too specific to task and can hardly be used for general-purpose discomfort prediction.

There is a need to propose a generalized approach of discomfort modeling. Within a European search project named REAL MAN (see Lestrelin & Trabot, 2005, for a general presentation of the project), a general discomfort modeling approach has been proposed. The basic idea is to evaluate the discomfort of a movement by assessing the discomfort from the smallest entity, that is, at the level of each degree of freedom (DoF) for each joint to the whole body movement. The approach requires:

The identification of the static discomfort function of each degree of freedom in terms of angle and force level

The investigation of how to combine these low-level discomfort functions to evaluate the discomfort of a static posture involving a set of DoFs

The extension of a static discomfort function into a dynamic one

More explicitly, a discomfort model for each DoF of each joint D_i was proposed in this project. It was based on the assumption that the discomfort is mainly dependent on its joint angle α_i relative to its maximum range of motion α_{imax} as well as on its joint moment (torque) M_i relative to its maximum moment M_{imax}. This can be expressed as

$$D_i = D_i\,(M_i/M_{imax}, \alpha_i/\alpha_{imax}) \tag{25.1}$$

Furthermore, it is assumed that joint angles and joint moments affect the discomfort *independently*, thus allowing the separation of discomfort function D_i into two functions, one only dependent on joint angle called angle discomfort function k_i, and the other only dependent on joint torque called torque discomfort function f_i,

$$D_i = f_i\,(M_i/M_{imax}, \alpha_i/\alpha_{imax}) = k_i\,(\alpha_i/\alpha_{imax})f_i(M_i/M_{imax}) \tag{25.2}$$

In addition, it is also hypothesized that f_i's may be the same for all DoFs, or at least for a group of DoFs,

$$D_i = D_i\,(M_i/M_{imax}, \alpha_i/\alpha_{imax}) = k_i\,(\alpha_i/\alpha_{imax})f(M_i/M_{imax}) \tag{25.3}$$

Although M_{imax} itself is joint angle dependent, the torque discomfort function f_i may not be dependent on joint angle but only on the ratio M_i/M_{imax}. Besides, Equation 25.3 states that a generic form of torque discomfort function may exist for at least a group of joint DoFs. Specific experiments were designed to test these assumptions and to identify joint angle and strength-based discomfort functions (Zacher & Bubb, 2004).

The second problem is how to evaluate discomfort of a static posture once discomfort function is known for each DoF implied in the posture. The aim is to evaluate the discomfort feeling of a static body posture from the individual DoF discomfort functions D_i. If we assume that a body part discomfort is a linear combination of D_i's, then

$$D_{body} = \Sigma c_i D_i + c_0 \tag{25.4}$$

where c_i is the weighting coefficient associated to each DoF involved in the body part.

Finally, once static joint and body discomfort functions are identified, one has to extend them to dynamic models, which can be used for evaluating the discomfort of a joint or a body part motion. The dynamic discomfort model may take a form like, for instance, for one DoF,

$$D_i^{dynamic} = F_i\left(\int D_i^{static}(M_i(t)/M_{imax}, \alpha_i(t)/\alpha_{imax})dt\right) \tag{25.5}$$

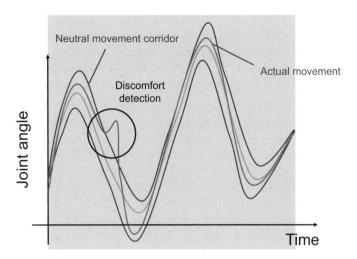

FIGURE 25.2   Neutral movement concept.

For the movement of a body part, the associated discomfort may be a combination of discomfort feelings of all DoFs involved in the motion, which can take a form like Equation 25.4.

The success of such an approach depends on the availability of the basic data on the discomfort perception in terms of joint angle and joint force. Bubb and his colleagues (Bubb, 2003; Zacher & Bubb, 2004) attempted to identify joint angle and torque-related discomfort functions. Kee and Karwowski (2001, 2003) defined joint angle iso-comfort functions and proposed a ranking system for evaluating joint motion discomfort. Interestingly, a similar approach was proposed by Chung et al. (2005) for postural classification and whole-body postural stress prediction from the postural classification of body parts.

Dufour and Wang (2005) proposed the concept of neutral movement for identifying joint discomfort function and applied it to the analysis of the discomfort of car ingress/egress movement. It is an extension of the concept of neutral or least discomfort posture, which was used for postural comfort evaluation (fig. 25.2). Ideally, the neutral movement for a person should be the one that generates the least discomfort for a task. The basic idea is to define a corridor for each joint angle, which reflects the intra- and inter-individual variability of the neutral movements. Any deviation from the neutral movement corridor due to environment constraints may reveal a possible discomfort perceived during the movement. Clearly, discomfort evaluation depends on how the neutral movement is defined. Monnier et al. (2006) implemented the neutral movement concept in a motion simulation and evaluation tool, named RPx (see also Monnier et al., 2009 in this handbook). The discomfort of ingress/egress movements for the light truck was evaluated. Only joint angle variation was considered in the discomfort function for each DoF D_i in Equation 25.1. In this study, 16 different light truck configurations were tested. The movements corresponding to the configurations rated as the lowest discomfort were chosen as neutral movements. The experimental data were used to adjust the coefficients required in Equation 25.4 for the key frame postures. The global movement discomfort was then evaluated from these key frame postures. Although the concept of neutral movement itself cannot explain why such a movement is preferred by subjects, it can be considered a practical computer-aided design tool that reconstructs the judgments of subjects.

25.4 Concluding Remarks

In this paper, the methods for evaluating motion-related discomfort have been summarized. The methods proposed for postural stressfulness analysis, such as OWAS and RULA, are not well adapted for digital human tools due to the fact that they were initially developed for qualitative field observation in industry. Currently available discomfort models are either design-parameter-based or biomechanical models specific to studied tasks. Although some interesting generic discomfort modeling approaches have been proposed in recent years, like the one based on the concept of neutral movement, further research efforts in motion-related discomfort modeling are highly required especially for:

Developing a more rigorous discomfort rating method
Having more detailed biomechanical data, like joint range of motion and joint strength
Developing more sophisticated musculoskeletal biomechanical models capable of predicting not
    only joint forces but also muscular forces and other soft tissue loads

Clearly, good predictive discomfort models should rely on reliable discomfort rating methods. The research community in (dis)comfort is more and more aware of the limitations of direct estimation rating scales.

More realistic data of joint range of motion (ROM) are still missing. The effects of coupling between joint axes and clothing are not well quantified. This is also true for joint moment strength, despite a large amount of human strength data for material handling. Maximum joint torques are even more difficult to obtain than joint ROMs, because of a high number of trials that a subject has to execute. In fact, they are joint position dependent and force direction dependent.

Until now, discomfort, induced by internal biomechanical constraints that affect the human musculoskeletal system, has not adequately been taken into account in digital human modeling. We believe that discomfort modeling requires a detailed muscular activities simulation. Though computation of muscular forces from kinematical data is a very challenging task, recent advances in modeling of the musculoskeletal system allow researchers to quantify muscle forces and the effect of tendon transfer (see a review by Van Sint Jan, 2005).

References

Ausejo, S., Wang, X., 2009. Motion capture and human motion reconstruction. In: Duffy, V.S. (Ed.), *Handbook of Digital Human Modeling*. Taylor & Francis, Boca Raton, FL.

Branton, P., 1969. Behavior, body mechanics and discomfort. *Ergonomics*, Vol. 12, No. 2, 316–327.

Bubb, H., 2003. Research for a strength based discomfort model of posture and movement. *International Ergonomics Association XVth Triennial Congress*, August 24–29, 2003, Seoul.

Chevalot N., Wang X., 2004. Experimental investigation of the discomfort of arm reaching movements in a seated position. *SAE International Conference and Exposition of Digital Human Modeling for Design and Engineering*, June 15–17, Oakland University, Rochester, MI. SAE paper 2004-01-2141.

Chung, M.K., Lee, I., Kee, D., 2005. Quantitative postural load assessment for whole body manual tasks based on perceived discomfort. *Ergonomics*, Vol. 48, No. 5, 492–505.

Corlett, E.N., Bishop, R.P., 1976. A technique for assessing postural discomfort. *Ergonomics*, Vol. 19, No. 2, 175–182.

De Looze, M.P., Kuijt-Evers, L.F.M., Van Dieën, J., 2003. Sitting comfort and discomfort and relationships with objective measures. *Ergonomics*, Vol. 46, No. 10, 985–997.

Dickerson, C.R., Martin, B.J., Chaffin, D.B., 2006. The relationship between shoulder torques and the perception of muscular effort in loaded reaches. *Ergonomics*, Vol. 49, No. 11, 1036–1051.

Dufour, F., Wang X., 2005. Discomfort assessment of car ingress/egress motions using the concept of neutral movement. *SAE International Conference and Exposition of Digital Human Modeling for Design and Engineering*, June 15–17, Iowa City, IA. SAE paper 2005-01-2706.

Genaidy, A.M., Al-shedi, A.A., Karwowski, W., 1994. Postural stress analysis in industry. *Applied Ergonomics*, Vol. 25, No. 2, 77–87.

Giacomin J., Quattrocolo S., 1997. An analysis of human comfort when entering and exiting the rear seat of an automobile. *Applied Ergonomics*, Vol. 28, No. 5/6, 397–406.

Han, S.H., Song M., Kwahk, J., 1999. A systematic method for analyzing magnitude estimation data. *International Journal of Industrial Ergonomics*, Vol. 23, 513–524.

Helander, M.G., 2003. Forget about ergonomics in chair design? Focus on aesthetics and comfort. *Ergonomics*, Vol. 46, 1306–1319.

Helander, M.G. and Zhang, L. 1997. Field studies of comfort and discomfort in sitting. *Ergonomics*, Vol. 40, No. 9, 895–915.

Hignett, S. and McAtamney, L., 2000. Rapid entire body assessment (REBA). *Applied Ergonomics*, Vol. 31, 201–205.

Jung, E.S., Choe, J., 1996. Human reach posture prediction based on psychophysical discomfort. *International Journal of Industrial Ergonomics*, Vol. 18, 173–179.

Karhu, O., Kansi, P., Kuorinka, I., 1977. Correcting working postures in industry: a practical method for analysis. *Applied Ergonomics*, Vol. 8, No. 4, 199–201.

Kee, D., Karwowski, W., 2001a. The boundaries for joint angles of isocomfort for sitting and standing males based on perceived comfort of static joint postures. *Ergonomics*, Vol. 44, No. 6, 614–648.

Kee, D., Karwowski, W., 2001b. LUBA: An assessment technique for postural loading on the upper body based on joint motion discomfort and maximum holding time. *Applied Ergonomics*, Vol. 32, 357–366.

Kee, D., Karwowski, W., 2003. Ranking systems for evaluation of joint and joint motion stressfulness based on perceived discomforts. *Applied Ergonomics*, 34, 167–176.

Keyserling, W.M., 1990. Computer-aided posture analysis of the trunk, neck, shoulders and lower extremities. In: Karwowski, W., Genaidy, A.M., Asfour, S.S. (Eds.), *Computer-Aided Ergonomics*. Taylor & Francis, London, pp. 261–272.

Laring, J., Forsman, M., Kadefors, R., Örtengren, R., 2002. MTM-based ergonomic workload analysis. *International Journal of Industrial Ergonomics*, Vol. 30, 135–148.

Lestrelin, D., Trabot, J., 2005. The REAL MAN project: Objectives, results and possible follow-up. *SAE International Conference and Exposition of Digital Human Modeling for Design and Engineering*, June 15–17, Iowa City, IA. SAE paper 2005-01-2682.

Maurin, M., 2003. Measures and measurement from data—advanced mathematical basements. *Proceeding of the Young Researchers Seminar ECTRI-FERSI*, Lyon, INRETS, December 2003.

Maurin, M., 2006. An original comfort/discomfort quantification in a bi-variate controlled experiment: Application to the discomfort evaluation of seated arm reach. *SAE International Conference and Exposition of Digital Human Modeling for Design and Engineering*, July, Lyon, France. SAE paper 2006-01-2347.

McAtamney, L., Corlett, E.N., 1993. RULA: a survey method for the investigation of work-related upper limb disorders. *Applied Ergonomics*, Vol. 24, No. 2, 91–99.

Miedema, M.C., Douwes, M., Dul, J., 1997. Recommended maximum holding times for prevention of discomfort of static standing postures. *International Journal of Industrial Ergonomics*, Vol. 19, 9–18.

Monnier, G., Renard, F., Chameroy, A., Wang, X., Trasbot, J. 2006. Motion simulation approach integrated into a design engineering process. *SAE International Conference and Exposition of Digital Human Modeling for Design and Engineering*, July, Lyon, France. SAE paper 2006-01-2359.

Monnier, G., Wang, X., Trasbot, J., 2009. A motion simulation tool for automotive interior design. In: Duffy, V.S. (Ed.), *Handbook of Digital Human Modeling*. Taylor & Francis, Boca Raton, FL.

Monnier, G., Wang, X., Verriest, J.P., Goujon, S., 2003. Simulation of complex and specific task-oriented movements—application to the automotive seat belt reaching. *SAE International Conference and Exposition of Digital Human Modeling for Design and Engineering*, June, Montreal, Canada. SAE paper 2003-01-2225.

Park, W., Chaffin, D.B., Martin, B.J., 2004. Toward memory-based human motion simulation: development and validation of a motion modification algorithm. *IEEE Transactions on Systems, Man, and Cybernetics—Part A: Systems and Humans*, Vol. 34, No. 3, 376–386.

Shackel, B., Chidsey, K.D., Shipley, P. 1969. The assessment of chair comfort. *Ergonomics*, Vol. 12, 269–306.

Shen, W., Parsons, K.C., 1997. Validity and reliability of rating scales for seated pressure discomfort. *International Journal of Industrial Ergonomics*, Vol. 20, 441–461.

Van Sint Jan, S., 2005. Introducing anatomical and physiological accuracy in computerized anthropometry for increasing the clinical usefulness of modelling systems. *Critical Reviews in Physical and Rehabilitation Medicine*, Vol. 17, No. 4, 249–274.

Wang, X., Le Breton-Gadegbeku, L., Bouzon, L., 2004. Biomechanical evaluation of the comfort of automobile clutch pedal operation. *International Journal of Industrial Ergonomics*, Vol. 34, 209–221.

Wang, X., Maurin, M., 2003. Discomfort/comfort assessment using an ordered category scale. *International Ergonomics Association XVth Triennial Congress*, August 24–29, 2003, Seoul.

Zacher, I., Bubb, H., 2004. Strength based discomfort model of posture and movement. *SAE Digital Human Modeling for Design and Engineering Symposium Proceedings*, June 15–17, 2004, Oakland University, Rochester, MI. SAE paper 2004-01-2139.

Zhang, L., Helander, M.G., Drury, C.G., 1996. Identifying factors of comfort and discomfort in sitting. *Human Factors*, Vol. 38, No. 3, 377–389.

26

Optimization in Design: A DHM Perspective

Matthew B. Parkinson

26.1 Introduction

Humans are highly variable on many functional measures that are related to artifact design. The wide ranges of adult standing height, hip breadth, and other body dimensions (called "anthropometry") are readily observed and often considered quantitatively in design. However, variability in human perception, behavior, and performance can be equally or more important than anthropometric variability—but these factors are less commonly considered in a quantitative manner. Human adaptability diminishes, but does not eliminate, the impact of inter-individual variability on artifact performance. The ubiquity of "one-size-fits-all" is a testament to this adaptability, but is not a prescription for good design, particularly in cases where performance is important and people interact with the artifact through multiple interfaces.

Designing for people requires the quantitative consideration of all relevant aspects of human variability. Simultaneous consideration of variability in both the users and the product or environment being designed is extremely complex and generally requires the assistance of computers. While rule-of-thumb guidelines or iterative trial-and-error are common approaches to design, the nature of the problem makes it ideal for the algorithms and tools that formal optimization provides.

A survey of the literature indicates that the words "optimization" and "optimal" have been used somewhat casually. Any iterative design process has been considered "optimization" and its results pronounced "optimal." For the purposes of this chapter the words will be used more particularly. Optimization is an iterative, algorithmically controlled process by which design variables are systematically varied and the performance of the resulting design assessed. This process continues until some stopping criteria (e.g., a maximum number of iterations or function evaluations is reached or some mathematical conditions are met) are achieved. The resulting design is considered optimal. Optimal may be local or global depending on whether the design is guaranteed to be the best possible design for the entire design space, or merely the best in a reduced region of that space. The algorithms that control the optimization process are generally classified by the method employed for conducting the iterations and improving the design (e.g., derivative-based, genetic, etc.) and whether the resulting design is a local or global optimum.

One of the primary advantages to using optimization in this context is the facility it provides for the simultaneous consideration of coupled systems. For example, mechanical design performance can be considered at the same time as the design ergonomics. This capability has become increasingly important as other principles of concurrent engineering have been adopted into design processes (Resnick, 1996). Without optimization the two are often decoupled or broken into hierarchical systems to simplify analysis (Helander & Lin, 1998), which may not produce optimal results (Burns & Vicente, 2000).

Optimization has been used to solve many different kinds of problems involving digital human models. These can be divided into these general classes: motion modeling and posture prediction; environment and task design; product design and model development.

26.2 Motion Modeling and Posture Prediction

Essential to digital human modeling (DHM) is the prediction of realistic postures and motions. Abdel-Malek et al. (2001) give a brief but excellent summary of several types of posture prediction. Traditional efforts in this area involve the creation of data-driven models that predict motions or postures based on the observation of real people performing similar tasks. Increased computing power has facilitated other approaches where the body is modeled as kinematic chains and optimization is used to find the joint angles that satisfy some criteria. For example, Kothiyal et al. (1992) predicted postures by minimizing the force at the L4/L5 intervertebral joint subject to constraints on the forces allowed at other joints in the body. Abdel-Malek et al. (2001) and subsequent papers have expanded on this approach by examining more complex objective functions and constraints and by extending the method to predicting entire motions rather than single postures. This usually relies on solving a multi-objective optimization problem. Since different weighting of the objectives results in different motion outcomes, their selection is a critical and problematic process. Also, since the method does not rely on actual data, there is no guarantee that the predicted motion has any relation to what might be observed in the real world. Despite these limitations the approach has found acceptance because of its strengths: It is readily adapted to new environments and tasks.

26.3 Environment and Task Design

Digital human models are frequently used in the design of tasks and environments. Since the nature of many tasks is highly dependent upon the environment in which the task is conducted, the two are often considered simultaneously. For example, consider a simple task in which a heavy part must be lifted and moved from a shelf in one location to a shelf in another. A single shelf height might be ideal for one individual, but what shelf height maximizes the number of people within a population who can safely and repetitively perform the task? This problem has traditionally been solved by evaluating a couple of representative figure models (called "mannequins" or "boundary mannequins") in the key postures within the task (Bittner, 2000; HFES 300 Committee, 2004). The environment is adjusted and evaluated iteratively until a good solution is obtained. Without optimization this process can be tedious and time-consuming. This limits the ergonomist or designer in both the number of figure models utilized and the quantity and variety of environmental variables that are modified.

The inclusion of optimization in the design process allows simultaneous consideration of the large amount of variability in the way tasks are performed, the people who do them, and the ways in which the work cell can be configured. For example, using genetic algorithms, Carnahan and Redfern (1998) applied the National Institute for Occupational Safety and Health (NIOSH) guidelines for manual lifting to task design to achieve a family of solutions. In another example that integrates motion and posture prediction, Abdel-Malek et al. (2004) used optimization to design environments that minimized measures such as the joint loading for an individual operator.

26.4 Layout Design

Layout design is related to the design of tasks and environments. It is considered separately here because it is usually posed as a different kind of optimization problem and solved using different algorithms. In the layout design problem there are a finite number of components that need to be optimally placed within a finite space (fig. 26.1). Sometimes this space is very well defined, but it can also be variable and subject to modification by the optimization algorithm. The application of optimization to the vehicle packaging problem has been demonstrated for both trucks (Parkinson et al., 2005) and cars (Parkinson & Reed, 2006). This includes the selection of locations for the seat, steering wheel, pedals, and other components, subject to boundary constraints (such as floor height, roof height, firewall position, and cab length). Within cab dimension limits and other constraints, the vehicle interior is engineered to maximize the accommodation of the design population, where accommodation means that a person is able to perform all required tasks while seated in a comfortable position. However, even among accommodated individuals, a vehicle usually provides a wide range of performance on other important measures, such as headroom and exterior vision.

26.5 Product Design

In addition to the design of environments within which people work and play, optimization and digital human models can influence the design of the products we use. Selecting the ideal shape and balance of a hand tool, for example, might involve the simultaneous consideration of biomechanical, anthropometric,

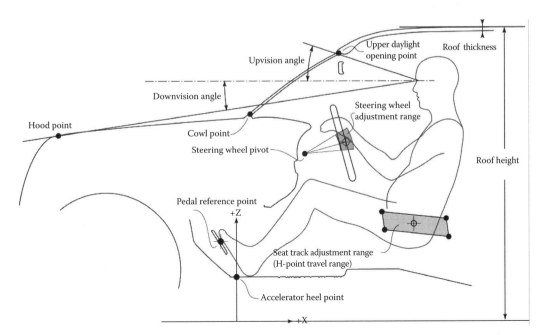

FIGURE 26.1 Dimensions and reference points used in the optimization of passenger car interior packaging. (From Parkinson & Reed, 2006.)

and tool performance issues. In a study of bicycle gearing, Cho et al. (1999) used optimization to determine the best possible system of gears given several design criteria.

Their objective was to quantify the optimal pedaling rates for a given output power by changing the number of gears and the gear ratios. The problem was constrained mechanically by gear geometries and by physiological measures including heart rate and muscle activation.

26.6 Model Development

The scope of digital human modeling has moved beyond mannequins and ergonomic lookup tables to include statistics-based models of human capability, preference, and performance. In some situations, optimization can be an invaluable tool in their development. For example, in Sasena et al. (2005) optimization was used to adaptively control an experiment designed to gather reach difficulty data. Participants were asked to perform a push-button task at a variety of locations within their right-hand reach envelope. Given the size of the space involved and the finite attention span of participants, it was important that data be gathered in the most efficient manner possible.

A statistical model of each participant's reach difficulty assessment was built in real-time and optimization was used to select which point the participant should evaluate next. The optimization problem was to determine what location in space, if sampled, would minimize the uncertainty in the model—and should therefore be selected next. Figure 26.2 shows how the variance within the model decreased as sample points were added (the variance at a sampled point is 0). Using this approach, the desired fidelity for a single participant was achieved in as few as 60 reach trials. The result was an invaluable design tool: a population model that included both the maximum reach envelope and reach difficulty for sub-maximal seated reaches.

The remainder of this chapter details how optimization problems incorporating human variability are posed, how that variability is modeled, how the optimization problems are solved, and the limitations of current methodologies.

26.7 Solving the Problem

A general approach to optimization in design using digital human models is outlined in Figure 26.3. First, the population of users, operators, or workers must be defined and modeled. This could be as rudimentary as selecting a pair of mannequins to represent the anthropometric extremes in the user population (the problems inherent in this approach are discussed in a coming section). Some authors have suggested using 3 or 17 mannequins (Bittner, 2000). With advanced models that include the residual variance, more mannequins might be used (Parkinson & Reed, 2006, used 1,000 mannequins). At a minimum, the gender mix and anthropometric variability of the user population should be considered.

After the user population has been specified, the objective function must be selected. This is the function or metric that will be maximized or minimized in the optimization process. It is an assessment of

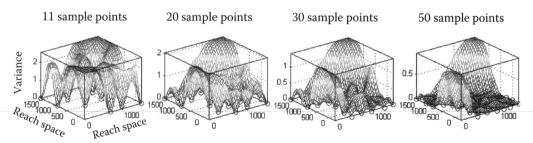

FIGURE 26.2 The use of optimization enabled a dramatic reduction in model variance with the addition of relatively small numbers of sample points. (From Sasena et al., 2005.)

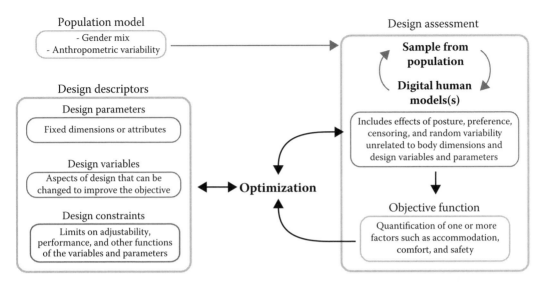

FIGURE 26.3 Schematic of a typical optimization methodology, showing submodels and information flow. The optimization algorithm coordinates the systematic modification of the design descriptors to improve the functionality as assessed by the digital human model or models.

how well a given design meets the needs of the user population. While this is often a single measure, such as accommodation, it can be a combination of measures, each assigned a weight toward the overall function. For example, it could be some combination of safety, accommodation, and cost. In the vehicle packaging example described above, the objective function might be to minimize the number of drivers who are unable to sit in their preferred location—in other words, to minimize the number of disaccommodated drivers.

In order for the objective function to be calculated, the candidate design must be specified. There are three types of design descriptors: design parameters, design variables, and design constraints. Design parameters are the dimensions or attributes that are fixed. In the vehicle packaging problem this might be the vehicle geometry. Design variables are the aspects of the design that will be changed to improve the objective function. In the packaging problem these might be the location of the seat and steering wheel adjustment ranges (fig. 26.1). The optimization programmatically changes the values of the design variables to create a new design, which is then evaluated to calculate the objective function. Based on this new calculated value, the design variables are modified and the process continues until the stopping criteria are met.

The final types of design descriptors are design constraints. These are limits on performance metrics like stress or strain, and on adjustability and other functions of the variables and parameters. Design constraints in the packaging problem might include Federal Motor Vehicle Safety Standards, vehicle performance measures, or simple geometric constraints such as the relative positions of the seat and steering wheel. Although limits on the ranges that the design variables can take are often instituted to speed algorithm convergence or restrict the models to areas for which they are validated, these limits are typically considered differently from design constraints involving more than one variable.

In order to calculate the value of the objective function, the design is evaluated for its ability to meet the specified needs of the population. This is done in one of two ways. In the first, a virtual assessment is conducted for each member of the population and the results of the assessments are combined in some way to obtain a representative metric (e.g., 92% accommodation or average accommodation score of "6"). In the second, the design is assessed using a population model, and the accommodation level is inferred directly. In either case, the design assessment provides an opportunity to calculate the objective function and to determine whether compliance with the constraints is maintained. The user models can include many kinds of effects: posture; preference; censoring (how drivers might behave when they are disaccommodated); and the random variability that is unrelated to anthropometry and design variables and parameters. This process is described in the following example.

26.8 Vehicle Packaging Example

The vehicle is represented by a set of parameter values, constraints, and design variables. The categorization of cab features varies across design situations. Usually some features of the cab are fixed by the desire to use an existing chassis or other components. The design variables can include features of the cab architecture, such as cowl height and roof height, but often will be restricted to component locations (fig. 26.1). In this example, the design variables define the locations of the seat (x_{seat}, z_{seat}) and steering wheel (x_{sw}, z_{sw}) adjustment ranges (fig. 26.4). The ranges of the design variables may be constrained by vehicle specifications and other considerations. For example, common design constraints limit the overall vehicle height and cab fore-aft length.

The fitness of a particular vector of design variable values is evaluated by a virtual fitting trial in which a population of drivers is postured. This population can be obtained in a number of ways, but in this example there was random sampling of 1,000 sets of anthropometric data from the specified (e.g., 1988 U.S. Army) population. These became 1,000 virtual drivers or mannequins used to evaluate candidate designs. The preferred steering wheel and seat positions for each driver are predicted as a function of body dimensions and vehicle interior geometry. This is done using the cascade modeling approach (Reed et al., 2002) applied to data from the laboratory and in vehicle studies of truck-driver posture (Reed et al., 2000a; Jahns et al., 2001). First, the preferred steering wheel position is predicted from the driver's stature. Next, the driver's seat position is predicted using a regression model that takes into account leg length and body mass index. Finally, torso posture is predicted from body dimensions, steering wheel position, and seat position, taking into account whether the headroom is restrictive to the individual driver.

Importantly, each measure (steering wheel, seat, hip, and eye positions) is predicted by both the anthropometry detailed above and a random component that describes each individual driver's deviation from the mean. This accounts for the residual variance in posture and component-location preference that is unrelated to body dimensions and is obtained by random sampling from appropriate distributions. For example, preferred steering wheel position is only weakly related to body dimensions,

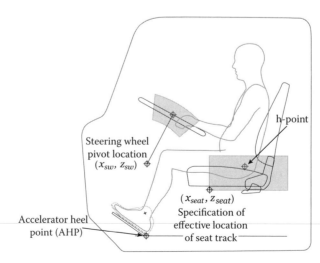

FIGURE 26.4 Cross section of a 2D truck packaging problem. The horizontal and vertical locations of the seat and steering wheel adjustment ranges are the design variables.

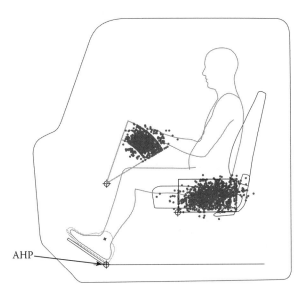

FIGURE 26.5 The results of a virtual fitting trial with 1,000 drivers (only one of the virtual drivers is shown). The preferred seat and steering wheel locations for the entire cadre are shown relative to the location of the vehicle geometry and truck and steering wheel adjustment ranges.

so including the residual variance is critical to ensuring an adequate adjustment range. To improve the efficiency of the simulation, the vector of random posture components is sampled once for each sampled vector of anthropometric variables, yielding a virtual driver vector, characterized by body dimensions and preferences relative to other drivers with the same dimensions. Figure 26.5 shows the results of a virtual fitting trial. The preferred seat and steering wheel locations are shown, relative to the cab geometry and component locations, for each of the 1,000 virtual drivers.

26.9 Modeling Performance

To this point, the models used to quantify the objective function and compliance with constraints have not been discussed. Good models are vital to successful optimization. In addition to the usual requirement of accuracy, the models must also be robust. Once computers have begun the optimization process, there is typically little interaction with the operator until it is complete. It is possible that the algorithm will choose to evaluate regions of the design space that are different from those that are typically utilized. Designers need to ensure that the models will be viable across a broad range of inputs.

Within digital human modeling, the models upon which the optimization routines operate typically relate population attributes and design attributes to quantify some level of performance. They can include factors like posture or preference and design performance measures as a function of body size. They might also consider the effects of restrictions due to component locations (censoring) and random variability unrelated to body dimensions and design variables and parameters.

For purposes of physical accommodation, human variability can usefully be partitioned into the variability related to body dimensions (anthropometric variability) and behavioral (e.g., postural) variability. For example, the height of a driver's eyes above the seat is related to both torso length and torso posture. The most common approach to representing anthropometric variability is the use of mannequins or templates that represent people at dimensional extremes. The (often implicit) rationale for using only a few "boundary mannequins" is that designs accommodating the anthropometric extremes (for example a woman who is 5th percentile by stature and a man who is 95th percentile by stature) will also accommodate people with less-extreme dimensions. Many contemporary research publications approach design problems in this manner, including methods for optimizing workspaces and controls in aircraft cockpits (Chung & Park, 2004; Hamza et al., 2004; Berson, 2002). The selection of anthropometric extreme cases has been extended to the use of many boundary mannequins selected with consideration of anthropometric covariance (HFES 300 Committee, 2004). For example, the A-CADRE family of 17 mannequins represents much of the multivariate anthropometric variability in an adult population (Bittner, 2000). Boundary mannequin sampling approaches are commonly used with DHM software to create figure models. None of these approaches, however, considers postural variability.

Any use of mannequins in design requires that they be postured in realistic ways. Posturing is often performed manually by the designer, but a number of approaches to posturing mannequins have been developed (Faraway et al., 1999; Chaffin et al., 2000; Reed et al., 2002; Abdel-Malek et al., 2001; Loczi, 2000). In the vehicle packaging example, driver postures are related by statistical models to data gathered from real drivers in a variety of laboratory and vehicle configurations. Mannequin posturing algorithms are usually deterministic, giving a single posture for a particular combination of mannequin body dimensions and task constraints. However, people who have the same body dimensions often perform tasks (driving in this example) with substantially different postures (Flannagan et al., 1998; Reed et al., 2000a). As a consequence, purely deterministic models, even with ideally accurate posture prediction, are insufficient for quantitative assessment of accommodation (Porter et al., 2004; Reed & Flannagan, 2000; Reed et al., 2000b; Parkinson & Reed, 2006).

The effects of postural variability that is unrelated to body dimensions must be taken into account in design. In vehicle packaging, the SAE Recommended Practices for driver packaging accomplish this by the use of unified statistical models that encompass population variability in both body dimensions and postural behavior. For example, the eyellipse (SAE J941) approximates the distribution of driver eye locations in vehicle space as a three-dimensional normal distribution. Because it models eye location directly while taking into account the combined effects of vehicle dimensions, anthropometric variability among drivers, and postural variability, the eyellipse has been one of the most elegant and effective tools ever developed for human factors analysis (Roe, 1993).

26.10 Posing the Problem

Optimization problems typically consist of an objective function, constraints, and design variables. Functions or models relate the design variables to the objective and constraints. Although most software now allows the objective to be specified as either a minimization or a maximization problem, most of the algorithms are actually developed to minimize the objective. In fact, many software tools automatically convert the user-specified maximization problem into a minimization problem for internal calculations. The standard form for an optimization problem is called negative-null form and looks like this:

$$
\begin{aligned}
\text{minimize} \quad & f(\mathbf{x}) \\
\text{subject to} \quad & h(\mathbf{x}) = 0 \\
& g(\mathbf{x}) \leq 0
\end{aligned}
\tag{26.1}
$$

where x is the vector of design variables, h are the vector-valued functions comprising the equality constraints, and g are the vector-valued functions comprising the inequality constraints.

In the vehicle packaging problem described in previous sections, the objective might be to minimize disaccommodation (i.e., minimize the number of people who are not able to sit in their preferred manner within the vehicle). For a fixed vehicle geometry, this could be achieved by finding the best possible location for the seat and steering wheel adjustment ranges. Their x and z locations, (x_{seat}, z_{seat}) and (x_{sw}, z_{sw}), comprise the vector of design variables, x. Constraints govern the location of the adjustment ranges. For example, the vertical location of the steering wheel needs to be above the vertical location of the seat. Similarly, the horizontal location of the seat needs to be aft of the horizontal location of the steering wheel. Finally, the size of the design space can be reduced by providing some limits on the values the design variables can take (they might be restricted to realistic numbers that lie within the cab, for example). In traditional form the optimization problem has become:

$$
\begin{aligned}
\text{minimize} \quad & f(\mathbf{x}) \\
\text{subject to} \quad & z_{seat} - z_{sw} \leq 0 \\
& x_{sw} - x_{seat} \leq 0 \\
& 0 \leq x_{sw} \leq 1000 \\
& 0 \leq z_{sw} \leq 1000 \\
& 0 \leq x_{seat} \leq 1000 \\
& 0 \leq z_{seat} \leq 1000
\end{aligned}
\tag{26.2}
$$

where $f(\mathbf{x})$ is a quantification of disaccommodation as a function of the location of the seat and steering wheel adjustment ranges. Notice there are no equality or h constraints. This is by design. Equality constraints are much more difficult than inequality constraints for most optimization algorithms to satisfy so they are generally avoided when possible.

Once the designer is able to define the problem in this manner, it can readily be implemented within an appropriate optimization environment—provided the equations calculating the objectives and constraints are also ready for implementation.

It is important to note that there are many, many ways in which to look at a design problem. For example, rather than minimizing disaccommodation and determining the best location for adjustment ranges of a fixed size, the problem could have been inverted. In other words, a related problem might be:

> Minimize   *magnitude of adjustment ranges*
> Subject to   *95% of drivers are accommodated*
>   *Seat and steering wheel are inside the car*
> By changing   *size of seat adjustment range*   (26.3)
>   *Size of steering wheel adjustment range*
>   *Location of seat adjustment range*
>   *Location of steering wheel adjustment range*

In this formulation, the answer (assuming there is one that meets the criteria) will accommodate the specified percentage of the population (95%). In contrast, using the formulation given in Equation 26.2 the accommodation level would have been the highest possible given the limitations imposed by the constraints. Each formulation has its advantages and might be used in different circumstances. For a detailed example of these issues, see Parkinson and Reed (2006) and Parkinson et al. (2005).

26.11 Limitations and Ongoing Work

The accuracy and utility of incorporating DHM into design are limited by the validity of the underlying models. Better models are needed to describe both postural and subjective responses to censoring of various kinds, particularly for censoring multiple degrees of freedom. In the cab optimization problem, improved cost functions are also needed for safety-related measures such as exterior vision. An additional limitation lies in the complexity of the problems that can be solved using optimization. It is frequently difficult for designers to understand all the intricacies of the results, putting them in the uncomfortable position of having to rely on the validity of the answers given them. This is particularly problematic because most optimization of complex problems will result in local, rather than global, optima (Burns & Vicente, 2000) and the designer must decide if the result is "good enough" or whether additional refinement and computation are required. In any case, all results should be carefully examined for validity.

The outlined approach can differ in important ways from standard industry practice. The inclusion of residual variance within the models, which the use of optimization facilitates, is a relatively new concept and an uncommon practice. In the packaging example presented here, the stochastic posturing methods explicitly consider residual variance in posture that is unrelated to body dimensions and cab geometry. This allows more accurate quantification of population accommodation and design fitness. In contrast to the current SAE tools, the new method is explicitly multivariate and allows simultaneous considerations of multiple design features while maintaining the quantitative rigor that is the primary strength of the SAE models. Moreover, the current implementation spans a larger range of potential design variables than current SAE tools and can be readily expanded to encompass more. Finally, the new methods allow unambiguous inclusion of both subjective (e.g., comfort) and objective (e.g., safety) metrics in cab optimization.

In addition to improved models, future work will examine this methodology in the broader context of designing for human variability. For example, since the entire population is currently sampled multiple times during each iteration of the optimization, the computational expense of complex problems can increase rapidly. At any particular step in optimization using digital human models, only a subset of the current population may be contributing to changes in the objective function. Optimizing for carefully selected subsets of the population prior to evaluation with a large group may provide the best balance between speed and accuracy. Other benefits from the methodology, such as exploring the design space for alternative designs and understanding design trade-offs, are also under investigation.

This is an active area of research. Current and future work involves the improvement of modeling strategies and the application of robust multi-objective optimization strategies to solve these problems. Recent advances in optimization, such as the inclusion of probabilistic constraints, are also being introduced into the problem formulation—which may be one way to treat multi-objective problems.

26.12 Review Questions

What are four ways in which optimization is used in design with digital human models?

What is the difference between an iterative design process and optimization?

From an optimization perspective, what is the difference between design parameters and design variables?

From an optimization perspective, what is the difference between constraints and objectives?

26.13 Other Resources

26.13.1 Optimization Software

Microsoft Excel Solver is built into Excel and is developed by Frontline Systems, Inc. More information at http://www.solver.com

MATLAB® and the MATLAB® Optimization Toolbox, http://www.mathworks.com

Optimization toolkit and summary of free and commercial packages, maintained by the Optimal Design Laboratory at the University of Michigan, http://www.optimaldesign.org/software.html

iSIGHT software, developed by Engineous, http://www.engineous.com

26.13.2 Online Resources

The OPEN Design Laboratory at Penn State University, http://www.opendesignlab.org

Text, "Principles of Optimal Design" by Papalambros and Wilde. 2000. Cambridge University Press. Online materials, including homework and solutions, at http://www.optimaldesign.org.

Designing for Human Variability, http://www.dfhv.org

References

Abdel-Malek, K., Yu, W., Mi, Z., Tanbour, E., and Jaber, M. (2001). Posture prediction versus inverse kinematics. Proceedings of the ASME International Design Engineering Technical Conferences. Pittsburgh, PA.

Abdel-Malek, K., Yu, W., Yang, J., and Nebel, K. (2004). A mathematical method for ergonomic based design: Placement. *International Journal of Industrial Ergonomics*, 34(5):375–394.

Berson, B. L. (2002). Modernization of a military aircraft cockpit. *Ergonomics in Design*, 10(3):11–16.

Bittner, A. C. (2000). A-CADRE: Advanced family of manikins for workstation design. *Proceedings of Human Factors and Ergonomics Society*, pages 774–777. Long Beach, CA.

Burns, C. M., and Vicente, K. J. (2000). Participant-observer study of ergonomics in engineering design: How constraints drive design process. *Applied Ergonomics*, 31(1):73–82.

Carnahan, B. J., and Redfern, M. S. (1998). Application of genetic algorithms to the design of lifting tasks. *International Journal of Industrial Ergonomics*, 21(2):145–158.

Chaffin, D., Faraway, J., Zhang, X., and Woolley, C. (2000). Stature, age, and gender effects on reach motion postures. *Human Factors*, 42(3):408–420.

Cho, C. K., Yun, M. H., Yoon, C. S., and Lee, M.W. (1999). Ergonomic study on the optimal gear ratio for a multi-speed bicycle. *International Journal of Industrial Ergonomics*, 23(1-2):95–100.

Chung, S. J., and Park, M. Y. (2004). Three-dimensional analysis of a driver-passenger vehicle interface. *Human Factors and Ergonomics in Manufacturing*, 14(3):269–284.

Faraway, J., Zhang, X., and Chaffin, D. (1999). Rectifying postures reconstructed from joint angles to meet constraints. *Journal of Biomechanics*, 32(7):733–736.

Flannagan, C. A. C., Manary, M. A., Schneider, L. W., and Reed, M. P. (1998). Improved seating accommodation model with application to different user populations. *Proceedings of the SAE International Congress & Exposition*, volume 1358, pages 43–50. SAE, Warrendale, PA.

Hamza, K., Hossoy, I., Reyes-Luna, J. F., and Papalambros, P. Y. (2004). Combined maximization of interior comfort and frontal crashworthiness in preliminary vehicle design. *International Journal of Vehicle Design*, 35(3):167–185.

Helander, M. G., and Lin, L. (1998). Optimal sequence in product design. *Proceedings of the Human Factors and Ergonomics Society*, 1:569–573.

HFES 300 Committee (2004). Guidelines for Using Anthropometric Data in Product Design. *Human Factors and Ergonomics Society*, Santa Monica, CA.

Jahns, S. K., Reed, M. P., and Hardee, H. L. (2001). Methods for in-vehicle measurement of truck driver postures. SAE Technical Paper 2001-01-2821.

Kothiyal, K. P., Mazumdar, J., and Noone, G. (1992). Biomechanical model for optimal postures in manual lifting tasks. *International Journal of Industrial Ergonomics*, 10(3):241–255.

Loczi, J. (2000). Application of the 3-D CAD manikin ramsis to heavy truck design. *Proceedings of the Human Factors and Ergonomics Society*, pages 832–835. Human Factors and Ergonomics Society.

Parkinson, M., and Reed, M. (2006). Optimizing vehicle occupant packaging. *SAE Transactions: Journal of Passenger Cars–Mechanical Systems*, 115.

Parkinson, M., Reed, M., Kokkolaras, M., and Papalambros, P. (2005). Robust truck cabin layout optimization using advanced driver variance models. *Proceedings of the ASME International Design Engineering Technical Conferences*. Long Beach, CA.

Parkinson, M., Reed, M., Kokkolaras, M., and Papalambros, P. (in press). Optimizing truck cab layout for driver accommodation. *ASME Journal of Mechanical Design*.

Porter, J. M., Case, K., Marshall, R., Gyi, D., and Sims, R. E. (2004). Beyond Jack and Jill: Designing for individuals using HADRIAN. *International Journal of Industrial Ergonomics*, 33(3):249–264.

Reed, M. P., and Flannagan, C. A. C. (2000). Anthropometric and postural variability: Limitations of the boundary manikin approach. Technical Paper 2000-01-2172. *SAE Transactions: Journal of Passenger Cars—Mechanical Systems*, 109.

Reed, M. P., Lehto, M. M., and Schneider, L. W. (2000a). Methods for laboratory investigation of truck and bus driver postures. Technical Paper 2000-01-3405. *SAE Transactions: Journal of Commercial Vehicles*, 109.

Reed, M. P., Manary, M. A., Flannagan, C. A. C., and Schneider, L. W. (2000b). Comparison of methods for predicting automobile driver posture. Technical Paper 2000-01-2172. *SAE Transactions: Journal of Passenger Cars—Mechanical Systems*, 109.

Reed, M. P., Manary, M. A., Flannagan, C. A. C., and Schneider, L. W. (2002). A statistical method for predicting automobile driving posture. *Human Factors*, 44(4):557–568.

Resnick, M. L. (1996). Concurrent ergonomics: a proactive approach. *Computers & Industrial Engineering*, 31(1-2):479–482.

Roe, R. W. (1993). Occupant packaging. In Peacock, B., and Karwowski, W., editors, *Automotive Ergonomics*, pages 11–42. Taylor & Francis, London.

Sasena, M. J., Parkinson, M., Reed, M. P., Papalambros, P. Y., and Goovaerts, P. (2005). Improving an ergonomics testing procedure via approximation-based adaptive experimental design. *Journal of Mechanical Design, Transactions of the ASME*, 127(5):1006–1013.

27

Ergonomics for Computer Usage

Paris F. Stringfellow,
Sajay Sadasivan, and
Anand K. Gramopadhye

27.1 The Computer Workstation

Like the species itself, human work has also gone through evolutionary change. Work we do today is very different from the work our bodies were biologically designed for. Our distant ancestors once spent a majority of their time hunting, foraging, and constantly moving. Over millennia, the human body adapted to the demanding conditions of this lifestyle as evident in our body structures. Ancient life involved substantial physiological demands—demands that have changed with today's work requirements. Now, instead of foraging for berries, we sort incoming documents; instead of traveling enormous distances to reach new hunting grounds, we sit for hours in a single location working with man's greatest achievement since the wheel—the computer. Yes, our lives have changed exponentially since Man 1.0. Our daily jobs now constrain the body that evolution equipped us with and require us to continually adapt to the evolution of human work.

27.2 And Along Came the Computer

We have come a long way since the Turing Machine. When the personal computer entered the mainstream in the late 1980s, life as we knew it would change forever. The modern office is equipped with a variety of technologies from wireless printers to handheld PDAs, the most prominent of which is the standard desktop PC. A recent U.S. census revealed that 76 million employed U.S. adults use a computer at their place of work (Cheeseman et al., 2003), an increase of 17% since the last census taken in 2001, and reports indicate that this trend is steadily growing. This piece of technology has single-handedly redefined the traditional work environment. However, as with all technological progress, there are unintended consequences. The popularity of the computer, both in the home and in the office, has resulted in new physiological strains on how we do our work.

27.3 Office Ergonomics: Benefits for People, Benefits for Business

With this increase in computer use, so too has arisen a new strain of work-related musculoskeletal disorders (MSDs), including carpal tunnel syndrome, back disorders, and so on. In 1999, Liberty Mutual reported that hand, wrist, and shoulder injuries were a fast-growing source of disability in the American workforce, rising in large part from the increase in computer usage in the workplace (Liberty Mutual, 1999). Musculoskeletal pain and discomfort, together with eyestrain, make up at least half of all subjective complaints of computer operators (Helander, 1988). Each year, estimated costs resulting from computer-based MSDs range in the millions, while employees are suffering from lost pay, and sometimes lifelong afflictions. Fortunately, the potential impacts of extended computer use are being realized, and much work and research has been done related to the ergonomics of computer workstations.

One major contributor, Antonio Grieco, looked at postural fixity and the impacts of sitting postures during computer-based work (Grieco, 1986). Etienne Grandjean also took a novel look at the visual aspects in VDT workstation design (Grandjean, 1987). Currently, the National Standard for Human Factors Engineering of Visual Display Terminal Workstations (ANSI/HFES 100-1988) provides standard ergonomic workstation measurements based on individual anthropometric data. Further, a number of books have been dedicated to the subject including Kroemer and Kroemer's *Office Ergonomics* published in 2001, *Ergonomics in Computerized Offices* by Etienne Grandjean published in 1986, and Dan MacLeod's *The Office Ergonomics Kit* from 1998.

27.4 What Influences Office Ergonomics?

Designing for the office requires the consideration of a number of different elements, from job rotation schedules to choosing the right chair. It can be broken up into four subsequent areas of focus: the

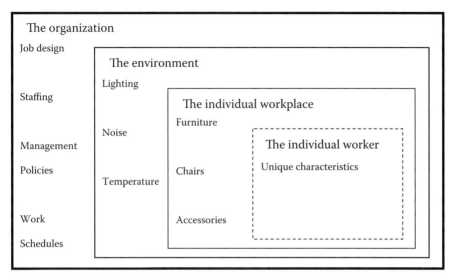

FIGURE 27.1 The organization of office ergonomics design.

organizational structure, environmental factors, the individual workstation, and the individual worker (WISHA, 2002), and is illustrated in Figure 27.1.

The organizational component looks at job design, staffing levels and procedures, as well as work schedules as they impact the worker. Here, items that are controlled through administrative procedures, such as scheduled break times, are considered. Within the organizational structure of the workplace, the office environment plays a part in affecting the ergonomic quality of the worker. Environmental elements such as monitor glare from uncovered windows or excessive noise from a busy hallway can influence both worker productivity and comfort. When environmental conditions cannot be controlled universally, personal adjustments can be made to the individual workplace. At this level, elements such as furniture and accessories can be added and customized to the worker's personal needs. Although much can be done to enhance the organization, environment, and workspace surrounding the worker, the worker's behavior toward ergonomics can also be improved through training and experience. All of these individual components can work together to impact the ergonomic quality of an office. This chapter considers the interaction of these levels as we begin to design and evaluate ergonomics for the office and for computer usage.

The objectives of this chapter are to provide a general overview of the ergonomic challenges associated with computer workstation use and to serve as a resource for ergonomic evaluation techniques with a special emphasis on computer-based human modeling approaches.

27.5 Office-Related CTDs

27.5.1 The Risks of Computer Usage

Workplace injuries are a major source of insurance claims today. In fact, repetitive motion claims make up 6.10% of all U.S. workplace accident claims (Michael, 2001). However, not all workplace injuries consist of slips and falls or loss of limbs. Many injuries occur during simple computer use and can have devastating impacts on workers' long-term health and livelihoods. For a significant period of time, injuries incurred during typewriter and computer use were not recognized as work-related injuries simply because they were not immediately recognizable or they could not be traced back to a single point of impact and therefore were not validated as a workplace claim. As a result, many office workers were forced to relinquish their jobs and were left with significant lifelong pain and even crippling effects.

It was not until the early 1980s that the ergonomic dangers of computer usage began to emerge. Now, we recognize the influence of such job demands on worker safety and classify injuries incurred as a result of long-term misusage, including computer work, as cumulative trauma disorders (CTDs). CTDs develop from long-term use, misuse, or overuse of joints, ligaments, muscles, tendons, cartilage, or spinal discs. Typically the disorder does not stem from one single acute or traumatic incident, but rather from the application of continual strain and pressure over time. Risk factors for CTDs include repetition, static loading or sustained exertions, awkward postures, mechanical contact stress, force, vibrations, and extreme temperatures. In effect, the dangers of these syndromes are twofold in that their origins can be difficult to identify as the injury results from long-term repeated actions that may not be harmful when performed infrequently.

27.6 Computer-Related Injuries: Their Causes, Their Fixes

The U.S. Department of Labor indicated that major causes of CTDs in the workforce are inflexible workstation designs, competitive work environments, poor education and training on proper workstation design, technological advances (including computer usage), and increased use of repetitive motions (U.S. Department of Labor, 1991). However, proper computer workstation design can help minimize the occurrence of CDTs incurred from this type of work. Some of the most common computer-workstation-related CTDs are discussed below.

Neck, shoulder, and back problems can be caused by stresses due to sitting without proper back support. They can also result from improper heights of the workstation relative to the body, which are often created by poor chair height adjustment. Lack of flexibility in workstation design can also contribute to these problems. Tendinitis is an inflammation of a tendon, usually near a joint or bony surface. Often, tendinitis caused by office work occurs in the lower arm/hand or shoulder regions. Carpal tunnel syndrome is a particular injury that often begins as tendinitis in the wrist or hand area and becomes progressively worse, leading to other problems. The thumb and first two or three fingers are affected due to compression of the median nerve. Repetitive motions such as typing extensively with the wrists in a bent position can contribute to the onset of carpal tunnel syndrome.

27.7 OSHA Algorithm for Treatment of CTDs

The Occupational Safety and Health Administration (OSHA) is a government organization that ensures the safety and well-being of U.S. workers through development of regulations and standards. OSHA provides industry-specific and task-specific guidelines for ergonomic conditions in a number of areas, and has recently published such guidelines for poultry processing and nursing home industries. In the OSHA Guidelines for Meatpacking publication, an algorithm is presented for the treatment of CTDs. This algorithm contains an iterative approach to addressing and minimizing the sources for CTDs. OSHA has also dedicated a significant amount of resources toward identifying the ergonomic risks associated with long-term computer use. They provide an interactive online toolkit that consists of guidelines and checklists for use in individual computer workstations (for more information, see http://www.osha.gov/SLTC/etools/computerworkstations).

27.8 Good Practices of Workspace Design

Most computer workstation ergonomic assessment tools will contain evaluation components for chairs and furniture, keyboards, mice, visual display terminals (VDTs), environmental elements such as lighting, temperature, and noise and organizational elements. Recommended settings will be based on empirical research findings and will be a function of the individual's unique physiological requirements and anthropometric dimensions.

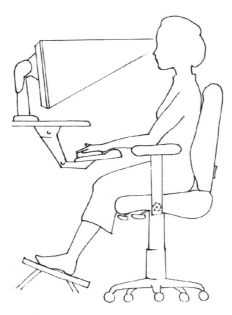

FIGURE 27.2 Neutral computer workstation posture.

27.8.1 Chairs and Furniture

To avoid and minimize potential CTDs during extended computer use, workers must maintain a neutral body position. Such a position requires the support of appropriate equipment placement and furniture. A neutral body position is illustrated in Figure 27.2. Since each worker maintains different anthropometric dimensions, no single computer workstation is appropriate for all users. Instead, workstations should be made adjustable so that their dimensions can be customized to the dimensional requirements of the worker.

Although appropriate workspace design is essential to maintaining healthy body posture, there are demonstrated benefits of periodic movement and repositioning while working. Stretch breaks and adjustability in the design of the workstation can facilitate such activity.

27.8.2 Keyboards, Mice, and Visual Display Terminals

The design and position of the keyboard, mouse, and the VDT are crucial elements to the computer workstation. As these components often serve as an extension of the workstation furniture; they too must be designed and positioned in such a way that ergonomic strain is minimized. The keyboard should be placed in a position that facilitates at least a 90° angle between the forearm and the upper arm. Further, it should support the wrists and promote a straight line from the elbow to the knuckle of each hand. Since much of the repetitive motion incurred during computer use comes from the frequent switching between the keyboard and the mouse, the position of the mouse should be close to and on the same surface as the keyboard. Finally, the size, resolution, brightness, reflectivity, and position of the VDT play a significant role in supporting the user's posture and comfort. The size, resolution, and position of the display need to be balanced to ensure that the user can ultimately view the information on the screen from an appropriate distance. A standard recommended distance for positioning the VDT from the user's eyes is 18 to 30 inches. Additionally, the brightness and the amount of glare reflected by the screen will contribute to the worker's eyestrain, which should be minimized.

27.8.3 Environmental Elements

Other factors that can have an impact on the ergonomic quality of a workstation include environmental elements such as lighting, temperature, and noise.

27.8.4 Lighting

In a typical office setting, many workers conduct a variety of office-type tasks in addition to purely computer work. These peripheral tasks, especially those that involve fine work and reading, usually require high contrast and subsequently, a significant amount of light to perform. On the other hand, computer monitors such as CRTs (cathode ray tubes) and LCDs (liquid crystal displays) are backlit and therefore emit their own light source for providing contrast. In fact, surrounding light can even have a negative effect on the ability to see the monitor as screen glare can be debilitating. The lighting requirements for the noncomputer tasks must be balanced against those lighting requirements for conducting computer work. Essentially, screen glare must be minimized while still allowing for adequate contrast on the noncomputer-related tasks.

27.8.5 Temperature

Temperature also is a significant factor in the workspace environment. As many can attest, a cold workspace can be detrimental to worker productivity and comfort. Typing, especially, which leaves the limbs exposed from the body's core, can result in numb fingers if the air is too cold. This condition can be exacerbated if the forearms are raised to less than a 90° angle from the upper arm. Likewise, extreme heat is dangerous for any individual and may cause unnecessary fatigue, stress, and in some cases heat stroke. The recommended temperature for performing standard computer work is between 68° and 79° F.

27.8.6 Noise

The effects of noise exposure on both short-term and long-term hearing are pronounced. The amount to which noise will degrade hearing is a function of the sound's loudness or intensity, measured in dBA, and its duration. OSHA mandates that workers not be exposed to more than 90 dBA for over 8 hours. While the standard office setting typically maintains a noise level of around 56 dBA, care should be taken to ensure that the level is not exceeded. In addition to physical injury in the form of hearing loss, excessive noise can be particularly distracting during computer use where cognitive demands are usually high (OSHA, Part 1910).

27.9 Organizational Elements

Other factors influential to ergonomic well-being in the office are often classified as *organizational* factors as they relate to the external conditions of the workspace such as managerial policies, deadlines, and psychosocial issues. These circumstances can have a significant effect on workers' stress levels and ability to adequately perform their jobs. One study determined that stress due to psychosocial issues resulted in a loss of productivity.

27.10 Ergonomic Evaluation of the Computer Workstation

To assess how well the above workspace design principles were met in a given workstation, an ergonomic evaluation of the computer workstation must be conducted. Currently, there is no standard or widely recognized assessment technique that outweighs the others. Most evaluation techniques are determined based on the experience level of their audience or user base, on how comprehensive the method is and based on how well the method meets the objectives set forth by management. There are a number of resources available that provide standards and regulations for ergonomic computer workstation policies. Consequently, many evaluation techniques are based on the guidelines provided by these organizations. Ergonomic evaluation techniques also differ by type of data collected, either objective or subjective. There are benefits and trade-offs to each, which will be discussed in more detail below.

27.11 Evaluation Methods

When performing an evaluation, there are two types of data that can be collected: objective data and subjective data. Objective data can come from checklists, comparing anthropometric and layout measurements, photographs or video recordings, counting activity frequency, collecting muscle activity through electromyography (EMG), and taking environmental measurements such as sound reading, light measurements, and temperature readings. This type of data can provide an actual and unbiased evaluation of ergonomic conditions. However, as most workstations are different, it is not always possible to capture a complete assessment of the workspace's ergonomic quality. Objective data collection can be time consuming and often requires some expertise in ergonomics to obtain and analyze. Further, most objective data do not provide an accurate assessment of the worker's personal level of discomfort. Thus, it can be beneficial to collect subjective ergonomic data through body discomfort questionnaires and through expert opinion. Subjective data are typically quick and easy to collect; however, the results can be different for each situation or individual and sometimes standardization across workstations can be a challenge. It is usually a good idea to include some of both data types in an ergonomic evaluation plan in order to utilize the strong points of both.

27.12 Evaluation Tools

Although any ergonomics evaluation can be conducted using basic methods such as comparing anthropometric dimensions to workstation measurements, these techniques are often time consuming and require some expert knowledge. Some tools have been developed for use by both novice and expert evaluators, which can help focus the evaluation to problem areas and do so in a time-efficient manner. These tools include, but are not limited to, checklists, subjective questionnaires, and computer modeling programs.

27.13 Checklists

One of the most basic and easy-to-use office ergonomics evaluation tools is a simple checklist. Typically, checklists help identify any significant ergonomic deficiencies existing in the workspace. They will tell you what is wrong or missing, but not necessarily *how* to fix it. Also, many checklists do not consider the degree of discomfort that an individual is experiencing, the duration or frequency with which the worker is performing a task, or the *actual* posture that the worker is adopting. Therefore, a fair degree of prior ergonomic knowledge is needed to effectively use a checklist. A good checklist should be comprehensive and have components that address the work area environment, overall posture, seating, monitor setup, and arm and wrist postures. When the checklist is completed, areas that are lacking ergonomic quality will be identified. Addressing these issues usually requires additional information about the situation to determine whether or not these unmet checkmarks are acceptable. For example, if the checklist indicates that the worker must reach extensively to access the telephone, this does not necessarily mean that the workspace configuration must be redesigned to accommodate this function. First, consider the frequency with which the telephone is typically accessed and then look at the severity of the reaching action. One who uses the telephone 2 to 3 times a day has very different needs from those who constantly take calls.

There are a number of acceptable office ergonomics checklists available for use. Try to choose one that incorporates more information about the situation than simply "yes, the workstation does have this item" or "no, the workstation does not have this item." Some checklists contain places for indicating the severity of the failure, such as low, medium, or high. Below is a list of recommended office ergonomics checklists.

- Ergonomics checklist—computer workstation:
  Dan McLeod, The office Ergonomics kit (Boca Raton, FL: CRC Press).
- OSHA computer wokstation checklist:
  http://www.osha.gov/SLTC/etools/computerworkstations/checklist.html.

27.14 Discomfort Questionnaires

Another way to evaluate the ergonomic effectiveness of an existing computer workstation is to ask users how comfortable they are. Worker discomfort questionnaires provide a detailed approach toward quantitatively assessing worker pain. The questionnaire typically contains an image of the body (see fig. 27.3) labeled with main body parts, and the user is required to indicate a level of pain associated with each body part. A standard Likert pain scale ranges from 1 to 5, where 1 = no pain and 5 = severe pain. To implement, workers complete the entire questionnaire at different times throughout the day. Analysis over time can then reveal areas subject to chronic pain and areas that experience discomfort as the workday progresses. These areas of discomfort can then point to components in the workspace design that may be contributing to the discomfort.

This approach can be applied to workstation design of both individuals and groups. When evaluating the ergonomic design of multiple workstations, say in a large office, all users can complete the questionnaires and aggregate discomfort ratings can be calculated. From these findings, large-scale ergonomic deficiencies can be identified as well as those for individual needs.

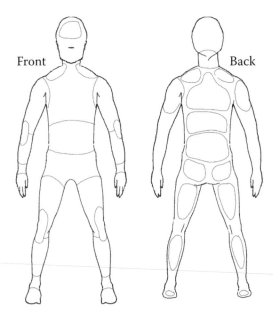

FIGURE 27.3 Standard body discomfort diagram.

27.15 Computer Modeling Programs

Finally, computer modeling programs can be used to both design and evaluate the ergonomic quality of computer workstations. Although little is available specifically for use in ergonomic evaluations of computer workstations, some human modeling systems that are available contain the flexibility to allow for computer workstation evaluation. With such tools, we are able to explore both virtual workplace design and virtual prototyping.

27.15.1 RULA

One of the most basic human modeling tools available for ergonomic evaluation is the Rapid Upper Limb Assessment, which was developed in 1993 by Dr. Lynn McAtamney and Professor E. Nigel Corlett who are ergonomists at the University of Nottingham in England (McAtamney & Corlett, 1993). The method assigns a score to the worker's working habits based on posture and physical workloads. Although originally developed as a worksheet, this approach has been modified for use online. A widely used online version of RULA can be found at http://www.rula.co.uk. Researchers have also attempted to adapt RULA specifically for computer use (Lueder, 1996), although these adaptations have yet to be fully validated.

27.15.2 3D Human Modeling Tools

Jack is a well-known human modeling tool developed by UGS Tecnomatix that is used in many industries to assess worker performance in a virtual environment or using a virtual product. Digital human models, which are based on anthropometric data, are integrated into virtual workstations where they perform tasks just as an actual worker would do. In turn, the software collects data on the virtual worker's performance and ergonomic quality. This system provides an excellent resource for understanding how a human worker will interact with a workstation design, and when used with the Task Analysis Toolkit add-on available through UGS Tecnomatix, it provides feedback on the ergonomics of the workstation relative to the task requirements.

One other 3D ergonomic analysis simulation tool includes Deneb/ERGO, which was purchased by Dassault Systemes in 1997, developer of CATIA CAD software, to later become Safework; ManneQuin-PRO by NexGen Ergonomics, Inc.; and RAMSIS by Human Solutions. Like Jack, Safework, ManneQuinPRO, and RAMSIS also use 3D human models based on anthropometric data to identify potential ergonomic flaws in a design. These types of modeling tools are particularly useful to industry not only for their ergonomic capabilities, but as a tool that significantly reduces rework once products hit the production floor.

27.15.3 Comprehensive Ergonomic Evaluation Software

Originally designed for field evaluations, the Pocket Ergo software is a digital evaluation tool that attempts to consolidate the many ergonomic principles into a single comprehensive utility. This system is available online or for PDAs. Unlike some of the other computer-based tools, this system does not afford customized workspace analysis, but rather it provides the evaluator with readymade generic assessment techniques followed by recommendations for individual design components.

The ErgoMaster system utilizes a combination of approaches for evaluating activities and workstations. While it does not make use of virtual human models, it uses actual photographs of the worker performing activities in the real workspace to assess ergonomic quality. Using the photograph provided by the evaluator, the system identifies poor postures and unsafe activities. The system then provides recommendations for improvement based upon approved ergonomic standards and guidelines. ErgoMaster also contains a number of standard evaluation checklists and tools similar to those in the Pocket Ergo system.

27.16 Integrated Performance Modeling Environment (IPME)

IPME, by Micro Analysis and Design Systems, considers other human factors issues than simply ergonomics. This program looks at the human's performance capabilities as part of a larger mechanical system and considers demands on the worker's visual, auditory, cognitive, and psychomotor resources.

27.17 DHM as Design Tool: Advantages and Disadvantages

Although many of the available ergonomic design tools are not specifically intended for use with the computer workstation alone, they have the capability to adapt to this situation and provide adequate means of assessment for both existing and prospective computer workstation designs. Determining which type of ergonomic evaluation tool to use depends on the context and goals of the evaluation itself. Traditional, stand-alone tools, such as RULA, can be used when there is a specific need to be addressed or if repetitive motion of the upper arms is suspected. For a more holistic evaluation of an existing workstation, one of the comprehensive ergonomic evaluation software systems may suffice. If a new computer workstation is under design and the ergonomic quality of this design must be determined prior to production, then the 3D human modeling tools coupled with other more specific tools should be used.

27.18 Designing the Office of the Future

The traditional and digital human modeling techniques discussed in this review provide a fairly sufficient and comprehensive package for designing computer workstations. As more and more of our daily task demands require us to rely on computer technologies both in our professional environments and at home, we must consider the dynamic nature of the technology itself. The designs that are developed and evaluated should facilitate future integration of new technologies as they become a regular part of our daily work center. Additionally, the designs should recognize the growing multimedia demands of this workspace. Finally, they should accommodate the needs of an aging workforce. Consequently, many of the tools discussed here may have to be revisited to account for such changes.

27.18.1 Growing Multimedia Demands

Online telecommuting, video conferencing, wireless communication, technology interfacing (e.g., mobile phones, digital cameras, PDAs, etc.), speech recognition, touch screens, eye tracking, and biometric identification are just some of the growing trends in multimedia resources for computing. These and other technologies are finding their way into the standard computer workstation. It is more important than ever that our workstations be able to integrate these multiple technologies into the human workflow efficiently and effectively.

27.18.2 Flexibility and Adaptability

The workstation of the future must be adaptive to the growing multimedia demands and flexible enough to accommodate these new technologies without great expense or disruption to the workplace. As technologies and the methods we use to evaluate those technologies change, workplace design must leave room to adapt to new currents in the work environment and also in the field of ergonomics.

27.18.3 Preparing for an Aging Population

By 2010, more than 51% of the workforce is expected to be 40 or older, a 33% increase since 1980, while the portion of the workforce aged 25 to 39 will decline 5.7% (Bureau of Labor Statistics, 2001). An older workforce also yields a change in the physiological capabilities of workers, such as decreased sight, hearing, and some motor skills (U.S. Department of Commerce, 2002). Consequently, by designing for this group, the ergonomic requirements for workstations will become skewed. Future workplace design and the tools that evaluate them must have the awareness of these physiological changes and be adaptable and flexible enough to accommodate the needs of an aging workforce.

References

American National Standards Institute and Human Factors and Ergonomics Society, "National Standard for Human Factors Engineering of Visual Display Terminal Workstations." ANSI/HFES (100-1988).

Cheeseman, J.D., Janus, A., Davis, J. *Computer and Internet Use in the United States: 2003, Special Studies* (U.S. Census of Population and Housing), Government Printing Office, 2003.

Fullerton, H.N., Toossi, M. "Labor Force Projections to 2010: Steady Growth and Changing Composition." *Bureau of Labor Statistics Monthly Labor Review Online*, 124(11), November 2001. http://www.bls.gov/opub/mlr/2001/11/contents.htm (accessed April 26, 2007).

Grandjean, E. *Ergonomics in Computerized Offices*. Taylor & Francis, 1987.

Grieco, A. "Sitting posture: an old problem and a new one." *Ergonomics,* 29(3): 345–62, 1986.

Helander, M. *Handbook of Human-Computer Interaction*. North-Holland, 1988.

Kroemer, K.H.E., Kroemer, A.D. *Office Ergonomics*. Taylor & Francis, 2001.

Liberty Mutual Research Center for Safety and Health. *From Research to Reality* (Annual Report). Liberty Mutual Research Center for Safety and Health, 1999.

Lueder, R. "A Proposed RULA for Computer Users. Proceedings of the Ergonomics Summer Workshop," UC Berkeley Center for Occupational & Environmental Health Continuing Education Program, San Francisco, August 8–9, 1996.

MacLeod, D. *The Office Ergonomics Kit*. Lewis Publishers, 1998.

McAtamney, L., Corlett, E.N. "RULA: a survey method for the investigation of work-related upper limb disorders," *Applied Ergonomics*, 24: 91–99, 1993.

Michael, R. "Liberty Mutual Liberty Mutual Releases Workplace Injury and Cost Data." *Ergonomics Today*, August 17, 2001. http://www.ergoweb.com/news/detail.cfm?id=395/ (accessed April 12, 2007).

Occupational Safety and Health Administration. "Occupational Noise Exposure." *Occupational Safety and Health Standards: Occupational Health and Control,* PART 1910.95.

U.S. Department of Commerce, *A Nation Online*, Economic and Statistics Administration/National Telecommunications and Information Administration, February 2002, based on CPS of September 2001.

U.S. Department of Labor. "Ergonomics: The study of work." OSHA Part 3125 (1991); 1–19.

WISHA Services Division, *Office Ergonomics: Practical Solutions for a Safer Workplace,* Washington State Department of Labor and Industries, 2002.

28

Workload Assessment Predictability for Digital Human Models

Jerzy Grobelny,
Rafal Michalski, and
Waldemar Karwowski

28.1 Introduction

Physical loading on the human motor system at work is a direct consequence of workstation and work-place conditions, as well as work system organization. The most efficient way to shape the various work system components would be to address them at the design stage, when functional requirements and system limitations can be considered most effectively. It is much more difficult to correct any limitations after the systems, especially large or complex systems, have already been designed (Karwowski, 2005). First attempts to model the human body and its relevant characteristics for system design purposes date as early as the 1970s (i.e., Ryan et al., 1970; Ryan, 1971; Kroemer, 1973; Bonney et al., 1974). Rapid advances in personal computers facilitated applications of computer-aided design (CAD) techniques and stimulated development of a variety of concepts and frameworks for modeling the human body. Such models include, for example, COMBIMAN (Kroemer, 1973; Evans, 1976; McDaniel, 1990), Sammie (Bonney et al., 1974, 1982; Case et al., 1990; Porter, 1992; Feeney et al., 2000), Apolin (Grobelny, 1988; Grobelny et al., 1992), Crew Chief (McDaniel, 1990), Ergoshape (Launis & Lehtelä, 1990, 1992), Heiner (Schaub & Rohmert, 1990), Safework (Fortin et al., 1990), Tadaps (Westerink et al., 1990), Werner (Kloke, 1990), Apolinex (Grobelny & Karwowski, 1994, 2000), Ramsis (Seidl, 1994, 1997; Bubb, 1998; Marach & Bubb, 2000; Bubb et al., 2006), Jack (Badler et al., 1995; UGS, 2004), Human (Sengupta & Das, 1997), Anthropos (Lippmann, 2000; Bauer et al., 2000; IST, 2002), and others. A comparison of such systems was discussed by Dooley (1982), Rothwell (1985), Karwowski et al. (1990), Porter et al. (1992), Hanson (2000), Wolfer (2000), Chaffin (2001), and Laurenceau (2001).

The main component of many computer-based models of the human body is the human mannequin generator that allows creation of virtual representations of the human body, with user-defined body segments according to the published anthropometric data (i.e., Pheasant, 1991). Typically, computer software generates desired models by scaling the human body according to the given percentile of body height. In some cases, it is also possible to specify additional parameters such as gender, nationality, or psychosomatic body type. Available systems differ from each other with respect to their modeling features, including precision of anatomical body representation, simulation of the effects of clothing, or the number of available parameters that define differences in the human body due to relevant or desired characteristics (Kee & Karwowski, 2002). In addition, some systems also allow for additional functions, such as the field of vision that aims to describe the effects of differing body postures at work (i.e., Anthropos, Apolinex, COMBIMAN, Safework, Sammie, Ramsis, Jack). Other useful design applications include simulated descriptors of physical workload, for example static force moments at different body joints calculated for a given body posture with due consideration of the eternal loadings and environmental conditions (i.e., Anthropos, Apolinex, COMBIMAN, Ergoshape, Safework, Sammie, Tadaps). Despite great progress in human body simulation, there are many outstanding issues that need to be resolved (Chaffin, 2005). Furthermore, limited validation of the available systems with respect to human body representation implies the need to exercise due care in reaching conclusions or making generalizations based on system-generated simulation results (Ruiter, 2000).

This chapter focuses on the applications of the Anthropos system operating in the 3D Studio Max environment, for analysis of static postural loading and comparison of results with estimates of perceived physical workload. The main objective of the study was to compare the results of the Anthropos ErgoMAX human body modeling system with the results of subjective estimates made by a group of industrial workers in one company. The results of this case study constitute a basis for discussing the advantages and disadvantages of applying human models for assessing physical workload. Some recommendations have also been proposed to evaluate the predictability of workload assessments carried out by the DHM software in a more systematic way.

28.2 A Case Study

28.2.1 Company Profile

This study was carried out in a branch of an international company, located in Poland and operating in the automobile sector. The company specialized, among other things, in manufacturing various rearview systems for cars—mostly internal and external mirrors. The enterprise was one of the leaders in this area with a production of 18 million mirrors a year, and with extensive experience of more than 30 years. The Polish branch employed more than 250 workers in 2006.

28.2.2 Subjects

Thirty-four employees from 12 different workstations took part in the study. In total, there were 10 females (29%) and 24 males (71%). The distribution of workers performing different tasks is shown in Table 28.1.

Eighteen of the participants were married (53%). One worker had primary education, five had vocational education, three had higher education, and the rest (25, 74%) were graduates of secondary schools (high school or equivalent). Other personal characteristics of the participating workers are presented in Table 28.2.

28.2.3 Workload Assessment by Means of Digital Human Models

28.2.3.1 Methods

The Anthropos ErgoMAX 6.0.2, working in the 3D Studio Max 6.0 environment, was used to assess workload due to working body postures. Anthropos provides for the following indices of postural

TABLE 28.1 Characteristics of Work Assignments

No.	Work Position	No. of Employees	Gender	
			Men	Women
1	Painter	3	2	1
2	Fitter (I)	4	2	2
3	Fitter (II)	5	3	2
4	Polisher (I)	2	2	0
5	Polisher (II)	2	0	2
6	Polisher (III)	2	2	0
7	Polisher (IV)	3	3	0
8	Presser	3	1	2
9	Technician	2	2	0
10	Forklift truck operator	2	2	0
11	Plastics deliverer	2	1	1
12	Stockroom deliverer	4	4	0

discomfort due to body postures at work (IST, 2002): discomfort; posture (angular joint position within motion limits); joint resistance; joint torque; normal forces (*x, y, z,* and vector); and difference of current posture angles to NASA neutral posture where the body experiences zero load (NASA, 1995). For all these indicators it is possible to take into consideration the support of selected body segments, as well as assignments of additional weights to hands, feet, and a head. The values of all percentage measures in Anthropos have been classified into three zones according to the increasing level of postural loading: green zone for values below 70%, yellow zone for values between 71% and 90%, and red zone for all values greater than 90%.

28.2.3.2 Procedures

All potential subjects were informed about the goals and study procedures, and had volunteered to participate in the study. The most prevalent body postures at all work positions were identified based on interviews with participating workers, direct on-site observations, and analysis of video recordings of all workers. Such body postures were then simulated using the Anthropos system, for both a 50th-percentile male and a 50th-percentile female. The adult mannequin with normal somatic type (50% value) representing the middle-European population was used for this purpose. Figure 28.1 depicts examples of the simulated work postures.

For the purpose of this project, the percentage index of postural discomfort was selected from those available in the Anthropos indices of postural loading. Such an index takes into consideration the following loading factors: joint angles (posture); resistance; force; and torque, which are estimated using

TABLE 28.2 Detailed Subject Characteristics

Subject Data	Mean	Standard Deviation	Minimum	Maximum
Age (years)	27.8	6.45	20	48
Weight (kg)	72.2	14.1	49	105
Height (cm)	173	10.1	150	187
Overall work experience (years)	7.82	6.89	0.13	28
Seniority in the company (years)	1.88	2.74	0.08	12

FIGURE 28.1 Examples of body positions simulated in Anthropos software.

empirically based proportions. More information about the calculation of this index can be found in IST (1998) or Deisinger et al. (2000). It should be noted that the percentage index of postural discomfort is a static measure, and as such, it does not consider the dynamics of human motions. The level of discomfort is calculated for each body segment under a given body posture. The percentage values of discomfort were used to define an average discomfort score for arms and legs, as well as an average discomfort score for the whole body. All calculations for male and female workers were done separately for representative body postures for all work positions.

For any given work position, the average postural discomfort score was defined as the arithmetic average of all products of the average value of discomfort and weights related to time exposure to a given posture during the entire workshift and by taking into account all representative body postures. The overall postural discomfort for a given work position was calculated as the average value of the discomfort scores for male and female workers.

28.2.3.3 Results: Perceived Body Discomfort

Detailed results of the discomfort scores for all work positions are presented in Table 28.3. The highest values of average discomfort scores were attributed to plastics delivery operations (87%), and technician work. The forklift truck operator exhibited the lowest score of postural discomfort. In general, the perceived discomfort scores were higher for the arms than for the legs. The reverse trend was observed in only three cases—the forklift truck operator, plastics delivery operations, and stockroom delivery operations.

28.2.4 Subjective Workload Evaluation

Numerous subjective tools for workload assessment focused on mental work tasks and psychological workload of employees—Cooper-Harper Scale (Cooper & Harper, 1969), Bedford Scale (Roscoe, 1987; Roscoe & Ellis, 1990), Workload Profile (Tsang & Velazquez, 1996), Multiple Resources Questionnaire (Boles & Adair, 2001), Integrated Workload Scale (Pickup et al., 2005), and Subjective Workload Assessment Technique (Reid & Nygren, 1988). Subjective methods were also used for a comprehensive workload evaluation. In this area, NASA Task Load Index (Hart & Staveland, 1988) is widely used. One of the latest proposals that take into account the multi-dimensional nature of workload was presented by Jung and Jung (2001), and Michalski and Grobelny (2007).

TABLE 28.3   Mean Discomfort Values for All Evaluated Positions

Position	Discomfort (%)		
	Arms	Legs	Average
Painter	70.4	44.4	57.4
Fitter (I)	61.4	51.4	56.4
Fitter (II)	59.1	46.1	52.6
Polisher (I)	84.3	54.6	69.4
Polisher (II)	61.3	43.7	52.5
Polisher (III)	56.7	43.7	50.2
Polisher (IV)	63.1	43.8	53.5
Presser	66.5	50.3	58.4
Technician	90.5	69.7	80.1
Forklift truck operator	20.1	48.0	34.0
Plastics deliverer	81.2	92.8	87.0
Stockroom deliverer	42.2	57.9	50.0
Mean	63.1	53.9	58.5
Standard deviation	18.9	14.4	14.3
Mean standard error	5.5	4.2	4.1

The major advantage of the methods presented by Michalski and Grobelny (2007)—the Overall Workload Level (OWL) and Subjective Overall Workload Assessment (SOWA)—is an application of the Analytic Hierarchy Process (AHP) developed by Saaty (1980, 1994, 1996). In the AHP method, by means of pair-wise comparisons of available variants and some additional calculations, one can obtain a vector of weights that allows for setting the hierarchy of importance of the analyzed items. A big advantage of the AHP is the possibility of controlling the ratings conformance during the pair-wise comparisons by calculating the inconsistency ratio (IR). One of the main drawbacks of this tool is the rapidly growing number of comparisons and increasing number of decision variants. Additional assessments of the virtues and constraints of various subjective methods of workload evaluation can be found, for example, in the works of Vidulich and Tsang (1986), Rubio et al. (2004), Phillips and Boles (2004), and Barriera-Viruet et al. (2006). An in-depth discussion of the subjective measures applied in ergonomics was presented by Annett (2002) and was followed by a series of commentaries.

28.2.4.1 Methods

In this work, the SOWA method (Michalski & Grobelny, 2007) and supportive software were employed to evaluate subjective workload levels. The method was based on the OWL (Jung & Jung, 2001) with some properties taken from other subjective techniques. The workload is evaluated in four fundamental dimensions: manual material handling (MMH); material work environment (MWE); body posture and movement (BPM); and mental demand environment (MDE). Each of these dimensions is characterized by several attributes. The detailed structure together with the full description of the SOWA method and supportive software can be found in the work of Michalski and Grobelny (2007).

28.2.4.2 Apparati

Specialized software supporting the SOWA technique was applied to improve the performance of questionnaire data input, and to allow for clear and comprehensible presentation of results. The computer application was based on the Microsoft Access database and was developed in the Microsoft Visual Basic 6.0 environment. All necessary information regarding these analyses, including personal details, parameters ratings, and comparisons together with inconsistency ratios, were compiled in a database file.

28.2.4.3 Procedures

A required number of assessment questionnaires were generated and printed out by means of the SOWA software. The generated reports consisted of four parts: a personal details survey, a workload attributes assessment form, and two pair-wise comparison forms. At first, subjects compared parameters within the confines of the individual workload dimensions. The same procedure was then used for all other workload dimensions. In the generated questionnaires, the order of parameters, parameters comparisons, and dimensions comparisons was set randomly. The assessment of the subjective workload was administered during work hours with groups of several employees.

The human subjects, who were all volunteers, were informed about the purpose and scope of the study. They were also instructed as to how to fill out the questionnaires, and they were assured of anonymity regarding their answers. The typical procedure for an individual group lasted 15 to 20 minutes. All subjects were given an opportunity to ask questions and received appropriate explanations and assistance any time during the study. The collected questionnaire data were entered into the software and then analyzed. The weights as well as overall perceived workload assessment index values (OPWAI) and inconsistency ratios were computed according to the AHP technique.

28.2.5 Results

The average value of the obtained workload index was 74.2%. The largest value of the OPWAI was registered for the stockroom delivery operators, which amounted to 91% of the maximum possible rate. A very high score was also received for the polisher (I) at 89.7%, while the lowest mean value of OPWAI were obtained for forklift truck operators at 58.6%. Taking into consideration the values calculated for different dimensions, it can be noted that the highest shares in the total value of OPWAI were attributed to the body posture and movement for the polisher (II) at 60.4%. In turn, the MWE component constituted a considerable part of OPWAI for the polisher (II) (44.9%) and painter (43.4%). In the case of stockroom delivery operations, the biggest value was observed for MMD (45.3%). The lowest value of the analyzed dimensions was obtained for polisher (I)—the relative share amounted to 1.6%.

In general, the highest scores were registered in the BMP factor, whereas the least important factor in all work positions was related to mental demands. Data about the overall workload index and mean IR values for all analyzed work positions are presented in Table 28.4. Because the applied software provides weighted scores for all attributes evaluated by an employee, it is possible to make further in-depth analyses within the confines of the given dimension. The most interesting dimension in this respect relates to body posture and movement (see Table 28.5).

28.2.6 Comparison of Results and Discussion

Figure 28.2 depicts the values of average discomfort scores with subjective average estimates of BPM for all work positions. In all cases, except for the polisher (II), the subjective estimates were much lower than indicators of discomfort generated by the Anthropos system. The largest differences were noted for the stockroom delivery workers, where subjective assessments of workload measured by BPM (11%) were nearly fivefold lower than the one derived from digital human modeling with an average score of discomfort of 50%.

A correlation analysis was performed in order to verify the results of the Anthropos analysis with subjective worker assessments of discomfort at each of the investigated work positions. Table 28.6 shows the correlation coefficients between the average values of postural discomfort (for arms, legs, and average) and main subjective workload indices (OPWAI MMH, MWE, BPM, MDE). The only statistically significant ($p < 0.05$) correlation was found between the reported arms discomfort and material work environment. The correlations between MWE and BPM and the average discomfort, as well as between BPM and arms discomfort, showed trends at the $p < 0.1$ level.

TABLE 28.4 Overall Perceived Workload Assessment Indices, Weighted Scores for Four Main Dimensions, and the Average Inconsistency Ratio for All Evaluated Positions

Position	OPWAI (%)	IR	[1] MMH (%)	[2] MWE (%)	[3] BPM (%)	[4] MDE (%)
Painter	75.8	0.568	4.2	43.4	23.2	5.0
Fitter (I)	68.2	0.251	7.4	13.1	40.1	7.6
Fitter (II)	64.7	0.214	12.0	14.2	31.6	6.9
Polisher (I)	89.7	0.198	1.6	44.9	40.0	3.2
Polisher (II)	75.1	0.642	3.5	6.9	60.4	4.3
Polisher (III)	85.4	0.952	22.1	12.3	39.2	11.8
Polisher (IV)	73.7	0.272	6.4	17.6	28.9	20.8
Presser	59.0	0.253	17.2	18.0	20.9	2.9
Technician	80.4	0.164	10.7	30.9	34.1	4.7
Forklift truck operator	58.6	0.312	22.5	6.3	15.6	14.3
Plastics deliverer	68.6	0.365	11.3	6.7	41.5	9.0
Stockroom deliverer	90.6	0.205	45.3	17.1	11.2	17.0
Mean	74.2	0.366	13.7	19.3	32.2	9.0
Standard deviation	10.9	0.237	12.1	13.4	13.5	5.8
Mean standard error	21.4	0.106	4.0	5.6	9.3	2.6

In general, it can be assumed that the anticipated positive relationship between the subjective scores of perceived workload due to posture at work and the average discomfort scores calculated with the help of Anthropos was only partially confirmed through the reported correlation coefficients. The low values of correlations between the indices of postural body discomfort and mental work demands were expected. Similarly, insignificant correlations were observed between the discomfort scores and OPWAI. Most likely this was due to the fact that the OPWAI value reflects a global loading on the human body, while

TABLE 28.5 Detailed Perceived Workload Assessment Weighted Scores for the BPM Dimension

Position	Decomposition of the BPM Dimension (%)				
	Standing	Stooping	Squatting & Kneeling	Twisting	IR
Painter	46.2	13.3	21.7	3.4	1.097
Fitter (I)	48.3	18.3	2.8	3.7	0.080
Fitter (II)	44.1	4.7	4.1	12.4	0.319
Polisher (I)	71.3	4.6	0.5	11.6	0.250
Polisher (II)	68.4	11.5	2.6	2.5	0.438
Polisher (III)	50.4	20.1	5.9	16.5	1.336
Polisher (IV)	71.2	8.1	3	5.9	0.250
Presser	52.2	18.7	0.6	2.8	0.179
Technician	27.4	27.8	27.9	7.6	0.147
Forklift truck operator	2.4	37.8	11.3	10.7	0.402
Plastics deliverer	42.6	18.5	22.3	7	0.210
Stockroom deliverer	35.2	29.6	14.1	10.2	0.051
Mean	46.6	17.8	9.7	7.9	0.397
Standard deviation	19.6	10.2	9.6	4.5	0.403
Mean standard error	5.7	2.9	2.8	1.3	0.116

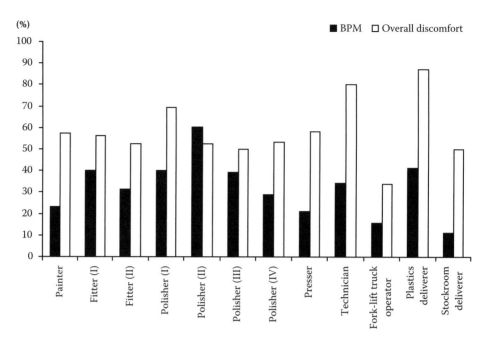

FIGURE 28.2   BPM assessment values and average discomfort indicators for all examined work positions.

the average values of discomfort reflect only the effects of static postural loading. While statistically insignificant, there was relatively high negative correlation (r = −0.493, p = 0.103) between MMH and arms discomfort. One possible explanation of this result is that manual material handlers are typically involved in dynamic work activities, which do not require supporting static body postures for a long period of time (Kee & Karwowski, 2001b).

The study results also revealed a positive linear relationship between work environment conditions and discomfort scores for arms and the average discomfort value. Such a relationship may indicate that those respondents who negatively assessed the quality of their working conditions also scored higher on the perceived postural loading at work. Therefore, poor postures at work can also negatively influence the overall subjective impression of working conditions.

In order to further explore results of this study, the correlation coefficients between the discomfort scores and the attributes of the subjective assessments of workload in the context of body postures and motions were also calculated (see Table 28.7). The highest correlation coefficient was observed between discomfort of the legs and squatting and kneeling posture (r = 0.616; p < 0.05). While statistically insignificant (p < 0.1), there was relatively high negative correlation between arms discomfort and standing and stooping posture, indicating less loading on arms during stooping at work. On the other hand, a

TABLE 28.6   Correlation Matrix between Calculated Discomfort Values and Subjective Workload Dimensions (p values in parentheses)

	OPWAI	MMH	MWE	BPM	MDE
Arms discomfort	0.291	−0.493	0.578	0.559	−0.058
	(0.359)	(0.103)	(0.049)	(0.059)	(0.857)
Legs discomfort	0.034	0.001	0.231	0.339	−0.029
	(0.916)	(0.998)	(0.471)	(0.281)	(0.929)
Average discomfort	0.210	−0.327	0.500	0.543	−0.053
	(0.512)	(0.300)	(0.098)	(0.068)	(0.869)

TABLE 28.7   Correlation Matrix between Calculated Discomfort Values and Components of a Body Posture and Movement Dimension (p values in parentheses)

	Standing	Stooping	Squatting & Kneeling	Twisting
Arms discomfort	0.505	−0.507	0.248	−0.372
	(0.094)	(0.093)	(0.436)	(0.233)
Legs discomfort	−0.232	0.239	0.616	−0.036
	(0.469)	(0.455)	(0.033)	(0.911)
Average discomfort	0.218	−0.216	0.476	−0.266
	(0.497)	(0.501)	(0.118)	(0.404)

positive correlation between arm discomfort and standing posture reflects that most workers engaged in standing postures performed tasks with a high level of static arm loading.

It should also be noted that the values of average IR indices were quite high (see Tables 28.4 and 28.5). This indicates the effect of inconsistent results of pair-wise comparisons and the final values of subjective assessment scores by the workers. If the subjective questionnaires were collected from a larger number of employees, the high values of IR for the individual workers would be less important, and would probably result in larger values of correlation coefficients (and also higher statistical significance of results). Moreover, it is also possible that IR values resulted from difficulties in distinguishing between different dimension variants because, for example, a given group of factors was found to be of no importance to the workers' perceived workload.

28.3   Conclusions

Despite their apparent shortcomings, the contemporary information systems that make possible digital modeling of humans at work are very useful in preventing basic errors at the system design stage. Furthermore, the tools and mannequins such as Anthropos provide very useful models of the human body characterized by relevant statistical parameters and high-quality visualization capability. Application of such systems can lead to better understanding of work system incompatibilities (Karwowski, 2005), and should help in developing comprehensive models of virtual working environments. Such models can also help in prevention efforts aimed at reducing the onset of work-related musculoskeletal disorders, especially those that are linked to poor workstation design and inadequate workspace organization.

The case study presented in this chapter is an example of applying the Anthropos system and SOWA software in order to simulate and analyze the extent of static loading due to different body postures at work in a single production company and a limited population of subjects. Therefore, the presented analysis does not constitute comprehensive verification of the usefulness and predictability of this applied human body modeling system. Such verification would require significantly enlarging the pool of subjects and performing similar studies at different companies. Furthermore, other DHM computer systems should also be considered.

Despite the above limitations, the results indicate that software modules for assessment of the effect of static postural loading on the human body in the context of virtual mannequins can be very useful as supplementary methods for physical workload analysis. Unfortunately, the high cost of virtual modeling environments (such as 3D Studio Max and Anthropos) may limit practical applications of these systems in the workplace. Another obstacle in this quest is the relative complexity of software generating digital human mannequins, as well as the complexity of the available human-computer interfaces. The clear benefit of the proposed approach is the ability to simulate static postural loading of workers for those workplaces and workstations that are still at the design stage. Finally, it should be noted that statistically significant correlations between subjective discomfort scores and some of the workload attributes make it possible to replace expensive analysis that utilizes virtual mannequins by a relatively simple and fast questionnaire for physical workload assessment.

Acknowledgments

The authors would like to acknowledge the useful comments and assistance provided by Dr. V. Duffy of Purdue University in preparing the final version of this chapter.

References

Annett J., 2002, Subjective rating scales: science or art?, *Ergonomics*, 45, 966–987.

Badler N.I., Becket W.M., Webber B.L., 1995, Simulation and analysis of complex human tasks for manufacturing [in:] *Proceedings of SPIE—Int. Soc. Opt. Eng.* 2596, 225–233.

Barriera-Viruet H., Sobeih T.M., Daraiseh N., Salem S., 2006, Questionnaires vs observational and direct measurements: a systematic review, *Theoretical Issues in Ergonomics Science*, 7, 261–284.

Bauer W., Lippmann R., Rössler A., 2000, Virtual human models in product development [in:] *Ergonomic Software Tools in Product and Workplace Design*, K. Landau (ed.), pp. 114–120, Stuttgart, Verlag ERGON GmbH.

Boles D.B., Adair L.P., 2001, The Multiple Resources Questionnaire (MRQ) [in:] *Proceedings of the Human Factors and Ergonomics Society*, 45, 1790–1794.

Bonney M.C., Case K., Porter J.M., 1982, User needs in computerized man models [in:] *Anthropometry and Biomechanics. Theory and Application*, R. Easterby, K.H.E Kroemer, D.E. Chaffin (eds.), pp. 97–101, New York, Plenum Press.

Bonney M.C., Case K., Hughes B.J., Kennedy D.N., Williams R.W., 1974, Using Sammie for computer-aided workplace and task design, Society of Automotive Engineers paper 740270, SAE Congress.

Bubb H., 1998, From measurement to design and evaluation. The idea of the CAD-tool Ramsis [in:] *Proceedings of the Workshop on 3d Anthropometry and Industrial Products Design*, Paris.

Bubb H., Engstler F., Fritzsche F., Mergl C., Sabbah O., Schaefer P., Zacher I., 2006, The development of Ramsis in past and future as an example for the cooperation between industry and university, *International Journal of Human Factors Modelling and Simulation*, 1, 140–157.

Case K., Porter J.M., Bonney M.C., 1990, Sammie: a man and workplace modelling system [in:] *Computer Aided Ergonomics*, W. Karwowski, A.M. Genaidy, S. Asfour (eds.), pp. 31–56, London, Taylor & Francis.

Chaffin D.B., 2001, *Digital Human Modeling for Vehicle and Workplace Design*. Society of Automotive Engineers, Inc. Warrendale, PA.

Chaffin D.B., 2005, Improving digital human modelling for proactive ergonomics in design, *Ergonomics*, 48, 78–491.

Cooper G.E., Harper R.P., 1969, The use of pilot ratings in the evaluation of aircraft handling qualities, NASA Ames Technical Report NASA TN-D-5153, Moffett Field, CA, NASA Ames Research Center.

Deisinger J., Breining R., Rößler A., 2000, Ergonaut: a tool for ergonomic analyses in virtual environments [in:] *Virtual Environments 2000, Proceedings of the 6th Eurographics Workshop on Virtual Environments*, Amsterdam, the Netherlands, J.D. Mulder, R. van Liere (eds.), Wien, New York, Springer.

Dooley M., 1982, Anthropometric modelling programmes—A survey, *IEEE Computer Graphics and Applications*, 2, 17–25.

Evans S.M., 1976, User's guide for the programs of COMBIMAN (Computerized Biomechanical Man-Model), AMRL Technical Report 76–117, Wright-Patterson AFB, OH: Aerospace Medical Research Laboratory, DTIC No. A038 323.

Feeney R., Summerskill S., Porter M., Freer M., 2000, Designing for disabled people using a 3D human modelling CAD system [in:] *Ergonomic Software Tools in Product and Workplace Design*, K. Landau (ed.), pp. 195–203, Stuttgart, Verlag ERGON GmbH.

Fortin C., Gilbert R., Beuter A., Laurent F., Schiettekatte J., Carrier R., Dechamplain B., 1990, Safework: a microcomputer-aided workstation design and analysis. New advances and future developments [in:] *Computer Aided Ergonomics*, W. Karwowski, A.M. Genaidy, S. Asfour (eds.), pp. 157–180, London, Taylor & Francis.

Grobelny J., 1988, Including anthropometry into the AutoCAD-microcomputer system for aiding engineering drafting [in:] *Trends in Ergonomics/Human Factors*, V, F. Aghazadeh (ed.), pp. 77–82, Amsterdam, North-Holland.

Grobelny J., Cysewski P., Karwowski W., Zurada J., 1992, Apolin: A 3-dimensional ergonomic design and analysis system [in:] *Computer Applications in Ergonomics, Occupational Safety and Health*, M. Mattila, W. Karwowski (eds.), pp. 129–135, Amsterdam, Elsevier.

Grobelny J., Karwowski W., 1994, A computer aided system for ergonomic design and analysis for Auto-Cad user [in:] *Proceedings of the 12th Triennial Congress of the International Ergonomics Association: Occupational Health and Safety*, pp. 302–303, Human Factors Association of Canada, Toronto, Canada.

Grobelny J., Karwowski W., 2000, Apolinex: A human model and computer-aided approach for ergonomic workplace design in open CAD environment [in:] *Ergonomic Software Tools in Product and Workplace Design*, K. Landau (ed.), pp. 121–131, Stuttgart, Verlag ERGON GmbH.

Hanson L., 2000, Computerized tools for human simulation and ergonomic evaluation of car interiors [in]: *Proceedings of the XIVth Triennial Congress of the International Ergonomics Association and 44th Annual Meeting of the Human Factors and Ergonomics Association, Ergonomics for the New Millennium*, San Diego, CA.

Hart S.G., Staveland L.E., 1988, Development of NASA-TLX (Task Load Index): Results of empirical and theoretical research [in:] *Human Mental Workload*, P.A. Hancock, N. Meshkati (eds.), pp. 139–183, Amsterdam, Elsevier.

IST, 1998, Anthropos 5 das Menschmodell der IST GmbH. Manual 2 Interaktionen, Germany, IST GmbH.

IST, 2002, Anthropos ErgoMAX—User guide, Germany, IST GmbH.

Jung H.S., Jung H.-S., 2001, Establishment of overall workload assessment technique for various tasks and workplaces, *International Journal of Industrial Ergonomics*, 28, 341–353.

Karwowski W., Genaidy A.M., Asfour S. (Eds.), 1990, *Computer Aided Ergonomics*, London, Taylor & Francis.

Karwowski W., 2005, Ergonomics and human factors: the paradigms for science, engineering, design, technology, and management of human-compatible systems, *Ergonomics*, 48 (5), 436–463.

Kee D., and Karwowski W., 2001, LUBA: An assessment technique for postural loading based on joint motion discomfort and maximum holding time, *Applied Ergonomics*, 32 (4), 357–366.

Kee D., and Karwowski W., 2001, The boundaries for joint angles of isocomfort for sitting and standing males based on perceived comfort of static joint postures, *Ergonomics*, 44 (6), 614–648.

Kee D., and Karwowski, 2002, Analytically derived three-dimensional reach volumes in a sitting posture for upper and lower body, *Human Factors*, 44 (4), 530–544.

Kloke W.B., 1990, Werner: a personal computer implementation of an extensible anthropometric workplace design tool [in:] *Computer Aided Ergonomics*, W. Karwowski, A.M. Genaidy, S. Asfour (eds.), pp. 57–67, London, Taylor & Francis.

Kroemer K.H.E., 1973, Combiman—COMputerized BIomechanical MAN model, Technical Report, AMRL-TR-72-16, Wright-Patterson Air Force Base, OH, Aerospace Medical Research Laboratory.

Launis M., Lehtelä J., 1990, Man models in the ergonomic design of workplaces with the computer [in:] *Computer Aided Ergonomics*, W. Karwowski, A.M. Genaidy, S. Asfour (eds.), pp. 68–79, London, Taylor & Francis.

Launis M., Lehtelä J., 1992, Ergoshape—a design oriented ergonomic tool for AutoCAD [in:] *Computer Applications in Ergonomics, Occupational Safety and Health*, M. Mattila, W. Karwowski (eds.), pp. 121–128, Amsterdam, Elsevier.

Laurenceau T., 2001, Benchmarking digital human modeling tools for industrial designers [in:] *Proceedings of the 2001 IDSA Conference on Design Education, Designing Your Life*.

Lippman R., 2000, Anthropos quo vadis? Anthropos human modeling past and future [in:] *Ergonomic Software Tools in Product and Workplace Design*, K. Landau (ed.), pp. 156–168, Stuttgart, Verlag ERGON GmbH.

Marach A., Bubb H., 2000, Development of a force-dependent posture prediction model for the CAD human model Ramsis [in:] *Ergonomic Software Tools in Product and Workplace Design*, K. Landau (ed.), pp. 105–113, Stuttgart, Verlag ERGON GmbH.

McDaniel J.W., 1990, Models for ergonomic analysis and design: Combiman and Crew Chief [in:] *Computer Aided Ergonomics*, W. Karwowski, A.M. Genaidy, S. Asfour (eds.), pp. 138–156, London, Taylor & Francis.

Michalski R., Grobelny J., 2007, Computer–aided subjective assessment of factors disturbing the occupational human performance, *Occupational Ergonomics*, 7.

NASA, 1995, Man-Systems Integration Standards, NASA-STD-3000, Vol. I, Rev. B., NASA Johnson Space Center, Houston, TX.

Pheasant S., 1991, *Bodyspace*, London, Taylor & Francis.

Phillips J.B., Boles D.B., 2004, Multiple Resources Questionnaire and Workload Profile: Application of competing models to subjective workload measurement [in:] *Proceedings of the Human Factors and Ergonomics Society*, 48, 1963–1967.

Pickup L., Wilson J.R., Norris B.J., Mitchell L., Morrisroe G., 2005, The Integrated Workload Scale (IWS): A new self-report tool to assess railway signaller workload, *Applied Ergonomics*, 36, 681–693.

Porter J.M., 1992, Man models and computer-aided ergonomics [in:] *Computer Applications in Ergonomics, Occupational Safety and Health*, M. Mattila, W. Karwowski (eds.), pp. 13–20, Amsterdam, Elsevier.

Porter J.M., Case K., Freer M.T., Bonney M.C., 1992, Computer aided ergonomics design of automobiles [in:] *Automotive Ergonomics*, B. Peacock, W. Karwowski (eds.), pp. 11–42, London, Taylor & Francis.

Reid G.B., Nygren T.E., 1988, The subjective workload assessment technique: A scaling procedure for measuring mental workload [in:] *Human Mental Workload*, P.A. Hancock, N. Meshkati (eds.), pp. 185–218, Amsterdam, Elsevier.

Roscoe A.H., 1987, The practical assessment of pilot workload, AGARD-AG-282, Neuilly Sur Seine, France, Advisory Group for Aerospace Research and Development.

Roscoe A.H., Ellis G.A., 1990, A subjective rating scale assessing pilot workload in flight. A decade of practical use. Royal Aerospace Establishment, Technical Report 90019, Farnborough, UK, Royal Aerospace Establishment.

Rothwell P.L., 1985, Use of man-modelling CAD systems by the ergonomist [in:] *People and Computers: Designing the Interface*, P. Johnson, S. Cook (eds.), pp. 199–208, Cambridge, UK, Cambridge University Press.

Rubio S., Díaz E., Martín J., Puente J.M., 2004, Evaluation of subjective mental workload: A comparison of SWAT, NASA-TLX, and Workload Profile Methods. *Applied Psychology: An International Review*, 53, 61–86.

Ruiter I.A., 2000, Anthropometric man-models, handle with care [in:] *Ergonomic Software Tools in Product and Workplace Design*, K. Landau (ed.), pp. 94–99, Stuttgart, Verlag ERGON GmbH.

Ryan P.W., Springer W.E., Hlastala M.P., 1970, Cockpit geometry evaluation, Janair report 700202, Boeing Military Airplane Systems Division.

Ryan P.W., 1971, Cockpit geometry evaluation, Phase II, vol 2, Joint Army-Navy Aircraft Instrument Research Report 7012313, The Boeing Company, Seattle, WA.

Saaty T.L., 1980, *The Analytic Hierarchy Process*, New York, McGraw-Hill.

Saaty T.L., 1994, Fundamentals of decision making and priority theory with the analytic hierarchy process, Pittsburgh, RWS Publications.

Saaty T.L. 1996, *The Analytic Hierarchy Process*, Pittsburgh, RWS Publications.

Schaub K., Rohmert W., 1990, Heiner helps to improve and evaluate ergonomic design [in:] *Proceedings to the 21st International Symposium on Automotive Technology and Automation*, vol 2, pp. 999–1016.

Seidl A., 1994, Das Menschmodell Ramsis—Analyse, Synthese und Simulation dreidimensionaler Körperhantlungen des Menschen, Dissertation an der Technischen Universität, München.

Seidl A., 1997, RAMSIS—a new CAD-tool for ergonomic analysis of vehicles developed for the German automotive industry, Automotive Concurrent/Simultaneous Engineering SAE, Special Publications, 1233, 51–57.

Sengupta A.K., Das B., 1997, Human: an AutoCAD based three dimensional anthropometric human model for workstation design, *International Journal of Industrial Ergonomics*, 19, 345–352.

Tsang P.S., Velazquez V.L., 1996, Diagnosticity and multidimensional subjective workload ratings, *Ergonomics*, 39, 358–381.

UGS, 2004, Jack, available on-line at http://www.ugs.com/products/tecnomatix/human_performance/jack/, accessed February 2007.

Vidulich M.A., Tsang P.S., 1986, Techniques of subjective workload assessment: A comparison of SWAT and NASA-Bipolar methods, *Ergonomics*, 29, 1385–1398.

Westerink J., Tragter H., van der Star A., Rookmaaker D.P., 1990, Tadaps: A three-dimensional CAD man model [in:] *Computer Aided Ergonomics*, W. Karwowski, A.M. Genaidy, S. Asfour (eds.), pp. 90–103, London, Taylor & Francis.

Wolfer B., 2000, Man modeling and human movement simulation in 3D—development trends and prospects for application in ergonomics [in:] *Ergonomic Software Tools in Product and Workplace Design*, K. Landau (ed.), pp. 100–104, Stuttgart, Verlag ERGON GmbH.

29

Human Modeling and Simulation

Jingzhou Yang

29.1 Introduction

Human modeling and simulation has attracted increasing attention from academia and industries in recent years. Various digital human models can be brought into the digital environment at the early design stage to check human factors issues, reducing the number of design iterations and saving money and time, while improving quality and safety. In this chapter we describe the different aspects of modeling human kinematics, summarize model-based and data-driven approaches, and review methods for workspace evaluation. We present a methodology for workspace zone differentiation for the solid zone or a surface in the 3D space. Finally, we provide a quick overview of the predictive dynamics such as motion prediction, walking, and running.

With the advances in computer hardware, computer graphics, and simulation techniques, the need to realistically represent humans within a digital environment has gained significant momentum. This need arises as industries continue to expand their use of digital humans in workspace/workstation design and prototype evaluation. For these purposes, the digital humans must have a realistic human appearance and mimic what real humans do in order to provide useful information about the workspace design or digital prototype under evaluation.

This chapter comprises four sections. The first section describes the basic concepts of modeling human kinematics, including the skeletal model, appearance, the musculoskeletal model, open-loop and closed-loop systems, and the Denavit–Hartenberg (DH) method.

The second section lists the general advantages and disadvantages for model-based and data-driven approaches. The data-driven method can provide only finite scenario solutions, while the model-based method can provide solutions for infinite scenarios.

The third section begins with an overview of studies on workspace evaluation and then reports a mathematical-based methodology for digital human workspace and zone differentiation. Generally, two approaches are used: experiment-based and mathematical-based methods. In this chapter, we present the Jacobian-rank deficiency method. The reach envelope is obtained by determining the singular surfaces from the augmented Jacobian matrix of the articulated linkage. An optimization-based zone differentiation method shows not only the reachability but also the discomfort levels inside the workspace.

The fourth section provides a general idea of the predictive dynamics that are used in digital human modeling and simulation. Predictive dynamics can give real-time or nearly real-time solutions. This approach is also task-based, meaning that different cost functions and constraints govern different tasks.

29.2 Modeling Human Kinematics

29.2.1 Human Modeling Approach

29.2.1.1 Skeletal Model

The human skeleton is comprised of 206 bones that are strong, light tubes, rods, and plates. Bones are linked by joints—some fixed fibrous, others mobile—with ligaments uniting bone ends buffered by shock-absorbent cartilage. To establish a skeletal model for biomechanically modeling human anatomy, researchers have implemented conventions for representing segmental links and joints. Human anatomy can be represented as a sequence of rigid bodies, or links, connected by joints. Of course, this serial linkage could be an arm, a leg, a finger, a wrist, or any other segment of the human body. Joints in the human body vary in shape, function, and form. The complexity offered by each joint must also be modeled, to the extent possible, to enable a correct simulation of the motion. The degree to which a model replicates the actual physical model is called the level of fidelity. It is important to distinguish the difference between a rigid body and a flexible body. A rigid body is one that cannot deform (e.g., we typically consider bone to be nondeforming). A flexible body (or deformable object) is one that undergoes relatively large strains when subjected to a load (e.g., soft tissue). Generally, the human skeletal model considers only the rigid motion. Indeed, for ergonomic design considerations, rigid body motion is adequate to address most problems.

Perhaps the most important element of a joint is its function, which may vary according to the joint's location and physiology. The physiology becomes important when we discuss the loading conditions of a joint. In terms of kinematics, we shall address the function in terms of the number of degrees of freedom associated with its overall movement.

For example, consider the elbow joint, which is considered a hinge or a one degree-of-freedom (DOF) rotational joint (e.g., the hinge of a door) because it allows for flexion and extension in the sagittal plane (fig. 29.1) as the radius and ulna rotate about the humerus. We shall represent it as a revolute joint about one axis that has no other motions (i.e., one DOF). Therefore, we can now say that the elbow is characterized by one DOF and is represented as a cylindrical rotational joint also shown in Figure 29.1.

On the other hand, consider the shoulder complex (fig. 29.2). The glenohumeral joint (shoulder joint) is a multi-axial (ball and socket) synovial joint between the head of the humerus (5) and the glenoid cavity (6). There is a 4 to 1 incongruency between the large round head of the humerus and the shallow glenoid cavity. A ring of fibrocartilage attaches to the margin of the glenoid cavity, forming the glenoid labrum. This serves to form a slightly deeper glenoid fossa for articulation with the head of the humerus.

A number of methods can be used to model this complex joint (fig. 29.3). One such method (Maurel, 1999) is to consider the shoulder girdle (considering bones in pairs) as four joints that can be

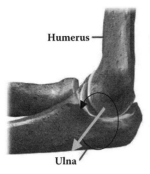
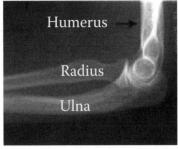

FIGURE 29.1 A one-DoF elbow.

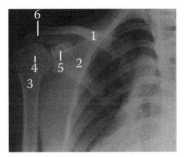

FIGURE 29.2 The shoulder joint (1. Clavicle; 2. Body of scapula; 3. Surgical neck of humerus; 4. Anatomical neck of humerus; 5. Coracoid process; 6. Acromion).

distinguished as the sterno-clavicular joint, which articulates the clavicle by its proximal end onto the sternum; the acromio-clavicular joint, which articulates the scapula by its acromion onto the distal end of the clavicle; the scapulo-thoracic joint, which allows the scapula to glide on the thorax; and the gleno-humeral joint, which allows the humeral head to rotate in the glenoid fossa of the scapula.

Another method takes into consideration the final gross movement of the joint (Yang, 2003; Yang et al., 2004a, 2005a), as abduction/adduction (about the anteroposterior axis of the shoulder joint), flexion/extension, and transverse flexion/extension (about the mediolateral axis of the shoulder joint). Note that

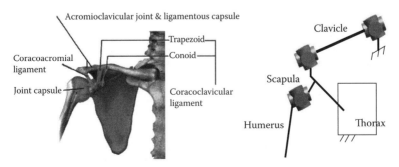

FIGURE 29.3 A model of the shoulder complex.

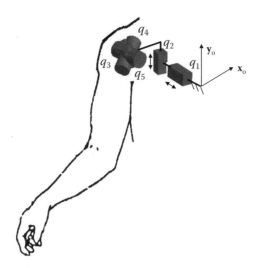

FIGURE 29.4 Model of the shoulder complex as three revolute and two prismatic DoFs.

these motions provide for three rotational degrees of freedom having their axes intersecting at one point. This gives rise to the effect of a spherical joint typically associated with the shoulder joint (fig. 29.4). In addition, the upward/downward rotation of the scapula gives rise to two substantial translational degrees of freedom and a total of five DOFs in the shoulder complex. This model allows for consideration of the coupling between some of the joints, as is the case in the shoulder where muscles extend over more than one segment. When muscles are used to lift the arm in a rotational motion, a translational motion of the shoulder occurs unwittingly.

Considering the whole human body, we obtain a skeletal model for Santos (developed by the Virtual Soldier Research group at the University of Iowa) shown in Figure 29.5 (Yang et al., 2006a).

29.2.1.2 Musculoskeletal Model

Muscles are one important component in the human system. Muscles are considered actuators for generating motion; they should be modeled when building a biomechanic model. When attaching the muscles to the skeletal model, two important factors should be considered. The first is the shape of the muscle and where it originates and inserts on the skeleton. Since the muscles are working in bones that are arranged in a lever-like structure, determining the correct attachment points is critical because there is a direct correlation between these and the torque that the associated muscle can create. The second factor that is significant when attaching the muscles to the skeletal model is the number of action lines used to approximate the force generated by a muscle.

Several digital human models have incorporated muscles. LifeMOD has 118 muscles that are automatically generated and attached to the bones at anatomical landmarks in Figure 29.6. This model has been used for rehabilitation research, motorcycle crash analysis, gait analysis, golf biomechanics, and so on.

Another musculoskeletal model is Anybody's model, shown in Figure 29.7 (Rasmussen et al., 2004). This system is used for biomechanical research, ergonomics design, computer-aided surgery, sports and fitness, and design of rehabilitation technology.

In the Santos system, a new muscle-wrapping algorithm has been developed (Patrick, 2005). There are 134 action lines; so far, each has one or more wrapping obstacles, and the complete musculoskeletal

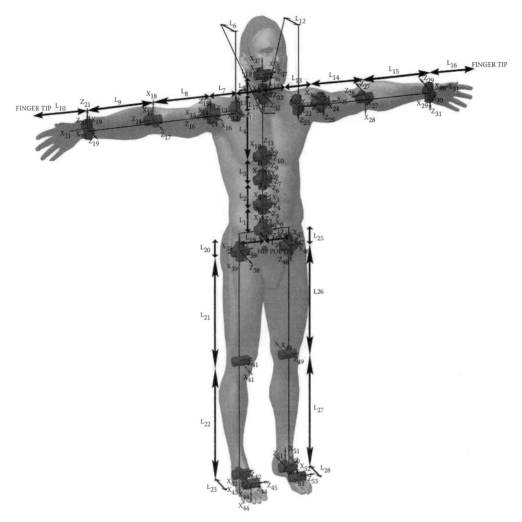

FIGURE 29.5 Santos model.

model is still under development. Furthermore, the bones are jointed into an appropriate kinematic skeleton, and interaction was added so the user has the ability to manipulate any of the joints. The result is a skeleton with real-time interaction as well as live muscle wrapping and sliding, all running at 30+ fps.

29.2.1.3 Appearance

Appearance is an important aspect of digital humans because the user trusts the results more if the digital humans look like real humans. So far the avatars in most digital human model systems are cartoon-like characters. Santos is the only model that has a realistic humanlike appearance and skin that can deform in a way that corresponds to the body segment movement. To visually simulate the elasticity of human skin as the joints are exercised, the amount of movement in the skin around a particular point must be defined when that point in the skin moves during joint rotation. This is done with a traditional

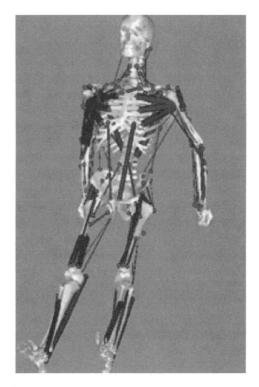

FIGURE 29.6   LifeMOD model.

animation technique called *skin weighting* (Choi, 2002), which addresses the aesthetic issue that would otherwise cause a 3D model to tear or break at the joints when rotated. Typically, this technique is accomplished subjectively through interactive tools—much like using a can of spray paint—that allow 8-bit gray-level values to represent how specific regions of the skin are anchored to specific joints. The

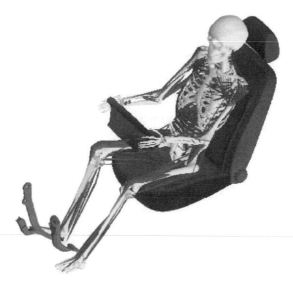

FIGURE 29.7   Musculoskeletal model of Anybody.

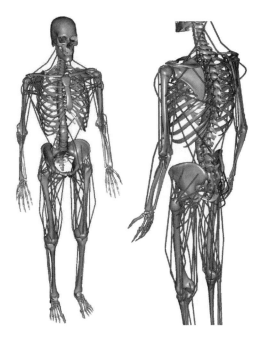

FIGURE 29.8 Santos musculoskeletal model.

higher the gray-level value, the greater the effect a given anchor has on that region. Figure 29.9 shows the skin weight associated with an anchor below the first spine joint.

Note the high gray-level values around the groin area, which cause the geometry in that skin region to be completely immoveable. Alternatively, the decreasing gray-level values above the groin area allow the skin increasing elasticity the further it is from the groin region.

29.2.1.4 Open-Loop versus Closed-Loop Chains

A kinematic chain consists of rigid links that are connected with joints, allowing the relative motion of neighboring links. Kinematic chains are categorized as open-loop or closed-loop chains. Open-loop

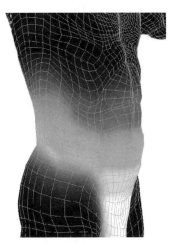

FIGURE 29.9 Skin with adjusted weight.

chains are defined as chains in which each member is simply jointed to a single previous-link neighbor and a single subsequent-link neighbor. The first link has no previous link, and the last link has no subsequent link; such series chains have one less joint than the number of links. Alternatively, it can be defined as a kinematic chain in which every link is connected to any other link by one and only one distinct path. In the human model, there are five open-loop chains starting from the hip point: to the right hand, left hand, head, left leg, and right leg, respectively.

A closed-loop chain is defined as a chain in which every link in the kinematic chain is connected to any other link by at least two distinct paths. A specific example of a closed-loop chain is a parallel robot. Human models are special mechanisms because in most scenarios they are considered open-loop chains; however, special cases such as the human swing motion where when the operator is trying to get in a cab of a heavy vehicle, his legs attach on the stairs and the hands hold the handles, and then he swings into the cab. This is one of a few special cases wherein the human model is considered a closed-loop chain since both hands and legs are fixed on the vehicle and the body swings. In digital human modeling, we transfer the closed-loop chain to an open-loop chain problem to solve.

29.2.1.5 Denavit–Hartenberg Method

To describe the translational and rotational relationships between adjacent links of the open kinematic chain, the Denavit and Hartenberg (1955) notation has been used because of its strength in handling large numbers of degrees of freedom and because of its ability to systematically enable kinematic and dynamic analyses. The DH notation is used to systematically establish a coordinate system (body-attached frame) to each link of an articulated chain in robotics (Asada & Slotine, 1986). The DH notation uses a minimum number of parameters to completely describe the kinematic relationship, and the relative location of the two frames can be completely determined by four parameters. Indeed, the Denavit–Hartenberg representation method has been demonstrated to yield an effective method for modeling humans (Jung et al., 1995; Yang et al., 2004b; Abdel-Malek et al., 2001).

Consider Figure 29.10 where two consecutive joints are shown.

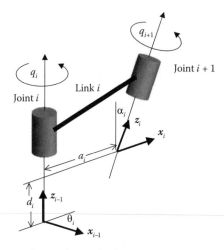

FIGURE 29.10   Defining the joint reference frames for the D-H representation.

The four parameters (depicted in fig. 29.10) are:

θ_i is the joint angle, measured from the x_{i-1} to the x_i axis about the z_{i-1} (right-hand rule applies). For a prismatic joint, θ_i is a constant. It is basically the angle rotation of one link with respect to another about the z_{i-1} axis.

d_i is the distance from the origin of the coordinate frame $(i-1)$ to the intersection of the z_{i-1} axis with the x_i axis along the z_{i-1} axis. For a revolute joint, d_i is a constant. It is basically the distance translated by one link with respect to another along the z_{i-1} axis.

a_i is the offset distance from the intersection of the z_{i-1} axis with the x_i axis to the origin of the frame i along x_i axis (shortest distance between the z_{i-1} and z_i axis).

a_i is the offset angle from the z_{i-1} axis to the z_i axis about the x_i axis (right-hand rule).

The four transformation matrices are

$$\mathbf{T}_{ai} = \begin{bmatrix} 1 & 0 & 0 & 0 \\ 0 & \cos \alpha_i & -\sin \alpha_i & 0 \\ 0 & \sin \alpha_i & \cos \alpha_i & 0 \\ 0 & 0 & 0 & 1 \end{bmatrix} \tag{29.1}$$

$$\mathbf{T}_{bi} = \begin{bmatrix} 1 & 0 & 0 & a_i \\ 0 & 1 & 0 & 0 \\ 0 & 0 & 1 & 0 \\ 0 & 0 & 0 & 1 \end{bmatrix} \tag{29.2}$$

$$\mathbf{T}_{ci} = \begin{bmatrix} \cos \theta_i & -\sin \theta_i & 0 & 0 \\ \sin \theta_i & \cos \theta_i & 0 & 0 \\ 0 & 0 & 1 & 0 \\ 0 & 0 & 0 & 1 \end{bmatrix} \tag{29.3}$$

$$\mathbf{T}_{di} = \begin{bmatrix} 1 & 0 & 0 & 0 \\ 0 & 1 & 0 & 0 \\ 0 & 0 & 1 & d_i \\ 0 & 0 & 0 & 1 \end{bmatrix} \tag{29.4}$$

The overall Denavit–Hartenberg coordinate transformation matrix from the frame i coordinate system relative to the frame $i-1$ coordinate system is then given by:

$$\mathbf{T}_{i-1}^{i} = \mathbf{T}_{ci}\mathbf{T}_{di}\mathbf{T}_{bi}\mathbf{T}_{ai} = \begin{bmatrix} \cos \theta_i & -\cos \alpha_i \sin \theta_i & \sin \alpha_i \sin \theta_i & a_i \cos \alpha_i \\ \sin \theta_i & \cos \alpha_i \cos \theta_i & -\sin \alpha_i \cos \theta_i & a_i \sin \alpha_i \\ 0 & \sin \alpha_i & \cos \alpha_i & d_i \\ 0 & 0 & 0 & 1 \end{bmatrix} \tag{29.5}$$

Similarly, for an n-joint manipulator, the global joint and end-effector frames using Equation 29.5 are restated using the n-homogeneous transformation matrices $\mathbf{T}_0^1, \mathbf{T}_1^2, \mathbf{T}_2^3, \mathbf{T}_3^4, \cdots, \mathbf{T}_{n-1}^n$. The transformation

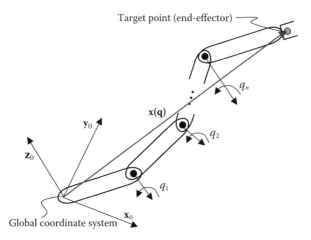

FIGURE 29.11   General kinematic model.

matrix from the end-effector frame to the global frame is then obtained by pre-multiplying each matrix in the series as:

$$\mathbf{T}_0^n = \mathbf{T}_0^1\, \mathbf{T}_1^2\, \mathbf{T}_2^3\, \mathbf{T}_3^4 \cdots \mathbf{T}_{n-1}^n \tag{29.6}$$

where the resulting transformation matrix $\mathbf{T}_0^n = \begin{bmatrix} \mathbf{R}_0^n & \mathbf{P}_0^n \\ \mathbf{0}_{1\times3} & 1 \end{bmatrix}$ contains the (3×3) $\mathbf{R}_0^n$ rotation matrix

and the (3×1) $\mathbf{P}_0^n$ position vector for joint n. Then the position vector of the point of interest is defined as

$$\begin{bmatrix} \mathbf{x}(\mathbf{q}) \\ 1 \end{bmatrix} = \mathbf{T}_0^n \begin{bmatrix} \mathbf{v}^n \\ 1 \end{bmatrix} \tag{29.7}$$

and the vector $\mathbf{x}(\mathbf{q})$ shown in Figure 29.11. From the above analysis, we can systematically represent the coordinates of the end-effector of a serial chain with respect to its global frame.

29.3 Model-Based versus Data-Driven Approaches

The data-driven approach is based on empirical task observation and statistical correlation. It starts with observing a subject performing the task and collecting motion capture data, then statistically analyzes the data to develop a mathematical model. However, the model-based approach is based on physiological study and empirical validation and starts with identifying key human performance measures that affect posture or motion, then determines the physiological factors and uses an optimization technique to model posture or motion, and finally uses an experiment to validate the model. While data-driven prediction of human motion has been attempted for the past several decades, it is now clear that the depth and breadth of the problem are too large and complex for simply recording postures/motions and playing them back again. This section addresses the difficulties and provides insight into the new and exciting field of predictive dynamics as an example.

Predictive dynamics is a term coined to characterize a new methodology for predicting human motion while considering dynamics and the environment. Whereas kinematics is the study of motion (position, velocity, and acceleration) without forces and torques, dynamics is the study of motion with all external and internal forces taken into consideration.

For every motion affected by physics, there are laws that govern that motion. These laws have undergone the test of time, have sent people to the moon, and have been implemented into every computer that governs dynamic systems. Equations that represent motion are called the equations of motion (EoM). For large redundant systems such as the human body, these equations become very sophisticated nonlinear differential equations subject to algebraic constraints; hence the term often used to represent these equations is *differential algebraic equations* (DAEs).

For a sophisticated system of segments, such as the human body which is made up of joints and rigid links, the formulation for multi-body dynamics becomes large and complex. Solving the consequent system of equations is almost impossible. Indeed, for structural systems with a limited number of degrees of freedom, numerical integrators have been developed to solve the problem iteratively. For high degrees of freedom, however, numerical integrators come to a halt.

We have focused on a method that employs optimization with dynamics to predict human motion. Recent results have demonstrated significant promise for resolving the problem of predicting human motion while considering external forces, obstacles, physiology, and most importantly the equations of motion (Kim et al., 2006; Yang et al., 2007a and b). This method, which we call *predictive dynamics*, provides a way to address the issue of predicting human motion in a general manner. Our recent results have shown that this method is applicable to gait prediction, lifting movements, pushing and pulling movements, climbing, and many others.

Thousands of experiments are typically done to capture a few motions. These motions are then compiled into large tables with many parameters. The data are then analyzed and modeled as a nonlinear or functional regression model that should, in principle, predict motion. There are many obvious problems with this method:

- Difficulties in collecting the data for varying anthropometries. This includes the changes of mass properties, moments of inertia, muscle performance, and many other parameters for each person.
- Difficulties in managing a large number of parameters in a functional or nonlinear regression algorithm. A large number of parameters results in a sophisticated and less accurate model.
- Difficulties in predicting postures and motions for reaches that have obstacles. For each obstacle, the experiments must be repeated.
- Difficulties in predicting motions where dynamics (external forces and loads) play an important role.

After the apple fell from the tree onto Newton's head, he proceeded to measure a few more, came up with the general theory, and finally devised a rigorous mathematical formulation for all falling objects and, furthermore, for all objects in motion. He did not measure every apple on every tree to come up with a theory that works and is the fundamental basis for all motion in our universe.

The idea of recording every motion for thousands of people and for thousands of different scenarios does provide a good way to study motion and to validate motion predicted with various methods. However, it has no value for the prediction of motions beyond static postures.

In the section "Dynamics at a Glance," we will provide a brief overview of the predictive dynamics formulation.

29.4 Digital Humans for Workspace Evaluation

Significant effort has been made in human workspace evaluation in recent years. In this section, we first review the literature, then illustrate the closed-form formulation for human workspace analysis, and finally, demonstrate a novel method for determining and visualizing zone differentiation of workspace.

29.4.1 Literature Review

Workspace is defined as the volume within which all the points can be reached by a reference point of the mechanism, for example, the center of the shoulder. Workspace properties can represent an important criterion in the evaluation, programming, and design of mechanisms, robots, and similar devices.

The earliest studies on the subject of manipulator performance in terms of workspace were conducted by Vinagradov et al. (1971), who introduced the term *service sphere*. A study of the relationship between kinematic geometry and manipulator performance including workspace was presented by Roth (1975). A numerical approach to this relationship was formulated and solved by Kumar and Waldron (1981) via tracing boundary surfaces of a workspace. Tsai and Soni (1981) studied accessible regions of planar manipulators, and Gupta and Roth (1982) studied the effect of hand size on workspace analysis. Other studies on the subject of manipulator workspaces were published by Gupta (1986), Sugimoto and Duffy (1982), Davidson and Hunt (1987), Yang and Lee (1983), Emiris (1993), Bulca et al. (1999), Cavusoglu et al. (2001), Monsarrat and Gosselin (2001), Ceccarelli and Ottaviano (2002), Di Gregorio and Zanforlin (2003), Yang and Abdel-Malek (2004a), and Abdel-Malek and Yang (2006a).

Human workspace is one of the most important areas in human factors and ergonomics. It refers to determining the workspace of human upper and lower extremities and body segments. Complete identification of human workspace is important to:

- Understand neural strategies allowing the positioning and orienting of the hand or foot during voluntary reaching movements, especially that of the human upper and lower extremities
- Quantify the full functional potential of a joint
- Study ergonomic postures and path trajectories

Generally, there are three methods for determining human workspace. The first method involves experimental-based methods using a large number of subjects to visualize human reach (Chaffin, 2002; Ottaviano et al., 2001; Badler, 1997; Reed et al., 2003; and Parkinson et al., 2003). The second is the voxel-based method (Troy & Guerin, 2004), and the third is the closed-form solution (Lenarcic & Umek, 1994; Yang, 2003; Sharma et al., 2004; Abdel-Malek et al., 2001, 2006a; Yang et al., 2005a and b). Abdel-Malek et al. (2006b) provided a complete review of swept volumes.

29.4.2 Reach Envelope

This section presents the closed-form formulation for human workspace determination. Examples of upper extremity, lower extremity, and fingers will illustrate the method.

29.4.2.1 Methodologies

The vector function $\mathbf{x}(\mathbf{q})$ characterizes the set of all points inside and on the boundary of the reach envelope generated by an anatomical landmark, typically a fingertip. The objective is to visualize this vector function, which consists of many parameters, and to better understand the motion governed by $\mathbf{x}(\mathbf{q})$. At a specified position in space given by $P(x_p, y_p, z_p)$, Equation 29.7 can be written as a constraint function:

$$\Omega(\mathbf{q}) = \begin{bmatrix} x(\mathbf{q}) - x_p \\ y(\mathbf{q}) - y_p \\ z(\mathbf{q}) - z_p \end{bmatrix} = \mathbf{0} \qquad (29.8)$$

Ranges of motion (Norkin & White, 1995) are imposed in terms of inequality constraints in the form of $q_i^L \leq q_i \leq q_i^U$, where $i = 1, \ldots n$ are transformed into equalities by introducing a new set of generalized coordinates $\boldsymbol{\lambda} = [\lambda_1 \ldots \lambda_n]^T$ such that

$$q_i = \left((q_i^L + q_i^U)/2\right) + \left((q_i^U - q_i^L)/2\right)\sin\lambda_i \qquad i = 1, \ldots, n \tag{29.9}$$

where the new variable λ_i is inherently constrained by the sine function and does not change the dimensionality of the problem. In order to include the effect of the ranges of motion, augmentation of the constraint equation $\Omega(\mathbf{q})$ with the parameterized ranges of motion is proposed, such that

$$\mathbf{H}(\mathbf{q}^*) = \begin{bmatrix} x(\mathbf{q}) - x_p \\ y(\mathbf{q}) - y_p \\ z(\mathbf{q}) - z_p \\ q_i - a_i - b_i \sin\lambda_i \end{bmatrix} = \mathbf{0} \qquad i = 1, \ldots, n \tag{29.10}$$

where $\mathbf{q}^* = [\mathbf{q}^T \quad \boldsymbol{\lambda}^T]^T$ is the vector of all generalized coordinates. Note that although n—new variables (λ_i) have been added, n—equations have also been added to the constraint vector function without losing the dimensionality of the problem.

The Jacobian (after the German mathematician Karl Gustav Jacob Jacobi) of the constraint function $\mathbf{H}(\mathbf{q}^*)$ at a point $\mathbf{q}^{*0}$ is the $(3 + n) \times 2n$ matrix

$$\mathbf{H}_{\mathbf{q}^*} = \partial\mathbf{H}/\partial\mathbf{q}^* \tag{29.11}$$

where the subscript denotes a derivative. With the modified formulation that includes ranges of motion, the Jacobian is expanded to

$$\mathbf{H}_{\mathbf{q}^*} = \begin{bmatrix} \mathbf{x}_\mathbf{q} & \mathbf{0} \\ \hline \mathbf{I} & \mathbf{q}_\lambda \end{bmatrix} \tag{29.12}$$

where $\mathbf{q}_\lambda = \partial\mathbf{q}/\partial\mathbf{q}$, $\mathbf{x}_\mathbf{q} = \partial\mathbf{x}/\partial\mathbf{q}$, $\mathbf{0}$ is a $(3 \times n)$ zero matrix, $\mathbf{I}$ is the identity matrix, and

$$\mathbf{x}_\mathbf{q} = \begin{bmatrix} x_{q_1} & x_{q_2} & \cdots & x_{q_n} \\ y_{q_1} & y_{q_2} & \cdots & y_{q_n} \\ z_{q_1} & z_{q_2} & \cdots & z_{q_n} \end{bmatrix} \tag{29.13}$$

$$\mathbf{q}_\lambda = \begin{bmatrix} -((q_1^U - q_1^L)/2)\cos\lambda_1 & 0 & \cdots & 0 \\ 0 & -((q_2^U - q_2^L)/2)\cos\lambda_2 & \cdots & 0 \\ 0 & 0 & \cdots & 0 \\ 0 & 0 & \cdots & -((q_n^U - q_n^L)/2)\cos\lambda_n \end{bmatrix} \tag{29.14}$$

Because the Jacobian is not square, rank deficiency criteria were developed. Before addressing these criteria, however, it is important to show why the singularity of the Jacobian has a direct effect on the control. The differentiation of Equation 29.7 with respect to time yields the velocity of the fingertip $\dot{\mathbf{x}}$ as

$$\dot{\mathbf{x}} = \mathbf{x_q}\dot{\mathbf{q}} \tag{29.15}$$

where $\dot{\mathbf{q}}$ is the vector of joint velocities. Given a specified path trajectory (i.e., $\dot{\mathbf{x}}$), the calculation of $\dot{\mathbf{q}}$ (i.e., joint velocities) requires computing an inverse of the Jacobian $\mathbf{x_q}$. For a singular Jacobian, it is not possible to compute the required velocities. These cases are typically associated with a kinematic configuration of the upper extremity that does not admit motion in a particular direction and requires a change in the arm's posture in order to execute the path. Because the Jacobian is not square, we define these singular sets as a subset of the workspace in which the Jacobian of the augmented constraint function of Equation 29.11 is *row rank deficient*; that is, the barriers are defined by *W and characterized by

$$\partial W \subset \left\{ \text{Rank } \mathbf{H_q}^* < k, \text{ for some } \mathbf{q}^* \text{ with } \mathbf{H}(\mathbf{q}^*) = \mathbf{0} \right\} \tag{29.16}$$

where k is at least $(3 + n - 1)$. Imposition of the rank deficiency condition can be implemented using a variety of methods; perhaps the most computationally efficient is the repeated elimination of square sub-Jacobians until several nonlinear equations are determined.

29.4.2.2 Upper Extremity

A restrained driver is modeled as a four-DOF model (Yang et al., 2005b). Figure 29.12 shows the model, and Figure 29.13 shows the reach envelope of this model.

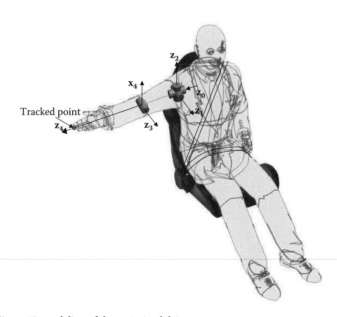

FIGURE 29.12 Kinematic modeling of the restrained driver.

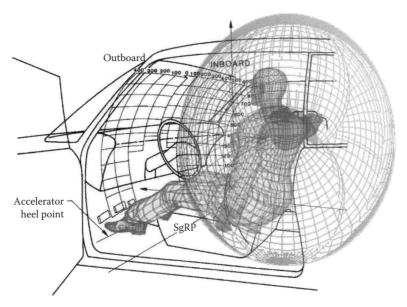

FIGURE 29.13   Cross section of reach envelope for the restrained driver.

29.4.2.3 Lower Extremity

The study of the toe's reach envelope with respect to the hip is a key for better ergonomic design of foot-operated mechanisms. For example, it is useful in the design of vehicle pedals. For the purpose of this study, a six-DOF kinematic model for the foot segment is used (Yang et al., 2006b). The point of interest in this case is the middle toe, and the reference point is taken at the right hip joint. Figure 29.14 shows the model.

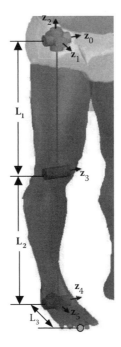

FIGURE 29.14   Leg segment model.

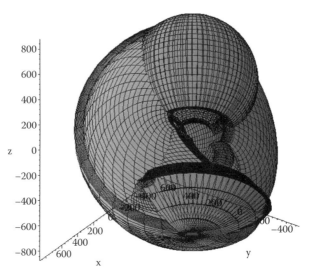

FIGURE 29.15   Reach envelope of the toe with respect to the hip.

Using the methodology presented in the previous section, we find the singular surfaces shown in Figure 29.13. When this reach envelope is plotted together with the human, it is shown in Figure 29.14.

29.4.2.4 Fingers

Consider the workspace of a point located at the tip of the index finger as shown in Figure 29.17a (Abdel-Malek et al., 2001). The kinematic motion of the finger is modeled as four revolute joints, two of which intersect and are shown in Figure 29.17b. The reach envelope is shown in Figure 29.18.

FIGURE 29.16   Reach envelope of lower extremity in the virtual environment.

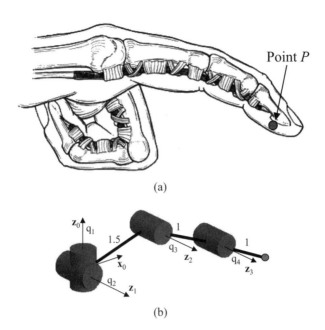

(a)

(b)

FIGURE 29.17 (a) A schematic of a finger. (b) Kinematic modeling of the finger as four revolute joints.

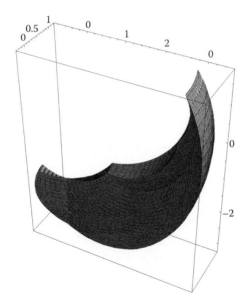

FIGURE 29.18 Workspace of the index finger.

29.4.3 Workspace Zone Differentiation

Reach envelope is not enough for designers because it tells only whether a point in the space is reachable or not. The designer really wants to know which area is best for placing the control or button. Nobody has investigated this, to our knowledge. This section presents an optimization-based approach for differentiating and visualizing the zones inside the workspace.

29.4.3.1 Methodology

The methodology of zone differentiation is based on a multi-objective optimization (MOO) posture prediction (Yang et al., 2004b and 2006a). We sample the points covering the workspace. For each point, we run the MOO posture prediction and have the feedback of objective function values. Here, the objective functions are also called human performance measures and include discomfort, effort, joint displacement, potential energy, and so on. After we normalize these human performance values, each voxel is assigned a binary number that lies between 00 and FF, which reflects a gray-level value. Hence, a grayscale picture can be visualized for each plane, where black represents a point with a maximum objective value, and white represents a point with a minimum objective value. Therefore, the result is a set of pictures that reflect the zone-differentiated workspace of each slice inside the sampled cube created earlier. A 3D view of the zone-differentiated workspace is created by assembling the resultant slices and visualizing them all at the same time (Yang et al., 2008).

29.4.3.2 Implementation

We implement two types of zone differentiation. The first one is solid zone differentiation, which is colored within a 3D cloud spherical shape. The second is surface zone differentiation. For example, if we have a panel inside a car, we can run this type of zone differentiation. Figure 29.19 shows the solid zone differentiation, which helps the user design products. Figure 29.20 shows the surface zone differentiation, which helps the user analyze the discomfort level of the given design.

FIGURE 29.19   Solid zone differentiation.

FIGURE 29.20   Surface zone differentiation.

29.5 Dynamics at a Glance

Predictive dynamics works based on a special optimization formulation. Consider a general optimization problem, for which there are three main ingredients: (1) a set of design variables, which in our case are the joint profiles (i.e., joint angles as a function of time) and the torque profiles at each joint; (2) multiple cost functions to be optimized, which are human performance measures that represent functions that are important to accomplishing the motion (e.g., energy, speed, joint torque); and (3) constraints on the motion (e.g., collision avoidance, joint ranges of motion). This general optimization problem is readily solved using existing optimization codes. The field of optimization is mature, and many such codes exist, have been verified and validated, and have been tested with many different complex problems.

We are interested in seeing how human motion is predicted for scenarios that involve dynamic influences including but not limited to external loads, obstacles, and running. In general, we consider any case where a human segment is undergoing motion that warrants the consideration of masses and moments of inertia. Predictive dynamics can incorporate such general cases.

Based on the general process discussed above, advanced dynamics simulation software is being developed to simulate any task. The structure for this environment is shown in Figure 29.21. At the core of this environment are general-purpose modules that can be used to study various dynamics activities. These consist of the optimization module, the B-spline module, the DH module, the inverse dynamics module, the cost function modules, and the constraint modules. The module for the DH method allows the user to build any form of skeleton. Using these modules, a variety of tasks, such as walking, can be modeled by modifying the performance measure(s) and constraints. In fact, this is the foundation for what we call *task-based motion prediction*, which suggests that what motivates human motion depends on the task being completed.

One of the predictive dynamics is gait prediction. Most of the work that has been conducted with gait/walking analysis is data-driven, meaning that it depends on prerecorded motion-capture data or EMG data. Using predictive dynamics, we have devised a new physics-based approach for modeling gait and balance that does not require prerecorded data. The process of taking a step is decomposed as shown in Figure 29.22. The process of walking is then decomposed into different steps (fig. 29.23), which are then simulated with separate optimization problems. In the case of walking, the performance measure is the total required joint torque throughout the body. There are five types of constraints for this

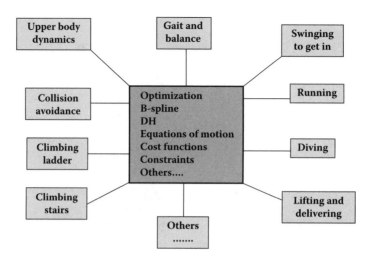

FIGURE 29.21 Structure of dynamics simulation.

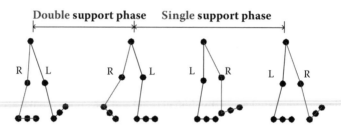

FIGURE 29.22 Two phases in a step (R: right leg, L: left leg).

FIGURE 29.23 Long-distance walking.

task: joint limits as discussed above; torque limits for each joint; restriction of feet from penetrating the ground; zero moment point (ZMP); and the specification of a series of foot-strike points.

There are other predictive dynamics examples at Virtual Soldier Research, such as human running simulation, grenade throwing simulation, and human diving simulation.

29.6 Summary

This chapter presents the basic concepts of human modeling and simulation, including the skeletal model, musculoskeletal model, appearance, open- and closed-loop systems, and DH method. We compare the data-driven approach with the model-based method, and a closed-form formulation for human workspace is demonstrated with different body segments. We also introduce the zone differentiation concept for digital human modeling and simulation, a new method that can improve design efficiency and bring the ergonomics evaluation into the early design stage. We discuss reach envelope and zone differentiation, which are useful tools for ergonomic design, and finally, we provide a brief overview of predictive dynamics in human modeling and simulation.

References

Abdel-Malek, K., Yang, J., Brand, R., and Tanbour, E. (2001). Towards understanding the workspace of the upper extremities. *SAE Transactions-Journal of Passenger Cars: Mechanical Systems*, Vol. 110, Section 6: 2198–2206.

Abdel-Malek, K. and Yang, J. (2006a). Workspace boundaries of serial manipulators using manifold stratification. *International Journal of Advanced Manufacturing Technology*, Vol. 28, No. 11–12: 1211–1229.

Abdel-Malek, K., Yang, J., and Blackmore, D. (2006b). Swept volumes: foundations, perspectives, and applications. *International Journal of Shape Modeling*, Vol. 12, No. 1: 87–127.

Asada, H. and Slotine, J.J.E. (1986). *Robot Analysis and Control*. New York: Wiley.

Badler, N.I. (1997). Real-time virtual humans. *Proceedings of the 1997 5th Pacific Conference on Computer Graphics and Applications*. Oct 13–16, 1997, pp. 4–13.

Biomechanics Research Group, Inc. (2005). *LifeMod Manual*.

Bulca, F., Angeles, J., and Zsombor-Murray, P.J. (1999). On the workspace determination of spherical serial and platform mechanisms. *Mechanism and Machine Theory*, Vol. 34, No. 3: 497–512.

Cavusoglu, M.C., Villanueva, I., and Tendick, F. (2001). Workspace analysis of robotic manipulators for a teleoperated suturing task. *IEEE International Conference on Intelligent Robots and Systems*, Vol. 4: 2234–2239.

Ceccarelli, M. and Ottaviano, E. (2002). A workspace evaluation of an eclipse robot. *Robotica*, Vol. 20, No. 3: 299–313.

Chaffin, D.B. (2002). On simulating human reach motions for ergonomics analysis. *Human Factors and Ergonomics in Manufacturing*, Vol. 12, No. 3: 235–247.

Choi, J. (2002). *Maya Character Animation*. Sybex Books.

Davidson, J.K. and Hunt, K.H. (1987). Rigid body location and robot workspace: some alternative manipulator forms. *ASME Journal of Mechanisms, Transmissions, and Automation in Design*, Vol. 109, No. 2: 224–232.

Denavit, J. and Hartenberg, R.S. (1955). A kinematic notation for lower-pair mechanisms based on matrices. *Journal of Applied Mechanics*, Vol. 77: 215–221.

Di Gregorio, R. and Zanforlin, R. (2003). Workspace analytic determination of two similar translational parallel manipulators. *Robotica*, Vol. 21, No. 5: 555–566.

Emiris, D.M. (1993). Workspace analysis of realistic elbow and dual-elbow robot. *Mechanisms and Machine Theory*, Vol. 28, No. 3: 375–396.

Gupta, K.C. (1986). On the nature of robot workspace. *International Journal of Robotics Research*, Vol. 5, No. 2: 112–121.

Gupta, K.G. and Roth, B. (1982). Design considerations for manipulator workspace. *ASME Journal of Mechanical Design*, Vol. 104, No. 4: 704–711.

Jung, E.S., Kee, D. and Chung, M.K. (1995). Upper body reach posture prediction for ergonomic evaluation models. *International Journal of Industrial Ergonomics*, Vol. 16: 95–107.

Kim, J., Abdel-Malek, K., Yang, J., and Marler, T. (2006). Prediction and analysis of human motion dynamics performing various tasks. *International Journal of Human Factors Modelling and Simulation*, Vol. 1, No. 1: 69–94.

Kumar, A. and Waldron, K.J. (1981). The workspace of a mechanical manipulator. *ASME Journal of Mechanical Design*, Vol. 103: 665–672.

Lenarcic, J. and Umek, A. (1994). Simple model of human arm reachable workspace. *IEEE Transactions on Systems, Man, and Cybernetics*, Vol. 24, No. 8: 1239–1246.

Maurel, W. (1999). 3D Modelling of the Human Upper Limb Including the Biomechanics of Joints, Muscles and Soft Tissues. Ph.D. Dissertation, Ecole Polytechnique Federale de Lausanne, France.

Monsarrat, B. and Gosselin, C.M. (2001). Singularity analysis of a three-leg six-degree-of-freedom parallel platform mechanism based on grassmann line geometry. *International Journal of Robotics Research*, Vol. 20, No. 4: 312–326.

Norkin, C.C. and White, D.J. (1995). *Measurement of joint motion: a guide to goniometry,* 2nd edition. Pennsylvania: F.A. Davis Company.

Ottaviano, E., Lanni, C., and Ceccarelli, M. (2001). Experimental determination of workspace characteristics of human arms. *9th International Conference on Control and Automation MED 2001,* Dubrovnik. CD Proceedings, paper n.017.

Parkinson, M.B., Reed, M.P., and Klinkenberger, A.L. (2003). Assessing the validity of kinematically generated reach envelopes. *SAE Digital Human Modeling Conference.*

Patrick, A. (2005) Development of a 3D Model of the Human Arm for Real-Time Interaction and Muscle Activation Prediction. M.S. Thesis, The University of Iowa.

Rasmussen, J., Damsgaard, M., Christensen, S.T., de Zee, M., Dahlquist, J., and Dhang, N. (2004). Musculoskeletal modeling by inverse dynamics. *The 14th European Society of Biomechanics Conference*, Hertogenbosch, The Netherlands.

Reed, M.P., Parkinson, M., and Chaffin, D.B. (2003). A new approach to modeling driver reach. SAE Technical Paper 2003-01-0587.

Roth, B. (1975). Performance evaluation of manipulators from a kinematic viewpoint. *NBS Special Publications,* Vol. 459: 39–61.

Sharma, G., Badescu, M., Dubey, A., Mavroidis, C., Sessa, T., Tomassone, M.S., and Yarmush, M.L. (2004). Kinematics and workspace analysis of protein based nano-motors. *2004 ASME Mechanisms and Robotics Conference, 2004 ASME Design Technical Conferences*, Salt Lake City, UT.

Sugimoto, K. and Duffy, J. (1982). Determination of extreme distances of a robot hand. Part 2: Robot arms with special geometry. *ASME Journal of Mechanical Design*, Vol. 104: 704–712.

Tsai, Y.C. and Soni, A.H. (1981). Accessible region and synthesis of robot arm. *ASME Journal of Mechanical Design*, Vol. 103: 803–811.

Troy, J. and Guerin, J. (2004). Human swept volumes. *SAE Technical Paper* 2004-01-2190.

Venema, S. and Hannaford, B. (2001). A probabilistic representation of human workspace for use in the design of human interface mechanisms. *IEEE/ASME Transactions on Mechatronics*, Vol. 6, No. 3.

Vinagradov, I., et al. (1971). Details of kinematics of manipulators with the method of volumes (in Russian). *Mexanika Mashin*, No. 27–28: 5–16.

Yang, D.C.H. and Lee, T.W. (1983). On the workspace of mechanical manipulators. *Journal of Mechanisms, Transmission and Automation Design*, Vol. 105: 62–69.

Yang, J., Marler, R.T., Kim, H., Arora, J., and Abdel-Malek, K. (2004b). Multi-objective optimization for upper body posture prediction, *10th AIAA/ISSMO Multidisciplinary Analysis and Optimization Conference,* Aug. 30–Sept. 1, 2004, Albany, NY.

Yang, J. (2003). Swept Volumes: Theory and Implementations. Ph.D. Dissertation, The University of Iowa.

Yang, J. and Abdel-Malek, K. (2004a). Singularities of manipulators with non-unilateral constraints. *Robotica*, Vol. 23, No. 5: 543–553.

Yang, J., Abdel-Malek, K., and Nebel, K. (2005a). The reach envelope of a 9 degree of freedom model of the upper extremity. *International Journal of Robotics and Automation,* Vol. 20, No. 4: 240–259.

Yang, J., Abdel-Malek, K., and Nebel, K. (2005b). On the determination of driver reach and barriers. *International Journal of Vehicle Design*, Vol. 37, No. 4: 253–273.

Yang, J., Man, X., Xiang, Y., Kim, H., Patrick, A., Swan, C., Abdel-Malek, K., Arora, J., and Nebel, K. (2007b). Newly developed functionalities for the virtual-human Santos. *SAE 2007 World Congress*, Cobo Center, Detroit, MI.

Yang, J., Marler, R.T., Beck, S., Abdel-Malek, K., and Kim, J. (2006a). Real-time optimal reach-posture prediction in a new interactive virtual environment. *Journal of Computer Science and Technology,* Vol. 21, No. 2: 189–198.

Yang, J., Sinokrot, T., and Abdel-Malek, K. (submitted 2006b). Swept volumes for the virtual human Santos. *International Journal of Robotics and Automation.*

Yang, J., Kim, J., Abdel-Malek, K., Marler, T., Beck, S., and Kopp, G. (2007a). *Computer-Aided Design,* Vol. 39, 2007, 548–558.

Yang, J., Sinokrot, T., and Abdel-Malek, K. (2008). Workspace zone differentiation analysis and visualization for virtual humans. *Ergonomics,* Vol. 51, Issue 3, 2008, 395–413.

30

Statistical Methods for Human Motion Modeling

Julian J. Faraway

30.1 Introduction

Human motion is inherently variable while statistics involves the study of variability. So it is natural that statistics should be applied to the study of human motion. Of course, statistical methods are widely used in ergonomics and human factors research, but they are limited in a certain way. Consider quantities such as the time taken to complete a task, the angle formed at the knee in a static posture, and the maximum velocity of the hand during a reach. These are all univariate, scalar quantities for which there is a large repertoire of statistical methods, many of which are available in standard statistical software. This chapter assumes the reader has some knowledge and experience with these textbook statistical methods, but we want to present here some more advanced techniques particularly appropriate for modeling motion.

Suppose we capture the motion of a subject performing some motions. Consider, for example, the included angle formed at the elbow. This angle will vary during the motion. Although we could model certain scalar characteristics of this curve such as its final or maximum value, it is more natural to model the whole curve, viewing it as a function. We shall use some more recently developed statistical methods called *functional data analysis,* which are designed to model data that are functions as opposed to just scalars or vectors. There is a growing literature on functional data analysis for univariate curve data.

A good general place to start is Ramsay and Silverman (2005), while Ramsay and Silverman (2002) provides a collection of analyses of functional data.

The type of data we propose to model is typically collected with motion capture equipment. We will not concern ourselves with the specifics of how this is done, but we shall assume that we have data on the 3D trajectories of some collection of markers on the human body. Typically, these are collected at some frequency, so we do not exactly view the whole curve, rather a collection of points along that curve. We shall assume that the frequency is sufficiently high that we can reasonably approximate the curve. Anything more than 25 Hz should be sufficient for our purposes.

There are some practical difficulties with motion capture data that we need to make allowances for. In particular, markers can get obscured so that some portions of some observed curves may be missing and so our statistical methods need to be able to accommodate this.

Human motion modeling requires that we model several different types of curve. The simplest example is the univariate curve such as the elbow included angle. We will present methods here capable of modeling this type of response. The motion of the hand forms a trajectory in 3D which we will also need to model. Although this could be viewed as three univariate curves, we shall see that it is better to model such a curve holistically. The motion of the head forms an orientation trajectory as it rotates during the motion. Again, this could be modeled in terms of its angular components, but is better modeled as a whole. We will also need to combine the various models for components of the body into a coherent whole. We will present methods for achieving this.

Of course, researchers have been modeling human motion without the use of statistics. However, the statistical approach does bring certain advantages. It is inherently data driven and relies less on prior theories about how people should move but instead allows the data to inform our knowledge. We shall return to this issue of the relative advantages and disadvantages of this approach at the end of this chapter.

30.2 Univariate Curves

Let us start with an example. The HuMoSim laboratory (www.humosim.org) at the University of Michigan has accumulated a large database of motions collected from subjects performing a variety of tasks in industrial and vehicle settings. The data we present here come from a study of the motions of 19 subjects in a standing position delivering cylinders to a collection of shelves positioned around the subject. Consider the right elbow included angle during the reach to place the cylinder on a high shelf in front of the subject as seen in the first panel of Figure 30.1. These particular curves are free of missing values and outliers. Unusual measures can be produced by failures in the motion capture and post-processing. It is

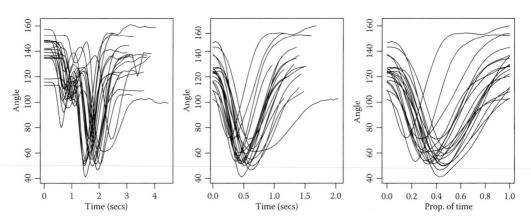

FIGURE 30.1 Right elbow included angle of 19 standing subjects delivering a cylinder to a high shelf.

important to develop screening procedures to detect such cases based on the typical ranges of response expected.

30.3 Registration

The curves we see in the first panel of Figure 30.1 do have some commonality in their shapes, but we can see that they are not well aligned. Before we can proceed with the analysis, we must *register* the curves so that a cross-sectional analysis is possible. Some more advanced ideas on how to register curves are presented in Ramsay and Li (1998), but we will present a simpler approach here. In this example, we can see stationary parts to the curve both before motion starts and after motion ends. We would like to remove these because they vary in length between the curves.

Actually determining when motion begins and ends is not straightforward. Subjects are never entirely stationery. Furthermore, some parts of the body start to move before others. In this example, the technician processing the data made a judgment of starting and ending times. We use these cut points to plot the trimmed data as seen in the second panel of Figure 30.1. Even here, we might doubt that the trimming has been done completely correctly. One might hope to develop a purely automatic way of performing this trimming, but in practice we have found this difficult to do in a completely reliable way, so some supervision will be necessary.

Even after the curves have been trimmed, we can see that they are of different lengths because the subjects took varying amounts of time to perform the motion. In the third panel of Figure 30.1, we see curves rescaled to a [0,1] interval representing the proportion of time rather than the clock time. The actual times should be saved and modeled separately. It is possible that the shape of the curve changes as the task is performed faster so we might consider the time to complete as a possible predictor.

At this point, we have a collection of curves, $y_i(t)$ for $i = 1, \ldots, n$ measured at set of points $t_{ij} \in [0,1]$ where $j = 1, \ldots, m_i$ and m_i is the number of observed points on the ith curve.

30.4 Curve Representation

Ideally one would like to model the curves directly, but in principle at least, these are infinite dimensional and so this is impractical. One could discretize to a fine grid as seen in Faraway (1997), but it is better to take advantage of the fact that the curves are not as high dimensional as they first appear and that they can be well approximated in a lower dimensional space.

We will approximate each curve using a set of cubic B-splines: $S_k(t)$ for $k = 1, \ldots, s$. The number of B-splines that used s should be kept as small as possible while still retaining a good fit to the data. A basis of eight cubic B-splines is shown in the first panel of Figure 30.2.

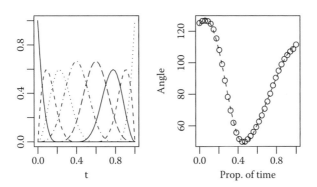

FIGURE 30.2 Basis of eight cubic B-splines is shown on the left. The fit to the first subject is shown on the right.

We aim to approximate $y_i(t)$ with $\sum_{k=1}^{s} y_{ik}S_k(t)$ by minimizing the integrated least squares criterion:

$$\int_0^1 (y_i(t) - \sum_{k=1}^{s} y_{ik}S_k(t))^2 \, dt$$

An example of this is shown in the second panel of Figure 30.2. We see that the fit is quite good.

As we shall see later, there is some variation even within the curves collected on the same subject performing the same action so that a perfect fit is not necessary. In this example, eight basis functions are more than sufficient. Thus, we reduce the curve data to just the eight coefficients. This has the effect of reducing the dimension of the data, which might be useful for storage purposes, but more importantly allows the application of standard statistical techniques. We now essentially have a vector response, namely the vector of fitted coefficients y_{ik} to which standard methods of multivariate analysis can be applied. See, for example, Johnson and Wichern (2002) for the more general methodology.

Other basis functions might be considered. We used cubic order splines because this allows the computation of continuous first- and second-order derivatives, which can be useful if velocity and acceleration are needed. It would not be sensible to use a polynomial basis as these tend to be numerically unstable. A Fourier basis is discouraged because these have non-local support. Bezier splines are a viable alternative as we shall see in the section on trajectory modeling.

30.5 Model Fitting

We now wish to relate the curve responses to the available predictors. We might be interested in how variables such as the height of the shelf or the height of the subject affect the shape of the curve. Suppose we collect all p predictors for curve i in the vector x_i where typically the first component is one in order to represent an intercept term. We then consider a functional regression model of the form:

$$y_i(t) = x_i^T \beta(t) + \varepsilon_i(t)$$

This is similar to the standard regression model except now the response, the (vector of) coefficients $\beta(t)$ and the error term $\varepsilon_i(t)$ are now functions. The coefficient function for a given predictor will now represent the effect on the response of that predictor over the whole duration of the reach. We cannot, in practice, fit this model exactly and must resort to an approximation. The corresponding model for the B-spline coefficients is:

$$Y = XB + E$$

where $Y \equiv y_{ik}$ is the $n \times s$ matrix of the B-spline coefficients. X is an $n \times p$ matrix of the predictors, B is a $p \times s$ matrix of regression coefficients, and E is an $n \times s$ matrix of errors. We can fit this model using least squares:

$$\hat{B} = (X^TX)^{-1}X^TY$$

Typically, users of these models want to do one of two things with a regression model. They want to understand how the predictors affect the response or they want to predict future observations. In human motion modeling both these applications are relevant.

30.6 Effects of Predictors

How does the height of the subject affect the way the reach is performed? To answer this kind of question, we need to recover estimates of the regression coefficient functions. First, we specify a fine grid of points t_j in $[0,1]$ for $j = 1, \ldots, g$ and then compute an estimate of the length p vector of coefficients at each grid point $\hat{\beta}(t_j)$:

$$\hat{\beta}(t_j) = \hat{B}S(t_j) \qquad \forall j$$

where $S(t_j)$ is the length s vector of B-splines evaluated at t_j. The pointwise standard error of these coefficient functions may be obtained from:

$$se(\hat{\beta}_i(t_j)) = \sqrt{(S^T(t_j)\hat{\Sigma}S(t_j))(X^TX)_{ii}^{-1}}$$

where:

$$\hat{\Sigma} = \frac{\hat{E}^T\hat{E}}{n-p} \qquad \text{where} \qquad \hat{E} = Y - X\hat{B}$$

Consider the curves presented in Figure 30.1. We collected the heights of these subjects and consider a model of the form:

$$y_i(t) = \beta_0(t) + \text{height}_i\beta_1(t) + \varepsilon_i(t)$$

where the heights are centered at their mean value so that $\beta_0(t)$ represents the mean response at the average height. We follow the method of analysis described above and display the results in Figure 30.3.

We see the estimated angle curve for a person of mean height on the left panel of Figure 30.3. We have added approximate 95% confidence bands using:

$$\hat{\beta}(t) \pm 2se(\hat{\beta}(t))$$

We see how the angle decreases in the middle of the motion before extending more at the end of the motion. We show the coefficient function for height in the panel on the right of Figure 30.3. At the beginning of the motion, we see that the function is positive, indicating that the taller subjects show a generally larger angle at the beginning of the motion. The function takes a value close to one at this point, indicating one degree extra for each one extra centimeter in height. Toward the end of the motion, the effect becomes negative, showing that taller people do not need to extend their arms as much to reach the high shelf. The 95% confidence bands show that the effect is significant since the

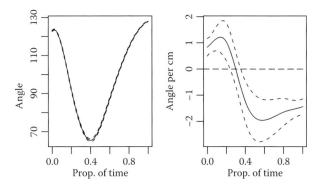

FIGURE 30.3 Coefficient functions for the height predictor model. The intercept function is shown on the left, and the coefficient function for height is shown on the right. 95% confidence bands are shown as dotted lines.

bands are well separated from the zero line shown on the plot. Note that the bands are widest in the middle of plot where the uncertainty about the effect of height is greatest.

30.7 Prediction

Suppose we wish to predict the response for a new value of the predictors given by the vector x_0. We can use:

$$\hat{y}_0(t) = x_0^T \hat{\beta}(t)$$

We can estimate the pointwise standard error of this prediction by

$$se(\hat{y}_0(t)) = \sqrt{(S^T(t)\hat{\Sigma}S(t))x_0^T(X^TX)^{-1}x_0}$$

To illustrate this, consider a new short subject who is 150 cm in height. We show the predicted response along with 95% pointwise confidence bands in the first panel of Figure 30.4. The confidence bands are for the mean response for a large number of subjects with 150 cm stature. The response of a given individual might be expected to vary more than these bands might suggest. It is possible to modify the bands by replacing the expression $x_0^T(X^TX)^{-1}x_0$ by $1 + x_0^T(X^TX)^{-1}x_0$, but since the bands are pointwise in nature, they do not give a full picture of the likely extent of the variation. Instead, it may be better to simulate responses in the following manner. Generate $\varepsilon^* \sim N(0, \hat{\Sigma})$ and produce:

$$\hat{y}_0^*(t) = S^T(t)(x_0^T\hat{B} + \varepsilon^*)$$

Repeat this as many times as desired. We show 19 simulated curves in the second panel of Figure 30.4. This can be compared to the original data as seen in Figure 30.1. Of course, we are simulating 19 people of the same height rather than 19 people of different heights, but the qualitative nature of the plots is similar. Notice that the simulated data convey the horizontal variation in the position of the minimum response much better than the pointwise confidence bands. The simulation approach can also be useful for answering other questions about the characteristics of motion. For example, we might ask how often the angle exceeds some specified level. To estimate the probability of such an event, we can simulate a large number of curves and record the proportion in which the event occurs. Calculating such a probability mathematically is far more difficult.

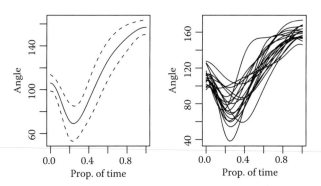

FIGURE 30.4 Predicted response for a subject 150 cm in height. Predicted mean response along with 95% confidence bands are shown on the left, and 19 simulated responses are shown on the right

Prediction of new responses involves extrapolation, both quantitative and qualitative. In our simple example, where the only predictor is height, we might not expect reasonable results if we try to predict the motion of a subject who is only 1 meter tall—in fact, the subject may not be able to reach the shelf at all. In this example, it is fairly easy to suggest bounds on the predictor space beyond which prediction is not recommended. However, in examples with multiple predictors, it is much more difficult because the predictors are likely to be correlated and so independent bounds will not work. This is a practical problem in the use of such models because it is difficult to place reasonable bounds on the predictor space. Make the bounds too stringent and the model will not be generally useful, but make them too lax and the model will produce poor predictions. There is no simple solution to this.

It is also important to recognize that the data were collected under certain qualitative conditions that may not be duplicated in the future situation. For example, the future subject may be stressed or tired, conditions not present in the original data. It is difficult to know whether the predictions of the model will be valid in such circumstances.

30.8 Variation within and between Subjects

When the same person repeats a motion multiple times, there will be some natural variation in the response. Now suppose multiple people with the same external characteristics of interest such as height, age, and gender repeat the motion. There will be even more variation between these subjects. This is of particular interest when considering the potential accuracy of predictions. Clearly there is no hope of predicting the response with an accuracy smaller than the within-subject variation, but more important, there is also no prospect of doing better than the between-subject variation. This imposes a bound on how well any model can perform. We are also interested in studying this variation for its own sake. Are motions for some types of individuals or tasks likely to be more variable than others?

We can illustrate this effect with data from the experiment. Six of the subjects repeat the reach to the shelf. We depict the curves from these six pairs in Figure 30.5. We see that the pairs of reaches tend to be much more similar within a pair rather than between a pair. We can fit the mean to each pair and then compute the residual SD—this is depicted as the solid line in the second panel of Figure 30.5. We can compute the overall pointwise SD for the 12 curves as the dotted line. We can see that variation within subjects is much smaller. We can also attempt to estimate the between-subject SD by regressing out the effect of height and computing the residual SD as shown by the dashed line. Now, there may well be other predictors than height that have a significant effect on the motion, and if we were we able to include these our estimate of the between-subject SD would decrease. The SD curve here is thus just an upper bound. Nevertheless, we can see that it falls somewhere between the within-subject and overall SD.

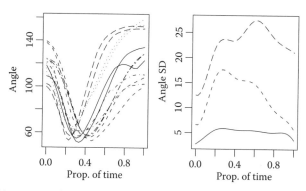

FIGURE 30.5 Six subjects repeat the reach. Each pair is shown with a different line type on the left. On the right, we show the pointwise SD, within subject (solid line), between subjects (dotted line), and overall (dashed line).

We can attempt to model this variation with a model of the form:

$$y_i(t) = x_i^T \beta(t) + s_{g(i)}(t) + \varepsilon_i(t)$$

where $s_{g(i)}(t)$ is the subject effect corresponding to the individual that generated the ith curve. This is the random effect of this individual that is not explained by the predictors x_i and can be modeled in a manner similar to $\varepsilon(t)$ except with its own covariance matrix. Although this model incorporates the unexplainable subject effect, it is fairly simplistic because it assumes that this effect is constantly additive across all input conditions. More complex models are possible and have been investigated in Faraway and Hu (2001).

30.9 Modifications to the Response

Bounded responses: In practice, most responses are physically bounded. For example, the elbow included angle discussed above has physical limits. One approach is to ignore these limits in the model fitting, but truncate the predictions toward the bound if the limit is exceeded. This approach is simple but somewhat unsatisfactory because it may result in predictions where the curve rides the bound for a substantial portion of the motion. This can give rise to unrealistic looking motions. Furthermore, extending a joint to its bound is usually uncomfortable and subjects will tend to avoid it. An alternative approach is to use the logit transformation. Suppose there are bounds $[b_l, b_u]$ on the response, then we use the transformed curve:

$$y'(t) = \log \frac{p(t)}{1 - p(t)} \qquad \text{where} \qquad p(t) = \frac{y(t) - b_l}{b_u - b_l}$$

We then model $y'(t)$ using the methods described above, backtransforming when the analysis is complete. The transformation approaches its limits asymptotically and so has the effect of avoiding the hard limits of the bounds. Complications arise when the bounds differ from subject to subject or depend on other aspects of the motion, such as the angles of shoulder whose bounds depend on the orientation of the arm.

Angular responses: Univariate curves in human motion modeling often represent an angular response. The methods described above work well when the typical range of this angle is relatively small, but will fail if observed angles lie on a wide arc where low values of the response tend to meet with higher values around the circle. Statistical methods for modeling such angular responses are discussed in Fisher (1993) although adaptations directly to functional angular responses are problematic. One simple but effective alternative to modeling an angular response $\theta(t)$ is to transform it to the pair $(\sin\theta(t), \cos\theta(t))$ and then model each of these components separately. When complete, these may then be transformed back to the angular response by using arctan transformation.

Even so, modeling such angular responses is difficult and it is best to avoid it if possible. Most local angles (i.e., angles between adjacent body segments) have a relatively small range and so this issue does not arise. However, global angles (i.e., angles measured with respect to a global coordinate frame) can have a very wide range. All things being equal, it is better to model the body in terms of the local angles.

Tied responses: In some applications, the user of the prediction model may want to fix the initial and/or final value of the curve. For example, the user may want to specify the initial posture of the subject or, indeed, may be chaining motions together so that new motion must take its initial frame from the final frame of the previous motion. Consider the case where both endpoints are specified. There are two possible approaches. One is to modify the unconstrained prediction to match the constraints. Let the specified endpoints be (s_0, s_1), then the modified prediction of $\hat{y}(t)$ becomes:

$$\hat{y}(t) - [(1 - t)(\hat{y}(0) - s_0) + t(\hat{y}(1) - s_1)]$$

The alternative is to model a constrained response, that is instead of modeling data $y(t)$, we model:

$$y(t) - [(1 - t)y(0) + ty(1)]$$

which constrains the curves to be zero at the endpoints. We modify the B-spline basis to remove the first and last basis functions and the effect of constraints is thus achieved since these are the only basis functions that are non-zero at the endpoints. We then backtransform the predictions from this model by adding the straight line joining s_0 and s_1.

The first approach is more convenient if we do not know whether the constraints will be imposed because only one model will be needed. However, the second approach has some advantages. In the example above, the effect of the height predictor was such that taller subjects start higher and end lower. By using the transformation on the response, the effect of the height predictor would be largely removed and a simpler model would result. Of course, this relies on the user subsequently setting sensible endpoints given the height of the subject. The case of only one tied endpoint is more problematic. Versions of the two approaches presented for the doubly tied curve can be applied. However, there are two important issues to consider.

If the user specifies an endpoint very different from those appearing in the data, the problem of extrapolation will occur in a more severe form than the doubly tied case where we need only model deviations from a straight line. Also, we need to decide whether, say, the effects of the initial position last throughout the motion and affect the final position or whether the initial effect wears off during the motion. In our example, the object must be placed on the shelf and the posture necessary for achieving this may not be very sensitive to the initial posture.

30.10 Trajectories in 3D

Consider the trajectory of the hand during a reach. We could consider the three univariate functions in the coordinate directions and model these using the techniques above, but there are some disadvantages to this approach. Typically, the coordinate frame is somewhat arbitrary so we would not wish our model to depend on this choice. We would prefer that it be invariant to rotations of this frame. Furthermore, the motion in each of the coordinate directions is quite dependent and it makes little sense to model it independently. Therefore we prefer an approach that models the curve holistically.

Consider the example discussed previously of 19 subjects reaching to a high shelf. Plots of these trajectories are shown in Figure 30.6. We can see a commonality in the pattern in that the hand loops

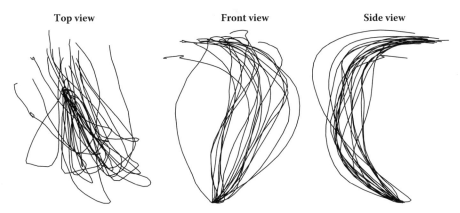

FIGURE 30.6 Three views of the trajectories of 19 subjects reaching to a high shelf. The trajectories have been aligned so that they start from the same origin.

back and to the right when delivering the object to the shelf. We would like to understand if the shape of the curve is related to predictors such as height, and we would like to model the variation. Incidentally, the reason that we model the wrist location rather than the hand itself is the hand will also tend to rotate as well as translate during the motion, which can complicate the appearance of the trajectory. In many applications, the endpoints of the trajectory are constrained. We want the hand to move from a given location to place an object in another location. Therefore, we will consider only the case of trajectories constrained at both endpoints. If this is not the case, the endpoints themselves will need to be predicted.

The method we propose is based on work described in Faraway et al. (2007) and uses Bezier curves. The model for the curve takes the form $\sum_{i=0}^{m} P_i B_i^m(t)$ where $B_i^m(t)$ are the Bernstein polynomials defined on [0,1] by:

$$B_i^m(t) = \binom{m}{i} t^i (1-t)^{(m-i)}$$

The P_i are called control points. P_0 coincides with the starting point of the curve at $t = 0$, while P_m lies at the endpoint at $t = 1$. For 3D curves, the control points have three dimensions. The interior control points determine the shape of the curve. In our experience, relatively few control points are required; we obtained satisfactory results with $m = 3$. The fitting of a 3D trajectory is shown in Figure 30.7.

Bezier curves have several desirable properties that have popularized their use in graphic design. Details can be found in texts such as Prautzsch et al. (2002). In particular, they are invariant to rotations and translations of the coordinate system because the shape of the curve is determined by the relative position of the control points. Also, the line segments $P_0 P_1$ and $P_{m-1} P_m$ are tangential to the curve at the start and end respectively. Thus they represent the direction of take-off and approach for the hand. This is particularly useful when modeling a task where the hand must reach in to perform a task where the approach is constrained by the surroundings or, say, where an object must be gripped in a particular way. The length of these line segments control how far the influence of the initial and final periods of the reach extend into the middle period of the reach.

The estimated control points P_i take the place of the B-spline coefficients y_{ik} above. We can then apply essentially the same approach to determining the effect of the predictor and/or generating new predictions.

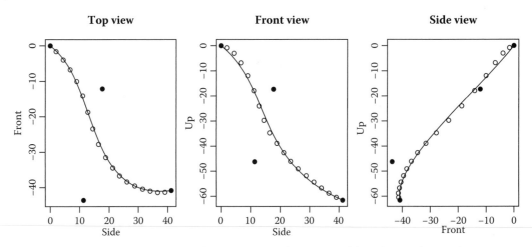

FIGURE 30.7 Three orthogonal views of a Bezier curve with two internal ($m = 3$) control points fit to some 3D trajectory data. The solid dots are the control points.

30.11 Orientation Trajectories

The orientation of a rigid body in 3D can be represented in various ways. Rotation matrices and Euler angles are both commonly used. This is discussed in books such as Zatsiorsky (1998). However, there are drawbacks to these representations, particularly with respect to the application of statistical methods. Rotation matrices use nine parameters to represent just the three degrees of freedom necessary to describe an orientation, while Euler angles are susceptible to the problem of gimbal lock and are non-commutative.

Quaternions provide a compact and elegant representation of orientation. We can think of a quaternion as a generalization of a complex number written as:

$$q = ix + jy + kz + w \equiv [v,w]$$

where $w, x, y, z \in$ IR and the imaginary numbers, i, j, k satisfy $i^2 = j^2 = k^2 = i\,jk = -1$. An orientation can be represented by a quaternion with vector of unit length. See, for example, Dunn and Parberry (2002) for an introduction in the context of human motion applications.

It is difficult to perform statistical methods directly on the quaternions. For example, the arithmetic average of two rotations represented as quaternions is not usually itself a rotation. An alternative approach is to perform a tangent mapping on the quaternions and compute the statistics in the tangent space. Orientation trajectories get mapped to 3D trajectories to which the methods of the previous section can be directly applied. We illustrate this in the first panel of Figure 30.8.

The *logarithm map* goes from the unit quaternion **q** to the tangent space at the identity quaternion:

$$logmap(\mathbf{q}) = \mathbf{v}/sinc(\theta/2)$$

where **v** is the vector part of **q** and $sinc(x) = \sin x/x$. The *exponential map* inverts the logarithm map and goes from a vector **v** in the tangent space to the space of unit quaternions with origin at the identity quaternion [**0**,1] can be obtained by:

$$exmap(v) = \begin{cases} [\mathbf{0},1] & \text{if } \mathbf{v} = \mathbf{0} \\ [\hat{\mathbf{v}} \sin\theta/2, \cos\theta/2] & \text{otherwise} \end{cases}$$

where $\theta = \|\mathbf{v}\|$ and $\hat{\mathbf{v}} = \mathbf{v}/\theta$. For more details on these maps, see Grassia (1998).

The mapping is nonlinear and the distortions become greater the further the orientation is from the origin as can be visualized in the first panel of Figure 30.8. We can recenter the orientation data around the origin, but for longer trajectories, the problem will be unavoidable. There is, however, a way to greatly reduce this distortion. The *slerp*, introduced by Shoemake (1985), is the geodesic joining the two endpoint orientations and is thus the equivalent of the straight line trajectory between the endpoints in the 3D case. We can compute the difference between this observed trajectory and the slerp (where the progress along the slerp trajectory is at the same speed as the observed trajectory). We call this the *slerp residual*, which is itself an orientation trajectory, but one that begins and ends at the origin. For most observed motions, the deviation from the slerp is not very large and so the distortion due to the exponential map is not substantial.

To summarize, the problem of modeling orientation trajectories can be reduced to that of modeling 3D trajectories with the use of quaternions combined with the residual slerp. More details may be found in Faraway & Choe (2008).

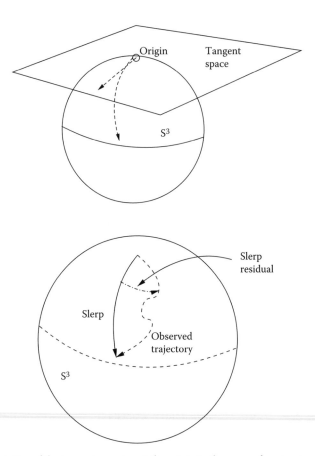

FIGURE 30.8   A depiction of the tangent mapping at the origin to the space of quaternions is shown on the left. In reality, orientations lie on the surface of a four-dimensional unit sphere S^3, while the tangent space has three dimensions. Geodesics, starting from the origin, map to straight lines in the tangent space. The slerp residual is the rotation from the slerp to the observed trajectory and is shown on the right. The slerp residual is itself an orientation trajectory.

30.12 Combining Elemental Models of Motion

We have described some basic building blocks of motion, but these need to be combined into a coherent whole. The body parts do not move independently, so we must link models for the components that respect the constraints imposed by the skeleton. Some constraints are simple—for example, the distance between the elbow and wrist is fixed. Other constraints involve joint angle limits.

The body linkage can be represented using a series of kinematic chains. Consider a chain with one end fixed at an origin. Using forward kinematics, we can use models for the angles at each link of the chain to represent the motion. Unfortunately, it is difficult to control the location of the other end of the chain, such as placing a hand. We might adjust the initially predicted chain using methods such as those described in Faraway et al. (1999), but this is inelegant and may involve substantial distortions.

The inverse kinematic approach fixes the endpoint and then uses a variety of techniques to place the connecting links. Some non-statistical approaches involve optimizing some criterion, such as comfort, to make this choice—a recent example of this approach is Marler et al. (2005). The key to making the statistical approach of modeling the components of the linkage work effectively is to choose the parameterization well. In particular, we want the same number of parameters as degrees of freedom in the linkage, and we want to be able to model those parameters as independently as possible.

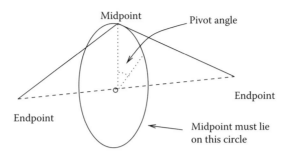

FIGURE 30.9 The pivot angle describes the location of the midpoint on the circle of its possible positions.

To illustrate this, consider a two-link chain in three dimensions, like the shoulder, elbow, and wrist linkage. Let the shoulder be fixed and let us require that the wrist begin and end in specified locations. This linkage is four degrees of freedom away from the endpoints. If we parameterize this in terms of the two angles, each describing the position of the upper and lower arm, it will not be so easy to use the angular response models and yet satisfy the endpoint constraints.

Alternatively, we can use the 3D trajectory model to describe the motion of the wrist between its specified endpoints. There remains only one parameter necessary to completely model the linkage. Since the elbow is constrained to lie on a circle whose center lies on and is orthogonal to an axis joining the endpoints at any given point in time. We call this angle the *pivot angle*. Such an angle was used by Korein (1985), and Wang (1999a and b). The angle is illustrated in Figure 30.9. The advantage of this parameterization is that the component parts can be modeled independently of each other while automatically satisfying the constraints.

This type of idea can be extended to more complex linkages. See Tolani et al. (2000), Faraway (2004), and Reed et al. (2006), for examples. A further consideration is that different parts of the body start and stop moving at different times. Models for these time offsets need to be integrated into the combined motion model.

30.13 Discussion

Ultimately, users of digital human models need some assurance that they are a reliable basis for making a decision. Does the model predict what people actually do? Is it valid? In this sense, the statistical approach has an inherent advantage over methods that do not directly use data collected from people in motion. We can at least fit the observed data reasonably well. However, any type of model must be able to predict new motion successfully.

Measuring success is not straightforward. Given a predicted motion and an observed motion, how do we measure how close they are and determine whether that is close enough? For such a multivariate object as motion, the range of possible types of metrics we could use is very large. However, there is some circumstantial evidence that some kinds of error are more noticeable or matter more than others (see Reed et al., 2005).

It is also important to understand that there are bounds on success. Previously we discussed that there is a certain inherent variability in human motion. We cannot expect to predict any closer than this. If we can come close to this bound, then we have done well. The other limitation to bear in mind is that human motion models have a large number of potential input values as we move about the task and subject space. Even the largest databases of motion lay quite sparsely in such a space. So even if we are able to demonstrate success across a large number of test cases, there is always the possibility of an unexpected bad prediction. All this means that any decent model validation would need to use a database comparable in magnitude to the one on which the model was based. Just running a few test cases is far from sufficient.

We have presented the basic building blocks for statistical models of human motion. These models have been used in practice to aid design decisions. The ability to easily see the effect of changes in the anthropometry of the subject or the dimensions of the task space make this approach useful. In particular, the ability to get more than just a single predicted motion, but also some idea of the variability concerning that motion, is particularly valuable.

References

Dunn, F., and I. Parberry (2002). 3D Math Primer for Graphics and Game Development. Wordware Publishing.

Faraway, J. (1997). Regression analysis for a functional response. *Technometrics* 39, 254–261.

Faraway, J. (2004). Human animation using nonparametric regression. *Journal of Graphical and Computational Statistics* 13, 537–553.

Faraway, J., S. Choe (2008). Modeling orientation trajectories. *Statistical Modelling* 8, in press.

Faraway, J., and J. Hu (2001). Modeling variability in reaching motions. SAE Technical Paper 2001-01-2094.

Faraway, J., M. Reed, and J. Wang (2007). Modeling 3D trajectories using Bezier curves with application to hand motion. *Applied Statistics* 56, 571–585.

Faraway, J., X. Zhang, and D. Chaffin (1999). Rectifying postures reconstructed from joint angles to meet constraints. *Journal of Biomechanics* 32, 733–736.

Fisher, N. (1993). *Statistical Analysis of Circular Data*. Cambridge University Press.

Grassia, F. (1998). Practical parameterization of rotations using the exponential map. *Journal of Graphics Tools* 3, 29–48.

Johnson, R., and D. Wichern (2002). *Applied Multivariate Statistical Analysis* (5th ed.). Upper Saddle River, NJ: Prentice Hall.

Korein, J. (1985). *A Geometric Investigation of Reach*. Cambridge, MA: MIT.

Marler, T., S. Rahmatalla, M. Shanahan, and K. Abdel-Malek (2005). A new discomfort function for optimization-based posture prediction. SAE Technical Paper 2005-01-2680.

Prautzsch, H., W. Boehm, and M. Paluszny (2002). *Bezier and B-Spline Techniques*. New York: Springer.

Ramsay, J., and X. Li (1998). Curve registration. *Journal of the Royal Statistical Society*, Series B 60, 351–363.

Ramsay, J., and B. Silverman (2005). *Functional Data Analysis* (2nd ed.). New York: Springer.

Ramsay, J. O., and B. W. Silverman (2002). *Applied Functional Data Analysis*. New York: Springer.

Reed, M., J. Faraway, and D. Chaffin (2005). Critical features in human motion simulation. Proceedings of the 49th Human Factors and Ergonomics Society Annual Meeting, Santa Monica, CA.

Reed, M., J. Faraway, D. Chaffin, and B. Martin (2006). The humosim ergonomics framework: A new approach to digital human simulation for ergonomic analysis. SAE Technical Paper 2006-01-2365.

Shoemake, K. (1985). Animating rotation with quaternion curves. *ACM SIGGRAPH* 19, 245–254.

Tolani, D., A. Goswami, and N. Badler (2000). Real-time inverse kinematic techniques for anthropomorphic limbs. *Graphical Models* 62, 353–388.

Wang, X. (1999a). A behavior-based inverse kinematics algorithm to predict arm prehension postures for computer-aided ergonomic evaluation. *Journal of Biomechanics* 32, 453–460.

Wang, X. (1999b). Three-dimensional kinematic analysis of influence of hand orientation and joint limits on the control of arm postures and movements. *Biological Cybernetics* 80, 449–463.

Zatsiorsky, V. (1998). *Kinematics of Human Motion*. Champaign, IL: Human Kinetics.

31

A Motion Simulation Tool for Automotive Interior Design

Gilles Monnier, Xuguang
Wang, and Jules Trasbot

31.1 Background of Ergonomic Simulation

Nowadays, digital human models (DHM) are widely used for ergonomic evaluation of products and workplaces. They actually enable a first evaluation of the interaction between future users and their environment at a very early stage of the product (or workplace) design. These models can be decomposed into a wide variety of functionalities such as cognitive assessment, strength prediction, body shape representation, discomfort evaluation, or posture and motion simulation. This last functionality is discussed in this chapter.

In order to define precisely what is intended by motion simulation, the term *simulation* is opposed to *animation*. The difference was initially defined by Badler et al. (1993):

- In animation, the goal is to describe the motion, and the animator usually imagines the desired motion before beginning the animation process. For this reason, there is only one motion to be simulated: the one that is expected by the animator.
- On the other hand, simulation can be considered an automation of the animation process. The system generates the motion based on user's input but the user knows less about what motion should result. In addition, the simulation process should not only reflect one individual's most probable behavior (or the one expected by the user) but the behavior of a target population of hundreds, thousands, or even millions of individuals.

This chapter presents a motion simulation approach integrated into a software named RPx. RPx is a project launched by Renault, Inrets, and Altran after a European project named REALMAN. The objective of the RPx project was to integrate the concepts and methodologies from the REALMAN project into a software used by the designers at Renault.

This chapter therefore reviews motion simulation approaches in order to define the objectives of the RPx motion simulation approach. The principles of this approach are then presented and applications for car interior design are illustrated.

31.2 Review of Motion Simulation Approaches

31.2.1 Inverse Kinematics

The main objective of an inverse kinematics (IK) algorithm is to calculate the posture of the digital mannequin from the position of the mannequin's end-effectors defined in task space (e.g., the index fingertip on the button, the heel on the heel point...). Upon the wide variety of IK algorithms described in the literature, one of the most popular techniques is the pseudo-inverse-based IK algorithm first introduced in the robotics field (Whitney, 1969), which aims to control the position of the end-effector of a robot.

Moreover, the pseudo-inverse IK algorithm was extended to follow the positions of controlled end-effectors while minimizing other performance criterion (Liegeois, 1977), such as avoiding joint limits (Liegeois, 1977), avoiding collision with the environment (Maciejeski & Klein, 1985), imitating a referential motion (Choi & Ko, 2000), defining multiple hierarchical constraints (Baerlocher & Boulic, 2004), or optimizing strength (Seitz et al., 2005). However, one must be aware that the posture generated by inverse kinematics is only a feasible solution among an infinite number of possible ones, depending on the performance criteria used for optimization.

31.2.2 Per Key Methods

A first motion simulation approach named the per key method, frequently used in the computer animation domain (Gleicher, 2001), is to consider that a motion is a succession of key postures, where intermediate frames can be calculated by interpolation between the key frames.

Applied to ergonomic simulations, this method takes advantage of motion capture systems that enable the capture of large number of motions. In order to simulate a new motion, a referential motion, close to the one to simulate, is retrieved from the database and adapted at relevant key frames. The intermediate frames are then interpolated while conserving the temporal characteristics of the referential motion in the angular domain (continuity, velocity, duration, etc.) between the key frames (Bindiganavale & Badler, 1998; Park et al., 2000; Zhang, 2002; Park et al., 2004).

However, this approach is limited to the tasks where key frames can be defined for the whole body such as reach movements. The main drawback is that the in-between constraints are not controlled at all by per key methods.

31.2.3 Per Frame Methods

Distinct from per key methods, per frame motion editing methods apply inverse kinematics to each frame of the motion. Such a per frame approach was for example used by Aydin and Nakajima (1999) for generating an automatic grasping motion that follows a pre-recorded trajectory and avoids joint limits.

Choi and Ko (2000) also proposed a similar algorithm. Their objective is to modify the recorded motion online to a differently sized person. Monnier et al. (2003, 2006) extended this approach by using a command table, which helps for control how to fulfill the geometric constraints for simulating complex motions such as seat belt handling and car ingress/egress movements.

Compared to per key methods, per frame methods guarantee constraints all along the motion but are computationally more expensive.

31.2.4 Motion Blending and Functional Regression

Compared to the methods introduced previously, motion blending needs a database of characteristic motions and consists in interpolating between their parameters in order to produce new motions (Multon et al., 1999).

Unuma et al. (1995) used Fourier decomposition to interpolate or extrapolate between two experimental human gait motions. For example, one might want to create a more or less tired walk. The objective of the motion blending method is to create the new motion by interpolating between a normal and a tired walk. As no quantitative parameters describe notions such as tiredness, two captured motions are necessary: a normal and a tired one. The qualitative parameters describing each motion are then obtained for each angular trajectory using Fourier series decomposition. Based on these quantitative parameters, it is then possible to create a new motion by interpolating from one motion to another.

Similar to motion blending in the way it combines a set of motion captured in different conditions, functional regression methods (Faraway, 1997, 2000) predict time variant curves such as joint angle from controlled variables defined in an experiment such as stature and target position for a reaching task, for example.

Although these methods seem very powerful with a low computational cost, they suffer from three main drawbacks. At first, it is difficult to apply such a method to the movement that is affected by a high number of parameters, as the size of the data to be collected must be very large if all the effects have to be studied. A second drawback is that, in some cases, interpolation along a variable is not appropriate, as the posture change may not be continuous. As an example that can be easily observed, when the hands are lowered under a threshold, one may adopt a squatting posture from a standing position. The interpolation along the hand's height may result in an unrealistic posture. The conclusion is that the movements over which the regression/blending are made have to be homogeneous in motion control: The same strategy has to be employed and the effects of a covariate have to be continuous on the performed movements. Finally, it should be noted that the motion constraints to be fulfilled are not explicitly defined in such methods. For example, it cannot be guaranteed that, in a walking sequence, the foot will stand on the floor and not slip on it. For this reason, it will usually be necessary to re-modify this new motion in order to fill the constraints of the task (Faraway et al., 1999).

31.2.5 Modular-Based Approach

A new approach proposed by Reed et al. in 2006 can be defined as modular-based as functionalities of their motion simulation approach are decomposed into small, individual modules. For example, one module handles task-oriented head and eye movement, another handles balance, another one hand trajectory, and so on. Each module is behavior based—or knowledge based—as it relies on motion analysis findings published in the literature. For example, the gaze module is based on a study published by Kim and Martin (2002), whereas the hand trajectory module is based on research published by Faraway (2000).

The main advantage of this approach is that it enables the combination of findings from different studies. For example, reaching studies, combined with walking studies and gazing studies, enable the simulation of a mannequin reaching an object at a given location and then carrying it to another location.

On the other hand, the central problem of this modular-based approach is how to ensure the natural coordination of these modules, as human movements are highly task dependent and multiple solutions exist due to the kinematic redundancy of the body. This reduces its application to simple and well-defined combined tasks: walk and reach, walk and carry a weight, and so on.

31.3 Objectives of RPx

From this short review, one can see that different motion simulation methods have been proposed in recent years. But none of them enable the simulation of complex tasks such as car ingress or egress.

After a recent European project (REAL MAN, IST 2000-29357, 2001–2004), which proposed and demonstrated some promising concepts in the area of motion simulation, a program called RPx (Realman Program eXtension) was launched by Renault, INRETS, and Altran to develop a simulation tool enabling the integration of motion simulation into the car design process. In addition, an important issue addressed by the RPx motion simulation approach is motion variability. In fact, even under the same conditions, people do realize very different motions of the same task. The proposed approach has therefore to preserve the variability observed experimentally. In this sense, several motions are to be simulated and evaluated rather than only one most probable motion.

31.3.1 Principles

The motion simulation approach developed in the RPx project is data based—the data contain the variability. It is also knowledge based, as the motion control rules, gathered from detailed motion analysis, are used to define how a motion can be modified to a new scenario in a command table (Monnier et al., 2003; Monnier, 2004).

31.3.2 Collecting Data

As for all data-based simulation approaches, basic data are motions performed experimentally by subjects in different conditions. The trajectories of surface markers are measured, and it is then necessary to calculate the joint angles of the digital mannequin at each frame of the motion using a motion reconstruction procedure (see Chapter 38).

Particular attention should be paid to the choice of controlled variables and to the experimental design for optimizing the number of tested configurations. In addition, one has to make sure that the motion control strategy used in experimental conditions corresponds to the one used in real life.

31.3.3 Gathering Knowledge

Once the data are collected, it is necessary to define the motion control rules that will drive the motion simulation process. These rules are based on the identification of:

- Key frames of each motion
- Strategies used by the subjects
- Constraints of each strategy
- How the body fulfills these constraints

This knowledge can either be gathered by motion analysis or defined by an expert who knows how people behave. Their definition by an expert makes it possible to define the influence of parameters not tested experimentally. However, simulation control rules defined by experts need to be validated experimentally.

31.3.4 Key Frames

Key frames of the motion correspond to particular events in the motion describing how a person interacts with the environment. Most frequently, these key frames are associated with one effector (the left foot, the pelvis, the head, etc.). Different types of events (or key frame) can be defined:

- When the effector leaves an object. For example, in a car ingress task, when the left foot leaves the ground.

- When the effector reaches an object. For example, in a car ingress task, when the left foot reaches the floor.
- When the effector avoids colliding with an object. For example, in a car ingress task, the frame at which the left foot avoids the rocker panel is defined when the foot is above the rocker panel.
- When the effector enters an environment. For example, in a car ingress task, when the pelvis is inside the car (is below the cant rail).
- When the hand is approaching or leaving an object. For example, in a reach to a hand brake, the hand starts closing onto the hand brake a few moments before reaching it.

31.3.5 Motion Strategies

Motion strategies allow differentiating different ways of performing the same task. For each motion captured experimentally, it is necessary to identify which strategy is used by each subject. This identification can be time consuming when a high number of motions were captured. Therefore, some automatic identification methods have to be defined. They can be based on key frame orders or by searching for a particular sub-motion sequence. For example, on a light truck ingress task, two strategies were observed when entering the torso (head first or pelvis first) and two others when climbing up the footboard (using the left foot or the right foot). In the first example, the strategy was automatically identified by looking at the order of the key frames: which body segment is entering the car first, the head or the pelvis. In the second example, the strategy was automatically identified by looking at which foot steps toward the footboard.

31.3.6 Constraints

In RPx, geometrical constraints of a task are taken into account. They define the influence of the environment on the motion. Two different types of constraints are defined.

The first are contact or collision avoidance relations between body elements (e.g., the foot, the head, etc.) and the environment (e.g., the ground, the sill, etc.). Such constraints can be defined over the whole motion or over a period of the motion. For example, in a car egress task, it is defined that the left foot has to be in contact with the floor from the beginning to the instant that the foot leaves the floor. These time periods are defined using the key frames identified for each motion.

The second type of constraints is defined on effector trajectories. For example, the x coordinate of the left foot pedal point can be controlled over the whole motion or over part of the motion. In a car egress task, its coordinate can be fixed to the clutch pedal position at the beginning of the motion and constrained relatively to the front pillar when getting out of the car. The shape of the in-between trajectory will be the same as that of the referential motion. Such constraints can be defined over position and orientation trajectory.

As constraints change with motion control strategy, they have to be defined for each motion strategy of a given task. The choice of the constraints has to be made by an expert who can use expertise as well as information coming from a motion analysis study.

31.3.7 Control of the Degrees of Freedom

Once the constraints are defined, it is necessary to identify how the different degrees of freedom (DOFs) are affected by the constraints due to the kinematic redundancy of the human body. Such rules are specified in a so-called command table. Of course, a first try could be that all DOFs are affected by all constraints. However, it is often more relevant to divide the whole body into separate body chains for a better control of motion simulation.

31.3.8 Simulating Motions

Based on data and control rules gathered during analysis, a motion simulation process can be started. It consists of two steps: (1) selecting motions from the database and (2) adapting the referential motion to the new conditions.

In this paper, only the issue of virtual experiment is addressed. The concept of a virtual experiment is to simulate the motions of the subjects who participated in the experiment: If 20 subjects took part in the experiment, they will "virtually" come back in order to evaluate a new configuration. Compared to simulating a population, composed of a high number of new randomly generated mannequins (over 100 for instance), it is simpler as the strategies and behaviors of the subject are known.

31.3.9 Selecting Motions

The very first step of the RPx motion simulation approach is to select motions in the database. For a virtual experiment, the most similar motion experimentally realized by a subject is selected. This similarity is defined as a weighted distance between the experimental conditions and the ones to be simulated. It therefore reflects how close are the conditions in which it was realized with respect to the conditions to be simulated.

31.3.10 Motion Adaptation

Once a referential motion is generated, it must be adapted to the constraints of the new simulation scenario. As the constraints are easier to define in the task space rather than in the joint angle domain, the effector trajectories are first adapted to the new constraints. In the RPx program, effectors are controlled not only in position (X, Y, and Z) but also in orientation. These trajectories are modified over key frames and conserve the shape of the trajectory.

The position of an effector at a given key frame is defined as a control point. The main attribute of a control point is the relation between the position of the effector and geometrical and anthropometrical parameters. The shape of the trajectory of the referential motion is conserved between key frames using a spline interpolation method.

As an illustration, in the egress database, six effectors are controlled: the two feet, the pelvis, the head, and the two hands. All are controlled in position, and the pelvis is also controlled in orientation. Regarding the vertical coordinate (Z position) of the left foot only, it is controlled at six key frames (fig. 31.1): (1) at the starting frame, (2) when the left foot leaves the car floor, (3) when the left foot is at its maximum height,

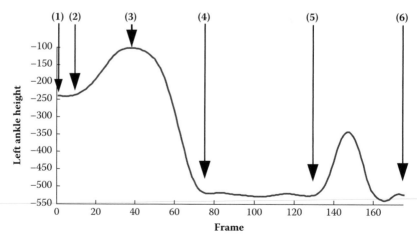

FIGURE 31.1 Key frames selected to control the vertical coordinate (Z) of the left foot trajectory during a car egress task. (From Monnier, G., et al., *A Motion Simulation Approach Integrated into a Design Engineering Process*, SAE International, Warrendale, PA, 2006. With permission.)

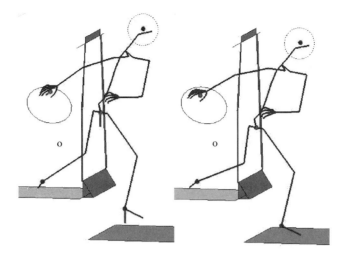

FIGURE 31.2 Constraints applied to the mannequin on the egress task (left) and result obtained with the inverse kinematic solver (right). Particular care is taken to avoid head collision into the cant rail. Constraints are indicated with red lines and head collision avoidance is indicated with a circle around the head center of gravity. (From Monnier, G., et al., *A Motion Simulation Approach Integrated into a Design Engineering Process*, SAE International, Warrendale, PA, 2006. With permission.)

(4) when the left foot reaches the ground, (5) at the beginning of the step outside of the car, (6) at the end of the step (and final frame). The control points of the left foot Z position are related to the vehicle floor, rocker panel, and the ground. Regarding the other effectors, they are controlled over the same key frames and over other key frames.

In the ingress/egress tasks, several geometrical parameters influence the trajectories of the effectors. The control rules gathered during the analysis of the experiments allowed us to take into account the effects of the heights of the floor and ground, the positions of the rocker panel, cant rail, front pillar, B-pillar, and steering wheel.

However, this approach allows the integration of any additional knowledge that might be gathered in other experimental conditions.

The final process consists in modifying the joint angles of the referential motion by inverse kinematics so that the effectors reach the modified trajectories (Choi & Ko, 2000). In the RPx project, the concept of chain was introduced in order to define which degrees of freedom are modified by which trajectories. At each frame, the posture is adapted to the trajectory constraints while remaining as close as possible to the referential posture. Multiple constraints are controlled in a hierarchic way (Baerlocher & Bouclic, 2004).

Figure 31.2 shows an example of how multiple constraints are controlled in an egress motion. The degrees of freedom of the lower limb (pelvis position and orientation, hip and knee angles) are modified according to the trajectories of the ankles and pelvis. Then, the spine angles are modified according to the head constraints; and finally, the arm angles are adapted to the hand trajectories.

Compared to the motion modification method proposed by Park et al. in 2002, this method handles constraints that are defined at intermediate frames.

31.4 Application Examples

Within the RPx project, most of the tasks performed by a car driver were captured and integrated into the software. These tasks are:

- Car ingress (front seat) and egress (front and rear seat)
- Light truck ingress and egress

- Reaches toward central rearview mirror, joystick for external left rear mirror, gearshift lever, right glove box, hand brake, central dashboard and roof push button, central dashboard rotary button
- Manipulation of seat commands located at different locations on the seat and using different grasping hand postures

The motion data of these tasks are structured so as to be used by RPx software for car interior design. Some examples will be shown here to illustrate how useful such a motion simulation tool is for car design engineers.

31.4.1 Light Truck Ingress

As a first example of application, a light truck ingress task is analyzed. The design questions were:

What is the optimal position of the cant rail?
Is a footboard located 480 mm above the ground too high? Who is affected by such a high position?

Figure 31.3 shows the referential and simulated motions of one mannequin. On the top row, one can see the referential motion. This motion has to be adapted to the new scenario (same mannequin and new geometry):

- At the beginning of the motion, both feet have to be on the ground
- When reaching the footboard, the left foot has to be on it

FIGURE 31.3 Referential (top row) and simulated (bottom row) motions of an ingress motion in a light truck. The referential motion is applied to the mannequin to be simulated.

- Both hands have to be in contact with the steering wheel
- When sitting, the feet have to be on the floor

No constraints were imposed on the head with reference to the cant rail and the captured motions selected as referential were free of constraint on the head, too (none of the subjects expressed any complaints about being bothered by the cant rail in this referential configuration).

On the bottom row, one can see that the simulated motion is correctly adapted to the scenario.

The optimal position of the cant rail can be defined as the minimum height so that any head collision should be avoided for any drivers. The volume swept by the head during ingress and egress motions needs to be evaluated.

Figure 31.4 shows the volume swept by one and by the 20 subjects who took part in the experiment— as in this case, a virtual experiment was defined.

Regarding the impact of the footboard height over the motion, it can be evaluated upon the left hip flexion when the left foot is on the footboard as it is the main degree of freedom adapted by a footboard height modification. Here, the hip flexion values are compared to a situation defined as comfortable. For this ingress task, a footboard height of 420 mm, being the lowest value among the existing vehicles of this type on the market, is considered here as comfortable.

Figure 31.5 shows that people with short legs have to flex their hip 7 more degrees for a footboard at 480 mm than for a footboard at 420 mm (10% increase). This difference is reduced to about 3 degrees for people with long legs. In order to evaluate if the footboard height is acceptable or not, a threshold of 5 degrees is chosen. The application of this criterion indicates that people with legs shorter than 820 mm will find this tested configuration unacceptable, which represents 58% of the French population.

31.4.2 Lateral Seat Command Reaching

Another application example concerns a reach toward a seat command located on the left side of the seat and backward. Such commands are usually used for adjusting the seat back inclination.

The design question is described as follows. Lateral cushions are necessary to maintain a lateral support so as to increase driving posture comfort. However, lateral cushions may lead to additional difficulty for drivers to reach a seat backward command. But how will reach motion be modified with this additional lateral cushion? Is the additional difficulty acceptable?

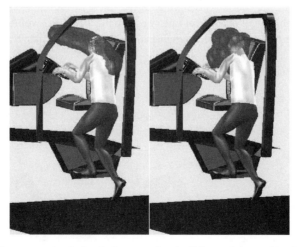

FIGURE 31.4 Left: Volume swept by the head when entering the light truck (one subject). Right: Volume swept by 20 mannequins when entering the light truck (one frame only: head center of gravity below the cant rail).

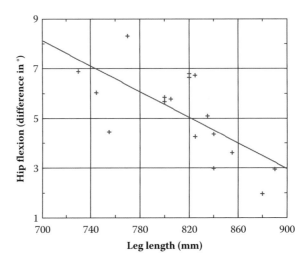

FIGURE 31.5 Virtual experiment of ingress in a light truck. Left hip flexion difference for two footboard heights: 420 mm and 480 mm with respect to the leg lengths.

As for this specific task, and due to the additional lateral cushion, people might have to bend forward in order to reach this command. The situation is simulated and illustrated with one subject in Figure 31.6.

Here, five end-effectors are controlled: the pelvis, the feet, and the hands. And an additional constraint is applied so that the left arm avoids the collision into the seat.

Due to the additional lateral cushion, the torso has to bend up to 24 degrees whereas without this cushion, it only has to bend up to 17 degrees. The lateral cushion therefore necessitates an additional 7 degrees (+41%) torso flexion, which probably generates more discomfort.

This increase of torso flexion has to be compared with the ergonomic criteria defined by the car manufacturer. Here, it might be worth changing the location of the seat command or reducing the lateral cushion thickness.

For this task, it is also necessary for car designers to evaluate the clearance required by the arm (upper arm and lower arm) when manipulating this command, as the room around the seat is usually quite small. Reach motions for the lateral seat command were therefore simulated for a sample of 50 mannequins in order to evaluate the overall volume swept by the arms. This volume, superposed with the door panel and B-pillar of a future vehicle, enables the designer to evaluate the collisions between the arm and the car.

FIGURE 31.6 Driving posture (left), referential (center) and simulated (right) motion for a reach toward a command on the seat. Compared to the referential motion, an additional constraint is applied so that the left arm avoids the collision into the lateral cushion of the seat.

FIGURE 31.7 Evaluation of the torso flexion increase from the seating posture. Left: referential motion colliding with the lateral cushion. Right: lateral cushion avoided.

Figure 31.8 shows the comparison of the future vehicle with five other existing competitors whose levels of feature were previously evaluated by the car manufacturer. This comparison ranks the future vehicle with the same level of collision feature as vehicle #5.

31.4.3 Foot Path during an Egress Motion from the Rear Seat

A third example of application is about rear seat egress. The objective of the design engineer is to evaluate the necessary room that will allow most users to exit without colliding with the environment (door panel and rear seat foot member).

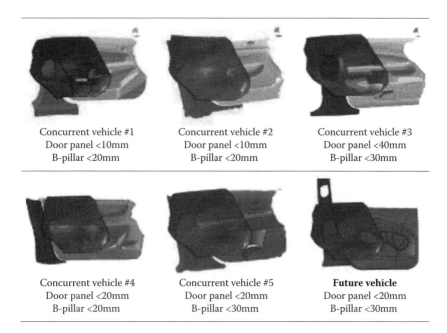

Concurrent vehicle #1	Concurrent vehicle #2	Concurrent vehicle #3
Door panel <10mm	Door panel <10mm	Door panel <40mm
B-pillar <20mm	B-pillar <20mm	B-pillar <30mm

Concurrent vehicle #4	Concurrent vehicle #5	**Future vehicle**
Door panel <20mm	Door panel <20mm	Door panel <20mm
B-pillar <20mm	B-pillar <30mm	B-pillar <30mm

FIGURE 31.8 Volume swept by the arm of a target population of 50 mannequins when reaching for a lateral seat command in order to evaluate the level of feature of the command reaching task. The collision between the left arm and the door panel and B-pillar geometries for different vehicles (already on the market or future vehicle) are shown. The seat is not represented. Only the B-pillar, door panel, and volume swept by the arms are represented. Lateral view from the driver's right side.

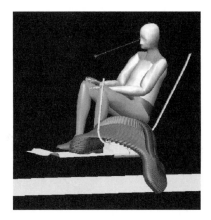

FIGURE 31.9   Constraints defined for an egress rear seat task: floor, ground, B-pillar, and sill.

For this, egress motions of 50 mannequins, a representative sample of the target population, were simulated. Only the floor, the ground, the B-pillar, and the sill were defined and were taken into account in the simulations (see fig. 31.9). The global volume swept by the left foot of the mannequins is generated for the evaluation of the collisions with the door panel and the rear seat foot member.

This volume therefore indicates to the designers which parts are to be optimized. Of course, a compromise between the ease of egress and the door storage capacity has to be found here.

31.5 Discussion and Conclusion

This chapter has given a brief summary of existing motion simulation approaches and presented in detail the one developed in the RPx project. Thanks to motion control rules, different constraints can be easily defined and fulfilled on different effectors in a generic way, allowing the simulation of simple reach movements as well as complex multi-sequenced ingress/egress ones using a unique simulation software.

The main advantages of this simulation approach are:

- No software development is necessary when including an additional task.
- Thanks to the virtual experiment and the automation of simulating a large number of mannequins, inter-individual motion variability is taken into account for optimizing the design.

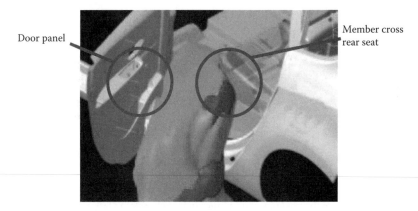

FIGURE 31.10   Volume swept by the left foot of a target population and by a 95th percentile height subject. Areas of collision are detected with the door panel and with the member cross rear seat.

The influence of parameters not tested experimentally over effector constraints can be defined by an expert. As a result, it is possible, in a certain range, to simulate motions for a new set of parameters and evaluate their influence over the motion.

The major drawbacks of this approach are:

- As all data-based simulation approaches, a new task cannot be simulated without running new experimentations.
- It is time consuming to gather the motion control rules for the simulation of complex motions. A detailed analysis of captured motions is needed.

Through the examples showed in this chapter, it is clear that tools like RPx software can assist car design engineers in vehicle packaging. It is now used in the everyday work of ergonomic specialists at Renault to evaluate future car configurations.

References

Ausejo, S., Wang, X. (2009). Motion capture and human motion reconstruction. In V.S. Duffy (Ed.), *Handbook of Digital Human Modeling*. Boca Raton, FL: Taylor & Francis.

Aydin, Y., Nakajima, M. (1999). Database guided computer animation of human grasping using forward and inverse kinematics. *Computers & Graphics*, Vol 23, pp. 145–154.

Badler, N.I., Phillips, C.B., Webber, B.L. (1993). *Simulating Humans: Computer Graphics Animation and Control*. New York: Oxford University Press.

Baerlocher, P., Boulic, R. (2004). An inverse kinematic architecture enforcing an arbitrary number of strict priority levels. *The Visual Computer*, Vol 20(6), pp. 402–417.

Bindiganavale, R., Badler, N.I. (1998). Motion abstraction and mapping with spatial constraints. Modelling and motion capture techniques for virtual environments, international workshop, *CAPTECH'98*, pp. 70–82.

Choi, K-J., Ko, K-S. (2000). Online motion retargetting. *Journal of Visualization and Computer Animation*, Vol 11, pp. 223–235.

Faraway, J.J. (1997). Regression analysis for a functional response. *Technometrics*, Vol 39(3), pp. 254–261.

Faraway, J.J. (2000). Modeling reach motions using functional regression analysis. Technical Paper 2000-01-2175. SAE International, Warrendale, PA.

Faraway, J.J., Zhang, X., Chaffin, D.B. (1999). Rectifying postures reconstructed from joint angles to meet constraints. *Journal of Biomechanics*, Vol 32, pp. 733–736.

Flash, T., Hogan, N. (1985). The coordination of arm movements: An experimentally confirmed mathematical model. *Journal of Neuroscience*, Vol 5, pp. 1688–1703.

Gleicher, M. (2001). Comparing constraint-based motion editing methods. *Graphical Models*, Vol 63, pp. 107–134.

Kim, K.H., Martin, B.J. (2002). Visual and postural constraints in coordinated movements of the head in hand reaching tasks. Proceedings of the 46th Human Factors and Ergonomics Society Conference, HFES, Santa Monica, CA.

Liegeois, A. (1977). Automatic supervisory control of the configuration and behavior of multibody mechanisms. *IEEE Transactions on Systems, Man, and Cybernetics*, Vol 7(12), pp. 868–871.

Maciejeski, A.A., Klein, C.A. (1985). Obstacle avoidance for kinematically redundant manipulators in dynamically varying environments. *International Journal of Robotics Research*, Vol 4(3), pp. 109–117.

Monnier, G. (2004). Simulation de mouvements humains complexes et prédiction de l'inconfort associé —application à l'évaluation ergonomique du bouclage de la ceinture de sécurité. Ph.D. Dissertation. INSA Lyon (in English).

Monnier, G., Renard, F., Chameroy, A., Wang, X., Trasbot, J. (2006). A motion simulation approach integrated into a design engineering process. Technical Paper 2006-01-2359. SAE International, Warrendale, PA.

Monnier, G., Wang, X., Verriest, J.-P., Goujon, S. (2003). Simulation of complex and specific task-orientated movements: Application to the automotive seat belt reaching. *SAE Transactions*, Vol 112(7), pp. 715–721.

Multon, F., France, L., Cani-Gascuel, M.-P., Debunne, G. (1999). Computer animation of human walking: a survey. *Journal of Visualization and Computer Animation*, Vol 10, pp. 39–54.

Park, W., Chaffin, D.B., Martin, B.J. (2000). Development of an angle-time-based dynamic motion modification method. Technical Paper 2000-01-2176. SAE International, Warrendale, PA.

Park, W., Chaffin, D.B., Martin, B.J. (2004). Toward memory-based human motion simulation: Development and validation of a motion modification algorithm. *IEEE Transactions on Systems, Man, and Cybernetics, Part A: Systems and Humans*, Vol 34(3), pp. 376–386.

Reed, M.P., Faraway, J., Chaffin, D.B., Martin, B.J. (2006). The HUMOSIM ergonomics framework: A new approach to digital human simulation for ergonomic analysis. Technical Paper 2006-01-2365. SAE International, Warrendale, PA.

Seitz, T., Recluta, D., Zimmermann, D., Wirsching, H.-J. (2005). FOCOPP—an approach for a human posture prediction model using internal/external forces and discomfort. Technical Paper 2005-01-2694. SAE International, Warrendale, PA.

Uno, Y., Kawato, M., Suzuki, R. (1989). Formation and control of optimal trajectory in human multijoint arm movement—minimum torque-change model, *Biological Cybernetics*, Vol. 61, pp. 89–101.

Unuma, M., Anjyo, K., Takeuchi, R. (1995). Fourier principles for emotion-based human figure animation. In R. Cook (Ed.), *SIGGRAPH 95 Conference Proceedings, Annual Conference Series*, pp. 91–96.

Whitney, D.E. (1969). Resolved motion rate control of manipulators and human prostheses. *IEEE Transactions on Man-Machine Systems*, Vol 10(2), pp. 47–53.

Wiley, D., Hahn, J. (1997). Interpolation synthesis of articulated figure motion. *IEEE Computer Graphics & Applications*, Vol 17(6), pp. 39–45.

Zhang, X. (2002). Deformation of angle profiles in forward kinematics for nullifying end-point offset while preserving movement properties. *Journal of Biomechanical Engineering*, Vol 124, pp. 490–495.

32

Human Performance: Evaluating the Cognitive Aspects

Brian F. Gore

32.1 Introduction

Modeling human cognition, and understanding the manner in which humans use information, are becoming increasingly important as system designers develop automation to support the human operators. Tasks that were traditionally manual and physical in nature are being replaced with tasks that are cognitive in nature. This is exemplified in the aviation community where automation is being adopted to increase efficiency and safety. It is also becoming more important in surface transportation where increasing levels of automation are placing the human into a role of supervising the automobile's performance and manually controlling the vehicle intermittently (e.g., adaptive cruise control, autonomous cruise control and vehicle guidance; Sheridan, 1992; Seppelt & Lee, 2007). Furthermore, nuclear power plant design, medical system design and operation, UAV and other telerobotic operations and manufacturing systems are all increasingly placing the human into similar supervisory roles (Sheridan & Ferrell, 1974; Miller, 2000; Zhai & Milgram, 1991, 1992; Sheridan, 1992; Boring et al., 2006). In these environments, it is important to model both the physical human as well as human cognition. Incorrectly modeling any of these performance factors, or ignoring one in favor of the other, may lead to incorrect predictions. It is important that the design community, including the digital human modeling community, include appropriately verified and validated human physical geometric models and human cognitive models. Focusing on human physical geometric models alone, while rather salient and very important for specific anthropometric and biomechanic considerations, is insufficient for fully representing the human as they solely generate a physical geometric representation of a given model (either the human form, or other physical feature in an environment) but ignore important cognitive representations such as perception, attention, decision making, and memory. On the other hand, focusing solely on the cognitive models is in itself insufficient because human cognition interacts with an outside world in a closed-loop manner. Furthermore, human cognition is often not directly observable, and integrating physical models with cognitive models provides an opportunity to visualize the complex interaction among the various interacting components in a modeled human's environment providing a system designer or an analyst the ability to see situations where the design as proposed may not be adequate for the operational conditions.

This chapter reviews cognitive and digital human models (physical models), the synergistic advantages of integrating the two together, the process taken to develop and validate an integrated model, the resultant need for model transparency due to the complexity of these integrated models, and the benefits afforded by comprehensive visualizations.

32.2 Models of Human Performance

Human performance models may encompass many forms—from the purely cognitive models built from empirical research and theories of human processes (attention, perception, decision making, response times, response characteristics), to digital human models, or physical models of human anthropometry, biomechanics, posture, movement, bones, and anatomy (most of which have been discussed in the current volume).

32.2.1 Cognitive Models

Cognitive models, also known as human process models, are computational representations of the human processes that are required to complete a goal behavior in a specific context. These analytical models are rooted in theory and are drawn from psychological literature that describes the range of underlying human processes. These cognitive (process) models predict the human's performance within the system. This includes the manner that humans attend and perceive information, learn and decide what to do with that information and carry out the response to this information. Cognitive models have been represented as an information processing (IP) system with an input, central processing, and output component.

IP models began in the early 1950s with a combination of Shannon and Weaver's (1949) mathematical theory of information and Wiener's (1950, 1980) ideas on feedback in controlled systems. IP models solidified the abstract concept surrounding information by fostering the notion that human behavior can be represented in mechanistic terms (input, processing, and output). Broadbent (1958) further formalized this IP approach to human perception and memory by developing block diagram analyses of information flow. His qualitative ideas laid the groundwork for a programming analogy founded on the idea of discovering algorithms to represent the manner in which humans process information. This approach spawned models of visual search and identification (Sternberg, 1969), short- and long-term memory (Baddeley & Hitch, 1974), reaction time underlying simple decision processes (Smith, 1968), and movement control (Baron et al., 1970) among others. These models tended to be of isolated psychological functions.

Card, Moran, and Newell (1983) continued this line of research describing the information-processing system as a set of memories linked together with principles that drive the memories. Their information-processing system, known as the model human processor (MHP), is divided into three segments: the perceptual system, the motor system, and the cognitive system. Each of these systems has its own set of memories and processes that work to buffer, encode, decay, and act on the information retrieved from the memory and process structure. MHP was applied at the low level task of keyboard input, basic reading, and responding to auditory input (Card et al., 1983).

A similar but more recent model to the IP and the MHP has been presented by Wickens and Hollands (2000)* who represent the human operator in terms of a closed-loop IP system that takes input information through sensory and perception processors then sends this information through attention and memory filters to a response and a response-selection architecture. All of the models in this IP representation are from empirical human performance data collection efforts that explored the effects of information being provided to the human. A representation of the IP model as presented by Wickens and Hollands (2000) can be seen in Figure 32.1, which shows a number of different models required to successfully predict human cognition.

Most integrated models follow some semblance of a modified stage model process to represent human information processing (Wickens, 1992). The modified stage model is a modification proposed by Wickens (1992) to the stage model (Pew & Mavor, 1998). The modified stage models internally transform the sensation and perception information from the external stimuli to enable computational cognitive processing of the world elements. The modified stage models also contain structures to represent the working and the long-term memory of the simulated operators. The cognitive functioning within the modified stage model contains a number of functions including situation awareness, mental models, multitasking, learning, and decision-making models. The modified stage model also incorporates a representation of the motor behavior (i.e., the functions performed by the neuromotor system to carry out the activities of the simulated operator).

Many different approaches have been taken to computationally represent human cognition, three of which include the Atomic Component of Thought-Rationale (ACT-R),[†] State, Operator and Results (Soar), and the Executive Process Interactive Control (EPIC) architectures.

ACT-R is a modeling approach that incorporates a unified theory of cognition using a production-rule cognitive architecture. It is referred to as a unified theory of cognition because it is designed to predict human behavior by processing information and generating intelligent behavior using two distinct memory structures (declarative and procedural) to guide goal performance and learning. ACT-R combines a goal-directed production system with a sub-symbolic activation calculus that tunes itself to

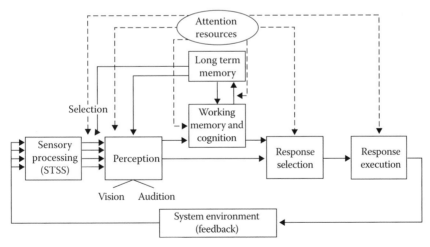

FIGURE 32.1 Wickens and Hollands (2000) Information Processing Model. (From Wickens, C. D., and J. G. Hollands, *Engineering Psychology and Human Performance*, 3rd ed., Pearson Education, Upper Saddle River, NJ, 2000. With permission.)

---

*Published originally by Wicknes (1992) as a composite model built from a number of previous investigators (Broadbent, 1958; Smith, 1968; Sternberg, 1969; Welford, 1976).

†ACT-R has been augmented to include a model of visual perception and motor movement (RPM) to the model's predictive capabilities (Lebiere, 2002).

the structure using Bayesian learning mechanisms (Lebiere & Anderson, 1993; Lebiere, 2002). Lebiere reports that ACT-R is aimed at the atomic level of thought and that the individual cognitive, perceptual, and motor acts that take place do so at the sub-second level. The ACT-R theory can accommodate various empirically determined modules (theories) of human cognition within its declarative/procedural architecture (Anderson, 1993; Anderson & Lebiere, 1998; Lebiere, 2002). The ACT theory holds that the behavior is organized according to a single production rule and as such possesses the potential for serial-defined bottlenecks to occur in generating its performance. Pew (2007) suggests that the ACT-R software is the world's most widely used cognitive architecture.

Soar is a parallel-matching, parallel-firing, rule-based system that aims to represent both procedural and declarative knowledge and is used to model all capabilities of an artificially intelligent system and for modeling human cognition and behavior (Newell, 1990). Soar is an example of a modified stage model focusing primarily on working memory but has been used to model human central-processing capabilities such as learning, problem solving, planning, search, natural language, and other human-computer interaction tasks. Soar can represent all ranges of behavior from highly routine to difficult behaviors and does so through a series of rule-testing operations (Newell, 1990). Soar assumes that human behavior can be modeled as the operation of a cognitive processor and a set of perceptual and motor processes all operating in parallel and maintains a single framework for all tasks and subtasks, permanent and temporal knowledge, goals, and a single learning mechanism (Pew & Mavor, 1998).

EPIC is an architecture developed to represent executive processes that control other processes during multiple task performance. This architecture is based on the EPIC spawned by Kieras and Meyer in the mid-1990s (Kieras & Meyer, 1997). EPIC is a production-rule cognitive processor that contains parallel perceptual and motor processors. It possesses fixed architectural properties in terms of components, pathways, and most time parameters while also possessing task-dependent properties of cognitive processor production rules, perceptual recoding, and response requirements and styles. It is currently able to model system performance but does not embody a theory of human performance (Pew & Mavor, 1998). EPIC approaches behavior representation from a low level, bottom-up fashion beginning with perceptual processors for input and output along the visual, auditory, and tactile dimensions. The input processors are symbolically coded changes in sensory properties. The visual-processing dimension is the higher fidelity representation in EPIC. For example, EPIC represents foveal, parafoveal, and peripheral vision. The output processors are items in modality-specific partitions of working memory. Many of the models that have been developed in these various cognitive architectures represent human performance and human cognitive processes at different levels of abstraction (granularity, detail) and have been validated at these varying levels of abstraction, generally focusing their validation efforts on a single dimension of performance that is determined based on the goals of the model being developed.

32.2.2 Digital Human Models

Digital human models (DHMs) are used to visualize the geometric form of humans in a particular context. For example, the Jack (Badler et al., 1993), Delmia/Safework Human Builder Simulation (Green & Charland, 2006), or the SANTOS (Abdel-Malek et al., 2006) software tools allow the placement of a human's anthropometry inside a crew station to evaluate the design of the interior of the crew station relative to a range of anthropometries (e.g., the 5th percentile female to the 95th percentile male).* Some of the more comprehensive digital mannequin models contain biomechanic/kinematic information as well to constrain the physical actions of the digital mannequin (e.g., Zhang et al., 1999; Chaffin, 2001, 2002; Farrell & Marler, 2004). Using models in this manner enables a designer to visualize the interaction that the human form will have with a computer-aided engineering/design (CAE/D) prototype of a cockpit or other operator position. Physical models can also be created at varying levels of fidelity

---

*Jack is a trademark of Siemens, Inc., Human Builder is a trademark of Delmia/Safework, and SANTOS is a trademark of the virtual research program at the University of Iowa.

depending on the needs of the analyst. Physical models do not necessarily behave in a closed-loop manner within the environment in which they are placed. They can be independent and when they are created as independent, entities behave according to strict rules of behavior programmed by an analyst. For example, the DHM will be scripted to reach for 5 seconds irrespective of other competing tasks and priorities.

32.2.3 Integrated Models

Cognitive and physical modeling approaches are often complementary, and can be integrated to enable predictions of complex, dynamic, human behaviors of operator-environment interactions. Integrated models such as the structure presented in Figure 32.2 combine a number of individual process models of operator performance into a coordinated representation of interacting micro-models of human perceptual, cognitive, and motor system representations within the context of an environment, thus incorporating the high-level behaviors that are characteristic of human performance. DHMs can feed physical constraints to cognitive process models that in turn feed information to the DHMs in a closed-loop fashion. Integrated modeling approaches focus on micro-models of human performance that feed-forward and feed-back to other constituent models in the human system depending on the contextual environment that surrounds the virtual operator.

For example, consider the following scenario. An astronaut is required to complete two tasks concurrently—reaching a switch with the left hand and pulling a lever with the right hand. The switch is located within the reach envelope of the anthropometry, but when the switch is being pressed the lever falls out of the reach envelope. An open-loop DHM that used sequential ordering of the tasks would reveal satisfactory performance. However, closed-loop integrated models that test performance in a realistic procedural manner by assessing task interactions and parallel task performance would reveal a limitation. This requires a feedback loop that in turn works to make the behaviors predictive in nature. The feedback loop contains success clauses that suggest to the analyst whether the activities are possible and when tasks become interrupted. Only with an integrated human performance modeling approach can the analyst see that the simulated astronaut cannot physically multitask in this manner. Either the left hand will reach its goal or the right hand will reach its goal but it is unlikely that both will attain their goals. The integrated approach implements a procedural set and knowledge of the environment

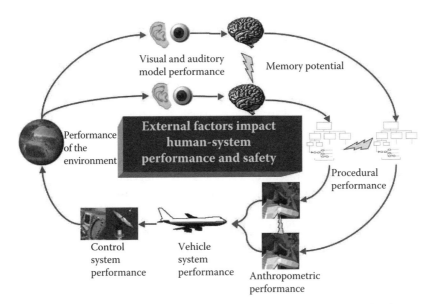

FIGURE 32.2 Interactive structure of a number of embedded models inside an integrated architecture.

that is triggered by changes in the environment but constrains the behaviors by the anthropometry. In physical models alone, the mannequin will carry out the reach task until it reaches the limit of its anthropometry (i.e., no recognition of meeting a goal unless the goal is to reach the end of the envelope). However, integrating the cognitive model with the digital mannequin will permit the digital mannequin to visually locate the target location, detect the target, recognize the target, reach the target, stop its reach once it arrives at the target, and feed success information (in the form of success clauses) to the cognitive model once it reaches its target location. The cognitive model will then complete a series of scheduling behaviors to complete the various tasks required by the specific scenario. Allowing models to operate in a closed-loop fashion increases the credibility that can be given to a model's performance prediction because feedback from the operational environment, often a necessary consideration in design, is being considered.

The output measures of interest for integrated HPMs have traditionally included task demands, workload, task load, information load, attention demands, stress, procedural timing measures, and the human's contributions to system errors. These measures have been used to identify when, where, and how often errors occurred within a specific job design and combined with the load measures could be used to re-organize procedures to reduce time and load demands (Siegel & Wolfe, 1969; Smith & Corker, 1993; Shively et al., 2000; Gore & Corker, 1999; Gore et al., 2001; Gore, 2002; Corker et al., 2003; Gore & Smith, 2006; Foyle & Hooey, 2008).

32.3 Integrated Model Example: Man-Machine Integration Design and Analysis System (MIDAS)

The Man-Machine Integration Design and Analysis System (MIDAS) is a 3D rapid-prototyping human performance modeling and simulation environment that facilitates the design, visualization, and computational evaluation of complex man-machine system concepts.[*] MIDAS aims to reduce design cycle time, support quantitative predictions of human-system effectiveness, and improve the design of crew stations and their associated operating procedures. MIDAS combines continuous-control, discrete-control, and critical decision-making models to represent the internal models and cognitive function of the human operator in complex control systems. It involves a critical coupling among humans and machines in a shifting and context-sensitive function. MIDAS links models of human anthropometry, biomechanics, and human cognition together with an environment to determine whether the human can perform various procedural sequences to a criterion level with new technological concepts. MIDAS' first principled approach to modeling human performance is an approach that is based on computational models of the mechanisms that underlie and cause human behavior within the context of human-system performance. The basic human perceptual and attentional processes, together with working and long-term memory models, action selection architectures, and physical representations of the human operator and environment models, have been validated. The interaction of these models produces behavior that closely approximates human behavior.

MIDAS accomplishes integrated behavioral modeling by linking a virtual human (a physical anthropometric mannequin model) to a computational cognitive structure representing human capabilities and limitations, and to a series of procedures, and places this virtual human within commercially available CAE/D databases to graphically represent and ergonomically assess the human with the physical entities in an environment (considered earlier in the current volume). Physical CAE/D component agents include external environmental influences such as terrain, weather, time of day, crew station component models, and other equipment models. The cognitive components are composed of a perceptual mechanism (visual and auditory), an attentional mechanism, a memory structure, a decision

---

[*]The MIDAS research program began in the fall of 1984 (at the time termed A^{3}I) with plans to develop the first fully integrated HPM linking together cognitive and performance models.

maker, and a response selection architecture. The human cognitive, perceptual, and motor process models associated with performance describe, within their limits of accuracy, the responses that can be expected of the human operator for safe operation of advanced automated technologies. Attention demands are represented by Wickens' multiple resource principle and incorporate the modified TAWL index for quantifying attention (McCracken & Aldrich, 1984; Miller, 2000, 2003). Combining attention demands along the input (visual, auditory), central cognitive processing (spatial, verbal), and output (psychomotor, verbal) resources in the manner that MIDAS does accomplishes the goal of developing a measure of attention demands. The complex interplay among bottom-up and top-down processes enables the emergence of unforeseen, and nonprogrammed behaviors.

MIDAS outputs include dynamic visual representations of the simulation environment, timelines, task lists, cognitive loads along six resource channels, actual/perceived situation awareness, and human error vulnerability and human performance quality. In addition, MIDAS incorporates functions that simulate the effects of stressors on skilled performance through workload and timing excesses. When the cumulative demands of concurrent tasks exceed a threshold of seven, the operator is assumed to be at greater risk for shedding tasks, or reduced performance levels, thereby leaving the operator vulnerable to error.

MIDAS has been used successfully in aerospace and military domains to visualize and evaluate human performance with conceptual designs (Corker & Smith, 1993; Corker et al., 1994; Gore & Corker, 2000; Gore & Smith, 2006). A significantly improved version of the MIDAS was released in 2006. MIDAS maintains its philosophy of being an integrated human performance modeling software tool that generates computational models of human operators performing complex tasks in simulated operational environments to predict operator workload, situation awareness, and task-performance timelines. MIDAS continues to combine a 3D model of the environment and of the crew station, an ergonomically correct human mannequin, and models of human cognition (attention, perception, decision making) and behavior. The new release of MIDAS, MIDAS 4.0,* has resulted in streamlined and improved code as it was transitioned from a Unix-based (SGI) development environment to a freely available Windows PC-based Microsoft .NET platform. This effort has resulted in a significantly improved user interface, model transparency, model applicability to new contextual environments (generalizability), and enabled MIDAS to use a host of new CAD formats (e.g., Pro-E and JT formats).

32.4 Integrated Model Development Process

Integrated models have arisen as viable options to improve design due to decreases in computer costs, increases in representative results, and increases in model validity. They are especially valuable because the computational predictions can be generated early in the design phase of a product, system, or technology to formulate procedures, training requirements, and to identify system vulnerabilities and where potential human-system errors are likely to arise. The model development process allows the designer to fully examine many aspects of human-system performance with new technologies to explore potential risks brought to system performance by the human operator. Often this can be accomplished before the notional technology exists for human-in-the-loop (HITL) testing (Gore, 2000). This method possesses cost and efficiency advantages over waiting for the concept to be fully designed and used in practice (characteristic of HITL tests). Using HPMs in this manner is advantageous because risks to the human operator and costs associated with system experimentation are greatly reduced: no experimenters, no subjects and no testing time (NASA, 1989; Gore, 2000).

---

*MIDAS 4.0 was fully redesigned/recoded by Alion Science and Technology's Micro Analysis and Design Operations. More details can be found at http://hsi.arc.nasa.gov/groups/midas.

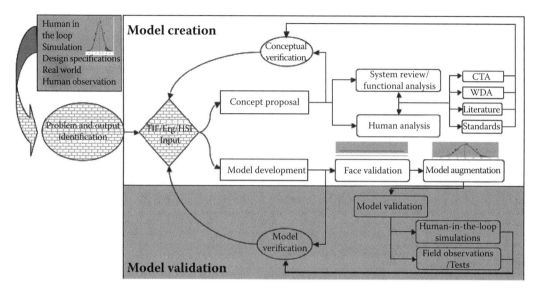

FIGURE 32.3   Model development process.

When models are used for complex system integration issues, the model development process generally follows an iterative design philosophy that collaboratively leverages off of empirical human data (i.e., either HITL simulations or real-time measurements) and concurrently feeds information to HITL simulation processes. A representation of a model development process is outlined in Figure 32.3.

Figure 32.3 outlines a number of iterative model development stages. This iterative process involves a model creation stage (the upper/white portion in the figure) and a model validation stage (the lower/dark portion in the figure).

32.4.1   Model Creation Stage

The first step in the model creation stage is to recognize the human performance area requiring examination/study, to identify concepts that could benefit the human-system performance, and to verify the concept (shown in the upper/lighter portion of fig. 32.3). The model creation stage then progresses into a stage where the system is reviewed and functionally analyzed, then is combined with human tasks as identified through task analysis (Kirwan & Ainsworth, 1992), cognitive task analyses (CTA; Diaper, 1989; Zachary et al., 1998; Klein, 2000), work domain analyses (Sanderson et al., 1999), literature reviews, or evaluation of guidelines and standards among other forms of analyses, and the model is operationally validated.

This model creation process will be described by way of an aviation system development effort. Consider an advanced automation concept on the flight deck of a commercial aircraft. The system concept is reviewed and functionally analyzed to identify the pilot's tasks required to carry out the cockpit duties with advanced technologies (the upper/lighter right portion of fig. 32.3). A series of structured approaches including the task and cognitive task analyses are completed. The task analysis is a process whereby the tasks required to safely flying the aircraft with the automation are analyzed, documented, and outlined (Kirwan & Ainsworth, 1992). The task analysis is a methodology covering a range of techniques to describe and, in some cases, evaluate the human-machine and human-human interaction in systems. It is often described as the study of what an operator (or team) is required to do in terms of actions or cognitive processes to achieve a system state. It is often a hierarchical decomposition of how a goal-directed task is accomplished, including a detailed description of activities, task and element durations, task frequency, task allocation, task complexity, environmental conditions, necessary clothing and equipment, and any other unique factors involved in, or required for, one or more people to perform a given task (Kirwan & Ainsworth, 1992).

The CTA identifies all of the critical cognitive tasks that the operator is required to perform with the automation (Diaper, 1989; Zachary et al., 1998). CTA is a family of methods and tools for gaining access to the mental processes that organize and give meaning to observable behavior. CTA methods describe the cognitive processes that underlie the performance of tasks and the cognitive skills needed to respond adeptly to complex situations. Knowledge is elicited through in-depth interviews and observations about cognitive events, structures, or models. Often the people who provide this information are *subject matter experts* (SMEs)—people who have demonstrated high levels of skill and knowledge in the domain of interest (Klein, 2000). A specialized form of task analysis is the cognitive work analysis/work domain analysis that identifies the tasks within their contextual domain (Sanderson et al., 1999; Vicente, 1999). The cognitive work analysis (CWA) framework focuses on identifying the constraints that shape behavior rather than trying to predict behavior itself describing five classes of constraints: work domain, control tasks, strategies, social-organizational, and worker competencies. CWA can be viewed as a complement to traditional task analysis as it adds the capability for designing for the unanticipated by describing the constraints on behavior rather than solely describing the behavior. These all feed into a concept-verification phase, where the research concept is verified by a human-system engineer, and preparations are made to implement the results from the task analyses into a model form.

The model development process varies with each modeling tool and is exemplified by the MIDAS process as presented next. MIDAS, like all models, requires cognitive triggers and a series of procedures to complete a goal activity. In creating a task/procedural model (a model of the task flow that the operator will be required to perform in the scenario), the human performance modeler decides on the time/duration that is important, and identifies a procedural sequence from one of the outlined procedure identification processes in Figure 32.3 (cognitive task, task or work domain analysis, or a literature or standards review). The integrated model will select human performance times from a combination of lookup tables, user input data, and unconstrained performance representations to produce an integrated HPM prediction of human-system performance.

Once procedures have been identified, the analyst uses a number of task identification tools to quantify and specify the loads associated with each task. MIDAS uses a validated and generally accepted task-load specification structure to develop the cognitive task model developed by the U.S. Army, termed the Task Analysis/Workload (TAWL) index or Modified TAWL (McCracken & Aldrich, 1984; Hamilton et al., 1991; Hamilton et al., 1994; Mitchell, 2000, 2003). The TAWL and the Modified TAWL are representations of a series of general tasks with loads along each of four channels (note, this can be augmented to five or six channels depending on the theory to which the user ascribes)—the visual, auditory, central cognitive processing, and psychomotor channels (VACP). The analyst enters empirically determined values (e.g., visual loads – 0.0 = no visual activity, 1.0 = visually detect/register, 3.7 = discriminate, etc.) into a spreadsheet-like structure that contains a number of task primitives that in turn get called by the relevant environmental triggers in a specific simulation of interest. The empirically determined values along this scale are based on U.S. military personnel in the Army Light Helicopter Experimental (LHX) Program (McCracken & Aldrich, 1984). The Modified TAWL has been augmented to include additional load characteristics and increased cognitive modeling requirements (Hamilton et al., 1991; Hamilton et al., 1994; Mitchell, 2000, 2003).

The initial HPM is implemented with the verified tasks and a face validation effort is completed (still in the upper/lighter portion of fig. 32.3). Verification is the process of determining whether a simulation model and its associated data behave as intended by the model developer/analyst (Sargent, 1980). One common approach to insure face validity is to ask SMEs to make judgments about the content or behavior of an integrated model (Campbell & Bolton, 2005). The subjective judgments are optimal when they follow formal standardized and structured stages (e.g., questionnaires, interviews), and when they are made by multiple SMEs who were not involved in creating the model.

Given, however, that judgments are subjective, they are subject to many biases in considering the model performance. This qualitative approach essentially tests the model's face validity, thus ensuring that simulation results are consistent with the expected system behavior (Law & Kelton, 2000). This qualitative validation effort was undertaken on MIDAS' cognitive performance when a simulated astronaut completed tasks according to one of two schedules of performance (reported in Gore & Smith, 2006). The simulation was developed with input from one astronaut SME and validated by an independent astronaut SME. The series of simulations predicted that the simulated astronaut possessed an increased vulnerability for engaging in erroneous performance under the parallel task due to increased cognitive loading, as compared to performance predicted under the sequential task schedule. Another qualitative approach to validation can involve the model developer viewing a videotape of the environmental situation, or generating a timeline sequence of the events completed by the human, and validating the timing of the events against the timeline as output from the HPM (NASA, 1989; DMSO, 1991; Deutsch & Pew, 2008; Lebiere et al., 2008). This approach is considered a qualitative approach to validation as opposed to verification because it uses subjective judgment of the observable behaviors produced by the human operator cross-referenced to the observable behaviors produced by the model.

32.4.2 Model Validation Stage

The second aspect of the iterative model development process is the model validation stage (the lower/darker portion of fig. 32.3). The validation stage evaluates system performance and often involves prototype design and implementation phases, which lead to HITL data acquisition and field observations and data comparisons. Generally, the more time spent on the scenario creation stage (the upper/lighter portion of fig. 32.3), the less time will be spent in the model validation phase (the lower/darker portion of fig. 32.3). HITL simulations are more costly and time consuming and it is very advantageous to pinpoint the potential problem areas through a modeling exercise (besides learning about the process that is undertaken when using the systems being developed in the field).

As Figure 32.3 alludes to, model verification and validation are critical steps that must be undertaken to establish model credibility. Validation is the degree to which a model or simulation and its associated data are accurate representations of the real world from the perspective of the intended uses of the model or simulation (Sargent, 1980; Balci, 1998; Law & Kelton, 2000). Simulation models are created with a specific purpose in mind (criterion), and it is against this criterion that model accuracy is judged. Model validation is an increasingly difficult goal to achieve given the complexities of the embedded models within an integrated HPM. The interactive nature and corresponding assumptions built within the integrated HPMs (that have the potential of occasionally contradicting one another) make validation increasingly difficult. A contrarian's view holds that validating these models of higher complexity is simply not possible, and the field has simply not been able to effectively validate these models of higher complexity even when simulations have been developed for the sole purpose of validation (Glenn et al., 2004). However, a review of the literature on validation efforts with these integrated HPMs suggests that the field of human performance modeling recognizes the importance of validation and is making a concerted effort to validate their models. Many of the integrated HPMs consider that integrating a validated model will result in a validated integrated HPM (Glenn et al., 2004). While this may be an acceptable conclusion to draw on some occasions, it is not necessarily the case for all conditions for two primary reasons. First, models sensitive to one context may not be sensitive to other contexts unless the model is designed with flexibility as one of its goals—something that had not been regularly done with existing integrated models. Second, integrated models possess many interacting submodels and this interaction may result in the model not being applicable to the new environment. An explanation of two commonly used validation approaches will be described followed by an explanation of some of the validation efforts undertaken on the MIDAS architecture.

Quantitative validation approaches are traditionally statistical in nature and measure the degree to which a model's data are similar to an empirically collected set of data. A quantitative test for a model's validity is the degree to which the model's output statistically differs from HITL data. The recommended statistical tests used to measure the similarity between the datasets are goodness-of-fit tests, such as regression, to assess trend consistency, completing analysis of variances (ANOVAs) on a comparable human and model dataset separately to assess trend consistency (see Campbell & Bolton, 2005, for dangers associated with the ANOVA procedure for model validation), root mean squared (RMS) scaled deviations to assess the exact match, and the chi-square goodness-of-fit approach (Campbell & Bolton, 2005).

For example, the original version of MIDAS has been validated on numerous occasions starting with a quantitative validation effort on process control operator task performance (Hartzell, 1990).* MIDAS was subsequently quantitatively validated with an empirical simulation of a Boeing 757 crews' response times to descent clearances from air traffic control in a variety of scenarios (Corker et al., 1994). This quantitative validation effort produced accurate predictions of flight crew response to the descent clearances issued by the air traffic controllers. In 2000, more formal quantitative validation efforts were undertaken on MIDAS behaviors and workload relative to experimental manipulations in an advanced air traffic concept (Corker et al., 2000; Gore & Corker, 2000). These MIDAS behaviors were validated by correlating the timing of tasks and workload for 50 simulation runs as a function of the rules of separation guiding aircraft in the National Airspace System, to the HITL flight deck performance measures collected in a simulation completed at Embry Riddle University. The results from the study indicate that as the flight deck became increasingly involved in self-separation, the controller task-loading increased and performance parameters (communication time, frequency, points of closest approach, and efficiency) varied systematically with the type of control (traditional ground-based control, traditional control but with all aircraft flying direct to, 20% free flight, and 80% free flight). An increasing relationship among the flight crew workload trends, as evidenced by the communication load, is reflected due to the reduction in the level of ground control. Air MIDAS, a branch of MIDAS for aviation modeling, quantitatively validated its visual fixation sequence model in a synthetic vision system model development effort (see Corker et al., 2003, for a full description of the quantitative validation effort).

Validation is always a challenge to the field of modeling, particularly when modeling human cognition, because human cognition is rarely directly observable and is generally only inferred. It must be ensured that the quantitative validation approach uses meaningful model parameters upon which to judge the model's performance (Roberts & Pashler, 2000). Solely using the goodness-of-fit as a measure to determine model acceptability can be misleading, and can be meaningless. As Roberts and Pashler (2000) point out, a theory can fit too much; it can be closely fit by a similarly flexible theory making very different assumptions; and it could be using an incorrect assumption while correctly fitting the data. For these reasons, both qualitative and quantitative measures should be used iteratively and constantly throughout the model development process to determine model validity. The challenge of creating valid models increases with model complexity.

32.5 Interpreting Complex Integrated Models

Integrating physical models with cognitive models provides an opportunity to visualize the complex interactions among the various components in a modeled human's environment, providing a system designer or analyst the ability to identify situations where the design as proposed may not be adequate for the operational conditions. Model transparency is a very large challenge for models that include representations of human cognition since cognition is an internal mechanism, thereby making cognition difficult to directly observe. *Model transparency* used in the present chapter refers to the ability of the model developer/user to comprehend the relationship that exists among the models being used

---

*MIDAS has progressed through many development stages over the years. The most recent information on MIDAS can be found at http://hsi.arc.nasa.gov/groups/midas.

in the simulation, the performance of the model in the simulation, which models are triggering in the model architecture, and whether the model is behaving as the model developer would expect. Gluck and Pew (2005) refer to this as runtime interpretability. Other researchers have referred to this with different terminology, including model traceability, model behavior visibility, model verifiability, model validity, and model interpretability (NASA, 1989; Napierky et al., 2004; Gluck & Pew, 2005; Foyle & Hooey, 2008).

Model transparency therefore brings in a number of key elements from model verification and validation of the model's performance leading ultimately to establishing model credibility. When models are transparent, the user can comprehend the model performance, thereby increasing their trust and confidence that the output from the model is in line with human performance. Conversely, when models are not transparent, the user cannot comprehend the model's performance. The user will not have confidence, nor is there trust that the model performed according to expectations. When models are developed that integrate a number of sub-models together as shown by the MIDAS software earlier, the transparency of the model is paramount because it becomes very difficult to determine which model is active at which time in a scenario (NASA, 1989; Deutsch & Pew, 2003). Without this insight into the model, and an accurate understanding of the assumptions embedded in the model, results may be overstated. For the field to advance, it is critical that a comprehensive understanding of the mechanisms operating in the model is formalized, and that there is sufficient transparency in the model's operation. This formalization and transparency will increase the likelihood that assumptions are properly identified and noted in the model's performance and that the correct model will be chosen and used for the specific application.

The MIDAS software team has invested significant resources in increasing the transparency of the MIDAS software by providing a graphic user interface (GUI; shown in fig. 32.4) that guides an analyst in the scenario creation process, and that provides feedback to the analyst (which models are performing, what the models are doing, and when the models are triggered), a capability that was absent in previous versions. The upper left portion of the GUI in Figure 32.4 shows the procedure development window,

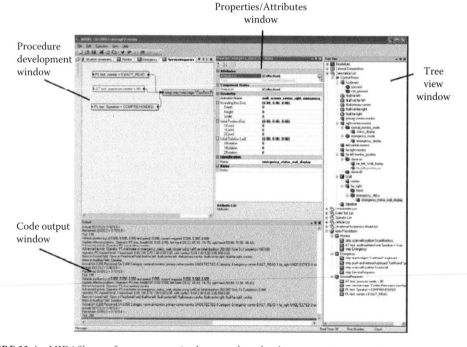

FIGURE 32.4   MIDAS's use of transparency in the procedure development environment.

FIGURE 32.5   Visualization of MIDAS's use of Jack with vision cones driven by empirical model.

which shows the procedures that have been programmed by the analyst that will be triggered by the environment. A number of tabs exist within this upper left portion of the GUI and can be selected by the analyst depending on the progress in the model development process. For example, analysts can switch to a different tab to show the model's procedures at differing layers/depths, and the analyst can also select output tabs to demonstrate the workload and situation awareness runtime output. The upper middle portion of the GUI demonstrates the properties window, which outlines the logic that drives the models (attributes, geometry, states of the CADs, and the CAD geometry's location according to an XYZ plane). The right side of the GUI demonstrates the interaction among all of the models desired by the analyst in the simulation, known as the tree view. The lower left portion of the GUI demonstrates the code output window to allow the analyst to evaluate the performance of the code.

By integrating a DHM along with the cognitive sub-models, insights about cognitive processes can be inferred from the observable behaviors. As an example, Figure 32.5 shows how MIDAS visualizes the modeled operator's visual attention utilizing a visual cone driven by empirical data (the reader is directed to Gore & Corker, 2000, for a list of empirical data contained within MIDAS) that indicates how the modeled operator visually samples the world. One can infer from the empirical models driving the visual cone, to what the modeled operator is attending. Another visualization technique within MIDAS is the presentation of dynamic timeline information, coupled with synchronized human performance output (i.e., workload and situation awareness) that demonstrates the cognitive tasks that are active at any given time in the simulation (figs. 32.6 and 32.7). These measures have recently been collected in a multi-crew aeronautics application environment. It has proven useful in understanding and verifying procedural sequences and their impact on the modeled operator's performance (Gore & Smith, 2006).

32.6 Conclusion

DHMs are especially useful for visualizing the geometric form of humans in a particular context as shown in many examples throughout this volume. Cognitive models are especially useful for representing

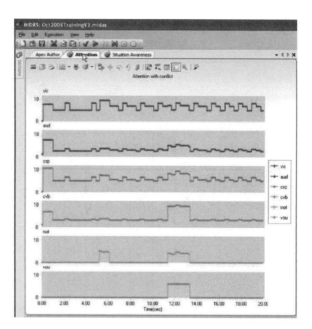

FIGURE 32.6   MIDAS workload visualization window for model validation.

human processes that are required to complete a goal behavior in a particular context. When integrated, cognitive and physical modeling approaches are synergistic approaches that enable predictions of complex, dynamic, human behaviors of operator-environment interactions. The closed-loop nature of the relationship between cognitive models and DHMs, as exemplified by the integrated modeling approach, results in more representative human-system performance than either of the approaches alone. Com-

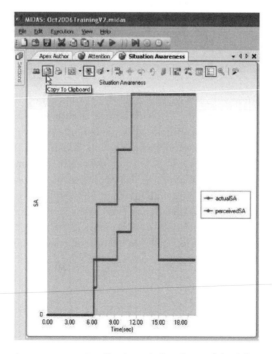

FIGURE 32.7   MIDAS situation awareness visualization window for model validation.

bining models in this fashion adds complexity, however, and this complexity highlights the need for improved model transparency and validation approaches.

Acknowledgments

The research presented herein was funded by, and supports, the NASA NGATS ATM-Airportal Project. The technical monitor of NASA Ames Research Center Grant # NNA06CB16A is Dr. Jessica Nowinski. The opinions expressed are those of the author and do not necessarily reflect the opinions of NASA, of the Federal government, or of SJSU. The author would like to thank all reviewers of the current document for their provocative thoughts regarding human performance modeling.

References

Abdel-Malek, K., J. Yang, R.T. Marler, S. Beck, A. Mathai, X. Zhou, A. Patrick, & J. Arora. 2006. Towards a new generation of virtual humans: Santos. *International Journal of Human Factors Modeling and Simulation*, Vol. 1, No. 1, pp. 2–39.

Anderson, J. R. 1993. *Rules of the mind*. Hillsdale, NJ: Lawrence Erlbaum.

Anderson, J.R., & C. Lebiere. 1998. *Atomic components of thought*. Hillsdale, NJ: Lawrence Erlbaum.

Baddeley, A.D., & G. Hitch. 1974. Working memory. In *The psychology of learning and motivation: Advances in research and theory*, Vol. 8, ed. G.H. Bower, pp. 47–89. New York: Academic Press.

Badler, N.I., C.B. Phillips, & B.L. Webber. 1993. *Simulating humans: Computer graphics, animation, and control*. Oxford: Oxford University Press.

Balci, O. 1998. Verification, validation, and testing. In *The handbook of simulation*, ed. J. Banks, pp. 335–393. New York: John Wiley & Sons.

Baron, S., D.L. Kleinman, & W.H. Levison. 1970. An optimal control model of human response. Part 2: Prediction of human performance in a complex task, *Automatica*, Vol. 6, No. 3, pp. 371–383.

Boring, R.L., D.D. Dudenhoeffer, B.P. Hallbert, & B.F. Gore. 2006. Virtual power plant control room and crew modeling using MIDAS. *Proceedings of the joint Halden reactor project and CSNI special experts' group on human and organisational factors workshop on future control station designs and human performance issues in nuclear power plants*, May 8–10, 2006, Halden, Norway, pp. 1–5:5.3.

Broadbent, D.E. 1958. *Perception and communication*. London: Pergamon.

Campbell, G.E & A.E. Bolton. 2005. HBR validation: interpreting lessons learned from multiple academic disciplines, applied communities, and the AMBR project. In *Modeling human behavior with integrated cognitive architectures: Comparison, evaluation and validation*, ed. K.A. Gluck & R.W. Pew, pp. 365–395. Hillsdale, NJ: Lawrence Erlbaum.

Card, S.K., T.P. Moran, & A. Newell. 1983. *The psychology of human computer interaction*. Hillsdale, NJ: Lawrence Erlbaum.

Chaffin, D.B. (Ed.). 2001. *Digital human modeling for vehicle and workplace design*. Warrendale, PA: Society of Automotive Engineers.

Chaffin, D.B. 2002. On simulating human reach motions for ergonomics analyses. *Human Factors and Ergonomics in Manufacturing*, Vol. 12, No. 3, pp. 235–247.

Corker, K., B.F. Gore, K. Fleming, & J. Lane. 2000. Free flight and the context of control: Experiments and modeling to determine the impact of distributed air-ground air traffic management on safety and procedures. *Proceedings of the 3rd USA-Europe Air Traffic Management R & D Seminar*, Naples, Italy: USA-Europe Air Traffic Management.

Corker, K.M., B.F. Gore, E. Guneratne, A. Jadhav, & S. Verma. 2003. *SJSU/NASA coordination of Air MIDAS safety development human error modeling: NASA aviation safety program. Integration of Air MIDAS human visual model requirement and validation of human performance model for assessment of safety risk reduction through the implementation of SVS technologies*, (Interim Report and Deliverable NASA Contract Task Order No. NCC2-1307), Moffett Field, CA: San Jose State University.

Corker, K., S. Lozito, & G.M. Pisanich. 1994. Flight crew performance in automated air traffic management. In *Proceedings of 21st biennial conference of the Western European Association for Aviation Psychology*, Dublin, Ireland.

Corker, K., & B. Smith. 1993. An architecture and model for cognitive engineering simulation analysis: Application to advanced aviation automation. *Proceedings of the AIAA computing in aerospace 9 conference*, October, San Diego, CA.

Defense Modeling and Simulation (DMSO). 1991. Validation of human behavior representations: RPG special topic. Washington, DC: U.S. Department of Defense.

Deutsch, S. 1998. Interdisciplinary foundations for multiple-task human performance modeling in OMAR. *Proceedings of the twentieth annual meeting of the Cognitive Science Society*, Madison, WI.

Deutsch, S., & R. Pew. 2003. Modeling the NASA SVS part-task scenarios in D-OMAR. (BBN Report No. 8399). Cambridge, MA: BBN Technologies.

Deutsch, S., & R. Pew. 2008. D-OMAR: An architecture for modeling multi-task behaviors. In *Human performance modeling in aviation,* ed. D.C. Foyle & B.L. Hooey. Boca Raton, FL: CRC Press/Taylor & Francis.

Diaper, D. 1989. *Task analysis for human computer interaction.* Chichester: Ellis Horwood.

Farrell, K., & R.T. Marler. 2004. Optimization-based kinematic models for human posture, University of Iowa, Virtual Soldier Research Program, Technical Report No. VSR-04.11.

Foyle, D.C., & B.L. Hooey (eds.). 2008. *Human performance modeling in aviation.* Boca Raton, FL: CRC Press/Taylor & Francis.

Glenn, F., K. Neville, J. Stokes, & J. Ryder. 2004. Validation and calibration of human performance models to support simulation-based acquisition. In *Proceeding of the winter simulation conference.* Vol. 1, ed. R. Ingalls, M. Rosetti, J. Smith, & B. Peters, pp. 1533–1540.

Gluck, K.A., & R.W. Pew (eds.). 2005. *Modeling human behavior with integrated cognitive architectures: Comparison, evaluation and validation.* Hillsdale, NJ: Lawrence Erlbaum.

Gore, B.F. 2000. The study of distributed cognition in free flight: A human performance modeling tool structural comparison. *Proceedings of the third annual SAE international conference and exposition—digital human modeling for design and engineering*, Warrendale, PA: SAE Inc.

Gore, B.F. 2002. An emergent behavior model of complex human-system performance: An aviation surface related application. *VDI Berichte 1675,* pp. 313–328. Düsseldorf, Germany: VDI Verl.

Gore, B.F., & K. Corker. 2000a. System interaction in free flight: A modeling tool cross- comparison. *SAE Transactions—Journal of Aerospace*, Vol. 108, No. 1, pp. 409–424, Warrendale, PA: SAE Inc.

Gore, B.F., & K.M. Corker. 2000b. A systems-engineering approach to behavioral prediction of an advanced air traffic management concept. *19th annual digital avionics systems conference (DASC): Entering the second century of powered flight, 1,* 4B3/1-4B3/8.

Gore, B.F., & K.M. Corker. 2001. Human error modeling predictions: Increasing occupational safety using human performance modeling tools. In *Computer-aided ergonomics and safety (CAES) 2001 conference proceedings,* ed. B. Das, W. Karwowski, P. Modelo, & M. Mattila, July 28-August 4, Maui, Hawaii.

Gore, B.F., & J.D. Smith, 2006. Risk assessment and human performance modeling: The need for an integrated approach. *International Journal of Human Factors of Modeling and Simulation*, Vol. 1, No. 1, pp. 119–139.

Gore, B.F., S. Verma, K. Corker, A. Jadhav, & E. Guneratne. 2004. Chapter 5: Human performance modeling predictions in reduced visibility operation with and without the use of synthetic vision system operations. In *Proceedings of the 2003 conference on human performance modeling of approach and landing with augmented displays* (NASA/CP-2003-212267), ed. D.C. Foyle, A. Goodman, & B.L. Hooey, pp. 119–142. Moffett Field, CA: NASA.

Gore, B.F., S. Verma, A. Jadhav, R. Delnegro, & K.M. Corker. 2002. *Human error modeling predictions: Air MIDAS human performance modeling of T-NASA.* NASA Ames Research Center Contract No. 21-1307-2344. CY01 Final Report.

Green, R.F., & J. Charland. 2006. Human modeling in the product lifecycle management of the Boeing 787 Dreamliner. In *Proceedings of the annual SAE international conference and exposition— digital human modeling for design and engineering*, Paper No. 2006-01-2315, Warrendale, PA: SAE Inc.

Hamilton, D.B., C.R. Bierbaum, & L.A. Fullford. 1991. Task analysis/workload (TAWL) user's guide, version 4 (Research Project 91-11). Alexandria, VA: U.S. Army Research Institute for the Behavioral and Social Sciences (AD A241 861).

Hamilton, D.B., C.R. Bierbaum, & D.M. McAnulty. 1994. Operator task analysis and workload prediction model of the AH-64D mission volume II: Appendices A through F (Contract No. MDA90-92-D-0025). Army Research Laboratory. Anacapa Sciences, Inc., Fort Rucker, AL.

Hartzell, E.J. 1990. Army-NASA Aircrew/Aircraft Integration Program (A3I). Executive summary. Moffett Field, CA: NASA-Ames Research Center.

Kieras, D.E., & D.E. Meyer. 1997. An overview of the EPIC architecture for cognition and performance with application to human-computer interaction. *Human-Computer Interaction*, Vol. 12, No. 4, pp. 391–438.

Kirwan, B., & L.K. Ainsworth. 1992. *A guide to task analysis*. Washington, DC: Taylor & Francis.

Klein, G. 2000. Using cognitive task analysis to build a cognitive model. *Proceedings of the IEA 2000/ HFES 2000 congress*, Vol. 1, pp. 596–599. Santa Monica, CA: HFES.

Laughery, R.L. 1999. Using discrete event simulation to model human performance in complex systems. In *Proceedings of the 31st conference on winter simulation: Simulation—a bridge to the future.* Vol. 1, ed. P.A. Farrington, H.B. Nembhard, D.T. Sturrock, & G.W. Evans, pp. 815–820.

Law, A.M., & D.W. Kelton. 2000. *Simulation, modeling and analysis* (3rd ed.). New York: McGraw Hill.

Lebiere, C. 2002. Modeling group decision making in the ACT-R cognitive architecture. In *Proceedings of the 2002 computational social and organizational science (CASOS)*. June 21–23, Pittsburgh, PA.

Lebiere, C., & J.R. Anderson. 1993. A connectionist implementation of the ACT-R production system. In *Proceedings of the 15th Annual Conference of the Cognitive Science Society*, pp. 635–640.

Lebiere, C., R. Archer, B. Best, & D. Schunk. 2008. Modeling pilot performance with an integrated task network and cognitive architecture approach. In *Human performance modeling in aviation*, ed. D.C. Foyle & B.L. Hooey. Boca Raton, FL: CRC Press/Taylor & Francis.

McCracken, J.H., & T.B. Aldrich. 1984. Analysis of selected LHX mission functions: Implications for operator workload and system automation goals. (Technical note ASI 479-024-84(b)), Fort Rucker, AL: Anacapa Sciences, Inc.

Miller, C. 2000. The human factor in complexity. In *Automation, control and complexity: New developments and directions*, ed. T. Samad & J. Weyrauch. New York: John Wiley.

Mitchell, D.K. 2000. Mental workload and ARL workload modeling tools. *Report No. ARL TN-161.* Aberdeen Proving Ground, MD: U.S. Army Research Laboratory.

Mitchell, D.K. 2003. Advanced improved performance research integration tool (IMPRINT) Vetronics technology test bed model development. *Report No. ARL-TN-0208.* Aberdeen Proving Ground, MD: U.S. Army Research Laboratory.

Napierski, D.P., A.S. Young, & K.A. Harper. 2004. Towards a common ontology for improving traceability of human behavior models. In the *Proceedings of the 9th annual behavioral representation in modeling and simulation conference.*

National Aeronautics and Space Administration (NASA). 1989. Human performance models for computer aided engineering. In *Panel on pilot performance models in a computer-aided design facility*, ed. J.I. Elkind, S.K. Card, J. Hochberg, & B.M. Huey. Washington, DC: National Academy Press.

Newell, A. 1990. *Unified theories of cognition*. Cambridge, MA: Harvard University Press.

Pew, R.W. 2007. Some history of human performance modeling. In *Integrated models of cognitive systems*, ed. W. Gray. Oxford: Oxford University Press.

Pew, R.W., & A.S. Mavor. 1998. *Modeling human and organizational behavior: Application to military simulation*. Washington, DC: National Academy of Science.

Roberts, S., & H. Pashler. 2000. How persuasive is a good fit? A comment on theory testing. *Psychological Review*, Vol. 107, No. 2, pp. 358–367.

Sanderson, P., N. Naikar, G. Lintern, & S. Gross. 1999. Use of Cognitive Work Analysis across the system lifecycle: Requirements to decommissioning. *Proceedings of the Human Factors and Ergonomics Society's 43rd annual meeting.* Santa Monica, CA: Human Factors and Ergonomics Society.

Sargent, R.G. 1980. Validation of simulation models. *Proceedings of the 11th conference on winter simulation*, pp. 497–503, December 3–5, 1979. San Diego, CA.

Seppelt, B.J. & J.D. Lee. 2007. Making adaptive cruise control (ACC) limits visible. *International Journal of Human-Computer Studies*, Vol. 65, pp. 192–205.

Shannon, C.E., & W. Weaver. 1949. *Mathematical theory of cognition.* Urbana and Chicago: University of Illinois Press.

Sheridan, T.B. 1992. *Telerobotics, automation, and human supervisory control.* Cambridge, MA: MIT Press.

Sheridan, T.B., & W.R. Ferrell. 1974. *Man-machine systems.* Cambridge, MA: MIT Press.

Shively, R.J., M.D. Burdick, M. Brickner, I. Nadler, & J. Silbiger. 2000. Comparison of a manned helicopter simulation to a computer-based human performance model. *Proceedings of the American Helicopter Society 56th annual forum,* Virginia Beach, VA.

Siegel, A.I., & J.J. Wolf. 1969. *Man-machine simulation models: psychosocial and performance interaction.* New York: Wiley.

Smith, E. 1968. Choice reaction time: An analysis of the major theoretical positions. *Psychological Bulletin*, Vol. 69, pp. 77–110.

Sternberg, S. 1969. The discovery of processing stages: Extensions of Donder's method. In *Attention and performance II,* ed. W.G. Koster., pp. 276–315. Amsterdam.

Welford, A.T. 1976. *Skilled performance.* Glenview, IL: Scott Foresman.

Wickens, C.D. 1992. *Engineering psychology and human performance* (2nd ed.). New York: Harper Collins.

Wickens, C.D., & J.G. Hollands. 2000. *Engineering psychology and human performance* (3rd ed.). Upper Saddle River, NJ: Pearson Education.

Wiener, N. 1950, 1980. *The use of human beings: cybernetics and society.* New York: Avon Books.

Vicente, K.J. 1999. *Cognitive work analysis: Toward safe, productive, and healthy computer-based work.* Mahwah, NJ: Lawrence Erlbaum.

Zachary, W. 2005. Chapter 5: A COGNET/iGEN cognitive model that mimicks human performance and learning in a simulated work environment. In *Modeling human behavior with integrated cognitive architectures: Comparison, evaluation and validation,* ed. K.A. Gluck & R.W. Pew, pp. 113–175. Hillsdale, NJ: Lawrence Erlbaum.

Zachary, W.W., J.M. Ryder, & J.H. Hicinbothom. 1998. Cognitive task analysis and modeling of decision making in complex environments. In *Decision making under stress: Implications for training and simulation,* ed. J. Cannon-Bowers & E. Salas. Washington, DC: American Psychological Association.

Zhai, S., & P. Milgram. 1991. A telerobotic virtual control system. In *Proceedings of SPIE Volume 1612: Cooperative intelligent robotics in space II,* ed. W. Stoney. Boston: SPIE.

Zhai, S., & P. Milgram. 1992. Human-robot synergism and virtual telerobotic control. In *Proceedings of the annual meeting of Human Factors Association of Canada.* Canada: Human Factors and Ergonomics Society.

Zhang, X., A.D. Kuo, & D.B. Chaffin. 1999. Optimization-based differential kinematic modeling exhibits a velocity-control strategy for dynamic posture determination in seated reaching movements. *Journal of Biomechanics,* Vol. 31, pp. 1035–1042.

33

Instrumentation in Support of Dynamic Digital Human Modeling

Douglas R. Morr,
John F. Wiechel, and
Sandra A. Metzler

33.1 Introduction

Scientific studies of human movement traces their origins back to the 17th century and include work up through the early 20th century. By the 19th century it had been realized that individual angular joint motion along with the translations of individual segments and the whole-body mass was necessary to fully characterize human motion. This hierarchical viewpoint allows the human body to be visualized as a system of joints and limbs with both active and passive loading members. It has, however, only been within the last century that true measurement, and therefore quantification, has been achieved.

Kinematics is the area of science that deals with motion, while kinetics deals with the results of such motion in terms of forces and accelerations. The instrumentation used to quantify the kinematics of human movement, along with those designed to measure the kinetics that exist during this movement, involve multiple areas of science and technology. With the advent of the computer, and the powerful capabilities of today's computing techniques, it has become more and more desirable to be able to create reliable, robust, and valid digital models of portions and even the entire human body. The kinematic and kinetic measurements of humans performing and being exposed to a wide variety of activities and events are a necessity for understanding both normal and abnormal function, and therefore for the development of such models. In order to achieve this, proper measurements of humans must be attained. This means that it is important to understand, at least in a broad sense, the instrumentation that is used in support of digital human modeling efforts.

This chapter is divided into three major sections. The first section discusses instrumentation and methods used to measure, at least directly or primarily, kinematic variables. The second section covers instrumentation employed to quantify kinetic quantities, and the third details those methods that are combined kinematic and kinetic measurement devices.

33.2 Kinematic Measurement

Accurately measuring motion is essential in any scientifically sound method of analyzing dynamic human activity. This then becomes equally, if not more, important, when attempting to create a reliable digital representation of a human, or some portion of the human body. By accurately measuring the motion of body segments in 3D space, as well as the angular motion of those segments and the resulting joint motion, kinematic comparisons can be made between living human tests and their surrogate digital models. This is usually accomplished by establishing movement corridors from volunteer or postmortem human subject (PMHS) tests, and then comparing model results to these corridors. As technology has advanced so have the complexities and techniques used to obtain kinematic data of humans, including stereometric methods, sound-based methods, electromagnetic systems, radiographic techniques, goniometric techniques, and inertial-sensing-based systems.

33.3 Stereometric Methods

At the simplest and earliest level, capturing dynamic human movement has involved some type of film technology using stereometric methodologies. This involves estimating the position of an object using multiple 2D images, or stereophotogrammetry. Photogrammetry involves obtaining reliable information about physical objects and their environments through the processes of reading, recording, and interpreting photographic images and patterns of electromagnetic radiant energy. The distinct advantage of these methods is that they allow for the inclusion of the entire human body and therefore enable the simultaneous viewing of relative motion between various body segments. The systems that are used to accomplish this task for human motion are generically referred to as motion capture systems. Regardless of the technology utilized, motion capture systems are designed to determine the instantaneous position and orientation of a system of axes that are considered to be fixed on, or at the joints of, the bones in order to reconstruct the kinematics of that human's movements. The bones are assumed to be rigid objects and a fundamental assumption is that these axes do not move relative to the bones.

33.3.1 Photographic

Prior to the advent of motion picture cameras, dynamic human movement was captured on still photographs using multiple exposure stroboscopic photography. If only one camera was utilized, this of

FIGURE 33.1 Early photographic example of human motion capture from Eadweard Muybridge, "Animal Locomotion," 1887. (From Muybridge, E., *Complete Animal and Human Locomotion*, Dover Publishers, New York, 1887. With permission.)

course could only capture planar motion. Therefore, in order to capture 3D motion multiple still cameras were utilized, along with strobic light sources. In order to obtain meaningful photographs, very precise placement and timing was required, followed by painstaking analysis of each set of photographs relative to each other. Without computer hardware and software to aid in this step, the kinematic data that resulted were often very course temporally, often contained large amounts of error spatially, and were usually discontinuous. The discontinuities arise due to body segments not constantly being visualized even when multiple cameras were used. An early example of such pictures is from the work of Eadweard Muybridge in the late 19th century (fig. 33.1). Muybridge is credited with being the first person to photographically analyze human motion in 1887 using a serial photographic technique. Also from the late 19th century is the first analysis of human and animal motion using a photographic gun by Etienne-Jules Marey (fig. 33.2). Marey was the first to use Muybridge's photographic techniques scientifically to correlate ground reaction forces with movement, and his chronophotograph became the first motion film camera in 1888. For his efforts Marey is considered by many to be the father of modern motion analysis.

With the use of computers, many of the temporal and spatial errors that occur when using photographic techniques can be overcome. However, photographs are just still pictures of one point in time. This is why once video cameras became more readily available and cost effective, kinematic studies obviously chose to use this technology.

33.3.2 Cinematographic

In today's human motion data acquisition systems, or motion capture systems, some type of camera system is usually employed as a direction sensor for incoming radiation. These systems have the common goal of measuring an object's position and orientation in physical space and recording this information in a computer-usable format. When these cameras are video based, whether that be high-speed digital or normal-speed analog, the method of data capture is referred to as cinematographic. Due to

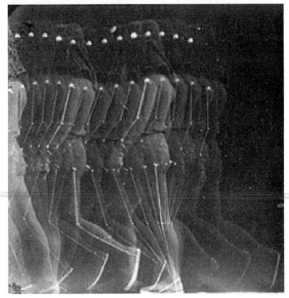

FIGURE 33.2 Early photographic examples of planar body segment tracking from Etienne-Jules Marey. (From Marey, E., *Le Mouvement*, Masson, Paris, 1894. With permission.)

the smooth and highly nonplanar nature of the human anatomy, it is nearly impossible to derive axial rotation from video images alone; instead special markers are placed on specific locations on the body. These markers are then used to reconstruct the human body kinematics by correlating common tracking points on the tracked objects in each image and then using this information with the knowledge of the relationship between each of the images and the camera parameters.

The markers used can be active or passive, depending on the nature of the measurement system. Passive systems utilize markers that reflect light back to the sensor. Active systems, on the other hand, have markers with a light source. Both active and passive systems are capable of recording the most complex human motions with the least deterioration of data quality. However, regardless of the type of marker used, when used on live volunteers, these markers must be placed on the skin or clothing of the volunteer and therefore may have inherent error due to relative motion of the marker. In order to handle the data collected, software is provided with these systems to aid in camera calibration, data collection, landmark identification and reconstruction, and other post-processing tasks. These systems often require the use of pattern recognition techniques, thereby requiring high-computational resources. Also, a skilled operator still must supervise the data sorting, which can take 1 to 2 minutes per second of simple data, or up to 30 minutes per second of complicated data. Some systems can have relatively

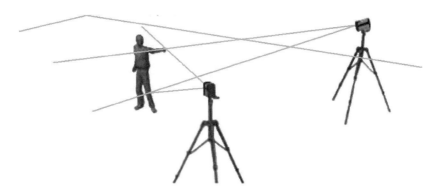

FIGURE 33.3 Multi-camera environment generic setup (Qualisys Motion Capture Systems). (From Qualisys Motion Capture Systems, www.qualisys.com. With permission.)

low spatio-temporal resolution. A generic example of how a lab space may be set up for video motion capture is shown in Figure 33.3.

33.3.3 Optoelectric

Optoelectric systems have been utilized for human movement studies since the early 1970s, and are among the most common today. These systems utilize video cameras to track the image coordinates of retroreflective markers (fig. 33.4), small lightbulbs, or photodiode arrays. Stroboscopic illumination using infrared LEDs or visible light is often used in conjunction with regular-speed cameras to avoid the blurring effect that can be caused by fast movement. In these systems, an array of cameras is used along with a multitude of markers, and the human subject must perform tasks in a calibrated volume (fig. 33.5).

FIGURE 33.4 Optoelectric markers on human volunteer (Vicon). (From Vicon Motion Systems, www.vicon.com. With permission.)

FIGURE 33.5 Typical camera setups for video-based systems. (From Qualisys Motion Capture Systems, www. qualisys.com. With permission.)

Passive systems are sensitive to background light, and sunlight and strong incandescent light typically are avoided. Also, due to the passive nature of the markers there can often be correspondence issues and occlusion (line of sight). However, the currently available passive systems can have less than 2.0 mm RMS errors during dynamic testing, high-speed cameras running at 120 to 1000 Hz, and normally are capable of high resolution (1000 × 1000 pixels). Active systems, on the other hand, can be utilized in many environments, including sunlight, while still maintaining relatively high speed and resolution, and do not have correspondence problems.

33.3.4 Optomechanical

Another approach to quantifying human motion is through the application of principles of scanning mirrors. This method was first introduced in the early 1970s and involves the projection of a V-shaped light image onto a rotating mirror. The mirror then sweeps the reflected image through space and the legs of the V repeatedly hit small photodetectors affixed as landmarks on a subject's body. When the photocurrent pulse occurs, it is converted into 2D direction information using splitting optics, diffractive gratings, and optoelectronic circuitry. Again, software is relied upon to reconstruct and quantify human motion. These types of systems have the advantages of the use of passive landmarks, high spatio-temporal resolution, and the landmark coordinates are available real-time. The disadvantage of such systems is that the number of landmarks that can be used is usually limited, and the limited depth range due to the fixed stereobase.

33.3.5 Lateral Effect Photodetectors

The lateral photoeffect can be used to determine the position of a light image on a semiconductor photo-detector (fig. 33.6). A photocurrent is produced by the incident light, and is divided between the lateral

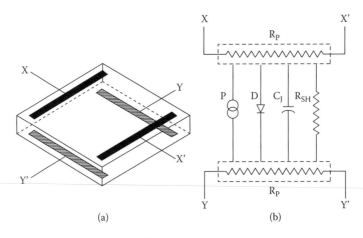

(a)   (b)

FIGURE 33.6 Typical two-dimensional photodiode.

contacts of the detector. If a fully reverse-biased detector is used, there is a linear relationship between the position of a light spot and the signal currents. In this method, infrared LEDs are used as landmarks on the subject. These LEDs are operated using time-division multiplexing in order to compensate for the influence of background light. This enables the system to be used, at least on a limited basis, in daylight. This use of the method also has the advantages of a much higher spatial resolution and automatic marker identification. A high temporal resolution can be achieved, but this is offset by a lower temporal resolution when there are large numbers of markers. The disadvantages of this system are the use of wired markers, making dynamic movement cumbersome and often challenging, and the system's sensitivity to marker light reflections.

33.3.6 Röntgenographic

In a fashion similar to the optical systems, Röntgenographic systems utilize electromagnetic radiation, but no lenses. The Röntgen is a unit of measurement of ionizing radiation in air, named after German physicist Wilhelm Röntgen. Selvik introduced this motion analysis methodology in 1983. In this type of system, the perspective center is formed by the Röntgen source and the human subject is placed between the perspective center and the recording device. The Röntgen absorption pattern is generally recorded on a photographic plate, so essentially the human subject performs activities inside a large camera. Digitizing the relevant features is required prior to any subsequent processing. In order to enhance the recognition of landmarks, small tantalum markers can be implanted. This allows for very high spatial resolution on the order of $+/-$ 0.01 mm; however, the measurement frequencies are relatively low in real-time. Blurred marker images can occur due to the finite apertures of the Röntgen foci making it difficult to accurately determine the centroid of the image. For this reason, automatic centroid detection systems are usually employed. Also, pattern recognition techniques are usually needed to enable clear identification of the passive, nondistinct, markers.

33.4 Sound-Based Methods

Goldman and Nadler were among the first to use a sound-based method to measure displacements, velocities, and accelerations of a body segment in 1956. They attached an emitter to a subject, oscillator-generated ultrasonic waves that were transmitted by the emitter, and microphones picked up the waves. The frequency of waves received by each microphone varied from the emitted frequency according to the Doppler effect. This, in turn, was proportional to velocity. In 1983 Fleischer and Lange measured hand motion using three pairs of transmitters and receivers and measured the time delay from transmission to reception to determine the location of each receiver from the transmitter. By 1990, Hsiao and Keyserling had developed a 3D ultrasonic system using eight receivers set around a subject. The signal generated from each transmitter was received by at least three receivers. A computer algorithm was used to determine position of transmitters mounted at body joints, both in their spatial coordinates and joint angles.

In general, acoustic systems typically have multiple transmitters that trigger a measurable high-frequency ultrasonic pulse. Receivers, microphones, measure the time taken for the sound to travel and triangulate the position of the receivers relative to the transmitters. The transmitters and receivers can either be placed on a body segment or fixed in the control volume. These systems are inexpensive and relatively easy to set up and use. However, since a clear line of sight must be maintained, there can be occlusion problems. Also, there is a limit to the range and number of sensors, cables have to be on body, and acoustic interference can occur. The accuracy, update rate, and range of these systems are limited by the physics of sound, with accuracies usually below that of other kinematic motion capture systems. Finally, transmitters or receivers are not at joint centers, therefore there can be relative motion errors.

33.5 Electromagnetic Systems

Electromagnetic systems typically have sensors placed on the body to measure the low-frequency magnetic fields generated by a transmitter source. The transmitter source is usually constructed of three perpendicular coils that emit a magnetic field when a current is applied. The current is applied sequentially, creating three mutually perpendicular fields during each measurement cycle. The measured strength of the fields is proportional to the distance of each coil from the field emitter assembly. The position and orientation of each sensor are then calculated from the nine measured field values. These systems have the advantages of being able to provide real-time data at up to 100 Hz without occlusion of markers and without correspondence issues. Other strengths include lower number of markers needed to track full body motion (fig. 33.7), clean data, and low processing times. Accuracy has been reported up to 0.1 inches and 0.1 degrees, but many other commercially available systems have accuracies that are not as good as video-based systems. However, since magnetic fields decrease in power rapidly as the distance from the source increases, ferrous materials within the control volume (fig. 33.8) can easily disturb them. Also, electromagnetic systems often have a more limited workspace (ranging generally between 3 and 20 feet), heavier sensors and wires on the body, and are sensitive to electromagnetic interference.

33.6 Radiographic Techniques

Various radiographic technologies are available to many researchers, including X-rays, magnetic resonance imaging (MRI), fluoroscopy, and computer-aided tomography just to name a few. Traditionally radiographical techniques are used in human motion analysis to measure bone position by identifying either anatomical landmarks or implanted markers and calculating relative positions over time during

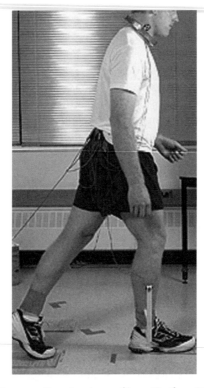

FIGURE 33.7   Example of electromagnetic system in use. (From Northern Digital Industries, www.ndigital.com. With permission.)

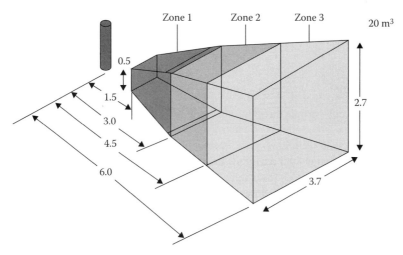

FIGURE 33.8 Typical calibrated volume of electromagnetic system. (From Northern Digital Industries, www.ndigital.com. With permission.)

planar motion. The impetus for use of these techniques can be to avoid skin motion artifacts by directly measuring bone movement and/or to visualize actual internal kinematics. A key advantage of these techniques is that any location on the bone's surface can be tracked, not just that of markers. Unfortunately, the need to have at least some markers affixed to the bones can be quite painful to living subjects, and there are obvious health concerns with radiation exposure. Also, these technologies are all designed for static or quasi-static studies. Even dynamic MRI and CT has too low of frame rates and the environment is too restrictive for most dynamic weight-bearing activities. Other issues with these types of methodologies arise with bone marker placement. Bone markers placed too closely and/or with small inter-marker angles can adversely affect rotational accuracy. However, these systems are highly accurate and precise in determining the absolute and relative location of anatomic structures. For example, muscle origin and insertion points can be accurately identified on CT, and therefore the length changes of soft tissue structures can be estimated quite well. High accuracy and precision of soft tissue changes can be critical with biologic tissues that have a strong rate dependence.

33.6.1 X-Ray

Stereo (biplanar) radiographic imaging enables accurate quantitative 3D motion assessment along with direct visualization of bone motion. This uses two X-ray sources (generators) and two image intensifiers optically coupled to high-speed video. Radio-opaque markers, usually tantalum due to its biocompatibility and high radio density, are placed on bones. A minimum of three markers is needed per bone. Video radiographs have poorer resolution than film. They require invasive procedures and radiation exposure.

33.6.2 MRI

MRI depicts soft tissues and bony structures non-invasively, and has been used for both static and dynamic joint imaging by comparing spatial and angular relationships between bones in select slices of joints in various positions. MRI provides a wealth of image data, excellent detail of anatomic structure in virtually any plane. Loss of signal and image distortion occur in the presence of metal. It is not really 3D-only planar, has long imaging times (shortened by increasing thickness, but loose detail and small variations are magnified by slice thickness). Cine MRI sequencing allows imaging during active movement, but currently the method is restricted to only one slice. Not until open MRI was weight-bearing activity even possible, but field strength lower in open MRI causing lower spatial resolution.

33.6.3 Fluoroscopy

Conventional fluoroscopy permits direct visualization of bone motion, but is limited to 2D assessment and is prone to parallax error and motion blurring,

Alternative fluoroscopic techniques have been applied based on Röntgen stereophotogrammetric methodologies: This is a sequential technique using biplanar views, and relying on a library of previously recorded 3D images to match the original fluoroscopic images using normalized contour matching to obtain coordinates. Radiation dosage, image distortion, low frame rate (8 frame/sec), and the need for complete 3D geometry are limitations, but accuracy is within +/− 1 mm.

33.6.4 Computer-Aided Tomography

CT used to create detailed and subject-specific 3D models of skeletal systems. It allows precise definition of local anatomically sound coordinate systems along with precise marker location determination. This is a computationally cumbersome requirement of such methods. CT can be used in conjunction with other kinematic measurement methods to determine discrete kinematics. It intrinsically allows relationships of anatomic geometry in local coordinate systems.

33.7 Goniometric Systems/Methods

All human joints have six degrees of freedom; therefore, they need three translational and three rotational measurements to be fully characterized. However, this is not always required or even useful depending on the application. Any joint motion can be described by relative motion between rigid members attached at that joint, so either the positions of the bones with respect to one another or their relative displacements can be analyzed. Positional data are usually measured, while displacements are usually calculated.

In its simplest form, goniometry is the direct measurement of joint angles. In this light, goniometers are angle-measuring devices. Goniometers are used to provide joint angle data to kinematic algorithms used to determine body posture. They can be either rigid (fig. 33.9) or flexible (fig. 33.10), but always are attached to or worn by the subject being measured. Most clinical goniometers are mechanical devices where the angular measurement is directly read from the device itself, while electrogoniometers (fig. 33.11) are electromechanical devices providing continuous data during movements. Electrogoniometers are essentially an exoskeleton attached to the subject. There is a general correspondence of the goniometer's joints to the joints of the user, and the angles recorded are used to estimate the joint angle portion of kinematics. Regardless of the type of goniometers used a number of issues can arise. The soft tissue of the body allows the position of the linkages relative to the body to change as motion occurs. Also, alignment with the body joints is difficult, especially highly multiple degree-of-freedom joints such as the shoulder. This can lead to cross-talk errors due to the goniometer axes not being aligned with the anatomical joint. Also, due to the wide anatomical and morphological differences in subjects, the systems need to be recalibrated for each user.

33.7.1 Quasi-Static Goniometry

Quasi-static, or clinical, goniometry (fig. 33.9) is the most basic form of joint angle measurement. A clinical goniometer usually is made of two rigid, nearly identical pieces that can rotate relative to one

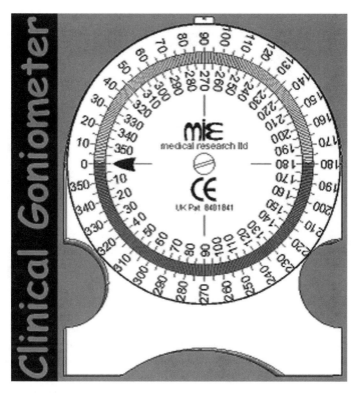

FIGURE 33.9 Example of a clinical goniometer from MIE Medical Research. (From MIE Medical Research, www.mie-uk.com. With permission.)

another about one common hinge point. Either one or both of the moving pieces are marked with angles that can be read directly. The device is manually held or temporarily attached to two segments of a joint; then the subject performs some dictated motion, and the angles viewed are manually recorded. This is very useful in the clinical setting to have a quantification of joint range of motion and is inexpensive and easy to use. However, it is obviously not well suited for dynamic activities.

FIGURE 33.10 Example of a flexible electrogoniometer from Noraxon. (From Noraxon, www.noraxon.com. With permission.)

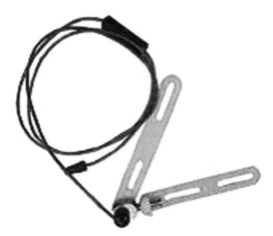

FIGURE 33.11 Example of an electro-mechanical goniometer from Noraxon. (From Noraxon, www.noraxon. com. With permission.)

33.7.2 Electrogoniometers

An electrogoniometer adds the capability of instrumented measurement to the clinical goniometer using a rotary displacement-measuring device such as a rotary potentiometer. Since each joint has multiple degrees of freedom, a number, or all of which may be of interest, more than one electrogoniometer is usually used at each joint creating a mechanical exoskeleton that is attached to the body of the subject (fig. 33.12). This allows for true real-time measurements without range limits, or occlusion and correspondence problems. Processing times are very fast, but movement can be restricted and sensor positions are fixed which can make the devices cumbersome and heavy to wear. Also, since these are electromechanical devices, mechanical components have a finite life, and electronics, including cabling, can wear out or be damaged through use.

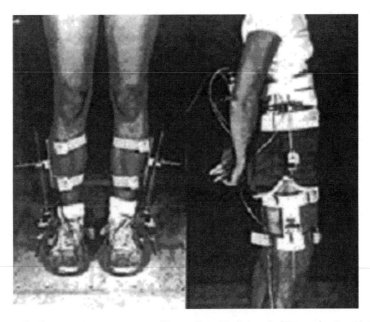

FIGURE 33.12 Early electrogoniometer system. (From Chao, Y., *Journal of Biomechanics*, 13, 989–1006, 1980. With permission.)

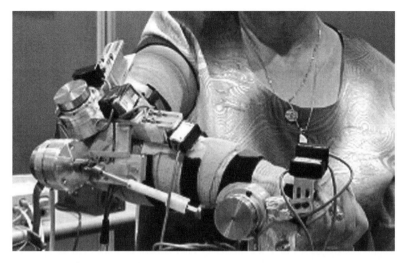

FIGURE 33.13 Example of an ISLD with inertial sensors. (From Brunetti, F., et al., "A new platform based on IEEE 802.15.4 wireless inertial sensors for motion caption and assessment," *Proceedings of the 28th International Conference of the IEEE/Engineering in Medicine and Biology Society*, IEEE, New York, 2006. With permission.)

One of the first electrogoniometers was created by Karpovich in 1960 to study joint motion during walking, but this device was limited to planar motion. In 1972 Kinzel measured human motion in 3D space and provided the mathematical formulation for treating and interpreting data from up to six degrees-of-freedom electrogoniometers. This device used a system of rotary and linear potentiometers on an external frame. A device such as this has since been referred to as an Instrumented Spatial Linkage Device (ISLD). ISLDs measure both rotational and translational motion using various measurement instruments. This allows more of the degrees of freedom of each joint to be measured and higher accuracy of these measurements. Today these systems are still limited to primarily planar motion and are best suited for quasi-static activities, although some systems have been developed that have been successfully used in full ROM dynamic testing (fig. 33.13).

33.7.3 Rotary Potentiometers

Potentiometers used in electrogoniometry are usually resistive-type voltage-dividing instruments. Resistive potentiometers have an electrical resistance element with a movable contact, called a wiper. In the case of rotary potentiometers (fig. 33.14), the contact motion is rotational with a range of a few degrees

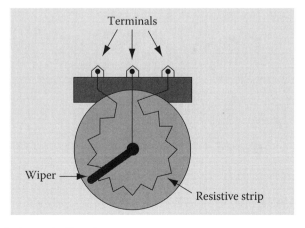

FIGURE 33.14 Typical rotary potentiometer.

to many full rotations. In order to measure a signal, the resistance element is provided an excitation voltage, usually DC, and the output is the voltage resulting from the potential difference at the location of the wiper contact with the resistance element. Rotary potentiometers are designed so that the output voltage has a linear relationship with the input angular displacement. The resolution of these devices can be very small, depending on how the device is made, and varies inversely with the diameter of the instrument. A larger diameter instrument allows for greater resolution. Dynamically, potentiometers are zero-order instruments; however, the inertia and friction of the potentiometer's moving parts do impose some mechanical loading.

33.7.4 Optical Encoders

All optical displacement measuring instruments have the distinct advantage of noncontacting operation with negligible force exerted on the measured object. Encoders have the distinct advantage of digital output without requiring the use of secondary electronics. Optical encoders utilize a small plastic disc that has been etched with graduated marks, or a code pattern, that rotates in response to input rotational motion (fig. 33.15). The optics inside the encoder usually consist of some type of light source mounted on one side of the rotary disc and a photodetector mounted opposing the light source on the other side of the disc. The photodetector records the movement of the disc as interruptions of the light source. This is then translated into the predetermined binary equivalent, which in turn can be converted to its equivalent rotational displacement. Since these devices are digital, the resolution of the device is limited by the number of bits it uses to represent motion. However, the size and spacing of the code pattern also dictates resolution. The resolution is generally presented as the number of counts per revolution, which increases as the size of the instrument increases. Resolutions of many thousands of counts per revolution are readily available (with 360 counts per revolution being equivalent to 1 degree resolution).

33.7.5 Linear Transducers

Multiple methods of linear displacement devices have been used in human motion studies, including linear potentiometers, string potentiometers, and linear variable-differential transformers (LVDT). Linear potentiometers operate in the same manner as rotary potentiometers; however, the contact motion is linear and its stroke length limits the range of the instrument (fig. 33.16). The available stroke lengths range from sub-inch to more than 20 inches, while the resolution can be as small as 0.001 inches.

A string potentiometer (string pot) is a special type of rotary potentiometer that has a flexible wire that winds around a spool. The wiper is connected to the spool, and as the wire is pulled into or out of the instrument the spool rotates thus changing the resistance. This output voltage has been calibrated to vary with the linear displacement of the string rather than the rotational displacement of the wiper. LVDTs consist of a sinusoidally excited primary coil and two secondary coils with the same frequency,

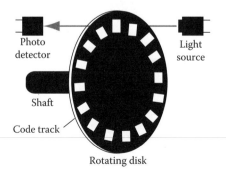

FIGURE 33.15 Typical optical encoder. (From Society Conference, National Instruments, www.ni.com. With permission.)

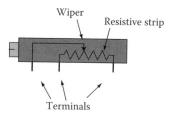

FIGURE 33.16 Typical linear potentiometer.

but amplitudes vary depending upon the location of the movable central core (fig. 33.17). This core element is iron and moves linearly through the armature. By observing the changes in the two secondary coil amplitudes, the core's position within the armature can be determined, thereby determining the displacement. LVDTs full stroke ranges are much less than that of potentiometers, on the order of a few inches, but due to the measurement method used the resolution is infinitesimal.

33.8 Inertial Sensing Systems

Another method of measuring human motion employs the use of inertial measuring instruments, or accelerometers. This has been made practical with advances in miniaturized and micro-machined sensor technologies, such as micro-electromechanical (MEM). Accelerometers measure linear accelerations directly, and then these data are integrated to get velocity and again integrated to get displacements. Due to inherent error progation that occurs with the integration process, and drift that occurs due to a non-zero fluctuating offset, long-term displacement tracking using accelerometers alone is impractical. Although accelerometers have been used to measure body segment inclination and in human activity of daily living assessment, alone they provide a poor quality estimate of inclination during large acceleration movements. Only when the acceleration measured is small compared to gravity can an accelerometer be used accurately as an inclinometer. This is because accelerometers have fluctuating offset due to their relative position and sensitivity to gravity. Their small size, low power requirements, relatively low expense, robustness, ease of attachment to the human body, ability to provide accurate readings without inherent latency, lack of occlusion issues, and the ability to have unsupervised monitoring of human motion make their use attractive to many.

In order to use kinematic data for estimating joint loading using inverse dynamics, both the orientation and the angular velocity must be known. In order to do this, rate gyroscopes, simply referred to as gyros by most, are used to measure angular velocity. The gyro data can then be integrated to find angular displacement. Inertial measurement units (IMUs) are devices used to measure the orientation of human body segments with a combination of three single-axis (or one three-axis) accelerometers and three single-axis gyroscopes. Again, the integration process itself introduces drift errors due to the non-zero offset of both instruments. These errors increase over time. A Kalman filter has been used to take into account the spectra of the signals from accelerometers and gyroscopes as well as the fluctuating

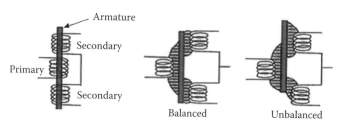

FIGURE 33.17 Typical LVDT operation.

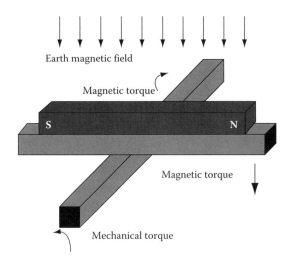

FIGURE 33.18   Typical operation of a magnetometer.

offsets to reduce the overall error in estimating the body segment orientations. In designing this type of filter, each data source is appropriately weighted using knowledge about the signal characteristics based on their theoretical models in order to optimally utilize the data from all sensors. Using such a filter it is reported that the drift can be reduced to 0.5 deg/second.

The errors encountered when deriving linear and angular displacements from inertial sensors can be minimized, and compensation for gravity effects can be implemented with use of additional complementary sensors. An example of such a system is an attitude and heading reference system (AHRS). In this system a combination of rate gyroscopes, accelerometers, and magnetometers (fig. 33.18) are used as complements to eliminate drift by continuous correction of the orientation obtained from angular rate sensor data.

The magnetometer senses absolute 2D orientation using the earth's magnetic field, similar to a compass. The combination of data from all three devices simultaneously allows for accelerometer orientation

FIGURE 33.19   Example of a full-body inertial sensing system. (From Xsens Motion Technology, www.xsens. com. With permission.)

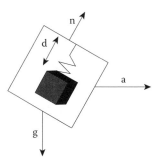

FIGURE 33.20 Single-axis accelerometer.

correction and when combined with the Kalman filter lower integration drift errors. With the use of MEM technology, these systems are becoming more readily available and much more convenient to the end user including full body suits with multiple AHRSs that allow for fully dynamic motion (fig. 33.19).

33.8.1 Accelerometers

Single-axis accelerometers have a mass suspended by a spring in a housing (fig. 33.20). Accelerations will cause movement of the mass. Then the acceleration can be measured according to Hooke's law and Newton's second law. Multi-axis accelerometers simply apply this theory to perpendicular axes (fig. 33.21). MEM accelerometers contain silicon beams that deform in the presence of acceleration. This deformation changes the capacitance, which is in turn processed as a voltage corresponding to the applied acceleration.

33.8.2 Gyroscopic Instruments

Gyroscopes are instruments used to measure angular motion by sensing the Coriolis forces when rotations are applied. There are three basic types: (1) mechanical, (2) optical, and (3) magnetohydrodynamic (MHD). Mechanical gyroscopes operate on the basis of conservation of angular momentum by sensing the change in direction of angular momentum. These types of gyroscopes use gimbals or lasers and are therefore not suitable for human motion analysis due to large size and cost. Micro-electromechanical machined (MEMS) inertial sensors have become more available. These use a vibrating resonator that, when rotated, is subjected to the Coriolis effect that causes secondary vibration orthogonal to the original vibrating direction (fig. 33.22). Sensing this secondary vibration allows detection of the rate

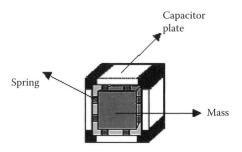

FIGURE 33.21 3-axis accelerometer.

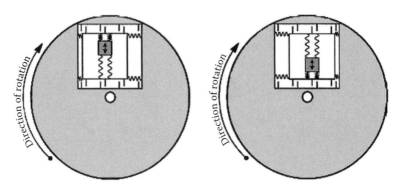

FIGURE 33.22   MEM rate gyroscope operation. (From Analog Devices, www.analog.com. With permission.)

of turn. With an MHD angular rate sensor angular motion about the sensitive axis results in a relative velocity difference between the highly conductive fluid proof mass and the static magnetic field applied normally (fig. 33.23). This relative velocity difference produces an electric potential that is measured. A mechanical gyro can measure constant input rates (including DC), while MHD devices cannot measure DC and usually have bandwidths of 1 to 1,000 Hz. These are ideal for human movement studies due to their low power requirements, small size, and low cost.

33.9 Kinetic Measurement

Biomechanical models are often used to take the kinematic measurement data and determine some degree of kinetic variables. Two fundamental types of models exist: linked segment and distributed. A linked segment model uses external exposure data (both kinematic and kinetic) and anthropometric data to calculate net joint moments and forces. The body is represented as a set of articulated links in a kinetic chain; intersegmental moments and forces are calculated from forces measured either up or down the chain. In a distributed model, electromyographical (EMG) data are additionally used to apportion the net forces and moments to muscles and passive structures. Most models that have been developed and used extensively have been 2D quasi-static, but as computing power and technology have advanced so too have the models to now include 3D dynamic models as well. Biomechanical

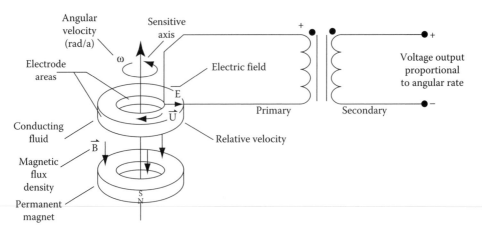

FIGURE 33.23   MHD angular rate sensor. (From Applied Technology Associates, www.aptec.com. With permission.)

models are useful to assess internal exposures, but are only estimates to external exposures. The most precise method to determine internal exposure is through direct measurement. This is not always practical or even safe internally, but often can be incorporated externally during testing. Direct measurement of external loading is currently the only way to accurately assess the forces exerted on the human body.

33.10 External Loads

33.10.1 Load Cells and Dynamometers

A dynamometer is generally a device for measuring mechanical force or torque, and is often used interchangeably with the terms load transducer and load cell. The force applied to a body is measured by measuring its effect on that body. Many types of devices exist that accomplish this task from simple balances to intricate semiconductors. Balances operate on the principle that an object's mass can be measured by balancing it against a known mass in the presence of gravity. The physical effect of an applied load can also be measured using the elastic deformation of a body. Almost any of the available methods can be used for static or quasi-static measurements; however, when dynamic measurements are to be taken, then the method of measuring deformations must be employed.

Elastic transducers can be modeled ideally as a mass supported by a spring and damper, and are therefore second-order systems. The system natural frequency is proportional to the ratio of the spring stiffness to the mass, while the sensitivity is inversely proportional to the spring stiffness. Both the sensitivity and the natural frequency of a sensor are important to match to the application, but when considering using an elastic transducer this essentially dictates the stiffness of the device. The simplest elastic transducers measure one load component, while up to six-component transducers can be obtained. Six-component transducers measure the three orthogonal forces as well as the moment about each axis.

The technology utilized to measure the deformation under an applied load drives the differences between various load cells. Strain-gauge transducers apply either metal-foil or semiconductor gauges directly to the deformable member to measure the strain caused by a deformation (fig. 33.24). Each gauge has a resistance that changes as the gauge's length is changed. Therefore, when the gauge is attached to a deformable body, any deformation of the body also causes deformation (i.e., a resistance change) of the strain gauge. By applying a constant input current to the gauge, any change in gauge length is seen as a change in the voltage produced across that gauge proportional to the strain induced. The gauges are designed to be sensitive only to length changes along a principal direction and are wired in a Wheatstone bridge configuration (fig. 33.25). This configuration allows isolated measurement of a single load component, along with compensation for any temperature-related deformations. The design trade-off with use of strain-gauged transducers involves the sensitivity to natural frequency inverse proportionality. In order to compensate for this inherent incompatibility, secondary electronics are normally employed that allow for stiffer transducers that are sensitive due to the improved signal-to-noise ratio characteristics.

Piezoelectric materials, such as quartz, can also be utilized in load transducers. Piezoelectric transducers utilize materials that produce an electrical charge that is directionally dependent when deformed (fig. 33.26). When the material is placed between metal electrodes and a force is applied to the material, a charge is produced that is proportional to the material's deformation. This charge is then also directly proportional to the applied force, and the output voltage can be calibrated to measure the force applied. Through careful orientation of the material in the transducer body, individual components of the load can be isolated. The advantage of piezoelectric transducers is their high sensitivity with high stiffness. The disadvantage is that they are not well suited for static or quasi-static measurements due to the charge produced under any constant load decaying to zero. These systems are first-order systems; therefore, in order to compensate for the inherent charge decay under constant load, the time constant must

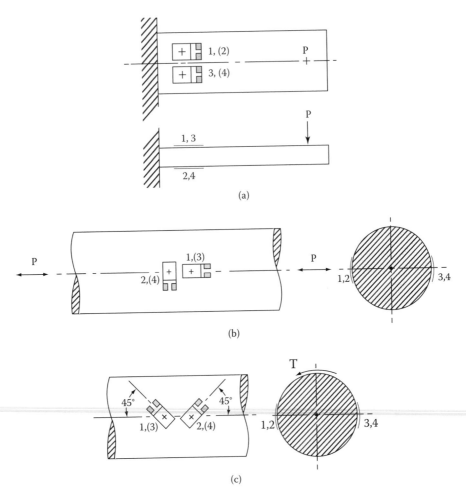

FIGURE 33.24 Typical strain gauge load cell designs: (a) cantilever beam, (b) axial load, (c) torque. (From Vishay Measurements Group, www.vishay.com. With permission.)

be increased. The time constant is equal to the resistance times of the capacitance. The capacitance is inherent to the material, while increasing the total resistance in the system in secondary electronics can increase the resistance and therefore the time constant. However, increasing the resistance also acts to decrease the sensitivity of the transducer. Therefore, when using piezoelectric transducers the trade-off is between duration of adequate static measurement and sensitivity.

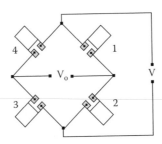

FIGURE 33.25 Wheatstone bridge.

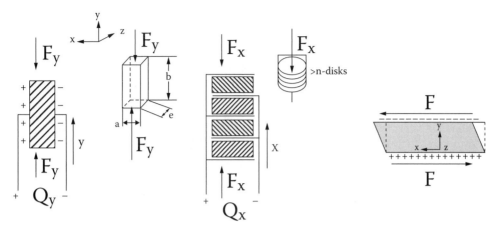

FIGURE 33.26 Various types of piezoelectric transducers. (From Kistler Instruments, www.kistler.com. With permission.)

33.11 Internal Loads

The *in vivo* measurement of internal loads involves inherent technical as well as practical and even ethical issues. Early work completed by Rydell in 1966 using strain-gauged hip prostheses, and Frankel in 1971 using an instrumented hip nail with biotelemetry to record proximal femur loads showed the remarkable potential of performing *in vivo* measurements along with the sometimes extreme difficulties. There is also often a high cost-to-benefit ratio in developing such measurement capabilities. However, such devices have been used to show peak forces throughout the musculoskeletal system including the joints, bones, muscles, and ligaments during a variety of tasks. They have also been used to assess or validate theoretical joint loading modeling efforts, provide accurate data to assist in orthopedic decision making, obtain loading and boundary conditions for implant-to-bone interface fixation assessment, and evaluate bone stress and strain states during remodeling and repair.

33.11.1 Stress-Strain Measurement

The stress and strain can be measured in rigid deformable bodies, such as bone, as well as soft tissues such as muscle, tendons, and ligaments. The stress is a measure of the internal forces present, while the strain is a measure of the deformation of the body. Unfortunately, since there are such large differences between the tissue properties and the anatomic geometric constraints, different types of transducers usually have to be developed for each specific measurement desired.

Muscles, tendons, and ligaments are soft tissue that only carry a tensile force. Therefore, the transducers developed for use with these tissues either involve clamping or suturing directly to the tissue of interest. The output of the measurement devices is then carried outside of the skin using fine wires. The application of such measurements has been limited primarily to isometric activities, and the use *in vivo* has been mainly limited to intra-operative settings where there is already an invasive procedure taking place.

Strain-gauge techniques have been employed to measure the strain in bone *in vivo*. Although theoretically simple, applying strain gauges directly to bone is a technically challenging task. Unfortunately, in order for the gauges to adhere well enough to the bone to measure the true deformations present, the surface of the bone must be defatted and dried. The chemicals used to do this cause osseous necrosis and result in a relatively short amount of time available to take any reliable measurements prior to gauge bond failure. In this light, this method has almost entirely been restricted to short-term animal studies. Instrumented plates and screws have also been utilized to measure load sharing between bones across joints. Use of plates and screws allows longer-term measurements including the changes in loading

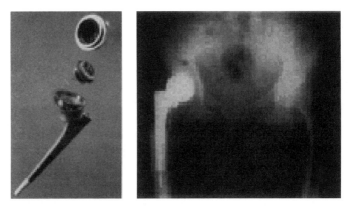

FIGURE 33.27   Instrumented hip prosthesis. (From Hodge, W., et al., *Proceedings of the National Academy of Science—Biophysics*, 83, 2879–83, 1986. With permission.)

conditions as bone remodeling occurs. In addition telemetry systems can be applied that preclude the need for connective wires.

33.11.2 Joint Force Instruments

Through the work of Rydell in the 1960s and Frankel in the early 1970s, *in vivo* joint contact forces were measured in the hip. Later in the 1980s Kilvington and Goodman instrumented a total hip prosthesis and took single-axis *in vivo* measurements for over a month postoperatively. In 1984 Bergmann showed that long-term instrumented prostheses work well *in vivo* in animals. Many others have since developed and tested *in vivo* instrumentation at the hip, knee, and shoulder among other areas (figs. 33.27 and 33.28).

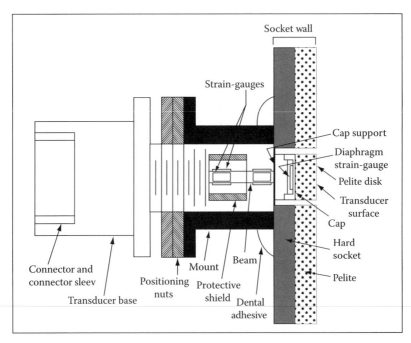

FIGURE 33.28   Example of instrumented knee for below-knee prosthetic limb. (From Sanders, J., C. Daly, and E. Burgess, *Journal of Rehabilitation Research*, 29(4), 1–8, 1992. With permission.)

33.11.3 Intra-Abdominal Instruments

The issues with directly measuring loads in the spine have led to indirect methods such as intra-abdominal pressure measurement. The basic theory behind these measurements is that the spinal pressures can be directly inferred from the measured intra-abdominal pressures. In the 1950s multiple researchers found that intragastric pressures increased when trunk flexion moment increased (Davis, 1956; Bartelink, 1957). These researchers used rubber balloons to measure intra-abdominal pressure. A pressure-sensitive wireless radio pill was introduced in the early 1960s that simply requires the subject to swallow the pill. A strain-gauged pressure transducer catheter used to measure internal abdominal pressure that is either swallowed or inserted intrarectally has also been used. By the early 1970s Stubbs used these measurements in the workplace and along with Davis has attempted to correlate the measured intra-abdominal pressures to safe levels of manual forces. These methods are safe and simple with little discomfort, but the pressure can change voluntarily simply by contracting abdominal muscles. Moreover, the relationship between measured pressure and spinal loading still is uncertain except when sagittal plane symmetry can be assumed.

33.11.4 Intradiscal Instruments

Intradiscal pressure measurement is the most direct and reliable method to assess spinal loading, and has been used in labs for many postures and activities. However, it is also the least practical and is invasive, risky to the subject, and has only been used in very controlled lab settings and primarily *in vitro* (fig. 33.29). In 1960 Nachemson inserted a membrane-covered liquid-filled needle connected to a pressure transducer into the intervertebral disc and measured spinal pressures while the disc was loaded compressively. Nachemson found the pressure to be approximately the same in all three principal stress directions so the nucleus was hydrostatic. He also found that the stresses were about 1.5 times the

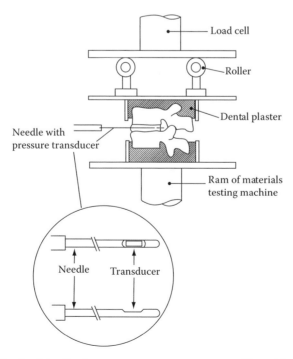

FIGURE 33.29 Example of an intradiscal pressure measurement test setup. (From Adams, M. A., D. S. McNally, and P. Dolan, *Journal of Bone and Joint Surgery*, 78-B(6), 965, 1996. With permission.)

applied load per unit of cross-sectional area. Additional experiments throughout the 1970s used a semi-conductor strain-gauge transducer needle to compare intradiscal pressures under various loading conditions including torsion, compression, tension, bending, and shear. This technique has allowed for the validation of biomechanical models in calculating the internal disc loading during dynamic events.

33.11.5 Muscle Force

There is no generally applicable method available to directly measure that muscles are acting and the magnitude of each muscle's exertion during human movement. Many have used modeling to estimate or predict muscle patterns from calculated net joint forces and moments; however, it is statically not possible to determine an individual muscle's contribution from these calculations due to the infinite number of muscle forces with the same net result. Although there is no method currently available to determine exact muscle force contribution from different synergistically contracting muscles, electromyography has been used for over a century as a tool to evaluate and estimate both individual and group muscle activity.

33.11.6 Electromyography (EMG)

Muscles are responsible for active movement; therefore, any full understanding of forces causing or contributing to movements must include muscle contributions. In 1792 Galvani showed that electricity could initiate muscle contractions. Marey became the first to record muscle activity in 1890 and introduced the term *electromyography* (EMG). EMG was originally used to search for abnormal action potentials (fibrillation or fasciculation) and to test for delays in nerve conduction velocity. Inman introduced modern EMG in 1944 along with 3D force and energy measurements in studying normal and amputee walking. Since that time EMG has been used extensively to evaluate posture and postural strain, among other specific applications.

The principles of EMG are a combination of muscle physiology and neurology. Muscles contract in response to motor neuron impulses. The impulses cause depolarization of the motor end-plate membrane, resulting in muscle action potentials and activation of muscle fibers. Highly sensitive EMG electrodes can measure these potentials, and therefore the muscle activation signal. This signal is then representative of muscle tension. So, EMG signals are used to predict muscle tension from myoelectric signals based on the theory that increase in muscle tension is proportional to myoelectric activity. However, only when a muscle acts isometrically has the relationship between EMG signal and muscle tension been shown to be linear. Using secondary signal conditioning equipment (analog) or software (digital), the raw EMG voltage output can then be analyzed more meaningfully.

Two basic quantification routines are generally used in EMG studies: continuous mean voltage EMG and "true" integrated EMG. The continuous mean voltage EMG (AvgEMG) is essentially equal to the RMS value times a constant coefficient. The integrated EMG (iEMG) is the area within the potential changes, which is equivalent to the integration of the average signal. The iEMG is usually expressed as an electrical output. When analyzing EMG signals the level of excitation is considered to be constant when the signal fluctuates randomly about a stable value. Researchers have developed relationships between iEMG and various biomechanical quantities to characterize mechanical muscle performance. However, these relationships have only been validated under very limited experimental conditions, and even under these conditions they are not a measure of muscle force. EMG has also been used to evaluate muscle fatigue with the theory that amplitude of signal increases and frequency decreases with fatigue, but these relationships are also widely unvalidated.

Even though true muscle force cannot be measured with EMG, and the relationships of EMG data to muscle tension are not always predictable, without EMG data only the net moment across a joint is determined. Two basic ways of attaining EMG data are used: surface (fig. 33.30) and fine wire (figs. 33.31 and 33.32). With surface EMG electrodes that are contained in pads are adhered on the skin of a subject over the muscle or group of muscles of interest. Fine-wire EMG involves directly placing the electrode

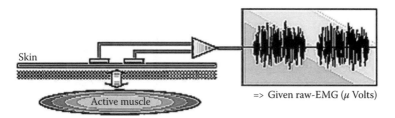

FIGURE 33.30 Typical surface EMG setup and raw output. (From Konrad, P., *The ABC of EMG*, Noraxon, Scottsdale, AZ, 2005. With permission.)

into the muscle fibers, and is therefore somewhat invasive. However, if the muscles are small, deep, or part of a larger group, then fine-wire EMG must be used. Otherwise it is difficult to compare relative muscle activity between two different postures due to changes in muscle length and surface electrode shift. Regardless of whether surface or fine-wire EMG is used, normally a bipolar electrode technique is employed. Signal errors can arise due to cross-talk between muscles and electromagnetic fields. The signals can also be affected by the electrode geometry, the distance between active fibers and the electrode, and the location of the muscle of interest. Finally, due to the very small (few micros to few millivolts) and low-power signal, there are inherent noise problems with data collection that must be overcome with secondary electronics.

33.12 Combined Methods

Most of the methods and instruments described thus far can be, and often are, used in combination with each other or other data in order to determine both kinematic and kinetic variables of interest. Some examples have already been discussed, including using biomechanical models to predict internal loading from kinematic studies, and inertial sensing systems used to estimate both kinematic and kinetic variables. There are a couple of other specific examples that should be discussed individually: kinesiological EMG (kEMG) and arthrometers.

33.12.1 Kinesiological Electromyography

A specific area of EMG analysis used to determine relationships of muscle activation signal (EMG) to joint movement and overall body motion is called kinesiological EMG (kEMG) (fig. 33.33). These relationships are needed to determine a single muscle's contribution to total joint reaction. This method usually requires use of fine-wire EMG electrodes, thereby requiring more channels of data collection than surface EMG and expertly trained medical staff. Since the internal kinetic reactions during movement are desired, kEMG also requires accurate external force and movement data. Even though the contribution of passive stretch on muscle tension is not measured, kEMG has been used quite successfully to analyze neuromuscular control.

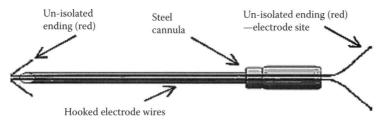

FIGURE 33.31 Typical fine wire EMG. (From Konrad, P., *The ABC of EMG*, Noraxon, Scottsdale, AZ, 2005. With permission.)

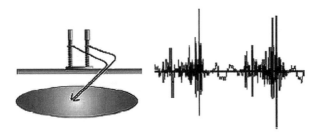

FIGURE 33.32 Typical fine wire application and output. (From Konrad, P., *The ABC of EMG*, Noraxon, Scottsdale, AZ, 2005. With permission.).

33.12.2 Arthrometers

Arthrometers are instrumented devices specifically designed to load and measure movement in response to this applied load at a particular joint in the body. Therefore, arthrometers measure both kinetics and kinematics simultaneously. Arthrometers have been shown to be highly accurate and robust during quasi-static planar *in vitro* tests, but they tend to be very cumbersome and limiting *in vivo*. These devices inherently limit and to some extent control the range of motion of a joint, and therefore the data collected often does not reflect what happens during normal dynamic full range of motion activities. Most commonly these devices have been developed and used for the ankle, but others have worked on developing knee, elbow, finger, and even shoulder arthrometers.

Ankle arthrometers are usually used to quantify anterior-posterior load-displacement and inversion-eversion laxity characteristics (fig. 33.34). Three rotational and three translational axes are measured using an instrumented spatial linkage or electrogoniometer system. The foot is typically clamped in

FIGURE 33.33 Application of kinesiological EMG. (From Konrad, P., *The ABC of EMG*, Noraxon, Scottsdale, AZ, 2005. With permission.)

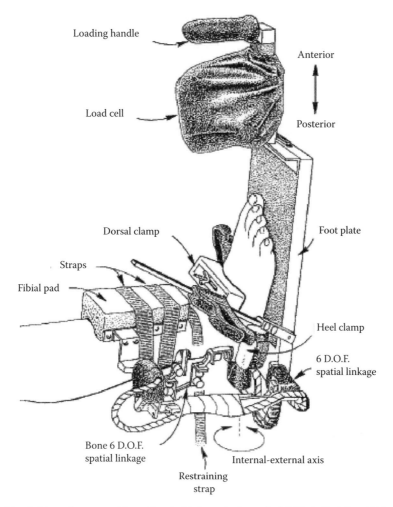

FIGURE 33.34 Ankle arthrometer. (From Kovaleski, J., et al., *Journal of Athletic Training*, 37(4), 467–74, 2002. With permission.)

frame with a reference point that is established and secured somewhere above the foot, usually on the anterior aspect of the tibia. Force is applied through some type of loading device attached to the footplate that in turn loads the foot. The force and/or torque applied are then directly measured using load cells, and the ISLD measures the relative movement between the footplate and the reference point.

References

Aarås, A., & Ro, O. (1997). Electromyography (EMG): methodology and application in occupational health. *International Journal of Industrial Ergonomics*, 20, 207–214.

Andersson, G., & Örtengren, R. (1974). Myoelectric back muscle activity during sitting. *Scandinavian Journal of Rehabilitation Medicine*, 3(Suppl.), 73.

Andersson, G., Örtengren, R., & Nachemson, A. (1977a). Quantitative electromyographic studies of back muscle activity related to posture and loading. *Orthopaedic Clinical North America*, 8, 85–96.

Andersson, G., Örtengren, R., & Nachemson, A. (1977b). Intradiskal pressure, intra-abdominal pressure and myoelectric back muscle activity related to posture and loading. *Clinical Orthopaedics*, 129, 156–164.

Andrews, B., & Harwin, W. (1984). An online multichannel electromagnetic goniometer for measuring hand movement. In *Proceedings 2nd International Conference of Rehabilitation Engineering,* 511–512.

Appoldt, F., Bennett, L., & Contini, R. (1970). Tangential pressure measurement in above-knee suction sockets. *Bulletin on Prosthetic Research,* 10(13), 70–86.

Bachmann, E., McGhee, R., Yun, X., & Zyda, M. (2001). Inertial and magnetic posture tracking for inserting humans into networked virtual environments. *In Proceedings of the ACM Symposium on Virtual Reality Software and Technology,* 9–16.

Barnes, S.Z., Morr, D.R., Oggero, E., Pagnacco, G., & Berme N. (1997). The realization of a haptic (force feedback) interface device for the purpose of angioplasty surgery simulation. In Benghuzzi and Bajpai (Eds.), *Biomedical Sciences Instrumentation* (pp. 19–24), Instrument Society of America, Research Triangle, NC.

Barshman, B., & Durrant-Whyte, H. (1995). Inertial navigational systems for mobile robots. *IEEE Transactions Robotics and Automation,* 11, 328–342.

Bartelink, D. (1957). The role of abdominal pressure in relieving the pressure on the lumber intervertebral discs. *Journal of Bone and Joint Surgery,* 39B, 718–725.

Bartsch, A.J., Bolte IV, J.H., Litsky, A.S., Herriott, R.G., McFadden, J.D. (2006). Application of anthropomorphic test device crash test kinetics to post mortem human subject lower extremity testing. Society of Automotive Engineers, 2006-01-0251.

Bartsch, A. J., Morr, D. R., Tanner, C. B., Wiechel, J. F. (2005). Using MADYMO to determine occupant kinematics for accident reconstruction purposes. Presented at the 6th MADYMO Users' Meeting of the Americas, Novi, MI.

Baselli, G., Legnani, G., Franco, P., Brognoli, F., Marras, A., Quaranta, F., & Zappa, B. (2001). Assessment of inertial and gravitational inputs to the vestibular system. *Journal of Biomechanics,* 34, 821–826.

Bergmann, G., Graichen, F., & Rohlmann, A. (1993). Hip joint loading during walking and running measured in two patients. *Journal of Biomechanics,* 26(8), 969–990.

Bergmann, G., Graichen, F., Siraky, J., Jendrzynski, H., & Rohlmann, A. (1988). Multichannel strain gauge telemetry for orthopaedic implants. *Journal of Biomechanics,* 21, 169–176.

Berme, N., & Cappozzo, A. (Eds.) (1990). *Biomechanics of Human Movement.* Worthington, OH: Bertec Corporation.

Berme, N., & Morr, D. (1999). Dynamometers. In J. Webster (Ed.), *Wiley Encyclopedia of Electrical and Electronics Engineering* (pp. 100–107). New York: John Wiley & Sons.

Bigland, B., & Lippold, O. (1954). The relation between force, velocity, and integrated electrical activity in human muscles. *Journal of Physiology,* London, 123, 214–244.

Bolte IV, J.H., Hines, M.H., Herriott, R.G., McFadden, J.D., & Donnelly, B.R. (2003). Shoulder impact response and injury due to lateral and oblique loading. *Stapp Car Crash Conference Journal,* 47, 35–54.

Bolte IV, J.H., Hines, M.H., McFadden, J.D., & Saul, R.A. (2000). Shoulder response characteristics and injury due to lateral glenohumeral joint impacts. *Stapp Car Crash Conference Journal,* 44, 261–280.

Boonstra, M., van der Slikke, R., & Keijsers, N. (2006). The accuracy of measuring the kinematics of rising from a chair with accelerometers and gyroscopes. *Journal of Biomechanics,* 39, 354–358.

Bouisset, S. (1973). EMG and force in normal motor activities. In J.E. Desmedt (Ed.), *Electromyography and Clinical Neurology* (pp. 547–583). Karger, Basel.

Brown, R., Davy, D., Heiple, K., Kotzer, G., Heiple, K. Jr., Barilla, J., Goldberg, V., & Burstein, A. (1985). In vivo load measurements on a total hip prosthesis. *Transactions of the 31st ORS Annual Meeting,* 10, 283.

Bull, A., & Amis, A. (1998). Knee joint motion: Description and measurement. *Proceedings of the IMechE,* 212(H), 357–372.

Burdorf, A., & Laan, J. (1991). Comparison of methods for the assessment of postural load on the back. *Scandinavian Journal of Work Environmental Health*, 17, 425–429.

Cappozzo, A., Catani, F., Della Croce, U., & Leardini, A. (1995). Position and orientation of bones during movement: anatomical frame definition and determination. *Clinical Biomechanics*, 10, 171–178.

Cappozzo, A., Leo, T., & Pedotti, A. (1975). A general computing method for the analysis of human locomotion. *Journal of Biomechanics*, 8, 307–320.

Carlson, C., Mann, R., & Harris, W. (2004). A look at the prosthesis-cartilage interface: Design of a hip prosthesis containing pressure transducers. *Journal of Biomedical Materials Research*, 8(4), 261–269.

Carter, D., Vasu, R., Spengler, D., & Dueland, R. (1981). Stress fields in the unplated and plated canine femur calculated from in vivo strain measurements. *Journal of Biomechanics*, 14, 63–70.

Cavagna, G. (1975). Force platforms as ergometers. *Journal of Applied Physiology*, 39, 174–179.

Chaffin, D. (1969). A computerized biomechanical model—development of and use in studying gross body actions. *Journal of Biomechanics*, 2, 429–441.

Chaffin, D., & Anderson, G. (1991). *Occupational Biomechanics*. New York: John Wiley.

Chao, E. (1978). Experimental methods for biomechanical measurement of joint kinematics. In *CRC Handbook of Bioengineering in Medicine and Biology*, Vol. 1 (pp. 385–411). West Palm Beach, FL: CRC Press.

Chao, Y. (1980). Justification of triaxial goniometer for the measurement of joint rotation. *Journal of Biomechanics*, 13, 989–1006.

Clarys, J. (1994). Electrology and localized electization revisited. *Journal of Electromyography and Kinesiology*, 4, 5–14.

Convery, P., & Buis, A. (1998). Conventional patellar-tendon-bearing socket/stump interface dynamic pressure distributions recorded during the prosthetic stance phase of gait of a trans-tibial amputee. *Prosthetics and Orthopaedics International*, 22(3), 193–198.

Cooney, W., An, K., & Chao, E. (1986). Direct measurement of tendon forces in the hand. *Transactions of the 32nd ORS Annual Meeting*, 11, 53.

Cunningham, B., Kotani, Y., McNully, P., Cappuccino, A., & McAfee, P. (1997). The effect of spinal destabilization and instrumentation on lumbar intradiscal pressure: An in vitro biomechanical analysis. *Spine*, 22(22), 2655–2663.

Davis, P. (1956). Variations of the intra-abdominal pressure during weight lifting in various postures. *Journal of Anatomy*, 90, 601.

Davis, P. (1981). The use of intra-abdominal pressure in evaluating stresses on the lumbar spine, *Spine*, 6, 90–92.

Davis, R., Ounpuu, S., Tyburski, D., & Gage, J. (1991). A gait analysis data collection and reduction technique. *Human Movement Science*, 10, 575–587.

Davis, P., & Troup, J. (1964). Pressures in the trunk cavities when pulling, pushing and lifting. *Ergonomics*, 7, 465–474.

Demes, B., Qin, Y.-X., Stern, J., Larson, S., & Clinton, R. (2001). Patterns of strain in the macaque tibia during functional activity. *American Journal of Physical Anthropology*, 116(4), 257–265.

Doeblin, E.O. (1990). *Measurement Systems Application and Design*, 4th Edition. New York: McGraw-Hill.

Eberhart, H. (1947). *Fundamental Studies of Human Locomotion and Other Information Relating to Design of Artificial Limbs*. Berkeley: University of California.

Edixhoven, P., Huiskes, R., De Graaf, R., Van Rens, T., & Slooff, T. (1987). Accuracy and reproducibility of instrumented knee-drawer tests. *Journal of Orthopaedic Research*, 5(3), 378–387.

Eie, N., & When, P. (1962). Measurements of the intra-abdominal pressure in relation to weight bearing of the lumbo-sacral spine. *Journal of the Oslo City Hospital*, 12, 205.

Elliot, B., & Blanksby, B. (1976). A cinematographic analysis of overground and treadmill running by males and females. *Medical Science in Sports and Exercise*, 8, 84–87.

Farris, M.V., Wiechel, J.F., & Guenther, D.A . (1987). Dynamic response of the part 572 neck. Proceedings, Canadian Multidisciplinary Road Safety Conference V, Calgary, Alberta, Canada, June 1–3.

Fleischer, A., & Becker, G. (1986). Free hand-movements during the performance of a complex task. *Ergonomics*, 29, 49–63.

Fleischer, A., & Lange, W. (1983). Analysis of hand movements during the performance of positioning tasks. *Ergonomics*, 26, 555–564.

Foxlin, E. (1996). Inertial head-tracker sensor fusion by a complementary separate-bias Kalman filter. Proceedings of the VRAIS.

Frankel, V., Burstein, A., Brown, R., & Lygre, L. (1971). Biotelemetry from the upper end of the femur. *Journal of Bone and Joint Surgery*, 53A, 1023.

Frigo, C. (1990). Three-dimensional model for studying the dynamic loads on the spine during lifting. *Clinical Biomechanics*, 5, 143–152.

Furnée, E. (1967). Hybrid instrumentation in prosthetics research. *7th ICMBE*, Stockholm, 446.

Giansanti, D., & Maccioni, G. (2005). Comparison of three different kinematic sensor assemblies for locomotion study. *Physiological Measurement*, 26, 689–705.

Goldman, J., & Nadler, G. (1956). Electronics for measuring human motions. *Science*, 124, 807–810.

Gottlieb, G., & Agarwal, G. (1971). Dynamic relationship between isometric muscle tension and the electromyogram in man. *Journal of Applied Physiology*, 30, 345–351.

Granata, K., & Bennett, B. (2005). Low-back biomechanics and static stability during isometric pushing. *Human Factors*, 47(3), 536–549.

Greenwald, A., & Haynes, D. (1972). Weight-bearing areas in the human hip joint. *Journal of Bone and Joint Surgery*, 54B, 157–163.

Hamilton, M., Wiechel, J.F., & Guenther, D.A. (1986). Development of a Child Lateral Thoracic Impactor. Passenger Comfort, Convenience and Safety: Test Tools and Procedures, Publication No. P-174, Society of Automotive Engineers.

Hanavan, E. (1964). A mathematical model of the human body. Report No. AMRL-TR-102, Wright-Patterson Air Force Base, OH.

Heinrich, I., Otun, E., & Anderson, J. (1988). Reproducibility of surface electromyogram and intra-abdominal pressure for use in ambulatory monitoring. *Ergonomics*, 31, 1821–1835.

Heller, M., Bergmann, G., Deuretzbacher, G., Durselen, L., Pohl, M., Claes, L., Haas, N., & Duda, G. (2001). Musculo-skeletal loading conditions at the hip during walking and stair climbing. *Journal of Biomechanics*, 34(7), 883–893.

Hines, M.H., Schmalbrock, P., Baker III, P.B., & Bolte IV, J.H. (2001). Comparison of autopsy, x-ray, and M.R.I. findings following a low speed impact to the shoulder. *Proceedings, Association for the Advancement of Automotive Medicine*, 45, 215–240.

Hodge, W., Carlson, K., Fijan, R., Burgess, R., Riley, P., Harris, W., & Mann, R. (1989). Contact pressures from an instrumented hip endoprosthesis. *Journal of Bone and Joint Surgery*, 71(9), 1378–1386.

Hodge, W., Fijan, R., Carlson, K., Burgess, R., Harris, W., & Mann, R. (1986). Contact pressures in the human hip joint measured in vivo. *Proceedings of the National Academy of Science—Biophysics*, 83, 2879–2883.

Hof, A. (1984). EMG and muscle force: an introduction. *Human Movement Science*, 3, 119–153.

Inman, V., Ralston, H., deSaunder, J., Feinstein, B., & Wright, E. (1952). Relation of human electromyogram to muscle tension. *EEG Clinical Neurology*, 4, 187–194.

Inman, V., Ralston, H., & Todd, F. (1981). *Human Walking*. Baltimore: Williams & Wilkins.

Jasty, M., Lew, W., & Lewis, J. (1982). In vivo ligament forces in the normal knee using buckling transducers. *Transactions of the 28th ORS Annual Meeting*, 7, 241.

Kabada, M., Ramakrishnan, H., & Wooten, M. (1990). Measurement of lower extremity kinematics during level walking. *Journal of Orthopaedic Research*, 8, 383–392.

Karpovich, P., Herden, E., & Asa, M. (1960). Electrogoniometric study of joints. *United States Armed Forces Medical Journal*, 11, 424–450.

Kattan, A., & Nadler, G. (1969). Equations of hand motion path for work space design. *Human Factors*, 11, 123–130.

Kinzel, G., Hall, A., & Hillberry, B. (1972a). Measurement of the total motion between two body segments. I. Analytical development. *Journal of Biomechanics*, 5(1), 93–105.

Kinzel, G., Hillberry, B., Hall, A., Sickle, A., & Harvey, W. (1972b). Measurement of the total motion between two body segments. II. Description and Application. *Journal of Biomechanics*, 5, 283–293.

Kirstukas, S., Lewis, J., & Erdman, A. (1992a). 6R instrumented spatial linkages for anatomical joint motion measurement—part 1: Design. *Journal of Biomechanical Engineering*, 114(1), 92–100.

Kirstukas, S., Lewis, J., & Erdman, A. (1992b). Instrumented spatial linkages for anatomical joint motion measurement—part 2: Calibration. *Journal of Biomechanical Engineering*, 114(1), 101–110.

Kleissen, R., Buurke, J., Harlaar, J., & Zilvold, G. (1998). Electromyography in the biomechanical analysis of human movement and its clinical application. *Gait and Posture*, 8, 143–158.

Koh, S.W., Cavanaugh, J.M., Mason, M.J., Peterson, S.A., Marth, D.R., Rouhana, S.W., & Bolte IV, J.H. (2005). Shoulder injury and response due to lateral glenohumeral joint impact: An analysis of combined data. *Stapp Car Crash Conference Journal*, 49, 291–322.

Konrad, P. (2005). *The ABC of EMG: A Practical Introduction to Kinesiological Electromyography*. Noraxon, USA.

Kovaleski, J., Hollis, M., Heitman, R., Gurchiek, L., & Pearsall, A. (2002). Assessment of ankle-subtalar-joint-complex laxity using an instrumented ankle arthrometer: an experimental cadaveric investigation. *Journal of Athletic Training*, 37(4), 467–474.

Kummer, F., Lyon, T., & Zuckerman, J. (1996). Development of a telemeterized shoulder prosthesis. *Clinical Orthopaedics and Related Research*, 330, 31–34.

Lanyon, L. (1976). The measurement of bone strain in vivo. *Acta Orthopaedics Belgium*, 42, Supplement 1, 98.

Lee, S.G., Wiechel, J.F., Morr, D.R., Ott, K.A., & Guenther, D.A. (2006). Response of neck muscles to rear impact in the presence of bracing. SAE 2006-01-2369, *Proceedings of the 2006 Digital Human Modeling for Design and Engineering Conference*, Society of Automotive Engineers.

Leskinen, T. (1985). Comparison of static and dynamic biomechanical models. *Ergonomics*, 28, 286–291.

Li, G., & Buckle, P. (1999). Current techniques for assessing physical exposure to work-related musculoskeletal risks, with emphasis on posture-based methods. *Ergonomics*, 42(5), 674–695.

Liu, W., Siegler, S., & Techner, L. (2001). Quantitative measurement of ankle passive flexibility using an arthrometer on sprained ankles. *Clinical Biomechanics*, 16, 237–244.

Luinge, H., & Veltnik, P. (2004). Inclination measurement of human movement using a 3-D accelerometer with autocalibration. *IEEE Transactions on Neural Systems and Rehabilitation Engineering*, 12(1), 112–121.

Luinge, H., & Veltnik, P. (2005). Measuring orientation of human body segments using miniature gyroscopes and accelerometers. *Medical and Biological Engineering and Computing*, 43, 273–282.

Marey, E. (1873). *La Machine Animale: Locomotion Terrestre*. Paris: Balliere.

Marey, E. (1885). *Development de la Methode Graphique par l'Emploi de la Photographic*.

Marey, E. (1894). *Le Mouvement*. Paris: Masson.

Marras, W., Fathallah, F., Miller, R., Davis, S., & Mirka, G. (1992). Accuracy of a three-dimensional lumbar motion monitor for recording dynamic trunk motion characteristics. *International Journal of Industrial Ergonomics*, 9, 75–87.

Mathie, M., Celler, N., Lovell, N., & Coster, A. (2004). Classification of basic daily movements using a triaxial accelerometer. *Medical and Biological Engineering and Computing*, 42, 679–687.

Mayagoitia, R., Nene, A., & Veltnik, P. (2002). Accelerometer and rate gyroscope measurement of kinematics: An inexpensive alternative to optical motion analysis systems. *Journal of Biomechanics*, 35, 537–542.

McGill, S., & Norman, R. (1987). Reassessment of the role of intra-abdominal pressure in spinal compression. *Ergonomics*, 30, 1565–1588.

McNally, D., Adams, M., & Goodship, A. (1992). Development and validation of a new transducer for intradiscal pressure measurement. *Journal of Biomedical Engineering*, 14(6), 495–498.

Moreno, J., Rocon, E., Ruiz, A., Brunetti, F., & Pons, J. (2005). Design of an inertial measurement unit for gait kinematics sensing in a lower leg orthosis. *Eurosensor XIX*.

Morr, D.R. (1996). Design of a Quantitative System for Assessing Static Human Balance. Thesis for Master of Science Degree, Ohio State University, OH.

Morr, D., Wiechel, J., & Bartsch, A. (2006a). Head impacts associated with daily living activities: Diagnosis versus injury potential. In *Proceedings of the 5th World Congress of Biomechanics*.

Morr, D., Wiechel, J., & Bartsch, A. (2006b). Utilization of MADYMO to determine and verify occupant kinematics, kinetics and injury mechanisms during a real world collision. In *Proceedings of the 5th World Congress of Biomechanics*.

Morr, D.R., Wiechel, J.F., Tanner, C.B., & Bartsch, A.J. (2006). Review of the measurement of linear center of gravity and angular head accelerations. In *Proceedings of the 5th World Congress of Biomechanics*.

Morris, J. (1973). Accelerometry, a technique for the measurement of human body movement. *Journal of Biomechanics*, 6, 729–736.

Morris, J., Lucas, D., & Bresler, B. (1961). Role of the trunk in stability of the spine. *Journal of Bone and Joint Surgery*, 43A, 327.

Morrison, J. (1968). Bioengineering analysis of force actions transmitted by the knee joint. *Biomedical Engineering*, 3, 164–170.

Muybridge, E. (1887). *Complete Human and Animal Locomotion*. Dover Publishers.

Nachemson, A. (1960). Lumbar intradiscal pressure. *Acta Orthopaedics Scandinavia*, 43(Supp. 1), 1–104.

Nachemson, A., & Elfstrom, G. (1970). Intravital dynamic pressure measurements in lumbar discs. *Scandinavian Journal of Rehabilitation Medicine*, Suppl. 1, 3–38.

Nachemson, A., & Morris, J. (1964). In vivo measurements of intradiscal pressure. *Journal of Bone and Joint Surgery*, 46A, 1077–1092.

Nordin, M., Elfström, G., & Dahlquist, P. (1984). Intra-adominal pressure measurements using a wireless radio pill and two wire-connected pressure transducers: A comparison. *Scandinavian Journal of Rehabilitation Medicine*, 16, 139–146.

Nunamaker, D., & Perren, S. (1976). Force measurements in screw fixation. *Journal of Biomechanics*, 9, 669–675.

O'Brien, C., & Paradise, M. (1976). The development of a portable non-invasive system for analyzing human movement. In the *Proceedings of the 6th Congress of the International Ergonomics Association* (pp. 390–392).

O'Connor, J., Goodship, C., Rubin, C., & Lanyon, L. (1981). The effect of externally applied loads on bone remodeling in the radius of sheep. In A.F. Stokes (Ed.), *Mechanical Factors and the Skeleton* (pp. 83–90), London: John Libbey.

Padgaonkar, A., Krieger, K., & King, A. (1975). Measurement of angular acceleration of a rigid body using linear accelerometers. *Journal of Applied Mechanics*, 42(3), 552–556.

Pagnacco, G., Oggero, E., Morr, D.R., Barnes, S.Z., & Berme, N. (1997). The mechanics of drop landing on a flat surface—a preliminary study. In Benghuzzi and Bajpai (Eds.), *Biomedical Sciences Instrumentation* (pp. 53–58), Instrument Society of America, Research Triangle, NC.

Pagnacco, G., Oggero, E., Morr, D.R., & Berme, N. (1998a). High resolution acquisition of force plate data. *Proceedings of the NACOB*.

Pagnacco, G., Oggero, E., Morr, D.R., & Berme, N. (1998b). Oversampling data acquisition to improve resolution of digitized signals. *Biomedical Sciences Instrumentation*, 34, 137–142.

Pagnacco, G., Oggero, E., Morr, D.R., & Berme, N. (1998c). Probability of single contact during gait. In *Proceedings of the NACOB*.

Pagnacco, G., Oggero, E., Morr, D.R., & Berme, N. (1998d). Average power spectral density of physiological tremor in normal subjects. *Biomedical Sciences Instrumentation*, 34.

Pagnacco, G., Oggero, E., Morr, D.R., & Berme, N. (1998e). Tremor in normal subjects: Spectrum characteristics. In *Proceedings of the NACOB*.

Pagnacco, G., Oggero, E., Morr, D.R., & Berme, N. (1998f). Probability of valid data acquisition using currently available force plates. *Biomedical Sciences Instrumentation*, 34, 392–397.

Patel, V., Hall, K., Ries, M., Lotz, J., Ozhinsky, E., Lindsey, C., Lu, Y., & Majumdar, S. (2004). A three-dimensional MRI analysis of knee kinematics. *Journal of Orthopaedic Research*, 22, 283–292.

Paul, J. (1974). Quantitative analysis of locomotion using television. In World Congress ISPO.

Paul, J.P. (1967). Forces transmitted by joints in the human body. *Proceedings of the Institution of Mechanical Engineers*, 181(3J), 8.

Pfau, T., Witte, T., & Wilson, A. (2005). A method for deriving displacement data during cyclical movement using an inertial sensor. *Journal of Experimental Biology*, 208, 2503–2514.

Polga, D., Beaubien, B., Kallemeier, P., Schellhas, K., Lew, W., Butterman, G., & Wood, K. (2004). Measurement of in vivo intradiscal pressure in healthy thoracic intervertebral discs. *Spine*, 29(12), 1320–1324.

Raab, F., Blood, E., Steiner, T., & Jones, H. (1979). Magnetic position and orientation tracking system. *IEEE Transactions Aerospace Electronic Systems*, AES15(5), 709–718.

Ralston, H., Todd, F., & Inman, V. (1976). Comparison of electrical activity and duration of tension in the human rectus femoris muscle. *Electromyography in Clinical Neurophysiology*, 16, 277–286.

Rhoad, R., Klimkiewicz, J., Williams, G., Kesmodel, S., Udupa, J., Kneeland, J., & Iannotti, J. (1998). A new in vivo technique for three-dimensional shoulder kinematics analysis. *Skeletal Radiology*, 27, 92–97.

Roetenberg, D. (2006). Inertial and magnetic sensing of human motion. Ph.D. Thesis, University of Twente.

Rydell, N. (1966). Forces acting on the femoral head-prosthesis: a study on strain gauge supplied prostheses in living persons. *Acta Orthopedica Scandinavia Supplement*, 88, 1–132.

Sanders, J., Daly, C., & Burgess, E. (1992). Interface shear stresses during ambulation with a below-knee prosthetic limb. *Journal of Rehabilitation Research*, 29(4), 1–8.

Sato, K., Kikuchi, S., & Yonezawa, T. (1999). In vivo intradiscal pressure measurement in healthy individuals and in patients with ongoing back problems. *Spine*, 24(23), 2468.

Selvik, G. (1990). Roentgen stereophotogrammetric analysis. *Acta Radiology*, 31, 113–126.

Selvik, G., Alberius, P., & Aronson, A. (1983). A Roentgen stereophotogrammetric system: Construction, calibration and technical accuracy. *Acta Radiology Diagnostics*, 24(4), 34–352.

Sheehan, F., Zajac, F., & Drace, J. (1998). Using cine phase contrast magnetic resonance imaging to noninvasively study in vivo knee dynamics. *Journal of Biomechanics*, 31, 21–26.

Siegler, S., Wang, D., Plasha, E., & Berman, A. (1994). Technique for in vivo measurement of the three-dimensional kinematics and laxity characteristics of the ankle joint complex. *Journal of Orthopaedic Research*, 12, 421–431.

Stansfield, B., Nicol, A., Paul, J., Kelly, I., Graichen, F., & Bergmann, G. (2003). Direct comparison of calculated hip joint contact forces with those measured using instrumented implants: An evaluation of a three-dimensional mathematical model of the lower limb. *Journal of Biomechanics*, 36(7), 920–936.

Sutherland, D. (2001). The evolution of clinical gait analysis—part I: Kinesiological EMG. *Gait and Posture*, 14, 61–70.

Sutherland, D. (2002). The evolution of clinical gait analysis—part II: Kinematics. *Gait and Posture*, 16, 159–179.

Tanner, C.B., Wiechel, J.F., & Guenther, D.A. (2001). Vehicle and occupant response in heavy truck to passenger car sideswipe impacts. Paper 2001-01-0900, Accident Reconstruction-Crash Analysis, SP-1572, 251–265, Society of Automotive Engineers.

Tashman, S., & Anderst, W. (2003). In-vivo measurement of dynamic joint motion using high speed biplane radiography and CT: Application to canine ACL deficiency. *Transactions of the ASME*, 125, 238–245.

Taylor, S., Gorjon, J., & Walker, P. (1999). An instrumented prosthesis for knee joint force measurement in vivo. *IEE Colloquium—Innovative Pressure, Force and Flow Measurements*, 89, 6.

Taylor, S., Walker, P., Perry, J., Cannon, S., & Woledge, R. (1998). The forces in the distal femur and the knee during walking and other activities measured by telemetry. *Journal of Arthroplasty*, 13(4), 428–437.

Thies, S., Tresadern, P., Kenney, L., Howard, D., Goulermas, J., Smith, C., & Rigby, J. (2006). Comparison of linear accelerations from three measurements during reach and grasp. Centre for Rehabilitation and Human Performance Research, University of Salford.

Tranberg, R., & Karlsson, D. (1998). The relative skin movement of the foot: a 2-D Roentgen photogrammetry study. *Clinical Biomechanics*, 13, 71–76.

Van der Beek, A., & Frings-Dresen, M. (1998). Assessment of mechanical exposure in ergonomic epidemiology. *Occupational and Environmental Medicine*, 55, 291–299.

vanDijk, R., Huiskes, R., & Selvik, G. (1979). Roentgen stereophotogrammetric methods for the evaluation of the three dimensional kinematic behaviour and cruciate ligament length patterns of the human knee joint. *Journal of Biomechanics*, 12, 727–731.

Van Sint Jan, S., Salvia, P., Hilal, I., Sholukha, V., Rooze, M., & Clapworthy, G. (2002). Registration of 6-DOF electrogoniometry and CT medical imaging for 3D joint modeling. *Journal of Biomechanics*, 35(11), 1475–1484.

Veltnik, P., Bussmann, H., & DVries, W. (1996). Detection of static and dynamic activities using uniaxial accelerometers. *IEEE Transactions on Rehabilitation Engineering*, 4, 375–385.

Watson, B., Ross, B., & Kay, A. (1962). Telemetering from within the body using pressure-sensitive radio pill. *Gut*, 3, 181.

Wiechel, J., & Bolte, J. (2006). Response of reclined post mortem human subjects to frontal impact. Paper 2006-01-0674, Society of Automotive Engineers.

Wiechel, J.F., & Guenther, D.A. (1989). Two-dimensional thoracic modeling considerations. SAE Transactions, 890605, *Journal of Passenger Cars*, 98(6), 700–717.

Wiechel, J.F., MacLaughlin, T.F., & Guenther, D.A. (1993). Head impact reconstruction—HIC validation and pedestrian injury risk. Paper 930895, Society of Automotive Engineers.

Wilke, H.-J., Neef, P., Caimi, M., Hoogland, T., & Lutz, C. (1999). New in vivo measurements of pressures in the intervertebral disc in daily life. *Spine*, 24(8), 755–762.

Winter, D. (1972). Television-computer analysis of kinematics of human gait. *Computational Biomedical Research*, 5, 498–504.

Winter, D., Quanbury, A., & Reimer, G. (1976). Analysis of instantaneous energy of normal gait. *Journal of Biomechanics*, 9, 253–257.

Woods, J., & Bigland-Ritchie, B. (1983). Linear and nonlinear surface EMG/force relationships in human muscles. *American Journal of Physical Medicine*, 62(6), 287–292.

Zahalak, G., Duffy, J., Stewart, P., Hawley, R., & Paslay, P. (1976). Partially activated human skeletal muscle: An experimental investigation of force, velocity, and EMG. *Journal of Applied Mechanics*, 98(1), 81–86.

Zhou, H., Huosheng, H., & Tao, Y. (2006). Inertial measurements of upper limb motion. International Federation for Medical and Biological Engineering.

Zhu, R., & Zhou, Z. (2004). A real-time articulated human motion tracking using tri-axis inertial/magnetic sensors package. *IEEE Transactions Neural Systems Rehabilitation Engineering*, 12, 295–302.

Zuniga, E., & Simons, D. (1969). Nonlinear relationship between averaged electomyogram potential and muscle tension in normal subjects. *Archives of Physical Medicine*, 50, 613–620.

34

Instrumentation for Evaluating Effective Human-Computer System Design

Jack T. Dennerlein and
Peter W. Johnson

34.1 Introduction

Desktop and laptop computer systems have become ubiquitous in home, school, and work environments. These systems' fundamental components are a display monitor, a keyboard, and pointing device whose use is dependent upon the appropriate use and adjustment of furniture systems. These components can affect the efficiency of the human-computer system. The proper component configuration and use can help maintain the health and performance of the computer operator. Therefore, effective human-computer systems design for applied ergonomics and human factors requires quantifying aspects of several interface components, including performance, usage patterns, postural congruency, applied forces, and muscle load. Quantifying these interface components brings together several instrumentation technologies, including embedded software, electrogoniometers, inclinometers, strain gauges, and psychophysical instruments to quantify factors pertaining to human-computer interaction. These instrumentation technologies have been utilized in several laboratory and a few field studies that have explored and contributed to effective human-computer system designs.

34.2 Defining Effective Human-Computer System Design

Effective human-computer system design takes on many different meanings depending upon the audience. Our definition is based on fundamental ergonomic definitions: matching task demands with users' capabilities with the goal of optimizing performance and worker well-being. Performance includes time to completion and error reduction. Performance (and production cost) has been a major driver in the current design of input devices; however, a major public health issue for workers' well-being would include measuring risk factors for computer-related illnesses, specifically musculoskeletal disorders (MSDs). MSDs can be highly prevalent among computer users (Gerr et al., 2002, 2004, 2006). Hence our definitions of effective human-computer system design incorporate both optimizing performance (both through software and hardware) and reducing risk for injury among users (primarily through hardware).

Prevalence of symptoms among computer users has been reported to be as high as 50% among computer workers (Gerr et al., 2002) and as high as 60% among college students (Schlossberg et al., 2004; Hupert et al., 2004). Epidemiological studies have linked many external exposures with health outcomes. Physical factors defined by these studies include repetition, force, awkward posture, direct pressure, and vibration (NRC/IOM, 2001). Using a computer includes many of these risk factors. These risk factors drive many of the measures of effective computer system design.

Epidemiological studies have consistently linked chronic musculoskeletal disorders (MSDs) to prolonged use of computers (Gerr et al., 2006). For example, Faucett and Rempel (1994) and Bergqvist et al. (1995a and b) link an increase in the number of hours spent working on VDTs with an increased risk of upper extremity musculoskeletal disorders. Matias et al. (1998) also found that the percentage of the day spent working at a VDT was a leading factor for the existence of MSDs.

Many studies have examined postural aspects associated with MSDs (Punnett & van der Beek, 2000). Sauter et al. (1991) reported arm discomfort increased with increases in keyboard height above elbow level, a surrogate for posture. A large-scale, longitudinal epidemiological study demonstrated that important relationships exist between certain postural aspects of an individual sitting at a workstation and the development of MSDs (Marcus et al., 2002).

Studies have linked forceful repetitive exertions to upper extremity musculoskeletal disorders (e.g., Luopajärvi et al., 1979; McCormack et al., 1990). For the submaximal conditions associated with computer work, there is evidence that force plays a role (Wahlström, 2005). In a cross-sectional laboratory study, Feuerstein et al. (1997) observed that people with symptoms of musculoskeletal disorders of the upper extremity applied larger forces to the keyboard than people without any musculoskeletal disorder of the upper extremity. Similarly, Pascarelli and Kella (1993) reported that 26% of the injured computer users in their study had a forceful keying style. Marcus et al. (2002) also observed higher incidence for those who type on a keyboard requiring more activation force.

34.3 Integrated Usage Monitors

Since duration is the most consistent risk factor associated with computer-related MSD, measuring usage via integrated software is a promising new technology for effective computer interaction (Gerr et al., 2006). Integrated usage monitors are software programs loaded on a personal computer that runs in the background environment of the operating system software. The program measures pointing device movement and the timing and duration of keyboard and pointing device events. These can be stored in a file locally on the host computer and then through the Internet collected at a central location (e.g., Ijmker et al., 2006). Total usage time can be estimated from the input devices by assuming that interactions such as viewing and reading the computer monitor (brief input device inactivity or idle periods) are bounded in time by input device usage. Usage time is then the sum of all of these input device activity periods along with the bounded idle periods (usually idle periods less than 30 seconds; Blangsted et al.,

2004). Usage monitors are available commercially. We have observed a specific relationship between daily computer usage and the report of MSD symptoms among college students (Chang et al., 2007). Increased usage within a 24-hour computer period was related to increased reporting of MSDs.

34.3.1 Keyboard and Typing Speeds

Since usage monitors record the timing of keystrokes, the rate of text entry into a computer can be calculated easily from the usage monitor's data; however, the calculations need to incorporate and account for errors, delete, and backspace keys (http://www.yorku.ca/mack/RN-TextEntrySpeed.html). Taking the inverse of the period between keystrokes as recorded by a monitor provides an instantaneous typing speed. In addition, the duration of individual keystrokes can be measured with millisecond accuracy (Chang et al., 2004), which may change with exposure to use and muscle fatigue (Chang, 2008).

34.3.2 Pointing Device Performance

For the mouse movement, the monitor can measure distance and time for each movement of the pointing device. The monitor can then evaluate many performance criteria for the pointing device (ISO, 1998). For example, performance can be quantified by information capacity theory proposed by Fitts (1954). Fitts' law (Equation 34.1) was derived and modeled after the motor performance when subjects performed reciprocal tapping tasks (see fig. 34.1) and is given by:

$$MT = a + b \log_2 (A/W) \qquad (34.1)$$

where MT is the movement time, A is the distance from the starting point to the center of the target, W is the width of the target measured along the axis of motion, and a and b are empirical constants typically determined by fitting a straight line to measured data. Fitts proposed that the term $\log_2 (A/W)$ defined the index of difficulty for the task, ID. The ISO 9241 Part 9 (ISO, 1998) recommends at least five different types of movements be tested when evaluating pointing devices. Fitts also defined the ratio of the ID/MT as the index of performance for a task (IP). The units for ID and IP are bits and bits per second. The ISO 9241 Part 9 (ISO, 1998) calls this index of performance the throughput of a pointing device. (Dennerlein & Yang, 2001; Dennerlein & Dimarino, 2006) saw an increase (improvement) in the index of performance with the addition of force-feedback mouse.

The Fitts index of performance as stated above determined the users' performance based on the accuracy of the task requirements, rather than the true accuracy of the user. Douglas, et al. (1999) describe an effective target width as $W_e = 4.33 * SD$, where SD is the standard deviation of the true locations of the

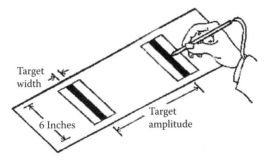

FIGURE 34.1 The one-dimensional reciprocal tapping paradigm (Fitts, 1954), where the target amplitude is A and the target width is W. (From Fitts, P. M., *Journal of Experimental Psychology*, 47, 381–91, 1954. With permission.)

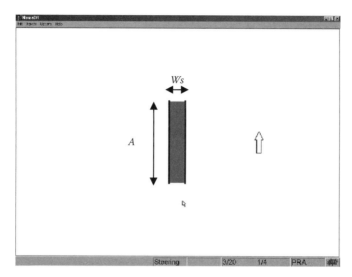

FIGURE 34.2   Fitts Law for steering task. The cursor must be moved from one end of the tunnel to the other end, amplitude (length) A, staying within the tunnel's accuracy (width) W_s.

taps about their mean location. Thus, Douglas et al. (1999) propose an effective index of performance that relies on the measured speed (MT) and the measured accuracy (W_e) of the movement. Dennerlein and DiMarino (2006) observed that force feedback in a pointing device improves the performance of a user by reducing the effective target width.

Pointing device tasks are not limited to point-and-click but often include tracking a path and maintaining the cursor within a certain boundary during the tracking tasks. As such this is known as steering the cursor along a path. Accot and Zhai (1997) proposed a modified Fitts Law to measure the performance of steering tasks. They proposed a measure of an index of difficulty similar to Fitts (1954). The index of difficulty for a simple tunnel steering task (fig. 34.2) is simply ID = A/W_s where ID is the index of difficulty, A is the amplitude or length of the required movement through the tunnel and the W_s is the width of the tunnel. Dennerlein et al. (2000) evaluated the performance of the steering task with and without force feedback mouse. Similar to Fitts's task, results from the tunnel tasks indicate task times are proportional to the ID.

34.4 Posture Congruency

With respect to ergonomics, the goal for effective human interface design is to promote natural and neutral postures that reduce or minimize the internal tissue loads of the muscles, tendons, connective tissues, and fluid pressures in body compartments (e.g., the hydrostatic pressure in the carpal tunnel). Measuring and reporting postures require standard nomenclature and definitions. Two organizations provide standardization for joint angle measurements, The American Academy of Orthopaedic Surgeons (Greene & Heckman, 1994) and the International Society of Biomechanics (www.isbweb.org).

Measurement of posture involves two concepts: relative joint angle and orientation of a musculoskeletal segment in 3D space. Relative joint angle is the orientation of a distal segment relative to its proximal and adjacent segment. Since neutral postures typically refer to the orientation of a distal segment relative to its proximal segment, relative segment/joint angles are the most often reported measure used when evaluating ergonomic interface design. When joint torques are needed, then segments orientations relative to ground are needed to calculate the effects of gravity and accelerations of the segments (Winter, 2005).

34.4.1 Relative Joint Angles Using Goniometers

Goniometers are used extensively to measure relative joint angles. Manual goniometers come in various sizes and shapes depending on the joint angle of interest (fig. 34.3). They are used extensively throughout the physical and occupational therapy fields to assess range of motion of joints for assessing progress during rehabilitation.

For computer assessment, manual goniometers have been used to measure several joint angles ranging from the fingers to the neck, back, hip, and knees. For example, in order to characterize working postures in a group of computer operators, Ortiz et al. (1997) relied on manual goniometers to measure the posture of the whole upper extremity, head and neck, and torso angles. Through using these measurement techniques, Marcus et al. (2002) described several postural factors that were associated with increased risk of developing computer-related musculoskeletal disorders. The major limitation of manual goniometers in many ergonomic studies, however, is that they do not capture dynamic motion and are difficult to use to measure range of motion of the joints during a task.

Electrogoniometers (fig. 34.4) provide instrumentation to measure both static (angles) and dynamic components (velocities and accelerations) of posture. Serina et al. (1999) demonstrated that dynamic accelerations of the wrist measured with electrogoniometers are quite high, of the order that have been deemed as high risk for MSDs (Schoenmarklin et al., 1994).

Electrogoniometers utilize two types of technology, either optical or strain-gauge-based sensors, to measure the bend in a media spanning the joint of interest. Optical goniometers are useful for small

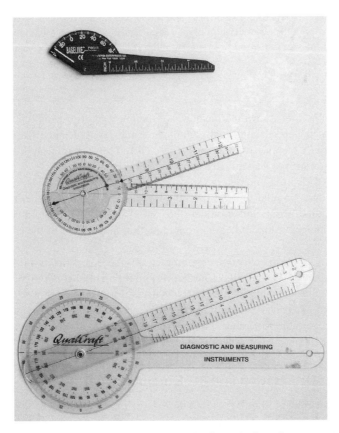

FIGURE 34.3   Manual goniometers for small joints such as the finger (top), medium-size joints such as the wrist (middle), and large joints such as the shoulder, hip, and knee.

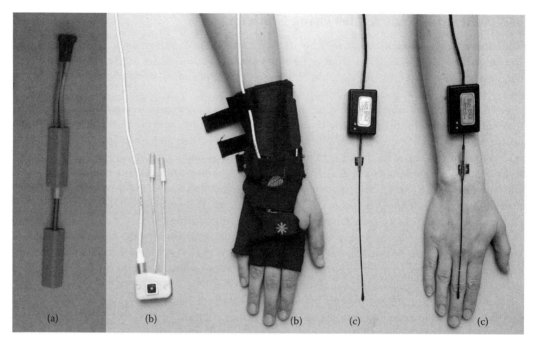

FIGURE 34.4 Electrogoniometers goniometers for the wrist (a and b) and the finger (c). The first system (a) is a bi-axial electrogoniometer (Biometrics, Gwent, UK) and the second (b) has two sensors that fit into the glove to keep the sensors aligned with the two wrist joint centers of movement (Greenleaf, Mountain View, CA). The shape sensor (c) on the right (Measurand, http://www.measurand.com/) utilizes fiber-optic technology to measure the finger angles.

joints and have been used to study the kinematics of the finger joints during typing and tapping (Jindrich et al., 2004a and b; Balakrishnan et al., 2006). These authors used Shape Sensors manufactured by Measurand Inc. (http://www.measurand.com/). These optical sensors are usually designed to measure one degree of freedom, such as flexion/extension of a joint.

Strain-gauge-based electrogoniometers can measure one or two axes of movement about a joint. Therefore, they are quite popular to measure the joint angles of joints with multiple degrees of freedom, such as the thumb, wrist, and shoulder. Jonsson et al. (2007) utilized a two-axes goniometer to measure the flexion/extension and abduction/adduction movements of the thumb during mobile phone use.

An important issue for the use of two-axes goniometers is the factors that influence the measurement and the congruency between the joint axes of rotation and the alignment of the goniometers. When the two items are not well matched, measurement errors occur. Jonsson and Johnson (2001) demonstrated that two different goniometric systems can provide different results when movements are not near the neutral position (fig. 34.5). The difference can be attributed to how the two designs align differently with the wrist's axes of rotation.

34.4.2 Segment Orientation and 3D Kinematics

There are several instrument technologies used for measuring orientation of body segments relative to an external reference frame. These include inclinometers, electromagnetic motion analysis systems, and optical-marker tracking motion analysis systems. Most of these technologies are easily interfaced with a computer and can record single postural measurements or can continuously record data from the sensors for kinematic and dynamic analysis.

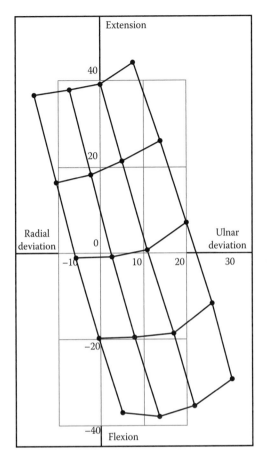
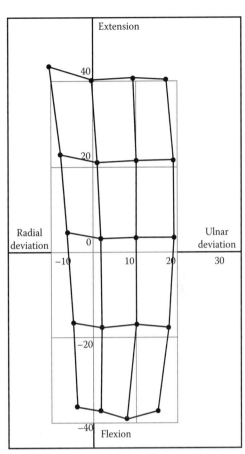

FIGURE 34.5 Error between the measured and actual wrist angles for two differently designed two-axes goniometers. (Reprinted from Jonsson and Johnson (2002) with permission.) The system on the right was designed to align with the axes of rotation for the wrist and hence has better accuracy. (From Jonsson, P., and P. W. Johnson, *Applied Ergonomics*, 32, 599–607, 2001. With permission.)

Inclinometers provide a measure of a body segment's orientation relative to gravity. Since gravity can create torques around joints supporting the segment, nonvertical orientations are considered to create larger torques about that joint. Inclinometers usually measure just one general angle relative to gravity; however, multiple inclinometers can be configured to get multiple angles relative to gravity. Dennerlein and Johnson (2006a and b) used a system consisting of two inclinometers mounted orthogonally in the horizontal plane to measure the flexion and abduction of the upper arm relative to gravity. The virtual corset (MicroStrain, Williston, Vermont) also uses two orthogonal inclinometers to measure trunk flexion and lateral bending of workers (Johnson et al., 2007). Inclinometers are also used to measure head angles when working at the computer (Airaksinen et al., 2005).

Electromagnetic sensor systems, using the power and strength of a local magnetic field induced a transmitter (e.g., Flock of Birds, Ascension Technology, Burlington, Vermont) can measure 3D joint position and motion. The x, y, z and three-orientation angles of the sensor are measured in reference to the reference frame of the transmitter. Through mounting multiple sensors on the body, the orientation and position of the segment can be recorded (e.g., Dennerlein and Johnson, 2006a and b for the upper arm, and Assink et al., 2005 for the head).

Finally, the position of either infrared (active) or reflective (passive) markers placed on the body can be tracked through a camera-based system. There are several commercially available products (e.g., Motional Analysis, Santa Rosa, CA, and Northern Digital (NDI) Waterloo, Ontario) that track the x,

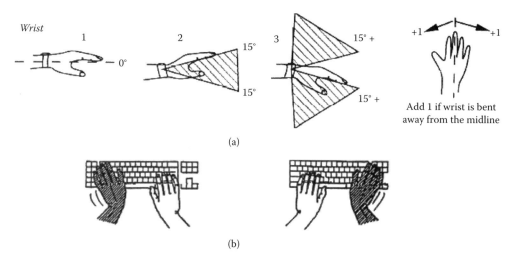

(a)

(b)

FIGURE 34.6 Two figures from two observational tools (a is from RULA and b is from the Berkeley Checklist) to assess wrist posture for human-computer interactions. The modified RULA figure assess elbow. (Reprinted from McAtamney and Corlett (1993) and Janowitz et al. (2002) with permission.)

y, z position of a single marker through the use of a series of cameras. Through triangulation, the position of each marker can be detected by the location on the camera's image. If three markers are placed on a segment, then the three markers can be used to determine the orientation of the body relative to the reference frame of the cameras. Winter (2005) explains the kinematic transformation for marker systems quite well. For computer interface, Dennerlein et al. (2007) explore the dynamics and the joint powers for the whole upper extremity while tapping on a computer keyswitch. Active marker systems have the advantage of reporting back to the camera in a known sequence whereas passive markers have to be identified and defined by the user before their images can be analyzed with the motion capture software.

34.4.3 Observational Methods to Measure Postural Congruency

In addition to directly measuring postural congruency, two methods exist for assessing general postural aspects of the human-computer interaction, a modified Rapid Upper Limb Assessment (RULA) for computer users (McAtamney & Corlett, 1993; Lueder, 1996) and the Berkeley Computer Check List (Janowitz et al., 2002). Both provide scores that increase with increased observations of awkward (less neutral and natural) postures (fig. 34.6). The RULA was originally developed for use among industrial workers, but a modified version for computer use among office workers has been developed. In contrast, the Computer Use Checklist was developed specifically for use among office workers. Both observational assessments were designed to be administered quickly and easily in an office environment to identify ergonomic interventions. However, the specific capabilities of these tools to measure the postural risk for MSDs and performance have not yet been formally tested.

34.5 Measuring Applied Forces and Muscle Efforts

Forces applied to the keyboard can be measured two ways: by measuring the forces at the key cap (Rempel et al., 1994) and by measuring the forces between the keyboard and the work surface, underneath the keyboard (Armstrong et al., 1994; Martin et al., 1996; Gerard et al., 1996, 1999). Individual key cap transducers have the advantage of measuring the applied force as close to the fingertip as possible

FIGURE 34.7 A force platform to measure keyboarding forces underneath the keyboard and a mouse with force sensors along with electrogoniometers to measure wrist posture as utilized in Dennerlein and Johnson (2006a and b).

and also allow the measurement of the force specific to each individual finger. The key cap transducer is custom produced for each keyswitch design. The post that connects the key cap to the keyswitch is replaced with a new post with a strain-gauge base load cell. Rempel et al. (1994) described the highly transient fingertip forces with three phases: (1) a brief key compression phase lasting some 10 to 50 ms; (2) an impact phase as the fingertip traveling with the key cap collides with the end of the key travel; and (3) the continual increase of force and then decrease of force and keyswitch release that last some 50 to 100 ms.

The advantages of measuring forces underneath the keyboard using a force platform include the ability to use the measurement system with almost any type of keyboard (rather than being limited to instrumenting specific keys on a keyboard) and capturing all the forces applied to the keyboard as a whole (fig. 34.7). A keyboard force platform is created by placing off-the-shelf transducers between two metal plates and then setting the keyboard on top of this system. These keyboard force platforms, however, do not measure several artifacts of keyboard dynamics as well as resting of the fingers and hands not associated with activation of the keyboard. The keyboard force platform typically suffers from low frequency force vibrations due to the plates themselves and/or the keyboard vibrating when a key is activated. As a result, some low-frequency components of the individual fingertip forces may be lost when filtering of the force signals to remove this primarily unpreventable artifact. Nonetheless, important information about the applied forces has been discerned demonstrating relationships between keyboard design (Gerard et al., 1996, 1999) and symptomatic populations (Feuerstein et al., 1997).

34.5.1 Applied Forces to Computer Mice

Recent studies have shown that musculoskeletal disorders may be associated with computer mouse use (Johnson et al., 2000) and that these apparent mouse-related disorders are increasing (Jensen et al., 1998; Wahlström, 2005). In order to understand the relationship between mouse use and the increasing prevalence of computer-related musculoskeletal disorders, the biomechanical components (e.g., finger force, wrist, forearm, and shoulder postures) and patterns of mouse operation must be assessed. For researchers, measuring the forces and the pattern of the forces applied to the mouse can provide biomechanical and physiological information related to computer mouse operation.

The physical challenges of developing force-sensing computer mice include developing/adapting force-sensing technologies to fit inside the mouse, not significantly altering the look and feel of the mouse, and maintaining the measurement accuracy of the force-sensing technology. The first force-sensing mouse was made by Johnson et al. (1993) and consisted of a computer mouse retrofitted with a computer-machined aluminum frame that consisted of bending elements instrumented with strain gauges.

Using this instrumented mouse, Johnson et al. (2000) measured the forces applied to the mouse in a group of 16 subjects at their actual workstations while they performed their regular work. Their results demonstrated that mouse use accounted for 24% of the time subjects spent working in front of the computer and the mean forces applied to the mouse were low, averaging 0.6% MVC (0.43 N) and 0.8% MVC (0.35 N) for the sides and button respectively. Peak forces were higher, ranging between 0.9% to 4.3% MVC (0.87 to 1.77 N) for the sides and 1.4% to 6.2% MBV (1.05 to 1.76 N) for the button. During mouse use, males and females applied the same absolute (Newton) force; however, the relative (%MVC) force applied by females was slightly higher. With virtually every force measure, the ability to measure and/or detect differences was increased when the mouse forces were compared using relative (%MVC) forces rather than absolute (Newton) forces. This demonstrates the importance of collecting both absolute and relative force measures. An undetermined issue is whether absolute or relative force exposure is a better measure of exposure and a predictor of disease and disorder.

34.5.2 Measuring Muscle Effort

Another measure used to quantify computer interaction is the level of required muscle activity to interact with the computer. Muscle activity is usually measured with electromyography (EMG), which has extensive use through the computer interaction community (Kumar & Mital, 1996; Basimajian & De Luca, 1985). EMG measures can provide relative efforts of muscles (e.g., Sommerich et al., 2001, 2002), the required motor control (e.g., Kuo et al., 2006) and the pattern of muscle activity through EMG gap analysis (e.g., Aarås et al., 1997; Blangsted et al., 2004b).

Under the right conditions (e.g., static contractions), the amount of the muscle's electrical activity can be related to the force exerted by the muscle. During computer work, the extensor muscles have been shown to work at higher relative force levels compared to the flexor muscles (Fernström et al., 1994, 1997). The extensor muscles appear to be a more common site of injury relative to the flexor muscles (Rose, 1991). The increased load on the extensor muscles may partially explain why these muscles are a common site of injury. So to more fully understand computer interaction, not only is it important to measure the flexor muscles that are responsible for holding and/or activating computer input devices, but to study and measure the loads in the extensor muscles that often provide a stabilizing role in work activities.

Finally, patterns of muscle activity can be analyzed to estimate how hard the muscle is working and how often and how much time the muscle is provided with opportunities to rest. Periods of muscle rest are important in order to allow rest and recovery and minimize muscle fatigue. With electromyography, periods of muscle rest can be identified by low-level or silent periods in the electrical activity of the muscle. These low-level or silent periods in the muscle's electrical activity are referred to as EMG gaps (Veiersted et al., 1993). It has become evident from GAP analyses, that at least in the shoulder muscles there are very few EMG gaps (rest periods) during computer mouse work (Aarås et al., 1997).

Standard methods exist in measuring and reporting EMG findings. *The Journal of Electromyography and Kinesiology* provides specific guidelines for publishing papers with EMG data. In addition, an international effort created specific recommendations on the use of bipolar surface electromyography electrodes, the SENIAM (http://www.seniam.org/), the Surface ElectroMyoGraphy for the Non-Invasive Assessment of Muscles.

34.6 Psychophysics: Measuring Perceptions and Preferences

An essential component in the effective design of human-computer systems is the measurement of the user's subjective perceptions and usability preferences of a system or product. Three major areas of subjective measurements in HCI usability research are: (1) satisfaction (the user's experience and attitude toward the system or product), (2) effectiveness (the level of comprehension and perception of outcome), and (3) task difficulty (the duration, mental workload, and efficiency of completing the tasks) (Hornbaek, 2006; ISO, 1998).

Satisfaction, which includes comfort, preference, ease-of-use, and attitude, is most often measured through self-reported questionnaires. Depending on the type and characteristics of the measure, these questionnaires are most commonly in the form of five-, seven-, or nine-point Likert scale (for ordinal measures), visual-analogue scale (VAS; for continuous measures), bipolar questionnaires (for dichotomous or continuous measures), rank-order (for preference), or a combination of the above (Helander et al., 1992; Hornbaek, 2006; Kroemer & Grandjean, 1997). Although usually custom-designed depending on the system or product, satisfaction should be measured through validated questionnaires, since the use of standardized usability measures allows for the comparisons between different designs of systems and products (Carey, 1991).

Measures of effectiveness include accuracy, completeness, outcome quality, and the user's ability to identify and predict the outcome and function of the system or product (Hornbaek, 2006; ISO, 1998). Effectiveness is often assessed by self-administered open- and closed-ended questionnaires and/or Likert-type scales on the user's perception of comprehension and outcome (McLoone et al., 2004).

Task difficulty, which includes effort, usage patterns, and duration of workload, can be measured through self-administered questionnaires in the various forms of Likert-type scales or VAS that assess the users' subjective experience of duration, perception of task difficulty, and mental workload. Another common subjective scale on the ratings of perceived exertions (RPE) includes the Borg CR10-scale and Borg RPE-scale (Borg, 1982). The Borg CR10-scale is used to assess perceived fatigue and exertion in various body segments (e.g., hands, wrists, arms shoulders, neck, back) and the RPE scale is calibrated to heart rate and used to measure whole-body exertions. It is important that the correct scale and scale procedures are used to ensure meaningful results (Borg, 1998).

Although the criteria on subjective perceptions and usability preferences should be established prior to user testing of the system or product (Carey, 1991), psychophysical measurements during the testing phase should be conducted prior, during, and after using the system or product. However, in a recent review study, a major challenge in HCI usability testing is that 93% of the studies measured satisfaction via questionnaires after using the system (Hornbaek, 2006). Therefore, user's subjective perceptions would then be formulated (and biased) by the user's level of (mis)comprehension of the questions, and consequently the change in the subjective perceptions from before to after using the system or product cannot be captured. Pre-usability questionnaires provide useful information that cannot be gathered after using the system or product. For example, one metric of pre-usability includes approachability, an important parameter especially for marketing. Approachability measures the level of perception of challenge for new systems and products based upon initial appearance; higher approachability ratings indicate a perception of a lower challenge in order to use the product, in which challenge is considered a negative consequence.

Collecting subjective ratings during the actual use of the system or product also provides informative data on subjective perceptions. For example, through an instantaneous self-assessment (ISA) technique, Tattersall and Foord (1996) demonstrated a method of measuring workload and satisfaction ratings during instead of after using the system.

An important component of the human-computer interface is the visual display, which is the primary unit for information transfer to the human. Primary evaluations methods for the visual display monitor are thus based on functional visual capabilities and discomfort. Functional visual capability is best described as the capability of correctly identifying letters and directions of the parallel lines of the letter E at various levels of contrast. For example, Sheedy et al. (2005) used low contrast (20%) visual acuity tests exploring disability glare from surrounding luminance.

References

Aaras, A., Fostervold, K. I., Ro, O., Thoresen, M., Larsen, S., 1997. Postural load during VDU work: a comparison between various work postures. *Ergonomics* 40, 1255–68.

Accot, J., Zhai, S., 1997. Beyond Fitts' Law: Models for Trajectory-Based HCI Tasks. In *Proceedings of the ACM Conference on Human Factors in Computing Systems—CHI '97*. New York: ACM, pp. 295–302.

Airaksinen, M. K., Kankaanpaa, M., Aranko, O., Leinonen, V., Arokoski, J. P., Airaksinen, D., 2005. Wireless on-line electromyography in recording neck muscle function: A pilot study. *Pathophysiol* 12, 303–6.

Armstrong, T. J., Foulke, J. A., Martin, B. J., Gerson, J., Rempel, D. M., 1994. Investigation of applied forces in alphanumeric keyboard work. *Am Ind Hyg Assoc J* 55, 30–5.

Assink, N., Bergman, G. J., Knoester, B., Winters, J. C., Dijkstra, P. U., Postema, K., 2005. Interobserver reliability of neck-mobility measurement by means of the flock-of-birds electromagnetic tracking system. *J Manipulative Physiol Ther* 28, 408–13.

Balakrishnan, A. D., Jindrich, D. L., Dennerlein, J. T., 2006. Keyswitch orientation can reduce finger joint torques during tapping on a computer keyswitch. *Hum Factors* 48, 121–9.

Basmajian, J. V., de Luca, C. J., 1985. *Muscles Alive: Their Functions Revealed by Electromyography*. Williams and Wilkins, Hagerstown, MD.

Bergqvist, U., Wolgast, E., Nilsson, B., Voss, M., 1995a. Musculoskeletal disorders among visual display terminal workers: individual, ergonomic, and work organizational factors. *Ergonomics* 38, 763–76.

Bergqvist, U., Wolgast, E., Nilsson, B., Voss, M., 1995b. The influence of VDT work on musculoskeletal disorders. *Ergonomics* 38, 754–62.

Blangsted, A. K., Hansen, K., Jensen, C., 2004. Validation of a commercial software package for quantification of computer use. *Int J of Ind Ergon* 34, 237–41.

Blangsted, A. K., Sogaard, K., Christensen, H., Sjogaard, G., 2004. The effect of physical and psychosocial loads on the trapezius muscle activity during computer keying tasks and rest periods. *Eur J Appl Physiol* 91, 253–8.

Borg, G., 1982. A category scale with ratio properties for intermodal and interindividual comparisons. In: H. G. Geissler, P. Petzold, H. F. J. M. Buffart, Y. M. Zabrodin (Eds), *Psychophysical Judgement and the Process of Perception*. VEB Deutscher Verlag der Wissenschaften, Berlin, pp. 25–34.

Borg, G., 1998. Borg's Perceived Exertion and Pain Scales. *Human Kinetics*, Champaign, IL.

Carey, J. M., 1991. *Human Factors in Information Systems: An Organizational Perspective*. Ablex Publishing, Norwood, NJ.

Chang, C. H., 2008. Validating computer usage monitors for exposure assessment of computer workers. Sc.D., Harvard School of Public Health, Boston, MA.

Chang, C. H., Amick, B. C., 3rd, Menendez, C. C., Katz, J. N., Johnson, P. W., Robertson, M., Dennerlein, J. T., 2007. Daily computer usage correlated with undergraduate students' musculoskeletal symptoms. *Am J Ind Med* 50, 481–88.

Chang, C. H., Wang, J. D., Luh, J. J., Hwang, Y. H., 2004. Development of a monitoring system for keyboard users' performance. *Ergonomics* 47, 1571–81.

Dennerlein, J. T., Becker, T., Johnson, P., Reynolds, C., Picard, R., 2003. Frustrating computers users increases exposure to physical factors, 15th Triennial Congress of the International Ergonomics Association (IEA 2003), Seoul, South Korea.

Dennerlein, J. T., DiMarino, M. H., 2006. Forearm electromyographic changes with the use of a haptic force-feedback computer mouse. *Hum Factors* 48, 130–41.

Dennerlein, J. T., Johnson, P. W., 2006a. Changes in upper extremity biomechanics across different mouse positions in a computer workstation. *Ergonomics* 49, 1456–69.

Dennerlein, J. T., Johnson, P. W., 2006b. Different computer tasks affect the exposure of the upper extremity to biomechanical risk factors. *Ergonomics* 49, 45–61.

Dennerlein, J. T., Kingma, I., Visser, B., van Dieen, J. H., 2007. The contribution of the wrist, elbow and shoulder joints to single-finger tapping. *J Biomech* 40, 3013–22.

Dennerlein, J. T., Martin, D. B., Hasser, C., 2000. Force-feedback improves performance for steering and combined steering-targeting tasks, Conference on Human Factors in Computing Systems, CHI-2000, The Hague, The Netherlands.

Dennerlein, J. T., Yang, M. C., 2001. Haptic force-feedback devices for the office computer: performance and musculoskeletal loading issues. *Hum Factors* 43, 278–86.

Douglas, S. A., Kirkpatrick, A. E., MacKenzie, I. S., 1999. Testing pointing device performance and user assessment with the ISO 9241, Part 9 standard, ACM Conference on Human Factors in Computing Systems—CHI '99, New York.

Faucett, J., Rempel, D., 1994. VDT-related musculoskeletal symptoms: interactions between work posture and psychosocial work factors. *Am J Ind Med* 26, 597–612.

Fernström, E., Ericson, M. O., 1997. Computer mouse or Trackpoint—effects on muscular load and operator experience. *Appl Ergon* 28, 347–54.

Fernstrom, E., Ericson, M. O., Malker, H., 1994. Electromyographic activity during typewriter and keyboard use. *Ergonomics* 37, 477–84.

Feuerstein, M., Armstrong, T., Hickey, P., Lincoln, A., 1997. Computer keyboard force and upper extremity symptoms. *J Occup Environ Med* 39, 1144–53.

Fitts, P. M., 1954. The information capacity of human motor systems in controlling the amplitude of a movement. *J Exp Psychol* 47, 381–91.

Gerard, M. J., Armstrong, T. J., Foulke, J. A., Martin, B. J., 1996. Effects of key stiffness on force and the development of fatigue while typing. *Am Ind Hyg Assoc J* 57, 849–54.

Gerard, M. J., Armstrong, T. J., Franzblau, A., Martin, B. J., Rempel, D. M., 1999. The effects of keyswitch stiffness on typing force, finger electromyography, and subjective discomfort. *Am Ind Hyg Assoc J* 60, 762–69.

Gerr, F., Marcus, M., Ensor, C., Kleinbaum, D., Cohen, S., Edwards, A., Gentry, E., Ortiz, D. J., Monteilh, C., 2002. A prospective study of computer users: I. Study design and incidence of musculoskeletal symptoms and disorders. *Am J Ind Med* 41, 221–35.

Gerr, F., Marcus, M., Monteilh, C., 2004. Epidemiology of musculoskeletal disorders among computer users: lesson learned from the role of posture and keyboard use. *J Electromyogr Kinesiol* 14, 25–31.

Gerr, F., Monteilh, C. P., Marcus, M., 2006. Keyboard use and musculoskeletal outcomes among computer users. *J Occup Rehabil* 16, 265–77.

Greene, W. B., Heckman, J. D., 1994. *Clinical Measurement of Joint Motion.* Amer Acad of Orthopaedic Surgeons, Rosemont, IL.

Helander, M. G., Landauer, T. K., Prabhu, P. V., 1992. *Handbook of Human-Computer Interaction* (2nd ed.). North-Holland, Amsterdam, The Netherlands.

Hornbaek, K., 2006. Current practice in measuring usability: Challenges to usability studies and research. *Int J Human-Computer Studies* 64, 79–102.

Hupert, N., Amick, B. C., Fossel, A. H., Coley, C. M., Robertson, M. M., Katz, J. N., 2004. Upper extremity musculoskeletal symptoms and functional impairment associated with computer use among college students. *Work* 23, 85–93.

Ijmker, S., Huysmans, M., Blatter, B. M., van der Beek, A. J., van Mechelen, W., Bongers, P. M., 2006. Should office workers spend fewer hours at their computer? A systematic review of the literature. *Occup Environ Med* 64, 211–22.

ISO. 1998. Ergonomic requirements for office work with visual display terminals (VDTs). Part 11: guidance on usability. International Organization for Standardization, (ISO 9241-11: 1998).

Janowitz, I., Stern, A., Morelli, D., Rempel, D., 2002. Validation and field testing of an ergonomic computer use checklist and guidebook. 46th Annual Conference of the Human Factors & Ergonomics Society, Baltimore, MD.

Jensen, C., Borg, V., Finsen, L., Hansen, K., Juul-Kristensen, B., Christensen, H., 1998. Job demands, muscle activity and musculoskeletal symptoms in relation to work with the computer mouse. *Scand J Work Environ Health* 24(5), 418–24.

Jindrich, D. L., Balakrishnan, A. D., Dennerlein, J. T., 2004a. Effects of keyswitch design and finger posture on finger joint kinematics and dynamics during tapping on computer keyswitches. *Clin Biomech* (Bristol, Avon) 19, 600–608.

Jindrich, D. L., Balakrishnan, A. D., Dennerlein, J. T., 2004b. Finger joint impedance during tapping on a computer keyswitch. *J Biomech* 37, 1589–96.

Johnson, P. W., Hagberg, M., Hjelm, E. W., Rempel, D., 2000. Measuring and characterizing force exposures during computer mouse use. *Scand J Work Environ Health* 26, 398–405.

Johnson, P., Ploger, J., Trask, C., Village, J., Chow, Y., Koehoorn, M. Teschke, K., 2007. Longitudinal exposure assessments of low back posture in five heavy industries in British Columbia. *Proceedings of the Sixth International Scientific Conference on Prevention of Work-Related Musculoskeletal Disorders*. PREMUS, Boston, MA.

Johnson, P., Tal, R., Smutz P., Rempel D., 1993. Computer mouse designed to measure finger forces during operation. *Proceedings of the IEEE Engineering in Medicine and Biology Society 15th Annual International Conference*. IEEE EMBS, San Diego, CA. 1404–1405.

Jonsson, P., Johnson, P. W., 2001. Comparison of measurement accuracy between two types of wrist goniometer systems. *Appl Ergon* 32, 599–607.

Jonsson, P., Johnson, P. W., Hagberg, M., 2007. Accuracy and feasibility of using an electrogoniometer for measuring simple thumb movements. *Ergonomics* 50, 647–59.

Kroemer, K. H. E., Grandjean, E., 1997. *Fitting the Task to the Human*. Taylor & Francis, Bristol, PA.

Kumar, S., Mital, A., 1996. *Electromyography in Ergonomics*. CRC Press, Boca Raton, FL.

Kuo, P. L., Lee, D. L., Jindrich, D. L., Dennerlein, J. T., 2006. Finger joint coordination during tapping. *J Biomech* 39, 2934–42.

Lueder, R., 1996. A proposed RULA for computer users, Ergonomics Summer Workshop, San Francisco, CA.

Luopajarvi, T., Kuorinka, I., Virolainen, M., Holmberg, M., 1979. Prevalence of tenosynovitis and other injuries of the upper extremities in repetitive work. *Scand J Work Environ Health* 5 (suppl 3), 48–55.

Marcus, M., Gerr, F., Monteilh, C., Ortiz, D. J., Gentry, E., Cohen, S., Edwards, A., Ensor, C., Kleinbaum, D., 2002. A prospective study of computer users: II. Postural risk factors for musculoskeletal symptoms and disorders. *Am J Ind Med* 41, 236–49.

Martin, B. J., Armstrong, T. J., Foulke, J. A., Natarajan, S., Klinenberg, E., Serina, E., Rempel, D., 1996. Keyboard reaction force and finger flexor electromyograms during computer keyboard work. *Hum Factors* 38, 654–64.

Matias, A. C., Salvendy, G., Kuczek, T., 1998. Predictive models of carpal tunnel syndrome causation among VDT operators. *Ergonomics* 41, 213–26.

McAtamney, L., Corlett, N. E., 1993. RULA: a survey method for the investigation of work-related upper limb disorders. *Appl Ergon* 24, 91–99.

McCormack, R. R., Jr., Inman, R. D., Wells, A., Berntsen, C., Imbus, H. R., 1990. Prevalence of tendinitis and related disorders of the upper extremity in a manufacturing workforce. *J Rheumatol* 17, 958–64.

McLoone, H., Hegg, C., Johnson, P., 2004. Evaluation of Microsoft's comfort curve keyboard. *Proceedings of the Human Factors and Ergonomics Society's 49th Annual Meeting*, Santa Monica, CA, pp. 1359–1363.

NRC/IOM, 2001. *Musculoskeletal Disorders and the Workplaces: Low Back and Upper Extremities.* National Academy Press, Washington, DC.

Ortiz, D. J., Marcus, M., Gerr, F., Jones, W., Cohen, S., 1997. Measurement variability in upper extremity posture among VDT users. *Appl Ergon* 28, 139–43.

Pascarelli, E. F., Kella, J. J., 1993. Soft-tissue injuries related to use of the computer keyboard. A clinical study of 53 severely injured persons. *J Occup Med* 35, 522–32.

Punnett, L., van der Beek, A. J., 2000. A comparison of approaches to modeling the relationship between ergonomic exposures and upper extremity disorders. *Am J Ind Med* 37, 645–55.

Rempel, D., Dennerlein, J., Mote, C. D., Jr., Armstrong, T., 1994. A method of measuring fingertip loading during keyboard use. *J Biomech* 27, 1101–4.

Rose, M. J., 1991. Keyboard operating posture and actuation force: Implications for muscle over-use. *Appl Ergon* 22, 198–203.

Sauter, S. L., Schleifer, L. M., Knutson, S. J., 1991. Work posture, workstation design, and musculoskeletal discomfort in a VDT data entry task. *Hum Factors* 33, 151–67.

Schlossberg, E. B., Morrow, S., Llosa, A. E., Mamary, E., Dietrich, P., Rempel, D. M., 2004. Upper extremity pain and computer use among engineering graduate students. *Am J Ind Med* 46, 297–303.

Schoenmarklin, R. W., Marras, W. S., Leurgans, S. E., 1994. Industrial wrist motions and incidence of hand/wrist cumulative trauma disorders. *Ergonomics* 37, 1449–59.

Serina, E. R., Tal, R., Rempel, D., 1999. Wrist and forearm postures and motions during typing. *Ergonomics* 42, 938–51.

Sheedy, J. E., Smith, R., Hayes, J., 2005. Visual effects of the luminance surrounding a computer display. *Ergonomics* 48, 1114–28.

Sommerich, C. M., Joines, S. M., Psihogios, J. P., 2001. Effects of computer monitor viewing angle and related factors on strain, performance, and preference outcomes. *Hum Factors* 43, 39–55.

Sommerich, C., Starr, H., Smith, C. A., Shivers, C., 2002. Effects of notebook computer configuration and task on user biomechanics, productivity, and comfort. *Int J Ind Ergon* 30, 7–31.

Tattersall, A. J., Foord, P. S., 1996. An experimental evaluation of instantaneous self-assessment as a measure of workload. *Ergonomics* 39, 740–48.

Veiersted, K .B., Westgaard R. H., Andersen P., 1993. Electromyographic evaluation of muscular work pattern as a predictor of trapezius myalgia. *Scand J Work Environ Health*, 19, 284–90.

Wahlstrom, J., 2005. Ergonomics, musculoskeletal disorders and computer work. *Occup Med* (Lond) 55(3), 168–76.

Wahlstrom, J., Lindegard, A., Ahlborg, G., Jr., Ekman, A., Hagberg, M., 2003. Perceived muscular tension, emotional stress, psychological demands and physical load during VDU work. *Int Arch Occup Environ Health* 76, 584–90.

Winter, D. A., 2005. *Biomechanics and Motor Control of Human Movement.* Wiley, New York.

35

The Psychophysiology of Emotion, Arousal, and Personality: Methods and Models

Wolfram Boucsein and
Richard W. Backs

35.1 Psychophysiological Measures

The term *psychophysiology* refers to the measurement of physiological phenomena as they relate to behavior in the broadest sense (Andreassi, 2000). A basic assumption in psychophysiology is that behavioral, cognitive, emotional, and social phenomena all have concomitant physiological processes (Hugdahl, 1995). Psychophysiological measures are not only a complement to introspective (subjective) and behavioral measures (e.g., movements); they have the unique advantage of constituting measures of the underlying psychological processes that cannot be faked, because psychophysiological responses are normally not under voluntary control.

Psychophysiology has traditionally focused on arousal and emotion. Within the last two or three decades, the interaction between cognitive and emotional processes has gained much importance in the wake of the cognitive neuroscience movement (Kosselyn & Andersen, 1992). More recently, the use of psychophysiology has become an important issue in engineering psychology (Backs & Boucsein, 2000). This chapter will give a brief outline of psychophysiological methods and theory that are applicable for digital human modeling (DHM) of emotion, arousal, and personality of virtual agents.

The majority of psychophysiological responses are measured by sensors placed on the body surface. An important issue in psychophysiological recording is the avoidance of artifacts resulting from noise caused by electromagnetic fields, speech, or body movements. Signal amplification, filtering processes, and analog/digital conversion (e.g., sampling rate) may exert additional influences on the recorded signal (Luczak & Göbel, 2000). Therefore, the use of psychophysiological methods requires training in both recording and evaluation techniques. As a rule of thumb, sampling rates of 100 Hz and more are required for each signal. Hence, a huge amount of data will result from long-term recording, which is typical for field applications. However, pre-processing during data acquisition can reduce the amount of collected data considerably.

Most psychophysiological measures need some parameterization before statistical evaluation. This can be a standard procedure such as fast Fourier transformation. In many instances, however, customized software has to be applied, for example, for getting amplitudes and recovery times from an electrodermal response (Boucsein, 1992, Appendix). During interactive use of psychophysiological methods, which will be necessary for most instances of DHM, online parameterization is needed. Such evaluation is challenging for psychophysiology, but it is feasible (e.g., Boucsein et al., 2005).

The following sections give a very brief description of five major groups of psychophysiological measures that have been used in engineering psychophysiology (Boucsein & Backs, 2000). We exclude measures of the endocrine system here, since their use in DHM is not foreseeable at present. More details on psychophysiological recording and parameterization can be found in Stern, Ray, and Quigley (2001).

35.2 Measures of Central Nervous System Activity

Spontaneous changes in the electroencephalogram (EEG) are observed with changing mental demands such as a decrease of alpha activity (8–12 Hz) and concomitant increase of beta activity (>13 Hz) that occur with increased mental load. They may also indicate different states of vigilance or attention. Increased frontal theta band activity (4–7 Hz) has been seen during periods of increased mental load. As a more refined measure of stimulus-directed central-processing capacity, an event related potential (ERP), the most frequently discussed component of which is the P3 or P300, can be taken from the EEG by averaging procedures. Averaging procedures also apply to the contingent negative variation (CNV) or the *Bereitschaftspotential* (readiness potential) used for determining response-directed central processing. Additional event-averaged EEG procedures that may be used in ergonomics are the lateralized readiness potential (LRP; Hohnsbein et al., 1995) and the eye-fixation-related potential (EFRP; Yagi, 1995). However, all these procedures require a highly controlled situation with access to responses to single stimuli. Furthermore, EEG variables in general have not been used much in field studies because they are prone to artifacts (cf. Gevins et al., 1995), although their use will undoubtedly increase with advances in so-called dry electrode technology (e.g., Taheri et al., 1994). Brain imaging has not been used in the field for technical reasons (Kodama et al., 1997), but this too may change, at least for imaging the cortex, with progress in functional near-infrared optical imaging (Hoshi, 2003).

35.3 Measures of Cardiorespiratory Activity

The heart is innervated by both sympathetic and parasympathetic fibers from the ANS. Cardiac activity is a rather complex phenomenon and psychophysiologists have used various measures, of which the most frequently applied is heart rate (HR). It is used as an indicator of physical as well as mental load (Mulder et al., 2000). More refined analyses of evaluating cardiac activity use measures of heart rate variability (HRV), such as the power of the 0.1 Hz component and respiratory sinus arrhythmia, as very sensitive indicators of cognitive activity. Although tonic HR changes may indicate the need for energy supply in response preparation, phasic HR patterns can be used to determine stimulus-directed central processing activity. Orienting responses consist of an HR deceleration, followed by acceleration, while defensive or startle responses cause an immediate HR acceleration (Turpin, 1985).

Unpleasant pictures produce the greatest initial deceleration, while pleasant pictures prompt the greatest peak acceleration (Winton et al., 1984). Recording respiration without wearing a mask, which may interfere with performance, is possible by using a thermistor in the nostrils or a strain gauge around the chest. However, these devices do not provide more than respiration rate, which is mainly used to control for respiratory artifacts in various psychophysiological measures, and much less seldom used as a measure itself (Wientjes, 1992). Peripheral blood flow can be recorded by plethysmographic techniques, preferably via infrared light reflection from a finger or the earlobe. A decrease in finger pulse volume is a very sensitive measure of orienting and defensive responses (Sokolov, 1963) and indicates both mental and emotional stress (see Section 35.9). Blood pressure (BP) is frequently recorded as an indicator of physical or mental stress, although problems with its continuous recording by means of the *Penaz* method still persist (Wesseling et al., 1986).

35.4 Measures of Electrodermal Activity

Electrodermal activity (EDA) is mainly dependent on sweat gland activity and is normally recorded from palmar skin. As a pure measure of sympathetic ANS activity, EDA is an exception in psychophysiological recording. Phasic EDA is measured as skin conductance response (SCR) or skin resistance response (SRR). Amplitudes of electrodermal responses (EDRs) reflect the amount of affective or emotional arousal elicited by a stimulus or situation. Orienting and habituation can be easily seen in EDR amplitudes. Both amplitude and recovery of an EDR have been demonstrated to be sensitive for certain aspects of central information processing (Boucsein, 1992), and may be used in a manner similar to ERPs as indicators of mental activity. The frequency of spontaneous electrodermal changes (called frequency of nonspecific SCRs or NS.SCR frequency) is a valid indicator of emotional strain (Boucsein, 2005) and has shown specific sensitivity during human-computer interaction (Boucsein, 2000).

35.5 Measures of Somatomotor Activity

Eye movements and eyeblink rate are sensitive indicators of mental and sometimes emotional activity. Eyeblink rate has also been used to detect fatigue in various fields of application (Stern et al., 1994). Pupillary dilation is a sensitive measure of both mental and emotional strain (see Section 35.9). Its application to field situations poses a problem principally because of difficulty in maintaining the eye in the field of view. Changes in light intensity pose no major problem as long as the amount of light impinging on the eye is monitored and corrections are made.

The electromyogram (EMG) can be used to evaluate the activity of certain muscle groups caused by physical work as well as by mental and emotional strain. EMG may be used to quantify specific kinds of strain such as neck muscle tension, especially in computerized work (Boucsein, 2000). Recording facial muscle activity consists of a fine-tuned methodology to evaluate emotions in applied contexts (Boucsein et al., 2002). The zygomaticus major reflects an increase in hedonic valence, while the levator labii superioris indicates unpleasant feelings. The corrugator supercilii is activated in case of critical evaluation of a stimulus or a situation.

35.6 Measures of Body Temperature

Core temperature, rectal or oral, is frequently used in addition to heart rate, especially in field studies on night and shift work, since it displays marked circadian variations and can be recorded easily. Body temperature is also increased during physical workload. However, it is not suitable for short-term changes. Finger temperature measured by thermistors may be used as an indicator for emotional stress (Levenson et al., 1990).

35.7 Modeling of Psychophysiological Systems

Despite great technological progress in psychophysiological recording and data evaluation that has been made during the past decades, there is only limited progress in the development of psychophysiological theory. Psychophysiologists with a few exceptions have not contributed to a theoretical approach integrating their measures (e.g., Boucsein, 1992; Kahneman, 1973; Pribram & McGuinness, 1975). One exception is Berntson and colleagues (Berntson et al., 1991; Berntson et al., 1994) whose autonomic space model for ANS responses may prove useful for DHM. For example, Backs (2001) has shown how Wickens' (2002) model of attention may map to the autonomic space for heart rate.

One of the reasons for this lack of progress is the nontheoretical multivariate approach frequently used in psychophysiological research. In addition, psychophysiologists have always been occupied with the development of refined techniques for data acquisition and evaluation. As a consequence, rather simple theoretical frameworks are still in use. The following sections will provide suitable theoretical frameworks for arousal, emotion/stress, and personality, which might be used in the process of making virtual agents more humanlike.

35.8 The Psychophysiology of Arousal

The description of any behavior requires at minimum a goal toward which it is directed and a measure of its intensity (Duffy, 1972). Intensity of behavior is related to an unspecific hypothetical excitation process in the central nervous system (CNS), which is called *arousal* or *activation*. It is regarded as a basic process that optimizes the information-processing flow from perception to behavior. Hence, arousal theories tried to explain how the brain organizes the allocation of resources to its subsystems and functional units, dependent on internal and external demands. With the advent of EEG measures, a widely accepted method for quantification of arousal became available (Lindsley, 1951). In particular, EEG desynchronization (reduction of alpha and increase of beta activity) was considered to reflect an increase in general arousal, the source of which was located in the reticular formation (the reticular activation system, RAS). Various measures of the ANS (HR, blood pressure, electrodermal activity, and temperature) and general muscle tension as an indicator from the somatomotor system were added (Duffy, 1957).

One of the major theoretical notions that came with arousal research was the inverted-U relationship between arousal and performance (Malmo, 1962). Its essence is that an increase from low to moderate arousal results in an increase of performance, whereas a further increase of arousal will result in performance decrement. During the years to come, psychophysiological responses were not always in line with the hypothesis of a single arousal dimension. Contrary to the predictions of an unspecific arousal theory, Lacey (1967) demonstrated that the pattern of psychophysiological responses varies with stimulus properties. In particular, HR and EDA sometimes changed in the same but sometimes in different directions, which was referred to by Lacey as *directional fractionation*. Starting with Lacey's work, one-dimensional arousal theories were considered too simplistic for adequate neurophysiological modeling. Instead, arousal proved to be a complex and multidimensional phenomenon that can be best taken care of by modeling different neurophysiological systems for different kinds of arousal. Therefore, we promote a four-arousal model to be used as a theoretical framework that combines structures relevant for both arousal and emotion.

Figure 35.1 depicts a model of the four-arousal systems and their neurophysiological underpinnings proposed by Boucsein (1992, 2006). The model integrates the two-arousal system of Routtenberg (1968), the three-arousal system of Pribram and McGuinness (1975), the three-arousal system of Fowles (1980), the behavioral inhibition system after Gray (1982), and the circuits between basal ganglia and frontal cortex after DeLong, Georgopoulos, and Crutcher (1983). The horizontal arrangement of Figure 35.1 corresponds to the structural and functional hierarchy within the brain, whereas its vertical arrangement follows a sequence from stimulation through information processing to response preparation.

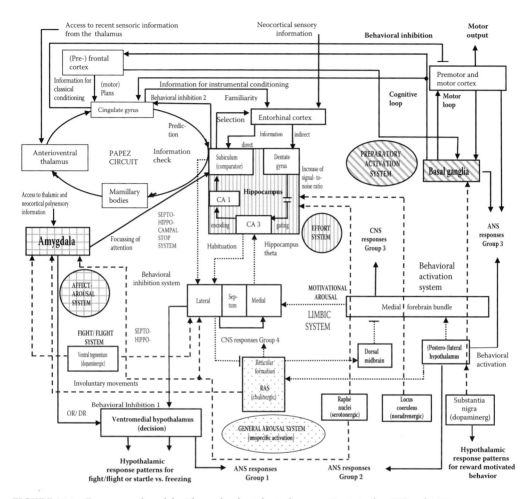

FIGURE 35.1 Four-arousal model with psychophysiological concomitants in the CNS and ANS.

Arousal System 1 (checkered background) is centered around the amygdala and is labeled *affect arousal* system. It is responsible for focusing attention and for the elicitation of orienting and defensive responses (see Sokolov, 1963), but also other immediate reactions such as flight/fight or freezing, the patterns of which are available in the ventromedial hypothalamus (lower left in fig. 35.1).

Arousal System 2 (vertically striped background) is centered around the hippocampus and is labeled *effort* system. It has the ability to connect or disconnect input and output. The physiological patterns generated by this system can be regarded as concomitants of central information processing. Arousal System 2 is also related to the basic hippocampal circuit involved in the behavioral inhibition system proposed by Gray (1982).

Arousal System 3 (oblique striped background) is centered around the basal ganglia and is labeled *preparatory activation* system. Its activation results in an increased readiness of brain areas involved in somatomotor activity. It has close relationships to the behavioral activation system described by Fowles (1980).

In the middle column, below the *effort system*, there is a fourth arousal system, which has been postulated as early as in the 1950s, but also by Fowles (1980): Arousal System 4 (dotted background), which consists of a *general arousal system* in the reticular formation, therefore labeled reticular activation system (RAS). It features a reciprocal relationship to the hippocampal *effort system* proposed by Routtenberg (1968) as described below.

The effort system plays a central role in arousal. It corresponds to Gray's (1982) behavioral inhibition system (BIS), which is activated by noradrenergic and serotonergic pathways from the locus ceruleus and the raphe nuclei, respectively. In case of stress-relevant stimuli, these pathways increase the signal-to-noise ratio in the indirect information flow within the hippocampus. As a consequence, the subiculum is alerted and the information is processed in the Papez circuit. Here, very recent (i.e., not yet cortically processed) information is available via the anterioventral thalamus. In addition, access is provided to motor plans and conditioned behavior patterns via the prefrontal cortex and the cognitive loop, respectively. If the system can predict what will happen next or what response will be appropriate, the subiculum will not impose an inhibition of behavior. If not, the BIS is activated via the cingulate gyrus and motor/premotor areas, and/or via the lateral septal area and the ventromedial hypothalamus.

Fowles (1980) postulated a behavioral activation system (BAS) that complements the BIS. The BAS is triggered by positive expectations, the presence of positive or the absence of negative reinforcement. The BAS is triggered by mesolimbic dopaminergic pathways. The dopaminergic system supplies the basal ganglia as well, which—together with the premotoric cortical areas—elicit the preparatory electrodermal activity 2 (Boucsein, 1992). This kind of EDA belongs to the autonomic concomitants of voluntarily initiated motor acts.

Routtenberg (1968) regarded the dopaminergic and noradrenergic stimulation of the limbic system through the medial forebrain bundle (lower right part of fig. 35.1) as a motivationally determined second arousal system besides general arousal, labeled Arousal System 4 here. It is responsible for maintaining vegetative functions and for initiating reward-directed behavior, thus corresponding to Panksepp's (1982) foraging-expectancy command system. Its output triggers posterolateral hypothalamic areas to elicit so-called ergotropic responses (e.g., tonic increases in HR). Neocortical concomitants of this kind of motivational arousal are the CNV that appears prior to a signaled imperative stimulus, or the *Bereitschaftspotential* (readiness potential), which appears prior to a motor output. The activity of the motivational arousal system is indicated by theta rhythm generated in the medial septal area, which is, however, not recordable with surface electrodes.

The typical indicator for general arousal is neocortical desynchronization, that is, alpha activity in the EEG is replaced by beta activity. According to Routtenberg (1968), motivational and general arousals have the property to inhibit each other, as depicted by means of dotted lines in Figure 35.1. The reticular formation inhibits the limbic system via the dorsal midbrain. In turn, the limbic system inhibits reticular activity via medial septal, hippocampal, and lateral septal areas; a pathway which at least partly resembles *behavioral inhibition 1* as proposed by Gray (1982). The reciprocal relationship of both systems is regarded as enabling selective and adaptive information processing, together with an adequate somatic regulation. The posterolateral hypothalamus has been proposed to facilitate both reticular and limbic arousal (Routtenberg, 1968).

The ventral tegmental area in the midbrain, which is the origin of the mesolimbic dopaminergic system, delivers dopamine to the amygdala and to septal areas (lower left part of Figure 35.1). The mesolimbic dopaminergic system facilitates involuntary movements that are part of the fight/flight system and of orienting and defensive responses, all of which are elicited by patterns stored in the ventromedial hypothalamus. Here, the group 1 ANS concomitants originate, such as the sweat-eliciting sympathetic pathway, which is responsible for electrodermal activity 1 (Boucsein, 1992, fig. 6). Since the ventromedial hypothalamus has been regarded as the origin of BIS-related response patterns (Gray, 1982), the increase of EDA during BIS activity as described by Fowles (1980) has its neurophysiological origin here. As a first approach, a marked increase in EDA can be indicative of BIS activity, whereas a gradual HR increase points to BAS activity. Cholinergic fibers that originate in the reticular formation facilitate the affect arousal system via the amygdala, together with serotonergic pathways from the raphe nuclei in the midbrain. The amygdala also acts on the hippocampal comparator (i.e., the subiculum) to increase and focus attention.

Table 35.1 lists the typical psychophysiological concomitants for each of the four arousal systems modeled in Figure 35.1 (Boucsein & Backs, 2000; Boucsein, 2001, 2006). They are labeled group 1 through 4 and comprise both autonomic and central nervous system responses.

TABLE 35.1 Psychophysiological Measures Having Specific Indicator Functions for the Groups of Responses of the Autonomic and Central Nervous Systems

Group 1 responses Affect arousal system and flight/fight system	Increase of frequency and sum of amplitudes of the emotionally negatively tuned non-stimulus specific electrodermal responses (electrodermal activity type 1; ANS) Phasic changes in heart rate (ANS): - Heart rate decrease followed by an increase as components of an orienting response - Heart rate increase without previous decrease as component of a defensive response

Group 2 responses
  Effort system and
  behavioral inhibition system
	Decrease of heart rate variability (ANS)
	Increase of recovery times of electrodermal responses (ANS)
	Increase of P300 amplitude in the evoked potential calculated from the electroencephalogram (CNS)
	Increase of frontal theta activity in the spontaneous electroencephalogram (CNS)

Group 3 responses
  Preparatory activation system and
  behavioral activation system
	Moderate increase of the tonic heart rate (ANS)
	Increase of the amplitude of the preparatory non-stimulus specific electrodermal responses (electrodermal activity type 2; ANS)
	Contingent negative variation and *Bereitschaftspotential* in the evoked potential calculated from the electroencephalogram (CNS)

Group 4 responses
  General arousal system
  (reticular activation system)
	Increase of the sympathicus-driven responses of the autonomic nervous system (ANS):
	- Marked increase of heart rate, blood pressure, and tonic electrodermal activity
	Desynchronization in the electroencephalogram (alpha blockade; CNS)

If a situation changes or certain stimulation occurs, affect arousal will show up in group 1 measures as frequency or amplitude of the electrodermal response (EDA 1) and phasic HR changes. Attention will be shifted toward the new stimulus, supported by involuntary somatomotor responses such as head- or eye movements (not shown in Table 35.1). The preparatory activation system will provide an increased readiness of brain areas involved in eliciting intended somatomotor actions. This will increase brain negativity as can be seen in the CNV (McGuinness & Pribram, 1980) and the *Bereitschaftspotential*, but also other group 3 responses such as a moderate increase of tonic HR (Fowles, 1980) and an increase of EDA 2 amplitudes (Boucsein, 1992).

Dependent on novelty, stress, or an increased emotional load, the normally rather straightforward chain of situation-reaction relationships can be modified by the effort system. It has the property to disconnect Arousal Systems 1 and 3 in order to prevent immediate action and facilitate deliberate analysis performed by certain cortical-subcortical brain circuits (Gray, 1982), including the so-called Papez circuit (see fig. 35.1, upper left). The ongoing central information processing will be reflected by a decrease in HRV, an increase of the ERP P300 component, enhanced frontal theta activity in the EEG, and an increase of EDR recovery times, shown as group 2 responses in Table 35.1.

Group 4 responses indicate the amount of general arousal, which is seen in various measures of the ANS (HR, blood pressure, and tonic EDA) and in EEG desynchronization (beta replaces alpha activity, which is often called alpha blocking). Dependent on the strength of stimulation or of the subject's intentions, group 4 responses may dominate the whole psychophysiological occurrence.

Although far from covering all possible psychophysiological relationships in arousal and emotion, the model proposed here may help provide a framework for generating refined hypotheses regarding the action of different arousal and emotional processes on physiological outcomes. It is crucial for such a neurophysiologically based approach to be supported with empirical results from the field. Boucsein and Backs (2000) have exemplified this with respect to workload and stress in the field of ergonomics, giving a summary of the sensitivity and reliability of all psychophysiological measures in this field. A summary of these and additional results is given in Table 35.2.

TABLE 35.2 Psychophysiological Measures in the Order of Their Usability in Human-Machine Interactions

Measure	Usability	Reliability	Arousal System Specificity	Application Field Specificity
Heart rate (tonic) from the electrocardiogram	Easy to record, a universal but rather unspecific indicator, not very sensitive for other kinds of strain except physical	0.94 (20 min; d) 0.59 (2 W; b)	Increase (marked): physical strain (general arousal) Increase (moderate): mental strain (preparatory activation) Decrease (moderate): fatigue	Strong physical activity (1); movement during flight task (2, 3); time pressure in HCI (4); fear/anger > happiness, disgust (5)
Heart rate variability (0.1 Hz-band; mean square of successive differences) between interbeat intervals from the electrocardiogram	Easy to record, but requires a relatively long recording period, a very sensitive and specific indicator	0.66 (2 D; g)	Decrease (marked): mental strain (effort) Decrease (moderate): fatigue Increase: relaxation, recovery	Decrease with time pressure in HCI (4); decrease during mental workload in complex task environments (6,7); during night shift work (8); increase after frequently interspersed breaks (9)
Electromyogram from muscles of thorax and limbs	Easy to record, but in the lower range of dynamics relatively easy to disturb	0.79 (20 min; d) 0.49 (2 W; b)	Increase: muscular (physical) strain, increased psychological strain (increased stress level)	Strength of muscular work (1), tension during computer breakdown (9) and noisy work (10)
Eyelid closure frequency (eye blink frequency)	Easy to record, does not require direct bodily contact (recordable by a video system)	0.87 (20 min; d)	Increase: visual strain, hyper-vigilance, emotional strain (affect arousal) Decrease: increased information uptake (effort)	Increase during complicated maneuvers in flying (3) and driving tasks (7) Decrease during surgery film compared to neutral film (11)
Eye movements (saccades and fixations)	Do not require direct bodily contact (recordable by a video system)	0.77 (2-4 M; e)	Increase: mental strain (effort) and fatigue	Increased saccadic duration with time on task (12); fixation of emotional pictures > neutral (13)
Frequency of non-specific electrodermal responses (phasic skin conductance changes)	If both hands are needed, recording from foot possible, artifact prone, but a very sensitive and specific indicator	0.81 (20 min; d) 0.62 (1 Y; c)	Increase: negatively toned emotion (affect arousal), emotional strain Decrease: hypovigilance with increased risk for sleep	Increase with driving demands (7) and long waiting times under time pressure in HCI (4) but decreases if time pressure is absent (14) and after night work (8); fast > slow news on a handheld PC (19)
Amplitude of non-specific electrodermal responses (phasic skin conductance changes)	If both hands are needed, recording from foot possible, artifact prone, but a very sensitive and specific indicator	0.72 (2 W; b)	Increase (marked): negatively toned emotions (affect arousal), emotional strain. Increase (moderate): increased cognitive activity (preparatory activation)	Increase with aversiveness of events (15), user-hostility of HCI (16). Indicates the intensity dimension of emotions (17); obstructive > conducive games (18)
Heart rate (phasic) from the electrocardiogram	Stimulus-locked, a rather specific indicator	0.72 (8 W; h)	Increase without previous decrease: defensive response Decrease with subsequent increase: orienting response	Aversive stimuli increase HR (20), negative stimuli initial decrease (21), greater mid-interval increase for pleasant stimuli (22)
Blood pressure: Riva-Rocci (discontinuously), Penaz (continuously)	Not easily continuously recordable, artifact prone	0.63 (1 W; f) 0.56 (2 W; b)	Increase (marked): physical strain (general arousal) Increase (moderate): mental strain (effort)	Systolic: high demands (23) and time pressure in HCI (4) Diastolic: angry > happy, sad, fearful, relaxed faces (24)

Measure	Practical notes	Retest reliability	Psychophysiological interpretation	Application field
Pulse volume amplitude	Restriction of one hand, artifact prone, but sensitive	0.84 (20 min; d) 0.68 (2 W; b)	Increase of the amplitude modulation: orienting response	Initial OR and persisting interest during interacting with a new product (25)
Electromyogram from facial muscles	Clearly visible facial electrodes can be obtrusive in real life	0.55 (2 W; b)	Indicators of positive or negative emotions, without requirement of visible changes in facial expressions	Differentiating between pleasant, unpleasant, and disgusting stimuli (26–28), including interactions with new products (25)
Body core temperature	Slow reactivity, rectal sensor uncomfortable to wear		Global indicator of general arousal, used for circadian rhythm and jet lag	Increase during physical and environmental strain (29); re-entrainment after time shift (30)
Skin conductance level	If both hands are needed, recording from the inner aspect of the foot possible	0.59 (2 W; b) 0.60 − 0.66 (3–5 W; c)	Increase: high strain Decrease: low strain (high or low general arousal)	Increase during prolonged waiting during HCI (4); challenge during computer game (31)
Skin temperature	Easy to record, but slow reactivity	0.61 (2 W; b)	Decrease: high strain	Decrease in emotional stressful HCI (32); increased in positive emotions (18)
Respiration rate	Easy to record, but not very sensitive	0.70 (1 W; f) 0.85 (2 W; b)	Increase: physical strain (general arousal)	Increase with demands (33) and time pressure during HCI (4); gets instable during boredom (32)
Alpha activity from the electroencephalogram	Artifact prone in field settings, not online calculable	0.92 (20 min; d)	Decrease: mental strain (effort) Increase: hypovigilance, decrease of general arousal	Decreases with workload during flight (34) and increases during extended night flying (35)
Theta activity from the electroencephalogram	Artifact prone in field settings		Increase: mental strain (effort), hypovigilance	Increases with cognitive demands during flight performance (3); increases in pleasant states (36)
Combined indices from different bands of the electroencephalogram	Artifact prone in field settings, not online calculable		Recording of general arousal for the control of hypovigilance	Indices from alpha, beta, and theta related to hypovigilance (37) and pleasantness/ unpleasantness (38)
Amplitude of P300 in the evoked potential taken from the electroencephalogram	Artifact prone in field settings, not online calculable, requires secondary task or additional stimulation	0.62 (2 Y; a)	Measure of an increased need for information processing (effort)	Indirect indicator of effort used in primary task (39) Specific patterns elicited by emotional pictures (40)

Source: Sources for reliability: (a) Segalowitz & Barnes (1993); (b) Waters et al. (1987); (c) Boucsein (1992); (d) Fahrenberg et al. (1979); (g) Amara & Wolfe (1998); (h) Rau et al. (1996). Intervals for retest reliability: D = days, W = weeks, M = months, Y = years. Sources for application field specificity: (1) Vitalis et al. (1994); (2) Roscoe (1993); (3) Hankins & Wilson (1998); (4) Boucsein (2000); (5) Cacioppo et al. (2000); (6) Veltman & Gaillard (1998); (7) Richter et al. (1998); (8) Boucsein & Ottmann (1996); (9) Boucsein & Thum (1997); (10) Hanson et al. (1993); (11) Palomba et al. (2000); (12) McGregor & Stern (1996); (13) Nummenmaa et al. (2006); (14) Kuhmann et al. (1987); (15) Backs & Grings (1985); (16) Muter et al. (1993); (17) Johnsen et al. (1995); (18) Rimm-Kaufman & Kagan (1996); (19) Kallinen & Ravaja (2005); (20) Klorman et al. (1977); (21) Simons et al. (1999); (22) Lang et al. (1993); (23) Rose & Fogg (1993); (24) Schwartz et al. (1981); (25) Boucsein et al. (2002); (26) Larsen et al. (2003); (27) Weyers et al. (2006); (28) Schienle et al. (2001); (29) Romet & Frim (1987); (30) Gander et al. (1989); (31) Mandryk et al. (2006); (32) Ohsuga et al. (2001); (33) Brookings et al. (1996); (34) Sterman & Mann (1995); (35) Gundel et al. (1995); (36) Sammler et al. (2007); (37) Scerbo et al. (2000); (38) Min et al. (2005); (39) Sirevaag et al. (1993); (40) Diedrich et al. (1997).

35.9 The Psychophysiology of Emotion and Stress

Relating psychophysiological measures to emotional states dates back to James and Lange in 1884/1885. According to the so-called James–Lange theory, emotion-relevant stimuli elicit ANS and somatomotor changes, which in turn trigger the experience of emotion. Although being counterintuitive, their theory shed light on the importance of physiological changes in the periphery for emotional processes. Their theory was challenged by Cannon and Bard in the early 1900s. Their notion that subjective and physiological aspects of emotions parallel each other is still valid. Although the Papez circuit (see upper left of fig. 35.1) that had been proposed in 1937 is no longer regarded as central for generating and maintaining emotions, the limbic system to which the Papez circuit belongs (see center of fig. 35.1) is still discussed as being of major importance in the realm of emotions.

The field of emotion suffers because it lacks a generally accepted definition. After reviewing the relevant literature, Kleinginna and Kleinginna (1981) described emotions as:

[A] complex set of interactions among subjective and objective factors, mediated by neural/hormonal systems, which can (a) give rise to affective experiences such as feelings of arousal, pleasure/displeasure; (b) generate cognitive processes such as emotionally relevant perceptual effects, appraisals, labeling processes; (c) activate widespread physiological adjustments to the arousing conditions; and (d) lead to behavior that is often, but not always, expressive, goal directed, and adaptive. (p. 335)

Such a definition stresses not only the necessity of psychophysiological response in emotion research; it also fosters a dimensional approach taking regard of at least the two aspects of arousal and valence (i.e., pleasure/displeasure).

Figure 35.2 shows such a dimensional system (Larsen & Diener, 1992), where single emotions (or affective states) are located in the quadrant of a circle, therefore labeled circumplex model by Russell (1980). The dimensionality of the emotional space has been proposed to be from only two to three (Schlosberg, 1954) and up to four orthogonal axes (Osgood, 1966; Sokolov & Boucsein, 2000). Because the arousal-relaxation and the pleasantness-unpleasantness (i.e., valence) axes were present in all systems, they will be used here to exemplify how psychophysiological measures can be applied in emotion detection.

In order to discover specific somato-visceral indicators for the valence and the arousal dimension, Greenwald, Cook, and Lang (1989) used emotion-eliciting slides as stimuli and found evidence for differential facial EMG activity and HR response peak acceleration being specific for valence, whereas EDA appeared more closely related to arousal. A similar specific sensitivity for cardiac and electrodermal measures was obtained by Johnsen, Thayer, and Hugdahl (1995), who presented faces with standardized emotional expressions (Ekman & Friesen, 1976). However, an almost opposite autonomic patterning has been found when subjects were asked to voluntarily produce facial expressions: Levenson, Ekman, and Friesen (1990) recorded ANS and EMG measures in both actors and control subjects during the expression of anger, fear, sadness, and disgust as negative emotions, and happiness and surprise as positive emotions. Emotional valence was best represented by EDA, being most pronounced during the expression of negative emotions, while HR was only sensitive to disgust. Stemmler (1989) induced the three emotions fear, anger, and pleasure, extracting various parameters from ANS and EMG recordings. Measures that significantly differentiated among the three emotions were subjected to a discriminant analysis, which resulted in different profiles for fear and anger but no specific one for pleasure. Moreover, this result was restricted to the real induction of emotion, whereas the profiles did not differ at all under imagery conditions. In a series of studies in which consumers were confronted with tactile and olfactory properties of cosmetic products and fabric care, Eisfeld et al. (2007) successfully applied multivariate psychophysiological recording (EDA, HR, HRV, pulse volume amplitude, and EMG from facial muscles) to get objective measures for emotions elicited by cosmetic products.

In general, ANS-based psychophysiology does not yet reveal a dimensional structure of emotion space that compares to the aforementioned two to four dimension models. Instead, psychophysiological

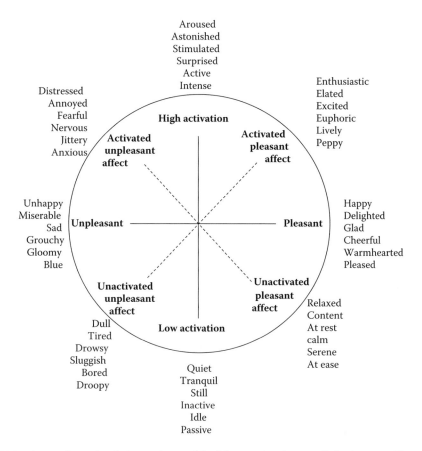

FIGURE 35.2 A two-dimensional circumplex model of the emotional space. (After Larsen and Diener, 1992, fig. 2.1.)

patterns have been frequently related to discrete emotions such as fear, anger, and joy (Levenson, 1988). The search for distinct somato-visceral response patterns indicating a specific emotion is further complicated by the need to keep the arousal level for the different emotion-eliciting situations comparable. Since most ANS measures undergo characteristic changes if general arousal is varied, the strength of emotions has to be controlled if ANS concomitants of different emotions are compared to each other.

Although measures of the ANS constitute the domain for psychophysiological research into emotions, EEG has been successfully used as well. Positive emotions have been found to increase beta activity (13 Hz and above) in the left frontal cortex, while negative emotions were more likely to activate the right frontal cortex (Davidson, 1993). EEG-derived measures may also be useful in determining the intensity of a negative emotional state (Davidson & Fox, 1982).

The use of the term *stress* in psychology comprises a wide range of phenomena, reaching from stimulus properties (mostly overstimulation) via coping mechanisms (e.g., cognitive appraisal) to typical response patterns. For the stimulus side, the term *stressor* is in use, while (although not very familiar to most North Americans) the term *strain* has been introduced in Europe for the impact of stress (Weiner, 1982). Most researchers use the term *stress* to characterize a state of high general arousal with a negative emotional tone. In the field of engineering psychophysiology, the concept of *workload* is often used instead of stress (Boucsein & Backs, 2000). Psychophysiological measures are very popular in stress research. However, other than being concerned with intense response reactions, the psychophysiology of stress is not principally different from what we discussed in the context of arousal and emotion.

35.10 The Psychophysiology of Personality

There have been various attempts to relate personality to ANS responses. Both Western (e.g., Eysenck, 1967) and Eastern (e.g., Teplov & Nebylitsyn, 1971) approaches assumed personality related to influences of higher cortical processes on the subcortical control of autonomic systems. However, these systems were based on a one-dimensional arousal concept, and therefore could not account for the complex interactions between various arousal systems shown in Figure 35.1.

Psychophysiological concomitants of broad personality characteristics such as the "big five" (Costa & McCrae, 1992) have been modeled for extraversion and neuroticism by Eysenck (1967) and have been modified by Gray (1972). Excitability of the limbic system, which was held responsible for processing emotion-relevant stimulus characteristics, formed the basis for neuroticism (i.e., emotional lability vs. stability), while influences of the reticular arousal system on cortical information processing served as basic neurophysiological concept in explaining behavioral differences between extraverts and introverts. According to Eysenck (1967), introverts are more easily conditioned, because their increased activity in a suggested RAS-cortex loop facilitates consolidation of learned material. Extraverts, on the other hand, are thought to elicit cortical inhibition faster, thus providing a kind of protection against strong stimulation. This brings extraverts close to sensation seekers (Eysenck & Zuckerman, 1978). Furthermore, introverts should be more generally aroused than extraverts in an average stimulus intensity range, whereas the opposite holds for higher stimulus intensities, due to an increased cortical (transmarginal) inhibition in extraverts. The Eysenck group repeatedly confirmed the proposed greater reactivity of introverts to stimuli of medium intensity (Eysenck & Eysenck, 1985), especially in electrodermal measures (Stelmack, 1981). The failure to confirm Eysenck's theoretical notions outside his own group was explained by Eysenck (1994) by the complexity of interactions between personality and situational characteristics. Because the empirical support for a greater general arousal in introverts was rather

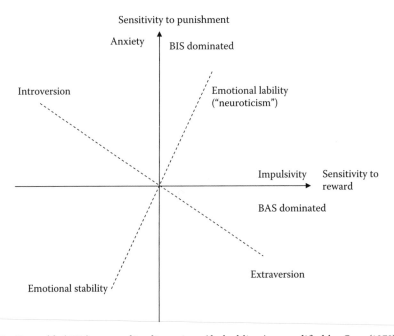

FIGURE 35.3 Eysenck's (1967) personality dimensions (dashed lines) as modified by Gray (1973) (solid lines) together with their behavioral (sensitivity to reward vs. punishment) and neurophysiological correlates (BAS vs. BIS). Note: the lability/stability axis is non-orthogonal, indicating the lack of independence between the original personality dimensions to the extraversion/introversion axis. (After Boucsein, 1992, fig. 49.)

weak, Gray (1973) proposed a modification of Eysenck's original notion, starting from the reward and punishment systems in the brain, which correspond to the BAS and BIS depicted in Figure 35.1. Instead of a higher conditionability of introverts as proposed by Eysenck, Gray suggested introverts were more susceptible to punishment and frustrating non-reward. Extraverts, on the other hand, were regarded as more prone to positive reward.

The modification of Eysenck's original notion proposed by Gray (1973) is shown in Figure 35.3, together with the identification of susceptibility to positive reinforcement with impulsivity and proneness to negative reinforcement with anxiety added by Gray (1981). Boucsein (1992) proposed a differential sensitivity of EDA and HR for these neurophysiologically based personality dimensions. An increase in EDA can be regarded as a specific indicator for anxiety (punishment), while an increased HR will more likely reflect impulsivity (reward). Gomez and McLaren (1997) partially confirmed this hypothesis, using a go/no-go discrimination task with monetary reward/punishment. While EDA was diminished under reward, subjects did not show an increase in HR in this condition. Arnett and Newman (2000) were also not able to show an HR increase during monetary reward for fast responses in a choice reaction task. It remains questionable whether generalized personality traits can be identified by a single psychophysiological measure. Instead, multivariate approaches should be favored in this domain in the same manner as in arousal research (Fahrenberg, 1987).

35.11 Summary

This brief overview of psychophysiological methods and theory was focused upon the measures and models that we believe will be most useful for consideration by the DHM community. As the realism of virtual agents continues to progress, we urge the DHM community to consider a psychophysiology for the agent as an important feature of humanlike behavior. Further, we believe that knowledge of the psychophysiology of the human who is interacting with the virtual agent will also be needed to shape the agent's response. Closing the loop between the human and the agent at multiple levels (i.e., cognitive, behavioral, psychophysiological) will ultimately be a necessity to have truly lifelike communication. In our companion chapter (see Chapter 16), we offer some thoughts on how psychophysiological measures can be employed to improve the naturalness of human-agent interactions.

References

Amara, C. E., & Wolfe, L. A. (1998). Reliability of noninvasive methods to measure cardiac autonomic function. *Canadian Journal of Applied Physiology, 23*, 396–408.

Andreassi, J. L. (2000). *Psychophysiology: Human behavior and physiological response*, 4th ed. Hillsdale, NJ: Erlbaum.

Arnett, P. A., & Newman, J. P. (2000). Gray's three-arousal model: an empirical investigation. *Personality and Individual Differences, 28*, 1171–1189.

Backs, R. W. (2001). An autonomic space approach to the psychophysiological assessment of mental workload. In P. A. Hancock, & P. A. Desmond (Eds.), *Stress, workload, and fatigue* (pp. 279–289). Mahwah, NJ: Lawrence Erlbaum.

Backs, R. W., & Boucsein, W. (2000). *Engineering psychophysiology. Issues and applications*. Mahwah, NJ: Lawrence Erlbaum.

Backs, R. W., & Grings, W. W. (1985). Effects of UCS probability on the contingent negative variation and electrodermal response during long ISI conditioning. *Psychophysiology, 22*, 268–275.

Berntson, G. G., Cacioppo, J. T., & Quigley, K. S. (1991). Autonomic determinism: The modes of autonomic control, the doctrine of autonomic space, and the laws of autonomic constraint. *Psychological Review, 98*, 459–487.

Berntson, G. G., Cacioppo, J. T., Quigley, K. S., & Fabro, V. T. (1994). Autonomic space and psychophysiological response. *Psychophysiology, 31,* 44–61.

Boucsein, W. (1992). *Electrodermal activity.* New York: Plenum Press.

Boucsein, W. (2000). The use of psychophysiology for evaluating stress-strain processes in human-computer interaction. In R. W. Backs & W. Boucsein (Eds.), *Engineering psychophysiology. Issues and applications* (pp. 289–309). Mahwah, NJ: Lawrence Erlbaum.

Boucsein, W. (2001). Psychophysiological methods. In W. Karwowski (Ed.), *International encyclopedia of ergonomics and human factors* (Vol. 3, pp. 1889–1895). London: Taylor & Francis.

Boucsein, W. (2005). Electrodermal measurement. In N. Stanton, A. Hedge, K. Brookhuis, E. Salas, & H. Hendrick (Eds.), *Handbook of human factors and ergonomics methods* (pp. 18-1–18-8). London: CRC Press.

Boucsein, W. (2006). Psychophysiologische Methoden in der Ingenieurpsychologie. In B. Zimolong, & U. Konradt (Eds.), *Enzyklopädie der Psychologie. Themenbereich D Praxisgebiete. Serie III Wirtschafts-, Organisations- und Arbeitspsychologie. Band 2 Ingenieurpsychologie* (pp. 317–358). Göttingen: Hogrefe.

Boucsein, W., & Backs, R. W. (2000). Engineering psychophysiology as a discipline: Historical and theoretical aspects. In R. W. Backs & W. Boucsein (Eds.), *Engineering psychophysiology. Issues and applications* (pp. 3–30). Mahwah, NJ: Lawrence Erlbaum.

Boucsein, W., Haarmann, A., & Schaefer, F. (2005). The usability of cardiovascular and electrodermal measures for adaptive automation during a simulated IFR flight mission. *Psychophysiology, 42 (Suppl. 1),* S26.

Boucsein, W., & Ottmann, W. (1996). Psychophysiological stress effects from the combination of night-shift work and noise. *Biological Psychology, 42,* 301–322.

Boucsein, W., Schaefer, F., Kefel, M., Busch, P., & Eisfeld, W. (2002). Objective emotional assessment of tactile hair properties and their modulation by different product worlds. *International Journal of Cosmetic Science, 24,* 135–150.

Boucsein, W., & Thum, M. (1997). Design of work/rest schedules for computer work based on psychophysiological recovery measures. *International Journal of Industrial Ergonomics, 20,* 51–57.

Brookings, J. B., Wilson, G. F., & Swain, C. R. (1996). Psychophysiological responses to changes in workload during simulated air traffic control. *Biological Psychology, 42,* 361–377.

Cacioppo, J. T., Berntson, G. G., Larsen, J. T., Poehlmann, K. M., & Ito, T. A. (2000). The psychophysiology of emotion. In R. Lewis & J. M. Haviland-Jones (Eds.), *The handbook of emotion* (2nd ed., pp. 173–191). New York: Guilford Press.

Costa, P. T., & McCrae, R. R. (1992). Four ways five factors are basic. *Personality and Individual Differences, 13,* 653–665.

Davidson, R. J. (1993).The neuropsychology of emotion and affective style. In R. Lewis & J. M. Haviland (Eds.), *Handbook of emotions* (pp. 143–154). New York: Guilford.

Davidson, R. J., & Fox, N. A. (1982). Asymmetrical brain activity discriminates between positive and negative stimuli in human infants. *Science, 218,* 1235–1237.

DeLong, M. R., Georgopoulos, A. P., & Crutcher, M. D. (1983). Cortico-basal ganglia relations and coding of motor performance. In J. Massion, J. Paillard, W. Schultz, & M. Wiesendanger (Eds.), *Experimental Brain Research, Vol. 49* (pp. 30–40). Heidelberg: Springer.

Diedrich, O., Naumann, E., Maier, S., & Becker, G. (1997). A frontal positive slow wave in the ERP associated with emotional slides. *Journal of Psychophysiology, 11,* 71–84.

Duffy, E. (1957). The psychological significance of the concept of "arousal" or "activation." *Psychological Review, 64,* 265–275.

Duffy, E. (1972). Activation. In N. S. Greenfield, & R. A. Sternbach (Eds.), *Handbook of psychophysiology* (pp. 577–622). New York: Holt.

Eisfeld, W., Wachter, R., Schaefer, F., & Boucsein, W. (2007). Objective emotional assessment of perceivable wellness effects. *Cosmetics & Toiletries, 122,* 63–72.

Ekman, P., & Friesen, W. V. (1976). *Pictures of facial affect.* Palo Alto: Consulting Psychologists Press.

Eysenck, H. J. (1967). *The biological basis of personality.* Springfield: Thomas.

Eysenck, H. J. (1994). Personality. Biological foundations. In P. A. Vernon (Ed.), *The neuropsychology of individual differences.* New York: Academic Press.

Eysenck, H. J., & Eysenck, M. W. (1985). *Personality and individual differences.* New York: Plenum Press.

Eysenck, S., & Zuckerman, M. (1978). The relationship between sensation-seeking and Eysenck's dimensions of personality. *British Journal of Psychology, 69,* 483–487.

Fahrenberg, J. (1987). Concepts of activation and arousal in the theory of emotionality (neuroticism): A multivariate conceptualization. In J. Strelau & H. J. Eysenck (Eds.), *Personality dimensions and arousal* (pp. 99–120). New York: Plenum Press.

Fahrenberg, J., Walschburger, P., Foerster, F., Myrtek, M., & Müller, W. (1979). *Psychophysiologische Aktivierungsforschung.* Munich: Minerva Publikation.

Fahrenberg, J., Walschburger, P., Foerster, F., Myrtek, M., & Müller, W. (1983). An evaluation of trait, state, and reaction aspects of activation processes. *Psychophysiology, 20,* 188–195.

Fowles, D. C. (1980). The three arousal model: Implications of Gray's two-factor learning theory for heart rate, electrodermal activity, and psychopathy. *Psychophysiology, 17,* 87–104.

Gander, P. H., Myhre, G., Graeber, C. R., Andersen, H. T., & Lauber, J. K. (1989). Adjustment of sleep and the circadian temperature rhythm after flights across nine time zones. *Aviation, Space, and Environmental Medicine, 60,* 733–743.

Gevins, A., Leong, H., Du, R., Smith, M. E., Le, J., DuRousseau, D., Zhang, J., & Libove, J. (1995). Towards measurement of brain function in operational environments. *Biological Psychology, 40,* 169–186.

Gomez, R., & McLaren, S. (1997). The effects of reward, heart rate and skin conductance level during instrumental learning. *Personality and Individual Differences, 23,* 305–316.

Gray, J. A. (1972). The psychophysiological basis of introversion-extraversion: A modification of Eysenck's theory. In V. D. Nebylitsyn & J. A. Gray (Eds.), *The biological bases of individual behaviour.* New York: McGraw Hill.

Gray, J. A. (1973). Causal theories of personality and how to test them. In J. R. Royce (Ed.), *Multivariate analysis and psychological theory* (pp. 409–463). New York: Academic Press.

Gray, J. A. (1981). A critique of Eysenck's theory of personality. In H. J. Eysenck (Ed.), *A model for personality* (pp. 246–276). New York: Springer.

Gray, J. A. (1982). *The neuropsychology of anxiety: An inquiry into the functions of the septo-hippocampal system.* Oxford: Clarendon Press.

Greenwald, M. K., Cook, E. W., & Lang, P. J. (1989). Affective judgement and psychophysiological response: Dimensional covariation in the evaluation of pictorial stimuli. *Journal of Psychophysiology, 3,* 51–64.

Gundel, A., Drescher, J., Maaß, H., Samel, A., & Vejvoda, M. (1995). Sleepiness of civil airline pilots during two consecutive night flights of extended duration. *Biological Psychology, 40,* 131–141.

Hankins, T. C., & Wilson, G. F. (1998). A comparison of heart rate, eye activity, EEG and subjective measures of pilot mental workload during flight. *Aviation, Space, and Environmental Medicine, 69,* 360–367.

Hanson, E. K. S., Schellekens, J. M. H., Veldman, J. B. P., & Mulder, L. J. M. (1993). Psychomotor and cardiovascular consequences of mental effort and noise. *Human Movement Science, 12,* 607–626.

Hohnsbein, J., Falkenstein, M., & Hoormann, J. (1995). Effects of attention and time-pressure on P300 subcomponents and implications for mental workload research. *Biological Psychology, 40,* 73–81.

Hoshi, Y. (2003). Functional near-infrared optical imaging: Utility and limitations in human brain mapping. *Psychophysiology, 40,* 511–520.

Hugdahl, K. (1995). *Psychophysiology. The mind-body perspective.* Cambridge, MA: Harvard University Press.

Hustmyer, F. E. Jr. & Burdick, J. A. (1965). Consistency and test-retest reliability of spontaneous autonomic nervous system activity and eye movements. *Perceptual and Motor Skills, 20,* 1225–1228.

Johnsen, B. H., Thayer, J. F., & Hugdahl, K. (1995). Affective judgment of the Ekman faces: A dimensional approach. *Journal of Psychophysiology, 9,* 193–202.

Kahneman, D. (1973). *Attention and effort.* Englewood Cliffs, NJ: Prentice-Hall.

Kallinen, K., & Ravaja, N. (2005). Effects of the rate of computer-mediated speech on emotion-related subjective and physiological responses. *Behaviour & Information Technology, 24,* 365–373.

Kleinginna, P. R., & Kleinginna, A. M. (1981). A categorized list of emotion definitions, with suggestions for a consensual definition. *Motivation & Emotion, 5,* 345–379.

Klorman, R., Weissberg, R. P., & Wiesenfeld, A. R. (1977). Individual differences in fear and autonomic reactions to affective stimulation. *Psychophysiology, 14,* 45–51.

Kodama, H., Yoshida, T., Yamauchi, Y., Takahashi, A., & Echigo, J. (1997). Application of functional MRI to ergonomics. *International Ergonomics Association 13th Triennial Congress, Tampere, Finland,* 1–3.

Kosselyn, S. M., & Andersen, R. A. (1992). *Frontiers in cognitive neuroscience.* Cambridge, MA: MIT Press.

Kuhmann, W., Boucsein, W., Schaefer, F., & Alexander, J. (1987). Experimental investigation of psychophysiological stress reactions induced by different system response times in human-computer interaction. *Ergonomics, 30,* 933–943.

Lacey, J. I. (1967). Somatic response patterning and stress: Some revisions of activation theory. In M. H. Appley, & R. Trumbull (Eds.), *Psychological stress: Issues in research* (pp. 14–37). New York: Appleton-Century-Crofts.

Lang, P. J., Greenwald, M. K., Bradley, M. M., & Hamm, A. O. (1993). Looking at pictures: Affective, facial, visceral, and behavioral reactions. *Psychophysiology, 30,* 261–273.

Larsen, R. J., & Diener, E. (1992). Promises and problems with the circumplex model of emotion. In M. S. Clark (Ed.), *Review of personality and social psychology annual, Vol. 13: Emotion* (pp. 25–59). Thousand Oaks, CA: Sage.

Larsen, J. T., Norris, C. J., & Cacioppo, J. T. (2003). Effects of positive and negative affect on electromyographic activity over zygomaticus major and corrugator supercilii. *Psychophysiology, 40,* 776–785.

Levenson, R. W. (1988). Emotion and the autonomic nervous system: A prospectus for research on autonomic specificity. In H. L. Wagner (Ed.), *Social psychophysiology and emotion: Theory and clinical applications* (pp. 17–42). New York: John Wiley & Sons.

Levenson, R. W., Ekman, P., & Friesen, W. V. (1990). Voluntary facial action generates emotion-specific autonomic nervous system activity. *Psychophysiology, 27,* 363–384.

Lindsley, D. B. (1951). Emotion. In S. S. Stevens (Ed.), *Handbook of experimental psychology* (pp. 473–516). New York: John Wiley & Sons.

Luczak, H., & Göbel, M. (2000). Signal processing and analysis in application. In R. W. Backs & W. Boucsein (Eds.), *Engineering psychophysiology: Issues and applications* (pp. 79–110). Mahwah, NJ: Lawrence Erlbaum.

Malmo, R. B. (1962). Activation. In A. J. Bachrach (Ed.), *Experimental foundations of clinical psychology* (pp. 386–422). New York: Basic Books.

Mandryk, R. L., Inkpen, K. M., & Calvert, T. W. (2006). Using psychophysiological techniques to measure user experience with entertainment technologies. *Behaviour & Information Technology, 25,* 141–158.

McGregor, D. K., & Stern, J. A. (1996). Time on task and blink effects on saccade duration. *Ergonomics, 39,* 649–660.

McGuinness, D., & Pribram, K. (1980). The neuropsychology of attention: Emotional and motivational controls. In M. C. Wittrock (Ed.), *The brain and psychology* (pp. 95–139). New York: Academic Press.

Min, Y. K., Chung, S. C., & Min, B. C. (2005). Physiological evaluation on emotional change induced by imagination. *Applied Psychophysiology and Biofeedback, 30,* 137–150.

Mulder, G., Mulder, L. J. M., Meijman, T. F., Veldman, J. B. P., & van Roon, A. M. (2000). A psycho-physiological approach to working conditions. In R. W. Backs & W. Boucsein (Eds.), *Engineering psychophysiology. Issues and applications* (139–159). Mahwah, NJ: Lawrence Erlbaum Associates.

Muter, P., Furedy, J. J., Vincent, A., & Pelcowitz, T. (1993). User-hostile systems and patterns of psycho-physiological activity. *Computers in Human Behavior, 9,* 105–111.

Nummenmaa, L., Hyönä, J., & Calvo, M. G. (2006). Eye movement assessment of selective attentional capture by emotional pictures. *Emotion, 6,* 257–268.

Ohsuga, M., Shimomo, F., & Genno, H. (2001). Assessment of phasic work stress using autonomic indices. *International Journal of Psychophysiology, 40,* 211–220.

Osgood, C. E. (1966). Dimensionality of the semantic space for communication via facial expressions. *Scandinavian Journal of Psychology, 7,* 1–30.

Palomba, D., Sarlo, M., Angrilli, A., Mini, A., & Stegagno, L. (2000). Cardiac responses associated with affective processing of unpleasant film stimuli. *International Journal of Psychophysiology, 36,* 45–57.

Panksepp, J. (1982). Toward a general psychobiological theory of emotions. *The Behavioral and Brain Sciences, 5,* 407–467.

Pribram, K. H., & McGuinness, D. (1975). Arousal, activation, and effort in the control of attention. *Psychological Review, 82,* 116–149.

Rau, H., Furedy, J. J. & Elbert, T. (1996). PRES- and orthostatic-induced heart-rate changes as markers of labile hypertension: Magnitude and reliability measures. *Biological Psychology, 42,* 105–115.

Richter, P., Wagner, T., Heger, R., & Weise, G. (1998). Psychophysiological analysis of mental load during driving on rural roads—a quasi-experimental field study. *Ergonomics, 41,* 593–609.

Rimm-Kaufman, S. E., & Kagan, J. (1996). The psychological significance of changes in skin temperature. *Motivation and Emotion, 20,* 63–78.

Romet, T. T., & Frim, J. (1987). Physiological responses to fire fighting activities. *European Journal of Applied Physiology, 56,* 633–638.

Roscoe, A. (1993). Heart rate as a psychophysiological measure for in-flight workload assessment. *Ergonomics, 36,* 1055–1062.

Rose, R. M., & Fogg, L. F. (1993). Definition of a responder: Analysis of behavioral, cardiovascular, and endocrine responses to varied workload in air traffic controllers. *Psychosomatic Medicine, 55,* 325–338.

Routtenberg, A. (1968). The two-arousal hypothesis: Reticular formation and limbic system. *Psychological Review, 75,* 51–80.

Russell, J. A. (1980). A circumplex model of affect. *Journal of Personality and Social Psychology, 39,* 1161–1178.

Sammler, D., Grigutsch, M., Fritz, T., & Koelsch, S. (2007). Music and emotion: Electrophysiological correlates of the processing of pleasant and unpleasant music. *Psychophysiology, 44,* 293–304.

Scerbo, M. W., Freeman, F. G., & Mikulka, P. J. (2000). A biocybernetic system for adaptive automation. In R. W. Backs & W. Boucsein (Eds.), *Engineering psychophysiology. Issues and applications* (pp. 241–253). Mahwah, NJ: Lawrence Erlbaum.

Schienle, A., Stark, R., & Vaitl, D. (2001). Evaluative conditioning: A possible explanation for the acquisition of disgust responses. *Learning and Motivation, 32,* 65–83.

Schlosberg, H. (1954). Three dimensions of emotions. *Psychological Review, 61,* 81–88.

Schwartz, G. E., Weinberger, D. A., & Singer, J. A. (1981). Cardiovascular differentiation of happiness, sadness, anger, and fear following imagery and exercise. *Psychosomatic Medicine, 43,* 343–364.

Segalowitz, S. J., & Barnes, K. L. (1993). The reliability of ERP components in the auditory oddball paradigm. *Psychophysiology, 30,* 451–459.

Simons, R. F., Detenber, B. H., Roedema, T. M., & Reiss, J. E. (1999). Emotion processing in three systems: The medium and the message. *Psychophysiology, 36,* 619–627.

Sirevaag, E. J., Kramer, A. F., Wickens, C. D., Reisweber, M., Strayer, D. L., & Grenell, J. F. (1993). Assessment of pilot performance and mental workload in rotary wing aircraft. *Ergonomics, 36,* 1121–1140.

Sokolov, E. N. (1963). Higher nervous function: The orienting reflex. *Annual Review of Physiology, 25,* 545–580.

Sokolov, E. N., & Boucsein, W. (2000). A psychophysiological model of emotion space. *Integrative Physiological and Behavioral Science, 35,* 81–119.

Stelmack, R. M. (1981). The psychophysiology of extraversion and neuroticism. In H. J. Eysenck (Ed.), *A model for personality.* New York: Springer.

Stemmler, G. (1989). The autonomic differentiation of emotions revisited: Convergent and discriminant validation. *Psychophysiology, 26,* 617–632.

Sterman, M. B., & Mann, C. A. (1995). Concepts and applications of EEG analysis in aviation performance evaluation. *Biological Psychology, 40,* 115–130.

Stern, J. A., Boyer, D., & Schroeder, D. (1994). Blink rate: A possible measure of fatigue. *Human Factors, 36,* 285–297.

Stern, R. M., Ray, W. J., & Quigley, K. S. (2001). *Psychophysiological recording.* New York: Oxford University Press.

Taheri, B. A., Knight, R. T., & Smith, R. L. (1994). A dry electrode for EEG recording. *Electroencephalography and Clinical Neurophysiology, 90,* 376–383.

Teplov, B. M., & Nebylitsin, V. D. (1971). Eigenschaften und Typen des Nervensystems. In T. Kussmann, & H. Kölling (Eds.), *Biologie und Verhalten.* Bern: Huber.

Turpin, G. (1985). Ambulatory psychophysiological monitoring: Techniques and applications. In D. Papakostopoulos, S. Butler, & I. Martin (Eds.), *Experimental and clinical neuropsychophysiology* (pp. 700–728). London: Croom Helm.

Veltman, J. A., & Gaillard, A. W. K. (1998). Physiological workload reactions to increasing levels of task difficulty. *Ergonomics, 41,* 656–669.

Vitalis, A., Pournaras, N. D., Jeffrey, G. B., Tsagarakis, G., Monastiriotis, G., & Kavvadias, S. (1994). Heart rate strain indices in Greek steelworkers. *Ergonomics, 37,* 845–850.

Waters, W. F., Williamson, D. A., Bernard, B. A., Blouin, D. C., & Faulstich, M. E. (1987). Test-retest reliability of psychophysiological assessment. *Behavioral Research and Therapy, 25,* 213–221.

Weiner, J. S. (1982). The Ergonomics Society—the Society's lecture 1982: The measurement of human workload. *Ergonomics, 25,* 953–965.

Wesseling, K. H., Settels, J. J., & de Wit, B. (1986). The measurement of continuous finger arterial pressure noninvasively in stationary subjects. In T. H. Schmidt, T. M. Dembrowski, & G. Blümchen (Eds.), *Biological and psychological factors in cardiovascular disease* (pp. 355–375). Berlin: Springer.

Weyers, P., Mühlberger, A., Hefele, C., & Pauli, P. (2006). Electromyographic responses to static and dynamic avatar emotional facial expressions. *Psychophysiology, 43,* 450–453.

Wickens, C. D. (2002). Multiple resources and performance prediction. *Theoretical Issues in Ergonomics Sciences, 3,* 159–177.

Wientjes, C. J. E. (1992). Respiration in psychophysiology: Methods and applications. *Biological Psychology, 34,* 179–203.

Winton, W. M., Putnam, L. E., & Krauss, R. M. (1984). Facial and autonomic manifestations of the dimensional structure of emotion. *Journal of Experimental Social Psychology, 20,* 195–216.

Yagi, A. (1995). Eye fixation-related potential as an index of visual function. In T. Kikuch, H. Sakuma, I. Saito, & K. Tsuboi (Eds.), *Biobehavioral Self-Regulation* (pp. 177–181). Tokyo: Springer.

36

Biometrics

Yingzi Du

36.1 Introduction

Biometrics attempts to automatically identify or verify a person using physical, biological, and behavior characteristics, including face, iris, fingerprints, hand geometry, voice, and so on. Compared to the traditional identification and verification methods (such as paper ID, plastic ID card, or password), biometrics is more convenient for users, reduces fraud, and is more secure. It is becoming an important ally of security, intelligence, law enforcement, and e-commerce. In the Executive Summary of the 9/11 Commission Report (2004), the following necessary improvements were recommended by the commission members in "Protect against and Prepare for Terrorist Attacks":

> Address problems of screening people with biometric identifiers across agencies and governments, including our border and transportation systems, by designing comprehensive screening systems that address common problems and sets common standards.
>
> Quickly complete a biometric entry-exit screening system, one that also speeds qualified travelers.

The applications of biometrics are mainly for verification, identification, and watchlists.

A verification process asks the user to input a username or ID number, and verifies the person against information stored in the database. For example, if you claim you are "Joe," the biometric system will try to pull out Joe's data from the system and compare it with yours. The output of the biometric system will be "yes, it is Joe" or "no, it is not Joe."

TABLE 36.1 Tasks of the Biometrics

	Matching	Solve Question	Difficulty Level
Verification	One-to-One	"Are you who you say you are?"	Hard
Identification	One-to-Many	"Who are you?"	Harder
Watchlist	A few-to-Many	"Are you a bad guy?"	Hardest

In an identification process, the user only needs to present the biometric data. The system will match it against all the biometric data in the database. For example, you only need to submit your data to the system, and the system will compare your data with every data in the database. The output of the biometric system will be, "yes, it is x," or "no, x is not in our database."

Watchlist refers to a small list of most-wanted criminals or terrorists. The watchlist function is very similar to the identification process. However, those on the watchlist do not want to be identified. Therefore, they will not cooperate with the biometric system. For example, there is a terrorist who wants to sneak through a security check. He will try his best to disguise himself and not be identified.

Table 36.1 compares these three functions of biometrics. Only verification is one-to-one matching, and requires the users to input their username or ID number. The watchlist is the most challenging because this group of people will try their best to sneak through any identification system.

Table 36.2 shows a list of biometrics we are using to identify/verify a person. The physiological biometrics are what we were born with as a person, such as the face, fingerprint, iris, hand geometry, ear, and DNA. The behavioral biometrics are what we developed over these years as a habit, such as voice, gait, signature, and keystroke dynamics. In general, physiological biometric traits are more stable than behavioral biometric traits.

In this chapter, we will describe a typical biometric system, and introduce fingerprint, face, iris, voice, hand geometry, ear, gait, vein, footprint, and soft biometrics recognition technologies. We especially discuss and compare different kinds of fingerprint, face, and iris recognition systems. In addition, we discuss the strengths and weaknesses of each biometric technology and how to combine multiple biometric systems to achieve better performance. In the end, we discuss and analyze the challenges and trends of biometrics.

36.2 A Typical Biometric System

A biometric system usually includes two subsystems: the biometric enrollment system (fig. 36.1a) and biometric matching system (fig. 36.1b).

For a biometric trait to be recognized and identified by the biometric system, an enrollment step will be necessary. This step usually takes multiple signals (images) from the same biometric trait of a person. The biometric system generates the templates of the signals (images) and fuses them to get a higher quality signal (image). A template is then generated from this signal and saved in the database. In this step, the user cooperation is usually necessary.

TABLE 36.2 List of Biometrics

Physiological	Behavioral
Face	Voice
Fingerprint	Gait
Iris	Keystroke Dynamics
Hand Geometry	Odor
Ear	Signature/Handwriting
Vein	
Footprint	
Retina	

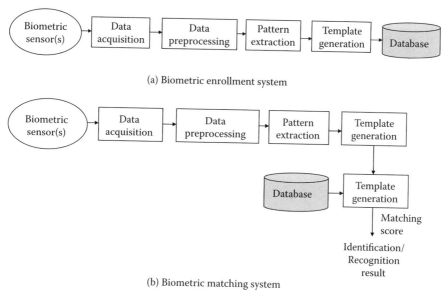

(a) Biometric enrollment system

(b) Biometric matching system

FIGURE 36.1 The diagram of a typical biometric system.

The biometric enrollment system includes the biometric sensor(s) module, the data acquisition module, the data preprocessing module, the pattern extraction module, the template generation module, and the biometric database module.

The biometric sensor(s) module sense and acquire data from biometric traits. Examples of the sensor are: digital camera (for face), fingerprint scanner (for fingerprint), recorder (for voice), and near-infrared (NIR) camera (for iris). The sensor can be triggered automatically when the biometric traits (identification applications) or username/ID number (verification applications) are presented.

The data acquisition module interprets the biometric data into digital signals (images) and right formats for future processing.

The data preprocessing module processes these signals to reduce the noise and extract the biometric data from the background. For example, for face recognition, the face will be detected and extracted in this module. In addition, the extracted faces will be normalized to have same sizes for further processing. The accuracy of data preprocessing is very important to the system performance.

The pattern extraction module uses mathematical models to analyze the special patterns of individual biometric traits and extract the distinctive features. These features could be some frequency characteristics or spatial correlations. The amount of extractable distinctive features will depend on biometric trait characteristics, and the mathematical models used for feature extraction. For example, if hand shape is used as a biometric trait for human identification, the amount of extractable distinctive feature is not high. There are a lot of hands that may have a similar shape. As a result, the recognition accuracy using hand shape will not be high. If iris is used as a biometric trait for human identification, the amount of extractable distinctive feature is high. The accuracy will be largely affected by the mathematical models used for feature extraction.

The template generation module will encode/quantize the extracted features into templates. The goal of this module is to reduce unnecessary redundancy in the data and to improve the efficiency in matching step.

The templates will be saved in the biometric database for further identification/recognition. Sometimes, for security purpose, the templates will first be encrypted before being saved in the database.

Compared to the biometric enrollment system, the biometric matching system adds the pattern matching module. In the biometric matching system, the newly sensed biometric data will be first processed similarly as the enrollment data and the system will generate the pattern templates from the data. The pattern matching module compares the newly generated templates with those in the biometric database, calculates matching scores, and does identification/recognition based on the matching scores. If the matching score is higher than the predetermined threshold, the system identifies/verifies it.

The false acceptance rate (FAR) and the false rejection rate (FRR) are used to measure if the biometric system is reliable (Woodward et al., 2002). A biometric system that generates high score of either FAR or FRR is not reliable and cannot be used.

The FAR measures the percentage of incorrect identification:

$$FAR(\%) = \frac{\text{Number of false acceptance}}{\text{Total number of acceptances by the system}} \times 100\% \tag{36.1}$$

The FRR measures the percentage of incorrect rejection:

$$FRR(\%) = \frac{\text{Number of false rejections}}{\text{Total number of rejections by the system}} \times 100\% \tag{36.2}$$

Receiver operating characteristic (ROC) curve is often used to measure the accuracy of the system performance. An ROC curve is a plot of FAR against FRR. An ROC curve can give a balanced view of FAR and FRR. To compare two ROC curves, equal error rate (ERR) is often used as an important parameter. ERR is defined as the decision threshold of a system set so that the proportion of false rejections will be approximately equal to the proportion of false acceptances (Woodward et al., 2002).

For a biometric system, if the output is not an identification result (one-to-many match) but a recognition result (few-to-many match), a cumulative math characteristic (CMC) curve is often used to measure the accuracy. The CMC curve is a plot of the accuracy percentage of the system in determining whether a test image is correctly identified in the top n matches, as n is varied. CMC curve is popularly used in analyzing face recognition system performance.

In addition, a biometric system performance will be judged by its failure to enroll rate (FTER). The FTER measures the percentage of unsuccessful enrollment. The unsuccessful enrollment could be because the acquired samples do not contain a sufficient quality to create a template. The FTER is defined as:

$$FTER(\%) = \frac{\text{Number of unsuccessful enrollments}}{\text{Total number of enrollment attempts}} \times 100\% \tag{36.3}$$

FTER is considered to be an important parameter for user acceptance.

36.3 Fingerprint Recognition

Using fingerprint as a way to identify a person began in ancient history. In 1000 BCE, ancient Chinese and Babylonian civilizations used fingerprints to sign legal documents. In modern society, fingerprint has been a great aid to law enforcement and criminal identification. It is a well established and accepted method for personnel identification. Every finger has ridges (raised minute) and valleys on the fingertips. The ridges and valleys display a number of characteristics, which are called minutiae. The minutiae do not change naturally during a person's life. Based on the core shape of the minutiae, fingerprints can be divided into three major types: arches, loops, and whorls. It is not difficult for human eyes to categorize fingerprints. But it would be very challenging for human beings to correctly recognize a fingerprint among thousands of fingerprint images. Automatic fingerprint recognition methods are playing critical roles in identification.

Over the decades, many different algorithms have been designed and developed for analyzing fingerprint patterns and matching the unknown fingerprint with those inside the database. In general, the fingerprint patterns are analyzed in three different levels.

Global level: The algorithm looks at the overall shape of a fingerprint. For example, in this level, the algorithm will decide the type of the fingerprint (arch, loop, or whorl), and then use the overall directions of the ridges/valleys for matching. To identify a fingerprint in this level, it requires major parts of a fingerprint for recognition. The center of part of the fingerprint must be available for recognition. This has been very challenging in criminal identification and is inconvenient in daily use of fingerprint identification. The accuracy of the algorithm is often not high.

Local level: The algorithm looks at the minutiae details, such as the ridge termination, bifurcation, dot, or crossover. A popular approach in this level is to find interest points. When matching, the interest points from two fingerprints will be calculated to find the correlation. If the majority of interest points can be matched, the two fingerprints will be identified as the same. In this level, the center part of the fingerprint is often not used to extract the interest points. To identify a fingerprint in this level, a small amount of available fingerprints will be good enough as far as there has enough overlapping between two fingerprint images.

Very-fine level: The algorithm looks at finger sweat pores. It uses the position and shape of the pores for identification. A high-resolution fingerprint sensor is required in this level. It is more like skin recognition than fingerprint recognition. Due to the high requirement for the fingerprint scanner in this level, it is not used often. In addition, it will be very impossible to have latent fingerprints with such high resolution. Therefore, the applications of the algorithms in this level have been very restricted.

A variety of fingerprint recognition systems are built based on analyzing fingerprint patterns in local levels. Based on difference sensors used in fingerprint identification, the systems can be divided into three major categories: optical, solid-state, and ultrasound. There are four kinds of solid-state sensors: capacity, thermal, electrical field, and piezoelectric (pressure). Table 36.3 lists current fingerprint sensors, their principles to take fingerprint images, their features, and applications.

Currently, optical sensor and capacity sensor are most popularly used. The optical sensors can take very high quality fingerprint images, which are critical for robust and accurate identification. They have been widely used by law enforcement, such as FBI (background check), Department of Justice (criminal identification), and Homeland Security (US-VISIT program). The fingerprint images in FBI databases are acquired from optical sensors.

The capacity sensors are cheap and compact, but the resolution is not very high. These kinds of sensors are usually integrated with consumer electronics, such as cell phones, handheld devices, and notebooks.

36.4 Face Recognition

Faces are defined as "the front part of the human head, including the chin, mouth, nose, cheeks, eyes, and usually the forehead" by Webster's Dictionary. It would be very difficult to describe a face using simple shapes or patterns, which creates the challenge for a computer to recognize a face automatically. Face recognition is defined as "given still or video images of a scene, identify or verify one or more persons in the scene using a stored database of faces" (source: http://face-rec.org).

Face recognition is how a person normally practices personal identification. Humans are very good at recognizing faces. However, faces will change with age, and can be easily changed with daily makeup. A lot of people are doing plastic surgery to their faces, which challenges face recognition.

In 1971, Goldstein et al. used 21 specific subjective markers (such as hair color and lip thickness) to identify a person. In late 1980s, Kohonen used a simple neural net to perform face recognition for aligned and normalized face recognition. In 1988, Kirby and Sirovich applied principle component analysis (PCA) to face recognition. This is the milestone in face recognition. Since then, many researchers

TABLE 36.3   List of Fingerprint Sensors

Sensor Type		Brief Description	Features	Applications
Optical		Use light reflections of a fingerprint to take the digital image.	1. Generate high-resolution images 2. Sensitive to the dirt, grease, and surface condition of the finger 3. Not good for dry fingers	Law enforcement, gate/door access, computer access, and e-commerce, good for large database identification
Solid-State				
	Capacity	Generate the digital image based on the capacitance of each detailed location of a finger. Ridges have greater capacitance than valleys.	1. Cheap 2. Compact size 3. Easy to integrate with other devices/systems 4. Medium resolution 5. Sensitive to dirt, grease, and surface condition of the finger	Internet authentication, e-commerce, and personal computer access. (Capacity sensor is most popular used in this category.)
	Thermal	Generates the digital image based on temperature differentials between ridges and valleys.	The sensor may get hot easily.	It can be only used for small database identification.
	Electrical Field	Measures the electric field beyond the surface layer of the skin, where the fingerprint begins.	1. Not sensitive to dry, worn, or dirty skin 2. Proper ground is necessary for this sensor	
	Piezoelectric (pressure)	Generates the current when encountering the pressure from the finger.	The resolution is low.	
Ultrasound		Generate a digital image based on the sound echo signals of the fingerprint. The contact scattering from ridges and valleys are different.	1. Expensive 2. Not sensitive to any dirt, grease, etc. 3. Resolution is not high	Not popular because it is not cost effective.

and engineers have designed and developed a number of approaches for a computer to recognize faces automatically. Each method has its pros and cons. In general, the face characteristics are analyzed in six different approaches.

Holographic approach: The algorithm recognizes the face as a whole face. This is similar to how people recognize a person. It uses some mathematical model to extract the features of the entire face for template generation. Eigenface is the most famous approach in this level. The templates in the database are generated faces (eigenfaces).

Local feature level: The algorithm looks at main patterns of the face, such as the distance between the eyes, the distance between the eyes to the mouth, and the size of the nose. In this level, the accurate location of the eyes is usually very critical, because the distance between the eyes and the size of the eyeballs are usually used as the scalar and axis. The algorithms in the level will be very sensitive to expression, which can dramatically change the face characteristics.

Hybrid approach: A combination of holographic approach and local feature level approach. An example of this approach is to segment key parts of the face (such as eyes, mouth, and nose) for recognition separately. Then combine those matching scores with eigenface approach for recognition/identification.

Skin level: The algorithm looks at the skin pores, and reflections. High resolution is important for this level. In this level, location of eyes and tip of nose are usually used to register two faces for recognition/identification. Similar to local feature level approach, this kind of algorithm will be sensitive to expression.

3D shape: The algorithm looks at the face as a 3D object with curves. In this level, specially designed or arranged camera(s) are necessary to take the 3D image of a face. Nose shape will be most important in 3D face recognition. Currently, the accuracy of 3D face recognition is restricted by the resolution accuracy of 3D face cameras.

Multi-spectra analysis: The algorithm analyzes the multiple spectral face images. At this level, multiple wavelengths are used to take face images in different frequencies. In addition to normal (visual frequency) charge-couple device (CCD) camera, infrared (IR) camera is often used to get an IR image of the face.

In general, we classify face recognition systems into three categories based on the camera system used: 2D face recognition system; 3D face recognition system; and multi-spectral face recognition system. Table 36.4 is a brief description and comparison of these face recognition systems.

National Institute of Standards and Technology (NIST) has organized the face recognition grand challenge (FRGC). NIST provided a number of faces (indoor and outdoor) for research groups to test and evaluate their own algorithms. The evaluation results show a lot of improvement in face recognition algorithms in recent years. With passport style photos, the top algorithms can achieve over 99% accuracy. But face recognition accuracy can be significantly affected by shadows and complicated background. As a result, outdoor front face image recognition rate is still far from satisfactory. In the FRGC workshop, it is reported that a combination of 3D and 2D in face recognition can improve the accuracy.

TABLE 36.4 Face Recognition System

System	Camera Type	Features	Applications
2D			
Holographic approach		1. Low-resolution requirement. 2. Very efficient in generating templates and searching database. 3. Not sensitive to camera type. 4. Sensitive to illumination variance	1. Surveillance 2. Law enforcement 3. Gate/door control
Local feature level	Webcam, digital camera	1. Very high requirement in finding the accurate positions of eyes. 2. Low accuracy.	Usually used with other methods to achieve higher accuracy
Hybrid approach		1. Combination of holographic approach and local feature level approach. 2. The accuracy in registration and segmentation will affect the recognition accuracy dramatically.	Some publications in literature. Not popular in commercialized system yet.
Skin level	High-resolution webcam or digital camera	1. Capable in discriminating identical twins with high-resolution image and moderate makeup. 2. High-resolution requirement. 3. Sensitive to the image resolution. 4. Sensitive to different type of camera.	Good for small database application in high security scenario
3D	3D camera	1. Insensitive to illumination. 2. Sensitive to eyeglasses. 3. Low processing speed. 4. Require more memory and more storage. 5. The recognition accuracy is restricted by the camera resolution and 3D acquisition accuracy.	Price is an issue for its application. The recognition accuracy is not high and is not good for large database application.*
Multi-spectra	Multi-spectra camera	1. Insensitive to makeup. 2. Easily detect the faked face, such as rubber face or dummy face.	1. Good for liveness test. 2. Surveillance. 3. Price is still an issue. 4. Accuracy is also an issue.*

* It is a pretty new area. More research and improvements are necessary in this approach.

36.5 Iris Recognition

The iris is the pigmented area of the eye surrounding the pupil that contracts and expands to control the amount of light entering the eye. It gives color to the eye (Forrester et al., 2001). Every iris exhibits a distinct pattern that not only differs from person to person, but from left eye to right eye. Iris recognition systems use different methods to exploit this high level of distinction for the purpose of recognizing a person.

In 1987, ophthalmologists Flom and Safir first stated that the iris is very unique for each person and remains unchanged after the first year of human life, and they applied a patent for a manual process for iris identification. It is not a realistic approach for iris recognition but it does give the foundation for iris recognition. In 1994, Daugman invented a first automatic iris recognition system. The iris has proven to be the most stable and reliable means of biometric identification (Daugman, 2004). Figure 36.2 shows the diagram of iris recognition.

To obtain a clear iris image, a near infrared (NIR) camera with enough sensitivity in 800 nm is often used (www.iridian.com). In the visible light wavelengths, it is difficult to get the detailed iris patterns from the dark color eyes. But under the NIR light, very fine details from the iris (even dark ones) are visible. The acquired iris image by NIR camera is in grayscale. To perform iris recognition, the iris area should be first extracted from the obtained image. This step is called preprocessing. In the preprocessing step, the system detects the pupillary and limbic boundaries, eyelashes, and eyelids. In addition, the system will find locations of glare or other possible noise in the iris image.

After preprocessing, the iris patterns are extracted from the image. Then the system transfers the iris to the polar axis for pattern analysis and extraction. Usually, normalization is performed in this step to reduce the effect of pupil contraction or dilation. The normalized iris patterns are sent to pattern analysis and extraction. There are a number of different approaches to analyze the iris patterns and generate the iris templates. According to the various template generation methods used, these algorithms can be grouped into:

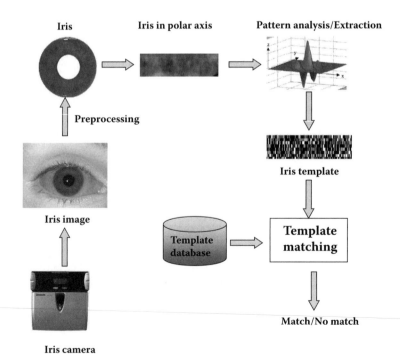

FIGURE 36.2   Iris recognition system.

Frequency domain methods: Using a waveform that is bounded in both frequency and duration to extract information. For example, Daugman used the phase measure of 2D Gabor wavelets as the iris code. The phase is quantized to four values (2 bits) and the iris code is 256 bytes long (Daugman, 2004). This method has been adopted by Iridian Technologies and currently is used in commercial iris recognition systems.

Spatial domain filter-based method: Analysis of the iris patterns in the spatial domain. For example, Wildes et al. decompose the iris region by constructing a Laplacian pyramid from an iris image. Recently, the authors designed a local texture analysis algorithm to calculate the local texture patterns (LTP) of iris images to generate a one-dimensional iris template (Du et al., 2006). This approach relaxed the requirement of a significant portion of the iris for identification and recognition.

The spatial domain filter-based methods and the frequency domain methods are related. The spatial domain-based methods often rely on a frequency-based decomposition of the iris image.

Other approaches: Some researchers have used independent component analysis (ICA) or principle component analysis (PCA) to analyze iris patterns.

Currently, there are five general categories of commercialized iris recognition systems: PC iris recognition system; walkup iris recognition system; standalone/portable iris recognition system; low-cost mobile personal iris recognition system; and remote iris recognition system (Table 36.5).

36.6 Voice Recognition

The vibration of the vocal cords, as well as the positions, shapes, and sizes of the various articulator changes over time to reflect the sound being produced (Rabiner et al., 1993). Speech is produced as a sequence of sounds. Information embedded in speech can be divided into three categories: linguistic (words in the speech); paralinguistic (the way of delivery); and nonlinguistic information (facial expression, hand gesture, and the speaker properties). Voice is a way for humans to identify a person. Even before a baby is born, the baby can distinguish mom's voice in the womb. It is claimed that "fetus' heart rate accelerates when mother reads, lowers when stranger reads" (www.webmd.com/content/article/64/72506.htm).

In 1936, the first machine to generate synthetic human voice from text was invented by Bell Lab. But voice recognition is much harder than voice synthetics. In the 1970s, the first product of voice recognition using speaker-dependent, small vocabulary, and discrete speech system was invented by Threshold Technology. In the late 1990s, it became possible to do real-time voice recognition.

Voice recognition in English is most studied because more researchers have worked on English speakers than any other languages. In English, the vowel sounds are perhaps the most interesting class of sounds. The English voice recognition systems rely heavily on vowel recognition to achieve high performance. But some other languages may not have such characteristics. Therefore, voice recognition can be very language dependent.

Applying voice recognition techniques to identify a person is called speaker identification. Compared to traditional voice recognition, the speaker identification does not care much about the content of the voice, but the person who is speaking. However, speaker recognition is very challenging because voice is a behavior biometrics, which is not as stable as physiological biometrics. It can be easily changed due to age, colds, cough, allergies, mood, accent, or language.

36.7 Other Biometrics

36.7.1 Hand Geometry Recognition

Hand geometry (or hand recognition) is used to identify a person based on the shape of one's hand. It usually measures the length of fingers, the shape of the palm, and thickness of the hand. Hand shape is not as unique as fingerprint or iris. It is very possible for two persons to have the same hand shape.

TABLE 36.5   Iris Recognition Systems

Typical System	Brief Description	Database	Applications
PC Iris Recognition System	Use a low-end iris camera, such as the Panasonic Authenticam. The iris camera is linked directly with a PC, commonly with a USB connection.	Small (<100 people)	Computer access control, Internet authentication and e-commerce applications.
Walkup Iris Recognition System	Use a server-client distribution system and comprise multiple iris cameras, such as the LG IrisAcess System. The enrollment is done within a secure environment with a separate enrollment iris camera. The identification is performed in a less secured environment, such as outside of a door or gate that leads to a secure area.	Large (>5000 persons)	Government, financial institutions, key research areas and other corporations for door/gate access control and positive identification.
Standalone/Portable Iris Recognition System [1]	A portable iris recognition system is a fully self-contained iris enrollment and recognition system. It is a stand-alone handheld device, such as PIER from Securimetrics Inc.	Small–medium (100–1000)	Iris recognition in the field for homeland security, law enforcement, or positive identification.
Low-Cost Mobile Personal Iris Recognition System [2]	Low-cost NIR camera that can be easily integrated with cell phone, laptop, or other devices/systems.	Very small database	For personal e-commerce, cell phone, and laptop log-in.
Remote Iris Recognition System (Sarnoff IOM) [3]	It can identify a person using iris over 10 feet distance. The system has much stronger NIR illuminators and the iris camera has been mounted with telescope-like optical lenses.	Small–medium (100–1000)	For security with minimal user cooperation.

[1] PIER Version 2.2. http://www.securimetrics.com.
[2] http://www.mobilewhack.com/reviews/oki_iris_recognition_for_mobile_devices.html.
[3] Sarnoff Iris on the Move (IOM), http://www.sarnoff.com/products/iris-on-the-move.

Usually, the left hand will have a similar hand shape as the right hand. Therefore, hand geometry can be only used for verification for small database. And it can not be used in high-security facilities. But hand geometry recognition has been popularly used in facilities as a way for access control. This is because hand geometry is not as invasive as fingerprint recognition or iris recognition. It is relatively easily to use and is well accepted by users.

36.7.2 Ear Recognition

Ear recognition uses the shape of the ear to identify a person. In 1989, Iannarelli found each person has unique ear shapes that can be used for personal identification. Ear recognition is based on distinctive ear shape and the structure of the cartilaginous, projecting portion of the outer ear. Different from the face, ears will not dramatically change from teenage to adult. In some states, for a minor to get a driver's license, the person is required to take a 45-degree photo with one ear visible. This is because while the face may dramatically change within several years the ear shape won't change. Recently, some researchers use 3D cameras to take ear shape for ear recognition and it is reported that this can improve the recognition accuracy. The challenges of ear recognition are: ears are not directly visible for many people (especially people with long hair) and it is hard to acquire a frontal ear image. Currently, ear recognition accuracy still needs improvement for large database applications.

36.7.3 Vein Recognition

Vein recognition is used to identify a person by recognizing the blood vessel patterns. Usually infrared light is used to detect vein vessel patterns from a hand or a finger. Vein patterns are developed before birth, and remain stable throughout a person's life. They are unique for a person, even identical twins. A person's left hand and right hand have different vein structures. It is not intrusive and has become popular in Asia.

36.7.4 Gait Recognition

Gait recognition is used to identify a person by the way one walks. Gait is a behavior biometric and therefore is not stable in a person's life. It can be changed due to age, mood, stuff a person is carrying, and road situation. But gait is a good way to remotely identify a person.

36.7.5 Footprint Recognition

Footprint recognition is used to identify a person based on one's footprint characteristics. It is similar to fingerprint recognition that ridges and valleys in the foot are used as characteristics in recognition. For an adult, a large scanner will be necessary to scan the full footprint. In addition, it will be inconvenient and invasive to have a person take off the shoes to take a footprint scan. However, footprint recognition is popularly used to identify a newborn baby. The fingerprint of a new baby is too small to have a decent identification. But a baby's foot is not very big and can be easily scanned by a small scanner for recognition. The accuracy of automatic footprint recognition is not well studied in the literature due to the limitation of the applications.

36.8 Soft Biometrics

Soft biometrics are the biometric traits that are not very unique to each person and can be easily changed even in a daily life (Jain et al., 2004). Examples of soft biometrics are weight and height. Weight can be easily changed by diet or simply by clothing. Height can be changed by age and footwear. Soft biometrics cannot be used alone for positive human identification. But it can provide some useful supplemental information for human identification. It is often used in criminal identification as additional evidence.

36.9 Multimodal Biometrics

A multimodal biometric system uses more than one physiological or behavioral characteristic for human identification. A unimodal biometric system uses only one physiological or behavioral characteristic for human identification. Each unimodal biometric system has its pros and cons.

36.9.1 Unimodal Biometric Systems

Table 36.6 compared different kinds of biometric technologies in terms of performance, cost, and their applications.

All of the above biometrics can be used for verification purposes. However, only fingerprint, iris, and vein can be used for reliable identification. Face and voice recognition systems do not need user cooperation, which are very suitable for surveillance.

Fingerprint and iris recognition systems are very reliable and accurate. They can be used for large database applications. Each has its advantages: Fingerprint recognition system is cheap; iris recognition system can achieve extremely high accuracy rate (1 in 131,000 error rate) [20]; vein recognition system is nonintrusive to the users, and is well accepted by Asian people.

36.9.2 Multimodal Biometric Systems

Figure 36.3 shows the block diagram of a multimodal biometric system, which integrates N (N>1) different types of unimodal biometric systems. Each unimodal biometric system will generate its matching scores. The multimodal biometric system will fuse these matching scores to make a final decision: match or non-match. Proper data fusion rules are very important for the success of multimodal biometric systems.

The multimodal biometric systems can improve the performance of unimodal biometric systems. For example, a face recognition system or a voice recognition system do not have high accuracy rate. But a multimodal biometrics system which is composed of a face recognition system and a voice recognition system could possibly achieve much higher accuracy. A face recognition system combined with an iris recognition system could improve the accuracy rate while keeping the ability of surveillance.

The multimodal biometric systems can reduce the vulnerability. A single biometric system might be easily fooled. However, it would be much more difficult to fake multiple biometrics traits to fool the system.

TABLE 36.6   Comparison of Biometric Systems

	Accuracy	Reliability	Stable	Intrusive	User Co-op	Large Population
Fingerprint	High	High	Yes	Yes	Yes	Yes
Face	Medium	Medium	No	Somewhat	No	No
Iris	Very High	Very High	Yes	Somewhat	Yes	Yes
Voice	Low	Low	No	No	No	No
Hand Geometry	Low	Low	Yes	No	No	No
Ear Shape	Medium	Medium	Yes	No	Yes	No
Vein	Medium	High	Yes	No	Yes	N/A

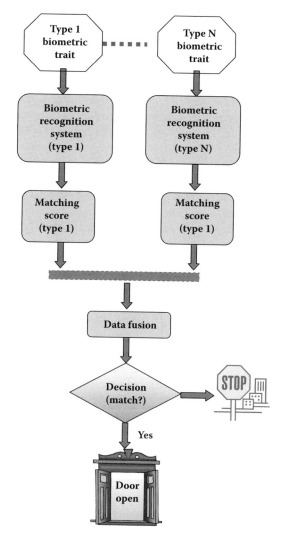

FIGURE 36.3 Diagram of multimodal biometric system.

However, multimodal biometric systems will ask the users to submit multiple biometric traits. User acceptance and convenience will be very important and should be seriously considered when designing a multimodal biometric system.

36.10 Trends

Any technology is a two-edged sword. While it can be properly used to improve the quality of life, misuse can do harm. It draws a great concern from the general public and government agencies about how to use and protect the biometric data. Biometric data are different from any ID number or password. It is a kind of identity for a person that is very difficult to be changed. For example, if a credit card number is stolen from the Internet, it will only take a little effort for the bank to reissue another one. But if a person's iris data were stolen, it would be impossible for a person to change the eye.

Privacy is another issue. While government agencies want to get more information from the general public for good purposes such as law enforcement, it is very possible that some of the information can be misused by a small group of bad people in or out of the government agencies. In addition, each person has a human right to privacy. Will collecting the biometric data be too intrusive?

36.10.1 Standardize

Currently, there are many of kinds of biometric systems using different interfaces, data formats, measurements, and outputs. This creates a tremendous problem for integrators and data exchanges. For example, fingerprints connected by different law enforcement centers may have different formats due to different fingerprint recognition systems used. It would be very hard to perform fingerprint recognition across different centers. As a result, this created the problem in criminal record share or cooperation among different law enforcement centers in criminal investigation.

In addition, each biometric system provider would claim its system has the best performance based on internal tests and evaluation methods. How could ender users choose the biometric system best for them? It would be important to standardize these systems, and have a standard performance test for each system. NIST (National Institute of Standards and Technology) has played a great role in standardizing biometric systems/technologies. International standards in biometrics will be important for developing compatible biometric systems. With the requirements of the US-VISIT program for some European countries to provide biometric passports, the international standardized biometric passport will be a key for the success of this program.

36.10.2 Less Vulnerable

The designers and operators of the biometric systems should consider possible attacks and reduce vulnerability of the system. These vulnerabilities include (but are not limited to) spoof, hacking and system attacking, and insider threat.

Spoof is used to fool the system using artificial or forged biometric trait(s)/templates. For example, some old versions of voice recognition could be fooled by using recorded voice. Liveness will be a key in preventing spoof. A liveness test determines whether the biometric traits are from a living person rather than an artificial or lifeless person. The liveness test is very important to prevent spoof. For example, the natural papillary response (changing pupil size in response to changes in illumination) can be used to confirm the liveness of an iris. The motion of a person could be used to confirm the liveness of a face. The response of a finger to a low electronic currency could be used to test the liveness of a finger.

36.10.3 Hacking and System Attack

Hacking is a big problem for any networked system. The criminals may hack the biometric system and alternate the biometric data in the database. It is very important for a biometrics system to encrypt the biometric data and personal information of the user. It is also important for the security of the network if the biometrics system uses server-client architecture.

36.10.4 Insider Threat

This is a human issue. Ethics education should be enforced. Companies/government agencies should be more careful about people who have the privilege to access the database or important information in the biometric system.

36.10.5 More Choices

In the near future, there will be more kinds of biometric systems available to fit different needs. It will be cheaper to integrate some biometric systems with consumer electronics, such as cell phones, handheld

PCs, and notebooks. There will be more choices in high-end biometric systems to be deployed in highly secured facilities. In addition, the multimodal biometric system may enable more choices for users in presenting biometric data.

36.11 Conclusion

Biometrics will be more popular and will be used more often in our daily lives to verify and identify a person automatically, because it is more convenient for users, reduces fraud, and is more secure. It is becoming an important ally of security, intelligence, law enforcement, and e-commerce.

In this chapter, we introduced the fingerprint, face, iris, voice, hand geometry, ear, vein, footprint, and soft biometrics recognition technologies. When used for human identification, each kind of biometric trait has its own advantages and disadvantages. Combining multiple biometric traits for human identification can improve overall accuracy. Therefore, multimodal biometrics will be a better choice.

In the near future, we expect to get more robust and less vulnerable biometric systems. At the same time, the smaller, cheaper biometric system will be preferred. With the advance of the technology and system, biometrics will become more and more popular for authentication and surveillance applications.

References

Daugman, J. (1994); Biometric Personal Identification System Based on Iris Analysis, United States Patent No. 5,291,560, Washington D.C.: U.S. Government Printing Office.

Daugman, J. (2004); "How Iris Recognition Works," *IEEE Transaction on Circuits and Systems for Video Technology*, 14(1), 21–30.

Du, Y.; Ives, R. W.; Etter, D.M.; and Welch, T.B. (2006); "Use of One-Dimensional Iris Signatures to Rank Iris Pattern Similarities," *Optical Engineering*, 45(3), 037201-1~10.

Flom, L.; and Safir, A. (1997); Iris Recognition System. United States Patent No. 4,641,349, Washington D.C.: U.S. Government Printing Office.

Forrester, J.; Dick, A.; McMenamin, P.; and Lee, W. (2001); *The Eye: Basic Sciences in Practice*, London: W B Saunders.

Jain, A. K.; Dass, S. D.; and Nandakumar, K.; "Soft Biometric Traits for Personal Recognition System," *Proceedings of International Conference on Biometric Authentication*, LNCS 3072, pp. 731–738, Hong Kong, China, July 2004.

Maltoni, D.; Maio, D.; Jain, A.K.; and Prabhakar, S. (2003); *Handbook of Fingerprint Recognition*, New York: Springer.

Rabiner, L.; Juang, B-H. (1993); *Fundamentals of Speech Recognition*, Englewood Cliffs, NJ: PTR Prentice-Hall Inc.

Woodward, J.D.; Orlans, N.M.; Higgins, P.T. (2002); *Biometrics*, Berkeley, CA: The McGraw-Hill Company.

37

Data Mining and Its Applications in Digital Human Modeling

Yan Liu

37.1 Introduction

Rapid growth of data acquisition systems, such as motion trackers, eye trackers, and brain imaging tools, has enabled the collection of large volumes of data on human behavior, performance, and psychophysical measures. In fact, the sheer volume of data has far exceeded humans' abilities to capture and analyze it without using powerful tools. Therefore, unless advanced knowledge discovery techniques are developed, researchers risk losing much valuable information from the data they have collected and stored. Data mining (DM) has become an active area of research and development in recent years. The emergence of DM as a new technology is due to the fast development and wide applications of database technologies. Many organizations already have the capability of collecting and managing massive quantities of data. DM techniques can take this evolutionary process beyond retrospective data access and management to prospective and proactive information delivery.

This chapter aims to provide a brief introduction to DM and illustrate how DM techniques can be applied in digital human modeling (DHM).

37.2 Data Mining Process

Many people view DM as a synonym for another widely used term, knowledge discovery in databases (KDD). A widely accepted definition of KDD was given in Fayyad et al. (1996): KDD is the nontrivial process of identifying valid, novel, potentially useful, and ultimately understandable patterns in data. DM is truly an interdisciplinary field in that it lies at the interface of database management, machine learning, pattern recognition, statistics, visualization, and many others.

Based on practical and real-world experiences, the CRoss Industry Standard Process for Data Mining (CRISP-DM, CRISP-DM Special Interest Group, 2000), shown in Figure 37.1, has been defined by a consortium of companies that used data mining from the days of its infancy. According to this definition, six main tasks are involved in DM: business understanding, data understanding, data preparation, modeling, evaluation, and deployment. As depicted in Figure 37.1, DM is not a one-way process; rather, it involves many iterations and may contain loops between any two steps.

37.2.1 Business Understanding

The first important step in the whole DM process is to understand the objectives to conduct DM, that is, the problems that need to be addressed. After the problems and objectives are determined, it is advisable to define the important performance criteria on which the success of the DM process can be judged. Such criteria can be objective or subjective. Another important task is to assess the current situations of the organization. Some questions need to be asked at this stage, such as how much background knowledge the organization has about the problem, where the data source is, how many resources are available for this project, and so forth.

37.2.2 Data Understanding

After the problem and a rough plan for its solution have been set up, the next step is to collect data that contain information and are relevant to the defined DM problem, either from scratch if no suitable data are available or from existing databases. After the data are acquired, they need to be described, defining the volume of data and identities and meanings of individual attributes, and describing the initial format of the data. Although data exploration/survey is not a required step, it is usually very helpful for understanding the data, such as identifying the distributions of attributes, relations among a small number of attributes, and results of simple aggregations. Statistical analysis, data visualization, and database queries can be helpful tools during this step. It is also important to verify the quality of the collected data, examining whether there are incomplete data, missing values, errors, and so on.

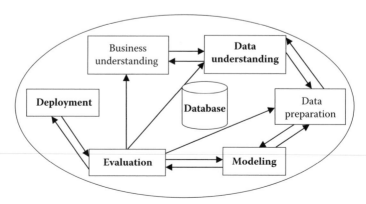

FIGURE 37.1   Cross-industry standard process for data mining. (From CRISP-DM Special Interest Group, *CRISP 1.0 Process and User Guide*, www.crisp-dm.org, 2000. With permission.)

37.2.3 Data Preparation

Data preparation is an extremely important but often neglected step in the DM process. The old saying "garbage in, garbage out" is particularly applicable to the typical DM projects in which large datasets are collected via some automatic methods such as sensors and web surveys. Major tasks in data preparation include data selection, data cleaning, data reduction, new data construction, and data formatting.

We need to select the data to be used for analysis using criteria including the relevance of the data to the data-mining goals and technical constraints such as limits on data volume and data types. Incomplete, noisy, and inconsistent data are common characteristics of real-world databases; therefore, data have to be cleansed in order to improve the efficiency and ease of the mining process. Examples of data-cleaning tasks are removing errors in data and filling in missing data with default values or estimates by modeling. The data construction step represents constructive operations on the selected data, such as generating new attributes or new records, merging tables, and transforming data. Data reduction is applied in DM to obtain a reduced representation of the dataset that is much smaller in volume, yet retains the important information. The new data construction step represents constructive operations on the selected data, for example, generating new attributes or new records, merging tables, and data transformation. The last data preparation step is data formatting, which involves syntactic modifications to the data without changing meaning. This modification may be required by the particular modeling tool chosen in the DM process.

37.2.4 Modeling

Modeling is the crucial step in which DM algorithms are applied in order to extract data patterns. The first step in the modeling task is to select appropriate DM modeling techniques based on the defined DM tasks and data characteristics. Some of the most commonly applied DM modeling techniques are described in the section "Data Mining Modeling Techniques." Each technique has its own pros and cons, and no single technique provides the best performance for all types of tasks. Therefore, a multi-strategy approach might be needed in dealing with real complex systems. Before we actually build a model, we need to generate a procedure or mechanism to test the model's quality and validity. For example, in classification and prediction tasks, it is common to use classification and prediction accuracy as quality measures for derived models. Next, the selected DM modeling techniques are run on the prepared dataset to build models. After the models are derived, they need to be assessed according to the users' domain knowledge and predefined success criteria and test design.

37.2.5 Evaluation

In addition to being evaluated in terms of their performance measures, such as accuracy, as described in the previous task, the models need to be evaluated with respect to the defined project objectives. Even if the models appear satisfactory, it may still be necessary to do a more thorough review of the data-mining engagement in order to determine if any important factor or task has been overlooked.

37.2.6 Deployment

If all the previous steps are satisfactory and the models fulfill the project objectives, the DM results can be deployed in the problem domain. Monitoring and maintenance are important issues if the data mining result becomes part of the day-to-day business and its environment. A final report is usually generated, explaining the deliverables and results produced and summarizing important points in the project and experiences gained in the project.

37.3 Data-Mining Modeling Techniques

Based on functionalities and application purposes, DM modeling techniques can be classified as concept/class description, association analysis, classification and prediction, cluster analysis, and evolution analysis (Han & Kamber, 2004).

37.3.1 Concept/Class Description

Data can be associated with concepts or classes. For instance, concepts of human operators include experienced operators and inexperienced operators, and classes of their workload include high workload, medium workload, and low workload. The task of concept/class description attempts to describe individual classes and concepts in summarized, concise, and yet precise terms. This task can be conducted via data characterization and data discrimination.

Data characterization refers to the process of summarizing the general characteristics or features of a target class of data, such as finding the characteristics of the tasks that result in high workload to operators. Simple descriptive statistical analysis (such as mean, median, variance, and quantities) and graphic displays (such as histograms, scatterplots, and quartile plots) can provide the first insights in the data. To best apply DM techniques, they should be integrated with data warehouses. A data warehouse is a subject-oriented, integrated, nonvolatile, and time-variant collection of data in support of management's decisions (Inmon, 2002). To facilitate complex analyses and visualization, the data in a data warehouse are typically modeled by a multidimensional database structure, where each dimension corresponds to an attribute or a set of attributes in the schema, and each cell stores the value of some aggregate measure. By providing multidimensional data views and the precomputation of summarized data, data warehouse systems are well suited for online analytical processing (OLAP). Typical OLAP operations include roll-up (increasing the level of aggregation) and drill-down (decreasing the level of aggregation or increasing detail) along one or more dimension hierarchies, slice-and-dice (selection and projection), and pivot (re-orienting the multidimensional view of data) (Chaudhuri & Dayal, 1997).

Data discrimination refers to the process of comparing the features of the target class with one or a set of contrasting classes. An example of the data discrimination operation is to compare the profiles (such as age, education, and workforce training) of operators with excellent performance and those with poor performance. The methods used for data discrimination are similar to those used for data characterization.

37.3.2 Association Analysis

Association rules algorithms are motivated by identifying rules that capture increased frequency of an item (or an attribute value) or a collection of items (Webb, 2000). An association rule is expressed in the form of $X \Rightarrow Y$, where X and Y are called *antecedent* and *consequent* of the association rule, respectively. An n-item set is a combination of n single items. These rules are computed from data, and, unlike the if-then rules of logic, association rules are probabilistic in nature.

One of the major problems with association rules mining is the overwhelming number of rules that can be generated, many of which may not be of interest. To address this issue, two important measures are often used to constrain derived rules. One measure is called the support of a rule, which is defined as the probability that a randomly selected data record will contain all items in both the antecedent and consequent of the rule. The other measure is the confidence of a rule, which is defined as the conditional probability that a randomly selected data record contains all items in the consequent of the rule given all items in the antecedent of the rule are present. The process of discovering association rules generally can be decomposed into two phases (Agrawal et al., 1993): Generate all n-item sets ($n \geq 2$) whose support values meet the specified minimum support threshold. These item sets are called large item sets. Then, from the large item sets, generate all rules that satisfy the specified minimum confidence threshold.

Association rules that satisfy the minimum support and confidence thresholds are not necessarily interesting. For instance, a rule with high confidence should not be confused with high correlation or high causality between the antecedent and consequence of the rule (Brijs et al., 2003). As a result, a number of other interestingness measures have been proposed, such as lift (Chen et al., 1996), chi-square test for correlation (Brin et al., 1997), collective strength (Aggarwal & Yu, 2001), and subjective measures like conformity and unexpectedness (Liu et al., 2000).

Association rules algorithms were developed originally for mining large item sets and their associated association rules, yet they have been extended to other applications, such as multidimensional association rules (Kamber et al., 1997), quantitative association rules (Srikant & Agrawal, 1996), and clustering association rules (Lent et al., 1997).

37.3.3 Classification and Prediction

Classification is the process of deriving a model that describes and distinguishes data classes or categories, so that the model can be used to predict the class of objects whose class labels are unknown. If the target attribute is a continuous variable, then a prediction model needs to be derived (Han & Kamber, 2004). Varieties of classification and prediction methods have been proposed, among which the most commonly applied include decision tree, artificial neural network (ANN), support vector machine (SVM), Bayesian classification, and regression. The first four techniques are briefly discussed in this section.

37.3.4 Decision Tree

As its name implies, a decision tree can be viewed as a model in the form of a tree structure. If the target attribute is categorical, then a classification tree is generated. On the other hand, if the target attribute is continuous, then a regression tree is generated (Breiman et al., 1984). A decision tree consists of decision nodes or leaf nodes. Each decision node has splits, testing the values of some function of its corresponding attribute, and each branch from the decision node corresponds to a distinct outcome of the test. Each leaf node gives the predicted class label or value of the target attribute. Decision tree is one of the most popular DM modeling techniques because a decision tree is easy to understand and can be constructed relatively quickly compared to other methods such as ANN and SVM. In addition, no prior assumptions about the data need to be made. Nevertheless, decision tree modeling also suffers from some problems. It is not flexible at modeling complex relationships in data. In addition, trees created from numeric datasets can be quite complex.

Although a variety of classification decision tree algorithms have been developed with different capabilities and requirements, most are variations of a core learning algorithm that employs a "greedy" top-down search through the space of possible decision trees, the most commonly used method for tree construction today (Murthy, 1998). The most crucial issues in decision tree modeling are selecting the splits and determining the size of the tree. Many goodness measures of splits that indicate their predictive power have been proposed. The three most popular are information gain (entropy) (Quinlan, 1993), Gini index of diversity (Breiman et al., 1984), and χ^2 statistics (Kass, 1980). An overly large tree is difficult for humans to understand and hence deteriorate its interpretability. In addition, a large tree can become fragmented, having many leaf nodes with few cases. Such leaf nodes are likely to be influenced by noise and thus cause overfitting problems when the tree captures both the target features and noise (Breslow & Aha, 1997). Pruning is a method for obtaining the right-sized trees (Murthy, 1998).

37.3.5 Artificial Neural Network

An artificial neural network (ANN) is an information-processing paradigm that is inspired by the way biological nervous systems process information. It is composed of a large number of highly interconnected processing elements (called artificial neurons) working in unison to solve specific problems.

Therefore, ANN is also referred to as connectionist learning. Two major advantages of neural networks are their capabilities of modeling complex nonlinear functions and their high tolerance to noisy data. However, ANNs also have some disadvantages. First, it can take a very long time to train an ANN when there are many hidden nodes. Second, it is difficult for a human to interpret the symbolic meaning of the weights and hidden neurons in the network. Therefore, an ANN is like a black-box without analytical basis. For this reason, ANNs are less preferred in many DM tasks where understanding the results is important.

Feedforward ANNs do not have connections between neurons that form directed cycles. Backpropagation, which is a multi-layer feedforward ANN, is probably the most popular type of ANN. It contains an input layer, one or more hidden layers, and one output layer. Backpropagation learning involves two steps. First, inputs are propagated forward from the input layer through the hidden layer to the output layer. Then, errors are backpropagated using a method of gradient descent to search for a set of weights that fits the training data so as to minimize the mean squared error (MSE) between the network's prediction and true target value. Radial basis function (RBF) ANNs are another commonly used type of ANNs. They use RBFs that have built into a distance criterion with respect to a center, as a replacement for the sigmoidal hidden layer activation characteristic in the backpropagation. RBF ANNs have the advantage of not being stuck in local minima in the same way as the backpropagation.

Contrary to feedforward ANNs, recurrent ANNs are models with bidirectional data flow, for instance, they propagate data from later layers to earlier ones. A simple recurrent ANN is a variation of the multi-layer feedforward ANN. It uses a three-layer network structure (an input layer, a hidden layer, and output layer), with the addition of a set of context neurons in the input layer. Through connections to the hidden layer, the context neurons always retain a copy of the previous values of the hidden neurons at each time step. Thus, the network can maintain a sort of state, allowing it to perform tasks such as sequence prediction that are beyond the power of a standard multilayer ANN. In a fully recurrent ANN, each neuron receives inputs from every other neuron in the network. The network is not arranged in layers. Typically, a subset of the neurons in the network receives external inputs in addition to the inputs from all the other neurons, and another disjunctive subset of neurons report their outputs externally as well as sending them to all the other neurons.

A number of other more advanced types of ANNs have been proposed recently, such as neural-fuzzy ANNs (Pal & Mitra, 1999), stochastic ANNs (Kappen, 2001), and associative ANNs (Tetko, 2002). A number of books have been dedicated to ANNs, such as Fausett (1994), Bishop (1995), and Haykin (1999).

37.3.6 Support Vector Machine

The basic idea behind a support vector machine (SVM) is to find the hyperplane or decision boundary in the feature space that separates data of one class from another with the largest margin (Cristianini & Shawe-Taylor, 2000). The hyperplane with the largest margin is called the maximum marginal hyperplane (MMH). Support vectors refer to data that are equally close to the MMH. SVMs have good generalization performance, because the complexity of the learned SVMs is characterized by the number of support vectors rather than the entire training data, and the SVMs always find the optimum solution. However, the most serious drawbacks of SVMs from a practical point of view are their high algorithmic complexity and extensive memory requirement in large-scale tasks (Horváth, 2003).

37.3.7 Bayesian Classification

Bayesian classification is based on Bayes' theorem (Bayes, 1958), which relates the conditional and marginal probability distribution of random variables. Bayesian classifiers can predict the probability that a given sample belongs to a particular class and assign the sample to the class having the highest posterior probability conditioned on the sample. The two most popular Bayesian classifiers are naïve Bayesian classifiers and Bayesian belief networks (Han & Kamber, 2004).

Naïve Bayesian classifiers assume that the effects of attributes on a given class are independent of one another; this assumption is made to simplify the computations involved. Studies have shown that depending on the nature of the probability model, naive Bayes classifiers, in spite of their naive design and apparently oversimplified independence assumptions, can achieve comparable performance to decision tree, support vector machine, and some selected network classifiers (Langley et al., 1992; Huang et al., 2003; Borro et al., 2006).

A Bayesian belief network is a directed acyclic graph model whose nodes represent random variables and arcs represent probabilistic dependence relations among the variables. The variables may be discrete or continuous, and they may correspond to actual attributes given in the data or to latent variables believed to form relationships. Besides, there is a conditional probability table (CPT) associated with each variable, which lists the conditional probability of the variable given the values of its parent(s) in the network. Once the network structure and the CPT for each variable are decided, the joint probability distribution of a particular combination of attribute values can be easily calculated (Heckerman, 1996; Friedman et al., 1997).

37.3.8 Cluster Analysis

Cluster analysis is an exploratory data analysis tool for solving classification problems. The goal of the cluster analysis is to partition data objects into clusters or groups, so that the objects within the same cluster are similar but objects in different clusters are dissimilar. The results of a cluster analysis may contribute to the definition of a formal classification scheme (e.g., taxonomy for related operators), indicate rules for assigning new cases to classes for identification and diagnostic purposes, or provide exemplars to represent classes. The major clustering algorithms can be generally classified into partitioning methods, hierarchical methods, density-based methods, grid-based methods, and model-based methods (Han & Kamber, 2004).

37.3.9 Partitioning Methods

Given a dataset of n data objects and the number of clusters, k, to be formed, a partitioning algorithm divides the data objects into k groups, where each group represents a cluster. The clusters are formed by optimizing an objective partitioning criterion, such as a dissimilarity function based on distance. The most well-known and commonly applied partitioning methods are k-means and k-mediods clustering.

In the k-means algorithm, cluster similarity is measured in terms of the mean value of the objects in a cluster, which can be thought of as the cluster's centroid. The k-means algorithm is composed of the following steps. First, k data objects are randomly selected, each representing the initial centroid of a cluster. Next, each of the remaining objects is assigned to the cluster to which it is the most similar. Then, the new mean of each cluster is recalculated. The second and third steps iterate until the criterion function converge. The square-error criterion, which is the sum of the square of the distance between each object and its cluster centroid, is usually used (Hartigan, 1975).

The k-mediods clustering was developed primarily to make the k-means clustering more robust to outliers. Whereas k-means clustering takes the mean value of the object in a cluster as a reference point, k-mediods clustering uses one object in each cluster as its representative. The partitioning method is then performed based on the principle of minimizing the sum of the distance between each object and the representative object of its associated cluster (Kaufman & Rousseeuw, 1990).

37.3.10 Hierarchical Methods

Hierarchical clustering creates clusters through hierarchical construction, which can be done in an agglomerative or a divisive form. Agglomerative hierarchical clustering starts by placing each object as an individual cluster and then merges these atomic clusters into larger and larger clusters, until all the objects are in a single cluster or a certain termination condition is met. Divisive hierarchical clustering does the reverse of agglomerative hierarchical clustering by starting with all objects in one cluster and then subdividing the cluster into smaller and smaller clusters, until each object forms a cluster on its own or certain termination condition is met. A tree structure called dendrogram is often used to represent the process of hierarchical clustering; it shows how objects are grouped together step by step (Kaufman & Rousseeuw, 1990).

Despite their simplicity, the hierarchical clustering methods often encounter problems regarding the selection of merging or splitting points. However, such a decision is very crucial because once the decision is executed, the later steps can neither undo what has been done nor perform object swapping between clusters. As a result, merging or splitting decisions, if not well chosen at some step, may lead to poor-quality clusters. Besides, the hierarchical clustering methods do not scale well because each splitting or merging decision requires the examination and evaluation of a number of objects or clusters in a large dataset. One potential approach to addressing these issues is to integrate hierarchical clustering with other non-hierarchical clustering methods, resulting in multiple-phase clustering (Zhang et al., 1996; Karypis et al., 1999).

37.3.11 Density-Based Methods

The main idea of the density-based methods is to find regions of high density (representing clusters) and regions of low density (representing noise), with the high-density regions being separated from low-density regions. These methods can make it easy to discover arbitrary clusters. Examples of the density-based methods include DBSCAN (density-based spatial clustering of applications with noise; Ester et al., 1996), OPTICS (ordering points to identify the clustering structure; Ankerst et al., 1999), and clustering based on density distribution functions called DENCLUE (Hinneburg & Keim, 1998).

37.3.12 Grid-Based Methods

Grid-based clustering methods use a grid mesh to partition the entire problem domain into a number of cells. Clustering is then performed to the cells instead of the database itself. Since the number of cells in the grid is usually much smaller than the number of the data objects, the processing speed can be significantly improved. Some interesting examples of the grid-based methods include statistical information grid (STING; Wang et al., 1997), WaveCluster, which clusters objects using a wavelet transform method (Sheikholeslami et al., 1998), and a combination of grid- and density-based approach called CLIQUE (Agrawal et al., 1998). However, for highly irregular or concentrated data distributions, a grid mesh with very fine granularity will be required in order to sustain a certain clustering quality. To address this issue, a grid-based algorithm that employs adaptive mesh refinement techniques was proposed recently in Liao et al. (2004).

37.3.13 Model-Based Methods

The basis of model-based clustering methods is to use certain models for clusters and attempt to optimize the fit between the data and the models. These are generally two major types of model-based methods: statistical approaches, such as expectation-maximization (EM) clustering and conceptual clustering, and ANN approaches like self-organizing feature maps (SOMs).

Expectation-maximization (EM) clustering can be viewed as an extension to the k-means paradigm. Instead of assigning cases to clusters to maximize the differences in means of different clusters, the EM algorithm computes probabilities of cluster memberships based on one or more probability distributions with the goal of maximizing the overall probability or likelihood of the data, given the clusters (Dempster et al., 1977; Bradley et al., 1998). The central idea behind conceptual clustering is to group together objects into classes in such a way that each class represents a descriptive concept, hence the clusters generated with this approach are more interpretable than those produced by conventional clustering techniques, which use a numeric similarity metric to group together similar objects (Fisher, 1987; Iba & Langley, 2001).

The SOMs (also known as the Kohonen feature maps) are one of the most well-known ANN algorithms for cluster analysis. The goal of SOMs is to produce low-dimensional representation of the training data while preserving the topological properties of the high-dimensional input space. This makes SOMs a useful tool for visualizing low-dimensional views of high-dimensional data. With SOMs, clustering is performed by having the output neurons compete for a training sample, and the neuron whose weight vector is the most similar to the input becomes activated (Haykin, 1999).

37.3.14 Evolution Analysis

The goal of evolution analysis is to model regularities or trends for objects whose behaviors change over time. Evolution analysis may involve other DM tasks, such as association analysis, classification and prediction, and cluster analysis, but its distinct features include time-series analysis and sequential pattern mining (Han & Kamber, 2004).

37.3.15 Time-Series Analysis

Time series is a sequence of data points, measured typically at successive times, spaced at (often uniform) time intervals. There are two main goals of time-series analysis: modeling time-series (i.e., identifying the nature of the phenomenon represented by the sequence of observations) and forecasting time series (i.e., predicting future values of the time-series variables). Statistical methods for time-series analysis have been studied extensively in statistics (Chatfield, 2003).

Most time-series patterns can be described in terms of two basic classes of components: trend and seasonality. The trend represents a general systematic linear or (most often) nonlinear component that changes over time and does not repeat or at least does not repeat within the time range captured by the sampled data. The seasonality, on the other hand, repeats itself in systematic intervals over time. Regression has been a popular tool for modeling time series, finding trends and outliers in the dataset. If the time-series data contain considerable error, the first step in the process of trend identification is smoothing, which usually involves some form of local averaging of data such that the nonsystematic components of individual observations cancel each other out. The most common technique is *moving average* smoothing, which replaces each element of the series by either the simple or weighted average of n surrounding elements, where n is the width of the smoothing window. To detect seasonal patterns, we can examine correlations between each ith elements of the series and $(i-k)$th element (where k is referred to as the lag) using autocorrelation analysis (Box et al., 1994).

There are several methods for time-series forecasting. The auto-regressive integrated moving average (ARIMA) is one of the most well-known techniques. It is powerful yet can be difficult to use, requiring a great deal of experience to accurately use and interpret its results (Bails & Peppers, 1980).

37.3.16 Sequential Pattern Mining

Sequential pattern mining is the discovery of frequently occurring ordered events or sequences as patterns, where the events are recorded with or without a concrete notion of time (Han & Kamber, 2004). Among the varieties of techniques for sequential pattern mining, only Markov models are discussed because they are the most widely used and relevant to digital human modeling.

A Markov model is a statistical model in which the system being modeled is assumed to be a Markov process. The key property of a Markov process, referred to as Markov property, is that the conditional probability distribution of the future states of the process, given its present state and all past states, depends only on the current state and not on any past states. A Markov model is defined by a set of states and a set of transitions between the states. Each transition has an associated transition probability (Grinstead & Snell, 1997).

When the states are directly visible to observers, the state transition probabilities are the only parameters. However, in the situations when the states are not directly observable but some variables influenced by the states are observable, a hidden Markov model (HMM) needs to be used. In addition to the (hidden) states and their transitions in the regular Markov process described above, a HMM includes a set of observable states and the probabilities of the observable states given the hidden states. There are three basic tasks of using an HMM. The first task is evaluation: Given the parameters of an HMM, compute the probability of a particular observed sequence. This problem can be solved using the forward-backward procedure (Baum & Petrie, 1966). The second task is decoding: Given the parameters of an HMM, find the most likely sequence of hidden states that could have produced a given observed sequence. This problem can be solved using the Viterbi algorithm (Viterbi, 1967). The third task is learning: Given a set of observed training sequences and a model, find the most likely model parameters that explain the sequences. This problem can be solved using the Baum-Welch algorithm (Baum et al., 1970).

37.4 Applications of DM in DHM

This section illustrates applications of DM techniques in three major tasks of DHM: mental workload assessment, human action recognition in image sequences, and ergonomics data analysis.

37.4.1 Mental Workload Assessment

Mental workload refers to the portion of operator information-processing capacity and resources required to meet system demands (O'Donnel & Eggemeier, 1986). Limitations in humans' ability to process and respond to information have become a limiting factor in many current complex dynamic environments. In such situations, therefore, accurate assessment of human mental workload is crucial in order to optimize the overall system performance. A number of measurement procedures have been used to classify and predict operator mental workload. In particular, ANNs have been widely employed to model workload by performing electroencephalogram (EEG) data classification. For example, in Mazaeva et al. (2001), a SOM network, which consisted of 30 neurons in the input layer (corresponding to six EEG channels with five bands each) and six neurons in the output layer (corresponding to six levels of mental workload ranging from very low to overload), was applied to cluster the training data into six categories. Given a testing dataset, the trained SOM achieved an overall classification accuracy of 89%. In another study by Wilson and Russell (2003), a backpropagation network using 43 physiological signals, which comprised six EEG channels and two electrooculographic (EOG) channels with five bands each plus three other peripheral features (ECG interbeat, EOG interblink, and respiration intervals), was generated to continuously monitor, in real time, the mental workload of seven subjects while they were performing the NASA multi-attribute task battery with two levels (low versus high) of task difficulty. The average classification accuracies of the backpropagation network were 82% and 86% for the low and high task difficulty, respectively.

37.4.2 Human Action Recognition in Image Sequences

The ability to recognize human activities by vision is crucial to designing machines that can interact intelligently with a human-inhabited environment. This domain is called human motion analysis, sometimes also referred to as Looking at People. It covers, among others, face recognition, hand gesture recognition, and whole-body tracking (Gavrila, 1999). There are three major tasks in the process of human motion analysis: human detection, human tracking, and understanding (including recognizing and describing) human actions in image sequences (Wang et al., 2003). This section illustrates applications of DM techniques for the recognition of human actions in image sequences.

Action recognition in human motion analysis can be thought of as a sequential data-matching problem. HMMs, as discussed in the section "Sequential Pattern Mining," can be applied to analyze sequential data with spatio-temporal variability. The use of HMMs involves two stages: training and classification. In the training stage, the number of states of an HMM must be specified, and the corresponding state transformation and output probabilities are optimized, so that the generated symbols can correspond to the observed image features within the examples of a specific movement class. Then, in the matching stage, the probability that a particular HMM could have generated the test symbol sequence corresponding to the observed image features is computed. HMMs have recently become popular tools for matching human motion patterns. For instance, Yamato et al. (1992) used HMM to recognize six different tennis strokes among three players. A method to model human daily-life actions was proposed in Mori et al. (2004), which used 3D motion data and associated each action with a continuous HMM, followed by hierarchical recognition. In Ren et al. (2005), HMMs were used to train a classifier to distinguish between natural and unnatural movement based on human-labeled data.

ANNs are another interesting yet less investigated approach to analyzing sequential data. For instance, an RBF network was developed to learn the correlation between facial feature motion patterns and human emotion (Rosenblum et al., 1994), in which the highest level of the network identified emotion, the middle level determined motions of the facial features, and the motion directions were recovered at the lowest level. A backpropagation network was used in Guo et al. (1994) to classify human motions represented by stick figure models into three categories: walking, running, and other motions.

37.4.3 Ergonomics Data Analysis

One of the main themes in digital human modeling is to develop human models that can be used to identify ergonomic problems such as the effects of heavy loads on the human musculoskeletal system or difficulties in reaching or seeing objects (Sundin, 2006). Classification and clustering are two typical types of problems in ergonomics data analysis that can be addressed with DM techniques. For example, in Zurada et al. (2004) various classification tools, including backpropagation ANN, logistic regression, decision trees, memory-based reasoning, and ensemble models, were explored to classify industrial jobs with respect to the risk of work-related low back disorders (LBDs). An incremental supervised method called clustering and classification algorithm-supervised (CCAS) was used in Ye et al. (2003) to distinguish human-generated sign language and to discover which signs were more difficult to differentiate in practice for computers as well as humans. In addition, in Abdali et al. (2004), four clustering techniques—partitioning, hierarchical, grid-based, and density-based—were applied to explore 3D body scans together with the relational anthropometric and demographic data contained in an integrated multimedia anthropometric database.

37.5 Summary

This chapter gave a brief introduction to the DM process and some commonly applied DM modeling techniques, along with illustrative examples of how DM can be applied in mental workload assessment, human action recognition in image sequences, and ergonomics data analysis.

Advances in high-quality measurement devices have enabled the collection of extensive data on human behavior, performance, and psychophysical measures. This creates opportunities and challenges for us to discover valuable information and knowledge about both physical and cognitive aspects of human beings. It is my hope that this chapter will serve to provide the basis for a better appreciation of DM and how it can be applied in the DHM research.

References

Abdali, O., Viktor, H., Paquet, E., & Rioux, M. (2004). Exploring anthropometric data through cluster analysis. *SAE 2004 Transactions Journal of Aerospace*, 113(1), 241–246.

Aggarwal, C.C., & Yu, P.S. (2001). Mining associations with the collective strength approach. *IEEE Transactions on Knowledge and Data Engineering*, 13(6), 863–873.

Agrawal, R., Gehrke, J., Gunopulos, D., & Raghavan, P. (1998). Automatic subspace clustering of high dimensional data for data mining applications. *Proceedings of 1998 ACM SIGMOD International Conference on Management of Data*, Seattle, WA, 94–105.

Agrawal, R., Imielinski, T., & Swami, A. (1993). Mining association rules between sets of items in very large databases. *Proceedings of the ACM SIGMOD Conference on Management of Data*, Washington, DC, 207–216.

Ankerst, M., Breunig, M., Kriegel, H-P., & Sander, J. (1999). OPTICS: ordering points to identify the clustering structure. *Proceedings of 1999 ACM SIGMOD International Conference on Management of Data*, Philadelphia, PA, 49–60.

Bails, D.G., & Peppers, L.C. (1982) *Business Fluctuations: Forecasting Techniques and Applications*, Englewood Cliffs, NJ: Prentice-Hall.

Baum, L.E., & Petrie, T. (1966). Statistical inference for probabilistic functions of finite state Markov chains. *Annals of Mathematical Statistics*, 37, 1554–1563.

Baum, L.E., Petrie, T., Soules, G., & Weiss, N. (1970). A maximization technique occurring in the statistical analysis of probabilistic functions of Markov chains. *Annals of Mathematical Statistics*, 41(1), 164–171.

Bayes, T. (1958). Studies in the history of probability and statistics: IX. Thomas Bayes's essay towards solving a problem in the doctrine of chances. *Biometrika*, 45, 296–315.

Bishop, C.M. (1995) *Neural Networks for Pattern Recognition*, Oxford: Oxford University Press.

Borro, L.C., Oliveira, S.R.M., Yamagishi, M.E.B., Mancini, A.L., Jardine, J.G., Mazoni, I., dos Santos, E.H., Higa, R.H., Kuser, P.R., & Neshich, G. (2006). Predicting enzyme class from protein structure using Bayesian classification. *Genetics and Molecular Research*, 5(1), 193–202.

Box, G., Jenkins, G.M., & Reinsel, G. (1994). *Time Series Analysis: Forecasting and Control (3rd Edition)*. Englewood Cliffs, NJ: Prentice Hall.

Bradley, P.S., Fayyad, U.M., & Reina, C.A. (1998). Scaling EM (Expectation Maximization) clustering to large databases. *Technical Report MSR-TR-98-35*, Microsoft Research.

Breiman, L., Friedman, J., Olshen, R., & Stone, C. (1984). *Classification and Regression Trees*. Wadsworth, CA: Pacific Grove.

Breslow, L. A., & Aha, D. W. (1997). Simplifying decision trees: A survey. *Knowledge Engineering Review*, 12(1), 1–40.

Brijs, T., Vanhoof, K., & Wets, G. (2003). Defining interestingness for association rules. *International Journal of Information Theories and Applications*, 10(4), 370–376.

Brin, S., Motwani, R., Ullman J., & Tsur, S. (1997). Dynamic itemset counting and implication rules for marked basket data. *SIGMOD Record*, 6(2), 255–264.

Chatfield, C. (2003). *The Analysis of Time Series: An Introduction* (6th Edition). Boca Raton, FL: Chapman and Hall/CRC.

Chaudhuri, S., & Dayal, U. (1997). An Overview of Data Warehousing and OLAP Technology, *ACM SIGMOD Record*, 26(1), 65–74.

Chen, M-S, Han, J., & Yu, P.S. (1996). Data mining: an overview from database perspective. *IEEE Transactions on Knowledge and Data Engineering*, 8(6), 866–883.

CRISP-DM Special Interest Group (2000). *CRISP 1.0 Process and User Guide*. Retrieved January 17, 2004 from http://www.crisp-dm.org.

Cristianini, N., & Shawe-Taylor, J. (2000). *An Introduction to Support Vector Machines and Other Kernel-based Learning Methods*. Cambridge, UK: Cambridge University Press.

Dempster, A.P., Laird, N.M., & Rubin, D.B. (1977). Maximum likelihood from incomplete data via the EM algorithm. *Journal of the Royal Statistical Society*, Series B, 39(1), 1–38.

Ester, M., Kriegel, H-P., Sander, J., & Xu, X. (1996). A density-based algorithm for discovering clusters in large spatial databases. *Proceedings of 1996 International Conference on Knowledge Discovery and Data Mining (KDD'96)*, Portland, OR, 226–231.

Fausett, L. (1994). *Fundamentals of Neural Networks*. New York: Prentice Hall.

Fayyad, U., Piatetsky-Shapiro, G., & Smyth, P. (1996). From data mining to knowledge discovery: An Overview. In U.M. Fayyad, G. Piatetsky-Shapiro, P. Smyth, & R. Uthurusamy (Eds.), *Advances in Knowledge Discovery and Data Mining* (pp. 1–36). Menlo Park, CA: AAAI Press.

Fisher, D. (1987). Knowledge acquisition via incremental conceptual clustering. *Machine Learning*, 2(2), 139–172.

Friedman, N., Geiger, D., & Goldszmidt, M. (1997). Bayesian network classifiers. *Machine Learning*, 29, 131–161.

Gavrila, D.M. (1999).The visual analysis of human movement: a survey. *Computer Vision and Image Understanding*, 73 (1), 82–98.

Grinstead, C.M., & Snell, J.L. (1997). *Introduction to Probability* (2nd Edition). Providence, RI: American Mathematical Society.

Guo, Y., Xu, G., & Tsuji, S. (1994). Understanding human motion patterns. *Proceedings of International Conference on Pattern Recognition*, 2(9), Jerusalem, Israel, 325–329.

Han, J., & Kamber, M. (2004). *Data Mining: Concepts and Techniques* (2nd Edition). San Francisco: Morgan Kaufmann.

Hartigan, J.A. (1975). *Clustering Algorithms*. New York: John Wiley & Sons.

Haykin, S. (1999). *Neural Networks* (2nd Edition). New York: Prentice Hall.

Heckerman, D. (1996). Bayesian networks for knowledge discovery. In U.M. Fayyad, G. Piatetsky-Shapiro, P. Smyth, & R. Uthurusamy (Eds.), *Advances in Knowledge Discovery and Data Mining* (pp. 273–305). Menlo Park, CA: AAAI Press.

Hinneburg, A., & Keim, D.A. (1998). An efficient approach to clustering in large multimedia databases with noise. *Proceedings of 1998 International Conference on Knowledge Discovery and Data Mining (KDD'98)*, New York, 58–65.

Horváth, G. (2003). Neural networks in measurement systems. In J.A.K., Suykens, G. Horváth, S. Basu, C. Micchelli, and J. Vandewalle (Eds.), *Advances in Learning Theory: Methods, Models and Applications, NATO-ASI Series in Computer and Systems Sciences*, (pp. 375–402). Amsterdam, The Netherlands: IOS Press.

Huang, J., Lu, J., & Ling, C.X. (2003). Comparing naïve Bayes, decision trees, and SVM with AUC and accuracy, *IEEE Computer Society*, 553–556.

Iba, W., & Langley, P. (2001). Unsupervised learning of probabilistic concept hierarchies. *Lecture Notes in Computer Science, Machine Learning and Its Applications*, 2049, 39–70.

Inmon, W.H. (2002). *Building the Data Warehouse (3rd Edition)*. New York: John Wiley & Sons.

Kamber, M., Han, J., & Chiang, J.Y. (1997). Metarule-guided mining of multi-dimensional association rules using data cubes. *Proceedings of 1997 International Conference on Knowledge Discovery and Data Mining (KDD'97)*, Newport Beach, CA, 207–210.

Kappen, H.J. (2001). An introduction to stochastic neural networks. In F. Moss and S. Gielen (Eds.), *Neuro-informatics and Neural Modelling* (Handbook of Biological Physics) (pp. 517–552). Amsterdam, The Netherlands: North-Holland Publishing.

Karypis, G., Han, E-H., & Kumar, V. (1999). CHAMELEON: a hierarchical clustering algorithm using dynamic modeling. *Computer*, 32, 68–75.

Kass, G.V. (1980). An exploratory technique for investigating large quantities of categorical data. *Applied Statistics*, 29, 119–127.

Kaufman, L., & Rousseeuw, P.J. (1990). *Finding Groups in Data: An Introduction to Cluster Analysis*. New York: John Wiley & Sons.

Langley, P., Iba, W., & Thompson, K. (1992). An analysis of Bayesian classifiers. *Proceedings of the Tenth National Conference on Artificial Intelligence*, San Jose, CA, 223–228.

Lent, B., Swami, A., & Widon, J. (1997). Clustering association rules, *Proceedings of 1997 International Conference on Data Engineering (ICDE'97)*, Birmingham, U.K., 220–244.

Liao, W-K, Liu, Y., & Choudhary, A. (2004). A grid-based clustering algorithm using adaptive mesh refinement. *Proceedings of the 7th Workshop on Mining Scientific and Engineering Datasets*, Lake Buena Vista, FL.

Liu, B., Hsu, W., Chen, S., & Ma, Y. (2000). Analyzing the subjective interestingness of association rules. *IEEE Intelligent Systems*, 15(5), 47–55.

Mazaeva, N., Ntuen, C., & Lebby, G. (2001). Self-Organizing Map (SOM) model for mental workload classification. *Joint 9th IFSA World Congress and 20th NAFIPA International Conference*, Vancouver, Canada, 1822–1825.

Mori, T., Segawa, Y., Shimosaka, M., & Sato, T. (2004). Hierarchical recognition of daily human actions based on continuous hidden markov models. *Proceedings of IEEE International Conference on Automatic Face and Gesture Recognition*, Seoul, Korea, 779–784.

Murthy, S. K. (1998). Automatic construction of decision trees from data: a multi-disciplinary survey. *Data Mining and Knowledge Discovery*, 2(4), 345–389.

O'Donnel, R.D., & Eggemeier, F.T. (1986). Workload assessment methodology. In K. Boff, L. Kaufinan, and J. P. Thomas (Eds.), *Handbook of Perception and Human Performance, v2: Cognitive Processes and Performance*, New York: Wiley Interscience.

Pal, S.K., & Mitra, S. (1999). *Neuro-Fuzzy Pattern Recognition: Methods in Soft Computing*. New York: John Wiley & Sons.

Quinlan, J.R. (1993). *C 4.5—Programs for Machine Learning*. San Francisco: Morgan Kaufmann.

Ren, L., Patrick, A., Elfros, A.A., Hodgins, J.K., & Rehg, J.M. (2005). A data-driven approach to quantifying natural human motion. *ACM Transactions on Graphics*, 24(3), 1090–1097.

Rosenblum, M., Yacoob, Y., & Davis, L. (1994). Human emotion recognition from motion using a radial basis function network architecture. *Proceedings of IEEE Workshop on Motion of Non-Rigid and Articulated Objects*. Austin, TX, 43–49.

Sheikholeslami, G., Chatterjee, S., & Zhang, A. (1998). WaveCluster: a multi-resolution clustering approach for very large spatial databases. *Proceedings of 1998 International Conference on Very Large Data Bases (VLDB'98)*, New York, 428–439.

Srikant, R., & Agrawal, R. (1996). Mining quantitative association rules in large relational tables. *Proceedings of 1996 ACM SIGMOD International Conference on Management of Data*, Montreal, Canada, 1–12.

Sundin, A., & Ortengren, R. (2006). Digital human modeling for CAE applications. In G. Salvendy (Ed.), *Handbook of Human Factors and Ergonomics* (pp. 1053–1078). New York: John Wiley & Sons.

Tetko, I.V. (2002). Introduction to associative neural networks. *Journal of Chemical Information Computer Sciences*, 42(3), 717–728.

Viterbi, A.J. (1967). Error bounds for convolutional codes and an asymptotically optimum decoding algorithm. *IEEE Transactions on Information Theory*, 13, 260–269.

Wang, L., Hu, W., & Tan, T. (2003). Recent developments in human motion analysis. *Pattern Recognition*, 36(3), 585–601.

Wang, W., Yang, J., & Muntz, R. (1997). STING: a statistical information grid approach to spatial data mining. *Proceedings of 1997 International Conference on Very Large Data Bases (VLDB'97)*, Athens, Greece, 186–195.

Webb, G.I. (2000). Efficient search for association rules. *Proceedings of 2000 International Conference on Knowledge Discovery and Data Mining (KDD-2000)*, Boston, 99–107.

Wilson, G.F., & Russell, C.A. (2003). Real-time assessment of mental workload using psychophysiological measures and artificial neural networks. *Human Factors*, 45(4), 635–643.

Yamato, J., Ohya, J., & Ishii, K. (1992). Recognizing human action in time sequential images using hidden markov model. *Proceedings of IEEE Conference on Computer Vision and Image Processing*, Champaign, IL, 379–385.

Ye, N., Li, X., & Farley, T. (2003). A data mining technique for discovering distinct patterns of hand signs: Implications in user training and computer interface design. *Ergonomics*, 46(1–3), 188–196.

Zhang, T., Ramakrishnan, R., & Livny, M. (1996). BIRCH: an efficient data clustering method for very large databases. *Proceedings of 1996 ACM SIGMOD International Conference on Management of Data*, Montreal, Canada, 103–114.

Zurada, J., Karowski, W., & Marras, W. (2004). Classification of jobs with risk of low back disorders by applying data mining techniques. *Occupational Ergonomics*, 4, 291–305.

38

Motion Capture and Human Motion Reconstruction

Sergio Ausejo and
Xuguang Wang

38.1 Introduction

Motion capture is becoming a popular technique for animating a virtual actor by a real person in the game industry. It is also used for biomechanical motion analysis in clinical and sports applications. In recent years, this technique is applied for analyzing and predicting complex human motions in ergonomic simulation by digital human figures. Indeed, a design engineer needs to simulate the motions of the future operators/users and to make ergonomic evaluation of a product in a very early phase. Due to limited knowledge on human motion control strategies, the data-based approach is the only one that is capable of simulating complex task-oriented movements today. For this, a well-structured motion database is needed (see for instance Park et al., 2004; Monnier et al., 2006; Wang et al., 2006). Realistic motion reconstruction is the first step for humanlike motion simulation. The quality of reconstructed motions directly affects the realism of simulated motions. As a large amount of motion data is necessary for simulating different human activities, a fast motion reconstruction method has to be developed. This chapter aims to give an overview of different motion reconstruction methods, especially for whole-body motion reconstruction from external markers trajectories.

Human motion reconstruction consists (fig. 38.1) of recording with a motion capture system the motion of a real subject at first. Then, the geometric dimensions (e.g., body segment length, body external contour) of a digital human model with a predefined kinematic linkage must be tailored to the real subject by estimating the subject-specific parameters. Finally, the motion of the digital human model can be reconstructed from the measured motion data using different inverse kinematics methods.

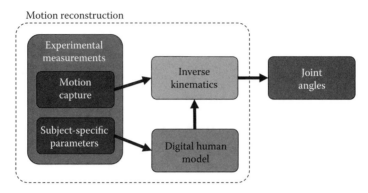

FIGURE 38.1   Human motion reconstruction process overview.

Motion capture consists of measuring the body position of a subject when performing a motion. By body position we mean the position and orientation of each body segment under analysis. There are different technologies for motion capture and all of them have advantages and limitations. Motion capture systems can be based on sensors located on the subject like accelerometers, acoustic transmitters, electromagnetic sensors, or inertial sensors. However, optoelectronic motion capture systems based on stereophotogrammetry are by far the most widespread in digital human modeling for ergonomic simulation. In this chapter, we consider only optoelectronic systems, although a brief review of other technologies is presented in Section 38.2.

Different digital human models can be used to reconstruct the recorded motion depending on the application. Clinical applications are usually interested only in a few body segments (e.g., the leg in gait analysis), which are usually assumed to move independently without any joint constraint. In this chapter, we consider only whole-body models like Jack, RAMSIS, Human Builder, Man3D, or ManneQuinPRO. All of these models have fixed joint centers and joint axes, which is a source of errors due to the complex nature of human joints, but it is a commonly accepted assumption for studying gross human motion. In order to adjust the digital human model to each subject, we need to estimate the subject-specific parameters, which are the skin surface, the joint parameters (joint centers and joint axes) in the corresponding local coordinate system (LCS), and marker coordinates in each LCS. We will refer to local marker coordinates as *model-marker coordinates* as opposed to *measured-marker coordinates*, which are the global marker coordinates measured by an optoelectronic system. The methods proposed to estimate the subject-specific parameters are reviewed in Section 38.3.

Finally, different inverse kinematics (IK) methods can be used to reconstruct the motion of a digital human model adjusted to the subject under investigation from the measured marker trajectories. In this chapter we use the term *IK method* to refer to any method that calculates joint angles of a digital human model from marker trajectories. Commercial digital human software tools and motion capture systems are usually equipped with IK methods for performing motion reconstruction. However, these IK methods are proprietary. Therefore, in this chapter we will only refer to those methods published in the literature. IK methods used in different application domains are reviewed in Section 38.4.

38.2 Data Collecting

In this section, we present the main motion capture systems that can be used for measuring human motion. The advantages and limitations of each system are briefly presented in Section 38.2.1. Additionally, in Section 38.2.2, as optoelectronic motion capture systems are by far the most widespread, we describe in more detail their main problems and current solutions proposed in the literature. The interested reader is referred to Allard et al. (1995), Medved (2001), and Chiari et al. (2005) for further details.

38.2.1 Motion Capture Systems

Goniometers are devices that measure directly the relative rotation between two body segments. They are usually a mechanical device with potentiometers attached to it for measuring joint angles or a flexible cable instrumented by strain gauges. Their main limitation for digital human applications is that they cannot be used for measuring the motion of the whole body because a goniometer is basically a one- or two-dimensional device and it is difficult to measure a three degrees-of-freedom (DOF) joint like the shoulder. Moreover, it is difficult to attach a goniometer to some joints. Other limitations are that they represent a mechanical constraint to the joint natural motion, specific goniometers for each joint are required, and alignment of the goniometer with the joint is difficult.

Accelerometers are used for measuring acceleration at a high frequency. They are able to measure acceleration in three mutual perpendicular directions and a double time integration of the acceleration signal yields the displacement of the sensor. Practical limitations are the need of the initial conditions for the sensor (i.e., initial position and velocity), the wiring associated with the sensor, and the low-frequency drift whose effect increases with time.

Acoustic systems are based on small acoustic transmitters located on the subject and an array of microphones (e.g., Zebris). As the speed of sound is known, the position of each source can be estimated. This technology can be disturbed by reflections of the sound and only a limited number of sensors can be used.

Electromagnetic systems use sensors that measure the magnetic fields generated by a source and from the measured strength of the field it is possible to estimate the position and orientation of the sensor (e.g., Ascension Technology Corporation or Polhemus). The main advantage of this technology is that the sensors can be always detected without any occlusion problem. However, it is affected by electromagnetic interferences from metallic objects that distort the sensor signal. Additionally, only a limited number of sensors can be used, the wiring associated with the sensors can hamper the subject motion, and the capture volume is small compared to optoelectronic systems. Nevertheless, they have been successfully used to measure the motion of the whole body (e.g., Bodenheimer et al., 1997; Molet et al., 1999).

Recently, new inertial sensors with reduced drift problems have been developed (e.g., Xsens). These sensors combine accelerometers, rate gyroscopes, and a magnetometer, which is sensitive to the earth's magnetic field and can be used to correct drift. Furthermore, the motion data are transmitted to a computer using wireless communication. An advantage of this inertial technology is that no external cameras, emitters, or markers are needed. However, the subject must wear a special suit with embedded sensors, wiring, and a power supply.

An emerging technology for motion capture is the so-called markerless systems. These systems are based on computer vision and they do not require any sensor on the subject. They estimate the posture of the subject directly from the 2D images by using computer vision techniques. Currently the accuracy of these systems is very low for digital human applications and they are used mainly on entertainment applications. However, if their accuracy is sufficiently improved for digital human applications, this technology may be the choice in the future. The interested reader is referred to Wang et al. (2003) and Hartley and Zisserman (2004) for further details.

Optoelectronic motion capture systems are nowadays the most widely used (e.g., Viconpeak, PhaseSpace, Optotrak, Motion Analysis Corporation, Elite, or Qualisys). They require several cameras to record the position of markers located on the skin of the subject. Then, from the 2D marker positions recorded at each frame, the 3D marker positions can be estimated using stereophotogrammetric methods (Medved, 2001). Typical values for the accuracy and precision are both in the range 1–5 mm, depending on the commercial motion capture system used (Chiari et al., 2005). There are two types of markers: passive and active. Passive markers are small retroreflective balls, which are used together with infrared stroboscopic illumination mounted around the lens of each camera. Active markers are light emitting diodes (LEDs) pulsed sequentially, so that the system can detect automatically each marker. This is an advantage over passive markers, which require complex methods for their identification.

Another advantage of active markers is that they allow for higher accuracies and sampling rates than passive markers. However, important advantages for passive markers are the absence of wires, batteries, and pulsing circuitry on the body of the subject.

38.2.2 Measurement Errors

It is usually assumed that the motion of each bone can be described by the markers on the body segment when an optoelectronic motion capture system is used. Unfortunately, marker trajectories contain two sources of inaccuracies (Cappozzo et al., 1996):

1. *Instrumental errors*, due to the motion capture system. These errors are systematic and random and they produce an apparent movement of the markers.
2. *Soft tissue artifact*, due to the real relative movement between markers and the underlying bone caused by passive and active soft tissues.

High-frequency content of instrumental errors can be dealt with by using smoothing methods (Woltring, 1995; Giakas & Baltzopoulos, 1997). However, soft tissue artifact, which has a similar frequency content as the body movement, cannot be removed by smoothing methods and it is recognized to be the bounding error source in the measurement of human motion (Alexander & Andriacchi, 2001; Leardini et al., 2005). The error introduced by the soft tissue artifact can be reduced at least in three ways:

1. Markers can be positioned in locations with minimum soft tissue artifact based on data reported on the literature. An excellent review on patterns and magnitudes of soft tissue artifact has been reported recently by Leardini et al. (2005).
2. Soft tissue artifact can be compensated provided it can be measured or modeled (Cappello et al., 1997; Ball & Pierrynowski, 1998; Lucchetti et al., 1998).
3. Soft tissue artifact can also be minimized by optimization using IK methods. This is the main topic of this chapter and it is reviewed extensively in Section 38.4.

Another common problem of optoelectronic systems is the *missing marker problem*. This problem appears when a marker is not visible from at least two cameras and then the 3D position of the marker cannot be calculated. This problem is usually handled by interpolating the gaps in data using cubic splines or other interpolation methods (Muijtjens et al., 1997). Other methods were also proposed, such as the one based on the assumption of a constant distance between markers (Desjardins et al., 2002) or the one based on the use of a simplified human model during the estimation of the marker trajectories from 2D camera data (Herda et al., 2001). An additional problem of optoelectronic systems is the impossibility of completely tracking the motion of some bones (e.g., vertebrae or scapula) using external markers fixed on the skin because of high relative skin/bone movement and an insufficient number of markers. An extensive survey of the main motion capture problems and the methods proposed to deal with them has been recently presented by Chiari et al. (2005) and Leardini et al. (2005).

38.3 Identification of Subject-Specific Parameters

Subject-specific parameters include the skin surface used for visualization, joint parameters (joint centers and joint axes) referred to the corresponding local coordinate system (LCS) of each segment, and model-marker coordinates in each LCS. Local model-marker coordinates can be calculated if the global positions of all the markers are known in a given posture and the position and orientation of each LCS can be estimated in the same posture. On the other hand, joint parameters are difficult to estimate and several methods have been proposed:

1. *Regression methods*, which estimate joint parameters from coordinates of anatomical landmarks using regression equations (Seidel et al., 1995; Meskers et al., 1998).

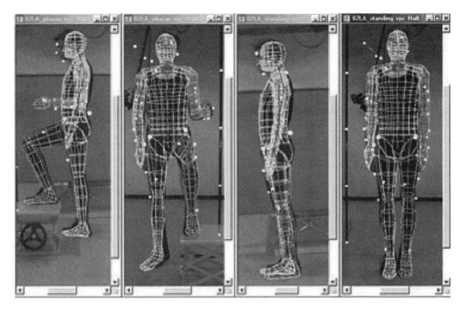

FIGURE 38.2 Superimposed RAMSIS model adjusted to a subject in two predefined postures. (From Ausejo, S., et al., "Robust Human Motion Reconstruction in the Presence of Missing Markers and the Absence of Markers for Some Body Segments," *Proceedings of the SAE 2006 Digital Human Modeling for Design and Engineering Conference* SAE International, Warrendale, PA, 2006. With permission.)

> 2. *Functional methods*, which estimate joint parameters from well-defined functional motion data (Halvorsen et al., 1999; Stokdijk et al., 2000; Gamage & Lasenby, 2002).

However, the previous methods are designed for specific joints in clinical and biomechanical applications and most of them require a large number of experimental measurements. Computer animation researchers have also proposed several methods for estimating joint parameters. These methods estimate joint parameters directly from the motion capture data (Bodenheimer et al., 1997; Ringer & Lasenby, 2002) and some of them estimate both joint parameters and kinematic structure of the digital model from the motion data (Silaghi et al., 1998; Kirk et al., 2005). A drawback of these methods is that the joint parameters depend on the quality of motion data.

Commercial digital human software tools (e.g., RAMSIS BodyBuilder) provide methods for creating an individualized human model from a set of anthropometric dimensions of a real subject. These methods are based on statistical techniques and large human databases. However, they do not provide the local model-marker coordinates.

Digital human applications require methods that allow estimating all subject-specific parameters of a whole-body model in a reasonable time. For example, the software PCMAN (Seitz et al., 2000) was used in the REALMAN project to adjust manually the RAMSIS model to a specific subject. The standard procedure requires a minimum of two simultaneous pictures in two predefined postures (fig. 38.2) and allows estimating all subject-specific parameters. However, this method is still time consuming (Wang et al., 2005) and more efficient methods are needed.

38.4 Motion Reconstruction by IK Methods

Inverse kinematics (IK) methods calculate at each frame the posture of a digital human model, which is tailored to a real subject, from the motion capture data (e.g., markers trajectories). Before presenting the different IK methods published in the literature, we will first introduce the types of motion reconstruction problems that can be found in real applications.

38.4.1 Definition of Motion Reconstruction Problems

We can classify motion reconstruction problems according to the amount of captured motion data into three groups: over-guided, under-guided, and exactly-guided problems. A motion reconstruction problem is over-guided when there is an excess of data to guide the motion of the digital human model (i.e., there is redundant information to define the posture of the model at each frame). Under-guided problems appear when the motion data are not enough to determine the position and orientation of each segment of the human model. A motion reconstruction problem is exactly-guided when the number of DOF of the model is equal to the number of measured-marker coordinates and the latter is enough to define the motion of the model. Exactly-guided problems rarely appear in real applications and they can be solved using the same IK methods as for over-guided problems.

38.5 Over-Guided Problems

Ergonomic, biomechanical, and clinical applications are typical examples where over-guided problems are found, as much more markers than strictly necessary are usually used to record the motion of the subject. IK methods for over-guided problems can be classified into two groups:

1. *Local IK methods*, which reconstruct the motion of each body segment independently without imposing joint constraints between body segments. This means that joints may dislocate in the reconstructed motion as joint integrity is not guaranteed.
2. *Global IK methods*, which reconstruct the motion of the whole body at once while guaranteeing joint constraints between body segments.

38.5.1 Local IK Methods

This type of method is used in applications in which assuming joint constrains (e.g., fixed joint center) is not appropriated and only a small number of body segments are studied. All these methods require a minimum of three non-collinear markers on each body segment and they consider each segment as independent. Notice that joint translations may appear in the reconstructed motion, which can be due to real joint translations and/or fictitious joint translations due to errors introduced by soft tissue artifact.

The first local IK methods (Apkarian et al., 1989; Kadaba et al., 1990) calculated the pose (position and orientation) of each body segment directly from the coordinates of three non-collinear markers on each segment without taking into account the errors introduced by the soft tissue artifact. Other local IK methods (Veldpaus et al., 1988; Söderkvist & Wedin, 1993; Challis, 1995; Carman & Milburn, 2006) minimize measurement errors in a least-squares sense at a body segment level. They are formulated as the following optimization problem:

$$\min\ f = \sum_{i=1}^{n} \left\| R \cdot x_i + d - y_i \right\|^2 \quad \text{subject to } R^T R = I;\ \det(R) = 1 \tag{38.1}$$

where x_i and y_i are the position vector of marker i in the reference coordinate system and a second coordinate system, respectively, and d is the optimal translation vector of the origin of the reference coordinate system referred to the second coordinate system.

Local IK methods can also be based on the minimization of the eigenvalue norm of the inertia tensor of a cluster of markers on the body segment (Alexander & Andriacchi, 2001) or on the extended Kalman filter, which enables the use of *a priori* information on the measurement noise and type of motion to tune the Kalman filter (Lasenby et al., 2001; Halvorsen et al., 2005).

38.5.2 Global IK Methods

Global IK methods for over-guided problems can be based on very different approaches but all of them need to solve a more or less complex optimization problem. Over-guided problems correspond to the case where more motion data are available than is strictly necessary to define the posture of the digital human model. Furthermore, due to measurement errors and the kinematic constraints, the digital human model cannot adopt a posture that matches all the measured markers. Global IK methods minimize the global measurement error which is defined as the sum of squared distances between the measured and model-marker positions. These methods are especially well-suited for reconstructing the motion of digital human models like Jack, RAMSIS, Safework, Man3D, or ManneQuinPRO, which impose joint constraints.

Lu and O'Connor (1999) proposed a global IK method based on the minimization of the weighted sum of squared distances between measured and model-marker 3D positions. Joint constraints were considered implicitly in the definition of the human model and they were automatically fulfilled because the model contained only open loops and relative coordinates were used. The method was formulated as the following optimization problem:

$$\min f(\mathbf{q}) = [\mathbf{P} - \mathbf{P}'(\mathbf{q})]^{\mathrm{T}} \mathbf{W} [\mathbf{P} - \mathbf{P}'(\mathbf{q})] \tag{38.2}$$

where q is the column vector of relative coordinates (e.g., joint angles), P is the column vector with the measured global coordinates of all markers in a given frame, P'(q) is a column vector with the global coordinates of all markers given by the human model as a function of the relative coordinates q, and W is a weighting matrix. P'(q) can be written as

$$\mathbf{P}'(\mathbf{q}) = \mathbf{T}(\mathbf{q})\mathbf{P}^* \tag{38.3}$$

where T(q) is the combined transformation matrix from local coordinate systems to the global coordinate system evaluated for a given vector q, and P* is a column vector with the coordinates of each marker in the corresponding local coordinate system. Given the measured-marker coordinates at a certain time, the goal is to find a vector of relative coordinates q such that the objective function (Equation 38.2) is minimized. This is a nonlinear unconstrained optimization problem that can be solved with several standard optimization algorithms. Lu and O'Connor applied their method to a lower limb model and Roux et al. (2002) applied the same method to an upper body model. Both papers concluded that the global IK method proposed by the authors estimate body segment poses more accurately than local IK methods. Other authors (Zhao & Badler, 1994; Bodenheimer et al., 1997; Riley et al., 2000) have also proposed similar global IK methods based on relative coordinates.

In motion reconstruction problems, relative coordinates are the most widely used because the main digital human models contain only open loops and these coordinates lead to a problem without explicit kinematic constraints. Note that although relative coordinates are independent for open-loop models, for closed-loop models they are dependent and require kinematic constraints for the model definition. Therefore, the previous methods are not valid for digital models that contain closed loops.

Other global IK methods valid for open- and closed-loop models are based on natural coordinates (Wang et al., 2005; Ausejo, 2006; Ausejo et al., 2006), which describe the position and orientation of bodies through the Cartesian components of points and vectors located at the mechanism joints. Natural coordinates (García de Jalón & Bayo, 1994) are dependent coordinates and they always require kinematic constraints for the definition of the human model. The kinematic constraints originate from rigid body conditions and explicit joint constraints, and they can be expressed in matrix form as follows:

$$\Phi(\mathbf{q}) = 0 \tag{38.4}$$

where q is the column vector of natural coordinates and each element $\phi_i(q)$ of $\Phi(q)$ is a kinematic constraint equation. The motion of the digital human model is defined by the driven coordinates z, which are a subset of q, and they can be expressed in matrix form as follows:

$$\mathbf{z} = \mathbf{Sq} \tag{38.5}$$

The value of the driven coordinates is defined by the driving constraints

$$\Psi(\mathbf{q},t) = \mathbf{Sq} - \mathbf{d}(t) = 0 \tag{38.6}$$

where $\mathbf{d}$ is a column vector whose ith element is a given function of time $g_i(t)$ that gives the value of corresponding driven coordinate. In a motion reconstruction problem z contains the model-marker coordinates and $\mathbf{d}$ contains the measured-marker coordinates. According to the definition of a global IK method, we have to satisfy the kinematic constraints, which include the joint constraints, and minimize the global measurement error, for instance, the quadratic error of the driving constraints. This is a nonlinear constrained optimization problem:

$$\underset{\mathbf{q} \in \mathfrak{R}^n}{\text{minimize}} \quad f(\mathbf{q}) = \frac{1}{2} \Psi^{\mathsf{T}}(\mathbf{q},t) \, \Psi(\mathbf{q},t)$$

$$\text{subject to:} \quad \Phi(\mathbf{q}) = 0 \tag{38.7}$$

The objective function is a quadratic function of the natural coordinates and the equality constraints are always linear or quadratic. Additionally the Jacobian matrix of the equality constraints is sparse with linear or constant terms. This means that efficient algorithms for sparse matrix factorization can be used within the optimization algorithm. Optimization problems formulated with relative coordinates require fewer variables than the equivalent problem formulated with natural coordinates. However, relative coordinates give highly nonlinear objective functions and equality constraints (only for models with closed loops). Unfortunately, an efficiency comparative between different global IK methods has not been published to date.

Baerlocher and Boulic (2004) proposed an efficient recursive IK algorithm that can handle multiple constraints. The constraints can be classified into different priority levels. The highest prioritized constraints are satisfied first. The solutions for the constraints of lower levels are searched in the null space of the Jacobian matrix defined from higher level tasks, implying that the lower lever constraint solution will not affect the higher level ones. Monnier et al. (2007) successfully adapted this method to the whole-body motion reconstruction by forcing model-based markers to follow the measured trajectories. Thanks to this priority-based resolution approach, the markers, trajectories with low confidence (easily missed or high skin movement) can be assigned with a lower priority level. As the method is an interactive one, the computational cost is very high.

Cerveri et al. (2003a and b, 2005) proposed a global IK method for estimating the motion of the whole body directly from 2D images recorded by the motion capture system instead of from the 3D marker trajectories. Their method is based on an extended Kalman filter and a human model with joint constraints. The posture of the human model, in terms of joint angles, is calculated by minimizing the distance between the markers measured on the 2D images recorded by multiple cameras and the back-projected markers of the human model. The authors demonstrated the robustness of the techniques against missing and phantom markers. However, the ability to cope also with the soft tissue artifact needs to be assessed further (Leardini et al., 2005).

38.6 Under-Guided Problems

Under-guided motion reconstruction problems appear when the motion capture data available are not enough to calculate the posture of the digital human model. All motion reconstruction methods for under-guided problems are based on the same principle: As the measured motion data are not enough to calculate the whole model posture, additional information is added in order to completely define the digital model posture at each frame.

Under-guided problems are usually found on virtual reality and computer graphics applications when a minimum number of sensors are used to measure and reconstruct the subject's motion in real time. Different motion capture systems are commonly used in these applications, including electromagnetic (Badler et al., 1993), optoelectronic (Grochow et al., 2004; Chai & Hodgins, 2005) and even markerless systems (Boulic et al., 2006). Some global IK methods use a prediction model of human motions, which is usually trained from a predefined motion database (Grochow et al., 2004; Chai & Hodgins, 2005; Liu et al., 2006). Then, the motion can be reconstructed from a reduced number of sensors using the prediction model. Their drawback is that the motions that can be reconstructed are restricted to those stored in the database. Another approach to compensate for the lack of motion data is to add constraints to the model (Boulic et al., 2006; Peinado et al., 2006), for example, a balance constraint to ensure the static equilibrium of model. The main goal of these methods is to produce believable motions rather than measuring accurately the real subject motion. Therefore, they are limited to applications where the accuracy of the reconstructed motion is not crucial.

Under-guided problems also appear in applications in which posture accuracy is important. One of the most common under-guided problems is associated with the spine. Digital human models usually include several segments to model the spine but the motion of these segments is difficult to measure using only external markers and the spine posture is not completely defined. Molet et al. (1999) proposed a method to estimate the posture of a seven-segment spine model using the orientation data provided by three electromagnetic sensors, while reconstructing the whole-body motion. Each sensor influenced only the motion of the neighboring joints according to tailored ratios. Recently, Ausejo et al. (2006) presented a method for reconstructing the spine posture of the RAMSIS model using a few markers on the spine while reconstructing the whole body motion (Figure 38.3a). A spine reference posture, which is defined in terms of joint angles, was used to complement the motion data provided by the markers and weighting factors were used to control the flexibility of each spine joint angle. Monnier et al. (2007) suggested the use of a coordination law between the joint angles of the spine. The four global spine movements were used to control all spine joint angles: torso global frontal and lateral flexion-extension, torso axial rotation, and torso elevation and compression (fig. 38.3b).

38.7 Concluding Remarks

Motion capture and reconstruction is the first step for many different digital human applications. In this chapter, currently available methods of motion capture and kinematic motion reconstruction are overviewed for digital human models. At present motion reconstruction methods are able to reconstruct the motion of a whole-body model with a good visual faithfulness (fig. 38.4). However, there is a need for more efficient, accurate, and less time-consuming methods. The following issues related to motion reconstruction requires to further investigation:

First, most currently available motion capture systems are expensive and require time-consuming data post-processing and a well-controlled laboratory setting. Their main limitation is the soft tissue artifact, which is recognized to be the bounding error source in the measurement of human motion (Alexander & Andriacchi, 2001; Leardini et al., 2005). Therefore, further improvements are needed in order to handle more efficiently the soft tissue artifact.

Second, the creation of an individualized digital human model is a difficult and time-consuming process. It involves two fundamental issues: the definition of the model itself and the estimation of

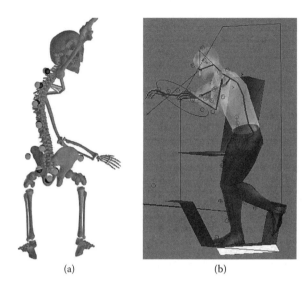

(a)   (b)

FIGURE 38.3 (a) Reconstructed posture of a generic reach movement (only markers on pelvis and spine are shown). The spine posture is not completely defined by the five markers on it and its posture was reconstructed using a referential posture. (From Ausejo, S., et al., "Robust Human Motion Reconstruction in the Presence of Missing Markers and the Absence of Markers for Some Body Segments," *Proceedings of the SAE 2006 Digital Human Modeling for Design and Engineering Conference* SAE International, Warrendale, PA, 2006. With permission.) (b) Car ingress motion reconstructed using the spine coordination law by Monnier et al. (2007). Only one marker (red circle) is positioned on the spine. (From Monnier, G., et al., "Coordination of Spine Degrees of Freedom during a Motion Reconstruction Process," *Proceedings of the SAE International Conference and Exposition of Digital Human Modeling for Design and Engineering*, SAE International, Warrendale, PA, 2007. With permission.)

subject-specific parameters. The most popular digital human models assume fixed joint centers and joint axes, which is a source of errors especially for some joints (e.g., hand or shoulder complex). Furthermore, there is not always an easy correspondence between the model joints and those of the real subject due to the complexity of human joints. Therefore, there is a special need for improving the model and for more efficient, accurate, and automatic methods for estimating subject-specific parameters.

Third, under-guided problems are still difficult to solve. As the posture of the digital human model is not completely defined in under-guided problems, the critical point is how to identify additional data or constraints and how to integrate the additional information into IK algorithms.

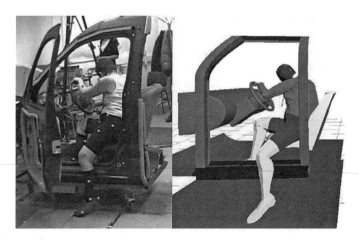

FIGURE 38.4 Real subject (left) and reconstructed posture of the digital human (right).

In this chapter, we have only reviewed methods for kinematic motion reconstruction in which forces are not taken into account. However, many applications such as ergonomic simulation will benefit from a dynamic motion reconstruction, which aims to reconstruct not only kinematic data of a motion but also its external contact forces for estimating internal joint torques. Dynamic data are necessary for better understanding motion control and perceived-motion-related discomfort (Wang, Chapter 25 in this handbook).

References

Alexander, E.J., Andriacchi, T.P., 2001. Correcting for deformation in skin-based marker systems. *Journal of Biomechanics* 34, 355–361.

Allard, P., Stokes, I.A.F., Blanchi, J.P., Eds. 1995. *Three-dimensional analysis of human movement.* Human Kinetics, Champaign, IL.

Apkarian, J., Nauman, S., Cairns, B., 1989. A three-dimensional kinematic and dynamic model of the lower limb. *Journal of Biomechanics* 22, 143–155.

Ausejo, S., 2006. A new robust motion reconstruction method based on optimisation with redundant constraints and natural coordinates. Ph.D. dissertation, University of Navarra, San Sebastian.

Ausejo, S., Suescun, A., Celigüeta, J.T., Wang, X., 2006. Robust human motion reconstruction in the presence of missing markers and the absence of markers for some body segments. In *Proceedings of the SAE 2006 digital human modeling for design and engineering conference.* Lyon, France.

Badler, N.I., Hollick, M.J., Granieri, J.P., 1993. Real-time control of a virtual human using minimal sensors. *Presence* 2(1), 82–86.

Baerlocher, P., Boulic, R., 2004. An inverse kinematic architecture enforcing an arbitrary number of strict priority levels. *The Visual Computer* 20(6), 402–417.

Ball, K.A., Pierrynowski, M.R., 1998. Modeling of the pliant surfaces of the thigh and leg during gait. In: Jacques, S.L. (Ed.), *Proceedings of SPIE laser-tissue interaction IX*, pp. 435–446.

Bodenheimer, B., Rose, C., Rosenthal, S., Pella, J., 1997. The process of motion capture: Dealing with the data. In Thalmann, D., van de Panne, M. (Eds.), *Computer animation and simulation '97.*

Boulic, R., Varona, J., Unzueta, L., Peinado, M., Suescun, Á., Perales, F., 2006. Evaluation of on-line analytic and numeric inverse kinematics approaches driven by partial vision input. *Virtual Reality* 10(1), 48–61.

Cappello, A., Cappozzo, A., La Palombara, P.F., Lucchetti, L., Leardini, A., 1997. Multiple anatomical landmark calibration for optimal bone pose estimation. *Human Movement Science* 16, 259–274.

Cappozzo, A., Catani, F., Leardini, A., Benedetti, M.G., Della Croce, U., 1996. Position and orientation in space of bones during movement: Experimental artefacts. *Clinical Biomechanics* 11(2), 90–100.

Carman, A.B., Milburn, P.D., 2006. Determining rigid body transformation parameters from ill-conditioned spatial marker co-ordinates. *Journal of Biomechanics* 39(10), 1778–1786.

Cerveri, P., Pedotti, A., Ferrigno, G., 2003a. Robust recovery of human motion from video using Kalman filters and virtual humans. *Human Movement Science* 22(3), 377–404.

Cerveri, P., Pedotti, A., Ferrigno, G., 2005. Kinematical models to reduce the effect of skin artifacts on marker-based human motion estimation. *Journal of Biomechanics* 38(11), 2228–2236.

Cerveri, P., Rabuffetti, M., Pedotti, A., Ferrigno, G., 2003b. Real-time human motion estimation through biomechanical models and nonlinear state-space filters. *Medical and Biological Engineering and Computing* 41(2), 109–123.

Chai, J., Hodgins, J.K., 2005. Performance animation from low-dimensional control signals. *ACM Transactions on Graphics* 24(3), 686–696.

Challis, J.H., 1995. A procedure for determining rigid body transformation parameters. *Journal of Biomechanics* 28(6), 733.

Chiari, L., Della Croce, U., Leardini, A., Cappozzo, A., 2005. Human movement analysis using stereophotogrammetry. Part 2: Instrumental errors. *Gait and Posture* 21(2), 197.

Desjardins, P., Plamondon, A., Nadeau, S., Delisle, A., 2002. Handling missing marker coordinates in 3D analysis. *Medical Engineering and Physics* 24(6), 437.

Gamage, S.S.H.U., Lasenby, J., 2002. New least squares solutions for estimating the average centre of rotation and the axis of rotation. *Journal of Biomechanics* 35, 87–93.

García de Jalón, J., Bayo, E., 1994. *Kinematics and dynamic simulation of multibody systems: The real-time challenge.* Springer-Verlag, New York.

Giakas, G., Baltzopoulos, V., 1997. A comparison of automatic filtering techniques applied to biomechanical walking data. *Journal of Biomechanics* 30(8), 847–850.

Grochow, K., Martin, S.L., Hertzmann, A., Popovic, Z., 2004. Style-based inverse kinematics. *ACM Transactions on Graphics* 23(3), 522–531.

Halvorsen, K., Lesser, M., Lundberg, A., 1999. A new method for estimating the axis of rotation and the center of rotation. *Journal of Biomechanics* 32, 1221–1227.

Halvorsen, K., Söderström, T., Stokes, V., Lanshammar, H., 2005. Using an extended Kalman filter for rigid body pose estimation. *Journal of Biomechanical Engineering* 127, 475–483.

Hartley, R., Zisserman, A., 2004. *Multiple view geometry in computer vision.* Cambridge University Press, Cambridge, UK.

Herda, L., Fua, P., Plankers, R., Boulic, R., Thalmann, D., 2001. Using skeleton-based tracking to increase the reliability of optical motion capture. *Human Movement Science* 20(3), 313.

Kadaba, M.P., Ramakrishnan, H.K., Wootten, M.E., 1990. Measurement of lower extremity kinematics during level walking. *Journal of Orthopaedic Research* 8, 383–392.

Kirk, A.G., O'Brien, J.F., Forsyth, D.A., 2005. Skeletal parameter estimation from optical motion capture data. In *Proceedings of the IEEE Computer Society conference on computer vision and pattern recognition.*

Lasenby, J., Gamage, S., Ringer, M., 2001. *Geometric algebra: Application studies.* Technical report. Department of Engineering, University of Cambridge, UK.

Leardini, A., Chiari, L., Della Croce, U., Cappozzo, A., 2005. Human movement analysis using stereophotogrammetry. Part 3: Soft tissue artifact assessment and compensation. *Gait and Posture* 21(2), 212.

Liu, G., Zhang, J., Wang, W., McMillan, L., 2006. Human motion estimation from a reduced marker set. In *Proceedings of the symposium on interactive 3D graphics and games.*

Lu, T.-W., O'Connor, J.J., 1999. Bone position estimation from skin marker co-ordinates using global optimisation with joint constraints. *Journal of Biomechanics* 32, 129–134.

Lucchetti, L., Cappozzo, A., Cappello, A., Della Croce, U., 1998. Skin movement artefact assessment and compensation in the estimation of knee-joint kinematics. *Journal of Biomechanics* 31, 977–984.

Medved, V., 2001. *Measurement of human locomotion.* CRC Press, Boca Raton, FL.

Meskers, C.G.M., van der Helm, F.C.T., Rozendaal, L.A., Rozing, P.M., 1998. In vivo estimation of the glenohumeral joint rotation center from scapular bony landmarks by linear regression. *Journal of Biomechanics* 31, 93–96.

Molet, T., Boulic, R., Rezzonico, S., Thalmann, D., 1999. Human motion capture driven by orientation measurements. *Presence* 8(2), 187–203.

Monnier, G., Renard, F., Chameroy, A., Wang, X., Trasbot, J., 2006. A motion simulation approach integrated into a design engineering process. SAE paper 2006-01-2359. In *Proceedings of the SAE international conference and exposition of digital human modeling for design and engineering.* Lyon, France.

Monnier, G., Wang, X., Beurier, G., Trasbot, J., 2007. Coordination of spine degrees of freedom during a motion reconstruction process. In *Proceedings of the SAE international conference and exposition of digital human modeling for design and engineering.* Seattle, WA.

Muijtjens, A.M.M., Roos, J.M.A., Arts, T., Hasman, A., Reneman, R.S., 1997. Tracking markers with missing data by lower rank approximation. *Journal of Biomechanics* 30(1), 95.

Park, W., Chaffin, D.B., Martin, B.J., 2004. Toward memory-based human motion simulation: development and validation of a motion modification algorithm. *IEEE Transactions on System, Man and Cybernetics—Part A: Systems and Humans* 34(3), 376–386.

Peinado, M., Meziat, D., Boulic, R., Raunhardt, R., 2006. Environment-aware postural control of virtual humans for real-time applications. In *Proceedings of the SAE 2006 digital human modeling for design and engineering conference*. Lyon, France.

Riley, M., Ude, A., Atkeson, C.G., 2000. Methods for motion generation and interaction with a humanoid robot: Case studies of dancing and catching. In *Proceedings of the AAAI and CMU workshop on interactive robotics and entertainment 2000*. Pittsburgh, PA.

Ringer, M., Lasenby, J., 2002. A procedure for automatically estimating model parameters in optical motion capture. In *Proceedings of the the British machine vision conference*. Cardiff, UK.

Roux, E., Bouilland, S., Godillon-Maquinghen, A.-P., Bouttens, D., 2002. Evaluation of the global optimisation method within the upper limb kinematics analysis. *Journal of Biomechanics* 35, 1279–1283.

Seidel, G.K., Marchinda, D.M., Dijkers, M., Soutas-Little, R.W., 1995. Hip joint center location from palpable bony landmarks: A cadaver study. *Journal of Biomechanics* 28(8), 995–998.

Seitz, T., Balzulat, J., Bubb, H., 2000. Anthropometry and measurement of posture and motion. International *Journal of Industrial Ergonomics* 25(4), 447–453.

Silaghi, M., Plaenkers, R., Boulic, R., Fua, P., Thalmann, D., 1998. Local and global skeleton fitting techniques for optical motion capture. In Magnenat Thalmann, N., Thalmann, D. (Eds.), *Modelling and motion capture techniques for virtual environments*, pp. 26–40. Lecture Notes in Artificial Intelligence, No. 1537. Springer, New York.

Söderkvist, I., Wedin, P., 1993. Determining the movements of the skeleton using well-configured markers. *Journal of Biomechanics* 26(12), 1473–1477.

Stokdijk, M., Nagels, J., Rozing, P.M., 2000. The glenohumeral joint rotation centre in vivo. *Journal of Biomechanics* 33, 1629–1636.

Veldpaus, F.E., Woltring, H.J., Dortmans, L.J.M.G., 1988. A least-squares algorithm for the equiform transformation from spatial marker co-ordinates. *Journal of Biomechanics* 21, 356–360.

Wang, L., Hu, W., Tan, T., 2003. Recent developments in human motion analysis. *Pattern Recognition* 36(3), 585–601.

Wang, X., Chevalot, N., Monnier, G., Ausejo, S., Suescun, Á., Celigüeta, J.T., 2005. Validation of a model-based motion reconstruction method developed in the REALMAN project. In *Proceedings of the SAE 2005 digital human modeling for design and engineering symposium*. Iowa City, IA.

Wang, X., Chevalot, N., Monnier, G., Trasbot, J., 2006. From motion capture to motion simulation: An in-vehicle reach motion database for car design. SAE paper 2006-01-2362. In *Proceedings of the SAE international conference and exposition of digital human modeling for design and engineering*. Lyon, France.

Woltring, H.J., 1995. Smoothing and differentiation techniques applied to 3-D data. In Allard, P., Stokes, I.A.F., Blanchi, J.P. (Eds.), *Three-dimensional analysis of human movement*, pp. 79–99. Human Kinetics, Champaign, IL.

Zhao, J., Badler, N.I., 1994. Inverse kinematics positioning using nonlinear programming for highly articulated figures. *ACM Transactions on Graphics* 13(4), 313–336.

39

The Use of Digital Human Models for Advanced Industrial Applications

Christina Godin
and Jim Chiang

39.1 The Value of Digital Human Modeling for Industrial Ergonomics

The field of ergonomics has gained considerable momentum in recent years, to the point where ergonomists are playing less of a reactive role and migrating into the design process. One of the key criteria for proactive ergonomics is efficiency during the product development process. In order to stay competitive in the consumer market, manufacturers are driven to shorten the development time for a new product, thereby responding to trends more quickly, as well as reducing cost and increasing the total number of products introduced in a given time period (Feyen et al., 2000). This goal has been accomplished largely due to the use of computer-aided design tools. The same holds true in ergonomics, where computer-generated environments and digital humans allow analyses to be performed without ever requiring physical data or prototypes. This trend toward computer-aided ergonomics has been observed in the military as well as a number of manufacturing industries, including the automotive, clothing, and aviation sectors (Yee & Nebel, 1999; Rigel et al., 2003; Dai et al., 2003; Blome et al., 2003; Doi & Haslegrave, 2003). A variety of human modeling tools have been introduced for the purpose of ergonomic analyses, such as RAMSIS, SafeWork, and Jack (Reed et al., 2003), as well as Dhaiba and Santos (in development). These software programs allow users to create virtual environments and generate human models within those environments. Once this is done, the user can manipulate the human model to interact with the surroundings as is predicted would be the case in a true physical environment. A variety of virtual analyses can be performed to predict the success of a product or workstation layout. Most software tools allow for clearance, posture, reach, line of sight, and strength predictions (Blome et al., 2003). Thus, testing that would normally be performed in the physical world is now done without ever building a part or using a true human being.

39.2 Identify Modeling Needs

Before making the decision to employ a digital human model within your organization, there are several points to consider:

What types of questions will you be addressing with digital human modeling (DHM)?

It is important to identify how you intend to utilize a human modeling software package and ensure that the questions you expect to answer can truly be addressed through DHM. For example, several human modeling tools include the ability to perform typical ergonomic assessments, such as low back, rest allowance, strength, reach, and clearance assessments. Reach and clearance assessments are built upon pre-existing anthropometric databases that have been incorporated into the digital human model. Human modeling packages are limited by the availability of assessment methods that have been published and made available in the literature. Most modeling packages have integrated pre-existing ergonomic assessment tools into the digital environment. The benefit is that assessments typically done via traditional methods can also be performed using a digital human model as the subject. However, if the goal is to expand your ergonomic assessment toolset far beyond what is readily available, you may find this a challenge. Digital human models should be seen as *enablers* or another avenue by which ergonomists can utilize their assessment tools.

Common reasons for incorporating digital human modeling into an ergonomics program:

There have been several stated reasons for adding human modeling capabilities to a new or already established ergonomics department or program. The more common rationales are stated below.

Integration into computer-aided design. Several corporations have adopted a product lifecycle management (PLM) solution, where their product is conceived, designed, and assembled using computer-aided design and manufacturing tools to aid the process. Digital human models allow ergonomists to work in the virtual world as well, responding to digital concepts and providing feedback concerning the predicted ergonomic risks of such a concept.

Realistic posture generation. Several DHM tools have incorporated anthropometric databases as well as accurate range of motion (ROM) profiles to ensure representative human movements. This allows the user to manipulate the model and create an infinite number of postures, which closely reflect the actual posture of interest. Furthermore, some digital humans have incorporated posture prediction research into their tools, which reduces the subjective nature of manually posturing the digital human.

For example, in traditional two dimensional (2D) and three dimensional (3D) biomechanical models that do not allow for high-fidelity posture generation, the postures that would be used to drive a biomechanical assessment are somewhat simplified. However, performing the same assessment using a digital human would allow the user to generate a closer representation of the actual posture through a manipulation of the multiple degrees of freedom within the human model.

Enhanced visualization. In general, good ergonomics can be associated with a cost saving due to decreased injury, increased quality, and several other factors. However, ergonomic changes generally require a financial commitment to make that change. Ergonomists will often need to present a rationale to their management to demonstrate the value and necessity of making the investment in ergonomics. Digital human models and their surrounding environment can be used to provide an accurate visual representation of the problem and value of the suggested resolution. Digital mockups and renderings of a situation can augment a business case.

Can DHM benefit other disciplines or areas of your business beyond ergonomics?

Absolutely. Some digital human models also include industrial engineering tools thus allowing the user to perform a predetermined time standards analyses within the digital world. Furthermore, digital human modeling can act as a training tool, using simulated tasks to demonstrate proper and improper performance techniques. Outside of manufacturing, DHM is commonly used for designing products for the end user, for example, the cockpit of an airplane or the interior of a new car or truck. For example, in the automotive industry, occupant packaging studies are conducted to assess the driver's seated position, the view through the window, as well as driver and passenger access to various equipment and controls (Rigel et al., 2003; Dai et al., 2003; Reed et al., 2003).

The evaluation of service and maintenance operations can also be performed using digital human modeling. It is not only important to determine if a product can be safely built and assembled but also disassembled after it goes to market. The feasibility of servicing a product, for example a car, can impact the cost of maintaining that product as well as the safety of the individuals working in the maintenance field.

39.3 Do the Benefits Offset the Cost?

Digital human modeling is a powerful tool and can greatly augment an ergonomics program. However, DHM solutions can be costly, may require user training for best results, and typically require an upgrade in hardware to get the most out of the DHM. For those who have identified DHM as a valuable asset to their work methods, this investment is relatively minor. Nevertheless, such factors should be considered prior to making the venture. If the tool is not going to be used regularly, there may be an alternative solution for your organization. Some companies will choose to outsource digital human modeling assessments if they are required infrequently, while others will use alternative assessment methods to derive ergonomic solutions.

Once you have identified that digital human modeling will be an asset to your current ergonomics program, your next endeavor will likely include a review of the currently available digital human modeling tool suites. While most commercial tools (albeit there are few) have similar capabilities, you will find differences in cost, function, and the ergonomic toolset embedded within each. The reader is encouraged to review Jack (UGS, Texas, USA), Safework (Dassault Systemes, Montreal, Canada), and RAMSIS (Human Solutions, Kaiserslautern, Germany, and Michigan, USA). Additional options include Dhaiba (Digital Human Research Center, Tokyo, Japan), AnyBody (AnyBody Technology, Aalborg, Denmark), and Santos (Virtual Soldier Research Group, University of Iowa, USA). The latter three options are primarily thought of as research-centric tools at this point in time and differ in function, philosophy, and capability from the more traditional human modeling options mentioned above.

39.4 Augmenting Digital Human Modeling with Motion Capture Technology

As computer-aided ergonomics becomes a regular part of the design process for a new product, the drive to improve the accuracy and validity of this technology will increase. Some researchers and manufacturers have already been faced with this challenge and have sought ways to enhance their virtual reality laboratories. One of the major limitations of simply manipulating digital humans is the inability to accurately predict the motion path of a human being. Digital humans are capable of assuming both static and dynamic postures, with dynamics being the most difficult and time consuming to achieve. Predicting a motion path can be accomplished using a variety of software packages, but the user must manually (or semi-manually) program this. Motion paths are usually generated in one of two ways: by chaining together motion blocks (e.g., in walking, leg lifting can be programmed to occur repeatedly to create locomotion) or by indicating a start and end point and allowing the software to predict the human's actions (Yee & Nebel, 1999). Both techniques can be tedious, especially when a series of complex actions must occur together. Motion capture technology is an alternative option for generating dynamic human motion within a virtual environment. This method alleviates the time-consuming task of programming a motion series for a digital human and also ensures that the movements will reflect true human actions.

A variety of collection devices has been used to track motion data. For example, ShapeTape (Measurand, Fredericton, Canada) has fiber-optic sensing arrays that track bending and twisting within the tape to capture movement. This can be placed on the limbs, torso, and head to sense human movement (Danisch & Lowery-Simpson, 2003). Optical systems utilize infrared cameras to locate an object or human in space. Reflective markers are placed on the human and positional data are captured via camera and stored in a PC. Optical systems do not require a tether to a central processor; however, this may occasionally result in lost data if the actions of the human block a camera's line of sight to a body marker (Yee & Nebel, 1999). This limitation can be avoided by ensuring that the appropriate number of cameras is used. For example, when studying passengers entering and exiting a vehicle, Rigel et al. (2003) found that quality motion data could not be obtained until the total number of cameras used in their study was increased to 14. This ensured that all 60 markers placed on the subject's body were sufficiently visible to the cameras during data collection.

Another common method of capturing motion is through electromagnetic devices which are used to track positional data in real time. The instrumentation required for this system is a central transmitter, which is connected to a central processing unit. This unit can have one or more sensors that are tracked in space based on its orientation to the electromagnetic field surrounding the transmitter (Maiteh, 2003). While magnetic tracking devices allow data to be concurrently processed and ensure a true representation of the movement patterns of the subject, there are limits to its application. In most cases, electromagnetic tracking devices are limited to laboratory settings given that interference with metallic objects or power sources will distort data. The magnetic fields from ferrous metal objects distort the transmitter field of the motion capture system, which ultimately causes the positional data to reorder refs.: 1999; 2001; 2003 drift, altering results (Yee & Nebel, 1999; Jayaram & Repp, 2001; Agnew et al., 2003).

Using an Ascension (Ascension Technology Corporation, Vermont, USA) electromagnetic motion capture system, integrated with the Jack human motion analysis software, Maiteh (2003) demonstrated how virtual ergonomic analyses can be performed in both proactive and reactive scenarios. When performed reactively, the author suggests that a video of the operation in question should be taken and a virtual environment built to replicate the true setting. An operator can interact with this environment in a motion capture lab in order to identify critical movements. These movements can then be analyzed using a software package such as Jack. If a job is assessed proactively, the analyzer can review the virtual environment that is being proposed and determine which variables require ergonomic consideration. A digital human can be programmed to follow the anticipated motion paths or more desirably a motion capture system can be used to generate true motions for analysis in a software program (Maiteh, 2003). Using a motion capture facility and Jack software, the author reviewed several scenarios for an automotive sub-assembly task and advocates the use of motion capture as a more accurate and time efficient method for job analyses.

39.5 Developing a Motion Capture/Virtual Reality (VR) Laboratory

Typically, implementing a VR lab would not be your first step into the world of human modeling. Rather, the addition of VR capability adds advanced capability to augment your existing human modeling toolset.

The cost of setting up your lab can vary widely depending on the equipment you choose and the types of operations you are trying to simulate. Picking appropriate hardware for your lab can be tricky. There is a vast array of devices available and understanding the benefits and limitations of each type of device is important. This section will give a brief overview of the types of equipment that are generally used with DHM software.

39.6 Tracking Systems

A tracking system is required to capture the movements of the subjects you will have in your lab. There are many types of tracking systems available on the market, but generally two types of systems are supported: magnetic and passive optical.

39.7 Optical, Active and Passive

Optical tracking systems can be divided into active and passive systems (see Figure 39.1). This designation is determined by the types of markers used in each system. In an active system, the markers require power and are thus considered active. In a passive system, the markers do not require power and are thus passive. While active optical systems have many advantages over passive systems, they do not tend to be used in VR applications with DHM software. As such this discussion will focus on passive optical tracking systems.

Optical systems use cameras that see spherical markers. Because each marker is spherical, it does not have any information about its orientation in space. To compensate for this, multiple markers are required on each body segment or object to generate information about both position and orientation.

The main disadvantage of optical systems is the line-of-sight requirement. Simply put, in order to create a 3D representation, multiple cameras must see the same marker from different angles. While two cameras for a given marker can supply sufficient data, three or more cameras are preferred. Optical systems tend to require a larger dedicated capture volume. In excess to the volume required for your actual study, additional space is needed to place cameras in appropriate locations and distances to achieve proper fields of views for all cameras (see Figure 39.2). Optical systems are also computationally intensive as 2D data must be processed from all cameras in order to generate the proper 3D representation. While more cameras provide more camera views and thus relieve some of the line-of-sight issues, they also increase the computing power required to handle this increased amount of data. Fortunately, computing speed has generally outpaced general computing needs so this becomes less of an issue with time. While it may be generally recommended to dedicate one fairly powerful PC to handle the optical system data, often it is possible for this same PC to also handle your DHM software. This is especially true with multiprocessor computers, which are quickly becoming more common.

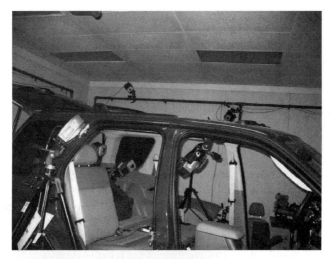

FIGURE 39.1   A lab using passive optical motion tracking methods to complete an ergonomic accessibility study.

39.8 Electromagnetic

Electromagnetic systems generate a magnetic field and then sense the position and orientation of the sensors within the field. They generally require less space than optical systems since camera field of view is not a consideration. An electromagnetic system senses both the position and orientation of each sensor; therefore, fewer sensors are required compared to markers for optical systems. Electromagnetic systems use active sensors so the subject must either be cabled to the tracking system, or wireless options can provide more movement freedom although a battery pack is required (see Table 39.1).

The main drawback of electromagnetic systems is metallic distortion. The use of metal objects within the tracking space leads to tracking errors. While these can sometimes be compensated for, this is not always the case. If you can get around this limitation, these systems can be a good choice.

FIGURE 39.2 Optical digital cameras. (Courtesy of Motion Analysis Corporation, www.motionanalysis.com.)

39.9 Other Types of Tracking Systems

There are other types of tracking systems available. Notably inertial tracking systems have appeared recently and some show good cost/performance ratios. As well some manufacturers are producing hybrid tracking systems by combining different tracking technologies into their products.

39.10 Data Gloves

Unlike a general tracking system, data gloves are specifically used for tracking hand postures (Figure 39.3). Some gloves may include sensors for tracking wrist posture, but all gloves will have sensors for tracking finger postures. The number of sensors will have a major influence on cost and usability. Other factors to consider are size, comfort, and intrusiveness to your application. Different glove manufacturers deal with sizing differently. Some produce gloves in various sizes while others take a one-size-fits-all approach. If you have a set target population in mind for your studies this may be an important consideration. The intrusiveness of a glove design may be a consideration. Data gloves are available with both closed and open fingertip. Open fingertip gloves provide more manual dexterity and may be easier to fit on a wider range of hand sizes. Lastly, the connector and material of the glove are also of consideration. Material feel, thickness, and cut will affect how the glove can be worn. Data connection to the glove varies as well. Some gloves have a connector. Other gloves have an integrated cable. Both approaches have advantages and disadvantages.

39.11 Head-Mounted Displays

Head-mounted displays (HMD) come in many varieties from light, simple monoscopic, eyeglass type to wide field of view, stereoscopic headpieces (Figure 39.4). Accordingly, these displays vary widely in price,

TABLE 39.1 Tracking System Comparison

	Relative Cost	Line-of-Sight	Metallic Interference
Passive optical	high	yes	no
Electromagnetic	low	no	yes

FIGURE 39.3 Wireless data glove kit. (Courtesy of 5DT, Inc., www.5dt.com.)

size, and weight. An HMD combined with a tracking system can provide an immersive experience for the visual system. This may allow the subjects to behave more naturally as they move through a virtual environment, although the weight, balance, and comfort of the HMD will influence this greatly.

39.12 Power and Communications

All hardware devices require power and a means to communicate with the computer. The main consideration here is if these will interfere with your application. For example, data gloves are available in both wired and wireless varieties. The wired type requires a cable from the glove to the control box. In some applications, like seated applications with limited movement, this may be acceptable. For applications involving full body motion through a volume, this cable may become tangled and limit movement. Going to a wireless option will relieve the need for this cable, but a battery is then required to power the glove and the wireless transmitter. The size, weight, and run-time of the battery then become the major considerations as well as how and where the battery pack will be worn. While this example focused on data gloves, these same considerations are valid for almost all types of hardware.

39.13 Cost Breakdown

Table 39.2 supplies general costs of VR hardware that you might consider for your lab. These costs span different vendors and levels of functionality and should be used to getting a general idea of the costs involved.

FIGURE 39.4 Head-mounted display. (Courtesy of 5DT, Inc., www.5dt.com.)

TABLE 39.2 Hardware Options and Pricing

Hardware Type	Cost Range
Electromagnetic tracking system	$25,000–$75,000
Passive optical tracking system	$80,000–$175,000
Data gloves, pair	$12,000–$25,000
Stereo head-mounted display	$25,000–$50,000

39.14 Hardware Evaluation

Researching equipment on the Internet is one thing, but there is nothing like using the equipment first-hand. Most often you should be able to arrange for a demonstration or evaluation of the products. This is strongly encouraged. Not only does it provide you with firsthand experience with the products, it will also give you an opportunity to evaluate the vendor company in general and how they sell and support their products.

39.15 Ongoing Costs

While the general cost of hardware has been addressed, keep in mind that there may be ongoing costs in the form of extended warranties and support contracts. It is important to inquire about these and evaluate whether you will need these.

39.16 Software Drivers

Aside from the need of a DHM package, there needs to be adequate software support for the hardware you wish to use. The availability, feature set, and ease of use of these software drivers are important decisions in the purchase of hardware devices. At the end of the day, if you have a piece of equipment with no usable way to use it with your DHM software, then likely you have made some poor choices. Drivers may be included as an add-on to your DHM package or they may be supplied by the hardware vendors themselves. In some cases you will need drivers supplied by both parties to enable the functionality you desire. Investigating these issues before purchasing can save you unexpected costs and headaches later. It may also be possible to create your own drivers. If so, you need to consider the technical skills required and the cost benefits of creating and maintaining your own drivers versus purchasing an off-the-shelf product.

39.17 Space Considerations

Lab space requirements depend on your applications and the type of equipment you have. For manufacturing operations, you would typically be interested in tracking a whole body as it moves through a volume. If you are looking at simulating an entire workstation, then the size of the lab needs to be able to accommodate this. Also, the tracking system itself may require additional space. For example, a magnetic system requires area for transceiver placement and an optical system requires space for camera placement. The space requirements for an optical system are more than those of a magnetic system. Let's say, for example, you wish to have a work area of 3 m^2 and wish to cover up to a height of 2.5 m. This volume should be able to capture the movements of a tall person and accommodate for some lateral movement and vertical reach. To accommodate a magnetic system, you would need to add approximately 1 m each side of the space to set up the transceivers. To set up a passive optical system to capture this volume would require approximately 6 m^2 of floor space with a ceiling height of 3.5 m. Optical system space requirements are also determined by the focal length of the camera lenses.

39.18 Prop Considerations

One of the big advantages of motion capture is that the movements and postures captured in your simulations are based on real people and real movements. This adds credibility to the validity of your data and your analysis. To achieve this goal of keeping the movements and postures real requires that the physical environment you set up in your lab adequately mimic the virtual world in which the simulation is taking place (Figure 39.5). This is the job of the props you set up for your motion capture session. While completely replicating the virtual world in your lab would yield the most realistic results, this would be time consuming, costly, and would also largely negate the need for doing the simulation. Rather you should use the simulation to enhance understanding of how the subject will interact with its environment. For example, if the work station has a seated operation, it would be not be practical to capture motions of your subject while pretending to sit and perform this operation. You would need to supply a seat for them to sit on of similar size. Also, if the subject is required to lean forward over a surface you would need to supply them with an appropriate hard point to lean over. Asking a person to pretend to lean over something would not provide valid results. Determining how many props you need in your simulation can be difficult at times. Also, props add complexity to capturing the motions. If you are using an optical system, each prop you place in the room creates a visual obstruction which you need to compensate for by either adding cameras and/or adjusting the position of cameras. If you are using a magnetic system you need to be careful of the choice of materials for the props, which may be difficult depending on the size and shape of the prop required. As a general guideline you will want to provide sufficient props to replicate the hard contact points and surfaces with which the human will interact.

39.19 Hardware Vendors

Below is a list of some of the more established hardware vendors. Keep in mind that this is by no means an exhaustive list.

Tracking systems
Ascension Technology Corporation, http://www.ascension-tech.com
Motion Analysis Corporation, http://www.motionanalysis.com
Vicon, http://www.vicon.com

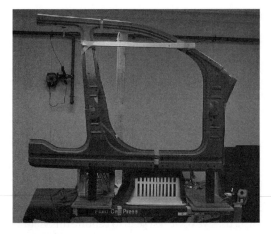

FIGURE 39.5   Props (left) used to replicate a virtual environment (right) while testing the reach to make an electrical connection to the centerline radio antenna in a vehicle.

Data gloves
Fifth Dimension Technologies, http://www.5dt.com
Immersion Corporation, http://www.immersion.com
Head-mounted displays
Fifth Dimension Technologies, http://www.5dt.com
nVis, Inc, http://www.nvisinc.com
Rockwell Collins, Inc., http://www.rockwellcollins.com/optronics
Virtual Research Systems, Inc., http://www.virtualresearch.com

39.20 Digital Human Modeling and Motion Capture for Training Applications

The use of motion capture and virtual reality environments has gained considerable momentum and credibility in recent years. In particular, digital human models and immersive virtual reality environments have proven to be effective training tools in a number of applications. For example, injured victims and medical personnel have been modeled to monitor the responses of medical staff (Chi et al., 1996, 1997). This technology has also extended into the world of sports, where virtual reality immersion has been tested to determine if it is a valid training tool for the game of handball (Bideau, 2003). The study consisted of two phases: once when a goalkeeper interacted with a real thrower, and once when the thrower was a digital human animated through motion capture. The goalkeeper's actions were compared between the true and the virtual conditions, and it was found that the virtual environment offered enough realism to generate the same outcome as a live training scenario. This study showed the success of using an advanced technology to improve the skills of athletes in handball, and may also serve to advance training in other sporting avenues as well.

Additional use of this technology can be seen at the Information Sciences Institute at the University of Southern California, where they have developed an animated human model for training. Steve (Soar Training Expert for Virtual Environment) is able to interact with students as well as monitor their progress, and offer guidance and assistance. Steve has been used in a variety of naval applications and offers both individual and team training, where the digital mannequin can even take on the role of an absent student (Johnson et al., 1997; Rickel & Johnson, 1999, 2000; Johnson & Rickel, 2000). In another application, the University of Pennsylvania (Dr. Norman Badler) and NASA have joined to develop integrated computer-based training and teaching tools. The research mandate is to improve the reliability of providing instruction to crews who have limited training, in particular for emergency situations. The group has developed a training system and set of instruction tools that are suitable for multiple applications, including medical and maintenance.

Interactive virtual worlds provide a powerful medium for experiential learning. The use of immersive technologies such as head-mounted displays, data gloves, and motion tracking systems can assist in providing environments that closely reflect real-world settings. This gives students the opportunity to learn by doing. This is particularly crucial in applications where students/trainees would not otherwise have exposure to a real-life setting. For example, consider a technician's access to the latest military vehicles. This equipment, like any other motorized vehicle, will require maintenance throughout its life cycle. However, given the vehicle expense and relatively small production numbers, maintenance crews are left with little to no opportunity to work on these machines in a learning environment prior to performing true maintenance operations. Furthermore, machinery and vehicles of this nature can often be found in remote locations and overseas where maintenance crews do not have ready access, again limiting exposure and practice opportunities. The responsibilities and expectations of a maintenance crew, including the safe operation and rapid repair of such vehicles, are extreme, especially given the lack of exposure to such performance conditions. However, through the use of immersive virtual reality, a toolset now exists to offer students exposure and practice with digital, 3D renderings of the machinery.

39.21 Refining the Use of Motion Capture and Digital Human Modeling for Ergonomics in Manufacturing

In recent years, a series of studies have been conducted through the University of Windsor (Windsor, Canada), McMaster University (Hamilton, Canada), and Ford Motor Company (Michigan, USA) to gain a better understanding of how DHM and motion capture can be used to solve ergonomic concerns. A summary for each study is presented below. Full manuscripts can be obtained by contacting the author or publisher.

Study 1: Assessing the Accuracy of Ergonomic Analyses When Human Anthropometry Is Scaled in a Virtual Environment (SAE, 2006: Godin, Chiang, Stephens, & Potvin)

This study addressed the effect of scaling subjects in a virtual reality environment when performing ergonomic evaluations for assembly automotive tasks. Ten male and 10 female automotive employees participated in this study. Subjects were selected to fit into one of four anthropometric groups (n=5/group); 5th percentile female (5F), 50th percentile female (50F), 50th percentile male (50M), or 95th percentile male (95M). Each subject was asked to perform three automotive assembly tasks while interacting with a digital rendering of a vehicle in virtual reality (Figure 39.6). The subjects were represented in virtual reality as a human mannequin (Classic Jack, UGS) whose actions were driven by their actual motions captured via motion tracking (EvaRT, MotionAnalysis). Each subject performed the tasks under four different conditions; in one condition, the subject appeared in true size, and in the three other conditions, they were scaled to appear as the size of the other three subject groups. Peak and cumulative low back loads, joint angles at the point of peak compression, and peak and cumulative resultant shoulder moments were output from the Task Analysis Tool Kit within Classic Jack. A repeated measures ANOVA with a Tukey's significance post-hoc test were used to identify differences within the data ($p < 0.05$). Results show that, for virtual assessments of peak and cumulative low back compression, scaling subjects between the range of the 50 F to the 95 M was deemed an acceptable practice. In terms of ergonomic assessments related to the shoulder, if limits are to be based on 5 F or 50 F individuals, subjects can be scaled anywhere within the range of 5 F to 50 F, without affecting the accuracy of the results

FIGURE 39.6 The actual assembly job (top), lab setup (middle), and virtual environments (bottom) for the three tasks described in the study summary.

and subsequent ergonomic decisions. If results will be based on 50 M or 95 M, it is acceptable to select subjects that fall within this range and scale them to the desired size. These recommendations are based on tasks typical of automotive assembly type tasks and are intended to act as a guideline when selecting subjects for ergonomic studies performed with motion capture and virtual reality integration.

Study 2: Measuring, Supporting Hand Forces for Common Automotive Assembly Applications (Godin, Cashaback, & Potvin)

Often, in automotive assembly, tasks will be performed where the worker is leaning on the vehicle with one hand for support while assembling a part using the other hand. This presents a challenge when predicting the demands of the task using a digital human modeling tool. It is largely unknown how much force a worker will translate through their supporting hand/arm and, consequently, the subsequent effect it may have on an overall assessment of joint moments. Furthermore, when motion capture is used in conjunction with a digital human model, subjects performing in the virtual environment are typically encouraged to perform the task naturally, as they would on a true assembly line. To closely reflect a true work task, props are typically used to guide the subjects as they interact with the digital rendering of the vehicle. However, given that subjects are not restricted in how they choose to perform the task, it is generally unknown, *a priori*, where individuals would choose to rest their hands. Thus, it is generally not possible to position force plates within the vehicle mockup in order to measure the forces in the supporting hand. Furthermore, if sites on a mockup are pre-selected for positioning force plates, workers may become biased toward leaning their hands at specific sites, which could impact the validity of the results of the motion capture study. In order to overcome some of these issues, a study was conducted using common automotive assembly tasks to estimate the forces applied by subjects when asked to perform the task in a lab setting using physical props to guide their actions.

The study was designed to determine the average leaning forces, as a percentage of body weight. Six male (height = 1.8 ± 0.05, mass = 88.7 ± 6.8 kg) and six female subjects (height = 1.6 ± 0.04, mass = 60.4 ± 17.3 kg) were recruited from a university environment.

Four conditions were tested, including: (1) Leaning across a table while bending and reaching forward. This was designed to simulate reaching across the engine compartment of a vehicle. (2) Leaning on a table while bending/twisting and reaching forward to simulating reaching under the instrument panel of a vehicle through the driver's side door. (3) Leaning on the side of a door frame when reaching over head and toward the center of the vehicle, where the overhead console would typically be located. (4) Leaning on the top edge of the door frame and reaching overhead and toward the center of the vehicle. Two locations were tested for overhead reaches made near the center of the vehicle, as assembly line workers use both techniques, depending on the task and the individual. Thus, it was necessary to provide an estimate for both techniques. Using physical props commonly seen in a motion capture laboratory, force plates were place in the predetermined locations listed above. Hand force data were collected using a 2.2 KN tri-axial load cell (XYZ Sensor, Sensor Development Inc., Michigan, USA). Force signals were A/D converted using a 12-bit analog to digital multifunction I/O board (National Instruments) that was attached to a PC-compatible computer. The signals were sampled at 1024 Hz and digitally filtered using a Butterworth filter with a cutoff of 2 Hz. Forces in all three dimensions were measured; however, only forces in the direction perpendicular to the plate surface (Z) were presented as the X and Y forces observed to be negligible. The force data were normalized as a percentage of body weight (% BW) and a repeated measures ANOVA ($p < 0.05$), with a $4 \times 2 \times 2$ factorial design, was conducted to determine if significant differences existed within the data.

There were no significant differences noted in the magnitude of the hand support forces, as a percentage of body weight, between males and females. Therefore, the following hand support forces were applied to both male and female subjects for scenario: (1) 12.1% BW for leaning forward, (2) 7.8% BW for leaning forward and twisted, (3) 5.5% BW with the hand on the side of door frame while reaching forward and overhead, (4) 8.0% BW with the hand on the top edge of a door frame while reaching forward and overhead.

These data provided an estimate of the load demands on the supporting hands during common automotive assembly postures, and can be used to better represent a true task when performing an ergonomic assessment using a digital human model.

Study 3: Ongoing Studies

A group of studies have been undertaken by Ford Motor Company and McMaster University to better define the methodology required for an ergonomics analysis completed using motion capture and a digital human model.

The first study is investigating within- and between-digital human model user variability in the manual posturing of human models. Users will complete three study repetitions on 12 different automotive assembly tasks. Dependent variables include joint angles and joint torques.

A second study was designed to evaluate the differences in posture when human models are manually postured versus when they are postured via motion capture. The goal of this study is to help define guidelines regarding when an ergonomist can accurately posture a human model manually and when to conduct the analysis using motion capture technology.

A third study, focusing on motion capture, is investigating posture differences between different subject types, including: (1) expert plant workers trained in the specific study tasks; (2) experienced plant workers not trained in the specific study tasks; (3) automotive engineering team members, not trained in vehicle assembly, but familiar with automotive manufacturing. The goal of this study is to help define if certain subject types are required during a motion capture study for each of the assembly tasks studied.

References

Agnew, M. J., D. M. Andrews, J. R. Potvin, and J. P. Callaghan. 2003. Dynamic 2-D measurements of cumulative spine loading using an electromagnetic tracking device. In *Proceedings of the 24th Association of Canadian Ergonomists conference* (London, Ontario).

Bideau, B., R. Kulpa, S. Menardais, L. Fradet, F. Multon, P. Delamarche, and B. Arnaldi. 2003. Real handball goalkeeper vs. virtual handball thrower. *Presence: Teleoperators and Virtual Environments* 12(4): 411–21.

Blome, M., T. Dukic, L. Hanson, and D. Hogberg. 2003. Simulation of human-vehicle interaction in vehicle design at Saab Automobile: Present and future. SAE paper number 2003-01-2129.

Chi, D. M., J. R. Clarke, B. L. Webber, and N. I. Badler. 1996. Casuality modeling for real-time medical training. *Presence: Teleoperators and Virtual Environments* 5(4): 359–66.

Chi, D. M., E. Kokkevis, O. Ogunyemi, R. Bindiganavale, M. J. Hollick, J. R. Clarke, B. L. Webber, and N. I. Badler. 1997. Simulated casualties and medics for emergency training. In *Medicine Meets Virtual Reality: Global Healthcare Grid*. K. S. Morgan, et al., eds. IOS Press, 1997, 486–94.

Dai, J., Y. Teng, and L. Oriet. 2003. Application of multi-parameter and boundary mannequin techniques in automotive assembly process. SAE paper number 2003-01-2198.

Danisch, L., and M. Lowery-Simpson. 2003. Portable, real-time shape capture. SAE paper number 2003-01-2175.

Doi, M., and C. M. Haslegrave. 2003. Evaluation of JACK for investigating postural behaviour at sewing machine workstations. SAE paper number 2003-01-2218.

Feyen, R., Y. Liu, D. Chaffin, G. Jimmerson, and B. Joseph. 2000. Computer-aided ergonomics: A case study of incorporating ergonomics analyses into workplace design. *Applied Ergonomics* 31: 291–300.

Godin, C. A., J. Cort, J. Cashaback, A. Stephens, and J. R. Potvin. 2008. An estimation of supporting hand forces for common automotive assembly tasks. SAE paper number 2008-01-1914.

Jayaram, U., and R. Repp. 2001. Calibrating virtual environments for engineering applications. In *Proceedings of the American Society of Manufacturing Engineering design engineering technical conference, 2001* (Pittsburgh, PA).

Johnson, W. L., J. Rickel, and J. C. Lester. 2000. Animated pedagogical agents: Face-to-face interaction in interactive learning environments. *International Journal of Artificial Intelligence in Education* 11: 47–78.

Johnson, W. L., J. Rickel, R. Stiles, and A. Munroe. 1998. Integrating pedagogical agents into virtual environments. *Presence: Teleoperators and Virtual Environments* 7(6): 523–46.

Maiteh, B. 2003. An application of digital human modeling and ergonomics analysis in workplace design. In *Proceedings of the American Society of Manufacturing Engineering international mechanical engineering congress, 2003* (Washington, DC).

Reed, M. P., M. B. Parkinson, and A. L. Klinkenberger. 2003. Assessing the validity of kinematically generated reach envelopes for simulations of vehicle operators. SAE paper number 2003-01-2216.

Rickel, J., and W. J. Johnson. 1999. Virtual humans for team training in virtual reality. In *Proceedings of the ninth world conference on AI in education*, 578–85.

Rickel, J., and W. J. Johnson. 2000. *Task-oriented collaboration with embodied agents in virtual worlds.* Cambridge, MA: MIT Press.

Rigel, S., E. Assman, and H. Bubb. 2003. Simulation of complex movement sequences in the product development of a car manufacturer. SAE paper number 2003-01-2194.

Yee, A., and K. J. Nebel. 1999. Digital humans in virtual environments. *American Society of Manufacturing Engineering* 5: 147–51.

Part 4

Applications

40

Digital Human Modeling in Automotive Product Applications

Lars Hanson, Dan
Högberg, and Arne Nåbo

40.1 Introduction

One approach toward successful business for companies that market, develop, and produce products is integrated product development (Andreasen & Hein, 2000). The fundamental idea is that by organizing cross-functional project teams to carry out product development activities more or less concurrently, the product development time and cost will decrease, and the product quality will increase. The concept promotes established, but continuously improved, design processes, the utilization of structured design methods, as well as the use of efficient design tools, such as computer-aided engineering (CAE) tools. The latter is apparent in modern industry where more and more design activities are carried out in a virtual world by the aid of computer-created products, environments, and situations.

This chapter deals with occupant packaging in the car design process, and more specifically, how digital human modeling (DHM) tools can support the consideration of physical ergonomics issues in the early virtual stages of the design process. Figure 40.1 illustrates a typical design setting for the matters discussed.

40.2 DHM in the Vehicle Design Process

It is worth noting that there is a lack of humans in Figure 40.1. As this is sales material, this is not surprising since the potential buyer is likely to be more interested in watching the car rather than any unknown person sitting in the car. That the car is designed so that the car fits the user comfortably

FIGURE 40.1   Example of product design context. (Courtesy of Saab Automobile.)

regardless of the user's age, abilities, or size is likely to be taken for granted. Indeed, car companies design their products with user variability and expectations in mind, but still with some limitations. Car companies often try to accommodate (i.e., make the car fit) 95% of the targeted user group. To verify this requirement is, however, not easy. As in all user-centered design problems there are three parameters that need to be considered in establishing the level of accommodation: the *product*, the *user*, and the *task* (Pheasant & Haslegrave, 2006). Without going into detail, the precise level of accommodation in a car interior is hard to define since it is a multivariate problem where several body dimensions affect the estimation, subjective issues are involved (e.g., comfort), and there are a variety of tasks that the driver and the passengers perform with or in the car. Moreover, there are variations between humans and also even within humans (e.g., the selection of different driving postures depending on the traffic situation). Hence, ergonomics in the vehicle design process has its complexities; particularly since other design disciplines often state conflicting requirements, meaning that the overall product development challenge is to realize a design with the best balance between all product requirements, including ergonomics.

Accepting these inherent difficulties, how can we deal with occupant packaging in the design process, particularly at the virtual stages? If Figure 40.1 were an illustration of the level we have reached when it comes to computer modeling in the design process, we would have problems. Fortunately, as this book clearly illustrates, we can also model humans. Digital human modeling enables ergonomics issues to be treated in the virtual car design process. Without these tools one would have to base occupant packaging solely on established, but not necessarily recent or flexible, standards (e.g., SAE standards; Roe, 1993), or postpone the ergonomics evaluation until physical prototypes are built, which often are made later in the design process (Porter et al., 1993). Prototypes may still be required, but by using human models in the early stages of the design process major deficiencies can be rectified in due time. If major ergonomics problems are detected late in the design process, the required design changes are likely to be refused due to time and cost. Basic mockups or flexible fixtures can work as an alternative or complement to human modeling tools for certain design problems. It is sensible to employ a pragmatic approach toward the tools and methods used in the design process, and utilize the ones that are most beneficial for illustrating, evaluating, and solving the design task.

There are three central elements of a design process: *exploration* (understanding the design problem), *generation* (finding possible solutions) and *evaluation* (assessing solutions) (Cross, 2000), typically performed iteratively in a design loop. As Högberg (2005) discusses, DHM tools can benefit all three elements of the design loop. However, the main function of DHM tools is to support ergonomics evaluation. Isaksson (1998) found that the requirements of evaluation methods change as the product development project

FIGURE 40.2 Anthropometric and posture variability.

matures. In the early stages of the development process, high flexibility, rapid modeling, and comparative analysis are important issues for evaluating a large number of concepts. When the form and dimensions of the vehicle begin to be clarified, the evaluation can then become more detailed. DHM tools particularly meet the requirements for evaluation of design proposals at early stages.

Ideally, for designers concerned with human-vehicle interaction issues, there would be computer models of the "entire" human, so that it would be possible to simulate both physical and cognitive ergonomics in the design process. Even though DHM tools for simulating cognitive issues are being developed, physical aspects are comparably easier to model in computers than the cognitive side, basically due to the complexity of the human brain. Still, issues like predicting realistic and individual postures and motions are a challenge within the physical DHM research and development area. An example of a DHM tool that includes both physical and cognitive models is SANTOS (Abdel-Malek et al., 2006).

A major issue to consider in occupant packaging is to ensure that the car suits people of a range of sizes, performing a number of tasks. Very simply, this can be expressed with the rule of thumb: "Let the small person reach and the tall person fit" (Helander, 1995). Figure 40.2 gives a hint of anthropometric variability and how it affects occupant packaging, such as when it comes to adjustability and reserving space in the car interior.

When designing a car interior, one needs to consider additional issues to the basic seated driving posture (e.g., to reserve space for bodily motions required or desired, and space for belongings, coffee mugs etc.). It may also be desirable to consider the subjective feeling of space. In addition to geometry issues, physical ergonomics also focuses on human strength and the mechanical exposure of the human as well as vision. Naturally, visual aspects are important in car driving (e.g., that all drivers can see important instruments on the control panel and have a good view of the surroundings through windows and mirrors).

40.2.1 Optimizing Reach, Fit, and Vision around a Seated Driver

Traditionally, vehicle designers use SAE — Society of Automotive Engineers — "two and a half" dimensional accommodation tools to design the cockpit, including the layout of a recommended seat position, hand reach envelopes, head contours, and eye ellipses (Roe, 1993). These accommodation tools and a human replica (fig. 40.3) were first available in plastic to serve as curves when using a pen and drawing table. Later they were integrated into CAD tools.

For the last few decades, three-dimensional DHM tools have complemented the traditional tools. In vehicle design, DHM tools are typically utilized to optimize comfort, fit, reach, and vision (Chaffin, 2001). A common design task when designing the cockpit is to optimize these parameters around a seated driver with an optimal posture. Figure 40.4 illustrates a typical application of DHM in cockpit

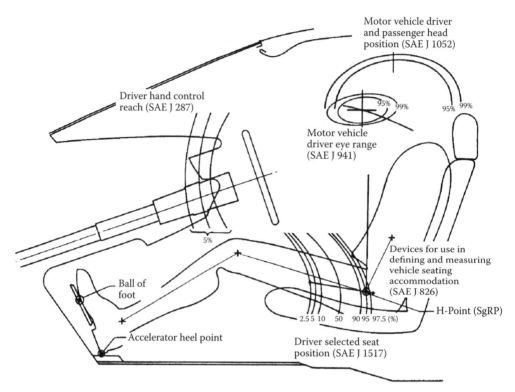

FIGURE 40.3 SAE accommodation design tools. (From Roe, R. W., "Occupant Packaging," in *Automotive Ergonomics*, Taylor & Francis, London, 11–42, 1993. With permission.)

design, in this case utilizing the DHM tool RAMSIS, which has been developed specially for the design of vehicle interiors (Seidl, 1997; Human Solutions, 2007).

It can be questioned if an optimal posture exists when drivers show great differences and variations in their applied posture. Sundström (2003) found that each driver applied a range of 2 to 4 postures, though one favorite posture dominated the overall seating behavior. The changes in posture could be correlated to events in the driving environment or handling of artifacts, tools, or equipment. Several studies, for instance, Rebiffe (1969), Grandjean (1980), Tilley and Dreyfuss (1993), as well as Porter and Gyi (1998), have examined driver postures and come up with recommended joint angle intervals for

FIGURE 40.4 DHM in cockpit design.

comfortable postures. The comfort intervals and comfort evaluation methods are questioned, but are considered suitable for comparing concepts (Nilsson, 1999). Others, for example, Andreoni et al. (2002) and Hanson et al. (2006b), use joint angle means. These mean angles can be used for comfort evaluation, and can be implemented in the DHM tool RAMSIS, especially its posture prediction model. As a default, RAMSIS uses the comfort intervals found by Krist (1994). Based on the comfort values and geometrical constraints, RAMSIS identifies the optimal driver posture. In the DHM tool JACK (Badler et al., 1993; UGS, 2007), the cascade prediction model is integrated. The model is based on laboratory experiments with American subjects in different specific vehicle package setups and focuses on eye and hip locations (Reed, 1998). There are also posture prediction models developed by vehicle manufacturers for internal purposes (e.g., Quattrocolo et al., 2002).

40.3 Work Process and Organizational Issues

As the functionality of today's DHM tools continuously develops, along with the areas of application, the efficient implementation of the tools in industrial development processes becomes an important issue, basically to achieve the best outcome from the tools. Companies regularly have a detailed description of their product development process, defining the things to do, when to do them, what methods are to be utilized at different phases, and so on. The use of DHM tools should be part of a company's formal design process (Sundin & Sjöberg, 2002), and the tools should be utilized together with other CAE tools and methods, such as the SAE mannequin and its accommodation guidelines. Later in the design process, when the project has matured, it is often beneficial to use physical mockups and prototypes in parallel with the virtual testing process, where DHMs work in cooperation with physical mannequins (e.g., the H-point machine; Roe, 1993), and real humans.

In order to acquire good results from DHM tools, it is recommended that a structured process of how to carry out human simulation analyses is established. Hanson et al. (2006a), Green (2000), and Ziolek and Kruihof (2000) have described different analysis procedures. In general, the processes cover the three parameters of accommodation mentioned earlier—the *product*, the *user*, and the *task*. In addition, in order to reduce the risk of misuse and misinterpretation, Ziolek and Nebel (2003) highlight analysis planning, including the definition of aims, as well as final documentation as parts of the process. Accordingly, the processes include major general steps such as mission statement, defining the physical environment, defining the task, defining how to represent the users, and then performing, interpreting, communicating, and storing the analysis. Crucial steps in the human simulation process are mannequin selection and the choice of evaluation method. Sundin and Sjöberg (2002) found, however, that objective ergonomics evaluation methods were seldom utilized, and that DHM tools often were used as visualization tools only.

Defining the mannequins so that they correctly represent the future users of the vehicle is frequently simplified into the use of a standard setup based on a few percentile values, or a combination of predefined key anthropometric variables. A target group can for certain design problems be represented by a few individuals. It is common within ergonomics to use 5th, 50th, and 95th percentile sizes according to stature. This descriptive statistical representation strategy works fine for one-dimensional design problems, for example, defining a doorway height (Dainoff et al., 2003). However, it is not suitable for multidimensional issues (Ziolek & Wawrow, 2004; Robinette & Hudson, 2006). For multidimensional problems other statistical approaches have been proposed (Bittner et al., 1987; Zehner et al., 1993; Bittner, 2000). A multidimensional boundary approach (Dainoff et al., 2003) is commonly recommended for vehicle cockpit design issues, for example when defining the seat adjustment range required to accommodate the user group (as will be illustrated in Case 1). Simply expressed, in the multidimensional boundary approach the dimensions for the mannequins (that are to represent the users) are acquired from selecting evenly spread points on the surface of a multidimensional ellipsoid. For some design problems, a distributed cases approach is appropriate, where representative cases within a boundary are defined or randomly selected (Dainoff et al., 2003). This approach has been adopted for mannequin definition (size and ability) in the design tool HADRIAN (based on the DHM tool SAMMIE), which is intended to aid designers to include users that are seldom considered in the design process (e.g., older and disabled people), together with able people, according to the inclusive design philosophy (Porter et al., 2004).

The following section illustrates a fabricated but realistic case of how a car manufacturer may work with DHM tools in the car design process.

40.4 Case 1: Seat Adjustment Range for a New Vehicle Platform

A large vehicle development project is launched where a number of car brands cooperate in order to develop a new common car platform. The mission is to create a platform suitable for a wide range of cars, from roadsters to sport utility vehicles.

David is responsible for seat-related issues in the platform project. After reading the newspapers and discussing with Cecilia, who is responsible for the platform's human factors characteristics, he is aware of the general anthropometric trends of people getting taller and more corpulent. With this in mind, David and Cecilia have decided to conduct a deeper analysis of the seat adjustment range required to ensure that the targeted population will be accommodated with the proposed design. They discuss using the in-house test group in the flexible buck (basically an adjustable car structure) and DHM tools to find a suitable adjustment range. Therefore, they contact Madeleine who is a human factors simulation engineer and responsible for the DHM tools RAMSIS and JACK at the car company. In the following section we follow David and Cecilia's first meeting with Madeleine where she guides David and Cecilia through the standardized simulation ordering process and discusses choice of method. After that we follow Madeleine's thoughts and actions when she generates mannequins, describes the physical environment, defines the task and runs the digital human modeling software. Finally we portray Madeleine delivering the result to David and Cecilia and storing the analysis report in the documentation database.

40.4.1 Identification and Background Information

Madeleine, David, and Cecilia meet to kick off the DHM analysis of an appropriate seat adjustment range for the platform. Their discussion is structured and follows the human simulation process proposed on the company's intranet system. The meeting starts with David and Cecilia describing the background: the new platform, the anthropometric trends, and their uncertainties about the appropriate seat adjustment range. Madeleine takes notes and fills in the blanks in the process protocol. The protocol is linked to a database and all employees within the company have access to the database. Madeleine summarizes the background and enters the data in the protocol. She expresses the analysis objective as follows: "To describe the appropriate seat adjustment range for European and American vehicles launched in 10 years, as well as the location of the seat adjustment range relative to the accelerator heel point." David and Cecilia agree upon this formulation. Madeleine adds her name and contact data as the one who performs the analysis, and David as the requester of the simulation. When all have a common picture of the aim, the desired output is discussed. David is interested in illustrating differences and similarities with the proposed adjustment range and existing car models, both for their own and competitor cars. Cecilia highlights the importance of comparison with SAE standards. She also suggests that the analysis should include comfort ratings as a complement to the hip-point locations predicted by the DHM tool. Madeleine confirms the desires by adding them to the protocol and promises to deliver tables and pictures for illustrative purposes. Finally, the group discusses desired completion date. The agreed date, two weeks from today, is entered into the protocol. Having this information entered in the protocol concludes the first of three stages in the human simulation process. Before David and Cecilia leave, Madeleine gathers the information required to perform the analysis properly, such as information about the targeted market segments and the physical environment as CAD geometry.

40.4.2 Method Description

Madeleine is now left alone and starts the analysis by searching the human simulation database on the intranet, hoping to find earlier similar studies to save time and gain from previous simulations. She uses "seat adjustment range" as a keyword. Unfortunately, she gets few hits. This is not surprising since DHM is a relatively new tool within the company, seat adjustment ranges are set when car platforms are designed, and platforms may last for decades. She finds one investigation about adjustable pedals and how they affect the seat adjustment range, which she found interesting. She also asks an experienced packaging engineer, who is responsible for SAE requirements, of how such studies have been performed previously. He tells her that the size and shape of the adjustment range and its location have been based on results from studies in the flexible buck and SAE seat adjustment curves, SAE J1100 (Roe, 1993). He also reminds Madeleine that ASPECT guidelines have been introduced and are now working in parallel with SAE recommendations (Schneider et al., 1999).

Madeleine continues the work by reading the market analysis report. In the report she finds that the premium mid-segment vehicle that David, Cecilia, and she chose as an example car is aimed at Australia, Western Europe (Germany, Holland, Great Britain, Scandinavia), and the USA. The vehicle will be produced as a sports sedan and wagon, and is projected to attract young people and families. Madeleine understands the underlying thoughts, because she knows that vehicle buyers are relatively brand faithful, and therefore it is strategic to attract younger people. With this fact in mind Madeleine starts to search for updated anthropometric data for the countries with expected large market shares for the future vehicle. She finds up-to-date American, Dutch, Swedish, and Australian anthropometric data. Based on this data, she generates representative mannequin descriptions. Madeleine uses five key anthropometric variables: sitting height, hip height, buttock-knee length, arm length, and waist circumference. Madeleine selects these body dimensions since she noticed that they were used in all reports about seat adjustment range found in the database. She determines anthropometric data for female and male mannequins for each country and ensures that the market share proportion is correct. She adjusts the anthropometric data for secular trends so that the mannequin sizes are likely to correspond to the customers' sizes when the car is eventually introduced on the market and throughout the car platform's market life cycle. All information is put in a table in the process protocol, and then entered in the DHM tool, JACK's principal component analysis module. The software defines anthropometric data of the boundary mannequins covering 95% of the population described in the table. Madeleine transfers the anthropometric information from JACK into mannequins in RAMSIS. By using this method, she has now created a set of mannequins representing the targeted customers in a multidimensional problem. In the analysis, she also has decided to include mannequins with the same anthropometrics as each person in the in-house test group, to facilitate comparison with the buck study. She also models David, Cecilia, and herself, to support discussions between the people involved in ordering and performing the analysis.

David, Cecilia, and Madeleine decided to use a mid-segment vehicle as an example in the investigation. Madeleine enters the PDM system and searches the database for the latest CAD geometry definition of the vehicle. She imports floor, pedals with their motion line, steering wheel with its adjustment area, proposed seat adjustment area, inner roof, and top of dashboard into RAMSIS. She tries to keep a minimal description since exchange between computer systems is time consuming and conversion errors sometimes occur. In addition, Madeleine believes that a simple description of the physical environment allows the viewer of the illustrations to focus on the ergonomics issues, in terms of posture and accommodation. Also, it indicates that the analysis has been performed at an early design stage with relatively rough assumptions, and that design corrections are both possible and recommended if required. Before the physical environment definition is ready she makes sure that the steering wheel adjustment areas are located for a quarter-to-three grip and moves the accelerator halfway down along the motion line.

The next step in the process is to define the task. This time it is a normal driving task, which Madeleine usually defines with the following constraints in RAMSIS (fig. 40.5). Since she does not want to

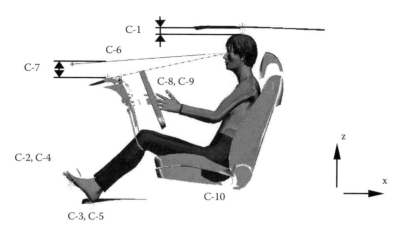

FIGURE 40.5   Constraints used in the case study.

restrict the seat adjustment area in the simulation (that is the aim of the study) she greatly enlarges that area so that it will be nonconstraining.

C-1:   Head clearance. Minimum 25-mm vertical distance between head top (vertex) and inner roof.
C-2:   Right pedal point on accelerator, pressed down halfway.
C-3:   Right heel point on floor.
C-4:   Left pedal point on foot support.
C-5:   Left heel point on floor.
C-6:   Line of sight. 5 degrees down from horizontal line.
C-7:   Line of sight clearance. Minimum 70 mm vertically between line of sight and top of instrument panel.
C-8:   Right grasp point within steering wheel adjustment area.
C-9:   Left grasp point within steering wheel adjustment area.
C-10: Head-point (hip point) within greatly extended seating adjustment area (nonconstraining in x and z direction).

With the constraints and the physical geometry defined, Madeleine enters the RAMSIS automatic analysis procedure. She inputs the mannequins and the constraints to be used. In the output section she describes that she is interested in receiving information about comfort and coordinates for the H-point. Before running the automatic analysis she remembers that she should change the settings so that the company's unique posture descriptions are used instead of the default RAMSIS values. When this is done she pushes the start button and takes a coffee break while the computer is working.

After the break the computer is ready and she has plenty of interesting output data to consider. However, she temporarily ignores the new information and instead starts to use the SAE packaging kit in RAMSIS. By entering the accelerator heel point and the desired vertical distance between the heel point and the seating reference point (SgRP) (fig. 40.3), that is, the H30 value (Roe, 1993), seat reference curves automatically appear. She now has a very interesting picture with SAE seat reference curves, marks for each mannequin's H-point, and the previously proposed seat adjustment area for the platform. Madeleine also arranges the comfort data in an illustrative table, showing mean and standard deviation values for the target population and the in-house test group (Table 40.1). Data for David and Cecilia are also presented, as are data from a previous analysis. As further benchmarking, Madeleine uses the kind of drawings that vehicle companies share with each other to find out competitors' design of the seat adjustment range. With this information in hand, Madeleine contacts David and Cecilia for a meeting to deliver the results.

TABLE 40.1   Comfort Data from the Analysis

	Target Population Mean (s.d.)	In-House Test Group Mean (s.d.)	David	Cecilia	Previous Car Model Mean (s.d.)	Concept Car with Flexible Pedals Mean (s.d.)
Overall	9.0 (2.2)	7.8 (1.8)	9.5	6.5	8.3 (1.9)	9.2 (2.1)
Neck	7.5 (1.9)	7.9 (2.0)	8.8	7.4	7.4 (1.6)	7.7 (2.0)
Arms	8.7 (1.3)	7.5 (1.1)	8.5	4.8	8.5 (1.2)	9.1 (1.4)
Back	9.5 (1.8)	8.1 (1.9)	10	7.1	9.1 (1.8)	9.9 (1.8)
Legs	6.9 (2.0)	6.9 (2.1)	7.1	4.6	7.1 (2.3)	8.8 (1.7)

40.4.3 Result and Closure of Simulation Mission

Madeleine has stored all pictures and tables generated in the analysis in the database. She uses the unique project number, and uses pictures and figures together with her own comments. Madeleine starts the presentation by showing David and Cecilia illustrations similar to Figure 40.6. She describes how the previously proposed seat adjustment range is shown by the thin line in Figure 40.6, and the adjustment range required according to the study to accommodate the target group (diamonds with thin lines) is represented by the thicker line. The diamonds with thicker lines represent H-points for the in-house test group, and filled diamonds represent David and Cecilia's H-points. The SAE seat adjustment range is presented with dotted lines.

They discuss the differences and conclude that the previously proposed seat adjustment area is sufficient, at least in length. The size and the shape are also comparable to competitors in the premium mid-segment vehicles. David and Cecilia are convinced that a horizontal extension of the adjustment range, which was their thought before seeing Madeleine's pictures, is not required. Instead, a decrease in length and increase in height seems preferable. However, David's view is that, even though obviously advantageous, a larger height adjustment range is likely to be complicated to realize and difficult to get approval for in the new platform. Also, David says that to decrease the seat adjustment range in length

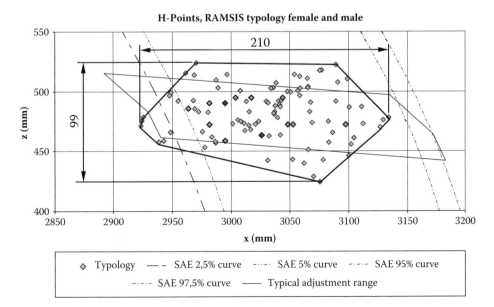

FIGURE 40.6   Illustration example presented at the case study meeting.

may not be an option in a conservative design environment. Based on the discussion, they together update the comments and conclusions about the seat adjustment range.

Cecilia who has short legs finds the comfort ratings interesting. Her ratings in Table 40.1 are lower than all others, which indicates lower comfort level. She finds this possible and she admits that she has problems in reaching the pedals without sitting too close to the steering wheel. Overall, the comfort ratings look promising, with high numbers close to the maximum of 10. However, the absolute comfort values are of minor interest. It is the value compared to different concepts that is interesting. In this case the seat adjustment range proposed by the DHM tool is almost as good as the concept car with adjustable pedals and better compared to the previous car model.

The issues that came up during the final discussion between Madeleine, David, and Cecilia are entered into the protocol, as a complement to the comments that Madeleine stored earlier about the analysis results. David and Cecilia thank Madeleine for a good job and write their name and date in the approval section in the protocol, which closes the digital human simulation and visualization mission. Before leaving they promise Madeleine that they will come back with information about H-point locations and comfort ratings from the tests with the in-house test group in the flexible buck, information that Madeleine can use for validation of the software.

After the meeting, Madeleine is satisfied with her job and the positive feedback. She thinks she showed the strengths of the DHM approach. She is glad that she was consulted since it indicates that the possibilities of the DHM tool are recognized and considered advantageous. Another positive thing is that the study was initiated at an early design stage where it is possible and comparatively easy to perform design changes when necessary. An extra bonus was that David and Cecilia promised to come back with information about H-point locations and comfort ratings from the tests with the in-house test group in the flexible buck, information that Madeleine can use for validation of the software—a validation she will use in her work and in convincing colleagues about the qualities of the DHM tool.

40.5 Case 2: Reachability Analysis for Cup Holder Location

In this second case, Madeleine initiates a DHM-based ergonomics analysis herself. In a couple of days she will attend a meeting where a relocation of the cup holder for a dashboard enhancement of the next year's premium car will be discussed. The company is organized so that all product characteristics have their own "advocates" in the development project (e.g., safety, climate comfort, drive quality, etc.). One of these areas is ergonomics and Madeleine's role at the meeting is to determine whether or not the proposed design change can be approved from an ergonomics point of view. She is a bit concerned that the reachability of the cup holder will be inferior to the current design, and might even be too poor for approval. She knows that even though the cup holder is not a primary control the ability for the driver to operate the cup holder and the cup located within it is important both for driving safety and convenience. It may also be one of those things a potential buyer tries out when sitting in the car for the first time in a car showroom.

Madeleine plans to use the DHM tool to study the reachability and visibility for all users. She could consult the in-house test group and run a trial in the flexible buck, but she thinks the DHM approach will be faster. Using the DHM tool may also be better since the in-house test group currently lacks a number of important anthropometric representatives, such as a very short female, with short legs and arms relative to sitting height. Her virtual test group is always available and easy to modify if required.

Madeleine uses the same human simulation process as described in Case 1, but in this case she both orders as well as carries out the analysis herself. Despite this, she thoroughly fills in all sections in the protocol to ensure that she remembers to consider all issues, and to have good documentation of the study for her own and others' benefit in the future.

Madeleine suspects that it might not be enough to do the cup holder study utilizing a short mannequin with short arms only. Since the seat is adjustable, it may well be another mannequin configuration that will be the limiting user. Since she has an established family of mannequins (i.e., her virtual test

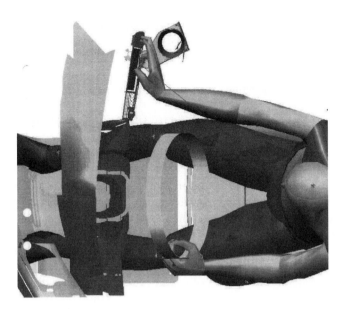

FIGURE 40.7   Image from the reachability study.

group), she decides to run the reachability study employing all mannequins. The DHM tool will perform the multiple simulations automatically so it will not require much additional time. She uses the posture prediction functionality with constraints similar to the previous case study, but in this case the H-points are located inside the existing seat adjustment area. The DHM tool helps Madeleine to judge whether all mannequins will be able to reach the cup holder and maneuver the adjacent controls (fig. 40.7).

Madeleine also uses the vision analysis function in the DHM tool to study if the symbols and buttons for the cup holder will be visible for all users. She uses both the view cone feature and the function to look through the mannequin's eyes, and produces illustrations similar to those shown in Figure 40.8.

The analysis results from the DHM tool indicate the cup holder to be within reach for all users. However, the vision analysis reveals that the symbols for the cup holder may be hidden for some drivers. Madeleine wants to discuss her findings before the formal project meeting and arranges a quick meeting with Cecilia and Martin, the leader of the enhancement project. They agree that Madeleine can approve

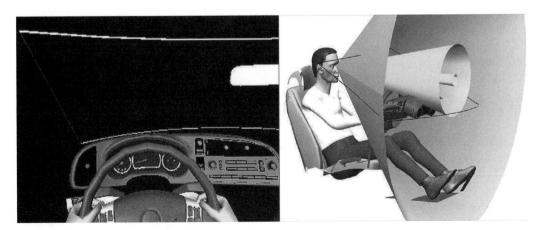

FIGURE 40.8   Vision analysis functionality.

the position of the cup holder, but that a redesign of the symbols will be proposed to ease readability for certain drivers. Madeleine documents everything in the protocol and starts to prepare for the meeting. Now she has good arguments and illustrations for her statements at the meeting. She also has some ideas of how to solve the problem with the visibility of the symbols, and she draws a simple sketch and some notes about it. After the formal meeting she will update the protocol so that the entire process is well documented.

After the meeting, it strikes Madeleine how well the DHM tool helped her to analyze the problem, and also how it inspired her to generate solutions. She has lately started to think about the way DHM tools could have a more widespread use. At present, Madeleine is the only one at the company who uses the tool within the product development process. The DHM tool's functionality and usability is getting better and better, and basic ergonomics analyses are quite easy to carry out well without requiring much knowledge of ergonomics. Of course, the basics about ergonomics are good to know, and are especially necessary so that one understands when to contact a specialist. However, for many studies a specialist may not be required as long as the person using the tool is aware of its basic functionality, limitations, and so on. Madeleine's outlook is that all who design anything in the car with a human interface, which indeed involves almost all products if one also considers assembly, service, disassembly, and so on, should benefit from using DHM tools, especially if they were integrated in the CAD tools used by the engineers and designers. Also, other tasks than those performed inside the car could well be simulated without much effort (e.g., loading the trunk, filling fuel, or checking oil level).

If Madeleine's scenario becomes reality, she thinks that the design of the cars would get even better than today, and the car development process would be faster and smoother. It would also allow her to focus her specialist competence on the more complicated tasks. Another idea is that all people who work with some sort of human simulation at the company today should collaborate more. She thinks, for example, of the human simulation activities performed at the production department when planning for the assembly of the cars. First, they could exchange experiences at a general level on how to perform human simulations quickly with good results, and second, some simulations are quite similar (e.g., the simulation of ingress and egress for a car owner, and a person in the assembly line getting in and out of the car).

Madeleine sometimes gets the feeling that physical ergonomics issues get low priority in the car design process. The cognitive side seems to get more attention and resources. She clearly understands the importance of considering cognitive ergonomics in car design when the driver's attention gets more and more distracted by all the functionalities added to the control panel, as well as the need to strive for usability and safety for all kinds of users and situations. She recently learned that usability comes quite low in the hierarchy of customer needs (Jordan, 2000). She thinks that managers and designers sometimes miss the message in the hierarchy model that good usability, including physical ergonomics issues like comfortable seating and good visibility, are prerequisites for creating the kinds of pleasures that come above usability in the hierarchy. Delivering pleasure without reducing usability is particularly important for creating long-term customer value, which Madeleine learned is the way to build the brand and earn customer loyalty.

40.6 Discussion

Customers ask for functional, safe, reliable, usable, and pleasurable vehicles. To deliver this, vehicle companies need to blend and balance all kinds of product characteristics in the vehicle development process. Some aspects will be closely related to ergonomics. Hence, vehicle companies consider occupant packaging and human vehicle interaction issues important. Perhaps these issues should be considered even more important? Porter and Porter (1998) argue for an "inside-out" approach in the vehicle development process, rather than an "outside-in" approach. They contend a clearer focus should be put on the humans inside the car, their accommodation, comfort, diversity, and so on, and they maintain that current

FIGURE 40.9 "The first sketch." (Courtesy of Saab Automobile.)

vehicle development processes typically let exterior styling prevail over ergonomics aspects. Certainly, exterior styling is extremely important for affection and pleasure, but the challenge is balancing these requirements. An indication of the "inside-out" approach is "The first sketch" (fig. 40.9), which was used by a car manufacturer to convey a common vision in the new car development project, and in the subsequent marketing campaign. The sketch clearly illustrates the importance of the human element, and that the vehicle is designed around a seated driver who fancies driving. The cockpit arrangement and vehicle characteristics are to be designed to the driver's physical and cognitive capacity. This, together with the expected functionality, safety, and reliability, and an attractive design, create altogether a pleasurable ride and ownership that keeps the customer smiling.

The previous two cases illustrate how DHM tools can assist in the vehicle design process for the evaluation of physical parameters such as accommodation, comfort, reach, and vision. Seat belt fit and clearance to maximize comfort and safety are other parameters that can be analyzed with the tools. Most of the DHM analyses performed today are based on a seated person who does not move. The seated position is essential, both for professional drivers who spend hours in the vehicle per day and for nonprofessionals who may only spend a few hours per week driving. However, sitting in the vehicle and performing small bodily motions while driving is not the only interaction performed with the car. A car owner also has to ingress the car to reach a seated position, and egress. The vehicle manufacturers prioritize ingress and egress and they now encourage DHM developments in this area. Responses on this request are, for instance, work from the RAMSIS group (Cherednichenko et al., 2006) and ANYBODY group (Rasmussen et al., 2006). Simulating, visualizing, and evaluating ingress and egress takes DHM to a new dimension where human motions are considered. This can be seen as a move from a static virtual world toward a dynamic virtual world.

As discussed in Case 2, the personnel at the assembly line also perform vehicle ingress and egress. Other tasks that are carried out by assembly personnel as well as by car owners include loading the trunk (e.g., the spare wheel) and filling windshield washer fluid and gas. These tasks are at present rarely analyzed with DHM tools at departments focusing on product applications (i.e., designing the vehicle). Instead these issues are more commonly analyzed at the departments focusing on production application, for instance, manual assembly in this case. There could be benefits from increasing the collaboration between these typically quite separate departments. As illustrated in the two cases, communication between disciplines is essential. Communication within disciplines is, of course, essential, too. Furthermore, there are a small number of DHM tool users at both departments. In order to increase communication and to achieve a critical mass of DHM tool users, an alternative might be to gather all virtual ergonomics work in a center. In such a center all human vehicle interactions during a vehicle's life cycle could be analyzed, including both the production and the use phase, and preferably even the disassembly phase.

As discussed by Sundin and Örtengren (2006), work procedures for DHM tool usage and the integration of the tools in design processes are important for an effective and efficient usage. Most vehicle companies have developed such work procedures, sometimes linked to a PDM (Product Data Management) system or similar. Researchers and people working within companies continuously improve evaluation methods. It is recommended that companies continuously update the DHM tools with new general and brand-specific information, to make sure that the most recent information is used. If possible, new evaluation methods developed are utilized for both virtual testing as well as for tests in physical prototypes. Also, by integrating evaluation methods that companies have confidence in, objective methods and visualizations can be used side by side. With such an up-to-date human modeling tool in their hands, the company will have a powerful design tool for human vehicle interaction evaluations. The DHM tools are powerful today. However, the tools' strength and impact on the vehicle design process are expected to grow significantly in the future.

References

Abdel-Malek, K., Yang, J., Marler, T., Beck, S., Mathai, A., Zhou, X., Patrick, A. and Arora, J. (2006). Towards a new generation of virtual humans. *International Journal of Human Factors Modelling and Simulation* 1(1): 2–39.

Andreasen, M. M. and Hein, L. (2000). *Integrated product development*. Lyngby, Denmark, Institute for Product Development.

Andreoni, G., Santambrogio, G. C., Rabuffetti, M. and Pedotti, A. (2002). Method for the analysis of posture and interface pressure of car drivers. *Applied Ergonomics* 33: 511–522.

Badler, N., Phillips, C. B. and Webber, B. L. (1993). *Simulating humans: computer graphics animation and control*. New York, Oxford University Press.

Bittner, A. C. (2000). *A-CADRE: Advanced family of mannequins for workstation design*. XIVth congress of IEA and 44th meeting of HFES, San Diego. 774–777.

Bittner, A. C., Glenn, F. A., Harris, R. M., Iavecchia, H. P. and Wherry, R. J. (1987). CADRE: A family of mannequins for workstation design. *Trends in ergonomics/human factors IV*. S. S. Ashfour, Ed., Elsevier Science Publishers: 733–740.

Chaffin, D. B. (2001). Introduction. *Digital human modeling for vehicle and workplace design*. D. B. Chaffin, Ed., Warrendale, Society of Automotive Engineers: 1–16.

Cherednichenko, A., Assmann, E. and Bubb, H. (2006). *Experimental study of human ingress movements for optimization of vehicle accessibility*. Proceedings of the IEA2006, 16th World Congress on Ergonomics, Maastricht, The Netherlands.

Cross, N. (2000). *Engineering design methods: strategies for product design*. Chichester, John Wiley & Sons.

Dainoff, M., Gordon, C., Robinette, K. M. and Strauss, M. (2003). *Guidelines for using anthropometric data in product design*. Santa Monica, Human Factors and Ergonomics Society.

Grandjean, E. (1980). Sitting posture of car drivers from the point of view of ergonomics. *Human factors in transport research, Part 1*. E. Grandjean, Ed. London, Taylor & Francis.

Green, R. F. (2000). *A generic process for human model analysis*. Warrendale, Society of Automotive Engineers. SAE Technical Paper 2000-01-2167.

Hanson, L., Blomé, M., Dukic, T. and Högberg, D. (2006a). Guide and documentation system to support digital human modeling applications. *International Journal of Industrial Ergonomics* 36: 17–24.

Hanson, L., Sperling, L. and Akselsson, R. (2006b). Preferred car driving posture using 3-D information. *International Journal of Vehicle Design* 42(1–2): 154–169.

Helander, M. G. (1995). *A guide to the ergonomics of manufacturing*. London, Taylor & Francis.

Högberg, D. (2005). *Ergonomics integration and user diversity in product design*. Department of Mechanical and Manufacturing Engineering, Loughborough University, UK. Doctoral thesis.

Human Solutions (2007). *Ergonomic simulation in automotive design*, Human Solutions, Kaiserslautern, Germany. Access: 25 January 2007. http://www.human-solutions.com/automotive_industry

Isaksson, O. (1998). *Computational support in product development. Applications from high temperature design and development.* Department of Mechanical Engineering. Luleå University of Technology, Luleå, Sweden. Doctoral thesis.

Jordan, P. W. (2000). *Designing pleasurable products: an introduction to the new human factors.* London, Taylor & Francis.

Krist, R. (1994). *Modellierung des sitzkomforts: Eine experimentelle studie (Modelling sit comfort: An experimental study).* Katholischen Universität Eichstätt, Germany. Doctoral thesis (in German).

Nilsson, G. (1999). *Validity of comfort assessment in RAMSIS.* Warrendale, Society of Automotive Engineers. SAE Technical Paper 1999-01-1900.

Pheasant, S. and Haslegrave, C. M. (2006). *Bodyspace: anthropometry, ergonomics and the design of work. 3rd ed.* Boca Raton, Taylor & Francis.

Porter, J. M., Case, K., Freer, M. T. and Bonney, M. C. (1993). Computer-aided ergonomics design of automobiles. *Automotive Ergonomics.* B. Peacock and W. Karwowski, Eds. London, Taylor & Francis: 43–77.

Porter, J. M., Case, K., Marshall, R., Gyi, D. E. and Sims, R. E. (2004). 'Beyond Jack and Jill': designing for individuals using HADRIAN. *International Journal of Industrial Ergonomics* 33: 249–264.

Porter, J. M. and Gyi, D. E. (1998). Exploring the optimum posture for driver comfort. *International Journal of Vehicle Design* 19: 255–266.

Porter, J. M. and Porter, S. C. (1998). Turning automotive design 'inside-out'. *International Journal of Vehicle Design* 19(4): 385–401.

Quattrocolo, S., Gario, R. and Pizzoni, R. (2002). *3D human and vehicle model for driver and occupants posture prediction and comfort evaluation.* Digital Human Modeling Conference, Munich, Germany. 485–492.

Rasmussen, J., Toerholm, S., Siebertz, K. and Rausch, J. (2006). *Posture and movement prediction by means of musculoskeletal optimization.* Warrendale, Society of Automotive Engineers. SAE Technical Paper 2006-01-2342.

Rebiffe, R. (1969). *An ergonomic study of arrangement of the driving positions in motorcars.* Symposium, London.

Reed, M. P. (1998). *Statistical and biomechanical prediction of automobile driving posture.* University of Michigan Transport and Research Institute. Ann Arbor, MI. Doctoral thesis.

Robinette, K. M. and Hudson, J. A. (2006). Anthropometry. *Handbook of human factors and ergonomics. 3rd ed.* G. Salvendy, Ed. Hoboken, John Wiley & Sons: 322–339.

Roe, R. W. (1993). Occupant packaging. *Automotive ergonomics.* B. Peacock and W. Karwowski, Eds. London, Taylor & Francis: 11–42.

Schneider, L. W., Reed, M. P., Roe, R. W., Manary, M. A., Flannagan, C. A. C., Hubbard, R. P. and Rupp, G. L. (1999). *ASPECT: The next-generation H-point machine and related vehicle and seat design and measurement tools.* Warrendale, Society of Automotive Engineers. SAE Technical Paper 990962, SAE Transactions: *Journal of Passenger Cars*, Vol. 108.

Seidl, A. (1997). *RAMSIS - A new CAD-tool for ergonomic analysis of vehicles developed for the German automotive industry.* Warrendale, Society of Automotive Engineers. SAE Technical Paper 970088.

Sundin, A. and Örtengren, R. (2006). Digital human modeling for CAE applications. *Handbook of human factors and ergonomics. 3rd ed.* G. Salvendy, Ed. Hoboken, John Wiley & Sons: 1053–1078.

Sundin, A. and Sjöberg, H. (2002). *How are computer mannequins used in Sweden?* Nordic Ergonomics Society Conference, Kålmården, Sweden. 745–750.

Sundström, J. (2003). *Contextual studies of truck drivers' sitting. Applying activity theory to studies of long-haul drivers' activities and postural changes during driving.* Department of Product and Production Development. Gothenburg, Chalmers University of Technology, Sweden. Licentiate thesis.

Tilley, A. and Dreyfuss, H. (1993). *The measure of man and woman. Human factors design.* New York, Whitney Library of Design.

UGS (2007). *Human performance: Jack*, UGS, Tecnomatix. Access: 25 January 2007. http://www.ugs.com/products/tecnomatix/human_performance/jack/

Zehner, G. F., Meindl, R. S. and Hudson, J. A. (1993). *A multivariate anthropometric method for crew station design: abridged.* Ohio, Wright-Patterson Air Force Base. AL-TR-1992-0164. 32.

Ziolek, S. A. and Kruithof, P. (2000). *Human modeling & simulation: a primer for practitioners.* XIVth Congress of IEA and 44th Meeting of HFES, San Diego. 825–827.

Ziolek, S. A. and Nebel, K. (2003). *Human modeling: controlling misuse and misinterpretation.* Warrendale, Society of Automotive Engineers. SAE Technical Paper 2003-01-2178.

Ziolek, S. A. and Wawrow, P. (2004). *Beyond percentiles: An examination of occupant anthropometry and seat design.* Warrendale, Society of Automotive Engineers. SAE Technical Paper 2004-01-0375.

41

Inclusive Design for the Mobility Impaired

J. Mark Porter, Russ
Marshall, Keith Case,
Diane E. Gyi,
Ruth E. Sims, and
Steve Summerskill

41.1 Overview

HADRIAN is a computer-based inclusive design tool developed initially to support the design of kitchen and shopping tasks. The tool is currently being expanded to include data on an individual's ability to undertake a variety of transport-related tasks, such as vehicle ingress/egress, coping with uneven surfaces, steps, street furniture, and complex pedestrian environments. A feature of the enhanced HADRIAN tool will be a journey planner that compares an individual's physical, cognitive, and emotional abilities with the demands that will placed upon that individual depending on the mode(s) of transport available and the route options.

41.2 Introduction

There are ethical, legislative, and financial reasons why products and services should be designed, wherever appropriate and possible, for the widest range of consumer ages, shapes, sizes, needs, preferences, abilities, and aspirations. For example, it is estimated by the World Health Organization that the world population of people aged 60+ years will increase from 600 million in 2000 to 1.2 billion by 2025, and 2 billion by 2050 (WHO, 2006, http://www.who.int/ageing/en/). The WHO website also states that in 2006 about two-thirds of all older people are living in the developing world; by 2025 it will be 75%. Furthermore, it notes that the very old (age 80+ years) is the fastest growing population group in the developed world. The UK Design Council website (http://www.designcouncil.org.uk) states that 50% of Britain's adult (16+) population will be aged 50 or over in 2020. It also points out that while 8.7 million people come under the remit of the Disability Discrimination Act 1995, fewer than 5% are wheelchair users. Collectively they have an estimated disposable income of £45-50 billion. This increasingly older population, together with the disabled population, will affect market forces and impact how designers consider their end-users (Vanderheiden & Tobias, 2000). While young and able people are often considered to be able to 'adapt' to a poor design, there is typically an associated human cost. For example, a poor posture that has to be maintained for prolonged periods will result in a high incidence of musculoskeletal troubles and possibly sickness absence. People who are older or disabled have less opportunity to adapt to a poor design. In many cases, they are effectively designed out and cannot use the product or service. The design for all or inclusive design philosophy aims to reduce, if not eliminate, such problems.

Inclusive design is an approach to design that seeks to ensure that mainstream products, services, and environments are accessible to the largest number of people. We are currently witnessing a shift from designers and planners treating disabled and older people as separate cases, requiring special design solutions, toward integrating them into the mainstream through a more inclusive approach to the design of buildings, public spaces, products, and services. Central to this inclusive approach is the challenge of understanding and quantifying the numbers of people adversely affected by decisions made during the specification and design process. This challenge has led to the creation and development of HADRIAN.

HADRIAN (Human Anthropometric Data Requirements Investigation & ANalysis) is our computer-based inclusive design tool that was created through the UK EPSRC 'EQUAL' (Extending Quality of Life) initiative. The tool is now being further developed through the EPSRC Sustainable Urban Environment (SUE) program. Accessibility and User Needs in Transport for Sustainable Urban Environments (AUNT-SUE) is a multi-disciplinary consortium of researchers from London Metropolitan University, Loughborough University, University College London, RNIB, London Borough of Camden and Hertfordshire County Council. The aim of the consortium is to improve our understanding of the needs, abilities, and preferences of people who experience transport-related exclusion in towns and cities. Better empathy with disadvantaged users and would-be users will be encouraged through an AUNT-SUE toolkit to support planners, designers, operators, user groups, and others working to make urban transport and street design more inclusive.

HADRIAN builds upon our 30+ years experience in developing the SAMMIE system and providing support to a large number of industrial, commercial, and government projects through SAMMIE CAD Ltd., a UK Ergonomics Society Registered Design Consultancy (see Porter et al., 1999). Figure 41.1 shows our current SAMMIE human modeling tool that runs on Windows XP. As expert users, we have also successfully used SAMMIE in a variety of inclusive design projects including the Brussels Tram 2000 project (see fig. 41.2). However, we feel there is a strong need to empower non-experts so that inclusive design becomes a fundamental part of design practice.

HADRIAN was therefore created to make a step-function change in the way that anthropometric data are used in the design and evaluation of products and services. Our approach has been to replace numerous separate tables of population percentiles (e.g., for arm reach, hip breadth, sitting height, and so on) with a holistic database of 100 individuals covering a wide range of shapes, sizes, and abilities (see fig. 41.3). This approach adds significantly to the richness of the data that a designer has access to. Additionally, by presenting the data as individual people, the designer has a much greater empathy with those they are designing for. To assist in making use of the data, there is a simple task analysis/synthesis tool that allows the designer to simulate various tasks with these individuals interacting with SAMMIE CAD models of equipment and products in the virtual environment. Each task element (e.g., reach a device, look at instructions) is automatically analyzed to identify if any of the individuals have a problem. If specific individuals experience problems, the designer can view these individuals' 'best attempts' to more fully understand the nature of the physical problems with a view to iteratively improving the design and making it more inclusive. Details of our methodology can be found in Porter et al. (2003, 2004a and b, 2006), Gyi et al. (2004) and Marshall et al. (2005).

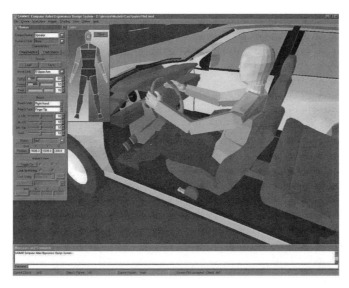

FIGURE 41.1 Latest version of SAMMIE runs on Windows XP (see http://www.sammiecad.com for further details of this human modeling system).

FIGURE 41.2 SAMMIE has been used in several multinational design projects working closely with designers from the concept stage onward.

Although there are a number of inclusive design tools/methods available, none of these include consideration for the emotional issues of users' lives in an explicit manner. We are now adding to HADRIAN's functionality by including additional data concerning an individual's emotional and cognitive dimensions.

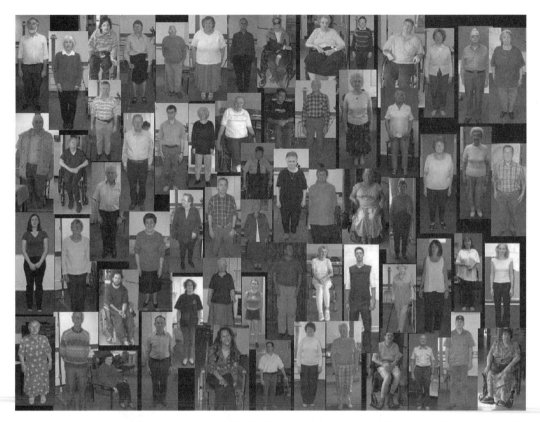

FIGURE 41.3 HADRIAN's database of 100 individuals covering a wide range of sizes and abilities. Each individual's data can be used as a design resource and to generate a digital human model of each individual using SAMMIE.

41.3 Problems with Percentiles

Design for all, or inclusive design, needs to move from a philosophical viewpoint to a central feature of design practice. Key to this is establishing empathy between designers, who usually start their careers while young, healthy, and able, and the people who would primarily benefit, who are often older, in poor health, and unable to achieve all the tasks they would like to with ease and confidence. Current design practice frequently involves using anthropometric and biomechanical databases that present percentile values for body size and strength, joint ranges for mobility, and so on. These numbers do not motivate the designer to vigorously explore design solutions that are more inclusive. In fact, the commonly accepted view for mainstream design is to cater to only the 5th to 95th percentile users of a product or service. This is designing for numbers, not people. Today, we believe that it is no longer acceptable to continue this approach of deliberate designing out of people who are in the top or bottom 5% of size or ability.

Percentiles are univariate in that they relate to only one dimension or variable. Knowing someone's stature does not provide an accurate estimation of their hip breadth, shoulder mobility, or grip strength because these variables are poorly correlated. Furthermore, people can be excluded from using a product or service because of a wide range of factors including their personal emotional and cognitive dimensions. For example, a person may be able to reach a control but not be able to manipulate it as required; be able to lift an item but not have the balance to carry it safely; be able to see a timetable but not be able to plan a journey route; be able to walk 100 meters but not be confident enough to cross a busy road; be able to climb stairs but not be willing to walk past a group of teenagers in an environment dominated by graffiti; and so on.

There is an important distinction between the specification of percentile values to be used for a particular design dimension and the percentage of the user population that will be accommodated in all respects. A common mistake made by many designers is to use the 5th percentile female stature and 95th percentile male stature mannequins to assess a workstation, assuming that if both of these mannequins can be accommodated, then so can 95% of the adult population. This is an incorrect assumption as it implies that those people designed out because either their sitting height, hip breadth, or leg length, for example, are greater than 95th percentile male values are all the same people. Similarly, all those with sitting eye height, arm length, or leg length smaller than 5th percentile female values are assumed to be the same individuals. As these dimensions are not strongly correlated then these assumptions are incorrect. A study of air crew selection standards and design criteria analysis reported by Roebuck, et al., (1975, p. 268) illustrates the problem perfectly as it was shown that nearly half of the air crew were designed out when the 5th to 95th percentile range was used on a large number of body dimensions (in this case 15 dimensions). Even limiting the number of dimensions to just 7 (sitting height, eye height-sitting, shoulder height-sitting, elbow rest height, knee height, forearm-hand length and buttock-leg length) designed out over 30% of the available air crew.

Clearly, the univariate nature of anthropometric percentiles is a significant limitation to success when being used by non-experts in the design process. Designers need to be able to consider multivariate issues right from the beginning of a design project; however, this is rarely achieved in practice until the project reaches the user trial stage with a full-size working prototype. By this stage it is often difficult to make substantial changes to the product. The next section describes how the HADRIAN inclusive design tool deals with multivariate issues and supports the designer during the early stages of the design process.

41.4 Our Approach to Inclusive Design

The HADRIAN inclusive design tool includes several features that are briefly described below.

41.4.1 Database of Individuals

Detailed information is provided on the size, shape, abilities, preferences, and concerns of a wide range of people, each presented as individual datasets. Data are collected using traditional anthropometers and a body scanner (see fig. 41.4 for examples of the output) while a variety of experimental rigs have been developed to assess 3D reach volumes, kitchen and shopping tasks, and vehicle ingress/egress (see fig. 41.5). Furthermore, a transport activities questionnaire is used to obtain extensive information about the nature of any problems experienced when making a journey, for example: physical barriers of all types, cognitive issues with information presentation or wayfinding, or emotional concerns with poor lighting, badly maintained or overcrowded/deserted environments, and so on.

This range of information is presented to the designer as a set of screen displays for each person, including video clips showing them undertaking a variety of tasks such as lifting a baking tray with oven gloves on, reaching to food stuffs on a low or high shelf, entering and leaving a bus/coach/tram/train (see fig. 41.6). These data are both informative and foster empathy between the designer and these future users/consumers. Basic emotional and cognitive data related to activities of daily living, such as shopping, cooking, and making a journey, are also presented. These screen displays are a very useful design resource and we have tried to make them easily navigable and to clearly show the extent of human variability that needs to be considered when practicing inclusive design. The anthropometric data taken from the body scans are used to create virtual SAMMIE human models for each individual in the database (see fig. 41.7), and these are used in the multivariate assessments described below.

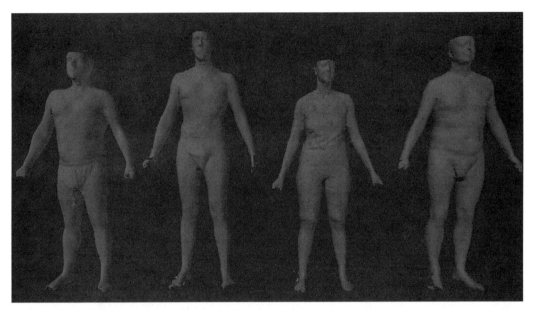

FIGURE 41.4 Scanned images for individuals that are used to collect anthropometric data and to construct the individual SAMMIE human models.

41.4.2 Task Analysis/Synthesis Tool

A simple task analysis/synthesis tool is provided whereby the designer can specify a series of task elements (such as look at, reach to, lift to, walk to, climb up) with reference to physical items in a CAD model of an existing product or an early prototype of a new design. The designer can specify parameters as appropriate (e.g., use thumb tip reach for inserting a card into a card reader; specifying a viewing distance for specific character sizes), although the tool will set default settings for each individual otherwise (e.g., preferred hand). The series of task elements can be easily constructed to define the critical aspects of interacting with a product or service. For example, when using an ATM (cash dispenser) the

FIGURE 41.5 Extracts from video clips for two individuals leaving the vehicle ingress/egress rig.

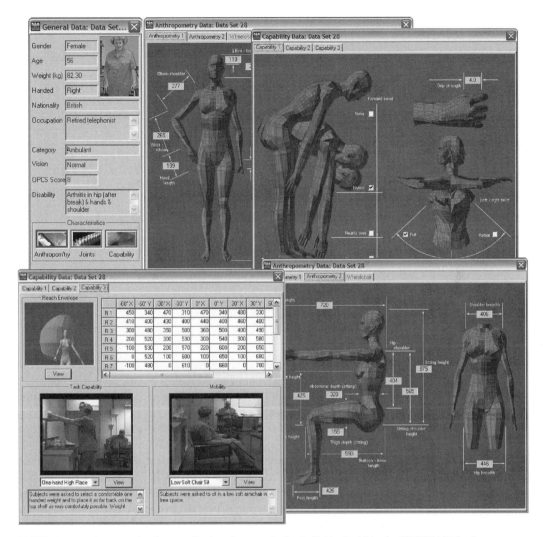

FIGURE 41.6 An example of screen displays for a particular individual within the HADRIAN database.

series would include: look at screen, look at card slot, reach to card slot, reach to screen buttons, look at keypad, reach to keypad, look at cash tray, and reach to cash tray.

41.4.3 Multivariate Assessment of Each Individual's Ability to Complete a Task

An automated analysis is then undertaken whereby each individual in the database is assessed in terms of ability to complete successfully each task element. This procedure deals with the multivariate nature of interactions with products, integrating the relevant physical, emotional, and cognitive issues. The physical issues are evaluated using the SAMMIE human models for each individual in the HADRIAN database to create a virtual simulation of the tasks, while the emotional and cognitive issues are assessed using personal lookup tables. The analysis presents the percentage of individuals within the database who are designed out from using a product or excluded from using a service, and the designer is encouraged to seek further details by scrolling through virtual simulations of the problem(s) encountered by

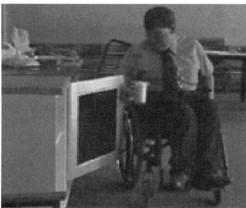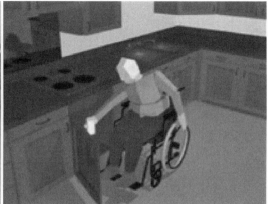

FIGURE 41.7   SAMMIE human models are constructed for each individual in the HADRIAN database.

each person. This visualization of the person and the problem encourages the designer to explore potential solutions by modifying the CAD model and repeating the analysis.

HADRIAN does not present people as anonymous numbers in percentile tables (where individuals are lost forever). Instead, individual datasets are presented describing each person in terms of body shape and size, abilities, coping behaviors, and emotional characteristics. The designer is, thereby, both more knowledgeable about individual differences and more motivated to make a difference due to the greater empathy with their problems in interacting with products and services. The interactive nature of the tool allows the designer to view video clips and other personal data for all individuals who are predicted to be designed out from using a product or excluded from using a service. The nature of their predicted problems, be they physical, emotional, cognitive, or a combination thereof, are highlighted for each specific task element. Some people may not be able to climb on board a bus without help, others may be able to get on board but not have the confidence to get to their seat before the bus moves off, and some may be discouraged from using the bus as they know it is too crowded or that the bus stop is in a dark and unpleasant location. In all such cases, people may decide not to use buses at all. The optimum solution to such problems involves both design (e.g., of the bus, bus stop/shelter, etc.) and operation (e.g., driver training, scheduling of bus service, etc.).

41.5 Data Collection Issues

Several issues have emerged during this research, some of which are discussed below.

41.5.1 Participant Selection

The complexity of carer interactions was considered too difficult to quantify and model within the current software tool. Consequently, the decision was made to exclude those who are not physically able to get themselves out of their house and onto the pavement to join the transport system. Clearly, such people still have needs and aspirations for transport usage, and a future study could exclusively explore the considerations involved with individuals and carers.

It was our intention to revisit as many of the 100 people who had previously participated in the original EQUAL funded project concerning kitchen and shopping tasks to see if they would be willing to participate again in the AUNT-SUE transport-related project. We hoped this would enable some study of the longitudinal aspects of age and disability, due to the number of years since the first study data collection took place. However, several of the original participants have sadly died in the interim and some others are no longer able to get outside without assistance. Furthermore, we have experienced problems contacting many of the remaining participants as their contact details are now out-of-date. As a consequence, we have had to find many new participants for the AUNT-SUE project.

All participants were required to complete a medical screening questionnaire before the trials. Typically, medical screening questionnaires are used to exclude just the type of participants that our research is looking for. It was important that we were well informed of the potential consequences of a participant's current medical condition so that appropriate changes could be taken to reduce any identified risk during the data collection trials. For example, those participants with vertigo/dizziness were not asked to bend low down and those who were epileptic were not body scanned.

41.5.2 Vehicle Ingress and Egress

When making a journey using a public transport vehicle, people encounter a variety of step heights and handle locations during ingress and egress. We designed a rig to assess participants' abilities in these situations, and decisions concerning which heights and handle positions to study were made after referencing the relevant public transport regulations and making field observations within the Midlands area of the United Kingdom. Train carriages and trams are covered by the Rail Vehicle Accessibility (Amendment) Regulations (2000). This states the maximum step height should be 200 mm with handrails placed internally, on either side of the external doorways, between 700 mm and 1200 mm above the floor. From our observations it was found that step heights into trains varied between 180 mm and 280 mm, with the one example of trams having no step at all. London Underground state that the maximum step height on their lines is 240 mm. Buses and coaches (carrying more than 22 people for public usage) are covered by the Public Service Vehicle Accessibility Regulations (2000). This states that the maximum step height from pavement to bus should be 250 mm, with the first handrail inside the bus being within 100 mm of the entrance and between 800 mm and 1100 mm above ground level. Observed step heights for buses were found to vary between 170 mm and 300 mm, and for coaches between 270 mm and 370 mm.

The rig, consisting of an entrance and exit, was designed to provide different door widths on each side: one side narrower to simulate an older-style bus, train, or coach entrance, and the other side wider to simulate the access of newer buses and trains (see fig. 41.5). The grab handles on each side can be placed in two positions; on the narrow side they can be set at 100 mm or 200 mm from the entrance to the 'vehicle', on the wider side they can be set at 300 mm or 400 mm from the entrance. The step heights can be varied from 150 mm, 250 mm, or 350 mm to reflect the worst-case scenario. There is a 100-mm horizontal gap between the ground and the vehicle on both sides to reflect the horizontal gap between pavement/platform and the body of the vehicle.

The design of the rig raised a number of ethical considerations. In order to keep the trials as safe as possible, participants completed the transport abilities questionnaire before attempting to use the rig, thereby giving advance information of what would be likely to cause problems. The initial rig setup was then adjusted according to these responses: able-bodied participants had both step heights set at the maximum 350 mm, with handles set at 200 mm and 400 mm, respectively. Less able participants had lower step heights and handle heights adjusted to their ability. When the rig was set correctly, participants were first asked to observe an experimenter demonstrating the task. Experimenters stood on both sides of the rig to offer assistance if required, and it was reinforced that participants should only attempt if they were happy to do so and they should take their time. When it came to stepping down participants were asked to first look at the required step and state whether they were happy to continue, before doing so in a controlled, safe manner. Anyone who felt unsure about the task was obviously free to stop, and steps could be removed if required during the trial.

41.5.3 Use of Whole Body Scanned Data

We are using a whole body scanner ([TC]² NX$_{12}$ Body Measurement System) to quickly collect body dimensions for use in constructing virtual human models of individual participants (see fig. 41.4). Participants undress in an enclosed private cubicle into lightweight and close-fitting clothing that is neutral to their skin tone, as high contrast with skin tone causes problems in attaining a complete scan. Once inside the scanner booth participants stand and then sit in standardized postures. The scan takes a matter of seconds, and then the person can exit and dress again.

Traditional external anthropometric data were collected to enable comparison with measurements from the scanner to see if the scanner made the process quicker and more accurate for both able-bodied and less able participants. This comparison is ongoing. We have concerns that less able participants might experience problems getting into the required positions, and there are issues concerning those in wheelchairs, as the scanner dislikes reflection or high contrast.

41.5.4 Emotional and Cognitive Dimensions

The latest version of HADRIAN now includes data on an individual's ability to undertake a variety of transport-related tasks, such as vehicle ingress/egress, coping with uneven surfaces, steps, street furniture, and complex pedestrian environments (see fig. 41.8). The latter includes capturing an individual's concerns about finding one's way during an unfamiliar journey, changing transport modes (e.g., from a bus to a train), crossing busy roads, walking past large groups of people and/or graffiti, and so on.

Our Transport Activities Questionnaire has been developed to get rich and detailed information regarding participants' physical abilities, and also to tap into their cognitive and emotional issues surrounding transport usage. Participants are asked questions concerning: their physical abilities, based on the Office of Population Censuses and Surveys scale (Martin et al., 1988); any problems encountered when using trains, buses, trams, London-style taxi cabs, and minicab taxis; their ability to walk distances, as well as issues surrounding taking luggage on the different transport modes; the types and frequency of journeys made; problems in using stairs, lifts, or escalators; and difficulties in understanding timetables and signs. The questionnaire also includes a request for information about problems experienced in the local area. Any local areas that participants identified as causing problems, when traveling, are visited by the experimenters to provide quantitative

FIGURE 41.8   Examples of transport tasks and environments that have physical, cognitive, and emotional dimensions for the traveler.

data to supplement the reports from the participants. For example, this may range from measuring the force required to open a heavy shop door, to assessing the cognitive and emotional issues at a transport node (e.g., changing from a bus to the train, involving crossing busy roads, walking through empty or crowded public spaces with poor street lighting). In short, the questionnaire aims to provide information concerning issues that may arise at any point during the whole journey process—from leaving home, traveling to a transport node, vehicle ingress/egress, changing transport modes, and arriving at the destination.

41.5.5 Testing and Validation of the Tool

The tool provides a database of physical, emotional, and cognitive information for 100 individuals, carefully selected to cover a very wide range of abilities. While such data can be obtained traditionally from user trials, this can only be done for existing designs or those at the full-size prototype stage. The HADRIAN database will allow access to this rich data at a much earlier stage in the process, allowing planners, designers, and operators of transport vehicles and systems to maximize social inclusion and access through the earliest consideration of the issues being faced by the users of public transport as they move to and from home, work, school, leisure destinations, and shopping. To validate the approach we plan to undertake some real-world case studies within the testbed areas identified by the AUNT-SUE consortium, or elsewhere if more suitable, where HADRIAN will be used by designers and planners to try out different options for a specific design problem, be it a ticket barrier, a train station, or access to the Olympics 2012. The tool should be able to identify the superior design option in terms of inclusivity, and to give direction as to how to improve it further, if appropriate. We also plan to compare the predictions made using HADRIAN for a specific virtual individual/object interaction with actual observations recorded during real-world usage with the same individual.

41.6 Development of a Personalized Journey Planner

A specific feature of the enhanced HADRIAN tool will be a journey planner that compares an individual's physical, emotional, and cognitive abilities with the demands that will placed upon that individual depending on the mode(s) of transport available and the route options. For example, some people do not use trains because of difficulties experienced getting on and off; some prefer not to walk through busy public areas or places with graffiti where they feel vulnerable; some experience problems finding their way on unfamiliar routes; some are unable to walk far or to climb steps with confidence; some are reluctant to cross busy road junctions; and so on. If a particular desired journey is unachievable or very difficult, either unaided or with support from others, then that person is likely to feel socially excluded. The prototype journey planner should allow people to predict problems that they may experience before deciding to make the journey. Hopefully, a suitable alternative route and choice of transport mode(s) can be identified using the planner such that the task demands fall within the person's abilities and preferences.

While the database only comprises 100 individuals, we envision a web-based planner could be made available. Members of the public would need to complete an online questionnaire to provide relevant personal data on their body size (i.e., clothing sizes), general health, abilities, and transport preferences. A major issue in such a planner would be compiling a database of the specific demands that would be placed on the traveler as a function of the exact geographic locations and the transport modes available for a particular journey. It is hoped that a pilot trial can be run in the testbed areas, with transport nodes, shopping areas, museums, theaters, cinemas, and restaurants providing data for their surrounding areas (i.e., distance from the nearest bus stops, train station, taxi rank, etc.). The data could include, for example, distances by foot, details of steps/lifts/escalators, performance of street lighting/signposting, quality of the pavement/street surface, and perceived safety (e.g., ratings from a sample of people covering a range of ages and abilities) and objective safety (data on thefts, accidents, incidents, etc.). Figure 41.9 shows an initial screen interface for the journey planner.

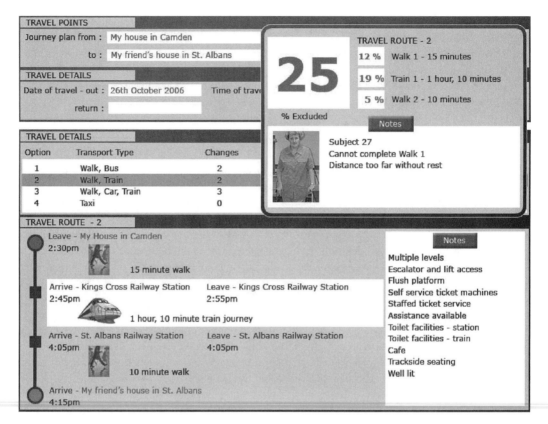

FIGURE 41.9 An initial screen interface for the prototype journey planner showing that 25% of the HADRIAN database would experience problems with Travel Option 2. A breakdown of problems for each stage of this journey is given, along with links to the individuals predicted to have problems. Such an analysis is useful for transport planners and designers. Alternatively, the planner can be used to evaluate proposed journeys just for one person (i.e., as a personalized journey planner).

Another issue for the personalized journey planner will be dealing with day-to-day transport problems that influence the selection of optimum routes and modes of travel, such as canceled or delayed trains, road construction, or equipment failure. Our vision for the future would be for the journey planner to receive up-to-date information from transport service providers, local communities, and individual travelers via the Internet. The modern-day mobile phone is now equipped with high resolution cameras and access to the Internet; combined with a global positioning system (GPS), this will allow travelers to take photos of transport problems they are personally experiencing and upload them with a text or voice explanation to a specialist website. This would enable real-time information to be accessed by the journey planner. To close the loop, travelers already on a planned journey could be contacted via their mobile phones with a revised journey plan to deal with these unforeseen problems.

41.7 Concluding Remarks

It is hoped that inclusive design tools such as HADRIAN will lead the way in integrating inclusive design within contemporary design practice. To have the greatest impact upon the quality of peoples' lives, it is important that such an approach is adopted by the majority of design professionals. Currently, inclusive design is practiced by only a few specialist designers/ergonomists who have access to the relevant data and have knowledge of appropriate methods. Furthermore, current practice primarily addresses just the physical issues. The HADRIAN approach provides the relevant data on people of all shapes, sizes, and abilities, plus the methodology to predict who will be designed out or excluded from using a particular product or service whether this is due to physical, emotional, or cognitive factors. We are not advocating that one design fits all, but the needs, abilities, preferences, concerns, and aspirations of all people who would want to use the product or service should be considered at the concept stage of design. This may result in a single inclusive solution or in a range of products and services to cater to various niche markets.

References

Gyi, D.E., Sims, R.E., Porter, J.M., Marshall, R., and Case, K., 2004, Representing older and disabled people in virtual user trials: data collection methods. *Applied Ergonomics*, 35, 443–451.

Marshall, R., Porter, J.M., Sims, R., Gyi, D., and Case, K., 2005, HADRIAN meets AUNT-SUE. *Proceedings of INCLUDE 2005*, Royal College of Art, London, UK (CD-ROM).

Martin, J., Meltzer, H., and Elliot, D., 1988, OPCS surveys of disability in Great Britain: The prevalence of disability among adults, (Office of Population Censuses and Surveys, Social Survey Division, HMSO).

Porter, J.M., Case, K., and Freer, M.T., 1999, Computer aided design and human models. In: *The Occupational Ergonomics Handbook*, W. Karwowski and W. Marras (Eds.), pp 479–500, CRC Press LLC, Orlando, Florida. ISBN 0-8493-2641-9.

Porter, J.M., Case, K., Marshall, R., Gyi, D.E., and Sims, R., 2004a, 'Beyond Jack & Jill': Designing for individuals within populations using HADRIAN. *International Journal of Industrial Ergonomics*, 33(3), 249–264.

Porter, J.M., Case, K., Marshall, R., Gyi, D.E., and Sims, R.E., 2006, Developing the HADRIAN inclusive design tool to provide a personal journey planner. In: *Contemporary Ergonomics 2006*, P.D. Bust (Ed.), pp 465–467. The Ergonomics Society Annual Conference—2006, Cambridge, UK. ISBN 0-4153-9818-5.

Porter, J.M., Marshall, R., Freer, M., and Case, K., 2004b, SAMMIE: A computer aided ergonomics design tool. In: *Working Postures & Movements: Tools for Evaluation and Engineering*, eds. N. Delleman, C. Haslegrave and D. Chaffin, pp 454–470, CRC Press LLC, Orlando, FL.

Porter, J.M., Marshall, R., Sims, R.E., Gyi, D.E., and Case, K., 2003, HADRIAN: a human modelling CAD tool to promote 'design for all'. *Proceedings of INCLUDE 2003: Inclusive Design for Society and Business*, CD-ROM, Volume 6, pp 222–228, March 2003, Royal College of Art, London, UK.

Public Service Vehicles Accessibility Regulations, 2000, Statutory Instrument No. 1970, Department for Transport, UK.

Roebuck, Jr, J.A., Kroemer, K.H.E., and Thomson, W.G., 1975, *Engineering Anthropometry Methods*, John Wiley, New York.

Rail Vehicle Accessibility (Amendment) Regulations, 2000, Statutory Instrument No. 3215, Department for Transport, UK.

Vanderheiden, G., and Tobias, J., 2000, Universal design of consumer products: current industry practice and perceptions. *Proceedings of the International Ergonomics Society/Human Factors and Ergonomics Society Congress 2000*, 19–22.

42

Digital Human Modeling Automotive Manufacturing Applications

Dan Lämkull, Roland
Örtengren, and
Lennart Malmsköld

42.1 Introduction

Any use of computer software to solve engineering problems is called computer-aided engineering, and has become one of the most important tools for simultaneous engineering. Fast, high-quality computer graphics now allow us to render lifelike images of people performing a multitude of tasks within various computer-aided engineering programs. Thus, it now is possible to position and move manikins to predict the performance capabilities of designated groups of people within a computer-rendered environment. Digital human modeling tools (DHM tools) have been introduced in industry to facilitate a faster and more cost-efficient design process. Most of the tool users are in the fields of automotive and aerospace engineering. The tools are applied in the design, modification, visualization, and analysis of human workplace layouts and product interactions.

This chapter covers the main reasons for the use of DHM tools in the automotive manufacturing industry. The focus is entirely on physical digital human modeling, not on cognitive digital human modeling. Important aspects regarding work process and organization are covered as well as how companies deal with some identified shortcomings.

The chapter also describes how fewer physical prototype vehicles have made it necessary to find new methods for training of operators adapted to the new conditions. Finally, an example of a simulation case from the order of the case to the result presentation is given.

42.2 The Main Reasons for the Use of DHM Tools in the Automotive Manufacturing Industry

In today's harsh business environment, it is necessary to focus on the ability to develop new products and processes with perfection in order to maintain competitiveness in the global market. International competition has become more intense as the world trade has increased with companies trying to conquer new markets. The number of world-scale competitors capable of delivering high-quality products to customers has increased, resulting in the challenge of today: to compete on previously unexplored and inaccessible market areas (Wheelwright & Clark, 1992). The challenge is to reduce the development cycle without sacrificing performance and quality (Cohen et al., 1996), and to maintain a sound and healthy work environment.

42.2.1 Time to Market

Time to market is a critical success factor in today's business environment. With the short-lived products of competitive global markets, most of the engineering has to be done before the start of production and simultaneous with product development (simultaneous engineering) (Rauglas, 1998). The consequences are that either the design of a new production system has to be without disabling features or the product has to be designed to fit into an existing production system. Both alternatives imply the need for adequate methods of analyses of production implications in the design process. Early involvement of many (e.g., manufacturing, purchasing, marketing) constituencies is essential for cycle time reduction and improvements in product innovation capabilities (Koufteros et al., 2001). In fact, the most cited reason for delays in product development projects in manufacturing systems is engineering change orders (Barkan, 1992; Millson et al., 1992). These occur when materials are unavailable or parameters about the product do not match manufacturing capabilities or customer expectations. This points to the lack of information or equivocality. Designers may find out that manufacturing cannot meet the specifications or that the specified material is unavailable or that customers are not satisfied. Time is wasted when physically separate functions need to communicate. The benefits are derived from fewer mismatches between product characteristics and existing process capabilities. These mismatches are often caused by the designer's misperceptions of factory capabilities (Langowitz, 1988). Manufacturing may also suggest ways to design the product for ease of manufacturing. Manufacturing may suggest how products can be designed with fewer parts, assembled or tested more easily, or accommodated to automated equipment (Boothroyd & Dewhurst, 1988).

Any use of computer software to solve engineering problems is called computer-aided engineering (CAE), and has become one of the most important tools for simultaneous engineering. With the improvement of graphics displays, engineering workstations, and graphics standards, CAE has come to mean the computer solution of engineering problems with the assistance of interactive computer graphics (Myklebust, 2001).

Computer-aided design (CAD) refers to the use of computers in converting the initial idea for a product into a detailed engineering design. The evolution of a design typically involves the creation of geometric models of the product, which can be manipulated, analyzed, and refined. In CAD, computer models and graphics replace the sketches and engineering drawings traditionally used to visualize products and communicate design information (Preston White & Richards, 2002). In the late 1950s, hardware became available that allowed the machining of 3D shapes out of blocks of wood or steel, and these shapes could then be used as stamps and dies for products such as the hood of a car. The exploration of the use of parametric curves and surfaces can be viewed as the origin of computer-aided geometric design (Farin, 1990). Groover (1984) states that: "Computer Aided Design (CAD) can be described as any activity that involves the use of the computer to create or modify an engineering design." This means that CAD can be used in many different ways and at different levels of complexity.

In line with other engineering software development, DHM tools have moved from a 2D to a 3D world. Fast, high-quality computer graphics now allow us to render lifelike images of people performing a multitude of tasks within various computer-aided design programs. Furthermore, our statistical description of various population attributes, such as size, shape, strength, and range of motion of a specific group, have become rather refined. Thus, it now is possible to position and move manikins to predict the performance capabilities of designated groups of people within a computer-rendered environment (Chaffin, 2001). DHM tools, such as 3DSSPP (Chaffin, 1969), Jack (Badler, 1993) and Ramsis (Seidl, 1997), have been introduced in industry to facilitate a faster and more cost-efficient design process. Most of the tool users are in the fields of automotive engineering, aerospace engineering, or industrial engineering (Landau, 2000; Chaffin, 2001). The tools are applied in the design, modification, visualization, and analysis of human workplace layouts or product interactions.

42.2.2 Musculoskeletal Disorders

The Swedish Work Environment Authority has performed work environment surveys in Sweden since 1989. The surveys are conducted every second year and the latest survey is from 2005 (Arbetsmiljöverket, 2005). In this survey, approximately 10,000 persons answered a questionnaire and 20,000 persons were interviewed. In the 2005 survey of work-related disorders, 8% of the men and 10% of the women replied that they had suffered discomfort for the past 12 months due to strenuous work postures. This means approximately 1.5 million employees of both sexes were working every day in strenuous work postures. During 2004 to 2005, some 370,000 employees had suffered discomfort from strenuous work postures. Upwards of 16,000 new musculoskeletal disorders[1] (MSDs) were reported in 2005 as work injuries. Furthermore, the Work Environment Surveys show assembly workers[2] to be a vulnerable group. More than 6 out of 10 women (62%) agree, partly or wholly, that they have strenuous work postures. A slightly smaller proportion of men agree, partly or wholly, that their work entails strenuous work postures (57%). Male assembly workers, and even more so vehicle assemblers, responded with many work-related disorders due to strenuous work postures (14%), heavy man handling (12%), and short, repetitive work operations (9%) in the year preceding the interview.

More than 600 million working days are lost due to work-related illness or injury each year in the European Union. Although precise figures do not exist, estimates from Member States of the economic costs of all work-related illness range from 2.6% to 3.8% of Gross National Product

---

[1] The U.S. Department of Labor defines a musculoskeletal disorder (MSD) as an injury or disorder of the muscles, nerves, tendons, joints, cartilage, or spinal discs. MSDs do not include disorders caused by slips, trips, falls, motor vehicle accidents, or similar accidents.

[2] Assembly workers are defined as workers who assemble parts and components, e.g. of machinery and vehicles, electrical and electronic equipment and metal, rubber, plastic and wooden products. There are appr. 16 000 female assembly workers and 46 500 male assembly workers in Sweden (2005).

(GNP) (European Agency for Safety and Health at Work, 1998). Available cost estimates of MSDs put the cost at between 0.5% and 2% of GNP[3] (European Agency for Safety and Health at Work, 2000). MSDs are a commonly reported work-related health problem by European workers: 30% (44 million European workers) complain of backache; 17% complain of muscular pains in their arms or legs; 45% report working in painful or tiring positions; 33% are required to handle heavy loads in their work (Paoli, 1999).

The only routinely collected national source of information about occupational injuries and illnesses of U.S. workers is the Annual Survey of Occupational Injuries and Illnesses conducted by the Bureau of Labor Statistics (BLS) of the U.S. Department of Labor. For cases involving days away from work, BLS reports that in 2001 (the last year of data available at the time), 522,528 cases (34%) were the result of overexertion or repetitive motion (BLS, 2003). The precise cost of occupational musculoskeletal disorders is not known. Estimates vary depending on the method used. A conservative estimate published by NIOSH is $13 billion annually (NIOSH, 1996). Others have estimated the cost at $20 billion annually (AFL-CIO, 1997). Regardless of the estimate used, the problem is large both in health and economic terms.

All these figures show how widespread MSDs are today, and justify the effort to design production systems to fit the workers. A sound and healthy working environment is not just something that the law demands, it is also profitable. It enhances employees' well-being, satisfaction, and performance and results in lower costs for sick-leave and less employee turnover (Toomingas et al., 2005). Many MSDs can be prevented using ergonomic interventions to modify work and workplaces based on assessment of risk factors (European Agency for Safety and Health at Work, 2000).

42.2.3 Relationships between Ergonomics and Manufacturing Quality: Productivity

Several studies have identified a relationship between ergonomically problematic work tasks and quality deficiencies to the extent that around 30% to 50% of all quality remarks are related to or directly due to ergonomic problems (Axelsson, 1994, 1995; Eklund, 1995; Hallberg, 1995; Sandström & Svensson, 1996; Eklund, 1997; Moestam Ahlström, 2002).

Product design processes have become increasingly complex and the number and complexity of the demands on the designers are increasing. These include demands on functionality, usability, cost, ease of assembly, service, reliability, logistics, ergonomics, appearance, coherence to international and national standards or laws, ecology and energy considerations, in addition to new, emerging aspects such as ethics, pleasure, trust, status, and image (Cushman & Rosenberg, 1991; Eklund, 1996; Jordan, 1996). Some of the major reasons for integrating ergonomics in product design are improved product usability, improved user performance, safer products, improved user comfort, enhanced user satisfaction, and differences among users (Cushman & Rosenberg, 1991). These aspects are also congruent with the definition of quality. Relevant considerations of ergonomics in product design include anthropometric, biomechanical, physiological, cognitive, environmental, and psychosocial aspects (Cushman & Rosenberg, 1991). There are also more precise ergonomics design recommendations that, among other things, would lead to improved quality in manufacturing (Helander & Nagamachi, 1992; Helander & Willén, 1996). It is recognized that the design of products has a decisive influence on (1) manufacturing costs (in terms of number of parts, manufacturing time), (2) quality output, and (3) ease of manufacturing (Willkrans & Norrblom, 1995). Improvements in working environment also contribute to a more efficient production in terms of fewer breakdowns, less production disturbances, and a reduced need for maintenance (Abrahamsson, 2000).

---

[3] The GNP for the European Union in 2000 was $7 892 billion (WTO, 2001).

42.3 Organizational and Work Process Aspects

DHM tools are complex and using them requires expertise in different fields; thus a wide spectrum of users with different backgrounds often works in cooperation with these tools (Lockett et al., 2005). The users come from several occupations, such as ergonomists, manufacturing engineers, designers, and simulation engineers. Focus of research within ergonomics simulation has primarily been related to improve the simulation tools with more enhanced functionalities resulting in better and more accurate posture and motion algorithms and biomechanical models (e.g., Ciavarro et al., 2004; Dickerson et al., 2004; van Hoof et al., 2004; Bhatti et al., 2005; Seitz et al., 2005; Wang et al., 2005). Considerably less research has been conducted to investigate the needs of organizations and end users and their experiences of using the DHM tools. In the commercial sector, marketing departments often serve as the primary means of determining what customers want in a product's human interfaces, or manufacturing engineers, managers, and ergonomists define what is needed in the work environment (Lockett et al., 2005). These mechanisms are imperfect, however, as described by Chaffin (2001).

Volvo Car Corporation (VCC) has made large investments in DHM tools for production engineering. These tools require substantial investments in time and money and are thus within reach only for mainly larger and resourceful companies. However, the users of the simulation tools have experienced unexpected difficulties. Work process and data retrieval have proven to be critical issues when using simulation tools (Ziolek & Kruithof, 2000; Ruiter, 2001; Ziolek & Nebel, 2003). Within VCC the experience is that the DHM tools are overly time consuming and the gathering of information for use in the simulation task takes too much time with the effect that too few simulations are made (Dukic et al., 2002; Blomé et al., 2003).

42.3.1 Improvements of the Use of DHM Tools

Volvo Car Corporation has used DHM tools within the manufacturing department for more than a decade. During this period, several activities have been formalized and implemented to improve the use of DHM tools. According to the simulation engineers, the two most important activities are the formalization of an ergonomic forum and a standardized web-based order and report form with predefined text fields.

42.3.2 Ergonomic Forum

An ergonomic forum is a meeting where all simulation engineers together with ergonomists and customers, and sometimes assemblers, carefully examine tasks (e.g., tasks carried out during the previous weeks) that have been identified as troublesome. By such an activity, the simulation engineers will learn from each other and experienced assembly workers can teach the simulation engineers how tasks are carried out in the plant(s). Another important activity, although rarely on the agenda, is to pay visits to the assembly plant(s). This is a most instructive activity where the simulation engineers can see how the tasks, once simulated, actually are carried out. Moreover, typical tasks could be observed and discussed. This is not always easily done since the plant(s) might geographically be situated far from the simulation engineers; however, in the case the plant(s) is located not too far away from the simulation engineers such visits should definitely be on the agenda.

42.3.3 Centralized Placement of Simulation Engineers

At Volvo Car Corporation all ergonomy simulation engineers are located together, which is most efficient because it facilitates communication among simulation engineers and they can learn from each other. They can easily discuss different software-related questions and also ask each other about cases they know others have done in the group. A decentralized location has advantages such as faster communications between the customers and the simulation engineers—obscurities could easily be discussed and solved "just by turning around and tapping the customer's shoulder" and, probably, such an arrangement also gives the simulation engineer a better insight in all ongoing car projects, leading to both better product and process knowledge. Despite these advantages, a centralized placement is chosen because the simulation engineers value the support from each other before a closer location to the customers. Furthermore, a geographical distance between the simulation engineers and their primary customers, the manufacturing engineers, forces a more careful order description. A careful, written order description facilitates a more efficient exchange of simulation orders between simulation engineers in case of sickness or heavy work load.

42.3.4 A Standardized Web-Based Order and Report Form

The simulation engineers criticize the quality of the simulation orders, since they rarely contain all information needed to successfully carry out a simulation. A working process such as described by Green (2000) would considerably contribute to a more robust and less troublesome simulation order process. Moreover, the improvements of the DHM tools integration into different CAD systems have increased the possibility of creating load lists containing the components describing the work cell and thus minimizing ambiguities in a simulation order.

A careful order description from customers including information of a wide range of issues decreases doubts of how a task should be carried out or how the immediate environment looks around the assembled car body. Examples are:

  A complete list of all CAD geometries (car components and equipments) necessary for the simulation, which gives information about, for instance, height above assembly line. The list also gives information about the assembly sequence and the components' version numbers and correct placements, provided that the data in the CAD-database is correct.
  Information of (hand) tools: weight, torque, one or two hands
  Weight of handled car component
  Distance between the car bodies: Is it possible to approach the car body from behind or from the front?
  If the operator is allowed to lean on the car body
  If the operator is allowed to enter the car body

The list could be filled with copious amounts of information. The most important thing is that the information gives an understanding of the task, which is the first activity in Green's (2000) suggestion of a DHM tool working process.

42.3.5 Digital Human in Operator Training

The methods for training assembly operators are shifting due to changes in product development during the last decade. Product development in the automotive industry has gone from product validations based on a fairly large number of physical prototypes to computer-based simulation technologies with only a small number of prototype vehicles produced. The result of this change is that the first vehicles in many cases are produced by components from serial tools in a much later stage, only a short period before the real production starts. Those vehicles are named pre-series vehicles, and they are, apart from the very first ones, produced in the existing manufacturing system. With the former conditions the prototype vehicles were used also for training purposes. Less prototype vehicles have made it necessary to find new methods for training operators adapted to the new conditions.

Learning new operational content can be described in two major phases: the cognitive phase and the motor phase, and the two phases are described as the learning curve (Dar-El et al., 1995). In early stages of learning, the cognitive system dominates the process. Focus for the assembly operator is understanding and following job instructions, remembering the sequence of operations, recognizing and understanding how the tools, equipment, or assembly parts should be handled, and so on.

However, as experience is gained through executing the task the learning process migrates from a cognitive focus to a motor focus.

Today computers are used for training operators by introducing them to the content of new operations, the shape of new components, assembly sequence of new operations, and so on. This type of training support is still in its infancy and research is ongoing but some large-scale examples exist. The main focus is in cognitive learning, and the idea is to use the computer for the main portion of cognitive learning and focus the few training occasions during the pre-series period when training can shift to motor training.

As digital humans are used for analyses and simulations by manufacturing engineering, the possibility to reuse this information for training purposes exists. The reason for doing an ergonomic analysis is in many cases due to a need for understanding how the conditions for the operator will look in the assembly operation. This understanding is also of interest when the initial training activity takes place. The operations that have been analyzed with a digital human can give much information to the operator, just by showing a specific pose. It gives information about the manufacturing engineer's or the ergonomist's intention regarding movement or position for the operator. This pose can be critical from an ergonomic viewpoint, and it can therefore act as a tool for preventing operators from developing wrong methods in the assembly work.

Figure 42.1 gives some examples of the difference in understanding how an operation shall be performed by using and not using a digital human as an illustration to the operation.

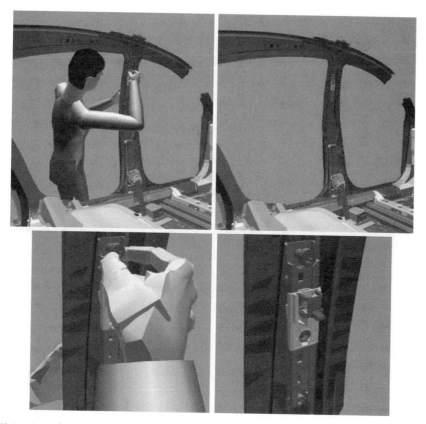

FIGURE 42.1 Assembly of screw for seat belt with and without a digital human model (macro and micro). (Courtesy of SAAB Automobile AB.)

42.4 Assessment Tools for Ergonomics

There has long been a need for practical, systematic, and simple methods for the identification and assessment of ergonomically hazardous jobs or situations. Employers, not the least of which, need aids of this kind for their continuous investigation of the risks entailed in the work activities, as stipulated in the rules on internal control of the working environment initiated by the companies (Munck-Ulfsfält et al., 2003) as well as by the authorities (Arbetarskyddsstyrelsen, 1998). By using such models, it should be possible to obtain an initial indication as to whether or not a certain job or operation entails physical loads dangerous to health, and in this way an initial documentary basis for remedial action.

42.4.1 Company Specific Assessment Tools

In addition to the manikin, the DHM tools include specific tools to assess working postures and physical workloads; however, these ergonomic methods do not seem to be frequently used. The DHM tools are mainly used for visualizations (Sundin & Sjöberg, 2004; Lockett et al., 2005), or are generally based around exploring the consequences of geometry (Neumann & Kazmierczak, 2005). Industry today is focused primarily on the evaluation of static postures, driven in part by the lack of human performance analysis tools available for dynamic motions, but also because the time investment to generate a simulation is relatively long (Raschke et al., 2005). Chaffin (2001) claims that the DHM tools are most often used to evaluate human "fit, clearance, and line of sight issues." According to Laitila (2005), most analyses done with DHM tools are related to reach ability and space demands, while less often workload analyses are done. Goutal (2000) claims that the performance of the assessment tools does not meet the requirements from the users and it seems that the weakness of the tools, although computerized, leads inevitably to the transfer of the assessments to physical mockups. Companies using DHM tools tend to rely on their uniquely designed assessment tools rather than using assessment tools already integrated in the DHM tools (Lämkull, 2005; Bäckstrand et al., 2006). The visualizations are frequently used to provide information about body posture, reach ability, field of view, and clearances for arms, hands, and tools, and serve as a basis for ergonomics evaluation and decision making.

42.4.2 Volvo Cars Web-Based Ergo Report Tool

Volvo Cars has developed a web-based ergo report tool. This tool gives all joint configurations of the manikin in the DHM tool according to the nomenclature used within the ergonomic community (flexion, extension, adduction, abduction, rotation, supination, pronation, etc.). The manikins' joint values are compared to Volvo Cars' standard (Munck-Ulfsfält et al., 1999) which is partly based on Petrén's (1968) maximum joint values, and gives a classification of the values by using a color chart (red = not allowed, yellow = semi-allowed, green = allowed). The report tool's limit values can be edited in case new limit values have to be inserted in the standard, for instance due to new laws or recommendations from the authorities. The report tool gives the joint values of neck, back, shoulders, elbows, hands, hips, knees, and feet (fig. 42.2). Additionally, the report tool also indicates whether the maximum recommended reach distance and hand position area/volume are violated or not (fig. 42.3).

42.5 A Simulation Case from Order to Final Presentation

The web-based order and report form displays a formalized process divided into three major sections: order, report and search.

Joint limits (deg)						
	Actual value	**Explanation**	**Allowed**	**Semi-allowed**	**Not allowed**	**Physical maximum**
Neck						
Flexion	6		0-15	>15-30	>30	50
Extension	0		0-5	>5-10	>10	50
Lateral flexion	4		0-15	>15-30	>30	60
Rotation	5		0-15	>15-30	>30	90
Back						
Flexion	72		0-20	>20-60	>60	100
Extension	0		0-5	>5-10	>10	40
Lateral flexion	10		0-10	>10-20	>20	45
Rotation	8		0-15	>15-30	>30	50

FIGURE 42.2 A part of Volvo Cars web-based ergo report tool (neck and back). The joint values of the manikin are displayed in the actual value-column. The column-fields are colored according to the joint value limits. In this example the flexion value for the back is exceeding the not allowed limit (>30 degrees), and is therefore colored red. (Courtesy of Volvo Car Corporation.)

42.5.1 Order Section

During the order stage, the simulation engineer collaborates closely with a customer. The main customers are manufacturing engineers without any tool experience. Figure 42.4 shows the order section of the web-based form. The order section identifies the simulation case and the customer, as well as specifying which car project the simulation case is related to. In the order section the following *required* information is given:

Title: An informative name of the simulation case; in this case Rod antenna assembly

Mail Address: A unique email address that identifies the customer

Life Cycle: In which life cycle is the product, two choices are available: running production or development

Area: In which part of the plant is the operation carried out that will be simulated: body in white, paint or final assembly

Build Phase: Which virtual build event is the simulation related to

Project: Which car project is the simulation case related to

Domain: Which engineering domain does the case belong to: chassi, exterior, interior or electric

Work Description: Ideally, a full task analysis is given that lists all sub-tasks involved from entering the work area and bringing necessary supplies to exiting. Details like what body positions are likely, and how the body could be supported should be considered. Personal communication is superior to written instructions, thus a lot of the work description is transmitted via

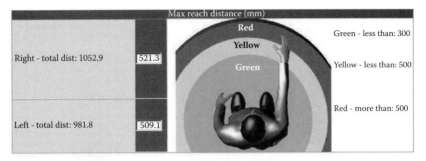

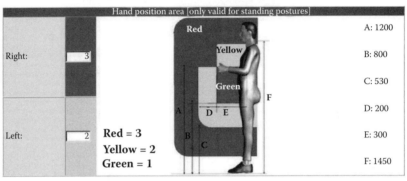

FIGURE 42.3 A part of Volvo Cars web-based ergo report tool (max. reach distance and hand position area/volume). In this example both hands are exceeding maximum recommended reach distance and the right hand is even violating recommended height distance. The ergo report tool should be used with care—a correct ergonomics assessment must be combined with several other characteristics such as weight of handled objects, frequencies, and duration. The manikin in the report tool is not occupying the same posture as the manikin in the reported case; it is just placed in the interface for illustrative purpose. (Courtesy of Volvo Car Corporation.)

personal communication directly between customers and simulation engineers. Nonetheless, a thoroughly written instruction is necessary to improve, or even enable, the documentation of the case and also makes it easier for the simulation engineers, or the customers, to change tasks between each other. It is also in this part of the order section the customer formulates the most important questions to be answered. It is possible for the customer to attach both a component list (containing the CAD components included in the case) and a so-called PII (Process and Inspection Instruction—the work description as it is presented to the assembly operators in the plant).

42.5.2 Report Section

In the report section the simulation results are presented. Figure 42.5 shows a screen dump of the report section. Common output formats are graphs, pictures, and text documents. The most common format is a PowerPoint document or a PDF with pictures showing work postures, field of views, hand and tool accessibilities, and comments about general findings and problems, if any (see fig. 42.6). The simulation engineer can easily attach all kinds of documents in the report section.

The interested parties discuss the results from the simulation case and if a further analysis is necessary with new conditions this is described in the solution/action/rework field. In this simulation case it was decided to change the antenna type from a rod antenna to a shark fin antenna. This was due to several reasons—reach and visual problems for the operators and risk of damaging the painted car body

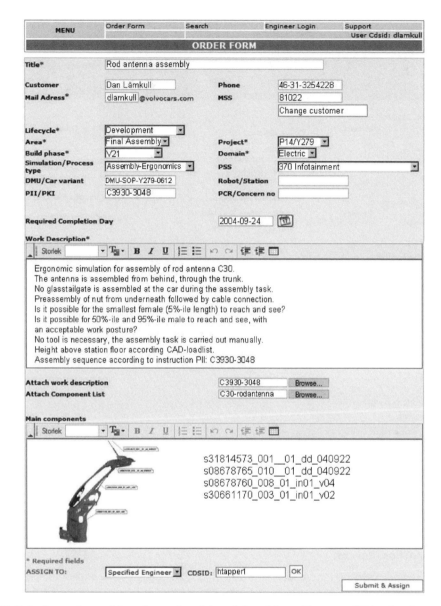

FIGURE 42.4 Screen dump of the order section of the web-based form. (Courtesy of Volvo Car Corporation.)

due to scratch marks from the connecting wires of the rod antenna (see fig. 42.7). The shark fin antenna enables a robotized assembly of the antenna, which avoids the reach and visual problems as well as the scratch marks problem.

42.5.3 Search Section

The simulation results are stored in a database and can be searched by using keywords such as part of the title, car project, area, build phase, customer name, and so on (see fig. 42.8). The search results are listed according to identification number, title, customer, build phase, completion day, status, result, area and assigned engineer.

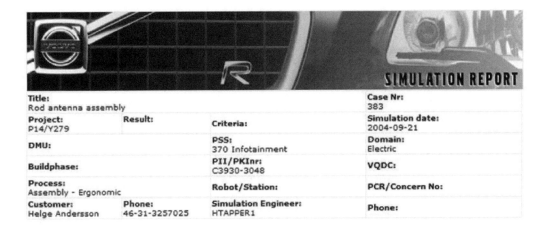

Title:			Case Nr:
Rod antenna assembly			383
Project:	**Result:**	**Criteria:**	**Simulation date:**
P14/Y279			2004-09-21
DMU:		**PSS:** 370 Infotainment	**Domain:** Electric
Buildphase:		**PII/PKInr:** C3930-3048	**VQDC:**
Process: Assembly - Ergonomic		**Robot/Station:**	**PCR/Concern No:**
Customer: Helge Andersson	**Phone:** 46-31-3257025	**Simulation Engineer:** HTAPPER1	**Phone:**

Work Description

Updated simulation of rod antenna assembly in new position.

Priority order:
- 5% female, assembly from top, preassembly of nut from underneath, cable connection. All done from the rear of the car only.
- 50% male, assembly from top, preassembly of nut from underneath, cable connection. All done from the rear of the car only.
- 95% male, assembly from top, preassembly of nut from underneath, cable connection. All done from the rear of the car only.

Attached work description files

No files included

Simulation result

See attachment.

Attached Simulation files

922_rod_antenna_w05_050112_1257.ppt

Solution/Action/Rework

Main component(s)

Attached Component files

C3D-rodantenna

Web link

http://www.manufacturing.volvocars.ford.com/manufacturing_engineering_81000/simreport/ShowPrint.asp?id=383

Case Opend From this Case

Check here

Reopend From:

FIGURE 42.5   Screen dump of the report section. (Courtesy of Volvo Car Corporation.)

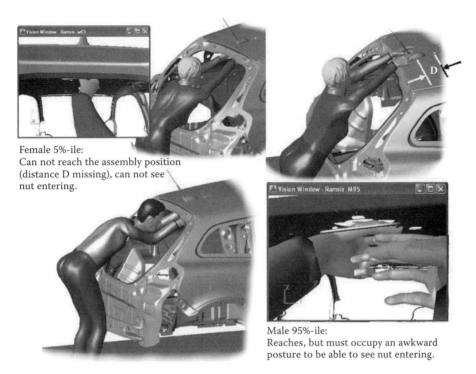

Female 5%-ile:
Can not reach the assembly position
(distance D missing), can not see
nut entering.

Male 95%-ile:
Reaches, but must occupy an awkward
posture to be able to see nut entering.

FIGURE 42.6 In the report section the simulation results are presented. The most common formats of result presentations are PowerPoint documents or PDFs with pictures showing work postures, field of views, hand and tool accessibilities, and comments about general findings and problems. (Courtesy of Volvo Car Corporation.)

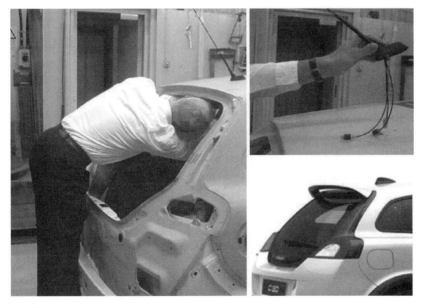

FIGURE 42.7 The rod antenna simulation case and its results led to a change of antenna type from a rod antenna to a shark fin antenna. This was due to several reasons—reach and/or visual problems for the operators and risk of damaging the painted car body due to scratch marks from the connecting wires of the rod antenna. The shark fin antenna enables a robotized assembly of the antenna, which avoids the reach and visual problems as well as the scratch marks problem. (Courtesy of Volvo Car Corporation.)

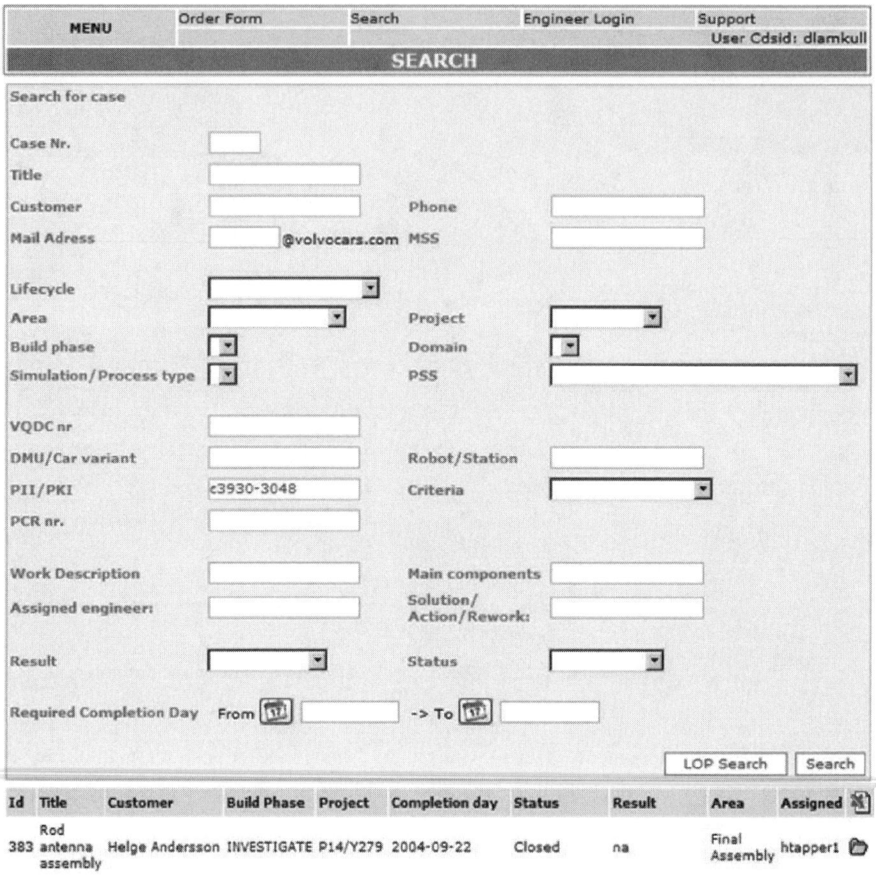

FIGURE 42.8 Screen dump of the search section. (Courtesy of Volvo Car Corporation.)

References

Abrahamsson, L. (2000). Production economics analysis of investment initiated to improve working environment. *Applied Ergonomics* 31, pp. 1–7.

AFL-CIO (1997). Stop the pain. The American Federation of Labor and Congress of Industrial Organizations (AFL-CIO). Washington, DC.

Arbetarskyddsstyrelsen (Swedish National Board of Occupational Safety and Health). (1998). AFS 1998:1. *Arbetarskyddsstyrelsens föreskrifter om belastningsergonomi* (*Ergonomics for the Prevention of Musculoskeletal Disorders*).

Arbetsmiljöverket. (2005). *Musculoskeletal Ergonomics Statistics*. Published in June 2006.

Axelsson, J. (1994). Ergonomic aspects on design and quality. In: *Proceedings of the 12th Triennial Congress of the International Ergonomics Association*, Toronto, Canada, 4, pp. 18–21.

Axelsson, J. (1995). The use of some ergonomic methods as tools in quality improvement. In: *Proceedings of the 13th International Conference on Production Research*, Tel Aviv, Israel, Freund Publishing House, pp. 721–723.

Bäckstrand, G., Högberg, D., De Vin, L.J., Case, K., and Piamonte, P. (2006). Ergonomics Analysis in a Virtual Environment. In: *Proceedings of the International Manufacturing Conference*, IMC 23, University of Ulster, Jordanstown, Ireland, pp. 165–172.

Badler, N. (1993). *Computer Graphics Animation and Control: Simulating Humans.* New York, Oxford University Press.

Barkan, P. (1992). Productivity in the process of product development—an engineering perspective. In: *Integrating Design for Manufacturing for Competitive Advantage.* Susman, G.I. (Ed.). New York, Oxford University Press, pp. 56–68.

Bhatti, M.A., Han, R.P.S., and Vignes, R. (2005). Muscle forces and fatigue in a digital human environment. In: *Proceedings of the 2005 SAE Digital Human Modelling for Design and Engineering Symposium,* Iowa City, IA. SAE 2005-01-2712.

Blomé, M., Dukic, T., Hanson, L., and Högberg, D. (2003). Simulation of human-car interaction at Saab automobile: Present and future. In: *Proceedings of the 2003 SAE Digital Human Modelling for Design and Engineering Conference and Exposition,* Montreal, Canada, SAE 2003-01-2192.

BLS. (2003). Number of nonfatal occupational injuries and illnesses with days away from work involving musculoskeletal disorders by selected worker and case characteristics, 2001. U.S. Department of Labor (Published in March 2003).

Boothroyd, G., and Dewhurst, P. (1988). Product design for manufacture and assembly. *Manufacturing Engineering* 4, pp. 42–46.

Chaffin, D.B. (1969). A computerized biomechanical model: development and use in studying gross body actions. *Journal of Biomechanics* 2, pp. 429–441.

Chaffin, D.B. (2001). *Digital Human Modelling for Vehicle and Workplace Design.* Warrendale, PA, Society of Automotive Engineers.

Ciavarro, G.L., Tramonte, A., Fusca, M., Santambrogio, G.C., and Andreoni, G. (2004). Evaluation of 3D kinematic model of the spine for ergonomic analysis. In: *Proceedings of the 2004 SAE Digital Human Modelling for Design and Engineering Symposium,* Rochester, MI. SAE 2004-01-2198.

Cohen, M.A, Eliashberg, J., and Ho, T.-H. (1996). New product development: The performance and time-to-market tradeoff. *Management Science* 42, pp. 173–186.

Cushman, W.H., and Rosenberg, D.J. (1991). A technique for assessing postural discomfort. *Ergonomics* 19, pp. 175–182.

Dar-El, E.M., Ayas, K., and Gilad, I. (1995). A dual-phase mode for the individual learning process in industrial tasks. *IIE Transactions* 27, pp. 265–271.

Dickerson, C.R, Rider, K.A., and Chaffin, D.B. (2004) Merging biomechanical models of the shoulder with digital human modelling. In: *Proceedings of the 2004 SAE Digital Human Modelling for Design and Engineering Symposium,* Rochester, MI. SAE 2004-01-2166.

Dukic, T., Rönnäng, M., Örtengren, R., Christmansson, M., and Davidsson, A. (2002). Virtual evaluation of human factors for assembly line work: A case study in an Automotive Industry. In: *Proceedings of the Digital Human Modelling Conference,* Munich, Germany. SEA 2002-07-0033.

Eklund, J. (1995). Relationships between ergonomics and quality in assembly work. *Applied Ergonomics* 26, pp. 15–20.

Eklund, J. (1996). Den svåra marknadsföringen av användarkvalitet. In: *Marknad för Ergonomi?* S. Franzén and J. Hedman (Eds.). Rådet för Arbetslivsforskning, Stockholm, Sweden.

Eklund, J. (1997). Ergonomics, quality and continuous improvement—Conceptual and empirical relationships in an industrial context. *Ergonomics* 40, pp. 982–1001.

European Agency for Safety and Health at Work. (1998). Economic impact of occupational safety and health in the Member States of the European Union. Agency report 1998.

European Agency for Safety and Health at Work. (2000). Work-related neck and upper limb musculoskeletal disorders. Agency report and factsheet WRULD: summary of Agency report, 2000.

Farin, G. (1990). *Curves and Surfaces for Computer Aided Geometric Design—A Practical Guide.* San Diego, CA, Academic Press.

Goutal, L. (2000). Ergonomics assessment for aircraft cockpit using the virtual mock-up. In: *Ergonomics Software Tools in Product and Workplace Design: A Review of Recent Developments in Human Modelling and Other Design Aids.* Landau, K. (Ed.). Stuttgart, Germany. Verlag ERGON GmbH.

Green, R. (2000). A generic process for human model analysis. In: *Proceedings of the Digital Human Modeling Conference*, Munich, Germany. SAE 2000-01-2167.

Groover, M.P. (1984). *CAD/CAM Computer Aided Design and Manufacturing*. Englewood Cliffs, NJ, Prentice-Hall.

Hallberg, Å. (1995). Samband mellan arbetsmiljö och kvalitet—en fallstudie. Avdelningen för industriell arbetsvetenskap. Linköping Institute of Technology at Linköping University. Report No. LiTH-IKP-R-847 (in Swedish).

Helander, M., and Nagamachi, M. (1992). *Design for Manufacturability*. London, Taylor & Francis.

Helander, M.G., and Willén, B.Å. (1996). Design for human assembly. In: *Handbook of Industrial Ergonomics*. Karwowski, W. and Marras, W. (Eds.). Boca Raton, FL, CRC Press.

van Hoof, J., van Markwijk, R., Verver, M., Furtado, R., and Pewinski, W. (2004). Numerical prediction of seating position in car seats. In: *Proceedings of the 2004 SAE Digital Human Modelling for Design and Engineering Symposium*, Rochester, MI. SAE 2004-01-2168.

Jordan, P.W. (1996). Displeasure and how to avoid it. In: *Contemporary Ergonomics*. Robertson, S.A. (Ed.). London, Taylor & Francis, pp. 56–61.

Koufteros, X., Vonderembse, M., and Doll, W. (2001). Concurrent engineering and its consequences. *Journal of Operations Management* 19, pp. 97–115.

Laitila, L. (2005). Datormannequinprogram som verktyg vid arbetsplatsutformning—en kritisk studie av programanvändning. Luleå tekniska Universitet. Institutionen för Arbetsvetenskap, Avdelningen för Industriell Produktionsmiljö. Licentiate thesis (in Swedish; abstract in English).

Lämkull, D. (2005). The daily use of mannequins within the manufacturing department at Volvo Car Corporation—working methodology, developments and wanted improvements. In: *Proceedings of the 37th Annual Conference of Nordic Ergonomics Society*, Oslo, Norway, pp. 86–90.

Landau, K. (2000). *Ergonomics Software Tools in Product and Workplace Design: A Review of Recent Developments in Human Modeling and Other Design Aids*. Landau, K. (Ed.). Stuttgart, Germany, ERGON GmbH.

Langowitz, N. (1988). An exploration of production problems in the initial commercial manufacture of products. Research Policy 17, North Holland, The Netherlands, Elsevier Science Publishers B.V, pp. 43–54.

Lockett, J.F., Assmann, E., Green, R., Reed, M.P., Rascke, R., and Verriest, J.-P. (2005). Digital human modeling research and development user needs panel. In: *Proceedings of the 2005 SAE Digital Human Modeling for Design and Engineering Symposium*, Iowa City, IA. SAE 2005-01-2745.

Millson, M.R., Raj, S.P., and Wilemon, D.A. (1992). A survey of major approaches for accelerating new product development. *Journal of Product Innovation Management* 9, pp. 53–69.

Moestam Ahlström, L. (2002). Prerequisites for Development of Products Designed for Efficient Assembly. Chair of Production Systems, Department of Production Engineering, Royal Institute of Technology, Stockholm, Sweden. Doctoral thesis.

Munck-Ulfsfält, U., Falck, A., and Forsberg, A. (1999). Requirement Specification for Load Ergonomics. Volvo Car Corporation, Göteborg, Sweden. Corporate document.

Munck-Ulfsfält, U., Falck, A., Forsberg, A., Dahlin, C., and Eriksson, A. (2003). Corporate ergonomics programme at Volvo Car Corporation. *Applied Ergonomics* 34, pp. 17–22.

Myklebust, A. (2001). Computer-aided engineering. In: AccessScience-McGraw-Hill, DOI 10.1036/1097-8542.153860.

Neumann, W.P., and Kazmierczak, K. (2005). Integrating Flow and Human Simulation to Predict Workload in Production Systems. In: *Proceedings of Nordic Ergonomics Society 37th Annual Conference*, Oslo, Norway, pp. 308–312.

NIOSH. (1996). National occupational research agenda. Department of Health and Human Services, Public Health Service, Centers for Disease Control and Prevention, National Institute for Occupational Safety and Health, DHHS (NIOSH). Cincinnati, OH. Publication No. 96-115.

Paoli, P. (1999). Data from the European survey on working conditions. European Foundation for the Improvement in Living and Working Conditions, unpublished article.

Petrén, T. (1968). *Lärobok i Anatomi: Rörelseapparaten* (Anatomy textbook). Nordiska bokhandeln, Stockholm, Sweden.

Preston White, K., and Richards, L.G. (2002). Computer-aided design and manufacturing. In: Access-Science-McGraw-Hill, DOI 10.1036/1097-8542.153810.

Raschke, U., Kuhlmann, H., and Hollick, M. (2005). On the design of a task based human simulation system. In: *Proceedings of the 2005 SAE Digital Human Modeling for Design and Engineering Symposium*, Iowa City, IA. SAE 2005-01-2702.

Rauglas, D. (1998). Concurrent industrial engineering: The marriage of predetermined time systems, ergonomics and work system design. Industrial Engineers IE Solutions. Conference Proceedings.

Ruiter, I.A. (2001). Development of a checklist for the use of anthropometric Man Models. In: *Proceedings of the Digital Human Modeling for Design and Engineering Conference and Exposition*, Arlington, VA. SEA 2001-01-2116.

Sandström, R., and Svensson, I. (1996). Konsekvenser av monteringsovänliga artiklar SAAB 900. In: *Ergonomi, Produktivitet, Kvalitet—Sex Praktikfall*. J. Eklund (Ed.). Linköping, Sweden, UniTryck, pp. 29–38.

Seidl, A. (1997) RAMSIS: A New CAD-Tool for Ergonomic Analysis of Vehicles Developed for the German Automotive Industry. Vol. 1233, Automotive Concurrent/Simultaneous Engineering, SAE, Special Publications, pp. 51–57.

Seitz, T., Recluta, D., Zimmermann, D., and Wirsching, H.-J. (2005). FOCOPP—An approach for a human posture prediction model using internal/external forces and discomfort. In: *Proceedings of the 2005 SAE Digital Human Modeling for Design and Engineering Symposium*, Iowa City, IA. SAE 2005-01-2694.

Sundin, A., and Sjöberg, H. (2004). Datormannequiner och ergonomi i produkt- och produktionsutveckling. Hur används mannequiner i Sverige och vilken nytta kan de göra? Arbetslivsrapport nummer: 2004:19. Göteborg, Sverige. (in Swedish; abstract in English).

Toomingas, A., Cohen, P., Jonsson, C., Kennedy, J., Mases, T., Norman, K., and Odefalk, A. (2005). A Sound Working Environment in Call and Contact Centres - Advice and Guidelines. Stockholm, Sweden.

Wang, Q., Xiang, Y.J., Kim, H.J., Arora, J.S., and Abdel-Malek, K. (2005). Alternative formulations for optimization-based digital human motion prediction. In: *Proceedings of the 2005 SAE Digital Human Modeling for Design and Engineering Symposium*, Iowa City, IA. SAE 2005-01-2691.

Wheelwright, S.C., and Clark, K.B. (1992). *Revolutionizing Product Development*. New York, Free Press.

Willkrans, R., and Norrblom, H. L. (1995). Design for Easy Assembly—Facts and Checklist. IVF, Gothenburg, Sweden.

Ziolek, S.A., and Kruithof Jr, P.C. (2000). Human modelling and simulation: A primer for practitioners. In: *Proceedings of the International Ergonomics Association Conference*, San Diego, USA.

Ziolek, S.A., and Nebel, K. (2003). Human modelling: controlling misuse and misinterpretation. In: *Proceedings of the 2003 SAE Digital Human Modeling for Design and Engineering Conference and Exposition*, Montreal, Canada. SAE 2003-01-2178.

43

Advanced Measurement Methods in Mining

Dean H. Ambrose

43.1 Introduction

Mining and processing earth's materials form the basic building blocks from which many technological advancements and products are made. Virtually all metallic and non-metallic products are derived from a substance found in our earth that we blend, mold, extrude, or pulverize into something useful. The mining industry is responsible for the first step of this process: extraction from the ground. Mining, particularly coal mining, poses significant risks to both humans and machines. As raw materials are forcibly broken or blasted apart, then roughly yet effectively excavated by large transport vehicles or by vertical and incline hoists, human operators must battle noise, dust, mud, and darkness while controlling and maintaining their machinery and keeping themselves and their crews safe from harm. Such a hazardous work environment poses extraordinary challenges to mine operators who must vigilantly recognize and control the perils of mining.

While mining is essential to the progress and productivity of our society, the early years of mining was characterized by miners' poor health, common accidents, and environmental impacts. Poor safety and health practices were costly in both personal injuries and property damage. Thanks to safety and environmental regulations and diligence from the mining community, mining safety records have significantly improved over the last 50 years, and progress in this area continues.

As digital human modeling (DHM) been more prevalently used by human factors and ergonomics professionals to analyze workplace hazards and improve workplace design (Brown, 1999; Badler et al., 2002; Chaffin, 2002; Määttä, 2003; Ferguson & Marras, 2004; Colombo & Cugini, 2005; Zhang & Chaffin, 2006), the National Institute for Occupational Safety and Health (NIOSH) Pittsburgh Research Laboratory (PRL) also saw the benefits to using DHM in their computer-simulated mining environments. This chapter features how researchers use DHM to assess and ultimately decrease the occupational risks and threats faced by underground coal mine machine operators.

43.2 Understanding the Mine Environment

The mining environment can be likened to a hostile workplace. Miners work with excessive noise levels in poorly illuminated areas, and may be exposed to toxic gases and dusts, excessive heat, humidity, and vibration. The machines are powerful, dangerous, and mobile with swinging, moving appendages and spinning, jagged, grinding cylinders (fig. 43.1). Miners confront roof falls or other wayward debris from the face or walls of the mine, inherent dangers of operating electrical equipment, malfunctioning machines, ignition and explosion hazards from gases and dust, mine fires, and sudden inundations of water or gas.

Mining is inherently uncomfortable. Prolonged stress from the mining environment often exceeds human tolerance levels. Stress factors often overlap and contribute to decreased alertness or performance and reduced productivity. Noise levels coupled with redundant machine tasks, for example, could induce fatigue or complacency and increase the chance of an injury or fatality.

FIGURE 43.1 Cutting cylinders on a continuous miner machine. (Photograph courtesy of Joy Mfg., Franklin, PA.)

Ongoing efforts strive for miners' protection, resulting in healthier miners and safer mines. The approaches taken by the mining industry have validity in any industry concerned with the health and safety of its workers. These are: engineering control, machine safety, industrial engineering, training, regulations, and unions.

Engineering control helps reduce exposure to potential hazards either by isolating the hazard or by removing it from the work environment. Here it includes mine design and equipment and workforce selection. *Machine safety* is a prominent concern inherent to the hazards of working in a mechanized environment, and includes automation and remote control technology. *Industrial engineering* refers to human factors and ergonomics, applying information from human characteristics, abilities, expectations, and common behaviors to the design of machines, procedures, and even the environments in which the miners operate. Adequate *training*, from initial job orientations to refresher training for mining veterans, must take into account new technological changes as the industry advances. Federal, state, and local *regulations* play an important role in controlling mine hazards, developing mandated training for the entire mining industry and a certification system for instructors. Mine *unions* have taken a major role to ensure that miners' safety and health remain as important as, if not more than, the daily production in mines.

43.3 "Measuring" in the Mine Environment

Accident statistics provided by the Mine Safety and Health Administration (MSHA) are extremely helpful in establishing trends of injuries and fatalities relating to specific mining machinery, job titles, accident and mine type, and so on. However, Smith (2006) finds that accident reports from MSHA are primarily focused on violations of mine laws, and lack helpful detail for researchers who are attempting to discern the root cause of injuries and fatalities. Furthermore, Ambrose (2003) states that MSHA accident investigation report narratives contain minimal information to facilitate studying interactions between machines and their operators.

Traditional mine research data are collected via field testing at an actual mine, or in a laboratory. It takes particular commitment for an active mine to work a research team's field tests into their production schedule. The research team must be qualified to operate within an active mining environment, and undergo mandatory mine training. They must also have personal safety and protection gear. In addition to the hazards and dangers identified earlier, the research team is still faced with variable and uncontrolled test conditions. They may have limited access to mine locations and personnel. Also, they may have no control over where or which machines or instruments are being used during their "time slot" for testing. In a laboratory test, researchers develop scaled or full-scale mockups of mine environments, equipment, and machinery. Laboratory tests (fig. 43.2) are useful for creating optimum test

FIGURE 43.2 Test subjects remove wire mesh from actual equipment under replicated mine seam height conditions.

conditions without time constraints; though they lack much of the realism of actual mine conditions, and experienced mining operators may not always be available for the tests. Both methods incur significant time and expense.

When DHM was first identified as a possible method of conducting mine research in the mid-1990s, an initial literature review by PRL found no previous DHM applications in the mining industry. An expanded search into the automobile, agricultural, factory floor, and sports industries provided much information. Early research endeavors (Ambrose, 1996, 1999, 2000a and b) reflect the learning curve in what is now a 10-year span of research incorporating DHM in safety and health for mining.

Aside from the problems associated with field tests in actual mines and laboratory tests, it should be noted that DHM has its own limitations. Researchers must be concerned with their organization's or industry's acceptance of DHM predictions, and the databases from which outcomes are derived. Decision makers within any industry must be able to justify their investments in new technology, and the consequences of such investments in terms of potential research results and their approval by stakeholders and customers. Embracing DHM also means researchers must be able to effectively communicate the results of their work for proactive uses, and show some measure of their research effectiveness. DHM is in competition, so to speak, with more tangible research results such as redesigned equipment, patents, or a quick turnaround response to a specific injurious or fatal incident. Nevertheless, DHM is becoming a recognized and proven research tool with potential to design products and systems, to study and apply human factors and ergonomic principles, and to serve as an effective vehicle for advocating safety and health research, as evidenced below.

43.4 Capturing Human Motion

PRL uses motion-capture technology in their simulation research methods. Motion capture technology (fig. 43.3) is a technique that uses the human body, or other object, as an input device. In most applications, sensors or optical markers are placed on the subject's body and the motion capture system monitors and records the subject's motions. If done properly, motion capture provides high-quality data more quickly and efficiently than traditional frame-by-frame animation techniques. Because the technology captures movement data directly from an actual subject, the data are much more accurate.

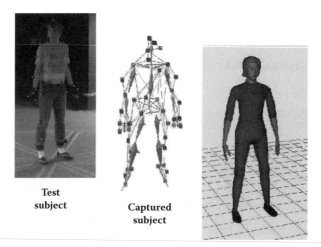

Test subject   **Captured subject**   **DHM environment**

FIGURE 43.3 Motion data are captured through sensors on the test subject and transferred to a DHM environment.

Motion, particularly human motion, is very subjective. There are subtleties in human motion that a motion tracking system will detect, but a human eye will not, using traditional key frame animation techniques. Motion capture systems provide an accurate, convenient, and quantitative assessment of functions by providing comparative or absolute motion measurements.

Specialized DHM simulation and analysis software sometimes is packaged with a motion capturing software module, providing an interface to a motion capture system. The result is a rather compelling research tool. Each sensor or reflective marker on a subject transfers data through a motion capture system into a DHM software environment. Both the position and orientation of each sensor are tracked and mapped to a virtual human with a predefined figure, enabling the virtual human to mimic the motions of the test subject, which can be saved for future analysis. If the connection between the motion capture system and the DHM software supports it, computer-generated objects, such as a virtual human, move in real time with the real-life subject. Therefore, investigators can observe real-time results on the virtual display, providing instant information that can significantly augment the experiment. Such DHM simulation has been used at PRL to conduct a detailed examination of lower back movement and muscle stress, virtual human joint information, muscle stress from hand loading mining material, and body part strength. This chapter explores these investigations in further detail in later sections.

PRL uses two primary types of whole-body motion capture systems, *electromagnetic* and *optical*. Both types are quickly becoming standards in the movie production, automobile, and military R&D communities for creating and studying animated motions of characters (and on the Hollywood side, for creating special effects). There are pros and cons to both types.

PRL's electromagnetic motion capture systems use tether sensors attached to the test subject, which then move within an electromagnetic field created by a transmitter. These sensors pick up changes in the electromagnetic field and the corresponding positions are fed to a controller connected to a computer workstation that processes the motions in real time. The electromagnetic system has no possibility of blocked or occluded markers, thereby allowing real-time motion processing and instantaneous playback of captured data.

Setting up an electromagnetic motion capture system isn't very complicated, but it is particularly sensitive to metal, including steel support columns or beams, steel-reinforced concrete floors, overhead light housings, metallic studs in walls, filing cabinets, and so on. In addition, electromagnetic systems work best in smaller performance areas using slower data rates, which limits capturing fast motions and motions in a broad space. Cable harnesses inherent with PRL's electromagnetic systems can pose a problem for more athletic test motions, and can unintentionally snag on stationary objects within the test field. Compared to an optical system, electromagnetic systems typically require less setup time and are more cost effective.

PRL's optical motion capture system uses reflective markers attached to a test subject, who is then digitally filmed with special high-resolution infrared cameras and an infrared light source. The digitized information feeds into a computer workstation that controls the optical tracking system and records the motions. The markers provide two-dimensional points for each camera, which the motion capture software translates to three-dimensional coordinates. Significant processing power is required to resolve two-dimensional camera data to three-dimensional motions data, possibly causing instant playback to be riddled with occluded markers. If test subjects get too close or markers overlap, it could confound the software. To combat these issues, a newer feature of motion capture software technology is biomechanical intelligence, which "knows" the human body and can compensate automatically when occlusion occurs. This allows real-time motion processing of captured data and instantaneous playback through DHM software.

Unlike their electromagnetic counterpart, optical systems allow unencumbered motion for a large number of markers in a fairly large space, and are unaffected by metals. For example, it's possible to track multiple subjects, each with 50 sensors attached, in a 900 square foot area, noting that area and accuracy are directly related to the number of, and positioning of, cameras and infrared light sources. Of course, setup time is typically more time consuming for optical systems, allowing for camera

FIGURE 43.4   Actual dual boom arm roof-bolting machine (Photograph courtesy of J. H. Fletcher & Co., Huntington, WV.)

positioning, calibrating all equipment, and masking unwanted infrared sources in the test area. Once setup is complete, the high-speed, high-resolution infrared cameras used in optical systems can accommodate high data rates such as 2,000 frames/second, allowing for extremely accurate capture of even high-speed motion. This has particular relevance to obtaining accurate samples of sudden motions, such as equipment operators avoiding moving machines or machine appendages. Motion capture is an effective validation and verification tool for digital human models. Woolley et al. (1999) suggest that human motion studies using dynamic biomechanical analyses or human motion simulation models should establish an empirical motion database. Efforts are underway. An Ambrose et al. (2005a) study used motions from movements of roof bolter appendages (fig. 43.4) and 12 human subjects with mine machinery experience to verify that the model's simulation predictions represent an accurate picture of the machine model and operator during roof bolting tasks. Valid random motions reported by Ambrose (2000b, 2001), and Volberg and Ambrose (2002) used the same 12 human subjects to study aspects of operator movements, the range of motion of operators, and variation in those movements. The same valid database of captured human motions contributed to the Ambrose et al. (2004) investigation of low back stress experienced by machine operators. The captured anthropometry of each of the 12 test subjects validated the database used by Ambrose and Cole (2005) to evaluate control interventions that reduced the severity of muscle recruitment and spine loads resulting from roof bolting in different work postures and seam heights.

   New human subject testing (Bartels et al., 2008) used 10 additional subjects to supply movements from experienced miners to help validate motions representing the sudden movements necessary to avoid

FIGURE 43.5   Test subject operating a roof-bolter boom arm assembly mockup.

moving mine machinery that can influence the industry standard for a specific machine's ground speed. This growing database will be integrated with DHM motion capture data and machine models to accurately simulate the operator working around a specific machine, and predict probable impact incidents through collision detection. Another new human subject test (Kwitowski & DuCarme, 2007) furnished movements from 12 experienced miners that will help validate motions that correctly mimic a machine operator controlling the horizontal motion of an appendage from a roof bolter machine (fig. 43.5). Further use of this database with DHM will examine low back stress experienced by these machine operators during this task.

43.5 DHM Technology Applications in Mining

This chapter has already established the unique and severe environmental conditions in which miners perform. Environmental stress and restrictions, particularly the restricted vertical workspace in many underground coal mines, make mining one of the most difficult industrial environments in which to make safety and health improvements. Studying the interaction between people and their environment, regardless of the industry, is essential to determining causal factors behind fatalities and injuries, and in developing controls to help prevent them. This section suggests methods for using DHM to understand and solve human-machine interactions.

Investigators must be discerning with DHM data. They must refrain from generating more data than needed, or read more from the database than what the model and simulations were designed to deliver. The model is only as good as the system it defines; certain parameters must be validated using real subjects and undergo further enhancements to streamline its efficiency. One must remember that DHM results are only predictions of outputs from events or conditions that *reflect* the real world through a virtual world. With real-world logistics—such as travel to mining sites and costs associated with experiments—no longer a factor, it's very easy to become overwhelmed with data because it's now possible to generate and track so *much* of it. Good research and intelligent conclusions that will benefit industry are served by planning simulations well, and judiciously using the right data for the right job.

43.6 Using DHM to Depict Motion and Behavior Variation

Early DHM software effectively portrayed simple movement behaviors and basic motions, but failed to capture the random nature of human motions or to depict path variance within human motions. The key to delivering motion variability in DHM involved the concept of *stochastic modeling*, but how was this to be accomplished in the DHM world? Ambrose (2001) reported a simple technique for representing and analyzing motion variations and hazardous events in a computer-simulated three-dimensional workplace using DHM software. Later, Ambrose (2004) discussed this same technique and detailed the code development for a model that demonstrated random human motion and behaviors. Now it was possible for researchers to study hazardous interactions in a virtual environment, in this case, unintentional contact between mining operators and mining machines.

In an underground coal mine, miners are at risk from being struck by the mining machinery due to the confined workspace. Miners who worked around roof-bolter machines were particularly susceptible to unintentional contact incidents, and researchers used DHM to increase miners' safety while working around this machine. In order to effectively study unintentional human-machine contact, the simulation had to account for random motion and behavior of the operator.

Basic motions and standard deviation data provided random parameters for the operator's movements in the random motion code. Each of the basic motions provided sets of three-dimensional (xyz) values. Results from motion envelope analyses by Bartels et al. (2001) provided standard deviation values. Applying standard deviation values to basic motions helped researchers devise "manipulation values" that used xyz-orientation angles and xyz-positional coordinates to define a set of final postures of

TABLE 43.1   Random Hand Motion Rules for a Roof Bolter Machine

			Operator's Hand Motion Type					
		Basic	Prominent vertical direction	Vertical with z direction	Vertical with x direction	Point to point	Leaning forward and backward	Starting position
Motion direction	orientation angles	x y z	x y z	x y z	x y z	x y z	x (random y) z	x y (random z)
	positional coordinates	x y z	x (random y) z	x (random y) (random z)	(random x) (random y) z	(random x) (random y) (random z)	x y z	(random x) y (random z)

the virtual human. Using positional x as an example, x is equal to multiplying the original x value from the basic motions by the standard deviation value from experiments on motion envelopes than minus a random number from zero to twice the standard deviation value for x. Tables imbedded into the code defined the standard deviation values, also called seed variables.

Guidelines were established (Table 43.1) that used manipulation values to cause random motion, which made it possible for investigators to program code correctly to recognize the motion's direction and how random changes affect basic motions. These guidelines also maintained directional integrity of the virtual human's intended or expected basic motions while providing random elements within those motions.

In contrast to random motions, random *behaviors* were easier to define. A random behavior is simply a series of human motions that mimic a specific action. Investigators can use statistical information about a job function to identify risky worker behavior with a model through a decision algorithm that formulates and determines what behavior to use during the simulation.

Results from this random motion and behavior study using DHM can provide increased awareness to workers and even potentially impact the engineering and design of the machinery. Finding the right software tool with all the features and capabilities to develop and execute customized computer code for virtual motions and behaviors was critical to this effort.

43.7 Using DHM to Determine Subject Response and Reach Envelopes

Due to a restricted workspace, miners are often forced into uncomfortable postures and encounter limited reach capabilities. DHM software has effectively depicted simple response and reaching movements. But the operator postures unique to operating machines in underground coal mines prompted the Bartels et al. (2001) study to measure human motion response times and motion envelopes in a restricted environment. Human motions were captured and recorded using a motion tracking system, as described earlier in this chapter. The following lessons learned from the Bartels et al. (2001) study have applications in any DHM model and simulation human response and motion envelope study.

43.7.1 Test Normal Subject Movements

To prepare for the motion capture, researchers set up a mock mining environment that included a wooden replica of a roof-bolter boom arm assembly (fig. 43.6). Test subjects were asked to position themselves around the roof bolter as they normally would for different job tasks, providing unique starting point data for each operator. These data were useful in analyzing collisions that occur between operators and an appendage from the roof bolter, and in examining a variety of reach envelopes from the test subjects in different work postures.

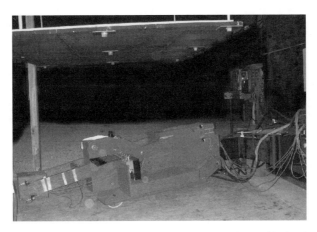

FIGURE 43.6 Mock mining environment includes a wooden replica of a roof-bolter boom arm assembly and adjustable roof to vary the seam height.

43.7.2 Exercise Repetition of Test Trials

Subjects repeated the test trials three times for each independent variable combination under investigation. In this study, variables included differing coal seam heights and work postures such as subjects kneeling on one or both knees, squatting versus stooping, and standing. Repeating the trials made a much more useful statistical analysis of the motions being captured. To help minimize fatigue, subjects were allowed to rest at least two minutes between repetitions.

43.7.3 Capture Quality Data from Test Subjects

Test subjects were instructed to follow standard, specific procedures relating to their job, operating a roof-bolter machine. If test subjects have difficulties following procedures, it's practical to dismiss those subjects or simply keep trying repetitions until they reach the desired number of successful trials, otherwise the data are not valid for the study's assumptions.

43.7.4 Structure Response Times

To capture accurate response times for all subjects, it's wise to create a common starting pose for all response time testing. Researchers staged each test subject relative to a specific point on the subject's body and to another specific point on the machine. Subjects were instructed to complete a roof-bolting task sequence, but when they received a verbal cue, they were to move as if to avoid being struck by a roof-bolter appendage. The timing of the verbal cue was random so the subject wouldn't be able to anticipate the action.

43.8 Virtual Human Vision

Virtual human vision is often overlooked when studying worker safety and health, perhaps because the ability to clearly see people, other machines, moving machine appendages, work area's characteristics, and environmental hazards seems obvious. Yet the inability to see people, objects, and hazards has contributed substantially to mining incidents and fatalities, and must be accurately accounted for by simulations. DHM software accounts for virtual human visual capabilities using a vision cone, which defines an area extending from the virtual human's eyes, or using a separate window that shows what the virtual human can see (fig. 43.7). Advancements in DHM software now allow vision tracking and

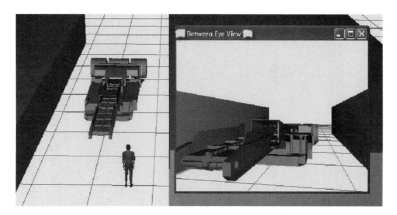

FIGURE 43.7   DHM mine environment with eye view of the machine operator inserted.

recording of objects seen, and not seen, by the virtual human. The following discussion describes how virtual vision was used to advance safety and health concerns using models and simulations in DHM software.

43.8.1 Vision Cone

When examining a database of collisions between a miner and mining machine appendages, the investigator needs to ensure the data accurately reflect the real world. Early investigation by Bartels et al. (2001) used a viewing area defined by an oval directrix model characterized by Humantech (1996). This viewing area looked like a cone extending from between the virtual human's eyes. Bartels et al. (2001, 2003) modified the cone to determine the optimal viewing area for unique lighting conditions found in underground mining environments. Surprisingly, a subject's vision cone was significantly reduced by the bill of a standard hard hat. Figure 43.8 shows the angular data of the viewing area for normal light, modified light (0.06 fL with cap lamp), and simulation (original area of the virtual operator). This study set the standard for virtual human vision within an underground mining environment while operating roof-bolter equipment.

43.8.2 Viewing Area

Investigators were also concerned with a virtual human's view, when considering operator response time to the event when the operator sees a moving object and gets out of the way of it. Using the modified viewing area already discussed, eight reference points on the viewing cone were identified (fig. 43.9). During a simulation, researchers recorded and calculated the distance between these reference points and a reference point on another object, such as a boom arm of a roof bolter. Bartels et al.'s (2001, 2003) experiments validated the premise that when one or more of the eight resulting distances was negative, the virtual operator couldn't see the boom arm. From this study, the DHM industry was better able to address vision tracking and recording of objects as seen by virtual humans.

43.8.3 Line of Sight

The mining industry was challenged to define an adequate field of vision for operators seated in cabs of earth-moving equipment, whose unique specifications simply were beyond conventional line of sight and conical section analysis techniques (Hella et al., 1991, 1996). Eger et al. (2004, 2005) used obscuration zone and coverage plane features in DHM as an effective tool for evaluating line of sight in earth-movers, yielding important enhancements to underground mining machine design. The technique can

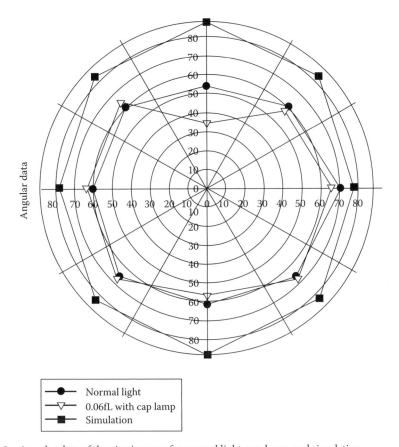

FIGURE 43.8 Angular data of the viewing area for normal light, cap lamp. and simulation.

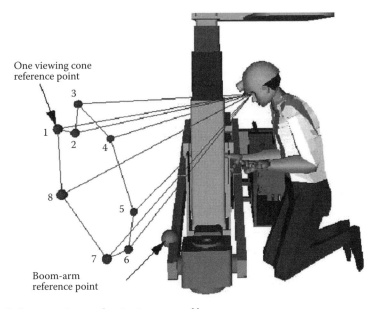

FIGURE 43.9 Reference points on the viewing cone and boom arm.

by used on any machine component, and involves using obscuration zones to show regions of space not visible to the operator. Coverage zones were generated with colored grids indicating visible and non-visible zones. Subsequent equipment enhancements provided greater visibility to operators within the cab of earth-moving machines.

43.8.4 Value Attention Locations

A very important visual aspect of mining machine operators is called value attention locations (VAL), key visual areas in which miners control and operate their machines. The mining industry also uses an educational aid called "red zones are no zones" to help remote-controlled machine operators understand which areas around their machines they should avoid, and to reduce the injury and fatality rates from miners making contact with moving machines. Bartels et al. (2007) used DHM vision tools and survey data to define the VAL for operator positions, the logic behind choosing various positions, and to define the operator's direct focus area during various tasks. The results to date provided preliminary recommendations for control interventions that enhance VAL, decreasing safety and health risks to mining equipment operators. In addition, researchers plan to use DHM illustrations to assist adopting new mining techniques and procedures, as well as applying technological advances to equipment to enhance VAL.

43.9 Using DHM to Characterize Machine Designs and Controls of Mining Equipment

As recently as the mid-1950s, one third of all coal produced in the United States was hand loaded. Physical demands on mine workers have been greatly reduced primarily due to advances in mechanization during the second half of the 20th century. The past decade has seen new mining technologies such as remote control, continuous haulage systems, and automated equipment. However, the physical demands on miners remain significant. Human-centered design principles have been minimal as new equipment and new technologies in the mining industry have surfaced, as recognized by Ambrose (2000a and b, 2003), Ambrose et al. (2005b), Cornelius and Turin (2001a and b), Cornelius et al. (2001c), and Steiner et al. (1998).

Operating equipment at the mine face, the point of coal extraction, is one of the most fundamental and risky elements of underground mining. It is performed in restricted workspaces with reduced visibility. Studying safety and health issues surrounding face equipment is extremely complex. Face equipment, such as roof bolters and continuous mining machines, are not only dangerous when performing tasks at the face, but also when moving from one task site to the next. These machines pose significant dangers to both the machine operators and their helpers, putting humans in awkward postures for tasks and requiring fast reactions to avoid being struck or pinched by moving equipment and machine appendages. Since it's neither feasible nor ethical to use human subjects to directly evaluate factors that precipitate such injuries in the field or in laboratory experiments, these machine design and control practice concerns are being addressed by researchers with the help of DHM.

43.9.1 Machine Appendage Speeds

Roof-bolter operators in underground coal mines have a high incidence rate of being struck by the roof bolter's drilling boom (fig. 43.10), often resulting in serious injury or fatality (MSHA, 1994). Ambrose (2003), and Ambrose et al. (2005a and b) used motion analysis data and DHM to analyze accident risks of miners working with roof bolters, manipulating key factors that influenced injuries including the speed of the roof-bolter boom, boom direction, vertical space constraints and work postures, operator location, operator sizes, and hand positioning behaviors when operating the roof bolter. The most

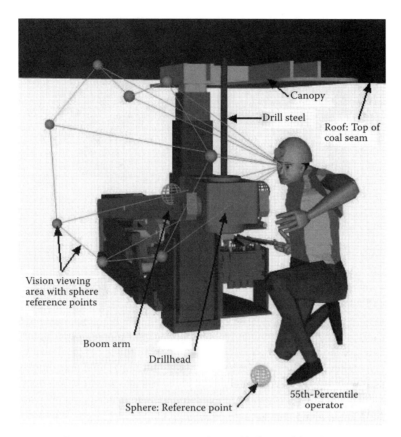

Canopy

Drill steel

Roof: Top of
coal seam

Vision viewing
area with sphere
reference points

Boom arm

Drillhead

Sphere: Reference point

55th-Percentile
operator

FIGURE 43.10 A view of a DHM mine environment with a roof-bolter model and virtual operator.

influential variable in causing a struck-by incident was boom speed. Operators working in a more restricted vertical workspace, in which an upright position was not possible, experienced greater struck-by risks than operators able to maintain an upright posture. Likewise, larger operators were at greater risk compared to average sized operators. This investigation identified the safest boom speed parameters for varying vertical workspaces and operator sizes.

43.9.2 Machine Tramming Speeds

In previous studies, Ambrose (2001, 2003) simulated an operator's behavior and machine motion to accurately predict and identify hazards, and used that information to form safe design parameters for mining equipment. Bartels et al. (2008) combined motion capture data and DHM simulations to gather struck-by and pinched-by data using collision detection features of DHM. This will determine safer workplace positions and safe tramming speeds for a continuous mining machine (fig. 43.11), one of the most basic and dangerous pieces of underground mining equipment. DHM simulations let researchers study multiple environments and virtual humans in differing scenarios that would be hazardous and cost prohibitive in field studies.

43.9.3 Mechanical versus Electronic Joystick Controls

In low-roof (also called low seam height) mining conditions, miners experience injuries from repetitive motions to their wrists, elbows, and shoulders while in different work postures such as kneeling

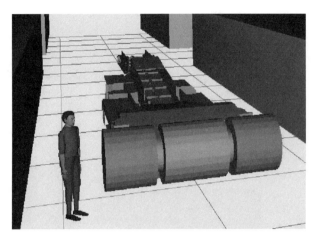

FIGURE 43.11   A DHM mine environment with continuous miner model and virtual operator.

on one or both knees. Exposure to these and similar stressors is a recipe for musculoskeletal disorders. Using DHM simulations, Ambrose et al. (2007) predicted joint moment and joint force effects to the right wrists, elbows, and shoulders of roof bolter operators while using electronic and mechanical joystick controls. As expected, electronic joystick controls significantly reduced joint movement and force compared to mechanical joystick controls. Using DHM in this study facilitated the estimation of upper extremity loads on equipment operators. Despite its findings, DHM data must still be validated in real-world situations through an epidemiological assessment of equipment operators in the field. For example, despite the physiological benefits of an electronically controlled joystick, it doesn't provide as much physical feedback as a mechanically controlled joystick. Some operators may prefer, or even need, the "feel" of the mechanical joystick feedback to safely and efficiently handle the machine.

43.9.4 Mining System Designs and Procedures

In Poland's mining industry, most mining accidents occur in anthracite coal mines, with the majority of fatalities occurring in underground coal mines around mining machines and equipment. Similarly in the U.S. mining industry, the majority of accidents occur in underground bituminous coal mines, with the balance occurring in surface mines. Dudek et al. (2005) used DHM simulations to help identify technical risks in standard Polish mining systems (longwall and roadheading) and devised methods to reduce or eliminate those risks. DHM simulations with mining machines were performed even before mining machine prototypes were manufactured, giving manufacturers the opportunity to make changes early in the design process, a significant economic advantage. By taking human factors into account, researchers and manufacturers developed a realistic picture of both operational conditions and human behavior. Potential machine operator mistakes were identified early enough to impact machine design and/or mining system procedures.

43.10 Ergonomics Analysis: Using DHM to Illustrate Material Handling in the Mine Environment

As previously illustrated, miners encounter physical demands and environmental restrictions that include confined workspaces, necessitating awkward postures, working in muddy or wet floor conditions, exposure to high levels of whole-body vibration, and performing significant heavy manual work. Back injuries from handling materials in underground mines continue to pose a major safety concern. Despite numerous mechanized aids, Patton et al. (2001) found that the number of materials-handling injuries remains the second most common injury in underground coal mines. Earlier studies by Gallagher et al. (1997a and b) developed recommendations for manual lifting tasks in underground mining, and examined the effects of posture on miner's back strength while kneeling and standing. The following DHM applications illustrate this enabling technology as an exceptional research tool that profoundly changed ergonomics analyses methods for studying low back issues in mine workers.

43.10.1 Lifting and Walking Stress Analysis

One of the major contributors to lower back disorders in any environment is simply lifting things. Proper body posture and keeping objects at the proper proximity to the body help reduce lower back strain when lifting or handling objects. PRL used DHM animations to train mine workers on proper lifting and carrying postures. The DHM carried an average load for a miner, represented by 40 pound wire-mesh, in a variety of typical underground mining scenarios. The objective was to make mine workers aware of proper body posture when lifting, and to instruct mine workers to keep loads close to their bodies when lifting or handling loads. A DHM's *watchdog* feature of the low back analysis was applied and changed colors to show when forces on the lower back increased (red) or decreased (yellow or green).

43.10.2 Spinal Load Analysis for Machine Operators

Ambrose et al. (2004) effectively used DHMs to evaluate the severity of muscle recruitment and spinal loads while operating a roof bolter in different work postures and mine seam heights. Researchers generated a database containing L4/L5 spinal joint and back muscles by processing captured motions from test subjects using DHM. They then analyzed the variance of the forward bending moment, for both standing and kneeling postures, using maximum values for spinal forces and moments, and estimated muscle forces from 10 trunk muscles. The results showed that an operator's forward bending moment increases significantly from a standing posture, while the compression force and trunk muscle activity were greater from the kneeling posture (fig. 43.12). DHMs in kneeling postures demonstrated an increased forward bending moment, twisting moment, compression force, and trunk muscle activity for lateral movements and extended torso in a 45-inch seam height versus a 60-inch seam height.

This research led the way for Ambrose and Cole (2005) to evaluate control interventions with DHMs that reduced the severity of muscle recruitment and spine loads for roof-bolter operators in different postures and seam heights. This study illustrates the benefits of using DHM to estimate spinal loads for equipment operators using what-if scenarios that contrast human motion in the workplace with control

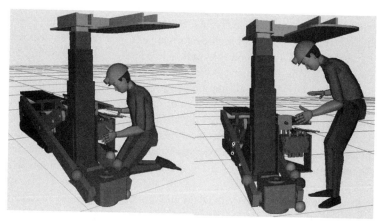

FIGURE 43.12 Illustration of forward bending comparing a standing work posture and a kneeling work posture.

interventions. The first database contained L4/L5 spinal joint and back muscles generated by processing captured motions from test subjects using DHM. A second database repeated this generation with job materials (drill bit, bolt, and wrench) that were one-half the normal weight. Finally, a third database repeated the generation once again using the bolting process that mimicked the subject performing tasks with full-weighted materials relocated to different positions. Comparing the resulting databases indicated that all work postures and seam heights benefited from a reduction in the weight of materials handled, and showed a significant decreased response force when the materials were relocated.

43.10.3 Reconstructing Material Handling Accidents

Winkler et al. (2005) first used DHM to visualize accidents involving handling materials in the mining industry, which were useful in developing "lessons learned" training materials. Understanding risk factors and dangerous situations, made more realistic by DHM visualizations, allows for both theoretical and realistic accident reconstruction and investigation. DHM makes it possible to exactly model human silhouettes, and reproduces detailed anthropometrical features of the humans involved in accidents.

43.11 Choosing the Right DHM Software and Motion Capture Hardware Systems

Finding the right software tool with all the features and capabilities to develop and execute computer code for virtual motions and behaviors is critical to any DHM research. Commercial DHM software tools provide a product with a virtual human modeling system for ergonomic analyses and work performance evaluations to help design and study mining systems with man-machine interactions. It would be prudent in early planning of using DHM that the correct research tools fit the immediate job and potential work for the future. The following points are for reference only, and do not take into account specific software uses, project scopes, or budget considerations.

Digital motion developed in DHM software is an approximation of actual human motions required to accomplish the task and usually results in rigid, robot-like movements. This animation is very time consuming, especially as you increase the number of virtual humans, and simultaneously, the number of joints involved. Conversely, motion capture systems deliver realistic motion data files generated quickly and easily from one motion capture session. Also, customizing the DHM user interface to implement graphical user interface (GUI) applications allows users to quickly develop and test user interface code without the need to write and compile complex code or work with abstruse "widget" libraries.

With increasing options, choosing which DHM and motion capture system will best suit one's needs is no easy task. Several factors will help determine the choice. Only some motion capture systems, for example, are real time. If one is looking for high-performance animation, the choices will narrow. If one needs a wireless system, that will limit choices as well. While motion capture systems don't come cheap, optical systems are definitely in the higher price range. Ensure that whichever motion capture system one chooses supports one's DHM—not all motion captures play well with all DHMs, and vice versa. Most high-quality DHM software will include a host of special features, but other features will add to the basic cost; plan well and choose soundly. All the DHM software and motion capture systems have their pros and cons, and there is no magic specification that measures the quality of data you can get from any one choice. Once the choices are narrowed, ask for sample data. Ask if one can take the DHM software and motion capture system for a test drive, that's even better. Some vendors will accommodate this request, so it's worth asking.

Add-on modules and devices available with DHM and motion capture could overwhelm the average researcher or scientist. Have an experienced vendor or representative give you a demonstration of these features, which are extremely helpful in understanding even the most challenging features. In addition, ask for references. Obtain a list of current and past users from the vendor. Customer feedback can be the most important decision-making aid, and gives one real-user perspective and *experience*—very helpful if you buy the same DHM or motion capture system.

Most vendors provide initial training when the product arrives, and technical support is usually free for the first year. Extending technical support is wise, and it's also recommended to seek training after one has a chance to "play" with the product, which may incur additional costs but in many cases is worth the expense. If one is able to attend additional training for the product, take specific problems, questions, and issues for discussion; use them as examples to practice new features. Take full advantage of the instructor, who should also play a consultant role to class participants during the training.

Maintaining yearly support and maintenance contracts usually includes updates to the vendor's software or firmware. Annual support and maintenance charges usually come as one package, but occasionally a vendor will offer different packages to cover limited support needs at a lower cost. There are plenty of support gurus out there. Look for one that provides the best support and explains issues and answers questions patiently and thoroughly within a reasonable response time via telephone, email, or in person.

Computer platforms and networking are important in the design of usage if one intends to share DHM and motion capture licenses and data files. Regardless, both systems require very large computer storage space and large data-streaming needs. It's recommended that one select scientific workstations with speed, storage, various media-recording capabilities, and various ports and slots to handle plug-ins and expansion needs.

Attending conferences that discuss and address DHM technology is an excellent resource to learn and share information. Usually, vendors of DHM software and motion capture systems attend conferences with exhibits, providing an opportunity to talk about needs and potential applications. In addition, conferences provide an exceptional opportunity to associate with long-time users of DHM and motion capture systems.

43.12 Looking to the Future

New generations of highly advanced virtual humans that reflect state-of-the-art technology are on the horizon. Whispers of what's to come are evidenced by Kim et al. (2005), Mi et al. (2004), Wang et al. (2005), Yang et al. (2004), and Zhou and Lu (2005). The next generation of DHM will come from current digital human model transformations and "newborn" DHMs. In addition to looking real, they will most likely include realistic gross movement and internal functions; autonomously predict postures and motions; monitor human performance indices including physiological and musculoskeletal quantities; provide valuable information pertaining to the efficiency and effectiveness with which a task is completed; exhibit cognitive behavior such as the ability to walk through a maze as the environment changes and discern which new path to take. These new DHMs will plug into motion capture systems, be easily inserted in a CAD environment, and deployed into vehicles, systems-maintenance settings, and hostile and hazardous areas.

As researchers use DHM systems to improve the safety and health of workers in hazardous environments, they should also look for improvements in DHM technology. Continued research incorporating the latest DHM advancements and features can only increase the effectiveness of DHM in improving, and ultimately saving, real human lives.

43.13 Mining Terms

Boom arm or roof-bolter arm: The arm is a roof-bolter appendage that vertically lifts the drill mast mounted to it and horizontally swings to adjust position for desired location of vertical lift.
Continuous haulage: A controlled mobile conveyor system that navigates in underground mines and moves coal from the working face to a main dumping point from which the coal is taken to the surface.

Continuous miner: A self-propelled machine that rips coal, metal, and nonmetal ores, rock, stone, or sand from the face and loads it onto conveyors or into shuttle cars in a continuous mining operation.

Entry: An entry is a horizontal mine passageway or room that is formed as a result of room-and-pillar mining operation. The passageway or room varies in height, width, and length, for example, 24 inches high, 16 feet wide, and 60 feet long or 15 feet high, 20 feet wide, and 120 feet long.

Face: The face is the working area in from the last open crosscut in an entry or a room. Crosscuts in room-and-pillar mining result from piercing of pillars at regular intervals for the purpose of haulage and ventilation.

Longwall: In longwall mining, a cutting head moves back and forth across a panel (longwall) of coal averaging 945 feet in width and 9,900 feet in length. The cut coal falls onto a flexible conveyor for removal. Longwall mining is done under hydraulic roof supports (shields) that are advanced as the seam is cut. The roof in the mined out areas falls as the shields advance.

Rib: In underground coal mines, it is the solid coal side of a passageway (entry).

Road header: An entry-boring machine, called a road header, which bores the entire section of the entry in one operation.

Roof: In underground coal mines, the rock immediately above a coal seam. Sometimes part of the coal is left for the roof.

Roof bolter: A machine to install roof support bolts in underground mine passageways or the one who operates this machinery.

Roof falls: This is when rock or coal falls from the roof into a mine passageway (entry).

Room-and-pillar: Most underground coal is mined by the room-and-pillar method, whereby rooms are cut into the coal bed leaving a series of pillars, or columns of coal, to help support the mine roof and control the flow of air. Generally, rooms are 16 to 30 feet wide and the pillars up to 100 feet wide. As mining advances, a grid-like pattern of rooms and pillars is formed.

Shuttle car: Diesel or electric-powered car in underground mine that transports materials from the working face to mine cars or conveyor belts.

Tramming: This term is used to define when the machine operator moves a self-propelled piece of equipment from place to place.

References

Ambrose DH [1996]. Unpublished white paper titled, Review of Jack modeling and simulation software for purchase recommendation, National Institute for Occupational Health and Safety-Pittsburgh Research Laboratory, April 1996.

Ambrose DH [1999]. Presentation titled, Random Jack Motion, Jack User Meeting, Ann Arbor, MI, May 11–12, 1999.

Ambrose DH [2000a]. Presentation titled, Research with Jack machine safety and operator health issues, Jack User Meeting, Ann Arbor, MI, October 31–November 1, 2000.

Ambrose DH [2000b]. A simulation approach analyzing random motion events between a machine and its operator. Ann Arbor, MI: Proceedings of the 2000 SAE Digital Human Modeling for Design & Engineering, number 2000-01-2160.

Ambrose DH [2001]. Random motion capture model for studying events between a machine and its operator. Seattle, WA: Proceedings of the Business and Industry Symposium of the 2001 Advanced Simulation Technologies Conference (SCS), April 22–26, 2001, pp. 127–134.

Ambrose DH [2003]. Machine injury prediction by simulation using digital humans. Montreal, Canada: June 20, 2003, Proceedings of the 2003 SAE Digital Human Modeling for Design & Engineering, number 2003-01-2190.

Ambrose DH [2004]. Developing random virtual human motions and risky work behaviors for studying anthropotechnical systems. Pittsburgh, PA: Department of Health and Human Services, CDC, NIOSH, IC 9468, NIOSH publication number 2004-130.

Ambrose DH, Bartels JR, Kwitowski, Helinski RF, Gallagher S and Battenhouse TR [2005a]. Mine roof bolting machine safety: A study of the drill boom vertical velocity. Pittsburgh, PA: Department of Health and Human Services, CDC, NIOSH, IC 9477, NIOSH publication number 2005-128.

Ambrose DH, Bartels JR, Kwitowski A, Helinski RF, Gallagher S and Battenhouse TR [2005b]. Computer simulations help determine safe vertical boom speeds for roof bolting in underground coal mines. *Journal of Safety Research*, Volume 36, Issue 4, pp. 387–397.

Ambrose DH, Burgess T and Cooper DP [2007]. Upper extremity force moment and force predictions when using joystick control. Seattle, WA: June 12–14, 2007, Proceedings of the 2007 SAE Digital Human Modeling for Design & Engineering, number 2007-01-2497.

Ambrose DH and Cole PC [2005]. Comparing estimated low back loads from control interventions for underground mine roof bolter operators. KOMTECH 2005 Conference, November 14–17, 2005, Zakopane, Poland.

Ambrose DH, Cole PC and Gallagher S [2004]. Estimating low back loads of underground mine roof bolter operators using digital human simulations. Oakland University, Rochester, MI: June 15–17, 2004, Proceedings of the 2004 SAE Digital Human Modeling for Design & Engineering, number 2004-012148.

Badler NI, Erignac CA and Liu Y [2002]. Virtual humans for validating maintenance procedures, *Communications of the ACM*, Volume 45, Issue 7, pp. 56–63.

Bartels JR, Ambrose DH and Gallagher S [2007]. Effect of operator position on the incidence of continuous mining machine/worker collisions. Baltimore, MD: October 1–5, 2007, Proceedings of the 51st Human Factors & Ergonomics Society Annual Meeting.

Bartels JR, Ambrose DH and Gallagher S [2008]. Analyzing factors influencing struck-by accidents of a moving mining machine by using motion capture and DHM simulations. Pittsburgh, PA: June 17–19, 2008, Proceedings of the 2008 SAE Digital Human Modeling for Design & Engineering, number 2008-01-1911.

Bartels JR, Ambrose DH and Wang R [2001]. Verification and validation of roof bolter simulation models for studying events between machine and its operator. Alexandria, VA: Proceedings of the 2001 SAE Digital Human Modeling for Design & Engineering, number 2001-01-2088.

Bartels JR, Kwitowski AJ and Ambrose DH [2003]. Verification of roof bolter simulation model. Montreal, Canada: June 20, 2003, Proceedings of the 2003 SAE Digital Human Modeling for Design & Engineering, number 2003-01-2217.

Brown AS [1999]. Role Models: Virtual People Take On the Job of Testing Complex Designs. *Mechanical Engineering*, Volume 121, Number 7, pp. 44–49.

Chaffin DB [2002]. On simulating human research motions for ergonomics analyses. *Human Factors and Ergonomics in Manufacturing*, Volume 12, Issue 3, pp. 235–247.

Colombo G and Cugini U [2005]. Virtual humans and prototypes to evaluate ergonomics and safety, *Journal of Engineering Design*, Volume 16, Number 2, pp. 195–203.

Cornelius KM and Turin FC [2001a]. A case study of roof bolting tasks to identify cumulative trauma exposure. http://www.cdc.gov/niosh/mining/pubs/downloadablepubs.htm

Cornelius KM and Turin FC [2001b]. Ergonomics considerations for reducing cumulative trauma exposure in underground mining. http://www.cdc.gov/niosh/mining/pubs/downloadablepubs.htm

Cornelius KM, Turin FC, Wiehagen WJ and Gallagher S [2001c]. An approach to identify jobs for ergonomic analysis. http://www.cdc.gov/niosh/mining/pubs/downloadablepubs.htm

Dudek M, Chuchnowski W and Tokarczyk J [2005]. Identification of technical risk factors by computer simulation. KOMTECH 2005 Conference, November 14–17, 2005, Zakopane, Poland.

Eger T, Jeffkins A, Dunn P, Bhattacherya and Djivre M. [2005]. Benefits of assessing line-of-sight from LHD vehicle in a virtual environment. *CIM Bulletin*, Volume 98, Number 1089.

Eger T, Salmoni A and Whissell R [2004]. Factors influencing load-haul-dump operator line of sight in underground mining. *Applied Ergonomics,* Volume 35, Number 2, pp. 93–103.

Ferguson SA and Marras WS [2005]. Workplace design guidelines for asymptomatic vs. low-back-injured workers, *Applied Ergonomics,* Volume 36, pp. 85–95.

Gallagher S and Bobick TG [1997b]. Effects of posture on back strength and lifting capacity. http://www.cdc.gov/niosh/mining/pubs/downloadablepubs.htm

Gallagher S, Hamrick CA and Love AC [1997a]. Biomechanical modeling of asymmetrical lifting tasks in constrained lifting postures. http://www.cdc.gov/niosh/mining/pubs/downloadablepubs.htm

Hella F, Schouller JF and Tisserand M [1996]. Development of a new method to evaluate driver visibility around trucks and industrial vehicles. In *Vision in Vehicles*—V (371–378). Amsterdam: Elsevier Science B.V.

Hella F, Tisserand M, Schouller JF and Engllert M [1991]. A new method for checking the driving visibility on hydraulic excavator. *International Journal of Industrial Ergonomics,* Volume 8, pp. 135–145.

Humantech [1996]. *Ergonomic design guidelines for engineers.* Ann Arbor, MI: Humantech.

Kim JH, Abdel-Malek K, Yang J, Farrell K and Nebel K [2005]. Optimization-based dynamic motion simulation and energy expenditure prediction for a digital human. Iowa City, IA: June 14–16, 2005, Proceedings of the 2005 SAE Digital Human Modeling for Design Human Modeling for Design & Engineering, number 2005-01-2717.

Kwitowski AJ and DuCarme JP [2007]. Human motion analysis of roof bolting postures involving boom swing. Baltimore, MD: October 1–5, 2007, Proceedings of the 51st Human Factors & Ergonomics Society Annual Meeting.

Määttä T [2003]. *Virtual environments in machinery safety analysis.* Espoo, Finland: VTT Publications.

Mi Z, Farrell K and Abdel-Malek K [2004]. Virtual environment for digital human simulation. Oakland University, Rochester, MI: June 15-17, 2004, Proceedings of the 2004 SAE Digital Human Modeling for Design Human Modeling for Design & Engineering, number 2004-01-2172.

MSHA [1994]. Coal mine safety and health roof bolting machine committee: report of findings, July 8, 1994. Arlington, VA: U.S. Department of Labor, Mine Safety and Health Administration, Coal Mine Safety and Health, Safety Division, pp. 1B28.

Patton PW, Stewart BS and Clark CC [2001]. Reducing materials handling injuries in underground mines. http://www.cdc.gov/niosh/mining/pubs/downloadablepubs.htm

Smith V [2006]. Study finds MSHA reports lack detail, usefulness, Daily Press—Associate Press writer, Morgantown, WV, July 25, 2006. (Newspaper article)

Steiner L, Cornelius K, Turin T and Stock D [1998]. Work sampling applied to a human factors analysis of miner positioning. http://www.cdc.gov/niosh/mining/pubs/downloadablepubs.htm

Volberg OS and Ambrose DH [2002]. Motion editing and reuse techniques and their role in studying events between a machine and operator. San Diego, CA: Proceedings at the Society of Computer Simulation International (SCS) Advance Simulation Technologies Conference 2002, Simulation Series Volume 34, Number 4, pp. 181–186.

Wang Q, Xiang J, Kim JH, Arora JS and Abdel-Malek K [2005]. Alternative formulations for optimization-based digital human motion prediction. Iowa City, IA: June 14–16, 2005, Proceedings of the 2005 SAE Digital Human Modeling for Design Human Modeling for Design & Engineering, number 2005-01-2691.

Winkler T, Michalak D, Bojara S and Jaszczyk L [2005]. Using of virtual technology in reconstruction of accidents in the mining industry. KOMTECH 2005 Conference, November 14–17, 2005, Zakopane, Poland.

Woolley C, Chaffin DB, Raschke U and Zhang X [1999]. Integration of electromagnetic and optical motion tracking devices for capturing human motion data. Hague, Netherlands: Proceedings of the 1999 SAE Digital Human Modeling for Design & Engineering, Number 1999-01-1911.

Yang J, Abdel-Malek K and Nebel K [2004]. Restrained and unrestrained drive reach barriers. Oakland University, Rochester, MI: June 15–17, 2004, Proceedings of the 2004 SAE Digital Human Modeling for Design Human Modeling for Design & Engineering, number 2004-01-2199.

Zhang X and Chaffin DB [2006]. Digital human modeling for computer-aided ergonomics. In *The occupational ergonomics handbook*, 2nd ed. Boca Raton, FL: CRC Press.

Zhou X and Lu J [2005]. Biomechanical analysis of skeletal muscle in an interactive digital human system. Iowa City, IA: June 14–16, 2005, Proceedings of the 2005 SAE Digital Human Modeling for Design Human Modeling for Design & Engineering, number 2005-01-2709.

44

Virtual Reality Training to Improve Human Performance

Sajay Sadasivan, Paris
F. Stringfellow, and
Anand K. Gramopadhye

44.1 Introduction

Reliable human task performance is important to the safe and seamless operation of industry, ranging from critical high-consequence industries like health care and aviation to productivity in classic manufacturing industries. Training has been shown to improve the reliability and productivity of human operators in both cognitive and psychomotor tasks in complex systems. Conventional training methods like classroom and on-the-job training have inherent limitations, ranging from the lack of exposure to critical scenarios to costs involved in training implementation. With advances in computer technology there has been a thrust to develop technology-based training simulators. Virtual reality (VR) technology took these initial forays to the next level as it allows for superior graphic realism and interactivity. This chapter discusses the application of virtual reality to training to improve human operator skills and reduce error in a variety of scenarios. An overview of factors influencing the design of VR training systems and an introduction to available technology are provided. In addition, a case study outlines the process for effective modeling of tasks for design and development of VR training simulators.

44.2 Need for Training

In the manufacturing industry, the effectiveness and the efficiency of processes depend largely on the skills of the workforce, which reflects in the quality of goods produced and the profitability of an organization. Wherever human operators are involved, there is a probability of errors. Research has shown that approximately 80% of aviation accidents are attributable to aircrew errors (Wiegmann & Shappell, 2001). Kohn et al. (1999) estimated that each year at least 44,000 deaths in American hospitals can be attributed to medical errors.

Training has long been identified as the most effective intervention to improve human performance whether that of an operator of heavy machinery or a surgeon using sophisticated, minimally invasive surgical instruments. Training is the primary intervention adopted for the acquisition of skills, such as in medicine, manufacturing, and aviation. With advances in technology, developments in training programs have progressed from classic on-the-job training scenarios to advanced synthetic simulators.

On-the-job training (apprenticeship) is currently the default training provided for improving task performance. This mode of training has its limitations. In the health care industry, there are ethical issues in training surgeons on patients, and in the aviation industry, on-the-job training is limited by the availability of aircraft to train on. Advanced physical simulators duplicating aircraft systems and manufacturing environments are expensive and their availability is limited. Surgical training using animals may be considered inhumane. Scenarios for the training of emergency response personnel such as firefighters and paramedics are difficult to simulate in the real world.

Simulated training provides a solution to many of these problems. In the aviation industry, simulators have been used extensively from the Link Trainer of 1929 to current full flight simulators that are capable of moving in all six degrees of freedom (DoF). In the medical field, simulator use has become popular with training students as it allows for extensive practice without mistakes having dire consequences to patients. In a full-scale simulation, training scenarios can be created for rare anomalies that would be almost impossible for trainees to encounter in a real-world, on-the-job training environment.

44.3 Application of Computer Technology to Training

Increased application of advanced technology in training can be expected with computer technology becoming cheaper. Over the past decades, numerous technology-based training devices have been offered to aid in improving efficiency and effectiveness of humans. These training delivery systems, such as computer-aided instruction, computer-based multimedia training, and intelligent tutoring systems, are extensively used in educational and industrial training settings. For example, Latorella et al. (1992) and Gramopadhye et al. (1993) have used low-fidelity inspection simulators with computer-generated images to develop off-line inspection training programs for inspection tasks. Despite their advantages, older PC-based low-fidelity simulators lack realism as most of these tools use only two-dimensional sectional images and do not provide the trainee with a holistic exposure to the task environment. Virtual reality allows a solution to this problem. With VR technology becoming cost effective and reliable, there is a push to apply this technology to training.

Due to advances in the commodity graphics market, the visual realism of the simulators has improved considerably. These training simulators vary in realism and degree of interaction, from a simple desktop point-and-click version to a fully immersive, virtual reality simulator. It was observed that participants who had prior training with training simulators performed better in virtual tasks than those who did not (Ferlitsch et al., 2002).

VR allows for detailed representation of complex scenarios that an operator might encounter in the real world. It allows for interactivity with the task environment and allows instructors to design training programs to cater to the adaptive needs of the trainee. A simulated VR training environment can allow active feedback on the process and performance characteristics of the task. VR allows a controlled environment to collect performance and process data that can be provided as feedback information. This environment can also be used to capture cognitive information that can be used for the modeling of experts performing specific tasks, which can be used for training purposes in designing the training systems as well as for providing novices with information on experts' strategy. In the next section, we provide a brief overview of training design and measurement.

44.4 Modeling Effective Training

Training systems can be effective in providing practice and exposure to complex scenarios. Virtual reality simulators provide graphic realism, and it is easy to be trapped in an optimistic conviction of their success in providing effective training. It is necessary to fall back on sound human factors research and empirically derived principles when designing training systems.

Cohen et al. (1979) point out that training success depends on the emphasis on learning safe behavior, ensuring the transfer of the learning through practice, and evaluating training effectiveness while providing feedback.

There is a wealth of information available in the design of training systems (Patrick, 1992; Gordon, 1994; Schneider, 1985). Some types of training that can be adopted either independently or in combination have been described below.

44.4.1 Pre-Training

Pre-training gives the trainee prior information about the objectives and scope of the training program and facilitates assimilation of new material. It has been prescribed that training should be imparted initially with a special overview that introduces general, simple and fundamental ideas before going into complex and specific concepts (Reigeluth & Stein, 1983).

44.4.2 Feedback Training

Feedback provided on task performance and the process adopted while the trainee attempts to perform the task lead to improving task performance and can be applied to learning facts, concepts, procedures, problem solving, cognitive strategies, and motor skills (Weiner, 1975).

44.4.3 Feedforward Training

It is often necessary to provide feedforward information or cue the trainee about what should be perceived. This information can be presented in the form of physical guidance, demonstrations, and verbal guidance. For example, in an aircraft inspection task, when a novice inspector tries to find defects in an airframe, the indications may not be obvious and the trainee must be initially provided information on what to look for and where to look.

44.4.4 Active Training

An active approach is recommended when keeping the trainee involved and to help with internalizing the material. The trainee makes an active response after each piece of new material is presented, allowing for interaction and feedback. Czaja and Drury (1981) used an active training approach, demonstrating its effectiveness in a complex inspection task.

44.4.5 Progressive Parts Training

Salvendy and Seymour (1973) successfully applied progressive parts training methodology to training industrial skills. In this type of training, small parts of the job are taught to criterion, then successively larger sequences until the entire task is taught. This method allows the trainee to understand each element separately as well as the links between them, the latter representing a higher level of skill.

44.4.6 Part-Task Training

Part-task training has been recommended if it is possible to break the task into components that draw on different working memory subsystems or if they can be independently broken off (Wickens & Hollands, 2000).

44.4.7 Schema Training

The trainee must be able to generalize the training to new experiences and situations. Exposure to controlled variability during training can lead to developing schema that will allow for a correct response in a novel situation.

44.4.8 Staggered Practice

Training with staggered practice sessions with gaps in the sequence that may be occupied by other cognitive tasks is desirable for strengthening knowledge acquisition, retention of the knowledge, and transfer to the actual task.

When designing training systems for a certain task, it is necessary to model the task effectively. The next section outlines a design process for the development of VR training simulators using the video borescope inspection process as an example.

44.5 Training Simulator Design Process

44.5.1 Case Study: Simulator for Visual Inspection Training

The first step in any design process is to understand the requirements for training. This would primarily involve documenting the processes and defining the tasks involved.

Determining the requirements for a training system deals with establishing human functional characteristics for performing the required process satisfactorily. This is done by identifying the tasks constituting the process and then identifying the skill and knowledge requirements for performing the job. This is done by observing the process, collecting detailed information on the task, identifying individual tasks elements and their interactions. The basic process of designing virtual reality training simulators is shown below.

    a. Understand the task
        i. Identify parameters that govern the task environment
        ii. Model the human performing the task
        iii. Develop an interaction design model for the different tasks
    b. Develop specifications for the training simulator
        i. The interface
        ii. Data capture and analysis
        iii. Tasks and scenarios
        iv. Measures to evaluate the training simulator
    c. Iterative design
        i. Identifying readily available VR technology

      ii. Prototyping
      iii. User testing
   d. Implementation and evaluation of training effectiveness

There are various knowledge elicitation tools available for understanding the task. They have their strengths and weaknesses based on the type of tasks being analyzed. Some methods elicit knowledge from experts (Ford & Sterman, 1997; Rush & Wallace, 1997), while others involve a basic interview process (Hendrick, 1997) such as teachback (Johnson & Johnson, 1987) and the think aloud protocol (Nielsen, 1997). Other more rigorous task analysis (Kirwan & Ainsworth, 1992) methods are Hierarchical Task Analysis (HTA) (Shepherd, 2000), Cognitive Task Analysis (CTA) (Klein, 2004), Operator Function Model (OFM) (Rubin et al., 1988), GOMS model (Card et al., 1983), and NGOMS (Kieras, 1988).

Cognitive Task Analysis (CTA) is used to identify the cognitive skills and processes of task performance that are needed for task performance and is beneficial in analyzing complex tasks to understand user behavior, while Hierarchical Task Analysis (HTA) describes the task in terms of a hierarchy of operations and plans based on structure diagram which is a part-whole decomposition of the overall task.

Good human factors design of training systems relies on understanding the system, modeling the task, iterative testing, and development. A basic design process of a virtual reality borescope inspection simulator (Vembar et al., 2005; Vembar et al., 2006; Sadasivan et al., 2006b) is described below. A video borescope is a nondestructive inspection tool used for the inspection of the internal sections of an aircraft engine for damage or flaws. It involves an inspector inserting a probe with a charge-coupled device (CCD) at its end into the engine. The camera output is presented to the inspector on a small LCD screen. The inspector has a handheld control that can be used to manipulate the articulation of the tip of the probe and freeze the image, control the zoom, and pan features. The inspector needs to have the psychomotor skills to guide the camera to the specific components to be inspected and also the cognitive skills to search for defects and to decide on their criticality.

44.5.2 Knowledge Elicitation

Initial explorations involved talk-throughs with the expert technicians giving an overview of the video borescope. This process was conducted with the investigator explaining the project and asking the technician to describe the borescope and explain its functionality. This was not a very structured interview. The investigator only asked questions to probe for explanations and to repeat the process to the expert to receive feedback verifying the understanding of the process. This helped in understanding the operations of the tool and its functionality. Additional documentation review of the video borescope operating manual was conducted.

This was followed by an introduction to the engine inspection process, using a similar process as described above. The inspection process is standardized and the description was consistent between the different technicians interviewed.

This was followed by a walk-through of the inspection process. The expert technician was instructed to give a demonstration of using the video borescope to perform an inspection task. Here the investigator would ask questions probing the details of each task and eliciting explanations to specific tasks. This process was videotaped for further reference. This was followed by asking the expert to perform a sample task while thinking aloud. The investigator demonstrated what he meant by thinking aloud and then the expert technician performed the task. An excerpt from this session is provided below.

Expert (AC) was thinking aloud while performing a borescope inspection task.

- Inserting probe.

AC: *I have to be careful while manipulating the probe .... If you don't, you stand a chance of locking it in the engine ... that's what I am trying to avoid ....*

- Manipulating probe.

AC: I'll zoom out so I can get a relationship of where I am at in the engine.

- Presses zoom button.

AC: I'll see if I can pan to the left a little.

- Moves articulation control to the left.

AC: I am looking at the leading edge of the stator right there (points at screen) ... I'll pull out just a little more.

- Pulls probe out a little until complete edge comes into view.

AC: I'll lock it right there (articulation) ... see if I can zoom in to inspect specific damage in the leading edge there (points at leading edge on screen) ... now that I have it in place and have the articulation locked I can zoom in .... that way I'll be able to get directly to the damage I was looking for ....

In addition, an actual borescope inspection task was observed on the hangar floor without any interference from the investigator. The investigator, in the presence of the expert, demonstrated using the tool to go through the steps involved according to understanding of the task after referring to notes. This allowed the verification of the information and allowed the expert to fill in any missing information.

44.5.3 Hierarchical Task Analysis

Hierarchical Task Analysis (HTA) is considered a broad approach to task analysis and hence this methodology was considered most advantageous to address the overall process of performing a video borescope inspection. Detailed knowledge elicitation for the HTA of the borescope inspection of the hot section of the Pratt & Whitney PT6 turboprop engine was conducted as described above. The information was then analyzed and documented. The two aspects of the video borescope inspection process, namely, the inspection skill and the skill in using the tool, are illustrated in the HTA along with the interaction between both aspects.

The information was analyzed using a task decomposition process by developing flowcharts of the process and by breaking the process into its subtasks for further analysis. A sample section of a task decomposition flowchart is shown in Figure 44.1. The analysis also included examining the relationship between the specific objectives, the activities required to perform them, and the different actions to be taken to perform the activities. A sample relationship is illustrated in Figure 44.2. This then led to generating the hierarchy. The partial HTA of the borescope inspection process is shown in Figure 44.3.

44.5.4 Developing Specifications for the Training Simulator

Studying the borescope inspection process can be broken down to two broad tasks: (1) examine the behavior, functionality and constraints of the video borescope, and (2) analyze the human interactions with the video borescope.

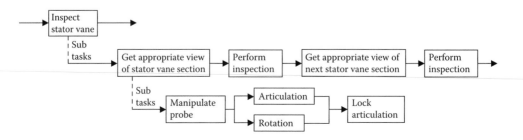

FIGURE 44.1   Task decomposition flowchart.

Objective	Activities	Actions
Stator vane inspection	Manipulate probe to appropriate section	Feed probe and control articulating tip till the specific stator vane comes into view
	Acquire suitable view	Control articulating tip and rotate probe to get appropriate view of section. Freeze articulation.
	Inspect section	Control digital zoom pan and rotate functions to inspect section for defects.
	Repeat process	Unlock articulation and articulate view to next section.

FIGURE 44.2 Relationship between objectives, activities, and actions.

Borescope inspection is analyzed using the task analysis with an aim to model the process and to determine the training needs for application to aircraft inspection. The major intermediate deliverables of this research will be: (1) specifications for the simulation of the borescope tool and (2) specifications for the development of the stimulus material to be used for applying the simulator to aviation inspection training.

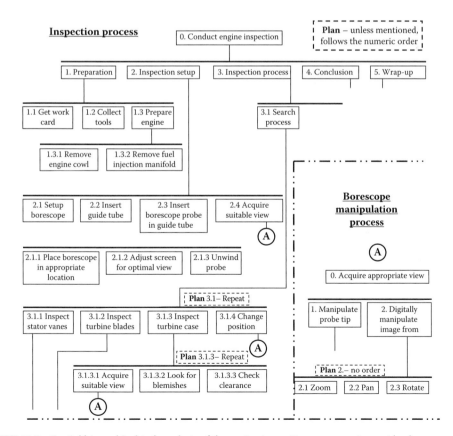

FIGURE 44.3 Partial hierarchical task analysis of the engine inspection process using a video borescope.

Once training needs are established and the engine components identified, we can start the development of the stimulus material. The engine components will be modeled. Defects will be identified for the training scenarios and textures will be developed to represent the defects, which can then be integrated with the model. This would be an iterative process involving presence studies with experienced borescope users to evaluate the quality and realism of the graphics.

The task analysis is the first step in developing specifications for the training simulator. This analysis reveals important notions critical to the future development of a borescope virtual training tool. For example, skilled coordination of the borescope probe tip is necessary to ensure an accurate inspection of an engine as well as to ensure that the probe is not damaged while inside the casing. Interviews with experienced borescope inspectors conveyed that the haptic feedback received as the probe moves throughout the internal casings greatly aids the inspector in the assessment of the probe's location and limitations. For example, as an inspector feeds the probe through the fuel manifold and into the engine casing, the resistance felt indicates that the probe has reached its maximum span. Another area highlighted by the task analysis is the importance of a clear and legible image through which to inspect and navigate. Once the actual inspection has begun, the inspector relies on the image produced by the probe tip to relay accurate conditions inside the engine; hence, the precision with which the image is relayed is critical to the inspector's ability to make correct and sometimes costly decisions.

44.5.5 Modeling the Video Borescope: Specifications for the Tool

It is necessary to clearly define the different elements of the tool to be modeled. The different constraints that need to be enforced need to be specified. A brief overview of some specifications is provided below.

The video borescope consists of a probe and an interface. The probe is inserted into the engine to be inspected and manipulated using the interface. The interface is made up of a display unit and a controller. The display unit renders the image obtained from the charge-coupled device (CCD) located at the tip of the video borescope probe. The controller has a device (e.g., mini joystick) to control the articulation of the tip of the probe and various buttons to control the image properties and so on.

The probe tip consists of the CCD attached to an articulating segment. The articulation of the borescope tip along two axes allows for a large number of camera positions, affording a large field of view. For simulation of the camera movement, we would have to specify the field of view of the static camera and then add the two movements that influence it, namely, the probe feed and the articulation of the probe tip (fig. 44.4). This would involve iteratively modeling a free camera initially with its limited field of view and then approximating the articulation and the probe feed into geometric constraints to guide its movement based on user interaction. In addition to the modeling the camera movement, orientation, and field of view, there is the capability of digital zoom and pan that would have to be modeled. The range of the zoom and the quality of the image displayed should be assessed.

The probe is a semi-flexible object that has constraints in movement below certain radii of curvature at different positions. The probe is also interacting with nondeformable objects (metal engine components) and it offers increasing frictional resistance to movement with increasing angles of contact with the metal components until it resists forward movement completely. The material properties of the probe will have to be specified to define constraints for the virtual reality model. Specifications for the interface also include the different controls that were used during the task. The display constraints include the illumination levels (variable) and resolution and size of displays commonly used.

44.5.6 Identifying Readily Available VR Technology

Once we have the specifications and the basic design, we would need to identify virtual reality technology to be used for the development of the simulator. It is always beneficial to consider available off-the-shelf solutions before building custom devices.

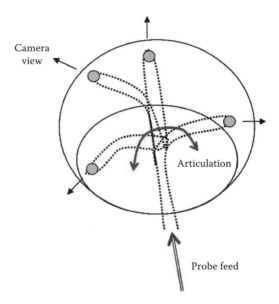

FIGURE 44.4   Borescope probe tip articulation.

44.5.6.1 Current Virtual Reality Technology

VR hardware can be broadly classified into three categories: display devices, interaction devices, and spatial orientation systems. Over the years a variety of high- and low-fidelity interface devices have been available. It is necessary to evaluate the devices to assess their suitability and feasibility for the task.

44.5.6.1.1 Display Devices

There are primarily two types of display devices: visual display systems and haptic interfaces.

Visual displays provide graphic information and are used to present a virtual scene to the user. They range from single user displays to multi-user displays and can be simple LCD desktop monitor or projector-based displays or high-fidelity immersive environments. A few of these displays are discussed here.

Head-mounted display (HMD) (fig. 44.5) consists of a head gear (helmet) with usually two displays—one for each eye—and has the potential to provide a different image to each eye and hence can easily achieve stereoscopic vision. It is usually integrated with a head movement tracker.

Cave Automatic Virtual Environment (CAVE) consists of multiple rear projection screens ranging from three to six (complete immersion). A pair of images is projected, one for each eye, based on the users' interaction. Using shutter glasses synchronized with the images, stereoscopic vision is achieved and users can see 3D objects and walk around them.

Window VR (fig. 44.6) consists of a touchscreen LCD panel suspended in space. Its position and orientation is tracked and it displays a section of the virtual environment to the user wherever it is pointed like a window to the virtual world.

Shutter glasses like CrystalEyes from StereoGraphics or the IR Pro from 3DTV are worn like regular eyeglasses, and they have small LCD screens (one for each eye) that operate like shutters (shutting off one eye at a time) synchronized with a display system like a regular desktop monitor or a projection system. They enable 3D stereo vision.

44.5.6.1.2 Interaction Devices

There are a variety of interaction devices available that can be used to specify input commands for movement in the virtual environment and the manipulation of objects.

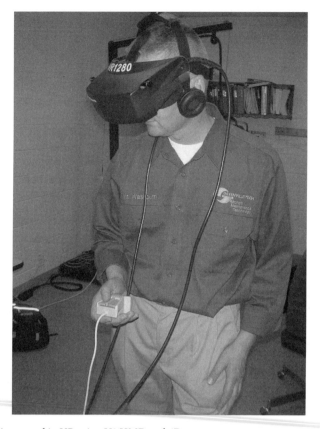

FIGURE 44.5   User immersed in VR using V8 HMD and 6D mouse.

6D mouse (fig. 44.7) from Ascension Technologies is a handheld device that can provide position and orientation coordinates on six degrees of movement.

SpaceNavigator, SpaceMouse, and the Spaceball from 3Dconnexion are devices that allow for multiple axis interaction for motion control in VR.

FIGURE 44.6   Window VR from Virtual Research.

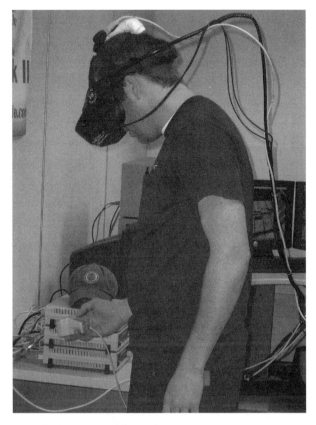

FIGURE 44.7   User immersed in VR using 6D mouse for interaction.

Data Glove (fig. 44.8) from 5DT and the CyberGlove from Immersion 3D are glove-based interaction devices that can detect finger and hand movement for interaction with virtual environments.

44.5.6.1.3 Haptic Devices

Haptics refers to technology that interfaces the user via the sense of touch to provide a sense of force, weight, texture, or other physical properties. Haptic or force feedback allows for a greater degree of

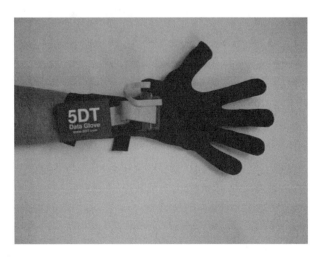

FIGURE 44.8   5DT glove.

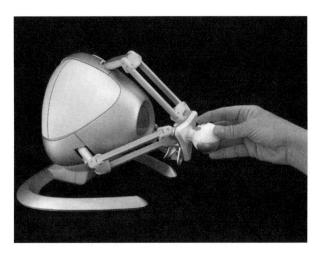

FIGURE 44.9   Novint falcon haptic device.

immersion in engineering design tasks, such as part assembly, or delicate manipulation tasks like in surgery or borescope inspection. Haptic feedback can be provided with available off-the-shelf devices as listed below or by using custom-built devices to suit the task.

- CyberGrasp and CyberTouch from Immersion 3D
- 3-DoF Omega and 6-DoF Delta from Force Dimension
- Novint Falcon (fig. 44.9) from Novint Technologies lets users feel weight, shape, texture, dimension, and force effects.

The Sensable Technologies PHANTOM product line of haptic devices makes it possible for users to touch and manipulate virtual objects.

44.5.6.2 Spatial Orientation Systems

Spatial orientation tracking is necessary to provide seamless virtual display responses to the user's actions. An error in tracking or calibration can lead to loss of presence.

One of the most popular tracking technologies is the Flock-of-Birds (fig. 44.10) from Ascension.

FIGURE 44.10   Flock of birds tracker.

44.6 Virtual Reality Assessment

Virtual reality is most applicably defined as "immersive, interactive, multi-sensory, viewer-centered, three-dimensional computer-generated environments and the combination of technologies required to build them" (Cruz-Neira, 1993). As this definition suggests, creating a virtual environment (VE) requires immersing humans into a world completely generated by a computer. The human user becomes a real participant in this world, interacting with and manipulating virtual objects. For virtual environments, presence, the subjective experience of being in one place or environment, even when physically situated in another, becomes the most important criterion. The success of using VR as a tool for training and job aiding, therefore, should be highly dependent on the degree of presence experienced by the users of the virtual reality environment.

The degree of immersion has classically been measured by presence. Traditionally, presence had been defined as the psychological perception of being transported to or existing in a virtual environment (Sheridan, 1992; Witmer & Singer, 1998). Witmer and Singer (1998) define immersion as "a psychological state characterized by perceiving oneself to be enveloped by, included in, and interacting with an environment that provides a continuous stream of stimuli and experiences" and also state that VR environments that are geared toward providing a greater sense of immersion will produce higher levels of presence. A different point of view is taken by Slater and Wilbur (1997) who define immersion as "the extent to which computer displays are capable of delivering an inclusive, extensive, surrounding, and vivid illusion of reality to the senses of the VE participant." They are of the opinion that presence is determined by the extent to which VR hardware is capable of recreating the physiological sensations of the real world in the virtual world.

Different assessment parameters are used for evaluating the effectiveness of VR. Sadowski and Stanney (2002) provide a detailed list of variables that influence presence. Some of the factors affecting performance in the simulator are presence, visual realism, haptics, and user and interaction constraints. Slater and Usoh (1993) describe system factors as external factors that relate to how well the system replicates the real-world equivalent. The relative importance of these factors in a training scenario is task dependent (Stanney et al., 1998). In some of the tasks, increased visual realism leads to an increased expectation of behavioral realism (Garau et al., 2003). Studies conducted to evaluate the aircraft inspection simulations found a high degree of "presence" and correlation between the real world and the simulated environment (Vora et al., 2001). Most of the studies have failed to follow up on the impact of simulator training on performance in the real world. Due to the non-uniform experiences of the users, as well as the subjective nature of the responses, there have been some controversies regarding the appropriateness of using presence in predicting simulator effectiveness and user performance (Slater, 1999; Usoh et al., 2000). Though presence is a popular parameter for evaluating virtual environments, it has been suggested that the evaluation of virtual environments should not be limited to the subjective feeling of being transported to a different scenario (Casati & Pasquinelli, 2005). The literature suggests various other parameters to be evaluated. It has been shown that there is a trade-off between the temporal and visual fidelity in virtual simulators (Watson et al., 1998). Additionally, it would be beneficial to measure the transfer effects of training in virtual worlds and their subsequent effect on real-world performance. Though haptic interfaces are gaining acceptance and becoming pervasive in training simulators, there is a lack of research in modeling human perception of haptics (Sheridan, 2000). Haptic illusions have been experienced by participants in virtual environments that have been attributed to multimodal interaction among the senses and motor systems (Biocca et al., 2002). This makes it critical to evaluate different modalities of interaction including perception and motor functions. The exploitation of the sensorimotor loop enables the user to act and to perceive the consequences of that action so as to create connections between movements and perceptual experiences (Casati & Pasquinelli, 2005). The development of VR simulators has kept up with the advances in technology and there is a tendency to fall into the trap of gold plating. This results in the technology being expensive and its benefits are lost as the colleges and small aviation maintenance firms can't afford to implement such solutions. There is a need for a selective

FIGURE 44.11   U.S. Navy personnel using a VR parachute trainer.

fidelity (Schricker & Lippe, 2004; Andrews et al., 1996) approach to find the trade-offs in using VR simulators for training. Experimentally evaluating different parameters with the users along with an iterative design process will lead to more effective training simulators suitable for specific task.

44.7 Applications of Virtual Reality for Training

Virtual reality has been extensively used as a training tool by the military, like the parachute training simulator shown in Figure 44.11 or the heavy vehicle simulator in Figure 44.12. In the civilian world, it has been used in the aviation industry, health care industry, and in manufacturing. We will present some of the applications of VR for training and discuss them briefly.

44.8 Applications of VR in Aviation

In the aviation industry, VR simulators have been used for pilot training for a couple of years now. Very realistic high-fidelity simulators have been built to allow pilots to practice scenarios, like landing an aircraft in a fog, with minimal risk compared to training in a similar scenario in the real world.

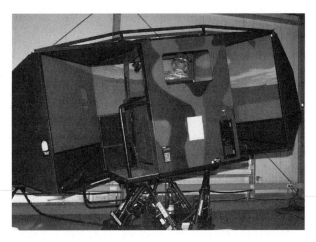

FIGURE 44.12   Heavy vehicle simulator.

FIGURE 44.13 Boeing 747 simulator.

Figures 44.13 and 44.14 show a Boeing 747 simulator at the Crew-Vehicle Systems Research Facility, a division of NASA Ames SimLabs. This flight simulator has six DoF motion capability. Another facility at the NASA Ames SimLabs is the FutureFlight Central (fig. 44.15), a virtual air traffic control/air traffic management training and planning facility.

In the field of aircraft maintenance training, virtual reality simulators have been developed for inspection training (Sadasivan et al., 2007; Stringfellow et al., 2006; Sadasivan et al., 2006a) like the 3D cargobay simulator (fig. 44.16) at Clemson University. Past research has shown that participants experienced a high level of presence in the simulator. Sadasivan et al. (2005) also showed that feedforward search strategy information can be taught. In this study, the search strategy of an expert inspector was provided to novice inspectors in the form of a static coded display in the virtual environment that was developed based on the analysis of eye tracking information collected from an expert inspector performing the inspection task in VR. There has also been an attempt to combine the benefits of on-the-job training and virtual reality training (Duchowski et al., 2004; Sadasivan et al., 2005a; Mehta et al., 2005) in the development of the collaborative virtual environment (CVE), which allows an instructor and trainees to be immersed simultaneously in the same virtual environment. Here, avatars are used to represent the user in the virtual environment, and the movements of the user are mapped to the movements of the avatar. Figure 44.17 shows the trainee's view of the trainer performing an inspection task. The eye movement of the trainer is dynamically plotted on the scenario to provide visual deictic reference. Deictic references

FIGURE 44.14   Boeing 747 cockpit simulator.

are associated with pointing and verbal expressions such as "look at this." By enabling deictic reference in a CVE, effective communication can be established in the immersive virtual environment.

44.9 Applications of VR in Health Care and Medicine

Virtual reality has extensive training applications in the health care industry, where human errors can be lethal. Providing training to novice resident surgeons on human patients is risky and unethical and sacrificing animals for practicing procedures is considered inhumane. Virtual reality simulators can allow extensive practice with active feedback. They would allow students to experience rare scenarios and infrequently occurring conditions. There are many surgical procedures that have been simulated using virtual reality for training, and some examples have been provided below.

LapMentor is a high-fidelity simulator for providing training in laparoscopic skills, and transfer effects studies have shown improved performance (Andreatta et al., 2006). LapSimGyn is a virtual

FIGURE 44.15   Air Traffic Control tower simulator NASA FutureFlight Central.

FIGURE 44.16 3D cargobay inspection training simulator.

reality simulator developed for training gynecologic laparoscopic skills (Larsen et al., 2006). Moorthy et al. (2003) evaluated a VR bronchoscopy training simulator and found that it can be an effective training tool. Nakao et al. (2002) developed an interactive VR simulation of a beating heart for cardiovascular surgery simulations. The SIMENDO is a VR simulator developed to train hand-eye coordination in endoscopic surgery (Verdaasdonk et al., 2006). Tsai et al. (2001) have developed an interactive virtual reality orthopedic surgery simulator to be used for training surgical skills in using instruments to operate on rigid anatomic structures and also for planning and rehearsal of operations. NES (Voss et al., 2000), a nasal endoscopy training simulator, is a VR training system for training in endonasal sinus surgery. O'Toole et al. (1999) evaluated a surgical simulator with force feedback designed to provide training and assessment of suturing techniques. Other applications of VR have been in training emergency response personnel like the simulation of a stretcher evacuation in an off-shore gas platform (Hubbold & Keates, 2000).

FIGURE 44.17 View of trainer in a CVE.

44.10 Applications of VR in Manufacturing

In the manufacturing industry also, VR has had extensive applications in training. Lin et al. (2002) developed a virtual Computer Numerical Control (VR-CNC) milling machine training simulator and also provided architecture for the development of VR-based training systems. Virtual reality simulators have been developed for welder training for shipbuilding (Porter et al., 2006).

   In this chapter we have presented an overview of the application of virtual reality for training and a basic methodology for modeling tasks for the development of VR training simulators. As technology becomes cheaper, it is expected that VR simulators for training will become pervasive in a variety of application areas.

References

Andreatta, P.B., Woodrum, D.T., Birkmeyer, J.D., Yellamanchilli, R.K., Doherty, G.M., Gauger, P.G., and Minter, R.M., 2006, Laparoscopic skills are improved with LapMentor training: results of a randomized, double-blinded study. *Annals of Surgery*, 243(6), 854–860.

Andrews, D.H., Carroll, L.A., and Bell, H.H., 1996, The Future of Selective Fidelity in Training Devices. Armstrong Lab Wright-Patterson AFB OH Human Resources Directorate Final Report No. ADA316902.

Biocca, F., Inoue, Y., Polinsky, H., Lee, A., and Tang, A., 2002, Visual cues and virtual touch: Role of visual stimuli and intersensory integration in cross-modal haptic illusions and the sense of presence. In Proceedings of *Presence 2002*. Porto Portugal: Fernando Pessoa University.

Card, S.K., Moran, T.P., and Newell, A.L., 1983, *The Psychology of Human Computer Interaction*. Hillsdale, NJ: Erlbaum.

Casati, R., and Pasquinelli, E., 2005, Is the subjective feel of "presence" an uninteresting goal? *Journal of Visual Languages and Computing* 16, 428–441.

Cohen, A., Smith, M.J., and Anger, W.K., 1979, Self-protective measures against workplace hazards, *Journal of Safety Research*, 11(3), 121–131.

Cruz-Neira, C., 1993, Virtual reality overview. ACM SIGGRAPH '93 course notes: applied virtual reality. ACM SIGGRAPH '93 Conference, Anaheim, CA.

Czaja, S.J., and Drury, C.G., 1981, Training programs for inspection. *Human Factors*, 23(4), 473–484.

Duchowski, A.T., Cournio, N., Cumming, B., McCallum, D., Gramopadhye, A.K., Greenstein, J.S., Sadasivan, S., and Tyrell, R.A., 2004, Visual deictic reference in a collaborative virtual environment. In: Proceedings of Eye Tracking Research & Applications (ETRA) Conference, March 22–24, San Antonio, TX, ACM, pp 35–40.

Ferlitsch, A., Glauninger, P., Gupper, A., Schillinger, M., Haefner, M., Gangl, A., and Schoefl, R., 2002, Virtual endoscopy simulation for training of gastrointestinal endoscopy. *Endoscopy*, 34(9), 698–702.

Ford, D. N., and J. D. Sterman, 1997, Expert knowledge elicitation to improve mental and formal models. *System Dynamics Review* 14(4), 309–340.

Garau, M., Slater, M., Vinayagamoorthy, V., Brogni, A., Steed, A., and Sasse, M.A., 2003, The impact of avatar realism and eye gaze control on perceived quality of communication in a shared immersive virtual environment. In: Proceedings of the SIGCHI Conference on Human Factors in Computing Systems, pp. 529–536. ACM.

Gordon, S.E., 1994, *Systematic Training Program Design: Maximizing Effectiveness and Minimizing Liability*. Upper Saddle River, NJ: Prentice Hall.

Gramopadhye, A.K., Drury, C.G., and Sharit, J., 1993, Training for decision making in aircraft inspection. In: Proceedings of the Human Factors and Ergonomics Society 37th Annual Meeting, Seattle, WA, pp. 1267–1272.

Hendrick, H.W., 1997, Organizational design and macroergonomics. In: *Handbook of Human Factors and Ergonomics,* 2nd ed. G. Salvendy, Ed. New York: John Wiley.

Hubbold, R., and Keates, M., 2000, Real-time simulation of a stretcher evacuation in a large-scale virtual environment. *Computer Graphics Forum,* 19, 123–134.

Johnson, L., and Johnson, N., 1987, Knowledge elicitation involving teachback interviewing. In: *Knowledge Elicitation for Expert Systems: A Practical Handbook.* A. Kidd, Ed. New York: Plenum Press, 91–108.

Kieras, D., 1988, Towards a practical GOMS model methodology for user interface design. In: *Handbook of Human-Computer Interaction.* M. Helander, Ed., pp. 135–158. Amsterdam: North-Holland.

Kirwan, B., and Ainsworth, L.K., Eds., 1992, *A Guide to Task Analysis.* London: Taylor & Francis.

Klein, G., 2004, Cognitive task analyses in the ATC environment: Putting CTA to work. Presentation given on May 19, 2004 to the FAA Human Factors Research and Engineering Division.

Kohn, L., Corrigan, J.M., and Donaldson, M.S., 1999, *To Err Is Human—Building A Safer Health System.* Washington, DC: National Academy Press.

Larsen, C.R., Grantcharov, T., Aggarwal, R., Tully, A., Sorensen, J.L., Dalsgaard, T., and Ottesen, B., 2006, Objective assessment of gynecologic laparoscopic skills using the LapSimGyn virtual reality simulator. *Surgical Endoscopy* 20, 1460–1466.

Latorella, K.A., Gramopadhye, A.K., Prabhu, P.V., Drury, C.G., Smith, M.A., and Shanahan, D.E., 1992, Computer-simulated aircraft inspection tasks for off-line experimentation. In: Proceedings of the Human Factors Society 36th Annual Meeting, Santa Monica, CA: The Human Factors Society, pp. 92–96.

Lin, F., Ye, L., Duffy, V.G., and Su, C.-J., 2002, Developing virtual environments for industrial training, *Information Sciences,* 140(1), 153–170.

Mehta, P., Sadasivan, S., Greenstein, J.S., Duchowski, A.T., and Gramopadhye, A.K., 2005, Evaluating different display techniques for search strategy training in a collaborative virtual aircraft inspection environment. In: Proceedings of Human Factors and Ergonomics Society Annual Conference, 2005, Orlando, FL.

Moorthy, K., Smith, S., Brown, T., Bann, S., and Darzi, A., 2003, Evaluation of virtual reality bronchoscopy as a learning and assessment Tool, *Respiration,* 70, 195–199.

Nakao, M., Oyama, H., Komori, M., Matsuda, T., Sakaguchi, G., Komeda, M., Takahashi, T., 2002, Haptic reproduction and interactive visualization of a beating heart for cardiovascular surgery simulation, *International Journal of Medical Informatics,* 68(1–3), 155–163.

Nielsen, J., 1997, Usability testing. In *Handbook of Human Factors and Ergonomics,* 2nd ed. G. Salvendy, Ed. New York: John Wiley.

O'Toole, R.V., Polayter, R.R., and Krummel T.M., 1999, Measuring and developing suturing technique with a virtual reality surgical simulator, *Journal of the American College of Surgeons,* 189, 114–127.

Patrick, J., 1992, *Training Research and Practice.* New York: Academic Press.

Porter, N.C., Cote, A.J., Gifford, T.D., and Lam, W., 2006, Virtual reality welder training, *Journal of Ship Production,* 22(3), 126–138.

Reigeluth, C.M., and Stein, R., 1983, The elaboration theory of instruction. In: *Instructional-Design Theories and Models: An Overview of Their Current Status.* C. M. Reigeluth, Ed. Hillsdale, NJ: Lawrence Erlbaum.

Rush, R., and Wallace, W., 1997, Elicitation of knowledge from multiple experts using network inference, *IEEE Transactions on Knowledge and Data Engineering,* 9(5), 688–696.

Rubin, K.S., Jones, P.M., and Mitchell, C.M., 1988, OFMspert: Inference of operator intentions in supervisory control using a blackboard architecture. *IEEE Transactions on Systems, Man, and Cybernetics,* 18(4), 618–637.

Sadasivan, S., Vembar, D., Washburn, C., and Gramopadhye, A.K., 2007, Evaluation of international devices for projector based virtual reality aircraft inspection traning environments. In: Proceedings of HCI International Annual Conference, July 22–27, 2007, Beijing, P.R. China.

Sadasivan, S., Vembar, D., Stringfellow, P.F., and Gramopadhye, A.K., 2006a, Comparison of Window-VR and HMD display techniques in virtual reality inspection environments. In: Proceedings of the International Ergonomics Association Conference, July 10–15, 2006, Maastricht, The Netherlands.

Sadasivan, S., Vember D., Stringfellow, P.F., and Gramopadhye, A.K., 2006b, Evaluation of interaction devices for NDI training in VR: gamepad vs. joystick. In: Proceedings of Human Factors and Ergonomics Society Annual Conference, 2006, San Francisco, CA.

Sadasivan, S., Rele, R.S., Greenstein, J.S., Duchowski, A.T., and Gramopadhye, A.K., 2005a, Stimulating on-the-job training using a collaborative virtual environment with head slaved visual deitic reference. In: Proceedings of HCI International Annual Conference, July 22–27, 2005, Las Vegas, NV.

Sadasivan, S., Greenstein, J.S., Gramopadhye, A.K., and Duchowski, A.T., 2005, Use of eye movements as feedforward training for a synthetic aircraft inspection task, CHI '05, April 2–7, 2005, Portland, OR, ACM.

Sadowski, W., and Stanney, K.M., 2002, Measuring and managing presence in virtual environments. *Handbook of Virtual Environments: Design, Implementation, and Applications.* Mahwah, NJ: Lawrence Erlbaum, pp. 791–806.

Salvendy, G., and Seymour, D.W., 1973, *Prediction and Development of Industrial Work Performance.* Toronto: John Wiley.

Schneider, W., 1985, Training high performance skills: Fallacies and guidelines. *Human Factors*, 27(3), 285–300.

Schricker, B.C., and Lippe, S.R., 2004, Using the high level architecture to implement selective-fidelity. In: Proceedings of the 37th Annual Symposium on Simulation April 18–22, 2004. IEEE Computer Society, Washington, DC, p. 246.

Shepherd, A., 2000, *Hierarchical Task Analysis.* New York: Taylor & Francis.

Sheridan, T., 1992, Musings on telepresence and virtual presence. *Presence: Teleoperators and Virtual Environments*, 1(1), 120–125.

Sheridan, T.B., 2000, Interaction, imagination and immersion some research needs. Proceedings of the ACM symposium on Virtual reality software and technology, October 22–25, Seoul, Korea.

Slater, M., 1999, Measuring Presence: A response to the witmer and singer presence questionnaire. *Presence: Teleoperators and Virtual Environments*, 8(5), 560–565.

Slater, M., and Usoh, M., 1993, Representations systems, perceptual position, and presence in Virtual Environments. *Presence: Teleoperators and Virtual Environments,* 2(3), 221–233.

Slater, M., and Wilbur, S., 1997, A framework for immersive virtual environments (FIVE): Speculations on the role of presence in virtual environments. *Presence: Teleoperators and Virtual Environments,* 6(6), 603–616.

Stanney, K.M., Mourant, R.R., and Kennedy, R.S., 1998, Human factors issues in virtual environments: A review of literature. *Presence,* 7(4), 327–351.

Stringfellow, P.F., Sadasivan, S., and Gramopadhye, A.K., 2006, Computer-based nondestructive inspection training and its potential in aviation maintainence. In: Proceedings of the Internationl Ergonomics Association Conference, July 10–15, 2006, Maastricht, The Netherlands.

Tsai, M-D, Hsieh, M-S, and Jou, S-B., 2001, Virtual reality orthopedic surgery simulator. *Computers in Biology and Medicine,* 31, 333–351.

Usoh, M., Catena, E., Arman, S., and Slater, M., 2000, Using presence questionnaires in reality. *Presence: Teleoperators and Virtual Environments,* 9(5), 497–503.

Vembar, D., Sadasivan S., Stringfellow, P.F., and Duchowski, A.T., 2006, Gamepad vs. keyboard: An evaluation of interaction devices for NDI training in VR. In: Proceedings of the International Ergonomics Association Conference, July 10–15, 2006, Maastricht, The Netherlands.

Vembar, D., Sadasivan, S., Duchowski, A.T., Stringfellow, P.F., and Gramopadhye, A.K., 2005, Design of a virtual borescope: A presence study. In: Proceedings of HCI International Annual Conference, July 22–27, 2005, Las Vegas, NV.

Verdaasdonk, E.G., Stassen, L.P., Monteny, L.J., and Dankelman, J., 2006, Validation of a new basic virtual reality simulator for training of basic endoscopic skills: the SIMENDO. *Surgical Endoscopy*, 20, 511–518.

Vora, J., Nair, S., Gramopadhye, A.K., Melloy, B.J., Meldin, E., Duchowski, A.T., Kanki, B.G., 2001, Using virtual reality technology to improve aircraft inspection performance: presence and performance measurement studies. In: Proceedings of the Human Factors and Ergonomic Society 45th Annual Meeting, pp. 1867–1871.

Voss, G., Ecke, U., Müller, W.K., Bockholt, U., and Mann, W., 2000, How to become the "High Score Cyber Surgeon"—endoscopic training using the nasal endoscopy simulator (NES). In: Proceedings of Computer Assisted Radiology and Surgery (CARS) 2000, pp. 290–293.

Watson, B.A., Walker, N., Ribarsky, W.R., and Spaulding, V., 1998, The effects of variation in system responsiveness on user performance in virtual environments. *Human Factors*, Special Section on Virtual Environments, 40(3), 403–414.

Weiner, E.L., 1975, Individual and group differences in inspection. In: *Human Reliability in Quality Control*. C. G. Drury and J.G., Eds. Fox. London: Taylor & Francis.

Wickens, C.D., and Hollands, J., 2000, *Engineering Psychology and Human Performance*. Upper Saddle River, NJ: Prentice Hall.

Wiegmann, D.A., and Shappell, S.A., 2001, A human error analysis of commercial aviation accidents using the human factors analysis and classification system (HFACS) DOT/FAA/AM-01/3. Washington, DC, FAA.

Witmer, B., and Singer, M., 1998, Measuring presence in virtual environments: A presence questionnaire. *Presence: Teleoperators and Virtual Environments*, 7(3), 225–240.

45

Lab Testing and Field Testing in Digital Human Modeling

Tania Dukic and
Lars Hanson

45.1 Introduction

Digital human modeling is used to verify human-machine interaction requirements for designing products and workplaces. Digital human modeling tools are essential, consisting of three major parts. First, a mannequin can be a replica of the human being with respect to the design of biomechanical design and motion behavior. Second, a CAD tool analyzes products and/or production environments. Third, ergonomic evaluation tools can analyze human products or human production environment interaction. Digital human modeling is a relatively new tool, and users ask for improvements in human motion behavior control and ergonomic evaluation methods. In order to know how human motions are carried out in certain situations, experiments in laboratories and in the field are conducted. Human health and quality of performance are also registered during these motion experiments. Information about human motions is implemented in digital human modeling tools as posture/motion databases that allow users to predict mannequin motions in the CAD environments. Linked to motion patterns are information about human health and performance quality, which can used as a human product or production environment interaction evaluation tool.

The different parts of the digital human model are unequally developed. Little work is done to develop methods for predicting a mannequin's vision behavior and integration of evaluation methods adapted to humans' vision capability. There is a need to conduct lab and field testing to develop such aspects further. The present chapter consists of three parts. First, a state of the art of digital models' performance on vision evaluation is presented. Second, three series of experiments are presented, one lab and two field experiments performed according to a standardized method to collect data on humans' visual behavior. All experiments come from applied research projects in collaboration with automotive companies (Dukic et al., 2005, 2006; Han Kim et al. 2006). Third, we discuss different ways to implement those data within actual digital human modeling tools.

45.2 Vision Module: Visual Evaluation

Among the first commercialized digital humans (Bonney et al., 1974), ADAPS incorporated a view module (Hoekstra, 1993). Users were able to direct the digital human's line of sight by pointing a mouse cursor at the target where the digital human was to direct its view. It was also possible to display the field of view of the digital human by using the "field of view" module. In the field of view, users were able to see through the digital human's eyes and different viewpoints could be displayed (Hoekstra, 1993). In today's digital human programs, new algorithms exist for displaying the field of view more dynamically. It is possible to remain within the digital human's field of view when performing a task and to obtain pictures from its environment. The digital human's user can "look around" and redefine a new center for the field of view, the algorithm recalculates a corresponding head and neck posture, and displays results. Usually, the scene can also be seen through the eyes of the digital human displayed in a special window. Visual fields characteristics are differently represented by various digital models and can be customized. The majority of the studies reported in the literature perform vision analysis in a similar way. A digital human, with relevant anthropometry, is placed in a workplace and an appropriate working posture is chosen, usually based on the posture prediction database or the judgment of the user. The fields of view are then displayed in order to visualize what the digital human is able to see. Vision analysis is applied in different fields: industrial (Sundin et al., 2000; Che Doi & Haselgrave, 2003), aerospace (Goutal, 2000; Nelson, 2001), and automotive (Rix et al., 2000; Bowman, 2001; Hudelmaier, 2002). The automotive studies attempt to verify whether visibility conditions inside and outside the vehicle are both feasible and in accordance with requirements. For assembly work, visual analysis takes the assembly sequence into account in order to check whether any visual obstruction will occur in a given sequence. In many cases, the digital human's user tries to test different computer mannequins characterized by different anthropometrics to represent the product or production environment target group. Documentation concerning vision analysis is composed of visualizations of what the digital human sees. Reflective zones can be part of this documentation, primarily for automotive industry applications (e.g., to check rear mirror vision). Field of view is the most frequently used option for evaluation of operator vision in a certain workplace or during the use of a product. However, the human vision is more complex and there is a need for further development beside the field of view simulation.

45.3 Standardization

If a question could not be answered by any other existing methods and available digital models, there is a gap that can be filled by collecting new data through experiments performed either in a lab or in a field study. Figure 45.1 illustrates the working method to follow in order to collect data in a standardized way. Collecting data to use for new models requires a standardized method. The standardized way of collecting data is illustrated through the case study and the field study below; however, it could be summarized as a general scientific method to collect data where some rules need to be followed such as: control of variables to be measured, describe the study so that it is possible to be reproduced by someone else, and so on. This standardized methodology could be summarized as:

- Define a research question to be answered
- "Operationalize" the research question
- Define dependent and independent variables
- Choose a relevant environment in which to perform the study (lab or field environment)
- Collect data

- Perform statistical analysis
- Model
- Answer the question
- Develop a model

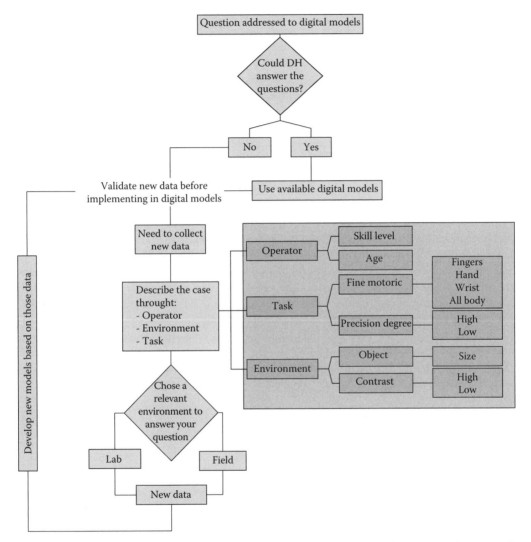

FIGURE 45.1 Process to collect new data to implement into digital models, from the question to be answered to validation of new data implemented into digital model.

The choice of lab or field study is often a consequence of the nature of the research question to be answered. Some issues are per nature not possible to study in a lab and some others are not able to be performed in a field study. This choice does sometimes not exist at all.

45.4 Field Study

45.4.1 The Gap in DHM

Digital human modeling tools are currently able to predict vision cones, generate SAE eye ellipse, and allow a user to see through the mannequin's eyes. When driving and operating a control in a center panel, the driver performs two manual tasks simultaneously guided by two visual tasks. The primary task, driving, requires almost 100% visual focus on the road and road environment and the hands on the steering wheel for manual control of the vehicle. The secondary task, to manually push a button with a finger, requires some visual attention for detection of button position. Determining how long a driver can keep the eyes from the road to perform secondary tasks without affecting driving quality (e.g., Zwahlen, 1998) would be an example in which existing digital human modeling tools do not include a way to control a digital model with constraints such as (1) primary vision task is to focus on road and road environment, (2) primary manual task is to control car's position with steering wheel, (3) secondary vision task is to detect control button, (4) secondary manual task is to push control button. It is not yet possible to perform simulation and analysis of two simultaneous tasks, and task description is not divided into manual and visual behavior. Furthermore, it is not possible to predict the visual time required to perform a task. Therefore, there is a need for a prediction tool that allows two simultaneous tasks described in manual and visual demand. Development of a prediction tool that can generate information about fixation time off road is a first start—information that can be compared with existing visual guidelines on maximum time off road.

Furthermore, for transportation, many young as well as older people drive vehicles. Aging affects vision, and most sensory functions (Fozard, 1990). The ability to accommodate visual input decreases with age and older people need more light than younger people to perform the same task (Olson, 1993). The near point distance (i.e., the closest point at which a person is able to accommodate) becomes more distant with age, and visual acuity decreases (Sekular et al., 1982; Shaheen & Niemeier, 2001). Furthermore, a decline can be observed in the size of the visual field (Brug, 1968). To predict visual time off road when driving and operating a button in central stack for a certain task, it is therefore necessary to consider the age of the person. In the digital human modeling tools today mannequin age is not a factor obviously considered. The mannequins' anthropometrics, including length, weight, and joint range motion can be modified to replicate the human of interest. Joint range motion is as affected by aging as vision, mainly due to joint diseases that are common in older ages. By modifying joint motion, aging could be considered. Vision cones and other vision analysis tools can not be modified in a similar way. Because there are several senses that are affected by aging, these changes can be added into the parameter and coupled with related parameters.

Two field experiments were conducted to investigate the visual time off road required when operating a push button in the center panel. Also, we investigated how the visual time off road was affected by the push button position and the driver's age (Dukic, 2005, 2006).

45.4.2 Methodology

45.4.2.1 Equipment

An instrumented vehicle was used in both experiments. The steering wheel deviations were recorded during the drive as a measure on driving quality. Car data were recorded in a vehicle information file with a sampling rate of 10 Hz. Eye movements were recorded with a head-mounted eye-tracking system, SMI iView, using infrared cameras to capture the positions of the pupil and the corneal reflex at 50 Hz. The video signals from the eye tracking system, eye, and scene with eye focus indicator were passed on to a video splitter and recorded on videotape. All video signals were sampled at 25 Hz. A manually operated control unit was connected to the video system and the recording unit for the car data. The control unit synchronized the two systems by placing an indication in both the video file and the car data file. In both experiments the control layout of the center panel was modified to allow comparison of horizontal and vertical eye movements and to cover a wider range of controls position. Such a wide range of control positions is facilitated by modeling. Furthermore, a modified center panel makes the data generic and not vehicle-brand specific.

45.4.3 Subjects

Two different strategies were used to recruit subjects to the two field experiments. In the first experiment, the subjects were invited to respond to an advertisement flyer distributed around campus. In second experiment, the subjects answered a personal invitation. Subjects and their addresses were randomly drawn from the Swedish car owner register. In both experiments the following two criteria were used. (1) The subject must be a resident in Lund municipality. This was to minimize subjects' time spent on the experiment. Time is valuable for everybody and lack of time is an excuse frequently mentioned for not participating. (2) The subject must have been in possession of a valid driving licence for at least 5 years. This criterion is due to the belief that experienced drivers require less time to get used to a new vehicle. To minimize the learning curve was of high importance. In the second field experiment, age was an additional criterion due to the purpose of the experiment, comparing younger to older drivers.

45.4.4 Procedure

During the field experiments, the driver's primary task was to drive as usual on a two-lane highway. The secondary task was to push a button on the panel when the experiment supervisor so instructed. Subjects were informed about the experimental procedure and signed a written consent before entering the car. They were asked to drive normally and to obey traffic rules. In the car, subjects were asked to adjust the seat and steering wheel to comfortable positions, and to adjust the mirrors to facilitate side and rear views. The subjects were accompanied in the car by an experiment supervisor who sat in the passenger seat and by an eye-tracking operator sitting in the backseat. Each subject started the experiment with a training period to become accustomed to the car and to the secondary tasks. After the training period, the eye-tracking equipment was mounted on the subject and was calibrated. During the data acquisition stage, the experiment supervisor orally gave instructions to the subjects to push a button, and simultaneously the supervisor pushed a button on the manual control unit, which entered a cue into the vehicle information file and the eye-tracker video. The experiment supervisor pushed the same button when the button task was completed (i.e., the subject's right hand had returned to the steering wheel, gear stick, or knee). After the drive, the subjects were asked to mark their perceived safety for each of the eight buttons on a 100-mm-long visual analogue scale with anchor points "very unsafe" and "very safe." The subjects were also asked to complete a questionnaire about their perception of the experimental situation, their driving behavior, and the ecology of the experiment.

45.4.5 Field Experiment Results

In the field study it was shown that control position and age affect visual behavior. Older drivers had significantly longer visual time off road than younger drivers for all button locations. Button location had a significant effect on visual time off road. Not surprisingly, a control further away from the primary focus point gets more visual time off road compared to a control closer to the primary focus point. When comparing controls with the same distance from the primary focus point, it was shown that horizontal eye movements are faster compared to vertical movements. However, this statement does not hold for a control close to the gear stick, which is far from the primary focus point. In a study it was shown that this control position almost requires low or sometimes no visual time off road. Three potential explanations for this difference are:

1. Discrimination level: To push a button on the center stack, subjects were required to find and choose the correct button among other alternatives placed relatively close together. The button close to the gear stick was alone with no other button alternative nearby. The number of alternatives to choose among affects decision time (Hick, 1952).
2. Driver's perception of risk: If subjects look at the control button near the gearstick, they cannot use their peripheral vision to observe the traffic environment. This may imply that subjects think

looking at the control button there is too dangerous, because they lose all visual contact with the traffic situation, as well as their motion detection capacity. Thus, they try to keep this visual time off road to a minimum.

3. Motor skill: Drivers were very quick in developing a good skill level in performing the arm movement needed to push a control button close to the gearstick, a movement similar to that needed for reaching for the gear stick, which is a frequent movement in driving. Moreover, the degree of freedom in performing this movement was very limited due to the interior design (i.e., seat design limits and guides for the arm movement). Due to the ease of movement, the need for visual guidance was limited.

45.5 Laboratory Study

45.5.1 The Gap in DHM

Current knowledge in ergonomics is rather extensive with respect to what kind of physical loads can lead to musculoskeletal disorders (Bernard, 1997). However, the knowledge on the effects of visual demand on manual task performance is not well documented (Delleman, 1992; Li & Haslegrave, 1999) but is considered important. Helander and Furtado (1992) said the following in *Design for Assembly* (p.182): "Visibility and visual feedback play a vital role in assembly. Visual feedback occurs simultaneously with motions such as 'reach,' 'move,' 'position,' and so forth." (In MTM it is not included in the analysis.) In the manufacturing process, one requirement is that all features should be fully visible and provide visual feedback. Hidden features complicate assembly and are costly. Thus, according to the literature, there seems to be an obvious need for practical ergonomic knowledge to evaluate visual demand during the performance of a manual task. Several studies have shown a clear relation between increased visual demand and poor working posture (Laville, 1985; Delleman, 1992; Haslegrave, 1994; Li & Haslegrave, 1999; Wartenberg et al., 2004). One example of a cause of an inappropriate working posture is when an operator has to crouch to see a component through a window in a machine (Haslegrave, 1994). It seems that poor compliance with visual demand in industry is responsible for many of the poor working postures observed.

The study aimed at collecting data to understand the effect of both visual and manual demands on upper body movements and visual behavior. This issue is very important for manual assembly.

45.5.2 Methodology

45.5.2.1 Equipment

Each target is composed of a small cylindrical button protruding to the right (diameter: 2.5 cm; length: 3.8 cm) from a board placed perpendicularly to the frontal plane (fig. 45.1a). The targets were placed at seated shoulder height at 70% (close) or 120% (far) of the extended arm and hand length of each subject. The targets were aligned with either the mid-sagittal plane (medial) or the right shoulder (lateral). Upper body motions were recorded by an optical motion-tracking system (PC Reflex) with six cameras. Eye motions were recorded using an infrared camera-based tracking system (SMI IView) placed on a helmet. Both systems were synchronized and body and eye movements were sampled at a frequency of 50 Hz.

45.5.3 Subjects

Eight male and one female volunteers participated in the experiment (mean age: $24 \pm$ SD 2.1). All were native English speakers, right-handed, with normal or corrected to normal vision.

45.5.4 Procedure

A preparation phase, before measurements, was needed to place the markers on the subject, set the eye-tracker, and calibrate both systems. From a neutral seated position on a stool, at x cm distance from the panel (depending on arm length), with the right hand on the knee and gaze at a reference fixation point in the mid-sagittal plane (fig. 45.2a), the subject was asked to either reach for the target with the right index finger (low-visual, high-manual demand condition), read a word placed just above the target (high-visual, low-manual demand condition), or simultaneously read and reach for the target (high-visual and high-manual demand condition). For each trial, the subject was orally informed of the task condition to be performed. For the high-visual condition, a randomly selected five-letter word was presented. The motion was initiated by the experiment leader's verbal signal and performed at comfortable speed, and no constraints were provided about movement time. After they performed the task, the subjects were asked to come back in the initial position with their hands on knees and the eyes back to the fixation point indicated before (marker on the top of the panel). Only the right side was recorded. For each target position and task condition, either two or six replications were performed for lateral or medial target configurations. The order of target positions and task conditions was randomized and balanced across subjects. A total of 48 trials were performed by each subject.

45.5.5 Lab Experimental Results

The results show that the manual and visual demands modify the movements associated with the manual and visual target acquisition, respectively. The observed differences between conditions suggest an interaction and cooperation between the visual and manual functions.

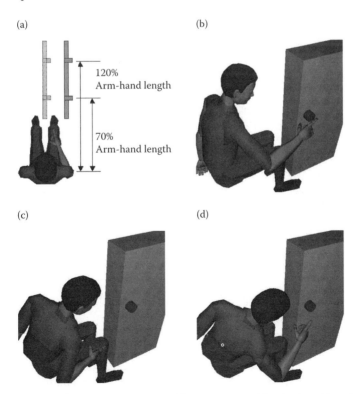

FIGURE 45.2 (a) Target configuration in a top view. The targets were either in the mid-sagittal plane or in the plane perpendicular to the right shoulder of each subject. (b–d) A read-only, reach-only, and simultaneous read and reach task, respectively (Han Kim et al., 2006).

Manual demand, represented by target distance, obviously influences torso flexion. High-manual demand also influenced the gaze on target duration. Although the origin of the longer fixations on the far target may not be clearly identified, it may be due to the small visual angles of the target (read task condition) and/or requirement for feedback control of hand movements (reach task condition). Visual demand influenced the torso flexion angle. The read-only task resulted in a larger torso angle compared to the reach-only task. It is suggested that the orientation of the visual target (word) facing to the right induces a large torso flexion, which is required to place the head and eyes in an adequate position for visual target acquisition. The effects of visual and manual demands interact with each other. Manual demand modulates the activities of visual function, as demonstrated by gaze-on-target duration changes with target distance in the read-only condition. When read and reach tasks are performed simultaneously, torso movements and gaze-on-target duration were comparable to those in read-only tasks. Hence in the context of the task conditions presented in this study, visual demand has a predominant effect on both visual and manual target acquisition movements.

Results of this study indicate that tasks to be simulated should be described in terms of both visual and manual demand, which is not the case today. It is only described manually to position the digital human. Modeling approaches should take into account the spatial and temporal interactions of the multiple systems responsible for specific sensorimotor functions in order to produce realistic movements and postures.

45.6 Implementing New Data Inside Digital Models, or How to Close the Loop?

This part refers to the feedback loop in Figure 45.1 from the step where new data have been collected to their implementation inside available digital models. This step is not an easy one, and there are several ways to proceed, depending on the type of data, in order to develop new models.

The results of both field and lab studies are clear: Vision is a parameter to consider. However, one would like to avoid collecting data again each time a similar question is raised since experiments are expensive to perform. It is preferable to incorporate results inside digital models in order to predict and evaluate visual demand in similar and other situations. An implementation would hopefully lead to the use of studies on visual issues that were considered earlier in the design process, and the number of vision-related field experiments could be reduced. A further approach would be to add a visual demand function link to the code-based motion MTM prediction tool in Jack. The actual MTM function in Jack works in the following way: The operator defines the human to perform the task and chooses a task such as "get, go, touch" and defines as well the object to perform the task on. A visual demand function link to the MTM code will position the digital model by adjusting the distance between the head and the hands, the position of the torso and the visual time spent on target based on the data recorded. Specifically, torso flexion angles (manual function) and gaze-on-target time (visual function) were measured to determine the effect of different task conditions (i.e., reach, read, and simultaneous read and reach tasks) on movement/posture organization. This new function would fine-tuned to the final position of the digital model to perform the task.

The experiences gained in the field study presented here could be implemented in a digital human model as a prediction for visual time off road considering control position, vehicle characteristics, and human characteristics. In order to model time off road, the neuro-fuzzy modeling technique has been used. The adaptive neuro-fuzzy interference system in MATLAB® was used (Jang et al., 1997) for defining and tuning the Gaussian-formed membership functions of the Sugeno type system. Tuning was done 20 times with a combination of back-propagation and least squares methods in order to minimize the model error. A MATLAB® program was designed to test all combinations of parameters describing control position, vehicle characteristics, and human characteristics. Giving

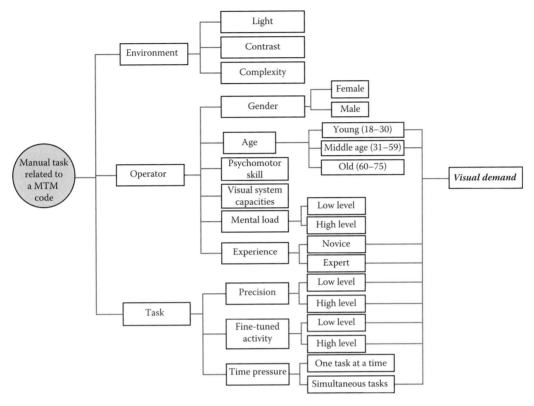

FIGURE 45.3   Relation between the manual task and the visual demand.

the lowest error, age, stature, and control position (y and z coordinates) were found as the variables suitable as input to a visual time off road prediction. A model was trained and evaluated. The performance of the visual tool and the modeled trend were acceptable. The root mean square deviation of the visual demand model with regard to the training data was 190 ms and 361 ms with regard to the testing data.

The model developed is possible to integrate in digital human modeling tools or use as a stand-alone module. Model input parameters such as stature and control position are parameters existing or possible to calculate in the tools. Age is a new variable, not considered in existing tools. However, it is found in both field experiments that age has an effect on visual behavior. Therefore, the age variable must be added manually until it is implemented. We propose that age be added in the digital human modeling tools as a mannequin design variable similar to anthropometrics. A variable that affects vision parameters, such as field of view, accommodation, and other parameters, is age, which also affects joint range, and so age should be a considered variable.

With new vision prediction models used in combination with existing posture and motion prediction tools as well as new evaluation models integrated or as stand alone, it is possible to analyze vision more thoroughly. For evaluation the driver's time off road tool, a not yet designed driver safety perception model, could be combined with existing view cones, and see-through eyes techniques (fig. 45.4). A combination of evaluation methods giving hard facts as time spent off road, driver perception, and illustrations in the form of view lines and view cones seems highly effective for accurate analysis.

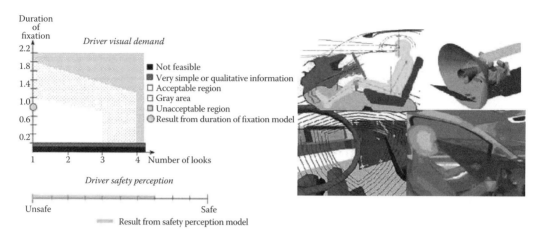

FIGURE 45.4   Output from the visual time off road model (circle) showed in guidelines from Zwahlen et al. (1988) and output from a safety perception (left), together with vision lines, view cone, and see-through eye technique.

45.7 Conclusions

Lab and field studies are necessary to deliver new data and models to digital human modeling. Following the standardized method for performing lab and field studies and being aware that implementation in digital humans require an extra step in contrast to traditional ergonomic research, namely a modeling step as described in Figure 45.1. The end product of traditional ergonomic research is frequently a new pool of data. A good contact with digital human retailers is an assumption for successful implementation.

References

Bernard, B.P. (1997). Musculoskeletal disorders and workplace factors: a critical review of epidemiologic evidence for work related musculoskeletal disorders of the neck, upper extremities, and low back. US Department of Health and Human Service: National Institute for Occupational Safety and Health, Cincinnati, OH.

Bonney, M.C., Case, K., Hughes, B.J., Kennedy, D.N., and Williams, R.W. (1974). *Using SAMMIE for computer-aided workplace and work task design.* Automotive Engineering Congress, Detroit, MI, SAE.

Brug, A. (1968). Lateral visual field as related to age and sex. *Journal of Applied Psychology* 52:10–15.

Bowman, D. (2001). Using digital human modeling in a virtual heavy vehicle development environment. *Digital Human Modeling for Vehicle and Workplace Design.* D.B. Chaffin, (Ed.). Warrendale, PA, SAE: 77–100.

Che Doi, M.A. and Haselgrave, C. (2003). *Evaluation of Jack for Investigating Postural Behavior at Sewing Machine Workstations.* Digital Human Modeling Conference, Montréal, Canada.

Delleman, N.J. (1992). Visual determinants of working posture. *Computer Application in Ergonomics, Occupational Safety and Health.* M.a.K. Mattila, W, (Ed.). Elsevier Science Publishers.

Dukic, T., Hanson, L., and Falkmer, T. (2006). Effect of drivers' ages and control location on secondary task performance during driving. *Ergonomics* 49(1): 78–92.

Dukic, T., Hanson, L., Wartenberg, C., and Holmqvist, K. (2005). Effect of button location on driver's visual behavior and safety perception. *Ergonomics* 48(4): 399–410.

Fozard, J.L. (1990). Vision and hearing in aging. *Handbook of the Psychology of Aging.* J.E. Birren and K.W. Schiare, (Eds.). New York, Academic Press, Inc.

Goutal, L. (2000). Ergonomics assessment of aircraft cockpit using the virtual mock-up. *Ergonomic Software Tools for Product and Workplace Design—A Review of Recent Developments in Human Modeling and Other Design Aids.* K. Landau, (Ed.). Stuttgart, ERGON.

Han Kim, K., Martin, B.J., Dukic, T., and Hanson, L. (2006). The role of visual and manual demand in movement and posture organization. In: Proceedings of the 2006 SAE Digital Human Modeling for Design and Engineering Conference and Exhibition, Lyon, France, July 4–6, 2006, SAE 2006-01-2312.

Haslegrave, C.M. (1994). "What do you mean by a 'working posture'?" *Ergonomics* 37(4): 781–799.

Helander, M. and Furtado, D. (1992). Product design for manual assembly. *Design for Manufacturability. A System Approach to Concurrent eEngineering and Ergonomics.* M. Helander and M. Nagamachi, (Ed.). London, Taylor & Francis: 171–188.

Hicks, W.E. (1952). On the rate of gain of information. *Quarterly Journal of Experimental Psychology* 4: 11–26.

Hoekstra, P.N. (1993). "Seeing what you are doing" with computer aided anthropometric assessment. In: Proceedings of the IEA World Conference on Ergonomics of Materials Handling and Information, Processing at Work. Warsaw, Poland, 14–17 June 1993. London, Taylor & Francis, 613–616.

Hudelmaier, J. (2002). Analysing driver's view in motor vehicles. In: Proceedings of the 2002 SAE Digital Human Modeling Conference, Munich, Germany, June 2002.

Jang, J.S.R., Sun, C.T. and Mizutani, E. (1997). *Neuro-Fuzzy and Soft Computing. A Computational Approach to Learning and Machine Intelligence.* Upper Saddle River, NJ, Prentice-Hall.

Laville, A. (1985). Postural stress in high-speed precision work. *Ergonomics* 28(1): 229–236.

Li, G. and Haslegrave, C.M. (1999). Seated work postures for manual, visual and combined tasks. *Ergonomics* 42(8): 1060–1086.

Olson, P.L. (1993). Vision and perception. *Automotive Ergonomics.* B. Peacock and W. Karwowski, (Eds.). London, Taylor & Francis: 161–183.

Nelson, C. (2001). Anthropometric analyses of crew interfaces and component accessibility for the international space station. *Digital Human Modeling for Vehicle and Workplace Design.* D.B. Chaffin, (Ed.). SAE. R-276.

Rix, J., Heidger, A., Helmstädter, C., Quester, R. and Ringhof, T. (2000). Integration of the virtual human in CA design review. *Ergonomic Software Tools for Product and Workplace Design—A Review of Recent Developments in Human Modeling and Other Design Aids.* K. Landau, (Ed.). Stuttgart, Germany, Verlag ERGON GmbH: 183–194.

Sekular, R., Kline, D., and Dismukes, K. (1982). *Aging and Human Visual Function.* New York, Alan R. Liss, Inc.

Shaheen, S.A. and Niemeier, D.A. (2001). Integrating vehicle design and human factors: minimizing elderly driving constraints. *Transportation Research, Part C* 9: 155–174.

Sundin, S., Christmansson, M. and Örtengren, R. (2000). Use of a computer mannequin in participatory design of assembly workstations. *Ergonomic Software Tools for Product and Workplace Design—A Review of Recent Developments in Human Modeling and Other Design Aids.* K. Landau, (Ed.). Stuttgart, ERGON: 204–213.

Wartenberg, C., Dukic, T., Falck, A. and Hallbeck, S. (2004). "The effect of assembly tolerance on performance of a tape application task: A pilot study." *International Journal of Industrial Ergonomics* 33: 369–379.

Zwahlen, H.T., Adams, C.C., Jr. and Debald, D.P. (1988). Safety aspects of CRT touch panel controls in automobiles. *Vision in Vehicle II.* A.G. Gale, M.H. Freeman, C.M. Haslegrave, P.A. Smith and S.P. Taylor, (Eds.). North Holland, Elsevier Science B.V. 2: 335–344.

46

Networking Human Performance Models to Substantiate Human-Systems Integration

Jennifer McGovern
Narkevicius, Timothy
M. Bagnall, Robert
A. Sargent, and
John E. Owen

46.1 Introduction

New systems, whether defined as implementation of new technology or as integration of legacy systems, are still populated by people. Those people may be remotely placed with respect to the hardware and software, but they continue to contribute their capabilities and limitations to the overarching system. Success of these systems is still related to meeting the operational capabilities and depends on the performance of the interrelated humans at expected levels. Achieving the desired operational capabilities is particularity important to the developers of complex systems. These complex systems are used by the military, power generation, and in other large organization or infrastructure entities including transportation. These all require large complex organizational implementation and technical solutions. In recent years, the human systems integration (HSI) approach has begun to be adopted as a user-centered approach for the design and implementation of advanced technology and automated systems. The HSI approach links human-based, hardware-based, and software-based domains through the systems engineering process. HSI has identified software toolsets and processes to support the challenging decision-making aspects of materiel acquisition.

HSI is enjoying a rise in popularity. There are two primary reasons. The military, currently the primary users of the HSI approach, are experiencing an unprecedented number of large, high-value, technically challenging programs in development. Concurrently, there has been a change in the zeitgeist of the intellectual elite as continued growth in technology has not delivered desired results. Thought leaders are beginning to understand the role people play in technology (and organizational) systems in new ways.

In the military acquisition of complex technological systems, cross-program integration is becoming a more significant issue. This has recently been reflected in programs such as the Navy's Next Generation Carrier (CVN-21). This program has a Manpower Key Performance Parameter (KPP) of 20% to 25% reduction of the ship's crew. The Littoral Combat Ship has worked with a very low number of personnel as reflected in its Manpower KPP. This Manpower figure has been a technical challenge for a number of human-based domains (including human factors engineering, habitability, and personnel) when taken together with the modular work structure proposed to meet the very flexible concept of operations.

To define the requirements of humans as a fundamental system component, it is essential to understand the inherent "capacity" of user populations and their typical operational environment. This is more than the basic anthropometrics or cognitive capability of the average member of the user population (Booher, 2003; Chapanis, 1996) and includes detailed descriptions and understanding of the target audience of users and maintainers and explicit understanding of the knowledge, skills, and abilities (KSAs) of the people that will be operating and maintaining the system as well as an understanding of the work that must be performed. A number of authors have explored the definition and use of occupational information as well as the effects of organizational structure, business processes, and work structure (Cook, 1996; Kubeck, 1995; Peterson et al., 1999; Sheridan, 2002; and Wilson & Corlett, 1995 are some examples of these diverse disciplines).

The human-based domains include not only human factors engineering but also manpower, personnel, training, system safety, habitability, personnel survivability, and environment, safety, and occupational health (ESOH) (Department of Defense, 2003). These domains may be distributed organizationally in a number of ways depending on the organization practicing HSI, but they are all present in some form in all currently established HSI programs worldwide (reference USAF/USN/USA/MOD/CF websites). These more diverse data must be included in SE and trade space analyses to ensure that the system will perform as envisioned and specified in the operational environment with the prescribed user population. Additionally, the team structure must also be addressed as many complex automated systems result in overt or covert changes to the organizational structure and business rules. These organizational changes can affect the work to be performed and must be considered as part of the overarching design. They must be reflected in the information architecture of the system as well, especially if there are automation or decision-making support elements to be included in the system.

It is essential to fully understand both the work to be performed by the system and the context in which that work will occur when users operate the system. Complex advanced technology systems, especially automated mission systems, have made many technological advances toward being more responsive or appropriate to the humans with whom they interact. However, the effect of the implementation of the system on the work performed by the component humans is not well understood or accounted for in design of those systems. The work the humans perform (including workflow) must be defined. That definition must be utilized locally in the human factors of the design and globally in the overall systems design and organizational structure. In addition, it is important to socialize that definition of the work to be performed into the organization and among the claimants of that work. These business processes, organizational structures, and occupational work must also be factored into the systems thinking and design. This includes eliciting the explicit and implicit information flows necessary to perform the mission and to support the human decision-making processes.

To achieve the goal of successfully integrating humans into the SE of automated systems, especially for decision making, it is essential to achieve HSI. This requires actually integrating the human domains and applying the products of that successful integration to the design of automated decision-making systems. Modeling and simulation (M&S) tools provide an excellent workspace to achieve these trade-offs, both across HSI domains and among competing engineering solutions.

46.2 Human Systems Integration Overview

Complex systems require careful engineering to deliver successful performance of the desired capability. This engineering is difficult when restricted to the development and integration of hardware and software; however, the challenges increase when the integration of the human begins and performance of the system again changes.

At the start of the new millennium, a metacognition began to take shape with respect to the integration of humans and hardware/software systems. That is, system designers began to understand that a well-designed system capitalized on the human element of the system. Too many systems had been successful in test and evaluation only to fail in operation because users were not able to optimize their own performance to ensure that the system exhibited the desired performance. Although the U.S. Army had developed an HSI process in the mid-1980s, in the early 2000s others began to join them. Defining a human-centered design policy for materiel acquisition in the Navy and the Air Force began. These HSI programs and others share mandating the implementation of HSI policy in acquisition. A formal HSI policy ensures that systems engineers treat the user and maintainer as the integral components of the system that they are. This inclusion reduces life cycle costs and contributes to optimizing system performance. HSI policies enumerate "best practices" for HSI practitioners to integrate seamlessly into SE for acquisition. These best practices include use of modeling and simulation to support successful analysis, development, and implementation of programs throughout the life cycle.

HSI programs must have organizational backing to ensure that the technical and management aspects of HSI are institutionalized. These management and technical processes must be set out in the organizational policies. These policies also ensure that HSI tools and analyses are integrated into the systems engineering process. This ensures that the human considerations are included as design drivers rather than as consequences of poor or poorly implemented hardware and software design.

One way to substantiate HSI in policy is to have tenets to which all programs must adhere. These program management and technical tenets have been developed and exercised in numerous programs. The following tenets have been adapted from the MANPRINT philosophy in the 1980s.

Initiating HSI considerations early in acquisition: During functional area analysis and initial capability definition.

Recognition of HSI issues and planning for resolution: Including an audit trail and an HSI management plan.

Explicit and consistent performance requirements: Tracked in a requirements traceability matrix.

An integrated technical process: Bringing together domain analyses and integrating them with systems engineering.

Proactive trade-offs during design: Utilizing a target audience description (TAD) and human task network modeling tools.

HSI as a factor in contracting and source selection: To ensure that industry partners understand the importance of HSI.

HSI milestone assessments: To ensure that HSI is continually and rigorously addressed.

As reflected in these tenets, proactive trade-offs must be made throughout the design process. Complex systems (including families of systems and systems of systems) are best developed following the iterative systems engineering process. HSI tools must be valid and effective and must work with other engineering tools in the SE process. Descriptions of the users, their work, their likely sources for recruiting, their physical and mental capabilities, the environment in which the work must be performed and any inherent hazards in that environment, and the habitability of that environment must be captured in a target audience description, which provides a complete and captured snapshot of users to be designed and planned for throughout the life cycle of the system. The tools must allow these snapshots to be generated quickly in a form useable in the individual domains and in the HSI trade space to perform the human-based integration. Further, these tools must allow sensible human-based solution sets to be integrated with the larger systems engineering trade space in which all the engineering disciplines are represented in hardware and software solution sets. The ability of the system developer to support these types of analyses should be included in the selection of a vendor to develop the required solution.

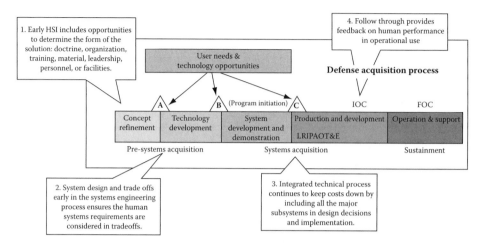

FIGURE 46.1   Systems Engineering Acquisition process with HSI Insertions. (From DoD Directive 5000.1, 2003. With permission.)

46.3 Systems Engineering

Systems engineering practice calls for rigorous definition of requirements. These requirements (Kossiakoff & Sweet, 2003; Martin, 1997) by logical extension include those fundamentals that detail the relationship between the human parts of the system and the more traditionally thought of hardware and software parts of the system. Human factors is the specialty engineering discipline that focuses on the human component of the system, including capabilities and limitations (e.g., sensation, cognitive abilities, ergonomics, and anthropometry). Collaboration between SE and human factors results in an understanding of the resources and requirements of the humans using the system as it was designed to be used. This includes the operation and maintenance of the system. This inclusion of the human as a fundamental system component is part of the trade space in any system design and an essential part for advanced mission systems, especially those with decision-aiding or automation systems. The challenge, however, is to understand this portion of the trade space by balancing "human-friendly" solutions with successful hardware solutions and software implementations. Figure 46.1 illustrates the major phases and iterative nature of the systems engineering and acquisition process (Department of Defense, 2003). While this process is appropriate to the design, development, and deployment of complex systems, systems of systems, and families of systems, performance expectations are not necessarily met when systems developed using this process are subsequently used in operational scenarios. Clearly, HSI will bring a great deal of improvement to SE by ensuring the most contentious element of the system is considered throughout the SE process. Research has shown that modeling and simulation (M&S) are effective tools for development and analysis throughout these phases. Further, M&S brings continuity for future programs, allowing development of baselines even for "clean sheet" programs.

46.4 Modeling: An Enabling Toolset

There are many types of models available from simple representations of concepts on paper through complex models of large segments of behavior. This includes both models of human behavior and models of hardware and software. These various tools allow the exercise and prediction of the behavior of the system or systems represented in the model. Each engineering and specialty engineering discipline has its own models that reflect the known areas of concern based on the experience of the discipline. Unfortunately these models do not represent a complete picture of the system or the environments in which the system must perform. This narrow scope may result in a lack of insight that can result in misrepresentation of the performance of the system, inadequate representation of the capabilities of the elements of the system, or inappropriate representation of the operational environment. At best, the model may be misleading. At worst, the model results in a picture that is irretrievably wrong.

Many advances have been made over the past several decades in the simulation field to improve interoperability among computer simulations. With the completion of the high-level architecture (HLA) baseline definition in August 1996, the Defense Modeling and Simulation Office built a strong foundation of standards to promote simulation reuse and interoperability. This ensured that different models could work to generate data together and share those data.

However, even with these strides in standardization and data sharing, in the human-based domains, semantic challenges persist in inhibiting simulation integration. The data themselves are the problem: on one hand, the domains use many of the same words to represent different concepts, and on the other hand the domains use different words to represent the same concepts. This semantic confusion has many outcomes but the challenge for this forum is the difficulty in collecting and using model data in light of these overlaps. A further challenge arises from the lack of interface between the domains in other data arenas, rendering modeling a more difficult endeavor. This lack of data transition or sharing could be mitigated through linked models, allowing for the proactive design and implementation trade-offs previously discussed. Tools that permit rapid, flexible analysis would further the individual domains by allowing a more complete exploration of the potential solutions to which individual domains could contribute while simultaneously expanding the capability of each domain to quickly modify inputs based on changes dealt by other domains to the solution trade space.

Modeling provides a forum for each HSI domain to present all the possible inputs to the solution space. Modeling provides the workspace in which different domains can come together and try out solutions in a rapid, inexpensive, bounded way. In this fashion, potential design alternatives can be explored. The effects of various solutions can be determined by focusing on the overall mission of the system rather than on the human elements of that system. The outcomes contribute directly to defining better, more diverse solutions. These more robust solutions contribute more to the overarching system solution implementation.

This multiplication of effect can only result from simulations that influence one another. To achieve this, the data exchanged must be standardized to ensure that participating models are able to use the data. Further, the standardization of data and their supporting lexicons ensures that the analysts are able to make sense of the ensuing data from the models, regardless of the model selected to generate final output. This requires interoperability of the models and simulations.

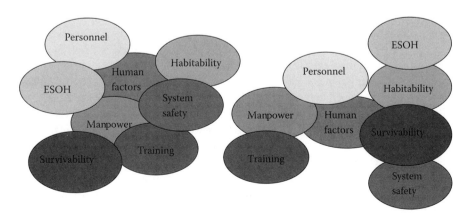

FIGURE 46.2 Two Notional Solutions Sets where overlap indicates degree of integration across domains. Note the notional differences between the two potential solutions.

Simulations must be integrated primarily because the distributed simulation has a superior ability to model a certain environment than any of the simulations independently. A distributed simulation consists of a mosaic of simulations with each providing a niche service to complement the whole. Some applications excel in modeling human performance, while others excel in modeling terrain. No matter what strength an application has, it is best to have it focus on what it does best in the distributed simulation. It is generally unwise to use a single simulation application with one strength for all modeling purposes when a distributed simulation can use the best of all strengths. Leveraging the strengths of different models further enhances the robustness of and utility of the data that result.

The fundamental reason conflicts in data sharing exist stems from this lexical discrepancy. For example, the definition of the term *function* (of functional analysis) from the *Handbook of Human Factors* (Salvendy, 2006) can be viewed from many perspectives. It acts as a logical unit of behavior of a human or machine component that is necessary to accomplish the mission of the system. When comparing independently conducted functional analyses, what ultimately surfaces is semantic differences arising from integration phases, or from the interpretation of a logical unit of behavior. What one team of analysts considers a function may be considered an inappropriate level of detail by another team. This discrepancy stems partly from the various goals for performing a functional analysis (or any other analysis in the human-based domains). Because not all functional analyses have the same goal or purpose, contradictions will exist.

Modeling is an excellent methodology from which to study the effects of the HSI disciplines on one another. It also allows examination of the effects of the integrated domains interacting with other potential elements of the system (hardware and software). Because these models represent both the system and the environment, it is essential that all of the system be represented—the hardware, the software, and the humans that work with the system. These users will be operators, maintainers, supervisors, suppliers, and other support personnel, all of whom interact with the system in some way. By failing to account for this major system element, predictions from the models are flawed at best and unusable at worst.

Much progress has been made to M&S software tools over the past decade. Software engineers have designed methods where disjointed M&S tools can communicate with each other to take advantage of the strengths of each tool. With the capability of integrating M&S tools, systems engineers can now analyze a more complete system model during the design process. Process models that study materiel acquisition can now include the effects of human performance to the system.

46.5 SEAPRINT Case Study

In 2004, the SEAPRINT team undertook a study to analyze the aircraft fueling system proposed for the CVN-21 aircraft carrier. The primary technical objective of the study was to perform a sensitivity analysis on how fueling affects aircraft launch and recover rates. Since an aircraft carrier is designed to provide power projection, delivering a portable military aircraft base, a key metric of its operational performance is the volume of aircraft launches and recoveries.

The impetus for the study came about from the pre-existence of a Navy engineering process model, FOCUS, which analyzed launch and recovery rates for the CVN-21. However, this model essentially ignored the human impact on system performance. During the study, we learned that the process model was using a simplifying assumption that all aircraft get fueled in the exact same amount of time. The model assumed a perfect, invariant human operator. Using an incomplete model of the operating environment such as this is prone to cause inaccurate results and faulty conclusions about the aircraft carrier's performance. In an effort to improve the accuracy of the results and the conclusions drawn from the process model, the SEAPRINT team conducted a task analysis of the aircraft fueling crew. This included the tasks of all personnel on the deck of the aircraft carrier that supported the refueling of aircraft.

Following the task analysis, the SEAPRINT team developed a human performance model (HPM) of the aircraft refueling crew using the Army's Improved Performance Research Integration Tool

(IMPRINT). Meanwhile, software engineers representing both modeling environments collaborated to facilitate the integration of the two tools. Communication between the two modeling environments was made possible using RockLobster, an available software communication protocol developed by NAVAIR in Lakehurst, New Jersey. Like other software communication protocols, RockLobster employs the typical features of a client-server architecture.

46.6 Imprint-Focus Client-Server Architecture

In a client-server architecture, the "client" makes requests for services from the "server." Figure 46.3 shows the fundamental properties of a client-server architecture from an IMPRINT-FOCUS perspective where FOCUS was the client and IMPRINT was the server. After establishing a communication protocol, in this case RockLobster, any SE process model can potentially request services from any HPM.

The client-server interface also allows process models to select only those entities for which a higher fidelity human behavior representation is desired. The process model is not required to perform HPM processing for all of its entities—only those entities selected for subscription will receive the higher fidelity human behavior representations. For example, within the context of the SEAPRINT case study, we could have chosen only a subset of the total refueling teams to subscribe for higher fidelity human behavior modeling. By providing HPM processing only where needed, the process model will run more efficiently and the extent of the analysis can be focused at the desired level.

In a client-server interface, the process model uses a subscription process to initiate HPM services. When the simulation commences, the process model sends a request for subscription to a server via data interactions. One or more behavior servers may respond to the request with a data interaction that specifies it can accept a new client. The process model then selects one of the available services and completes the subscription process by replying with a data interaction to the selected server. The behavior server, in turn, confirms the subscription with another data interaction.

Figure 46.4 shows the variables communicated between the FOCUS process model and the IMPRINT HPM. The integrated simulation begins with a subscription process whereby FOCUS subscribes to services of the IMPRINT model. After the simulation has commenced, FOCUS will call upon IMPRINT to provide a fueling time for an aircraft whenever an aircraft needs refueling. At this time, FOCUS pauses execution and sends five variables describing the nature of the aircraft requiring fuel and the number of fueling teams currently active to IMPRINT. When IMPRINT receives these five variables, it

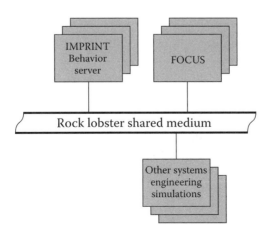

FIGURE 46.3 Client-server architecture.

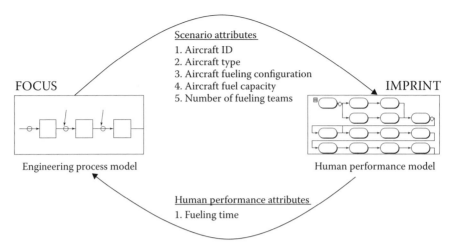

FIGURE 46.4   FOCUS-IMPRINT variable communication.

calculates a fueling time by running the SEAPRINT fuel system HPM. After IMPRINT has calculated a fueling time, it sends the information over to FOCUS which continues its execution until the next fueling request.

46.7 Integrated Simulation Results

This case study linked indirectly related models (one of a system and the other of the people operating another, related system). Data collection continues as the project continues, but preliminary results include success in linking an HPM to an engineering process model.

Linking the IMPRINT model of below-decks fuel preparation and above-decks fueling model is not a breakthrough. IMPRINT has been linked to process models previously. Linking IMPRINT to the FOCUS model was, unexpectedly, in a kluged retrofit. However, a more elegant linkage between the models would not significantly alter the outcome.

This demonstration of linking an HPM with a process model is kluged but effective. The outcome and continued data collection illustrate that adding a representation of human operators to the process model yields improved results from the process model. This improvement is reflected in the usefulness of the inflected data.

The models themselves are unremarkable. The resulting data, however, are remarkable because of their application. IMPRINT provided data to FOCUS that the process model did not account for: the performance of the human subsystem. The injection of accurate human performance representation into the FOCUS process model allows system developers to generate more accurate data that reflects the much more likely outcome of system use in real operational usage. This improvement in the model data can return better design that results in systems that work as advertised when introduced to fleet users.

The introduction of FOCUS data into the IMPRINT model also yields benefits for the human performance model. The input of system data introduces opportunity for more different HSI trade-offs to be modeled. Changes in performance due to manpower, personnel, training, or human factors changes can be modeled in the human performance model and then introduced in the process model. These proactive trade-offs are at the heart of successful HSI.

46.7.1 The Way Forward

The integration of human performance modeling tools with engineering process modeling tools is essential to ensure that the human subsystem is represented throughout the SE process. The HSI approach provides the overarching process to support that integration. Incorporation of modeling tools also provides a means of collecting data of a system in operation and performing trade-offs across the entire trade space.

While modeling tools continue to develop within individual domains of systems engineering and human systems, they remain monolithic and therefore ungeneralizable. The case study presented illustrates the viability of linking disparate models to more fully explore systems under development.

The alignment of different "faces" or aspects of the same event in different models is not new. However, linking a human performance model that is one part of a process to a process model of a different facet of the same process allows exploration of more in depth trade-off questions. In addition, the processes and policies developed in concert with these models provide a platform to successfully include modeling and simulation tools to support complete human systems integration throughout the engineering phases of acquisition.

The success of linking disparate models is a clear indication that modeling tools linked together provide a robust, effective platform for generating solutions for systems engineering. The need for data standardization remains equally clear. Establishing data standards for human behavior models will allow diverse models, such as human figure models and human behavior models, to work together with system process models (of both hardware and software) and agent-based simulations.

References

Booher, H.R. (Ed.). (2003). *Handbook of Human Systems Integration*. Hoboken, NJ: John Wiley & Sons, Inc. Wiley Interscience.

Chapanis, A. (1996). *Human Factors in Systems Engineering*. Hoboken, NJ: John Wiley & Sons, Inc. Wiley Series in Systems Engineering and Management, Andrew Sage, Series Editor.

Cook, M.A. (1996). *Building Enterprise Information Architectures: Reengineering Information Systems*. Upper Saddle River, NJ: Prentice Hall PTR.

DoD Directive 5000.1. (2003). "The Defense Acquisition System," May 12.

Kossiakoff, A. & Sweet, W.N. (2003). *Systems Engineering Principles and Practice*. Hoboken, NJ. John Wiley & Sons, Inc. Wiley Series in Systems Engineering and Management. Andrew Sage, Series Editor.

Kubeck, L.C. (1995). *Techniques for Business Process Redesign: Tying It All Together*. New York: John Wiley & Sons, Inc. A Wiley-QED Publication.

LaVine, N.D. & Peters, S.D. (2001). "An Advanced Software Architecture for Behavioral Representation within Computer Generated Forces." Final Report, Defense Modeling and Simulation Office, Alexandria, Virginia, December.

Martin, J.N. (1997). *Systems Engineering Guidebook: A Process for Developing Systems and Products*. Boca Raton, FL: CRC Press.

Peterson, N.G., Mumford, M.D., Borman, W.C., Jeanneret, P.R., & Fleishman, E.A. (1999). *An Occupational Information System for the 21st Century: The Development of O*NET*. Washington, DC: American Psychological Association.

Salvendy, G. (2006). *Handbook of Human Factors and Ergonomics* (3rd ed.). Hoboken, NJ: John Wiley & Sons, Inc. Wiley Interscience.

Sheridan, T.B. (2002). *Humans and Automation: System Design and Research Issues*. Hoboken, NJ: John Wiley & Sons, Inc. Wiley Series in Systems Engineering and Management, Andrew Sage, Series Editor. And HFES Issues in Human Factors and Ergonomics Series, Vol. 3, Supervising Editor: David Meister.

Wilson, J.R. & Corlett, E.N. (Eds.). (1995). *Evaluation of Human Work: A Practical Ergonomics Methodology* (2nd ed.). Philadelphia, PA: Taylor & Francis, Inc.

47

Digital Modeling of Behaviors and Interactions in Teams

Barrett S. Caldwell

47.1 Introduction

A frequent question raised in the study of digital human modeling of teams is similar to that raised in the study of real human groups or teams: What is the value added of studying teams that cannot be achieved by studying individuals? In essence, this question addresses both the issues of appropriate unit of analysis, and emergent properties. As computer power continues to improve, it becomes increasingly possible to model behaviors at increasingly fine grain sizes. However, when analyzing system design or performance issues, the ability to analyze teams is not only a philosophical one, but a practical issue of efficient evaluation, analysis, and improvement of the system at the appropriate level of investigation.

Task coordination and information flow among team members is a critical aspect of human task performance. Modern complex systems (such as air traffic coordination, emergency response systems, process control plants, or space flight operations), spanning large physical distances and multiple technical subsystem functions, require substantial synchronization and monitoring in order to diagnose, respond, and recover from adverse events. In such systems, it is difficult to imagine a single operator with the cognitive capabilities to attend to, integrate, or effectively respond to the number of potential situations and contingencies that will emerge in ongoing operations. Thus, it is critical to evaluate, and thus understand and improve, the behaviors and interactions of teams and groups conducting coordinated, distributed tasks (Tambe, 1997).

Some examples of team coordination have demonstrated effective problem solving and exceptional recovery from catastrophic system degradation, such as the actions of astronauts and flight controllers during Apollo 13 and other human spaceflight missions (Kranz, 2000). However, this same task environment has seen breakdowns in effective coordination or information flow leading to loss of life and endangering an entire spaceflight program (CAIB, 2003). In ways that are still being uncovered and addressed, team behaviors during and after Hurricane Katrina are highlighting the challenges of effective translation from planning through diagnosis through response and contingency adaptation (Select Bipartisan Committee, 2006). Effective preparation and responsible management of complex civil infrastructures in the face of many opportunities for degradation require a vastly improved capability to design, model, and enhance team behaviors and interactions during routine, non-routine, and emergency situations.

47.2 Definitions

For the purposes of this chapter, I will make assumptions regarding size and function of groups and teams. In the psychological literature, groups and teams may be used interchangeably, especially as distinguished from individual processes and those of larger collections of humans (which are more frequently studied by sociology and anthropology) (Caldwell, 1990/1991; Deal & Kennedy, 1982; Fine, 1979; Kiesler, 1997; Shweder, 1991). Some researchers, however, will focus on the term *team* to indicate groups with specific intentional organization, patterns of interaction, shared goals, and mutually understood roles for behavior (Sundstrom, et al., 1990; Swezey & Salas, 1992). For the purposes of this chapter, this task-focused, performance-oriented definition of teams will be used.

Although there are debates about the minimum size of groups or teams (Shaw, 1981), this chapter assumes a minimum size of three. It is at this size that multiple interaction pathways and information exchanges are first possible. Ideal and maximum size estimates depend on the focus of the researcher, and the tasks and interaction dynamics of the members (Bales & Slater, 1955; McGrath, 1984; Shaw, 1981). Small group and team researchers tend to use a maximum size guideline of 20 members, because interaction patterns and coordination of task roles and functions becomes difficult for most people to manage at larger sizes. Groups, with their lack of defined goal and role structures, may not be as strictly determined by size. However, some researchers do focus on small groups to maintain an emphasis on manageable sizes and numbers of direct relationships and interpersonal dynamics.

Digital modeling of teams can easily be seen as an extension or variation of the multi-agent systems literature. However, there is substantial confusion and conflict over the nature of the concept of agent and what aspects of role or activity are necessary or sufficient to characterize agent behavior, singly or in cooperation (Weiss, 1999). Authors such as Tambe (1997) and Josslyn (1999) do address these concerns, and emphasize that intelligent agents and humans in complex socio-technical systems require awareness of, autonomy in, and complexity of interactions with the task performing system. Even in studies of human organizational behavior, there are numerous terms and definitions for describing the nature of shared goals, values, or norms affecting behavior (Hofstede, 1998). The question for efforts in creating effective digital agents, as well as developing realistic mathematical descriptions of real human behaviors, becomes one of what factors, characteristics, and aspects of performance and action need to be modeled.

47.3 What Needs to Be Modeled?

One of the most essential points to be made is that modeling group behavior as a collection of rational actors with immediate, reliable, and logically consistent access to information is neither useful nor desirable in attempting to create valid models of team task coordination. The history of decision-making literature has already identified substantial limits to human information collection, prioritization, weighting, and integration for decisions and performance. Both physiological and psychological mechanisms are responsible for the phenomenon of bounded rationality, and the limits of human processing at multiple timescales (Newell, 1990).

When looking at team behaviors, a simple aggregation of individual behaviors or decision-weighting utility models is not sufficient. One assumption that can be problematic is that individual weighting functions in a multi-attribute decision framework differ only by value. However, each person operating in a socio-technical task context can have a potentially infinite list of factors and criteria that are implicitly assigned zero weight. Given different organizational, social, and individual priorities and experiences, it is unlikely that different people working together will automatically select from the same set of criteria to assign any weight at all (Green, 2002)—some actors will find some criteria important, while others will not consider them as part of the set of criteria requiring weighting. Thus, a concept of socio-technical bounded rationality must explore the issues that arise when distributed task performers must agree on what is important, or what information and goals are shared, and not simply processes of coordination of information and tasks that are already well defined and understood. These issues are not absolute, but require re-evaluation with each distinct configuration of team membership. Thus, the issues of compatible and coordinated understandings ("shared mental models") and processes of developing and enhancing well-integrated team functioning must be included in the modeling of team behaviors and interactions in complex task settings (Bradley, 1992; McGrath & Hollingshead, 1994; Orasanu, 1990; Sundstrom et al., 1990).

These distinctions of task types and priorities for performance reflect a broader taxonomy of team tasks (McGrath, 1984; Sundstrom et al., 1990). McGrath's taxonomy includes four major task types: making decisions, performing tasks, resolving conflicts, and solving problems. Sundstrom and colleagues also emphasize cycles of activities, where the same team conducts multiple performances over time and therefore develop shared understandings of each other's role as well as operational histories that can be referenced as a shorthand for a rich set of meanings, expectations, and intentions. A great deal of agent research and laboratory studies of human team performance have focused on a subset of team tasks, primarily group decision making and decision support, and economic-based competitions (Billard, 1996; Brown, Armstrong, Swierenga, & Wellend, 1990; Green, 2002; McGrath & Hollingshead, 1994; Stasser, Vaughan, & Stewart, 2000; Weiss, 1999).

However, standing teams conducting critical tasks in time- or resource-constrained contexts must simultaneously address a full taxonomy of task interactions (Cooke, et al., 2001; Josslyn, 1999). Ongoing team dynamics, including coordination of multiple information flow channels and human-computer interactions for distributed tasks, require a combination of task performance, team interaction, and data coordination flow emphases known as *taskwork, teamwork,* and *pathwork* (Caldwell, 2005a; Garrett & Caldwell, 2002b). Researchers may assume that team structures and information flow patterns will represent a command and control hierarchy (Cuevas, et al., 2004; Jentsch, Barnett, et al., 1999; Josslyn, 1999; Swezey & Salas, 1992) based on a human-machine supervisory control paradigm (Sheridan, 1988, 1992), and in most military scenarios, this is an appropriate assumption. However, there are several other structures that may emerge, based on different group norms, goals, and personality factors, as well as situation constraints.

For instance, settings such as NASA Mission Control work on a shifting expertise model, with the assumption that the complexities of the task (managing a spacecraft during routine or non-routine operations) require coordination of multiple experts with distinct areas of domain specialization. This model, known as Distributed Supervisory Coordination (Caldwell, 2005b, 2006), suggests that different configurations of information flow and team maintenance are required compared to a more structured command and control model. Recent advances in agent research, particularly swarm behaviors, have demonstrated that *no* structured supervisory coordination or control model is *required* to achieve coordinated performance. Thus, an effective DHM for team behavior must recognize that the very structure and organization of information flow, task coordination, or expertise in the team must be available for modeling, based on the task and situation context as well as the characteristics of team members (both alone and in combination). A primary requirement for team DHM, therefore, includes an understanding of model parameters that define and characterize interactions between team members, and not simply attributes of individual team members themselves.

47.4 Model Parameters for Group and Team DHM

One of the oldest forms of digital modeling of teams and groups comes from role-playing games such as Dungeons and Dragons (D&D). The process of research and development devoted to behavioral DHM in the computer game industry lags behind the popularity of such games, and thus reflects a common pattern of popular usage driving research priorities (Fulton, 2002). In its original (tabletop, paper-based, human-only) forms, D&D-style games characterized individual players and "non-player characters" (generated characters not controlled by one of the human game players) by "character statistics" or stats. These stats represented both individual characteristics (strength, stamina, and intelligence as the ability to learn skills) and variables that influenced more social interactions (charisma, intelligence as the ability to influence people, race and guild allegiances) (SRD, 2007). Some skills could even be used to affect whether or not a specific character's actions would result in social interactions at all (such as stealth, which influenced whether a thief could pass by another character or steal items without being detected).

These character stats were used by the person in charge of the game (Dungeon Master, or DM) to determine probabilities of successful actions in a complex turn-taking, role-playing imagined environment. As D&D games evolved into computer-based applications, player stats became the most effective means for developing more sophisticated and rich patterns of interactions between and among human players and computer-generated characters (SRD, 2007). As players individually and cooperatively interact with the game and each other, they gain experience and grow in various stat values. Significant work has been conducted in the computer game industry to determine how to use psychological theory to adjust reward contingencies to support and encourage more extensive game playing (Hopson, 2004).

However, a recent development in massively multiplayer online role-playing games (MMORPGs) is the ability to purchase in the "real world" a character with particular character stats that have been achieved through another player's experience (Thompson, 2007). The very nature of MMORPGs is that interactions between characters are based both on character stats and player actions (communications, strategic choices, use of tools or weapons). The practice of purchasing characters has generated negative responses from other experienced players, as the player using the purchased character does not exhibit the shared experience and relative expertise that others expect based on the level of game activity required to achieve the stat levels. Thus, the model of an experienced character in an MMORPG not only represents the development of character stats, but the exposure of the player to the cultural socialization and expertise gained from the team-based playing experience. Player expertise, therefore, is described as a multidimensional construct including individual stat capabilities, interpersonal stat capabilities, and cultural experience within the specific social context of interest (Caldwell, 2005b). The concepts of modeling cultural patterns of interaction or specific social contexts are relatively little studied.

47.5 Modeling Cultural Interactions

Cross-cultural research by Hofstede has highlighted a number of presumed cultural dimensions of interpersonal behavior (Hofstede, 1983, 1998). These dimensions, emphasizing characteristics such as the importance of individuality versus group affiliation, respect and deference for authority, or strictness of gender role differences, are supposed to represent important aspects of national differences in employees, even when working in a single organization. The use of such dimensions has substantial power, due to the attempt by this research to develop more universal scales of human interaction. However, collections of individuals from a variety of national origins, primary and secondary socializations, and professional as well as non-professional affiliations, are likely to differentially represent or characterize different patterns of group identity and reference. In other words, someone may reflect their professional subculture quite strongly, such as a fact-based, merit-oriented engineer who does not find emphasizing social history or family relations important. However, there would be more socialization or national origin conflict for such a person as a South Asian female than an Anglo American male. Thus, any individual may find varying levels of comfort and consistency between multiple cultural and subcultural associations, which would then influence the likelihood to respond in particular ways in one group compared to another.

As described, the assumptions of command and control structures and information flow patterns used in some agent literature (Josslyn, 1999) are actually assumptions about a particular style of cultural interaction rather than a logically required aspect of team performance. The nature of task coordination does have both structural and functional implications on information and other resource coordination, as well as the use of team member expertise as a particular type of resource available. Modeling the performance of the team as an emerging unit of analysis does require an examination of how structural or functional aspects of the team affect expertise sharing and coordination among team members.

47.6 Expertise, Resource Coordination, and Team Performance

In an ongoing research program to understand team performance in distributed coordination task environments, the author has led an investigation into different modules describing information flow and team coordination. There are four types of modules involved, based on analyses of NASA mission control teams, large research groups, and others (Caldwell & Ghosh, 2003; Garrett & Caldwell, 2002a). The modules can be briefly described as follows (Caldwell, 2005a):

Asking: Novices (or team members acting outside of their area of expertise) bring queries to the team; queries are then available to be answered by one or more experts, depending on complexity, comprehensibility, expert availability, and query understanding.

Learning: Members are socialized as part of an expert team, and use existing experts and reference sources to develop expertise in a particular area while learning about the structure and processes of the team.

Sharing: A mixed group of novices and experts interacts using shared information tools (such as a discussion list or chat room) to exchange information, perspectives, and social affiliation, in addition to specfic task-oriented discussions.

Solving: Members of the expert team are responsible for monitoring and troubleshooting problems and are focused on effective task performance to maintain system functioning.

The author's research lab has developed initial model simulations of the *Asking* module using discrete event simulation tools (AutoMod) approximating a multi-server queuing process (Caldwell & Ghosh, 2003; Ghosh & Caldwell, 2006). Unlike a rational agent model of novices interacting with experts, this *Asking* module includes several probabilistic parameters that describe plausible (but inefficient or non-rational) real-world behaviors. For instance, novices and experts are modeled using expertise parameters in a number of knowledge domains. In this module, the probability of a novice receiving an appropriate (understood) answer from an expert *decreases* as the difference in expertise between the novice and expert *increases*. In this situation, experts frequently forget their accumulated expertise, and how difficult it would be for a novice to understand an explanation without that expertise. In addition, a novice may ask a question expecting a simple answer, when in fact a deep knowledge of the subject matter may reflect an understanding that the question is actually not simple to answer. Novices who are "clueless" (unaware of the expert community or the subject focus of their question) may not approach the relevant expert to ask a question. Finally, an available expert may simply wait and then not respond, without providing feedback or reasons. Analysis of the expert network simulations can examine the throughput of a stream of novices who come to ask questions. Measures of a network's performance are described in terms of fractions of satisfied and dissatisfied novices, the length of queues, and the effects of different strategies for directing novices to experts.

Data collection from real human interactions in a weblog were analyzed to develop model parameters and simulation seeds for models of interactions in a *Sharing* module (Ghosh & Caldwell, 2006). Differences in participation in the weblog included rules for posting and linking to other weblogs (thus providing additional sources of topics for information sharing and knowledge development over time), as well as achievement of critical mass of participation-based different types of users with heavy, medium, and light rates of participation. In addition, the real weblog data indicated differences in user participation based on thread topic, which could be clustered in one of a few domain areas (interest areas, which may or may not correspond to expertise areas defined in the *Asking* module). AutoMod simulations are able to determine differences in total weblog activity based on modifying probabilities of topic domain matching to user interest areas, participation and posting rates of different types of users, or relative persistence of topic participation based on number of interested users.

It is expected that improved team interaction models evolving from these first modules will also help define and enable multi-scale simulations of *Learning* and *Solving* modules of team interactions. However, one significant challenge is that these two modules may operate over very different timescales. *Learning* processes that enable a novice to become an expert may take months or years; *Solving* problems among expert teams may address task performances over minutes. In addition, both the *Learning* and *Solving* modules involve interacting with a dynamic external set of information sources, and not just the interactions between the human members of the expert team (and any novices interacting with the experts). When combining internal team coordination and external information access and flow processes, issues of coordination efficiency over time take on additional value.

47.7 Information Flow Efficiency and Coordinated Task Performance

One issue in coordinated task performance, especially in strictly hierarchical command and control structures (Josslyn, 1999) is the issue of effective transfer of information among members of the team. As the number of team members and the number of coordination links between members increases, there is an increasing opportunity for information loss or distortion (Artman, 2000; Kling et al., 2001; Stasser, et al., 1995; Stasser et al., 2000). Based on information theory, we can describe the efficiency (reliable transmission of signal without loss or equivocation) of communication between sender and receiver as

$$S_r = e_c * S_s$$

In a multi-hop transmission network, the cumulative efficiency for j hops would be calculated as

$$S_r = prod(k = 1, j) e_{ck} * S_s$$

Traditional considerations of network transmission (as well as social psychology perspectives on group coordination losses) would thus attempt to maximize each e_{ck} and minimize the number or hops j from sender to receiver.

In a multidimensional distributed expertise network, the types of signals, and types of expertise, exchanged may be technical, social, or a combination of the two (Watts & Monk, 1998). Based on the author's work on distributed expertise, we must also consider that an individual may have different levels of expertise (and thus reliability of understanding and effectively transmitting information) in each of n dimensions (Garrett & Caldwell, 2002a). Thus the equations above become vector representations of signals S_{sn}, and efficiency must be managed as the reliable transmission of the vector representation of the signal through intermediaries with m dimensions of transmission capability, or:

$$S_{rm} = vector(m) prod(k = 1, j) e_{ck(i=1,m)} \cdot vector(n) S_{s(i=1,n)}$$

Any signal of dimension n sent through an intermediary of m < n will then be reduced to a signal of dimension m, as the intermediary has no capability to relay (100% loss) signal components in dimensions m + 1 … n. Information richness and social information processing studies of information technology also use this concept to describe the loss of social information through information technologies (Caldwell, et al., 1995; Kiesler, et al., 1984; Rice, 1992). In this sense, the communication channel has transmission capability of dimension m < n where n includes the range of non-verbal or other social cues. Team coordination in agent networks usually assume relatively low dimensions of communication channel capability, based on focused and limited mission goals and structured task requirements (Josslyn, 1999; Kempf, et al., 2000; Tambe, 1997).

Task coordination and performance in a distributed expertise network, however, must consider two additional points. Over the course of a performance task, the signal S about the state of the world is actually a function of time, requiring ongoing sampling and possible updates on the part of the expert (Sheridan, 1992). The communication network processes described above do not consider the complication of signals that change over time, or the need to transmit the signal to the receiver S_r that is accurate (i.e., updated) to the time of receipt, and not simply at the time of transmission (Caldwell, 2000). In addition, an expert may in fact improve the transmission of a signal in a coordinated expertise network, by adding expertise and interpretive capability to a signal received by the expert that was not included in the capability of the sender (Watts & Monk, 1998). (One example of this concept is described in the capability of an expert to detect and resolve problems in a complex engineering system based on integrating user complaints, where no user has sufficient understanding or awareness of the system to identify the root problem or perform the detection or problem resolution singlehandedly.)

We can no longer assume the form of the equation above, because any expert operating during hop j may have efficiency greater than 1 in some dimension of the signal transmission, and may even increase signal in some dimension from 0 to some value. In addition, we must have the capability that experts can update signals $S_{rn,t}$ so that they reflect the state of the world at receipt and relay time t, not the state of the world at original send time t_0. Numerous examples exist of real-world team behaviors that exhibit these characteristics. An appropriate mathematical formalism for modeling such coordinated expert behaviors over time still requires substantial development. Thus, team coordination and knowledge generation processes cannot be derived simply by extensions of reliable lossless transmission of signals from classical information theory or Markov analysis. Most importantly, a vector product of efficiencies cannot be used to define or measure the coordination gains and performance enhancements of a distributed expert network performing knowledge sharing and information coordination tasks in a dynamic task environment (since any signals that degrade to zero in any dimension cannot be recovered in a multiplication operation).

47.8 Future Research Needs and Summary

One issue for future modeling research is whether the traditional size limits on effective team size and organization can be verified by models of team size and resulting patterns of coordination. Limits of coordination or communication of historical concern in studies of face-to-face groups can be addressed in some ways through the uses of information technology and adapted group processes (Caldwell, 2003; Cooke et al., 2001; McGrath & Hollingshead, 1994). However, the adoption of new tools and interaction patterns creates different constraints on group and team functions, which could be effectively modeled in terms of knowledge coordination, task and situation awareness communication, and timing constraints for effective information flow and synchronized performance.

Ongoing research questions in the study of human teams address team communication patterns in both timing and process, and how teams respond to non-routine events. Given the immature state of the art in understanding how high-performing teams differ from lower performing teams in critical non-routine task performance, it is unclear that the group dynamics literature is able to provide definitive guidance for DHM in teams. There is some promise that a combination of software simulation and human research studies will help to develop a more robust understanding of team performance.

Since human performance research is often time and resource intensive, it is impractical to conduct exhaustive analyses of all possible DMH or team interaction parameters and parameter values. Thus, simulation can be used to determine, through sensitivity analysis, which combinations of team interaction parameters are likely to yield qualitative shifts in performance. The impacts of these parameter combinations can then be further addressed and (dis)confirmed through focused human performance studies.

DHM of team performance is an area of research that is still in its very early stages of development. In order to more fully understand the properties, processes, and responses of teams to performance demands and task events, it is first important to recognize that actual teams have a variety of structural, organizational, and experiential characteristics. Teams must manage task dynamics and changing situational constraints based on ongoing events. Performance challenges and demands for team innovations or responses may be due to internal task performances by team members, or external events, and interactions between them (Helmreich & Merritt, 2000). In addition, teams working in complex environments must also manage the use of information and other resources in a distributed task context, using a variety of technologies and flow pathways (Jasek & Jones, 2001; Kling et al., 2001). Finally, even structured teams have interpersonal team dynamics that affect the effective flow of information and coordination of task performance. Structured command and control hierarchies, however, are not reflective of the full range of effective task performing units. Some teams are skilled in recognizing and shifting information flows and coordination processes according to real-time changes in required expertise, information, or task goals. In all cases, though, it is important for DHM efforts to have the capacity to model a broad set of task and team parameters. Some existing tools have been developed to model agents conducting routine or highly structured tasks in fairly constrained task scenarios, and some attempt to define optimal performance given particular *a priori* assumptions regarding task settings and goal weightings. Additional work is required to effectively model and describe the complex interplay of cultural, socio-technical, and operational factors that affect how teams do manage robust operations and event responses in dynamic tasks.

References

Artman, H. (2000). Team situation assessment and information distribution. *Ergonomics, 43*(8), 1111–1128.

Bales, R. F., & Slater, P. E. (1955). Role differentiation in small decision-making groups. In: T. Parsons (Ed.), *Family, Socialization, and Interaction Process* (pp. 259–306). Glencoe, IL: The Free Press.

Billard, E. A. (1996). Stability of adaptive search in multi-level games under delayed information. *IEEE Transactions on Systems, Man, and Cybernetics—Part A: Systems and Humans, 26*(2), 231–240.

Bradley, G. (1992). Computers and human communication in the organization—A psychosocial perspective on the individual and the society in change—Research in progress. Paper presented at the Proceedings of the Human Factors Society 36th Annual Meeting, Santa Monica, CA.

Brown, C. E., Armstrong, H. G., Swierenga, S. J., & Wellend, A. R. (1990). Human-computer "friendship"; a metaphor for intelligent system design. Paper presented at the Proceedings of the Human Factors Society 34th Annual Meeting, Santa Monica, CA.

CAIB. (2003). Columbia Accident Investigation Board Report Vol. 1. Washington, DC: Columbia Accident Investigation Board Report/NASA.

Caldwell, B. S. (1990/1991). Social Processes of Isolated Groups of U.S. National Park Rangers (Doctoral Dissertation No. 91-02, 63). Ann Arbor, MI: University Microfilms.

Caldwell, B. S. (2000). Information and communication technology needs for distributed communication and coordination during expedition-class space flight. *Aviation, Space, and Environmental Medicine, 71*(9, Supp), A6–A10.

Caldwell, B. S. (2003). Distributed supervisory coordination with multiple operators and remote systems. Paper presented at the IEEE International Conference on Systems, Man and Cybernetics, Washington, DC.

Caldwell, B. S. (2005a). Analysis and modeling of information flow and distributed expertise in space-related operations. *Acta Astronautica, 56*, 996–1004.

Caldwell, B. S. (2005b). Multi-team dynamics and distributed expertise in mission operations. *Aviation, Space, and Environmental Medicine, 76*(6), Sec II, B145–B153.

Caldwell, B. S. (2006). Issues of task and temporal coordination in distributed expert teams. Paper presented at the Proceedings of the 16th World Congress of the International Ergonomics Association, July 10–14, 2006, Maastricht, The Netherlands.

Caldwell, B. S., & Ghosh, S. K. (2003). Describing and modeling information flow in expert communities. Paper presented at the Seventh International Conference on Human Factors in Organizational Design and Management, Aachen, Germany.

Caldwell, B. S., Uang, S.-T., & Taha, L. H. (1995). Appropriateness of communications media use in organizations: Situation requirements and media characteristics. *Behaviour and Information Technology, 14*(4), 199–207.

Cooke, N. J., Kiekel, P. A., & Helm, E. E. (2001). Measuring team knowledge during skill acquisition of a complex task. *International Journal of Cognitive Ergonomics, 5*(3), 297–315.

Cuevas, H. M., Fiore, S. M., Salas, E., & Bowers, C. A. (2004). Virtual teams as sociotechnical systems. In: S. H. Godar & P. Ferris (Eds.), *Virtual and Collaborative Teams: Process, Technologies, and Practice*. Hershey, PA: Idea Group Publishing.

Deal, T. E., & Kennedy, A. A. (1982). *Corporate Cultures: The Rites and Rituals of Corporate Life*. Reading, MA: Addison-Wesley.

Fine, G. A. (1979). Small groups and culture creation: The idioculture of little league baseball teams. *American Sociological Review, 44*, 733–745.

Fulton, B. (2002). Beyond psychological theory: Getting data that improve games. Paper presented at the Game Developer's Conference 2002 Proceedings, March, 2002, San Jose, CA.

Garrett, S. K., & Caldwell, B. S. (2002a). Describing functional requirements for knowledge sharing communities. *Behaviour and Information Technology, 21*(5), 359–364.

Garrett, S. K., & Caldwell, B. S. (2002b). Mission control knowledge synchronization: operations to reference performance cycles. Paper presented at the Proceedings of the Human Factors and Ergonomics Society 46th Annual Meeting, Santa Monica, CA.

Ghosh, S., & Caldwell, B. S. (2006). Usability and probabilistic modeling for information sharing in distributed communities. Paper presented at the Proceedings of the Human Factors and Ergonomics Society 50th Annual Meeting, San Francisco.

Green, K. C. (2002). Forecasting decisions in conflict situations: a comparison of game theory, role-playing, and unaided judgement. *International Journal of Forecasting, 18*, 321–344.

Helmreich, R. L., & Merritt, A. C. (2000). Safety and error management: The role of crew resource management. In: B. J. Hayward & A. R. Lowe (Eds.), *Aviation Resource Management* (pp. 107–119). Aldershot, UK: Ashgate.

Hofstede, G. (1983). National cultures in four dimensions: A research-based theory of cultural differences among nations. *International Studies of Management and Organization, 13*(1–2), 46–74.

Hofstede, G. (1998). Attitudes, values and organizational culture: Disentangling the concepts. *Organizational Studies, 19*(3), 477–492.

Hopson, J. (2004). Behavioral game design. Paper presented at the Game Developer's Conference. Retrieved February 22, 2007, from http://mgsuserresearch.com/publications/default.htm.

Jasek, C. A., & Jones, P. M. (2001). Cooperative support for distributed supervisory control. In: G. M. Olson, T. W. Malone & J. B. Smith (Eds.), *Coordination Theory and Collaboration Technology* (pp. 311–339). Mahwah, NJ: Lawrence Erlbaum.

Jentsch, F., Barnett, J., Bowers, C. A., & Salas, E. (1999). Who is flying this plane anyway? What mishaps tell us about crew member role assignment and air crew situation awareness. *Human Factors, 41*(1), 1–14.

Josslyn, C. (1999). Semiotic Agent Models for Simulating Socio-Technical Organizations (DS Project, PSL/NMSU). Los Alamos, NM: Los Alamos National Laboratory.

Kempf, K., Uzsoy, R., Smith, S., & Gary, K. (2000). Evaluation and comparison of production schedules. *Computers in Industry, 42*, 203–220.

Kiesler, S. (Ed.). (1997). *Culture of the Internet.* Mahwah, NJ: Lawrence Erlbaum.

Kiesler, S., Siegel, J., & McGuire, T. W. (1984). Social psychological aspects of computer-mediated communication. *American Psychologist, 39*(10), 1123–1134.

Kling, R., Kraemer, K. L., Allen, J. P., Bakos, Y., Gurbaxani, V., & Elliot, M. (2001). Transforming coordination: The promise and problems of information technology in coordination. In: G. M. Olson, T. W. Malone & J. B. Smith (Eds.), *Coordination Theory and Collaboration Technology* (pp. 507–533). Mahwah, NJ: Lawrence Erlbaum.

Kranz, G. (2000). *Failure is Not an Option: Mission Control from Mercury to Apollo 13 and Beyond.* New York: Berkeley Books.

McGrath, J. E. (1984). *Groups: Interaction and Performance.* Englewood Cliffs, NJ: Prentice-Hall.

McGrath, J. E., & Hollingshead, A. B. (1994). *Groups Interacting With Technology.* Newbury Park, CA: Sage Publications.

Newell, A. (1990). *Unified Theories of Cognition.* Cambridge, MA: Harvard University Press.

Orasanu, J. M. (1990). Shared Mental Models and Crew Decision Making (No. CSL Report 46). Princeton NJ: Princeton University, Cognitive Science Laboratory.

Rice, R. E. (1992). Task analyzability, use of new media, and effectiveness: A multi-site exploration of media richness. *Organization Science, 3*(4), 475–500.

Select Bipartisan Committee (2006). A Failure of Initiative: Final Report of the Select Bipartisan Committee to Investigate the Preparation for and Response to Hurricane Katrina (US House of Representatives Report). Washington, DC: US Government Printing Office.

Shaw, M. E. (1981). *Group Dynamics: The Psychology of Small Group Behavior* (3rd ed.). New York: McGraw Hill Book Company.

Sheridan, T. B. (1988). Human and computer roles in supervisory control and telerobotics: Musings about function, language and hierarchy. In: L. P. Goodstein, H. B. Anderson & S. E. Olsen (Eds.), *Tasks, Errors, and Mental Models* (pp. 149–160). London: Taylor & Francis.

Sheridan, T. B. (1992). *Telerobotics, Automation, and Human Supervisory Control.* Cambridge, MA: MIT Press.

Shweder, R. A. (1991). *Thinking Through Cultures.* Cambridge, MA: Harvard University Press.

SRD. (2007). Revised (v3.5) System Reference Document: Basics and Ability Scores. Retrieved March 6, 2007, from http://www.wizards.com/d20/files/v35/Basics.rtf

Stasser, G., Stewart, D. D., & Wittenbaum, G. M. (1995). Expert roles and information exchange during discussion: The importance of knowing who knows what. *Journal of Experimental Social Psychology, 31*(3), 244–265.

Stasser, G., Vaughan, S. I., & Stewart, D. D. (2000). Pooling unshared information: The benefits of knowing how access to information is distributed among group members. *Organizational Behavior and Human Decision Processes, 82*(1), 102–116.

Sundstrom, E., DeMeuse, K. P., & Futrell, D. (1990). Work teams: applications and effectiveness. *American Psychologist, 45*(2), 120–133.

Swezey, R. W., & Salas, E. (Eds.). (1992). *Teams: Their Training and Performance.* Norwood, NJ: ABLEX.

Tambe, M. (1997). Towards flexible teamwork. *Journal of Artificial Intelligence Research, 7*, 83–124.

Thompson, C. (2007, February). Mr. know-it-all. *Wired, 15,* 42, 44.

Watts, L. A., & Monk, A. F. (1998). Reasoning about tasks, activities, and technology to support collaboration. *Ergonomics, 41*(11), 1583–1606.

Weiss, G. (Ed.). (1999). *Multiagent Systems: A Modern Approach to Distributed Artificial Intelligence.* Cambridge, MA: MIT Press.

48

Digital Human Modeling for Palpatory Medical Training with Haptic Feedback

Robert L. Williams II,
John N. Howell, and
Robert R. Conatser, Jr.

48.1 Introduction

This chapter discusses work in palpatory diagnosis training for schools of medicine and allied health. Palpatory diagnosis involves identifying medical problems via touch. First we present a literature review to establish the state of the art in palpatory diagnosis using virtual reality and then we present a case study: the Virtual Haptic Back project at Ohio University. Haptics indicates the human sense of touch, and haptic interfaces provide force and touch feedback from virtual environments. Our digital human modeling includes both realistic 3D surface models of the back geometry and accurate compliance measurement of human tissue in vivo.

48.2 Literature Review

Haptics, the science of touch, is being applied in medical virtual reality environments to increase realism and training effectiveness. In the medical field, most haptics/biomedical references relate to surgery, including suturing, endoscopy, intubation, injections, and patient rehabilitation, all of which require significant tactile skills. The Immersion Corporation (www.immersion.com) has developed haptic interfaces for injection training and sinus surgery simulation. Virtual reality (VR) with haptic feedback is also currently of interest in the dental field for simulation of drilling and of other dental procedures (e.g., Yau et al., 2006; www.sys-consulting.co.uk/web/projects/project_view.jsp?code=haptics). The remainder of this literature review will focus on a less-developed, but promising area, applying haptics and virtual reality to non-invasive, non-surgical palpatory training for medical diagnoses.

In the Stanford Visible Female project (Heinrichs et al., 2000), a 3D stereoscopic visualization of the female pelvis has been developed from numerous slices of 2D pelvis data. Haptic feedback was enabled via the PHANToM haptic interface, allowing the user to interact with and feel the virtual model. A virtual reality-based simulator prototype for the diagnosis of prostate cancer has been developed using the PHANToM haptic interface (Burdea et al., 1999). Another tumor palpation VR simulation was developed by Langrana (1997). Crossan et al. (2001) are using a PHANToM haptic interface in an equine ovary palpation simulator, for pregnancy determination in veterinary training. The Virtual Haptic Back is under development at Ohio University to augment the palpatory training of osteopathic medical students (Howell et al., 2006). This project has implemented a combined graphical and haptic model of a human back on a PC, using two PHANToM 3.0 interfaces for haptic feedback.

Temkin et al. (2002) are augmenting the Visible Human Project of the National Library of Medicine with a haptic device to improve the teaching of anatomy via touch. Basdogan et al. (2001) have presented a haptic simulator for bile duct diagnosis. For actual patients, Stalfors et al. (2001) have developed a method to model head/neck cancer conditions graphically and haptically, enabling doctors to palpate remotely, thus using telemedicine for remote diagnosis and monitoring. Riener et al. (2004) have presented a haptic knee joint simulator for clinical knee joint evaluation training. McLaughlin et al. (2004) have developed a haptic simulation of the female clinical breast examination since breast cancer is a major health problem for women in the United States. Kevin et al. (2001) report a system wherein a pressure transducer records the real-world forces in an expert's palpatory examination of a patient's abdomen for later playback via a haptic interface in a virtual model, for student training purposes.

48.3 Case Study: The Virtual Haptic Back (VHB)

48.3.1 VHB Project Motivation and Overview

The initial stage of learning palpatory diagnosis is a challenge for many osteopathic medical students. The Virtual Haptic Back (VHB) is being developed as an aid in the teaching and learning of palpatory diagnosis. It simulates the contours of human backs and the compliances (reciprocal of stiffness) over their surfaces, and allows these to be felt through the haptic interfaces. Regions of abnormal tissue texture are simulated with altered surface compliance.

The Virtual Haptic Back (VHB) was initially developed for training students of osteopathic medicine, but will be applicable in related areas, such as physical medicine and rehabilitation, physical therapy, chiropractic and massage therapy. The VHB augments traditional training in the difficult art of palpatory diagnosis (identifying medical problems via touch). Via two PHANToM 3.0 haptic interfaces, the student can explore a realistic virtual human back with accurate graphical and haptic (force and touch feedback) representations (fig. 48.1). Realistic somatic dysfunctions of different difficulty levels are programmed in random locations for the student to find by touch. The VHB can be used for student practice and as a repeatable, objective evaluation tool to track student progress.

The model consists of the contour of a back plus the compliances of the surface. For the initial version, the contour was modified from the Visible Female dataset. Subsequent contours have been obtained from living subjects with 3D photography. The compliance values were initially chosen to match the subjective feel of a back, as determined by osteopathic specialists in manual medicine. They were spot checked against compliance measurements made on actual human backs, using the PHANToM 3.0, equipped with a modified probe 2 cm in diameter, which assessed displacement as a function of force applied in graded steps up to 6 N.

FIGURE 48.1   Medical student practicing palpation with the virtual haptic back.

VHB users feel the virtual back with two fingers (or a finger and a thumb) placed into the thimble-like receptacles at the ends of the PHANToM haptic interfaces. Behind the virtual haptic back by approximately 15 cm is a full-sized image of the back displayed on a 23-inch flat-screen monitor (fig. 48.1). Two dots (L and R) on the screen indicate to the user visually where the left and right fingers are with respect to the haptic back (see fig. 48.2). In this way, the user is able to bring the fingers directly to the center of the back in order to begin palpation.

In the model used for testing, the back was programmed (C++ in the General Haptic Open Software Toolkit, GHOST SDK, with OpenGL for graphics) to have a uniform compliance except for one 2.5- by 3.0-cm region. The entire region of testing was a rectangle superimposed on the graphics image of the back 13.5 cm wide and 22 cm high; it encompassed thoracic segments T5–T10. The compliance of the abnormal region, which ranged from 2.45 to 0.972 mm/N, was made to blend smoothly into the compliance of the surrounding areas (2.52 mm/N) with a hyperbolic tangent function. Subjects typically moved their fingers along the back searching for regional differences in feel, and then went back to explore the region or regions they suspect might be abnormal. When they had decided which area was abnormal, they pressed a foot switch. A recorded voice provided immediate feedback as to whether their choice was correct or not.

In discriminating between two different linear compliances, applying greater force causes increasingly greater differences in displacement. This led users to press harder if they were having difficulty detecting the abnormal region. This was undesirable for two reasons. Application of force levels over 6 N caused the PHANToM electric motors to overheat. Second, application of high forces is inappropriate clinically, both because of potential patient discomfort and because palpatory information from superficial soft tissues can be lost by applying too much force. We did the following in order to discourage users from pushing too hard: (1) When they applied unacceptably high forces, automated voice feedback warned them not to press so hard. (2) A visual gauge in the lower right of the screen monitored their force levels, enabling users to see when they were approaching unacceptable levels. (3) More important, the programmed compliance difference between the abnormal area and the surrounding areas was multiplied by a hyperbolic tangent function that made the difference gradually disappear with increasing displacements between 8 and 16 mm. Thus, the differences were maximum in a desirable range of force

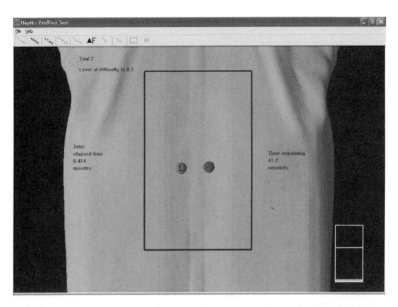

FIGURE 48.2 Graphic image of VHB during the pre- and post-tests. Dots marked L and R indicate the position of the two palpating fingers. The large rectangle indicates the region where abnormal compliance can be found. Trial number and difficulty level appear at upper left. Total time elapsed in the test appears at left; time remaining in the present trial appears at right. The box in the lower right is a force indicator, which rises to the level of the horizontal line before a voice warns against using so much force.

application, about 3 N in the normal regions. Based on preliminary measurements, this force level falls within the range of forces typically exerted by fingers in clinical palpatory diagnosis. Figure 48.3 shows the compliance functions programmed.

Table 48.1 shows the range of compliance values used in the current VHB model. The Weber fraction W expressed as a percentage is defined to be:

$$W = \frac{C_b - C_x}{C_b} \times 100\%$$

where $C_b = 2.52$ mm/N is the background compliance used for normal back properties and C_x is the abnormal area compliance, given in the table. We assign arbitrary difficulty levels as given in Table 48.1.

TABLE 48.1   VHB Abnormal Compliance Values Used

Difficulty Level	Compliance C_x (mm/N)	Weber Fraction W (%)
0.00	0.97	61.5
0.25	1.19	52.8
0.50	1.51	40.1
0.70	1.98	21.4
0.75	2.14	15.1
0.80	2.25	10.7
0.85	2.29	9.1
0.90	2.35	6.7
0.95	2.45	2.8
0.99	2.50	0.8

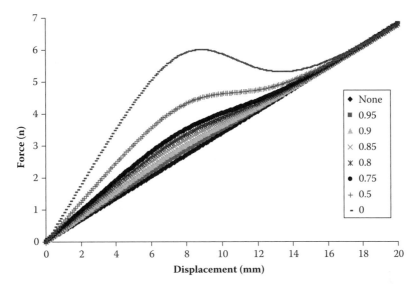

FIGURE 48.3   Relation between force and displacement at different difficulty levels. The straight line indicates background stiffness (the reciprocal of compliance). Increasing deviations from background make the task progressively easier. The deviations disappear at high displacements produced by application of high forces. The functions are implemented to simulate clinical palpatory situations.

48.3.2 Hardware and Software

For details regarding our VHB hardware and software, please see: http://www.ent.ohiou.edu/~bobw/html/VHB/VHB.html. This includes information on system specifications, haptic interfaces, our 3D viewing options (Ji et al., 2006), the playback feature (Williams et al., 2004), multi-point collision detection, our 3D camera for measuring the 3D contour of human backs, virtual motion testing (Chen et al., 2006), and our measurement of human tissue compliance *in vivo* (Williams et al., 2007).

48.3.3 Training Methods

The volunteer subjects (N = 21) were first-year osteopathic medical students within the first 3 months of their palpatory training. During their first session in the lab, they were given an opportunity to familiarize themselves with the haptic interfaces, practicing 10 to 15 minutes identifying regions of abnormal compliance until they were comfortable with the task. During these initial familiarization sessions, a transparency function was activated, permitting the user to see the skeletal elements beneath the skin (fig. 48.4). This feature was turned off during the pre- and post-training sessions and during the training sessions in between the tests.

48.3.4 The Pre- and Post-Tests

Following the practice phase, subjects took a test in which they had to locate the regions of abnormal compliance presented in successive trials. The locations varied randomly between sessions. The abnormalities could be on the left or on the right and at any one of six vertebral levels (T5–T10). Five different levels of difficulty (i.e., magnitude of compliance differences) were presented, starting with the easiest and progressing step-wise to the most difficult (1.51, 1.98, 2.25, 2.35, and 2.45 mm/N, compared to the background compliance of 2.52 mm/N). At each difficulty level there were two trials. Each trial was completed in 1 minute; time remaining in each was presented on the screen. Midway through the test, the program paused, giving the user an opportunity to take the fingers out of the apparatus and rest the arms, before finishing the test. This test was administered again at the end of the two-week practice sequence as a post-test in order to determine the improvement in performance resulting from the practice.

48.3.5 The Practice Sequence

Following the pre-test, subjects carried out the first of eight practice sessions which were completed over a 2-week period. Subjects were permitted to do the practice sessions at their own convenience, but no more than one session per day. The total time of each practice session was limited to 15 minutes. Although the default setting of the program started at the easiest level (greatest compliance difference), subjects could at any time pick any level of difficulty on which to work. Most tended to start with the easier levels and progress to the harder levels. In the practice sessions, when subjects made an incorrect diagnosis (i.e., incorrect localization), the recorded voice told them of their error and the program displayed a box around the correct area (fig. 48.4) on the screen with the transparency function turned on. Subjects could then go back and feel the abnormality before going on to the next trial. Subjects could also choose to pause the program at any time in order to rest their arms.

48.3.6 Data Analysis

Results from the pre- and post-tests were analyzed for each difficulty level by t-test. Results from the practice sessions were analyzed by repeated measures ANOVA.

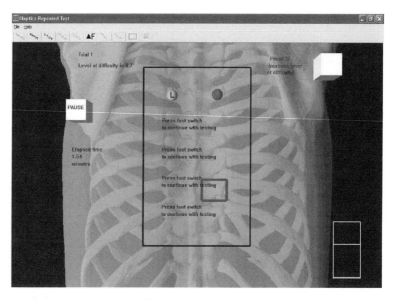

FIGURE 48.4 Appearance of the screen following a wrong answer in the practice sessions. The transparency function is activated to reveal underlying bone. The small green box indicates the actual location of the abnormal area. The user can practice palpating this area before going on to the next trial. By touching the upper left box with an L or R finger dot, the user can pause the program; by touching the upper right box, the user can alter the difficulty level of the next trial. These boxes can be accessed at any time during the practice sessions.

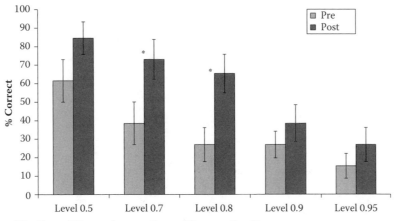

*Significant difference between Pre- and Post-test P < .05

FIGURE 48.5 Comparison of percent correct responses in tests before and after the practice sessions as a function of difficulty level of the task. The difficulty levels from easiest to hardest (left to right) correspond to compliance values of 1.51, 1.98, 2.25, 2.35, and 2.45. Baseline compliance value was 2.52. In this and subsequent figures, vertical bars represent standard errors of the mean values.

48.3.7 Evaluation Results

Significant differences in performance accuracy between the pre- and post-tests were seen only at difficulty levels of 0.7 and 0.8, corresponding to compliance values of 1.98 and 2.25, respectively (fig. 48.5). At the easiest levels, performance was better than at the harder levels, especially in the pre-test. A trend toward better performance with practice at these levels might have reached statistical significance with a bigger N. Even easier levels that were included in the practice sessions were not included in the pre- and post-tests. At the harder levels, performance dropped off to what are probably chance levels (see Discussion) and at these levels no significant pre- to post-practice performance differences were observed.

Performance monitored in the practice sessions revealed gradual improvement over the eight sessions revealed at difficulty levels of 0.7, 0.75, and 0.8 (compliance values of 1.98, 2.14, and 2.25, respectively). This is illustrated in Figure 48.6 for the difficulty level of 0.75.

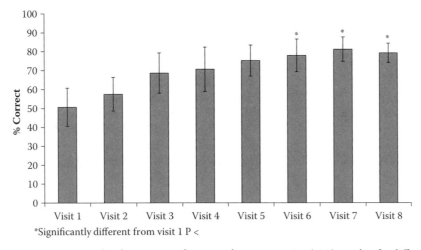

*Significantly different from visit 1 P <

FIGURE 48.6 Percent correct localizations as a function of practice session (visit) number for difficulty level of 0.75 (compliance value of 2.14). Performance in the last three sessions was significantly (P < 0.05) better than that in the first session.

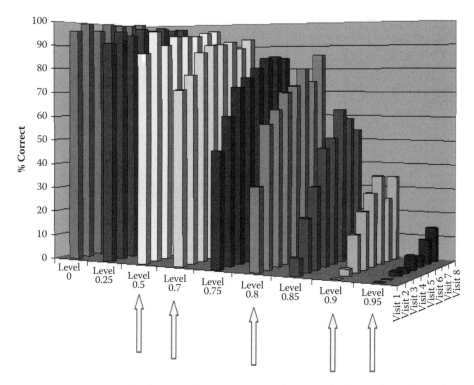

FIGURE 48.7 Percent correct localizations during practice as a function of both visit (session) number and difficulty level (designated in units of compliance). As the difficulty level increases, moving from left to right on the graphed surface, performance falls off. At the difficulty levels at which performance falls off, improvement can be seen with increasing visit number. The arrows indicate difficulty levels used in the pre- and post-tests.

Cumulative averaged results for all difficulty levels, all practice sessions, and all subjects are shown in the 3D plot of Figure 48.7. It emphasizes that performance falls off as the difficulty level rises. The rise in performance as a function of visit number is also apparent in the range of difficulties at which performance falls off. At the easiest levels, to the left, users did very well even in the first practice session. At the hardest level, at the far right, accuracy improved with successive practice sessions, but remained near chance level throughout. The most dramatic improvement is seen at intermediate levels of difficulty where users performed poorly during the first practice sessions, but did progressively better in successive sessions.

Upon finishing the training and the post-test, subjects were asked three questions. The first question was "Do you think this practice with the haptic back will be of help to you in the development of your palpatory skills in OMM lab?" Of the 21 subjects, 17 marked "yes," 4 marked "maybe," and no one marked "no." The second question was, "Do you think further practice with the haptic back would be of help to you in the development of your palpatory skills?" Twelve subjects answered "yes," 8 marked "maybe," and 1 subject marked "no." They were also asked to rate the realism of the simulation on a 0 to 10 scale, with 0 being unrealistic and 10 being very realistic. The mean value reported was 6.5.

48.4 Discussion

48.4.1 VHB Capabilities and Limitations

The VHB combines graphics and haptics into the simulation of the human back and is beginning to find applications as a training tool in medical education and as a research tool in the study of touch. The data presented here relate to both of these applications.

A simulation is only as good as the data upon which the simulation model is based. The intent of the VHB is to simulate the process of palpatory diagnosis in which a practitioner of manual medicine uses fingers and hands to sense the mechanical properties of the patient's body surface. The VHB model, using the PHANToM, is limited in that it simulates only the gross contours and the compliance of the back normal to the surface. The haptic interfaces do not permit the user to feel fine contours detectable only by the mechanoreceptors of the skin, and the model is devoid of any thermal component. Shear forces are also currently not simulated. The force feedback provided by the haptic interfaces simulates primarily the proprioceptive component of palpation.

In principle, the most accurate force feedback simulation would be based on high-resolution compliance measurements over the entire surface of the back. Although current work in our laboratory is directed toward that goal, the compliance model used in this study was largely based on feedback from practicing physicians who specialize in manual medicine. This evaluation by practitioners seems vital in that, because of the incompleteness of the model with respect to sensory modalities, it is conceivable that the most perfectly simulated compliance characteristics would not provide the best simulation of the palpatory experience. Only the practitioners can tell us that, and, unless the model passes this test, it will not be accepted by them as an effective aid in teaching/learning palpatory diagnosis.

48.4.2 Compliance Detection Results

A standard measure in the analysis of sensory systems is that of the limit of detection, the just noticeable differences (JND) (Gescheider, 1997). In some sensory systems, such as auditory and visual, two questions are of interest: (1) the lowest signal level that can be detected; and (2) the smallest change in signal intensity expressed as a fraction of the absolute intensity value, known as the Weber fraction. In the case of compliance detection, only the latter has meaning. Although our experimental setup was not optimally designed for the precise determination of the Weber fraction, our results do yield an average value of 11% after training.

Using a device that permitted control of the compliance and of the total displacement used in a pinching movement between the thumb and index finger, Tan et al. (1995) demonstrated that compliance detection measured against a baseline compliance of 4 mm/N averaged a Weber fraction of 8% when displacement in all trials was the same. The authors argue that under these conditions the subjects have information about the terminal force at the end of the movement and what was measured was the ability to discriminate force, rather than compliance. The similarity of their value with previously measured Weber fraction for force measurement, 7% (Pang et al., 1991) supported their argument. They then repeated the experiment, but varied the displacement randomly during the trials. Without the terminal force cues, compliance detection then decreased to 22%.

Using a PHANToM 1.5 haptic interface, Dhruv and Tendick (2000) found Weber fractions in the range of 14% to 25%, depending on the baseline level of compliance against which the differences were detected. With a baseline of compliance of 8 mm/N the detection threshold was 14%, but with a baseline of compliance of 2 mm/N the detection threshold was 25%. A subsequent study with the PHANToM yielded a detection threshold of 8% to 12% (De Gersem et al., 2003, cited in De Gersem et al., 2005).

The classical method of determining the Weber fraction, done in these studies, presents the subject sequentially with two surfaces, objects, or situations. The subject indicates which has the higher compliance. This is repeated many times. Since there are two choices only, chance score is 50%. Typically, the threshold value for detection is taken as 75%—halfway between chance and completely reliable detection. In our study the chance value was somewhere between 8.3% and 16.7%, depending on whether or not the user was touching one or two back regions when the foot switch was depressed. This is consistent with performance at the hardest level in the pre- and post-tests (fig. 48.5). Taking 20% as chance levels, 60% correct identifications would then be halfway between chance and 100%. A mean of 60% was achieved in the pre-test at a difficulty level of 0.5, which corresponds to a Weber fraction of 40%. In the post-test, subjects achieved 60% correct at a difficulty level of 0.8. This corresponds to a Weber fraction of 11%. This figure agrees well with the Weber fraction of 8% to 12% reported by De Gersem et al. (2003).

The improvement in the Weber fraction from 40% to 11% between pre- and post-tests may overestimate the training effect somewhat, in that part of the learning may really represent familiarization with the haptic apparatus. This is suggested by the fact that performance improved significantly between the pre-test and the first practice test session. At the 0.7 difficulty level pre-test performance was about 35% while performance during the first practice session at that difficulty level was already 70%, similar to the post-test value (fig. 48.5). The gradual improvement in performance seen at somewhat harder levels of difficulty, 0.75 (fig. 48.6) and 0.80 (not shown), suggests that, in addition to the rapid learning from familiarization, a slower learning process also took place.

Further work will be necessary to determine what the limits are of learning in this context and what the absolute physiological limits of compliance detection might be. It will be interesting to see if, through more extensive practice than used in this study, the Weber fraction can be brought down to that of the force detection results reported by others in the literature.

Clinical palpatory diagnosis training undoubtedly involves improvement in both the ability to feel small compliance differences and the ability to impart meaning to what is felt. The improvement in performance with practice on the VHB is likely to reflect primarily the former, but the complexity and realism of the model may require some of the latter. Strong positive feedback from medical student subjects suggests that experience with the VHB may be also helpful in both of these aspects.

The data discussed were collected for N=21 medical students during winter 2006. In fall 2006, we improved the realism of our model and the precision of our training protocol and repeated the study with the entire incoming class of first-year medical students at Ohio University. Data from 89 students were included in the study. These data, as yet unpublished, indicate the same trends as shown in the study with the 21 volunteer subjects discussed in this chapter, with regard to improvement in accuracy (Weber fraction) and speed with practice on the VHB.

48.5 Summary

We presented a literature review detailing the state-of-the-art in palpatory diagnosis training with virtual environments and haptics augmentation. This area requires digital human modeling in 3D surface models and tissue compliance. We then presented a case study of the VHB, a haptic simulation of human back for medical and related palpatory diagnosis training. We used the VHB to assess the limit of human compliance detection and to explore the effects of training. Compliance detection values obtained are similar to those reported by other investigators and an eight-session training period over 2 weeks significantly improves performance.

Acknowledgments

The authors gratefully acknowledge grant funding from the Osteopathic Heritage Foundation of Columbus, OH, for supporting this work. We also thank members of the Virtual Haptic Back Team and Interdisciplinary Institute for Neuromusculoskeletal Research for project work. We acknowledge expert palpatory physicians David Eland, D.O., and Janet Burns, D.O. The authors thank David Noyes for hardware construction and intellectual contributions.

References

Basdogan, C., C. Ho, and M.A. Srinivasan, 2001, "Virtual Environments for Medical Training: Graphical and Haptic Simulation of Common Bile Duct Exploration," *IEEE/ASME Transactions on Mechatronics*, 6(3): 267–285.

Burdea, G., G. Patounakis, and V. Popescu, 1999, "Virtual Reality-Based Training for the Diagnosis of Prostate Cancer," *IEEE Transactions on Biomedical Engineering*, 46(10): 1253–1260.

Chen, M.-Y., R.L. Williams II, R.R. Conatser Jr., and J.N. Howell, 2006, "The Virtual Movable Human Upper Body for Palpatory Diagnostic Training," *SAE Digital Human Modeling Conference*, Paper #06DHM-5, July 4–6, Lyon, France.

Crossan, A., S. Brewster, S. Reid, and D. Mellor, 2001, "Comparison of Simulated Ovary Training over Different Skill Levels," *Eurohaptics* 2001: 17–21.

De Gersem, G., H. Van Brussel, and F. Tendick, 2003, "A New Optimization Function for Force Feedback in Teleoperation," *Proceedings of the International Conference on Computer Assisted Radiology and Surgery* (CARS), London, UK, June: 1354.

De Gersem, G., H. Van Brussel, and F. Tendick, 2005, "Reliable and Enhanced Stiffness Perception in Soft-tissue Telemanipulation," *International Journal of Robotics Research*, 24: 805–822.

Dhruv, N., and F. Tendick, 2000, "Frequency Dependence of Compliance Contrast Detection," *Proceedings of the ASME Dynamic Systems and Control Division,* DSC-Vol. 69-2.

Gescheider, G.A., 1997, *Psychophysics: The Fundamentals*, 3rd ed., Lawrence Erlbaum Associates, Mahwah, NJ, 3.

Heinrichs, W.L., S. Srivastava, J. Brown, J.-C. Latombe, K. Montgomery, B. Temkin, and P. Dev, 2000, "A Steroscopic Palpable and Deformable Model: Lucy 2.5," *Third Visible Human Conference*, Bethesda, MD.

Howell, J.N., R.L. Williams II, J.M. Burns, D.C. Eland, and R.R. Conatser Jr., 2006, "The Virtual Haptic Back: Detection of Compliance Differences," *SAE Digital Human Modeling Conference*, Paper #06DHM-27, July 4–6, Lyon, France.

Ji, W., R.L. Williams II, J.N. Howell, and R.R. Conatser Jr., 2006, "3D Stereo Evaluation for the Virtual Haptic Back Project," *14th Symposium on Haptic Interfaces for Virtual Environments and Teleoperator Systems*, IEEE-VR2006, March 25–26, Arlington, VA.

Kevin, C., T. Kesavadas, and J, Mayrose, 2001, "Development of the Virtual Human Abdomen: Algorithms and Methodologies," *Proceedings of ACM-SIGGRAPH*: 170.

Langrana, N., 1997, "Human Performance Using Virtual Reality Tumor Palpation Simulation," *Computers and Graphics*, 21(4): 451–458.

McLaughlin, M., I. Cohen, M. Desbrun, L. Hovanessian, M. Jordan-Marsh, S. Narayanan, G. Sukhatme, and P. Georgiou, 2004, "Haptic Simulator for Training in Clinical Breast Examination," NSF Report. http://imsc.usc.edu/research/project/hapticsim/hapticsim_nsf.pdf.

Pang, X.-D., H.Z. Tan, and N.I. Durlach, 1991, "Manual Discrimination of Force Using Active Finger Motion," *Perception and Psychophysics*, 49(6): 531–540.

Riener, R., M. Frey, T. Proll, F. Regenfelder, and R. Burgkart, 2004, "Phantom-Based Multimodal Interactions for Medical Education and Training: the Munich Knee Joint Simulator," *IEEE Transactions on Information Technology in Biomedicine*, 8(2): 208–216.

Stalfors, J., T. Kling-Petersen, M. Rydmark, and T. Westin, 2001, "Haptic Palpation of Head and Neck Cancer Patients: Implication for Education and Telemedicine," *Studies in Health Technology and Informatics*, 81: 471–474.

Tan, H.Z., N.I. Durlach, G.L. Beauregard, and M.A. Srinivasan, 1995, "Manual Discrimination of Compliance Using Active Pinch Grasp: The Roles of Force and Work Cues," *Perception and Psychophysics*, 57: 495–510.

Temkin, B., E. Acosta, P. Hatfield, E. Onal, and A. Tong, 2002, "Web-Based Three-Dimensional Virtual Body Structures: W3D-VBS," *Journal of the American Medical Information Association*, 9(5): 554–556.

Williams II, R.L., W. Ji, J.N. Howell, and R.R. Conatser Jr., 2007, "In Vivo Measurement of Human Tissue Compliance," submitted, SAE Digital Human Modeling Conference, Seattle, WA.

Williams II, R.L., M. Srivastava, R.R. Conatser Jr., and J.N. Howell, 2004, "Implementation and Evaluation of a Haptic Playback System," *Haptics-e Journal*, IEEE Robotics and Automation Society, 3(3): 1–6.

Yau, H.T., L.S. Tsou, M.J. Tsai, 2006, "Octree-Based Virtual Dental Training System with a Haptic Device," *Computer-Aided Design and Applications*, 3(1–4): 415–424.

49

Health Care Delivery and Simulation

Kathryn Rapala and
Julie Cowan Novak

49.1 Introduction

In the wake of the landmark 1999 Institute of Medicine report *To Err Is Human: Building a Safer Health System,* international attention has been focused on the safety of our patients. Since then, health care practitioners, administrators, patients, industry, and academia have put time, energy, intellect, and dollars into reducing the number of patients who die or are injured due to health care errors.

It is difficult to discern if this significant effort has reduced health care error. Leape and Berwick, in their article "Five Years after To Err Is Human: What Have We Learned?" note that although incremental gains in patient safety have been made, barriers remain:

> The combination of complexity, professional fragmentation, and a tradition of individualism, enhanced by a well-entrenched hierarchical authority structure and diffuse accountability, forms a daunting barrier to creating the habits and beliefs of common purpose, teamwork, and individual accountability for successful interdependence that a safe culture requires. (Leape & Berwick, 2005)

It is time to revisit our patient safety interventions for effectiveness, and to eliminate those barriers.

49.2 High-Reliability Organizations

49.2.1 Reliability and Health Care

Reliability is the measurable capability of a process or procedure to perform its intended function in the required time under commonly and uncommonly occurring conditions (Berwick & Nolan, 2003). High-reliability organizations are those that, even with highly complex and dangerous activities, have very few errors. Examples of high-reliability organizations include nuclear power plants, aircraft carriers, and air traffic control systems.

Health care is even more complex than the examples listed above. An aircraft carrier focuses on landing one aircraft at a time. Manufacturers are able to standardize parts and processes on the assembly line. In contrast, health care practitioners manage the arrivals, transfers, and departures of highly individual patients with very different needs on any given day. For the individual practitioner, the practice, process, and products are factored into patient needs, resulting in extremely complex care. The challenge for all of us, as health care professionals, is to convert the health care environment into high-reliability organizations.

49.3 Organizational Elements Essential to Reach Reliability

It is one thing to study high-reliability organizations, but quite another to introduce reliability into practice. Gaba stated that there are four elements present in every high-reliability organization. These elements, which may be used as a framework for practice, include:

Systems, structures, and procedures conducive to safety and reliability are in place.
A culture of safety permeates the organization.
Safety and reliability are examined prospectively for all the organization's activities; organizational learning by retrospective analysis of accidents and incidents is aggressively pursued.
Intensive training of personnel and teams takes place during routine operations, drills, and simulations. (Gaba, 2003)

Aspects of all of these elements are present in every health care service and academic setting. Health care organizations wishing to decrease error need to address every element in depth, implementing the latest and best evidence in each area. This chapter focuses on the training and simulation in health care.

49.4 Simulation

Simulation may be defined as a "technique, not a technology, to replace or amplify real experiences with guided experiences, often immersive in nature, that evoke or replicate substantial aspects of the real world in a fully interactive fashion" (Gaba, 2007). Like airline industries that put their pilots and crews through teamwork and simulation exercises, simulation training and exercises have been developed and used in health care in the military and with cardiopulmonary resuscitation (CPR) certification. Over the past two decades, simulation in health care has begun to be used more broadly across multiple settings.

There are four major types of simulation and training. The first is using equipment or props to perform a task. Examples include anchoring a Foley catheter, mock codes, or starting an intravenous line. A second type of simulation is one of a particular procedure or scenario, using high-fidelity simulated patients or simulated portions of patients. These patients may be used in a variety of circumstances and scenarios, such as performing a cardiac catheterization or simulating a process such as an operating room experience. This type of simulation may be done individually or with many configurations of teams. Simulations may also be computer based. Much like video games, these two-dimensional simulations can teach the operation of equipment or any number of processes or procedures. Finally, simulations may be drills, such as tabletop drills used in public health for natural disaster preparedness or avian influenza readiness.

49.5 Simulation Conceptual Framework

Simulation is a relatively new health care application. In the 1960s, the Laerdal Company released a series of mannequins that could be used in resuscitation efforts. In the late 1960s, the computer-controlled simulators were developed, with the high-fidelity simulators we know today developed in the mid-1990s (Hunt et al., 2006). Since then much work, in both dedicated simulation centers and individual organizations, has been done to integrate simulation and health care.

With the diversity of simulation applications, the professions using and producing simulation techniques, scenarios, and devices, along with the speed of simulation science application, it is not surprising that there is not a wealth of accepted simulation conceptual frameworks. Kneebone writes that this diversity is a strength that encourages creativity and choice, but there is a danger that without a coherent theory, fragmentation and lack of direction will ensue (Kneebone, 2006).

With that caution, following are three applicable but different simulation conceptual frameworks. These conceptual frameworks, although untested, do not appear mutually exclusive. In fact, Kneebone's simulation and boundary areas may be seen as a high-range theory, Gaba's dimensions as a mid-level theory, and Jeffries' framework as a way to design education.

49.5.1 Kneebone: Simulation and Boundary Areas

Kneebone states most growth occurs in "boundary zones," or those areas where different domains or disciplines interact and cross-fertilize. Work occurs exclusively within individual disciplines, but collaborative work in the boundary zones is what produces new ideas and growth (Kneebone, 2006). Kneebone further posits that seven disciplines—and he cautions that these are based on his personal experience—are relevant to simulation. These simulation domains are: clinical practice; simulator technology; education; communication; psychology, sociology and human factors; health care policy; and performance arts. A barrier to the boundary zone concept is "total internal reflection"; that is, a discipline is essentially in its own silo. Within this silo, the discipline looks only inward and is unable to look at the total landscape. Kneebone states that awareness of boundary zones is not enough, and that rigor needs to ensure collaboration in simulation.

49.5.2 Gaba: 11 Dimensions of Simulation

Gaba states that current and future simulation applications—and these applications are diverse in nature—can be categorized by 11 simulation dimensions. These dimensions are the:

Purpose and aims of the simulation activity
Unit of participation in the simulation
Experience level of simulation participants
Health care domain in which the simulation is applied
Health care discipline of personnel participating in the simulation
Type of knowledge, skill, attitudes, or behavior addressed in the simulation
Age of the patient being simulated
Technology applicable or required for simulation
Site of simulation participation
Extent of direct participation in simulation
Feedback method accompanying simulation

The dimensions additionally, according to Gaba, represent potential areas for research, both across the dimensions, and in assessing the outcomes of simulation across the dimensions (Gaba, 2007).

49.5.3 Jeffries: Designing, Implementing, and Evaluating Simulations

Jeffries proposes a framework of a slightly different type: an educational framework to design, implement, and evaluate simulations. Core to this model are teacher factors, student factors, and education practices. The expertise and skills of teachers interact with those of the student, in particular the student program, level, and age. Educational practices—active learning, feedback, student/faculty interaction, collaboration, high expectations, diverse learning, and time on task—completes this core unit. An intervention, known as a design characteristic and simulation, is introduced to this core unit. These design characteristics and simulations may consist of objectives, fidelity or realism, complexity, cues or debriefing. The effect of the design characteristics and simulation produces the outcomes, which are learning or knowledge, skill performance, learner satisfaction, critical thinking, and self-confidence (Jeffries, 2005).

49.6 Communication and Teamwork

Moving from the old model of "see one, do one," simulation and training provides learning and practice on virtual patients rather than real patients. Additionally, the concept of teamwork training is very important to health care. Sentinel event data, both internal and external, indicate that ineffective communication is a root cause of almost every adverse event (The Joint Commission, 2007). The study "Silence Kills" showed that 84% of physicians and 62% of nurses and other clinical-care providers have seen coworkers take shortcuts that could be dangerous to patients, yet fewer than 10% of physicians, nurses, and other clinical staff directly confront their colleagues about their concerns, and one in five physicians said they have seen harm come to patients as a result (VitalSmarts, 2005). This idea of "organizational silence" refers to the notion of a collective-level phenomenon of saying or doing very little in response to significant problems that face an organization (Henrikson & Dayton, 2006). Simulation provides a mechanism to teach these concepts.

Learning how to communicate and work in teams not only raises awareness, but also teaches essential skills as a provider develops from novice to expert (Larew et al.). Simulation and teamwork training also can mimic work complexity. Traditional training teaches a perfect procedure in a perfect environment, which happens infrequently in the real environment. The human factors present in work environments are an area that has great potential for simulation research.

49.7 Cost

Cost is a factor in all simulations and is relatively easy to measure compared to other parameters; however, the literature does not reflect a great degree of information in this regard. Cost may depend on the target population, purpose of the simulation, technology used, and how organizations incorporate simulation into work. Cost varies per type of simulation used: It is lower for simple skills-based simulations, and higher for high-fidelity team simulation. One must also consider time spent off the unit for health care providers and staffing of the simulation (Gaba, 2007). Additional considerations of cost include space for the simulation and time and expertise needed for scenario and simulation development and programming.

Now that health care simulation is frequently implemented in academia and practice, a need exists to study the impact of technology on costs. Finding only one study that related to the cost of simulation construction rather than operating costs over time, a discounted cash flow methodology was used to evaluate a regional simulation center that was created by two nursing programs and a hospital. This analysis evaluated the difference between three separate entities as compared to one center with common equipment, facility, and faculty costs. Although faculty savings were significant, investment costs and revenue assumptions were higher—overall there were no savings. The model did not take into account intermediate and long-term outcomes (Harlow & Sportman, 2007).

49.8 Outcomes

Despite the spread of simulation in health care practice, there is not a great deal of outcomes research that speaks to its effectiveness. This may be because the benefits of simulation are difficult to measure, and are more long-term in nature than an immediate intervention. Gaba notes that there are no comparable studies available from high-reliability organizations in non-health-care industries (Gaba, 2007).

In the nursing education literature, Ravert identified nine qualitative studies in which the effectiveness of computer-based simulator education was measured. Data from these studies indicated that simulation did enhance education (Ravert, 2002).

It is clear that the study of simulation outcomes has many research opportunities. Teamwork and communication, work complexity and human factors, in addition to the more traditional competency and education areas, are areas in which various interventions can be simulated and tested.

49.9 Case Study: Purdue University School of Nursing

An Institute of Medicine (IOM) multidisciplinary summit was convened to address educators' roles in patient care error prevention in 2002. *Health Professions Education: A Bridge to Quality* was published as a result (Institute of Medicine, 2003). The report noted:

> Although there is a spotlight on the serious mismatch between what we know to be good quality care and the care that is actually delivered, students and health professionals have few opportunities to avail themselves of coursework and other educational interventions that would aid them in analyzing the root causes of errors and other quality problems and in designing system-wide fixes.

A simulation learning environment is one way in which the Purdue University School of Nursing has bridged this gap.

The Purdue School of Nursing simulation lab began as a vision in the late 1980s as a simulated intensive care unit (ICU) in the Purdue Center for Nursing Education (CNE). This simulated ICU included a

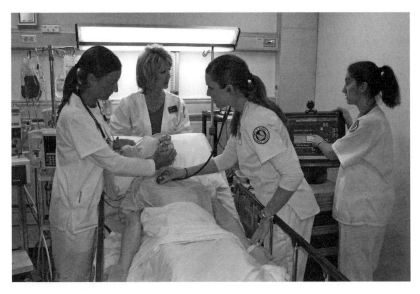

FIGURE 49.1 Purdue CNE students practice clinical tasks on a mannequin.

mechanical ventilator, an electrocardiogram, and a training mannequin. A faculty member was charged with providing medical surgical nursing students with a simulated critical care experience prior to the actual critical care experience. Nursing students were guided through monitor placement, the mechanics and physiological basis of ventilation, and mock codes. This "ICU on Campus" day was very well received by students. And while an early Purdue study focus on the reduction of anxiety of students who underwent simulation training did not show a reduction in student anxiety, students embraced the program and found it so valuable that the simulation experience is now firmly embedded in the curriculum (Erhler & Rudman, 1993). In the present day, instead of anxiety, measures are student performance, measuring human factors, and other patient safety-driven measures.

The Purdue Simulation Program has grown and matured over time. Initial funding of the simulated ICU was from a Helen Fuld grant; other funds from the Purdue Presidents fund, a HRSA grant, and a recent Fuld grant in 2006 have provided state-of-the-art simulator updates. Starting from a simple simulating mannequin in the 1990s, the Purdue simulation family now consists of an adult, a child, an infant, and a pregnant female.

With this sophisticated equipment, the objective of the Purdue CNE simulation laboratory is for students to learn skills, both in tasks and in teamwork, and practice as much as necessary in a low stress environment. Critical thinking skills and collaboration are encouraged. To address the nursing shortage, student admissions has increased from 100 to 170 students per class; simulation time allows campus rotations prior to hospital and agency placements. With faculty and clinical rotations at a premium, simulation increases educational efficiency and quality.

Scenarios based on practice are created for the students by expert faculty, with input from health care providers and the literature. During a clinical experience in the health care setting, students may not experience many critical aspects of patient care. Simulation bridges that gap. For example, a student on an obstetrical clinical rotation may not, due to timing, see a delivery. However, the pregnant simulated patient has provided over 300 "deliveries," illustrating many aspects of obstetrics that students would not have experienced otherwise.

Simulation is a key component of the Purdue doctor of nursing practice program (DNP). The DNP program addresses the complexity of the health care system: the information, knowledge, and technology explosion; escalating costs; and the need for a systems approach to create new models of care and solve existing health care dilemmas (Wall, Novak, & Wilkerson, 2005). Due to the complexity of health care, and the need to grow the program due to the nursing shortage at all levels and in most settings, a program was designed to include the purchase of simulators to support the undergraduate and graduate programs. The DNP program was funded, including simulation, with a Helen Fuld grant for $2.49 million. DNP students not only use the simulators, design scenarios, and laboratories based on practice and evidence, but look for gaps in product design. In collaboration with mechanical engineering students, several prototypes have been designed for health care improvement. Today, the simulation lab is more than a "learning lab" of old. Rather, the Purdue simulation lab is a Mecca for collaboration, design, observation, interaction, and immersion.

Kneebone writes that many see simulation as a means to practice clinical tasks, but the true value of simulation occurs when the real world is paralleled (Kneebone, 2006). The latter is what the Purdue CNE simulation aspires to achieve. Simulation is one of those new care models that has almost unlimited application not only for practitioners but for engineers and other experts. The 20-year Purdue simulation lab has matured as the uses for simulation, which is a fairly young process in health care, have evolved along with it.

49.10 Conclusion

The Purdue School of Nursing case study is representative of simulation history. Simulation is well grounded in rationale and is intuitively a good thing—one practices scenarios or procedures to increase competency and safety. Despite the lack of research to provide an evidence base for simulation, organizations are rapidly adopting the technology to address patient safety issues at a fundamental level. Simulation is providing opportunities not only to the students and practitioners using the simulation, but also to researchers, industry, and other health care partners.

Acknowledgments

The authors gratefully acknowledge grant support from the U.S. Department of Health and Human Services, Health Resources and Services Administration (#1 DO9HPO5304-01-00), "Rural Advanced Practice Nursing: Post-BSN to MS/DNP," and from the Helene Fuld Health Trust, "The Doctor of Nursing Practice: Reengineering Healthcare."

The authors acknowledge Michael Criswell, MSN, RN, CCNS, who provided historical context for the Purdue School of Nursing simulation case study.

References

Berwick, D., & Nolan, T. (2003). *High reliability health care*. Retrieved October 2004, from http://www.ihi.org/IHI/Topics/Reliability/ReliabilityGeneral/EmergingContent/HighReliabilityHealthCarePresentation.htm

Gaba, D. (2003) Safety first: ensuring quality care in the intensely productive environment—the HRO model. *Anesthesia Patient Safety Foundation Newsletter 18*(1):1–16.

Gaba, D. (2007). The future vision of simulation in healthcare. *Simulation in Healthcare 2*(2): 126–135.

Harlow, K., & Sportman, S. (2007). An economic analysis of patient simulators for clinical training in nursing education. *Nursing Economics 25*(1): 24–29.

Henriksen, K., & Dayton, E. (2006). Organizational silence and hidden threats to patient safety. *Health Services Research 41*(4p2): 1539–1554.

Hunt, E., Nelson, K., & Shilkofski, N. (2006) Simulation in medicine: Addressing patient safety and improving the interface between healthcare providers and medical technology. *Biomedical Instrumentation & Technology 40*: 399–404.

Institute of Medicine (2003). Greiner A, Knebel, E. *Health professions education: a bridge to quality*. Washington DC: National Academy Press.

Jeffries, P. (2005). A framework for designing, implementing and evaluating simulations used as teaching strategies in nursing. *Nursing Education Perspective 26*(2): 96–103.

Kneebone, R. (2006). Crossing the line: Simulation and boundary areas. *Simulation in Healthcare 1*(3): 160–163.

Kohn, L., Corrigan, J., & Donaldson, M. (2000). *To err is human: Building a safer health system*. Washington, DC: National Academy Press.

Larew, C., Lessans, S., Spunt, D., Foster, D., & Covington, B. Innovations in clinical simulation: Application of Benner's theory in an interactive patient care simulation. *Nursing Education Perspectives. 27*(1): 16–21.

Leape, L., & Berwick, D. (2005) Five years after *To err is human*: What have we learned? *Journal of the American Medical Association 292* (19), 2384–2390.

Novak, J. (2006). The Doctor of Nursing Practice: Reengineering Healthcare, Helen Fuld Healthcare Trust.

Novak, J., Wall, B., & Edwards, N. (2006). U.S. Department of Health and Human Services, Health Resources and Services Administration (#1 DO9HPO5304-01-00), "Rural Advanced Practice Nursing: Post-BSN to MS/DNP"

Rapala, K., & Novak, J.C. (2007). Integrating patient safety into curriculum—the Purdue Doctor of Nursing Practice. *Patient Safety Quality Healthcare 4*: 16–18, 20–23.

Ravert, P. (2002). An integrative review of computer-based simulation in the education process. *CIN: Computer, Informatics, Nursing 20*(5): 203–208.

The Joint Commission. (2007). *Sentinel Event.* Retrieved April 9, 2007 from http://www.jointcommission.org/

VitalSmarts. (2005). *Silence Kills: The 7 Crucial Conversations in Healthcare.* Retrieved May 24, 2007 from http://www.vitalsmarts.com.

Wall, B., Novak, J., and Wilkerson, S. (2005). The doctor of nursing practice: Reengineering healthcare. *Journal of Nursing Education 44*(9): 396–403.

50

Modeling Human Physical Capability: Joint Strength and Range of Motion

Laura Frey Law, Ting Xia,
and Andrea Laake

50.1 Introduction

Digital human modeling (DHM) has seen success in industries such as vehicle design and workplace improvement. More recently, DHM has shown great potential for applications such as injury analysis and prevention, orthopedics, rehabilitation, and sports performance simulation. [1] While these applications span a variety of areas, they all depend on our knowledge of the human musculoskeletal system. In other words, to implement a DHM capable of predicting realistic human motion and performance, the capabilities and limitations of the human musculoskeletal system must be imposed on the model.

50.2 What Makes a Digital Model Human?

A digital model could be created that represents a superhuman with little resemblance to any living individual. This, however, is not likely to be considered a digital human, but rather a digital model of a robotic or humanlike entity. Humans inherently have great inter-individual variability—making the modeling of humans a complex process. Indeed, specific applications may have different goals, requiring different representations of human capabilities. While there is a continuum of human classifications, commonly recognized in anthropometry as normal percentiles of individual size (e.g., 5th through

95th percentile female), we often consider discrete facets to simplify and encompass the wide range possible (5th, 50th and 95th percentiles). Similarly, DHM could categorize human capability into broad categories: the ideal human, the impaired human, and the realistic human. Each could have its unique application and advantages.

50.2.1 Ideal Digital Human

The *ideal* digital human may be a representation of a highly trained, healthy individual such as a professional athlete. This ideal model might be useful for modeling various training and competitive applications: optimizing figure skating routines with complex jumps, designing protective gear for football players, or racquet design for elite tennis players. No one ideal human exists of course, and thus would have to be tuned based on the specific application.

50.2.2 Impaired Digital Human

This approach could be used to predict how various impairments impact function or how preventative health approaches could benefit society in general. This is not to imply that there exists an ideal size or shape, but rather to represent individuals with specific impairments: limited joint range of motion (ROM), strength, or cardiovascular endurance. For example, digital human modeling may one day be able to demonstrate whether strength and aerobic conditioning may benefit the progression and/or prevention of musculoskeletal disorders. Numerous epidemiological studies support that being overweight and unfit have numerous negative consequences, but no one has yet to demonstrate actual prevention through modeling.

50.2.3 Realistic Digital Human

The *realistic* DHM must be able to accurately represent the majority of the population's capabilities: size, strength, endurance, ROM, and so on. This DHM category is most likely to have the greatest usefulness for ergonomic applications, as a wide range of individuals span the workforce. This chapter will discuss methods of representing realistic physiological capabilities for a digital human model, with the primary focus on musculoskeletal strength.

50.3 Overview of Muscle Mechanics

Muscles are the force generating components within the body that provide a means of mobility and functionality. The essential force producing cells making up muscle are the muscle fibers. These in turn are composed of compartments called sarcomeres, which contain the myofilaments referred to as thick and thin filaments. The most commonly accepted muscle contraction mechanism is known as the sliding filament theory [2] by which muscle shortening occurs due to sliding of these filaments relative to one another. This theory postulates that the thick filaments (consisting of primarily myosin proteins) have projections known as cross-bridges that can bind to the adjacent thin filaments (consisting of primarily actin proteins). After binding, these heads rotate, similar to a rowing stroke, causing a sliding motion of the thin filaments. The thin filaments on either side of the sarcomere are brought closer to the center, making the sarcomere shorten. As this occurs throughout a muscle, tension is produced and transmitted to the skeleton through tendons. Ultimately, if the muscle force generated is greater than an opposing force, joint movement will occur.

Because the sarcomere is the basic force production unit, muscle force capability is proportional to the number of sarcomeres in parallel. This explains why muscle cross-sectional area is relevant to maximal muscle strength. Each sarcomere can only shorten a finite amount. The more sarcomeres in series in a muscle, the greater the rate at which total shortening can occur. Since the cross-bridge cycling rate is relatively constant, ultimately this impacts maximum shortening velocity of a muscle (greater total shortening over the same time period). Thus, long, thin muscles are well suited for generating speed and movement, whereas short, fat muscles are better suited for generating force.

50.3.1 Force-Velocity Relationship

It has been long understood that muscles can produce greater force when contracting slowly than at faster velocities. Muscle force decays in a curvilinear manner as shortening velocity increases (fig. 50.1). This was first described by Hill in 1938, and is described by Hill's equation: $(P + a)*v = b*(Po − P)$; where P is muscle force, Po is the maximum static force, v is muscle contraction velocity, and a and b are constants. The underlying mechanism for this phenomenon is theorized to be a result of muscle cross-bridge cycling rate. [3] The speed of the cross-bridge cycling is believed to be constant, unable to adapt with increasing speeds of muscle shortening. Therefore at higher contraction velocities, there would be fewer myofilament cross-bridges attached at any given moment in time, resulting in less total force produced.

While the Hill relationship has become a widely used principle in describing muscle force characteristics, this single phenomenon does not directly correspond to the net torque-angular position relationship at the joint level. The complex arrangement of muscles, and their corresponding moment arms that vary with position, complicate this relationship. Indeed, not all muscles contributing to joint rotation will be shortening at the same linear velocity despite a single angular velocity (moment arm length dependent).

Muscles can shorten during force production (concentric contraction), which corresponds to the right half of the force-velocity curve (fig. 50.1). When muscle-shortening velocity is zero, an isometric (static) contraction occurs. However, muscles can lengthen during contractions as well (eccentric contractions). This occurs when the external force is greater than the muscle force generated. Functionally, this occurs when lowering a weight, as the volitional muscle torque is less than the moment due to the external object. Eccentric contractions can generate more force than isometric (constant length) contractions, which in turn generate more force than concentric (shortening) contractions. The underlying mechanism of increased force resulting from eccentric contractions is hypothesized to be due to the stretching of the active cross-bridges and passive connecting filaments.

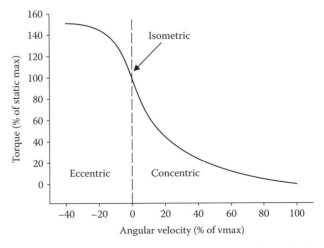

FIGURE 50.1 Normative force-velocity relationship (concentric portion of curve) demonstrating decreasing force with increasing velocity. The eccentric portion of the curve (lengthening contraction) is based on empirical data replicated over the years since Hill's original work.

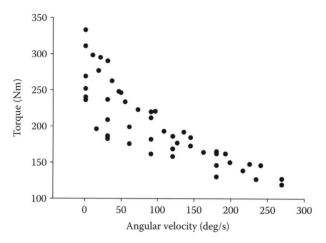

FIGURE 50.2   Male knee extension torque-velocity data from 14 studies. Note the decay in peak torque with increasing angular velocity, similar to Hill's force-velocity relationship.

Torque-velocity data for joints such as the elbow [4–6] and the knee [7–9] are abundant in the literature. Figure 50.2 plots knee extension torque-velocity data compiled from 14 studies, showing only data collected from males. [4, 7, 8, 10-20] These curves grossly follow the force-velocity relationship originally modeled by Hill (fig. 50.1).

In addition to the physiological factors influencing muscle force production, several inter-individual factors can be important. Training status clearly impacts muscle-force-generating capability. Repetitive resistive exercise results in muscle hypertrophy, or increased cross-sectional area [21] whereas immobilization and inactivity result in significant atrophy, or loss of muscle cross-sectional area. [22] Females produce on average two-thirds the peak force of men, [23] but this can vary from 35% to 85% depending on the task [24] (see fig. 50.3). Age is another potential influence, where strength typically peaks between 20 and 40 years of age. [25] With increasing age, reduction in muscle oxidative capacity, [26] degeneration of muscle fiber, [27] decrease in activity level, [28] and endocrine changes [29] are believed to contribute to decreases in muscle strength. The effects of anthropometry on muscle strength are somewhat inconsistent. Attempts have been made to predict muscle strength from height, weight, limb

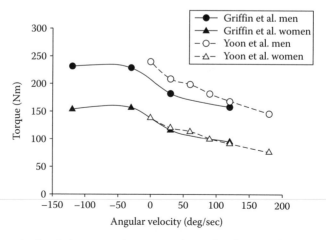

FIGURE 50.3   Male versus female knee extension torque-velocity data from Griffin et al. [4] and Yoon et al. [7]. Note that men were consistently higher than women in both studies.

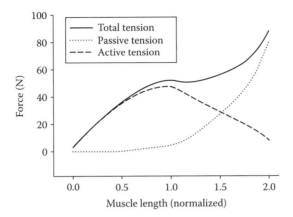

FIGURE 50.4 Hypothetical length-tension curve involving total muscle tension (solid line), passive tension (dotted line), and active tension due to cross-bridges (dashed line) relative to optimal muscle length (1.0).

volume or girth, and other factors (i.e., allometric scaling). [30] To date, the single best predictor for muscle strength is fat-free body mass (FFM), which reflects total muscle mass. [25] Indeed, the effects of gender, age, and training on strength can be partially explained by their corresponding differences in muscle mass. [31–33]

50.3.2 Length-Tension Relationship

Muscle force production is dependent on muscle length, commonly referred to as the length-tension relationship. In isolated muscle fibers, the relative overlap between the force-producing myofilaments, myosin (thick filaments) and actin (thin filaments), directly influences active force production. In addition, passive force production (due to connecting filaments within muscle) is non-existent at shorter lengths, but increases in an exponential manner as muscle is stretched beyond its ideal overlap (fig. 50.4). This again has been more clearly delineated in single muscle fiber preparations than in intact human joints. However, static joint torque has been shown to vary with joint angle. [34]

When coupled, these two muscle force relationships—force-velocity and length-tension—result in a relatively complex three-dimensional (3D) force-velocity-length relationship. Interactions between velocity and position have not been fully characterized, but rather the torque-angular velocity-position relationship has been extrapolated from knowledge of isolated force-velocity and length-tension relationships. [35]

50.4 Modeling Muscle Force

The origins of modern mathematical muscle modeling are credited to Hill's force-velocity relation, (F + a)(v + b) = constant, first published in 1938. Hill first proposed a simple phenomenological model to describe muscle shortening, composed of a contractile element (a motor) and a series elastic element (a spring) in 1938. [36, 37] The Hill model is difficult to use to predict isometric force, [38] as it was developed to model the force-velocity relationship. Since that time, different modeling approaches have been developed to estimate muscle forces for a variety of applications. Huxley's biophysical (cross-bridge) model of muscle, based primarily on the calcium dynamics associated with muscle contraction, was developed in 1957. [2, 37] Traditional engineering systems control theory has also been used to model muscle force as a simple second-order linear system [39] or with added nonlinearities. [40] Each of these modeling approaches has been utilized in isolation and/or in combination for many successive model variations since their introduction. Despite extensive work in the field of muscle modeling over the past 65 years, there is yet to be a single universally accepted model. The best model may be dependent on the specific application or target population. [41]

50.5 Modeling Joint Torque

Animal studies have greatly advanced our knowledge of basic muscle biomechanics, driving our ability to develop reasonable isolated muscle models. In humans, however, we cannot easily measure single muscle forces to validate these models. We can simply measure net torque produced about a joint. As many synergistic muscles (agonists) often work together to produce joint torque, and opposing muscles (antagonists) may concomitantly be activated to stabilize the joint, it is difficult to estimate single muscle forces. Considering simulation-time for DHM, a muscle-based model incorporating up to hundreds of individual muscles may not always be practical. Perhaps a better approach to DHM whole-body strength assessment is to focus on joint-space strength rather than individual muscles.

50.6 Joint Strength

While the term *strength* is commonly used, it is not so readily defined. Practically, an individual's strength is dependent on the ability to generate torque about a joint, but may inherently include the ability to sustain that torque as needed (e.g., endurance). For clarity, we will use the term *strength* to refer to the maximum torque-generating capability about a specific joint and consider endurance and fatigue separately.

50.7 Normative Strength Determination

There are different methods for testing strength, including static and dynamic methodologies. Isometric contractions can be used to determine static maximum strength at various joint angles with no joint movement. This type of testing is often used to determine joint torque-angle curves. While this relationship is largely analogous to the muscle length-tension relationship, additional complexity arises from the muscle moment arms varying throughout the ROM. [42] The testing procedure for isometric strength is relatively simple, has been implemented since the 1970s, [43–45] and has been well summarized previously. [42] One of the early attempts to document the static joint torque-angle relationship involved the trunk and upper extremity. In 10 men and 10 women, peak elbow strength occurred mid-range at ~90° of elbow flexion. [34] Similarly, isometric trunk strength in seated subjects (59 males, 43 females) was greatest in a neutral sitting posture and decreased as the posture deviated from neutral. [46]

Isometric strength has also been examined extensively at the whole-body level in tasks such as lifting, pulling, and pushing. The relationship between whole-body strength and body posture is complex due to the involvement of multiple joints, two-joint musculature, as well as tremendous redundancy of the musculoskeletal system. Still, it was demonstrated that body posture had significant effects on whole-body strength. [42] From the ergonomics perspective, the critical issue regarding whole-body strength is different from that of individual joint strength. For individual joints, the load-strength ratio may be useful to predict fatigue and risk of injury at that specific location. Whole-body models may be used to estimate where the "weakest" link is located (i.e., largest load to joint strength ratio). For instance, the critical region may be the low back when lifting with a bent forward posture, but during a squat lift it may be the knees. Computerized strength prediction programs that evaluate individual joint load were first implemented in 1968 by Chaffin. [47] Using a seven-link planar model, later known as the 2D Static Strength Prediction Program (2D SSPP, University of Michigan), static lifting strength in the sagittal plane was predicted by comparing the loads of six major body joints to their corresponding static

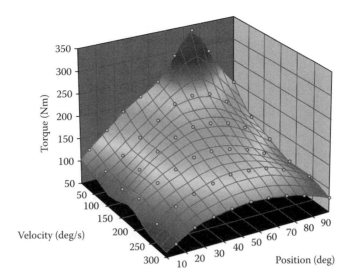

FIGURE 50.5　Knee extension torque-angular velocity-position 3D surface for a single subject. Note that peak torque occurs mid-range and decreases with increasing velocity, as expected.

strength limits. The 3D version of the program (3D SSPP, University of Michigan) was further developed by Garg and Chaffin in 1975 [48] and continues to be a commonly used tool for strength assessment in DHM today.

While these static models have proven to be highly useful, advances in DHM to include dynamic motion predictions have escalated the need for similar dynamic strength representations. Large errors are expected from static DHM under high-speed and high-frequency conditions such as running, jumping, and vibration. Increasingly, data is becoming available to develop normative databases of major dynamic joint strength. Quantitative dynamic testing can be performed using isokinetic or isotonic testing. Maximum isotonic testing involves repeated lifting of a constant load to determine a maximum lift for one or more repetition maximums (e.g., 1RM, 3RM). A single value for strength (maximum load) is obtained but there is no consideration for the velocity at which the motion is performed. Isokinetic testing determines dynamic maximum muscle strength at constant velocities. The torque produced by a muscle group is measured throughout the joint ROM while moving at a fixed angular velocity. While this is not a functional movement, we do not move with the same velocity throughout a motion in daily life; this type of testing is well controlled. For the purpose of DHM, this can estimate maximum torque capability at a variety of velocities. Using a combination of isometric and isokinetic strength values, we can develop realistic limits of static and dynamic strength capability for humans, including both the angle-specific and velocity-specific influences on torque production. Figure 50.5 provides an example of a single subject's maximum knee extension torque as a function of joint position and velocity. Disadvantages of this approach may include: static strength does not necessarily correlate with dynamic strength; joint strength in one plane of motion may not be representative of strength in another plane of motion; and the ability to generate maximum exertion is influenced by psychological and pathological factors.

DHM can incorporate dynamic strength representations in several ways. While using actual database-driven approaches is certainly feasible, modeling the 3D dynamic torque surface analytically has the added advantage of being able to directly program this information into DHM. Non-uniform rational B-splines (NURBS), an industry standard tool for the presentation and design of geometry, is one method that can be used to model these strength surfaces. [49] Figure 50.6 provides an example of a NURBS representation of mean female knee extension peak torque (N = 9) as a function of angular velocity and joint position. [50]

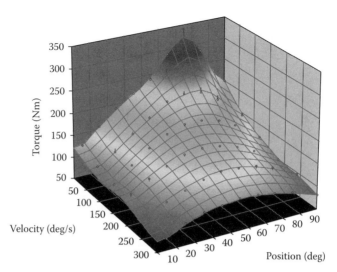

FIGURE 50.6 Non-uniform rational B-splines (NURBS) representation of the single subject knee extension torque (see fig. 50.5) as a function of angular velocity and joint angle.

50.8 Joint Range of Motion

ROM and joint mobility in humans have been well studied over the years, with a classic study on ROM by Barter et al. in 1957. [51] There are many individual differences that can potentially affect ROM. Females may have slightly greater ROM than males on average; [52] children can have larger ROM than adults, but normal joint mobility doesn't change substantially once maturation is reached with increasing age; [53] obesity can cause decreases in available ROM most likely due to soft-tissue approximation; however, anthropometric variations do not have consistent effects on ROM. [54] In contrast to the small influence of personal factors on ROM, several biomechanical factors can have a significant impact. Muscles that span two joints may cause changes in maximum ROM at one joint that is dependent upon the position of the adjacent joint (the two-joint muscle effect). For instance, knee extension can be reduced by up to 34° when the hip is also flexed. [24] Additionally, ligaments that provide joint stability can alter maximum ROM depending on their biomechanical position, and ultimately alter muscle tension. For instance, maximum glenohumeral abduction (raising your arm out to the side) is reduced when the arm is internally rotated versus in a neutral or externally rotated position. Pathological conditions resulting in altered ROM (both reduced and exaggerated) can also occur, but are beyond the scope of this chapter. Currently, simplified representations of joint ROM are used in digital human modeling—considering maximum normal joint ROM for isolated joints. As models increase in fidelity, DHM may increasingly represent these additional inter-individual and biomechanical factors that influence ROM.

50.9 Medical Applications

While the application of DHM for ergonomic purposes is becoming increasingly common, its application in other health areas is lagging. The potential, however, for DHM to have a significant impact on advancing health is immense. [1] It has been estimated that approximately 3 million people in the United States are using ambulatory aids, such as wheelchairs, canes, and walkers. Further, numerous individuals experience temporary or chronic functional impairments as a result of disease or injury (e.g., knee osteoarthritis, low back pain). DHM may be useful to better understand the influences of muscle weakness, decreased ROM, and pain on functional ability. For example, a 17-link DHM has been used to investigate the kinematic changes during reaching and lifting associated with low back pain. [55, 56] Further, the French HANDIMAN project is aimed at modeling car ingress and egress motions for young and older adult populations with and without joint replacements. [57] HADRIAN, a predictive DHM and design tool developed in the United Kingdom, is aimed at predicting reach and lift capabilities in older and disabled individuals. [58]

DHM potential also extends to applications associated with physical rehabilitation. For example, gait analysis is a tool used for assessment of orthopedic and neuromuscular disorders. Piazza and Delp [59] simulated abnormal activation patterns of the lower extremity muscles using software more recently known as SIMM (MusculoGraphics, Inc., Santa Rosa, CA) and were able to reproduce a clinically observed stiff-knee gait. DHM has been also used to assist patients in regaining their upper-body physical strength and mobility following a stroke using a home-based haptic telerehabilitation regimen. [60] In brief, the telerehabilitation system consisted of a patient interface and a therapist interface connected by the Internet. The patient interface allowed patients to practice upper extremity motions using video games such as driving simulation. Concurrently, the patient's motion and force data were recorded using haptic/force-feedback devices and were sent to the therapist. The therapist interface consisted of a patient DHM based on JACK (Motion Analysis Corp., Santa Rosa, CA), which allowed playback of the patient's activities to monitor the rehabilitation progress.

Attempts have also been made to model surgical procedures on a digital patient. Free and Delp [61] built a 3D hip model using SIMM to investigate the effects of trochanteric transfer in total hip replacement. The authors demonstrated that the trochanteric transfer had limited effect on the moment arm of primary hip abductor muscles. It was concluded that restoration of muscle length was a higher priority for such a surgical procedure. A knee model using SIMM demonstrated that a small error in the tilting angle during knee prosthesis implantation caused a substantial change in knee kinematics. [62]

To date, most DHMs are associated with the human musculoskeletal system. Attempts are being done to build a virtual human that contains other physiological systems and allows researchers to investigate the human body as a single complex system. The process of building a virtual physiologic human has been initiated in Europe. [63] Further, Santos, a DHM created at the University of Iowa, is incorporating muscle strength, fatigue, energy expenditure, and heart rate predictions. [1] The future of DHM will continue to grow in model fidelity and prediction accuracy. The targeted applications will likely continue to expand as our ability to adequately represent human physiology improves.

References

1. Abdel-Malek K, Yang J, Marler T, Beck S, Mathai A, Zhou X, Patrick A, Arora J, Towards a new generation of virtual humans. *Int J Hum Factors Modeling Simulation* 2006, 1, (1), 2–39.
2. Huxley AF, Muscle structure and theories of contraction. *Prog in Biophysics Biophy Chem* 1957, 7, 255–318.
3. Landesberg A, Sideman S, Force-velocity relationship and biochemical-to-mechanical energy conversion by the sarcomere. *Am J Physiol Heart Circ Physiol* 2000, 278, (4), H1274–84.
4. Griffin JW, Tooms RE, vander Zwaag R, Bertorini TE, O'Toole ML, Eccentric muscle performance of elbow and knee muscle groups in untrained men and women. *Med Sci Sports Exerc* 1993, 25, (8), 936–44.
5. Farthing JP, Chilibeck PD, The effects of eccentric and concentric training at different velocities on muscle hypertrophy. *Eur J Appl Physiol* 2003, 89, (6), 578–86.
6. Gallagher MA, Cuomo F, Polonsky L, Berliner K, Zuckerman JD, Effects of age, testing speed, and arm dominance on isokinetic strength of the elbow. *J Shoulder Elbow Surg* 1997, 6, (4), 340–46.
7. Yoon TS, Park DS, Kang SW, Chun SI, Shin JS, Isometric and isokinetic torque curves at the knee joint. *Yonsei Med J* 1991, 32, (1), 33–43.

8. Westing SH, Seger JY, Karlson E, Ekblom B, Eccentric and concentric torque-velocity characteristics of the quadriceps femoris in man. *Eur J Appl Physiol Occup Physiol* 1988, 58, (1-2), 100–104.

9. Arnold BL, Perrin DH, Kahler DM, Gansneder BM, Gieck JH, A trend analysis of the in vivo quadriceps femoris angle-specific torque-velocity relationship. *J Orthop Sports Phys Ther* 1997, 25, (5), 316–22.

10. Taylor NA, Cotter JD, Stanley SN, Marshall RN, Functional torque-velocity and power-velocity characteristics of elite athletes. *Eur J Appl Physiol Occup Physiol* 1991, 62, (2), 116–21.

11. Wickiewicz TL, Roy RR, Powell PL, Perrine JJ, Edgerton VR, Muscle architecture and force-velocity relationships in humans. *J Appl Physiol* 1984, 57, (2), 435–43.

12. Seger JY, Thorstensson A, Electrically evoked eccentric and concentric torque-velocity relationships in human knee extensor muscles. *Acta Physiol Scand* 2000, 169, (1), 63–69.

13. Esselman PC, de Lateur BJ, Alquist AD, Questad KA, Giaconi RM, Torque development in isokinetic training. *Arch Phys Med Rehabil* 1991, 72, (10), 723–28.

14. Froese EA, Houston ME, Torque-velocity characteristics and muscle fiber type in human vastus lateralis. *J Appl Physiol* 1985, 59, (2), 309–14.

15. MacIntosh BR, Herzog W, Suter E, Wiley JP, Sokolosky J, Human skeletal muscle fibre types and force: velocity properties. *Eur J Appl Physiol Occup Physiol* 1993, 67, (6), 499–506.

16. Caiozzo VJ, Perrine JJ, Edgerton VR, Training-induced alterations of the in vivo force-velocity relationship of human muscle. *J Appl Physiol* 1981, 51, (3), 750–54.

17. Weir JP, Evans SA, Housh ML, The effect of extraneous movements on peak torque and constant joint angle torque-velocity curves. *J Orthop Sports Phys Ther* 1996, 23, (5), 302–8.

18. Folland JP, Irish CS, Roberts JC, Tarr JE, Jones DA, Fatigue is not a necessary stimulus for strength gains during resistance training. *Br J Sports Med* 2002, 36, (5), 370–73; discussion 374.

19. Prietto CA, Caiozzo VJ, The in vivo force-velocity relationship of the knee flexors and extensors. *Am J Sports Med* 1989, 17, (5), 607–11.

20. Thorstensson A, Grimby G, Karlsson J, Force-velocity relations and fiber composition in human knee extensor muscles. *J Appl Physiol* 1976, 40, (1), 12–16.

21. Cureton K, Collins M, Hill D, McElhannon FJ, Muscle hypertrophy in men and women. *Med Sci Sports Exerc* 1988, 20, (4), 338–44.

22. Appell H, Skeletal muscle atrophy during immobilization. *Int J Sports Med* 1986, 7, (1), 1–5.

23. Roebuck J, Kroemer K, Thomson W, *Engineering Anthropometry Methods.* Wiley-Interscience: New York, 1975.

24. Webb Associates, *Anthropometric Source Book, Vol. 1, NASA Ref. 1024.* National Aeronautics and Space Administration: 1978; Vol. 1.

25. McArdle W, Katch F, Katch V, *Exercise Physiology: Energy, Nutrition and Human Performance.* 5th ed. Lippincott Williams & Wilkins: Philadelphia, PA, 2001.

26. Cartee G, Influence of age on skeletal muscle glucose transport and glycogen metabolism. *Med Sci Sports Exerc* 1994, 26, (5), 577–85.

27. Doherty T, Vandervoort A, Taylor A, Brown W, Effects of motor unit losses on strength in older men and women. *J Appl Physiol* 1993, 74, (2), 868–74.

28. Nygard C, Luopajarvi T, Ilmarinen J, Musculoskeletal capacity and its changes among aging municipal employees in different work categories. *Scand J Work Environ Health* 1991, 17, (Suppl 1), 110–17.

29. Lamberts S, van den Beld A, van der Lely A, The endocrinology of aging. *Science* 1997, 278, (5337), 419–24.

30. Vanderburgh P, Katch F, Schoenleber J, Balabinis C, Elliott R, Multivariate allometric scaling of men's world indoor rowing championship performance. *Med Sci Sports Exerc* 1996, 28, (5), 626–30.

31. Heyward V, Johannes-Ellis S, Romer J, Gender differences in strength. *Res Q Exerc Sport* 1986, 57, 154–59.

32. Bishop P, Cureton K, Collins M, Sex difference in muscular strength in equally-trained men and women. *Ergonomics* 1987, 30, (4), 675–87.
33. Castro M, McCann D, Shaffrath J, Adams W, Peak torque per unit cross-sectional area differs between strength-trained and untrained young adults. *Med Sci Sports Exerc* 1995, 27, (3), 397–403.
34. Schanne F, A three dimensional hand force capability model for the seated operator. University of Michigan, Ann Arbor, MI, 1972.
35. Winter D, *Biomechanics and Motor Control of Human Movement*. 2nd ed. Wiley-Interscience: Waterloo, Ontario, Canada, 1990.
36. Hill VA, The heat of shortening and the dynamic constants of muscle. *Proceedings of the Royal Society of London B* 1938, 126, 136–95.
37. Epstein M, Herzog W, *Theoretical Models of Skeletal Muscle*. John Wiley & Sons: Chichester, England, 1998, 238.
38. Phillips CA, Repperger DW, Neidhard-Doll AT, Reynolds DB, Biomimetic model of skeletal muscle isometric contraction: I. an energetic-viscoelastic model for the skeletal muscle isometric force twitch. *Computers in Biology and Medicine* 2004, 34, 307–22.
39. Close CM, Frederick DK, *Modeling and Analysis of Dynamic Systems*. 2nd ed. John Wiley & Sons: New York, 1995, 681.
40. Bobet J, Stein RB, A simple model of force generation by skeletal muscle during dynamic isometric contractions. *IEEE Transactions on Biomedical Engineering* 1998, 45, (8), 1010–16.
41. Frey Law LA, Shields RK, Predicting human chronically paralyzed muscle force: a comparison of three mathematical models. *J Appl Physiol* 2006, 100, (3), 1027–36.
42. Chaffin D, Andersson G, Martin B, *Occupational Biomechanics*. 4th ed. Wiley & Sons: New York, NY, 2006.
43. Caldwell L, Chaffin D, Dukes-Dobos F, Kroemer K, Laubach L, Snook S, Wasserman D, A proposed standard procedure for static muscle strength testing. *Am Ind Hyg Assoc J* 1974, 35, (4), 201–6.
44. Chaffin D, Ergonomics guide for the assessment of human static strength. *Am Ind Hyg Assoc J* 1975, 36, (7), 505–11.
45. Kroemer K, Kroemer H, Kroemer-Elbert K, *Ergonomics: How to Design for Ease and Efficiency*. Prentice-Hall: Englewoods Cliffs, NJ, 1994.
46. Kumar S, Isolated planar trunk strengths measurement in normals: part III—results and database. *Int J Ind Ergon* 1996, 17, (2), 103–11.
47. Chaffin D, A computerized biomechanical model-development of and use in studying gross body actions. *J Biomech* 1969, 2, (4), 429–41.
48. Garg A, Chaffin D, A biomechanical computerized simulation of human strength. *AIIE Trans* 1975, 7, (1), 1–15.
49. Castro MJ, Apple DF, Jr., Rogers S, Dudley GA, Influence of complete spinal cord injury on skeletal muscle mechanics within the first 6 months of injury. *European Journal of Applied Physiology* 2000, 81, (1-2), 128–31.
50. Laake A, Frey Law LA, In *Modeling 3D Knee Torque Surfaces for Males and Females*. American Society for Biomechanics, Stanford, CA, August 2007.
51. Barter J, Emmanuel I, Truett B, *A Statistical Evaluation of Joint Range Data*; WADC-TR-57-311; Wright-Patterson Air Force Base, OH, 1957.
52. Snelkinoff E, Grigorowitsch M, The movement of joints as a secondary sex and constitutional charateristics. *Zeitschrift für Konstitutionslehre* 1931, 15, (6), 679–93.
53. Salter N, Darcus H, The amplitude of forearm and of humeral rotation. *J Anat* 1953, 87, 407–18.
54. Laubach L, Body composition in relation to muscle strength and range of joint motion. *J Sports Med Phys Fitness* 1969, 9, (2), 89–97.
55. Kim K, Martin B, Chaffin D, Woolley C, In *Modeling of Effort Perception in Lifting and Reaching Tasks*, SAE Digital Human Modeling Conference, Arlington, VA, June 26–28, 2001.

56. Chaffin D, On simulating human reach motions for ergonomics analyses. *Hum Factor Ergon Man* 2002, 12, (3), 1–13.

57. Chateauroux E, Wang X, Pudlo P, Ait El Menceur M, Difficulties of elderly and motor impaired people when getting in and out of a car, *11th International Conference on Mobility and Transport for Elderly and Disabled Persons*, June 19–21, 2007.

58. Gyi D, Sims R, Porter J, Marshall R, Case K, Representing older and disabled people in virtual user trials: data collection methods. *Appl Ergon* 2004, 35, (5), 443–51.

59. Piazza S, Delp S, The influence of muscles on knee flexion during the swing phase of gait. *J Biomech* 1996, 29, (6), 723–33.

60. Jadhav C, Nair P, Krovi V, Individualized interactive home-based haptic telerehabilitation. *IEEE MultiMedia* 2006, 13, (3).

61. Free S, Delp S, Trochanteric transfer in total hip replacement: effects on the moment arms and force-generating capacities of the hip abductors. *J Orthop Res* 1996, 14, (2), 245–50.

62. Piazza S, Delp S, Stulberg S, Stern S, Posterior tilting of the tibial component decreases femoral roll-back in posterior-substituting knee replacement: a computer simulation study. *J Orthop Res* 1998, 16, (2), 264–70.

63. STEP, http://www.europhysiome.org (May 24, 2007).

Part 5

Current Implementation and the Future of Digital Human Modeling

Impact of Digital Human Modeling on Military Human-Systems Integration and Impact of the Military on Digital Human Modeling

John F. Lockett III
and Susan Archer

51.1 Introduction

Military organizations have driven much of the research and development in ergonomics and human factors engineering and as a result have realized significant benefits. During the past three decades, these organizations made significant investments in development of digital human models as tools to enable human systems integration. Use of these models has resulted in earlier identification of usability issues, increased opportunity to resolve those issues less expensively, and more effective and safer systems. In this chapter, we review these developments and discuss their application via case studies. Although military investment is certainly not limited to the United States Department of Defense (DoD), given the basis of our experience, this is our focus. Additional discussion of these developments and human systems integration can be found in, for example, Booher (1990, 2003) and Salvendy (2006).

Digital human models represent a wide range of human physical, mental, and social attributes and behaviors. Many of the models and the analysis tools in which they have been embedded characterize multiple attributes and so are not easily categorized. However, the following categories are useful in describing the predominant attributes represented by the models most commonly used for human systems integration (HSI). They are human figure models, task models, cognitive models, information flow models, and integrated or hybrid models.

51.2 Human Figure Models

Digital human models that include a visualization of the human body are often called human figure models. While many such models have been developed for animation purposes, those that are based on empirical physiological data are of interest for human systems integration, particularly human engineering. The empirical data may be anthropometric, such as body size, body shape, and range of motion. This empirical foundation supports analyses of design accommodation of a population rather than one figure size and shape. The tools are used for workspace design and process evaluation. Factors analyzed using the tools include size and layout of workspaces, entry and exit into those workspaces, accessibility of maintenance areas, ability to reach and see controls and displays, and field of view. Many of the three-dimensional, computer-aided design tools based on these models also incorporate biomechanics models to help evaluate lifts and predict posture. See Chapter 8 in this volume for a discussion of biomechanics models. Chaffin (2001) provides a history of these models, early case studies, and lessons learned. Some models and tools that fit this category are Jack (http://www.plm.automation.siemens.com/en_us/products/tecnomatix/assembly_planning/jack/index.shtml; Badler, Erignac, & Liu, 2002), Santos (http://www.digital-humans.org/research.htm; Abdel-Malek et al., 2006), RAMSIS (http://www.human-solutions.com/automotive_industry/ramsis_en.php; Bubb et al., 2006), Safework (http://www.safework.com/safework_pro/sw_pro.html; http://www.dtic.mil/dticasd/ddsm/srch/ddsm0130.html)—the first two were developed with significant financial and technical support from the U.S. DoD, the third with support from the German automotive industry, and the last with Canadian Ministry of Defense sponsorship.

51.3 Task Models

One of the most commonly used human modeling methods is task modeling. The purpose of a task model is to build a computer-based replica of a process. Once validated in order to ensure its accuracy, the computer model can then be used to predict the likely effect of changes to the process on performance measures such as time and accuracy.

Typically, task models are developed by methodically decomposing a large and complex process into individual elements. This is often referred to as a reductionist approach (Laughery, Lebiere, & Archer, 2006) and is a transparent method of developing a system description. These elements are then connected into a network that represents process flow. This network can become quite complex (see fig. 51.1), and many of the connections can be implicit (i.e., be controlled by program logic rather than explicitly drawn as a network connection).

These models can trace their pedigree to the 1960s, when they became popular as industrial engineering tools for achieving process improvements (Law & Kelton, 1991). Their popularity is largely due to the fact that they are easy to understand, their outputs are directly traceable to task analysis data, and they support a broad range of analysis fidelity. A primary strength of task models is that they can vary greatly in terms of the level of specificity (i.e., the extent to which the process has been decomposed) they achieve. This enables the modeler to tune the model complexity to the precise question that the model must answer.

Primarily, task models are meant to be predictive in that they are used to extend what we know about a process to predict how that process will be affected by a change of some type. The change may be simple (e.g., an element of the process is redesigned to take less time or fewer resources) or it may be very complex (e.g., elements of the task are reassigned to distributed team members, requiring new communication methods and protocols as well as substantive modifications to other parts of the process).

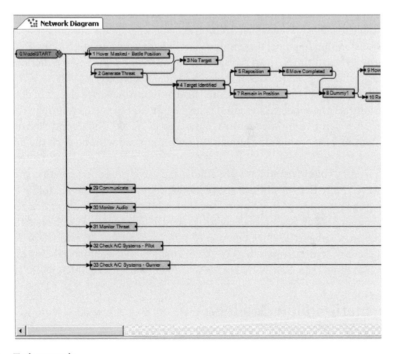

FIGURE 51.1 Task network.

Task models have been developed to estimate performance time for a sequence of tasks (Bowman, 2004), task loading and task allocation (Middlebrooks et al., 1999), to estimate the likelihood of error within a process (Archer, Keller, Archer, & Scott-Nash, 1999; Leiden et al., 2001), and to evaluate the effects of culture on operations (Mui et al., 2003). Regardless of the desired predictive capability, task models always result in providing the developer with a thorough understanding of the target process itself. Simply the process of building a description that is of sufficient detail to support a model provides an important and valuable knowledge base. Analysts often state that this process supplied outputs that were as valuable as any analysis efforts they conducted (Whicker & Sigelman, 1991). Examples of task modeling tools developed by the military are the Improved Performance Research, Integration Tool (IMPRINT) (Allender et al., 1995) and the Integrated Performance Modeling Environment (IPME) (Belyavin, Woodward, Nguyen, Robel, & Woolworth, 2005).

51.4 Cognitive Models

As military systems become more complex, the human performance modeling community has shown increased interest in cognitive models. As a general class, cognitive models are used to predict cognition based on perception of stimulus and memory (including knowledge). In contrast with task models, which are often theoretically agnostic, cognitive models are designed to instantiate specific theories of how humans perceive and process information. Also, in contrast with task models, the fidelity of cognitive models is necessarily quite high. These models process details about the human's current environment, determine what elements the human is likely to perceive and then what the human is likely to do (based on what lurks in the short- and long-term memory, as well as the information processing mechanism). In order for this to work well, the simulated memory stacks must be maintained as the human performs, requiring very fine slices of the task environment as the simulated scenario progresses.

Cognitive models are used to predict human decision making with a significant level of detail. Outputs can include decision-making time, the likelihood of a correct decision, and, if the model is implemented with sufficient fidelity, even which decision is made when many options are available.

In terms of decision-making time, these outputs are calculated using predictions of information processing and memory recall times. In well-respected cognitive models, these predictions are supported by empirical data collected over decades of research (Anderson & Lebiere, 1998).

To accomplish these predictions, cognitive models require two separate classes of data. The first class is the information that the modeled human has as part of experience or training. This knowledge is often referred to as declarative knowledge and is simulated as "facts." The second class of data represents the dynamic environment in which the human is performing and within which the human is making decisions.

Until fairly recently, cognitive models were almost exclusively used in academic and research environments. However, more than any other type of model, cognitive models are rapidly evolving. Recent investment is being brought to bear on improving the usability of these tools, which removes a significant barrier in applying them. Examples of cognitive models are Adaptive Control of Thought-Rational (ACT-R) (Anderson & Lebiere, 1998), CODAGE (Cognitive Decision AGEnt) (Kant & Thiriot, 2006), and COGNet (Zachary, Ryder, Santarelli & Weiland, 2000) as well as models that predict Recognition Primed Decision Making (Fan, Sun, McNeese, & Yen, 2005; Warwick & Hutchins, 2004).

51.5 Information Flow Models

Information flow models are used to predict the workload required to manage and process information within a unit, and they often are used to predict the level at which information flow and routing support situation awareness and decision making.

Most information flow models are strongly related to task models, but instead of the nodes in the model representing a process and the entities flowing through the nodes representing the humans and agents involved in performing the process, the nodes in the information flow model represent information-processing steps, and the entities are messages.

Information flow models must link data elements to decision-making needs (Gacy & Dahn, 2001). Additionally, they must be able to represent information quality and how time in the system or latency impacts information quality. This is necessary in order to predict the impact of the information availability and recency on system performance. The challenge, then, is to settle upon a level of abstraction that balances the need for the model to be predictive and have face validity, with the need for the model to be generalized (Estell, 2001). This typically drives the modeler to some sort of information taxonomy in which the content of the message is represented by the type of information it delivers. So messages about threats that contain information about location, heading, and type are represented as a "type" and messages that contain friendly force status reports are another "type." These types are then linked to properties such as source, recency, information decay rate, and reliability. The message load, or frequency, can be used to drive workload predictions and the message types and properties can be used to predict situation awareness and decision-making performance (Plott, Quesada, Kilduff, Swoboda, & Allender, 2004).

Information flow models have been used to assess how well an organization is designed to share information, manage their expect inflow of message traffic, and make timely decisions (Swoboda, Kilduff, & Katz, 2005). They have also been used to assess the efficiency of track management techniques, such as those found in air traffic control systems (Leiden, Kamienski, & Kopardekar, 2005), command, control, and communication: techniques for reliable assessment of concept execution (C3TRACE), and information flow (Wu, Huberman, Adamic, & Tyler, 2004).

51.6 Integrated and Hybrid Models and Tools

As the various modeling tools have been applied to real military problems, we have learned that each has its strengths and few complex problems can be completely solved by the application of a single type of tool. This has led to interest in hybrid tools. We define hybrid tools as disparate modeling types that are applied cooperatively to a common analysis problem and that contribute different information to accomplish a joint solution. As defined, we see three different types of hybrid models and tools, each distinguished by the method through which communication occurs.

The first type of hybrid model is one in which separate model types are brought to bear serially on a shared problem, often communicating through a shared database and an open architecture, as shown in Figure 51.2 and described in Plott, Hamilton, and Laughery (2003).

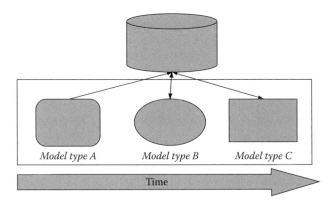

FIGURE 51.2 Serial hybrid model.

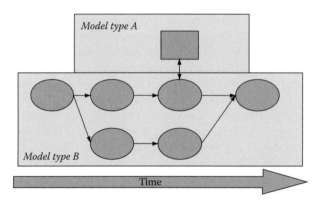

FIGURE 51.3   Hierarchical hybrid model.

An example of an analysis challenge that benefits from this approach is the challenge of selecting and laying out controls and displays in a crew station (Walters, Bzostek, & Li, 2005). Anthropometric models are clearly needed to perform this work. They are necessary to predict reach envelopes for the population of body sizes, to analyze visual cones, and to assess how various postures will impact control and display access. However, it is also necessary to evaluate how the controls and displays are used to perform the human's missions. Specifically, it is important to know the frequency with which each control will be used so that the most often used controls are within easy reach and the most often used displays are central to the human's view. Additionally, the sequence in which they will be used is needed in order to provide a layout that supports effective scan patterns. These factors are appropriate for analysis by task models, so that a large set of potential missions and task sequences can be assessed automatically, with results representing more than a single path through a process. All of these models need to share a description of task type linked to control and display locations and human capabilities and limitations.

The second type of hybrid model is shown in Figure 51.3.

In this type, one model is clearly the primary model (and is in charge of system execution control parameters like the clock), and some of the processing is outsourced to a different model type. An example of this is found in recent work where task and cognitive models have been used together (Archer, Lebiere, & Warwick, 2005). These hybrids have been available since 2003 (Archer & Lebiere, 2003) and are gaining attention because they allow a modeler to use task models to assess the efficiency and performance of a process and they support special, high-fidelity cognitive modeling of decision-making tasks. The advantage of this class of hybrid is that it plays to each model's strengths, and it avoids burdening the analyst with the details required by cognitive models for routine or automated tasks. This type of hybrid can be thought of as two models that exchange control, depending upon which tasks are in the execution queue.

A third example of a hybrid model is a combination of models that execute in parallel, sharing information and running with a shared time-management system (fig. 51.4). These are clearly the most sophisticated of the three hybrid types, requiring the most "middle-ware" (i.e., custom software to advance the clock and arbitrate the order in which information is passed between models, which model owns the data, and which model has the right to modify the data).

This type of model is often found in distributed simulation environments, and can be composed of models that have very different concepts of time (Archer & LaVine, 2000). That is, one model can be clock driven and progress in regular clock "ticks" while another can be event driven, progressing in irregular intervals dictated by the events that are occurring in the model and in the environment. The primary challenge with these models is to communicate efficiently, so that the entire hybrid does not execute too slowly to be practical.

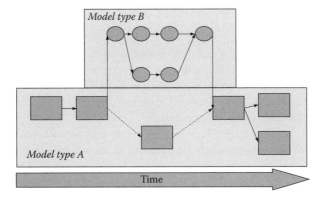

FIGURE 51.4 Parallel or concurrent hybrid model.

It is important to note that any of these three architectures could be used by models of the same type to communicate. Examples of integrated and hybrid modeling tools are the Graph-Based Interface Language (GRBIL) (Archer & Lebiere, 2005), Crew Station Design Tool (CSDT) (Walters, Bzostek, & Li, 2005), Man-Machine Integration Design Analysis System (MIDAS) (Gore & Jarvis, 2005), and the anthropometrics, vehicle, and biodymanics software tool (AVB-DYN) (Liang et al., 2005).

51.7 Analysis

All of the models and tools discussed have been applied to military human systems integration problems, and in the process we have learned much about better approaches for conducting the analysis and factors to consider when selecting a model or tool. Some of these lessons are specific to a category or model and will be highlighted in the case studies but those that are generally applicable are discussed here.

51.8 Analysis Approach

Perhaps the most important lesson learned for conducting a model-based analysis is to approach it just as one would a well-designed empirical study—that is, with a clearly defined hypothesis, independent (i.e., what is different between designs) and dependent (i.e., what is measured) variables, and a data analysis plan. Rather than diving in and building a model of the system of interest, we have found that explicitly stating the question you are attempting to answer prior to beginning the model development is important to input data collection, a necessary level of resolution, required output, and ultimately making the analysis more focused, efficient to conduct, and easier to defend. In the case studies that follow, we give examples of analysis issues and the approach taken to investigate them. All of the case studies were completed within a year and generally less than six months.

51.9 Application Issues

As is the case with most projects, time and resources drive the analysis approach decisions. If an analysis approach is technically sound but too expensive to implement or will be completed too late to support design decisions, its value is lost. Factors to consider when tailoring an approach to available time and resources include the following.

- Assess the expertise of the analyst and time needed to gain proficiency with tools and software required for the project. Projects may fail when an analyst is not trained in the use of a tool and is not given the time or funds necessary to learn to use it. Often the complexity of the tool is blamed rather than poor project planning. If time and resources are limited, then a less rigorous or complex approach may be more successful.
- Determine the availability of input data and subject matter experts for generating input data. If properly formatted and annotated data are not available, then the approach must require those data that are available or sufficient time and resources must be budgeted to collect and format them. Some tools and algorithms help generate input for other tools, but if this approach is taken, mismatches in underlying assumptions and constraints between tools must be assessed carefully to ensure a valid analysis.
- Assess the importance of verification, validation, and accreditation (i.e., the formal process of considering evidence supporting the valid use of a model to answer specific questions) to acceptance of the results. In some cases, only formally accredited models may be used. This is often the case when models or simulations are proposed to replace or augment testing. In other cases, generally accepted methods may be used to bound a problem and to identify issues and risks. This is more common in analyses supporting concept development and early design trade-offs.
- Ensure that the output of the analysis supports other analyses, decisions, and tools that will be used in the overall design project. Compatibility in terms of underlying data assumptions, data format required as input to other methods, and electronic format should be considered.

Lockett and Powers (2003) detail these and other factors to consider when planning for analysis and selecting tools to use to execute it.

51.10 Case Studies

51.10.1 Number of Operators and Workload

51.10.1.1 Tactical UAV Crew Size

Task modeling has been used for many years to predict the number of operators needed to successfully perform missions in new or modified crew stations. While a variety of task models exist, they all share a common output. This is the time required to perform a linked sequence of tasks, and the order in which they were performed. This is the simplest level of task modeling, and is useful in studies of process engineering and efficiency. Additionally, comparing the time available to complete a process to the time predicted through modeling is a straightforward method of assessing operator workload (Siegel & Wolf, 1969).

In 2002, Walters, Huber, French, and Barnes used the Micro Saint task network modeling tool to predict how crew size and rotation schedule affected operator workload and performance during the control of a Tactical Unmanned Aerial Vehicle (TUAV). Their work is a good example of how the basic capabilities of a task model can evaluate interesting crew size concepts.

As with all task models, the basic elements of the TUAV model task were the tasks,* the logic and branches that linked the tasks together into a flow, or process, and task-level performance time and accuracy that drove the predictions of crew performance. As is typical with most task models, the task-level time and accuracy estimates for the TUAV model were represented as a distribution of possible performance, rather

---

* Note that the nodes in the task model can be assigned to non-human team members, although they are typically referred to as "tasks."

than a point estimate. In this way, the analysts described a range of likely performance as well as the logic and variables that affected the range. Specifically, the level of fatigue of an operator has an impact on how quickly and accurately the operator can analyze the imagery collected by the TUAV. Then, as the model runs, it can stochastically sample from the specified range. In such cases, the model should be run many times in order to ensure that the range has been sufficiently explored and the output of the model will, in turn, be experienced as a range of possible results.

Task time and accuracy estimates can come from a range of sources, spanning empirical data collection through elicitation of estimates from subject matter experts. In some cases, a combination of these techniques is used, and the model itself can be used to explore the sensitivity of the model's predictions to uncertainty of a task's performance estimate. In the TUAV, modeling effort task data were collected directly through a combination of subject matter expert (SME) interviews, questionnaires, system requirements, the operational model summary and mission profile for the Shadow 200 and from previous TUAV models and studies (Barnes, Knapp, Tillman, Walters, & Velicki, 2000).

The models evaluated a combination of search type, emergency events, weather, and terrain. They sampled a large distribution of possible crew sizes (from 8 to 15) working 2-, 3-, 4-, or 6-hour rotation schedules. The TUAV model simulates 12- and 18-hour missions (over a 24-hour period) during 15 different conditions for 3 consecutive days. During missions, there are times when two aerial vehicles (AV) are in flight: when one AV is observing the targets and one AV is flying to assume control of the search.

The conclusions from executing the models and interviewing SMEs (during 12- and 18-hour missions) indicate that reducing the number of skilled personnel below six in the operations center could result in more vehicle mishaps during emergencies, an increase in the time to search for targets, and a decrease in the number of targets detected. More importantly, increasing crew size above this level resulted in only slight performance gains in target detection and target search time. This work demonstrated that task modeling was an effective tool for identifying the most efficient trade-offs between performance and crew size. Additionally, the Micro Saint task models could be parameterized such that they could be used to evaluate a wide range of parameters (e.g., weather, terrain, mission length) without modifying the network itself. This enabled the analysts to quickly investigate a broad set of independent variables to evaluate the effect on system performance.

51.11 Light Helicopter Experimental (LHX) and NBC Fox Operator Workload Analyses

Often, task models are used to predict the optimal allocation of tasks to crew members. The literature supports many methods of task allocation from a very simple strategy of assigning each worker only one task at a time (Archer, Keller, Archer, & Scott-Nash, 1999; Archer, Lebiere, & Warwick, 2005) to using predictions of cognitive workload and workload capacity (Sarno & Wickens, 1992; Aldrich, Craddock, & McCracken, 1984) in which tasks are rated on how much effort they require from the human operator of several types (e.g., cognitive, perceptual, motor). The approach that is chosen should be dictated by the question the analyst wants to answer. A simple approach might be appropriate to answer the question of whether you need four or five diesel mechanics in a maintenance organization to achieve a particular level of operational readiness. However, a more complex approach would probably be necessary to predict how long it will take an operator to notice a new target on a cluttered interface.

There are several examples of task modeling that have been used to predict how workload and the techniques humans use to manage it impact total system performance. One notable example is discussed in Laughery, Lebiere, and Archer (2006), which describes the use of task modeling of operator workload to determine whether a one-man Scout Helicopter (the LHX) was feasible. The central design issue was workload: Could one individual reasonably be expected to perform all of the tasks required within the available time?

Based on a review of the literature, Drews, Laughery, Kramme, and Archer (1985) concluded that the most promising theory of operator workload that was consistent with task network modeling was the multiple resource theory proposed by Wickens (e.g., Wickens, Sandry, & Vidulich, 1983). Simply stated, the multiple resource theory suggests that humans have not one information-processing resource that can only be tapped singly but several different resources that can be tapped simultaneously. Depending upon the nature of the information-processing tasks required of a human, these resources would have to process information sequentially (if different tasks require the same types of resources) or possibly in parallel (if different tasks require different types of resources).

While the multiple resource theory provided a theoretical basis for addressing the issues, the particular approach that was selected for this work was developed by McCracken and Aldrich (1984). This approach provided a simplified representation of the concept of multiple human information-processing resources. Using this technique, each operator activity in a task network was characterized by the workload demand the human experiences in four resources: auditory, visual, cognitive, and psychomotor. McCracken and Aldrich provided benchmark scales for rating tasks by the amount of resource-related effort required to perform each task.

During the task network simulation, the model of the operator may predict that he or she is required to perform several tasks simultaneously as defined by the task network. The task network model evaluates total attentional demands for each of the four channels (visual, auditory, psychomotor, and cognitive) by simply summing the attentional demands across all tasks that are being performed simultaneously. By examining the points in the mission at which these attentional demands are high, Drews, Laughery, Kramme, and Archer (1985) were able to assess the mission segments for which operator workload would be excessive. Using this approach, it was evident that a one-man cockpit was going to be quite difficult to achieve, even with the most optimistic projections of pilot-supporting technologies.

A similar approach was used by McMahon, Spencer, and Thornton (1995). In that project, the U.S. Army was working to modify a vehicle that had been designed to accommodate a four-person crew. This modified vehicle became the NBC Fox. The design team had attempted to modify the design for a three-person crew, but this redesign had resulted in forcing the operator to switch positions repeatedly in order to do the job, resulting in safety problems and unacceptable performance. Two task models were developed to examine alternative task allocations, resulting in substantially reduced workload and improved performance.

51.12 The Future Combat System (FCS)

The U.S. Army Research Laboratory has conducted a series of analyses aimed at answering crew size, level of automation and function allocation questions for future combat systems (Mitchell, Samms, Henthorn, & Wojciechowski, 2003; Mitchell, Samms, & Henthorn, 2004; Mitchell, 2005). These analyses employed primarily the IMPRINT task modeling tool and incorporated task analysis, SME interview, workload theory and research concerning driver distraction, situation awareness, trust in automation, and human information processing. These analyses were very effective in focusing the consideration of limits of human ability to handle simultaneous tasks and high workload, the maturity of advanced technologies to relieve these demands, and the resulting trade-offs to mission capability and performance. Ultimately, the analyses resulted in changes to system requirements documents (increasing the number of crew expected to operate specific vehicles for a given technology maturity) and the design of some of the vehicles—a significant accomplishment.

The analyses were particularly successful due to the approach used to conduct them. To gather the information needed to implement the approach, analysts read existing requirements documents, talked to users of predecessor systems, and interviewed design engineers to fully understand the mission and operating conditions for the vehicle (e.g., heat and extended hours of operation contributing to fatigue), proposed training, tactics, and procedure, proposed technologies requiring user interaction (e.g., control of unmanned vehicles), including how it would operate if not technically mature (i.e., what must the human do in the event the technology fails), and assess the knowledge skills and abilities of proposed operators. Only after these steps were complete could an analysis question be stated explicitly and the rest of the approach executed. A typical analysis question for the FCS vehicles was of the form "determine the best allocation of functions among crew members of the X system when performing a specific mission type using a specific level of enabling technology" (e.g., an automated driving system). Prior to model building, independent variables were selected, and levels were chosen and modeled separately as design options. Typical independent variables were number of operators/crew and allocation of mission functions (e.g., driving, scanning for targets, communicating with higher headquarters) to those operators. Variables held constant across models were explicitly declared and reported as assumptions. Dependent variables or measures were determined to compare the output from each model to evaluate the merit of each design option in terms of human and system performance. Typical dependent measures were maximum workload value per operator, number of times workload exceeded a recommended threshold, and tasks dropped most frequently during high workload. Separate models were built to represent each experimental condition (i.e., design option). The models were run, and output from each was compared statistically. The quantitative comparison results were critical to analysis success, but equally important was discussing them relative to mission capability resulting from the design options. The ARL HRED analysts considered this in their selection of the dependent variables. They were able to relate operator workload with errors and dropped tasks and thus potentially decreased mission performance.

As with models of task accuracy, the most powerful task models predict the effects of operator overload. Research indicates that operators may select from a large set of possible strategies to manage excessive workload (Hart, 1989). Specifically, they may drop a task, or try to perform all tasks at the same time with some performance penalty, or perhaps another operator would be assigned one or more of the ongoing tasks. As these management strategies play out, they may cause the task flow model to "flex" dynamically and new behaviors may emerge as new paths through the process may be created by the new operator assignments.

51.13 Brigade Staffing

More recent work has demonstrated that task models are very tolerant of a range of fidelity. In 2006, task models were used to estimate manpower requirements for the Battalion and Brigade S-1 unit as a result of the U.S. Army's transformation effort. This effort was unique in that much of the workload is driven by routine, day-to-day work. Unscheduled events can either interrupt this routine work or can augment it, depending upon the priority and the availability of qualified personnel. The purpose of the study was to determine whether specified manning concepts, in which shifts are manned by a team of personnel of various ranks and specialties, was sufficient for performing all the work. And if not, to identify the processes that were not likely to be performed to completion. A secondary purpose of the study was to assess the level of fatigue the unit would experience during sustained deployment, home operations, and conflict.

The model that was developed for this effort simulated manning requirements from a macro perspective to determine the adequacy of a proposed manning level, and is based on a product developed for the Navy called Total Crew Model (TCM) (Wetteland, Miller, French, O'Brien, & Spooner, 2000). The inputs into the model consist of routine workload and scenario events (such as training, maintenance, casualty operations, and days off). Schedules were developed to show what a typical day might look like for each person. In addition to the normal schedules, events were also created to represent extra duties—activities that occur at different times on different days throughout the time of the model (i.e., training or maintenance). The numbers and specialties of people required and desired to perform the tasks were also necessary inputs. Finally, there is a trump matrix that specifies priorities between combinations of competing tasks.

After the simulation runs, outputs are provided that assess the impact of sustained operations. The output generated by the model includes whether the normal work was completed (normal work being the hours of work taken from the pilot study data), the number of hours each soldier worked, and a task breakout for each soldier showing how much time was spent on each task. The model also provided data on fatigue levels for each MOS over the month and indicated whether all extra duties were completed.

In this study, special attention was given to enabling the modeler to visualize how well the manning concept was working as the simulation scenario progressed. Figure 51.5 shows the completed animation at the end of the model run. The bars in each cell of the grid indicate the amount of work that needs to be completed by each person in each area of work. In the output, green bars indicate the work that has been completed while red is used to indicate remaining work. In this example, not everyone was able to complete the work. At the bottom of the model is the percentage of work completed and the average number of hours worked by each individual over the course of the model.

As shown, the manning levels that were not sufficient to complete all the work looked specifically at the load of "normal work." PT, training, and maintenance were all separate duties at a higher priority than "normal work" and thus these activities were completed before "normal work" activities. During the PSDR pilot study, each soldier who was found to be unable to complete work under the model consistently worked longer than 40 hour weeks.

The model was run using several different manning levels to demonstrate the versatility of the model and the ability to experiment with manning levels and the addition of extra duties. Figure 51.6 shows the results of the model and illustrates that the results can be used to compare workload across the team.

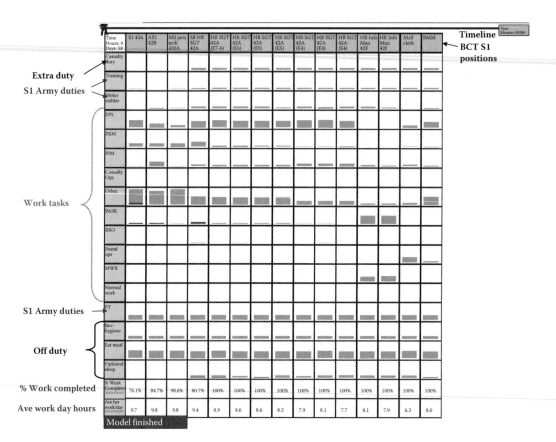

FIGURE 51.5 Completed animation screen shot.

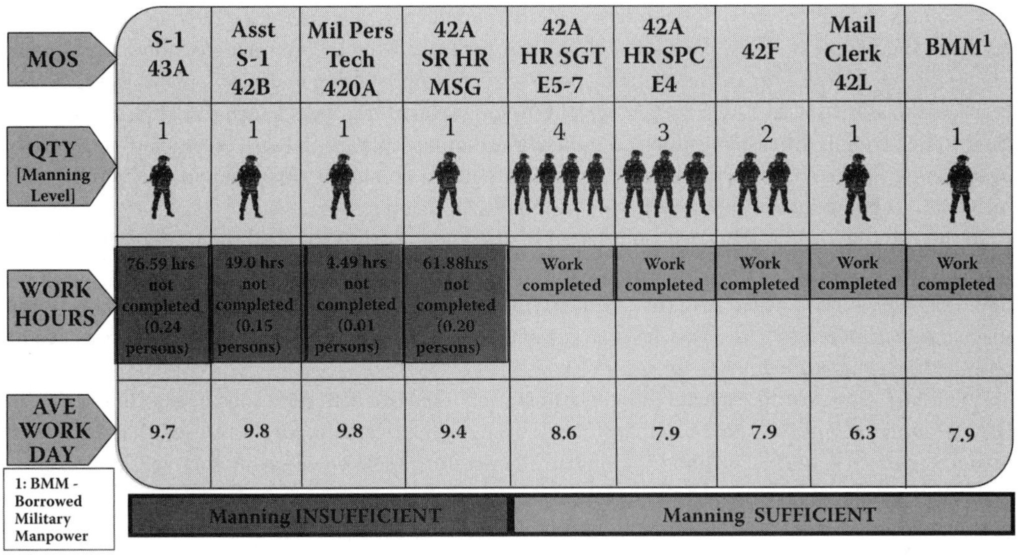

MOS	S-1 43A	Asst S-1 42B	Mil Pers Tech 420A	42A SR HR MSG	42A HR SGT E5-7	42A HR SPC E4	42F	Mail Clerk 42L	BMM[1]
QTY [Manning Level]	1	1	1	1	4	3	2	1	1
WORK HOURS	76.59 hrs not completed (0.24 persons)	49.0 hrs not completed (0.15 persons)	4.49 hrs not completed (0.01 persons)	61.88hrs not completed (0.20 persons)	Work completed	Work completed	Work completed	Work completed	Work completed
AVE WORK DAY	9.7	9.8	9.8	9.4	8.6	7.9	7.9	6.3	7.9

1: BMM - Borrowed Military Manpower

Manning INSUFFICIENT   Manning SUFFICIENT

FIGURE 51.6 Manning results, fully staffed.

As an additional output, the model predicted average fatigue levels for each crew member for the eight-week scenario. In the modeled scenario, each soldier experiences normal levels of fatigue roughly 90% of the time. Fatigue levels are a bit higher when soldiers have a 12-hour duty assignment overnight, which happens once a month, but the number of hours each person has to recover from the overnight duty work allows the fatigue levels to remain quite low. Soldiers experience bad fatigue less than 2% of the time.

In addition to seeing what the workload looks like for a fully staffed Brigade S-1 section, the model was adjusted to show how workload changes with different manning levels and additional work, illustrating the value of modeling as a powerful tool to examine "what-if" staffing scenarios.

51.14 Information Flow

51.14.1 Command and Control Case Study

The military command and control process is certain to change given the introduction of new information technology and new organizational structures. To predict how these changes will impact system performance, the Army Research Laboratory (ARL) Human Research and Engineering Directorate (HRED) sponsored the development of the Command, Control, and Communication: Techniques for Reliable Assessment of Concept Execution (C3TRACE) information flow modeling environment in which one can develop multiple concept models for organizations of any size, staffed by any number of people, performing any number of functions and tasks, and under various communication and information loads. C3TRACE predicts utilization, errors, message throughput, and decision-making performance.

In order to preserve a level of abstraction, C3TRACE is based on an information-driven decision architecture (Wojciechowski, Wojcik, Archer, & Dittman, 2001) based on the U.S. Army's accelerated decision-making process (Military Intelligence Officer Advanced Course, 1996) and designed to support prediction of how well the information flow and the organization's decision-making process supported timely and effective decision making. This architecture essentially consists of a taxonomy of the information elements (e.g., friendly force location, enemy location, etc.), each of which is attached to attributes such as recency, initial quality, priority, and decay rate. As messages come into the organization, their content is described through the taxonomy, causing the messages to follow a prescribed "route" through the C2 organization, being processed and contributing to decision making as they proceed. As this occurs, C3TRACE keeps track of how long it takes messages to be processed, and whether their route through the system, which accounts for task interruptions and task incompletion due to the competing priorities and workload, is likely to lead to decision making that is supported by timely and relevant data being shared appropriately within the C2 decision makers.

In the end, the quality of a decision is based on the match between the information received by the decision maker and the information required to make a decision and also by how much the "value" of the information has decayed over time. This technique can help to identify system and organizational inefficiencies, bottlenecks, or obstacles relevant to the high quality and recent information required for effective decision making.

C3TRACE was used to evaluate differences between alternate configurations of the Unit of Action (UA) Mounted Combat System Company Headquarters. The two configurations used the same information technology, and each configuration had 10 operators and 3 vehicles, but the types of personnel and vehicles differed as shown in Table 51.1.

Information flow models of both configurations were driven by a 96-hour scenario, consisting of nearly 9,000 messages. The routing of individual messages did vary between configurations to support the manpower assignments to specific vehicle types (e.g., when the CO moved from a C2V to an MCS vehicle, the modelers assumed that the CO would spend more time actually fighting the battle than directing it, requiring the XO to process the overflow).

C3TRACE outputs in this study included operator utilization and performance, completed versus dropped messages, mental workload levels, and decision quality. Based on the scenario duration and message volume modeled, the results showed differences among configurations in terms of operator utilization, operator performance, and workload peaks.

The CO's utilization was lower in Configuraton #2 due to a redistribution of responsibilities. In Configuration #2, the CO handled messages that concerned fighting the battle rather than directing it and the Robotics NCO processed all messages involved with controlling the unmanned assets. The XO assumed some of the CO responsibilities and also handled the messages from the platoons that concerned status, situation reports, and readiness condition. This resulted in an elevated utilization rate for the XO (as compared to the Deputy CO). The CO experienced fewer interruptions and fewer dropped tasks in Configuration #2 due to the off-loading of battle-directing tasks. In contrast, the 1SG was much

TABLE 51.1   Results of C3TRACE Evaluations

	Configuration 1	Configuration 2
Vehicle 1	Command and Control Vehicle (C2V) #1 Commanding Officer (CO) Fires Non-Commissioned Officer (NCO) Vehicle Commander (VC) NCO Chemical Driver NCO	Mounted Combat System CO Gunner Crew Chief
Vehicle 2	C2V #2 Deputy CO C4SIG NCO VC NCO Driver	C2V Executive Officer (XO) Fires NCO Robotics Enlisted, VC Chemical C4SIG Driver Enlisted
Vehicle 3	Future Tactical Truck System–Utility (FTTS-U) First Sergeant (1SG) NCO Supply Sgt NCO	FTTS-U First Sergeant (1SG) NCO Supply Sgt NCO

more heavily utilized in Configuration #2. Decision quality did not show a difference between configurations in this particular study.

This study demonstrated that information flow models for people assigned to an organization, the tasks and functions they perform, and the communications pattern within and outside of the organization, are all represented, and could facilitate what-if evaluations of numerous concepts without the need for live exercises or experiments (Plott et al., 2004). This capability will save resources and will provide an evaluation of many more concepts than could be accomplished by human-in-the-loop experiments alone.

51.15 CART Case Study

Beginning in 1999, the Air Force Research Laboratory/Human Effectiveness Directorate (AFRL/HECI) began an effort to provide a modeling environment in which human performance models were able to interact with engineering level, constructive, and first principle human performance models. To accomplish this goal, AFRL chose to expand the ARL HRED's Improved Performance Research Integration Tool (IMPRINT) (Allender et al., 1995). IMPRINT is government software consisting of a set of automated aids to assist analysts in conducting human performance analyses. IMPRINT provides the means for estimating manpower, personnel, and training (MPT) requirements and constraints for new weapon systems very early in the acquisition process. This required expansion of IMPRINT to enable it to communicate during run-time with other types of models via common protocols and to better simulate goal-oriented human performance. This effort became known as the Combat Automation Requirements Testbed (CART) project, and led to the development of an expanded version of IMPRINT that can be used to evaluate crew system and cockpit design for determining the impact of human factors and performance issues on the requirements generation process. These capabilities help analysts more fully understand the implication of human performance on total system performance, and to use that understanding to impact system requirements and to assess future training needs, and to use this understanding to affect system design very early in the acquisition process (Martin, Hoagland, Brett, & LaVine, 2001).

An example of the type of analysis performed using this expanded capability is the evaluation of the performance requirements for a new radar system to be used by a combat pilot and investigation of how the capabilities of the radar system would affect the performance of the pilot flying a reconnaissance mission during which the pilot is to locate and identify a potential target. If the pilot detects a threat during this planned mission, then the pilot will probably interrupt the planned mission and evade the threat. The time at which the threat is detected will affect the point in the mission that the pilot begins the maneuver and the time available to successfully evade. Similarly, if the pilot detects a high payoff target within range, he may interrupt the mission to attack the target. Depending upon the success of that maneuver, the planned mission might be resumed, aborted, or restarted.

In order to develop a tool that is capable of representing this behavior, many complex issues had to be resolved. First, a task model was used to represent the mission, and separate, largely independent task models were used to represent the goals. Triggering and prioritization mechanisms were implemented such that the goals would fire when the ongoing scenario dictated it, and they would properly interrupt and resume, representing the pilot switching goals as the pilot's environment changes. Most importantly, these task models had to exchange variable values and model control parameters with other currently executing models. In this way, variables that control the triggering conditions (e.g., threat availability) can be established through other models, such as radar or sensor models. These capabilities are provided through a straightforward architecture (see fig. 51.7).

The success of the architecture shown in Figure 51.7 rests upon the ability to send and receive variables from the CART human performance simulation model to other simulations. To achieve this, CART HPM acts as a federate within a high level architecture (HLA)–compliant federation. In order for this interaction to occur, middleware was developed. It is through this middleware that the goal is triggered and simulated human performance is communicated.

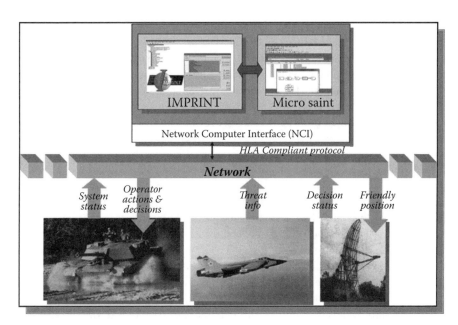

FIGURE 51.7   System architecture.

The most interesting aspect of this work is in the manner in which two disparate model types communicate and keep track of time. The CART models are discrete event simulations, whereas they may be communicating with continuous or real-time simulation models. For these models to properly interact, they can send and receive time-stamp-ordered (TSO) events. The CART IMPRINT models use a single-step, zero look-ahead time management scheme. This ensures that the models do not miss any object attribute changes or interactions that are received between time advancements. Because of the zero look-ahead, the CART models do not have to consider out-of-order event processing. In other words, the models do not have to cycle back in time to recover TSO events.

Hoagland et al. (2001) describe how the IMPRINT CART models were validated against human-in-the-loop data collected in a flight simulator performing a time-critical targeting mission. In this work, the models not only predicted the likelihood that the human pilot would identify and prosecute targets, but were also used to investigate new tactics, techniques, and procedures that could contribute to improved performance.

Perhaps most importantly, this work broke new ground for providing ways in which human performance modelers could interact with much larger "war-game" style models. This work demonstrated that models of human performance could exchange data with larger simulation federations, and that they could be used as "servers" for human performance estimates in order to more realistically represent human behavior in the other models. The project demonstrated that this could be done efficiently, and a human performance model could be designed that could maintain pace with the real-time distributed systems.

51.16 Maintenance Manpower

51.16.1 JBPDS Case Study

Modeling has also been successfully used to predict the impact of various maintenance concepts on system readiness. In 2005, ARL HRED conducted a maintenance manpower analysis of the Joint Biological Point Detection System (JBPDS) using the maintenance manpower modeling capabilities in IMPRINT.

The goal of this effort was to review the maintenance manning concept to ensure that the system could achieve its targeted operational readiness.

The JBPDS can detect and identify up to 10 biological warfare agents (e.g., bacteria, rickettsia, viruses, and toxins) and collects agent samples for subsequent laboratory analysis. It is a multi-service program with each service having a slightly different concept of operations. The Army planned to use the system housed in an HMMWV shelter, in a military trailer, and on the IAV, the Marines and Air Force plan to use it within a shelter system, on a trailer, and in a man portable mode, the Navy plans to use it on various ships and at ports of embarkation and debarkation.

The Army was considering a maintenance concept that involved using a signals support system specialist at the company level to service the 35 JBPDS units distributed across the corps sector. They were interested in predicting the utilization (or "busyness") of that soldier, and the resulting operational readiness of the JBPDS.

To conduct this study, ARL analysts imported data from the Logistics Support Analysis (LSA) Report describing the types of maintenance tasks to be performed, by whom, for how long, how frequently, and under what conditions. Next, they collected information regarding the mission scenario. This included information about how intensely the system would be used and the amount of time available between missions to repair any broken or worn components. Scenario data also include shift length for the repair team, spare parts availability, combat damage, and so on. This information is typically found in an operational mission profile or operational mission summary.

These data were combined and executed using the stochastic model within the define equipment portion of the IMPRINT tool (Archer, Gosakan, Shorter, & Lockett, 2005). This model simulates the maintenance requirements caused by accruing usage on the system components according to the scenario information. In very simple terms, as usage accrues, the amount of wear on each component is compared to the reliability of that component. When the model predicts a need for maintenance, it triggers a repair task for the maintainers, which is performed when the appropriate maintainers and any needed spares are available. This actually occurs in a sophisticated queuing model that manages competing requests for resources and the variability inherent in all the parameters. The result of the analysis is a predicted maintenance man-hour requirement and operational readiness based on the equipment characteristics, the scenario, and the maintenance concept.

The results of the work indicated that although the support system specialist would be considered a company expert, the model predicted that due to the high reliability of the system, he would actually only perform about 40 hours of total repair time on each JBPDS. Even when the model was adjusted to account for environmental stressors, such as the need to work in biochemical protective gear, extreme weather, and sometimes noisy environments, this soldier appeared to be spending much more time driving between the systems distributed across what could be very difficult and unfamiliar terrain than actually performing maintenance tasks.

The maintenance modeling performed by ARL HRED caused another maintenance concept to emerge. Specifically, one of the four chemical operations specialists that are already working at the unit could perform the remove and replace actions associated with failed LRUs, and using the organic support vehicle, evacuate them to the BDC. One of these soldiers and one support vehicle could gather up other failed units nearby on the way to the BDC repair site. This would alleviate the need for the company-level signals support specialist, reducing total life cycle cost.

In addition to the maintenance man-hour requirements and soldier utilization outputs, the model enabled the analysts to assess how sensitive JBPDS operational readiness was to spare availability. This output was extremely useful to the program manager, as he could then optimize his sparing concept.

This work illustrates one of the primary benefits of employing these modeling techniques. They are very useful in providing the analyst visibility into not just how the system is likely to be perform, but *why* it is performing in that way, enabling the analyst to identify alternative concepts that may not have been obvious before the analysis.

51.17 Decision Making

51.17.1 GRBIL Case Study

Hybrid tools that unite the strengths of task and cognitive models are beginning to become available. Archer and Lebiere (2003) have developed a system design and evaluation tool, GRBIL, in which developers can easily design and evaluate system interfaces. The system allows users to graphically define a system interface, walk through a set of operator goals for using the interface, and automatically generate a cognitive model of a system operator (built in the ACT-R cognitive architecture described in Anderson & Lebiere, 1998), providing metrics of time on task and potential errors. In addition, the system allows for the easy incorporation of dynamic models of the external world in order to evaluate interfaces that involve continually changing environments.

The architecture for GRBIL consists of four components:

- The GRBIL interface: Used to construct other interfaces and to specify an operator's goals and tasks for using that interface.
- The cognitive modeling system: Implemented in the ACT-R cognitive modeling architecture, and responsible for generating predictions of human performance.
- A dynamic environment model: Responsible for modeling environments external to the interface that may be changing and whose changes will affect users' performance with the interface. The current system uses the IMPRINT task network modeling architecture. The system has been designed so that different environment models may be easily "plugged" into the system.
- A software hub: Mediates communication between the other components.

The first step in constructing an interface in GRBIL is to design the physical layout of the interface. This is done in a similar fashion to many modern interface layout applications using WYSIWYG (what you see is what you get) drag and drop functionality. Once added to a window, a control may then be customized further (size, background image, etc.) from a menu. The second step in designing an interface is to provide a description of what actions each control will be capable of and what the desired effect of each action will be. This is done for each control in GRBIL via an "Event Actions" menu for each control. Using this process of adding interface windows, placing controls on those interfaces and then describing the effects of using those controls on the state of the interface, a GRBIL user can describe the functionality of an entire user interface (see fig. 51.8).

The next step in setting up an interface for evaluation is to define the interface tasks that an operator of the system might wish to perform. This is accomplished by stepping through and recording a series of actions (i.e., button clicks, moving of objects, etc.). The recorded series of actions is referred to as a "goal state" in the GRBIL terminology. Many separate goal states can be defined. Once recorded, a goal state may be executed by a cognitive model. It should be noted that while the recorded actions may list the exact steps to perform the interface tasks correctly every time, the ACT-R cognitive model has parameters in its architecture that can account for and model mistakes that could be made, steps that could be forgotten, and latencies in performing those tasks.

In GRBIL, the cognitive model interacting with the system is a representation of the knowledge that an average user would bring to bear on a new task. In particular, it represents as procedural knowledge the skills needed for manipulating the interface controls defined in the GRBIL toolbox. It also has the necessary skills to navigate the interface. When retrieving a control step that is not immediately available, the cognitive model is able to reason about how to navigate the interface using declarative knowledge to put itself in a position where that control is available. By default, it will do that using the knowledge of the interface that is automatically downloaded into the model before it is run. That knowledge takes the form of declarative memory chunks for each control that describes its position in the interface (e.g., which window or subwindow that it is a part of) and which actions are associated with it.

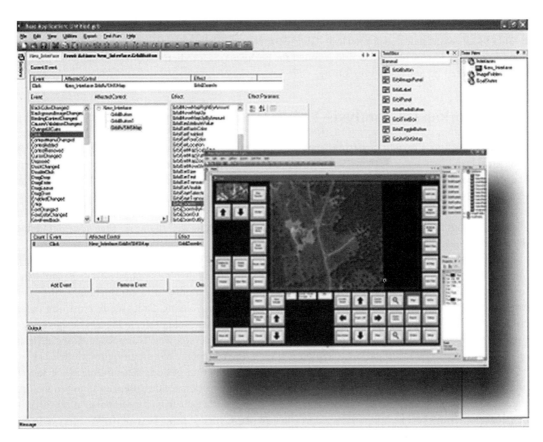

FIGURE 51.8   User interface constructed in GRBIL.

A number of characteristics of ACT-R are relevant to the performance of the model. First of all, ACT-R is limited in how fast it can perform its actions, especially external actions such as perceptual scanning and manipulation of the interface. In a dynamic, real-time environment such as robotic control, this can give rise to errors as the model is not able to keep pace with the demands of the task in the same way that a human operator would be unable to keep pace given the current system interface. Another source of errors is memory retrieval; ACT-R can skip steps or retrieve them in the wrong order (Anderson, Bothell, Lebiere, & Matessa, 1998) in the same way that a human might. Finally, performance can vary as a function of individual differences in working memory, psychomotor speed, or individual strategies, all of which can be represented in a constrained manner in the cognitive model.

To test and validate the predictive capabilities of ACT-R models generated from a GRBIL description, a detailed model of an unmanned vehicle operator control unit (OCU) was developed. The OCU is used by operators to control unmanned vehicles in the field. The interface for this OCU is quite complex, with several modes of operation and layers of buttons and controls. To validate the model's predictions of errors, latencies, and other performance measures, the project team has begun to collect performance data from human operators performing interface tasks such as mission planning and execution on the actual robotic OCU. Two participants familiar with the OCU interface were asked to set up the routes for two vehicles, which involves using the interface to indicate waypoints on a map that each vehicle should pass through and monitoring the vehicles' progress through the waypoints.

Timing predictions of the ACT-R model have matched preliminary data to some degree of accuracy and the model predicts potential human errors that could be made using the OCU to plan the vehicle routes (Archer & Lebiere, 2005), indicating that this is a promising hybrid architecture that will provide useful outputs to graphical user interface (GUI) designers. GRBIL research is continuing with the addition of more GUI elements that would provide for performance predictions on a broader set of tasks, such as map reading and spatial reasoning.

51.18 Workspace Analysis Using Human Figure Models

Workspaces include a variety of settings such as a crew station in a tank or submarine, the pilot's seat and controls in a helicopter or plane, a set of controls in an operations center, the driver's compartment of an automobile, or a position on an assembly line. Workspaces also include the areas necessary for technicians or service personnel to perform maintenance tasks. Although many workspaces are stationary, some are located on moving platforms such as ships or armored vehicles crossing rough terrain. In all cases, human figure modeling is a useful tool for evaluating how well the workspace accommodates the physical characteristics of its intended users.

The first step in performing a workspace analysis is to identify the critical actions that the human must be able to perform in the workspace. Typically these actions involve reaches with arms and legs, seeing specific locations within and outside the workspace, the ability to fit comfortably within the workspace, and entering or exiting the workspace. If an analyst is having difficulty determining these critical actions, we have found it useful to imagine which tests one would perform if every design under consideration could be built and used by every size and shape person who is supposed to use it. In a sense, human figure modeling allows us to do just that. The advantage of human figure models is that we can evaluate the critical tasks across the range and combination of body sizes found in the target user population and quantitatively compare the results. Often, workspace designers and human factors engineers evaluate reach, fit, and vision issues one dimension at a time (e.g., functional arm reach), but this does not always correspond to how a user must perform critical tasks in real life. Workspace tasks sometimes require multiple constraints to be satisfied simultaneously (e.g., reaching for a control while maintaining eye contact and keeping a foot on a pedal). Tests and analyses that consider these unidimensionally may optimize one while suboptimizing the rest.

Human models may also be used for analyzing the design and use of workspaces from a biomechanical perspective. Issues such as load carrying, weight distribution, vehicle rollovers, lifting, emergency ejection, and crash simulation have all been investigated using models. These models and their application are discussed in Chapter 8 of this volume.

51.19 Future Combat System (FCS): A Common Approach

Given the size and complexity of the FCS program and the number of workstations to be designed, it was critical that a common analytical approach be used for human engineering analysis of those workstations. In previous Army acquisition programs, differences in data, analytical tools, and methodology led to difficulty in timely, accurate, and decisive evaluation of vehicles. To address this for FCS, a common approach was developed to provide all the data and guidance necessary to make FCS operator and maintainer accommodation design decisions but eliminate sources of disagreement concerning validity or ability to compare results across the FCS team (lead system integrator, government, and platform developers). The approach is documented in Lockett, Kozycki, Gordon, and Bellandi (2005). The sources of variability among approaches considered were:

- Source of the anthropometric (body measurement) data used to perform accommodation analyses (i.e., which survey, database, or standard)
- Projection of anthropometric data to fielding date of the system to include accounting for secular and demographic trends

- Statistical method of sampling anthropometric data for sizing mannequins and testing for accommodation (univariate vs. multivariate)
- Procedure for sizing human figure model mannequins against anthropometric data
- Method for considering clothing and equipment worn and used by operators and maintainers
- Human figure modeling software used to perform analyses (i.e., software package)
- Sources, currency, and format of computer-aided design (CAD) files representing the vehicle designs
- Techniques for using human figure models to evaluate accommodation, considering factors such as positioning mannequins and specific types of accommodation (e.g., reaching, line of sight, head clearance)

The resulting common approach attempts to address these sources of variability. The basic steps of the approach are as follows.

1. Derive a common set of anthropometric data for use in analyses:
   a. Determine key accommodation analyses for FCS vehicles and the critical anthropometric dimensions associated with those analyses;
   b. Project anthropometric data from the most recent, 1988, Army anthropometric survey (ANSUR 88) to the year 2015 (at the time, the fielding year for FCS) for each of the critical dimensions;
   c. Derive dimensions to use in sizing mannequins using multivariate rather than univariate statistics.
2. Build clothed and equipped human figure model mannequins from the anthropometric data:
   a. Build boundary mannequin sets in human figure modeling tools;
   b. Add digitized clothing and equipment;
   c. Validate the combined mannequin and clothing model against 3D scan data;
   d. Distribute mannequins to FCS team.
3. Obtain CAD files of vehicles and optimize for use with human figure modeling tools such as Jack.
4. Perform analysis:
   a. Place mannequins in CAD file;
   b. Check for static and dynamic accommodation;
   c. Develop recommendations for addressing accommodation issues identified.

Even though the FCS program is ongoing, this common analytical approach has already proven effective in focusing discussion around design decisions affecting physical accommodation rather than wasting time trying to resolve differences among data sources, methodology, and models used.

51.20 Advanced Amphibious Assault Vehicle

Early in the design of a new amphibious assault vehicle (a very large armored military vehicle), before any physical models or prototypes existed, human figure models were used to evaluate a key vision requirement. The requirement involved being able to see a person, called a ground guide, from the waist up when the guide was standing a specific distance from the front of the vehicle. The ground guide would be viewed from the commander's crew station located in a turret on top of the vehicle. To evaluate whether or not current vehicle designs met the requirement, computer-aided design (CAD) representations of the vehicle designs were ported into the Jack tool. The most extreme case for the vision requirement was determined to be a ground guide of short stature. In human engineering terms this translated as a Jack human figure with a female 5th percentile waist (omphalion) height for the male and female target user population. This measurement is closest to the top of the pelvis segment on a Jack

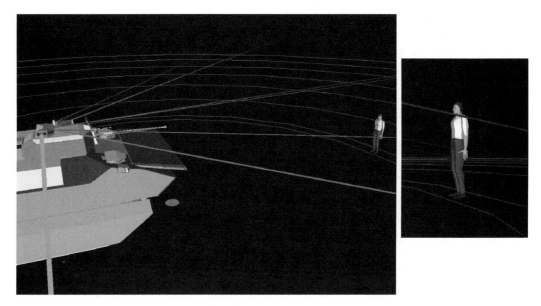

FIGURE 51.9 Analysis of vision requirement for armored vehicle.

figure. Note that female body dimensions are not always smaller than those of males. Figure 51.9 depicts analysis of the vision requirement for one of the vehicle designs.

The analysis revealed that the most promising vehicle design would not meet the vision requirement—the ground guide could only be seen from the shoulders up. As a result of using human figure modeling, the design team was able to meet as a group, then propose and evaluate design modifications to achieve the vision requirement. This concurrent engineering saved considerable time. The vehicle turret was moved forward to avoid costly redesign of the vehicle's bow mechanism. Making these changes early in development provided the program more design options and cheaper solutions versus finding out about problems meeting the requirement when prototype vehicles were being evaluated. At that later stage, more aspects of the design are by necessity set or are more expensive to rework.

51.21 Comanche Helicopter

In the early 1980s, the U.S. Army initiated a program known as light helicopter experimental (LHX) to seek a replacement for its fleet of Kiowa scout and Cobra light attack helicopters. LHX became the Comanche program in the early 1990s.

The original anthropometric accommodation requirement for the crew stations was bracketed around the central 90% of the Army male soldier population, including all mission-essential combat gear and protective clothing ensembles for the pilot and copilot. In 1993, the Army changed its policy about women aviators and permitted them to fly combat missions. This policy change resulted in new opportunities for women to qualify in previously male-only aircraft including the Comanche. As a consequence, the Comanche crew station requirements were changed from accommodating only a male soldier population to the current requirement of accommodating the central 90% of an equally weighted population of Army male and female soldiers.

Accommodating the change in requirement presented several problems. The crew stations had already been designed, only two prototypes existed, and access to a pool of pilots representing the full range of the target population was limited. To address these problems, Kozycki and Gordon (2002) applied multivariate methods to anthropometric data and human figure modeling to determine the extent of

FIGURE 51.10 Small female in Comanche cockpit, modeled using Jack.

crew station modifications necessary to accommodate the new requirement. They conducted princi-pal components analyses of the critical body dimensions for cockpit accommodation. Based on the results, a set of Jack figures representing the combinations of body dimension on the boundary of the population accommodation envelope was created. Pilot clothing and equipment were added to the Jack figures, and each was positioned in the cockpit and used to evaluate accommodation against 10 criteria. These criteria were agreed upon ahead of time and failure to meet all of them meant that the particular figure was not accommodated. The criteria concerned clearances between the pilot and cockpit features, the ability of the pilot to reach and see critical controls, and the ability of the pilot to sit in a posture with the eyes positioned to see at an acceptable angle outside the cockpit (i.e., at the design eye line). Figure 51.10 shows a smaller female successfully gripping the collective control but without sufficient extended reach to grasp engine control levers.

Although the Comanche program was eventually cancelled, this analysis revealed that the existing designs presented accommodation problems for not just smaller females but also larger males. Based on the analysis, several design changes were recommended to accommodate up to 40% more of the target population. These changes were implemented prior to cancellation of the Comanche program.

Motion capture technology can be used to position digital human figure models. Kennedy, Durbin, Faughn, Kozycki, and Nebel (2004) conducted an ingress and egress study of the Comanche crew sta-tion. Although the study involved live test participants entering and exiting a high-fidelity mockup of the crew station, an optical motion capture system was used to record the postures of the participants during the study. Later, these data were used to position Jack human figure models sized to the partici-pants' body dimensions. The resulting simulation allowed the researchers to identify and investigate clearance problems by viewing the model from many different angles for individual postures.

51.22 Stryker

During procurement and development of the Stryker family of armored vehicles, human figure mod-eling was used by ARL in innovative ways. It helped define accommodation requirements and then confirm claims made by vendors about their ability to meet those requirements. For example, because the acquisition strategy for the vehicles was to modify existing armored vehicles, the Army needed to ensure that at least one candidate vehicle had the potential to carry the number of soldiers they would specify in their contract requirements. ARL HRED used basic measurements of likely candidate vehicles to create Jack models representing the maximum space available. ARL then positioned different numbers of large male soldiers (in terms of critical dimensions such as bideltoid breadth, buttock-knee length, chest depth, and seated stature) to determine if the proposed requirement was feasible. Several candidates failed but enough had the potential to succeed to justify the requirement. Another acquisition innovation entailed requiring vendors to provide evidence in their proposals, in the form of human figure models and analyses, that they would be able to meet the soldier accommodation requirement. The usual case is for the vendor to state that it will meet the requirement to physically accommodate the central 90% of the target users as the program progresses through the design process. Then as the program progresses and difficult design trade-offs must be made, the accommodation requirement may get compromised. By requiring analysis evidence to support claims made, the Army was able to quantitatively evaluate each bidder's ability to meet requirements during source selection, before a winner was selected.

After a vendor was selected for Stryker, ARL used human figure modeling and worked with the former Military Traffic Management Command (MTMC) to investigate possible solutions for a loadmaster to perform safety checks on Stryker vehicles when being airlifted in a C-130 Hercules cargo aircraft. To perform safety checks such as ensuring the vehicle is tied down securely, the loadmaster must travel from one end of the vehicle to the other while it is loaded on the C-130. Unfortunately, there is very little space in which the loadmaster can move. At the time of the analysis, there was insufficient space for the loadmaster to travel through the interior of the vehicle or underneath it. Normally, the loadmaster travels between the side of the vehicle and the cargo hold wall of the aircraft. However, using Jack, ARL showed that this safety aisle was too narrow to meet Federal Aviation Administration guidelines. To help the Army develop a case for a waiver to these guidelines, ARL used Jack to investigate the feasibility of the loadmaster traversing the top of Stryker vehicles. For each type of Stryker, ARL positioned figures at key positions along the traversal path as shown in Figure 51.11.

The figures were sized to represent large male loadmasters wearing essential equipment. Key positions included those with minimal clearance between the top of the Stryker and the roof of the aircraft. ARL was able to measure how much space was available, to identify potential safety issues, and to flag obstacles or obstructions. Along with the measurements, the safety issues and obstacles were forwarded

FIGURE 51.11   Loadmaster traversing top of Stryker vehicle.

to the Stryker program for possible redesign. Based on this information, and then some redesign and confirmation via live testing, the Air Force ultimately granted a waiver to the FAA guidelines.

51.23 Army Airborne Command and Control System (A2C2S)

The Army Airborne Command and Control System (A2C2S) is a command and control system designed to be installed in a Blackhawk utility helicopter. The A2C2S product manager requested that the ARL perform an evaluation of the emergency egress characteristics of the A2C2S. ARL used a combination of human figure modeling and egress testing for the evaluation. The evaluation was conducted by Havir and Kozycki (2006). They used the Jack human figure model to perform a detailed analysis of all egress routes to identify whether the larger end of the male soldier population wearing clothing and equipment could fit through the egress routes, and to identify design characteristics of the A2C2S that enhance or degrade the soldier's ability to egress the aircraft. The addition of clothing and equipment to the models was critical to representing the clearances required by soldiers. Figure 51.12 shows one of the many egress paths evaluated using Jack.

A live test of emergency egress was used to validate the results of the model, verify that the egress could meet the time requirements, and identify additional safety concerns that may be encountered during actual egress trials. The results of the egress modeling identified some minor shortcomings with the egress characteristics of the A2C2S. The results of the egress testing validated the modeling that was performed. In addition, all egress trials successfully met or exceeded the time standard for emergency egress. The results and recommendations from the modeling and testing were provided to the A2C2S product manager for implementation.

51.24 Future Challenges and Development

DHM technologies have progressed significantly in the last few years, fueling important advances in the application of models to improve system performance, reduce cost, and improve safety. Despite this progress, it is beneficial to conduct an ongoing analysis of how DHM might continue the progression of ideas and value.

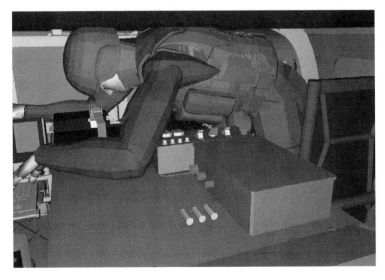

FIGURE 51.12   Evaluation of egress path.

The first place to look is at user requirements to determine whether they are adequately addressed with current capabilities. Lockett et al. (2005) identified an interesting combination of requirements gathered across commercial, academic, and military DHM users. These requirements included increased fidelity in predictive capabilities (e.g., better anthropometric and posture predictions for human figure models) and new measures of performance (e.g., comfort). They also included a set of requirements that, when met, would reduce barriers to usage. These include methods to (1) facilitate data sharing, (2) ease the development of hybrid and integrated models, (3) match user interfaces to user knowledge and skills, and (4) enable users to ensure their results are valid. Many of these requirements are being addressed by the tool development community; however, none of the requirements has yet been fully satisfied and they remain as opportunities for improvement. Similar and additional opportunities were identified in a study conducted by the U.S. National Academy of Sciences, National Research Council, Committee on Human-System Design Support for Changing Technology (Pew & Mavor, 2007).

In the remainder of this chapter, we provide some specific recommendations for further work that overlap the Lockett et al. and Pew and Mavor findings that we believe are within the short- to mid-term reach of the DHM tools we discuss in this chapter.

Predictive human performance models that account for differences in perception, information dissemination, and decision-making processes across a diverse team are a challenging next step for our field. Such models would enable us to better predict process improvements and potential decision-making biases in multicultural groups and would provide a basis for influencing the design of systems intended to be used by coalition teams. If we could predict these differences with validity, we could plan, train, and design for them, improving situational understanding and ultimately improving team decision making.

Also, we discuss hybrid models in this chapter. It is clear that the recent advancements in hybrid models will broaden the user community. As an example, users of anthropometric modeling tools will now be able to connect to information-processing models, enabling them to assess the impact of decision making on the physical motions of the human in the working environment, leading to insight into how decision-making errors might require new control or display layouts for error recovery. Now that the fundamental challenges of developing protocols and middleware through which disparate tools can communicate have been met, the next step will be to selectively implement this integration, with an eye toward developing more type III hybrid models (i.e., those in which models are permitted to execute in parallel, communicating as needed with a shared time management system). We would like to see further research and development in this area, as it would capitalize on the current capabilities and reduce duplication of effort. Such work requires prior negotiation and agreement on the definitions of tasks and performance measures across user communities. This is arduous and unglamorous work, but is necessary for advancement.

Next, in order to provide powerful capabilities to trade off system capabilities across multiple HSI domains (manpower, personnel, training, safety, etc.), we must develop quantitative algorithms of how each domain impacts performance, cost, and safety. This process is discussed in Chapter 46 in this volume; however, it is important to note that this challenge can only be met through an investment in fundamental research. This research must be carefully planned and orchestrated so that it can lead to findings that are generalizable through reliance on a shared task taxonomy. It is clear that life cycle cost is a well-understood system measure and that providing a direct method to link the HSI domains to performance and ultimately to cost would provide a powerful analysis capability.

Finally, in this chapter we discussed recent work extensions of DHMs into applications involving multiple systems (armored vehicles, aircraft, dismounted units) performing shared missions. Existing tools are also being applied to problems associated with predicting performance of distributed and collocated teams. One of the many challenges in this work is to ensure that the tools are scalable, that potential efficiencies are realized when they are scaled, and that we are predicting and reporting meaningful metrics for teams and across multiple systems. It is unrealistic to simply introduce multiple entities into a single model to represent multiple systems. This solution overlooks how systems communicate, the additional workload required when crews must share data and work together, and the impact of command structure on overall performance. Therefore, an additional layer must be imposed on system-specific models to represent these aspects of the mission. Providing this capability in a reasonably automated fashion is certainly possible but does not yet exist.

Clearly, DHM has a successful track record for influencing system design to benefit the warfighter. The opportunities for DHM to continue this work are evident and should continue to provide value to the military.

References

Abdel-Malek, K., Yang, J., Marler, R.T., Beck, S., Mathai, A., Zhou, X., Patrick, A., & Arora, J. (2006). Towards a new generation of virtual humans: Santos. *International Journal of Human Factors Modeling and Simulation*, 1(1), 2–39.

Aldrich, T.B., Craddock, W., & McCracken, J.H. (1984). *A computer analysis to predict crew workload during LHX scout-attack missions* (Vol. 1, MDA903-81-C-0504/ASI479-054-I-84[B]). Fort Rucker, AL: U.S. Army Research Institute Field Unit.

Allender, L., Kelley, T., Salvi, L., Lockett, J., Headley, D.B., Promisel, D., Mitchell, D., Richer, C., & Feng, T. (1995). Verification, validation, and accreditation of a soldier-system modeling tool. *Proceedings of the 39th Human Factors and Ergonomics Society Meeting*. Santa Monica, CA: Human Factors and Ergonomics Society.

Anderson, J.R., Bothell, D., Lebiere, C., & Matessa, M. (1998). An integrated theory of list memory. *Journal of Memory and Language*, 38, 341–380.

Anderson, J.R., & Lebiere, C. (1998). *The atomic components of thought*. Mahwah, NJ: Lawrence Erlbaum.

Archer, S.G., Gosakan, M., Shorter, P., & Lockett, J. (2005). New capabilities of the Army's maintenance manpower modeling tool. *Journal of the International Test and Evaluation Association*, 26(1), 19–26.

Archer, S.G., & LaVine, N. (2000). Modeling architecture to support goal oriented human performance. *Proceedings of the Interservice-Industry Training and Simulation Education Conference (IITSEC)*. Orlando, FL: National Training and Simulation Association.

Archer, R., Keller, J., Archer, S., & Scott-Nash, S. (1999). Discrete event simulation as a risk analysis tool for remote afterloading brachytherapy *(NUREG/CR-5362, 1 & 2)*. U.S. Regulatory Commission Office of Nuclear Regulatory Research.

Archer, R.D., & Lebiere, C. (2003). Integration of task network and cognitive models to evaluate system designs. *Proceedings from Advanced Simulation Technologies Conference*. San Diego, CA: Society for Modeling and Simulation International.

Archer, R.D., Lebiere, C., & Warwick, W. (2005). Design and evaluation of interfaces using the GRaph-Based Interface Language (GRBIL) tool. *Human Systems Integration Symposium*. Arlington, VA: American Society of Naval Engineers.

Badler, N., Erignac, C., & Liu, Y. (2002). Virtual humans for validating maintenance procedures. *Communications of the ACM*, July 2002, pp. 57–63.

Barnes, M., Knapp, B., Tillman, B., Walters, B., & Velicki, D. (2000). Crew systems analysis of Unmanned Aerial Vehicle (UAV) future job and tasking environments (ARL-TR-2081). Aberdeen Proving Ground, MD: U.S. Army Research Laboratory.

Belyavin, A., Woodward, A., Nguyen, D., Robel, G., & Woolworth, J. (2005). Development of a novel model of pilot control behavior in balked landings. In *American Institute of Aeronautics and Astronautics (AIAA) Conference Proceedings* (Vol. 10). Reston, VA: American Institute of Aeronautics and Astronautics.

Booher, H.R. (Ed.). (1990). *MANPRINT: An approach to systems integration*. New York: Van Nostrand Reinhold.

Booher, H.R. (Ed.). (2003). *Handbook of human systems integration.* Hoboken, NJ: John Wiley & Sons.

Bowman, E. (2004). Modeling the process and organization to support effects based planning. *Military Operations Research Society Symposium.* Monterey, CA: Military Operations Research Society.

Bubb, H., Engstler, F., Fritzsche, F., Mergl, C., Sabbah, O., Schaefer, P., & Iris Zacher. (2006). The development of RAMSIS in past and future as an example for the cooperation between industry and university. *International Journal of Human Factors Modeling and Simulation,* 1(1), 140–157.

Chaffin, D. (Ed.). (2001). *Digital human modeling for vehicle and workplace design.* Warrendale, PA: Society of Automotive Engineers.

Drews, C., Laughery, R.R., Kramme, K., & Archer R. (June 1985). *LHX cockpits: Micro SAINT simulation study and results.* Dallas, TX: Report prepared for Texas Instruments Equipment Group.

Estell, R. (December 2001). Abstraction and reality—a fundamental dilemma. *Phalanx,* 34, 4, 14, 39.

Fan, X., Sun, S., McNeese, M., & Yen, J. (2005). Extending recognition-primed decision model for human-agent collaboration. *Proceedings of the Fourth International Joint Conference on Autonomous Agents and Multi Agent Systems (AAMAS 2005).* The Netherlands.

Gacy, A.M., & Dahn, D. (2002). An information taxonomy combining military course of action development, crew factors and information perception. *Proceedings of the Military, Government & Aerospace Simulation Symposium: Advanced Simulation Technologies Conference.* San Diego, CA: Society for Modeling and Simulation International.

Gore, B.F., & Jarvis, P. (2005). Modeling the complexities of human performance. *2005 IEEE International Conference on Systems, Man, and Cybernetics.* New York: IEEE Systems, Man and Cybernetics Society.

Hart, S.G. (1989). Crew workload management strategies: A critical factor in system performance. *Fifth International Symposium on Aviation Psychology,* Columbus, OH.

Havir, T., & Kozycki, R. (2006). An assessment of the emergency egress characteristics of the U.S. Army Airborne Command and Control System (A2C2S). (ARL-MR-0635). Aberdeen Proving Ground, MD: U.S. Army Research Laboratory.

Hoagland, D.G., Martin, E.A., Anesgart, M., Brett, B.E., LaVine, N.D., & Archer, S.G. (2001). Representing goal-oriented human performance in constructive simulations: Validation of a model performing complex time-critical-target missions. *Simulation Interoperability Workshop.* Orlando, FL: Simulation Interoperability Standards Organization.

Huey, B.M., & Wickens, C.D. (Eds.). (1993). *Workload transition: Implications for individual and team performance.* Washington, DC: National Academy Press.

Kant, J-D., & Thiriot, S. (2006). Modeling one human decision maker with a multi-agent system: The CODAGE approach. *Proceedings of the Fifth International Joint Conference on Autonomous Agents and Multiagent Systems.* Hakodate, Japan.

Kennedy, J., Durbin, D., Faughn, J., Kozycki, R., & Nebel, K. (2004). Evaluation of an army aviator's ability to conduct ingress and egress of the RAH-66 Comanche crew station while wearing the Air Warrior Ensemble. (ARL-TR-3404). Aberdeen Proving Ground, MD: U.S. Army Research Laboratory.

Knapp, B., Archer, S., Archer, R., & Walters, B. (1999). Innovative approaches to modeling: An application in missile defense. *Proceedings of the Society for Computer Simulation Conference.* San Diego, CA: Society of Computer Simulation.

Kozycki, R., & Gordon, C. (2002). Applying human figure modeling tools to the RAH-66 Comanche crew station design. *Proceedings of the 2002 SAE Digital Human Modeling Conference.* Warrendale, PA: Society of Automotive Engineers.

Laughery, R., Lebiere, C., & Archer, S.G. (2006). Modeling human performance in complex systems. In G. Salvendy (Ed.), *The handbook of human factors* (3rd ed.), pp. 967–996. Hoboken, NJ: John Wiley & Sons.

Law, A.M., & Kelton, W.D. (1991). *Simulation modeling and analysis* (2nd ed.). New York: McGraw-Hill.

Leiden, K., Kamienski, J., & Kopardekar, P. (September 2005). Human performance modeling to predict controller workload analysis methodology to predict increases in airspace capacity (AIAA-2005-7378). *Proceedings of the Aviation Technology, Integration, and Operations Forum.* Reston, VA: American Institute of Aeronautics and Astronautics.

Leiden, K., Laughery, K.R., Keller, J., French, J., Warwick, W., & Wood, S. (2001). *A review of human performance models for the prediction of human error.* NASA System-Wide Accident Prevention Program. Moffett Field, CA: NASA Ames Research Center.

Liang, C., Magdaleno, R., Lee, D., Klyde, D., Allen, W.R., Overmeyer, K., & Rider, K. (2005). Validation of a biodynamic model for the assessment of human operator performance in a vibration environment. *Proceedings of the SAE Digital Human Modeling for Design and Engineering Conference,* Iowa City, Iowa. Warrendale, PA: Society of Automotive Engineers.

Lockett, J.F., Assmann, E., Green, R., Reed, M.P., & Raschke, U. (2005). Digital human modeling research and development user needs panel. *SAE Transactions,* 2005. Warrendale, PA: Society of Automotive Engineers.

Lockett, J., Kozycki, R., Gordon, C., & Bellandi, E. (2005). Human figure modeling analysis approach for the army's future combat systems. *Proceedings of the 2005 SAE World Congress,* Detroit, MI. Warrendale, PA: Society of Automotive Engineers.

Lockett, J.F., & Powers, J. (2003). Human factors engineering. In H.R. Booher (Ed.) *Handbook of human systems integration.* Hoboken, NJ: John Wiley & Sons.

Martin, E.A., Hoagland, D.G., Brett, B.E., & LaVine, N.D. (2001). Developing and integrating advanced human performance models with mission-level simulations: Progress and lessons learned. *Simulation Interoperability Workshop.* Orlando, FL: Simulation Interoperability Standards Organization.

McCracken, J.H., & Aldrich, T.B. (1984). *Analyses of selected LHX mission functions: Implications for operator workload and system automation goals* (Technical Note ASI479-024-84). Fort Rucker, AL: U.S. Army Research Institute Aviation Research and Development Activity.

McMahon, R., Spencer, M., & Thornton, A. (1995). A quick response approach to assessing the operation performance of the XM93E1 NBCRS through the use of modeling and validation testing. Presented at Military Operations Research Society Symposium: Annapolis, MD.

Middlebrooks, S.E., Knapp, B.G., Barnette, B.D., Bird, C.A., Johnson, J.M., Kilduff, P.W., Schipani, S.P., Swoboda, J.C., Wojciechowski, J.Q., Tillman, B.W., Ensing, A.R., Archer, S.G., Archer, R.D., & Plott, B.M. (1999). CoHOST (computer modeling of human operator system tasks) computer simulation models to investigate human performance task and workload conditions in a U.S. Army heavy maneuver battalion tactical operations center. *(ARL-TR-1994).* Aberdeen Proving Ground, MD: Human Research and Engineering Directorate, U.S. Army Research Laboratory.

Mitchell, D.K. (2005). Soldier workload analysis of the MCS platoon's use of unmanned assets. (ARL-TR-3476). Aberdeen Proving Ground, MD: U.S. Army Research Laboratory.

Mitchell, D.K., Samms, C.L., Henthorn, T., & Wojciechowski, J.Q. (2003). Trade study: Two-versus three-soldier crew for the Mounted Combat System (MCS) and other Future Combat System platforms. (ARL-TR-3026). Aberdeen Proving Ground, MD: U.S. Army Research Laboratory.

Mitchell, D.K., Samms, C.L., & Henthorn, T. (2004). Workload analysis of two- versus three-soldier crew for the Non-Line-of-Sight Cannon (NLOS-C) system. (ARL-TR-3406). Aberdeen Proving Ground, MD: U.S. Army Research Laboratory.

Mui, R.C., LaVine, N.D., Bagnall, T., Sargent, R.A., Goodin, J.R., & Ramos, R. (2003). A method for incorporating cultural effects into a synthetic battlespace. *Behavioral Representation in Modeling and Simulation Conference.* Orlando, FL: Simulation Interoperability Standards Organization.

Pew, R.W., & Mavor, A.S. (Eds.). (2007). *Human-system integration in the system development process: A new look.* Washington, DC: The National Academies Press.

Plott, B., Hamilton, A., & Laughery, K.R. (2003). Linking human performance and anthropometric models through an open architecture (2003-01-2203). *Proceedings of the Society for Automotive*

Engineers, Digital Human Modeling for Design and Engineering. Warrendale, PA: Society for Automotive Engineers.

Plott, B., Quesada, S., Kilduff, P., Swoboda, J., & Allender, L. (2004). Using an information-driven decision-making human performance tool to assess U.S. Army command, control, and communication issues. *Proceedings of Human Factors and Ergonomics Society 48th Annual Meeting.* Santa Monica, CA: Human Factors and Ergonomics Society.

Salvendy, G. (Ed.). (2006). *Handbook of human factors and ergonomics.* New York: Lawrence Erlbaum.

Sarno, K., & Wickens, C.D. (1992). *The role of multiple resources in predicting time-sharing efficiency: An evaluation of three workload models in a multiple task setting* (ARL-91-3/NASA A31-91-1). Moffett Field, CA: NASA AMES Research Center.

Siegel, A.I., & Wolf, J.A. (1969). *Man-machines simulation models.* New York: Wiley-Interscience.

Swoboda, J., Kilduff, P., & Katz, J. (2005). A platoon level model of communication flow and the effects on soldier performance. *Proceedings of Human Factors and Ergonomics Society 49th Annual Meeting.* Santa Monica, CA: Human Factors and Ergonomics Society.

Walters, B., Bzostek, J., & Li, J. (2005). Integrating human performance and anthropometric modeling in the crew station design tool. *Proceedings of the Digital Human Modeling for Design and Engineering Symposium* (Paper No. 05DHM-24). Iowa City, IA: Society of Automotive Engineers.

Walters, B., Huber, S., French, J., & Barnes, M. J. (2002). Using simulation models to analyze the effects of crew size and crew fatigue on the control of tactical unmanned aerial vehicles (TUAVs) (ARL-CR-0483). Aberdeen Proving Ground, MD: U.S. Army Research Laboratory.

Warwick, W., & Hutchins, S. (2004). Initial comparisons between a "naturalistic" model of decision making and human performance data. *Proceedings of the 13th Conference on Behavior Representation in Modeling and Simulation*, Simulation Interoperability Standards Organization.

Wetteland, C.R., Miller, J.L., French, J., O'Brien, K., & Spooner, D.J. (2000). The human simulation: Resolving manning issues onboard DD21. *Proceedings of the 2000 Winter Simulation Conference.* San Diego, CA: Society for Computer Simulation International.

Whicker, M.L., & Sigelman, L. (1991). *Computer simulation applications: An introduction.* Applied Social Research Methods Series, 25. Newbury Park, CA: Sage.

Wickens, C.D., Sandry, D.L., & Vidulich, M. (1983). Compatibility and resource competition between modalities of input, central processing, and output. *Human Factors*, 25(2), 227–248.

Wojciechowski, J.K., & Archer, S.G. (2002). Human performance modeling—art or science? *Proceedings of the Military, Government, and Aerospace Simulation Conference: Advanced Simulation Technologies Conference.* San Diego, CA: Society for Computer Simulation International.

Wojciechowski, J.Q., Wojcik, T., Archer, S., & Dittman, S. (2001). Information-driven decision-making human performance modeling. *Proceedings of the Military, Government, and Aerospace Simulation Symposium: Advanced Simulation Technologies Conference.* San Diego, CA: Society for Computer Simulation International.

Wu, F., Huberman, B.A., Adamic, L.A., & Tyler, J.R. (2004). Information flow in social groups. *Physica A*, 337, 327–335.

Zachary, W., Ryder, J., Santarelli, T., & Weiland, M. (2000). Applications for executable cognitive models: A case study approach. *Proceedings of International Ergonomics Association 14th Triennial Congress and Human Factors and Ergonomics Society 44th Annual Meeting.* San Diego, CA: Human Factors and Ergonomics Society.

52

Advanced Human Modeling in Support of Military Systems

Pierre Meunier,
Brad Cain, and
Mark Morrissey

52.1 Introduction

A key issue in executing a simulation-based acquisition (SBA) strategy is the concept of process integration. An identified key factor in the widespread deployment of emerging technologies has been the ability of individual technologies (such as Dassault Systemes' V5 Human Modeling solutions, commonly known as SAFEWORK, and Alion Science's Integrated Performance Modelling Environment, IPME) to become accepted, integrated components within a digital design and manufacturing product life cycle management strategy. This is illustrated in the huge consolidation the CAD/CAM/PDM industry has experienced over the past few years. The world's biggest CAD companies now realize that it is not merely enough to create product content in their CAD software, but it is equally important to manage the product data (in their PDM solutions) and to simulate how the product will be built, assembled, operated, maintained, and decommissioned from both a product and resource (i.e., human operator) perspective. As such, digital human modeling is now closely integrated into the digital design process.

Defense research agencies have been investing in toolsets to support simulation-based acquisitions to ensure that they are able to examine competing designs in a like-for-like simulation environment. In addition, SBA techniques are being deployed more frequently to provide added value in terms of process (time and money) as well as functionality. Because of this, defense contractors are becoming more reliant on simulation technology to conform to the requirements of defense procurement agencies. In many recent large procurement activities, the use of such technology was not only desirable, but also mandated. The added benefit of simulation for contractors is that it enables them to reduce the inherent financial and technical risk by understanding their proposed designs in greater detail at the bid stage, not to mention the sales and marketing potential of the end product.

This chapter will describe the integration of advanced digital human modeling tools that can be used not only within a product life cycle management strategy to provide value added for process and functionality, but also by defense agencies to predict system performance (including the human) in an SBA context.

52.2 Digital Human Modeling in a Product Life Cycle Management Environment

The domain of digital human modeling has evolved significantly over the last 10 years. The value proposition associated with digital human models (DHMs) has been generally accepted by the human factors (HF) user community since the initial commercial availability of DHMs afforded a pragmatic, while not always intuitive, computer-based methodology for analyzing human-product-system interaction. However, while early commercially available DHMs offered *alternative* approaches to performing HF analysis, they did not always offer a *better* or more *affordable* approach to performing HF and human systems integration (HSI) analysis within the product life cycle.

HF and HSI professionals have long been aware of the potential strategic importance of HF/HSI knowhow to effective system design, but have been hindered by a lack of appropriate, generally accepted tools that encapsulate HF/HSI data throughout the design life cycle. It is in this context that DHMs possess a significant and increasingly important role within a holistic, simulation-based approach to product design, manufacture, and acquisition. Indeed, human modeling has moved beyond physical form models to include behavioral, cognitive, and performance models that can be used to explore the constraints and capabilities of the entire system, including its operator.

52.3 DHM Integration in PLM Tools

As manufacturing organizations around the world continue to design, develop, manufacture, and *acquire* machines, vehicles, products, and systems that are capable of performing better, faster, and longer, an increasingly important design consideration is to ensure that these technological innovations are being designed from the perspective of the person who will actually gets be building and using them, be it a shop floor worker on an assembly line or a pilot of a multimillion-dollar fighter aircraft. DHMs, fully integrated into PLM environments, provide such organizations with a suite of human simulation and HF tools specifically geared toward understanding, and optimizing, the relationship between humans and the products they manufacture, install, operate, and maintain within the context of a highly dynamic, ever changing, digital design and definition process.

Organizations are increasingly viewing DHMs as one of the most versatile and scalable components of a PLM, or simpler design-centric, solution. This is, in part, driven by the relevance of DHMs to a wide range of applications. For example, the scalability of a DHM to a specific design activity, such as vehicle occupant accommodation, can be demonstrated by the fact that Formula 1 teams are using individual mannequins that accurately represent the anthropometry of individual Formula 1 drivers, whereas volume automobile manufacturers are using complex anthropometry accommodation techniques to simulate a global target audience mass-market for their vehicles. In the same light, defense and military organizations are using DHMs as a mechanism to ensure that equipment suppliers bidding for contracts are technically compliant with HF/HSI program requirements in an open, standardized, and quantifiable manner.

In addition to being scalable, DHMs are also highly versatile. From their origins as tools to address military operability scenarios (aircraft, armored fighting vehicles, ship bridge design, etc.), DHMs are now actively used in a multitude of product life cycle domains, including the aerospace, automotive, plant design, heavy engineering, shipbuilding and electrical goods industries, with the greatest number of users now focused on digital factory applications. To ensure competitiveness in the global marketplace, manufacturing organizations are becoming increasingly aware of the value in designing manufacturing facilities specifically geared to the skills, and limitations, of the local workforce. As a result, the digital factory is now a reality; in the automotive domain, work cell layouts, line balancing, process verification, body-in-white, and final assembly scenarios are commonly analyzed using commercial-off-the-shelf (COTS) digital manufacturing tools. However, the digital approach to production performance would not be viable without effective DHMs, as they allow manufacturing organizations to analyze scenarios to ensure that employees are capable of the required tasks in a safe and sustainable—ergonomically efficient—fashion.

52.4 Expanding User Base of DHMs

As DHMs became more sophisticated in terms of functionality, detailed analysis was necessary to determine specifically how a DHM can best serve the needs of design, manufacturing, and equipment-acquiring organizations. As a result of such analysis, end-user organizations and DHM software vendors alike have made, and continue to make, significant investments to understand not just *what* should be included in an effective DHM, but also *when*, *where*, and *how* DHMs can provide a tangible benefit to the way in which products are designed, manufactured, used, and acquired.

In recognition of the varying requirements of an organization, it has been imperative that DHM vendors design and package their portfolios to ensure that the appropriate amount of technology and information is presented to each user. In today's enterprise simulation systems, it is not rare to find such diverse occupations as simulation engineers, process planning specialists, maintenance engineers, manufacturing and assembly engineers, as well as human factors engineers and biomechanists actively using DHMs as part of their regular tasks. DHM vendors have, for the most part, responded by providing a flexible product packaging paradigm to meet the needs of the "new users" who possess a markedly different skill set than traditional users of DHMs.

52.5 Enterprise-Wide Human Engineering

The progression of DHMs from niche, expensive, difficult-to-use tools for a highly skilled, yet niche, user community, to a more widely adopted, integral, and intuitive component in the digital product design and definition process, has emerged out of necessity. This transformation occurred not just to ensure a sustainable business model for DHM vendors, but to ensure that DHMs reach their full potential within an organization, provide a clear advantage for end-user organizations, and provide a suitable vehicle to propagate an organization's HF and HSI know-how throughout the design life cycle.

One of the major benefits associated with the integration of DHMs into PLM solutions is that it has effectively empowered the key HF/HSI data owners to provide information to necessary stakeholders throughout the enterprise. For example, domain experts might author content such as DHM attribute data (anthropometry, comfort, posture, etc.), and then store that data on enterprise product data management (PDM) solutions for retrieval and use by all appropriate stakeholders who have the need to consume HF/HSI data in a *controlled and configured* manner.

Another benefit lies in the ability to use DHMs to integrate specific HF/HSI data as a design paradigm, or a set of rules, into the design process, thereby allowing such data to become pervasive in the design process. Today's flexible PLM systems permit an organization's knowledge to become parametrically linked throughout the design process. In such a scenario, the role of the HF/HSI engineer is to ensure that the rules assigned to the DHM rule base (which may include the specification of what mannequin libraries to use, what powertools must be used in the context of a specific vehicle type, what standards must be checked while using a DHM) are appropriate for any specific equipment type, program, or country. For example, Dassault Systemes' PLM portfolio includes the concept of knowledgeware that permits HF and HSI specialists to capture their know-how in an *expert* capacity by generating DHM content such as mannequin libraries, anthropometry, vision data, and functional envelopes, and share it in an *advisory* capacity within the PLM ecosystem.

HF/HSI experts can build up and share corporate HF/HSI know-how in rule bases, and leverage that knowledge across the enterprise to ensure design compliance with established HF/HSI standards. These rule bases can capture and automate knowledge processes such as best practices, application processes, and design validation and corrections. For example, best practices associated with the use of DMHs can be captured, and automated as a process wizard. Users define as many generic HF/HSI rules and checks specifications as needed for a project (i.e., DHM anthropometry, posture data, comfort data, etc.) between all design objects: from mechanical features such as forces and torques on components, to manufacturing or maintainability scenario rules (such as space envelopes to operate a tool efficiently).

These expert rules and checks are embedded in a unique design, and can then be used to automatically monitor the actions of any designer throughout the company. As the design is created or changed, the system uses the expert rules and checks to ensure compliance to corporate standards defined by the HF/HSI engineer while assisting the designer. An early example of this approach was produced by Beevis et al. and focused on large complex systems such as ships (Beevis et al., 2000).

Once DHM know-how is captured and stored, the rule can be made widely available across the enterprise, and also externally to supplier organizations and customer focus groups. From a non-expert perspective, DHM users have simple access to this high-level *validated* data, thereby ensuring, in a permanent way, the consistency of their design with corporate HF/HSI and DHM standards and best practices. This results in a reduction of risk and an increase in productivity as the user quickly converges toward an optimized design through this process.

By accelerating the exploration of design alternatives with regard to rules, knowledgeware-type approaches assist users in making better decisions and reaching optimal designs in a shorter amount of time. Also, and equally as important, they permit experts to convert implicit practices into explicit knowledge, thus automating design generation and reducing the risk and cost of repetitive tasks. Making previously implicit DHM practices explicit and leveraging them through knowledge design documents is an efficient way to transform specific know-how into corporate knowledge and disseminate it to those involved in the design process. Thus, by including the best practices of the company while working on a design and explaining the what, why, and how, a designer, or a design engineer, is able to expand the corporation's knowledge capital, while ensuring the consistency of design.

52.6 Systems Design

While considering the use of DHMs in the context of the design life cycle, it is possible to infer a distinction between the human being as simply a *resource* within product design, operability, manufacturing, training, and decommissioning activities, and the human being as a *component* of the system whose characteristics—physical, physiological, and psychological—need to be included in order to assess and optimize total system performance.

Consideration of human beings as simply system *resources* that possess a number of physical attributes and limitations that can be modeled and simulated has been the main approach to the use of DHMs to date. The tasks allocated to the *resources* can be as diverse as assembling a component on a moving assembly line, driving a vehicle, or performing an extra-vehicular space walk on the International Space Station, but these are not sufficient to deal with issues found in larger, more complex systems.

The type of issues found in those areas can best be modeled through a combination of physical, physiological, and cognitive characteristics integrated in a wider, more holistic, simulation environment.

Scientific research is advancing in all of these areas, and computer models of many phenomena are being produced as a consequence of improved personal computer capabilities. These models are beginning to find their way into commercial software that practitioners can begin to apply to design.

52.7 System Performance Modeling

In addition to the physical aspects of design, there is a parallel process of analysis and evaluation for cognitive fit. The concepts of human-centered design and human performance modeling and simulation (M&S) are critical factors to the success of complex system design. The emergence of powerful modeling frameworks promises a future in which modeling and simulation are a routine part of the design process for human-machine systems. The Canadian Department of National Defense research branch, Defense Research and Development Canada (DRDC), is developing integrated simulation solutions to support HSI with a focus on human factors and performance prediction. This is one example of several similar endeavors to advance the utility of human engineering tools for answering client questions.

52.8 Empirical Methods

Traditionally, integration of human capabilities into the design came after many of the hardware and software choices had been made. The human factors engineer (HFE) studied mission analyses and the workspace layout to assess what the human tasks were and how information flowed to or from the operator. These studies used storyboards, cardboard mockups, or part-task simulators to evaluate the design concepts; more complete mockups, could also be used to develop training schedules to teach new operators how to use the system. With this information in hand, the HFE could begin assessment of the mental demands and sources of failure or error with the system.

Advantages of this approach are that you can merge the physical and cognitive fit assessments in a limited sense, relying on the expertise of subject matter experts to extrapolate to in-service use. The insight afforded by simulation and modeling engineers (SMEs), both operational and HFE analysts, provides valuable guidance that can identify many critical design flaws.

This process is not without its disadvantages, however. Typically, these evaluations are labor-intensive so exposure to a broad operational SME population is rare, limiting the generalizability to the end-user population with all their individual differences. This is further complicated as selected SMEs are typically expert users while the end-users may span a range of expertise from novice to expert. If the SMEs cannot make the extrapolation to the user population from their own experience, the assessment will be biased toward the expert user. Finally, the time and cost of assessing designs in a variety of plausible scenarios are usually prohibitive, meaning that the HFE analyst must often extrapolate observed results into untested application areas, drawing on their own expertise to note differences and similarities that could moderate predicted performance.

52.9 Modeling Goal-Directed Behavior

Modeling the behavioral and cognitive activities of the operator provides the analyst with a means of mitigating some of the disadvantages of the empirical method. There is an adage that has been adapted from a number of sources that conveys a misleading message for analyses of systems:

> Everyone believes an experiment, except the experimenter; no one believes a model, except the modeler.

This belief stems from observers failing to recognize the difficulties of extrapolating performance derived from empirical studies into the field and that simulations with models are another form of experimentation but with fewer, or rather different, restrictions. Experimentation and analytical simulation should not be considered competitive or redundant; rather they should be considered complementary and supportive.

52.10 Basis for Modeling

Some of the processes used by HFE specialists as part of their traditional assessment can provide a suitable base upon which to build these models, such as hierarchical task analysis and information flow and cognitive processing studies. Simple models are often the best as they are easier to understand and develop, but the complex nature of the typical operator's role in today's work environment often requires more complex models of goal-directed behavior.

Analyzing the duties of personnel performing various functions has been a standard human engineering tool for some time, but the formal hierarchical task analysis (HTA) approach proposed by Annett and his colleagues set the standard (Annett & Duncan, 1967, 1971; Annett, 2003; Schraagen et al., 2000; Shepherd, 2000). Many of the task analysis tools in use today either elaborate on, or use a subset of, the HTA steps, such as cognitive task analysis (which focuses more on the cognitive components of the operator functions), although the cognitive work analysis (CWA) approaches provide a distinct, complementary perspective for analyzing problem spaces in design.

Advocates of CWA feel that HTA is too prescriptive when defining what operators do and that HTA does not adequately consider the context in which the operator performs the functions, leading to misinterpretations of what operators do in practice. Further, there may be several methods of achieving the same goal, but incorporating all solutions in a model is very demanding even if all such solutions can be identified. Ecological interface design (EID; Vicente, 1999), an approach to CWA, seeks to expand the analysis to include a broader perspective of the problem context and domain to better address these concerns. The EID approach seeks to produce design concepts that are more adaptable, supporting the operator, particularly in unforeseen situations, constraining the range of options by environmental and contextual factors.

Many of the CWA concerns are valid and parallel some artificial intelligence concerns about "just needing a larger rule base" to capture intelligent behavior. Other aspects seem to suggest that CWA and HTA approaches can coexist with complementary strengths. Nevertheless, the task analysis approach has been the dominant application within the Canadian Department of National Defense, and most CWA applications have been in research activities. The expectation is that these approaches will become increasingly prominent in acquisition and design to evaluate concepts and establish confidence in proposals before building costly prototypes.

Annett and colleagues took a control-systems approach to extend psychological methods for analyzing operator functions. HTA was originally developed for training, but subsequently HTA and its derivatives found widespread application in design and risk analysis. HTA takes a human-centered perspective, deconstructing an operator's activities into sets of operator goals, methods for achieving goals, and the sources of information used to recognize when the state of the world differs or is converging on the goals. High-level goals are decomposed into subgoals and submethods that are more accurately described until a cost-benefit criterion is reached. Each level of the decomposition represents or contains the methods used to control the state of the world and drive it toward the operator's desired goal state.

The HTA formalism is particularly useful for modeling complex behaviors in task network tools such as IPME (http://www.maad.com/index.pl/ipme). Commercial tools exist for conducting HTA, such as Task Architect (http://www.taskarchitect.com) and the HTA Tool (http://www.hfidtc.com) that provide a standard interface for collecting task data that can be customized to the users' requirements. Such tools typically collect only a subset of the information required for subsequent predictive modeling exercises; however, they nevertheless provide easy-to-use interfaces for rapid data collection. The IPME interface provides a more complete means of collecting the data required for building models, although it is more difficult to use in the field when talking with SMEs.

52.11 IPME

The Integrated Performance Modeling Environment (IPME) is a task-network-based tool for modeling human task performance. It is being developed by Alion Science and Technology's Micro Analysis and Design Division under direction from the UK QinetiQ Ltd. and DRDC Toronto. IPME was originally conceived as a stand-alone, integrated human performance modeling tool with component models that may be used to represent operator activities, the work environment, and personnel characteristics of the crew performing the tasks. IPME has the ability to communicate with other applications through high-level architecture (HLA) and TCP SOCKETS communications protocols, a feature that provides it scalability and flexibility as well as expanding its stand-alone capabilities to participate in distributed simulations.

IPME is built around the MicroSaint discrete event simulation engine, with added models of human workload and task scheduling to capture operator task execution in complex, multitasking environments. Task characteristics can interact with the workload model as well as external moderators (such as thermal physiology or fatigue models) to simulate the interaction among task execution, physiological or cognitive states given knowledge of these interactions from the relevant human sciences. IPME can readily adapt to changing scenarios through changes in the environment model and can generate team members with different capabilities and attributes.

IPME provides a means to model goal-directed behavioral characteristics of HTA using the hierarchical task network to model an operator's functional responsibilities and tasks. This approach creates a modular structure that can be elaborated as required, supporting model reuse. Traditionally, function and task analyses for various domains are developed separately to meet different objectives. This goal-directed, modular hierarchy framework is being developed so that these objectives can be met with a common model that evolves over the life cycle of concept development, design, and training.

Operator state and individual differences are modeling aspects that are seldom represented in HFE models used for acquisition, although they are acknowledged by clients as key components of overall system performance variability. Heat strain, fatigue, and other stressors all lead to changes in the operator state that can in turn affect performance; personality traits and experience can lead to different goals or methods of achieving the same goals. While stochastic approaches can introduce some of the variability observed in practice, they don't address the underlying causes, and while they make one simulation look good, they can fail to adequately capture the human variability in another application. Human modeling tools such as IPME provide the analyst with mechanisms for integrating validated models of operator state, traits, and biases that then moderate task execution to begin to reflect the individual differences observed in practice. Additionally, this allows the analyst to include propensity for errors as functions of the operator state to test the predicted performance sensitivity to expected field conditions.

Development of validated component models is receiving increased interest in the broader human modeling community, drawing on the wealth of knowledge generated by traditional studies within many of the human sciences R&D organizations and academia. Development of IPME has led to a product that provides considerable flexibility for integrating component models in a modular architecture that allows the user to include relevant aspects that contribute to observed performance variability. More diverse investigations will be possible as scientific knowledge of human behavior, cognition, and performance is incorporated into computational models, taking human models beyond scripted rule sets into formal system controllers, processing information and responding in a plausible fashion.

52.12 Toward a More Complete Human Model

In the process of designing complex systems, whether it be a command center or an air, land, or sea vehicle, there are many unknown quantities that should be resolved long before fielding or the beginning of production. Resolving all of these unknowns can be difficult, time-consuming, and expensive.

What is required is a valid set of tools that can be used to assess the effects of design parameters on mission performance and enable the designers to make informed design trade-off decisions. The knowledge used can be derived from previous experience, empirical studies, or analytical predictions of the performance implications. Indeed, in many cases, all three knowledge sources are desirable and they should be considered complementary rather than competitive, and they can be completed in an iterative manner. The objective is to develop sufficient confidence in the design of the product that manufacturing can proceed. This has led to design philosophies, such as Spiral Development (Boehm, 2000), that embrace uncertainty and plan for iterative development, while entailing more traditional development processes as components of each design or development iteration.

In this context, techniques that were traditionally stand-alone activities are increasingly becoming integrated processes. By the same token, DHMs that were used independently are now linked in a synergistic manner.

52.13 INSIGHT

INSIGHT, or the IPME/Safework Graphical Human Task environment, was developed to provide a holistic environment in which a user can view simulated humans operating in their proposed, or existing, workplace (Zhang, 2005). IPME captures the tasks that are allotted to humans and their relationships and timing, while Safework (http://www.safework.com) provides detailed anthropometry, postural information, and vision within the physical constraints imposed by the workspace. INSIGHT provides a standardized communications interface linking the behavioral IPME model with the Safework anthropometric model to act out the scenario for all to see. Figure 52.1 shows a conceptual layout INSIGHT.

IPME operator models can be built to react to external events from other simulations or operators, as well as generate external events to drive those agents. In this way, operator models can support analysis, either purely constructive or with Human In The Loop, as well as virtual team training, filling the role of missing crew members performing tasks and interacting with operators in a natural manner. What IPME lacks in these applications is an embodiment of the operator conducting the tasks as well as the ability to interact with the physical environment that Safework can provide.

INSIGHT couples IPME and Safework through a dynamic, bidirectional interface. It allows the analyst to animate operator actions and apply the power of multiple modeling environments while maintaining their independence. In this synchronized communication strategy, IPME sends relevant information to Safework if there is a task executing that involves physical behavior. The Safework mannequin performs those actions within the physical constraints of the workspace, while IPME checks feedback from Safework concerning the completion of the task.

Multiple tasks can be performed concurrently, according to the IPME task scheduler. Tasks can be interrupted, delayed, resumed, or shed, with appropriate responses from Safework. If Safework detects an inability to perform a task, say a failed reach or an obstructed view, this feedback to IPME can be used to change the course of action, with associated consequences (e.g., penalty in task performance).

There is a visualization advantage to IPME modelers with this link, because developing and debugging models typically involves observing the order of task execution, variable values and changes as functions of time, and so on. The output of performance models is typically a collection of numerical results. These data must then be reduced, tabulated, graphed, interpreted, and presented to the client in an intelligible manner. Linking the performance model with Safework makes a compelling demonstration, assisting in the explanation of results to clients that typically do not have a modeling background that puts the numbers into context.

But the association between the two models goes beyond simply visualizing IPME models. By making the link bidirectional, the anthropometric model can affect the behavioral model by indicating a reach failure, for example, that in turn results in contingency branching for successive tasks. For a visual task,

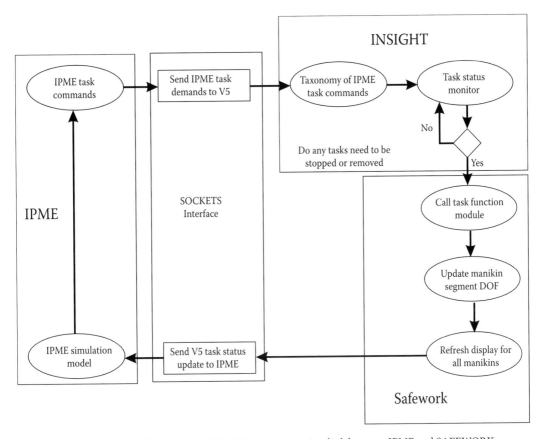

FIGURE 52.1 Conceptual layout of the INSIGHT communication link between IPME and SAFEWORK.

the position of the mannequin's eyes relative to its environment will dictate whether a target is visible or invisible due to an obstruction. Only with the proper combination of valid DHMs operating in accurately constructed 3D environments will designers of complex systems begin to reap the benefits of the wealth of HF knowledge currently available to build better products, and do so in a most cost-effective manner.

References

Annett, J. (2003). Hierarchical task analysis, In *Handbook of cognitive task design*, E. Hollnagel (Ed.), 17–35, Mahwah, NJ: Lawrence Erlbaum.

Annett, J., and Duncan, K. D. (1967). Task analysis and training design. *Occupational Psychology*, 41.

Annett, J., Duncan, K. D., Stammers, R. B., and Gray, M. J. (1971). *Task Analysis*, London: Her Majesty's Stationery Office.

Beevis, D., Davidson, P., Webb, R., and Coutu, E. (2000). Software support for sharing and tracking human factors issues during ship design. In *Human factors in ship design and operation,* Downsview, ON: Defence and Civil Institute of Environmental Medicine.

Boehm, B. (2000). Spiral development: Experience, principles, and refinements (CMU/SEI-2000-SR-008), Pittsburgh, PA: Carnegie Mellon Software Engineering Institute.

Schraagen, J. M. C., Ruisseau, J. I., Graff, M., Annett, J., Strub, M. H., Sheppard, C., Chipman, S. E., Shalin, V. L., and Shute, V. L. (2000). Cognitive task analysis (RTO Technical Report 24), North Atlantic Treaty Organization, Research and Technology Organization, AC/323(HFM)TP/16.

Shepherd, A. (2000). HTA as a framework for task analysis, In *Task analysis*, J. Annett and N. A. Stanton (Eds.), 7–24, London: Taylor & Francis.

Vicente, K. J. (1999). *Cognitive work analysis: Toward safe, productive, and healthy computer-based work*, Mahwah, NJ: Lawrence Erlbaum.

Zhang, L. (2005). Integrated Performance Modeling Environment (IPME)/Safework Graphical Human Task Environment (INSIGHT) (DRDC CR 2005-069). Safework (2000) Inc.

53

Digital Human Modeling and Scanner-Based Anthropometry

Jun-Ming Lu and
Mao-Jiun Wang

53.1 Introduction

Digital human models exhibit a variety of forms. Some use two-dimensional figure templates, while others are represented by three-dimensional geometry. No matter what form is taken, the basic requirement for anthropometric data holds. It defines body sizes of the digital human model, which reflects its similarity to the real human. Traditional anthropometric methods take lots of time and effort for data collection. Nowadays, 3D scanning technologies open opportunities for collecting anthropometric data more rapidly in a more convenient manner. In order to ensure the validity and reliability of data collection, the importance of anthropometry standards should be highlighted as well.

53.2 Anthropometric Methods

Anthropometric methods can be classified into direct methods and indirect methods (Roebuck, 1995). Direct methods use instruments such as calipers and measuring tapes to make direct contact with the human body. Sometimes a grid board and blocks are used to assist with positioning. It is tedious work and may involve human errors. In addition, the available anthropometric data are limited to a number of one-dimensional dimensions. Therefore, indirect methods were developed to solve these problems. In the 1990s, three-dimensional scanning technology made it possible to capture the whole-body data at once and has become the mainstream of indirect methods. Human interventions are less, and two-dimensional and three-dimensional data are available for building more realistic digital human models.

53.3 3D Scanning Technology for Anthropometry

The Loughborough Anthropometric Shadow Scanner (LASS) was the first application of 3D scanning technology for providing information on 3D size and shape of the human body (Jones et al., 1989). However, it was not comparable with traditional anthropometric methods in terms of reliability. From 1992 to 1994, HQL conducted a large-scale anthropometric survey of Japanese people with both traditional methods and whole-body scanners. At that time, scanner-based anthropometry was not reliable enough to be used as the main method for data collection. Therefore, traditional methods were taken for most measurements. From 1998 to 2002, a multinational anthropometric survey, called CAESAR (Civilian American and European Surface Anthropometry Resource), was undertaken in the United States, the Netherlands, and Italy (Daanen & van de Water, 1998). In addition to measurements, it was the first anthropometric survey to provide 3D human models. At present, the database it provides remains the most important one for the applications in digital human modeling. With the encouragement of these studies, more and more countries were trying to use 3D scanners to conduct national surveys. SizeUK, which began in 2001, was the first sizing survey ever to use body scanners as the principal means of capturing measurements (Treleaven, 2004). Following the model of SizeUK, scanner-based national surveys were then conducted in the United States and France. More recently, a growing number of countries endeavored to use 3D scanners to conduct national surveys, including China, South Korea, and Taiwan (Fan et al., 2004).

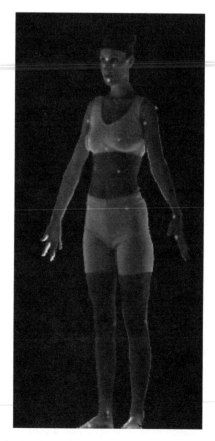

FIGURE 53.1 A female scan in the CAESAR project. (From Robinette, K. M., and H. Daanen, "Lessons Learned from CAESAR," *Proceedings of IEA*, 2003. With permission.)

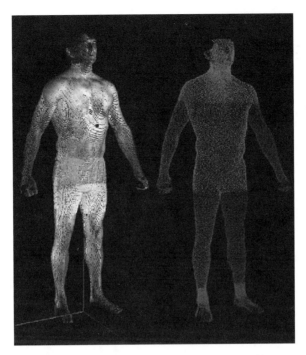

FIGURE 53.2 A male scan in SizeUK. (From Treleaven, P., *IEEE Spectrum*, 41(4), 28–31, 2004. With permission.)

53.4 Scanning and Preparation Works

Before analyzing the scanning data, it is critical to ensure the quality of the outputs. Since the 3D scanner is an optical instrument, it is very sensitive to light conditions and the geometric nature of the object being scanned. Therefore, strict settings of light and development of measurement attire for scanning will be very helpful (D'Apuzzo, 2005; Min et al., 2003). Also, due to the long duration of scanning, movement artifact is also an important issue to be addressed. Asking subjects to adopt standard postures and hold their breaths while scanning can help to minimize the effects of movements (Daanen et al., 1997). For international use, ISO (2005b) has presented the standards of scanning postures and measurement attires (fig. 53.3). In this way, the quality of scanning outputs can be ensured, and thus the accuracy of measurements can be improved. After scanning, the 3D body scanner generates more than 100,000 points in three dimensions from the surface of the human body. However, the scanning data provide very little information for applications. Therefore, landmarking is needed to categorize and summarize the scanning data.

53.5 Landmarking

Before collecting anthropometric data, key sites that define body segments and sizes, called landmarks, should be marked. Landmarks are mostly anatomical points. Therefore, palpation works are needed. Traditionally, with direct methods, this was done by human hands based on their tactile senses. Since there are differences between each person, some standards were developed to reduce human errors, such as ISO 8559 (1989) and ISO 7250 (1996). For the use of 3D scanning technology, human interventions are less obstructive, but the lack of standards brings about some problems.

Since the scanning data are very complicated, relationships of body parts in lengths have been widely used to narrow the range for searching landmarks (Pargas, 1998; Wang et al., 2003). For example, while locating the armpits, the range of search can be narrowed according to the proportion of the height

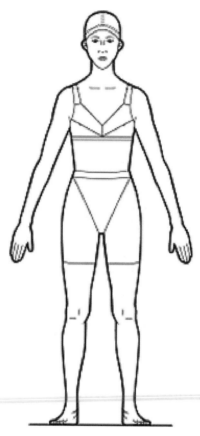

FIGURE 53.3 Standing posture and the measurement attire. (From International Organization for Standardization, *ISO 20685*, 2005. With permission.)

from the bustline to the floor to the stature. In this way, the success rate can be up to 100% (Pargas, 1998), and the time needed for searches can be greatly reduced. Traditionally, markers or stickers were placed on the surface of the human body to make their positions clear. Then they can be easily found on the scanning image with human eyes (Burnsides et al., 2001). With little human intervention, the landmarks can be located on the computer screen. However, in addition to the time needed for landmark identification, the human intervention may involve errors due to the inconsistency between different persons. Further, by using the color information that scanners provide, this work can be done automatically with computers (Wang et al., 2006). After putting on the markers manually, the staff can just wait till the computer completes data processing. If this method is combined with the use of the length relationships of body parts (fig. 53.4), the recognition rate can be rather high.

Nevertheless, despite the higher recognition rate, the pre-marking process requires additional effort and time. Moreover, the direct contact between the subjects and the staff may reduce the willingness of being scanned. For these reasons, developing marker-free methods has been an important issue to researchers. Recently, analyzing the geometry of the human body has become a major manner for landmarking without pre-marking. For example, the reconstruction of curves and surfaces can help to locate the landmarks consistently (Dekker et al., 1999; Buxton et al., 2000). Further, by analyzing the silhouette of the human body, properties such as gradient can be directly used for extraction (Douros et al., 1999). Once the key points are selected as control points, the curves or surfaces can be reconstructed to form the digital human model. Besides, the characteristics of the reconstructed curves or surfaces can be applied for data collection, which enhances its reliability and decreases its complexity. However, the reconstructed curves or surfaces may be a little different from the original data. Therefore,

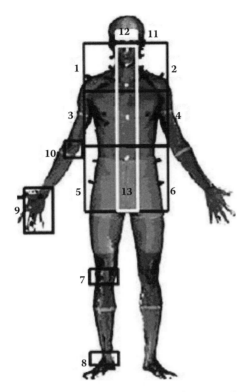

FIGURE 53.4 Automatic landmark identification by using color information. (From Wang, M.-J. J., et al., *International Journal of Advanced Manufacturing Technology*, 32, 109–15, 2006. With permission.)

if higher accuracy is required, this method would be better for digital human modeling, rather than for anthropometric data collection.

Further, based on a comprehensive database of 3D human models, the technique of template mapping makes it possible to extract landmarks very rapidly (Allen et al., 2003). Besides, this method ensures that the extracted landmarks will not be too far away from the actual positions, which yields a higher accuracy. The concept is to construct several standard human models with different sizes in advance. Thus, when a new scan is considered, the standard model most similar to the new scan in terms of body

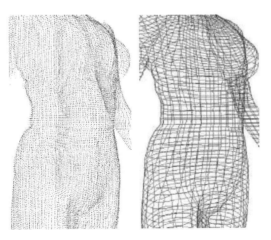

FIGURE 53.5 From original scanning data to reconstructed surfaces. (From Buxton, B., et al., "Reconstruction and Interpretation of 3D Whole Body Surface Images," *Proceedings of Scanning*, 2000. With permission.)

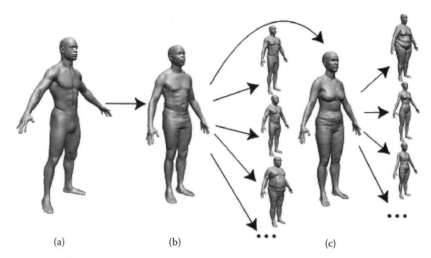

(a)   (b)   (c)

FIGURE 53.6 Standard human models for transformation. (From Allen, B., B. Curless, and Z. Popović, *ACM Transactions on Graphics*, 22(3), 587–94, 2003. With permission.)

sizes will be chosen. Finally, the selected standard model will be transformed to represent the new scan. As the model is transformed, the landmarks on the standard human model are also changed to new positions. Therefore, the time for landmarking decrease, and the results can be very consistent. However, this method requires a comprehensive database of 3D human models, or the mapped human model may not be similar to the real human. As a result, the positions of the landmarks may be incorrect.

In order to reserve the original geometry of the human body, Lu and Wang (2007) proposed a method for automated landmarking based on original scanning data. Pre-marking is not necessary, and the subjects only have to be scanned once in a standard posture. After body segmentation, the relationships of the lengths of body parts are applied for initial searches. Due to the diversity of body shapes, it is unlikely to use a single method to identify all the landmarks. Thus, there are four algorithms developed for different kinds of landmarks, including silhouette analysis (fig. 53.7), minimum circumference determination, gray-scale detection (fig. 53.8), and human-body contour plots (fig. 53.9).

53.6 Data Collection

Based on the landmarks, anthropometric data can be collected from 3D scanning data by using some approximation methods. One-dimensional data such as linear distances and circumferences can be obtained rapidly. In addition, two-dimensional and three-dimensional data such as surface areas and

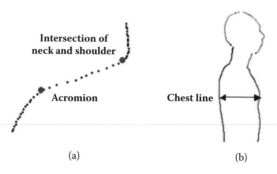

Intersection of
neck and shoulder

Acromion   Chest line

(a)   (b)

FIGURE 53.7 Silhouette analysis: (a) acromion and intersection of neck and shoulder, (b) chest line.

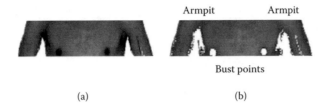

FIGURE 53.8 Illustration of gray-scale detection: (a) original scan image, (b) points with smaller gray-scale values are detected as the possible landmarks.

volumes are available, which provides more detailed description of the shape of human body. Figure 53.10 is an example of anthropometric data collection from 3D scanning data.

For the use of 3D scanners for anthropometric data collection, validity and reliability are important issues. Therefore, some discussion on this issue will be provided in the following sections.

53.7 Validity

Validity is here defined as the difference between scan-derived and manual measurements. The comparison of validity among related studies is illustrated in Figure 53.11. The vertical lines from left to right are the acceptable criteria from different standards, including ISO 8559 (1989), tailors' criteria (Bradtmiller & Gross, 1999), ISO 20685 (2005), ANSUR (Gordon et al., 1989), and grading for garments (Bradtmiller & Gross, 1999). All the horizontal lines are the differences between scan-derived and manual measurements. It is obvious that almost all the differences are larger than the acceptable threshold for traditional anthropometry (ANSUR). Therefore, improvements are needed.

53.8 Reliability

Reliability is here defined as the difference between repeated scan-derived measurements. The comparison is illustrated in Figure 53.12. The vertical lines are the same as those in Figure 53.11. Nevertheless, all the horizontal lines are the differences between repeated scan-derived measurements. As we can see that all the differences are smaller than the acceptable threshold for traditional anthropometry (ANSUR), it seems that the scan-derived measurements can be more consistent than the traditional anthropometry.

53.9 Sources of Errors

As previously mentioned, the scanner-derived measurements are not yet comparable with the traditional measurements. The sources of errors can be classified as hardware and software limitation. Hardware limitations, such as the number of cameras and the speed of scanning, will affect the quality of

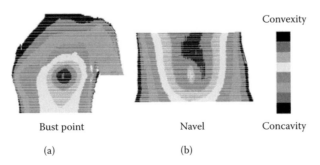

FIGURE 53.9 Human-body contour plots for locating: (a) bust point (b) navel.

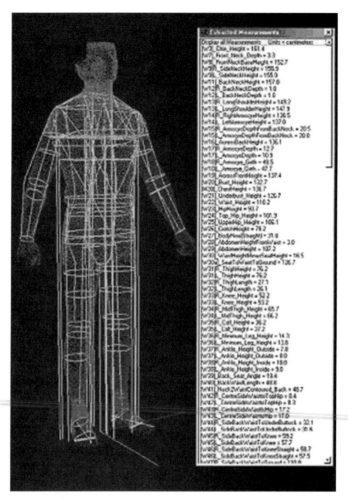

FIGURE 53.10   Anthropometric data collection from 3D scanning data. (From Treleaven, P., *IEEE Spectrum*, 41(4), 28–31, 2004. With permission.)

scans. The fewer the number of cameras, the more likely it is that missing areas may occur (Van Stralen, 2003). Besides, if the time for scanning can be reduced, the problems of breaths and movement will be solved. As for software limitations, poor definitions of landmarks and dimensions may greatly influence the results of data collection. Therefore, anthropometry standards should be developed and updated.

53.10 Anthropometry Standards

Existing anthropometry standards include ISO 15535, ISO 15536-1, and ISO 20685. ISO 15535 (2003) provides information for traditional anthropometry, while ISO 20685 (2005) provides anthropometry information using 3D scanning technology. Both standards only deal with collecting one-dimensional data on real humans. As for ISO 15536-1 (2005), the computer mannequins are considered. Therefore, it can be directly applied to digital human modeling.

For future developments, more detailed definitions for anthropometric properties of digital human modeling should be highlighted. Roebuck (2004) proposed "Anthropometric Dimension Delimitations" for digital human modeling, which can provide better understanding. In addition, standards for validation should be noted (Kouchi et al., 2004).

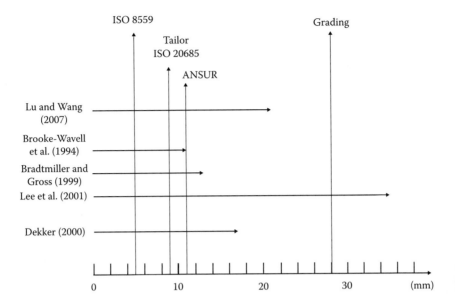

FIGURE 53.11 Comparison of validity. (Adapted from Dekker, L., "3D Human Body Modeling from Range Data," Ph.D. diss., 2000.)

53.11 Concluding Remarks

With 3D scanning technology, it is now more convenient to build anthropometric properties in digital human models. However, there are hardware and software limitations. Therefore, improvements of data processing, as well as the development of anthropometry standards, are needed for better use.

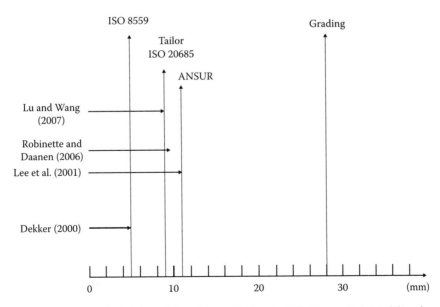

FIGURE 53.12 Comparison of reliability. (Adapted from Dekker, L., "3D Human Body Modeling from Range Data," Ph.D. diss., 2000.)

References

Allen, B., Curless, B., and Popović, Z. (2003). The space of human body shapes: Reconstruction and parameterization from range scans. *ACM Transactions on Graphics*, 22(3), 587–594.

Bradtmiller, B., and Gross, M.E. (1999). 3D whole body scans: Measurement extraction software validation. *Proceedings of International Conference on Digital Human Modeling* (paper number: 199-01-1892), The Hague.

Brooke-Wavell, K., Jones, P.R.M., and West, G.M. (1994) Reliability and repeatability of 3-D body scanner (LASS) measurements compared to anthropometry. *Annals of Human Biology*, 21(6), 571–577.

Burnsides, D., Boehmer, M., and Robinette, K. (2001). 3-D landmark detection and identification in the CAESAR project. *Proceedings of the Third International Conference on 3-D Digital Imaging and Modeling* (pp. 393–398), Quebec City.

Buxton B., Dekker L., Douros I., and Vassilev T. (2000). Reconstruction and interpretation of 3D whole body surface images, *Proceedings of Scanning 2000*, Paris.

Campagne Nationale de Mensuration 2003–2004, Available from http://www.ifth.org/mensuration.

Daanen, H.A.M., Brunsman, M.A., and Robinette, K.M. (1997). Reducing movement artifacts in whole body scanning. *Proceedings of International Conference on Recent Advances in 3-D Digital Imaging and Modeling* (pp. 262–265), Ottawa.

Daanen, H.A.M., and van de Water, J. (1998). Whole body scanners. *Displays*, 19(3), 111–120.

D'Apuzzo, N. (2005). Digitization of the human body in the present-day economy. *Proceedings of Videometrics VIII* (pp. 252–259), San Jose.

Dekker, L. (2000). 3D Human body modeling from range data. Ph.D. dissertation, University College London.

Dekker, L., Douros, I., Buxton, B.F., and Treleaven, P. (1999). Building symbolic information for 3D human body modeling from range data. *Proceedings of the Third International Conference on 3-D Digital Imaging and Modeling* (pp. 388–397).

Douros, I., Dekker, L., and Buxton, B. (1999). Reconstruction of the surface of the human body from 3D scanner data using B-splines. *SPIE Proceedings* (vol. 3640, pp. 234–245), San Jose.

Fan, J., Yu, W., & Hunter, L. (2004). *Clothing Appearance and Fit: Science and Technology*, Woodhead Publishing Ltd.

Gordon, C.C., Bradtmiller, B., Clauser, C.E., Churchill, T., McConville, J.T., Tebbetts, I., and Walker, R.A. (1989). 1987–1988 Anthropometric Survey of U.S. Army Personnel: Methods and Summary Statistics. Technical Report NATICK/TR-89-044, U.S. Army Natick Research, Development and Engineering Center, Natick, Massachusetts.

International Organization for Standardization. (1989). ISO 8559: Garment construction and anthropometric surveys—Body dimensions.

International Organization for Standardization. (2003). ISO 15535: General requirements for establishing anthropometric databases.

International Organization for Standardization. (2005a). ISO 15536-1: Ergonomics—Computer manikins and body templates – Part 1: General requirements.

International Organization for Standardization. (2005b). ISO: 20685: 3D Scanning methodologies for internationally compatible anthropometric databases.

Jones, P.R.M., West, G.M., Harris, D.H., and Read, J.B. (1989). The Loughborough Anthropometric Shadow Scanner (LASS). *Endeavor*, 13 (new series), 162–168.

Kouchi, M., Mochimaru M., and Higuchi, M. (2004). A validation method for digital human anthropometry: Towards the standardization of validation and verification. *Proceedings of Digital Human Modelling Symposium* 2004.

Lee, H.Y., Huang, K.W., and Jou, G.T. (2001). The verification of measurement extracted from 3D (three-dimensional) scanned against manual data. *Proceedings of the 6th Asian Textile Conference*, Hong Kong.

Lu, J.M., and Wang, M.J.J. (2007). Automatic landmarking from 3D human body scanning data. *IEA 2006 Conference Proceedings*, Maastricht.

Min, I.S., Nam, Y.J., Choi, K.M., Yun M.H., and Jung E.S. (2003). Development of measurement attire for three-dimensional anthropometric measurement survey. *IEA 2003 Conference Proceedings*, Seoul.

Pargas, R.P. (1998). Automatic measurement extraction from 3D scan data. DLA-ARN T2P5 Project Report for US Defense Logistics Agency.

Pargas, R.P., Staples, N.J., and Davis, J.S. (1997). Automatic measurement extraction for apparel from a three-dimensional body scan. *Optics and Lasers in Engineering*, 28, 157–172.

Research Institute of Human Engineering for Quality Life (HQL), Available from http://www.hql.or.jp.

Robinette, K.M., and Daanen, H. (2003). Lessons learned from CAESAR: A 3-D anthropometric survey. *Proceedings of IEA* 2003, Seoul.

Robinette, K.M., and Daanen, H.A.M. (2006). Precision of the CAESAR scan-extracted measurements. *Applied Ergonomics*, 37, 259–265.

Roebuck Jr., J.A. (1995). *Anthropometric Methods: Designing to Fit the Human Body*. Human Factors and Ergonomics Society, Santa Monica.

Roebuck Jr., J.A. (2004). Developing a Dictionary of dimension delimitation for digital human modeling. *Proceedings of Digital Human Modelling Symposium* 2004.

Treleaven, P. (2004). Sizing us up. *IEEE Spectrum*, 41, 28–31.

Van Stralen, M. (2003). Towards a human specific hole filling tool box for 3D body scans. Master's thesis, Institute of Information and Computing Sciences, Utrecht University.

Wang, C.C.L., Chang, T.K.K., and Yuen, M.M.F. (2003). From laser-scanning data to feature human model: A system based on fuzzy logic concept. *Computer-Aided Design*, 35, 241–253.

Wang, M.-J.J., Wu, W.-Y., Lin, K.-C., Yang, S.-N., and Lu, J.-M. (2006). Automated anthropometric data collection from three-dimensional digital human models. *International Journal of Advanced Manufacturing Technology*, 32(1-2), 109–115.

54

Digital Human Modeling Packages

Zhizhong Li

54.1 Introduction

Time-to-market and cost of a product have become critical for success in the global competitive market. During the last two decades, there has been significant technological progress to speed up the process and reduce the cost of product development, leading to a digital product development era. Various computer-aided tools have been practiced covering the whole product life cycle. Many computer tools are not simply computerization of manual/paper work to improve efficiency and quality, but greatly extend engineering computation and analysis capability, or even open up a field of engineering analysis. Digital human modeling (DHM) extends engineering analysis from functional mechanism analysis to human-centered analysis for compatibility, efficiency, comfort, and safety considerations.

After its emergence in the 1960s, much effort has been expended on DHM, resulting in many DHM packages with powerful functions. There are already several commercial DHM packages that have been approved for industry. Most DHM packages only represent physical characteristics of the complex human being by digital figures and embedded biomechanical disciplines, while some DHM packages also digitalize more complex cognitive aspects of the human being to some extent. A few DHM packages provide the possibility for end users to extend the functionality to their application or research purposes by computer language programming.

Although DHM is still far less popular than many computer-aided tools such as computer-aided design (CAD) and computer-aided manufacturing (CAM), more and more companies have realized its importance. Traditionally, the success of a product highly depends on its functionality; in today's consumer market, success also highly depends on its human factors considerations—how easy it is to learn and use, how comfortable is it to use for a long duration, how well does the product match the user's size, what are its safety and health problems, and so on. Subjective judgment is untrusting. Physical prototype testing is costly and time-consuming. An accurate analysis tool, DHM, is thus needed to evaluate the design at the early design phase.

To those expensive products with complex structure or confined space, such as aircrafts, maintainability needs to be carefully checked. Can a maintenance worker access the particular component? Can the worker replace it? How difficult will it be to do so? What can the worker see at this position? To answer these questions accurately, DHM offers virtually the only technological assistance before a physical prototype is available.

Many design decisions have a significant influence on safety to manually operated/driven products such as control rooms, cars, aircrafts, and so on. All design alternatives should be evaluated with DHM packages since safety performance depends on how a design fits the physical, psychological, or even behavioral characteristics of the target population.

Occupational safety and health (OSH) is now really a social issue for a company to earn respect and thus for its products to be welcomed. Actually, many good companies look at their OSH commitment and performance as one of their key competition advantages. DHM packages are very helpful for these companies to design/evaluate workplaces and jobs especially during the manufacturing planning stage. Very commonly, these companies also benefit from the analysis that they can use DHM models to train their workers.

Many companies have found DHM packages to be very helpful in training, communication, and product demonstration. With computer mannequin figures or animations, the instruction, communication, and demonstration become very effective. Unlike the traditional videos recorded from an actor operating a real product (or prototype), DHM packages can conveniently provide multiple views (e.g., worksite overall view and hand detail view of an operation).

It has been proved that DHM technology can save time and cost significantly. It can be expected that in the near future, DHM will become a standard tool for product engineers and safety engineers. Moreover, DHM has been regarded as an essential method for safety research. In many situations, like car accidents and fighter plane emergency ejection, we cannot ask human subjects to undergo obvious dangers, and physical experiments are always too expensive. DHM provides a low-cost, flexible, and adequately accurate solution for the research related to these situations.

In this chapter, DHM applications are introduced with examples. Then the selection and evaluation of DHM packages are discussed, aiming at providing a practical guide for those companies interested in adopting DHM technology. Then, some DHM packages are briefly introduced. Finally, limitation and current trends of DHM are discussed.

54.2 DHM Applications

DHM technology can be applied in many areas. This section will illustrate the industrial applications in design aiding and validation, job analysis, biomechanical analysis, maintainability analysis, and training/demonstration, but exclude DHM applications in research. In the following introduction of industrial applications, some specific DHM packages are cited. Note that the author does not imply that these specific DHM packages are the best for these applications.

54.2.1 Design Aid and Validation

As Duffy (2005) pointed out, DHM bridges computer-aided engineering design, human factors engineering, and applied ergonomics. It provides an easy way to include human considerations in engineering decisions. With DHM technology, design and manufacturing problems can be uncovered at the earlier design phase and the development cycle can be cut down dramatically.

Before DHM technology became feasible, physical human models were adopted in engineering design for a very long time. In the garment industry, 3D physical mannequins are still used. In many other industries, 2D articulated physical sheet mannequins were widely used for engineering design in earlier days. As a typical example, 2D articulated mannequins were put on an engineering drawing of a car cabin to guide component arrangement or to reach validation. There are still some standards providing technical details (e.g., dimensions and joint ranges of motion) of 2D articulated mannequins, such as GB/T 14779-93 "Functional design requirements of mannequins in sitting posture" and GB/T 14759-1995 "Mannequins design and application requirements." Later on, computerized 2D mannequins were used in computer-aided drawing software. These earlier digital human models mainly helped design engineers on anthropometric checking. With the development of the computer industry, 3D mannequins became available and more human characteristics were modeled into the digital mannequins—gradually, modern DHM technology was developed.

DHM has been widely applied in the design of vehicle interiors, aircraft cockpits, passenger compartments, and workplaces. The benefits are so clear that most if not all big car manufacturers are using DHM technology. Zygmont reported in 2000:

> Until recently, Bavarian Motor Works AG (BMW, Munich, Germany) ran through four physical mockups of car interiors, carefully evaluating each in its turn for ergonomic concerns. Now BMW builds only one final interior prototype for a last test of all its human-factor decisions regarding such niceties as seat firmness, the positions of controls for easy reach, the placement of gauges for quick reading, the view through windows, and even the sight line to mirrors.

The secret of the change at BMW was the application of a DHM method. An image at http://www.human-solutions.com/bilddatenbank/Ramsis/S2_LZW_klein.jpg (accessed on February 12, 2007) well illustrates the application of a DHM package, RAMSIS, in a car cab design.

Similar to the vehicle industry, the aircraft industry embraces DHM technology especially in cockpit design and in maintainability evaluation. Most aircraft manufacturers are using DHM packages, either their own or commercial. For example, Boeing has its own DHM package called BHMS (Boeing Human Modeling System) for airplane design. As a typical application of BHMS, reach envelope can be generated as pilot reach requirements for the design of a cockpit control layout (Boeing, 2007a).

DHM is also applied in the design of living facilities. More and more people care about whether their living facilities are friendly to them. To these individual-centered design problems, population data-based design is being replaced by personalized design. DHM makes it possible to be sure that the design favors the users' living activities. UGS reported a case of living space design for comfort and safety by Osaka Gas Co. using Jack software (UGS, 2007a).

For the design of many products, even 3D static human models can be very helpful. Liu et al. (2007) reported their toolkit under Unigraphics environment for the preliminary design of a helmet shell using a 3D head model generated from the 3D head scan of a representative user of an intended population group (figs. 54.1 and 54.2). This toolkit can not only shorten the helmet development cycle, but also significantly reduce the weight and enhance the stability of the helmet in a case study.

Various applications of DHM packages were introduced in a book edited by Chaffin (2001). Actually, wherever products or facilities will be used or operated by human beings, DHM can always provide design assistance.

FIGURE 54.1   A static 3D head model.

54.2.2 Job/Biomechanical Analysis

Industrial ergonomics is moving away from a reactive approach, in which jobs that cause injuries are modified, to a proactive approach that emphasizes assessing each job for feasibility and safety as the workplace and processes are designed (Reed, 2006). Under the current global competition situation, OSH becomes more and more influential to a company's success. DHM strongly supports the proactive

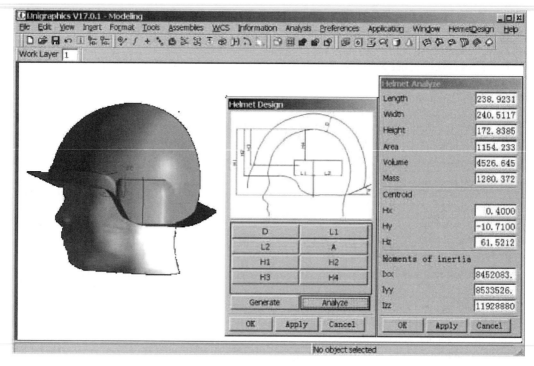

FIGURE 54.2   Helmet shell design based on 3D head model. (From Liu, H., Z. Z. Li, and L. Zheng, "Rapid Preliminary Helmet Shell Design Based on 3D Anthropometric Head Data," *Journal of Engineering Design*, accepted 2007. With permission.)

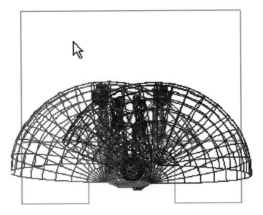

FIGURE 54.3 Work area check for a printed circuit board testing workstation (Wang, 2006).

approach. With DHM packages, a job can be evaluated from maximum lifting weight limit, maximum force to exert, how awkward the work posture can be, how difficult it is to reach a target, how much energy will be consumed, or even how long rest time should be arranged after work, and so on. Figures 54.3 and 54.4 show a work area and posture check for a PCB (printed circuit board) testing workstation with Mannequin Pro (Wang, 2006).

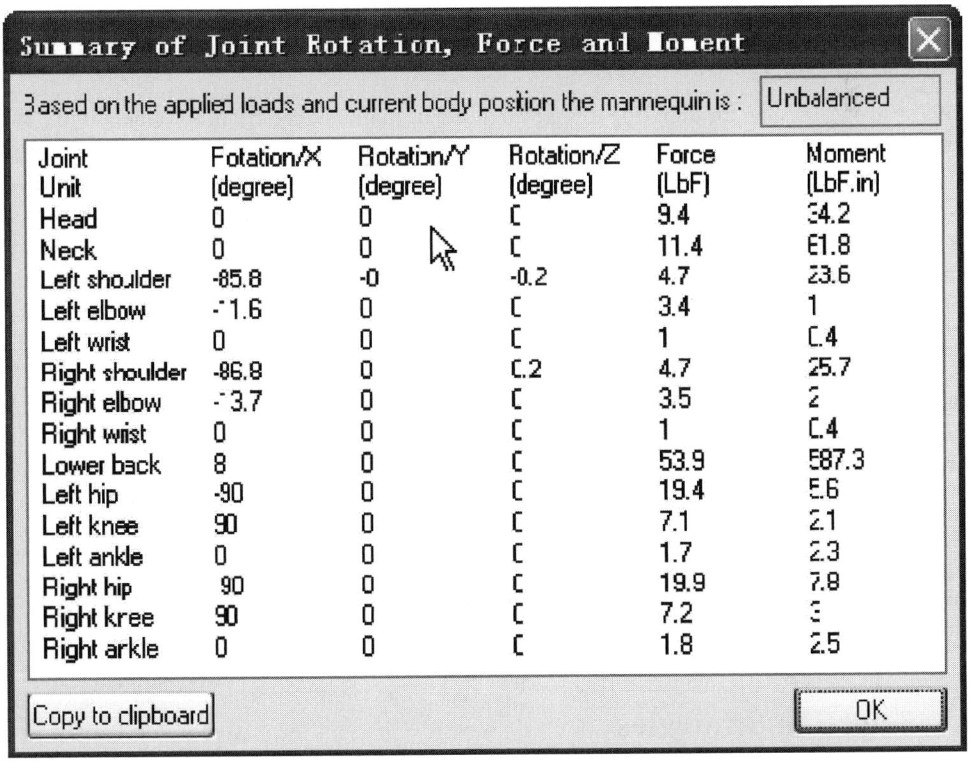

Summary of Joint Rotation, Force and Moment

Based on the applied loads and current body position the mannequin is : **Unbalanced**

Joint	Rotation/X	Rotation/Y	Rotation/Z	Force	Moment
Unit	(degree)	(degree)	(degree)	(LbF)	(LbF.in)
Head	0	0	0	9.4	34.2
Neck	0	0	0	11.4	61.8
Left shoulder	-85.8	-0	-0.2	4.7	23.6
Left elbow	-1.6	0	0	3.4	1
Left wrist	0	0	0	1	0.4
Right shoulder	-86.8	0	0.2	4.7	25.7
Right elbow	-3.7	0	0	3.5	2
Right wrist	0	0	0	1	0.4
Lower back	8	0	0	53.9	587.3
Left hip	-90	0	0	19.4	5.6
Left knee	90	0	0	7.1	2.1
Left ankle	0	0	0	1.7	2.3
Right hip	90	0	0	19.9	7.8
Right knee	90	0	0	7.2	3
Right ankle	0	0	0	1.8	2.5

Copy to clipboard   OK

FIGURE 54.4 Posture check for a printed circuit board testing workstation (Wang, 2006).

54.2.3 Maintainability

A big application area of DHM is maintainability analysis and maintenance training. As pointed out in the Introduction, to those expensive products with complex structure or confined space, DHM is virtually the only technology for maintainability evaluation before a physical prototype is available. Maintainability evaluation is a must in the area of engineering analysis to guarantee the success of these products. Because of the high cost of a physical prototype, today's companies try to replace the traditional physical mockups with digital mockups and adopt DHM technology to evaluate maintainability and usability of their complex products by visualizing and analyzing the actions required to assemble, maintain, and operate the products.

Abshire and Barron (1998) reported application of computer-based mockups (COMOK), including human models in Lockheed Martin Tactical Aircraft System's (LMTAS) F-16, F-22, and the Joint Strike Fighter (JSF) programs in the area of maintainability and human factors. With the imported CAD models, the DHM package, Jack (formerly from Transom Technologies, now UGS), performed Virtual Maintenance (VM). The following benefits were expected: reduced costs by phasing out expensive metal mockups, shortened schedules, standardized analyses, greater exchange and reuse of information, and integration of multiple databases.

According to UGS (2007b), maintenance is a particularly costly activity in the aerospace industry, accounting for 30% of the total life cycle cost of an aircraft; Jack has been used to evaluate designs from maintenance accessibility, part removal and replacement, and manual task timing by aerospace organizations including GE Aircraft Engines, Lockheed Martin, Short Brothers, and the U.S. Air Force. A piece of video at http://www.ugsitalia.it/products/efactory/jack/docs/f-22.avi (accessed on March 8, 2007) shows Jack replacing a component for an F22 fighter. In another example, physical accessibility to bolt with torque wrench was checked with BHMS software (Boeing, 2007b). It can be seen that DHM technology does contribute a lot to guarantee the maintainability of complex products.

54.2.4 Training/Demonstration

There are strong reasons for adopting digital methods in maintenance training (Dong et al., 2007): The traditional maintenance training approach with a physical model of the product is not satisfactory, while the quickly changing and intensely competitive market requires sufficient maintenance staff soon after or before a product is released to the market; the life of products has become shorter, and the knowledge of maintenance technicians must be updated more frequently; and it is unsafe and thus normally impossible for maintenance technicians to practice handling hazardous situations. Animated human figure models allow maintenance activity to be visualized (Ianni, 1995), like Figure 54.5 (Fu, 2005), thus DHM can be used to produce maintenance training animations even when a product is still under design, so that the maintenance staff can be ready to serve users.

Figure 54.6 shows a maintenance training system taking advantage of the animations generated by a DHM package, Jack (Fu, 2005). The customizable maintenance procedure is presented as a flow chart. By clicking on one box representing one maintenance step, the corresponding animation of this step will be played. It can be seen that the animation can show multiple views with different view scopes and directions. The systems can be upgraded to an Internet-based version so that the maintenance staff can be trained with the training materials updated at any moment.

Similarly, DHM can be used to generate realistic product or process demonstration without the need of physical models.

54.3 Selection Strategies

Success of DHM application depends on many issues, among which selection of a suitable DHM package is the most important. The selection work is not simply selecting one package from several options.

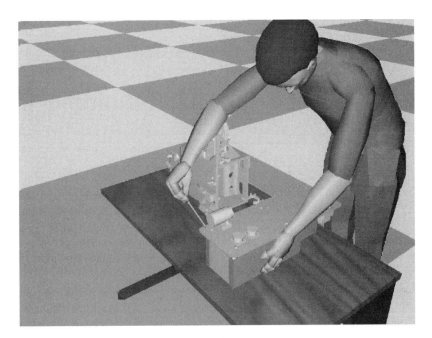

FIGURE 54.5   Maintenance action simulation (Fu, 2005).

Careful selection of DHM packages follows these strategies (see fig. 54.7): First, clearly define your application purpose, scope, and the interfaces with other related systems based on a fundamental understanding of DHM technology, then clarify available human and finance resources as the constraints of the selection decision. After that, collect information about the various DHM packages and evaluate them on the context of your application. You may start to contact DHM package providers. A checklist provided in the next section may be helpful for this step. If necessary, consult with experts, institutes, societies, associations, consultant services, or even some related government departments. Before you make a final decision, you may have to ask the DHM package providers to demonstrate their products

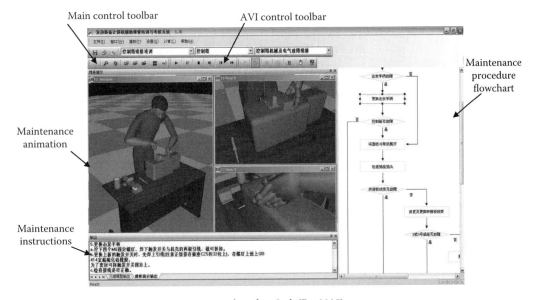

FIGURE 54.6   A maintenance training system based on Jack (Fu, 2005).

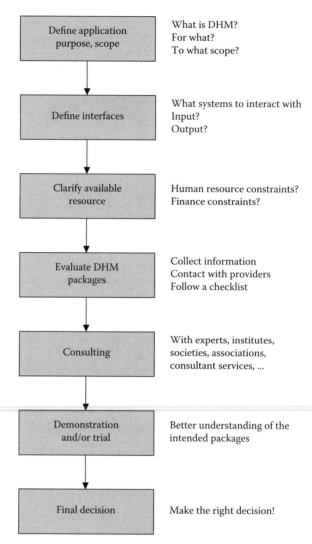

FIGURE 54.7 The recommended selection strategies.

for your evaluation. Normally, you can also ask for free trials of the packages for a limited period so that your researchers or engineers can get a deeper understanding. With these strategies, you may be able to make a right decision.

54.3.1 Application Purpose and Scope

When defining an application's purpose and scope, normally you are quite clear whether you want DHM packages for design aid, job analysis, maintainability analysis, training/demonstration, or other purposes. Besides that, you may also need to think about the following questions:

(1) What kind of human model is really needed for the application?

If you want to design a product to better fit the shape of a human body's surface, you need 3D static models of the interested human body surface, like the one used by Liu et al. (2007). Of course, you should carefully select the representative models for the intended population. However, if dynamic human activities are essential for consideration in product design, you need articulated human models. If it is a spatial problem, 3D models are required.

(2) What level of reality is required?

For most technical design problems, except shape and size, appearance (e.g., skin color and deformation, face expression, etc.) of the human model is not important; it just needs to look good. Wireframe models, rather than fully rendered models, are good enough in many cases. However, to examine the effect of clothes or the complex effect of clothes with an environment, high-level reality on appearance becomes very important.

(3) What functions should be provided?

The more complex functions a DHM package provides, the more expensive it is, and the more effort it requires to learn! You need to consider some future application extension, but think more about the success of the first step. A thin DHM package with easy to learn and use functions may be enough for your application. Unless you have qualified staff and abundant budget, try to avoid complex DHM packages.

(4) What is the scope for using the DHM package?

This question can be broken down into several sub-questions: What products does it need to cover? What job shops does it need to cover? What job types should it cover? … Answers to these questions may influence the answers to the above questions.

54.3.2 Integration with Other Systems

54.3.2.1 CAD Systems

DHM packages are often misunderstood as separate applications. That is true only in very simple cases. More often, DHM packages need to import geometrical models from geometry modeling systems such as CAD. Most DHM packages only provide very limited geometrical modeling support for constructing objects with regular shapes (e.g., box, cylinder, cone, etc.) in simple applications. After all, geometrical modeling is not the kernel function of a DHM package. Because most products are modeled in CAD systems, it is a fundamental requirement for the selected DHM package to integrate with the CAD systems currently used in the company. There are already several standard geometrical data format for data exchange between DHM and CAD systems, such as IGES, STEP, DXF, VRML, STL, and so on. In this chapter, design aid/validation of a DHM package by accepting models from geometrical modeling systems is called EXTERNAL design aid/validation. Note that some DHM packages have been fully integrated with CAD systems and can be launched under a CAD environment. This is called internal design aided/validation. It surely provides better support in design aid and validation.

If a DHM package cannot be launched directly in a CAD system, it can be helpful for the DHM package to export generated data, the user reach envelope as an example, for the CAD system as design references.

54.3.2.2 Motion Capturing Systems

Another important system for a DHM package to integrate with is motion capturing. Before human motion simulation in a DHM package, motion data should be prepared either by a motion capturing system or interactive setting. An interactive setting is inefficient and only suitable for evaluation of simple operations; human motion capturing can help prepare a great amount of motion data conveniently and accurately. Here, the performer for motion capturing should be strictly selected as a representative of the intended population. Integration between a DHM package and a motion capturing system smoothes data flow and enables instant ergonomics assessment of the operation tasks. Approaches to implement the integration can be categorized into two main groups: off-line (playback) methods, where data are exchanged by a medium file with a standard or specified format (such as C3D); and on-line (track) methods, where data are exchanged by direct communication between the two systems.

There are optical, magnetic, mechanical, sonic, biofeedback, electric field, inertial, video, and other motion capturing systems. Please refer to related chapters or other references.

54.3.2.3 Process Planning Systems

With increasingly public concern about OSH and the rise of the green manufacturing (GM) concept, manufacturers have recognized the importance of an OSH evaluation of manufacturing process plans. DHM provides a way to validate the ergonomic integrity of process plans and to reduce OSH risk during the planning stage. The validation requires a DHM package to receive data from and feedback evaluation results to the process planning system. Up to now, there are no standard data formats supporting this integration, although there are some standards of manufacturing process plans. The process plan information has to be manually interpreted and input into a DHM package, or a dedicated program should be coded in a DHM package to automate it for a special application.

54.3.2.4 Facility Layout Systems

Many layout systems are mainly oriented for mathematical optimization, focusing on minimization of transportation distance, time, and cost, even if they are visual systems with What You See Is What You Get (WYSIWYG) interfaces. However, the optimization of a facility layout should also be ergonomics-oriented, to improve efficiency, safety, and comfort of manual work. For those facility layout systems without DHM functionality, the ergonomics evaluation has to be finished in a DHM package. If the application purposes include facility layout evaluation, the selected DHM package should be able to accept layout results from the intended facility layout system.

54.3.2.5 Other Systems

The ergonomics evaluation results of product designs, process plans, or facility layouts may be submitted to a product data management (PDM) system or a product life cycle management (PLM) system for purposes like version management and information sharing within a product development team. The integration of a DHM package with such systems is commonly based on document sharing without strict format requirements. The only technical requirement is that the DHM package should be able to output its evaluation results into an easy-to-read file. If the DHM package should receive data from related systems through a PDM/PLM system, then a simple data interface between DHM and PDM/PLM systems should be implemented.

54.3.3 Current Resources

Available financial resources are an obvious constraint in making a decision in the selection of DHM packages, because the price of DHM packages varies significantly while normally quite high. Moreover, the supporting hardware and software (such as motion capturing systems) can also be expensive, or even more expensive than DHM packages.

If complex DHM packages are to be considered, another resource constraint is human resources. Without a qualified technical staff, complex DHM packages should not be considered; otherwise, the expected outcome may not be realized.

54.3.4 Package Evaluation

After you make clear your application purpose, scope, and constraints, you can start to collect technological and business information about DHM packages: Who are the providers? What are the functions? What are the prices? What are their market situations? What is the technical support? What are their success and failure stories? Much information can be found from the Internet. Contacting DHM package providers or sellers can normally get positive feedback. Then, carefully evaluate these packages on the context of your application with the aid of a checklist presented in the Appendix.

54.3.5 Consulting

It is often common to find it valuable to consult with related professional experts, research institutes, academic societies, industrial associations, or consultant services. If your application is OSH related, you can also get help from related government departments.

54.3.6 Package Demonstration/Trial Period

Very commonly DHM package providers/sellers would like to demonstrate their products with a focus on their support to your application. Take advantage of this as much as possible, because it is a good opportunity to get a better understanding of the packages. When you are able to narrow down your options, you may like to ask for free trials for a limited period to see whether the interested DHM packages do match your needs.

54.4 Evaluation of DHM Packages

The evaluation of DHM packages involves many aspects. By exploring these aspects, a template for the detailed evaluation and information collection is given in the appendix of this chapter. Here some additional remarks are presented for better understanding of some contents in the template:

1. The functionality of a DHM package highly depends on the structure of its human model.
2. The openness/extensibility of a complex DHM package can be very important. By providing script/language programming mechanisms and application programming interfaces (APIs), the user can fully explore potential benefits of the DHM packages. By programming the user can implement motion generation, constraints handling, interaction between objects, domain knowledge processing, and many other application-specific functions. Integration with other systems may also require programming in the DHM package.
3. A plentiful figure library can greatly improve the efficiency of scenario construction. Make sure to include the figure library when ordering a DHM package.

54.5 DHM Software

This section provides a short description of several DHM packages based on the author's limited knowledge and understanding. Commercial packages are the focus here, and so most research projects and military packages (please refer to Sundin & Ortengren, 2005, for examples) are excluded. Please note that the description does not constitute any endorsement by the author. Since software is updated often, all descriptions here should not be relied upon without further investigation.

54.5.1 AnyBody

(AnyBody Technology A/S, Aalborg, Denmark)

Provider: AnyBody Technology A/S

Web URL: http://www.anybodytech.com

AnyBody is unique among DHM packages as an open musculoskeletal modeling and simulation system. AnyBody was originated from the AnyBody research project (http://anybody.auc.dk/index.htm) at Aalborg University, Denmark, and is now owned by AnyBody Technology company. Bone geometries and muscular systems can be modeled with a script language, AnyScript, to establish a multi-body biomechanical system for muscle force calculation.

54.5.2 BHMS

Provider: The Boeing Company, Seattle, Washington

Web URL: http://www.boeing.com/assocproducts/hms

The Boeing Human Modeling System (BHMS) is a 3D DHM package for analysis of vision, distance, collision, population, reach envelopes, static volume envelope and swept volumes. It has been used in aircraft cockpit design and maintenance/assembly analysis, and even for extra-vehicular operations on space craft (zero gravity environment, suited human model).

54.5.3 Delmia Human

(DELMIA CORP., Auburn Hills, Michigan)

Provider: DELMIA CORP.

Web URL: http://www.delmia.com/solutions/html/addons.htm

http://www.delmia.com/gallery/pdf/DELMIA_V5Human.pdf

Description from the web: "DELMIA Human is a powerful human modeling tool used to create, validate, and simulate advanced user-defined digital human manikins, 'workers' in the DELMIA DPM environment for human interaction and worker process analysis early in manufacturing process."

54.5.4 DI-Guy

(Boston Dynamics, Cambridge, Massachusetts)

Provider: Boston Dynamics

Web URL: http://www.bostondynamics.com/content/sec.php?section=diguy

Among DHM packages, DI-Guy is also unique, for it was specially developed for real-time simulation with lifelike human characters for the U.S. Armed Forces. It provides photorealistic human models and an extensive library of behavior. Models and behavior for approximately 100 characters are initially included. New models can be added to the library following the OpenFlight standard. DI-Guy's expressive faces option can show a range of emotional expressions. API functions are provided with DI-Guy.

54.5.5 ErgoEASE

(EASE, Inc., Amherst, New Hampshire)

Provider: EASE, Inc.

Web URL: http://www.easeinc.com/solutions/ergoEASE.php

ErgoEASE is a comprehensive ergonomic tool for strength, fatigue and recovery, energy expenditure, RULA, repetitive motion, posture, and force analysis.

54.5.6 Jack

(UGS, Plano, Texas)

Provider: UGS

Web URL: http://www.ugs.com/products/tecnomatix/human_performance/jack

Originally, Jack was developed at the Center for Human Modeling and Simulation at the University of Pennsylvania, then known as EAI/Transom Jack, and is now available from UGS. Besides the full-featured stand-alone package, Jack is also integrated into UGS products NX (development product) and Teamcenter (PLM).

Jack's human models have 69 segments, 68 joints, a 17-segment spine, 16-segment hands, coupled shoulder/clavicle joints, and 135 degrees of freedom. Multiple population human models are available in Jack.

Jack has two ergonomics toolkits: Task Analysis Toolkit and Occupant Packaging Toolkit. Task Analysis Toolkit is helpful for optimal industrial task design, providing 10 functions including low back spinal force analysis, static strength prediction, NIOSH lifting analysis, predetermined time analysis, rapid upper limb assessment, metabolic energy expenditure, manual handling limit, fatigue/recovery time analysis, and working posture analysis; while Occupant Packaging Toolkit provides six human factors analysis tools that help you design vehicle interiors for optimal occupant comfort and performance, including SAE packaging guidelines, posture prediction, comfort assessment, advanced reach analysis, advanced anthropometry, and specialized part libraries. Besides the above toolkits, Jack has another toolkit for integration with motion capturing systems.

Jack can be programmed with JackScript, which is written in Python language. An external program coded in C/C++ language can be encapsulated to interact with Jack.

54.5.7 MADYMO

(TNO, JA Delft, The Netherlands)

Provider: TNO

Web URL: http://www.madymo.com

MADYMO is specialized for the automotive industry for occupant safety analysis and simulation with its generic multi-body and finite element capability. Key techniques include airbag simulation, bus rollover, rear impact safety, seating comfort, and side impact safety.

54.5.8 ManneQuinPro

(NexGen Ergonomics Inc., Pointe Claire/Montreal, Quebec, Canada)

Provider: NexGen Ergonomics Inc.

Web URL: http://www.nexgenergo.com/ergonomics/mqpro.html

ManneQuin is one of the most successful DHM packages in the world with thousands of users since its origin in 1990. It provides articulated human models covering 11 populations, adult and child, and three somatotypes. ManneQuinPro capabilities include analysis of field of view cones, reach envelopes and NIOSH lifting; automatic calculation of reaction force at feet; and biomechanical calculation of reaction joint force and torque due to applied load and body posture.

54.5.9 MIDAS

Provider: NASA

Web URL: http://humansystems.arc.nasa.gov/groups/midas/index.html

Man-machine Integration Design and Analysis System (MIDAS)

Description from the NASA web page:

> The Man-machine Integration Design and Analysis System (MIDAS) is a 3-D rapid prototyping human performance modeling and simulation environment that facilitates the design, visualization, and computational evaluation of complex man-machine system concepts in simulated operational environments.
>
> MIDAS links a virtual human, which consists of a physical anthropometric character, to a computational cognitive structure that represents human capabilities and limitations. The cognitive component is made up of a perceptual mechanism (visual and auditory), memory, a decision maker and a response selection architecture (Apex). The complex interplay among bottom-up and top-down processes enables the emergence of unforeseen and nonprogrammed behaviors.
>
> MIDAS outputs include dynamic visual representations of the simulation environment, timelines, task lists, cognitive loads along 6 resource channels, actual/perceived situation awareness, and human error vulnerability and human performance quality.

54.5.10 Poser

(e frontier, Inc., Scotts Valley, California)

Provider: e frontier, Inc.

Web URL: http://www.e-frontier.com/go/poser

Unlike most DHM packages for ergonomics applications, Poser provides realistic human models with facial expression, lighting effects, clothing, hair behavior (grown, styled, and blown), and so on.

54.5.11 RAMSIS

(HUMAN SOLUTIONS GmbH, Kaiserslautern, Germany)

Provider: HUMAN SOLUTIONS GmbH

Web URL: http://www.human-solutions.com/automotive_industry/ramsis_en.php

RAMSIS was designed in cooperation with the German automobile industry for the development of vehicles and cockpits. It was reported that more than 70% of the world's car manufacturers use RAMSIS for the ergonomic design of vehicle interiors.

RAMSIS human models have 53 joints and 104 degrees of joint freedom with databases for Germany, North and Central Europe, USA/Canada, Mexico, South America, Japan/Korea, China, and so on.

Besides some common ergonomics evaluation functions, RAMSIS provides other functions especially related to vehicles: simulation of the H-point, shoe models, children's data for different age groups from 9 months to 12 years, orthopaedic evaluation of spinal curvature, mirror visibility and mirror view, calculation and visualization of the belt run over the mannequin, and so on.

54.5.12 SAFEWORK Pro

(SAFEWORK Inc., Montreal, Quebec, Canada)

Provider: SAFEWORK Inc.

Web URL: http://www.safework.com

SAFEWORK Pro mannequins have 104 anthropometric variables, 99 segments, 148 degrees of freedom, and fully articulated spine, hand, shoulder, and hip models. Major modules of SAFEWORK Pro include vision analysis, postural analysis, comfort angle analysis, reach analysis, human activity analysis, task module, animation, collision detection, virtual reality, and clothing module. It can be used as a stand-alone application, but it has also been integrated into CATIA, DELMIA, and Pro/ENGINEER.

54.5.13 SAMMIE CAD

(SAMMIE CAD Limited, Leicestershire, England)

Provider: SAMMIE CAD Limited

Web URL: http://www.lboro.ac.uk/departments/cd/docs_dandt/research/ergonomics/sammie/home.htm

The SAMMIE system is a 3D DHM package providing the evaluation of fit, reach, vision, postural comfort, and mirrors (reflected glare). The human models have 23 body segments and 21 constrained joints.

54.5.14 Santos

Provider: The University of Iowa

Web URL: http://www.digital-humans.org/santos

Santos digital humans can be modeled with both physical and cognitive characteristics. Major issues to simulate include fatigue, endurance, cognition, situational awareness, behavior and prediction, crowd simulation, and so on.

54.5.15 THUMS

(Toyota Central R&D Labs., Inc., Aichi-gun, Aichi-ken, Japan)

Provider: Toyota Central R&D Labs, Inc.

Web URL: http://www.tytlabs.co.jp/english/tech/saf.html or http://www.dynamore.de/models/con_models_human.php?frame=ok

THUMS (Total Human Model for Safety) is a DHM package developed by Toyota Central R&D Labs, Toyota System Research, and Toyota Motor Company in conjunction with partners, mainly for vehicle safety study. Besides the skeletal structure, THUMS also models internal organs, muscles, skin, ligaments, and tendons of the human body, making it possible to predict the effects of various impacts on vehicle occupants and pedestrians.

Currently, a typical male occupant, a small female, a six-year-old child pedestrian, and a typical male pedestrian model are provided with THUMS.

54.5.16 Geometrical Modeling Systems

Common geometrical modeling systems can contribute to the modeling of human body surfaces either for direct design aid or for import into DHM packages. These systems include CAD and Reverse Engineering (RE) systems such as Unigraphics, CATIA, Pro/Engineer, SolidEdge, SolidWorks, AutoCAD, RapidForm, Imageware, Geomagic, and so on. CAD systems can be used to model complex geometrical entities. When a surface is to be modeled based on scanned point cloud, RE tools are often very helpful.

54.6 Limitations and Current Trends

Although great progress has been achieved and successful applications have been widely found, current DHM packages also have obvious limitations. Unrealistic appearance is an obvious issue for most DHM packages. Fortunately, this is not important for many industrial applications. On physical aspects, simulating a human as a multiple rigid body system is already very successful from the viewpoint of science, except that sometimes we may encounter improper reverse kinematical solutions. However, users find it is not easy to produce highly realistic motion animation. Moreover, even body surface deformation under pressure is seldom simulated by current DHM packages. The lack of realism is for computing performance consideration, and because of the conflict between high reality and complicated motion definition.

Vibration and other environment factors affect the human body, and human performance is seldom simulated. Except for the visual channel, other sensation channels are not well modeled. Human emotion, decision making, cognition, error, mental workload, communication, coordination, and teamwork have not adequately represented in current models. All these limitations can be attributed to our extremely limited understanding of internal working principles of the human.

Computing capability is increasing rapidly. In the near future, it may turn out to be no longer a problem, while better user interfaces for easy motion and task definition should be invented. With continuous effort on human performance modeling and simulation, more and more psychological and social models will be established for better prediction and simulation, although there is still a long way to go.

Acknowledgments

The chapter is partly supported by the National Natural Science Foundation of China (No. 70571045).

Appendix

Appendix. A template for the evaluation and information collection of DHM packages

Package name: _____ Provider: _____

Web: _____

Contact person: _____ Tel: _____ Fax: _____

E-mail: _____

1. Application domains:

☐ Internal design aid/validation ☐ External design aided/validation

☐ Job analysis ☐ Biomechanical analysis ☐ Maintainability analysis

☐ Training/demonstration ☐ Safety/injury assessment ☐ Entertainment

☐ Others: _____

2. Geometrical data integration: (write down version after name)

(1) Direct integration with geometrical modeling systems:

☐ Unigraphics _____ ☐ Pro/Engineer _____ ☐ CATIA _____ ☐ SolidEdge _____

☐ Solid Works _____ ☐ AutoCAD _____ ☐ Others: _____

(2) Input from an intermediate file:

☐ IGES _____ ☐ STEP _____ ☐ DXF _____ ☐ VRML _____ ☐ STL

☐ Others: _____

(3) Output to an intermediate file:

☐ IGES _____ ☐ STEP _____ ☐ DXF _____ ☐ VRML _____ ☐ STL

☐ Others: _____

3. Motion data integration: (write down version after name)

(1) Integration methods:

☐ Off-line (Playback) ☐ On-line (Track)

(2) Motion capturing systems:

☐ Vicon _____ ☐ MotionAnalysis _____ ☐ Optotrak _____

☐ PeakPerformance _____ ☐ Flock of Birds _____ ☐ Optotrak _____

☐ FullBodyTracker _____ ☐ Others: _____

(3) Motion animation systems: (write down version after name)

Input: ☐ FullBody/Tracker _____ ☐ Others: _____

Output: ☐ 3D Max _____ ☐ Others: _____

4. Simulation outputs:

(1) Motion related:

☐ Displacement ☐ Rotation/posture ☐ Speed ☐ Acceleration

☐ Force ☐ Moment ☐ Work ☐ Others: _____

(2) Task related:

☐ Vision ☐ Collision ☐ Reach ☐ Pass ☐ Maintenance/assembly

☐ Others: _____

(3) Performance related:

☐ Rotation/Posture ☐ Energy consumption ☐ Comfort ☐ Easiness

☐ Fatigue ☐ Safety ☐ Error ☐ Others: _____

(4) Behavior related:

☐ Perception (☐ Visual ☐ Auditory ☐ Haptical ☐ Olfactory ☐ Gustatory)

☐ Emotion ☐ Decision making ☐ Communication ☐ Learning

☐ Social Behavior ☐ Team work ☐ Others: _____

(5) Reality related:

☐ Surface deformation ☐ Face expression ☐ Breathing ☐ Perspiration

☐ Secondary motions (clothes, hair, ...) _____

☐ Others: _____

(6) Multimedia related:

☐ Static images ☐ Animations ☐ Others: _____

5. Human models:

(1) Model types:

☐ Physio-anatomical model ☐ Motion-mechanical model ☐ Psycho-cognitive model

(2) Anatomical systems:

Presentation:

☐ 2D

☐ 3D (☐ Stick ☐ wireframe ☐ Polyhedra ☐ Skeleton ☐ Enfleshed ☐ Dressed)

Skeletal systems: (DOF—Degrees of Freedom)

Whole body: Segments ＿＿＿＿＿ Joints ＿＿＿＿＿ DOFs ＿＿＿＿＿

Hand: Segments ＿＿＿＿＿ Joints ＿＿＿＿＿ DOFs ＿＿＿＿＿

Vertebral column: Segments ＿＿＿＿＿ Joints ＿＿＿＿＿ DOFs ＿＿＿＿＿

Other body components:

☐ Obstacle bypassing ☐ Tendon ☐ Ligament ☐ Skin ☐ Internal organs

☐ Others: ＿＿＿＿＿

(3) Psycho-cognitive models:

☐ Obstacle bypassing ☐ Collision avoidance

☐ Communication ☐ Team work ☐ Other Social behaviour

☐ Others: (list)

(4) Anthropometric setting: (ROM—Ranges of Motion)

☐ Dimensions ＿＿＿＿＿ (num) ☐ Joint ROMs ＿＿＿＿＿ (num)

☐ Inertial properties ＿＿＿＿＿ (num) ☐ Strength ＿＿＿＿＿ (num)

☐ Shape (☐ Interactive change ☐ 3D whole body scan import)

(5) Motion animation and control:

☐ Reverse kinematics ☐ Patterns and postures ＿＿＿＿＿ (num)

☐ Auto balancing ☐ Walk along path ☐ Collision detection

Hand motion: ☐ Auto grasp ☐ Attach to ☐ Constrained ☐ Others: ＿＿＿＿＿

Foot motion: ☐ Stand on ☐ Attach to ☐ Constrained ☐ Others: ＿＿＿＿＿

Leg and arm: ☐ Auto posture ☐ Others: ＿＿＿＿＿

Trunk motion: ☐ Auto move ☐ Auto posture ☐ Others: ＿＿＿＿＿

Eye motion: ☐ Object tracking ☐ Others: ＿＿＿＿＿

☐ Secondary motions: ＿＿＿＿＿

☐ Gravity effects ☐ nG effects ☐ Weightless effects

☐ Friction effects ☐ Load application

(6) Population databases:

☐ North America: Children Adult Aged Military ＿＿＿ ＿＿＿ ＿＿＿ ＿＿＿

☐ ＿＿＿: ＿＿＿ ＿＿＿ ＿＿＿ ＿＿＿ ＿＿＿ ＿＿＿ ＿＿＿ ＿＿＿

☐ ＿＿＿: ＿＿＿ ＿＿＿ ＿＿＿ ＿＿＿ ＿＿＿ ＿＿＿ ＿＿＿ ＿＿＿

☐ ＿＿＿: ＿＿＿ ＿＿＿ ＿＿＿ ＿＿＿ ＿＿＿ ＿＿＿ ＿＿＿ ＿＿＿

☐ South America: Children Adult Aged Military _____ _____ _____ _____

☐ _____: _____ _____ _____ _____ _____ _____

☐ _____: _____ _____ _____ _____ _____ _____

☐ _____: _____ _____ _____ _____ _____ _____

☐ Europe: Children Adult Aged Military _____ _____ _____ _____

☐ _____: _____ _____ _____ _____ _____ _____

☐ _____: _____ _____ _____ _____ _____ _____

☐ _____: _____ _____ _____ _____ _____ _____

☐ Asia: Children Adult Aged Military _____ _____ _____ _____

☐ _____: _____ _____ _____ _____ _____ _____

☐ _____: _____ _____ _____ _____ _____ _____

☐ _____: _____ _____ _____ _____ _____ _____

☐ Africa: Children Adult Aged Military _____ _____ _____ _____

☐ _____: _____ _____ _____ _____ _____ _____

☐ _____: _____ _____ _____ _____ _____ _____

☐ _____: _____ _____ _____ _____ _____ _____

☐ Oceania: Children Adult Aged Military _____ _____ _____ _____

☐ _____: _____ _____ _____ _____ _____ _____

☐ _____: _____ _____ _____ _____ _____ _____

☐ _____: _____ _____ _____ _____ _____ _____

(7) Animation methods:

☐ Interactive animation

☐ Motion data driven animation

☐ Programming animation

(8) Standard conformation:

☐ SAE (_____ _____ _____ _____)

☐ ISO: (☐ 15536 ☐ Others: _____)

☐ H-Anim

☐ Others: _____

6. Figure library:

☐ Facility library _____ (num) Note: _____

☐ Tool library _____ (num) Note: _____

7. Evaluation functions:

☐ Distance/reach/pass/accessibility analysis

☐ Manual handling limits   ☐ NIOSH lifting analysis   ☐ Low back spinal force analysis

☐ Static strength prediction   ☐ Metabolic energy expenditure

☐ Rapid Upper Limb Assessment (RULA)   ☐ Predetermined time analysis

☐ Fatigue/recovery time analysis   ☐ Working posture analysis

☐ Operation space analysis   ☐ Vision analysis   ☐ Injury analysis

☐ Population analysis   ☐ Personnel selection analysis

☐ Vibration analysis l   ☐ Noise analysis l   ☐ Illumination analysis

8. Openness/Extensibility/Programming support:

Programming language/Script: ☐ C/C++ ☐ Java ☐ VB ☐ Python ☐ Others: _____

☐ API library ☐ Project template/wizard

9. Complexity: ☐ C/C++   ☐ Medium ☐ C/C++

10. Usability (ease to learn/use): ☐ Easy ☐ Medium ☐ Difficult

11. Cost (purchase/supporting/use):

☐ Price/year _____ ☐ Price/license _____ ☐ Technical support _____ ☐ Training ___

☐ Hardware _____ ☐ Supporting software _____ ☐ Others: _____

12. Status and growth:

☐ User number _____ Growth by _____ Market share _____

Company profile: ☐ Very good ☐ Good ☐ Not bad ☐ Bad

Cooperation with respected companies: ☐ Very good ☐ Good ☐ Not bad ☐ Bad

Notes:

References

Abshire, K. J., and Barron, M. K. 1998. Virtual maintenance real-world applications within virtual environments. *IEEE Proceedings Annual Reliability and Maintainability Symposium*, 1998.

Boeing. 2007a. Case study 1—pilot reach accommodation. http://www.boeing.com/assocproducts/hms/case1.htm. (accessed on February 12, 2007)

Boeing. 2007b. Case study 2—maintainer accommodation. http://www.boeing.com/assocproducts/hms/case2.htm. (accessed on February 12, 2007)

Chaffin, D.B. (Ed.). 2001. *Digital human modeling for vehicle and workplace design*. Warrendale, PA: Society of Automotive Engineers.

Dong, W., Li, Z.Z., Yan, J. B., Wu, X. W., Yang, Y. H., and Zheng, L. 2007. Adaptive interaction in a 3D product structure browsing system for maintenance training. *Human Factors and Ergonomics in Manufacturing*, in press.

Duffy, V. 2005. Digital human modeling for applied ergonomics and human factors engineering. http://cyberg.wits.ac.za/cb2005/method2.htm. (accessed on February 12, 2007)

Fu K. 2005. A maintenance training system based on flow charts and simulation. Master's thesis, Tsinghua University.

Ianni J. D. 1995. *Maintenance simulation: Research and applications*, AL/HR-TP-1995-0019.

Liu, H., Li, Z.Z., and Zheng, L. 2007. Rapid preliminary helmet shell design based on 3D anthropometric head data. *Journal of Engineering Design*, in press.

Reed, M. P. 2006. Digital human modeling for ergonomics. http://mreed.umtri.umich.edu/mreed/research_dhm.html. (accessed on February 12, 2007)

Sundin, A., and Ortengren, R. 2006. Digital human modeling for CAE applications. In: *Handbook of human factors and ergonomics*, G. Salvendy (Ed.). New York: John Wiley & Sons, 1053–1078.

UGS. 2007a. Designing for comfort and safety: Tecnomatix helps Osaka Gas Co. develop friendly, personalized living spaces. http://www.ugs.com/about_us/success/docs/cs_osaka.pdf. (accessed on February 12, 2007)

UGS. 2007b. Jack at the movies. http://www.ugsitalia.it/products/efactory/jack/movies.shtml. (accessed on February 12, 2007)

Wang, Y.F. 2006. Ergonomics intervention for occupational safety in a PCB plant. Master's thesis, Tsinghua University.

Zygmont, J. 2000. Virtual humans save carmakers time and money. *Managing Automation*, no. 2. http://www.managingautomation.com/maonline/magazine/read/view/Virtual_Humans_Save_Carmakers_Time_and_Money_1403. (accessed on February 12, 2007)

55

Modeling and Augmenting Cognition: Supporting Optimal Individual Warfighter Human Performance

Dylan D. Schmorrow, Peter B. Muller, David A. Kobus, Karl Van Orden, Leah Reeves, and Kelly A. Rossi

55.1 Introduction

The human is a complex organism, rapidly adapting to many changes in the environment. How humans adapt is highly dependent upon their current mental and physical state. In addition, a plethora of technological advances have been developed over the last decade to enhance human performance and allow one to better interact with the environment. The interaction of human and technology while performing tasks has been termed the "human system" or the "soldier system" within the military context (Allender et al., 1995; Martinez-Lopez, 2006).

Many physical and psychological facets need to be considered to fully understand how the "system" interacts with the environment. Modeling the system as a whole, therefore, is not an easy task. A full and comprehensive analysis of the human system involves a number of aspects to consider that are internal or external to the individual performer. Internally, performer cognitive state, situational awareness, fatigue, experience, and physical readiness are some of the key human system components. External components such as the environment, biomechanics, human integration, and augmented cognitive technologies also need to be considered for a fully integrated human systems model. Thus, a full understanding of all aspects that affect the performance and well-being of an individual warfighter will allow for the development of flexible performance models, which could be used to assess the impact of a variety of technologies in any operational environment ranging from an infantryman in combat to an industrial worker performing a daily routine. Although this chapter specifically addresses the challenges faced by today's military regarding modeling the warfighter as a system, the issues may be considered universal regarding optimized human performance modeling--whether the human is in the military, is an industrial worker, or is an extreme athlete working with the various tools and equipment pertinent to the work environment.

The Department of Defense (DoD) has taken a significant interest in modeling human behavior of both individuals and teams during training and while they conduct operational tasks (Martinez-Lopez, 2006; Nicholson et al., 2005; Snow et al., 2006; Lawton et al., 2005). To become truly effective and useful, the many disparate models that currently exist need to be refined, validated, and integrated. To achieve such DoD goals, researchers must strive to create a model that is a valid representation of the real world from the perspective of the intended use of the model (U.S. Department of Defense, 2001). Achieving such a valid model is not a trivial task. For example, consider the real-world situation of today's warfighter—a true symbiosis of a technologically equipped system with the human element at its core. Collectively, the combination of these elements comprises the warfighter system and should, in theory, provide an integrated human-system solution for maximizing performance. However, in practice, this has not been the case. Individual elements have been added or subtracted in a systemized trial and error (and sometimes ad hoc) manner, with little regard for the overall integrated suite of tools for meeting mission objectives. This has resulted in what some defense engineers have termed the Christmas tree effect. New technologies serve as ornaments and are simply hung on the warfighter who serves as the tree (Miles, 2003). This is not the fault of the individual programs, which are doing their best to field individual optimal pieces of equipment to meet specific objectives. Rather, it is a larger problem because no one has the overall responsibility of looking at the entire warfighter system. The science of human factors and ergonomics needs to be applied for optimizing the warfighter's physical and cognitive performance while considering the interactive effects among brain, body, and combat gear. Yet to build a comprehensive model of an integrated human-system, new, expanded, or combined modeling approaches are needed. The intended use of such a model is to apply it to various warfighter tasks to optimize the warfighter system. Although there has been increased interest in simulation-based acquisition (SBA) in the last several years (Nicholson et al., 2005), the focus has been on major weapons systems and not the human as a system, as is the focus of this chapter.

55.2 Background

Ships, planes, and tanks are the big investment items for the military and, not surprisingly, they are robustly represented in DoD modeling and simulation programs. While these platforms have their own human modeling challenges, the key human-system interaction is between the human(s) and the operating platform(s). For the most part, this is a relatively static and well-defined interface with limited scope and usually a limited number of actions. The nature of modern warfare, with particular emphasis on smaller unit (e.g., 10 or less warfighters) operations in austere (and distributed/asymmetric) environments, necessitates a more robust understanding of the many facets of the human component and its relationship to overall system performance. Specifically, field commanders, planners, and systems developers need to know how warfighters will perform under the many conditions they might face in the field and how different types of equipment and ensembles would affect that performance. The optimal human-systems model needs to address a much larger scope—all human and technology capabilities, constraints, and the interactions of each with various operational environments. The need for such a model is apparent when one considers the combat performance requirement differences highlighted between the wars in Afghanistan and Iraq in which a single warfighter must be trained and ready to perform on a moment's notice. "Compared to how much money the United States spends on new fighter jets, submarines and other big-ticket weapons systems, it is severely 'under-investing' in technologies for unconventional warfare," says George Solhan of the Office of Naval Research (ONR) (Erwin, 2006). As infantry and irregular operations gain status from the operational community, funding opportunities will increase, and we will see a new emphasis and requirement for human-system modeling. Currently, there are 17 nations other than the United States with infantry modernization programs that have the potential to radically transform the infantryman system, yet few are currently applying integrated modeling and simulation (Mahon, 2006).

This chapter focuses on the importance of developing technologies and tools for applied human-system modeling. Modeling the human system not only takes into account the interaction of human and technology, but it also encompasses all other factors that affect human performance in a variety of environments. Many of these areas (e.g., physiology, biomechanics, kinesiology, augmented cognition, physics, and environmental science, etc.) have independently made some progress toward a better understanding of human performance. However, each is still being expanded to achieve a better understanding of human performance and, more specifically, understanding the interaction and integration of each of the model components. Currently, applied researchers are investigating various prototypes and technology developments that affect human performance. It is the integration of these investigations that will be used to build the models of the future.

55.3 Purpose

The purpose of this type of comprehensive, multifaceted modeling and simulation-based application of human system performance is to yield a more effective, maintainable, and survivable suite of combat capabilities for the individual warfighter. The model must encompass all aspects of cognition and decision making, physiology and ergonomics, and the technologies needed to integrate these components to support a fully optimized, capable human. As discussed herein, this approach augments conventional science and technology (S&T) approaches by including the impact of physiological demands and enhancements on cognition (e.g., via augmented cognition tools and techniques) (Schmorrow, 2005; Schmorrow & McBride, 2004; Schmorrow, Nicholson, Drexler, & Reeves, 2007; Schmorrow & Reeves, 2007; Schmorrow, Stanney, Wilson, & Young, 2006).

The goal of this type of state-of-the-science modeling is to optimize an individual in various environments using a range of solutions that are scalable across all potential operational environments. This research notion places dual emphasis on understanding human capabilities and limitations as they apply to the human performance arena by expanding the former and minimizing or mitigating the latter. The net result of this research area will aggressively expand the understanding of the human performance envelope.

The end-state would be a suite of tools, developed through merging the science of human factors with the engineering power of advanced anthropometric, behavioral, and cognitive modeling and simulation, that enables rapid and simultaneous prototyping and testing of multiple technology solutions. New models, methods, and technologies that support optimal individual mental, physical, and physiological readiness will improve the effectiveness of individuals in numerous situations and activities. The research and development areas for optimizing the human systems model may be classified into three separate components: *internal, external,* or *integrative.* Each component provides a lens through which independent capabilities can be explored, assessed, and improved. It is essential to note, however, that while each can stand alone as a conceptual topic, they are not truly independent and, in fact, their products need to be integrated or combined to generate solutions that span the broad requirements of human performance.

55.3 Infantry Modeling Challenges

55.3.1 Overview

The simplest fighting unit, the infantryman, is one of the most complex entities to model. New, integrated, and possibly iterative methods must be applied to optimize the combined warfighter system for a specific operational environment. The research areas identified in the following sections highlight the foundational science, strategies, and methodologies needed to drive the components, capabilities, and integration for optimizing human (warfighter) performance. As future warfighter-as-a-system modeling approaches should strive to do, the following sections address the need for technical and scientific collaboration across all three domains (internal, external, and integrative components) to ensure that all aspects of human performance have been fully investigated and understood.

55.3.2 Internal Components

Modern warfare presents the DoD with new challenges for individual and small unit (group/team) readiness. Smaller unit configurations place a greater burden on each individual to have near-optimal physical readiness and tactical proficiency. The emerging need for strategies to improve individual readiness in the areas of physical training, nutrition, fatigue management, and related factors is clear and has significant implications for the performance of warfighters in the future.

The physiological (cognitive and physical) makeup of an individual has a profound impact on performance. Sleep cycles, nutritional intake and adequacy, physical fitness, blood profiles, and genetics all affect one's physiology and thus overall performance. Likewise, education, genetics, intelligence, personality, and other dispositional factors also affect various aspects of cognitive abilities. While the impact of many of these components remains insufficiently modeled, their *combined* effects on individual health and performance are also virtually unknown. Furthermore, what little is known about the mitigating effects of training and self-management on physical and physiological viability has not been applied or demonstrated in various, complex operational environments.

There is tremendous evidence indicating that the physiological makeup of each individual, and the changes the body experiences on a daily basis, have a profound impact upon the individual's performance (Nieman et al., 1987; Zinker et al., 1990; Hackney et al., 1992; Friedlander et al., 2005). Nutritional intake varies daily, but long-term intake will have the most profound impact based on balance, macronutrient composition of one's diet, micronutrient deficiencies, and so forth. Physical activity can affect an individual's ability to adapt to an ever-changing environment, and daily, as well as long-term exercise and movement, will have a major impact. Fatigue related to sleep loss has been extensively studied and modeled (see Anderson & Horne, 2006; Chaiken, 2005; Harrison & Horne, 2000; Wertz, Ronda, Czeisler, & Wright, 2006), but its relationship to other physiological and psychological factors requires further investigation. Previous research has also demonstrated that stress affects human performance on complex tasks (Morgan et al., 2001, 2004). Recent findings indicate that individual stress reactivity may be predictive of cognitive performance but also related to resiliency against adverse combat operational stress reactions, including post-traumatic stress disorder (PTSD) (Carr et al., 2006). Modeling complex interactions between these and other factors influencing performance will be necessary to develop effective intervention and performance enhancement strategies.

It is evident there are a number of key physiological and psychological factors that affect human performance. These include physical fitness, nutrition, fatigue, cognitive abilities, and others. Some factors, like fatigue, have been well modeled, while others, such as stress reactivity, have not. The effect of some factors on performance lends themselves to modeling at the individual level (e.g., intelligence and cognitive abilities), while factors like fatigue and physical fitness can be modeled at the individual or small unit (team) level. It is important to note that these factors are not orthogonal and independent. Physical fitness, for example, can affect proneness to injuries, immune system health, and other factors.

A number of internal components must be considered as instrumental for the accurate development of a comprehensive human performance model. A brief overview of several of these components is presented next.

55.3.2.1 Physical Fitness

Physical training, the scientific validity of different training methods, and their applicability to individuals and performance are areas that need further research and investigation. Related to physical fitness is the concept of hardiness, which refers to everything from an individual's immune system health and resilience, to resistance to injuries and the physical ability to adapt to heat, cold, altitude, and other stressors. Recently, there has been an appreciation of the concept of functional fitness training for military personnel. Functional fitness effectively tailors strength, endurance, balance, and physical skill training to specific occupational categories in an attempt to maximize job-specific performance and minimize injuries. The U.S. Marine Corps has begun a process of evolving physical fitness training toward more functional exercise strategies (Doyle & McDaniel, 2006). The challenge will be to optimize the benefits of the functional fitness approach with its myriad of training options against the requirements for physical fitness in operational settings, while keeping injuries at acceptable levels and adhering to logistical constraints of deploying specialized training equipment.

55.3.2.2 Nutrition

The balance of calories, carbohydrates, proteins, and complex sugars continues to challenge military scientists and operational forces. Meals-ready-to-eat (MREs) are capable of meeting dietary requirements, but only if the entire package is consumed. Anecdotal evidence suggests that portions of the meal are often discarded by warfighters because of taste, or a host of other factors (e.g., dry, rely upon alternative foods from home). Nutritional problems can be exacerbated by factors such as altitude, which is known to suppress appetite. Finally, vitamins, minerals, and other nutrients and performance-enhancing additives have not been thoroughly investigated for their short-term performance and longer term health effects by the military.

55.3.2.3 Fatigue

The effects of sleep deprivation on performance have been extensively studied. Generally, tasks requiring sustained, continuous attention are most affected by fatigue. These include monotonous visual tracking (e.g., driving, lookout), vigilance tasks, and complex decision making. Current fatigue estimation models may benefit from studies addressing individual differences in sleep patterns. For example, Carr, Sausen, Taylor, and Drummond (2006) have found that habitually longer sleepers may be more resilient to sleep deprivation–related deficits in a verbal learning task than individuals requiring less sleep. Functional magnetic resonance imaging data from these participants indicated distinctly different cerebral activation patterns between long and short habitual sleepers under normal conditions. In addition to individual differences, complex task performance and team performance under dynamic situations are all areas in need of further research. Furthermore, the effects of sleeping aids and agents used to promote wakefulness will continue to require further attention. Additionally, closed-loop methodologies are being pursued by all branches of the military to help sustain vigilance performance.

55.3.2.4 Cognitive Ability and Intelligence

Innate cognitive abilities and intelligence affect skill acquisition, problem solving, and psychological resiliency. The types of tasks impacted include cognitive decision making, situation awareness (SA), multitasking, and task management. Warfighters of higher intelligence have been shown to have fewer stress-related mental health issues (Macklin et al., 1998; McNally & Shin, 1995; O'Toole, Marshall, Schureck, & Dobson, 1998). In the near future, smaller unit operational concepts may require greater emphasis on selection and classification of personnel assigned to specific unit positions. At a minimum, effective training strategies will need to account for a wide range of abilities and intelligence levels.

55.3.2.5 Personality and Stress Reactivity

Personality factors have been shown to be relevant for success in adjusting to military life and may become more important as smaller unit configurations are considered, developed, and deployed. The research literature clearly supports the notion that stress reactions affect performance on complex tasks (Morgan et al., 2001, 2004). Recent findings may indicate that individual stress reactivity relates not only to cognitive performance but also to resiliency against adverse combat operational stress reactions, including PTSD. Morgan et al. (2001) have shown that Special Forces soldiers have a greater response of Neuropeptide-Y (NPY, a stress-modulating peptide) to acute stress exposure when compared to non–Special Forces personnel. However, 24 hours after exposure to stress, NPY levels had returned to baseline levels in Special Forces soldiers, but were still elevated in non–Special Forces personnel. Although there are a number of potentially confounding factors, the fact that Special Forces personnel are found to be generally more resistant to PTSD may indicate that there may be fundamental neurophysiological mechanisms at work, and that individual stress reactivity may be worthy of further investigation as integrated human performance optimization strategies are developed.

55.3.2.6 Summary

Clearly, not all of the aforementioned internal components are currently understood nor readily modeled. However, each affects performance in some way and may become more important as warfighters, both on the ground and at sea, become composed of smaller unit sizes, have less redundancy with regard to personnel, and become more dependent on the overall performance of individual team members. Understanding how components such as fatigue, physical fitness, and nutrition affect performance is needed to develop strategies that mitigate their potentially deleterious effects. Doing so will work to "reduce the noise" they produce on modeled estimates of performance and result in warfighters who are better prepared to deal with both the internal and external stresses of military training and combat.

55.3.3 External Components

The way each person moves and reacts to physical surroundings determines the efficiency and effectiveness of each aspect of performance in various situations. Thus, capturing the physical and biomechanical aspects of the warfighter is an important focus of digital human modeling. The knowledge of not only the physics of human movement and locomotion but also the interaction with clothing, individual equipment, and manmade interfaces is also required. This interaction with the physical environment falls under the rubric of an external component. These components, external to the human body, may have a significant effect upon human performance. Physical activity can affect the biomechanical aptitude of an individual. However, clothing and tangible objects that interact with an individual may also affect one's biomechanical efficiency, as well as the extremes of climate and other types of weather conditions (e.g., snow, rain, etc.).

Several key external components that have a direct impact upon warfighter performance have been identified:

55.3.3.1 Temperature

Researchers have long been concerned with the effect of temperature (both high and low) upon human performance (Wing & Touchstone, 1963). The ability of a human to maintain core body temperature is directly influenced by the individual's acclimatization state (Wenger, 1988), aerobic fitness, and hydration level (Sawka & Pandolf, 1990). Aerobically fit persons who are heat acclimatized and fully hydrated have less body heat storage and perform optimally during exercise-heat stress.

55.3.3.2 Noise

When determining the effect of noise upon human performance, it is necessary to consider the nature of the task performed (Smith, 1989). The effect is also related to internal components, such as biological and psychological state, as well as other external components such as task complexity and the presence of other stressors. Tasks requiring continuous performance are especially affected by noise levels over 100 dB. This is a larger concern if the job requires a high level of sustained performance, such as with warfighters during complex, sustained operations, where complex or multitasking activities are much more likely to be disrupted by noise than are simple tasks during normal operations. In combat, there has also traditionally been a further extreme—the effects of blast overpressure (BOP), also known as high-energy impulse noise, which is a damaging outcome of explosive detonations and firing of weapons (Elsayed & Gorbunov, 2007). Exposure to BOP shock waves often results in injury predominantly to the hollow organ systems (e.g., auditory, respiratory, and gastrointestinal). With the threat of terrorist bombings and armed conflicts continuing to grow across the world, the dangers of BOP are now becoming global societal problems for both military and civilian populations and not just confined to military settings or societal anomalies, as has traditionally been the case.

55.3.3.3 Altitude

The effect of altitude on a warfighter's ability to perform optimally in combat is a component that has a direct and often very detrimental impact. When an individual is exposed to higher altitudes than the normal living environment, the warfighter may often experience both cognitive and physical impairments, but susceptibility varies considerably between individuals (Chapman et al., 1998). These symptoms are often termed acute mountain sickness (AMS), which is due to insufficient oxygen and fluid leakage from blood vessels (CDC, 2006). Symptoms such as headache, fatigue, nausea, loss of appetite, vomiting, and others will have an almost immediate impact on the warfighter's stamina, decision making, strength, fatigue, physiological factors, mood, cognitive/motor performance, and even vision (Shukitt-Hale et al., 1991; Karakucuk et al., 2004; Paone, 2006). Two major factors that influence an individual's response to altitude are the tasks performed and the amount of prior altitude exposure. At altitude, performance of aerobic tasks decreases, whereas performance of anaerobic tasks often stays the same (Burtscher et al., 2006). Also, previous exposure to altitude will assist in acclimatizing an individual and decreasing the likelihood of the above and other, often life-threatening, symptoms. Since the mid-1990s, the dogma of "live high, train low" has been accepted as the most effective method of preparing for environments at high altitudes. The landmark studies of Levine and Stray-Gundersen (1997) were instrumental in supporting this type of training theory for optimal performance at altitude. The amount of time of altitude exposure is also a factor when acclimatizing, as full acclimatization occurs over several months (Perry et al., 1992). Thus, the physically adept warfighter who has been predisposed to altitude for several months will be the most prepared to perform in these conditions.

55.3.3.4 Communication/Collaboration/Coordination

Technological advances in the information sciences have produced increased opportunities for quickly providing situational information to the human. The warfighter, in turn, needs to rapidly perceive, comprehend, and translate this information into actions. A critical problem is determining what information each warfighter needs and how to provide it in a way that can be understood and exploited quickly. To ensure situational superiority, the information must be successfully understood while the individual's cognitive capacity is being stressed by fatigue, heat, altitude, interruptions, and so on. This problem is further magnified when a team or multiple teams are required to act together on the same mission or objective. Thus, methods and processes need to be explored that enhance peer-to-peer communication/ collaboration/coordination, shared SA, and rapid decision making. For example, augmented cognition S&T, which utilizes physiological- and neurophysiological-based computational technologies explicitly designed to address human information-processing bottlenecks (e.g., attention, memory, decision making, etc.), is currently being explored in both military and non-military settings to extend the cognitive and decision-making capabilities of humans (Schmorrow, 2005; Schmorrow & McBride, 2004; Schmorrow, Nicholson, Drexler, & Reeves, 2007; Schmorrow & Reeves, 2007; Schmorrow, Stanney, Wilson, & Young, 2006). The future will thus bring advanced methods for understanding the interaction between the physical *and* cognitive demands on warfighters, as well as strategies to allow for the optimization of mission success (e.g., mitigation strategies to avoid/reduce physical *and* cognitive and overload). Assessment and mitigation strategies for SA, communication, and collaboration issues will be necessary for a complete understanding of individual cognition.

55.3.3.5 Chemical and Biological Hazards

Biological hazards pose a great threat to performance in combat, with various agents having numerous effects on the body physically, physiologically, and psychologically. Thus, identifying the gamut of biological agents that may be used and their effect on the warfighter is a complex task. Performance is affected in a variety of ways, based on the system targeted by the agent (e.g., respiratory, cardiovascular, gastrointestinal, nervous, etc.). Many symptoms are similar to common diseases and may be ambiguous if numerous chemical or biological substances are deployed at once, and some may be nonexistent with long-term effects, like neurocognitive impairment (Patel et al., 2003). Those such as burns, respiratory depression, neurological damage, and shock will prevent optimal physical functioning and decrease one's mental toughness, cognitive aptitude, and ability to make quick, effective decisions under stress. Casualty prediction models that estimate mortality and incapacitation from exposure to chemical/ biological agents are already in progress at the Armed Forces Radiobiological Research Institute (AFRRI) (Knudson et al., 2002). Long-term psychological effects have also been explored, as it is critical to the success of military units to understand the overall behavioral and psychological effects (Fullerton & Ursano, 1990). Another factor that will affect performance under exposure to chemical/biological hazards is the dose-toxicity relationship, increasing the effect as the amount of the toxin or agent rises (Martinez-Lopez, 2006). With the large array of different agents and their uncountable effects, classifying each chemical and biological hazard and its effect on performance will take time. However, much of the previous research will assist in identifying the numerous threats these hazards place on warfighters (be they physical, physiological, or psychological) and their effects on mission success.

55.3.3.6 Summary

Although there is abundant research exploring physical elements (e.g., heavy equipment loads; demanding physical tasks) and their effect on an individual's movement and their physical activity, the future holds much room for integration with other elements such as weather, climate, noise or other hazards, and the physics of the environment. To date, insufficient data are available for effective modeling of these factors and their *combined* effects on the warfighter(s). Consequently, automated performance measurement technologies (e.g., integrated behavioral, physiological, and neurophysiological data sensing that can be collected), based on metrics for combined biomechanical and cognitive assessment of the aforementioned internal and external factors' effects, are needed.

55.4 Integrative Components

Cognitive augmentation has become an increasingly important field for evaluating and affecting human performance. Creating synergy between humans and the systems they interact with is paramount to optimizing human performance and warfighter performance in particular (Schmorrow, 2005; Schmorrow et al., 2005, 2006; Kobus et al., 2006). This area of research requires knowledge and understanding of both internal and external components, thus leading to an integration of the two. Although numerous advancements have been made to understand how it affects an individual and the way one acts and reacts to the surroundings, further research will continue to shed light on the importance of augmented cognition to all aspects of human performance. Stress presents numerous issues to add to the way an individual performs. Mitigation of stressors through augmenting cognition is a necessity for understanding the human, as the effect of stress on the nervous system alters the performance outcome from one individual to the next.

Augmenting technologies are most often conceived of as working to improve the integration of the operator with the specific system being controlled. As such, these technologies have focused on methods of identifying the onset of fatigue, the focus of attention, and the manner in which automation interacts with the operator, based upon the current state, the nature of the problem, and the cognitive demands upon the operator. There appear to be two general areas of focus for the application of integrative, augmenting technologies for warfighters and the general population: (a) dismounted, or pedestrian individuals, and

(b) mounted, or seated, operators. For dismounted operators, augmenting technologies can run the gamut from indicators of simple life/death, actigraphy-based indicators of rest/sleep to sophisticated systems that estimate caloric expenditure (e.g., via GPS and actigraphy tracking of movement and motion) to head-mounted display and communication systems. Each subsequent layer of technological complexity presents challenging integration problems.

An example of one such integrative approach, as applied to the context of the dismounted ground infantry soldier, may be seen with the Army's recent augmented cognition efforts within their Future Force Warrior (FFW) program. The FFW will have more information available (via physical sensors, communications, etc.) than any dismounted soldier in history, and processing that information in dynamic and lethal environments requires careful management of limited attentional resources. Up to 380% performance improvement in a warfighter's attentional resources and decision-making capabilities were seen when closed-loop augmented cognition approaches were employed (see fig. 55.1) (Kobus et al., 2006; Schmorrow, Kruse, Reeves & Bolton, 2007). Such closed-loop approaches have also shown similar performance improvements and enhanced survivability for command and control system operators (e.g., stationary/mounted command center). These integrative approaches combine the availability of netted communications with mitigation strategies enabling sharing and real-time collaboration—approaches that enhance the kind of warfighter communication/collaboration/coordination and SA necessary to drive decisive actions in diverse and stressful operational environments (e.g., distributed, asymmetric/irregular warfare). Consequently, the tools and techniques utilized in these integrative approaches hold much promise for their applicability to the warfighter-system modeling approaches of the future, where key internal and external factors can be accounted for and their resulting combined effects on performance can be predicted. Further benefits of such modeling approaches include:

- Sufficient comprehension of individual cognition through an improved understanding of how to augment and enhance human attention, SA, and a broader domain of information processing and decision making in complex environments

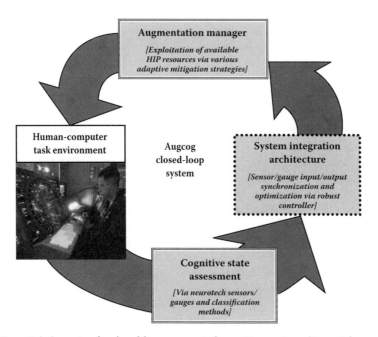

FIGURE 55.1 Essential elements of a closed-loop augmented cognition system. (From Schmorrow, D., et al., "Augmenting Cognition in HCI," in *Handbook of Human-Computer Interaction*, 3rd ed., Lawrence Erlbaum, Mahwah, NJ, 2007. With permission.)

- An advanced understanding of the interaction between physical environments and cognitive demands upon performance
- An improvement in the applied knowledge base of communications on individual decision making throughout various circumstances, with further support and focus on decision outcomes of individual warfighters and teams

55.5 Future Direction

There are two distinct paths for the future of human modeling approaches that consider the human as a fully integrated system. The first is the logical growth and extension of current efforts to include many of the areas discussed in this chapter. Imagine a model that shows the effects of fatigue, both physical and cognitive, based on lack of sleep, heavy workload, and other stressors and indicates differences in human performance in high heat conditions or in cold conditions at high altitudes. Such an outgrowth would naturally foster the aggregation of a number of factors that would allow a new level of equipment evaluation. Consider being able to evaluate the effects of adding 50 pounds of equipment to an individual warfighter. The extra equipment might increase chances of survival if the warfighter is hit by incoming fire, but it most likely would increase the chances of being hit because the extra weight makes the movement slower and less efficient.

The second path is in many ways the opposite of the first. It is the simplification and abstraction of the complex human model into an infantry-friendly planning tool. It is envisioned that such a tool could be provided to individual mission planners and combat leaders, so they could define mission profiles and current states and then obtain guidance from the modeling tool's algorithms regarding proper load out for a particular mission. The tool could draw on a myriad of sources to provide the user with unprecedented planning ability. It could suggest, for example, that there is only a 10% chance of rain in the next 48 hours, so there is no need for rain gear. It may even suggest that Corporal Jones is more suited to this mission than Corporal Smith according to the known (modeled parameters) differences in their physical and cognitive capabilities and limitations for the given mission requirements. The ultimate challenge will be making such a modeling tool simple to use yet powerful enough to retain the strengths of advanced digital human modeling techniques, such as those presented throughout the numerous other chapters in this volume.

References

Allender, L. (2000). Modeling human performance: Impacting system design, performance, and cost. In M. Chinni (Ed.), *Proceedings of the Military, Government, and Aerospace Simulation Symposium, 2000 Advanced Simulation Technologies Conference*, Washington, DC, 139–144.

Allender, L., Kelley, T.D., Salvi, L., Lockett, J., Headley, D.B., Promisel, D., Mitchell, D., Richer, C., & Feng, T. (1995). Verification, validation, and accreditation of a solider-system modeling tool. *Proceedings of the Human Factors and Erogonomics Society 39th Annual Meeting*, San Diego, CA, 1219–1223.

Anderson, C., & Horne, J.A. (2006). Sleepiness enhances distraction during a monotonous task. *Sleep, 29*(4), 573–576.

Burtscher, M., Faulhaber, M., Flatz, M., Likar, R., & Nachbauer, W. (2006). Effects of short-term acclimatization to altitude (3200 m) on aerobic and anaerobic exercise performance. *Int. J. Sports Med., 27*(8), 629–635.

Carr, W., Sausen, K., Taylor, M., & Drummond, S. (2006). Influence of individual differences and task difficulty on cerebral and behavioral responses during cognitive performance following total sleep deprivation. Presented at the 112th annual meeting of Association of Military Surgeons of the United States, San Antonio, TX.

CDC (Centers for Disease Control) (2006). Health Information for International Travel 2005–2006.

Chaiken, S. (2005). Verification and Analysis of the USAF/DoD Fatigue Model and Management Technology—Interim Rept. 1 Jan-30 Sep 2005. (AFRL Technical Report AFRL-HE-BR-TR-0162). Brooks Air Force Base: Air Force Research Laboratory.

Chapman, R.F., Stray-Gundersen, S., & Levine, B.D. (1998). Individual variation in response to altitude training. *J. Appl. Physiol., 85*(4), 1448–1456.

Crosby, M.E., Iding, M.K., & Chin, D.N. (2002). Research on task complexity as a foundation for augmented cognition. Presented at the 36th Hawaii International Conference on System Sciences.

Doyle, E., & McDaniel, L. (2006). *A concept for functional fitness.* Marine Corps Combat Development Command special report, November 2006.

Elsayed, N.M., & Gorbunov, N.V. (2007). Pulmonary biochemical and histological alterations after repeated low-level blast overpressure exposures. *Toxicol. Sci., 95*(1), 289–296.

Erwin, S.I. (2006). National Defense, Office of Naval Research Turns Attention to 'Irregular' Warfare, August. Retrieved September 1, 2007, from: http://www.nationaldefensemagazine.org/issues/2006/August/OfficeofNavalRe.htm.

Friedlander, A., Braun, B., Pollack, M., MacDonald, J., Fulco, C., Muza, S., Rock, P., Henderson, G., Horning, M., Brooks, G., Hoffman, A., & Cymerman, A. (2005). Three weeks of caloric restriction alters protein metabolism in normal weight, young men. *Am. J. Physiol.-Endocrinol. Med., 289*, E446–455.

Fullerton, C.S., & Ursano, R.J. (1990). Behavioral and psychological responses to chemical and biological warfare. *Mil. Med., 155*(2), 54–59.

Gore, B.F., & Corker, K.M. (2001). Human error modeling predictions: Increasing occupational safety using human performance modeling tools. Presented at the Computer-Aided Ergonomics and Safety Conference.

Hackney, A.C., Kelleher, D.L., Coyne, J.T., & Hodgdon, J.A. (1992). Military operations at moderate altitude: effects on physical performance. *Mil. Med, 157*(12), 625–629.

Harrison, Y., & Horne, J.A. (2000). The impact of sleep loss on decision making – a review. *J. Exp. Psychol. Appl., 6*, 236–249.

Karakucuk, S., Oner, A.O., Goktas, S., Siki, E., & Kose, O. (2004). Color vision changes in young subjects acutely exposed to 3,000 m altitude. *Aviat. Space Environ. Med., 75*(4), 364–6.

Knudson, G.B., Elliott, T.B., Brook, I., Shoemaker, M.O., Pastel, R.H., Lowy, R.J., King, G.L., Herzig, T.C., Landauer, M.R., Wilson, S.A., Peacock, S.J., Bouhaouala, S.S., Jackson, W.E., & Ledney, G.D. (2002). Nuclear, biological, and chemical combined injuries and countermeasures on the battlefield. *Mil. Med. 167* (2 Suppl.), 95–97.

Kobus, D., Brown, C., Morrison, J., Kollmorgen, G., Cornwall, R., & Schmorrow, D. (2006). DARPA improving warfighter information intake under stress—augmented cognition phase II: The concept validation experiment (CVE). Submitted for DTIC publication.

Lawton, C.R., Miller, D.P., & Campbell, J.E. (2005). *Human performance modeling for system of systems analytics: Soldier fatigue* (No. SAND2005-6569). Albuquerque: Sandia National Laboratories.

Levine, B.D., & Stray-Gundersen, J. (1997). "Living high—training low": effect of moderate-altitude acclimatization with low-altitude training on performance. *J. Appl. Physiol., 83*(1), 102–112.

Macklin, M.L., Metzger, L.J., Litz, B.T., McNally, R.J., Lasko, N.B., Orr, S.P., et al. (1998). Lower precombat intelligence is a risk factor for post-traumatic stress disorder. *J. Consulting Clin. Psychol., 66*, 323–326.

Mahon, T. (2006, September 11). Over the top? *Defense News*, p. 74.

Martinez-Lopez, L. (2006). *Biotechnology Enablers for the Soldier System of Systems.* Retrieved November 1, 2006, from National Academy of Engineering Web site, *The Bridge, 34*(3): http://www.nae.edu/NAE/bridgecom.nsf/weblinks/MKEZ-65RJZV?OpenDocument.

McNally, R.J., & Shin, L.M. (1995). Association of intelligence with severity of post-traumatic stress disorder symptoms in Vietnam combat veterans. *Am. J. Psychiatry, 152*(6), 936–938.

Miles, D. (2003). Military uniforms of the future. *American Forces News Service.*

Morgan, C.A., Southwick, S., Hazlett, G., Rasumusson, A., Hoyt, G., Zimolo, Z., & Charney, D. (2004). Relationships among plasma dehydroepiandrosterone sulfate and cortisol levels, symptoms of dissociation, and objective performance in humans exposed to acute stress. *Arch. Gen. Psychiatry, 61,* 819–825.

Morgan, C.A., Wang, S., Rasmusson, A., Hazlett, G., Anderson, G., & Charney, D.S. (2001), Relationship among plasma cortisol, catecholamines, neuropeptide Y, and human performance during exposure to uncontrollable stress. *Psychosom. Med., 63,* 412–422.

Nicholson, D.M., Lackey, S.J., Arnold, R., & Scott, K. (2005). Augmented cognition technologies applied to training: A roadmap for the future. In Dylan D. Schmorrow (Ed.), *Foundations of augmented cognition,* Mahwah, NJ: Lawrence Erlbaum, 931–940.

Nieman, D., Carlson, K., Brandstater, M., Naegele, R., & Blankenship, J. (1987). Running endurance in 27-h-fasted humans. *J. Appl. Physiol., 63,* 2502–2509.

O'Toole, B.I., Marshall, R.P., Schureck, R.J., & Dobson, M. (1998). Risk factors for posttraumatic stress disorder in Australian Vietnam veterans. *Australian & New Zealand J. Psychiatry, 32,* 21–31.

Paone. C. (2006). Scientists look to help soldiers overcome high altitude. Electronic Systems Center Public Affairs, October. Retrieved October 1, 2007, from: http://www.pentagon.mil/transformation/articles/2006-10/ta100606a.html.

Patel, M., Schier, J., Belson, M., Rubin, C., & Garbe, P. (2003). *Recognition of illness associated with exposure to chemical agents—United States, 2003.* MMWR Weekly, Div of Environmental Hazards and Health Effects. National Center for Environmental Health, CDC. 52(39), 938–940. Retrieved October 1, 2007, from: http://www.cdc.gov/mmwr/preview/mmwrhtml/mm5239a3.htm.

Perry, M.E., Browning, R.J., Jackson, R., & Meyer, J. (1992). The effect of intermediate altitude on the Army Physical Fitness Test. *Mil. Med., 157*(10), 523–526.

Sawka, M.N., & K.B. Pandolf (1990). Effects of body water loss on physiological function and exercise performance. In: C.V. Gisolfi and D.R. Lamb (Eds.). *Perspectives in exercise science and sports medicine* (pp. 1–38). Vol. 3, *Fluid homeostasis during exercise.* Indianapolis, IN: Benchmark Press.

Schmorrow, D.D. (Ed.) (2005). *Foundations of augmented cognition.* Mahwah, NJ: Lawrence Erlbaum.

Schmorrow, D., Kruse, A., Reeves, L.M., & Bolton, A. (2007). Augmenting cognition in HCI: 21st century adaptive system science and technology. In J. Jacko & A. Sears (Eds.) *Handbook of human-computer interaction* (3rd ed.). Mahwah, NJ: Lawrence Erlbaum.

Schmorrow, D., & McBride, D. (2004). Introduction. *Int. J. Human-Computer Interact., 17*(2), 127–130.

Schmorrow, D.D., Nicholson, D.M., Drexler, J.M., & Reeves, L.M. (2007). *Foundations of augmented cognition* (4th ed.). Arlington, VA: Strategic Analysis, Inc.

Schmorrow, D.D., & Reeves, L.M. (2007). *Foundations of augmented cognition* (3rd ed.). Heidelberg, Germany: Springer-Verlag.

Schmorrow, D.D., Stanney, K.M., & Reeves, L.M. (Eds.) (2006). *Foundations of augmented cognition* (2nd ed.). Arlington, VA: Strategic Analysis, Inc.

Schmorrow, D., Stanney, K.M., Wilson, G., & Young, P. (2006). Augmented cognition in human-system interaction. In G. Salvendy (Ed.), *Handbook of human factors and ergonomics* (3rd ed.). New York: John Wiley.

Schnell, T., Macuda, T., Poolman, P., Craig, G., Erdos, R., Carignan, S., et al. (2006). Toward the "cognitive cockpit": Flight test platforms and methods for monitoring pilot mental state. In D. Schmorrow, K. Stanney & L. Reeves (Eds.), *Foundations of augmented cognition* (2nd ed., pp. 268–278). Arlington, VA: Strategic Analysis, Inc.

Shukitt-Hale, B., Banderet, L.E., & Lieberman, H.R. (1991). Relationships between symptoms, moods, performance, and acute mountain sickness at 4,700 meters. *Aviat. Space Environ. Med. 62* (9 Pt 1): 865–869.

Smith, A. (1989). A review of the effects of noise on human performance. *Scand. J. Psychol., 30*(3), 185–206.

Snow, M.P., Barker, R.A., O'Neill, K.R., Offer, B.W., & Edwards, R.E. (2006). Augmented cognition in a prototype uninhabited combat air vehicle operator console. In D. Schmorrow, K. Stanney & L. Reeves (Eds.), *Foundations of augmented cognition* (2nd ed., pp. 279–288). Arlington, VA: Strategic Analysis, Inc.

Tremoulet, P., Barton, J., Craven, P., Gifford, A., Morizio, N., Belov, N., et al. (2006). Augmented cognition for tactical Tomahawk weapons control system operators. In D. Schmorrow, K. Stanney & L. Reeves (Eds.), *Foundations of augmented cognition* (2nd ed., pp. 313–318). Arlington, VA: Strategic Analysis, Inc.

U.S. Department of Defense (2001). VV&A recommended practices guide glossary. Washington, DC: Defense Modeling and Simulation Office. Retrieved October 6, 2006, from: http://vva.dmso.mil/Glossary/Glossary-pr.pdf.

Wenger, C.B. (1988). Human heat acclimatization. In: K.B. Pandolf, M.N. Sawka, and R.R. Gonzalez (Eds.). *Human performance physiology and environmental medicine at terrestrial extremes* (pp. 153–197). Indianapolis, IN: Benchmark Press.

Wertz, A. T., Ronda, J.M. Czeisler, C.A., & Wright, K.P. (2006). Effects of sleep inertia on cognition. *JAMA, 295*(2), 163–164.

Wing, J., & Touchstone, R.M. (1963). A bibliography of the effects of temperature on human performance. Final Report, Aerospace Medical Research Labs, Wright-Patterson AFB, Dayton, OH.

Young, P.M., Clegg, B.A., & Smith, C.A. (2004). Dynamic models of augmented cognition. *Int. J. Human-Computer Interact., 17*(2), 259–275.

Zinker, B., Britz, K., & Brooks, G. (1990). Effects of a 36-hour fast on human endurance and substrate utilization. *J. Appl. Physiol., 69,* 1849–1855.

56

Future Needs and Developments in Support of Computer-Based Human Models

Norman I. Badler and
Jan M. Allbeck

56.1 Introduction

Digital human models (DHM) have been developed to serve a variety of applications. Engineering human operator workplaces has driven DHM toward accurate models of human shape, articulation, and function. Biomechanics has developed empirical methods for obtaining human motion, torque, force, and energy expenditure that may be used as lookup tables or modeled in procedures or functions. Outside of human factors engineering, however, computer graphics has also been engaged in a relentless pursuit of improvements to human models: obtaining and modeling surface shape, creating realistic joint rotations and body segment deformations, and inventing modeling and rendering (drawing) techniques that affect appearance for maximal realism. Beyond surface appearance, moreover, psychological, psychosocial, and cognitive studies of human behavior are finding their way into control mechanisms for DHM that make them move as humans do, but also react, decide, behave, feel, and express emotions as real people might. Our intention in this chapter is to examine how some of these domains will influence, evolve, and challenge the DHM of the future.

During this discussion it is most helpful to keep in mind a "gold standard" for human modeling—namely, a real person. Because a DHM is a subset of the full set of human capabilities, it clarifies what we get and what we miss in considering some feature or other of a DHM. Obviously, the evaluation of DHM must consider an application's requirements and priorities that might require digital toolsets. For example, it may be rather superfluous to consider a psychosocial behavior model if one is only considering the reach volume of a cockpit design; but if one wants to simulate the possible behavior of an aircrew under off-nominal flight conditions, then psychosocial and team coordination factors may dominate performance.

While we advocate the development and deployment of more complete and more highly integrated human models, we do not press a point of view that one model will fit all purposes. If anything, we would like to see a software architecture with modules that can be selected as needed for population, capabilities, tasks, and analyses that fit together as "plug-ins" to a larger general framework. We believe this is a principal challenge and near term opportunity for the DHM community.

In this chapter we will first summarize the arsenal of tools that are available for DHM, mostly from "outside" the human factors community, look into how to access and control those tools relative to our "gold standard" and finally offer an ambitious pathway toward a DHM of the future. This cannot be an exhaustive survey; rather it tends to focus on illustrative examples and current system developments.

56.2 Tools for Future Human Factors Analyses

Tools for human factors analysis should include:

- Methods for obtaining baseline data on human shape, joint articulation, and joint limits
- Methods for acquiring empirical motion data from people in order to create predictive models for novel situations
- Methods for setting up and running analyses efficiently and effectively, with readily modified populations or workspace modifications
- Methods for digitally simulating the normal and exceptional tasks a human may be called on to perform, including sensing and reacting to others and the environment

No single human factors tool or system has this complete set of functionalities. Many approaches and directions have been discussed in this volume. We'll look at each of these categories in turn, checking ourselves against the gold standard and examining new or emergent concepts applicable to future DHM.

56.3 Skin Shape and Joint Motions

The computer animation and gaming communities understand the importance of visual appearance to a viewer's acceptance of the portrayed individual. Characters can even depart from realistic human proportions as long as the skin seems flexible and supple, muscles ripple and reshape with use, and joint angles show pose dependencies that make them move in "natural" ways. The DHM of today basically ignores the "squishiness" of the human body.

In some preliminary work we undertook a few years ago, we showed that a 10% to 15% change in reachable range occurred in the presence of obstacles if the obstacles were allowed to gently "penetrate" a lower arm surface mesh [1]. Although collisions (intersections) between polygonal skin meshes and a designed environment are readily computed, human skin is not rigid. Fixed meshes ignore the fact that the visible skin surface and the effective functional surface are not at all identical. The "iso-pain" surface displaces surface points toward the axis of a body segment to a distance where a subject starts to feel discomfort from the incursion and steady pressure. This iso-pain surface differs from the exterior skin shape especially where fat and muscle cover the deep skeletal structure: Consider the amount of surface displacement allowed in the abdomen, the upper thighs, or the buttocks. A visualization of the iso-pain surface will look very strange relative to the normal body surface, so the best realization for DHM might just consist of two nested surface layers: one representing the exterior skin for visual appearance and an internal iso-pain threshold surface for collision detection. The most advanced efforts that consider this effect are in seat design where force and pressure distributions are necessary to measure seated comfort [2]. Dynamic simulation for crash testing may also consider penetration depth as a component feeding injury assessment [3].

There are new approaches to body surface shape that explicitly include deformation in the body surface representation. These may be based on finite element methods [4, 5] or simpler and faster numerical methods using exemplar shapes for a set of given body poses [6]. The former admit better interaction with environment obstacles; the latter work well to give the look of muscle exertion or fat distribution relative to body pose.

A useful side effect of deformable surface modeling is that the surface actually responds to internal and external forces—a *physics-based* model. This means that the masses around a joint (say, the elbow) can deform the skin naturally as the joint flexes. Also, if an external force arises from a workplace obstacle, a manipulated object, a tool, or simple contact with a surface, the skin can respond appropriately. Moreover, a physics-based deformable model is essential for the correct modeling of the interaction of clothing with a body. Too many DHM appear to have painted-on tights and t-shirts. Modeling bulky, constrictive, or loose-fitting clothing requires a physics approach for deriving the resultant surface shape, functional joint limits, and skin pressure points [7]. To date, most of the research on clothing digital humans has focused on the cloth model rather than its impact on the underlying deformable body [8].

Even if we have physics-based deformation tools, we still need nominal body shape and joint articulations. The CAESAR [9] dataset has been a tremendous boon to the study of the space of contemporary body shape. Being generally limited to external skin shape in three canonical poses, it cannot fully describe body joint centers, limits of rotation, or multi-joint pose dependences. The latter, for example, occur when one moves the leg: The muscles coupling the hip and knees constrain the possible joint angle sets. There has been a notable effort to use the CAESAR data to anthropometrically and continuously scale body shapes by fitting a skeleton into the shapes based on surface landmarks [10]. Subsequently, the SCAPE project extended these methods to include deformable bodies and motion data obtained from non-CAESAR subjects [11]. The trick to managing scanned surface data for anthropometric scaling is to create a canonical skeleton and mesh such that all scanned datasets are resampled and remeshed to match it topologically; the remaining differences are purely geometric and subject to rescaling landmarks and numerically relaxing the points in between.

56.4 Human Motion Models

Once we have a scalable and deformable DHM, we need to make it move as a person does. The gold standard is a useful yardstick here, as real-time motion capture is an industry to itself, feeding the computer game and special effects communities. Motion capture has the distinct advantage of measuring the dependent and subtle motions of a real person across all body joints in a skeleton; but that is also its weakness. Since the skeleton itself cannot be tracked, external landmarks must be acquired. These can then be related to an estimated rigging (articulation) of a suitable body model. The errors introduced often manifest themselves as a lack of consistent relationship between the body and the environment, such as the floor. The consequent "foot-skate" is a very visually distracting artifact, and manual or semi-automated procedures must be used to clean up the motion and appropriately "lock" the feet to the floor [12].

The other aspect of motion capture that is both a boon and a bane is that ultimately it captures individual performance. The motion of one individual can be mapped to or retargeted to another with differing body proportions, but the methods cannot guarantee the physical correctness of the result. While suitable for computer animation, the validation of the resulting movements for human factors analyses is still under study [13]. Accordingly, there is much research activity in creating reusable motion libraries or databases that can be retargeted to novel human models as needed [14, 15, 16]. Setting aside the retargeting problem for the time being, there are still tough issues that need to be addressed if motion capture is to be used for motion simulation in human factors applications. The primary problem is the limitation of motion sensing technologies. The principal methods involve active or passive markers and sensing cameras. Active markers are LEDs that emit light; when seen and localized by the cameras, the 3D position of the marker can be established. Passive markers reflect light, which is then sensed.

In either case the subject must wear a body suit with carefully positioned markers, and even more importantly, the markers and cameras must enjoy an unobstructed line-of-sight. In practice, multiple cameras are used to obtain redundant data and minimize occlusions, but the limitations create additional constraints on the motions that can be performed and tracked.

In human factors analyses, the workplace is constraining; otherwise, there would be little to analyze, for example, in the suitability of design for an airplane cockpit, vehicle interior, factory work cell, or maintenance bay and in the processes of ingress and egress into tight spaces. As the workspace becomes cluttered or cramped, the analyses of fit and function become more important just as motion capture becomes technologically more difficult.

A new approach to obtaining real human motion data in such confining environments relies on wearable body suits such as the ShapeWrapII system [17] or inertia sensor suits [18]. By sensing joint angles indirectly from fiber optics and/or inertial sensors embedded in the suit, the body pose can be readily computed without any line-of-sight concerns. The technology is still new and expensive, and its accuracy is still being investigated. But it does bode well for obtaining empirical data on the tricky maneuvers people make in real situations. This is essential for applications with difficult body fits and reaches, such as vehicle power train maintenance and aircraft equipment bays.

An even more general motion capture system moves the environment outdoors and allows normal objects as props [19]. This method uses ultrasonic sources and sensors, inertial accelerometers, and ultimately GPS to sense unrestricted motions. Eventually, the state of computer vision may be such that video cameras can be used to completely disencumber the subject from the sensing technology. In addition, video still has other significant issues, requiring an unobstructed line of sight and sufficient light (visibility) to create a good image.

An unfortunate gap still exists no matter how we capture a subject's motion: We will still lack knowledge of *object* movements, effects, and interactions that people cause in the nearby environment. While computer vision techniques eventually may be robust enough to capture physical side-effects of human behavior in systems, there is no technologically simple way to do it otherwise. If we know that someone has grasped an object we can partially infer its motion: We would not know, for example, where along some graspable area the subject actually fixed her grip, and therefore cannot know the precise 3D configuration of the object. This problem will certainly need to be addressed by both DHM designers and roboticists [20].

56.5 Specifying the Desired Tasks and Analyses

Digital human models had their earliest engineering instantiation as substitutes for real people for evaluating virtual mockups. The prime motivator was reducing costs by building fewer full-scale prototypes. The primary analyses needed were fit and function for a representative target population. As DHM capabilities improved, however, the methods by which the designer or human factors engineer specified the workplace tasks for the digital model evolved as well. The earliest systems used batch-processing programs to run analyses on mainframe computers. As interactive graphical computing became available, more interactive systems emerged to give users more direct control over the body scaling, posing, and moving processes such as SAMMIE, Jack, and Safework/DELMIA [21].

When considered against our gold standard, the requirement that a human factors engineer manually insert, pose, and manipulate a DHM even within a well-designed interactive user interface is primitive. Were we using a live subject in a physical mockup, we would give him instructions on the tasks required, maybe show him a few visual examples of the tests we'd like, and ask for verbal reports on task comfort or exertion. In other words, we would use *instructions* to describe to the subject what we would like accomplished. Given his understanding of language, and his body's innate ability to conform physically to the workplace environment, we would quickly ascertain to what extent the instructions can be carried out.

Now, of course, we have added a number of new problems that need to be solved to realize movement toward the gold standard:

- The DHM must be "smart" about fitting into an environment and moving in ways that conform to or at least respect human propensities.
- The instructions must convey necessary information about tasks, goals, and measurements to be recorded.
- The instructions must be reusable across individuals in a population.
- The instructions should be authorable with no more effort than would be expended in a typical graphical user interface to a DHM. The instructions need not be arbitrary natural language utterances, but could be constrained to workable syntactic structure as long as the meanings (semantics) are well defined.

The first requirement places a direct burden on a DHM to automatically pose and move within a workplace given a set of constraints from task and physical geometry. Methods to accomplish this could include pose libraries (just choose the closest to the working configuration, e.g., seated operator), motion capture retargeting, or procedural techniques such as optimization to relax the body into a pose subject to biomechanical fitness measures such as minimum joint torques. The burden of posture and movement is shifted from the interactive user to the DHM itself. Many tools have been developed to aid this process, such as inverse kinematics [22], walking models [23, 24, 25], lifting models [26], collision avoidance [27], heuristic reach models [28], and procedural motion controllers [29]. These and other models potentiate prescriptive approaches to motion and task analyses: "Just do it."

The user, however, is still obligated to express what "it" actually is. Tasks consist of actions to be performed in a workplace and a binding of those actions to object components of the workplace. We have been engaged in building a parameterized action representation (PAR) for human task description [30, 31]. Unlike motion capture libraries that are limited to specific examples of joint poses over time, action parameterization implies flexibility to reuse core actions in widely varying contexts. For example, the action "walk to goal" is parameterized by "goal," thus leaving the details of how the walk is actually accomplished to the body receiving the instruction (how it executes a walk model) and a path planner that finds a feasible route (if one exists). Notice that a task instruction (as a PAR) may have many possible successful executions, or even failure.

Actions represented by PAR are at a higher descriptive level than motion capture joint angle data. PAR is more than just a macro capability of applying the same action to new figures. The PAR readily yields databases that provide reusable value for an analysis where the digital human, the workplace geometry, or the task measurements are simply re-specified as needed. Want a population accommodation analysis for a proposed work cell layout with a series of complex tasks? Set up the PARs for the task, tag (name) the critical reference points in the work cell, and apply the PAR to the desired population exemplars. Because the PARs are interpreted by the body motion generators, some individuals may have to take steps or use extensions to reach goals while others do not. Want to parametrically alter work cell dimensions to find the optimal performance timing for a particular individual? No problem: Just set up the task as a PAR and refer actions to the work cell tags. The DHM and its internal performance model that executes PAR will perform the task in each geometrical instantiation. We can even think of PAR as actively documenting a set of requirements, such as accessibility and placement of controls, which can be automatically checked during the design process [32].

A fruitful area requiring more study is the role of failure in PAR executions. Failure means that the requested task does not succeed for some reason, but in fact the reason is critical. Is the operator in the wrong place relative to the manipulated object? Is the problem in the digital human assigned capabilities? Is the problem in the workplace design? Is the problem in the PAR structure, such as a failure to list necessary pre-conditions? Is the problem external to the instruction, such as contact with a hazardous condition during task execution? There are rich sets of failure modes and failure analyses associated with particular systems; some of these could integrate well with a PAR framework to allow better semantic analyses of failure conditions.

There is already one commercial realization of some of the PAR features. TSB by UGS allows users to set up tasks, assign them to digital humans, and let them figure out how to coordinate and execute the task set [33]. Important components in this process are motion models that predict times for task execution (if known from MTM or Fitts' Law models), or compute time contextually (how long it takes to walk from here to there). TSB uses a straightforward user interface to avoid the pitfalls of using natural language directly as a prescriptive specification of tasks.

An alternative to prescribing motion via PAR or TSB would be for a user to demonstrate the motions required of the digital subject, and then have the digital model "do its best" to mimic the motion in the virtual workplace. There has been recent computer animation work on using low-dimensional motion inputs to drive high-quality, biomechanically reasonable, whole motion outputs [34, 35]. For human factors analyses, we must include not just motion, but motion in the context of the environment (collision avoidance) and task accomplishment (e.g., workspace reach goals).

Leaving aside the delicate issue of whether the human factors engineer should wear a motion capture suit in order to drive the smart digital avatar (an unlikely scenario), we turn instead to more fruitful territory if we consider the larger issue of generating instructions and verifying that those instructions result in reasonable digital human behaviors. That is, can we author instructions (for real or virtual people) in such a way that they can be used to both simulate a performance (and hence analyze useful variables) as well as compare a specific person's performance against a real subject matter expert's execution of the same task? If the task is a very specific one, such as lifting or reaching, such tools already exist. But if the task is more extensive and complex, such as engine part replacement or power system maintenance, the motions themselves may be less important than overall task timing, missteps, improper tool use, and so on.

Consider a maintenance task with, say, 10 steps. Each one can be represented as a PAR, as substeps to the global PAR task. A subject matter expert is motion captured doing the entire procedure and then the motion data are segmented and parameterized into the individual steps. The PARs then describe the tasks to be performed on the workplace objects, rather than the exact motions executed by the particular subject's performance. If we play the PAR through a DHM, we should see the task executed by the DHM's *own* performance model. This could even fail if the DMH can't reach, lift, or see crucial portions of the task specified by the PAR. If we motion capture another subject (perhaps a novice trainee) attempting the same complex task, we can compare that performance against the PARs derived from the expert performance to obtain a numerical measure of task difference. Ultimately, PARs intermediate the expert and novice performances to allow comparison at the task rather than the joint angle level. We believe this is critical in eventually establishing automated training and effectiveness evaluations for real users and maintainers of complex workplaces.

To move toward this goal, we created a system called RIVET (Rapid Interactive Visualization for Extensible Training) [36] that allows an operator to convert a subject's performance of a task into instruction "chunks" (fig. 56.1). The first stage is to acquire any sort of motion data possible, such as multiple video camera streams, audio, and even motion capture. These data streams are input to a user interface built on Apple Quicktime; the operator can delete useless segments, align the different media, and manually segment the streams into meaningful steps. The next module takes the aligned and segmented data streams and allows the operator to associate each one with an existing PAR from a database or to create a new one. At this point the performance has essentially been converted to PARs, but the original source material is accessible if needed, say, to show a video clip during a future training session. The final component takes the PARs and assembles them for any of several instruction presentation interfaces; we have experimented with a web-browser interface, instructions on a PDA, and even augmented reality presentations though a tiny head-mounted display. Thus a trainee, maintainer, or operator could receive appropriately segmented instructions during actual task execution at a level of detail that can vary according to skill or detail requests.

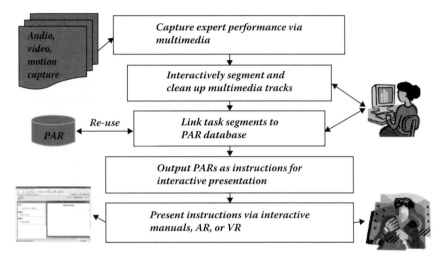

FIGURE 56.1   RIVET architecture for instruction segmentation, generation, and presentation.

Although there are still manual steps involved in converting a performance into instructions, the process is mostly straightforward and readily extensible given the accumulated PARs in the database. Automated motion segmentation methods are being actively researched [37]; automated PAR recognition is feasible [38] but requires further study.

56.6 Sensing and Reacting

Once we begin to look at human performance at the task level, new issues arise that transcend biomechanical models and simulation. For example, consider an automobile driver. The cockpit design may be perfect and the driver can fit and function flawlessly—unless she is talking on a cell phone [39]. Then task accomplishment (driving) is suddenly confounded with cognitive attention issues. Or perhaps a machine tool operator is emotionally agitated from a domestic dispute. How might these cognitive and psychological components be factored into a DHM's performance?

Consider first the role of attention in human behavior. DHM systems provide visual cones that characterize the operator's physical field of view, but not everything within the field of view is actually consciously perceived. In order to construct a computationally tractable model of human visual attention, we have adapted concepts from the psychology literature on attention factors and a human phenomenon called inattention blindness. There are complex visual processes at work in bringing current sensory information into the brain, but there is far more information flow than can be accommodated. Green's four factor model [40] posits that (sensory and cognitive) *conspicuity, mental workload, expectation*, and *capacity* influence what stimuli make it all the way into consciousness. In order to build a plausible computational model of attention, we constructed a simple analog [41] to the classic inattentional blindness experiments [42]. In these, a subject must count ball passes among like-color-shirted players, but in the middle of the activity a person in a gorilla suit walks on the scene, stands in the middle of the player circle, then walks off. In general, after the experiment ends and the subject reports the number of ball passes, half of the subjects report seeing nothing unusual! Of course they could have *seen* the gorilla, but it did not enter *consciousness*. The attention system filters stimuli according to an attentional set—in this case the subject's task of counting ball passes—which often eliminated the otherwise obvious gorilla [43]. In our analogous computer experiments, we could control for cognitive workload and vary the experiment parameters to create an approximate model of inattentional blindness [44]. By applying this model to a DHM engaged in multi-party conversation or driving behaviors, visual stimuli could often be perceived but not registered in consciousness. Thus a person engaged in a dialog might be distracted by a face popping up in a window, but if the discussion involved instead six people, the distractor might go unnoticed. Likewise, an object that appears ahead on the road might be seen or not by the driver model depending on the driver's cognitive workload: if on a cell phone, the object might well be missed. It is reasonable to imagine that with more comprehensive attention models, better (more realistic) behaviors in a complex workplace may be simulated. This could be crucial in applications such as air traffic control or nuclear power plant operation, where vigilance may be compromised by fatigue, boredom, or workload.

The MIDAS system developed over many years at NASA Ames has similar goals but different means. MIDAS uses empirically determined models of load based on cognitive, psychomotor, auditory, visual, and speech workload scales and a careful model of resource allocation and overload [45]. Given a mission, such as a helicopter raid, MIDAS can simulate the overall task and compute the instantaneous workload for each subtask. Overload time periods are triggers for control redesign or workload redistribution. Other related efforts are underway to combine psychomotor and cognitive models to vehicle driving [46].

There is a large literature on human performance moderators. Barry Silverman has surveyed and condensed a huge number of such studies into a simulation architecture called PMFserv [47]. The main ideas are that a person has goals, standards, and behaviors, and that an individual's selection of a behavior is based on a valenced (numerically computed) reaction to events and others in the immediate environment. How one acts is dependent upon how one feels (an emotional response) about the current situation. The performance moderators can act at the psychophysical level (increased fatigue, demand for nourishment) as well as the cognitive (choosing one action over another).

In a somewhat similar vein, but with a purpose more explicitly directed toward interactive virtual humans for training partners, researchers at the Institute for Creative Technologies at the University of Southern California construct virtual human models with emotions and an understanding of interpersonal relationships [48]. Building on classic emotion models [49] and user gaze and gesture interaction models [50], their human models attempt to stand in for real people in decision-making [51] or negotiation [52] settings.

At least three crucial components are needed to bring cognitive and psychosocial models into digital humans:

- They must perceive and understand the world around them as people do.
- They must have embedded models to process and react to those perceptions (or at least those perceptions that survive to consciousness).
- The virtual world they inhabit must behave like the real world.

These three problems are at the very heart of artificial intelligence (AI) research. Adding such components to a DHM is nothing less than adding a level of consciousness and reasoning and simulating the physical world and the behaviors of the engineered artifacts. Tools and methods abound in AI, but how are they to be incorporated into the DHM of the future? That's the topic we'll address next.

56.7 Putting Together the Next Generation of Digital Humans

It might appear by now that our goals greatly exceed any realistic software DHM tool likely to be constructed within the next few years. After all, we expect the DHM to reshape and move as if it were a supple and muscled individual, while it interacts with its environment in an intelligent and cognitive understanding of exactly what it is tasked to do. Even if we succeed in building this general DHM, will its very generality make it so software intensive and complex that no one or no corporate entity can risk investing in its development and maintenance?

In the 1990s, an analogous situation existed in computer animation software. Many individual production shops were developing local, proprietary, but often extensive systems to meet job and artistic needs. Software re-usability was low and programmers were expensive relative to art design talent. New techniques from published papers were quickly snatched up and repetitiously re-coded for local consumption. Consequently, new software architectures emerged such as 3DSMax and Alias (now Autodesk) Maya that allowed power users to add plug-ins into a core engine to address particular features. The plug-ins could be shared or sold, and only the plug-ins actually needed were used. Moreover, proprietary interests could still be satisfied through in-house plug-in development. Beyond plug-ins, users can create as-needed specialty code via scripting languages: Maya uses MEL scripting, but interpreted programming languages such as Python and Lua are likely to be the norm in the next generation of animation tools.

This general software architecture is an excellent model for future DHM systems. Digital human tool development tends to be dispersed in small research groups. There are only a few major vendors, and there is a small but dedicated user community, while U.S. government funding consists mostly of low-level grants. Such a distributed community seems ideal to tackle the challenge of building a next-generation DHM since individual expertise can be leveraged to support and benefit a broad community.

In order to create a plug-in software architecture for DHM, the developer, user, and application communities must first agree on a framework for the plug-ins that allows access to and modification of all human modeling, animation, and analysis features. This framework can be novel or may adapt an existing software tool such as Maya. The advantages to the former are emphases on problems of greatest interest to DHM such as joint motion models, functional reach space, and analysis of biomechanical workloads. The advantages to the latter, however, may outweigh the former, since an existing system avoids new system development time and costs, has an extensive existing user community, and offers a well-understood plug-in structure.

Consider how this new architecture for DHM, might work. A user licenses the core software engine (perhaps even open source) and obtains with it a basic set of capabilities and plug-ins. The core would support the articulation structure of the human body, a native 3D object representation for modeling body surfaces and workplace objects, and a generic graphical user interface. Basic plug-ins might include rigid skinning per body segment, joint limits, inverse kinematics, a nominal population, and perhaps a few key analysis tools. More extensive sets of plug-ins could be available when authored, for example, detailed hands, better coupled joint motion models, body mass distribution, deformable skin, CAESAR dataset, collision detection, collision avoidance, motion capture libraries, visual field handicaps, age-dependent joint mobility, procedural motion generators, dynamic simulation, psychophysical load models, visual perception filters, graphics rendering tools, 3D object model import and export, etc. The concept of mixing and matching to make the DHM help the human factors engineer best solve the application's simulation and analysis requirements is novel to the DHM community but not to related communities outside it.

It is worth noting that this general approach is designed to empower the application and human factors engineers, rather than argue for "more automation" to distance them from the process. Ultimately design suitability decisions should be made by people who can bring their best practices and judgments to bear on analytic data and simulations produced by computers. But the quality of such data can be vastly improved if the underlying models can be commandeered and expanded as needed.

Improvements in DHM capabilities will also be linked to a better annotation of the 3D workplace model environment. If the DHM is expected to reach for something, that object or part must be spatially referenced (by name "the pump" or by function "grasp the handle and lift"), the object's shape and the task should dictate a handshape, and its weight or extraction path may impose body pose constraints. Typical 3D object representations such as polygonal meshes hold geometry but not semantics: We see shape but not function. Our gold standard human has learned about the world through time, education, and experience. While we might train the DHM similarly, it is more efficient to have annotations added to the geometry to cue function. This information is generally known at design time, and often exists implicitly in the part naming hierarchy for a 3D model. But additional manual effort is necessary to provide the human-understandable tags that would be helpful to the user manipulating or commanding the digital model. We have called some of these "maintenance features" as opposed to mere "geometry features," since they are the points and places needed to manipulate an object rather than manufacture it. More automated annotation methods can sometimes be employed; for example, given a 3D model the set of "walkable spaces" can be computed directly [53]. It remains a general challenge to automatically annotate a given 3D model with semantically useful tags for manual manipulations. An avenue for future exploration could be based on a taxonomy of manual actions applied to objects [54]: by knowing the task to be performed (e.g., whether the action requires force or torque, one hand or two), the grip or contact places might be inferred directly from the geometry.

The plug-in architecture also assists in modeling the workspace inhabited by the digital human. At the simplest level, a physics plug-in can manage motions and support of mechanical models. The Havoc [55] and PhysX [56] engines are used commercially in games for this sort of functionality. For an engineered system, we need plug-ins that emulate the system's electrical, thermal, vibratory, or toxicity properties over time and usage. Often these data exist in some other large-scale engineering analysis tool, but there is no ready or convenient way to integrate them with a DHM [57]. The plug-in architecture would provide a steppingstone to system simulations without having to laboriously integrate disparate large engineering software systems. A small step in this direction uses lower resolution models to simplify the simulation [58].

Beyond the interactions of one or a small set of individuals in a workspace, we can imagine simulations at the level of entire factories, ships, or buildings. The key problems are to define the roles of the individuals, the tasks they need to (or could) perform, and the places where those tasks are done. There are no simple methodologies to describe the level of activity suitable for driving a set of digital humans "in the large." We have looked into commercial project management tools such as Microsoft Project as a substrate for organizing the myriad activities required, but it is not quite the right structure to admit low-level opportunity and variation. MicroAnalysis and Design have shown some progress in characterizing the roles and duty cycles of individuals on a ship [59] that could be elaborated (in the future, through PAR) into a flexible task set for each individual.

This new architecture for digital human systems presents an opportunity that should be discussed by the stakeholders in the DHM community. In the short term, it will require a consensus on the minimal level of support need in the core engine and a decision whether or not that engine is supplied or generated by an existing DHM vender or adapted from a broad-range product platform such as Maya. The plug-in architecture should be public and open source so that developers clearly understand how to replace and expand on core functionality.

An intriguing opportunity is presented by the plug-in architecture when viewed in the wider context of the Internet. Critical to many human factors analyses are data derived from experiments. Most of these studies end up published in the scientific literature and are, in that sense, open and public. In fact, however, such data are almost universally disconnected from a DHM environment. Suppose that a researcher could make charts, tables, and models accessible via the Internet and plug-in tools that "understood" the data and their field names in the context of the DHM core architecture "standard." Then that data could be used for analyses without having to wait for a vendor to explicitly incorporate it. Moreover, such data may be less well validated than national standards such as those managed by NIOSH, but they can still be useful with the appropriate cautions. A retrospective analysis and proceduralization of biomechanical data would be a mammoth undertaking, but if almost every researcher did a part, it would be possible to bring an immense amount of data online for the community to use. A similar model has been used in the genomic research community to link together disparate laboratories and their highly dissimilar databases into a common, searchable, useful resource [60, 61].

The evolution of a plug-in architecture, availability of web-based data resources, and higher user expectations and demands on digital human model appearance and function will drive feature integration. Digital human models have been created for specific applications for too long; consider the separate software systems currently used for special effects generation, engineering design evaluation, military simulation and training, computer game development, biomechanics analysis, robotics simulation, autonomous agent behavior, conversational agents, and crowd modeling. Ultimately they all model virtual people to variously generate, simulate, visualize, analyze, substitute for, and understand human performance. Our community is at a crossroads; no, actually it's more like an entrance to a superhighway. It's time to get on.

Acknowledgments

The support of NSF IIS-0200983, Office of Naval Research Virtual Technologies and Environments grant N0001 4-04-1-0259, U.S. Air Force AFRL/HEAL (AVIS-MS) and NASA 03-OBPR-01-0000-0147 are gratefully acknowledged. The opinions expressed here are solely those of the authors.

References

1. H. Shin, J. Allbeck, D. Elliott, N. Badler. "Modeling deformable human arm for constrained reach analysis." Digital Human Modeling Conference, June, 2002, Munich, Germany, VDI Berichte 1675, pp. 217–228 VDI-Gesellschaft Fahrzeugund Verkehrstechnik - Düsseldorf: VDI Verlag and SAE-International.
2. A. Hirao, S. Kitazaki, N. Yamazaki. "Development of a new driving posture focused on biomechanical loads." SAE Document Number: 2006-01-1302, SAE World Congress & Exhibition, Detroit, MI, 2006.
3. T. Jang, J. Lee, Y. Yoon. "Occupant behaviour simulation using DADS and design of seat belt for occupant safety." International Journal of Vehicle Design 21(4–5), 1999, pp. 402–423.
4. J. Teran, E. Sifakis, G. Irving, R. Fedkiw. "Robust quasistatic finite elements and flesh simulation." Proc. Eurographics/ACM SIGGRAPH Symposium on Computer Animation, 2005, pp. 181–190.
5. F. Cordier, N. Magnenat-Thalmann. "Integrated system for skin deformation." Proc. Computer Animation, 2000, pp. 2–8.
6. D. James, C. Twigg. "Skinning mesh animations." ACM Transactions on Graphics 24(3) (SIGGRAPH 2005), August, 2005, pp. 399–407.
7. K. Abdel-Malek, J. Yang, R. Marler, S. Beck, A. Mathai, X. Zhou, A. Patrick, J. Arora. "Towards a new generation of virtual humans: Santos." International Journal of Human Factors Modeling and Simulation 1(1), 2006, pp. 2–39.
8. P. Volino, N. Magnenat-Thalmann. *Virtual Clothing: Theory and Practice*. Springer, 2000.
9. http://store.sae.org/caesar.
10. B. Allen, B. Curless, Z. Popović. "The space of human body shapes: reconstruction and parameterization from range scans." ACM Transactions on Graphics 22(3) (SIGGRAPH 2003), July 2003, pp. 587–594.
11. D. Anguelov, P. Srinivasan, D. Koller, S. Thrun, J. Rodgers, J. Davis. "SCAPE: shape completion and animation of people." ACM Transactions on Graphics 24(3) (SIGGRAPH 2005), July 2005, pp. 408–416.
12. L. Kovar, J. Schreiner, M. Gleicher. "Footskate cleanup for motion capture editing." Proc. Eurographics/ACM Symposium on Computer Animation, 2002, pp. 97–104.
13. C. Godin, J. Chiang, A. Stephens, J. Potvin. "Assessing the accuracy of ergonomic analyses when human anthropometry is scaled in a virtual environment," SAE Digital Human Modeling for Design and Engineering Conference, July 2006, Lyon, France, Document Number: 2006-01-2319.
14. L. Kovar, M. Gleicher, F. Pighin. "Motion graphs." ACM Transactions on Graphics 21(3) (SIGGRAPH 2002), July 2002, pp. 473–482.
15. J. Lee, J. Chai, P. Reitsma, J. Hodgins, N. Pollard. "Interactive control of avatars animated with human motion data." ACM Transactions on Graphics 21(3) (SIGGRAPH 2002), July 2002, pp. 491–500.
16. O. Arikan, D. Forsyth. "Interactive motion generation from examples." ACM Transactions on Graphics 21(3) (ACM SIGGRAPH 2002), July 2002, pp. 483–490.
17. http://www.measurand.com/products/ShapeWrap.html.
18. http://www.xsens.com/moven; http://suit.innalabs.com/; http://animazoo.brightstarrdemo.co.uk.
19. D. Vlasic, R. Adelsberger, G. Vannucci, J. Barnwell, M. Gross, W. Matusik, J. Popović. "Practical motion capture in everyday surroundings." ACM Transactions on Graphics 26(3) (SIGGRAPH 2007), August 2007, in press.

20. G. Anthes. "I, coach: What's in store in robotics." Computerworld, May 21, 2007.

21. D. Chaffin, D. Thompson, C. Nelson, J. Ianni, P. Punte, D. Bowman. *Digital Human Modeling for Vehicle and WorkPlace Design.* Society of Automotive Engineers, 2001.

22. J. Zhao, N. Badler. "Inverse kinematics positioning using nonlinear programming for highly articulated figures." ACM Transactions on Graphics 13(4), 1994, pp. 313–336.

23. H. Ko, N. Badler. "Animating human locomotion in real-time using inverse dynamics, balance and comfort control." IEEE Computer Graphics and Applications 16(2), March 1996, pp. 50–59.

24. K. Yin, K. Loken, M. van de Panne. "SIMBICON: simple biped locomotion control." ACM Transactions on Graphics (SIGGRAPH 2007), 2007, in press.

25. F. Multon, L. France, M-P. Cani-Gascuel, G. Debunne. "Computer animation of human walking: A survey. Journal of Vizualization and Computer Animation 10, 1999, pp. 39–54.

26. J. Faraway. "Statistical modeling of reaching motions using functional regression with endpoint constraints." Journal of Visualization and Computer Animation 14, 2003, pp. 31–41.

27. K. Yamane, J. Kuffner, J. Hodgins. "Synthesizing animations of human manipulation tasks." ACM Transactions on Graphics 23(3) (SIGGRAPH 2004), August 2004, pp. 532–539.

28. L. Zhao, Y. Liu, N. Badler. "Applying empirical data on upper torso movement to real-time collision-free reach tasks." SAE Digital Human Modeling Conference, Iowa City, IA , 2005, SAE Transactions Journal of Passenger Cars—Mechanical Systems, paper 2005-01-2685.

29. P. Faloutsos, M. van de Panne, D. Terzopoulos. "Composable controllers for physics-based character animation." Proc. ACM SIGGRAPH 2001, pp. 251–260.

30. N. Badler, M. Palmer, R. Bindiganavale. "Animation control for real-time virtual humans." Comm. of the ACM 42(8), August 1999, pp. 65–73.

31. N. Badler, R. Bindiganavale, J. Allbeck, W. Schuler, L. Zhao, M. Palmer. "A parameterized action representation for virtual human agents." In J. Cassell, J. Sullivan, S. Prevost, and E. Churchill (eds.), *Embodied Conversational Agents*, MIT Press, 2000, pp. 256–284.

32. J. Allbeck, N. Badler. "Automated analysis of human factors requirements." SAE Digital Human Modeling Conf. Proceedings, 06DHM-49, Lyon, France, 2006.

33. U. Raschke, H. Kuhlmann, M. Hollick. "On the design of a task based human simulation system." Proceedings of the SAE Digital Human Modeling, Iowa City, IA, 2005.

34. A. Safonova, J. Hodgins, N. Pollard. "Synthesizing physically realistic human motion in low-dimensional, behavior-specific spaces." ACM Transactions on Graphics (SIGGRAPH 2004), 23(3), August 2004, pp. 514–521.

35. M. Dontcheva, G. Yngve, Z. Popović. "Layered acting for character animation." ACM Transactions on Graphics 22(3) (SIGGRAPH 2003), 2003, pp. 409–416.

36. N. Badler, J. Allbeck, A. Megahed, M. Whitmore. "RIVET: Rapid Interactive Visualization for Extensible Training." Habitation 2006. Orlando, FL. Feb. 5–8, 2006.

37. J. Barbič, A. Safonova, J.-Y. Pan, C. Faloutsos, J. Hodgins, N. Pollard. "Segmenting motion capture data into distinct behaviors." Proceedings of Graphics Interface, May 2004, pp. 185–194.

38. R. Bindiganavale, N. Badler. "Motion abstraction and mapping with spatial constraints." Workshop on Motion Capture Technology, Geneva, Switzerland, November 1998.

39. D. Strayer, F. Drews, W. Johnston. "Cell phone-induced failures of visual attention during simulated driving." Journal of Experimental Psychology: Applied, 9(1), 2003, pp. 23–32.

40. G. Green. "Inattentional blindness and conspicuity." http://www.visualexpert.com/Resources/inattentionalblindness.html, 2004.

41. E. Gu, C. Stocker, N. Badler. "Do you see what eyes see? Implementing inattentional blindness." Intelligent Virtual Agents (IVA), LNCR 3661, Springer-Verlag, 2005, pp. 178–190.

42. A. Mack, I. Rock. *Inattentional Blindness.* MIT Press, 1998.

43. S. Most, B. Scholl, E. Clifford, D. Simons. "What you see is what you set: Sustained inattentional blindness and the capture of awareness." Psychological Review 112, 2005, pp. 217–242.

44. E. Gu, N. Badler. "Visual attention and eye gaze during multipartite conversations with distractions." Intentional Virtual Agents, Marina del Rey, CA, 2006.

45. S. Tyler, C. Neukom, M. Logan, J. Shively. "The MIDAS human performance model." Proceedings of the Human Factors and Ergonomics Society 42nd Annual Meeting. Chicago, IL, 1998, pp. 320–325.

46. O. Tsimhoni, M. Reed. "The virtual driver: Integrating task planning and cognitive simulation with human movement models." SAE World Congress & Exhibition, Detroit, MI, SAE Document Number 2007-01-1766, April 2007.

47. B. Silverman, M. Johns, J. Cornwell, K. O'Brien. "Human behavior models for agents in simulators and games: Part I: Enabling science with PMFserv." Presence, 15(2), 2006, pp. 139–162.

48. J. Gratch, J. Rickel, E. André, N. Badler, J. Cassell, E. Petajan. "Creating interactive virtual humans: Some assembly required." IEEE Intelligent Systems, July/August 2002, pp. 54–63.

49. A. Ortony, G. Clore, A. Collins. *The Cognitive Structure of Emotions*. Cambridge University Press, 1988.

50. J. Cassell. "Nudge nudge wink wink: Elements of face-to-face conversation for embodied conversational agents." In J. Cassell et al. (eds.), *Embodied Conversational Agents*. MIT Press, 2000, pp. 1–27.

51. S. Marsella, J. Gratch. "Modeling the interplay of emotions and plans in multi-agent simulations." Proceedings of the 23rd Annual Conference of the Cognitive Science Society, Edinburgh, Scotland, 2001.

52. M. Core, D. Traum, H. Lane, W. Swartout, J. Gratch, M. van Lent, S. Marsella. "Teaching negotiation skills through practice and reflection with virtual humans." Simulation, 82(11), November 2006, pp. 685–701.

53. S. Bandi, D. Thalmann. "Space discretization for efficient human navigation." Proc. Eurographics, Computer Graphics Forum 17(3), 1998, pp. 195–206.

54. A. Bloomfield, Y. Deng, P. Rondot, J. Wampler, D. Harth, M. McManus, N. Badler. "A taxonomy and comparison of haptic actions for disassembly tasks." IEEE Virtual Reality Conf., Los Angeles, CA, March 2003, pp. 225–231.

55. http://www.havok.com.

56. http://www.ageia.com.

57. A. Sundin, R. Örtengren. "Digital Human Modeling for *CAE Applications*." *In Handbook of Human Factors and Ergonomics* (3rd ed.), G. Salvendy (ed.), John Wiley & Sons, 2006, pp. 1053–1078.

58. N. Badler, C. Erignac, Y. Liu. "Virtual humans for validating maintenance procedures." Communications of the ACM, 45(7), July 2002, pp. 56–63.

59. D. Schunk, S. Archer. "Evaluating workload capabilities with the Ship Manpower Analysis and Requirements Tools (SMART)." Proceedings of the Military, Government, and Aerospace Simulation (MGA 2001) Conference, M. Chinni (ed.), 2001, pp. 39–43.

60. S. Cohen-Boulakia, S. Davidson, C. Froidevaux. "A user-centric framework for accessing biological sources and tools." Lecture Notes in Computer Science, Volume 3615, Data Integration in the Life Sciences (DILS), 2005, pp. 3–18.

61. T. Green, G. Karvounarakis, N. Taylor, O. Biton, Z. Ives, V. Tannen. "Orchestra: Facilitating collaborative data sharing." Proceedings of ACM SIGMOD International Conference on Management of Data, 2007.

Index

Q